MATHEMATICAL STATISTICS

(*Twentieth Revised & Enlarged Edition*)

J.N. KAPUR

Emeritus Professor, J.N.U., New Delhi

and

H.C. SAXENA

Ex-Reader in Mathematics & Statistics
Ramjas College, Delhi

S. CHAND
PUBLISHING

S Chand And Company Limited

(ISO 9001 Certified Company)

S Chand And Company Limited

(ISO 9001 Certified Company)

Head Office: Block B-1, House No. D-1, Ground Floor, Mohan Co-operative Industrial Estate, New Delhi – 110 044 | Phone: 011-66672000

Registered Office: A-27, 2nd Floor, Mohan Co-operative Industrial Estate, New Delhi – 110 044 | Phone: 011-49731800

www.**schandpublishing.com**; e-mail: **info@schandpublishing.com**

Branches

Chennai : Ph: 23632120; chennai@schandpublishing.com
Guwahati : Ph: 2738811, 2735640; guwahati@schandpublishing.com
Hyderabad : Ph: 40186018; hyderabad@schandpublishing.com
Jalandhar : Ph: 4645630; jalandhar@schandpublishing.com
Kolkata : Ph: 23357458, 23353914; kolkata@schandpublishing.com
Lucknow : Ph: 4003633; lucknow@schandpublishing.com
Mumbai : Ph: 25000297; mumbai@schandpublishing.com
Patna : Ph: 2260011; patna@schandpublishing.com

First Edition 1960

Subsequent Editions and reprints 1961, 63, 67, 69, 71, 73, 76, 78, 79, 81, 82, 84, 86, 91, 93, 95, 97, 99, 2001, 2003, 2005, 2006, 2007, 2009, 2010, 2011, 2013, 2014, 2015, 2018 (Twice), 2019 (Twice), 2020

Reprint 2022

ISBN: 978-81-219-1246-4 **Product Code:** H6MST64STAT10ENAT0XR

PRINTED IN INDIA

By Vikas Publishing House Private Limited, Plot 20/4, Site-IV, Industrial Area Sahibabad, Ghaziabad – 201 010 and Published by S Chand And Company Limited, A-27, 2nd Floor, Mohan Co-operative Industrial Estate, New Delhi – 110 044.

PREFACE TO THE TWENTIETH EDITION

We thank all the students and teachers of Mathematics, Statistics and Economics all over India for receiving well our earlier editions of the book "Mathematical Statistics".

This new edition has been revised and updated as per the valuable suggestions of the users of the book. Latest topics with solved and unsolved examples taken from question papers of different universities have been incorporated at the proper places. Moreover, the chapter on "Non-Parametric Methods" has be rewritten to make it more useful to the readers.

We are grateful to Shri Ravindra Kumar Gupta, Managing Director, S. Chand & Co. Ltd. for the timely publication of this book in such a nice getup.

Suggestions for further improvement of the book are welcome.

AUTHORS

PREFACE TO THE FIRST EDITION

The present book is based on our experience, extending over a period of about fifteen years of teaching Mathematical Statistics to B.A. (Hons) and M.A. classes of the University Delhi. During this period, a number of excellent books on Mathematical Statistics have appeared, but, surprisingly enough, few have treated the subject on the model of the standard text books of Pure and Applied Mathematics. One well recognized practice in these text books is to provide illustrative examples, easy exercises, which are straight deductions from the theory and which throw light on the various important points in the theory, exercises of moderate difficulty for the average student and more difficult exercises to test the ingenuity of the brightest student. Books on Statistics have so far tended to give either very difficult exercises taken from research papers or only very easy exercises and, in general, these exercises are too few to give the students enough practice and confidence. The present book is intended to fill the gap by providing a book on the same model as the books in calculus, differential equations, statics, dynamics, etc. and is based on our conviction that such a book is not only possible and desirable, but is also absolutely necessary. We have given, we believe, enough exercises of all types and these exercises and the text form an integrated pattern. A special feature is the large number of exercises in the bodies of the chapters, more than fifty per cent of which have been specially constructed to illustrate the theory and are so designed that after doing these, the student should not only have a better grasp of the theory, but should also know the motivation for the various steps. Another important feature is that unlike the other standard books on the subject, answers have been provided to almost all the exercises, so that the student can always decide for himself or herself, whether he or she has proceeded rightly or not.

There is also a tendency in some recent books on Statistics, to slip over certain elementary, but basic topics, as well as to omit a large number of proofs, in order to cover as large a ground as possible in a small space. We have resisted this temptation, except in the last chapter on Testing of Hypotheses, Sequential Analysis and Distribution-free Methods. These topics are by their nature too wide to be covered completely in an introductory book, but we feel that a student of Statistics should be at least aware of the important basic concepts and results of these topics and, therefore, we have omitted a number of proofs in these chapters. In all the earlier chapters, however, we have given proofs of almost all the statements that have been made and very often more than one proof have been given. Quite often alternative proofs have been sketched in the exercises that follow the main proof.

While a great deal of emphasis has been placed on the Theory of Probability and its application in Mathematical Statistics, Descriptive Statistics and numerical problems on the same have also been given considerable space, specially in chapters on Measures of Location and Dispersion, Correlation and Regression, Fitting of Curves and Tests of Significance. Chapters 15 and 16 on Tests of Significance have been written, so as to be independent of chapters 13 and 14 on Exact Sampling Distributions, so that a student who is interested in these tests only and not on the derivations of the sampling distributions on which these tests are based can study these two chapters independently.

As for mathematical pre-requisites, only the knowledge of calculus up to the B.A. (Pass) standard is assumed, so that the book can be useful for undergraduate students, who have Mathematics as one of their elective or subsidiary subjects. However, in view of the important part played by Beta and Gamma functions of Mathematical Statistics, their important properties have been discussed in Appendix I.

Authors of a text book are always indebted to all the earlier books on the subject and as such we acknowledge with gratitude our debt to the authors and publishers of those books, which we have often consulted in the course of our teaching. We are deeply indebted to Professor R.A. Fisher and to Messrs Oliver and Boyd Ltd., Edinburgh for permission to reprint tables from Statistical Methods for Research Workers and to Prof. George W. Snedecor and Iowa State University Press, Iowa, U.S.A. for permission to reprint the tables from Statistical Methods (5th edition, 1956).

In a book of this nature, a number of misprints and mistakes are likely to have crept in and we shall be grateful for any notice of the same as well as for any suggestions for improvement which will be gratefully acknowledged and given full consideration by us for the future.

Delhi :
December, 1959.

J.N. KAPUR
H.C. SAXENA

CONTENTS

CHAPTER 1
INTRODUCTION

CHAPTER 2
FREQUENCY DISTRIBUTIONS AND MEASURES OF LOCATION

CHAPTER 3
MEASURES OF DISPERSIONS, SKEWNESS & KURTOSIS, MOMENTS OF FREQUENCY DISTRIBUTIONS

CHAPTER 4

THEORY OF PROBABILITY

CHAPTER 5

DISCRETE PROBABILITY DISTRIBUTIONS

CHAPTER 6

SPECIAL DISCRETE PROBABILITY DISTRIBUTIONS

CHAPTER 7

UNIVARIATE CONTINUOUS PROBABILITY DISTRIBUTIONS

CHAPTER 8

SPECIAL CONTINUOUS PROBABILITY DISTRIBUTIONS

CHAPTER 9

PRINCIPLE OF LEAST SQUARES, FITTING OF CURVES & ORTHOGONAL POLYNOMIALS

CHAPTER 10

CORRELATION AND REGRESSION

CHAPTER 11

MULTIPLE AND PARTIAL CORRELATION

CHAPTER 12

THEORY OF SAMPLING

CHAPTER 13

EXACT SAMPLING DISTRIBUTIONS

CHAPTER 14

EXACT SAMPLING DISTRIBUTION (*Continued*)

CHAPTER 15

TESTS OF SIGNIFICANCE BASED ON t, F AND Z DISTRIBUTIONS

CHAPTER 16

TESTS OF SIGNIFICANCE BASED ON THE CHI-SQUARE DISTRIBUTION

CHAPTER 17

STATISTICAL THEORY OF POINT ESTIMATION

CHAPTER 18

TESTING OF HYPOTHESES, SEQUENTIAL ANALYSIS, DISTRIBUTION-FREE METHODS, STATISTICAL DECISION THEORY

CHAPTER 19

ELEMENTS OF STOCHASTIC PROCESSES

CHAPTER 20

ORDER STATISTICS

APPENDICES

1

Introduction

1·1. Pure and Applied Statistics. Like Mathematics, the Science of Statistics may be classified into :

(*i*) *Pure Statistics* or *Mathematical Statistics* or the *Theory of Statistics* ; and

(*ii*) *Applied Statistics.*

In Pure Statistics, we develop the research tools which are then applied to specific problems in different fields, such as Physics and Chemistry, Anthropology and Biology, Agriculture and Animal Husbandry, Economics and Commerce, Education and Psychology, Medicine and Public Health, Industry and Insurance, etc. As in Mathematics, here also a theory of Statistics has been possible, since the same type of techniques are required in all the diverse fields of application. Again, as in Mathematics, Pure and Applied Statistics act and react and thus enrich each other. Pure Statistics continues to develop newer and stronger tools for Applied Statistics, while Applied Statistics continues to present new and challenging problems for Pure Statistics. Pure Statistics not only creates new tools but also tries to perfect existing ones by placing them on more and more rigorous foundations, and in this process, it depends much on recent developments in Pure Mathematics. At the same time, Applied Statistics dealing, as it does, with the application of statistical methods to specific problems, continues to find new uses for the existing tools and the new tools that are being created.

In this book, we shall be concerned mainly with the theory of Statistics, though we shall very often draw upon Applied Statistics for illustrating the theory. The Theory of Statistics is based on the *Theory of Probability* though it supplements that theory by certain additional axioms. We shall accordingly also study and in fact lay a great deal of emphasis on the Theory of Probability.

1·2. Definition and Characteristic Features of the Science of Statistics. Before embarking on the study of the Science of Statistics, it is desirable to define Statistics. In 1935, Dr. W.F. Willcox listed over a hundred definitions of Statistics and the list was even then far from being exhaustive. In fact, no definition can give a complete picture of a fast growing subject like Statistics, which is bound to outgrow any definition within which it may be sought to be encompassed, and as such, we shall not attempt to define Statistics. However, we give below some of the characteristic features of the Science of Statistics, in order to help the reader in understanding the nature of the subject.

(*i*) *Statistics is concerned with reduction of data or with obtaining correct facts from figures*. The data, which are collected or compiled in the course of an enquiry, are usually very voluminous and as such, it is often not possible to draw any conclusions from the data in their original form. It is, therefore, desirable to express all the relevant information contained in the unwieldy data by means of a comparatively few numerical values. The layman's concept of Statistics, as the science of collecting and tabulating numerical data, of representing data by means of charts and diagrams, of calculating the mean, median, mode, standard deviation, etc. of the data, though fifty years out-of-date, represents one important aspect of the reduction of data.

(*ii*) *Statistics deals with populations or aggregate of individuals, rather than with individuals*. Thus 'Milton was blind' is not a statistical statement, whereas 'The average height of an Indian is five feet' is a statistical statement. It should be noted that the kinetic theory of gases is a statistical theory, as its results apply to the population of gas molecules and not to any individual molecule.

(*iii*) *Statistics deals with variation,* for a completely homogeneous population will be specified in terms of one individual only.

(*iv*) *Statistics deals with only numerically specified populations*. Thus the population of words in a dictionary is not of statistical interest unless we want to know how many of the words begin with the various letters of the alphabet or with some other property of words capable of being expressed in numerical form. Qualitative characteristics like intelligence, beauty, goodness, etc. do not form part of statistical studies unless numerical equivalents are assigned to these in terms of, for example, intelligence quotients, beauty measurements, etc.

(*v*) *Statistics deals with populations which occur in nature and are subject to a large number of random forces*. Thus, tables of logarithms and of other mathematical functions, though they constitute numerical populations, are not of statistical interest, for they are determined by exact mathematical laws ; while populations of births and deaths, of imports and exports, of yield of crops, are of statistical interest, as they are affected by a large number of random causes.

(*vi*) *The logic used in statistical inference is inductive*. In statistical methods, we draw inference about the population from the information contained in a *sample*, which is only a part of the population and thus we pass from the particular to the general so that there is a chance of our conclusions being untrue.

(*vii*) *Statistical inferences are uncertain.* This is quite natural, since the logic used is inductive. The chief merit of statistical theory is that, under certain conditions, it enables us to get a more or less precise idea of the measure of this uncertainty in terms of probability statements. Thus though it may not be possible to say that a certain result will hold in all cases, we can with a certain measurable degree of confidence, say that the result is likely to hold in a known percentage of cases when the experiment or observation is repeated a large number of times.

(*viii*) *Statistical results might lead to fallacious conclusions, if quoted out of context or manipulated.* Statistical conclusions should not be based on incomplete information. Thus the argument that drinking is bad for longevity since 99% of the persons who drink beer die before reaching the age of 100 years is statistically defective, since we are not told what percentage of persons who do not drink beer die before reaching that age.

1·3. Main Divisions of Statistical Theory. Statistical theory can be classified into two main divisions :

(1) Design of Experiments, and

(2) Statistical Inference.

We discuss these briefly in the following paragraphs.

1·3·1. Design of Experiments. The subject of design of experiments whether in Science, Agriculture, Industry or Sample Survey tries to lay down the basic principles and designs for collecting data in the most economic and useful forms. We shall, however, be not concerned with this aspect of the theory in this book.

1·3·2. Statistical Inference. Statistical inference is the science of drawing valid conclusions from the given data, which are *assumed* to have been so collected as to satisfy the basic requirements of statistical theory. The problem of statistical inference may be formulated as follows :

The given data are regarded as a *sample* from some population and then valid inferences about the population are drawn from the information supplied by the sample. Thus the inference drawn from the sample is inductive in nature and has some degree of certainty. The first stage is of *specification*, that is, we try to specify the population from which the sample might have been drawn. In some cases, the theory of a particular problem may help us in specifying the population, but in most cases, experience, aided by the study of a large number of standard populations, is the only guide. The specification of the mathematical form of the population will, in general, involve a number of constants, called the *parameters* of the population. These parameters are to be *estimated* from the sample. In general, it would be possible to build a number of estimates from the sample called *statistics*, for the same population parameters. The *Theory of Estimation* consists in trying to find the *best estimators* for all the parameters. The possibility of discrimination arises, as due to the inductive nature of the inference, all estimators would be subject to error and the theory of sampling *distributions* helps us in evaluating these errors. The correct specification of the population and satisfactory estimation of the parameters, together with *measure of reliability* of these estimators, constitute the solution of the problems of statistical inference.

1·4. The Road Ahead. Purview of the Book

1·4·1. Chapters 1–4. Large amounts of numerical data are collected by various private and government agencies in the course of their normal work. Thus we have data of marks obtained by students in university examinations, number of students enrolled in different classes, amount of rainfall on different days of the year, amount of wheat produced in each district, number of railway passengers travelling in different months of the year, population in each town in a census year, the number of computers in use in India in different years and so on. Numerical data are also collected through astronomical observations, social and demographic surveys, national sample survey and planned experiment in laboratories.

The data collected are so huge that they cannot very often be stored in ordinary registers and have to be stored on magnetic tapes and discs of computers which can store huge amounts of data in a small space. A modern statistician has, therefore, to be familiar with the methods of storing and retrieving numerical information on computer systems.

After having collected the data, our first task is to tabulate these in a form in which these can be easily understood. Some of the methods for doing this are explained in chapter 2. The pictorial representations through histograms, bar charts and pie charts are extensively employed. In fact Japan celebrates every year a Statistics Day on which every school, every business firm and every government office reviews its progress in the previous year and displays it pictorially. We would strongly suggest to the Statistics students to collect data on various aspects of our national life (education, population, food production, energy availability, etc.) from various reports, for the last thirty or forty years and demonstrate pictorially the progress we have made and the targets we have still to attain in order to reach the standards of the developed countries.

Once we have got the data on any subject, we want to get an overall view before going into details. We ask questions like the following : What is the average amount ? What is the variation ? Are the data symmetrical ?

Is there more than one maximum ? Does the maximum represent a sharp peak ? We understand these questions by *developing measures of location, dispersion, skewness* and *kurtosis* in chapters 2 and 3 to answer these questions.

We now need a machinery to help us in drawing *valid conclusions* from the data collected. Since almost all phenomena we are interested in studying are subject to chance, we need the *theory of probability*. We develop this in chapter 4.

The theory of probability was first developed by mathematicians to answer problems posed by their gamble friends. Since in gambling all results are subject to chance, a theory of chance for explaining results obtained was needed. Later the theory was developed by demographers and actuaries to answer questions about insurance of human lives, the longevity of which, of course, depends on chance. More recently, the theory was developed by agriculturists and biologists to study chance variations in yields of crops and effects of medicines. In the present century, physicists, economists, psychologists, and management scientists have also made significant contributions. The mathematicians have abstracted from all this the work a *mathematical theory of probability* which is equally applicable to all these fields. In fact, mathematician have gone on making the theory more and more elegant, more and more rigorous and more and more abstract till it has become a branch of pure mathematics and its most advanced parts have very little relevance to day-to-day problems. We shall not study the more abstract theory here, but those students who like abstraction and generalisation, will find it a fascinating discipline.

The probability of an event gives a quantitative measure of the likelihood of the event happening. There are a number of approaches to the theory of probability *viz.* the *classical approach* due to the French mathematician P.S. Laplace (1749-1827), the *statistical approach* due to the German-American mathematician Richard Von Mises (1883-1953), the *axiomatic approach* due to the Russian mathematician A.N. Kolmogrov (1903—] and the *measure-theoretic approach* based on the *measure theory* developed by the French mathematician Henri Lebesgue (1875-1941) and others. We discuss all these in chapter 4, but we mention the last approach very briefly, since its value is in proving advanced theorems of the theory of probability, which are not needed in a first course.

1·4·2. Chapters 5–8. The most important concept needed is that of a *random variable* or *variate* or a *stochastic variable* or a *chance varible X, i.e.*, of a variable whose values depend on chance. Thus the number of points when a die is thrown, or the number of births in a month or the amounts of exports of a particular commodity, say steel, are random variables, since their values cannot be predicted exactly and only probabilistic statements can be made about them. The first two of these are *discrete random variables*, since they can take only a discrete set of values, while the third is a *continuous random variable*, since it can take all values in a certain range.

We first consider discrete random variables (also called *discrete random variates*) in chapters 5 and 6. Here a random variable X takes values, $x_1, x_2, \ldots\ldots, x_n$ with respective probabilities $p_1, p_2, \ldots\ldots, p_n$ where

$$0 \le p_i \le 1, i = 1, 2, \ldots, n \quad \sum_{i=1}^{n} p_i = 1$$

The values which the random variable X takes together with the probabilities with which these variate values are taken, constitute a *probability distribution*. An important concept associated with a random variable is that of the *expected value*. The expected value of X, defined as $\sum_{i=1}^{n} p_i x_i$ is also called the *mean* of X and is denoted by $\overline{X}$. We also find useful expected values of $f(X^r)$ and $f(X - \overline{X})^r$. In particular the expected value of X^r and $(X - \overline{X})^r$ *viz.*

$$\sum_{i=1}^{n} p_i x_i^r \text{ and } \sum_{i=1}^{n} p_i (x_i - \bar{x})^r$$

play an important role and are called *rth order moments about the origin and the mean* respectively. The second order moment *viz.*, $\sum_{i=1}^{n} p_i (x_i - \bar{x})^2$ is called the *variance* and plays an important role in the theory of statistics.

The moments in a sense characterise the probability distribution (*p.d.*) and their knowledge is up to a point, equivalent to the knowledge of the distributions. We, therefore, lay a great deal of emphasis on these moments. The expected values of e^{tx} and e^{itx}, where $i = \sqrt{-1}$ and t is real are also important and are called *moment-generating* (m.g.f.) and *characteristic function* of the probability distribution. They help us in finding the moments. In fact m.g.f. and c.f. may be regarded as alternative ways of defining the p.d., since the knowledge of either of these and specially the c.f. enables us, under certain conditions, to know the p.d.

There are two types of mathematical models for physical, biological and social phenomena *viz.*, *deterministic* and *stochastic*. In a determinstic model, given the value which the independent variable or variables take, we can predict the value which the dependent variable will take. Thus if a particle is thrown vertically upwards under gravity in vacuum, we can predict its height and velocity at any time. There is no uncertainty about these and the phenomenon is deterministic. On the other hand, if we talk of the enrollment in higher education in 1890 A.D. or the population of India in 2000 A.D., we can talk only of the probabilities certain enrollments or certain population sizes having certain values. There is an uncertainty associated with these. These are examples of stochastic phenomena.

For stochastic phenomena, we need *stochastic models* and these models have to be specified in terms of probability distributions. For making stochastic models, we need a large number of *standard probability distributions*, with which we can compare our observations and determine whether the distributions fit the data. The probability distributions will be called *discrete* or *continuous* probability distributions according as the random variate takes a discrete or a continuous set of values. In chapters 5 and 7, we discuss properties of discrete and continuous probability distributions respectively and in chapters 6 and 8, we study special theoretical discrete and continuous probability distributions.

For discrete distributions, we use summation of finite or infinite series and for continuous distributions we use the theory of integration due to German mathematician G.F.B. Riemann (1826-1866). We can treat both these distributions together, either by using the Reimann-Stielejes integrals due to Reimann and the Dutch mathematician T.J. Stielejes [1856-1894] or by using the unit function of the French mathematician C. Hermite [1822-1905] and the delta function of the British physicst P.M.A. Dirac [1902—], We discuss these approaches briefly in, section 7·7, but we still discuss the two types of distributions separately since we do not expect most of our readers to be familiar with the theory of Reimann-Stielejes integral or with the theory of unit and delta functions. In any case, unification can apply only to the proofs of certain theorems ; special distributions have to be discussed separately.

1·4·3. Chapters 9–11. In experimental work, the scientist studies the variation of one variable with respect to another. Thus he may go on changing the volume of a given quantity of gas and go on noting the corresponding pressure values or he may count the number of bacteria in agriculture at different times. He plots his observations in a two dimensional graph and he gets a set of points, which almost lie on a curve. The scientist believes in the regularity of behaviour of nature and as such he assumes that the departure of observed points from points on a smooth curve are due to *experimental errors* or *errors of observation*. He tries to *fit a curve* or *orthogonal polynomials* which is as close to the data-points as possible or which fits the data best in some sense. This is an important scientific problem and falls within the scope of our subject if we regard the errors as *random errors* or as due to chance. We discuss the problem of *fitting of curve* in chapter 9, where we make use of the important *principle of least square.* due to the great German mathematician C.F. Gauss [1777-1855] or Legendre.

The possibility or a relationship between two or more random variables leads us to the study of *multivariate probability distributions.* Here a number of variates are correlated and, in some sense, vary together. Thus heights and weights of individuals may vary together, not in a deterministic manner, but in a probabilistic manner. In anthropological work, when ancient skulls found in a certain region, may have to be studied to find whether they belong to the same race, we may have to study here simultaneous variation of even one hundred variates. We study bivariate distributions in detail and study some aspects of multivariate distributions in chapters 10 and 11.

The relationship between two random variates is measured by a *correlation coefficient* which is a dimensionless measure, lying between –1 and 1. For p variates, there will be p_{C2} pairwise correlation coefficients. The correlation between any two variables may be sometimes due to the influence of a third variable on both. It is, therefore, useful to study the *partial correlation coefficient* between x_1 and x_2 when the linear influences of other variates have been eliminated. We study these partial correlation coefficients in chapter 11. We are also interested in knowing whether knowledge of $X_2, X_3..., X_p$ can enable us to estimate X_1 reasonably well. We, therefore, find the correlation coefficient between the observed values of X_1 and the expected values of X_1 on the basis of a linear relationship between X_1 on the one hand and $X_2, X_3, \ldots, X_p$ on the other. This coefficient is called *multiple correlation coefficient* and is also studied in chapter 11.

1·4·4. Chapters 12–16. Having studied a large number of probability distributions in chapters 6 and 8 and also some characteristics of bivariate and multivariate distributions in chapters 10 and 11, we are now ready to build up the equipment for drawing valid conclusions from given data, *i.e.,* for *statistical inference.*

Our object is to study aggregates of populations, *e.g.*, we may want to know the total wheat production in the country or to know the quality of all the products made in a factory. It is not always possible to study every item of the population. This may not be feasible or may be too costly. We, therefore, resort to *sampling*. We take a *random sample* from this population and study its mean, variance, correlation etc. and from these we want to draw inferences about the population. Since we are going from the smaller sample to the bigger populations, the inference is *inductive* and has to be based on probability theory. The *theory of sampling* is given in chapter 12. In most cases, the population is assumed to be normal and as such the only unknown quantities are its two parameters, *viz.*, the mean and variance. We have to estimate their values. We set up *hypotheses* about these values and then see whether the data are consistent with the hypotheses. We study some of the simpler *tests of hypotheses* in chapter 12.

When the sample size is large, the tests developed in chapter 12 hold, but when *sample size is small*, as is usually the case in practice, we need *exact sampling distributions i.e.*, we want to know the probability distributions of sample statistics like the following, which are functions of sample observations $x_1, x_2, \ldots\ldots, x_n$.

$$\overline{x} = \frac{x_1 + x_2 + \ldots\ldots + x_n}{n}, \quad S^2 = \frac{1}{n} \sum_{i=1}^{n} (x_i - \overline{x})^2$$

$$t = \frac{\overline{x} \sim \mu}{S/\sqrt{n}}, \quad F = \frac{S_1^2}{S_2^2}, \quad \chi^2 = x_1^2 + x_2^2 + \ldots\ldots + x_n^2$$

$$z = \tfrac{1}{2} \log_e F, \quad r_{xy} = \frac{1}{N} \frac{\sum_{i=1}^{n} (x_i - \overline{x})(y_i - \overline{y})}{s_x s_y}$$

These exact sampling distributions are derived in chapters 13 and 14 and are used in chapters 15 and 16 to carry out the tests of significance *i.e.* to test whether the given samples could have come from the populations determined by the hypotheses we make.

We also give an introduction to analysis of variance and to the analysis of two simple experimental designs, *viz.*, *randomised block experiments* and *Latin-square designs*.

1·4·5. Chapters 17–19. Any given data can be regarded as a random sample from a given population. The specification of the popu- lation contains some parameters whose values have to be estimated, *e.g.*, to estimate the population means, we may use sample mean as an estimate. Which one we use : Arithmetic, Geometric or Harmonic Mean or Median or Mode ? For this we need criteria for good estimators. We give these in chapter 17 and we also give there a number of methods to obtain good estimators satisfying these criteria.

Instead of giving a *point estimate* for a population parameter, we may like to give an *interval* in which the population parameter may be expected to lie and we may state the *degree of confidence* with which the parameter is expected to be in that interval. This method of *confidence intervals* is discussed in chapter 18, where we also discuss the general theory of testing statistical hypotheses.

One of the objects of science of statistics is to aid in making *decisions under uncertainty*. An introduction to *statistical decision theory* is given in chapter 18. Another topic introduced in this chapter is *non parametric theory* which is concerned with drawing conclusions from sample populations, without making any assumption about the parametric form of the populations.

Chapter 19 gives an introduction to the *dynamic aspects of probability theory*. The probability distributions need not be static but can change with time or some other characteristics, *e.g.*, consider the number of persons in a queue and let $p_n(t)$ be the probability of there being n persons in the queue at time t. Here n can be 0, 1, 2, 3, ... and $\sum_{n=0}^{\alpha} p_n(t) = 1$. For a fixed t, we have a probability definition, but this p.d. goes on changing with time.

These *stochastic processes* occur in queuing theory, reliability theory, population growth, nuclear fission, cascade showers and in other hundreds of phenomena in science and engineering. Their study is a fascinating and rewarding aspect of statistical theory today.

1·4·6. The Road Ahead. This book is a first introduction to probability and mathematical statistics. There is an enchanting and fascinating world beyond this course.

The theory of probability and stochastic process can be very abstract for those whose interests are purely mathematical. At the same time it is most prolific in its applications and no branch of science or technology is unaffected by it.

We have just introduced statistical estimation, testing of hypotheses, statistical decision theory and non-parametric analysis in the last-but-one chapter. There is a huge literature in each field.

We have not even touched sequential analysis where we do not take a sample of fixed size, but we go on taking more and more observations till we reach conclusions of certain validity. This may even reduce sampling by half and it is valuable when sampling is costly.

The entire treatment of chapters 13—16 is generalized in *multivariate statistical analysis*, which is an important subject in its own right. We only introduce the object in an appendix.

For agriculture and industrial experiments, we need to design experiments so that we can find whether the new agriculture or industrial products are really different or what the influences of various factors on these products are. We have in this book seen how to analyse two simple designs. The analysis of more complex designs and the construction of these designs present interesting combinational and probabilitical problems.

The *sampling survey theory* is another interesting area which is important for agriculture and consumer surveys. Indian statisticians have made notable contributions to sampling theory, contribution of designs, multivariate analysis, stochastic processes and statistical inference. The valuable contributions of P.C. Mahalanobis, R.C. Bose, S.N. Roy, P.V. Sukhatme, C.R. Rao and a host of others to statistics have earned the respect of the whole world and have placed India high on the world map of Statistics.

We would like the students to consult other books and journals to widen their horizons throughout this course. We have given a bibliography for this purpose at the end.

The object of the study of statistics is not only to learn some manipulative abilities which no doubt are important ; it is also to get an insight into probabilistic and statistical thinking. The students are invited to study the succeeding pages with this aim in view.

1·5. Reference System. The chapters have been divided into numbered sections ; the numbering begins anew in each chapter. The chapter number is prefixed to the section number and separated from it by a dot. Thus section 4·2 refers to section 2 of chapter 4 and 4·2·1 refers to sub-section 1 of section 2 of chapter 4. The equations are numbered anew in each chapter and equation numbers are always enclosed in parentheses. Thus equation (6·5) refers to the fifth equation of the sixth chapter. Similarly, solved examples and figures in each chapter are numbered anew. Thus example 15·5 means the fifth solved example of chapter 15.

Exercises 1

1. Discuss the following :

(*a*) "The Science of Statistics is essentially a branch of Applied Mathematics and may be regarded as Mathematics applied to observational data." —R.A. Fisher*.

(*b*) "Statistics is the branch of scientific method which deals with the data obtained by counting or measuring the properties of populations of natural phenemena," M.G. Kendall**

(*c*) "Statistics may be regarded as (*i*) the study of population, (*ii*) the study of variation, (*iii*) the study of methods of reduction of data." —R.A. Fisher.

---

* Sir Ronald (Aylmer) Fisher (Feb. 17, 1890—July 29,1962), F.R.S., was one of the greatest statisticians of modern times.

** M.G. Kendall wrote the most monumental treatise on statistics in 1943.

2. Criticise the following arguments :

(*a*) The average depth of a river is 122 cm. and the average height of the members of a family is 125 cm. therefore the family can safely cross the river.

(*b*) 10% of the cases treated by a surgeon proved fatal. Out of the first ninety cases this year only five died, therefore out of the next ten cases, five must die.

(*c*) The average income of a factory proprietor and his employees is Rs. 5000 p.m. Therefore all the employees must be fabulously or handsomely paid.

(*d*) All the persons who have written to the firm about the medicine, have praised it. Therefore the medicine must be effective.

(*e*) On measuring ten Englishmen and ten Indians, it was observed that the average height of the Englishmen exceeds that of the Indians. Therefore the Englishmen are taller than the Indians.

(*f*) 99% of the members, who voted for the Army Estimates, were military officers. Therefore it was unfair to suppose that the voting was unbiased.

(*g*) The death rate in the American Navy during the Spanish American War was nine per thousand, while in the city of New York, it was sixteen per thousand. It was safer to be a sailor in American Navy than to live in New York city.

(*h*) From personal discussion with my friends, I can definitely say that most of the people in our country subscribe to the points of view which I hold.

(*i*) People's capitalism in America is a statistical myth in spite of most Americans holding shares in industry.

(*j*) In a country 20,000 vaccinated persons died of small-pox. Therefore vaccination is useless.

(*k*) The examination result of school *A* was 75% in a particular year. In the same year and at the same examination only 400 out of a total of 600 students were successful in school *B*. Therefore the teaching standard of *A* was decidedly better.

(*l*) The gross national product has increased many times since independence. Therefore every Indian is much better off than he was in 1947.

3. Comment on the following : (State with reason whether the following statements are true or false).

(*i*) A : According to this chart of standard heights and weights, you weigh too much.

B : No. According to this chart, I am not sufficiently tall.

(*ii*) A person asks the same set of 25 questions from all the members of a large joint family and also measures the size of their shoes. From this he deduces that there is a high degree of relationship between knowledge and size of the shoes.

(*iii*) In every country where the sale of contraceptives is allowed by the government, the birth-rate is declining. Therefore contraceptives must exert an influence on the birth-rate.

[Two unrelated factors may appear to be related, if they are both related to some third factor.]

(*iv*) A lady was told that for her height, her weight is above the average. She claimed that for her weight, she should have been taller. Which statement is correct and why ?

4. (*a*) Are the following statistical statements or not ? Give reasons.

(*i*) Indians are hard working.

(*ii*) The attraction between two particles varies inversely as the square of the distance between them.

(*iii*) The production of wheat per hectare in 1978 in a certain district was 100 quintals.

(*b*) Give some examples of statistical theories from Economics, Genetics and Physics.

(*c*) Out of 100 packets in a box, one is drawn at random and its weight is found to be 10 lbs. What inference, if any, can you draw about the population of packets ?

(*d*) In a city, 80% of road accidents involve male drivers, while only 20% involve female drivers. Therefore females are better drivers.

(*e*) Female workers in a factory are more punctual than male workers; since percentages of female and male persons coming on time are 90% and 50% respectively.

5. Do you agree with the following interpretations made on the basis of the facts given? Explain briefly your answer.

(*i*) The examination result in a college *A* was 70% in the year 1991. In the same year and at the same examination, only 500 out of 750 students were successful in college *B*. Hence the teaching standard in college *A* was better.

(*ii*) The number of deaths in military in the recent war was 15 of 1.000 while the number of deaths of civilian in the same period, was 18 per 1000. Hence it is safer to join military service than to lead a civil life.

2

Frequency Distribution and Measures of Location

2·1. Frequency Distributions. A frequency distribution is defined whenever (*i*) the values which the variate takes and (*ii*) the frequencies or the number of times each variate value is taken, are both specified.

Consider first an example where the marks obtained by a B.A/ B.Sc.—Part III class, of 60 students in a Statistics paper, the maximum marks allotted to the paper being 50, are :

22, 47, 9, 42, 31, 17, 13, 15, 18, 13, 2, 21, 27, 38, 15, 0, 33, 10, 34, 29, 26, 16, 25, 33, 36, 10, 24, 22, 26, 19, 14, 36, 18, 25, 21, 33, 35, 25, 18, 28, 25, 17, 38, 10, 3, 31, 24, 3, 12, 16, 33, 18, 26, 29, 27, 29, 28, 35, 26, 27.

Here the variable is the 'number of marks'. The data in the original form as above are called raw or *ungrouped data.* We would like to bring out certain salient features of this data. When arranged in ascending or descending order of magnitude, the data are said to be arranged in an *array*. Take each mark from the data and place a bar (1) or tally mark against the number when it occurs. Tally marks are recorded in batches of five, the fifth occurrence is shown by putting a cross tally (/) on the first four bars i.e. by (𝍸). This technique facilitates the counting of tally marks at the end. Counting the number of times each variate value occurs, we get the following Frequency Table of marks in Statistics of 60 students.

Table 2·1

Frequency Table of marks in Statistics of 60 students

Marks in Statistics	*Number of Students Tally marks*	*Cumulative Frequency*	*Marks in Statistics*	*Number of Students Tally Marks*	*Cumulative Frequency*
0	I = 1	1	24	II = 2	29
2	I = 1	2	25	IIII = 4	33
3	II = 2	4	26	IIII = 4	37
9	I = 1	5	27	III = 3	40

10	III = 3	8	28	II = 2	42
12	I = 1	9	29	III = 3	45
13	II = 2	11	31	II = 2	47
14	I = 1	12	33	IIII = 4	51
15	II = 2	14	34	I = 1	52
16	II = 2	16	35	II = 2	54
17	II = 2	18	36	II = 2	56
18	IIII = 4	22	38	II = 2	58
19	I = 1	23	42	I = 1	59
21	II = 2	25	47	I = 1	60
22	II = 2	27	–	—	
				60	

The frequency table given above is an improvement over the arrangement of figures in an array, as it presents a clearer picture about the data, but to get a still better picture about the structure of the frequency distribution, we re-classify the data into groups as follows :

Table 2·2

Frequency distribution of marks in Statistics of 60 students

Marks in Statistics (Class)		*Number of Students (Frequency)*	*Cumulative Frequency*
0—5	IIII	4	4
6—10	I	1	5
11—15	𝍸 II	7	12
16—20	𝍸 𝍸 I	11	23
21—25	𝍸 I	6	29
26—30	𝍸 𝍸 𝍸 I	16	45
31—35	𝍸 II	7	52
36—40	𝍸 I	6	58
41—45	I	1	59
46—50	I	1	60

This type of representation of frequencies is called a *grouped frequency distribution* or simply a grouped distribution. In the above example, 60 students have been divided into 10 groups *viz.*, 0–5, 6–10,......, 46–50. The groups are also known as *classes* and the boundary figures under the column heading 'marks' are called *class limits*. The figures on the left side of the classes are called *lower limits* while figures on the right are known as *upper limits*. Thus, for the class 21–25, 21 is the lower limit and 25 is the upper limit. The difference between the upper and lower limits of a class is called the *width* of the class or simply the *class interval*. The interval of the class 21–25 is 25–21, *i.e.*, 4. The variate value which lies mid-way between the lower and upper limits of a class is called its *mid-value* or *central value*, *e.g.*, the central value of the class 21–25 is 23. The number of individuals or observations in each class is called the *frequency* of that class or simply *class frequency*. The frequency of the class 21–25 is 6. The *cumulative frequency* corresponding to any variate value is the number of observations less than or equal to that value. The cumulative frequency corresponding to a class is the total number of observations less than or equal to the upper limit of the class, *i.e.*, it is the total of all the frequencies up to and including that class. The frequency corresponding to the upper limit of the class itself may, however, have to be left out of the total if, the class *a–b* is defined as '*a* and less than *b*'. In both the above tables, the cumulative frequencies are shown in the last column.

The following points should be kept in mind while tabulating or classifying an observed frequency distribution :

(*i*) For the sake of ease in computation and comparison, the class -intervals should, wherever possible, be of equal width. However, in some cases, *e.g.*, in the frequency distribution of the number of persons paying income-tax in different ranges of income or that of the number of children dying in different age-groups, it may be desirable to use unequal class -intervals due to greater concentration of observations in certain class-intervals.

(*ii*) The number of class-intervals should not ordinarily exceed 20 and should not, in general, be less than 10. With more than 20 classes, the computations become increasingly tedious, whereas with less than 10 classes, a great deal of accuracy may be lost.

(*iii*) The class -intervals should be defined with precision, *i.e.*, it should be made clear whether the class *a–b* means all values of the variate x for which $a \leq x < b$ or $a < x \leq b$. For continuous variates, it is desirable to take into account the accuracy of measurement. Thus if heights are measured correct to the nearest centimetre it is better to make class-intervals as 145·5–150·5 rather than 145–150.

(*iv*) The class -interval should be an integer as far as possible.

Note. A variable, which can take any numerical value within certain range is called a continuous variable. While dealing with a continuous variable, it is not possible to classify the data into groups of the type 0–5, 6–10, etc. Consider, for an example, the distribution of ages in years of students in a college. Clearly we cannot arrange the data in the age-groups 16–20, 21–25, etc. since there can be students having their ages between 20 and 21. Thus in case of a continuous variable, the data should be classified as 16–20, 20–24, 24–28, etc., it being understood that the students with ages less than 20 are included in the class 16–20. Thus a *continuous frequency distribution* has the form :

Class (Age in years)	*Frequency*
16—20	f_1
20—24	f_2
24—28	f_3
....	....

'Exclusive' and 'Inclusive' class -intervals

Class -intervals of the type $\{x : a \le x < b\} = (a, b)$ are called 'exclusive' since these classes exclude the upper limit of the class. In the above examples, the upper limits 20, 24 of the classes 16–20, 20–24 are excluded; class-intervals of the type $\{x : a \le x \le b\} = (a, b)$ are called *inclusive*, since they include the upper limit of the class. An example of this type of 'inclusive' class-intervals is given in table 2·2, page 15, where in the classes limits 5, 10, 15, of the respective classes are included in the classes 0–5, 6–10, 11-15. In order to ensure continuity of class-limits, we convert inclusive class-intervals into exclusive class-intervals after making an adjustment to the upper limits by using : adjustment $= \frac{1}{2}$ (lower-limit of succeeding class–upper-limit of this class). Thus classes would be 0·5–5·5, 5·5–10·5,... size of each class would be 5.

2·2. Graphical Representation of a Frequency Distribution. The graphical representation makes the unwieldy data readily intelligible and brings to light the salient features of the data at a glance. It makes visual comparison of data easier. It facilitates the comparison of two frequency distributions. It is also useful in locating certain *Partition Values*. (See §2·5·3).

The following types of graphs are used in representing frequency distributions :

1. Histogram. 2. Frequency Polygon and Frequency Curve 3. Cumulative Frequency Curve or the 'Ogive'.

2.2.1. Histogram. In drawing the histogram of a given grouped frequency distribution, we first mark off along the axis of x all the class-intervals on a suitable scale. With the class-intervals as bases, we draw rectangles with *areas* proportional to the frequencies of the class-intervals. For equal class-intervals, the heights of the rectangles will be proportional to the frequencies, while for unequal class-intervals, the heights will be proportional to the ratios of the frequencies to the widths of classes. Fig. 2·1 is the histogram of the following frequency distribution showing the number of boys on the registers of primary schools in a certain state :

Age (*in years*)	*No. of boys* (*in thousands*)	*Age* (*in years*)	*No. of boys* (*in thousands*)
2—5	150	13—14	496
5—11	3066	14—15	143
11—12	497	15—17	162
12—13	477		

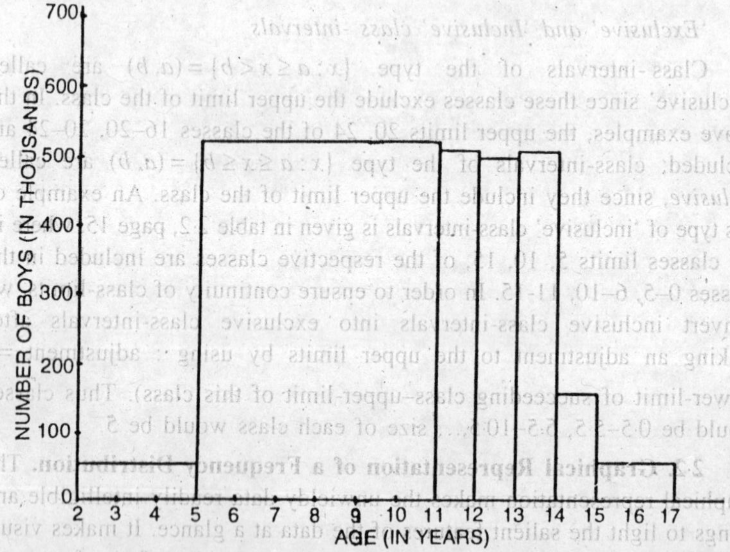

Fig. 2·1. Histogram of frequency distribution in the Table given above.

Here we first construct a rectangle with the class 2—5 as base, width $= 5 - 2 = 3$ and height equal to $\frac{150}{3}$ *i.e.* 50. Adjacent to this rectangle, we construct another rectangle with class 5–11, *i.e.* width $= 11 - 5 = 6$ as base and $\frac{3066}{6}$ *i.e.*, 511 as its height. Proceeding in this way, we get a set of rectangles constituting the histogram.

Notes 1. In case, the grouped frequency distribution is not continuous, it should first be converted into a continuous distribution by using *adjustment* as explained on page 17 and then the histogram be drawn for the adjusted grouped frequency distribution.

2. To draw the histogram for an ungrouped frequency distribution of a variate, we shall have to assume that the frequency corresponding to the variate value x is spread over the interval $x - \frac{1}{2}h$ to $x + \frac{1}{2}h$, where h is the jump from one value to the next value.

3. The histogram makes it appear as if the frequency in each class is uniformly distributed over that class which, in general, need not be true. The histogram is, however, particularly appropriate to cases in which the frequency changes rapidly or the class-intervals are not of the same width.

2·2·2. Frequency Polygon and Frequency Curve. For an ungrouped frequency distribution, the frequency polygon is obtained by plotting points with abscissae as the variate values and the ordinates as the corresponding frequencies and joining the plotted points by means of straight lines. For a grouped frequency distribution, the abscissae of the points are the mid-values of the class-intervals.

For equal class-intervals, the frequency polygon can be obtained by joining the middle points of the upper sides of the adjacent rectangles of the histogram by means of straight lines.

If the class-intervals are of small widths, the frequency polygon can be approximated by a smooth *frequency curve*. The frequency curve can be obtained by drawing a smooth free-hand curve through the vertices of the frequency polygon. In Fig. 2.2., the frequency polygon for the continuous frequency distribution given in table 2·2 has been shown in dotted lines.

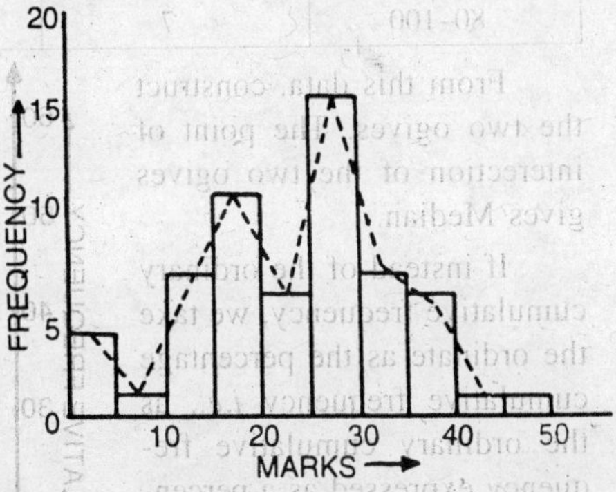

Fig. 2·2. Histogram and Frequency Polygon for the distribution given in Table 2·2, page 15.

2·2·3. Cumulative Frequency Curve or the 'Ogive'. We first form the cumulative frequency table and then obtain the curve by plotting the points. The curve so obtained is called a cumulative frequency curve or an ogive (pronounced ogive). There are two types of ogives.

(1) **Less than Ogive.** Plot the points with the *upper limits* of the classes as abscissae and the corresponding less than cumulative frequency as ordinates. The points are joined by a free-hand smooth curve to give less than *the cumulative frequency curve or the less than Ogive*. It is a rising curve. In Fig. 2·3, the ogive of the frequency distribution given in table 2·2 has been shown.

(*ii*) **More than Ogive.** Plot the points with the *lower limits* of the classes as abscissae and the corresponding more than cumulative frequency as ordinates. Join the points by a free-hand smooth curve to get 'the more than ogive'. It is a falling curve.

Draw the two ogives for the following frequency distribution of the weekly wages of ('less than' and 'more than') number of workers :

Weekly Wages	0–20	20–40	40–60	60–80	80–100
Number of Workers	41	51	64	38	7

Hence find the value of Median. (*B. Com., Delhi*)

[**Hint.** First make a cumulative frequency table as follows :–

Weekly Wages	*No. of Workers*	*Cumulative frequency*	
		less than	*more than*
0–20	41	41	201
20–40	51	41 + 51 = 92	194
40–60	64	92 + 64 = 156	156
60–80	38	156 + 38 = 194	92
80–100	7	194 + 7 = 201	41

From this data, construct the two ogives. The point of interection of the two ogives gives Median.

If instead of the ordinary cumulative frequency, we take the ordinate as the percentage cumulative frequency *i.e.*, as the ordinary cumulative frequency expressed as a percentage of the total frequency, we get the *percentage cumulative frequency curve* which is useful for comparing different frequency distributions as the frequency distributions are thus adjusted to uniform scale.

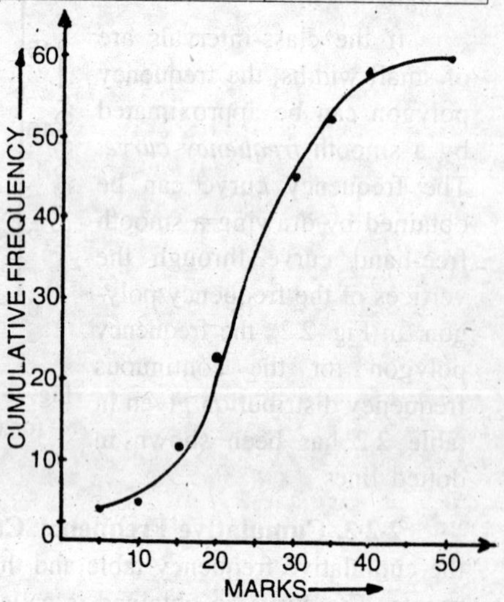

Fig. 2·3. (*a*) Ogive of the distribution given in Table 2·2

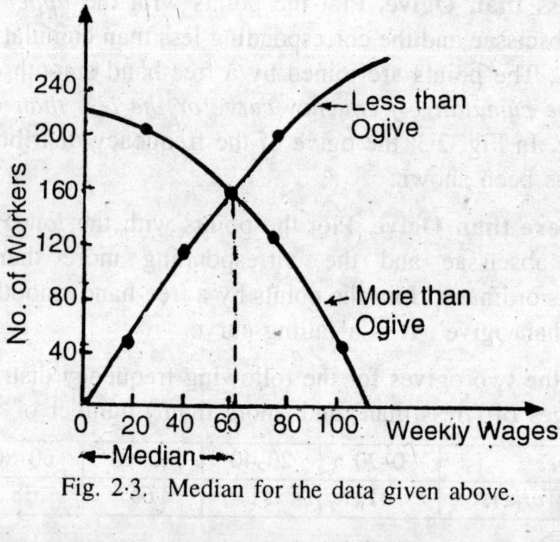

Fig. 2·3. Median for the data given above.

Note. Median is the abscissa of the point of intersection of the two ogives. In Fig. 2·3 (*b*), it is the point 'X' = 42·65.

2·3. Forms of Frequency Curves. The following are types of curves which we generally get in the graphical representation of frequency distributions.

(*a*) *Symmetrical Curve*. A frequency curve is said to be symmetrical if it can be folded along a vertical line (ordinate) so that the two halves of the figure coincide. An important type of a symmetrical curve is a bell-shaped curve which has got a single smooth hump in the middle and tails off' gradually at either end. It has a maximum at the centre the observations equidistant from the central maximum have the same frequency. Thus the following is a symmetrical frequency table.

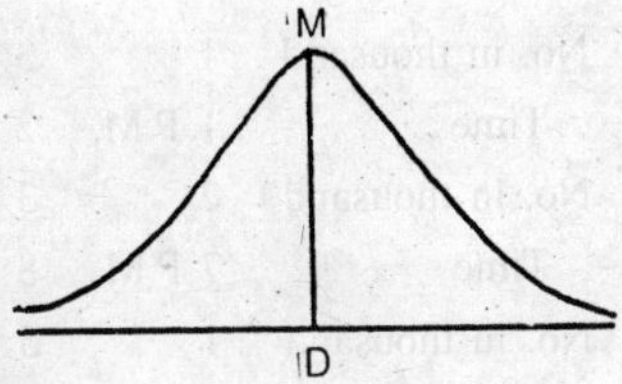

Fig. 2·4. Symmetrical curve

Variable :	1	2	3	4	5
Frequency :	3	5	9	5	3

(*b*) *Moderately Asymmetrical or Skew Curve*. A curve is said to be skew if there is no symmetry *i.e.* if it lacks in symmetry. In a skew curve, observations tend to pile up at one or the other end of the curve. Thus a skew curve may have a longer tail to positive (right) side in which case it is said be positive skew curve or to the negative (left) side in which case it is known as negative skew curve.

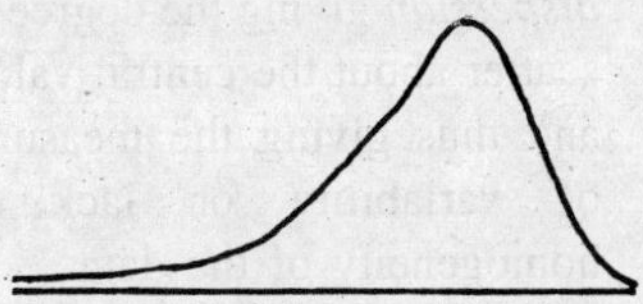

Fig. 2·5. (*a*) Negative skew curve.

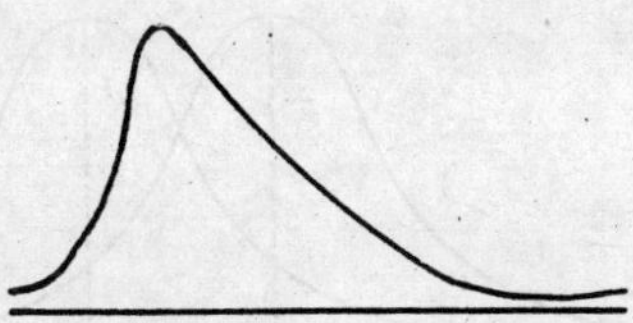

Fig. 2·5. (*b*) Positive skew curve.

(*c*) *Extremely Asymmetrical* (*or J-shaped*) *and U-shaped Curves*. Such curves are so called as they have the shape of a '*J*' or '*U*'.

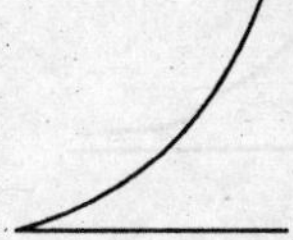

Fig. 2·6. (*a*) Negative J-shaped curve

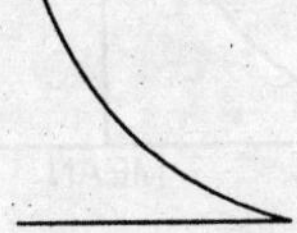

Fig. 2·6. (*b*) Positive J-shaped curve.

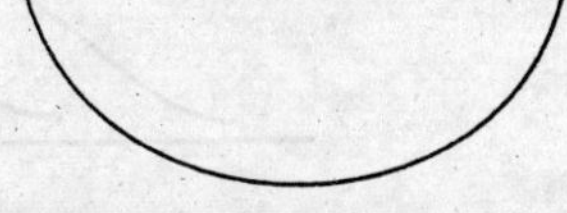

Fig. 2·7. *U*-shaped curve

The curve representing the distribution of assesses of income-tax will be a positive *J*-shaped.

U-shaped curve exhibits a maximum frequency at the ends of the range and a minimum towards the centre, as in Fig. 2·7.

Ex. The number of persons travelling by public buses at different times in the day in a capital city is given below :

Time	7 A.M.	8 A.M	9 A.M.	10 A.M.	11 A.M.	12 NOON
No. in thousand	1	3	5	9	3	1
Time	1 P.M.	2 P.M.	3 P.M.	4 P.M.	5 P.M.	6 P.M.
No. in thousand	2	3	5	9	5	1
Time	7 P.M.	8 P.M.	9 P.M.			
No. in thousand	1	5	2			

What is the form of the frequency curve between

(*i*) 7 A.M. and 10 A.M. (*ii*) 7 A.M. and 1 P.M. (*iii*) 10 A.M. and 5 P.M. (*iv*) 5 P.M. and 6 P.M. (*v*) 1 P.M. and 7 P.M (*vi*) 7 P.M., and 9 P.M. ?

2·4. Comparison of Frequency Distributions. In statistical work, it is frequently necessary to compare two frequency distributions. If they are of different types a precise comparison is difficult and is usually not required. If, however, they are of the same type a comparison can be made in terms of the values of the following four types of measure :

(*a*) *Measure of location or central tendency* gives the whereabouts of some central value, *e.g.*, the maximum and thus, to say, locate the distribution.

(*b*) *Measures of dispersion* giving the degree of scatter about the central value, and thus giving the measures of variability or lack of homogeneity of the data.

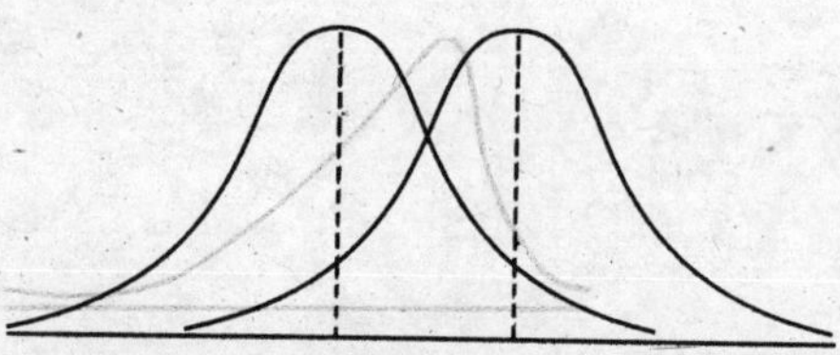

Fig. 2·8. (*a*). Difference in central value.

(*c*) *Measures of skewness* measuring the degree of departure from symmetry.

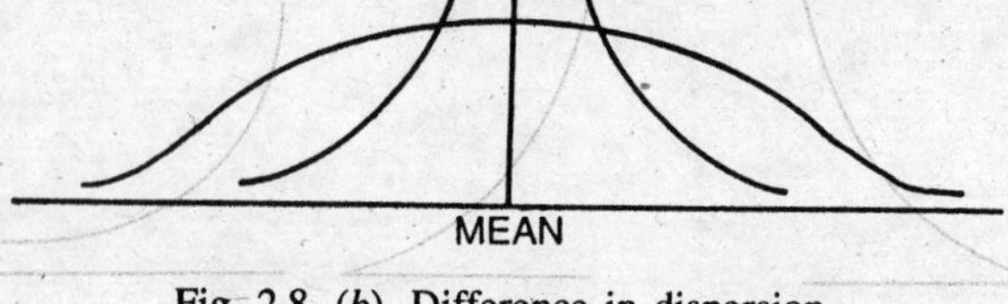

Fig. 2·8. (*b*). Difference in dispersion.

(*d*) *Measure of kurtosis*, measuring the degree of 'flatness' of the 'top' as compared with the 'normal' curve (defined in chapter 8).

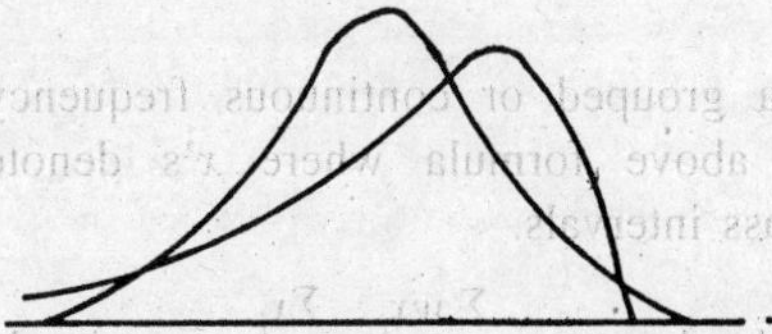

Fig. 2·8 (*c*). Difference in symmetry.

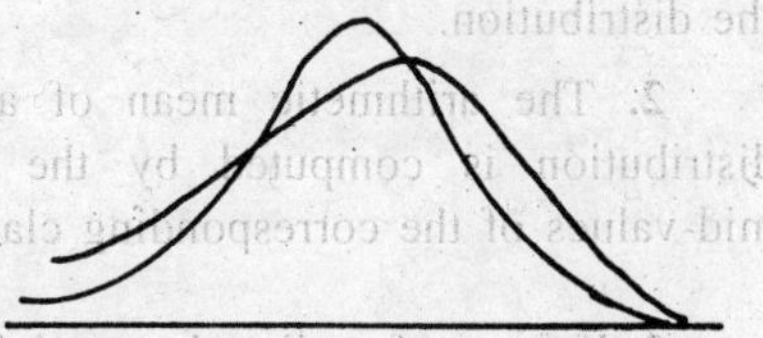

Fig. 2·8 (*d*). Difference in kurtosis.

The measures of dispersion, skewness and kurtosis will be discussed in the next chapter.

2·5. Measures of Location or Measures of Central Tendency. The various types of measures of location, in common use, are :

(1) Arithmetic Mean or simply the Mean.

(2) Geometric Mean. (3) Harmonic Mean.

(4) Median and Quartiles. (5) Mode.

2·5·1. Arithmetic Mean. Let the variate X take the values $x_1, x_2, \ldots\ldots, x_n$. The arithmetic mean of the set of observations x_i, $i = 1, 2, \ldots, n$ denoted by $\bar{x}$, is defined by

$$\bar{x} = \frac{x_1 + x_2 + \ldots\ldots + x_n}{n} = \frac{1}{n} \sum_{i=1}^{n} x_i \qquad \ldots(2 \cdot 1)$$

If the variate-values are not of equal 'importance', we may attach to them 'weights' $w_1, w_2, \ldots\ldots, w_n$ as measures of their importance, so that the weighted mean, $\bar{x}_w$, is defined by

$$\bar{x}_w = \frac{w_1 x_1 + w_2 x_2 + \ldots + w_n x_n}{w_1 + w_2 + \ldots + w_n} = \frac{\sum_{i=1}^{n} w_i x_i}{\sum_{i=1}^{n} w_i} \qquad \ldots(2 \cdot 2)$$

If the variate-value x_1 occurs f_1 times, x_2 occurs f_2 times and so on,

then $$\bar{x} = \frac{f_1 x_1 + f_2 x_2 + \ldots + f_n x_n}{f_1 + f_2 + \ldots + f_n} = \frac{1}{N} \sum_{i=1}^{n} f_i x_i \qquad \ldots(2 \cdot 3)$$

where $$N = f_1 + f_2 + \ldots\ldots + f_n = \sum_{i=1}^{n} f_i = \text{total frequency.}$$

This formula expresses that the arithmetic mean $\bar{x}$ is the weighted mean of the values $x_1, x_2, \ldots\ldots, x_n$ whose weights are the respective frequencies $f_1, f_2, \ldots\ldots, f_n$.

Notes 1. The arithmetic mean may be regarded as the x-co-ordinate of the centre of gravity of a plane of the same shape as the histogram of the distribution.

2. The arithmetic mean of a grouped or continuous frequency distribution is computed by the above formula where x's denote mid-values of the corresponding class intervals.

3. If $w_i = w$ for all i, then weighted mean $= \frac{\Sigma\, wx_i}{\Sigma\, w} = \frac{\Sigma x_i}{n} = \bar{x} =$ simple mean. However, if smaller weights are attached to smaller observations and larger weights to larger observations, then weighted mean has a higher value than the simple mean. In the reverse case, if the smaller weights are attached to larger items and also large weights attached to smaller items, then the weighted mean will be smaller in value than the simple mean.

Example 2·1 (*a*). *Find the arithmetic mean of the following marks obtained by 10 students in Statistics.*

52, 40, 70, 75, 43, 40, 35, 65, 48, 62

(*b*) *Given the following frequency distribution, calculate the mean.*

Monthly Wages (in Rs.)	*No. of Workers*	*Monthly Wages (in Rs.)*	*No. of Workers*
125—175	*2*	*375—425*	*4*
175—225	*22*	*425—475*	*6*
225—275	*19*	*475—525*	*1*
275—325	*14*	*525—575*	*1*
325—375	*3*		

(*a*) Here $\bar{x} = \frac{\Sigma x}{n} = (52 + 40 + 70 + 75 + 43 + 40 + 35 + 65 + 48 + 62) \div 10$

$$= \frac{530}{10} = 53 \text{ marks.}$$

(*b*) The calculation of Σfx is shown in the following table ;

Mid-value x	*No. of workers* f	*Product* fx	*Mid-value* x	*No. of workers* f	*Product* fx
150	2	300	450	6	2700
200	22	4400	500	1	500
250	19	4750	550	1	550
300	14	4200	Total	72	20050
350	3	1050			
400	4	1600			

Hence $\bar{x} = \frac{1}{N} \sum_{i=1}^{n} f_i x_i = \text{Rs. } \frac{20050}{72} = \text{Rs. } 278{\cdot}47$

Change of Origin and Scale. Shifting the origin to a point whose abscissa is a, the equation (2·3), page 23 takes the form

$$\bar{x} - a = \frac{1}{N} \sum_{i=1}^{n} f_i(x_i - a)$$

or,
$$\bar{x} = a + \frac{1}{N} \sum_{i=1}^{n} f_i d_i, \text{ where } d_i = x_i - a \qquad ...(2{\cdot}4)$$

This formula gives a Short Method for the computation of the arithmetic mean. The arbitrary number 'a' is called the provisional or assumed or working arithmetic mean.

In case of frequency tables having equal class-intervals, say of width h, it is sometimes convenient to use a different scale. We take a new unit $(1/h)$ times the original unit and the given distribution is changed to a new distribution, called 'derived' distribution.

Thus, if u be the new variate, we have

$$u = (x - a)/h \text{ or } x = a + h u \qquad ...(2{\cdot}5)$$

Multiply both sides by f and then sum to obtain $\Sigma fx = \Sigma af + \Sigma hfu$,

Dividing both sides by Σf, we get

$$\bar{x} = a + h \frac{\Sigma fu}{\Sigma f} = a + h\,\bar{u} \qquad ...(2{\cdot}6)$$

where $\bar{u}$ is the mean of the distribution of u.

Thus mean of the given distribution = arbitrary number plus h times the mean of the derived distribution.

Example 2·2. *Solve Example 2·1 (b), page 24 by applying formula (2·6).*

Let us take 300 as the provisional mean, *i.e.* a = 300 and h = 50 so that $u = (x - 300)/50$.

x	f	Derived distribution $u \left(= \frac{x-a}{h}\right)$	Product $f \times u$
150	2	− 3	−6
200	22	− 2	−44
250	19	− 1	−19
300	14	0	0
350	3	1	3
400	4	2	8

450	6	3	18
500	1	4	4
550	1	5	5
	$\Sigma f = 72$		$\Sigma fu = -31$

Hence $\bar{x} = a + h\,\bar{u} = 300 + 50 \times \dfrac{(-31)}{72} = \text{Rs. } 278{\cdot}47.$

Properties of the Arithmetic Mean.

I. *The sum of the deviations of values x_i, $i = 1, 2, ..., n$ from their mean $\bar{x}$ is zero, f_i being the frequency of x_i.*

Proof. We have $\sum_{i=1}^{n} f_i(x_i - \bar{x}) = \sum_{i=1}^{n} f_i x_i - \bar{x} \sum_{i=1}^{n} f_i = N\bar{x} - N\bar{x}$

$$= 0, \quad N = \sum_{i=1}^{n} f_i .$$

II. *The sum of squares of deviations of a set of values is minimum when taken about mean.*

Proof. Let x_i, $i = 1, 2, ...,n$ be the set of observations or values with frequencies $f_1, f_2, ...f_n$, and a be the arbitrary number from which the deviations are taken. Then let

$$S = \sum_{i=1}^{n} f_i\,(x_i - a)^2 .$$

We want to show that S is minimum when $a = \bar{x}$. Given x_i, S is a function of a only. $\quad\therefore\quad$ For S to be minimum,

$$\frac{dS}{da} = 0 \text{ and } \frac{d^2S}{da^2} > 0 \text{ for the value of } a \text{ given by } \frac{ds}{da} = 0$$

Now $\dfrac{dS}{da} = -2\,\Sigma f_i\,(x_i - a) = -2\;[\Sigma f_i\,x_i - a\,\Sigma f_i] = -2\,(N\bar{x} - aN)$

$$\frac{d^2S}{da^2} = -2\,(-N) = 2N > 0$$

Thus S is minimum when $a = \bar{x}$. Hence proved.

III. Mean of the combined distribution. *If $\bar{x}_1, \bar{x}_2, ...\bar{x}_k$, be the arithmetic means of k distributions with respective frequencies n_1, n_2,..., n_k, then the mean $\bar{x}$ of the whole distribution with frequency $N \left(= \sum_{i=1}^{k} n_i \right)$ is given by the formula* $\quad \bar{x} = \dfrac{1}{N} \sum_{i=1}^{k} n_i\,\bar{x}_i. \quad$ ...(2·7)

Proof. Let x_{ij} be an observation of the i-th distribution and let its frequency be f_{ij}, then $\bar{x}_i = \dfrac{\sum_j f_{ij} x_{ij}}{\sum_j f_{ij}} = \dfrac{\sum_j f_{ij} x_{ij}}{n_i}$

If $\bar{x}$ denotes the mean of all the observations, then

$$\bar{x} = \frac{\sum_i \sum_j f_{ij} x_{ij}}{\sum_i \sum_j f_{ij}} = \frac{\sum_{i=1}^{k} n_i \bar{x}_i}{\sum_{i=1}^{k} n_i} = \frac{1}{N} \sum_{i=1}^{k} n_i \bar{x}_i.$$

2·5·2. Geometric Mean. If the n non-zero and positive variate-values $x_1, x_2, \ldots\ldots, x_n$ occur $f_1, f_2, \ldots\ldots, f_n$ times respectively, then geometric mean G of the set of observations is defined by

$$G = \left[x_1^{f_1} \,.\, x_2^{f_2} \ldots x_n^{f_n} \right]^{\frac{1}{N}} = \left[\prod_{i=1}^{n} x_i^{f_i} \right]^{\frac{1}{N}}$$

i.e. $$\log G = \frac{1}{N} \sum_{i=1}^{n} (f_i \log x_i). \text{ where } N = \sum_{i=1}^{n} f_i \qquad \ldots(2\cdot 8)$$

Thus *the logarithm of the geometric mean is the weighted mean of the different values* $\log x_i$ *when weights are the frequencies* f_i.

Again, from (2·8), $$G = \text{Anti. log} \left[\frac{1}{N} \sum f_i \log x_i \right] \qquad \ldots(2\cdot 8(a))$$

G.M. of combined series. *If* G_1 *and* G_2 *are the geometric means of the two series with* n_1 *and* n_2 *observations respectively, then the geometric mean* G *of the combined series is given by*

$$\log G = \frac{n_1 \log G_1 + n_2 \log G_2}{n_1 + n_2}$$

Proof. Let x_{1i}, $(i = 1, 2, \ldots, n_1)$ and x_{2j}, $(j = 1, 2, \ldots, n_2)$ be n_1 and n_2 observations of the two series respectively. Then

$$G_1 = (x_{11} \,.\, x_{12} \ldots x_{1n_1})^{1/n_1} \Rightarrow \log G_1 = \frac{1}{n_1} \sum_{1}^{n_1} \log x_{1i}$$

$$G_2 = (x_{21} \,.\, x_{22} \ldots x_{2n_2})^{1/n_2} \Rightarrow \log G_2 = \frac{1}{n_2} \sum_{1}^{n_2} \log x_{2i}$$

Now $G = (x_{11}\, x_{22} \ldots x_{1n_1}\, x_{21}\, x_{22} \ldots x_{2n_2})^{1/(n_1 + n_2)}$, by def.

$$\Rightarrow \log G = \frac{1}{n_1 + n_2} \left[\log (x_{11}\, x_{12} \ldots x_{1n_1}\, x_{21}\, x_{22} \ldots x_{2n_2}) \right]$$

$$\Rightarrow G = \frac{1}{n_1 + n_2}\left[\sum_{i=1}^{n_1} \log x_{1i} + \sum_{1}^{n_2} \log x_{2j}\right] = \frac{n_1 \log G_1 + n_2 \log G_2}{n_1 + n_2}.$$

N.B. The result can be generalized to the case of n series.

2·5·3. Harmonic Mean. The harmonic mean, H, of n non-zero different variate-values x_i with frequencies f_i is given by

$$\frac{1}{H} = \frac{\sum_{i=1}^{n}\left(f_i \times \frac{1}{x_i}\right)}{\sum_{i=1}^{n} f_i} = \frac{1}{N}\sum_{i=1}^{n}\left(f_i \times \frac{1}{x_i}\right) \quad \ldots(2\cdot9)$$

Thus *the harmonic mean of the variate-values is the reciprocal of the arithmetic mean of their reciprocals.*

2·5·4. Median. Quartiles. *Partition Values and their graphical location* The *median* is defined as that value of the variate for which the total of the frequencies above (or below) that variate value is half the overall total frequency. Thus if the n values of the variate in an ungrouped frequency distribution are arranged in ascending (or descending) order of magnitude, the median is the middle value, if n is odd; it is taken to be the arithmetic mean of the two middle values if n is even, though in this case any value between the two middle values will satisfy the definition and this median will not be uniquely determined.

Partition values are the values of the variate which divide the total frequency into a number of equal parts. Some important partition values are *quartiles*, *deciles* and *percentiles.*

Quartiles are those variate values which divide the total frequency into four equal parts ; deciles and percentiles divide into ten and one hundred equal parts respectively. Suppose the values of the variate have been arranged in ascending order of magnitude, then the value of the quartile having the position between the lower extreme and the median, is the *lower* or *first quartile* and is denoted by Q_1 and that between the median and the upper extreme is the *upper* or *third quartile* and is denoted by Q_3. The median is thus one of the three quartiles and is sometimes denoted by Q_2. Also clearly median is the fifth decile D_5 and fiftieth percentile P_{50}. The difference between the upper and lower quartiles, *i.e.*, $Q_3 - Q_1$ is called the *inter-quartile range*. For a grouped frequency distribution, the quartiles, the deciles and the percentiles are given by

$$Q_i = l + \frac{\left(\frac{iN}{4} - C\right)}{f} \times h, \; i = 1, 2, 3,$$

$$\left.\begin{aligned} D_j &= l + \frac{\left(\dfrac{jN}{10} - C\right)}{f} \times h, j = 1, 2, ..., 9 \\ P_k &= l + \frac{\left(\dfrac{kN}{100} - C\right)}{f} \times h, k = 1, 2, ..., 99 \end{aligned}\right\}^{*} \quad ...(2{\cdot}10)$$

where l is the lower limit of the class in which the particular quartile lies, f is the frequency of this class. h is the width of this class, C is the cumulative frequency up to and including the class preceding the class in which the particular quartile lies and N is the total frequency.

Example 2·3. *Calculate the three quartiles for the following frequency distribution of the number of marks obtained by 49 students in a class.*

Marks group	*No. of students*	*Marks group*	*No. of students*
5—10	*5*	*25—30*	*5*
10—15	*6*	*30—35*	*4*
15—20	*15*	*35—40*	*2*
20—25	*10*	*40—45*	*2*

The following is the cumulative frequency table :

Class	*Frequency*	*Cumulative Frequency*
5—10	5	5
10—15	6	11
15—20	15	26
20—25	10	36
25—30	5	41
30—35	4	45
35—40	2	47
40—45	2	49
	Total 49	

To find Q_1, we have $\frac{N}{4} = \frac{49}{4}$ and this clearly lies in the class 15—20.

$\therefore \quad l = 15,\ h = 15 - 10 = 5,\ C = 11,\ f = 15.$

---

* The cumulative frequency up to the ith quartile Q_i is $iN/4$ and, therefore, $iN/4 - C$ is frequency between the lower limit of the quartile class and Q_i. *Assuming* that the frequency f in the quartile is distributed uniformly over the length h, which will give us this frequency $\frac{iN/4 - C}{f} \times h$ and hence Q_i is given by (2·10). Similarly, the formulas for D_j and P_j can be obtained.

Hence $Q_1 = 15 + \dfrac{\frac{49}{4} - 11}{15} \times 5 = 15{\cdot}47$ marks.

For median, $\dfrac{N}{2} = 24\frac{1}{2}$ and, therefore, median class is 15—20.

Hence median $= 15 + \dfrac{\frac{49}{2} - 11}{15} \times 5 = 19{\cdot}50$ marks.

To find Q_3, we have $\dfrac{3N}{4} = \dfrac{147}{4}$ and this lies in the class 25—30.

$\therefore\ l = 25,\ C = 36,\ f = 5,\ h = 5.$

Hence $Q_3 = 25 + \dfrac{\left(\frac{147}{4} - 36\right)}{5} \times 5 = 25{\cdot}75$ marks.

Graphical Location of Partition Values. For finding graphically the values of the median, quartiles and other partition values we must first draw the cumulative frequency curve or the *Ogive*. To determine the value of the median we mark a point along the *y*-axis corresponding to *N*/2. From this point, draw a line parallel to the *x*-axis and find the point where this line cuts the curve. Draw a line perpendicular to *x*-axis from this point. The distance between the origin and the foot of the perpendicular is the median.

For finding the value of Q_1 take a point on the *y*-axis corresponding to N/4 and then proceed as above.

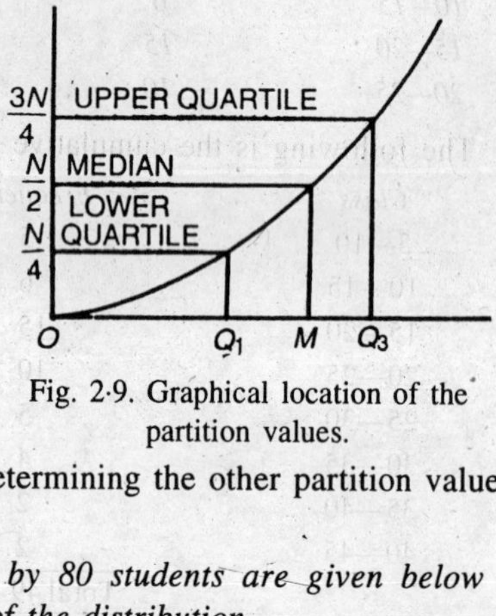

Fig. 2·9. Graphical location of the partition values.

The method is applicable for determining the other partition values also from the Ogive.

Example 2·4. *Marks obtained by 80 students are given below : Determine the value of the median of the distribution.*

Marks	*Frequency*	*Marks*	*Frequency*
0—10	*3*	*30—40*	*30*
10—20	*9*	*40—50*	*18*
20—30	*15*	*50—60*	*5*

The cumulative frequency table is as under :

Marks	*Frequency*	*Cumulative Frequency*	*Marks*	*Frequency*	*Cumulative Frequency*
0—10	3	3	30—40	30	57
10—20	9	12	40—50	18	75
20—30	15	27	50—60	5	80

Plot the points (10, 3) ; (20, 12) ; (30, 27) ; (40, 57) ; (50, 75) ; (60, 80). Join them by a free hand smooth curve to obtain the Ogive.

To determine the median, we have $N/2 = 40$.

From the point 40 on the y-axis, draw a straight line parallel to the axis of x cutting the Ogive at a point. Read the abscissa of this point. It is found out to be 34·3 marks (see Fig. 2·10).

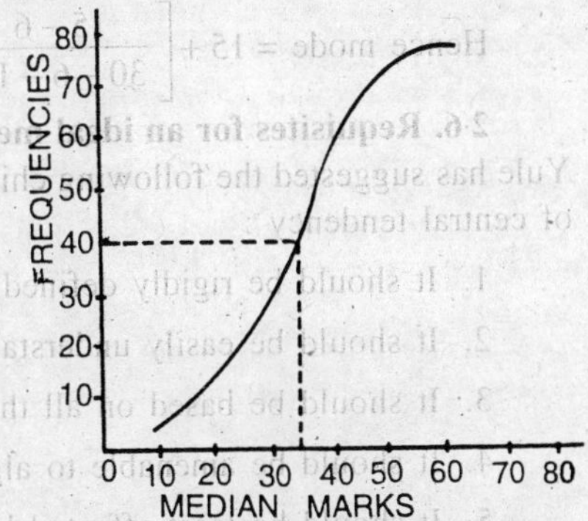

Fig. 2·10. Graphical location of the median of the distribution of marks obtained by 80 students.

Verification.

$$\text{Median} = 30 + \frac{40-27}{30} \times 10 \text{ marks}$$

$$= 30 + \frac{13}{3} = 34{\cdot}3 \text{ marks.}$$

Ex. 1. Determine graphically the quartiles of the distribution given in Example 2·4.

Ex. 2. Draw a graph showing the number of students getting more than 0, 10, 20, 30, 40, 50 marks. Show that its intersection with the Ogive determines the median. Explain also how this curve can be used to determine the quartiles and the deciles.

2·5·5. Mode. The mode or modal value of the distribution is that value of the variate for which the frequency is maximum.

For a frequency distribution, the mode is computed by the following formula :

$$^*\text{Mode} = l + \left[\frac{f_m - f_1}{2f_m - f_1 - f_2} \times h\right] \quad \ldots(2{\cdot}11)$$

where l is the lower limit of the modal class (*i.e.*, class having maximum frequency), f_m is the maximum frequency, f_1 and f_2 are the frequencies of the classes preceding and following the modal class.

Example 2·5. *Find the mode of the distribution given in Example 2·3. page 29.*

Clearly 15—20 is the modal class since its frequency in maximum.

* For computing the mode, we divide the modal class in the ratio $f_m - f_1 : f_m - f_2$ Another formula for calculating the mode is :

$$\text{Mode} = l + \frac{f_2}{f_1 + f_2} \times h$$

where f_1 and f_2 are the frequencies of the classes preceding and following the modal class.

$\therefore \quad l = 15, h = 5, f_1 = 6, f_2 = 10, f_m = 15,$

$$\text{Hence mode} = 15 + \left[\frac{15-6}{30-6-10} \times 5\right] = 18{\cdot}21 \text{ marks.}$$

2·6. Requisites for an ideal measure of central tendency. G. Udny Yule has suggested the following chief characteristics of an ideal measure of central tendency :

1. It should be rigidly defined.
2. It should be easily understandable and easy to calculate.
3. It should be based on all the observations.
4. It should be amenable to algebraic treatment.
5. It should be least affected by fluctuations of sampling.
6. It should not be affected much by the extreme values.

Exercises 2

1. Make a frequency table having grades of wages with class-interval of two rupees each from the following data of daily wages received by 30 labourers in a certain factory. Daily wages in rupees are :

14, 16, 16, 14, 22, 13, 15, 24, 12, 23, 14, 20, 17, 21, 22
18, 18, 19, 20, 17, 16, 15, 11, 12, 21, 20, 17, 18, 19, 23

Ans.

Wages in rupees	*No. of labourers*	*Wages in rupees*	*No. of labourers*
11—13	3	19—21	5
13—15	4	21—23	4
15—17	5	23—25	3
17—19	6		

2. The weights in grams of 50 apples picked out at random from a consignment are as follows :

106, 107, 76, 82, 109, 107, 115, 93, 187, 95, 123, 125, 111,
92, 86, 70, 126, 68, 130, 129, 139, 119, 115, 128, 100, 186,
84, 99, 113, 204, 111, 141, 136, 123, 90, 115, 98, 110, 78,
90, 107, 81, 131, 75, 84, 104, 110, 80, 118, 82.

Form the grouped frequency table by dividing the variate range into intervals of equal width, each corresponding to 20 grams, in such a way that the mid-value of the first class corresponds to 70 grams.

Ans.

Weight in grams	*No. of apples*	*Weight in gram*	*No. of apples*
60—80	5	140—160	1
80—100	14	160—180	0
100—120	17	180—200	2
120—140	10	200—220	1

3. (*a*) Show how the limit of classes are defined in grouping a given data.

(*b*) Mention the considerations which help you in deciding the class-limits and class-intervals ?

(*c*) What are the principles governing the choice of (1) the number of class-intervals, (2) the length of the class-intervals and (3) the mid-points of the class-intervals ?

4. Form an ordinary frequency table from each of the following tables :

(*a*) *Marks*		*No. of students*	(*b*) *Marks*		*No. of students*
Below	10	3	Above	0	30
"	20	8	"	10	26
"	30	17	"	20	21
"	40	20	"	30	14
"	50	22	"	40	10
			"	50	0

Ans. (*a*)

Marks	*No. of students*
0—10	3
10—20	5
20—30	9
30—40	3
40—50	2
Total	22

(*b*)

Marks exceeding		*Not exceeding*	*No. of students*
0	—	10	4
10	—	20	5
20	—	30	7
30	—	40	4
40	—	50	10
50 and above			0
		Total	30

5. (*a*) Write a short note on the use of graphical methods in Statistics.

(*b*) Write short notes on the following :

(*i*) Graphical representation of distributions. (*ii*) Histogram.

(*iii*) Ogive. (*iv*) Frequency polygon and frequency curve.

(*c*) Is there any relation between frequency polygon and frequency curve ?

6. Draw the histogram and frequency polygon for the data in Ex. 1. and 2, page 32.

7. Represent the following distribution by a frequency polygon :

Source	*Frequency*	*Source*	*Frequency*
90—99	2	50—59	14
80—89	12	40—49	3
70—79	22	30—39	1
60—69	20	20—29	1

8. The following table shows the frequency distribution for the number of students per teacher in 750 colleges and professional schools :

Students x	1	4	7	10	13	16	19	22	25	28
Frequency f	7	46	165	195	189	89	28	19	9	3

Draw a histogram and a frequency polygon.

[**Hint.** 1, 4, 7, etc., are the mid-points of the classes 0-2, 3-5, 6-8, etc.]

9. Draw a histogram representing the following frequency distribution :

Wages (Monthly)	*Frequency*	*Wages (Monthly)*	*Frequency*
Rs.		Rs.	
150	2	350	9
200	20	400	4
250	26	450	3
300	16		

(**Hint.** 150, 200,..., 450 are the mid-points of the classes 125–175, 175—225,...., 425—475]

10. Draw a histogram and a frequency polygon for the following distribution :

Degree of cloudiness	*Frequency*	*Degree of cloudiness*	*Frequency*
x	f	x	f
10	580	4	45
9	150	3	68
8	196	2	75
7	75	1	130
6	55	0	220
5	40		

[**Hint.** Take x as the central value and then plot.]

11. Draw a histogram for the data in Q. 13.

12. Draw the histogram and the frequency curve for the following distribution of 727 deaths from cholera (for various ages) classified according to age :

Age in years	*Number of deaths*	*Age in years*	*Number of deaths*
0—5	322	45—	6
5—	213	50—	7
10—	70	55—	5
15—	27	60—	0
20—	26	65—70	1
25—	17		Total 77
30—	12		
35—	11		
40—	10		

13. Draw a cumulative frequency curve for the following data :

Marks	*No. of students*	*Marks*	*No. of students*
0—4	4	12—16	8
4—8	6	16—20	4
8—12	10		

14. (*a*) Construct an Ogive from the following data relating to corn stalks (Height in feet). Locate graphically the median, quartiles and deciles :

Exceeding	*Not Exceeding*	*No. of corn stalks*
3	4	92
4	5	211
5	6	612
6	7	1802
7	8	2029
8	9	913
9	10	243

(*b*) Draw a cumulative graph of the following distribution showing the monthly wages of a group of workmen.

Wages (Rs.)	*No. of workmen*	*Wages (Rs.)*	*No. of workmen*
200	8	250	25
210	10	260	15
220	11	270	9
230	16	280	6
240	25		

15. The following marks were given to a batch of candidates :

66, 62, 45, 59, 32, 51, 56, 60, 51, 49,
25, 42, 54, 54, 58, 70, 43, 58, 50, 52,
38, 67, 50, 59, 48, 65, 71, 30, 46, 55,
82, 51, 63, 45, 53, 40, 36, 56, 70, 52,
67, 55, 57, 30, 63, 42, 74, 58, 44, 55.

Draw a cumulative frequency curve and from this curve, find the proportion of candidates receiving more than 50 marks.

16. Draw an Ogive curve showing the number of candidates obtaining marks higher than assigned value from the following data :

Marks Obtained	0–10	11–20	21–30	31–40	41–50	51–60	61–70	71–80
No. of candidates	33	108	221	439	529	495	322	153

17. The following table gives the frequency distribution of marks in two examination papers :

Class	*Frequency* A	B	*Class*	*Frequency* A	B
11—15	8	5	56—60	53	65
16—20	15	16	61—65	48	48
21—25	39	24	66—70	45	32
26—30	47	39	71—75	39	25
31—35	52	50	76—80	32	18
36—40	53	61	81—85	23	13
41—45	54	68	86—90	17	9
46—50	56	63	91—95	13	9
51—55	59	72	96—100	8	5

Draw the cumulative frequency curve of mark distribution and answer the following questions :

(*a*) For paper A, if 30% of the candidates are to be failed, what is the pass mark ?

(*b*) For paper B, if pass mark is 33, what per cent of candidates fail ?

[**Hint.** (*a*) Divide the ordinate of the highest point of the Ogive into 100 equal divisions. Through the 30th division, draw a horizontal line and find its intersection with the curve. The abscissa of this point is the required pass mark. [**Ans.** (*a*) 39, (*b*) 18]

18. Give one illustration each of the type of data for which you would expect the frequency curve to be :

(*i*) fairly symmetrical, (*ii*) positively skewed,

(*iii*) negatively skewed, (*iv*) J-shaped,

19. (*a*) What are the various measures of location of a distribution and to what purposes are they used ?

(*b*) In what circumstances would you consider the Arithmetic Mean, the Geometric Mean, the Harmonic Mean respectively, the most suitable statistic to describe the central tendency of a distribution ?

Ans. 19.(*b*). Which Average to be used ?

1. A.M.is preferred when all the items are to be taken into account.

2. G.M. is useful in averaging ratios and precentages and also in finding average rates of increase or decrease.

3. H.M. is important in problems in which variable-values are compared with a constant quantity of another variable *i.e.* time, distance covered within certain time, etc.

4. Median is the best average in open end grouped distributions.

5. Mode is used in problems involving expression of preferences where quantitative measurements are not possible.

20. The frequency distribution below gives the cost of production of sugarcane in different holdings obtain the arithmetic mean.

Class	*Frequency*	*Class*	*Frequency*
2—6	1	18—	52
6—	9	22—	36
10—	21	26—	19
14—	47	30—34	3

[**Ans.** 19·21]

21. The yields of grain (x k. gms). from 500 small plots are grouped into classes with a common class-interval (0·2 k. gms) in the table below, the values of x given being the mid-values of the class. Show that the mean of the distribution is 3·95 k. gms. and the quartiles 3·63 k. gms. and 3·28 k. gms.

x	f	x	f	x	f	x	f	x	f
2·8	4	3·4	47	4·0	88	4·6	35	5·2	4
3·0	15	3·6	63	4·2	60	4·8	10	—	—
3·2	20	3·8	78	4·4	59	5·0	8	—	—

22. Obtain the median and the mode for the following table :

No. of days absent	*No. of Students*	*No. of days absent*	*No. of students*
Less than 5	29	Less than 30	644
" " 10	224	" " 35	650
" " 15	465	" " 40	653
" " 20	582	" " 45	655
" " 25	634	"	

[**Ans.** median = 12·1 days, mode = 11·4 days]

23. Find the median and the two quartiles for the following table :

Marks	*No. of students*	*Marks*	*No. of students*
Below 10	15	Below 50	96
" 20	35	" 60	127
" 30	60	" 70	198
" 40	84	" 80	250

[**Ans.** median = 59·35 marks, $Q_1 = 31{\cdot}00$ marks, $Q_3 = 68{\cdot}52$ marks]

24. Find the median, lower and upper quartiles, 8th decile and 5th percentile for the following distribution of 45 workers :

Daily wages (in Rs.)	*No. of workers*	*Daily wages (in Rs.)*	*No. of workers*
1—2·99	6	9—10·99	21
3—4·99	53	11—12·99	16
5—6·99	85	13—14·99	4
7—8·99	56	15—16·99	4

[**Ans.** median = Rs. 6·49, Q_1 = Rs. 5·05, Q_3 = Rs. 8·40, D_8 = Rs. 8·86, P_{56} = Rs. 6·80

25. Find the median, lower and upper quartiles, 4th decile and 40th percentile for the following distribution :

Marks	No. of students	Marks	No. of students
0—4	10	14—18	5
4—8	12	18—20	8
8—12	18	20—25	4
12—14	7	25 and over	6

[**Ans.** median = 10·9 marks. $Q_1 = 6{\cdot}5$ marks, $Q_3 = 18{\cdot}12$ marks
$D_4 = 9{\cdot}33$ marks, $P_{40} = 9{\cdot}33$ marks.]

26. (*a*) Calculate the modal and the median wages of the following distribution of wages per thousand employees in a certain factory :

Daily wages (in Rs.)	*No. of employees*	*Daily wages (in Rs.)*	*No. of employees*
2—4	3	14—	204
4—	13	16—	139
6—	43	18—	69
8—	102	20—	25
10—	175	22—	6
12—	220	24--26	1

[**Ans.** Modal wages = Rs. 13·50, median wage = Rs. 13·50]

(*b*) The daily expenditure of 100 families is given as under :

Expenditure	0—10	10—20	20—30	30—40	40—50
No. of families	14	?	27	?	15

The median and mode for the distribution are Rs. 25 and Rs. 29 respectively. Calculate the missing frequencies. [**Ans.** 25, 26]

27. Determine graphically the values of the median and the two quartiles of the data given in exercises 14, 15 and 16 on page 35.

28. A distribution consists of 3 components with frequencies 45, 40 and 65 having their means 2, 2·5 and 2 respectively. Prove that the mean of the combined distribution is 2·13 approximately.

29. (*a*) Show that the arithmetic mean of the first *n* natural numbers is $\frac{1}{2}(n+1)$.

(*b*) Show that the weighted arithmetic mean of first *n* natural numbers when weights are equal to the corresponding numbers is equal to $\frac{1}{3}(2n+1)$.

(*c*) The arithmetic mean of *n* items of a series is $\bar{x}$. The sum of the first (n—1) items is *k*. Show that the *n*th item is $n\bar{x} - k$.

(*d*) The frequencies of values 0, 1, 2,....., *n* of a variable are proportional to

$$q^n, {}^nc_1 q^{n-1} p, {}^nc_2 q^{n-2} p^2, \ldots\ldots, p^n$$

where $p + q = 1$. Show that the mean is *np*.

30. If $x_1, x_2, \ldots\ldots, x_n$ are non-zero positive numbers, prove that $A \geq G$.

[**Hint.** Use the result $\dfrac{x_1 + x_2 + \ldots + x_n}{n} \geq (x_1 \,.\, x_2, \ldots, x_n)^{\frac{1}{n}}$, the equality sign holding when the x's are equal.]

31. Show that $\left[\sum_{i=1}^{n} w_i\right]\left[\sum_{i=1}^{n} w_i (x_i - \bar{x}_w)^2\right] = \sum_{i=1}^{n} \sum_{j>i} w_i w_j (x_i - x_j)^2$

where $\sum_{i=1}^{n} w_i \neq 0$ and $\bar{x}_w$ is the weighted mean of x_i's with weights w_i.

(*B. Sc. Hons. Sub, Delhi 1968*)

[**Hint.** R.H.S. $= \frac{1}{2} \sum_{1}^{n} \sum_{1}^{n} w_i \,.\, w_j .(x_i - x_j)^2 = \frac{1}{2} \sum_{1}^{n} \sum_{1}^{n} w_i w_j \{(x_i - \bar{x}_w) - (x_j - \bar{x}_w)\}^2$

= etc.]

32. Show that for J-shaped distribution with the maximum frequency towards the lower values of the variate, the median is nearer to Q_1 (lower quartile) than to Q_3 (upper quartile).

33. A distribution $x_1, x_2, \ldots\ldots, x_n$ with frequencies $f_1, f_2, \ldots\ldots f_n$ is transformed into the distribution $X_1, X_2, \ldots, X_n$ with the same corresponding frequencies of the relation $X_i = ax_i + b, i = 1, 2, \ldots\ldots, n$ where a and b are constants. Show that the mean, mode and median of the new distribution are given in terms of those of the first distribution by the same transformation.

(*B.A., Aligarh; B. Sc., Agra.*)

34. A variate takes values $a, ar, ar^2, \ldots\ldots, ar^{n-1}$ each with frequency unity. Show that the *A.M. A*, is $\dfrac{a(1-r^n)}{n(1-r)}$, the G.M. G., is $a.\ r^{\frac{n-1}{2}}$ and the *H.M. H*, is $\dfrac{an(1-r)r^{n-1}}{1-r^n}$. Prove that $AH = G^2$. Prove also that $A > H > G$ unless $n = 1$ when all the three means coincide.

35. A person travels n equal distances with velocities $v_1, v_2, \ldots\ldots, v_n$. Show that his average velocity (total distance/total time) cannot exceed $(v_1 + v_2 + \ldots + v_n)/n$. When will it be equal to this value ?

36. An aeroplane flies around a square the sides of which measure 100 km. each. The aeroplane covers at a speed of 400 km. p.h. the first-side, at 500 km. p.h. the second side, at 700 km. p.h. the third side, and at 600 km. p.h. the fourth side. Use the correct mean to find the average speed round the square.

[**Ans.** H.M. = 526·64 km. p.h.]

37. (A) Which type of average would be suitable for the following cases ? Give reasons :

(*a*) Size of agricultural holdings.

(*b*) Comparison of intelligence.

(*c*) Marks obtained in any examination.

(*d*) The average speed of a man travelling 1 k.m. at 5 km.p.h., 2 km. at 7 km. p.h., and $2\frac{1}{2}$ km. at 8 km.p.h.

(*e*) *Per capita* income in several countries.

(B) What measure do you suggest for each of the following distributions :

(*a*) Income of workers in a factory,

(*b*) Heights of students, and

(*c*) Number of petals of flowers ? **(B.A. Hons. Punjab.)**

[**Ans.** (A) (*a*) mode (*b*) median (*c*) mode (*d*) harmonic mean (*e*) weighted mean] (B) (*a*) A.M. (*b*) median (*c*) median

38. A population of values is symmetrically distributed about the constant k. Examine whether the following statements are correct :

(*a*) the mean is k.

(*b*) the mean coincides with the mode,

(*c*) the distribution cannot be bimodal,

(*d*) $\frac{1}{2}$ (upper quartile + lower quartile) is equal to k,

(*e*) the median coincides with mean.

[**Ans.** (*a*) Correct (*b*) Incorrect (*c*) Incorrect (*d*) Correct (*e*) Correct]

39. In a frequency table the upper boundary of each interval has a constant ratio to the lower boundary. Show that the geometric mean, G, may be expressed by the formula

$$\log G = x'_0 + \frac{C}{N} \sum_{i=1}^{n} f_i (i-1),\ N = \sum_{i=1}^{n} f_i.$$

where x'_0 is the logarithm of the mid-value of the first interval and C is the logarithm of the ratio between the upper and lower boundaries.

(B. Sc. Delhi 1972 ; B. Sc. Meerut, 1970)

40. If $a_1 = \frac{1}{n} \sum_{i=1}^{n} x_i,\ a_2 = \frac{1}{n} \sum_{i=2}^{n+1} x_i$ and $a_3 = \frac{1}{n} \sum_{i=3}^{n+2} x_i$, then show that

(*a*) $a_2 = a_1 + \frac{1}{n}(x_{n+1} - x_1)$ and

(*b*) $a_3 = a_2 + \frac{1}{n}(x_{n+2} - x_2)$. **(B.Sc., Calcutta)**

41. If G_x be the geometric mean of N x's and G_y be the geometric mean of N y's, then prove that (i) the geometric mean G of the $2N$ values is given by

$$G^2 = G_x \cdot G_y \,.$$

$$[\, G_x = (x_1 x_2 \ldots x_N)^{\frac{1}{N}},\ G_y = (y_1 y_2 \ldots y_N)^{\frac{1}{N}},\ G = (x_1 x_2 \ldots x_N y_1 y_2 \ldots y)^{\frac{1}{2N}} = \text{etc.}\,]$$

(*ii*) The geometric mean of N values (x_i/y_i) is G_x/G_y.

42. The following table shows the distribution of 100 families according to their expenditure per week. Number of families according to expenditure groups Rs. 10—20 and Rs. 30—40 are missing from the table. The median and mode are given to be Rs. 25 and Rs. 24. Calculate the missing frequencies and then arithmetic mean of the data.

Expenditure :	0—10	10—20	20—30	30—40	40—50
No. of families :	14	?	27	?	15

(B.Sc. Jiwaji 1971)

(Ans. 25, 24)

43. (*a*) Show that the arithmetic mean may be regarded as the x-coordinate of this centre of mass of a number of particles, masses $f_1, f_2,....., f_n$ placed at points on x-axis with x-coordinates $x_1, x_2,......., x_n$.

(*b*) What is the y-coordinate of the centre of mass of a plane of the same shape as the histogram of the distribution ?

(*c*) Show that $\overline{x}_w > \overline{x}$ if smaller weights are attached to smaller observations or larger weights are attached to larger observations.

44 Prove that for a discrete distribution :

$$\frac{f_1x_1 + f_2x_2 + ... + f_nx_n}{N} \geq \left(x_1^{f_1} x_2^{f_2} .... x_n^{f_n}\right)^{\frac{1}{N}} \text{ where } N = \sum_1^n f_i .$$

45. State or mention with reason in each case, the most appropriate diagram to represent the following data :–

(*i*) Distribution of a large number of candidates, according to the number of marks scored by each candidate in an examination.

(*ii*) the number of car accidents on a road during each of last fifty days.

46. Following is the age-distribution of 2,000 workers working in a factory :—

Age-group	20–25	25–30	30–35	35–40	40–45	45–50	50–55	55–60
No. of workers	250	380	325	287	220	268	150	120

(*a*) Find the number of workers to be retrenched from the lower-age groups, if it is desired to retrench the first 15% of the total workers from this group.

(*b*) Find the number of premature retirements if 10% of the total workers are to be retired from the highest age group.

Ans. (*a*) 250 from 20-25 and 50 from 25-30.

(*b*) 120 from 55-60 and 80 from 50-55.

47. The following table shows the distribution of monthly income for 580 persons.

[*monthly income (Rs.) less than*]	1000	1500	2000	2500	3000	3500	4000	4500	5000
[*Number of persons*]	53	133	254	394	500	542	560	571	576

Find (*a*) the number of persons having income between Rs. 1500 and Rs. 2500.

(*b*) the number of persons having income between Rs. 3500 and Rs. 4500.

Ans. (*a*) 261 (*b*) 29

48. The mean marks obtained by 300 candidates in Statistics is 46. The mean of the top 100 of them was found to be 70 and the mean of the last 100 was known to be 20. What is the mean of the remaining 100 candidates ?

Ans. 48 marks.

49. The mean weakly salary paid to all employees of a company was Rs. 5000. The mean weakly salaries paid to male and female employees were Rs. 5200 and Rs. 4200 respectively. Determine the percentage of males and females employed by the company. **Ans.** 80, 20.

50. In a factory employing 3,000 persons, 5 per cent earn less than Rs. 3 per hour, 580 earn from Rs. 3.01 to Rs. 4.50 per hour, 30 per cent earn from Rs. 4.51 to Rs. 6.00 per hour, 500 earn from Rs. 6.01 to Rs. 7.50 per hour, 20 per cent earn from Rs. 7.51 to Rs. 9.00 per hour, and the rest earn Rs. 9.01 or more per hour. What is the median wage ?

[**Hint.**

Earnings less than (in Rs.)	3	3·01–4·50	4·51–6·00	6·01–6·50	7·51–9·00	9·01 & above
No. of workers	150	580	900	500	600	270

Adjust the class-limits and use :

$$\text{Median} = 4{\cdot}505 + \frac{1{\cdot}5}{900}(1500 - 730) = \text{Rs. } 5{\cdot}79]$$

51. The mean age of 50 persons was found to be 32 years. Later was detected that the age 57 was misread as 27, age 60 was misread as 35 and age 28 was misread as 33. Find the corrected mean age. **Ans.** 33 years.

3

Measures of Dispersion, Skewness and Kurtosis. Moments of Frequency Distributions

3·1. Introduction. Dispersion. It should be clear that a measure of central tendency by itself can exhibit only one of the important characteristics of a distribution and, therefore, while studying a distribution, it is equally important to know how the variates are clustered around or scattered away from the point of central tendency. Thus if two groups of students have the same average marks, we may like to know whether one group consists of students of average and near-average intelligence and the other group is made up of a large number of very bright and very dull students. Consider the two distributions A and B :

A :	75	85	95	105	115	125
B :	10	20	30	70	190	280

Both the distributions have the same mean 100, yet they are different. In fact; in A the observations have range = 125–75 = 50 while in B, the range is 280 – 10 = 270. Thus, the two distributions widely differ from each other in their formations. Fig. 2·8(*b*) page 22 clearly shows that the two distributions centre round the same point, *viz.*, the mean, yet they differ in variation from the *mean*. Such variation is variously called as *dispersion*, *spread*, *scatter*, or *variability*, and we consider in the next section some of the measures that have been proposed to get a quantitative measure of this dispersion.

3·1·1. Characteristics of an ideal measure of Dispersion. The following are the requisites for an ideal measure of dispersion :

1. It should be rigidly defined.
2. It should be easily understandable and easy to calculate.
3. It should be based on all the observations.
4. It should be amenable to further mathematical treatment.
5. It should be least affected by *fluctuations of sampling*.

3·2. Measures of Dispersion. The measures of dispersion can be classified into three groups :

(*i*) Measures of the difference between representative variate values, such as the *range*,—the *interquartile* or the *interdecile range*.

(*ii*) Measures obtained from the deviations of every variate value from some central value, such as the *mean deviation from the mean* or the *mean deviation from the median* or the *standard deviation*.

(*iii*) Measures obtained from the variations of all the variates among themselves, such as the *mean difference*.

3·2·1. Range. Interquartile Difference. *Range* is the difference between the extreme values of the variate, that is, range is equal to $x_n - x_1$, where x_n and x_1 are the greatest and the least values of the variate. It is easily understood and computed, but depends exclusively on the two extreme values, while it is desirable that a measure of dispersion should depend on all the variate values. Nevertheless, the range is a very useful measure in industrial engineering work, specially in *statistical quality control* work.

The *interquartile range* is the difference between the upper and lower quartiles. The semi-interquartile range or the *quartile deviation*, Q, is half of it *i.e.*,

$$Q = \frac{Q_3 - Q_1}{2} \qquad ...(3\cdot1)$$

A dimensionless measure of dispersion based on quartile deviation is given by

$$\frac{Q_3 - Q_1}{Q_3 + Q_1}.$$

The *interdecile range* is the difference between the ninth and the first decile and contains eighty per cent of the total frequency, while the interquartile range contains fifty per cent. But these measures are easily calculated and give a fairly good idea of the spread of the distribution and are commonly used in descriptive statistics, but in advanced theory their use is rather rare because of the mathematical difficulty in handling them.

3·2·2. Average Deviation or Mean Deviation. It would be useless to take the sum of the deviations from the mean as a measure of dispersion since some deviations are positive and some are negative and their algebraic sum is zero. (Compare § 2·5·1).

However, if we take the mean of the absolute values of the deviations, we get a useful measure of dispersion called the mean deviation from the mean. Thus,

Mean deviation from the mean

$$\text{Mean deviation} = \frac{1}{N} \sum_{i=1}^{n} f_i \mid x_i - \bar{x} \mid \qquad \ldots(3{\cdot}2)$$

where $N = \sum_{i=1}^{n} f_i,\ \bar{x} = \frac{1}{N} \sum_{i=1}^{n} f_i x_i.$

Here $\bar{x}$ has to be replaced by median or mode if the mean deviation from median or mode is required.

The mean deviation is based on all the values of the variate and hence it is a better measure of dispersion. However, there is an element of artificiality in mean deviation since the deviations are not taken with their proper signs. The mean deviation is also difficult to handle mathematically.

Example 3·1. *Calculate the mean deviation (from the mean) of the following distribution :*

Marks	*0–10*	*10–20*	*20–30*	*30–40*	*40–50*
No. of Students	*5*	*8*	*15*	*16*	*6*

We first calculate mean by applying formula (2·6), page 25 and then apply formula (3·2), to find mean deviation.

Mid-value x	$u = \frac{x - 25}{10}$	f	fu	$x - \bar{x}$	$f \mid x - \bar{x} \mid$
5	–2	5	–10	–22	110
15	–1	8	–8	–12	96
25	0	15	0	–2	30
35	1	16	16	8	128
45	2	6	12	18	108
	Total	50	10		472

$$\bar{x} = \text{Mean} = a + \frac{h}{N} \Sigma f u = 25 + \left(\frac{10}{50} \times 10 \right) = 27 \text{ marks.}$$

$$\text{Mean deviation} = \frac{1}{N} \sum_{i=1}^{n} f_i \mid x_i - \bar{x} \mid = \frac{472}{50} = 9{\cdot}44 \text{ marks.}$$

Ex. Calculate mean deviation from mode of the above distribution.

3·2·3. Minimal Property of Average Deviation. *The sum of the absolute deviations from the median is a minimum.* In other words, *the mean deviation from median is less than that measured from any other value.*

Proof. Let A, B represent the smallest and largest observations and let O be the median. Let the frequencies in the intervals (AO), (OD), (D, C), (CB) be N, $f_1, f_2, N - f_1 - f_2 - \delta_1 - \delta_2$ respectively and let $\delta, \delta_1, \delta_2$, represent the frequencies at O, D and C, where

$OC = C$ and $OD = \frac{1}{2} C$. All these frequencies make a contribution to the excess of sum of absolute deviations from C over the sum of absolute deviations from O.

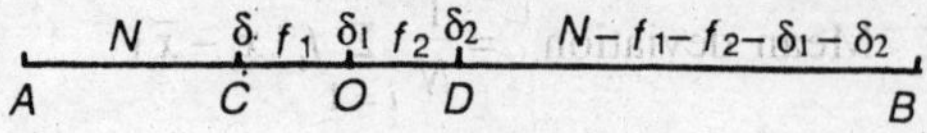

Fig. 3·1

The distance from C of any observation in OD exceeds its distance from O by a quantity which is less than C. The total contribution to the excess from all the observations in OD will therefore be $f_1 x$, where x will be less than C. Similarly the total contribution to this excess as made by all the observations f_2 in DC will be $f_2 y$ where y is less than C.

$$\text{The total excess} = NC + \delta C + f_1 x + \delta_1 D - f_2 y - \delta_2 C - (N - f_1 - f_2 - \delta_1 - \delta_2)C$$

$$= \delta C + f_1(x + C) + f_2(C - y) + \delta_1 C \geq O,$$

since $\quad O \leq x < C, O \leq y < C.$

Thus the mean deviation from $O \leq$ mean deviation from C.

Similarly the mean deviation from $O \leq$ mean deviation from C'.

Thus the mean deviation is least when measured from the median.

Alternative Proof.

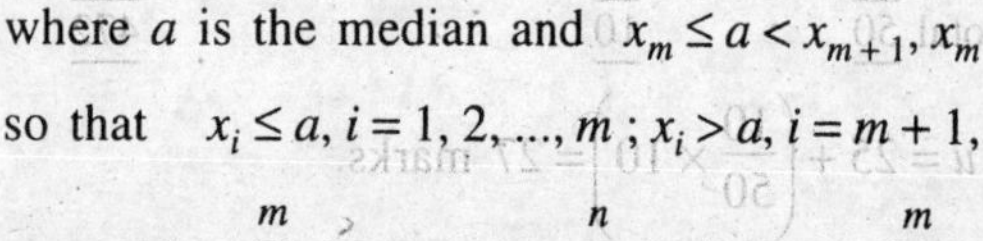

Fig. 3·2

$$S_a = \frac{1}{n} \sum_{i=1}^{n} f_i \,|\, x_i - a \,|, ;$$

$$S_a' = \frac{1}{n} \sum_{i=1}^{n} f_i \,|\, x_i - a' \,|,$$

where a is the median and $x_m \leq a < x_{m+1}, x_{m-1} \leq a' < x_m, \ldots$

so that $\quad x_i \leq a, i = 1, 2, \ldots, m \,;\, x_i > a, i = m+1, m+2, \ldots, n$

$$NS_a = \sum_{i=1}^{m} f_i \,|\, x_i - a \,| + \sum_{i=m+1}^{n} f_i \,|\, x_i - a \,| = \sum_{i=1}^{m} f_i (a - x_i) + \sum_{i=m+1}^{n} f_i (x_i - a)$$

$$= a\left[\sum_{i=1}^{m} f_i - \sum_{i=m+1}^{n} f_i\right] + \left[-\sum_{i=1}^{m} f_i x_i + \sum_{i=m+1}^{n} f_i x_i\right] = O + \left[-\sum_{i=1}^{m} f_i x_i + \sum_{i=m+1}^{n} f_i\right]$$

$$NS_{a'} = a'\left[\sum_{i=1}^{m-1} f_i - \sum_{i=m}^{n} f_i\right] + \left[\sum_{i=1}^{m-1} f_i x_i + \sum_{i=m}^{n} f_i x_i\right]$$

$$= a'\,[-2f_m] + \left[-\sum_{i=1}^{m} f_i x_i + \sum_{i=m+1}^{n} f_i x_i + 2f_m x_m\right]$$

$$\therefore \quad N(S_a' - S_a) = 2f_m (x_m - a') \geq O$$

Similarly $N(S_a'' - S_a) = 2f_{m+1}(a'' - x_{m+1}) \geq O$

$$N(S_a''' - S_a) = 2f_m(x_m - a''') + 2f_{m+1}(x_{m+1} - a''') \geq O$$

$$N(S_a^{iv} - S_a) = 2f_{m+1}(a^{iv} - x_{m+1}) + 2f_{m+2}(a^{iv} - x_{m+2}) \geq O$$

⋮

and so on.

This shows that the sum of absolute deviations is least when these deviations are measured from the median.

Note. A third proof of this important result is outlined below in the Ex. For a proof of this result for a continuous probability distribution, please see Example 7·7. And for a proof for the case where the probability density function has a finite number of points of discontinuity see Ex. 3 following Example 7·7. This will give the proof for a grouped frequency distribution.

Ex. 1. Show that the sum of absolute deviations from the point x is a function of x defined as follows :

$$S(x) = \begin{cases} N(\bar{x} - x), -\infty \leq x \leq x_1 \\ N(\bar{x} - x) + 2f_1(x - x_1), x_1 < x \leq x_2 \\ N(\bar{x} - x) + 2f_1(x - x_1) + 2f_2(x - x_1), x_2 < x \leq x_3 \\ ----- \\ N(\bar{x} - x), x \leq x_n \end{cases}$$

Ex. 2. Show that $S(x)$ is a continuous function of x with derivatives at all points except $x_1, x_2, \ldots, x_n$. Prove that

$$\frac{dS}{dx} = S'(x) = \begin{cases} -N & , -\infty < x < x_1 \\ -N + 2f_1 & , x_1 < x < x_2 \\ -N + 2f_1 + 2f_2 & , x_2 < x < x_3 \\ ---- \\ -N & , x < x_n \end{cases}$$

so that $S'(x)$ has jumps of $2f_1, 2f_2, 2f_3$ ..... at points x_1, x_2, x_3 ....

Ex. 3. Show that if $f_1 + f_2 + \ldots + f_k = \frac{1}{2}N$, then $S'(x) = O, x_k < x < x_{k+1}$

and the sum of absolute deviations from every point in (x_k, x_{k+1}) is the range and as such that every point can be regarded as the median.

Ex. 4. Show that if $f_1 + f_2 + \ldots + f_k = \frac{1}{2}N$ for any k, then $S(x)$ is minimum where k is the first integer for which $f_1 + f_2 + \ldots + f_k$ exceeds $\frac{1}{2}N$.

Ex. 5. A die is thrown 100 times and points 1, 2, 3, 4, 5, 6 occur with frequencies 15, 18, 14, 20, 16, 17. Draw the graph of $S(x)$ in this case.

3·2·4. Variance. Standard Deviation. Coefficient of variation. The *mean square deviation* of the variate x from any value a is defined as the mean value of the square of the deviations of x from a and is, thus given by

$$\mu'_2 = s^2 = \frac{1}{N}\sum_{i=1}^{n} f_i (x_i - a)^2, \; N = \sum_{i=1}^{n} f_i. \qquad ...(3\cdot3)$$

The positive square root of this quantity is called the *root-mean square deviation*, taken from a. In the particular case when deviations are taken from the mean $\bar{x}$, the mean-square deviation is called the *variance* of x and is denoted by μ_2 or σ_x^2 so that we have

$$\mu_2 = \sigma_x^2 = \frac{1}{N}\sum_{i=1}^{n} f_i\, (x_i - \bar{x})^2, N = \sum_{i=1}^{n} f_i \qquad ...(3\cdot4)$$

The standard deviation (*s.d.*) of x is the positive square root of the variance of x and is denoted by σ_x.

The *coefficient of variation* is defined as $100 \times$ (*s.d.* $\div$ mean). It should be noted that this ratio is dimensionless since *s.d.* and mean have the same dimension as that of the variate. Further if the standard deviation is expressed in cm. then variance will be (cm.)^2.

Other forms of the formula for Variance given in (3·4)

The formula for variance *viz.*, equation (3·4) can be put as :

I. $$\sigma_x^2 = \frac{1}{N}\sum_{i=1}^{n} f_i (x_i - \bar{x})^2 = \frac{1}{N}\sum_{i=1}^{n} f_i (x_i^2 - 2x_i\bar{x} + \bar{x}^2)$$

$$= \frac{1}{N}\sum_{i=1}^{n} f_i x_i^2 - 2\bar{x}\,.\,\frac{1}{N}\sum_{i=1}^{n} f_i x_i + \bar{x}^2\,.\,\frac{1}{N}\sum_{1}^{n} f_i$$

$$= \frac{1}{N}\sum_{i=1}^{n} f_i x_i^2 - 2\bar{x}\,.\,\bar{x} + \bar{x}^2\,.\,1$$

$$= \frac{1}{N}\sum_{i=1}^{n} f_i x_i^2 - \bar{x}^2$$

$$\sigma_x^2 = \frac{1}{N}\sum_{i=1}^{n} f_i x_i^2 - \left(\frac{1}{N}\sum_{i=1}^{n} f_i\; x_i\right)^2$$

II. $$\sigma_x^2 = \frac{1}{N}\sum_{i=1}^{n} f_i (x_i - \bar{x})^2 = \frac{1}{N}\sum_{i=1}^{n} f_i (\overline{x_i - a} - \overline{\bar{x} - a})^2, \; a \text{ is arbitrary.}$$

$$= \frac{1}{N}\sum_{i=1}^{n} f_i (x_i - a)^2 - 2.\frac{1}{N}\sum_{i=1}^{n} f_i (x_i - a).(\bar{x} - a) + (\bar{x} - a)^2\,.\,\frac{1}{N}\sum_{i=1}^{n} f_i$$

$$\sigma_x^2 = \frac{1}{N} \sum_{i=1}^{n} f_i (x_i - a)^2 - \left[\frac{1}{N} \sum_{i=1}^{n} f_i (x_i - a) \right]^2 \quad ...(3\cdot4)'$$

or $$\sigma_x^2 = s^2 - (\bar{x} - a)^2, \quad ... (3\cdot4)''$$

since $$\left[\frac{1}{N} \sum_{i=1}^{n} f_i x_i - \frac{1}{N} \sum_{i=1}^{n} f_i \cdot a \right]^2$$

$$= \bar{x} - \frac{1}{N} N \cdot a = \bar{x} - a.$$

3·2·5. Relation between Standard and Root-mean Square Deviations. If f is the frequency of the observation x_i, $i = 1, 2, ..., n$ then we have the relation from 13.4, between s and σ

$$s^2 = \frac{1}{N} \sum_{i=1}^{n} f_i (x_i - a)^2 = \sigma_x^2 + (\bar{x} - a)^2 \quad ...(3\cdot5)$$

Clearly s^2 will be least when $a = \bar{x}$.

Hence *the root-mean square deviation is least when deviations are measured from the mean i.e., the standard deviation is least possible root-mean square deviation.*

Aliter. $s^2 = \sigma_x^2 + (\bar{x} - a)^2$, s^2 is a function of a alone,

Now $$\frac{ds^2}{da} = -2(\bar{x} - a) \text{ and } \frac{d^2 s^2}{da^2} = +2 > 0$$

Hence s^2 is minimum when $a = \bar{x}$.

3·2·6. Effect of change of origin and scale on standard deviation

Theorem. The Standard deviation is unaltered by the change of origin but its value gets changed by the change of scale.

Proof. Let 'a' be taken as the new origin and scale be 'h' (positive or negative) times the original scale so that

$$u = \frac{x - a}{h} \Rightarrow x_i = a + hu_i \text{ and } \bar{x} = a + h\bar{u} \quad ...(3\cdot6)$$

Thus from (3·4), we have as s.d. is non-negative,

$$\sigma_x^2 = \frac{h^2}{N} \sum_{i=1}^{n} f_i (u_i - \bar{u})^2 = h^2 \sigma_u^2 \Rightarrow \sigma_x = h\sigma_u \quad ...(3\cdot7)$$

where σ_x^2 and σ_u^2 denote variances of x and u respectively. Thus standard deviation of x is independent of a and Standard deviation of x = h times the standard deviation of u, u = (x – a)/h.

Remark. The standard deviation is the best measure of dispersion, since it satisfies all the properties laid down for an ideal measure of dispersion. However, there is only one difficulty of extracting the square root. Further, the standard deviation remains unaffected whatever point be taken for finding deviations, it plays an important role in Statistics.

Example 3·2. *Find the mean and the variance of the distribution in which the values of X are 1, 2, ..., N, the frequency of each being unity.*

Here $\bar{x} = \dfrac{1+2+\ldots+N}{N} = \frac{1}{2}(N+1)$.

The mean square deviation from $x = 0$ is

$$s^2 = \frac{1^2+2^2+\ldots+N^2}{N} = \frac{1}{6}(N+1)(2N+1)$$

Hence $\sigma^2 = \frac{1}{6}(N+1)(2N+1) - \frac{1}{4}(N+1)^2$, on using (3·5), page 48.

$$= \frac{1}{12}(N^2-1).$$

Example 3·3. *Find the s.d. and the coefficient of variation from the following table giving wages of 230 persons :*

Wages in Rs.	*No. of Persons*	*Wages in Rs.*	*No. of Persons*
140–160	*12*	*220–240*	*50*
160–180	*18*	*240–260*	*45*
180–200	*35*	*260–280*	*20*
200–220	*42*	*280–300*	*8*

We first calculate the mean and then the standard deviation.

Wages (in Rs.)	*Mid-value x*	*No. of persons f*	$u = \dfrac{x-210}{20}$	u^2	$f \times u$	$f \times u^2$
140–160	150	12	–3	9	–36	108
160–180	170	18	–2	4	–36	72
180–200	190	35	–1	1	–35	35
200–220	210	42	0	0	0	0
220–240	230	50	+1	1	+50	50
240–260	250	45	+2	4	+90	180
260–280	270	20	+3	9	+60	180
280–300	290	8	+4	16	+32	128
Total		230			+125	753

Mean $= a + h\bar{u} = 210 + 20 \times \dfrac{125}{230} =$ Rs. 220·87, using (3·6)

Using (3·7), we have

$$S.D. = 20 \times \sqrt{\frac{753}{230} - \left(\frac{125}{230}\right)^2} = \text{Rs. } 34.52$$

$\therefore$ Coefficient of variation (C.V.) $= 100 \times \dfrac{34{\cdot}52}{220{\cdot}87} = 15{\cdot}63$, using (3.5)

Example 3·4. (*Variance of the combined set of group sets*). *The standard deviation of two sets containing n_1 and n_2 members are σ_1 and σ_2 respectively, being measured from their respective means m_1 and m_2. If the two sets are grouped together as one set of (n_1+n_2) members, show that the standard deviation, σ, of this set, measured from its mean is given by*

$$\sigma^2 = \frac{n_1\sigma_1^2 + n_2\sigma_2^2}{n_1+n_2} + \frac{n_1 n_2}{(n_1+n_2)^2} \cdot (m_1 - m_2)^2.$$

Let s_1^2 and s_2^2 be the mean square deviations of the two sets respectively having observations $x_{1i}, i = 1, 2, ..., n_1$ and $x_{2j}, j = 1, 2, ..., n_2$ then the mean square deviation, s^2 of the combined set is given by

$$s^2 = \frac{1}{n_1+n_2}\left[\sum_{i=1}^{n_1}(x_{1i}-a)^2 + \sum_{j=1}^{n_2}(x_{2j}-a)^2\right]$$

$$= \frac{n_1 s_1^2 + n_2 s_2^2}{n_1+n_2} \qquad \left[\because\ s_1^2 = \frac{1}{n_1}\sum_1^{n_1}(x_{1i}-a)^2\right]$$

$$= [n_1(\sigma_1^2 + d_1^2) + n_2(\sigma_2^2 + d_2^2)]/(n_1+n_2)\text{, using (3·5)}$$

or $$s^2 = \frac{n_1\sigma_1^2 + n_2\sigma_2^2}{n_1+n_2} + \frac{n_1 d_1^2 + n_2 d_2^2}{n_1+n_2} \qquad ...(3·8)$$

where $d_1 = m_1 - a$, $d_2 = m_2 - a$, a being an arbitrary number.

Let a be taken as the mean of the whole set, then

$$a = m = (n_1 m_1 + n_2 m_2)/(n_1+n_2),\ s^2 = \sigma^2.$$

$$d_1 = m_1 - a = m_1 - (n_1 m_1 + n_2 m_2)/(n_1+n_2)$$

$$= n_2(m_1 - m_2)/(n_1+n_2),$$

Similarly, $d_2 = n_1(m_2 - m_1)/(n_1+n_2)$.

Substituting the values of a, d_1 and d_2 in (3·8) and simplifying we get the required result.

Example 3·5. *Show that, for any discrete distribution, the standard deviation is not less than the mean deviation from the mean.*

(B.Sc. Gujarat 1971 ; M.Sc. Allahabad 1967 ; B.A. Hons., Delhi 1966 ; I.A.S. 1968, 1961)

We have to show that S.D. $\nless$ Mean deviation from mean

or Variance $\geq$ (mean deviation from mean)^2

i.e., $$\frac{1}{N}\sum_i f_i (x_i - \bar{x})^2 \geq \left[\frac{1}{N}\sum_i f_i \mid x_i - \bar{x} \mid\right]^2 \quad ...(1)$$

or $$(\sum_i f_i y_i^2)(\sum_i f_i) \geq (\sum_i f_i y_i)^2, \qquad \text{where } y_i = \mid x_i - \bar{x} \mid, N = \sum_i f_i$$

or $$\sum_i f_i^2 y_i^2 + \sum_{i<j} f_i f_j (y_i^2 + y_j^2) \geq \sum_i f_i^2 y_i^2 + 2\sum_{i<j} f_i f_j y_i y_j$$

or $$\sum_{i<j} f_i f_j (y_i - y_j)^2 \geq 0$$

which is always true. Hence the result follows.

Aliter. Let $\mid x_i - \bar{x} \mid = y_i$ so that (1) reduces to

$$\frac{1}{N}\sum_i f_i y_i^2 \geq \left(\frac{1}{N}\sum_i f_i y_i\right)^2$$

$$\Rightarrow \quad \frac{1}{N}\sum_i f_i y_i^2 - \left(\frac{1}{N}\sum_i f_i y_i\right)^2 \geq 0$$

i.e., $$\text{Var } y \geq 0 \quad ...(3{\cdot}9)$$

which is true. Hence the result is proved.

Note. The equality sign in (3·9) will hold when all the variate–values are equal.

Ex. Show that the variance σ^2 of a distribution with k component distributions is given by $N\sigma^2 = \sum_{i=1}^{k} n_i (\sigma_i^2 + d_i^2)$.

where m_i is the mean of the ith component with frequency n_i, m is the mean of the whole distributions, $d_i = m_i - m$ and N is the total frequency $\sum_{i=1}^{k} n_i$.

3·2·7. Mean Difference. For a frequency distribution the *coefficient of mean difference* is defined by

$$\Delta_1 = \frac{1}{N(N-1)}\sum_{j=1}^{n}\sum_{i=1}^{n} \mid x_i - x_j \mid f_i f_j$$

or by $$\Delta_1 = \frac{1}{N^2}\sum_{j=1}^{n}\sum_{i=1}^{n} \mid x_i - x_j \mid f_i f_j$$

depending on whether we exclude or include zero differences. The difference lies only in the denominator and is unimportant for large N.

This coefficient was first defined by Gini (1912) and has a certain theoretical attraction, since it depends on all the variate values and the difference of the values among themselves and not on deviation from some central value. It is, however, difficult to compute and the appearance of the absolute values makes it difficult to be handled mathematically.

The dimensionless measure corresponding to mean difference is *Gini's coefficient of concentration* which is defined by

$$G_1 = \frac{\Delta_1}{2\mu_1'}, \text{ where } \mu_1' \text{ is given by } (3{\cdot}3)$$

Ex. 1. Prove that $\frac{1}{N} \sum_{j=1}^{n} \sum_{\substack{i=1 \\ i \neq j}}^{n} (x_i - x_j) f_i f_j = 0.$

Ex. 2. Prove that $\frac{1}{N^2} \sum_{j=1}^{n} \sum_{i=1}^{n} (x_i - x_j)^2 f_i f_j = 2 \text{ var } x.$

Ex. 3. Find G_1 for the distribution in which the variate takes values $a, a+d, \ldots\ldots, a+\overline{n-1}\, d$, each with the same frequency.

3·3. Moments. We define the rth moment about any point a as μ_r' where

$$\mu_r' = \frac{1}{N} \cdot \sum_{i=1}^{n} f_i (x_i - a)^r, \ \sum_{i=1}^{n} f_i = N. \qquad \ldots(3{\cdot}10)$$

Thus the rth moment about any point a can be regarded as the arithmetic mean of the variate values $(x_i - a)^r$,

The rth moment about the mean $\overline{x}$ is written without a dash so that

$$\mu_r = \frac{1}{N} \sum_{i=1}^{n} f_i (x_i - \overline{x})^r, \ \sum_{i=1}^{n} f_i = N. \qquad \ldots(3{\cdot}11)$$

The moments about the mean are much important, because their values are independent of the origin of reference. [See 3.3.1]

Putting $r = 0, 1$ respectively in (3·10) and (3·11), we have

$$\mu_0' = \mu_0 = \frac{1}{N} \sum_{i=1}^{n} f_i = 1 \qquad \ldots(3{\cdot}12)$$

$$\mu_1' = \frac{1}{N} \sum_{i=1}^{n} f_i (x_i - a) = \overline{x} - a = d, \text{ say} \qquad \ldots(3{\cdot}13)$$

Thus *first moment about a* = mean − *a*.

Also $$\mu_1 = \frac{1}{N} \sum_{i=1}^{n} f_i (x_i - \bar{x}) = 0 \qquad ...(3\cdot14)$$

3·3·1. Moments about the mean in terms of Moments about any point and conversely.

We have $$\mu_r = \frac{1}{N} \sum_{i=1}^{n} f_i (x_i - \bar{x})^r = \frac{1}{N} \sum_{i=1}^{n} f_i (\overline{x_i - a} - \overline{\bar{x} - a})^r$$

$$= \frac{1}{N} \sum_{i=1}^{n} f_i (X_i - d)^r) \text{ where } X_i = x_i - a,\ d = \bar{x} - a = \mu_1'$$

$$= \frac{1}{N} \left[\sum_{i=1}^{n} f_i X_i^r - {}^rC_1 d \sum_{i=1}^{n} f_i X_i^{r-1} + {}^rC_2 d^2 \sum_{i=1}^{n} f_i X_i^{r-2} - \right.$$

$$\left. \ldots + {}^rC_{r-1} (-d)^{r-1} \sum_{i=1}^{n} f_i X_i + (-d)^r \sum_{i=1}^{n} f_i \right], \text{ using binomial expansion}$$

$$= \mu_r' - {}^rC_1 d \mu'_{r-1} + {}^rC_2 d^2 \mu'_{r-2} - \ldots + (-1)^{r-1} r d^{r-1} \mu_1' + (-1)^r d^r$$

or $$\mu_r = \mu_r' - {}^rC_1 \mu'_1 \mu'_{r-1} + {}^rC_2 \mu'^2_1 \mu'_{r-2} - \ldots + (-1)^{r-1} (r-1) \mu_1'^r \qquad ...(3\cdot15)$$

In particular, $$\mu_2 = \mu_2' - \mu_1'^2 \qquad ...(3\cdot16)$$

$$\mu_3 = \mu_3' - 3\mu_2'\mu_1' + 2\mu_1'^3 \qquad ...(3\cdot17)$$

$$\mu_4 = \mu_4' - 4\mu_3'\mu_1' + 6\mu_2'\mu_1'^2 - 3\mu_1'^4 \qquad ...(3\cdot18)$$

These three formulae should be committed to *memory*.

Note. In each of these relations, (*i*) the sums of the coefficients of various terms of the right hand side is equal to zero (*ii*) each term of the right hand side is of the same dimension as the term on the left and (*iii*) the number of terms on R.H.S = order of the moment on L.H.S.

Conversely, $$\mu_r' = \frac{1}{N} \sum_i f_i (x_i - a)^r = \frac{1}{N} \sum_i f_i (\overline{x_i - \bar{x}} + \overline{\bar{x} - a})^r$$

$$= \frac{1}{N} \sum_i f_i (x_i' + d)^r \text{ where } x_i' = x_i - \bar{x},\ d = \bar{x} - a$$

$$= \frac{1}{N} \left[\sum_i f_i x_i'^r + {}^rC_1 d \sum_i f_i x_i'^{r-1} + {}^rC_2 d^2 \sum_i f_i x_i'^{r-2} + \ldots \right.$$

$$\left. + {}^rC_{r-1} d^{r-1} \sum_i f_i x_i' + d^r \sum_i f_i \right]$$

$$\mu_r' = \mu_r + {}^rC_1\, \mu_{r-1}\, d + {}^rC_2\, \mu_{r-2}\, d^2 + \ldots + {}^rC_1\, \mu_2\, d^{r-2} + d^r.$$

In particular,

$$\mu_2' = \mu_2 + d^2,\ \mu_3' = \mu_3 + 3\, d\, \mu_2 + d^3,\ \mu_4' = \mu_4 + 4d\mu_3 + 6\, d^2\, \mu_2 + d^4.$$

Ex. Show that the sum of the coefficients of the various terms of the right hand side of (3·15) is equal to $(1-1)^r$ *i.e.*, zero.

3·3·2. Effect of Change of Origin and Scale on Moments.

Let us introduce a new variate u related to x by

$$u = \frac{x-a}{h} = \text{ so that } x - a = hu,\ \bar{x} - a = h\,\bar{u} \text{ and } x - \bar{x} = h\,(u - \bar{u}\,).$$

$$\therefore \qquad \mu_r' = \frac{1}{N} \sum_i f_i\,(x_i - a)^r = h^r \,.\, \frac{1}{N} \sum_i f_i\, u_i^r$$

and
$$\mu_r = \frac{1}{n} \sum_i f_i\,(x_i - \bar{x}\,)^r = h^r \,.\, \frac{1}{N} \sum_i f_i\,(u_i - \bar{u}\,)^r.$$

Thus the rth *moment of the variate x*

= h^r times the corresponding moment of the new variate u, where $u = (x - a)/h$.

In particular, taking $a = \bar{x}$ and $h = \sigma_x$ we find that the distribution of u has zero mean and unit variance and $(x - \bar{x}\,)/\sigma_x$ is said to be the form of the variate in *standard measure.*

Thus standardised random variate corresponding to $X = \dfrac{x - \bar{x}}{\bar{}_x} = \dfrac{x - \text{mean}}{\text{s.d.}}$

Note. The standardised r.v. is dimensionless and, as such, all its moments are also so. Further, since its mean is zero, its central moments coincide with the corresponding ordinary moments.

3·3·3. Sheppard's corrections to moments of grouped frequency distributions. In grouped frequency distributions, we had assumed that the frequencies are concentrated at the mid-points of the class-intervals. This assumption is, however, only an approximation. W.F. Sheppard proved that if

(*a*) the frequency distribution is continuous and,

(*b*) the frequency tapers off to zero in both directions, then corrections to the various moments are :

$$\left.\begin{aligned} \mu_1' \text{ (corrected)} &= \mu_1', \\ \mu_2 \text{ (corrected)} &= \mu_2 - \frac{h^2}{12}, \\ \mu_3 \text{ (corrected)} &= \mu_3, \\ \mu_4 \text{ (corrected)} &= \mu_4 - \frac{1}{2}h^2\mu_2 + \frac{7}{240}h^4 \end{aligned}\right\} \quad ...(3\cdot19)$$

where h is the width of each class- interval and the moments on the right side refer to the moments of the grouped frequency distribution.

Charlier Check

The following identities :

$$\Sigma f(x+1) = \Sigma fx + N$$

$$\Sigma f(x+1)^2 = \Sigma fx^2 + 2\Sigma fx + N$$

$$\Sigma f(x+1)^3 = \Sigma fx^3 + 3\Sigma fx^2 + 3\Sigma fx + N$$

$$\Sigma f(x+1)^4 = \Sigma fx^4 + 4\,\Sigma fx^3 + 6\Sigma fx^2 + 4\,\Sigma fx + N$$

are found useful in checking the values of Σfx, Σfx^2, etc. These identities are named after their discoverer.

3·4. Symmetrical and Skew Distributions. A distribution is said to be symmetrical when the frequencies are symmetrically distributed about the mean, that is, when the values of the variate equidistant from the mean have equal frequencies. For example, the following distribution is symmetrical about its mean 5.

x :	1	2	3	4	5	6	7	8	9
f :	3	4	6	9	10	9	6	4	3

It can be easily established that $\frac{1}{N}\sum_{i=1}^{n} f_i (x_i - \bar{x})^{2r+1} = 0$ since all the terms cancel in pairs, n being odd $= 2r + 1$, and $f_1 = f_n, f_2 = f_{n-1}$, etc.

$\therefore\ \mu_{2r+1} = 0\ \forall\ r$. In particular $\mu_3 = 0$. Hence for a symmetrical distribution,

$$\beta_1 = \mu_3^{\ 2}/\mu_2^{\ 3} = 0.$$

Thus β_1 is a measure of departure from symmetry.

Also for a symmetrical distribution, the mean, mode and median coincide. Further, in the case of such a distribution, median lies half-way between the two quartiles, *i.e.*, Q_3 – median = median – Q_1.

Skewness means lack in symmetry. In other words, for an unsymmetric distribution, it is the departure from symmetry. Skewness indicates whether the frequency curve is inclined more to one side than to the other, that is whether the curve has a longer tail on one side. Skewness is said to be *positive*, if the curve is more elongated to the right side, that is to say, if the mean of the distribution is to the right of or greater than the mode. In the reverse case, skewness is said to be *negative*. [See fig. 8·2·3]

For *distributions of moderate skewness*, there is an *empirical* relationship between the mean, the mode and the median, expressed by the approximation : (see Fig. 3·3 where M_o, M_d, M denote mode, median and mean respectively

$$\text{mean} - \text{mode} \cong 3\,(\text{mean} - \text{median}). \quad ...(3\cdot20)$$

where $\cong$ stands for, 'approximately equal to' so that mode = 3 median – 2 mean. ...(3.20*a*)

The object of measuring skewness is to estimate the direction in which and also the extent to which the distribution is distorted from the symmetrical distribution.

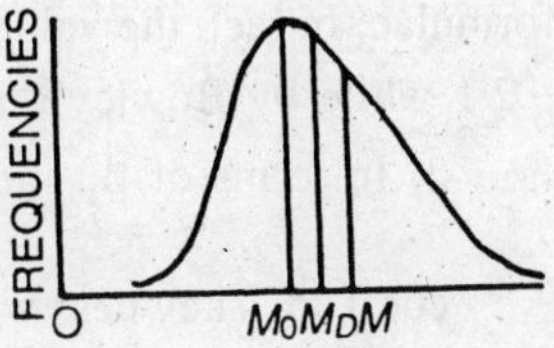

Fig. 3·3: Asymmetrical distribution.

Karl Parson's coefficient of skewness is given by the formula :

$$\text{Coeff. of skewness} = \frac{\text{mean} - \text{mode}}{\text{standard deviation}} \cong \frac{3\,(\text{mean} - \text{median})}{\text{standard deviation}}$$

This is clearly a pure number, since both the numerator and the denominator have the same dimension. Also, it is zero for a symmetrical distribution.

It can be shown that Karl Pearson's coefficient of skewness lies between –3 and +3.

If the standard deviation is not easily found, a rough measure of coefficient of skewness due to Prof. A. Bowley is given :

$$\text{Coeff. of skewness} = \frac{(Q_3 - \text{median}) - (\text{median} - Q_1)}{(Q_3 - \text{median}) + (\text{median} - Q_1)}$$

$$= \frac{Q_3 + Q_1 - 2\,\text{median}}{Q_3 - Q_1} = \frac{Q_3 + Q_1 - 2Q_2}{Q_3 - Q_1} \quad ...(3\cdot22)$$

This is evidently dimensionless and is zero for a symmetrical distribution.

Limits for Bowley's Coefficient of Skewness

Bowley's coefficient of skewness ranges from –1 to +1

Proof : From Algebra, we know that if a and b are positive real numbers, then

$$|a - b| \le |a + b| \quad \text{and} \therefore \quad \left|\frac{a-b}{a+b}\right| \le 1 \quad ...(A)$$

Now choose, $a = Q_3 - \text{median}$, $b = \text{median} - Q_1$, so that $a > 0$, $b > 0$.

$\therefore$ Hence using (A), $\left|\dfrac{(Q_3-\text{median})-(\text{median}-Q_1)}{(Q_3-\text{median})+(\text{median}+Q_1)}\right| \le 1$

In other words, $|\text{ skewness }| \le 1 \Rightarrow -1 \le \text{skewness} \le 1.$

Note 1. Skewness = 1 when median = Q_1 and

skewness = –1 when Q_3 = median.

2. Values of the coefficient of skewness due to Karl Pearson and Bowley cannot be compared, since they are based on entirely different formulae. In fact, the value of β_1 say by Pearson's method comes out to be – 2·1 while by Bowley's method, the value may come out to be – 0·8.

3. In terms of β_1 and β_2,

$$\text{coeff. of skewness} = \frac{\sqrt{\beta_1}\,(\beta_2+3)}{2\,(5\beta_2-6\beta_1-9)}$$

where β_1 and β_2 are defined in the next subsections 3.5. [Also see §8·10]

For a symmetrical distribution, $\beta_1 = 0$. $\therefore$ Coeff. of skewness = 0.

Example 3·6. *Find out a coefficient of dispersion based on quartile deviation and a measure of skewness from the following table giving the wages of 230 persons :*

Wages (in Rs.)	*No. of persons*	*c.f.*	*Wages (in Rs.)*	*No. of persons*	*c.f.*
140–160	*12*	*12*	*220–240*	*50*	*157*
160–180	*18*	*30*	*240–260*	*45*	*202*
180–200	*35*	*65*	*260–280*	*20*	*222*
200–220	*42*	*107*	*280–300*	*8*	*230*

Here N/2 = 115 and the 115th person has a wage in the class 220–240.

$$\text{Hence median} = Q_2 = 220 + \frac{115-107}{50}\times 20 = \text{Rs. } 223{\cdot}20$$

Similarly $$Q_1 = 180 + \frac{57{\cdot}5-30}{35}\times 20 = \text{Rs. } 195{\cdot}71$$

and $$Q_3 = 240 + \frac{172{\cdot}5-157}{45}\times 20 = \text{Rs. } 246{\cdot}88$$

By example 3·3 Mean = Rs. 220·87

Standard deviation = Rs. 34·52

Coefficient of dispersion based upon quartile deviation

$$\frac{Q_3-Q_1}{Q_3+Q_1} = \frac{51{\cdot}17}{442{\cdot}59} = 0{\cdot}1156$$

$$\text{Measure of skewness} = \frac{\text{mean}-\text{mode}}{\text{standard deviation}} = \frac{220{\cdot}87-233{\cdot}00}{34{\cdot}52}$$

$$= -0.3514$$

Second measure of skewness

$$= \frac{Q_3 + Q_1 - 2Q_2}{Q_3 - Q_1} = -\frac{3.81}{51.17} = -0.07446$$

Ex. Show that all measures of skewness are independent of both origin and scale.

3·5. Pearson's β and γ-Coefficients. Karl Pearson defined the four important coefficients as follows :

$$\left.\begin{aligned} \beta_1 &= \frac{\mu_2^{\ 2}}{\mu_2^{\ 3}}, & \gamma_1 &= \sqrt{\beta_1} \\ \beta_2 &= \frac{\mu_4}{\mu_2^{\ 2}}, & \gamma_2 &= \beta_2 - 3 \end{aligned}\right\} \quad ...(3{\cdot}23)$$

The values of the four coefficients depend upon the first four moments only. The values of the two coefficients β_1 and β_2 enable us to know whether the given distribution is symmetrical and is relatively more or less flat than the normal curve.

It is to be noted that these four coefficients are all dimensionless, that is to say there are independent of origin and of scale of measurement of the variate [since μ_n has the dimension (variate)^n hence the above quotients have dimension zero].

In the case of symmetrical distribution, all the moments of odd order about the mean are equal to zero, since the terms of the sum in μ_{2r+1} $= \frac{1}{N} \sum_{i=1}^{n} f_i (x_i - \bar{x})^{2r+1}$ cancel in pairs and hence β_1 is also equal to zero. Thus β_1 gives a *measure of departure for symmetry*, that is to say, *of skewness.*

Thus *Test for symmetry* of a distribution is $\beta_1 = 0$

The term *Kurtosis* due to Karl Pearson, is described as :

"Given two frequency distributions, which have the same variability, as measured by the standard deviation, they may be relatively more or less flat topped than the normal curve' (discussed in Chapter 8). A frequency curve may, in other words, be symmetrical, but it may fail to be equally flat, topped with the normal curve. The relative flatness of the top is called the *kurtosis* or 'convexity of curve' and can be measured by β_2." It will be proved in §8·1·2 that for a normal distribution, $\beta_2 = 3$. Hence for any distribution, the quantity $\beta_2 - 3$ is called *excess of kurtosis* or simply its *excess.* Curves with values of β_2 less than 3 are called *Platykurtic* (curves which are flat, like 'B') while those with values of β_2 greater than 3 are called *Leptokurtic*. (curves which are thin). In between these two types, the curve is called Mesokuric (Curve A in Fig. 3.4).

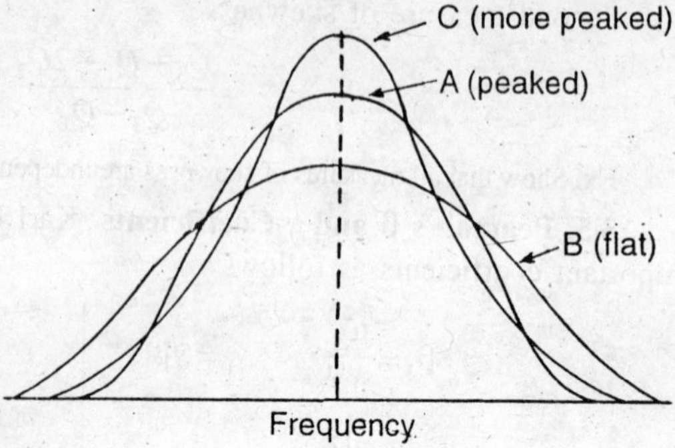

Fig. 3·4 Types of frequency curves

3·6. Factorial and Absolute Moments.

3·6·1. Factorial Moments. The *factorial moment* of order r about the origin $(x = 0)$ of a distribution is defined as

$$\mu'_{(r)} = \frac{1}{N} \sum_{i=1}^{n} f_i x_i^{(r)}, N = \sum_{i=1}^{n} f_i \qquad ...(3·24)$$

where $\qquad x^{(r)} = x(x-1)(x-2) \ldots (x-r+1).$

The factorial moment of order r about any point $x = a$ is defined by

$$\mu'_{(r)} = \frac{1}{N} \sum_{i=1}^{n} f_i (x_i - a)^{(r)}, \sum_{i=1}^{n} f_i = N$$

where $\qquad (x_i - a)^{(r)} = (x - a)(x - \overline{a+1}) \ldots (x - \overline{a+r-1}).$

Factorial moments in terms of ordinary moments and conversely

For $r = 1$, $\qquad \mu'_{(1)} = \frac{1}{N} \sum_{i=1}^{n} f_i x_i = \mu'_1 \qquad ...(1)$

For $r = 2$, $\qquad \mu'_{(2)} = \frac{1}{N} \sum_{i=1}^{n} f_i x_i^{(2)} = \frac{1}{N} \sum_{i=1}^{n} f_i x_i (x_i - 1)$

$$= \frac{1}{N} \sum_{i=1}^{n} f_i x_i^2 - \frac{1}{N} \sum_{i=1}^{n} f_i x_i = \mu_2' - \mu_1' \qquad ...(2)$$

For $r = 3$, $\qquad \mu_{(3)}' = \frac{1}{N} \sum_{i=1}^{n} f_i x_i^{(3)} = \frac{1}{N} \sum_{i=1}^{n} f_i (x_i)(x_i - 1)(x_i - 2)$

$$= \mu_3' - 3\,\mu_2' + 2\,\mu_1' \qquad ...(3)$$

$$\mu_{(4)}' = \frac{1}{N}\sum_{i=1}^{n} f_i\, x_i^{(4)} = \frac{1}{N}\sum_{i=1}^{n} f_i\, x_i (x_i - 1)(x_i - 2)(x_i - 3)$$

$$= \frac{1}{N}\sum_{i=1}^{n} f_i\, x_i\,(x_i^3 - 6x_i^2 + 11x_i - 6)$$

$$= \frac{1}{N}\sum_{i=1}^{n} f_i\, x_i^4 - 6\,.\,\frac{1}{N}\sum_{i=1}^{n} f_i\, x_i^3$$

$$+ 11\,.\,\frac{1}{N}\sum_{i=1}^{n} f_i\, x_i^2 - 6\,.\,\frac{1}{N}\sum_{i=1}^{n} f_i\, x_i$$

$$\mu_{(4)}' = \mu_4' - 6\,\mu_3' + 11\,\mu_2' - 6\,\mu_1' \qquad ...(4)$$

Conversely, we find *moments in terms of factorial moments.*

$$\mu_1' = \mu'_{(1)} \qquad \text{by (1)}$$

$$\mu_2' = \mu'_{(2)} + \mu'_{(1)} \qquad \text{by (2)}$$

$$\mu_3' = \mu'_{(3)} + 3\mu_2' - 2\mu_1' \qquad \text{by (3)}$$

$$= \mu'_{(3)} + 3[\mu'_{(2)} + \mu'_{(1)}] - 2\mu'_{(1)} = \mu'_{(3)} + 3\mu'_{(2)} + \mu'_{(1)}$$

$$\mu_4' = \mu'_{(4)} + 6\,\mu'_3 - 11\,\mu'_2 + 6\,\mu'_1 \qquad \text{by (4)}$$

$$= \mu'_{(4)} + 6\,[\,\mu'_{(3)} + 3\,\mu'_{(2)} + \mu'_{(1)}\,] - 11\,[\,\mu'_{(2)} + \mu'_{(1)}\,] + 6\,[\,\mu'_{(1)}\,]$$

$$= \mu'_{(4)} + 6\,\mu'_{(3)} + 7\,\mu'_{(2)} + \mu'_{(1)}.$$

3·6·2. Absolute moment. The *absolute moment* of order r about the origin ($x = 0$) of a distribution is defined as

$$\upsilon_r = \frac{1}{N}\sum_{i=1}^{n} f_i\,|\,x_i\,|^r,\ N = \sum_{i=1}^{n} f_i \qquad ...(3\cdot25)$$

Putting $r = 1$, we get that the mean deviation about any point is the first absolute moment about that point.

Ex. (*i*) Prove the following inequality for the absolute moments defined by (3·25) : $\upsilon_r^{2r} \le v^r_{r-1}\, v^r_{r+1}$

(*ii*) Replacing r successively by 1, 2, ..., r and multiplying all the inequalities thus formed, so that $v_r^{1/r} \le v_{r+1}^{\frac{1}{r+1}}$, $r = 1, 2, ...,$

[**Hint.** (*i*) Use the fact that $\frac{1}{N}\sum_{i=1}^{n} f_i\left[a\,|x_i|^{\frac{r-1}{2}}+b\,|x_i|^{\frac{r+1}{2}}\right]^2$ is non-negative.

In other words, the quadratic form in a and b *viz.*, $\nu_{r-1}a^2+2\nu_r ab+\nu_{r+1}b^2$ is non-negative, the condition for which is $\nu_{r-1}\nu_{r+1}-\nu_r^2\geq 0$],

3·7. Moments for a Bivariate Distribution. Let X and Y be two variates, *e.g.*, ages of bride and bridegroom ; heights and weights of students, quantity of manure applied and the corresponding yield obtained. Further, let f_i be the frequency of the pair (x_i, y_i), $i = 1, 2, ..., n$. Then we define the moment μ_{rs}' about the origin (0, 0) of order r in X and s in Y by

$$\mu_{rs}'=\frac{1}{N}\sum_{i=1}^{n} f_i\,x_i^r\,y_i^s, \quad \text{where} \quad N=\sum_{i=1}^{n} f_i \qquad ...(3\cdot26)$$

In particular,
$$\mu_{10}'=\frac{1}{N}\sum_{i=1}^{n} f_i\,x_i=\bar{x},$$
$$\mu_{01}'=\frac{1}{N}\sum_{i=1}^{n} f_i\,y_i=\bar{y}\,;$$
$$\mu_{11}'=\frac{1}{N}\sum_{i=1}^{n} f_i\,x_i y_i$$

Similarly, we define moment μ_{rs} about the mean $(\bar{x}, \bar{y})$ by

$$\mu_{rs}=\frac{1}{N}\sum_{i=1}^{n} f_i\,(x_i-\bar{x})^r(y_i-\bar{y})^s \qquad ...(3\cdot27)$$

In particular, $\mu_{10}=0, \mu_{01}=0$

$$\mu_{11}=\frac{1}{N}\sum_{i=1}^{n} f_i\,(x_i-\bar{x})(y_i-\bar{y}) \qquad ...(3\cdot28)$$

Also
$$\mu_{11}=\frac{1}{N}\sum_{i=1}^{n} f_i\,x_i\,y_i-\bar{x}_i\frac{1}{N}\sum_{i=1}^{n} f_i\,y_i-\bar{y}_i\frac{1}{N}\sum_{i=1}^{n} f_i\,x_i+\bar{x}\,\bar{y}$$
$$=\mu_{11}'-\bar{x}\,\bar{y} \qquad ...(3\cdot29)$$

This formula is very important in the study of *correlation* between two variates. μ_{11} is often called the *covariance* between the variates x and y and is denoted by cov (x, y).

Also
$$\mu_{20}=\sigma_x^2=\frac{1}{N}\sum_{i=1}^{n} f_i\,(x_i-\bar{x})^2=\mu_{20}'-\bar{x}^2 \qquad ...(3\cdot30)$$
$$\mu_{02}=\sigma_y^2=\frac{1}{N}\sum_{i=1}^{n} f_i\,(y_i-\bar{y})^2=\mu_{02}'-\bar{y}^2 \qquad ...(3\cdot31)$$

Exercises 3

1. (*a*) What do you understand by dispersion ? How is it measured ? What are different measures of dispersion ? Explain their uses.

(*b*) How does skewness differ from dispersion ? What is the object of measuring skewness ?

(*c*) Write short notes on :

(*i*) Variance, (*ii*) Quartile deviation, (*iii*) Coefficient of variation,

(*iv*) Skewness and kurtosis, (*v*) Positive and negative skewness.

(*d*) Show that for any frequency distribution :

(*i*) kurtosis is greater than unity.

(*ii*) coefficient of skewness is less than 1 numerically.

(*e*) Why do we, in general, calculate first four moments about the mean only and not higher moments ?

2. Calculate the arithmetic mean and the standard deviation of the following values of the world's annual gold output (in millions of pounds) for 20 different year :

94	95	96	93	87	79	73	69	68	67
78	82	83	89	95	103	108	117	130	97

[**Ans.** Mean = 90·15, standard deviation = 15·98836]

3. The following table gives the monthly wages of workers in a factory. Compute the standard deviation, quartile deviation, coefficient of variation and skewness.

Monthly wages (in Rs.)	*No. of workers,*	*Monthly wages (in Rs.)*	*No. of workers*
125–175	2	375–425	4
175–225	22	425–475	6
225–275	19	475–525	1
275–325	14	525–575	1
325–375	3		

[**Ans.** S.D. = Rs. 88·52, Q.D. = Rs. 51·46, C.V. = 31·8. Skewness = 0·7]

4. Find the mean and s.d. of the following series :

Expenditure	*No. of students*
Below Rs. 5	6
,, 10	16
,, 15	28
,, 20	38
,, 25	46

[**Ans.** Mean = Rs. 12·93, S.D. = Rs. 2]

5. The value of the arithmetic mean and the standard deviation of the following frequency distribution of a continuous variable derived by using both arbitrary working origin and scale are 135·3 lb. and 9·6 lb. respectively. Determine the actual class-intervals.

u :	– 4	–3	–2	–1	0	1	2	3	Total
f :	2	5	8	18	22	13	8	4	80

[**Hint.** Use $\bar{x} = a + h\bar{u},\ \sigma_x = h\sigma_u$]

6. For a frequency distribution of marks in History of 200 candidates (grouped in intervals 0–5, 5–10, ...) the mean and standard deviation (s.d.) were found to be 40 and 15. Later it was discovered that the score 43 was misread as 53 in obtaining the frequency distribution. Find the corrected mean and s.d. corresponding to the corrected frequency distribution.

[**Ans.** Mean = 39·95. Standard deviation = 14·975 approx.]

7. A student while calculating the mean and the standard deviation on 25 observations, obtained the following values :

mean = 56 cms. : standard deviation = 2 cms.

It was later discovered at the time of checking that he had wrongly copied down an observation as 64. What is the mean and s.d. if correct value is omitted ? [**Ans.** Mean = 55·67 cms., S.D. = 1·18 cms. approx.]

8. An analysis of the monthly wages paid to workers in two firms A and B, belonging to the same industry, gives the following results :

	Firm A	*Firm B*
Number of wage earners	586	648
Average monthly wages	Rs. 525	Rs. 475
Variance of the distribution of wage	10000	12100

(*a*) Which firm, A or B, pays out the larger amount as monthly wages ?

(*b*) In which firm, A or B is there greater variability in individual wages ?

(*c*) What are the measures of (*i*) average monthly wages and (*ii*) the variability in individual wages of all the workers in the two firms, A and B taken together ? (*I.A.S.*)

[**Ans.** (*a*) Firm B pays a larger amount as monthly wages.

(*b*) There is greater variability in individual wages in firm B

(*c*) (*i*) Combined arithmetic mean = Rs. 499.

(*ii*) Combined standard deviation of the two series = Rs. 105.

9. The scores of two golfers for 10 rounds each are :

A : 58, 59, 60, 54, 65, 66, 52, 75, 69, 52

B : 84, 56, 92, 65, 86, 78, 44, 54, 78, 68

Which may be regarded as the more consistent player ?

[**Hint.** Compare the coefficients of variation.]

[**Ans.** A is more consistent player]

10. Goals scored by two teams A and B in a football season were as follows :

No. of goals scored in a match	*Number of matches* A	B
0	27	17
1	9	9
2	8	6

3	5	5
4	4	4

By calculating the coefficient of variation in each case, find which team be considered more consistent.

[**Ans.** Team B]

11. (*a*) The means of two samples of sizes 50 and 100 respectively are 54·4 and 50·3 and the standard deviations are 8 and 7. Obtain the mean and standard deviation of the sample of size 150 obtained by combining the two samples. ***(B.Sc. Delhi)***

[**Ans.** Mean = 51·67, S.D. = 7·598 approx.]

(*b*) The first of two samples has 100 items with mean 15 and standard deviation 3. If the whole group has 250 items with mean 15·6 and standard deviation $\sqrt{13 \cdot 44}$, find the standard deviation of the second group. ***(B.A. Hons. Eco. Delhi) (B.Sc. Delhi 1972)***

[**Ans.** 4]

12. (*a*) Show that, if the variable takes the values 0, 1, 2, ..., n with frequencies proportional to the binomial coefficients, 1, ${}^nC_1, {}^nC_2, \ldots {}^nC_n$ respectively then the mean of the distribution is $\frac{1}{2}n$, the mean square deviation about origin is $\frac{1}{4}n\,(n+1)$ and the variance is $\frac{1}{4}n$.

(*b*) Show that, if the variable takes the values 0, 1, 2, ..., n with frequencies given by the terms of the binomial series $q^n, {}^nC_1 q^{n-1}p$, ${}^nC_2\, q^{n-2}p^2, \ldots, p^n$ where $p+q=1$, then the mean square deviation is n^2p^2+npq and the variance is npq.

13. Show that, if deviations are small compared with the mean M of a variable X, so that $\left(\dfrac{x}{M}\right)^3$ and higher powers of $\dfrac{x}{M}$ may be neglected, and σ is the s.d. of X,

(*a*) (*i*) $G = M\left(1 - \dfrac{1}{2}\dfrac{\sigma^2}{M^2}\right)$ (*ii*) $M^2 - G^2 = \sigma^2$ ***(B.A. Delhi 1971)***

(*iii*) $H = M\left(1 - \dfrac{\sigma^2}{M^2}\right)$ ***(M.Sc. Agra, B.A. Hons., Delhi)***

(*b*) mean $(\sqrt{x}) = \sqrt{M}\left(1 - \dfrac{\sigma^2}{8M^2}\right)$ approx.

(*c*) $V = \sqrt{\dfrac{2\,(M-G)}{M}}$ where V is the coefficient of variation.

14. From a sample of n observations, the arithmetic mean and variance are calculated. It is then found that one of the values, x_1, is in error, and should be replaced by x_1'. Show that the adjustment to the variance to correct this error is

$$\frac{1}{n}(x_1' - x_1)\left[(x_1' + x_1) - \frac{x_1' - x_1 + T}{n}\right]$$

where T is the total of the origin results.

(B.A. Stat. Hons. Delhi 1976 ; M.A. Agra; B.Sc. Kanpur 1974)

$$\left[\textbf{Hint.}\text{ Corrected variance} = \frac{1}{n}(x_1'^2 + x_2^2 + \ldots + x_n^2) - \left\{\frac{T - x_1 + x_1'}{n}\right\}^2\right]$$

15. Find the mean deviation from the mean and the standard deviation of the A.P. $a, a + d, a + 2d, ..., a + 2nd$ and prove that the latter is greater than the former. *(B.Sc. Delhi 1975)*

$$\left[\textbf{Ans.}\text{ Mean deviation} = d \times \frac{n(n+1)}{2n+1},\ \text{S.D.} = d \times \sqrt{\frac{n(n+1)}{3}}\right]$$

16. If in a series of measurements we obtain m_1 values of magnitude x_1, m_2 of magnitude x_2, and so on, and if $\bar{x}$ is the mean value of all the measurements, prove that the standard deviation is

$$\sqrt{\frac{\Sigma m_r (k - x_r)^2}{\Sigma m_r} - \delta^2}$$

where $\bar{x} = k + \delta$, and k is any constant. ***(M.A., Agra ; B.Sc. Bombay 1974)***

[**Hint.** $\sigma^2 = \Sigma m_r (x_r - \bar{x})^2 / \Sigma m_r,\ \bar{x} = \Sigma m_r x_r / \Sigma m_r$

$\therefore \quad \sigma^2 = \Sigma m_r (x_r - k + k - \bar{x})^2 / \Sigma m_r =$ etc.]

17. (*a*) The first three moments of a distribution about the value 2 of the variable are 1, 16 and –40. Show that the mean is 3, the variance = 15 and $\mu_3 = -86$. Also show that the first three moments about $x = 0$ are 3, 24 and 76. ***(B.Sc. Bombay ; B.Sc. Lucknow)***

(*b*) The first four moments of a distribution about the value 5 of the variable are 2, 20, 40 and 50. Show that the mean is 7, variance 16, $\mu_3 = -64$, $\mu_4 = 162$, $\beta_1 = 1$ and $\beta_2 = 0{\cdot}63$.

(B.Sc. Lucknow, B.Sc. Kanpur 1969, B.Sc. Utkal 1973, M.Sc. Aligarh)

(*c*) For a certain the distribution, the mean is 10, variance is 16, γ_1 is + 1 and β_2 is 4. Find the first four moments about the origin. ***(B.Sc. Delhi)***

[**Ans.** $\mu_1' = 10, \mu_2' = 1167\ \mu_3' = 1544$ and $\mu_4' = 82384$].

18. (*a*) Calculate the first four moments about the mean for the following data. Also calculate β_1 and β_2.

x :	1	2	3	4	5	6	7	8	9
f :	1	6	13	25	30	22	9	5	2

(*b*) In a certain distribution, the first four moments about the point 4 are $-1{\cdot}5, 17, -30$ and 108. Calculate β_1 and β_2 and state whether the distribution is leptokurtic or platykurtic.

(B.A. Delhi 1971 ; B.A. & B.Sc., Lucknow 1979)

[**Ans.** (*a*) $\mu_1' = -0{\cdot}09, \mu_2 = 2{\cdot}4873, \mu_3 = 0{\cdot}6789, \mu_4 = 18{\cdot}3358$ approx.,

$\beta_1 = 0{\cdot}0299$ approx., $\beta_2 = 2{\cdot}9627$

(*b*) $\mu_2 = 14{\cdot}75$, $\mu_3 = 39{\cdot}75$, $\mu_4 = 142{\cdot}31$, curve is platykurtic.]

19. Show that, if the class interval of a grouped distribution is less than one-third of the calculated standard deviation (S.D.), Sheppards' adjustment makes a difference of less than $\frac{1}{2}$% in the estimates of *S.D.* **(B.Sc. Agra)**

20. Show that for discrete distributions

(*i*) $\beta_2 \geq 1$. ***(M.Sc. Agra, B. Sc. Delhi 1973, B.Sc. Lucknow)***

(*ii*) $\beta_2 > \beta_1$. ***(B.A. Delhi 1971 ; B.Sc. Hons. Delhi 1977)***

$\left[\textbf{Hint.}\ (i)\ \text{We have to prove } \frac{1}{N}\sum_i f_i y_i^2 \geq \left[\frac{1}{N}\sum_i f_i y_i\right]^2\right.$

where $y_i = (x_i - \bar{x})^2$. *i.e.*, $\sigma_y^2 \geq 0$, which is true. Hence etc.]

21. Show that, for bivariate distributions, the moments about the mean are invariant for the change of origin but not for the change of scale.

22. For a distribution of 250 heights, calculations showed that the mean, standard deviation, β_1 and β_2 were 54 inches. 3 inches 0 and 3 inches respectively. It was however discovered on checking that the two items 64 and 50 in the original data were wrongly written in place of correct values 62 and 52 inches respectively. Calculate the correct frequency constants. ***(B.A. Hons. Delhi)***

[Ans. Mean remains unltered, *s.d.* $= 2{\cdot}968$, $\beta_1 = {\cdot}004$, $\beta_2 = 2{\cdot}815$]

23. In calculating the moments of a frequency distribution based on 100 observations the following results were obtained :

mean = 9, variance = 19, $\beta_1 = 0{\cdot}7$, $\beta_2 = 4$

But later on it was noticed that one observation 12 was read as 21. Obtain the correct value of the first four central moments. ***(I.A.S. 1963)***

[Ans. Corrected mean = $8{\cdot}91$, $\mu_2 = 17{\cdot}64$, $\mu_3 = 57{\cdot}05$, $\mu_4 = 1257{\cdot}15$, $\beta_1 = 0{\cdot}54$

$\beta_2 = 4{\cdot}04$]

24. (*a*) If *r* be the range and $S = \left[\frac{1}{n-1}\sum_{i=1}^{n}(x_i - \bar{x})^2\right]^{1/2}$ be the standard deviation

of a set of *n* observations $x_1, x_2, ..., x_n$, then show tnat $S \leq r\left(\frac{n}{n-1}\right)^{1/2}$

(B.A., Hons. Delhi 1967, 1982, B.Sc. Hons. Punjab, 1993)

[Hint. $r = x_i - x_k$ where $x_i = \max\ (x_1, x_2, ..., x_n)$, $x_k = \min.\ (x_1, ..., x_n)$)

(*b*) If *r* be the range and *s* the s.d. of a set of observations $x_1, x_2, ..., x_n$, prove that $s \leq r$.

[Hint. $x_i - \bar{x} \leq r \ \therefore\ s^2 = \frac{1}{N}\sum f_i (x_i - \bar{x})^2 \leq \frac{1}{N}\sum f_i r^2; ...\ s \leq r$

25. Prove that (*i*) $\mu_r' = \sum_{j=0}^{r} {}^{r}C_j\, \mu'_{r-j}\, \mu_1'^{\,j}$

(*ii*) $\mu_r = \sum_{j=0}^{r} \binom{r}{j} \mu'_{r-j} (-\mu'_1)^j$ **(B.Sc. Sardar Patel 1967)**

$$\left[\textbf{Hint. } (i)\ \mu_r'(a) = \frac{1}{N}\sum_{i=1}^{n} f_i (x_i - a)^r = \frac{1}{N}\sum_{1}^{n} f_i \left[(x_i - b) - (a - b)\right]^r\right.$$

$$\mu_r' = \frac{1}{N}\sum_{1}^{n} f_i\,[(x_i - b) - c]^r \quad \text{where } c = a - b.]$$

26. (*a*) If δ_r is the *r*th absolute moment about zero, use the mean value of

$$\left[u\,|x|^{\frac{r-1}{2}} + v\,|x|^{\frac{r+1}{2}}\right]^2.$$

to show that $\delta_r^{2r} < \delta_{r-1}^{\ r}\delta_{r+1}^{\ r}$

From this, derive the following inequalities :

(*i*) $\delta_r^{\ r+1} \le \delta_{r+1}^{\ r}$ (*ii*) $\delta_r^{\ 1/r} \le \delta_{r+1}^{\ 1/r+1}$ **(I.A.S.)**

(*b*) If λ and μ are positive integers, $\lambda < \mu$ and if β_i stands for the *i*th absolute moment of a distribution, show that $\beta_\lambda^{\ 1/\lambda} < \beta_\mu^{\ 1/\mu}$ **(I.A.S.)**

(*c*) Prove the following relation for Pearson's β-coefficients for skewness and kurtosis : $\beta_2 > \beta_1 + 1$ **(B.Sc. Delhi 1971 ; B.Sc. Hons. Gauhati 1968 ; B.A. Hons. Delhi 1977)**

(*d*) For any real *k*, prove that $\beta_2 \ge \beta_1 - (2k + k^2)$

Deduce that (*i*) $\beta_2 \ge \beta_1$ (*ii*) $\beta_2 \ge 1$.

[**Hint.** (*d*) Let $Y(t) = E[x^2 + tx + k\,\mu_2\,]^2 \ge 0 \quad \forall\ t,\ \mu_2 = Ex^2.$

$$= E\,[\,x^4 + t^2x^2 + k^2\mu_2^{\ 2} + 2\,tx^3 + 2\,k\,\mu_2\,x^2 + 2k\,\mu_2\,tx\,]$$

$$= t^2\mu_2 + 2\,t\,\mu_3 + (\,\mu_4 + k^2\mu_2^{\ 2} + 2k\mu_2^{\ 2})\text{ for all } t$$

Since $Y(t)$ is quadratic in *t*, $Y(t) \ge 0$ for all *t*

$\therefore\ (2\,\mu_3)^2 - 4\mu_2\,(\mu_4 + k^2\,\mu_2^{\ 2} + 2k\,\mu_2^{\ 2}) \le 0$ [*disc.* ≤ 0]

Divide both sides by $4\,\mu_2^{\ 3}\ (> 0)$ to get $\beta_1 - \beta_2 - (2k + k^2) \le 0$

$\therefore\quad \beta_2 \ge \beta_1 - (2k + k^2)$

Put $k = 0$ and $k = -1$ resp. to get the deductions.]

27. (a) Prove that if the *k*th moment of *X* exists, then all the moments of order less than *k* exist.

[**Hint.** Since $|x - a|^{k-1} \le |x - a|^k + 1$ for all *x*, it follows that

$$E\,|x - a|^{k-1} \le E\,|x - a|^k + 1.\text{ Hence etc.}$$

(*b*) Prove that Var $X = 0 \Rightarrow P[X = EX] = 1$.

[**Hint.** By hyp., Var $X = E[X - EX]^2 = 0$

$\therefore\ [X - E(X)]^2 = 0$ with probability 1

$\Rightarrow X - E(X) = 0$ with probability 1 *i.e.* $P\,[X = EX] = 1$.

4

Theory of Probability

4·1. Introduction. Probability in every day life.

The concept of probability is extremely important, as it has very extensive applications in the development of all physical sciences.

The* history of probability suggests that its theory developed with the study of games of chance, such as rolling a dice, drawing a card from a pack of cards, etc. Two French gamblers had once decided that any one person who will first get a 'particular point' will win the game, but for quite a long time that point could not be attained. This and similar problems were then posed to the great French mathematician Blaise Pascal, who after consulting another great French mathematician, gave the solution of the problems and then laid down a strong foundation of Probability. Later on, another French mathematician, Laplace, improved the definition of probability. Apart from games, uncertainty prevailed in every sphere of life.

Every day we have been coming across statements like the ones mentioned below. For example,

'Probably it will rain to -night'. It is quite likely that there will be a good yield of cereals this year and so on'. These statements indicate that even the men in the street has got some idea of probability. To put the word probability on a better footing, we define it. But before defining it, we have to explain a few terms, which we shall be making use in its context or in its definition.

4·1·1. Random Experiments or Observations

By a statistical or random experiment, we mean an action or an operation, which can produce some well -defined outcomes. There are experiments, which when repeated under essentially homogeneous and similar *i.e.* identical conditions, always produce the same result *i.e. unique* outcome or unique *events*. Such experiments are known as *deterministic*. There is another type of experiments which do not produce the same result

---

* For 'History of Probability'. see Tod Hunter treatise on Probability.

or outcome on every trial and are called random experiments. In other words, if a random experiment is repeated under essentially identical conditions, the results will vary at random, as it is impossible to predict about the performance of the experiment. Thus for example, if we toss an honest coin or throw an unbiased die, we may not get the same result or outcome even if we bear in mind the identical conditions. Clearly, we cannot say with certainty or cannot predict in advance what result or outcome is going to occur when the experiment is performed. It should be clear that the outcome has to be one of several possibilities associated with the random experiment. The totality of all the outcomes, of the random experiment is denoted by Greek alphabet Ω or English alphabet S and is called the sample space. Each outcome or element of this sample space is known as the sample point or *event* and is denoted by Greek alphabet ω. Thus the outcomes of a random experiment of throwing (*i*) a coin are 'head' (H) and 'tail' (T), (*ii*) two coins (or two tosses of a coin) are HH, HT, TH, TT.

4·2·1. Definitions of various terms used in the definition of probability.

Equally likely events. The possibilities or events are said to be *equally likely* when we have no reason to expect any one rather than the other. For example, as the result of drawing a card from a well -shuffled pack, any card may appear in the draw, so that the 52 different events are equally likely.

Mutually exclusive events. Events are called mutually exclusive (m.e.) or disjoint or incompatible, if the occurrence of one of them precludes the occurrence of all the others. On the contrary, events are compatible if it is possible for them to happen simultaneously. For instance in the rolling of two dice, the cases of the face marked 5 appearing on the dice and face 5 appearing on the other are compatible since in rolling two dice it is possible to get face 5 on each dice.

Exhaustive events. Events are said to be exhaustive when they include all possibilities. In throwing a coin, the turning up of head and of a tail are exhaustive events assuming of course that the coin cannot rest on its edge.

Favourable event. The trials which entail the happening of an event are said to be *favourable* to the event. For example, in the tossing of a die, the number of favourable events to the appearance of a multiple of 3 are two *viz.*, 3 and 6.

The following is the broad classification of the concepts used in probability :

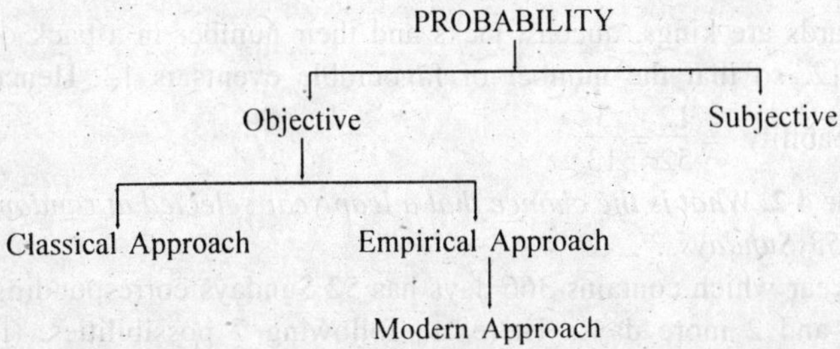

We shall now consider two definitions of probability.

(*i*) Mathematical or a *priori* or classical

(*ii*) Statistical or empirical

4·2·2. Mathematical (or a priori) definition of probability. *If there are n exhaustive, mutually exclusive and* **equally likely** *events, probability of the happening of A is defined as the ratio m/n, m is favourably to A.*

Thus probability is a concept which measures numerically the degree of certainty or uncertainty of the occurrence of an event.

Thus denoting the probability (or chance) of the happening of an event A by p, we have

$$P(A) = p = \frac{m}{n} \qquad \ldots(4\cdot1)$$

This definition of probability was given by James Bernoulli.

This gives the *numerical measure* of probability. Clearly, p is a positive number not greater than unity, so that $0 \le p \le 1$.

Since the number of cases in which the event A will not happen is n – m, the probability q that the event A will not happen is given by

$$P(\overline{A}) = q = \frac{n-m}{n} = 1 - \frac{m}{n} = 1 - p = 1 - P(A) \qquad \ldots(4\cdot2)$$

so that $p + q = 1,\ 0 \le q \le 1$.

An event A is *certain* to happen if all the trials are favourable to it and then the probability of its happening is unity, while if it is *certain* not to happen, its probability is zero.

Thus if $p = 0$, then the event is an impossible event while if $p = 1$, the event is *certain.*

Examples in which the number of favourable events can either be counted or enurmerated with the help of *Permutation and Combination* are given below :

Example 4·1. *Find the chance that if a card is drawn at random from an ordinary pack, it is one of the court cards.*

Court cards are kings, queens, jacks and their number in a pack of 52 cards is 12, so that the number of favourable events is 12. Hence required probability $=\frac{12}{52}=\frac{3}{13}$.

Example 4·2. *What is the chance that a leap year selected at random will contain 53 Sundays ?*

A leap year which contains 366 days has 52 Sundays corresponding to 52 weeks and 2 more days. There are following 7 possibilities. (i) Sunday and Monday (*ii*) Monday and Tuesday, (*iii*) Tuesday and Wednesday, (*iv*) Wednesday and Thursday (*v*) Thursday and Friday, (*vi*) Friday and Saturday, and (*vii*) Saturday and Sunday. Out of these 7 possibilities, there are two favourable outcomes *viz.*, (*i*) and (*vii*). Thus the required probability $=\frac{2}{7}$.

Example 4·3. *An urn contains 9 balls, two of which are red, three blue and four black. Three balls are drawn from the urn at random, that is every ball has an equal chance of being included in the three. What is the chance that*

(*i*) *three balls are of different colours ?*

(*ii*) *two balls are of the same colour and third of different ?*

(*iii*) *the balls are of the same colour ?*

(*i*) The number of ways in which three balls can be drawn from 9 balls is ${}^9C_3$ or 84, and these are equally likely, exhaustive and mutually exclusive cases.

A red ball can be drawn in 2 ways, a blue in 3 and a black in 4 ways, so that the total number of ways of drawing three differently coloured balls is $2\times3\times4=24$.

Therefore the required chance is $\frac{24}{84}$ or $\frac{2}{7}$.

(*ii*) Two blue balls can be drawn in ${}^3C_2$ ways and then a red or black ball in 6 ways so that the two blue balls and another coloured ball can be drawn in $6\times{}^3C_2$ or 18 ways. Two black balls can be drawn in ${}^4C_2$ ways and another coloured ball in 5 ways so that two black and another coloured ball can be drawn in $5\times{}^4C_2$ or 30 ways. Similarly, the number of ways in which two red balls and another coloured ball can be drawn in $7\times{}^2C_2$ or 7 ways. The total number of ways in which two balls of the same colour and a third of a different colour may be drawn is $18+30+7=55$.

Therefore the required chance is $\frac{55}{84}$.

(*iii*) Three blue balls can be drawn in 1 way and 3 black balls in $^4C_3$ or 4 ways so that the total number of ways of drawing three balls of the same colour is 5 and the chance is $\frac{5}{84}$.

Example 4·4. *In a hand at whist, what is the chance that the 4 kings are held by a specified player ?*

In a hand at whist, the specified player must have 13 cards including 4 kings. Thus he should have 9 cards of other type or types. 4 kings can be drawn in $^4C_4$ ways and 9 cards can be drawn out of the remaining 48 cards in $^{48}C_9$ ways. The total number of ways in which 4 kings and 9 cards of other type or types can be drawn in $^4C_4 \times {}^{48}C_9$ from a full pack of 52 cards. Any 13 cards can be drawn in $^{52}C_{13}$ ways.

Therefore the required chance $= \dfrac{^4C_4 \times {}^{48}C_9}{^{52}C_{13}} = \dfrac{11}{4165}$.

Example 4·5. *Each coefficient in the equation $ax^2+bx+c=0$ is determined by throwing an ordinary die. Find the probability that the given equation will have real roots.* **(*I.A.S. 1992 ; M.Sc., Bombay, 1992*)**

The equation $ax^2+bx+c=0$ will have its root real if $b^2 \geq 4ac$.

Now each coefficient of the given equation is determined by throwing an ordinary die so that each of the coefficients *a*, *b* or *c* can take the values 1 to 6.

Hence total number of possible outcomes or equally likely cases $=6\times6\times6=216$.

The number of favourable cases (satisfying $b^2 \geq 4ac$) are obtainable as follows :

ac	*a*	*c*	*4ac*	*b*(with $b^2 \geq 4ac$)	*No. of cases*
1	1	1	4	2,3,4,5,6	5
2	1 2	2 1	8	2,4,5,6	$2\times4=8$
3	1 3	3 1	12	4,5,6	$2\times3=6$
4	1 4 2	4 1 2	16	4,5,6	$3\times3=9$
5	1 5	5 1	20	5,6	$2\times2=4$

6	1, 6, 2, 3	6, 1, 3, 2	24	5,6	$4 \times 2 = 8$
8	2, 4	4, 2	32	6	$2 \times 1 = 2$
9	3	3	36	6	1
					Total 43

Clearly ac cannot take the values 7, 10, 11, 12 so as to satisfy $b^2 > 4ac$ and further none of the coefficients can be greater than 6.

Hence the required probability $= \dfrac{43}{216} = 0{\cdot}19907$

Ex. Find the probability that the roots of the equation in example 4·5 are equal. **Ans.** 3/216

Example 4·6. *If four squares are chosen at random on a chess-board, find the chance that they should be in a diagonal line.*

(*B.Sc. Stat. Hons. Delhi, 1988*)

A chess-board, has $8 \times 8 = 64$ equal squares (see Fig. 4·1) so that the number of equally likely cases of choosing 4 squares out of 64 squares will be ${}^{64}C_4 = (64 \times 63 \times 62 \times 61) / (4 \times 3 \times 2 \times 1)$.

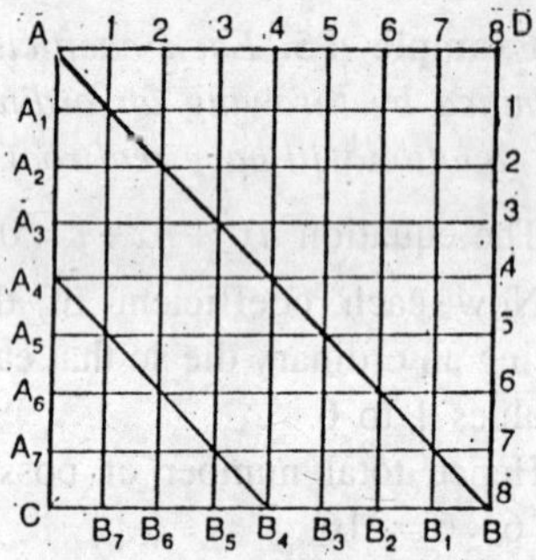

Fig. 4·1. Chess board with equal 64 squares

Now we want to find the numbers of ways in which 4 squares selected at random will be in a diagonal line parallel to AB.

Consider half of the chess-board parallel to ΔABC. Now clearly the number of ways in which 4 selected squares are along the lines $A_1 B_1$, $A_2 B_2$, $A_3 B_3$, $A_4 B_4$ are ${}^7C_4$, ${}^6C_4$, ${}^5C_4$, ${}^4C_4$ respectively and the numbers of ways for the 4 squares to be in line with AB will be ${}^8C_4$. Hence the required number of ways that 4 squares lie in a diagonal

$$= 2\,[\,2\,({}^7C_4 + {}^6C_4 + {}^5C_4 + {}^4C_4\,] + 2 \times {}^8C_4$$

$$= 4\,[\,35 + 15 + 5 + 1\,] + 2 \times 70 = 224 + 140 = 364$$

$$\therefore \text{ required probability } = \frac{364 \times 4 \times 3 \times 2 \times 1}{64 \times 63 \times 62 \times 61} = \frac{91}{158844}.$$

Example 4·7. Let $A_1, A_2, \ldots, A_n$ *be independent events and* $P(A_i) = p_i$, $i = 1, 2, \ldots, n$. *Further let p be the probability that none of the events* *occurs, then show that* $p \le e^{-\Sigma p_i}$ **(*M.Sc., Agra, 1987*)**

We have $p = P(\overline{A_1} \cap \overline{A_2} \cap .... \cap (\overline{A_n}) = P(\overline{A_1})\, P(\overline{A_2}) .....P(\overline{A_n})$

$= (1 - p_1)(1 - p_2) .....(1 - p_n)$

since $A_1, A_2, ..., A_n$ are independent, See cor. 1 §4·5, page 89.

$$= \prod_{i=1}^{n} (1 - p_i) \le \prod_{i=1}^{n} e^{-p_i}.$$

since $1 - x \le e^{-x}$ for $0 \le x \le 1$ which can be established by showing that for

$0 < x < 1,\ \log(1 - x)^{-1} = -\log(1 - x) = x + \dfrac{x^2}{2} + ... > x$

$\Rightarrow\ (1 - x)^{-1} > e^x\ \Rightarrow\ 1 - x < e^{-x}.$

Example 4·8. *What is the probability that at least two out of n people have the same birthday ? Assume 365 days in a year and that all days are equally likely.* **(*B.Sc. Delhi*)**

Since the birthday of a person can fall on any one of 365 days, the number of exhaustive cases for n persons

$= 365 \times 365 \times ....365\ n$ times

If the birthdays of all the n persons fall on different days of the year, then the number of favourable cases would be

$365(365 - 1)(365 - 2) .... (365 - \overline{n - 1})$

Hence required. probability =

[1– probability that no two persons have birthdays of the same day]

$$= 1 - \frac{365 \times (365 - 1) \times (365 - 2) \times ...... \times (365 - \overline{n - 1})}{365 \times 365 \times 365 \times ...... \times 365}$$

$$= 1 - \left(1 - \frac{1}{365}\right)\left(1 - \frac{2}{365}\right) .... \left(1 - \frac{n-1}{365}\right).$$

Example 4·9. *From an urn containing N_1 white and N_2 black balls ($N = N_1 + N_2$), balls are successively drawn without replacement one by one. What is the probability that i black balls will precede the first white ball ?*

Suppose all the balls are drawn one by one, without replacement and arranged in N different rooms. The total number of event-points is the number of ways in which N distinguished balls can be arranged in N rooms *i.e.*, N. Now the number of favourable event-points are given when the first i rooms are occupied by black balls, the $(i + 1)$ *th* room by a white ball and the last $N - i - 1$ rooms are filled by the remaining $N - i - 1$ balls in any manner whatsoever and therefore

$= N_2(N_2 - 1) .... (N_2 - i + 1) \cdot N(N - i - 1)\,!$

$$\therefore \quad \text{required probability} = \frac{N_1 N_2 (N_2 - 1) \ldots\ldots (N_2 - i + 1)}{N (N - 1) \ldots\ldots (N - i)}, i \geq 1$$

Example 4·10. *When n dice are thrown, find the probability that the sum of the points on the dice has a prescribed value s.*

The total number of equally likely events is 6^n while the number of favourable events is the number of events, which is the number of different sets of integers $(x_1, x_2, \ldots, x_n)$ such that $(x_1 + x_1 + \ldots + x_n) = s$ where $x_1, x_2, \ldots\ldots, x_n$ can take values 1, 2, ..., 6, though not necessarily in this order. But this number

$$= \text{coefficient of } x^s \text{ in the expansion of } (x + x^2 + \ldots\ldots + x^6)^n$$

$$= \quad " \quad " \quad " \quad " \quad " \quad " \quad [x(1 - x^6)/(1 - x)]^n$$

where $(1 - x^6)^n = \sum_{i=0}^{n} (-)^i \binom{n}{i} x^{6i}$, $(1 - x)^{-n} = \sum_{i=0}^{\infty} \binom{n+j-1}{n-1} j$

Hence $(x + x^2 + x^3 + x^4 + x^5 + x^6)^n$

$$= \sum_{i=0}^{n} \sum_{i=0}^{\infty} (-)^i \binom{n}{i} \binom{n+j-1}{n-1} x^{n+6i+j}$$

If $n + 6i + j = s, j = s - 6i - n$ and as $j \geq 0$, $i \leq (s - n)/6$,

Then the coeff. of $x^s = \Sigma (-)^i \binom{n}{i} \binom{s - 6i - n}{n - 1} = a$, say

where i ranges from 0 to the greatest integer $\leq (s - n)/6$.

$\therefore$ Required probability $= a / 6^n$.

Objections to a priori or classical or mathematical definition of probability are :

1. It involves the phrase 'equally likely' or 'equally probable'. What happens when events are not equally likely ? For example, if we have a coin known to be biased in favour of heads, that is, a head is more likely to appear than a tail, then what is the probability of a head ? The classical definition leaves us completely helpless here. Another example is that of finding the probability of a ceiling fan falling down, when it is tied with the hook in the ceiling.

2. The classical definition fails when the total number of possible outcomes is infinite. It also restricts value of probability to rational numbers only.

3. It may not be practically possible to enumerate all the equally likely cases. For example, it is practically not possible to count the hair in the head of a human being, when we are interested to find the probability of grey hair in the head. Or, it is not possible to count the number of leaves in a tree, when we are interested in finding the

probability of pale leaves in a tree.

4·2·3. Von Mises's Statistical (or Empirical) definition of probability. *If trials be repeated a great number of times, under essentially the same conditions, then the limit of the ratio of the number of times that an event happens to the total number of trials, as the number of trials increases indefinitely, is called the probability of the happening of the event. It is assumed that the ratio approaches a finite and a unique limit.*

Symbolically, if m be the number of times in which the event A happens in a series of n trials, then the probability, p, of the happening of this even t is given by

$$P(A) = p = \lim_{n \to \infty} \frac{m}{n} \qquad \ldots(4 \cdot 3)$$

provided that the limit is finite and unique.

Note. The two definitions of probability are apparently different. In the 'a priori' definition, it is the relative frequency of favourable events to the total number of events, while in empirical definition, it is the limit of the relative frequency of the happening of the event under consideration. However, in § 7·4 we shall see how the two definitions can be reconciled in same sense.

4·3. Notations. Let $A_1, A_2, \ldots\ldots, A_n$, be n events, then, in dealing with the theory of probability, we shall find it convenient to use the following notations :

$P(A_i)$ denotes the probability for the event A_i to materialize.

$P(\overline{A_i})$ denotes the probability for the event A_i *not* to materialize or complement event $\overline{A_i}$

$\overline{A_i}$ will be called the event opposite to A_i. Clearly A_i and $\overline{A_i}$ are exhaustive and mutually exclusive and (4·2) gives

$$P(A_i) + P(\overline{A_i}) = 1. \quad \therefore \quad P(\overline{A_i}) = 1 - P(A_i)$$

$P(A_i + A_j)$ denotes the probability of occurrence of *at least* one of the events A_i, A_j. We write $A_1 \cup A_2$ as $A_1 + A_2$ when $A_1 \cap A_2 = \phi$.

Similarly, $P(A_1 + A_2 + \ldots + A_n)$ denotes the probability of occurrence of *at least* one of the n events $A_1, A_2, \ldots, A_n$.

$P(A_i A_j)$ denotes the probability of the occurrence of both the events A_i and A_j. $A_i A_j$ will be called a *compound* event.

Many a time, $A_1 \cap A_2$ is written as $A_1 A_2$.

$P(A_1 \mid A_j)$ denotes the *conditional* probability for the event A_i to happen, when *it is known* that the event A_j has already happened.

$P(A_i \mid A_j A_k)$ denotes the conditional probability for the event A_i to happen, when *it is known* that A_j and A_k have already happened.

Two events are said to be *independent* (in probability sense) if the probability of happening of one event does not depend on the happening or not happening of the other event. For example, if from a bag containing a white and *b* black balls, we draw a ball and replace it before the next draw is made, the chance of the ball being white or black in the second draw is indepen- dent of the result of the first draw that is of the colour of the first ball.

De-Morgan rule : $(\bigcap_i A_i)^c = \bigcup_i A_i^c$; $(\bigcup_i A_i)^c = \bigcap_i A_i^c$

4·4. Additive Law of Probability or Theorem of Total Probability. *The probability that one of several* **mutually exclusive** *events $A_1, A_2, ..., A_n$, will happen, is the sum of the probabilities of the separate events.* In symbols,

$$P(A_1 + A_2 + \dots + A_n) = P(A_1) + P(A_2) + \dots + P(A_n)$$

Proof. Let N be the total number of exhaustive, mutually exclusive and equally likely events and out of these let $m_1, m_2, ..., m_n$ be respectively favourable to the events $A_1, A_2, \dots, A_n$. Since the n events are mutually exclusive these $m_1, m_2, ..., m_n$ events are quite distinct and non-overlapping. Therefore the total number of events which are favourable to either A_1 or A_2 or ...... or A_n is $m_1 + m_2 + \dots + m_n$.

Hence $$P(A_1 + A_2 + \dots + A_n) = \frac{m_1 + m_2 + \dots + m_n}{N} = \frac{m_1}{N} + \frac{m_2}{N} + \dots + \frac{m_n}{N}$$

$$= P(A_1) + P(A_2) + \dots + P(A_n) = \sum_{i=1}^{n} P(A_i)$$

which proves the theorem.

Cor. 1. If an event A consists of n mutually exclusive forms A_1, $A_2, \dots, A_n$ so that A happens whenever any one of these events happens and conversely, then $A = A_1 + A_2 + \dots + A_n$

and $$P(A) = P(A_1) + P(A_2) + \dots + P(A_n)$$

Cor. 2. If n mutually exclusive events $A_1, A_2, ..., A_n$ are exhaustive also, so that the probability of *at least* one of the n events to materialize is a certainty, then $P(A_1 + A_2 + \dots + A_n) = 1$

or $$P(A_1) + P(A_2) + \dots + P(A_n) = 1$$

Def. Two events A and B are said to be equivalent ($A = B$), if the happening or not happening of either implies the happening or not happening of the other.

Ex. 1. Prove that if A, B and C are any event of a sample space S.

(*i*) $A+B=B+A$ (*ii*) $AB=BA$

(*iii*) $A+(B+C)=(A+B)+C=A+B+C$ (*iv*) $\overline{A+B}=\overline{A}+\overline{B}$

(*v*) $\overline{AB}=\overline{A}+\overline{B}$ (*vi*) $A+A=A,\ AA=A$ (*vii*) $A(B+C)=AB+AC$

(*viii*) $S+A=S$ (*ix*) $SA=A$ (*x*) $\overline{S}=O, \overline{O}=S$

Ex. 2. Prove that (*i*) $P(A\overline{A})=0$ (*ii*) $P(A \mid B)=P(A)$ and $P(B \mid A)=P(B)$.

if A and B are independent events.

(*iii*) $P(AB)=0$ if A and B are mutually exclusive.

Ex. 3. Let A, B, C be three arbitrary events. Find expressions for the following events :

(*i*) Only A occurs ; (*ii*) Both A and B, but not C, occur ; (*iii*) All three events occur ; (*iv*) At least one event occurs ; (*v*) At least two events occur ; (*vi*) One and no more occurs ; (*vii*) Two and no more occur ; (*viii*) None occurs.

[**Ans.** (*i*) $A\overline{BC}$; (*ii*) $AB\overline{C}$; (*iii*) ABC ; (*iv*) $A+B+C$; (*v*) $AB+BC+CA$; (*vi*) $A\overline{BC}+\overline{A}B\overline{C}+\overline{AB}C$; (*vii*) $AB\overline{C}+CA\overline{B}+BC\overline{A}$; (*viii*) $\overline{A}\,\overline{B}\,\overline{C}$.

Ex. 4. Use the theorem of total probability to show that

(*i*) $P(A)=P(A.B)+P(A\overline{B})$ (*ii*) $P(A_1+A_2+\ldots+A_n)+P(\overline{A_1}\,\overline{A_2}\ldots\ldots\overline{A_n})=1$.

[use Ex. 1 (*v*) above]

(*iii*) $P(AB)+P(\overline{A}B)+P(A\overline{B})+P(\overline{A}\,\overline{B})=1$.

(*iv*) $P(A+B)=P(AB)+P(\overline{A}B)+P(\overline{B}A)$.

Ex. 5. If $A_1, A_2, \ldots, A_n$ are n events, then prove that

(*i*) $\overline{A_1+A_2+\ldots+A_n}=\overline{A_1}\,\overline{A_2}\ldots\ldots\overline{A_n}$;

(*ii*) $\overline{A_1 A_2 \ldots\ldots A_n}=\overline{A_1}+\overline{A_2}+\ldots\ldots+\overline{A_n}$.

4·4·1. Generalization of the Theorem of Total Probability.*

If the events $A_1, A_2, \ldots\ldots, A_n$, are not mutually exclusive, then the probability of the happening of at least one of the n events is given by $P(A_1+A_2+\ldots+A_n)$ and is

$$P\left(\bigcup_{i=1}^{n} A_i\right)=\sum_{i=1}^{n} P(A_i)-\sum_{\substack{i,j=1\\ i<j}}^{n} P(A_iA_j)+\ldots+(-1)^{n-1}P(A_1A_2\ldots A_n)$$

Consider first two events A and B only.

Since AB and $A\overline{B}$ are two exhaustive and mutually exclusive forms in which A can occur, we have

* The result of this theorem is known as Poineare's Formula after Henri Poineare (1854—1912), the great French mathematician who is called the last great universal list in mathematics, since he was almost the last great mathematician who made contributions to almost all branches of pure and applied mathematics. His contribution to probability theory was largely philosophical and pedagogic.

$$P(A) = P(AB) + P(A\bar{B})$$

Similarly, $P(B) = P(BA) + P(B\bar{A}) = P(AB) + P(\bar{A}B)$,

since the order of events is immaterial to the probability of the compound event. Adding the two results, we get

$$P(A) + P(B) = P(AB) + [P(A\bar{B}) + P(\bar{A}B) + P(AB)]$$

By the theorem of total probability the expression within the square brackets represents the probability $P(A + B)$ of the occurrence of at least one of the events A and B. $\therefore\ P(A) + P(B) = P(AB) + P(A + B)$

Hence $$P(A + B) = P(A) + P(B) - P(AB) \qquad ...(4{\cdot}4)$$

Let B mean the occurrence of at least one of the events A_1 and A_2 then (4·4) gives

$$\begin{aligned} P(A + A_1 + A_2) &= P(A) + P(A_1 + A_2) - P[A(A_1 + A_2)] \\ &= P(A) + P(A_1) + P(A_2) - P(A_1 A_2) \\ &\quad - P(AA_1) - P(AA_2) + P(AA_1A_2) \end{aligned}$$

Rearranging and modifying the notations slightly, we have

$$\begin{aligned} P(A_1 + A_2 + A_3) &= P(A_1) + P(A_2) + P(A_3) - P(A_1 A_2) - P(A_2 A_3) \\ &\quad - P(A_3 A_1) + P(A_1A_2A_3) \\ &= \sum_{i=1}^{3} P(A_i) - \sum_{\substack{i,j=1 \\ i<j}}^{3} P(A_i A_j) + P(A_1A_2A_3). \qquad ...(4{\cdot}5) \end{aligned}$$

The general law for n events, which may be proved by mathematical induction, is

$$\begin{aligned} P(A_1 + A_2 + \ldots + A_n) &= \sum_{1 \le i \le n} P(A_i) - \sum_{1 \le i < j \le n} P(A_i A_j) + \sum_{1 \le i < j < k \le n} P(A_i A_j A_k) \\ &\quad - \ldots + (-1)^{n-1} P(A_1A_2 \ldots A_n) \qquad ...(4{\cdot}6) \end{aligned}$$

where the second sum is over all combinations of the numbers 1, 2, ...,n taken two at a time, the third is over all combinations of the numbers taken three at a time, and so forth.

Alternative Proof of (4·4)

If n be the cardinal number of the given set A, then

$$n(A \cup B) = n(A) + n(B) - n(A \cap B)$$

Let N be the cardinal number of the sample space.

then we have $$\frac{n(A \cup B)}{N} = \frac{n(A)}{N} + \frac{n(B)}{N} - \frac{n(A \cap B)}{N}$$

$$\Rightarrow \quad P(A + B) = P(A) + P(B) - P(A \cap B)$$

Another Proof. (See Fig. 4·2)

$$P(A \cup B) = P[A \cup (\bar{A} \cap B)]$$
$$= P(A) + P(\bar{A} \cap B)$$
$$= P(A) + [P(\bar{A} \cap B) + P(A \cap B)] - P(A \cap B)$$
$$= P(A) + P(B) - P(A \cap B)$$

Cor. 1. If $A_1, A_2, \ldots\ldots, A_n$ are n mutually exclusive events, then

$$P(A_i A_j) = 0, \quad P(A_i A_j A_k) = 0, \quad P(A_1 A_2 \ldots\ldots A_n) = 0.$$

Consequently (4·6) reduces to

$$P(A_1 + A_2 + \ldots + A_n) = \sum_{i=1}^{n} P(A_i) = P(A_1) + P(A_2) + \ldots + P(A_n)$$

which is the theorem of total probability.

Cor. 2. $P(A_1 + A_2 + \ldots + A_n) \geq \sum_{i=1}^{n} P(A_i) - \sum_{\substack{i,j=1 \\ i<j}}^{n} P(A_i A_j)$

From (4·5). $P(A_1 + A_2 + A_3) \geq \sum_{i=1}^{3} P(A_i) - \sum_{\substack{i,j=1 \\ i<j}}^{3} P(A_i A_j)$

This result can be obtained by mathematical induction.

Note 1. If $P(A_i)$ be written as p_i, $P(A_i A_j)$ as p_{ij}, $P(A_i A_j A_k)$ as p_{ijk} and so on, then (4·6) can be written as

$$P(A_1 + A_2 + \ldots + A_n) = \Sigma p_i - \Sigma p_{ij} + \Sigma p_{ijk} - \ldots + (-1)^{n-1} p_{12\ldots n}$$
$$= S_1 - S_2 + \ldots + (-1)^{n-1} S_n. \qquad \ldots(4\cdot6')$$

where $S_1, S_2 \ldots\ldots$ denote the sum of the probabilities of events taken one, two,........ at a time.

Note 2. Consider any two events A and B. (Fig. 4·2) The events $A - AB$, AB, $B - AB$ are pairwise mutually exclusive and we have

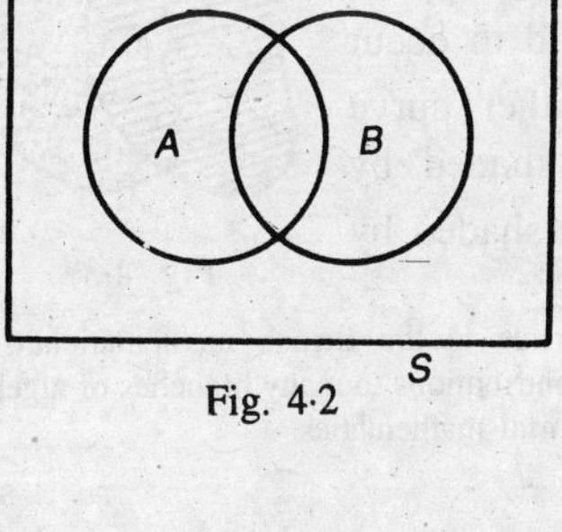

Fig. 4·2

$$A = (A - AB) + AB,$$
$$B = AB + (B - AB),$$
$$A + B = (A - AB) + AB + (B - AB)$$
$$\therefore \quad P(A) = P(A - AB) + P(AB),$$
$$P(B) = P(AB) + P(B - AB),$$
$$P(A + B) = P(A - AB) + P(AB) + P(B - AB).$$

Eliminating $P(A-AB)$ and $P(B-AB)$, we get

$$P(A+B)=P(A)+P(B)-P(AB). \qquad \ldots(4\cdot6'')$$

Ex. 1. (*a*) For any two events A_1 and A_2, prove that

(*i*) $P(A_1+A_2) \le P(A_1)+P(P(A_2)$

(*ii*) $P(A_1A_2) \ge 1-P(\overline{A_1})-P(\overline{A_2})$ **(*B.A. Hons. Delhi, 1988*)**

[**Hint.** (*ii*) Replace A by $\overline{A_1}$ and B by $\overline{A_2}$ and use $P(\overline{A_1}+\overline{A_2}) \le 1$ in [4·6]

(*b*) Arrange for any two events A_1 and A_2 the following probabilities in descending order of magnitude : $P(A_1)$, $P(A_1A_2)$, $P(A_1+A_2)$. and $P(A_1)+P(A_2)$.

(*B.A., Hons, Delhi; M.A., Statistics, Delhi*)

[**Ans.** $P(A_1A_2) \le P(A_1) \le P(A_1+A_2) \le P(A_1)+P(A_2)$]

Ex. 2. Use the theorem of Total probability to show that

(*i*) $P(A_1+A_2+A_3)=P(A_1\overline{A_2A_2}+P(A_2\overline{A_3A_1})+P(A_3\overline{A_2A_1})+P(A_1A_2\overline{A_3})$
$+P(A_2A_3\overline{A_1})+P(A_3A_1\overline{A_2})+P(A_1A_2A_3)$

(*ii*) $P(A_1)=P(A_1\overline{A_2}\overline{A_3})+P(A_1\overline{A_2}A_3)+P(A_1\overline{A_2}A_3)+P(A_1A_2A_3)$

(*iii*) $P(A_1A_2)=P(A_1A_2A_3)+P(A_1A_2\overline{A_3})$ Hence deduce that

$$P(A_1+A_2+A_3)=P(A_1)+P(A_2)+P(A_3)-P(A_1A_2)-P(A_2A_3)-P(A_3A_1)+P(A_1A_2A_3)$$

Ex. 3. Assuming the above theorem to be true for n events and less prove by using mathematical induction that it also holds for n+1 events.

Ex. 4. Prove the above theorem by the *method of induction and exclusion.**

[To find the *L.H.S.* of (4·6′), we have to find the number of distinct cases which are favourable to $A_1, A_2, \ldots, A_n$. The formula will be proved if we show that each case occurs once and once only on the *R.H.S.* of (4·6′). Consider a particular case. We can easily number the cases so that this case is favourable to $A_1, A_2, \ldots, A_m$ only. Then this case occurs ${}^mC_1$ times in S_1, ${}^mC_2$ times in $S_2, \ldots\ldots$, ${}^mC_m$ time in S_m and none at all in the remaining terms *i.e.*, it occurs ${}^mC_1-{}^mC_2+\ldots.+(-1)^{m\cdot1}\,{}^mC_m=1-(1-1)^m$ times on the *R.H.S.*]

4·4·2. Geometrical interpretation of the above theorem. Let us suppose that an event A_1 is said to occur when the *sample point* lies within a certain *sample space* bounded by a curve C_1. The space outside the curve will, of course, corresponds to A_1. Similarly, the event A_2 will be said to occur when the sample-point lies inside another curve C_2. Thus in Fig. 4·3 the whole area shaded by parallel lines corresponds to A_1, the area shaded by

Fig. 4·3

* The principle of inclusion and exclusion is due to the British mathematician J.J. Sylvester (1814—1894) who made important contributions to many branches of algebra. This principle is used extensively in combinatorial mathematics.

dots and parallel lines to $A_1 A_2$, and the dotted area to $A_2 A_1$. The shaded area shown by dots and parallel lines *i.e.*, the area common to the two curves corresponds to $A_1 A_2$. The entire area bounded by curves C_1 and C_2 corresponds to $A_1 + A_2$.

Area for $A_1 + A_2$

= Area for A_1 + Area for A_2 − Area for $A_1 A_2$.

Dividing both sides by the total number of sample points we get

$$P(A_1 + A_2) = P(A_1) + P(A_2) - P(A_1 A_2).$$

Similarly, Fig. 4·4 explains equation (4·5).

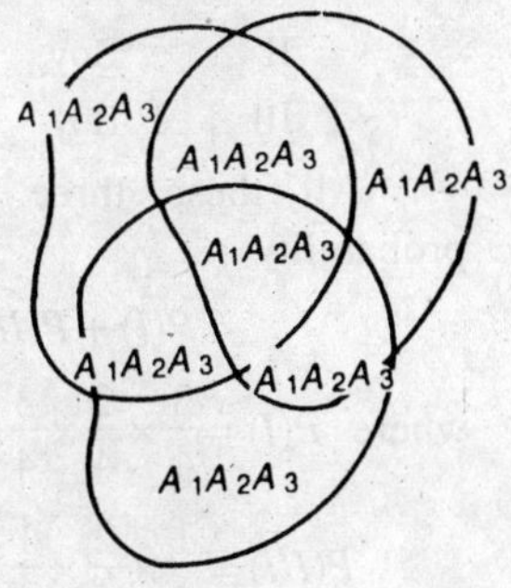

Fig. 4·4

4·4·3. *Boole's Inequality :

For n events $A_1, A_2, \ldots, A_n$, we have

(*i*) $P(A_1 \cap A_2 \cap \ldots \cap A_n) \geq \sum_{i=1}^{n} P(A_i) - (n-1)$

(*ii*) $P(A_1 \cup A_2 \cup \ldots \cup A_n) \leq \sum_{i=1}^{n} P(A_i)$

The equality sign holding, when events are mutually exclusive.

Proof. (*i*) Since $P(A + B) \leq 1$, we have $P(A) + P(B) - P(AB) \leq 1$

$\Rightarrow P(AB) \geq P(A) + P(B) - 1$. Let B stand for event $B \cap C$.

$\therefore P(ABC) \geq P(A) + P(BC) - 1 \geq P(A) + P(B) + P(C) - 2$

Generalizing, $P(A_1 A_1 \ldots A_n) \geq P(A_1) + P(A_2) + \ldots + P(A_n) - (n-1)$

(*ii*) Using this result for n events $\overline{A}_1, \overline{A}_2, \ldots, \overline{A}_n$, of we get

$$P(\overline{A}_1 \overline{A}_2 \ldots \overline{A}_n) \geq P(\overline{A}_1) + P(\overline{A}_2) + \ldots + P(\overline{A}_n) - (n-1)$$

$$\Rightarrow P(\overline{A_1 \cup A_2 \cup \ldots \cup A_n}) \geq [1 - P(A_1)] + [1 - P(A_2)] + \ldots + [1 - P(A_n)] - (n-1)$$

Hence $P(A_1) + P(A_2) + \ldots P(A_n) \geq 1 - P(\overline{A_1 \cup A_2 \cup \ldots \cup A_n})$

$$= P(A_1 + A_2 + \ldots + A_n)] = \sum_{i=1}^{n} P(A_i)$$

Ex. Give a geometrical interpretation of Boole's inequality.

Example 4·11. *Three groups of children contain respectively 3 girls and 1 boy, 2 girls and 2 boys and 1 girl and 3 boys. One child is selected at random from each group. Show that the chance that the three selected consist of 1 girl and 2 boys is 13/32.* ***(B.Sc. Nagpur, 1991)***

* Boole's inequality is known after the British mathematician George Boole (1815-1864), founder of *Boolean algebra* used in simplifying electric circuites for computers. He also made important contributions to symbolic logic through his book '*Laws of Thought*'.

There are three ways in which 1 girl and 2 boys can be selected *viz.*

I	Girl	Boy	Boy
II	Boy	Girl	Boy
III	Boy	Boy	Girl

All these three being mutually exclusive cases, the required probability

$$= P(I) + P(II) + P(III)$$

where $P(I) = \frac{3}{4} \times \frac{2}{4} \times \frac{3}{4} = \frac{9}{32}$, $P(II) = \frac{1}{4} \times \frac{2}{4} \times \frac{3}{4} = \frac{3}{32}$,

$$P(III) = \frac{1}{4} \times \frac{2}{4} \times \frac{1}{4} = \frac{1}{32}$$

$\therefore$ required probability $= \frac{9}{32} + \frac{3}{32} + \frac{1}{32} = \frac{13}{32}$.

Example 4·12. *A and B take turns in throwing two dice, the first to throw 9 being awarded the prize. Show that if A has the first throw, their chances of winning are in the ratio 9 : 8.* **(*B.Sc. Delhi, 1972*)**

9 can be made up in 4 ways *viz,* 3, 6 ; 4, 5 ; 5, 4 ; 6, 3 with two dice and thus the chance of throwing 9 is 4/36 or $\frac{1}{9}$ and the chance of not throwing is $1 - \frac{1}{9}$ or 8/9.

If *A* is to win, he should throw 9 in either the first, third or fifth or ...... throw.

The probability of *A*'s throwing 9 on the first throw is $\frac{1}{6}$, in third throw, it is $\frac{8}{9} \times \frac{8}{9} \times \frac{1}{9}$.. or $\left(\frac{8}{9}\right)^2 \times \frac{1}{9}$, in the fifth throw it is $\frac{8}{9} \times \frac{8}{9} \times \frac{8}{9} \times \frac{8}{9} \times \frac{1}{9}$ or $\left(\frac{8}{9}\right)^4 \times \frac{1}{9}$ and so on.

Hence $\quad p_A$ = *A*'s total chance of winning

$$= \frac{1}{9} + \left(\frac{8}{9}\right)^2 \times \frac{1}{9} + \left(\frac{8}{9}\right)^4 \times \frac{1}{9} + \text{......}$$

$$= \frac{1}{9} \Big/ \left[1 - \left(\frac{8}{9}\right)^2\right] = \frac{9}{17}$$

In other words, *odds* are 9 to 8 in *favour* of *A* or 8 to 9 *against A*. *B* can win in second, fourth, or sixth,....... throw.

$$\therefore \quad p_B = \frac{8}{9} \times \frac{1}{9} + \frac{8}{9} \times \frac{8}{9} \times \frac{8}{9} \times \frac{1}{9} + \text{......}$$

$$= \frac{8}{9} \times \left[\frac{1}{9} + \left(\frac{8}{9}\right)^2 \times \frac{1}{9} + \left(\frac{8}{9}\right)^4 \times \frac{1}{9} + \text{......}\right] = \frac{8}{9} \times \frac{9}{17} = \frac{8}{17}.$$

$\therefore$ *A*'s chance to *B*'s is in the ratio $\frac{9}{17} : \frac{8}{17}$ *i.e.*, 9 : 8.

Note. It will be observed that p_B could have been written down immediately from the relation $P_B - 1 - P_A$

Example 4·13. *Six cards are drawn with replacement from an ordinary deck. What is the probability that each of the four suits will be represented at least once among the six cards ?*

We shall first find the probability that all suits do not appear. Let A symbolize the appearance of all the suits and B the non-appearance of at least one of the suits. Since either A or B is certain to happen, $P(A+B)=1$ and since A and B are mutually exclusive, $P(A)=1-P(B)$.

To get $P(B)$, let us classify the possible outcomes favourable to B into four sets. B_1 is the set of all outcomes in which spades are absent, B_2 is the set for which hearts are absent, B_3 and B_4 are the sets for which diamonds and clubs are absent respectively.

These sets are overlapping : an outcome which consists of only spades and hearts falls in B_3 and in B_4. Clearly

$$P(B)=P(B_1+B_2+B_3+B_4)$$

$$=\sum_{i=1}^{4} P(B_i)-\sum_{1\le i<j\le 4} P(B_i B_j)+\sum_{1\le i<j<k\le 4} P(B_iB_jB_k)-P(B_1B_2B_3B_4) \quad ...(1)$$

The probability $P(B_1)$ that a spade will not appear in the six draws is $\left(\frac{3}{4}\right)^6$, and the value is the same for all B_i ; hence $\Sigma P(B_i)=4\left(\frac{3}{4}\right)^6$

The probability $P(B_1 B_2)$ that neither spades, nor hearts, will appear in the six draws is $\left(\frac{1}{2}\right)^6$ and is the same for all six pairs of the four suits taken two at a time ; hence $\Sigma P(B_i B_j)=6\left(\frac{1}{2}\right)^6$

Similarly, $\Sigma P(B_i B_j B_k)=4\left(\frac{1}{4}\right)^6$ and clearly $P(B_1B_2B_3B_4)=0$

Substituting the values of $\Sigma P(B_i), \Sigma P(B_iB_j), \Sigma P(B_iB_jB_k)$ etc. in (1), we get $P(B)=4\left(\frac{3}{4}\right)^6-6\left(\frac{1}{2}\right)^6+4\left(\frac{1}{4}\right)^6$

$\therefore$ Required probability $=P(A)=1-P(B)$

$$=1-4\left(\frac{3}{4}\right)^6+6\left(\frac{1}{2}\right)^6-4\left(\frac{1}{4}\right)^6=\cdot 38086$$

Example 4·14. *n letters to each of which corresponds an envelope, are placed in the envelopes at random. (i) What is the probability that no letter is placed in the right envelope ? (ii) What is the probability that exactly r letters are placed in the right envelopes ?* ***(I.A.S. 1993, M.A. Rajasthan 1975, M.A. Poona, 1970, M.A. Delhi 1966)***

Let A_i, $i=1, ,..., n$ denote the event that the nth letter is placed in the right envelope. Then the

Required probability $=P(\overline{A_1A_2.....A_n})=1-P(A_1+A_2+...+...+A_n)$

$$= 1 - \sum_{i=1}^{n} P(A_i) + \sum_{\substack{i=1 \\ i<j}}^{n} \sum_{j=1}^{n} P(A_iA_j) + \sum_{i=1}^{n} \sum_{j=1}^{n} \sum_{\substack{k=1 \\ i<j<k}}^{n} P(A_i A_j A_k)$$

$$+ \ldots + (-1)^n P(A_1A_2\ldots A_n)$$

Now $P(A_i) = \dfrac{1}{n}, i = 1, 2, ..., n$

$$P(A_i A_j) = P(A_i)\, P(A_j \mid A_i) = \frac{1}{n} \cdot \frac{1}{n-1}, \text{ for all } i, j, i<j$$

$$P(A_i A_j A_k) = P(A_i)\, P(A_j \mid A_i) P(A_k \mid A_i A_j)$$

$$= \frac{1}{n} \cdot \frac{1}{n-1} \cdot \frac{1}{n-2} \quad \text{for all } i, j, k, i<j<k$$

$$\ldots \quad \ldots \quad \ldots \quad \ldots \quad \ldots \quad \ldots$$

$$P(A_1 A_2 \ldots A_n) = \frac{1}{n} \cdot \frac{1}{n-1} \cdot \frac{1}{n-2} \ldots\ldots \frac{1}{2} \cdot 1$$

Hence the required probability

$$= 1 - {}^nC_1 \cdot \frac{1}{n} + {}^nC_2 \cdot \frac{1}{n(n-1)} + {}^nC_3 \frac{1}{n(n-1)(n-2)}$$

$$\ldots\ldots + (-1)^n \frac{1}{n(n-1)\ldots 3.2.1}$$

$$= \frac{1}{2!} - \frac{1}{3!} + \frac{1}{4!} - \ldots\ldots + (-1)^n \cdot \frac{1}{n!}$$

Aliter. (*i*) Suppose that u_n is the number of ways in which all letters go wrong. Consider any two letters. If these occupy each other's envelope the number of ways in which the remaining n—2 letters go wrong is u_{n-2} and there are n—1 ways in which two letters can be interchanged. But if one letter occupies another's place and not *vice versa* which can happen in $(n-1)$ ways there are u_{n-1} ways in which the others can go wrong. Hence we have the *difference equation*

$$u_n = (n-1)(u_{n-1} + u_{n-2})$$

Re-writing we have

$$u_n - nu_{n-1} = -(u_{n-1} - \overline{n-1}\, u_{n-3}) = \ldots \quad \ldots \quad \ldots$$

$$= (-1)^{n-2}\, (u_2 - 2u_1).$$

But $u_1 = 0$ and $u_2 = 1$

and thus $\quad u_n - nu_{n-1} = (-1)^n$

or $\quad \dfrac{u_n}{n!} - \dfrac{u_{n-1}}{(n-1)!} = \dfrac{(-1)^n}{n!}$

Putting $n = 2, 3, \ldots\ldots, n$ and adding, we get

$$u_n = n!\left\{\frac{1}{2!} - \frac{1}{3!} + \ldots + \frac{(-1)^n}{n!}\right\}$$

$$\text{Required probability} = \frac{u_n}{n!} = \frac{1}{2!} - \frac{1}{3!} + \ldots\ldots + \frac{(-1)^n}{n!}$$

i.e., the first (n—1) terms of $1 - e^{-1}$.

(*ii*) r letters can be chosen in ${}^nC_r$ ways and the rest arranged in u_{n-r} ways.

$$\therefore \text{ Required probability} = \frac{{}^nC_r\, u_{n-r}}{n!} = \frac{1}{r!} \cdot \frac{u_{n-r}}{(n-r)!}$$

$$= \frac{1}{r!}\left\{\frac{1}{2!} - \frac{1}{3!} + \ldots + \frac{(-1)^{n-r}}{(n-r)!}\right\}.$$

Example 4·15. Match or Rencontre Problem. *Two similar decks of n different cards are put into random order and matched against each other. If a card occupies the same position in both the decks we speak of a match (coincidence). Find the probability of atleast one match.*

Since we are interested in the rélative order in two decks we can take without loss of generality one deck to be in its natural order. Let A_i be the event that a match occurs at the i-th place. This means that card number 'i' is the i-th place while the other $(n-1)$ cards may be in any order.

$$\text{Thus} \qquad p_i = \frac{(n-1)!}{n!} = \frac{1}{n}.$$

Similarly, probability that cards numbered i and j are in their natural places while the remaining $(n-2)$ cards may be any arbitrary order is given by

$$p_{ij} = \frac{(n-2)\,1}{n!} = \frac{1}{n\,(n-1)}$$

Thus sum S_r of formula (4·6′) contains ${}^nC_r$ terms each of which is $\frac{(n-r)!}{n!}$ and therefore $S_r = {}^nC_r \cdot \frac{(n-r)!}{n!} = \frac{1}{r!}$, $r = 1, 2, \ldots\ldots$

$$\text{Hence } P(A_1 + A_2 + \ldots + A_n) = S_1 - S_2 + S_3 - \ldots\ldots \pm S_n$$

$$= 1 - \frac{1}{2!} + \frac{1}{3!} - \ldots\ldots \text{to } n \text{ terms.}$$

Ex. 1. Deduce the result of Ex. 4·15 from that of Ex. 4·14.

Ex. 2. Find the probability of exactly r matches in Ex. 4·15.

4·3·4. Theorem. *For any integer m with $1 \le m \le N$, the probability $P(m)$ that exactly m among the N events $A_1, A_2, \ldots\ldots, A_N$ occur simultaneously is given by*

$$P(m) = S_m - \binom{m+1}{m} S_{m+1} + \binom{m+2}{m} S_{m+2} - \ldots \pm \binom{N}{m} S_N$$

$$= \sum_{k=m}^{N} (-1)^{k-m} \binom{k}{m} S_k. \quad \text{where } S_m = \sum_{k=1}^{m} p_{ik1} p_{ik2} \cdots p_{ikn} \quad ...(i)$$

Proof. Let E be an arbitrary sample point, and suppose that it is contained in exactly n among the N events A_j.

Then $P(E)$ appears as a contribution to $P(_m)$ only if $n = m$.

Now $P\{E\}$ appears as a contribution to the section

$$S_i = \sum_{i=1}^{n} p_i, \quad S_2 = \sum_{\substack{i,j=1 \\ i<j}}^{n} p_{ij}, \ldots, S_n = \sum_{k=1}^{n} p_{ik1} \, p_{ik2} \cdots p_{ikn}$$

and not to $S_{n+1}, S_{n+2}, \ldots, S_N$.

Thus if $n < m$, $P\{E\}$ does not contribute to the right side of (i).

If $n = m$, then $P\{E\}$ appears in one and only 1 term of S_m,

If $n > m$, the contributions of $P\{E\}$ to the terms $S_m, S_{m+1}, \ldots$ $\ldots, S_n$, on the right of (i) cancel. Hence the total contribution of $P\{E\}$ to the right side of (i),

$$= \binom{n}{m} - \binom{m+1}{m}\binom{n}{m+1} + \binom{m+2}{m}\binom{n}{m+2} - \ldots$$

$$\ldots + (-1)^{n-m} \binom{n}{m}\binom{n}{n}$$

$$= \sum_{k=0}^{n-m} (-1)^k \binom{m+k}{k}\binom{n}{m+k}$$

$$= \sum_{k=0}^{n-m} (-1)^k \frac{(m+k)!}{k!\, m!} \cdot \frac{n!\,(n-m)!}{(m+k)!\,(n-m-k)!\,(n-m)!}$$

$$= \binom{n}{m} \sum_{k=0}^{n-m} (-1)^k \binom{n-m}{k}$$

$$= \binom{n}{m}\left[1 - \binom{n-m}{1} + \binom{n-m}{2} \ldots + (-1)^{n-m} \binom{n-m}{n-m}\right]$$

$$= \binom{n}{m} (1-1)^{n-m} = 0.$$

Cor. The probability P_m that at least m among the N events occur simultaneously is given by

$$P_m = P(_m) + P(_{m+1}) + P(_{m+2}) + \ldots + P(_N)$$

$$= S_m - \begin{pmatrix} m \\ m-1 \end{pmatrix} S_{m+1} + \begin{pmatrix} m+1 \\ m-1 \end{pmatrix} S_{m+2\ldots} \ldots \pm \begin{pmatrix} N-1 \\ m-1 \end{pmatrix} S_N$$

Ex. Prove that

(*i*) $S_m - (m+1) S_{m+1} \leq P_{(m)} \leq S_m$ (*ii*) $S_m - m\ S_{m+1} \leq P_{(m)} < S_m$.

(M.Sc. Bombay 1971 ; M.Sc. Gauhati ; M.Sc. Delhi)

4·5. Multiplicative Law of Probability or Theorem of Compound Probability. *The probability of the simultaneous occurrence of two events A and B is equal to the probability of A multiplied by the conditional probability of B given that A has occurred (or it is equal to the probability of B, multiplied by the conditional probability of A given that B has occurred). In symbols,*

$$P(AB) = P(A)P(B \mid A) = P(B)P(A \mid B).$$

Proof. Let N denote the total number of mutually exclusive and equally likely cases among which m cases are favourable to the event A. The cases favourable to both the events A and B are included in the m cases favourable to A. Let their number be m_1. The the probability $P(AB)$ that both the events A and B will happen is given by

$$P(AB) = \frac{m_1}{N} = \frac{m}{N} \cdot \frac{m_1}{m}.$$

The ratio m/N is the probability of A, that is $m/N = P(A)$.

Assuming the occurrence of A, there are only m equally likely cases left out of which m_1 are also favourable to B. Hence the ratio m_1/m represents the conditional probability $P(B \mid A)$ supposing that A has occurred.

Hence $$P(AB) = P(A)P(B \mid A)$$

Since the compound event AB involves A and B symmetrically, we shall also have $$P(AB) = P(B)P(A \mid B).$$

Cor. 1. If A and B are independent events then $P(B \mid A)$ is the same as $P(B)$. Thus for two *independent events* the theorem of compound probability becomes $P(AB) = P(A))P(B)$.

In general, *if the constituent events of a compound event are mutually independent, the probability of the compound event is equal to the product of the probabilities of the constituent events.*

In symbols, $P(A_1, A_2, \ldots A_n) = P(A_1)P(A_2)\ldots P(A_n)$

provided that $P(A_i) = (P(A_i \mid A_j \ldots A_p)$ for all $i, j, \ldots p$, $i \neq j \neq \ldots \neq p$.

Cor. 2. $$P(AB \mid C) = P(A \mid C)\, P(B \mid AC).$$

The above proof holds in this case provided N represents the total number of equally likely and mutually exclusive events favourable to C.

Cor. 3. $P(ABC) = P(A)P(B \mid A)P(C \mid AB)$

For $P(ABC) = P[A(BC)] = P(A)P(BC \mid A)$

$= P(A)P(B \mid A)P(C \mid AB)$ [using Cor. 2]

Cor. 4. $P(A_1, A_2 ..., A_n) = P(A_1)P(A_2 \mid A_1)P(A_3 \mid A_1 A_2)......$

$$P(A_n \mid A_1 A_2 ... A_{n-1})$$

$$= P\left(A_{i_1}\right) P\left(A_{i_2} \mid A_{i_1}\right) P\left(A_{i_3} \mid A_{i_1} A_{i_2}\right) ... P\left(A_{i_n} \mid A_{i_1} A_{i_2} .... A_{i_{n-1}}\right)$$

where $i_1, i_2, ..., i_n$ denote any permutation of the integers 1, 2,......, n.

Example 4·16. *A bag contains 4 red and 3 blue balls. Two drawings of 2 balls are made. Find the chance that the first drawing gives 2 red balls and the second drawing 2 blue balls*

(*a*) *if the balls are returned to the bag after the first draw,*

(*b*) *if the balls are not returned.*

(*a*) The number of ways in which any two balls may be drawn is ${}^7C_2$. The number of ways in which two red balls may be drawn is ${}^4C_2$. Therefore, the probability of drawing two red balls is ${}^4C_2/{}^7C_2 = \frac{2}{7}$.

Similarly, the probability of drawing two blue balls is ${}^3C_2/{}^7C_2 = \frac{1}{7}$.

Therefore, the chance of the compound event is $\frac{2}{7} \times \frac{1}{7} = \frac{2}{49}$.

(*b*) As before the chance of drawing 2 red balls is $\frac{2}{7}$. The balls are not returned to the bag and the number of ways in which further two balls may be drawn is ${}^5C_2$ so that the chance of drawing two blue balls is ${}^3C_2 / {}^5C_2 = \frac{3}{10}$.

Therefore the chance of compound event is $\frac{2}{7} \times \frac{3}{10} = \frac{3}{35}$.

Note. In (*a*) of this example the two events are independent but in (*b*) they are clearly not independent since the chance of drawing two blue balls is affected by the first draw.

Example 4·17. *The probability of n independent events are* $p_1, p_2, ..., p_n$. *Find an expression for the probability that at least one of the events will happen.*

Use the result to find the chance of obtaining at least one 6 in a throw of four dice.

For the rth event, the chance that it will not happen is given by $1 - p_r$, $r = 1, 2, ..., n$. The chance, therefore, that none of the events will happen is given by $(1 - p_1)\ (1 - p_2) .....(1 - p_n)$, and since it is certain that either no event will happen or that at least one event will happen the probability that at least one event will happen is given by $1-(1-p_1)(1-p_2).....(1-p_n)$.

The probability of obtaining a 6 in a throw of a dice is $\frac{1}{6}$ so that using the above result the chance of at least one 6 in a throw of four dice is given by $1-\left(\frac{5}{6}\right)^4=\frac{671}{1296}=0{\cdot}51775$.

Example 4·18. *For any three events A, B, and C, prove that*

(*i*) $P[(A+B) \mid C]=P(A \mid C)+P(B \mid C)-P[(AB) \mid C]$

(*ii*) $P(A\bar{B} \mid C)+P(AB \mid C)=P(A \mid C)$. **(B.A. Hons. Delhi)**

We have

(*i*) $P(A+B)=P(A)+P(B)-P(AB)$

Replace A and B by the compound events AC and BC respectively.

$\therefore \quad P(AC+BC)=P(AC)+P(BC)-P(ABC)$

Hence $$\frac{P(AC+BC)}{P(C)}=\frac{P(AC)}{P(C)}+\frac{P(BC)}{P(C)}-\frac{P(ABC)}{P(C)},\ P(C)\neq 0$$

$$\therefore \quad \frac{P[(A+B)C]}{P(C)}=P(A \mid C)+P(B \mid C)-P(AB \mid C)$$

$$\Rightarrow \quad P[(A+B) \mid C]=P(A \mid C)+P(B \mid C)-P(AB \mid C)$$

(*ii*) $$P(A\bar{B} \mid C)+P(AB \mid C)=\frac{P(A\bar{B}C)}{P(C)}+\frac{P(ABC)}{P(C)}$$

$$=\frac{P(A\bar{B}C)+P(ABC)}{P(C)}=\frac{P(AC)}{P(C)}=P(A \mid C).$$

Ex. 1. The probability that a teacher will give an unannounced test during any class meeting is $\frac{1}{5}$. If a student is absent twice, what is the probability that he will miss at least one test ? **[Ans.** $\frac{9}{25}$**]**

Ex. 2. *A* can solve 75% of the problems in this book and *B* can solve 70%. What is the probability that either *A* or *B* can solve a problem chosen at random? **(*B.A. Punjab*)**

[Ans. $\frac{27}{40}$**]**

Ex. 3. Let *A*, *B* and *C* be any three events. If $P(A \mid C)\geq P(B \mid C)$ and $P(A \mid \bar{C})\geq P(B \mid \bar{C})$. then prove $P(A)\geq P(B)$. **(*B.A. Hons. Stat. Delhi, 1984*)**

4·6. Bayes Theorem. *An event A can be explained by a set of n exhaustive and mutually exclusive hypotheses* $B_1, B_2, ..., B_n$. *Given*

(*i*) *'a priori' probabilities* $P(B_1), P(B_2),..., P(B_n)$; $P(B_i)\neq 0$, $i=1, 2, ...,n$ *corresponding to a total absence of a knowledge regarding the occurrence of A and* (*ii*) *conditional probabilities* $P(A \mid B_1)$, $P(A \mid B_2)$,...., $P(A \mid B_n)$

(*a*) *It is required to form the 'a posteriori' probabilities* $P(B_1 \mid A)$, $P(B_2 \mid A)$,......,$P(B_n \mid A)$.

(*b*) *Further, find the probabilities of materialisation of another event C, given the probabilities*

$$P(C \mid AB_1),\ P(C \mid AB_2).....,\ P(C \mid AB_n).$$

Proof. (*i*) By the theorem of compound probability,

$$P(AB_i) = P(B_i)P(A \mid B_i) = P(A)P(B_i \mid A)$$

$$\therefore \quad P(B_i \mid A) = \frac{P(B_i)P(A \mid B_i)}{P(A)} \quad ...(4{\cdot}7)$$

Since the event A can materialise in n mutually exclusive forms AB_1, AB_2.......,AB_n. we have by the theorem of total probability

$$P(A) = P(AB_1) + P(AB_2) + \ldots + P(AB_n)$$

$$= \sum_{i=1}^{n} P(B_i)P(A \mid B_i) \quad ...(4{\cdot}8)$$

$$\therefore \quad P(B_i \mid A) = \frac{P(B_i)P(A \mid B_i)}{\sum_{i=1}^{n} P(B_i)P(A \mid B_i)} \quad ...(4{\cdot}9)$$

From this result we can obtain the value of $P(B_i \mid A)$ given $P(A \mid B_i)$ and $P(B_i)$ for $i = 1, 2, n$

(*ii*) $$P(C \mid A) = \sum_{i=1}^{n} P(CB_i \mid A) = \sum_{i=1}^{n} P(B_i \mid A)P(C \mid AB_i)$$

$$P(C \mid A) = \frac{\sum_{i=1}^{n} P(B_i)\, P(A \mid B_i)P(C \mid AB_i)}{\sum_{i=1}^{n} P(B_i)P(A \mid B_i)}, \quad ...(4{\cdot}10)$$

Remarks. 1. The probabilities $P(B_i)$ is called the 'a priori' probabilities is because they exist before we have any information from the experiment itself.

2. The probabilities $P[B_i \mid A]$, $i = 1, 2,..., n$ are called the 'posteriori' probabilities.

3. In case the materialisation of B_i makes C independent of A, then $P(C \mid AB_i \mid = P \mid C \mid B_i)$

Note. The formula (4·9) was published in 1763 and was used extensively to decide between the hypotheses B_1, B_2..., B_n for explaining an event A which has occurred. Thus it was used in jurisprudence to decide between the various possible criminals for a crime. It was used in 'geophysics to assign probabilities in the various causes of an earthquake. The great French mathematician Laplace (1749-1827) who wrote a monumental book on probability about 1815 used it even to find the probability that the sun will rise tomorrow. The main problem in applying this theorem is over lack of knowledge of a 'prior' probability. The influence of Bayes thinking has been considerable and we have even a Bayesian school of statistical inference.

Example 4·19. *A bag A contains 2 white and 3 red balls and a bag B contains 4 white and 5 red balls. One ball is drawn at random from one of the bags and is found to be red. Find the probability that it was drawn from bag B.*

Let E_1 the event that the ball is drawn from the bag A, E_2 the event that it is drawn from the bag B and E be the event that the drawn ball is red. $\therefore\ P(E_1) = \frac{1}{2} = P(E_2)$.

Since the drawn ball is found to be red, we have to find $P(E_2 \mid E)$. Also $P(E \mid E_1) =$ the probability that the ball drawn is red if it is drawn from bag A $= \frac{3}{5}$. Similarly, $P(E \mid E_2) = \frac{5}{9}$

Hence by Bayes theorem, we have

$$P(E_1 \mid E) = \frac{P(E_2)\,P(E \mid E_2)}{P(E_1)\,P(E \mid E_1) + P(E_2)\,P(E \mid E_2)}$$

$$= \frac{\frac{1}{2} \times \frac{5}{9}}{\frac{1}{2} \times \frac{3}{5} + \frac{1}{2} \times \frac{5}{9}} = \frac{\frac{5}{9}}{\frac{27+25}{45}} = \frac{25}{52}.$$

Example 4·20. *A factory has 3 machines A, B and C, producing 1000, 2000 and 3000 bolts per day respectively. A produces 1% defective, B 1·5% and C 2% defective. A bolt is checked at random at the end of a day and is found to be defective. What is the probability that it came from machine A ?*

Let E_1, E_2, E_3 be the events that the bolt was chosen from machine A, B and C respectively. Let E be the event that it is defective.

Then $P(E_1) = \dfrac{1000}{1000 + 2000 + 3000} = \dfrac{1}{6}$, $P(E_2) = \dfrac{2000}{6000} = \dfrac{2}{6}$

$$P(E_3) = \frac{3000}{6000} = \frac{3}{6}$$

Also $P(E \mid E_1) = \frac{1}{100} = \cdot 01, \ P(E \mid E_2) = 0{\cdot}015, P(E \mid E_3) = 0{\cdot}02$

Hence by Bayes theorem,

$$P(E_1 \mid E) = \frac{P(E_1) \times P(E \mid E_1)}{P(E_1) \times P(E \mid E_1) + P(E_2)\, P(E \mid E_2) + P(E_3)\, P(E \mid E_3)},$$

$$= \frac{\frac{1}{6} \times \cdot 01}{\frac{1}{6} \times \cdot 01 + \frac{2}{6} \times \cdot 015 + \frac{2}{6} \times \cdot 02} = \frac{\cdot 01}{\cdot 10} = \frac{1}{10} = \cdot 1$$

Thus the probability of the defective bolt came from A is 1%. see also Ex. 58 and 59 page 122.

Example 4·21 *If a day is dry, the conditional probability that the next day will also be dry is p ; if a day is wet, the conditional probability that the next day will be dry is* p^1. *If* u_n *is the probability that the nth day will be dry. prove that* $u_n - (p - p^1)\, u_{n-1} - p^1 = 0, \quad (n \geq 2)$

If the first day is sure to be dry and $p = \frac{3}{4}, \ p^1 = \frac{1}{4}$, *find* u_n.

Let $A_1, A_2, \ldots\ldots A_{n-1}$ be the first, second, ... $(n-1)^{th}$ day is dry, wet respectively. Let $X - n\ th$ day be dry. We are given :

$$P(X \mid A_1) = p, \ P(X \mid A_2) = p^1, P(A_1) = u_{n-1}, P(X) = u_n,$$

Then $P(A_2) = 1 - P(A_1) = 1 - u_{n-1}$

But $P(X) = P(A_1)\, P(X \mid A_1) + P(A_2)\, P(X \mid A_2) + \ldots$

$\therefore \quad u_n = u_{n-1}\, p + (1 - u_{n-1})\, p^1 \quad$ or $\quad u_n - (p - p^1)\, u_{n-1} - p^1 = 0.$

If $p = 3/4, \ p^1 = 1/4$, then $u_n - \frac{1}{2}\, u_{n-1} - \frac{1}{4} = 0$

or $\quad 2u_n - 1 = \frac{1}{2}(2\, u_{n-1} - 1) = \frac{1}{2^2}(2u_{n-2} - 1) = \ldots\ldots$

$$= \frac{1}{2^{n-1}}(2u_1 - 1) = \frac{2 \times 1 - 1}{2^{n-1}} = \frac{1}{2^{n-1}}, \quad \therefore \ u_1 = 1$$

$$\therefore \quad u_n = \frac{1}{2} + \frac{2}{2^{n-1}}.$$

Ex. 1. The contents of three urns 1, 2, 3 are as follows :

1 white, 2 red, 3 black balls

2 white, 3 red, 1 black balls

3 white, 1 red, 2 black balls

An urn is chosen at random and from it two balls are drawn at random. The two balls are one red and one white. What is the probability that they come from the second urn ?

(*B.Sc. Hons. Cal. 1968*)

[**Hint.** $P[B_1] = P[B_2] = P\{B_3] = \frac{1}{3}$, $P[A \mid B_1] = \frac{2}{15}$, $P[A \mid B_2] = \frac{2}{5}$
$P[A \mid B_3] = \frac{1}{5}$, $P[B_1 \mid A] = \frac{3}{11}$, $P[B_2 \mid A] = \frac{6}{11}$, $P[B_3 \mid A] = \frac{4}{11}$.]

Ex. 2. From an urn containing 3 white and 5 black balls, 4 balls are transferred into an empty urn. From this urn 2 balls are taken and both happen to be white. What is the probability that the three balls taken from the same urn will be white ?

[**Hint.** $P(B_1) = \frac{3}{7}$, $P(B_2) = \frac{3}{7}$; $P(B_3) = \frac{1}{14}$; $P(A \mid B_2) = \frac{1}{6}$, $P(A \mid B_3) = \frac{3}{6}$,
$P(C \mid A B_2) = \frac{1}{2}$, $P(C \mid AB_3) = \frac{3}{4}$, $P(C \mid A) = \frac{1}{12}$.]

4·7·1. Set-theoretic Formulation of Probability Theory. The above discussion of the theory of probability is restrictive in the sense that it applies only when all the events under consideration are *equally likely*. We want to generalise it to the case when all the events are not necessarily equally likely. However the generalisation should include the earlier discussion as a special case.

The above discussion is also quite intuitive. We want to make it more rigorous and axiomatic. The definitions and axioms will, of course, be motivated by the results obtained above. In our discussion, we shall identify events with sets and as such some concepts from *set theory* will also be useful.

A set is a well-defined collection of objects, so that given any object, we can say whether it belongs to the set or not. Each object of the set is called its *element*. The set of all elements under consideration is called the *universal set* and the set which contains no element is called the *null set* or the empty set and is denoted by Greek alphabet φ (phi), read as 'Fie'. The *union* $A \cup B$ of the sets A and B is set of all those elements which belong to A or to B or to both. The *intersection* $A \cap B$ of the two sets A and B is the set of all the elements common to A and B. The *complement* $\bar{A}$ of a set A is the set of all those elements of the universal set which do not belong to A. Two sets A and B are said to be disjoint if they have no common element/elements *i.e.* if $A \cap B = \varphi$. A set A is said to be a *subset* of a set B if whenever an element x belongs to A, it also belongs to B and we write $A \subset B$ and $B \subset A$, the two sets are the same and we write $A = B$.

Definition of probability. When we perform an experiment, actual or conceptual, there may be a number of possible outcomes $e_1, e_2, \ldots, e_n$, say. We assume that these outcomes are *mutually exclusive i.e.* one of these outcomes is possible at a time and these are *exhaustive i.e.* one of these outcomes must occur. Such outcomes are called *elementary events* and the set of all such outcomes *i.e.* the set of all elementary events is called the *sample space*.

Thus if the experiment consists of throwing of a six-faced die, the possible outcomes are ; turning up of 1, 2, 3, 4, 5 or 6. These outcomes are mutually exclusive since only one of the numbers may turn up at a time, and they are exhaustive, since one of these must occur (we, of course, assume that the die will not stand on an edge !). Thus there are six elementary events and the sample space consists of 6 points viz. 1, 2, 3, 4, 5, 6.

When we throw up a coin, the sample space consists of two points viz. Head and Tail or H and T.

When we throw up two dice simultaneously, the sample space consists of 36 elementary events viz.

$$\begin{array}{cccccc} (1, 1), & (1, 2), & (1, 3), & (1, 4), & (1, 5), & (1, 6) \\ (2, 1), & (2, 2), & (2, 3), & (2, 4), & (2, 5), & (2, 6) \\ (3, 1), & (3, 2), & (3, 3), & (3, 4), & (3, 5), & (3, 6) \\ (4, 1), & (4, 2), & (4, 3), & (4, 4), & (4, 5), & (4, 6) \\ (5, 1), & (5, 2), & (5, 3), & (5, 4), & (5, 5), & (5, 6) \\ (6, 1), & (6, 2), & (6, 3), & (6, 4), & (6, 5), & (6, 6) \end{array}$$

where the elementary event (3, 2) means that the first die shows 3 and the second die shows 2.

With three dice, the sample space will consist of 216 elementary events or points viz. all ordered triplets of the form (i, j, k) where i, j, k can take the values 1 to 6.

Our experiments may also be conceptual *e.g.*, we may think of a rational number, so that an elementary event is any rational number. In this case our sample space is countably infinite or *denumerable i.e.* in this case we can set up a one-one correspondence between elements of the set of elementary events and the members of the set of natural numbers. If our experiment is thinking of a point on a line or in a plane, the sample space is non-countably infinite. We shall confine ourselves, however, mainly to finite sample spaces.

We now assign real numbers $p_1, p_2, \ldots\ldots, p_n$ to each of the elementary events $e_1, e_2, \ldots\ldots, e_n$ in such a way that

(i) $p_1 \geq 0$ $(i = 1, 2, \ldots\ldots, n)$ $\qquad$ (ii) $p_1 + p_2 + \ldots + p_n = 1$.

These numbers are called *probabilities of elementary events* and the assignment of these numbers satisfying these two conditions is called an *allowable* or *permissible assignment of probabilities.*

The assignment in which $p_1 = p = \ldots\ldots = p_n = 1/n$ *i.e.* in which all the elementary events are assigned the same probability is called the *natural assignment of probabilities.* We make this assignment when we have reasons to believe that all outcomes are "*equally likely*" *i.e.* when there is no reason to believe that any one outcome is more likely than any other *e.g.* if we have an 'unbiased die', we may assign the probability 1/6 to each of the 6 elementary events and when we have two unbiased die we may assign probability 1/36 to each of the 36 elementary events.

In the case of a biased die which is loaded on one side, we cannot use the natural assignment to probabilities. How do we then assign the probabilities to the elementary events ? One method to throw the die a large number of times and note the numbers of times it shows a fixed number say 2. If a die is thrown repeatedly a large number of times under more less identical conditions, it will be found that the ratio of number of times the die shows 2 to total number of times it is thrown becomes almost stationary at some number. This ratio may be assigned as the probability for the elementary event corresponding to throwing up of number 2. In the theory of probability, we shall however assume that such a permissible assignment of probabilities of elementary events is already given to us.

We now define an *event* as any subset on the sample space, and probability of an event E is the sum of the assigned probabilities to the elements of the events.

Thus if $E = e_{i_1} + e_{i_2} + ... + e_{i_m}$ then $P(E) = p_{i_1} + p_{i_2} + ... + p_{i_m}$

when $e_{i_1}, e_{i_2}, ..., e_{i_m}$ are the m elementary events which are elements of B.

There are 2^n subsets of the sample space S and each of these will have a probability associated with it. It is obvious that $P(\varphi) = 0$, $P(S) = 1$ and that probability of any other event will lie between 0 and 1.

In the case of natural assignment of probabilities, the probability of each elementary element is $1/n$ and $P(E) = \frac{m}{n} = \frac{N(E)}{N(S)}$

i.e. probability of an event is the number of elementary elements in E to the number of elementary events in S. This is sometimes expressed by defining probability or an event E as the ratio of the number of cases favourable to E to the total number of cases when the cases are exclusive, exhaustive and equally likely. The definition is useful only for the natural assignment of probabilities. This was the definition we used earlier.

For the natural assignment of probabilities, it is easily see that

P (Head in the case of a coin throwing) $= \frac{1}{2}$

P (Odd number turning up in the case of die) $= \frac{3}{6} = \frac{1}{2}$.

P (Sum of points being 7 in the case of two dice) $= \frac{6}{36} = \frac{1}{6}$.

Now suppose the assignment of probabilities for the die is,

$$p_1 = \frac{1}{12},\ p_2 = \frac{1}{12},\ p_3 = \frac{1}{12},\ p_4 = \frac{1}{3},\ p_5 = \frac{1}{12},\ p_6 = \frac{1}{3}$$

This is allowable since the sum of these numbers is unity and each one is a non-negative real number, then

$$P(\text{odd number}) = p_1 + p_3 + p_5 = \frac{1}{12} + \frac{1}{12} + \frac{1}{12} = \frac{1}{4}$$

$$P(\text{even number}) = p_2 + p_4 + p_6 = \frac{1}{12} + \frac{1}{3} + \frac{1}{3} = \frac{3}{4}.$$

Theorem of Total Probability.

$P(A \cup B) =$ sum of probabilities assigned to all elementary events in $A \cup B$. (Fig. (4·5))

Now the elementary events in $A \cup B$ can be obtained by taking all the elementary events in A and all the elementary events in B together, but in this process, we are considering all the elementary events in $A \cap B$ twice and therefore we remove all the elementary events in $A \cap B$ once, so that we get

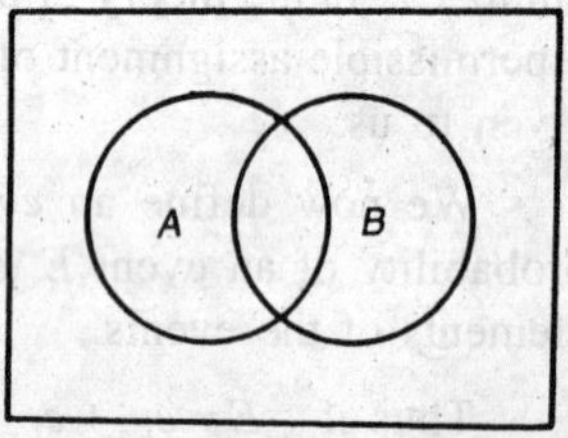

Fig. 4·5

$$P(A \cup B) = P(A) + P(B) - P(A \cap B)$$

In the same way, we get

$$P(A \cup B \cup C) = P(A) + P(B) + P(C) - P(A \cap B) - P(A \cap C) - P(B \cap C) + P(A \cap B \cap C).$$

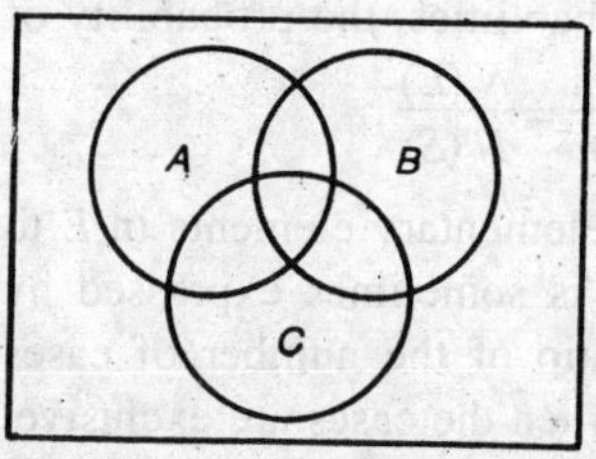

Fig. 4·6

since to get all the elementary events in $A \cup B \cup C$, (Fig. 4·6)we first combine together all the elementary events in A, B and C, but in this pairwise common events have been counted twice, so we remove them once, but in this removal, we have removed the elementary events of $A \cap B \cap C$, once too often, and so we include these events once more.

Two events are said to be mutually exclusive if they have no elementary event in common *i.e.* if the happening of one precludes the happening of the other *i.e.* if the corresponding subsets of S are disjoint.

Thus A and B are disjoint if $A \cap B = \varphi$. For two mutually exclusive events $P(A \cap B) = P(\varphi) = 0$

Similarly a number of events are said to be mutually exclusive if the corresponding sets are all disjoint.

For *mutually exclusive events*

$$P(A \cup B) = P(A) + P(B)$$

$$P(A \cup B \cup C) = P(A) + P(B) + P(C)$$

and in general, $P(A_1 \cup A_2 \cup \ldots\ldots \cup A_m) = P(A_1) + P(A_2) + \ldots\ldots + P(A_m)$

Theorem of Compound Probability. Suppose an event B is known to have happened then $B \neq \varphi$ or $P(B) \neq 0$.

We now define the conditional assignment of probability to an elementary event $\{e_i\}$ by

$$P(\{e_i\}) \mid B) = \frac{P(\{e_i\})}{P(B)} \qquad \text{if } e_i \in B$$

$$= 0 \qquad \text{if } e_i \notin B$$

This defines an allowable assignment of probabilities in S (Fig. 4·7), since

(*i*) $P(\{e_i\}) \geq 0$, $P(B) > 0$, $\dfrac{P(\{e_i\}) \geq 0}{P(B)}$

(*ii*) $P(\{e_1\}) \mid B) + P(\{e_2\} \mid B) + \ldots\ldots, + P(\{e_n\} \mid B) = \dfrac{\Sigma' P(\{e_i\})}{P(B)} = 1,$

where Σ' denotes summation over these elementary events which are elements of B (for other elementary events, the contribution is 0).

We now define the conditional probability $P(A \mid B)$ of the happening of A when B is known to have occurred as sum of the probabilities $P(\{e_i\} \mid B)$ over all the elements of A. However the contribution to the sum is 0, from all those elements of A which are not elements of B *i.e.* the contribution is only from the elements of $A \cap B$ so that we get $P(A \mid B) = \dfrac{P(A \cap B)}{P(B)}$ or $P(A \cap B) = P(B)P(A \mid B)$

Similarly $P(B \cap A) = P(A)P(B \mid A)$ But $P(A \cap B) = P(B \cap A)$ so that we get, $P(A \cap B) = P(B)\,P(A \mid B) = P(A)P(B \mid A)$

This is known as the *theorem of compound probability.*

Now an event A is said to be *independent* of an event B if

$$P(A \mid B) = P(A)$$

i.e., if the probability of happening of A is independent of the happening of B. For such an event, the theorem gives

$$P(A \cap B) = P(B)P(A \mid B) = P(B)P(A)$$

showing that $\qquad P(A) = P(A \mid B)$

i.e., B is independent of A.

Here we have assumed that $P(A) \neq 0$, $P(B) \neq 0$, that A and B are non-null sets or A and B are not impossible events. Thus if A is independent of B, then B is independent of A. We can speak of A and B as independent events.

4·7·2. Pairwise Independent Events. Two events A and B are defined to be stochastically independent if $P(A \cap B) = P(A)\,P(B)$

Theorem : Converse. *If for any two events A and B, $P(A \cap B) = P(A)P(B)$, then A and B are independent events.*

Proof. $P(A \cap B) = P(A)P(B \mid A) = P(B)\,P(A \mid B)$

Hence A and B are independent events.

For three events A, B, C, repeated use of the theorem of compound probability gives

$$P(A \cap B \cap C) = P[(A \cap B) \cap C] = P(A \cap B)P(C \mid A \cap B)$$
$$P(A \cap B \cap C) = P(A)\,P(B \mid A)P(C \mid A \cap B)$$

Since $A \cap B \cap C$ is unchanged by the change of order of the events, we get

$$\begin{aligned} P(A \cap B \cap C) &= P(A)P(B \mid A)P(C \mid A \cap B) = P(A)\,P(C \mid A)\,P(B \mid A \cap C) \\ &= P(B)P(C \mid B)P(A \mid B \cap C) = P(B)\,P(A \mid B)P(C \mid A \cap B) \\ &= P(C)P(B \mid C)P(A \mid B \cap C) = P(C)P(A \mid C)P(B \mid A \cap C) \end{aligned}$$

If A, B, C, are pairwise independent, then these give

$$\begin{aligned} P(A \cap B \cap C) &= P(A)P(B)P(C \mid A \cap B) = P(A)P(C)P(B \mid A \cap C) \\ &= P(B)P(C)P(A \mid B \cap C) \end{aligned}$$

Example 4·22. *Show by an example that if C is independent of A and B separately, it will not be necessarily independent of $A \cap B$.*

Consider the simultaneous throw of two die.

Let A be the event that the number on the first die is odd.

Let B be the event that the number on the second die is odd.

Let C be the event that the sum of numbers on the two die is odd.

It is easily seen by considering the 36 points in the sample space that, $P(A) = \frac{1}{2}$, $P(B) = \frac{1}{2}$, $P(C) = \frac{1}{2}$

$$P(A \cap B) = \frac{1}{4},\quad P(A \cap C) = \frac{1}{4},\quad P(B \cap C) = \frac{1}{4}\ ;\quad P(A \cap B \cap C) = 0$$

Since $P(A \cap C) = P(A)P(C)$, $P(B \cap C) = P(B)P(C)$,

C is independent of both A and B. Further, $P(C \mid A \cap B) = 0 \neq P(C)$ so that C is not independent of $A \cap B$.

Thus the result $P(A \cap B \cap C) = P(A)\,P(B)P(C)$ is not necessarily true for pairwise independent events. We require additional condition(s).

Conditions for Mutual independence of n Events. We define events $E_1, E_2, \ldots\ldots, E_n$ to be mutually independent if and only if, for

$i \neq j;\ i, j = 1, 2, \ldots\ldots, n \cdot i \neq j \neq k;\ i, j, k = 1, 2, \ldots\ldots, n$

we have
$$P(E_i \cap E_j) = P(E_i)P(E_j)$$
$$P(E_i \cap E_j \cap E_k) = P(E_i)P(E_j)P(E_k)$$
$$P(E_i \cap E_j \cap E_k \cap E_l) = P(E_i)P(E_j)P(E_k)P(E_l)$$
$$P(E_1 \cap E_2 \ldots\ldots \cap E_n) = P(E_1)P(E_2)\ldots\ldots P(E_n)$$

for all unequals, $i, j, k, l, \ldots\ldots$

The total number of conditions for mutual independence of n events is,

$$= {}^nC_2 + {}^nC_3 + \ldots\ldots + {}^nC_n = 2^n - n - 1.$$

Example 4·23. *Prove that if A, B and C are random events in a sample space and if A, B, C are pairwise independent and A is independent of* $(B \cup C)$, *then A, B and C are mutually independent.*

(*M.A. Delhi*)

Since A, B, C are pairwise independent, we have

$$P(A \cap B) = P(A)P(B); \quad P(B \cap C) = P(B)P(C); \quad P(A \cap C) = P(A)P(C)$$

and $$P[A \cap (B \cup C)] = P(A)P(B \cup C) \qquad \ldots(1)$$

Now $P[A \cap (B \cup C)] = P[(A \cap B) \cup (A \cap C)]$
$$= P(A \cap B) + P(A \cap C) - P[(A \cap B) \cap (A \cap C)]$$
$$= P(A)\,P(B) + P(A)\,P(C) - P(A \cap B \cap C)$$

Also $P(A)\,P(B \cup C) = P(A)[P(B) + P(C) - P(B \cap C)$
$$= P(A)\,P(B) + P(A)\,P(C) - P(A)\,P(B \cap C)$$

Hence using (1), we have $P(A)\,P(B) + P(A)\,P(C) - P(A \cap B \cap C)$
$$= P(A)\,P(B) + P(A)\,P(C) - P(A)\,P(B \cap C)$$

$$\Rightarrow \quad P(A \cap B \cap C) = P(A)\,P(B)\,P(C)$$

Hence the result is established.

Example 4·24 *Two fair dice are thrown independently. Three events A, B, and C are defined as follows :*

A : *Odd face with first dice.* B : *odd face with second dice.*

C : *sum of points on two dice is odd.*

Are the event A, B, and C mutually independent ?

(*B. Sc. Stat. Hons. Delhi, 1983*)

Since any one of the dice faces 1, 2, . . ., 6 can appear on any of the two dice, we have

$$P(A) = \frac{3 \times 6}{36} = \frac{1}{2}, P(B) = \frac{3 \times 6}{36} = \frac{1}{2} \quad \therefore \quad P(A)P(B) = \frac{1}{2} \times \frac{1}{2} = \frac{1}{4}$$

The sum of points on two dice will be odd if the number on one die is odd and on the other, it is even. We shall have the following 18 favourable cases for C viz.

(1, 2), (1, 4), (1, 6), (2, 1), (2, 3), (2, 5),...., (6, 1), (6, 3), (6, 5)

Thus $\quad P(C) = \frac{18}{36} = \frac{1}{2}$

For event, $A \cap B$, number of favourables cases are 9 viz.

(1, 1), (1, 3), (1,5), (3, 1), (3, 3), (3, 5), (5, 1), (5, 3), (5, 5)

Similarly for event $A \cap C$, number of favourable cases = 9

and for event $B \cap C$, number of favourable cases = 9

Also number of cases favourable to the event $A \cap B \cap C = 0$

Thus $P(A \cap B) = \frac{9}{36} = \frac{1}{4}, P(B \cap C) = \frac{9}{36} = \frac{1}{4}, P(B \cap C) = \frac{9}{36} = \frac{1}{4}$

$$\therefore \quad P(A \cap B) = \frac{1}{4} = P(A)\,P(B); \quad P(B \cap C) = P(B)\,P(C)$$

and $\quad P(C \cap A) = \frac{1}{4} = P(C)\,P(A)$

Also $\quad P(A \cap B \cap C) = P(\phi) = 0 \neq P(A)\,P(B)\,P(C)$

Hence the events A, B and C are pairwise independent, but not mutually independent.

Example 4·25 *An event A is known to be independent of the events B, $B \cup C$ and $B \cap C$. Show that it is also independent of C.*

Since A and B are independent, $P(A \cap B) = P(A)\,P(B)$

Similarly $\quad P[A \cap (B \cup C)] = P(A)\,P(B \cup C)$

$\quad P[A \cap (B \cap C)] = P(A)\,P(B \cap C)$

$$\therefore \quad P(A \cap B) + P(A \cap C) - P(A \cap B \cap C)$$
$$= P(A)[P(B) + P(C) - P(B \cap C)]$$

$$\Rightarrow \quad P(A \cap B) + P(A \cap C) - P(A \cap B \cap C)$$
$$= P(A)\,P(B) + P(A)\,P(C) - P(A)\,P(B \cap C)$$

$\Leftarrow \quad P(A \cap C) = P(A)P(C).$

Hence A and C are independent.

Ex. Prove or distrove: If A and B are independent events, then

$P(A \cap B \mid C) = P(A \mid C)\; P(B \mid C)$ ***(B.Sc. Math Hons. Delhi 1997)***

Hint. Take Example 4.24. Show that $P(A \cap B \mid C) = P\dfrac{(A \cap B \cap C)}{P(C)} = 0$

and $P(A \mid C)\, P(B \mid C) = \dfrac{1/4}{1/2} \cdot \dfrac{1/4}{1/2} = \dfrac{1}{4}$ $\therefore$ L.H.S. $\neq$ R.H.S.

4.8. Axiomatic Development of Probability Theory. In the axiomatic development of probability theory, probability is defined as a function defined on events (subsets of S) *i.e.*, it is a rule which associates to each event A certain real number $P(A)$ which satisfies the following three axioms.

Axiom I. $P(A) \geq 0$, *i.e.*, the probability of every event is non-negative.

Axiom II. $P(S) = 1$, *i.e.*, the probability of a certain event is unity.

Axiom III. $A_1, A_2, \ldots$ are a countable number of sub-events of S such that $A_1 \cup A_2 \cup A_2 \ldots = S$ and $A_i \cap A_j = \phi$, $(i \neq j)$,

then $P(A_i \cup A_j \cup A_k \cup \ldots) = P(A_i) + P(A_j) + P(A_k) + \ldots$

i.e., the probability of a union of disjoint events is the sum of probabilities of the events themselves.

From these axioms, we easily deduce the following theorems :

Theorem 1. $P(\phi) = 0$

Proof : $S \cup \phi = S$,

and therefore $P(S \cup \phi) \Rightarrow P(S) + P(\phi) = P(S)$, so that $P(\phi) = 0$

Theorem 2. $P(\bar{A}) = 1 - P(A)$.

Proof : $A \cup \bar{A} = S, A \cap \bar{A} = \phi$,

$$P[A \cup \bar{A}] = P(S) = 1 = P(A) + P(\bar{A})$$

so that $P(\bar{A}) = 1 - P(A)$.

Theorem 3. $A \subset B \Rightarrow P(A) \leq P(B)$

Proof : $A \subset B \Rightarrow A \cup (B \cap \bar{A}) = B \cup A = B$

Also $A \cap (B \cap \bar{A}) = \phi$ so that

$$P[A \cup (B \cap \bar{A})] = P(A) + P(B \cap \bar{A}) = P(B)$$

But $P(B \cap \bar{A}) \geq 0$, so that the result follows.

Theorem 4. $0 \leq P(A) \leq 1$ for any A in S

Proof : $A \subset S \Rightarrow P(A) \leq P(S) \Rightarrow P(A) \leq 1$

since from axiom $P(A) \geq 0$, it follows that $0 \leq P(A) \leq 1$.

Definition. The conditional probability of event A, given B, is defined $P(A \mid B) = \dfrac{P(A \cap B)}{P(B)}$ wherever $P(B) \neq 0$

We can build up a rigorous theory of probability based on the above axioms and definitions. Hence we define probability as a function satisfying certain axioms. The theorem of total probability is taken an axiom and the theorem of compound probability is taken as a definition.

This approach is similar to that of Euclidean geometry, where points, lines etc., are undefined terms, satisfying certain axioms and where we prove certain theorems based on these axioms. Geometry becomes practical when we give intuitive meaning to points and lines. In the same way, the theorem of this deductive theory of probability, will be useful when probability is given an intuitive meaning, as in the earlier sections of this chapter.

Example 4·26 *If A and B are events such that* $A \subset B$ *then* $P(A) \leq P(B)$.

Since $A \subset B$ we have from figure or otherwise $B = A \cup (B \cap \bar{A})$

$$\therefore \qquad P(B) = P(A) + P(B \cap \bar{A})$$

since A and $B \cap \bar{A}$ are mutually exclusive events.

Hence $P(B) \geq P(A)$ since $P(B \cap A) \geq 0$

Example 4·27 *If for two mutually exclusive events A and B*

$$P(B) = 2\,P(A) \text{ and } A \cup B = S, \textit{ find } P(A)$$

We have $\quad P(A) + P(B) = P(A \cup B) = P(S) = 1$

or $\qquad 3P(A) = 1 \quad [\because\ P(B) = 2P(A)] \quad \therefore\ P(A) = \frac{1}{3}\cdot$

Example 4·28. *Prove the following laws, assuming that in each case the conditional probabilities are defined.*

(*a*) $P(B) = 1$, *then* $P(A \mid B) = P(A)$.

(*b*) *If A and B are disjoint and* $P(B) > 0$, *then* $P(A \mid B) = 0$.

(*c*) *If* $A_1 \subset A_2$, *then* $P(A_1 \mid B) \leq P(A_2 \mid B)$

(*d*) $P(A_1 \cup A_2 \mid B) = P(A_1 \mid B) + P(A_2 \mid B) - P(A_1 \cap A_2 \mid B)$.

(*a*) Since $P(B) = 1$

$\therefore\ \ B = S =$ entire sample space and $P(S) = 1$

Now $P(A \mid B) = P(A \cap B)/P(B) = P(A \cap S)/P(S) = P(A)$.

(*b*) Since $A \cap B = \phi$, we have $P(A \cap A) = P(\phi) = 0$.

$\therefore\ \ P(A \mid B) = P(A \cap B)/P(B) = 0$.

(*c*) Since $A_1 \subset A_2$, we have $A_1 \cap B \subseteq A_2 \cap B$

$\therefore\ \ P(A_1 \cap B) \leq P(A_2 \cap B) \quad$ or $\quad P(B)P(A_1 \mid B) \leq P(B)P(A_2 \mid B)$.

Cancelling $P(B)$ which is not zero, we obtain the required result.

(*d*) $P(A_1 \cup A_2 \mid B) = P(B \cap (A_1 \cup A_2))/P(B) = P[BA_1 \cup BA_2]/P(B)$

$$= [P(BA_1) + P(BA_2) - P(A_1A_2B)]/P(B)$$

$$= P(A_1 \mid B) + P(A_2 \mid B) - P(A_1A_2 \mid B)$$

Ex. 1. *For any two events A and B, prove that*

$$P[(A \cap B) \cup (B \cap \bar{A})] = P(A) + P(B) - 2P(A \cap B).$$

Ex. 2. *Let A, B and G be three events associated with an experiment { A, B,*

C} *and let* $P(B) = \frac{3}{2} P(A), P(C) = \frac{1}{3} P(B)$, find $P(A)$ [**Ans.** $\frac{1}{3}$]

Ex. 3. *If A, B, C are mutually independent events, show that* $A \cup B$ *and C are also independent.*

4·9. Axiomatic Definition of Probability. The axiomatic approach was given in 1933 by A.N. Kolmogrov (1903—). Let *A* be the σ -field of events associated with outcomes of an experiment. Then, probability is defined on *A* as a real-valued function $P(\cdot)$ such that for any $A, A_1, A_2, \ldots . A_n \in A$

Axiom I. Non-negative, $P(A) \geq 0$.

Axiom II. Normed, $P(S) = 1$.

Axiom III. (*Axiom of finite or infinite Addition*). If A_1, A_2, ..., A_n, are pairwise : mutually exclusive, then

$$P(A_1 + A_2 + \ldots + A_n) = P(A_1) + P(A_2) + \ldots + P(A_n)$$

However, for further extension of the theory, we need axiom III which may be called the *extended axiom of addition.* It can be shown that this axiom can be replaced by the equivalent *axiom of continuity* which states that if a sequence of events $A_1, A_2, \ldots, A_n$ is product of all the events $\{A_n\}$ is an impossible event, *i.e.*

$$P(A_n) \to 0 \text{ when } n \to \infty$$

For further rigorous development of the theory, we need the concept of a *Borel field.* A class *C* of sets which is such that when $A_1, A_2 \in C$, then $A_1 \cup A_2, A_1 \cap A_2, \overline{A}_1$ also belong to *C* is said to constitute a *Boolean Field* which may be denoted by *F.* It is easily seen that if $A_1, A_2 \in C$, then $A_1 - A_2 = A_1 \cap \overline{A}_2$ also belongs to *C.*

A Borel Field need not contain union of all countable sequences of sets of *C.* A field which contains all such unions (and therefore countable intersection) is called a Borel Field or a σ– *field.* Given a Boolean Field *F,* there exists a numerical Borel Field containing *F* which is denoted by *B*(*F*) We can talk of a probability function on a Borel Field if we can specify it on the Boolean field generating it.

We can develop a calculus of probabilities on the larger space Ω of elementary events, a Borel or a σ– field *B* of sets (events), in Ω and a probability measure *P* on *B* the triplet (Ω,*B*,*P*) is called a probability space while (Ω,*B*) is called the *sample space.*

We can then develop probability theory as a branch of measure theory. The advantage is in mathematical sophistications and it helps in proving deeper theorems of probability theory. For our purpose, the three approaches viz. classical, statistical and axiomatic are quite enough.

Exercises 4

1. (*a*) Define independent and mutually exclusive events. Can two events be mutually exclusive and independent simultaneously ? **[Ans.** No**]**

(*b*) Define mutually exclusive events and state whether they can be regarded as a special case of dependent or independent events.

(*c*) Give the 'a priori' and statistical definition of probability. Which of these is more general and why ?

(*d*) In a report on consumer preference, it was given that out of 500 persons surveyed, 410 preferred variety A, 380 preferred variety B and 270 persons liked both.

Can the above report be correct ? Justify your answer.

[Use : $n(A \cup B) = n(A) + n(B) - n(A \cap B)$]

(*e*) Examine the correctness of the following report about the liking of consumers for two products *A* and *B*.

"50% liked both *A* and *B*, 35% liked *A* but not *B* and 25% liked *B* but not *A*."

2. (*a*) Discuss and crticise the following :

$P(A) = \frac{2}{3}$ $P(B) = \frac{1}{4}$ $P(C) = \frac{1}{6}$ for the probabilities of three mutually exclusive events *A*, *B* and *C* for which $P(A) + P(B) + P(C) = 1$.

(*b*) *A*, *B* and *C* are three independent events such that

$P(A\bar{B}\bar{C}) = \frac{1}{8}$, $P(AB\bar{C}) = \frac{1}{12}$ and $P(\bar{A}\bar{B}\bar{C}) = \frac{1}{12}$. Prove that $P(A) = \frac{1}{2}$ and $P(\overline{ABC}) = \frac{1}{4}$. **(*B. Sc. Bombay 1970*)**

(*c*) Prove or disprove ! $P(A \mid B) \geq P(A) \Rightarrow P(B \mid A) \geq P(B)$.

(*d*) Prove : $P(\bar{A} + \bar{B}) = 1 - P(AB)$, $P(\bar{A}\bar{B}) = 1 - P(A) - P(B) + P(AB)$

3. (*a*) Given $p_1 = P(A)$, $p_2 = P(B)$, $p_3 = P(A \cap B)$, express the following in terms of p_1, p_2, p_3 :

(*i*) $P(\bar{A} \cup \bar{B})$ (*ii*) $P(\bar{A} \cap B)$ (*iii*) $P(\bar{A} \cup B)$ (*iv*) $P(\bar{A} \cap \bar{B})$

(*v*) $P(A \cap \bar{B})$ (*vi*) $P(\bar{A} \mid \bar{B})$ (*vii*) $P(\bar{A} \cup \bar{B})$ (*viii*) $P[A \cup (\bar{A} \cap B)]$.

(*B.A., Hons., 1971*)

[Ans. (*i*) $1 - p_3$, (*ii*) $p_2 - p_3$, (*iii*) $1 - p_1 + p_3$, (*iv*) $1 - p_1 - p_2 + p_3$,

(*v*) $p_1 - p_3$ (*vi*) $\frac{1 - p_1 - p_2 + p_3}{1 - p_2}$, (*vii*) $1 - p_3$, (*viii*) $[p_1 + p_2 + p_3]$

(*b*) In terms of non-zerp probabilities $p_1 = P(A)$, $p_2 = P(B)$, $p_3 = P(AB)$, express

(*i*) $P(A + B)$, (*ii*) $P(A \mid B)$, (*iii*) $P(\bar{A}B)$, (*iv*) $P[\bar{A} \cap (A \cup B)]$,

(*v*) $P[A \cup (\bar{A} \cap B)]$, (*vi*) Condition that *A* and *B* are mutually exclusive.

(*vii*) Condition that $\bar{A}$ and $\bar{B}$ are mutually independent.

(*B.Sc. Meerut, 1970; B.A. Hons. Delhi, 1969*)

$$\left[\textbf{Ans. } (i)\ p_1+p_2-p_3,\quad (ii)\ \frac{p_3}{p_2},\quad (iii)\ p_2-p_3\quad (iv)\ p_2-p_3,\quad (v)\ p_1+p_2-p_3\quad (vi)\ p_3=0\quad (vii)\ 1-(p_1+p_2-p_3)=0\right]$$

4. (*a*) If A and B are any two events and the probability $P(B) \neq 1$, prove that

$$P(A \mid \overline{B}) = \frac{P(A) - P(A \cap B)}{1 - P(B)}$$

and hence deduce that $P(A \cap B) \geq P(A) + P(B) - 1$.

Also show that $P(A) >$ or $< P(A \mid B)$ according as $P(A \mid \overline{B}) >$ or $< P(A)$.

(*B.Sc. Stat. Hons. Delhi, 1989*)

(*b*) If A and B are mutually exclusive and $P(A \cup B) \neq o$, then prove that

$$P(A \mid A \cup B) = \frac{P(A)}{P(A) + P(B)}$$

(*c*) If $A \cup B \cup C = \phi$, then $P(B \cup C \mid A) = P(B \mid A) + P(C \mid A)$

(*d*) Prove :

(*i*) $P(A \cup B \mid C) = P(A \mid C) + P(B \mid C) - P(A \cap B \mid C)$

(*B.A. Hons. Delhi, 1984*)

(*ii*) $P(\overline{A}_1 \cup \overline{A}_2 \mid \overline{A}_2) = 1 - P(A_1)P(A_2 \mid A_1)P(A_3 \mid A_1A_2)$.

(*B. Sc. Bombay*)

(*e*) For three events A, B, C prove or disprove :

$P(A) > P(\overline{B}) \Rightarrow P(A \mid C) > P(B \mid C)$. **(*B.A. Hons. Delhi, 1983*)**

[Hint. (*a*) We have $P(A \mid \overline{B}) = P(A\overline{B})/P(\overline{B}) = [P(A) - P(AB)]/[1 - P(B)]$ ...(*i*)

Since $P(A \mid B) \leq 1$, we have $P(A) - P(AB) \leq 1 - P(B)$ ...(*ii*)

$\Rightarrow \quad P(AB) \geq P(A) + P(B) - 1$

Clearly we have $\dfrac{P(A \mid B)}{P(A)} = \dfrac{P(B \mid A)}{P(B)} = \dfrac{1 - P(B \mid A)}{1 - P(B)}$

Hence $P(A \mid B) >+ P(A)$

if $P(B \mid A) > 1 - P(B)$

i.e. if $P(B \mid A) < P(B)$ $\Big\}$ *i.e, if* $P(A \mid B) < P(A)$ ...(*iii*)

since $\dfrac{P(A \mid B)}{P(A)} = \dfrac{P(B \mid A)}{P(B)}$ · Hence the result.]

5. If A, B and C are three arbitrary events and we denote by

$S_1 = P(A) + P(B) + P(C)$

$S_2 = P(A \cap B) + P(B \cap C) + P(C \cap A)$, $S_3 = P(A \cap B \cap C)$

prove that exactly one of the three events occurs is given by $S_1 - 2S_2 + 3S_3$.

(*M. A. Statistics, Delhi*)

6. It is 8 : 5 against a person who is now 40 years old living till he is 70 and 4 : 3 against a person now 50 living till he is 70. Find the probability that at least one of these persons will be alive 30 years hence

[Ans. Required probability $= 1 - \dfrac{8}{13} \times \dfrac{4}{7} = \dfrac{59}{91} = 0{\cdot}64835$]

7. A and B are two independent witnesses (*i.e.*, there is no collusion between them) in a case. The probability that A will speak the truth is x, and the probability that B will speak the truth is y, A and B agree in a certain statement. Show that the probability that this statement is true is $\dfrac{xy}{1-x-y+2xy}$.

8. (*a*) Let p be the probability that a man aged x years will die in a year. Find the probability that out of n men $A_1, A_2, \ldots, A_n$, each aged x, A_1 will die and be the first to die. $\left[\textbf{Ans. } \dfrac{1}{n}[1-(1-p)^n]\right]$ (***M.A. Punjab 168***)

(*b*) Two dice are thrown. Let A and B be two events described as follows :

A : The event that sum of faces is odd.

B : the event that at least one ace occurs.

Find the probabilities of the events

(*i*) BA, (*ii*) $A+B$, (*iii*) $A\bar{B}$. (***B. Sc. Bombay 1967***)

$\left[\textbf{Ans. } (i)\ \dfrac{6}{36}\quad (ii)\ \dfrac{23}{36}\quad (iii)\ \dfrac{5}{36}\right]$

9. A and B throw alternately with a pair of dice. A wins if he throws 6 before B throws 7, and B if he throws 7 before A throws 6. If A begins, show that his chance of winning is $\dfrac{30}{61}$. (Huyghen's Problem)

(***B.A. Hons., Delhi 1960***)

10. Consider families with exactly two children. Let b and g stand for a boy and a girl respectively and first letter for the older child, we have four possibilities *bb, bg, gb* and *gg*. These are the four sample points, and we associate probability 1/4 with each.

(*a*) Given that the family has a boy, what is the probability that the other one is also a boy.

(*b*) Given that the family has the older child as a boy, what is the probability that the other one is also a boy ? $\left[\textbf{Ans. } (i)\ \dfrac{1}{3},\quad (ii)\ \dfrac{1}{2}\right]$

11. Three groups of children contain respectively 3 girls and 1 boy, 2 girls and 2 boys, 1 girl and 3 boys. One child is selected at random from each group. Show that the chance that the three selected consists of 1 girl and 2 boys is 13/32. (***I.F.S., 1997, B.Sc. Delhi 1978***)

12. Out of $(2n+1)$ tickets consecutively numbered, three are drawn at random. Find the chance that the numbers on them are in (*i*) Arithmetical Progression, (*ii*) Geometrical Progression if number of tickets is 101.

(***B. Sc. Delhi 1973, 75***)

Ans. (*i*) $\dfrac{3n}{4n^2-1}$, (*ii*) $\dfrac{53}{10\times 50\times 33} = 0{\cdot}0$[illegible]3

13. (*a*) A player tosses a coin and is to score one point for every head turned up and two for every tail. He is to play on until his score reaches or passes n. If p_n is the chance for attaining exactly n, show that

$p_n = \frac{1}{2}(p_{n-1} + p_{n-2})$ and hence find the value of p_n.

[Hint. $p_n - p_{n-1} = -\frac{1}{2}(p_{n-1} - p_{n-2})$] $\left[\textbf{Ans.}\ \frac{1}{2}\left\{2 + (-1)^n \frac{1}{2^n}\right\}\right]$

(*b*) There are *n* similar biased dice, such that the probability of obtaining a 6 with each one of them, is the same and equal to *p*. If all the dice are rolled once, show that p_n, the probability that an odd number of 6's is obtained, satisfies the difference equation

$$p_n + (2p - 1)\, p_{n-1} = p$$

and hence derive an explieit expression for p_n.

[Hint. $p_n - \frac{1}{2} = (1 - 2p)\left(p_{n-1} - \frac{1}{2}\right)$]. **[Ans.** $p_n = \frac{1}{2}[1 - (1 - 2p)^n]$]

14. Each of the *n* urns contains *a* white and *b* black balls. One ball is transferred from the first urn into the second, then one ball from the later into the third and so on. Finally one ball is taken from the last urn; what is the probability of its being white ? (***M. Sc. Punjab 1970***)

$\left[\textbf{Ans.}\ p_n = \frac{a}{a+b}\right]$

[**Hint.** If p_k be the probablity of drawing a white ball from the *k*th urn; then

$$p_k = \frac{a+1}{a+b+1}\, p_{k-1} + \frac{a}{a+b+1}(1 - p_{k-1}) = \frac{1}{a+b+1}\, p_{k-1} + \frac{a}{a+b+1}.$$

Putting $k = 2$ and using $p_1 = \frac{a}{a+b}$, we get $p_2 = \frac{a}{a+b}$. Proceeding in this manner and using the method of induction, we conclude that $p_n = \frac{a}{a+b}$]

15. (*a*) Four different objects 1, 2, 3, 4 are distributed at random on four places marked 1, 2, 3, 4. What is the probability that none of objects occupies the place corresponding to its number ?

(*b*) A person writes four letters and four envelopes. If the letters are placed in the envelopes at random, what is the chance that not more than one letter is placed in the correct envelope ?

[Hint. (*a*) See solved example 4·14, page 85.] **[Ans.** (*a*) $\frac{3}{8}$ (*b*) $\frac{17}{24}$]

16. (*a*) Prove that the following theorem in probability for the events E_1, E_2 and E_3 :

(*i*) $P(\overline{E_1} + \overline{E_2}) = 1 + P(E_1E_2) - P(E_1) - P(P_2)$.

(*ii*) $P(\overline{E_1} + \overline{E_2} + \overline{E_3}) = 1 - P(E_1)P(E_2 \mid E_1)P(E_3 \mid E_1E_2)$.

where $P(\overline{E})$ is the probability that *E* does not occur.

(*b*) If $A + B$ means the events *A* and $A\overline{B}$ and if $A - B$ means the event *AB* prove that

(*i*) $P[(A - AB) + B) = P(A + B)$ and

(*ii*) $\overline{P(A + B + C\,)} = P(\overline{ABC})$ (***M. A., Statistics, Delhi***) (*c*) For any two events A and B, find whether the following statements are true or not :

(*i*) $P(A \mid B) + P(\overline{A} \mid B) = 1$

(*ii*) $P(A \mid B) + P(A \mid \overline{B}) = 1.$ **(B.A., Hons., Delhi 1966)**

[**Ans.** (*i*) Correct (*ii*) Incoreect]

(*d*) Examine whether the following relations are correct.

(*i*) $(A \cup B) - C = A \cup (B - C)$

(*ii*) $\overline{(A \cup B \cup C)} = \overline{A}\ \overline{B}\ \overline{C}$ **(M.A., Statistics, Delhi, 1969)**

[**Ans.** (*i*) incorrect (*ii*) correct]

17. What is the probability that at least one of the players in a bridge game will get a complete suit of cards ?

(M.A. Statistics, Delhi; I.C.A.R, Delhi, 1974)

[**Hint.** Required probability $= P(A_1 + A_2 + A_3 + A_4)$

$$= \sum_i p_i - \sum_{ij}\sum p_{ij} + \sum_{ijk}\sum\sum p_{ijk} - p_{1234}$$

$$= \frac{4c_1 \times 4}{52c_{13}} - \frac{4c_2 \times 4 \times 3}{52c_{13} \times 39c_{13}} + \frac{4c_3 \times 4 \times 3 \times 2}{52c_{13} \times 39c_{13} \times 26c_{13}} - \frac{4c_4 \times 4 \times 3 \times 2 \times 1}{52c_{13} \times 39c_{13} \times 26c_{13} \times 13c_{13}}$$

$$= \frac{16 \cdot 13\,!\,39\,! - 72(13\,!)^2\,26\,! + 72(13\,!)^4}{52\,!}$$]

$$= 2{\cdot}52 \times 10^{-11}$$

18. In a lottery m tickets are drawn at a time out of the total number of n tickets and returned before the next drawing is made. Show that the probability that in k drawings each of the numbers 1, 2, 3, ..., n will appear at least once is given by $p_k = 1 - {}^nC_1\left(1 - \frac{m}{n}\right)^k + {}^nC_2\left(1 - \frac{m}{n}\right)^k\left(1 - \frac{m}{n-1}\right)^k - \ldots$

(M. Sc., Bombay; B.A., Hons., Delhi; B.A. Hons. Calcutta)

[**Hint.** $p_k = 1 - P$ (ticket numbered i does not appear in any one of the drawings.

$$P(A_1) = (n-m)^k / n^k\,,\ P(A_1A_2) = \binom{n-2}{m}^k \Big/ \binom{n}{m}^k = \left(\frac{n-m}{n}\right)^k\left(\frac{n-m-1}{n-1}\right)^k = \text{etc.}$$

19. We have k varieties of objects each variety consisting of the same number of objects. These objects are drawn one at a time and replaced before the next drawing. Show that the probability p_n, that n and no less drawings will be required to produce objects of all varieties is given by

$$p_n = \sum_{0}^{k-1} (-1)^r \binom{k-r}{r}\left(\frac{k-1-r}{k}\right)^{n-1}$$

i.e., $k^{n-1}p_n = (k-1)^{n-1} + {}^{k-1}c_1\,(k-2)^{n-1} + {}^{k-1}c_2(k-3)^{n-1} - \ldots$

(M.A. Stat. Delhi)

20. (*a*) Cards are dealt one by one from a well shuffled pack until ace appears. Show that the probability that exactly n cards are dealt *before* the first ace, is $\dfrac{(51-n)(50-n)(49-n)}{13 \cdot 51 \cdot 50 \cdot 49}$.

(*B. Sc. Delhi, 1992, B.A., Hons., Delhi, 1971*)

If cards continue to be dealt until a second ace appears, prove that the probability that exactly r cards are dealt in all *before* the second ace, is

$$\frac{r(41-r)(50-r)}{13 \cdot 17 \cdot 50 \cdot 49}$$

[**Hint.** (*i*) By hyp., there should be no ace in first n cards, so that the probability of such an event $= {}^{48}c_n / {}^{52}c_n$ and the remaining $(52-n)$ cards must contain one ace. Then required probability $= \dfrac{4}{52-n} \times \dfrac{48c_n}{52c_n} =$ etc.

(*ii*) Required probability $= 4c_1 \times \dfrac{48c_{r-1}}{52c_r} \times \dfrac{3}{52-r} =$ etc.]

(*b*) Cards are dalt one by one from an ordinary pack (without replacement) until two aces have appeared. Find the most probable number of cards to be turned up. **[Ans.** 18]

[**Hint.** The most probable number to be turned up will be the value of x which gives the chance of success at the *xth trial viz.*,

$$\frac{(x-1)(52-x)(51-x) \cdot 4 \cdot 3}{52 \cdot 51 \cdot 50 \cdot 49}$$

its greatest value. Hence etc.]

21. Consider n independent tosses of a coin which has probability p of falling heads. Let S be the number of tosses in which the coin falls head. Find the probability that S is even. **[Ans.** $\frac{1}{2}\{1+(q-p)^n\}$]

22. (*a*) Show that the chance of throwing 15 with 3 dice is $\dfrac{5}{108}$.

(*b*) A and B throw with 3 dice; if A throws 8, what is B's chance of throwing a higher number ?

(*c*) If $6n$ tickets numbered 0, 1, 2, ..., $6n-1$ are placed in a bag and three are drawn out, find the chance that the sum of the numbers is equal to $6n$.

(*d*) The probability of occurrence of an event A is 0·7, the probability of non-occurrence of B is 0·5 and that of at least one of A and B non-occurring is 0·6. Find the probability that at least one of A and B occurs.

[**Hint.** (*a*) The required probability is the coefficient of x^{15} in the expansion of $\dfrac{1}{6^3}(x^1+x^2+x^3+x^4+x^5+x^6)^3$, That is, in $\dfrac{1}{6^3} \cdot \dfrac{x^3(1-x^6)^3}{(1-x)^3}$. Hence etc.]

[Ans. (*b*) $\dfrac{7}{27}$. (*c*) $3n/(6n-1)(6n-2)$]

23. An urn contains a white and b black balls. Balls are drawn one by one until only those of the same colour are left. What is the probability that they are white ? (***M.A., Punjab 1960***)

[Reqd. prob. $= \left(\sum_{x=0}^{a-1} {}^{b+x-1}C_{b-1} \right) \div {}^{a+b}C_b$

$$= \sum_{0}^{a-1} \left({}^{b+x-1}C_x \right) \div {}^{a+b}C_b = {}^{a+b-1}C_{a-1} \div {}^{a+b}C_b = \frac{a}{a+b}.$$

For ${}^{b-1}c_0 + {}^{b}c_1 + {}^{b+1}c_2 + \ldots\ldots + {}^{b+a-2}C_{a-1}$

$$= \text{term independent of } x \text{ in } \left[(1+x)^{b-1} + \frac{(1+x)^b}{x} + \frac{(1+x)^{b+1}}{x^2} + \ldots \ldots + \frac{(1+x)^{b+a-2}}{x^a - a} \right]$$

$$= \text{,, ,, ,, in } (1+x)^{b-1} \left[\left(\frac{1+x}{x} \right)^a - 1 \right] \div \left[\left(\frac{1+x}{x} \right) - 1 \right]$$

$$= \text{,, ,, ,, in } (1+x)^{a+b-1} \div x^{a-1} = \binom{a+b-1}{a-1} = \binom{a+b-1}{b} \text{ where } \binom{n}{r} \equiv {}^nC_r.$$

24. Of three independent events, the prabability that the first only should happen is 1/4 ; the probability that the second should happen is 1/8, and the probability that the third only should happen is 1/12. Obtain the unconditional probability of the three events. [**Ans.** $\frac{1}{2}, \frac{1}{3}, \frac{1}{4}$]

(***B.A. Hons. Delhi. 1969***)

25. A coin is tossed $(m+n)$ times $(m > n)$. Show that the probability of getting m consecutive heads is $(n+3)/2^{m+2}$ and that of at least m consecutive heads is $(n+2)/2^{m+1}$ (***M. A., Madras, 1969***)

[**Hint.** The chance that the event has already occurred at the end of $(m+n-1)$ tosses is p_{n-1} and is only increased with an additional toss by the fact that an incomplete run of $(m-1)$ heads preceded by a tail may be followed by a head. The chance for this is $1/2^{m+1}$. We have $p_n = p_{n-1} + 1/2^{m+1}$. Write $n-1, n-2, \ldots.., 1$ for n and add using $p_0 = 1/2^m$].

Aliter. Let $(m+n)$ tosses be represented by $m+n$ rooms. Total no of event points $= 2^{m+n}$. Since $m > n$, only one run of m heads is possible which begins at the Ist or 2nd or . . . . .$(n+1)$*th* room. Except for the first & last cases, the runs just before & after the runs of m heads must be sealed off with tails in order that the runs of heads may not get longer. In the first case the runs just after & in the last case, the run just before the run of heads must be filled with a tail. Hence the event 'run of m heads' contains $(n-1)\,2^{n-2} + 2 \cdot 2^{n-1} = (n+3)\,2^{n+2}$ points and first result follows.

For second result, we have $p_m = (n+3)/2^{m+2}$, $p_{m+1} = (n+2)/2^{m+3}$,....,

$p_{m+n-2} = 5/2^{m+n}$. But $p_{m+n-1} = 2/2^{m+n} = 1/2^{m+n-1}$ and $p_{m+n} = 1/2^{m+n}$. Reqd. prob. $= p_m + p_{m+1} + \ldots\ldots + p_{m+n} =$ etc.

$$\left[\text{Put } x = 2 \text{ in } \frac{d}{dx}\{1 + x + x^2 + \ldots + x^{n+3}\} = \frac{x^{n+4} - 1}{x - 1}\right]$$

26. An urn contains N balls, p^N of which are white, the rest black. From it are drawn n_1 balls r_1 of which are found to be white. The balls are then replaced, and n_2 balls are now extracted among which r_2 are white balls, Show that the most probable value of p for such a sample is $p = (r_1 + r_2) \div (n_1 + n_2)$.

(**Hint.** Maximize the expression $f(p) = {}^{n_1}C_{r_1} p^{r_1} (1-p)^{n_1 - r_1} \times {}^{n_2}C_{r_2} p^{r_2} (1-p)^{n - r_2}$]

27. An urn contains N identical balls, black and white in unknown proportion, there being at least one white and one black ball is drawn out and replaced n times, the balls being mixed after each drawing with the result that first r white balls are seen. If all compositions of the population are equally likely, prove that the most probable value of R, the number of white balls in the urn is given by $R/N = r/n$.

28. (*a*) What is the probability that two integers chosen at random will be prime to each other ?

(*b*) Four possible integers are chosen at random. Find the probability of their having a common factor. Use II $(1 - p^{-4}) = 90/n^4, p = 2,3,5,\ldots$

(*c*) A hunter's chance of shooting an animal at a distance $r (> a)$ is a^2/r^2. He fired when $r = 2a$, and if he misses he reloads and fires when $r = 3a, 4a, \ldots\ldots$ If he misses at distance na, the animal escapes. What are the odds against the hunter ? **Ans.** $n + 1 : n - 1$

[**Hint.** (*a*) Let a and b be any two integers and r a prime number. The possible remainder are, $1, 2, \ldots\ldots r - 1$, then the chance of a being divisible by r is $1/r$ and same is that for b. The integers being chosen at random, the probability of both a and b having r as a factor is $1/r^2$ and therefore not being divisible by r is $1 - 1/r^2$. Reqd. prob. $= \prod_2^n \left(1 - \frac{1}{r^2}\right) = 6/r^2$

(*c*) The chance of escape of the animal in the end $= \prod_2 \left(1 - \frac{a^2}{i^2 a^2}\right)$

$$= \prod_2^n \left(1 - \frac{1}{i^2}\right) = \prod_{i=2}^{n} \frac{i-1}{i} \cdot \prod_{i=2}^{n} \frac{i+1}{i} = \frac{1}{2} \cdot \frac{n+1}{n} = \frac{n+1}{2n}$$

29. From a pack of 52 cards an even number of cards is drawn. Show that the probability of half of these cards being red is $\left[\frac{52!}{(26!)^2} - 1\right] / (2^{51} - 1)$

(*M. Sc. Meerut, 1968*)

30. (*a*) If each of n independent trials can result in one of the k incompatible events $E_1, E_2, ...+ E_k$ with respective probabilities $p_1, p_2,........., p_k$; $(p_1 + p_2 + ... + p_k = 1)$ show the probabilities to have n_1 events E_1, n_2 events $E_2,....., n_k$ events E_k where $n_1 + n_2 + ... + n_k = n$, is given by

$$Pn_1, n_2..... n_k = \frac{n\,!}{n_1\,!\, n_2\,! .... n_k\,!} p_1^{n_1} p_2^{n_2} .... p_k^{n_k}.$$

(*b*) A bag contains 5 white, 7 green, 12 red and 14 black balls. The balls are indistinguishable from each other except by colour. A ball is drawn and replaced after its colour has been noted, on ten occasions. If any ball is likely to be drawn as any other, show that the probability that of the ten balls seen 3 will be white, 3 green, 2 red and 2 black is

$$\frac{10\,!}{3\,!\,3\,!\,2\,!\,2\,!}\left(\frac{5}{38}\right)^3\left(\frac{7}{38}\right)^3\left(\frac{12}{38}\right)^2\left(\frac{14}{38}\right)^2$$

31. A bag contains n balls, k drawings are made in succession, and the ball on each occasion is found to be white, find the chance that the next drawing will give a white ball when the balls are not replaced after each drawing.

$$\left[\textbf{Ans.}\ \text{Reqd. prob} = 1 - {}^kC_1 \frac{(n-1)\,!}{n\,!} + {}^kC_2 \frac{(n-2)\,!}{n\,1} + ... + \frac{(-1)^k (n-k)\,!}{n\,!}\right]$$

32. In a bolt factory machines *A, B, C* manufacture respectively 25, 35 and 40 per cent of the total. Of their output 5, 4 and 2 per cent are defective bolts. A bolt is drawn at random from the product and is found defective. What are the probabilities that it was manufactured by machines *A, B,* or *C ?*

(M. A., Madras, 1960; M. A., Statistics, Delhi, 66)

$$\left[\textbf{Ans.}\ \frac{125}{345}, \frac{140}{345}, \frac{180}{345}\right]$$

33. The probability of the occurrence of two independent events is known to be each greater than $\frac{1}{2}$. It is given that probability of the first event occurring simultaneously with the second not occurring is 3/25. Also the probability of the second occurring simultaneously with the first not occurring is 8/25. Find the probabilities of the respective events.

[Hint. $P(A_1 \cap \overline{A}_2) = 3/25, P(\overline{A}_1 \cap A_2) = 8/25 \quad \therefore \quad P(A_1) - P(A_1)P(A_2) = 3/25, P(A_2) - P(A_1)\,P(A_2) = 8/25$, solve for $P(A_1)$ and $P(A_2)$]

34. (*a*) Three newspapers, *A, B, C,* are published in a certain city. It is estimated from a survery that of the adult population :

20% read *A,* 16% read *B,* 14% read *C,* 8% read both *A* and *B,* 5% read both *A* and *C,* 4% read both *B* and *C,* 2% read all three. What percentage reads at least one of the papers ? Of those that read at least one, what percentage reads both *A* and *B* ? (Use Bayes theorem)

(M.A. Delhi, 1968)

[Ans. (*i*) 35% (*ii*) 28%]

(*b*) In a bolt factory machines *A, B* and *C* manufacture respectively 25%, 35% and 40%. Of the total of their output 5, 4, 2 percent are defective bolts. *A* bolt is drawn at random from the product and is found to be defective. What are the probabilities that it was manufactured by machine *A*, *B* or *C*. (Use Baye's theorem).

(***M.A. Delhi, 1969, 68; B.Sc. Hons. Cal., 1968***)

$\left[\textbf{Ans. } \frac{23}{69}, \frac{28}{69}, \frac{36}{69}\right]$

35. There are two urns. In the first urn there are 3 white and 2 black balls and in the second 1 white and 4 black, From any arbitrarily chosen urn, one ball is taken. What is the probability for its being white if probability of choosing any ball from both the urns is 0·5 each ? [**Ans.** 0·4]

36. The probability that a family chosen at random has exactly n children is αp^n $(0 < p < 1)$. Suppose also that the chance of any child having blue eyes is a $(0 < a < 1)$, independently of the others. Prove that the probability of a family chosen at random having exactly k children with blue eyes is

$$\alpha p^k a^k / [1 - p(1 - a)]^{k+1}, k \geq 1 .$$

37. Let the probability p_n that a family has exactly n children be $\alpha p^n, n \geq 1$ and $p_0 = 1 - \alpha p(1 + p + p^2 + \ldots)$. Suppose that all sex distributions of n children have the same probability. Show that probability of family containing k boys is $2\alpha p^k / (2 - p)^{k+1}, k \geq 1$. (***M.A. Statistics, Delhi, 1964***)

Given that a family includes at least one boy, show that the probability that there are two or more boys is $p/(2 - p)$.

[**Hint.** Let E_j be the event that the number of children in a family is j and let A be the event that a family contains exactly k boys. Then $p(E_j) = p_j$, $j = 0, 1, \ldots$

Again, each child can have any of the two sex-distributions, male or female, the total number of possible distributions for a family to have j' children is 2^j.

$$\therefore \; P(A \mid E_j) = \binom{j}{k} / 2^j, j \geq k$$

$$\therefore \; P(A) = \sum_{j=k}^{\infty} p(E_j)\, P(A \mid E_j) = \sum_{j=k}^{\infty} p_j \,.\, P(A \mid E_j)$$

$$= \Sigma\, \alpha\, p^j \cdot \binom{j}{k} / 2^j = \alpha \sum_{j=k}^{\infty} \left(\frac{p}{2}\right)^j \binom{j}{k}$$

$$= \alpha \sum_{r=0}^{\infty} \binom{k+r}{k} \left(\frac{p}{2}\right)^{k+r}, \; ; j - k = r$$

$$= \alpha \left(\frac{p}{2}\right)^k \sum_{r=0}^{\infty} \binom{k+r}{r} \left(\frac{p}{2}\right)^r = \alpha \left(\frac{p}{2}\right)^p \Sigma (-1)^r \binom{\overline{-k+1}}{r} \left(\frac{p}{2}\right)^r$$

$$= \alpha \left(\frac{p}{2}\right)^k \left(1 - \frac{p}{2}\right)^{-(k+1)} = \text{etc.}]$$

(*ii*) $P(C \mid B) = P(C)/P(B) = \text{etc.}$

38. A man is given n keys of which one fits the lock. He tries them successively without replacement. Find the probability that at the rth trail the lock will open. Also determine the expected number of trials and their variance (*i*) if successful keys are not eliminated from further selection and (*ii*) if they are.

(*M.A. Statistics, Delhi*)

$$\left[\textbf{Ans. } (a)\ \frac{1}{n} \text{ or } \left(1-\frac{1}{n}\right)^{r-1}\left(\frac{1}{n}\right)\right.$$

$$\left.(b)\ (i)\ \mu = n,\ \sigma^2 = n(n-1),\ (ii)\ \mu = \frac{1}{2}(n+1),\ \sigma^2 = \frac{1}{12}(n^2-1)\right]$$

39. A and B are independent events both of which are neither certain nor possible. Prove that the events $\bar{A}$ and $\bar{B}$ are independent. **(*B.Sc., Delhi*)**

40. (*a*) If A and B are independent events, prove that (*i*) A and $\bar{B}$ *are independent,* (*ii*) $\bar{A}$ and B are independent and (*iii*) $\bar{A}$ and $\bar{B}$ are independent.

[**Hint** (*iii*) $P(\bar{A} \cap \bar{B}) = P(\overline{A \cup B}) = 1 - P(A \cup B)$

(*B.Sc., Poona, 1969; B.SC., Bombay*)

(*b*) A, B and C are three events such that A and B are independent and $P(C) = 0$. Show that A, B and C are independent.

41. The chance of success in each trial is p. If p_k is the probability that there are even number of success in k trials, prove that $p_k = p + p_{k-1}(1-2p)$

Deduce that $p_k = \frac{1}{2}[1 + (1-2p)^n]$. **(*B.Sc., Bombay, 1970*)**

[**Hint.** The event can happen in the following mutually exclusive ways;

(*i*) in $(n-1)$ trials even number of successes occur and then a failure with probability $(1-p)$ at the kth trial.

(*ii*) in $(n-1)$ trials, odd number of successes and then a success at the nth trial. hence $p_k = (1-p)p_{k-1} + p\,(1-p_{k-1}) = p + (1-2p)p_{k-1}$.

Also $p_k - \frac{1}{2} = [1 - 2p\,(p_{k-1} - \frac{1}{2})]$.

42. A group of $2N$ boys and $2N$ girls is divided into two equal groups. Find the probability 'p' that each group will be equally divided into boys and girls. Estimate 'p' using Stirling's formula. **(*M.A., Stat., Delhi*)**

$\left[\text{Use } n! \simeq \sqrt{2\pi}\, e^{-n}\, n^{n+\frac{1}{2}}\right]$ $\left[\textbf{Ans. } \dfrac{(2N!)^4}{(N!)^4(4N!)} \cong \sqrt{\dfrac{2}{\pi N}}\right]$

43. (*a*) What is the probability that at least two out of n people have same birth date ? **(*B.Sc., Delhi, 1971*)**

(*b*) What is the probability that

(*i*) the birthdays of twelve people will fall in twelve different calendar months (assume equal probabilities for the 12 months) ?

(*iii*) the birthdays of six people will fall in exactly two calendar months ?

(*B.Sc., Poona, 1965*)

$$\left[\textbf{Ans. } (a)\ 1-\left(1-\frac{1}{365}\right)\left(1-\frac{2}{365}\right)\cdots\left(1-\frac{n-1}{365}\right)\right.$$

$$\left.(b)\ (i)\ \frac{12}{12^{12}}\quad (ii)\ \frac{\binom{12}{2}(2^6-2)}{12^6}\right]$$

44. A and B have equal chances of winning a single game. A wants n games and B $n+1$ games to win a match. Show that the odds in favour of A are $1+P$ to $1-P$ where

$$P=\frac{2n\,!}{(n\,!)^2\,2^{2n}}$$ **(*M.Sc., Kerala, 1969*)**

[**Hint.** Probability of A winning at least n games out of $2n$ games

$$={}^{2n}c_n q^n p^n + {}^{2n}c_{n+1}\, q^{n-1} p^{n+1} + \ldots + {}^{2n}c_{2n} p^{2n} \text{ where } p=q=\frac{1}{2}.$$

$$=\frac{1}{2^{2n}}\left[\frac{1}{2}\{(1+1)^{2n}-{}^{2n}c_n\}\right]=\text{etc.}]$$

45. (*a*) Each coefficient of the equation $Ax^2+Bx+C=0$ is determined by the score thrown up in a throw of an unbiased six-faced die having number from 1 to 6. Obtain the probability that the equation has (*i*) equal and (*ii*) real roots. **(*M.A., Poona, 1965*)**

$$\left[\textbf{Ans. } (i)\ \frac{5}{216}\ (ii)\ \frac{43}{216}\right]$$

(*b*) The variate a and b are uniformly distributed in the intervals $0\le a\le 6$ and $0\le b\le 9$. Find the probability that the equation $x^2-ax+b=0$ has two real roots.

[**Hint.** $P(b\le a^2/4)=\int_{a=0}^{1}\int_{b=0}^{a^2/4}\frac{1}{6\times 9}\,da\,db$ $\left[\textbf{Ans.}\ \frac{1}{3}\right]$

(*c*) The variates a and b are uniformly distributed over the interval $(-1,1)$. Find the probability that the equation $x^2+2x+b=0$ has real roots.

$\left[\textbf{Hint. } P(a^2>b)=1-P(a^2<b)=\left[1-\int_{-1}^{+1}\left\{\int_{-\sqrt{b}}^{+\sqrt{b}}\frac{1}{4}\,da\right\}db=\text{etc.}\right]\right]$ $\left[\textbf{Ans.}\ \frac{1}{3}\right]$

(*d*) If the random variable Y is uniformly distributed over $(0,5)$, what is the probability that the roots of the equation $4x^2+4xY+Y+2=0$ are real ?

[**Hint.** $(4Y)^2-4\times 4\,(Y+2)\ge 0 \Rightarrow y^2-y-2\ge 0$ *i.e.* $y\ge 2$,

Hence $P(y\ge 2)=\int_2^5\frac{dy}{5}=\frac{3}{5}\Big]$. $\left[\textbf{Ans.}\ \frac{3}{5}\right]$

47. In a bridge game North and South have ten spades between them. Find the probability that either East and West has no spade (there are 13 spades in a pack of 52 cards and a player has 13 cards. The four players are designated by the position they occupy, *viz.* North, South, East and West).

(*B.Sc., Hons., Calcutta, 1968*)

$$\left[\textbf{Ans. } \frac{11}{50}, \frac{13}{50}\right]$$

48. Find the probability for a poker hand *i.e.,* a hand of five cards taken at random from a full pack of 52 cards to be a (*i*) royal flush, *i.e.* jack, queen, king and ace in a single suit, (*ii*) full house *i.e.* one pair and one triple of cards with equal face values, (*iii*) straight, *i.e.,* five cards in a sequence regardless of suit and (*iv*) four of a kind. **(*M. Sc. Gauhati, 1963*)**

$$\left[\textbf{Ans. } \text{Let } {}^{52}C_5 = q, \text{ then } (i) \frac{4}{q} \;\; (ii)\; 13 \cdot 12 \frac{4}{q} \;\; (iii) \frac{9 \cdot 4^5}{q} \;\; (iv)\; 13\; 12 \cdot \frac{4}{q}\right]$$

49. A and B engage themselves in a game each step of which consists in one of them winning a counter from the other. At beginning, A has a counters and B has b counters and in each successive step the probability of A winning a counter from B is p. The game is terminated when either of the two has all $a + b$ counters; what is the probability of A's winning when he has n counters ?

(*a*) Consider the two cases (*i*) $p \neq \frac{1}{2}$ (*ii*) $p = \frac{1}{2}$

(*b*) Consider the two cases (*i*) $p \neq q$ (*ii*) $p = q$ **(*Mysore, M. Sc.*)**

50. Let A B and C be any three events. If $P(A \cap \overline{B} \cap \overline{C}) = \frac{1}{4}$,

$P(\overline{A} \cap \overline{B} \cap C) = \frac{1}{4}$, $P(\overline{A} \cap B \cap \overline{C}) = \frac{1}{4}$ and $P(A \cup B \cup C) = \frac{1}{4}$. Verify whether A, B and C are (*i*) pairwise independent (*ii*) completely independent.

[**Hint.** Use Venn diagram.] **(*B. Sc., Hons., Calcutta, 1967*)**

51. If r *indistinguishable* balls are placed at random into n different cells, find the probability that the 1st cell contains r_1 balls, the 2nd cell r_2 balls,.... and the nth cell contains r_n balls, where $r_1 + r_2 + \ldots + r_n = r$) where each r_i can take values from 0 to r.

[**Hint.** We have $(n - 1)$ internal partitions and number of balls is $n + r - 1$. Mark the points $1, 2, \ldots, n + r - 1$ on the number axis. Any event may be obtained by choosing $n - 1$ out of $n + r - 1$ marked points, in which the partitions are inserted, the remaining r points being occupied by the r balls.

Total no of event-points $= \binom{n+r-1}{n-1} = \binom{n+r-1}{r}$

Thus the probability of the required event is $\begin{bmatrix} n+r-1 \\ r \end{bmatrix}^{-1}$

Note: In Physics, this result goes by the name of *Bose-Einstein Statistics.*

52. If r indistingiushable balls be distributed at random to n ($\geq r$) cells, such that a cell is either empty or occupied by a single ball, Find the probability that the 1st cell contains r_1 balls, . . . the nth cell contains r_n balls where

$r_1 + r_2 + \ldots + r_n = r$ and each $r_j = 0$ or 1.

Ans. Reqd. probability is $\begin{bmatrix} n \\ r \end{bmatrix}^{-1}$

In Physics, this assignment of probabilities is known as *'Fermi-Direc' Statistics.*

53. *De Mere'* Paradox. Show that the probability of at least one six in four throws of a die is $1-(5/6)^4 < 3{\cdot}5177$ and the probability of at least 'one pair of sixes' in 24 throws of two dice is $1-(35/36)^{24} < 0{\cdot}4914$. [A French gambler H. de Mere found by his long experience at gambling that the former event had greater probability than the latter. He referred to the French mathematician Blaise Pascal (1623–1662) who, in a letter dated 29th July, 1654, wrote to another great mathematician P. Fermat (1601–1665) about it. This led to the foundations of the theory of probability.

54. A hereditary characteristic (colour of hair, colour of eyes etc.,) is passed on through a pair of genes say *A, a.* A person is called a domainant, hybrid or recessive for this characteristic according as he has the pair, *AA, Aa, aa.* Let the proportion of dominants, hybrids and recessives in a population be u, $2v$, w where $u \geq 0, v \geq 0, w \geq 0, u + 2v + w = 1$

In random mating, a child gets one gene from the father and one from the mother and each of the pair of genes is passed on with probability $\frac{1}{2}$. Show that in the next generation the two proportions of dominants, hybrids and recessives are : $(u+v)^2 = p^2, 2(u+v)(v+w) = 2pq\,; (v+w)^2 = q^2$

Show that in the next and succeeding generations under random mating the proportions would always be $p^2, 2pq, q^2$

[This law is known as Hardy-Weinberg law (Weinberg was a genetist, but G.H. Hardy (1877-1947) was a leading English mathematician, well-known for his work on number theory and divergent series. In India, he is gratefully remembered for discovering the genius of Srinivasa Ramanujan (1887-1920). The greatest Indian mathematician of modern times.]

55. The sample space consists of the integers 1 to $2n$ which are assigned probabilities proportional to their logarithms. Find the probabilities and show that the conditional probability as getting integer 2, given that one even integer has been obtained is $\log 2/[\log 2 + \log n\,!]$

56. Suppose an isosceles triangle with two sides of length a if formed by randomly selecting the length of the third side from the set of all possible lengths. Find the probability that the triangle formed is obtuse.

[**Hint.** Length of third side must lie in the interval $(0, 2a)$. A right triangle will be formed if this side has length a $\sqrt{2}$ and an obtuse triangle will be formed if the length of this side lies in $[\sqrt{2}\, a, 2a]$

So the required probability $= (2a - \sqrt{2}a)/2a = \dfrac{2-\sqrt{2}}{\sqrt{2}} = 0{\cdot}293]$

57. In Bernoullian trials with $p = q = 1/2$, show that the probability p_n that two consecutive failures occur for the first time at $(n-1)$th and nth trial satisfies the difference equation $p_n = \frac{1}{2} \cdot p_{n-1} + \frac{1}{4} p_{n-2}, n > 2.$

Hence show that $$p_n = \frac{5-\sqrt{5}}{10}\left(1+\frac{\sqrt{5}}{4}\right)^n + \frac{5-\sqrt{5}}{10}\left(1-\frac{\sqrt{5}}{4}\right)^n.$$

58. Suppose that in answering a question in a multiple-choice test, an examinee knows the answer with probability p and he guesses with probability $1-p$. Assume that the probability of answering a question correctly is unity for an examinee who knows the answer and is $1/m$ for the examinee who guesses, where m is the number or multiple choice alternatives. Use Baye's theorem to show that the probability that an examinee knows and answers to a problem, given that it has been correctly answered, is $mp/[(1+(m-1)p.]$

59. Two players A and B start playing a series of games with Rs. a and b respectively. The stake is Re. 1 in a game and no game can be drawn. If the probability of A winning any game is a constant p, find the probability of A being ultimately ruined. (see section 19·2).

60. In answering a question of a multiple choice test, a student either knows the answer or he guesses. Let p be the probability that he knows the answer and $1-p$ the probability that he guesses. Assume that a student who guesses at the answer will be correct with probability $1/5$ where 5 is the number of multiple choice alternatives. What is the conditional probability that the student knew the answer to a question, given that he answered correctly.

(***B. Sc. Math Hons., Delhi 1985***)

[**Hint.** Let E be the event that the student gets the right answer. Let E_1 be the event that he knew the right answer, and E_2 be the event that he guesses the right answer

Then, $P(E_1) = p, P(E_2) = 1-p, P(E \mid E_2) = 1/5,$

$P(E \mid E_1) = p. \quad P(E_1 \mid E) = ?$] **Ans.** $\dfrac{p \times 1}{p \times 1 + (1-p) \times \frac{1}{5}} = \dfrac{5p}{4p+1}.$

61. In 1989, there were three candidates for the position of principal. Mr. Chatterji, Mr Ayangar and Dr. Singh— whose chances of getting the appointment are in the proportion 4 : 2 : 3 respectively. The probability that Mr. Chatterji, if selected, would introduce co-education in the college is 0·3. The probabilities of Mr. Ayangar and Dr. Singh doing the same, are respectively 0·5 and 0·8. What is the probability that there was co-education in the college in 1990 ? (***B. Sc. Statistics Hon. Delhi 1992***)

[**Hint.** Let A be the event of introducing the co-education , E_1, E_2, E_3 be the events that Mr. Chatterji, Ayangar and Dr. Singh are selected as principal.

Then $P(E_1) = \frac{4}{9}, P(E_2) = \frac{2}{9}, P(E_3) = \frac{3}{9}; P(E \mid E_1) = \frac{3}{10},$

$P(E \mid E_2) = \frac{5}{10}, PE \mid E_3) = \frac{8}{10}$

Now $P(E) = P[(E \cap E_1) \cup (E \cap E_2) \cup (E \cap E_3)]$

$= P(E \cap E_1) + P(E \cap E_2) + P(E \cap E_3)$

$$= \sum_{i=1}^{3} P(E_i)\, P(E/E_i) = \frac{4}{9}\cdot\frac{3}{10} + \frac{2}{9}\cdot\frac{5}{10} + \frac{3}{9}\cdot\frac{8}{10} = \frac{23}{45}. \Big]$$

62. A manufacturer firm produces steel pipes in three plants, with daily production volume of 500 ; 1,000 and 2,000 units respectively. According to past experience, it is known that the fraction of defective outputs produced by the these plants are respectively ·005; ·008 and ·010. If a pipe is selected from day's total production and found to be defective, find out (*i*) which plant the pipe cames (*ii*) what is the probability that it came from the first plant ?

[Hint. Let A_1, A_2 and A_3 be the events that a pipe is manufactured in plant 1, 2 or 3 respectively.]

$$\therefore \quad P(A_1) = \frac{500}{3500} = \frac{1}{7},\ p(A_2) = \frac{1000}{3500} = \frac{2}{7},\ P(A_3) = \frac{2000}{3500} = \frac{4}{7}$$

Let B be the event that a defective pipe is drawn. Then

$P(B \mid A_1) = \cdot 005,\ P(B \mid A_2) = \cdot 008,\ P(B \mid A_3) = \cdot 010$

Also $\quad P(A_1 \cap B) = P(A_1)\, P(B \mid A_1) = \cdot 005 \times \frac{1}{7},$

Similarly $\quad P(A_2 \cap B) = \cdot 008 \times \frac{2}{7},\ P(A_3 \cap B) = \cdot 010 \times \frac{4}{7}$

Baye's theorem gives

$$P(A_1 \mid B) = \frac{P(A_1 \cap B)}{P(A_1 \cap B) + P(A_2 \cap B) + P(A_3 \cap B)}$$

$$= \frac{0{\cdot}005/7}{(0{\cdot}005 \times \frac{1}{7}) + (0{\cdot}008 \times \frac{2}{7}) + (0{\cdot}010 \times 4/7)} = \frac{5}{61}$$

Similarly, $\quad P(A_2/B) = \frac{\cdot 016/7}{\cdot 061/7} = \frac{16}{61},\ P(A_3/B) = \frac{\cdot 040/7}{\cdot 061/7} = \frac{40}{41}$

Thus it is most probable that the defective pipe has been drawn from the output of the third plant.]

63. A tells the truth with probability p_1 and B tells the truth with probability p_2 independently. *A makes a statement, that 'B* says that *A* had lied'. Use Baye's theorem to show the probability that *A* told the truth is

$$\frac{p_1(1-p_1)}{p_1(1-p_2) + p_2(1-p_1)}.$$

Miscellaneous Exercises on Probability

1. An urn contains four tickets marked with numbers 111, 121, 211, 222 and one ticket in drawn at random. Let $A_i (i = 1, 2, 3)$ be the event that *i*th digit of the number of the ticket drawn is 1. Discuss the independence of the events A_1, A_2 and A_3. **[*B. Sc. Lucknow, 1971, B. Sc. Calicut, 1973*]**

[Hint. Find $P(A_1 \cap A_2), P(A_2)$, etc.

Also that $P(A_1 \cap A_2 \cap A_3) \neq P(A_1)\, P(A_2)\, (P(A_3))$]

2. Eight tickets numbered 111, 121, 122, 122, 211, 212, 221 are placed in a hat and stirred. One of them is then drawn at random. Show that the event *A*. "the first digit on the ticket drawn will be 1", *B* : "the second digit on the ticket drawn will be 1" are not pairwise independent although

$$P(A \cap B \cap C) = P(A)P(B)P(C)$$

(*B. Sc. Hons. Dacca, 1969*)

3. (*a*) If $P(A) = a$, $P(B) = b$, show that $P(A \mid B) > (a + b - 1)/b$

[*B. Sc., (S), Delhi, 1971*]

(*b*) If *A* and *B* are two mutually exclusive events, show that

$P(A \mid \overline{B}) = P(A)/[1 - P(B)]$ **(*B. Sc. Shivaji, 1971*)**

4. If *A* and *B* are two events and the probability $P(B) \neq 1$, prove that

$$P(A \mid \overline{B}) = [P(A) - P(A \cap B)]/[1 - P(B)]$$

Hence deduce that $P(A \cap B) > P(A) + P(B) - 1$

Also, show that $P(A) >$ or $< P(A \mid B)$ according as $P(A \mid \overline{B}) >$ or $< P(A)$

(*B. Sc. Hons. Delhi, 1972*)

5. A certain item is manufactured by three factories 1, 2 and 3. It is known that 1 turns out twice as many items as 2, and that 2 and 3 turn out the same number of items (during a specified production period). It is also known that 2 per cent of the items produced by 1 and 2 are defective, while 4 per cent of those manufactured by 3 are defective. All the items produced are put into one stockpile, and then one item is chosen at random. What is the probability that this item is defective ? **[Ans. ·025]**

6. For three independent events *A, B,* and *C*, the probability for *A* to occur is *a,* the probability that *A, B* and *C* will not occur is *b,* and the probability that at leas one of the three events will not occur is *C*. If *p* denotes the probability that *C* occurs but neither *A* nor *B* occurs, prove that *p* satisfies the quadratic equation $ap^2 + [(ab - (1 - a)(a + c - 1)]\, p + b(1 - a)\, 1 - c) = 0$

7. The sum of two positive quantities is constant and equal to 2*n*. Find the chance that the product of the two quantities is not less than 3/4 of their greater product. **[Ans. $\frac{1}{2}$]**

8. *A* and *B* are to play a series of games, the probability that any given game is won by *A*, won By *B* or drawn being respectively *p, q* and $1 - p{\cdot}q$ The series is won by the player who fist wins two consecutive games (even without the interruption of a drawn game) prove that $A^{'s}$ or $B^{'s}$ probabilities of winning the series are in the ratio $p^2(1 + q) : q^2(1 + p)$.

9. Prove that *A, B,* and *C* are random events in a sample space and if *A, B, C,* are pairwise independent and *A* is independent of $(B \cup C)$, than *A, B* and *C* are mutually independent. **[*M. Sc. Stat., Delhi, 1977*)**

10. The contents of urns I, II and III are as follows :

1 white, 2 black and 3 red ball,

2 white, 1 black and 1 red ball, and

4 white, 5 black and 3 red balls.

One urn is chosen at random and two balls drawn. They happen to be white and red. What is the probability that they come from urns, I, II or III ?

(B. Sc. Hons, Calcutta)

11. Those are $(N+1)$ identical urns marked 0, 1, ... , N each of which contains N, white and red balls. The kth urn contains k red and $N-k$ white balls, $k = 0, 1, \ldots\ldots, N$. An urn is chosen at random and n random drawings of a ball are made from it, the ball drawn being replaced after each draw. If the balls drawn are all red, show that the probability that the next drawing will also yield a red ball is approximately $(n+1)(n+2)$ when N is large.

(M. Sc. Stat., Delhi, 1973)

[**Hint.** Show that the required probability

$$= \frac{1}{N+1} \sum_{k=0}^{N} \left(\frac{k}{N}\right)^{n+1} \Big/ \frac{1}{N+1} \sum_{k=o}^{N} \left(\frac{k}{n}\right)^{n}$$

If N is large, then $\frac{1}{N} \sum_{n=0}^{N} \left(\frac{k}{N}\right)^{n} \approx \int_0^1 x^n dx = \frac{1}{n+1}$

and $\frac{1}{N} \sum_{k=0}^{N} \left(\frac{k}{n}\right)^{n+1} \approx \int_0^1 x^{n+1} dx = \frac{1}{n+2}$.

12. The outcome of an experiment are represented by points in the square board by $x = 0, y = 0, x = 2$ and $y = 2$ in the $x - y$ plane. If the probability is distributed uniformly, determine the probability that $x^2 + y^2 > 1$.

(B. Sc. Bombay, 1969)

$\Bigg[$ **Hint.** Reqd. probability $= \iint_A \frac{1}{4}\, dxdy - \iint_{A^c} \frac{1}{4}\, dxdy$ where A is the region for which $x^2 + y^2 > 1$ and $A^c =$ complement of A $\Bigg]$

13. A special die is prepared such that the probability of throwing 1, 2, 3, 4, 5 and 6 points are respectively

$$\frac{1-k}{6}, \frac{1+2k}{6}, \frac{1-k}{6}, \frac{1+k}{6}, \frac{1-2k}{6} \text{ and } \frac{1+k}{6}.$$

If two such dice are thrown, show that the probability of getting a sum equal to 9 is $\frac{1}{18}(1+k)(2-3k)$.

(B. Sc. Bombay, 1969)

14. An urn contains a white and b black balls and a series of drawings of one ball at a time is made the ball removed being returned to the urn immediately after the next drawing is made. If p_n denotes the probability that the nth ball drawn is black, show that $p_n = (b - p_{n-1})/(a+b-1)$. Hence find p_n.

(B. Sc. Hons. Calcutta, 1970)

15. The chance of success in each trial is p. If p_k is the probability that there are even number of success in k trials, prove that

$p_k = p + p_{k-1}(1-2p)$

Deduce that $p_k = \frac{1}{2}[1 + (1 - 2p)^k]$ *[B. Sc. Bombay, 1972)*

16. Suppose that each day the weather can be uniquely classified as 'fine' or 'bad'. Suppose further that the probability of having fine weather on the last day of a certain year is p_0 and we have the probability p that the weather on an arbitrary day will be of the same kind as on the preceding day. Let the probability of having fine weather on the nth day of the following year be p_n Show that

$$p_n = (2p - 1)p + (1 - p)$$

Deduce that

$$p_3 = (2p - 1)^3 (p_0 - \tfrac{1}{2}) + \tfrac{1}{2} \cdot$$

(I.S.I., 1970)

17. A certain mathematician always carries two match boxes, each time he wants a match stick, he selects a box at random. Inevitably a moment comes when he finds a box empty. Find the probability that at the moment the first box is empty, the second contains exactly r match sticks (assuming that each box contained N match sticks initially).

[**Ans.** ${}^{2n-r}c_n/2^{2n-r}$] *(B. Sc. Hons. Calcutta, 1968)*

18. Three players A, B and C agree to play a series of games, observing the following rules; two players participate in each game while third is idle, and the game is to be won by one of them. The loser in each game quits and his place in the next game is taken by the player who was idle. The player who succeeds in wining over both of his opponents without interruption wins the whole series of games.

Supposing the probability for each player to win a single game is $\frac{1}{3}$ and that the first game is played by A and B, find the probability for A, B and C respectively to win the whole series if the number of games is unlimited.

[**Ans.** 5/14, 5/14, 2/7] *(B. Sc. Shivaji, 1970)*

19. A drug manufactured by a company is tested for its toxic effect, Let B be the event that the drug is toxic and let A be the event that the test reveals that the drug is toxic. Let $P(A \mid B) = \theta = P(\bar{A} \mid \bar{B})$

and $P(B) = 1 - \theta, o < \theta, < 1$

Show that the probability that the drug is not toxic given that the test reveals that it is 'toxic' is free from θ. *(B. Sc. Bombay, 1971)*

20. Two-thirds of the students in a class are boys and the rest girls. It is known that the probability of a girl getting a first class is 0·25 and that of a boy is 0·28. Find the probability that a student chosen at random will get first class marks.

[**Ans.** 0·27]

(B. Sc., Bombay, 1976)

21. Two dice are rolled. Let A be the event that the sum of the points on the faces shown is odd, and B the event that there is at least one 3 shown. Find the probabilities of the events (*i*) $A \cup B$ (*ii*) $A \cap B$ (*iii*) $A - B$ and (*iv*) $(A \cap \bar{B}) \cup \bar{A}$.

(B.Sc. Gauhati, 1973) [**Ans.** (*i*) 23/36 (*ii*) 6/36 (*iii*) 11/36 (*iv*) 23/36]

22. If A,B and C are mutually independents, then show that the following events are also independent :

(*i*) A and $(B \cap C)$, (*ii*) B and $(A \cup C)$, (*iii*) $(A \cup B)$ and C,

(*iv*) $\bar{A}$ and $(B \cap \bar{C})$, and (*v*) A, B, and $\bar{C}$. **(*M.Sc. Vikram, 1972*)**

23. (*a*) An event A is known to be independent of the events B, $B \cup C$ and $B \cap C$. Show that it is also independent of C.

(*b*) If A, B, C are mutually independent events, then (*i*) $A \cup B$ and C are also independent (*ii*) A and $B + C$ are independent. (*iii*) $\bar{A}, \bar{B}, \bar{C}$ are mutually independent. **(*I.A.S. 1992*)**

(*c*) If C is independent of two mutually exclusive events A, B, then C is independent of $A \cup B$.

Hint. (*a*) $P(A \cap C) = P[A \cap \{(B \cup C) - (A \cap B) + A \cap (B \cap C)\}]$

$$= P[(A) \cap (B \cup C)] - P(A \cap B) + P[A \cap (B \cap C)]$$
$$= P(A)\,P(B \cup C) - P(A)\,P(B) + P(A)\,P(B \cap C)$$
$$= P(A)\,[P(B \cup C) - P(B) + P(B \cap C)]$$
$$= P(A)P(C). \text{ Hence etc.}$$

(*b*) (*ii*) $P\,[A(B + C)] = P(AB + AC) = P(AB) + P(AC) - P(ABC)$

$$= P(A)\,P(B) + P(A)\,P(C) - P(A)\,P(BC)$$
$$= P(A)[P(B) + P(C) - P(BC)] = P(A)\,P(B + C)$$

Thus A and $B + C$ are independent.

Also by theorem page's are independent.

Also $\bar{A}$ and $\overline{B + C}$ are also independent, so $\bar{A}$ and $\bar{B}\,\bar{C}$ are independent.

(*c*) We are given,

$$P(A \cap B) = o,\ P(C \cap A) = P(C)P(A),\ P[(AB) = P(A)P(B)$$
$$P[C \cap (A \cup B)] = P[C \cap (A + B - AB)] = P[C \cap (A + B)]$$
$$= P(C \cap A) + P(C \cap B) = P(C)P(A) + P(C)P(B)$$
$$= P(C)[P(A) + P(B)] = P(C)\,P(A \cup B)$$

24. One bag contains three identical cards, marked 1, 2, 3, and another contains two cards, marked 1, 2 Two cards are randomly chosen, one from each bag and the number observed. Denote the minimum of two by X and sum of the two by Y, Find the joint probability distribution of the random variable pair (X, Y). Find also the two marginal distributions and the conditional distribution of X when $Y \le 4$. **(*I.A.S. 1995*)**

25. What is the probability that in a company of 500 people, only one person will have birthday on New Year's 1st day. Assume that a year has 365 days.

[**Hint.** Here $n = 500, p = 1/365$ $\therefore$ $np = 500/365 = \mu$]

(**Ans.** $10 \simeq e^{-\mu}\mu = \cdot 348$)

26. A and B play a game of dice. A wins if he throws 11 with 3 dice before B throws 7 with 2 dice. B wins if he throws 7 before A throws 11. A starts the game and they throw alternately. What are the odds against A winning the game ultimately ? **(*I.A.S. 1995*)**

5

Discrete Probability Distributions

5·1·1. Definitions. Discrete Random Variables. *A variable, which can assume only a discrete set of real values and for which the value which the variable takes depends on chance, is called a discrete random variable or discrete stochastic variable or a discrete chance variable (d, r, v.).* Thus if we throw a perfect die, then X, the number of points on the die is a d.r.v. since (*i*) X takes only a set of discrete values viz., 1, 2, 3, 4, 5, 6 (*ii*) the value which it actually takes in an experiment depends on chance. In fact X can take values 1, 2, 3, 4, 5, 6, each with probability 1/6.

We may note that 'twice the number of points on a die' is also a random variable, which takes the values 2, 4, 6, 8, 10, 12 with probability 1/6 in each case. The 'square of number of points on a die' is also a random variable, which takes value 1, 4, 9, 16, 25, 36 with probability 1/6 in each case.

In fact if X is a d.r.v., which takes values $x_1, x_2, \ldots\ldots, x_n$ with probabilities $p_1, p_2, \ldots\ldots, p_n$ and if $f(X)$ is a single-valued function of X, then $f(X)$ takes values $f(x_1), f(x_2), \ldots\ldots, f(x_n)$ with probabilities $p_1, p_2 \ldots\ldots, p_n$ respectively.

We give below a few examples of a discrete r.v.:

(*i*) In throws of two perfect dice, the sum of points is a d.r.v. which takes the values 2, 3, 4, 5, 6, 7, 8, 9, 10, 11, 12 with respective probabilities 1/36, 2/36, 3/36, 4/36, 5/36, 6/36, 5/36, 4/36, 3/36, 2/36, 1/36 and the product of points on the two dice is a d.r.v. which takes the values 1, 2, 3, 4, 5, 6, 8, 9, 10, 12, 15, 16, 18, 20, 24, 25, 30, 36 with probabilities 1/36, 2/36, 2/36, 3/36, 2/36, 4/36, 2/36, 1/36, 2/36, 4/36, 2/36, 1/36, 2/36, 2/36, 2/36, 1/36, 2/36, 1/36.

(*ii*) In successive throws of a perfect coin, the number of throws before the first head appears takes the values 0, 1, 2, 3,......, n,...... with respective probabilities $\frac{1}{2}, \frac{1}{2^2}, \frac{1}{2^3}, \frac{1}{2^4}, \ldots\ldots, \frac{1}{2^{n+1}} \ldots.$

In the second example, we have a d.r.v. which takes an infinite set of values. The sum of the probabilities is to be unity.

It can be shown that if X and Y are discrete random variables, then $Y, CX, X \pm Y . f(X)+f(Y), \varphi(X, Y), |X|, \max[X, Y]$ and $\min [X, Y]$, $|X+Y|, |X-Y|$ are also discrete random variables where C is a constant.

In chapter 7, we shall consider continuous random variables where a continuous random variable can take all real values in certain interval (a, b) or even over the whole real line. In this chapter, we shall however consider discrete random variables only.

Thus to each outcome ω there corresponds a real number $X(\omega)$ which is defined for each point of the sample S *i.e.* for each $\omega \in S$. there corresponds an-dimensional random or *stochastic variable*.

Let S be the sample space associated with a given random experiment. A real-valued function defined on S and taking values in $R(-\infty, \infty)$ is called a one dimensional random variable. In general, a real -valued function, whose domain is S and whose range is a set *collection* of n -tuples of real numbers, is called an n-dimensional random variable.

Thus a random variable is a function $X(\omega)$ with domain S and range $R(-\infty, \infty)$ for which corresponding to every real number a, the event $\{\omega : x(\omega) \le a\} \varepsilon B$, where B is the σ – field of subsets in S ; Sometimes, we write $P(X \le a) = P\{\omega : x(\omega) \le a\}$

One dimensional *r.v.* is denoted by capitals $X, Y, Z, \ldots$.. and the values which the *r.v.* can assume are denoted by lower case letters $x, y, z, \ldots$..

Illustrations 1. Consider the random experiment of tossing of a coin with sides denoted by head 'H' and tail 'T' respectively.

Then $S = \{\omega_1, \omega_2\}$ where $\omega_1 = H, \omega_2 = T$

and $x(\omega) = \begin{cases} 1, & \text{if } \omega = H \\ 0, & \text{if } \omega = T \end{cases}$

outcome :	H	T
value of x :	1	0

2. Let a random experiment E be performed by throwing a coin and x be the number of heads. Then

Outcome :	HHH	HHT	HTH	THH	HTT	THT	TTH	TTT
Value of X :	3	2	2	2	1	1	1	0

or

outcome :	HHH	HHT or HTH or THH	HTT or THT or TTH	TTT
value of X :	3	2	1	0

Suppose X is a one -dimensional random variable which takes at most a countably infinite number of values $x_1, x_2, \ldots$.

Let each possible outcome x_i be associated with a number $p_i = P(x = x_i)$ $= p(x_i) =$ probability of the value x_i of the $r.v.X$.

The number $p(x_i)$ satisfies the following conditions :

(i) $p(x_i) \geq 0 \;\forall\; i$ (ii) $\sum_{i=1}^{\infty} p(x_i) = 1$

The function p satisfying the above two conditions is called the *probability mass function* of the *r.v.X*. The set $\{x_i, p(x_i)\}$ is known as the *probability distribution*, briefly denoted by p.d. of X.

The set of values which x takes is called the *spectrum* of the random variable.

Illustrations 1. Let the spectrum consist of a single point a, and $P(X = a) = 1$.

Then a is called the parameter of the *causal distribution* and X is causally distributed with parameter a.

2. In the Binomial distribution, the spectrum consists of the $(n + 1)$ points, 0, 1, 2, ..., n *i.e.* $x_i = i$ $(i = 0, 1, \ldots, n)$

and $p_i = \binom{n}{i} p^i (1-p)^{n-i}, \quad 0 < p < 1$

Also p_i's satisfy the condition

$$\sum_{i=0}^{n} p_i = \sum_{i=0}^{n} \binom{n}{i} p^i (1-p)^{n-i} = [p + (1-p)\}^n = 1.$$

5·1·2. Probability Mass Function, Probability Distribution.

The function $f(x)$ defined as follows :

$$f(x) = p_i \text{ when } x = x_i,\; i = 1, 2, \ldots, n; \; = 0 \text{ otherwise}$$

is called the *probability mass function* of the discrete r.v. X.

The probability distribution of X is specified when

(i) the values which X can take are given,

(ii) the probabilities of these variate values are given.

Thus the set of ordered pairs

$$\{(x_1, p_1), (x_2, p_2), \ldots, (x_n, p_n)\}$$

specifies the probability distribution (p.d.).

5·1·3. Distribution Function. Let the r.v. X take values $x_1, x_2, \ldots, x_n$ with probabilities $p_1, p_2, \ldots, p_n$ and let $x_1 < x_2 < x_3 < \ldots, < x_n$, then it is easily seen that

$P[X < x_1] = 0$, $P[X \leq x_1] = P[X < x_1] + P[X = x_1] = p_1$, $P[X < x_2] = p_1$,

$P(X \leq x_2) = [P(X < x_2) + P(X = x_2)] = p_1 + p_2$

$P[X < x_i] = p_1 + p_2 + \ldots + p_{i-1}$, $P(X) \leq x_i] = p_1 + p_2 + \ldots + p_i$

$P[X < x_n] = p_1 + p_2 + \ldots + p_{n-1}$, $P[X \leq x_n] = p_1 + p_2 + \ldots + p_n = 1.$

In fact for any real number x corresponding to a random variable X we can find $P[X < x]$ and $P[X \le x]$.

We now define the distribution function $F(x)$ of a random variable X defined in $(-\infty, \infty)$ as $F(x) = P[X \le x]$

We may recall that $F(a-o)$ denotes the limit of $F(x)$ as x approaches 'a' through values less that 'a' and $F(a+o)$ denotes the limit when x approaches 'a' through values greater than 'a'. It is easily seen that

$$\begin{aligned}
F(x) &= 0, && -a < x < x_1 \\
F(x) &= p_1, && x_1 \le x < x_2 \\
F(x) &= p_1 + p_2 && x_2 \le x < x_3 \\
&\dots\dots\dots\dots\dots\dots\dots\dots\dots\dots \\
F(x) &= p_1 + p_2 + \dots\dots + p_i && x_i \le x < x_{i+1} \\
&\dots\dots\dots\dots\dots\dots\dots\dots\dots\dots \\
F(x) &= 1 && x_n \le x < \infty
\end{aligned}$$

The graph of this for $x = 6$ is shown below :

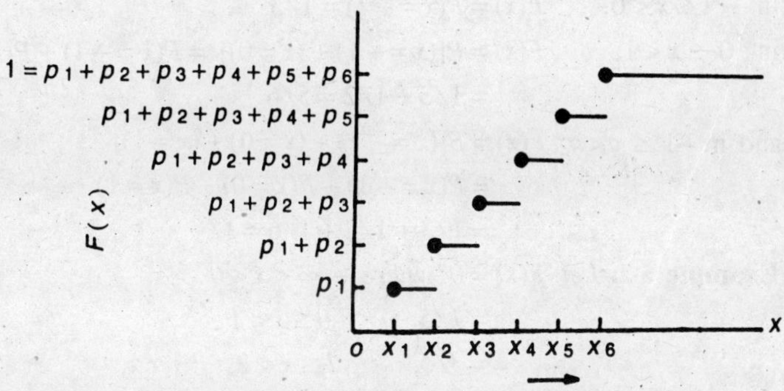

Fig. 5·1. Distribution function of a discrete random variate.

We may note the following *properties* of $F(x)$ defined in $(-\infty, \infty)$:

(*i*) $0 \le F(x) \le 1$

(*ii*) $F(x)$ is a monotonic increasing function of X.

i.e. $b \ge a \Rightarrow F(b) \ge F(a)$

(*iii*) $F(x)$ has n points of discontinuity at $x_1, x_2, \dots\dots, x_n$. At each of these points the function is continuous on the right, but is discontinuous on the left.

(*iv*) At the point x_i, $F(x_i + o) = F(x_i - o) + p_i$

where $F(x)$ has a jump or saltus of p_i at x_i.

(*v*) $F(-\infty) = \lim_{x \to -\infty} F(x) = 0, \; F(\infty) = \lim_{x \to \infty} F(x) = 1,$

(*vi*) $F(b) - F(a) = P[X \le b] - P[X \le a] = P[a < X \le b]$

(*vii*) $F(a < X < b) = P(a < X \le b) - P(X = b) = F(b) - F(a) - P(X = b)$

$$P[a \le X \le b] = P(a < X \le b) + P(X = a) = F(b) - F(a) + P(X = a)$$

$$P[a \le X < b] = P(a < X < b) + P(X = a)$$

$$= F(b) - F(a) - P(x = b) + P(X = a)$$

(*viii*) $F(x) \le F(y)$ if $x < y$

Remark. If $P(x = a) = 0 = P(x = b)$, Then all the events $a < X < b$, $a \le X < b$, $a < X \le b$, $a \le X \le b$ have the same probability $F(b) - F(a)$.

Ex.1. If *X, Y* are points in two perfect dice, find the probability function for max (*X, Y*), min (*X, Y*) and *X/Y*

Ex. 2. Plot $F(x)$ for each of the probability distributions of Ex. 1.

Example 5·1. *A random variable X assumes the values – 1, 0, 1 with probability 1/3, 1/2, 1/6. respectively. Determine the distribution of X.*

In $-\infty < x < -1$, $F(x) = P(x < -1) = 0$

in $-1 \le x < 0$, $F(x) = P(x = -1) = 1/3$

in $0 \le x < 1$, $F(x) = P[(x = -1) + (x = 0)] = P(x = -1) + P(x = 0)$

$$= 1/3 + 1/2 = 5/6$$

and in $-1 \le x < \infty$, $F(x) = P[(x = -1) + (x = 0) + (x = 1)]$

$$= P(x = -1) + P(x = 0) + P(x = 1)$$

$$= 1/3 + 1/2 + 1/6 = 1.$$

Example 5·2. *Let* $F(x) = 0$ *where* $-\infty < x < 0$

$$= 1/5, \quad 0 \le x < 1$$

$$= 3/5, \quad 1 \le x < 2$$

$$= 1, \quad 2 \le x < \infty$$

Determine the spectrum and the probability masses of the above distribution.

Here $F(x)$ is a step-function, the step-points being 0, 1, 3 which are the points of the spectrum.

Also $P(X = 0) = F(0) - F(0 - 0) = 1/5$

$$P(X = 1) = F(1) - F(1 - 0) = 3/5 - 1/5 = 2/5$$

$$P(X = 3) = F(3) - F(3 - 0) = 1 - 3/5 = 2/5.$$

Example 5·3. *A random variable X has the following probability function :*

$X = x$:	–2	–1	0	1	2	3
$p(x)$:	0·1	k	0·2	2k	0·3	3k

(*i*) *Find* k (*ii*) *Evaluate* $P(X < 2)$, $P(X \geq 2)$, $P(-2 < X < 2)$ (*iii*) *Find the minimum value of k such that* $P(X \leq 1) > \cdot 36$ (*iv*) *Determine the distribution function* $F(x)$ *of X.*

(*i*) $\sum_{x=-2}^{3} p(x) = 1$ gives $0\cdot1 + k + 0\cdot2 + 2k + 0\cdot3 + 3k = 1.$

Thus $k = 1/15.$

(*ii*) $P(X < 2) = P(X = -2) + P(X = -1) + P(X = 0) + P(X = 1)$

$= 0\cdot1 + k + 0\cdot2 + 2k = 3k + 0\cdot3 = 0\cdot5$

$P(X \geq 2) = P(X = 2) + P(X = 3) = 0\cdot3 + 3k = 0\cdot3 + 0\cdot2 = 0\cdot5$

Aliter. $P(X \geq 2) = 1 - P(X < 2) = 1 - 0\cdot5 = 0\cdot5$

$P(-2 < X < 2) = P(X = -1) + P(X = 0) + P(X = 1)$

$= k + 0\cdot2 + 2k = 3k + 0\cdot2 = 0\cdot2 + 0\cdot2 = 0\cdot4.$

(*iii*) $P(X \leq 1) = P(X = -2) + P(X = -1) + P(X = 0) + P(X = 1)$

$= 0\cdot1 + k + 0\cdot2 + 2k = 3k + 0\cdot3.$

By hyp., $P(X \leq 1) > \cdot36 \Rightarrow 3k + 0\cdot3 > \cdot36 \Rightarrow 3k > \cdot06 \Rightarrow k = \cdot02.$

(*iv*) $F(x) = 0, x < -2; = \frac{1}{10}, \; -2 \leq x < -1; \; = \frac{14}{15}, \; -1 \leq x < 3.$

Example 5·4. *Let X be a random variable such that*

$$P(X = -2) = P(X = -1) = P(X = 1) = P(X = 2) \text{ and}$$

$$P(X < 0) = P(X = 0) = P(X > 0).$$

Determine the probability mass function of X and the distribution of X.

X takes the values $-2, -1, 0, 1$ and 2 with the numbers $P(X = x_i)$ associated to these values.

X :	-2	-1	0	1	2
$p(x)$:	$\frac{1}{6}$	$\frac{1}{6}$	$\frac{1}{3}$	$\frac{1}{6}$	$\frac{1}{6}$

In $-\infty < x < -2$, $F(X) = 0$

in $-2 \leq x \leq -1$, $F(x) = P = [x = -2] = 1/6$

in $-1 \leq x < 0$, $F(x) = P[(x = -1) + (x = 0)]$

$= P(x = -1) + P(x = 0)$

$= 1/6 + 1/3 = 1/2$

in $0 \leq x < 1$ $F(x) = P[(x = 0) + (x = 1)] = P(x = 0) + P(x = 1)$

$= 1/3 + 1/6 = 1/2$

in $1 \leq x < 2$ $F(x) = P(x = 1) + P(x = 2) = 1/3$

and in $-2 \leq x < \infty$. $F(x) = P(x = -2) + P(x = -1)$

$+ P(x = 0) + P(x = 1) + P(x = 2)$

$$= 1/6 + 1/6 + 1/3 + 1/6 + 1/6 = 1.$$

Ex. A random variable assumes the values $-3, -2, -1, 0, 1, 2, 3$ such that

$$P(x = -3) = P(x = -2) = P(x = -1)$$
$$P(x = 1) = P(x = 2) = P(x = 3)$$

and $$P(x < 0) = P(x = 0) = P(x = 0)$$

obtain the probability mass function of X and its distribution function. Find also the probability mass function of $Y = 2x^2 + 3x + 4$

(*B.Sc. Poona, 1991*)

Example 5·5. *From a lot of 25 items, containing 5 defectives, a sample of 4 items was drawn at random (i) without replacement (ii) with replacement. Find the probability distribution of X, the number of defectives.*

(*i*) We draw a sample of 4 items from a lot of 25 items and this can be done in $\binom{25}{4}$ ways, which are equally likely. The number of ways in which out of 5, x defectives can be chosen and out of remaining 20 items, $4 - x$ non-defectives can be chosen $= \binom{5}{x} \times \binom{20}{4-x}$.

Consequently, $P(X = x) = \binom{5}{x} \times \binom{20}{4-x} \div \binom{25}{4}$.

(*ii*) When there is drawn a sample of 4 items with replacement, the number of ways $= \binom{4}{x}$.

Further the probability of choosing one defective out of 5 defectives is 1/5 so that the probability of choosing x defectives out of 5 is $(1/5)^x$. The probability of not choosing a defective is 4/5 so that the probability of not choosing 4—x non-defective items is $\left(\frac{4}{5}\right)^{4-x}$. Hence the probability of choosing x defective and $4 - x$ non-defective items is

$$P(X = x) = \binom{4}{x} \times \left(\frac{1}{5}\right)^x \left(\frac{4}{5}\right)^{4-x}$$

Example 5·6. *An experiment consists of three independent tosses of an honest coin.*

Let X = the number of heads,

Y = the number of consecutive occurrence of at least two heads. Determine the probability function of (*i*) *X* (*ii*) *Y* (*iii*) *XY* (*iv*) *X+Y*

Outcomes	*HHH*	*HHT*	*HTH*	*HTT*	*THH*	*THT*	*TTH*	*TTT*
X	3	2	2	1	2	1	1	0
Y	1	1	0	0	1	0	0	0
$XY(=U)$	3	2	0	0	2	0	0	0
$X+Y\ (=V)$	4	3	2	1	3	1	1	0

The sample space S is : {*HHH, HHT, HTH, HTT, THH, THT, TTH, TTT* } which consists of 8 elementary events.

(*i*) The tosses are independent and the spectrum of the *r.v.X* is 0, 1, 2, 3.

Now $P(0) = P(TTT) = 1/2 \cdot 1/2 \cdot 1/2 = 1/8$

$P(1) = P(HTTUTHTUTTH) = P(HTT) + P(THT) + P(TTH)$
$= 1/8 + 1/8 + 1/8 = 3/8$

$P(2) = P(HHTUHTHUTHH) = P(HHT) + P(HTH) + P(THH)$
$= 1/8 + 1/8 + 1/8 = 3/8$

$P(3) = P(HHH) = 1/2 \cdot 1/2 \cdot 1/2 = 1/8.$

The above results can be put in the tabular form as follows :

	Values of X (x) :	0	1	2	3
(*ii*)	Probabilities $p(x)$:	1/8	3/8	3/8	1/8
	Values of Y (y) :	0	1	[value 0 is occurring in Table 1. 5 times with each probability = 1/8]	
	Probabilities, p (y) :	5/8	3/8		
(*iii*)	Values of XY (u) :	0	2	3	
	Probability	5/8	2/8	1/8	
(*iv*)	Values of $X+Y = V$ (v) :	0	1	2	3	4
	Probabilities :	1/8	3/8	1/8	2/8	1/8

5·2. Mathematical Expectation Def. Let the *r.v.X.* take values $x_1, x_2, ..., x_n$ with probabilities $p_1, p_2, ..., p_n$. This means that if we perform a random experiment, X can have the value x_1 or x_2 ..... or x_n. Let us perform the experiment for a very large number of times under identical conditions. Out of these, let f_i give x_i, $(i = 1, 2, ..., n)$ where $f_1 + f_2 + \ldots + f_n = N$, then the mean of X is given by

$$\frac{f_1 x_1 + f_2 x_2 + \ldots + f_n x_n}{N} = \frac{f_1}{N} x_1 + \frac{f_2}{N} x_2 + \ldots + \frac{f_n}{N} x_n$$

Let $N \to \infty$, then using the statistical definition of probability , we get the limiting value for the mean as

$$\overline{X} = p_1 x_1 + p_2 x_2 + ... + p_n x_n = \sum_{i=1}^{n} p_i x_i.$$

This value is defined as the *mathematical expectation* of X and is denoted by $E(X)$. We might even call it the statistical expectation of X.

Thus $E(X)$ may be regarded as the limiting value of the average value of X realised in N random experiments as $N \to \infty$.

We also know that if $\varphi(X)$ is a single-valued function of X, then in N random experiments, $\varphi(X)$ will take the values $\varphi(x_1), \varphi(x_2)\ldots\ldots, \varphi(x_n)$ with frequencies $f_1, f_2, \ldots, f_n$ and the average value of $\varphi(X)$ in N experiments is

$$\frac{f_1\,\varphi(x_1) + f_2\,\varphi(x_2) + \ldots + f_n\,\varphi(x_n)}{N} = \frac{f_1}{N}\varphi(x_1) + \ldots + \frac{f_n}{N}\varphi(x_n)$$

and as $N \to \infty$, this approaches. $p_1\,\varphi(x_1) + p_2\,\varphi(x_2) + \ldots + p_n\,\varphi(x_n)$

which is defined as $E[\varphi(X)]$.

In particular, $\mu'_r(a) = E(X-a)^r$, $\mu_r = E[(X-\overline{X})^r]$, $\mu_2 = E[(X-\overline{X})]^2 = \sigma^2$.

Example 5·7. *What is the expected value of the number of points that will be obtained in a single throw with an ordinary dice ?*

Hence the variate X is the number of points on a dice. Its values are 1, 2, 3, 4, 5, 6 each having the probability $\frac{1}{6}$.

Hence $E(X) = p_1x_1 + p_2x_2 + \ldots\ldots + p_6x_6 = \frac{1}{6}\cdot 1 + \frac{1}{6}\cdot 2 + \ldots + \frac{1}{6}\cdot 6 = \frac{7}{2}$.

Example 5·8. *What is the expectation of the number of failures preceding the first success in an infinite series of independent trials with constant probability p of success in a trial ?*

(I.A.S. 1992 ; B.A. Hons. Delhi ; M.A. Madras ; M.A. Agra ; B.Sc. Hons. Gauhati, 1971 ; M.A. Bombay ; B.Sc. Hons. Sub.Statistics, Delhi)

If X denotes the number of failures preceding the first success we find that X takes the values 0 1 2 3 $p \quad qp \quad q^2p \quad q^3p$

with probabilities $p \quad qp \quad q^2p \quad q^3p$ .... where $q = 1 - p$..

$\therefore$ Probability density function $f(x) = q^r p$, $r = 0, 1, 2, 3, \ldots\ldots$

Hence $E(X) = 0\cdot p + 1\cdot qp + 2\cdot q^2p + 3\cdot q^3p + \ldots$

$$= pq(1 + 2q + 3q^2 + \ldots\ldots),\ q < 1 = pq(1-q)^{-2} = \frac{q}{p} = \frac{1}{p} - 1.$$

Example 5·9. *A person draws cards one by one from a pack until he draws all the aces. How many cards he may be expected to draw ?*

(Indian Statistical Institute, Calcutta)

Suppose in x drawings he draws all the aces. This means that in $(x-1)$ drawings he should draw 3 aces and the xth draw should be an ace. The probability of such an event is

$$\frac{{}^4C_3 \times {}^{48}C_{x-4}}{{}^{52}C_{x-1}} \times \frac{1}{52-x+1} = \frac{4(x-1)(x-2)(x-3)}{52 \times 51 \times 50 \times 49}$$

Hence required expectation $= \sum_{x=4}^{52} x. \frac{4(x-1)(x-2)(x-3)}{52.51.50.49}$

$$= \frac{4}{52.51.50.49} \times \left[\sum_{x=4}^{52} x^4 - 6 \sum_{x=4}^{52} x^3 + 11 \sum_{x=4}^{52} x^2 - 6 \sum_{x=4}^{52} x \right]$$

$$= \frac{68872440}{162430} = 42{\cdot}4.$$

Ex. 1. Show by an example that mathematical expectation need not be finite.

[**Hint.** Suppose the elementary probability law of the discontinuous random variable K is $p(k) = \frac{e^{-1}}{k\,!}, k = 0, 1, 2, ...\infty.$

$$\therefore \quad E(k\,!) = \sum_{k=0}^{\infty} \left[k\,! \cdot \frac{e^{-1}}{k\,!} \right] = e^{-1} \cdot \sum_{0}^{\infty} 1, \text{ which is not finite.}]$$

Another Example. Let $P[X = x] = 6/\pi^2 x^2, x = 1, 2, ..., \therefore E X = \sum_{1}^{\infty} \frac{6}{\pi^2 x} = \infty$

Ex. 2. Let a random variable X take the values $X_k = (-1)^k 2^k k^{-1}, k = 1, 2, 3, ...,$ with probabilities $p_k = 2^{-k}$, Find $E(X)$. [**Ans.** —log 2]

Ex. 3. Show that if X is a discrete random variable and a is a constant,

(*i*) $E(a)=a$; (*iii*) $E(aX)=a\ E(X)$;

(*ii*) $E[a\psi(X)] = a\,E[\psi(X)]$; (*iv*) $E(X - \bar{X}) = 0.$

where $\psi(X)$ is any function of X, where $\bar{X}$ is the mean of X

Ex. 4. Show that (*i*) $Var(aX) = a^2\ Var(X)$, a being a constant.

(*ii*) $\mu_r(aX) = a^r\ \mu_r(X)$, a being a constant.

Ex. 5. Two unbiased dice are thrown. Find the expected value of the sum of numbers of points drawn. [**Ans.** 7]

Ex. 6. Prove that moments about the mean are invariant for change of origin, but not for change of scale.

Ex. 7. If $P(X \le 0) = 1$ and if $E(X)$ exists, then $E(X) \le 0$.

Ex. 8. Prove that $|\ E(X)\ | \le E\ |\ X\ |$.

[**Hint.** $|\ E(X)\ | = |\ \Sigma x_i p_i\ | < \Sigma_i\ |\ x_i p_i\ | = \Sigma_i\ |\ x_i\ |\ p_i = E\,[\,|\ X\ |\ ]\,]$.

Aliter. Since $x \le |x|\ \therefore\ EX \le E|X|$ Also $-x < |x|\ \therefore\ E(-X \le E|X|$ $\Rightarrow -E(x) \le E\ |x|$ Hence etc.

Ex. 9. If X assumes only positive integral values, show that

$$EX = \sum_{n=1}^{\infty} P(x \ge n)$$

provided that the above series converges. (***B.A. Hons Maths Delhi, 1981***)

[**Hint.** $EX = 1\,P(x=1) + 2P(x=2) + 3\,P(x=3) + \ldots$

$= [P(x=1) + \ldots + P(x=n) + \ldots] + [P(x=2) + \ldots P(x=n) + \ldots] + \ldots$

$= P(x \geq 1) + P(x \geq 2) + \ldots = \sum_{n=1}^{\infty} P(x \geq n)$

Ex.10. If X assumes only positive values and has finite expectations, then

(*i*) $E(1/X) \geq \frac{1}{EX}$. (*ii*) $E(X^{1/2}) \leq [E(X)]^{1/2}$.

5·2·1. Expectation of a sum of Random Variables.

Theorem. *The expectation of the sum of two random discrete variables is equal to the sum of their expectations.*

In symbols, *if X and Y are random discrete variables, then*

$$E(X+Y) = E(X) + E(Y).$$

Proof. Let X and Y be two stochastic variates. Let X assume the m values $x_1, x_2, \ldots, x_m$ with probabilities $p_1, p_2, \ldots, p_n$ and Y assume the n values $y_1, y_2, \ldots, y_n$ with probabilities $p'_1, p'_2, \ldots p'_n$. Then the sum $X+Y$ is a stochastic variate which can take the mn values $x_i + y_j$, $(i = 1, 2, \ldots, m\,;\ j = 1, 2, \ldots, n)$. Let p_{ij} denote the probability of X assuming the value x_i, and at the same time, Y assuming the value y_j, *i.e.*, $p_{ij} = P(X = x_i,\ Y = y_j)$. As X assumes a definite value x_i, Y assumes one of the values $y_1, y_2, \ldots, y_n$ so that the sum $\sum_{j=1}^{n} p_{ij}$ represents the probability p_i of X assuming the value x_i, that is, $\sum_{j=1}^{n} p_{ij} = p_i$.

Similarly, $\sum_{i=1}^{m} p_{ij}$ represents the probability of Y assuming the value y_j that is $\sum_{i=1}^{m} p_{ij} = p_j$.

By definition,
$$E(X+Y) = \sum_{i=1}^{m} \sum_{j=1}^{n} p_{ij}\,(x_i + y_j)$$

$$= \sum_{i=1}^{m} \sum_{j=1}^{n} p_{ij}\,x_i + \sum_{i=1}^{m} \sum_{j=1}^{n} p_{ij}\,y_j$$

$$= \sum_{i=1}^{m} x_i \left(\sum_{j=1}^{n} p_{ij} \right) + \sum_{j=1}^{n} y_j \left(\sum_{i=1}^{m} p_{ij} \right)$$

$$= \sum_{i=1}^{m} x_i\,p_i + \sum_{j=1}^{n} y_j\,p'_j$$

Thus $E(X+Y) = E(X) + E(Y)$.

Cor. 1. Generalisation. The expectation of the sum of a number of discrete random variables in the sum of their expectations. In symbols,

$$E(X+Y+Z+\ldots) = E(X) + E(Y) + E(Z) + \ldots$$

For $E(X+Y+Z+\ldots) = E(X+\overline{Y+Z+\ldots}), = E(X) + E(\overline{Y+Z+\ldots}),$

$$= E(X) + E(Y) + E(Z+\ldots),$$

$$= E(X) + E(Y) + E(Z) + \ldots$$

Cor. 2. $E(aX+b) = aE(X) + b$, a and b being constants.

[**Hint.** $E(aX+b] = \sum_{i=1}^{n} p_i (ax_i + b) = a \sum_{i=1}^{n} p_i x_i + b = aEX + b)$]

Cor. 3. If X and Y are random variables, then

$$E(a\,X + bY) = aEX + bEY,$$

a and b being constants.

For $$E(aX+bY) = \sum_{i=1}^{m} \sum_{j=1}^{n} p_{ij} (ax_i + by_j)$$

$$= a \sum_{i=1}^{m} \sum_{j=1}^{n} p_{ij} x_i + b \sum_{i=1}^{m} \sum_{i=1}^{n} y_j p_{ij} = aEX + bEY.$$

Cor. 4. $E[a\,\psi_1(x) + b\psi_2(x)] = aE[\psi_1(x)] + bE[\psi_2(x)]$ where $\psi_1(x)$ and $\psi(x)$ are random functions of x, while a and b are constants.

Example 5·10. *What is the mathematical expectation of the sum of points on n dice ?*

Denoting by x_i the number of points on the ith dice, the sum of points on n dice will be $s = x_1 + x_2 + \ldots + x_n$

$$\therefore\ E(s) = E(x_1 + x_2 + \ldots + x_n) = E(x_1) + E(x_2) + \ldots + E(x_n).$$

But for every single dice, $E(x_i) = \frac{7}{2}$, $i = 1, 2, \ldots, n$. (See example 5·7, page 136)

Hence $$E(s) = \frac{7n}{2}.$$

Example 5·11. *Suppose that we have a series of n independent trials, the probability of success being p_i for the ith trial. Show that the mean number of successes is $\sum_{i=1}^{n} p_i$.* **(B.A., Hons., Delhi, 1955)**

Let us associate with ith trial a random variable X_i which has the value 1 in case of success and the value 0 in case of a failure. Then the number of successes in n trials is given by

$$m = X_1 + X_2 + \ldots + X_n, \text{ so that } E(m) = E(X_1) + E(X_2) + \ldots + E(X_n)$$

Since X_i can take the values 1 and 0 with probabilities p_i and $1 - p_i$, we have $E(X_i) = 1 \cdot p_i + 0(1 - p_i) = p_i$

Hence $E(m) = p_1 + p_2 + \ldots + p_n = \sum_{i=1}^{n} p_i.$

Cor. If $p_1 = p \ \forall i$, then $E(m) = np$.

Ex. 1. If X is a continuous random variable, show that (*i*) $Var(X) = E(X)^2 - [E(X)]^2$; (*ii*) $Var(aX) = a^2 . Var(X)$, (*iii*) $E(aX + bY) = aE(X) + bE(Y)$, a and b are constants. **(B.Sc. Delhi, 1971)**

Ex. 2. Thirteen cards are drawn simultaneously from a deck of 52. If faces count 1, face cards 10 and other according to denomination, find the expectation of the total score on the 13 cards. [Ans. 85]

Definition. *Two random variables are said to be independent if probability of either taking a particular value does not depend on what value the other variate takes.*

5·2·2. Expectation of product of independent stochastic variables.

Theorem. *The expectation of the product of two independent random variables is equal to the product of their expectations ; or in symbols, if X and Y are two independent random variables, then*

$$E(XY) = E(X)E(Y).$$

[Assume that X and Y are discrete]

Proof. Let the random variable X assume the m values $x_1, x_2, \ldots, x_m$ and the variable Y the n values $y_1, y_2, \ldots, y_n$. Let the probability of X assuming the value x_i be p_i and the probability of Y assuming the value y_j be p'_j. Since the variables X and Y are independent, by the theorem of compound probability, the probability that X assumes the value x_i and at the same time Y assumes the value y_j is $p_i p'_j$.

$$\text{Hence } E(XY) = \sum_{i=1}^{m} \sum_{j=1}^{n} p_i p'_j x_i x_j = \left(\sum_{i=1}^{m} p_i x_i \right) \left(\sum_{j=1}^{n} p'_j y_j \right)$$

$$= \sum_{i=1}^{m} p_i x_i E(Y) = E(Y) \sum_{i=1}^{m} p_i x_i = E(X) E(Y).$$

Generalization. This theorem can be extended to the case of several independent variables, so that if $X, Y, Z, \ldots$ are independent stochastic variables,

$$E(XYZ \ldots\ldots) = E(X) E(Y) E(Z) \ldots\ldots$$

Cor. $E[\varphi_1(x)\varphi_2(y)] = E[\varphi_1(x)] \, E[\varphi_2(y)]$ where X, Y are independent.

Ex. 1. Find the expected value of the product of points on n dice.

Ans. $(7/2)^n$

Ex. 2. Prove that if X, Y, are two random variables whose expectations exist and if $P(X \le Y) = 1$, then $E(X) \le E(Y)$.

Ex. 3. If $X_1, X_2, \ldots\ldots, X_n$ are independent in pairs, are they necessarily independent variable ? Support your answer by an example.

5·3. Covariance. Definition. If X and Y are two random variables with the respective expected values (or means) $\overline{X}$ $\overline{Y}$, then the covariance between X and Y is defined by the relation

$$\mu_{11} = Cov\,(X, Y) = E[(X - \overline{X})\,(Y - \overline{Y})] \qquad \ldots(5\cdot5)$$

Cor. If X and Y are independent random variables, then (5·5) reduces to

$$Cov\;(X, Y) = E(X - \overline{X})\;E(Y - \overline{Y}) = 0,$$

since $$E(X - \overline{X}) = E(X) - E(\overline{X}) = \overline{X} - \overline{X} = 0.$$

Thus the covariance of two independent random variables is equal to zero.

The *converse* of this result is *not necessarily true, i.e.*, if the covariance between two variables is zero, it is not necessary that they are independent. Thus let u and v be two independent random variables with same variance and let $X = u + v$, ; $Y = u - v$,

so that $$\overline{X} = \overline{u} + \overline{v}\ ;\quad \overline{Y} = \overline{u} - \overline{v}.$$

Now $$Cov\;(X, Y) = E(X - \overline{X})\,(Y - \overline{Y}) = E(u - \overline{u} + v - \overline{v})\,(\overline{u - \overline{u} - v - \overline{v}})$$

$$= E(u - \overline{u})^2 - E(v - \overline{v})^2 = \sigma_u^2 - \sigma_v^2 = 0,$$

but X and Y are not necessarily independent *e.g.*, if u and v denote the number of points on two dice, then their sum and difference X and Y are either both even or both odd and thus are dependent.

5·3·1. Correlation Coefficient, Definition. The ratio

$$r_{XY} = \frac{Cov\;(X, Y)}{\sqrt{[(Var\,(X)\;\;Var\,(Y)]}} = \frac{E(X - \overline{X})\,(Y - \overline{Y})}{\sqrt{E(X - \overline{X})^2}\;\sqrt{E(Y - \overline{Y})^2}} = \frac{\mu_{11}}{\sigma_X\,\sigma_Y} \qquad \ldots(5\cdot5')$$

is defined as the correlation coefficient between X and Y.

Obviously $r_{XY} = r_{YX}$ and both have the same sign as $Cov\;(X, Y)$.

Note. (5·5) can be put in the form :

$$Cov\;(X, Y) = E(XY - \overline{X}Y - \overline{Y}X + \overline{XY}) = E(XY) - E(\overline{X}Y) - E(\overline{Y}\,X) + E(\overline{XY})$$

$$= E(XY) - \overline{X}\,E(Y) - \overline{Y}\,E(X) + \overline{XY} = E(XY) - \overline{X}\,\overline{Y} - \overline{Y}\,\overline{X} + \overline{X}\,\overline{Y}$$

$$= E(XY) - E(X)\,E(Y) \qquad \ldots(5\cdot5'')$$

Ex. Show that, if a and b are constants,

(*i*) $Cov\,(aX, bY) = ab\;Cov\,(X, Y)$ (*ii*) $Cov\,(X + a,\; Y + b) = Cov\,(X, Y)$

(*iii*) $Cov(X,X) = Var(X)$. (*iv*) $Cov\left(\frac{X-\bar{X}}{\sigma_x}, \frac{Y-\bar{Y}}{\sigma_y}\right) = \frac{1}{\sigma_x\sigma_y} Cov(X,Y)$

(*v*) $r_{ax+b,\, cy+d} = r_{ax,\, cy} = \pm\, r_{xy}$ (*vi*) $Cov(X, X-Y) = Var\,X - Cov(X,Y)$.

according as a and c have the same or opposite signs.

5·3·2. Expectation and Variance of a Linear Combination of Random Variables.

Let $x_1, x_2, \ldots, x_n$, be n random variables with finite variances $\sigma_1^2, \sigma_2^2, \ldots, \sigma_n^2$, then we have to find the variance of u where

$$u = a_1x_1 + a_2x_2 + \ldots + a_nx_n.$$

a's being constants.

We have
$$E(u) = E(a_1x_1 + a_2x_2 + \ldots + a_nx_n)$$
$$= E(a_1x_1) + E(a_2x_2) + \ldots E(a_nx_n)$$
$$= a_1 E(x_1) + a_2E(x_2) + \ldots + a_nE(x_n)$$

$$\therefore \quad u - E(u) = a_1[x_1 - E(x_1)] + a_2[x_2 - E(x)] + \ldots + a_n[x_n - E(x_n)]$$

Squaring both sides and taking expected values, we get

$$Var(u) = a_1^2\, Var(x_1) + a_2^2\, Var(x_2) + \ldots + a_n^2\, Var(x_n)$$
$$+ 2a_1a_2\, Cov(x_1, x_2) + \ldots + 2a_{n-1}\, a_n\, Cov(x_{n-1}, x_n)$$
$$= \sum_{i=1}^{n} a_i^2\, Var(x_i) + 2 \sum_{\substack{i,j=1 \\ i<j}} a_ia_j\, Cov(x_i, x_j) \qquad \ldots(5\cdot6)$$

$$\Rightarrow \quad Var\,u = \sum_{i=1}^{n} a_i^2\sigma_i^2 + 2\sum_{\substack{i,j=1 \\ i<j}}^{n} a_ia_j\sigma_i\sigma_j r_{ij} \quad \text{using } (5\cdot5'),\ (5\cdot6)$$

where r_{ij} is correlation coefficient between x_i and x_j.

Cor. 1. Let $a_1 = 1,\ a_2 = 1,\ a_3 = a_4 = \ldots = a_n = 0$, so that

$$Var(x_1 + x_2) = Var(x_1) + Var(x_2) + 2\, Cov(x_1, x_2) \qquad \ldots(5\cdot7)$$

Cor. 2. Let $a_1 = 1,\ a_2 = -1,\ a_3 = a_4 = \ldots = a_n = 0$ so that

$$Var(x_1 - x_2) = Var(x_1) + Var(x_2) - 2\, Cov(x_1, x_1) \qquad \ldots(5\cdot7')$$

Cor. 3. Let x_1 and x_2 be independent chance variables, then

$$Cov(x_1, x_2) = 0,\ Var(x_1 \pm x_2) = Var(x_1) + Var(x_2) \qquad \ldots(5\cdot7'')$$

Cor. 4. Let $a_1 = a_2 = a_3 = \ldots = a_n = 1/n$,

so that
$$u = \sum_{1}^{n} a_i\, x_i = \frac{1}{n} \sum_{1}^{n} x_i = \bar{x}$$

Also let x_i's be independent variables and let $Var(x_i) = \sigma^2$, then

$$Var(\bar{x}) = \frac{\sigma^2}{n^2} + \frac{\sigma^2}{n^2} + \ldots\ldots \text{ upto } n \text{ times} = n\frac{\sigma^2}{n^2} = \frac{\sigma^2}{n}.$$

Cor. 5. *Var* $(aX+b) = a^2\, Var\, X$, *a and b being constants.*

Example 5·12. *Find the variance of the number of successes in a series of n independent trials in which the probability of success in the ith trial is* p_i.

Using the notation of §5·3, we have

$$Var(m) = Var(x_1 + x_2 + \ldots + x_n) = Var(x_1) + Var(x_2) + \ldots + Var(x_n)$$

x_i^2 takes the value 1 if there is a success and zero in case of failure

so that $$E(x_i^2) = 1.\ p_i + 0.\ (1-p_i) = p_i.$$

Hence the variance of the number of successes in the ith trial is

$$Var(x_i) = E[x_i - E(x_i)]^2 = E[x_i]^2 - 2E[x_i E(x_i)] + \{E(x_i)\}^2$$

$$= E(x_i^2) - \{E(x_i)\}^2 = p_i - p_i^2 = p_i q_i\ ;\ q_i = 1 - p_i$$

$$\therefore\ Var(m) = \sum_{i=1}^{n} p_i q_i \le \frac{n}{4} \quad \text{as } p_i\ (1-p_i) \le \frac{1}{4}$$

since $p(1-p)$ is maximum when $p = \frac{1}{2}$. so that max. value of $p\,(1-p) = \frac{1}{4}$.

Cor. If $p_i = p$ then $Var(m) = pq + pq + \ldots + pq = npq$.

Example 5·13. *A box contains* 2^n *tickets among which* ^nC_i *tickets bear the number i* $(i = 0, 1, 2, \ldots, n)$. *A group of m tickets is drawn. Let S denote the sum of their numbers. Find E(S) and Var S.*

(B.A., Hons. Calcutta ; M.A. Delhi ; M.Sc., Calcutta)

Let $x_1, x_2, \ldots\ldots, x_m$ be the chance variables representing the numbers on the first, second ...... m-th tickets.

Now x_i is a variable which takes the values 0, 1, 2,......, n with

respective probabilities. $$\frac{^nc_0}{2^n}, \frac{^nc_1}{2^n}, \ldots\ldots, \frac{^nc_n}{2^n}$$

$$\therefore \quad E(x_i) = \frac{1}{2^n}\,[\,1.\,^nc_1 + 2.\,^nc_2 + \ldots\ldots + n.\,^nc_n\,]$$

$$= \frac{n}{2^n}\,[\,1 + {}^{n-1}c_1 + {}^{n-1}c_2 + \ldots\ldots + {}^{n-1}c_{n-1}\,]$$

$$= \frac{n}{2^n}(1+1)^{n-1} = \frac{n}{2}.$$

Hence $E(S) = E(x_1 + x_2 + \dots + x_m) = \sum_{i=1}^{m} E(x_i) = \dfrac{mn}{2}$

$$E(x_i^2) = 1^2 \cdot \frac{{}^nc_1}{2^n} + 2^2 \cdot \frac{{}^nc_2}{2^n} + \dots + n^2 \cdot \frac{{}^nc_n}{2^n}$$

$$= \frac{n}{2^n}\left[(1+1)^{n-1} + (n-1)(1+1)^{n-2}\right]$$

$$= \frac{n}{2^n}\left[2^{n-1} + (n-1)2^{n-2}\right] = \frac{n(n+1)}{4}$$

$$E(x_i x_j) = \sum_{\substack{i,j \\ i<j}}^{n} \left[ij \, \frac{{}^nc_i}{2^n} \cdot \frac{{}^nc_j}{2^n - 1} \right] + \left[\sum_{i=1}^{n} i^2 \, \frac{{}^nc_i}{2^n} \cdot \frac{{}^nc_i - 1}{(2^n - 1)} \right]$$

$$= \sum_{j=1}^{n} \left[j \cdot \frac{{}^nc_j}{2^n - 1} \left(\sum_{i=1}^{n} \cdot i \, \frac{{}^nc_i}{2^n} - j \cdot \frac{{}^nc_j}{2^n} \right) + \sum_{i=1}^{n} \frac{i^2 ({}^nc_i)^2}{2^n (2^n - 1)} - \sum_{i=1}^{n} i^2 \frac{{}^nc_i}{2^2 (2^n - 1)} \right]$$

$$= \sum_{i} \left[j \cdot \frac{{}^nc_j}{2^n - 1} \left(\frac{n}{2} - j \cdot \frac{{}^nc_j}{2^n} \right) \right] + \sum_{j=1} j^2 \frac{({}^nc_j)^2}{2^n (2^n - 1)} - \sum_{j=1} \frac{j \, {}^nc_j}{2^n (2^n - 1)}$$

$$= \frac{1}{2^n - 1} \left[\frac{n}{2} \cdot n \cdot 2^{n-1} - \frac{1}{2^n} \sum_{j=1}^{n} j^2 \, {}^nc_j \right]$$

$$= \frac{1}{2^n - 1} \left[n^2 . 2^{n-2} - \frac{n(n+1)}{4} \right]$$

$$\therefore \quad E(S^2) = \sum_{i=1}^{m} E(x_i^2) + \sum_{\substack{j \; i \\ i<j}}\sum E(x_i x_j)$$

$$= \frac{m\,n(n+1)}{4} + \frac{m(m-1)\,n}{2^n - 1} \left[n.2^{n-2} - \frac{n+1}{4} \right].$$

Hence $\quad Var(S) = E(S^2) - [E(S)]^2 = \dfrac{mn}{4} - \dfrac{mn(m-1)}{4(2^n - 1)}$, on simplification.

Example 5·14. *Use relation that* $E(Ax^a + Bx^b + Cx^c)^2$ *is always*

≥ 0, x being a random variable and E denotes the mathematical expectation, show that

$$\begin{vmatrix} \mu_{2a} & \mu_{a+b} & \mu_{a+c} \\ \mu_{a+b} & \mu_{2b} & \mu_{b+c} \\ \mu_{a+c} & \mu_{b+c} & \mu_{2c} \end{vmatrix} \geq 0, \ \mu_n \text{ being the nth moment about mean.}$$

Hence or otherwise show that $\beta_2 - \beta_1 - 1 \geq 0$. Show also that $\beta_2 \geq 1$.

(B.A. Hons. 1978 ; M.Sc. Banaras; I.C.A.R., Delhi)

We are given that $E(Ax^a + Bx^b + Cx^c)^2$ is always ≥ 0 *i.e.*,

$$E[A^2x^{2a} + B^2x^{2b} + C^2x^{2c} + 2ABx^{a+b} + 2BCx^{b+c} + 2CAx^{c+a}] \text{ is always } \geq 0.$$

If x denotes deviations from the mean, this gives

$$A^2\mu_{2a} + B^2\mu_{2b} + C^2\mu_{2c} + 2AB\mu_{a+b} + 2BC\mu_{b+c} + 2CA\mu_{c+a} \text{ is always } \geq 0.$$

We know that the conditions that

$$a'x^2 + b'y^2 + c'z^2 + 2f'\ yz + 2g'\ zx + 2h'xy$$

is non-negative for all values of x, y and z are

$$(i)\ a' \geq 0, \qquad (ii) \begin{vmatrix} a' & h' \\ h' & b' \end{vmatrix} \geq 0, \qquad (iii) \begin{vmatrix} a' & h' & g' \\ h' & b' & f' \\ g' & f' & c' \end{vmatrix} \geq 0.$$

In the present case

$$a' = \mu_{2a},\ b' = \mu_{2b},\ c' = \mu_{2c},\ f' = \mu_{b+c},\ g' = \mu_{c+a},\ h' = \mu_{a+b}.$$

$\therefore$ We get the required result on substituting the values of a', b', c', f', g', h' in *(iii)*.

Putting in this result $a = 0$, $b = 1$, and $c = 2$ and remembering that $\mu_1 = 0$, we get

$$\begin{vmatrix} 1 & 0 & \mu_2 \\ 0 & \mu_2 & \mu_3 \\ \mu_2 & \mu_3 & \mu_4 \end{vmatrix} \geq 0$$

i.e.,

$$\mu_2\mu_4 - \mu_3^2 - \mu_2^3 \geq 0.$$

For β_2 to be finite, μ_2 should be ≥ 0.

Hence on dividing by μ_2^3, we have

$$\frac{\mu_4}{\mu_2^2} - \frac{\mu_3^2}{\mu_2^3} - 1 \geq 0 \text{ i.e., } \beta_2 - \beta_1 - 1 \geq 0$$

Further since $\beta_1 \geq 0$, we have $\beta_2 \geq 1$,

otherwise. From (*ii*) putting $a = 1$, $b = 1$ we get

$$\begin{vmatrix} \mu_2 & \mu_3 \\ \mu_3 & \mu_4 \end{vmatrix} \geq 0 \text{ or } \mu_4\mu_2 \geq \mu_3^2$$

or $\beta_2 \geq 1$.

Example 5·15. *Let the variable X have the distribution $P(X = 0) = P(X = 2) = p$, $P(X = 1) = 1 - 2p$, for $0 \leq p \leq \frac{1}{2}$. For what value of p is the Var X a maximum ?* **(B.Sc. Math Hons. 1985. 1987)**

The probability distribution is

X :	0	1	2
$P(x)$:	p	$1\text{-}2p$	p

$$\therefore \quad EX = \sum_{x=0}^{2} x\, P(X = x) = 0 \times p + 1 \times (1 - 2p) + 2 \times p = 1$$

$$\therefore \quad EX^2 = \sum_{0}^{2} x^2\, P\,(X = x) = 0^2 \times p + 1^2 \times (1 - 2p) + 2^2 \times p = 1 + 2p$$

$$\therefore \quad Var\,X = Ex^2 - (Ex)^2 = (1 + 2p) - 1 = 2p$$

$\therefore$ $Var\ X$ will be maximum when $p = 1/2$ and then $Var\ X = 1$.

Example 5·16. *An urn contains N tickets numbered 1, 2,....., N from which n tickets are drawn successively without replacement. Find the mean and variance of sum of the numbers on the tickets drawn.*

Define random variables $X_1, X_2, \ldots\ldots, X_n$ on the event space of n drawings by x_i the number on the ith ticket, $i = 1, 2, \ldots, n$

$$\therefore \quad S_n = X_1 + X_2 + \ldots\ldots + X_n$$

Now $X_1, X_2, \ldots, X_n$ all have the same distribution.

Now X_1 takes values 1, 2, ..., N each with probability $1/N$ so that

$$N(X_1) = \sum_{i=1}^{N} i/N = (N + 1)/2, \quad Var\,X_1 = EX_1^{\,2} - (EX_1)^2 = \Sigma i^2/N - (\Sigma i/N)^2$$

$$= \frac{(N + 1)\,(2N + 1)}{6} - \frac{(N + 1)^2}{4}$$

$$= (N^2 - 1)/12$$

$$\therefore \; ES_n = n\,(N + 1)/2$$

Now $P(X_1 = i, X_2 = j) = 1/N\;(N - 1)$ if $i \neq j$ and if $i = j$.

$$\text{Hence } E(X_1X_2) = \frac{1}{N(N-1)} \sum_{j=1}^{N} \sum_{i=1}^{N} ij - \sum_{i=1}^{N} i^2 = (N+1)\,(3N+2)/12$$

$Cov\;(x_1, x_2) = -\;(N + 1)/12 \;\therefore\; Cov\,(X_i, X_j) = -\,(N + 1)/12, i \neq j$

$$\therefore \; Var\,S_n = \frac{n\,(N^2 - 1)}{12}\left(1 - \frac{n-1}{N-1}\right).$$

Ex. 1. Show that

$$Cov\left(\sum_{i=1}^{n} a_i x_i, \sum_{j=1}^{n} b_j x_i\right) = \sum_{j=1}^{n} \sum_{i=1}^{n} a_i b_j\; Cov\,(x_i, x_j).$$

Ex. 2. Prove that $\mu'_{2a}\,\mu'_{2b} \geq (\mu'_{a+b})^2$ and $\mu_{2a}\,\mu_{2b} \geq (\mu_{a+b})^2$

Deduce (*i*) $\mu'_2 \geq \mu'^2_1$ (*ii*) $\beta_2 \geq 1$.

Example 5·17. *Let X_1, X_2,......., X_n be a sequence of mutually independent random variables with common distribution. Suppose X_k assumes only positive integral values and $E(X_k) = a > o$ exists, $k = 1, 2,......., n$. Let $S_n = X_1 + X_2 + ... + X_n$.*

(*i*) *Show that* $E\left(\dfrac{S_m}{S_n}\right) = \dfrac{m}{n}$ *for* $1 \leq m \leq n$

(*ii*) *Show that* $E\,(S_n^{-1})$ *exists and*

$$E\left(\frac{S_m}{S_n}\right) = 1 + (m - n)a\;\; E\,(S_n^{-1})\; for\; 1 \leq n \leq m$$

(*iii*) *Verify and use the inequality* $x + x^{-1} \geq 2, x > 0$ *to show that*

$$E\left(\frac{S_m}{S_n}\right) \geq \frac{m}{n}\; for\; m,\, n \geq 1$$

(*M.A. Stat., Delhi, 1976*)

(*i*) Since X_k's $k = 1, 2,..., n$ are independent *r.v.*, it follows that (X_k/S_n) are also independent where $S_n = X_1 + X_2 + ... X_n$. Also

$$E\left(\frac{X_1}{S_n}\right) = E\left(\frac{X_2}{S_n}\right) = ... = E\left(\frac{X_k}{S_n}\right) ... = E\left(\frac{X_n}{S_n}\right) = l, \text{ say} \quad ...(1)$$

$$\therefore\; E\left(\frac{X_1 + X_2 ... + X_n}{S_n}\right) = E\left(\frac{X_1}{S_n}\right) + ... + E\left(\frac{X_n}{S_n}\right) = nl$$

$$\Rightarrow \qquad E\left(\frac{X_k}{S_n}\right) = l = \frac{1}{n}\ E\left(\frac{X_1 + ... + X_n}{S_n}\right)$$

$$= \frac{1}{n}\ E\left(\frac{S_n}{S_n}\right) = \frac{1}{n}\cdot E(1) = \frac{1}{n}.$$

Hence
$$E\left(\frac{S_m}{S_n}\right) = E\left(\frac{X_1 + ... + X_m}{S_n}\right)$$

$$= E\left(\frac{X_1}{S_n}\right) + E\left(\frac{X_2}{S_n}\right) + ... + E\left(\frac{X_m}{S_n}\right)$$

$$\therefore \qquad E\left(\frac{S_m}{S_n}\right) = \frac{m}{n}.$$

(*ii*) Suppose X_k assumes only positive integral values *i.e.* $1 \le X_k \le \infty$

$$\therefore\ S_n = X_1 + X_2 + ... + X_n = \sum_{k=1}^{n} X_k \Rightarrow n \le S_n < \infty \Rightarrow 0 < S_n^{-1} \le \frac{1}{n}$$

$\therefore\ 0 < E\left(S_n^{-1}\right) \le \frac{1}{n}$ and hence $E\left(S_n^{-1}\right)$ exists.

Now
$$E\left(\frac{S_m}{S_n}\right) = E\left(\frac{S_n + X_{n+1} + ... + X_m}{S_n}\right),\ 1 \le n \le m$$

$$= E\left(1 + \frac{X_{n+1}}{S_n} + ... + \frac{X_m}{S_n}\right)$$

$$= E(1) + E\left(\frac{X_{n+1}}{S_n}\right) + ... + E\left(\frac{X_m}{S_n}\right)$$

$$= 1 + (m-n)\ E\left(\frac{X_i}{S_n}\right) \text{ for } n+1 \le i \le m$$

$$= 1 + (m-n)\ E(X_i) E\left(S_n^{-1}\right)$$

since X_i and S_n are mutually independent

$$= 1 + (m-n)\ a\ E(S_n^{-1}),\ 1 \le n \le m$$

(*iii*) The inequality $x + \frac{1}{x} \ge 2 \Rightarrow x^2 - 2x + 1 \ge 0 \Rightarrow (x-1)^2 \ge 0$ and is easily verified for $x > 0$.

From (*ii*) $1 + (m-n)a\ E\left(S_n^{-1}\right) \ge \frac{m}{n} \quad \Rightarrow E\left(S_n^{-1}\right) \ge \frac{1}{an}$

Putting $x = \frac{S_n}{an} > 0$ in the given inequality, we have

$$E\left(\frac{S_n}{an}\right) + E\left(an\, S_n^{-1}\right) \geq 2$$

$$\Rightarrow \quad \frac{1}{an} E(S_n) + an\, E(S_n^{-1}) \geq 2$$

$$\Rightarrow \quad \frac{1}{an}. an + an\, E(S_n^{-1}) \geq 2 \quad \because \; E(S_n) = n\, E(X_k) = na$$

$$\Rightarrow \quad E(S_n^{-1}) \geq \frac{1}{an}$$

Hence $\quad E\left(\frac{S_m}{S_n}\right) = mE\left(\frac{1}{S_n}\right) = m\; E(S_n^{-1})$ by (1)

$$\therefore \quad E\left(\frac{S_m}{S_n}\right) \geq m$$

5·3·3. Theorem. Cauchy- Schwartz Inequality.

If X, Y are random variables taking real values, then

$$[E(XY)]^2 \leq E(X^2) \,.\, E(Y^2)$$

Proof. Consider the expression $(X + tY)^2$ which is a function of real variable t. Since it is always non-negative for all real values of X, Y and t, it follows that $\quad E(X + tY)^2 \geq 0 \quad \forall \; t$

$$\Rightarrow \; E(X^2) + 2t\, E(XY) + t^2\, E(Y^2) \geq 0 \text{ for all } t$$

i.e. $\quad \varphi(t) = At^2 + Bt + C \;\geq 0$ for all t

where $\quad A = E(Y^2),\; B = 2\, E(XY),\; C = E(X^2)$.

Now $\varphi(t) \geq 0$ implies $B^2 - 4AC \leq 0$

$$\therefore \quad 4\,[E(XY)]^2 - 4\, E(X^2)\, E(Y^2) \leq 0 \;\Rightarrow\; [E(XY)]^2 \leq E(X^2)\, E(Y^2)$$

Cor. 1. The equality sign holds good if and only if

$$E(X + tY)^2 = 0 \;\Rightarrow\; P[(X + tY)^2 = 0] = 1$$

[See Ex 2. (b), page 68]

$$\Rightarrow \quad P\{X + tY = 0\} = 1 \;\Rightarrow\; P[Y = AX] = 1,\; A = -1/t.$$

$$\therefore \; x_i/y_i = \text{ constant } \forall i,\; i = 1, 2, ..., n$$

i.e., $\quad \frac{x_1}{y_i} = \frac{x_2}{y_2} = \; = \frac{x_n}{y_n} = k$, say.

Cor. 2. Replace X by $|X - E(X)|$ and Y by 1, then

$$E\,|X - EX|^2 \leq E\,(X - EX)^2 \;\therefore\; M.D. \leq S.D.$$

Example 5·18. *Let X and Y be two random variables having finite means. Prove or disprove :*

(*a*) $E[Min(X, Y)] \leq Min[EX, EY]$ (*b*) $E[Max(X, Y)] \geq Max[EX, EY]$

(*c*) $E[Min(X, Y) + Max(X, Y)] = EX + EY$

(B.A. Stat. Hons. Spl. Course, Delhi, 1989)

$$\left.\begin{aligned} \text{We know that} \quad & Min\,(X, Y) = \frac{1}{2}\,[(X+Y) - |X-Y|] \\ \text{and} \quad & Max\,(X, Y) = \frac{1}{2}\,[(X+Y) + |X-Y|] \end{aligned}\right\} \quad ...(i)$$

(*a*) $E\,[Min\,(X, Y)] = \frac{1}{2}\,E(X+Y) - E\,|X-Y|$

and Max $(X, Y) = \frac{1}{2}\,[(X+Y) + |(X-Y)|]$

Now $|E\,(X-Y)| \le E\,|X-Y| \;\Rightarrow\; -|E(X-Y)| \ge -E\,|X-Y|$

$$\Rightarrow\; -E\,|X-Y| \le -|E(X-Y)| = -|E(X) - E(Y)| \quad ...(ii)$$

From (*ii*), we get

$$E\,[\,Min\,(X, Y)\,] = \frac{1}{2}\,[E\,(X+Y) - E\,|X-Y|]$$

$$\le \frac{1}{2}\,[[EX + EY] - |EX - EY|], \text{ using } (ii)$$

$\Rightarrow E\,[\,Min\,(X, Y)\,] \ge Min\,[EX, EY]$, thus the given result is proved.

(*b*) $E\,[\,Max\,(X, Y)\,] = \frac{1}{2}\,[\,E\,(X+Y) + E\,|X-Y|\,]$

$$\ge \frac{1}{2}\,[E(X+Y) + |E\;(X-Y)|\,]$$

$$= \frac{1}{2}\,EX + \frac{1}{2}\,EY + \,|EX - EY| \quad \because\; |E(X-Y)| \le E\,|X-Y|$$

$$= Max\,[EX, EY]$$

$\therefore\; E\,|\,Max\,(X, Y)\,| \ge Max\,[E(X), E(Y)]$ thus the given result is proved.

(*c*) From (*i*), $Min\;(X, Y) + Max\,(X, Y) = X + Y$

$\Rightarrow E\,[Min\,(X, Y) + Max\,(X, Y)] = E(X+Y) = E(X) + E(Y).$

The given result is established.

5·4. Conditional Expectation. Conditional Variance.

Let X be a random variable which takes the values x_i with probability $p_i(Y)$ subject to the occurrence of the event Y, that is,

$$P_i\,(Y = y_j) = P(X = x_i \mid Y = y_j),\; i = 1, 2......, n$$

$$= P\{(X = x_i) \cap (Y = y_j)\}/P(Y = y_j)$$

Then the *conditional expectation* of X, given Y, is defined as

$$E(X \mid Y = y_j) = \frac{\Sigma\, x_i\;\; P(X = x_i\,;\;\, Y = y_j)}{P(Y = y_j)}$$

and the *conditional variance of* X, given Y, is defined as

$$Var\,(X \mid Y = y_j) = E[\,\{X - E\,\{(X \mid Y = y_j\}^2 \mid Y = y_j\,]$$

The *conditional variance* of Y given $X = x_i$ is defined as

$$Var\,(Y \mid X = x \mid = E[\,\{Y - E(Y \mid X = x_i\}^2 \mid X = x_i\,]$$

The *Conditional Covariance* may be defined similarly.

Theorem I. *The expected value of X is equal to the expectation of the conditional expectation of X, given Y. Symbolically,*

$$E(X) = E[E(X \mid Y)]$$

Proof. $E[E(X \mid Y)] = E[\sum_j E(X \mid Y = y_j)\,]$

$$= \sum_j \left[\Sigma_i\, x_i \frac{P\,(X = x_i \cap Y = y_j)}{P\,(Y = y_j)} \right] P\,(Y = y_j)$$

$$= \sum_i \sum_j x_i \frac{P(X = x_i \cap\; Y = y_j)}{p_j} = \sum_i x_i\; P(X = x_i) = E(X\,)$$

Theorem II. *The variance of X can be regarded as consisting of two parts, the expectation of the conditional variance and the variance of the conditional expectation.* Symbolically,

$$Var\,(X) = E\,[Var\,(X \mid Y)] + Var\,[E(X \mid Y\,\}]$$

Proof. $Var\,(X) = E[X - E(X\,)]^2$

$$= E[\overline{X - E(X \mid Y\,)} \;+\; \overline{E(X \mid Y\,) - E(X\,)}]^2$$

$$= E[X - E(X \mid Y)]^2 + E[E(X \mid Y\,) - E(X\,)]^2 \quad ...(5{\cdot}8)$$

the middle term being zero since it is equal to

$$E[X\,E(X \mid Y) - XE(X\,) + E(X).E(X \mid Y\,) - [E(X \mid Y\,)]^2 \quad ...(5{\cdot}9)$$

and $\quad E[XE(X \mid Y)] = E[E(X \mid Y).\;E(X \mid Y)]$

and similarly, $E[XE(X)] = E[E(X \mid Y).\;E(X)]$ by theorem I.

so that the terms in (5·9) cancel. Hence the result.

Theorem III. $Cov\;(X,\,Y) = E\;Cov\;(X,\,Y \mid Z\,) + Cov\;[E(X \mid Z\,)]\;\;E(Y \mid Z\,)$.

Proof. $Cov\,(X,\,Y) = E(X\,Y\,) - E(X)E(Y\,)$, X and Y measues from mean.

$$= EE(XY \mid Z\,) - EE(X \mid Z\,)EE\,(Y \mid Z\,). \text{ (by theorem I)}$$

$$= E\{Cov\,(X,\,Y \mid Z\,) + E(X \mid Z\,)E(Y \mid Z\,)\}$$
$$- EE(X \mid Z)\;EE(Y \mid Z\,)$$

$$= E\;Cov(X{,}Y \mid Z) + EE(X \mid Z)E(Y \mid Z)$$
$$- EE(X \mid Z)\;EE\,(Y \mid Z)$$

$$= E\;Cov\,(X,\,Y \mid Z) + Cov\,[E(X \mid Z)]E(Y \mid Z).$$

5·5. Moment Generating Function. The moment generating function (*m.g.f.*) *of the discrete probability distribution of a random* *discrete variable X about the point a is defined by*

$$M_{X=a}(t) = E[e^{t(x-a)}] = \sum_i e^{t(x_i - a)} p_i \qquad \ldots(5\cdot10)$$

where t is a real parameter.

The m.g.f. is a function of the real parameter t only. The subscript 'a' shows the point about which the m.g.f. is found out.

Expanding the exponential in (5·10), we get

$$M_{X=a}(t) = \left[\sum_i p_i + t \sum_i p_i(x_i - a) + \frac{t^2}{\lfloor 2} \sum_i p_i (x_i - a)^2 + \ldots\ldots \right.$$

$$\left. \ldots + \frac{t^r}{\lfloor r} \sum_i p_i(x_i - a)^r + \ldots \right]$$

$$= 1 + t\mu'_1 + \frac{t^2}{2\,!}\mu'_2 + \ldots\ldots + \frac{t^r}{r\,!}\mu'_r + \ldots \qquad \ldots(5\cdot11)$$

where μ'_r is the moment of order r about the point a.

Thus we see that the function $M_{X=a}(t)$ defined by (5·10) generates moments and that is why it is called the moment generating function.

From (5·11), we find $\mu'_r =$ coefficient of t^r/r in the expansion of $M_{X=a}(t)$ in ascending powers of t.

This gives a method of finding the moments. However, it may be found more convenient to evaluate the moments by the following formula obtained by differentiating the relation (5·11) r times with respect to t and then putting t=0. Thus

$$\mu'_r = \left| \frac{d^r}{dt^r} M_{X=a}(t) \right|_{t=0} \qquad \ldots(5\cdot12)$$

The equation (5·10) can be written as $M_{X=a}(t) = e^{-at} \;\; \Sigma e^{tx_i} p_i$

or
$$M_a(t) = e^{-at}\; M_0(t) \qquad \ldots(5\cdot13)$$

Thus *the m.g.f. about the point a is equal to* e^{-at} *times the m.g.f. about the origin*

Advantages and disadvantages of the m.g.f.

1. For many probability functions, the sum $\sum_i e^{tx_i} p_i$ may not exist for all real values of t. Thus if $f(x) = 6/\pi^2 x^2$, $x = 1, 2, 3.....$, then the *m.g.f.* does not exist. M.G.F. always exists for $t = 0$ since $M_{x=0}(0) = 1$.

2. Let us give an example to show that the *m.g.f.* of a discrete random variable may exist yet no moments even the mean does not exist.

Consider the discrete *r.v.X.* with *p.d.f.*

$$f(x) = \frac{1}{x(x+1)},\ x = 1, 2, \ldots = 0, \quad \text{otherwise}$$

$$\therefore\ E(X) = \sum_{1}^{\infty} x . \frac{1}{x(x+1)} = \sum_{1}^{\infty} \frac{1}{x+1} \text{ which is a divergent series so that}$$

the mean does not exist, consequently no moment of x exists. However, the *m.g.f.* is given by

$$m.g.f. \text{ about origin} = \Sigma\, e^{tx} f(x) = \sum_{x=1}^{\infty} e^{tx} \cdot \frac{1}{x(x+1)} = \sum_{x=1}^{\infty} \frac{Z^x}{x(x+1)},\ Z = e^t$$

$$= \frac{Z}{1 \cdot 2} + \frac{Z^2}{2 \cdot 3} + \frac{Z^3}{3 \cdot 4} + \ldots$$

$$= Z\left(1 - \frac{1}{2}\right) + Z^2\left(\frac{1}{2} - \frac{1}{3}\right) + \ldots$$

$$= \left(Z + \frac{Z^2}{2} + \ldots\right) - \frac{1}{Z}\left(\frac{Z^2}{2} + \frac{Z^3}{3} + \ldots\right)$$

$$= -\log(1 - Z) - \frac{1}{Z}[-\log(1 - Z) - Z],\ |Z| < 1$$

$$= 1 + \left(\frac{1}{Z} - 1\right)\log(1 - Z)$$

$$= 1 + (e^{-t} - 1)\log(1 - e^t),\ t < 0 \text{ as } |Z| < 1$$

$$\Rightarrow\ e^t < 1 \quad \Rightarrow \quad t < 0$$

Clearly for $t > 0$, $M_X(t)$ does not exist. For $t = 0$, $M_X(t) = M_0 = 1$.

3. Let us give an example to show that the *m.g.f.* of a random variable does not exist, except at one point, though all the moments exist.

Let X be a *r.v.* with *p.d.f.*

$$P(X = \pm 2^x) = \frac{e^{-1}}{2(x!)},\ x = 0, 1, 2, \ldots; = 0, \quad \text{otherwise}$$

$$M_X(t) = \sum_{0}^{\infty} (e^{t.2^x} + e^{-t.2^x}) \cdot \frac{1}{(2e)(x!)} = e^{-1} \sum_{0}^{\infty} \frac{\cosh(2^x t)}{x!}$$

It can be proved by D'Almbert ratio test that the series on the right side converges for $t = 0$. For $t > 0$ and $t < 0$, the series diverges and therefore does not generate moments.

$$\text{Again } E(x^r) = \sum_{x=0}^{\infty} (2^x)^r P(x = 2^x) = \frac{e^{-1}}{2} \sum_{x=0}^{\infty} \frac{(2^r)^x}{x!} = \frac{e^{-1}}{2} \cdot \exp(2^r).$$

Hence all the moments of X exist.

Note. In chapter 13, we shall see that the moments of Student's

t-distribution and F-distribution exist but the *m.g.f.* of the distributions does not exist.

5·5·1. Effect of change of Origin and Scale on M.G.F.

Let us introduce a new variate U related to X by

$$U = \frac{X-a}{h}, \quad \text{so that } X = a + hU$$

By definition, $M_0(t) = \sum_i e^{tXi} \; p_i = \sum_i e^{t(a+hUi)} \; p_i = e^{at} \; \sum_i e^{thUi} \; p_i$

Writing t/h for t on both sides, we get $\underset{x=0}{M}\left(\frac{t}{h}\right) = e^{at/h} M\,(u)$

$$\therefore \quad M.G.F. \text{ for } \frac{X-a}{h} = e^{-at/h} M_{X=0}\left(\frac{t}{h}\right). \qquad ...(5·14)$$

Cor. If $a = 0$, then $M.G.F.$ for $X/h = M_{x/h}(t) = M_x(t/h)$.

5·5·2. A Property of Moment Generating Function. *The moment generating function of the sum of two independent chance variables is the product of their respective moment generating functions.*

Symbolically, $\quad M_{x+y}(t) = M_x(t) \times M_y(t)$,

provided that x and y are independent random variables.

Proof. Let X and Y be two independent chance variables so that $X+Y$ is also a chance variable.

The *m.g.f.* of the sum $X+Y$ with respect to origin is

$$M_{x+y}(t) = E[e^{t(X+Y)}] = E(e^{tX} \cdot e^{tY}) = E(e^{tX}) \cdot E(e^{tY}), \qquad ...(5·15)$$

since X and Y are independent variables and so are e^{tX} and e^{tY}. $M_{x+y}(t) = M_x(t) \cdot M_y(t)$ Hence the theorem.

Exercise. Generalize the theorem and give its proof.

5·5·3. Factorial Moment Generating Function. The factorial moment generating function $w(t)$ of the discrete probability distribution of X about the origin is defined as

$$w(t) = E[(1+t)^X] \qquad ...(5·16)$$

Since $\quad (1+t)^X = \sum_{r=0}^{\infty} \frac{X^{(r)} t^r}{r\,!}, \; X^{(r)} = X(X-1)(X-2) \ldots (X-\overline{r-1})$

$$\therefore \qquad w(t) = 1 + t\,\mu'_{(1)} + \frac{t^2}{2\,!}\mu'_{(2)} + \cdots + \frac{t^r}{r\,!}\mu'_{(r)} + \cdots$$

we get $\qquad \mu_{(r)}' = \text{coefficient of } \frac{t^r}{r\,!}$ in the expansion of $w(t)$ in ascending power of t.

5·6. Cumulants. If the logarithm of the *m.g.f.* of a distribution can be expanded as a convergent series in powers of t *viz.*,

$$K_x(t) = \log_e M_x(t) \equiv k_1 t + k_2 \frac{t^2}{2!} + \ldots + \frac{k_r t^r}{r!} + \ldots \qquad \ldots(5{\cdot}17)$$

then the coefficients $k_1, k_2, \ldots$ are called the first cumulant, second cumulant,....of the distribution and $K(t)$ is called the *cumulative function.*

Differentiating r times both sides of the relation (5·17) with respect to t and putting $t = 0$, we have

$$k_r = \left[\frac{d^r}{dt^r} \log_e M_x(t) \right]_{t=0} = \left[\frac{d^r}{dt^r} K_x(t) \right]_{t=0} \qquad \ldots(5{\cdot}18)$$

5·6·1. Effect of Change of Origin and Scale on Cumulants.

Let $\quad U = \dfrac{X-a}{h}$, so that $x = a + hu$.

The cumulative function about $X = 0$ is given by

$$K_x(t) = \log_e M_x(t) = \log_e [\Sigma e^{tx_i} p_i] = \log_e [\Sigma e^{t(a+hu_i)} p_i]$$

$$= \log_e [e^{at} \Sigma e^{thu_i} p_i] = at + \log_e M_x(th)$$

or $$K_x(t) = at + k'_1(th) + k'_2 \frac{(th)^2}{2!} + \ldots + k'_r \frac{(th)^r}{r!} + \ldots$$

$$= [a + (k'_1 h)]\, t + (k'_2 h^2) \frac{t^2}{2!} + \ldots + (k'_r h^r) \frac{t^r}{r!} + \ldots$$

$\therefore \quad k_1 = a + hk'_1\, ;\ k_r = k'_r h^r \cdot r = 2, 3, \ldots;\ k'_r = r$th cumulant of u.

Thus *except the first cumulant, the r-th cumulant of the distribution of X is h^r times the r-th cumulant of the distribution U.* In other words, *cumulants are not invariant for the change of scale.* Also we notice that *all the cumulants except the first are invariant for the change of origin.*

Note. The cumulants are also called semi-invariants due to two very remarkable properties. In the first place, all cumulants except the first are independent of origin of reference. Secondly, if the scale of measurement of the variable is altered by multiplying all values by a constant h, the rth cumulant is multiplied by h^r.

5·6·2. Additive Property of Cumulants. *The r-th cumulant of the sum of independent stochastic variables is the sum of the r-th cumulants of the independent chance variables.*

Symbolically $\quad k(x_1 + x_2) = k(x_1) + k(x_2)$

Taking logarithms of both sides of (5·15), we get

$$\log_e(m.g.f. \text{ for } X + Y) = \log_e(m.f.g. \text{ for } X) + \log_e(m.g.f. \text{ for } Y).$$

Equating coefficients of like powers of t, we get the required result.

Ex. Generalize the theorem and give its proof.

5·7. Relations between Moments and Cumulants.

5·7·1. Cumulants in terms of moments. From (5·17) we have the identity

$$k_1 t + k_2 \frac{t^2}{2\,!} + k_3 \frac{t^3}{3\,!} + \ldots \equiv \log_e \left(1 + t\mu'_1 + \frac{t^2}{2\,!}\mu'_2 + \ldots\right) \qquad \ldots(5\cdot19)$$

$$= \left(t\mu_1' + \frac{t^2}{2\,!}\mu_2' + \frac{t^3}{3\,!}\mu_3' + \frac{t^4}{4\,!}\frac{\mu_4'}{4} + \ldots\right) - \frac{1}{2}\left(t\,\mu_1' + \frac{t^2}{2\,!}\mu_2' + \frac{t^3}{3\,!}\mu_3' + \ldots\right)$$

On comparing the coefficients of like powers of t, we find

$$\left.\begin{aligned} k_1 &= \mu_1',\; k_2 = \mu_2' - \mu_1'^2 = \mu_2, \\ k_3 &= \mu_3' - 3\mu_2'\mu_1' + 2\mu_1'^3 = \mu_3, \\ k_4 &= \mu_4' - 4\mu_3'\mu_1' + 3\mu_2'^2 + 12\mu_2'\mu_1'^2 - 6\mu_1'^4 = \mu_4 - 3\mu_2^2 \end{aligned}\right\} \qquad \ldots(5\cdot20)$$

These results can be *directly* obtained as follows :

From (5·13), $M_{\mu_1'}(t) = e^{-\mu_1' t} M_0(t) \;\therefore\; \log_e M_{\mu_1'}(t) = -\mu_1'(t) + \log_e M_0(t)$

or $$\log_e \left(1 + \mu_2 \frac{t^2}{2\,!} + \mu_3 \frac{t^3}{3\,!} + \ldots\ldots\right) = -\mu_1' t + k_1 t + k_2 \frac{t^2}{2\,!} + \ldots\ldots \qquad \ldots(5\cdot21)$$

Equating the coefficients of like powers of t after expanding the left hand member, we obtain the required results.

Example 5·19. Show that $\mu_r' = \sum_{j=1}^{r} {}^{r-1}c_{j-1}\mu'_{r-j}\, k_j$.

(*B.Sc. Shivaji, 1971 ; B.Sc. Hons. Gauhati*)

Differentiating equation (5·19) with respect to t and cross-multiplying, we get

$$\mu_1' + t\,\mu_2' + \ldots + \frac{t^{r-1}}{(r-1)\,!}\mu_r' + \ldots = \left[1 + t\mu_1' + \ldots + \frac{t^{r-j}}{(r-j)\,!}\mu'_{r-j} + \ldots\right]$$

$$\times \left[k_1 + k_2 t + \ldots + k_j \frac{t^{j-1}}{(j-1)\,!} + \ldots\right]$$

Comparing the coefficient of t^{r-1} on both sides we get the required result.

5·7·2. Moments in terms of Cumulants. From (5·20),

we get $\quad \mu_1' = k_1\,;\; \mu_2 = k_2\,;\; \mu_3 = k_3,$

$$\mu_4 = k_4 + 3\,\mu_2^2 = k_4 + 3\,k_2^2$$

$$\beta_1 = \mu_3^2/\mu_2^3 = k_3^2/k_2^3\,;\quad \beta_2 = \mu_4/\mu_2^2 = (k_4 + 3\,k_2^2)/k_2^2 = k_4/k_2^2 + 3$$

Aliter. $1+\mu_1{}'t+\mu_2{}'\frac{t^2}{2!}+\ldots+\mu_r{}'\frac{t^r}{r!}+\ldots\ldots$

$$=\exp\left[k_1t+\frac{k_2t^2}{2!}+\ldots\frac{k_rt^r}{r!}+\ldots\right], \text{ on using (5·19)}$$

$$=\exp[k_1t]\exp\left[\frac{k_2t^2}{2!}\right]\ldots\ldots\exp\left[\frac{k_rt^r}{r!}\right]\ldots\ldots$$

$$=\left[1+k_1t+k_1^2\frac{t^2}{2!}+\ldots\right]\left[1+k_2\frac{t^2}{2!}+\ldots\right]\left[1+k_r\frac{t^r}{r!}+\ldots\right]\ldots$$

Equating the coefficient of t^r on both sides, we have

$$\mu_r{}'=\sum_{m=0}^{r}\sum\frac{k_{p1}{}^{\pi}{}_1}{(p_1\,!)^{\pi 1}{}_{\pi 1}\,!}\cdot\frac{k_{p2}{}^{\pi}{}_2}{(p_2\,!)^{\pi 2}{}_{\pi 2}\,!}\ldots\ldots\frac{k_{pm}{}^{\pi m}}{(p_m\,!)^{\pi m}{}_{\pi m}\,!}\,r\,!$$

where the second summation extends over all over-negative values of π's such that $p_1\pi_1+p_2\pi_2+\ldots\ldots+p_m\pi_m=r.$

Example. Using the result of G.S. Mahajani and Ram Behari that the

$$\text{Exp. } [k_1t+k_2t^2+\ldots\ldots]=1+D_1\,t+\frac{D_2t^2}{2!}+\ldots\ldots+\frac{D_rt^r}{r!}+\ldots\ldots$$

where
$$D_r=\begin{vmatrix} k_1 & -1 & 0 & 0 & \ldots & 0\\ \frac{2k_2}{2!} & k_1 & -2 & 0 & \ldots & 0\\ \frac{3k_3}{3!} & \frac{2k_2}{2!} & k_1 & -3 & \ldots & 0\\ \ldots & \ldots & \ldots & \ldots & \ldots & \ldots\\ \frac{rk_r}{r!} & \frac{(r-1)\,k_{r-1}}{(r-1)\,!} & \ldots & \ldots & \ldots & k_1 \end{vmatrix} \quad \ldots(5·22)$$

prove that $\mu_1{}'=k_1,\ \mu_2{}'=k_1{}^2+k_2,\ \mu_3{}'=k_1{}^3+3k_1k_2+k_3,$

$$\mu_4{}'=k_1{}^4+6k_1{}^3k_2+4k_1k_3+3k_2{}^2+k_4\,. \quad \ldots(5·23)$$

Example 5·20. *If X is the length of run of either successes or failures, starting with the first trial, find the m.g.f. of X and hence deduce the value of mean of X. Also find Var X.*

$P(X=r)=P[(SS..SF)+(FF..FS)]$, S occurs r times while in second F occurs r times.

$$=P[SS..SF]+P[FF..FS]$$

$$=\{P(S)\}^r P(F)+\{P(F)\}^r P(S), \text{ since trials are independent.}$$

$$=p^rq+q^rp,\ r=1,2,3,$$

p = probability of a success S, q = probability of a failure $F, p+q=1$.

$$M_x(t) = \sum_{r=1}^{\infty} e^{tr} P(X=r)$$

$$= \Sigma e^{tr}(p^r q + q^r p) = q \sum_{1}^{\infty} (pe^t)^r + p \sum_{1}^{\infty} (qe^t)^r$$

$$= q \cdot pe^t \sum_{1}^{\infty} (pe^t)^{r-1} + p \cdot q e^t \sum_{1}^{\infty} (qe^t)^{r-1}$$

$$= p\, q\, e^t [(1 + pe^t + (pe^t)^2 + \ldots) + (1 + qe^t + (qe^t)^2 + \ldots)]$$

$$= p\, q\, e^t [(1 - pe^t)^{-1} + (1 - qe^t)^{-1}]$$

$$= p\, q\, e^t \left[\frac{1}{1-pe^t} + \frac{1}{1-qe^t}\right] = q\left[-1 + \frac{1}{1-pe^t}\right] + p\left[-1 + \frac{1}{1-qe^t}\right] \quad \ldots(1)$$

Differentiating w.r.t. t both sides of (1), we obtain

$$M_x'(t) = \frac{qpe^t}{(1-pe^t)^2} + \frac{pqe^t}{(1-qe^t)^2},$$

$$M_x''(t) = pq \frac{e^t(1+pe^t)}{(1-pe^t)^3} + \frac{pqe^t(1+qe^t)}{(1-qe^t)^2}$$

$$\therefore \quad \mu_1' = M_x'(0) = \frac{qp}{(1-p)^2} + \frac{pq}{(1-q)^2} = \frac{qp}{q^2} + \frac{pq}{p^2} = \frac{p}{q} + \frac{q}{p}.$$

$$\mu_2' = M_x''(0) = \frac{pq(1+p)}{(1-p)^3} + \frac{pq(1+q)}{(1-q)^3} = \frac{p(1+p)}{q^2} + \frac{q(1+q)}{p^2}$$

$$= \frac{p^2}{q^2} + \frac{q^2}{p^2} + \frac{p}{q^2} + \frac{q}{p^2}$$

$$\therefore \quad \mu_2 = \mu_2' - \mu_1'^2 = \frac{p^2}{q^2} + \frac{q^2}{p^2} + \frac{p}{q^2} + \frac{q}{p^2} - \left(\frac{p}{q} + \frac{q}{p}\right)^2 = \frac{p}{q^2} + \frac{q}{p^2} - 2.$$

5·8. Probability Generating Function. For a random variable X which takes integral values 0, 1, 2, 3,.......only, we define the probability generating function $P(s)$ by

$$P(s) = p_0 + p_1 s + p_2 s^2 + \ldots = \sum_{n=0}^{\infty} p_n s^n = E(s^X)$$

The coefficient of s^n in the expansion of P(s) in powers of s gives $P(X=n)$. This explains why $P(s)$ is called the probability generating function (p.g.f.). It is easily seen that $\left(\frac{\partial P}{\partial s}\right)_{s=1} = \sum_{n=0}^{\infty} np_n = \mu_1'$

$$\left(\frac{\partial^2 P}{\partial s^2}\right)_{s=1} = \sum_{n=0}^{\infty} n\,(n-1)\,p_n = \mu_2' - \mu_1'$$

so that $$\mu_2 = \mu_2' - \mu_1'2 = \left[\frac{\partial^2 P}{\partial s^2} + \frac{\partial P}{\partial s} - \left(\frac{\partial P}{\partial s}\right)^2\right]_{s=1}$$

For integral valued variates $P_x(e^t) = E(e^{tX}) = M_rG.F.$ for X.

Theorem. *The probability generating function of the sum of a number of independent random variables is equal to the product of their p.g.f.'s.*

Proof. We shall prove the result for two independent random variables though the proof is general.

Let X_1 and X_2 be two independent random variables. Then

$$P_{X_1+X_2}(s) = E(s^{X_1+X_2}) = E(s^{X_1})\,E(s^{X_2}),$$

since X_1 and X_2 are independent.

$$= P_{X_1}(s)\,P_{X_2}(s)$$

which proves the result.

In general, $$P_{X_1+X_2+\ldots+X_n}(s) = \prod_{i=1}^{n} P_{X_i}(s)$$

Cor. If the independent random variables $X_1, X_2, \ldots\ldots, X_s$ are all identically distributed and hence have the same *p.g.f.*, $P_x(s)$. Then

$$P_{X_1+X_2+\ldots+X_n}(s) = [P_X(s)]^n.$$

Example 5·21. *If X is the number scored with the throw of an unbiased die, find the probability generating function for X.*

Here $p_i = \frac{1}{6}$ for $i = 1, 2, \ldots\ldots, 6$

Hence the generating function $= \frac{1}{6}s + \frac{1}{6}s^2 + \ldots\ldots + \frac{1}{6}s^6 = \frac{1}{6}s(1-s^6)/(1-s)$.

Example 5·22. *Let X be a random variable with generating function p(s). Find generating function of (a) X + 1 (b) 2X. (c) Determine $P(X \le n)$.* **(M.A. Delhi 1969, 74)**

(*a*) We have $P_X(s) = \sum_{k=0}^{\infty} p_k s^k$

$\therefore$ Probability generating function of (X + 1)

$$= \sum_{k=0}^{\infty} p_k s^{k+1} = s \sum_{k=0}^{\infty} p_k s^k = sP(s).$$

(*b*) Probability generating function of $2X$

$$= \sum_{k=0}^{\infty} p_k s^{2k} = \sum_{k=0}^{\infty} p_k (s^2)^k = P(s^2).$$

(*c*) Let $P(X \le n) = q_n$ and $p(X = n) = p_n$

so that
$$q_n = P(X \le n) = P(X=0) + P(X=1) + \dots + P(X=n)$$
$$= p_0 + p_1 + \dots\dots + p_n$$

$$\therefore \quad Q(s) = \sum_{n=0}^{\infty} q_n s^n = \sum_{n=0}^{\infty} (p_0 + p_1 + \dots + p_n) s^n$$
$$= p_0 s^0 + (p_0 + p_1)\, s + (p_0 + p_1 + p_2) s^2 + \dots\dots$$
$$= p_0(1 + s + s^2 + \dots\dots) + p_1 s(1 + s + s^2 + \dots\dots)$$
$$+ p_2 s^2 (1 + s + s^2 + \dots\dots)$$
$$= p_0 \frac{1}{1-s} + p_1 \frac{s}{1-s} + p_2 \frac{s^2}{1-s} + \dots$$

$$Q(s) = \frac{1}{1-s} \sum_{n=0}^{\infty} p_n s^n = \frac{1}{1-s} \cdot P(s).$$

Ex. 1. Obtain the *p.g.f.* for the distributions

(*a*) $P(X = r) = {}^nC_r p^r (1-p)^{n-r}$, $r = 0, 1, 2, \dots\dots, n : 0 \le p \le 1$

(*b*) $P(X = r) = (1-p) p^{r-1}$, $r = 1, 2, \dots\dots, 0 \le p \le 1$

(*c*) $P(X = r) = \dfrac{e^{-m} m^r}{r\,!}$, $r = 0, 1, 2, \dots\dots,$ **(B.A. Hons. Stat. Delhi, 1984)**

(*d*) $P(X = r) = p(1-p)^r$, $r = 0, 1, 2, \dots\dots, 0 \le p \le 1$

(*e*) $P(x = r) = \dfrac{e^{-m} m^r}{(1 - e^{-m})\,(r\,!)}$, $r = 1, 2, \dots\dots$

Ans. (*d*) $P(s) = \dfrac{p}{1 - (1-p)s}$. (*e*) $P(s) = (e^{ms} - 1)/(e^m - 1)$

Ex. 2. If X and Y denote points on two perfect dice, find the *p.d.f.* and *m.g.f.* for X/Y. Can you find the *p.g.f.* for it ?

5·9. Moments of Bivariate Probability distributions. The mathematical expectation of a function of a two-dimensional random variable (X, Y) is given by $E[g(X, Y)] = \sum_i \sum_j g(x_i, y_j)\, P(X = x_i, Y = y_j)$

In particular, the (r, s) moments about origin of order r in X and s in Y and the *M.G.F.* about origin are defined by

$$\mu_{rs}' = E(X^r Y^s) = \sum_i \sum_j (x_i^r\, y_j^s) \cdot P(X = x_i, Y = y_j)$$

$$M.G.F. = E(e^{t_1 x + t_2 y}) = \sum_i \sum_j e^{t_1 x_i + t_2 y_j} P(X = x_i, Y = y_j)$$

Example 5·23. *Two ideal dice are thrown.* Let X_1 *be the score on the first die and* X_2 *the score on the second die. Let Y denote the maximum of* X_1 *and* X_2 *i.e. max* (X_1, X_2).

(*a*) *Write down the joint distribution of Y and* X_1.

(*b*) *Find* $E(Y)$, *Var* (Y) *and Cov* (Y, X_1).

(*M.Sc. Kerala 1977 ; M.Sc. Dibrugarh, 1969*)

Clearly X_1 and X_2 can each take the values 1 to 6 with probability 1/6.

$$P(Y = 1, X_1 = 1) = P(X_1 = 1, X_2 = 1) = P(X_1 = 1)P(X_2 = 1) = \frac{1}{6} \cdot \frac{1}{6} = \frac{1}{36}$$

$$P(Y = 2, X_1 = 1) = P(X_1 = 1, X_2 = 2) = P(X_1 = 1)P(X_2 = 2) = \frac{1}{6} \cdot \frac{1}{6} = \frac{1}{36}$$

$$P(X_1 = 2, Y = 1) = 0,\ P(X_1 = 2, Y = 2) = P(X_1 = 2, X_2 = 1) + P(X_1 = 2, X_2 = 2)$$

$$= \frac{1}{36} + \frac{1}{36} = \frac{2}{36}$$

and so on. We find the following joint probability table of X and Y.

$X_1 \downarrow$ \ $X_2 \rightarrow$	1	2	3	4	5	6	marginal total
1	1/36	1/36	1/36	1/36	1/36	1/36	6/36
2	0	2/36	1/36	1/36	1/36	1/36	6/36
3	0	0	3/36	1/36	1/36	1/36	6/36
4	0	0	0	4/36	1/36	1/36	6/36
5	0	0	0	0	5/36	1/36	6/36
6	0	0	0	0	0	6/36	6/36
marginal Total	1/36	3/36	5/36	7/36	9/36	11/36	1

$$E(Y) = 1 \cdot \frac{1}{36} + 2 \cdot \frac{3}{36} + 3 \cdot \frac{5}{36} + 4 \cdot \frac{7}{36} + 5 \cdot \frac{9}{36} + 6 \cdot \frac{11}{36}$$

$$= \frac{1}{36}(1 + 6 + 15 + 28 + 45 + 66) = \frac{161}{36} = 4{\cdot}47222$$

$$E(Y^2) = 1^2 \cdot \frac{1}{36} + 2^2 \cdot \frac{3}{36} + 3^2 \cdot \frac{5}{36} + 4^2 \cdot \frac{7}{36} + 5^2 \cdot \frac{9}{36} + 6^2 \cdot \frac{11}{36}$$

$$= \frac{791}{36} = 21{\cdot}97222$$

$$\text{Var } Y = E(Y^2) - [E(Y)]^2 = \frac{791}{36} - \left(\frac{161}{36}\right)^2 = \frac{2555}{1296} = 1{\cdot}97145$$

$$E(X_1) = \frac{6}{36}(1+2+3+4+5+6) = \frac{126}{36} = \frac{21}{6} = 3{\cdot}5$$

$$\text{Cov}(X_1, Y) = E(X_1 Y) - E(X_1) \cdot E(Y)$$

where $E(X_1 Y) = 1 \cdot \frac{1}{36} + 2 \cdot \frac{1}{36} + 3 \cdot \frac{1}{36} + 4 \cdot \frac{1}{36}$

$$+ 5 \cdot \frac{1}{36} + 6 \cdot \frac{1}{36} + 4 \cdot \frac{2}{36} + 6 \cdot \frac{1}{36} + 8 \cdot \frac{1}{36} + 10 \cdot \frac{1}{36} + 12 \cdot \frac{1}{36} + 9 \cdot \frac{3}{36} + 12 \cdot \frac{1}{36}$$

$$+ 15 \cdot \frac{1}{36} + 18 \cdot \frac{1}{36} + 16 \cdot \frac{4}{36} + 20 \cdot \frac{1}{36} + 24 \cdot \frac{1}{36} + 25 \cdot \frac{5}{36} + 30 \cdot \frac{1}{36} + 36 \cdot \frac{6}{36}$$

or $$E(X_1 Y) = \frac{616}{36} = 17{\cdot}11111$$

$$\text{Cov}(X_1, Y) = \frac{616}{36} - \frac{21}{6} \cdot \frac{161}{6} = -\frac{2765}{36} = -76{\cdot}80556$$

Ex. 1. Show that $\mu_{rs}' = \left(\frac{\partial \phi^{r+s}}{\partial t_1{}^r \partial t_2{}^s}\right)_{t_1 = t_2 = 0}$

Ex. 2. Show that $\sigma_x{}^2 = \left[\frac{\partial^2 \phi}{\partial t_1{}^2} - \left(\frac{\partial \phi}{\partial t_1}\right)^2\right]_{t_1 = t_2 = 0}$

$$\sigma_y{}^2 = \left[\frac{\partial^2 \phi}{\partial t_2{}^2} - \left(\frac{\partial \phi}{\partial t_2}\right)^2\right]_{t_1 = t_2 = 0}$$

$$\mu_{11} = \text{cov}(x, y) = \left[\frac{\partial^2 \phi}{\partial t_1 \, \partial t_2} - \left(\frac{\partial \phi}{\partial t_1}\right)\left(\frac{\partial \phi}{\partial t_2}\right)\right]_{t_1 = t_2 = 0}$$

Ex. 3. For integral-valued variables, if

$$P(u, v) = E(u^X v^Y) = \sum_{n=0}^{\infty} \sum_{m=0}^{\infty} p_{mn} u^m v^n,$$

prove that $$\bar{X} = \frac{\partial P}{\partial u}, \ \bar{Y} = \frac{\partial P}{\partial v}, \ \sigma_{X^2} = \frac{\partial^2 P}{\partial u^2} + \frac{\partial P}{\partial u} - \left(\frac{\partial P}{\partial u}\right)^2$$

where all derivatives are taken at $u = v = 1$.

5·10. Analogy between probability distributions and mechanical mass distributions.

There is a strict quantitative analogy between probability distribution of a discrete random variable and a mechanical system of a particle or total mass units rigidly attached to points in a straight line.

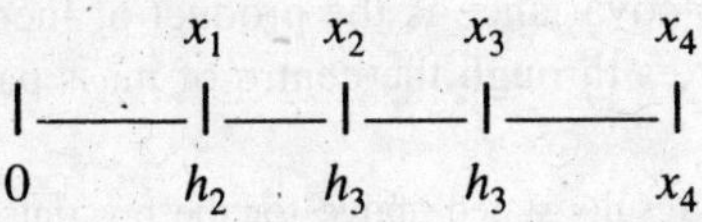

Fig. 5·2. Mechanical analogue of a p.d.

Let particles of masses $p_1, p_2, ...p_n$ of total mass units placed at points with abscissae $x_1, x_2, ...x_n$ on OX, then

$$\text{sum of masses} = p_1 + p_2 + ... + p_n = 1$$

sum of moments of masses about the origin

$$= p_1x_1 + p_2x_2 + p_3x_3 + ... + p_nx_n = \mu_1' = E(X)$$

Distance of centre of mass from the origin

$$= \frac{p_1x_1 + p_2x_2 + \cdots\cdots + p_nx_n}{p_1 + p_2 + \cdots\cdots + p_n} = \overline{X} = \mu_1'$$

Moment of inertia of the system of particles about OY

$$= p_1 x_1^2 + p_2x_2^2 + ... + p_nx_n^2 = \mu_2'$$

Moment of inertia of the system of particles about an axes parallel to OY through the centre of mass

$$= p_1(x_1 - \overline{x})^2 + p_2(x_2 - \overline{x})^2 + \cdots\cdots + p_n(x_n - \overline{x})^2 = \mu_2 = \sigma^2$$

Square of radius of gyration about OY

$$\frac{p_1x_1^2 + p_2x_2^2 + \cdots\cdots + p_nx_n^2}{p_1 + p_2 + \cdots\cdots + p_n} = \mu_2'$$

Radius of gyration about the parallel axes through the centre of mass

$$= \sqrt{\frac{p_1(x_1 - \overline{x})^2 + \cdots\cdots + p_n(x_n - \overline{x})^2}{p_1 + p_2 + \cdots\cdots + p_n}} = \sigma$$

Thus the mass of the *p.d.* determines the position of the centre of mass and its standard deviation gives the radius of gyration about the axis parallel to OY through the centre of mass. Also $\mu_2 = \mu_2' - \mu_1'^2$, so that the, M.I. about the axis through the C.M.

= M.I. about a parallel axis through the origin

– square of the distance between the parallel axis

This is easily recognized as the theorem of parallel axes.

It is similarly seen that for a bivariate probability distribution, the means $\overline{x}, \overline{y}$ determine the coordinates of the *c.m.* of n particles of masses $p_1, p_2, \cdots, p_n$ placed at points $(x_1, y_1), (x_1, y_2, \cdots\cdots\cdots, (x_n, y_n)$, the variances $\sigma x^2, \sigma y^2$ are the moments of inertia about axes through the *c.m.* parallel to

y and x axes and the covariance is the product of inertia of the system of particles about the axes through the centre of mass parallel to the axes of coordinate.

Ex. 1. Prove the results stated above for the bivariate distribution

Ex. 2. Interpret $\mu_{20} = \mu_{20}' - \overline{X}^2, \mu_{02} = \mu_{02}' - \overline{Y}^2, \mu_{11} = \mu_{11}' - \overline{XY}$ as theorems of parallel axes for moments and products of inertia.

Ex. 3. If $x_1 = x \cos\theta - y \sin\theta, y_1 = x \sin\theta + y \cos\theta$, use the results in mechanics to show that

$$\overline{x}_1 = \overline{x} \cos\theta - \overline{y} \sin\theta, \overline{y}_1 = \overline{x} \sin\theta + \overline{y} \cos\theta$$

$$\sigma_{x1}^2 = \sigma_x^2 \cos^2\theta + \sigma_y^2 \sin^2\theta - 2\mu_{11} \sin\theta \cos\theta$$

$$\sigma_{y1}^2 = \sigma_x^2 \sin^2\theta + \sigma_y^2 \cos^2\theta + 2\mu_{11} \sin\theta \cos\theta$$

$$\text{cov. } (x_1, y_1) = (\sigma_x^2 - \sigma_y^2) \sin\theta \cos\theta + \mu_{11} \cos 2\theta$$

Ex. 4. Verify the results of Ex. 3 directly without using analogy.

Ex. 5. Find $p_1, p_2, \cdots\cdots, p_n$ of a.d.r.v. with values $x_1, x_2, \cdots\cdots x_n$ has moments $\mu_1', \mu_2' \cdots\cdots, \mu'_{n-1}$.

Ex. 6. Show that for a.d.r.v.

$$\mu_k' = a_{1k} \mu_1' + a_{2k} \mu_2' + \ldots\ldots\ldots + a_{n-1,k} \mu_n', k = n+1, n+2, \cdots$$

where a's depend on $x_1, x_2, \cdots\cdots, x_n$ only.

Ex. 7. For a biased die, $\mu_1', \mu_2', \ldots\ldots \mu_5'$ are given show how you would find $p_1, p_2 \cdots\cdots, p_s$. Can you always find p's if μ's have any arbitrary values ?

Exercises 5

1. (*a*) Define (*i*) a discrete random variable, (*ii*) probability distribution, (*iii*) distribution function (*iv*) probability mass function (*v*) Spectrum (*vi*) degenerate random variable which degenerates at a point c *i.e.* $P(X = c) = 1$.

(*b*) State and prove the important properties of distribution function,

(*c*) Why is the distribution function called the 'step' function ?

(*d*) Let $x_1, x_2, \cdots\cdots$ be a countable number of points and $p(x_i)$ be the probability of x_i. Then show that, if F is the distribution of X,

$F(x_i) - F(x_{i-1}) = p(x_i)$.

(*e*) Show that $P(a < X \leq b) = P(X \leq b) - P(X \leq a) = F(b) - F(a)$

(B.A. Hons. Delhi, *1993*)

2. A random variable X has the probability function.

X :	0	1	2	3	4	5	6	7	8
$P(X = x) = p(x)$:	a	$3a$	$5a$	$7a$	$9a$	$11a$	$13a$	$15a$	$17a$

(*i*) Determine the value of a. (*ii*) Find $P(X < 2), P(X \geq 6), P(3 < X < 5)$

(*iii*) Find the minimum value of a such that $P(X \leq 3) > \cdot 5$.

(*iv*) What is the distribution function of X ?

$$\left[\textbf{Ans. } (i)\,\frac{1}{81}\quad (ii)\,\frac{4}{81},\frac{45}{81},\frac{1}{9}\quad (iii)\,\frac{1}{32}\right]$$

3. The spectrum of the random variable X consists of the points 1, 2, ...n, and $P(x=i)$ is proportional to $1/i(i+1)$. Determine the distribution function of X. Compute $P(3<x\le n)$ and $P(x>5)$.

[**Hint.** Let $P(x=i)=k/i(i+i)$ where k is determined by

$$1=\sum_{i=1}^{n}\frac{k}{i\,(i+1)}=k\sum_{i=1}^{n}\left(\frac{1}{i}-\frac{1}{i+1}\right)=k\left(1-\frac{1}{n+1}\right)=\frac{kn}{n+1}\quad\therefore\ k=\frac{n+1}{n}.$$

$$F(x)=\sum_{-\infty}^{i}P(X=x_i)=\sum_{1}^{i}\frac{k}{i\,(i+1)}=k\sum_{1}^{i}\left(\frac{1}{i}-\frac{1}{i+1}\right)=k\left(1-\frac{1}{i+1}\right)$$

$$=\frac{ki}{i+1}=\frac{(n+1)i}{n(i+1)},\ i\le x<i+1.$$

$$P(3<x\le n)=\sum_{4}^{n}p_i=k\sum_{4}^{n}\frac{1}{i(i+1)}=\frac{n+1}{n}\left[\frac{1}{4}-\frac{1}{n+1}\right]=\frac{(n-3)}{4\,n}.$$

$$P(x>5)=\sum_{6}^{n}p_i=\frac{n+1}{n}\left(\frac{1}{6}-\frac{1}{n+1}\right)=\frac{n-5}{6\,n}.$$

4. The distribution function $F(x)$ of a variate X is defined as follows :

$$\begin{aligned} F(x) &= A & -\infty<x<-1\\ &= B & -1\le x<0\\ &= C & 0\le x<2\\ &= D & 2\le x<\infty \end{aligned}$$

where A, B, C, D are constants. Determine the values of A, B, C, D, it being given that $P(X=0)=1/6$ and $P(X>1)=2/3$.

Ans. $A=F(-\infty)=0,\ \ D=F(\infty)=1,\ 1/6=P(X=0)=F(0)-F(-0)=C-B$, $\frac{2}{3}=P(X>1)=1-F(1)=1-C$. Thus $B=1/6$, $C=1/3$.

5. Suppose that the random variable X assumes three values 0, 1 and 2 with probabilities $\frac{1}{3},\frac{1}{6}$ and $\frac{1}{2}$ respectively. Obtain the distribution function of X.

(***B.Sc. Gujarat, 1992***)

6. The probability mass function of a random variable Y is zero except at the points $X=0, 1, 2$. At these points it has the values $p(0)=3c^3$, $p(1)=4c-10c^2$ and $p(2)=5c-1$ for some $c>0$.

(*i*) Determine the value of c.

(*ii*) Compute the probabilities : $P(X<2)$ and $P(1<X\le 2)$

(*iii*) Describe the distribution function and draw its graph.

(*iv*) Find the largest x such that $F(x)<\frac{1}{2}$

(*v*) Find the smallest x such that $F(x)\ge\frac{1}{3}$ (***B.Sc. Poona, 1983***)

Ans. (*i*) $\frac{1}{3}$, (*ii*) $\frac{1}{3}, \frac{2}{3}$, (*iv*) 1, (*v*) 1.

7. A number is chosen at random from each of the two sets 0, 1, 2, 3 and 0, 1, 2, 3. Find the probability distribution of the random variable denoting the sum of the numbers chosen.

Ans. Let X and Y be the two *r.v.* each taking the values 0, 1, 2, 3 with probability 1/4, then $Z = X + Y$

Values (Z)	:	0	1	2	3	4	5	6
Probabilities $p(Z)$	:	1/16	1/8	3/16	1/4	3/16	1/8	1/16

8. A coin is tossed four times. Let X denote the number of times a head is followed by a tail. Find the distribution of X.

Ans. Sample space : {*HHHH, HHHT. HHTH, HTHH, HTHT, ..., TTTT* }

X (number of outcomes) :	0	1	1	1	2, ..., 0	
Value of X (= x) :	0	1	2			[0 occurs 5 times and
Prob. $p(x)$:	5/16	10/16	1/16			$P(HHHH) = [P(H)]^4 = \frac{1}{2^4} = \frac{1}{16}$

9. Let A and B be two mutually exclusive events, then prove that

$$E(X \mid A \cup B) = \frac{P(A)\,E(X \mid A) + P(B)\,E(X \mid B)}{P(A \cup B)}.$$ State the result when $B = \overline{A}$.

[**Hint.** By def., $E(X \mid A \cup B) = \frac{1}{P(A \cup B)} \sum_{x_i \,\varepsilon\, A \cup B} x_i\, P(X = x_i)$

$$= \frac{1}{P(A \cup B)} \left[\sum_{x_i \,\varepsilon\, A} x_i\, P(X = x_i) + \sum_{x_i \,\varepsilon\, B} x_i\, P(X = x_i) \right],$$ as A and B are *m.e.*

$$= [\,P(A)\,E(X \mid A) + P(B)\,E(X \mid B)\,] / P(A \cup B) = \text{etc.}$$

When $B = \overline{A}$, the result reduces to

$$E(X \mid A \cup \overline{A}) = P(A)\,E(X \mid A) + P(\overline{A})\,E(X \mid \overline{A}).]$$

10. From a lot containing 12 items, 4 of which are defective, 5 are chosen at random. If X be the number of defectives found in the sample, write down

(*i*) the probability distribution of X.

(*ii*) $P(X \le 1)$, (*iii*) $P(1 < X < 3)$. (*iv*) Verify $\sum_{x=0}^{4} p(x) = 1$.

[Assume that the sample is drawn without replacement.]

[**Ans.** (*i*) $p(x) = \binom{4}{x}\binom{8}{5-x} \div \binom{12}{5}$, $x = 0, 1, 2, \ldots\ldots, 4$

(*ii*) $P(X \le 1) = P(X = 0) + P(X = 1) = \left[\binom{8}{5} + \binom{4}{1}\binom{8}{4}\right] \div \binom{12}{5} = \frac{336}{792} = \frac{14}{33}$.

(*iii*) $P(1 < X < 3) = P(X = 2) = \binom{4}{2}\binom{8}{3} \div \binom{12}{5} = \frac{14}{33}$]

11. Define a random variable and its distribution. Obtain the distribution function of the total number of heads occurring in three tosses of an unbiased coin.

(*B.Sc. Kerala 1970*)

12. Define the distribution function of a random variable. Let a coin with probability p for the occurrence of a head be tossed until a head appears. Let X denote the number of tails observed. Find $P(X = r)$ and hence obtain its distribution function. **(*B.Sc. Kerala, 1971*)**

[**Ans.** $P(X = r) = (1 - p)^r \, p$]

13. A random variable X has the following probability function :

x :	0	1	2	3
$p(x)$:	0·1	0·2	0·3	0·4

Find the probability function of Y and the mean of Y where $Y = x^2 + 1$.

y :	1	2	5	10
$p(y)$:	0·1	0·2	0·3	0·4

14. A can have any integral value from 0 to a–1, the chance of any value x varying as $a - x$, and B can have any integral value from 0 to $b - 1$, the chance of any value y varying as $b - y$ $(a > b)$. If A and B are independent, find $P(A = B)$.

[**Hint.** $$P(A = B) = \sum_{x=0}^{b} P[(A = x) \cap (B = x)] = \sum_{x=0}^{b} P(A = x)\, P(B = x)$$

where $P(A = x) = y_1 (a - x)$, y_1, being constant.

with $$1 = \sum_{x=0}^{a-1} P(A = x) = \sum_{x=0}^{a-1} y_1 (a - x) = y_1 \left[a \cdot a - \frac{a(a-1)}{2} \right] = y_1 \frac{a(a+1)}{2}$$

$$\Rightarrow \quad y_1 = 2/a(a+1) \quad \text{and,} \quad \therefore \quad P(A = X) = \frac{2(a - x)}{a(a+1)}$$

Similarly, $P(B = y) = y_2(b - y)$ with $1 = \sum_{y=0}^{b-1} y_2(b - y)$

giving $$P(B = Y) = \frac{2(b - y)}{b(b+1)}.$$]

15. An experiment consists of three independent tosses of a fair coin. Let X = the number of heads, Y = number of head runs, Z = length of head run, a head run being defined as consecutive occurrence of at least two heads, its length then being the number of heads occurring together in three tosses of the coin.

(*a*) Find the probability function of the random variables (*i*) X (*ii*) Y (*iii*) Z (*iv*) $X + Y$ (v) XY.

(*b*) Write down the probability table of (*i*) $X + Y = U$, (*ii*) $XY = V$.

(*B. Sc. Delhi 1970*)

[**Ans.**

Elementary Event	X	Y	Z	$X + Y$	XY
HHH	3	1	3	4	3
HHT	2	1	2	3	2
HTH	2	0	0	2	0
HTT	1	0	0	1	0
THH	2	1	2	3	2
THT	1	0	0	1	0

TTH	1	0	0	1	0
TTT	0	0	0	0	0

(*b*) (*i*) u : 0 1 2 3 4 (*ii*) v : 0 2 3

$p(u)$: $\frac{1}{8}$ $\frac{3}{8}$ $\frac{1}{8}$ $\frac{2}{8}$ $\frac{1}{8}$ $p(v)$: $\frac{5}{8}$ $\frac{2}{8}$ $\frac{1}{8}$

16. Define for a discrete random variable, the following terms :

(*i*) Mathematical Expectation. (***B.A. Hons., Delhi***)

(*ii*) Cumulative Function. (***B.A. Hons., Delhi***)

(*iii*) Moment Generating Function. (***B. Sc., Agra***)

(*iv*) Probability Generating Function. (***B.Sc. Shivaji 1970***)

Show how they are related to each other.

17. From an urn containing 3 red and 2 white balls, a man is to draw 2 balls at random without replacement, being promised 20 P. for each red ball he draws, and 10 P. for each white one. Find his expectation. [**Ans.** 32 P]

18. A man draws 2 balls from a bag containing 3 white and 5 black balls. If he is to receive Rs. 70 for every white ball which he draws and Rs. 7 for every black ball, find his expectation. [**Ans.** Rs. 61·25]

19. The face cards are discarded from two decks of cards ; then one card is drawn from each deck. What is the expectation of the sum of the numbers drawn ? [**Ans.** 11]

20. (*a*) On the average how many times must a die be thrown until one gets a 6 ? [**Ans.** 6]

(*b*) Four coins are tossed. What is the expectation of the number of heads ? [**Ans.** 2]

(*c*) A coin is tossed until a head appears. Find the expectation of the number of tosses. [**Ans.** 2]

(*d*) Two unbiased dice are thrown. Find the expected value of the sum of numbers of points on them. [**Ans.** 7]

21. In a lottery m tickets are drawn out of n tickets numbered from 1 to n. What is the expectation and dispersion of the sum of the numbers drawn ? (***M.A. Bombay***)

$$\left[\textbf{Ans. } \frac{m(n+1)}{2}, \frac{m(n^2-1)}{12}\left(1-\frac{m-1}{n-1}\right)\right]$$

22. A bag contains a coin of value M and a number of other coins whose aggregate value is m. A person draws one at a time till he draws the coin of value M. Find the value of his expectation.

$$\left[\textbf{Ans. } M+\frac{m}{2}\right]$$

23. Balls are taken one by one out of an urn containing a white and b black balls until the first white ball is drawn. Show the expectation of the number of black balls preceding the first white ball is $b/(a+1)$.

(***Indian Statistical Institute, Calcutta***)

Sol. Let x_1 denote the number of black balls preceding the first white ball. Let

x_i denote the number of black balls following the $(i-1)th$ white balls and preceding the *i-th* white ball. Let x_{a+1} denote the number of black balls left after all the white balls have been drawn.

Then evidently, $x_1 + x_2 + \cdots\cdots + x_{a+1} = b$

and therefore, $E(x_1) + E(x_2) + \cdots\cdots + E(x_{a+1}) = b.$

Now the probability of every sequence of balls *i.e.*, of every system of numbers $x_1, x_2, \cdots\cdots x_{a+1}$ is the same *viz.*, $\dfrac{a\,!\,b\,!}{(a+b)\,!}$.

Thus $\quad E(x_1) = E(x_2) = \cdots\cdots = E(x_{a+1}) = M$, say

Hence $\quad (a+1)M = b \;\therefore\; M = \dfrac{b}{a+1}$ **Ans.**

24. A box contains a white and b black balls ; c balls are drawn. Find the expectation of the number of white balls drawn.

(***M.A. Punjab Sept., 1956***)

Sol. Let a variable x_i be defined as follows :

$x_i = 1$ if i-th ball drawn is white, $i = 1, 2, \ldots.., c;$ $= 0$ if " " " " black.

Then the number of white balls s among the c balls drawn is given by $s = x_1 + x_2 + \cdots\cdots + x_c$

Now $\quad P(x_i = 1) = \dfrac{a}{a+b},\; P(x_i = 0) = \dfrac{b}{a+b}$

$$\therefore \quad E(x_i) = 1.\frac{a}{a+b} + 0.\frac{b}{a+b} = \frac{a}{a+b}$$

$$\therefore \quad E(s) = \frac{a}{a+b} + \frac{a}{a+b} + \cdots + \frac{a}{a+b} = \frac{ca}{a+b}.$$

25. An urn contains pN white and qN black balls, the total number of balls being N, $p + q = 1$. Balls are drawn one by one (without being returned to the urn) until a certain number n of balls is reached. What is the dispersion of the number of white ball drawn ? (***M.A., Statistics, Delhi, 1961 ; M.A., Delhi, 1959***)

Hint. Let $x_i = 1$ if the i-th ball drawn is white

$= 0$ " " " black then we have

$$E(x_i) = p,\; E(m) = np,\; E(x_i^2) = p,\; E(x_i - p)^2 = E(x_i^2) - p^2 = p - p^2 = pq,\; q = 1 - p$$

and $\quad E(x_i - p)(x_j - p) = E(x_i x_j) - p^2 = \dfrac{pN}{N} \cdot \dfrac{pN-1}{N-1} - p^2.$

$$= p \cdot \left[\frac{pN-1}{N-1} - p\right] = -\frac{pq}{N-1},\; i \neq j$$

$$\therefore \text{ Dispersion} \quad = E(m - np)^2 = E\left[\sum_{i=1}^{n} (x_i - p)^2 + 2 \sum_{i \neq j} (x_i - p)(x_j - p)\right]$$

$$= \sum_{t=1}^{n} E(x_i - p)^2 + 2 \sum_{i \neq j} E(x_i - p)(x_j - p)$$

$$= npq + 2.\frac{n(n-1)}{2}\left(-\frac{pq}{N-1}\right) = npq\,\frac{N-n}{N-1}$$

26. In a lottery containing n numbers (1, 2, ..., n), m numbers are drawn at a time. Let x_i represent the frequency of a specified number of N drawings.

Prove that $E(x_i) = Np$,

$$E(x_i - Np)^2 = Npq,\ E(x_i - Np)(x_j - Np) = E(x_i x_j) - N^2p^2 = N^2p(p' - p),\ i \neq j$$

where $p = \frac{m}{n}, q = 1 - p, p' = \frac{m-1}{n-1}$. **(*M. Sc., Madras*)**

[**Hint.** Let Y_{ij} denote the random variable associated with the *j-th* draw such that

$$y_{ji} = \begin{cases} 1 \text{ if number } i \text{ comes on } j\text{–th draw} \\ 0, \text{ otherwise} \end{cases}$$

Then $X_i = \sum_{i=j}^{N_i} Y_{ji}$ is the frequency of a specified number j in N drawings.

Then $E(X_i) = E[\sum_{j=1}^{N} Y_{ji}] = \sum_{j=1}^{N} E(Y_{ji}) = \sum_{j=1}^{N}\left[1.\frac{m}{n} + 0.\left(1 - \frac{m}{n}\right)\right] = \frac{m}{n}N = Np.$

Also $E(X_iX_j) = E\left[\left(\sum_{j=1}^{N} Y_{ji}\right)\left(\sum_{k=1}^{N} Y_{kj}\right)\right] = E\left[\sum_{k=1}^{N} Y_{ki}Y_{kj} + \sum_{\substack{k=1 \\ k\neq j}}^{N} Y_{ki}\ Y_{kj}\right]$

$$= \sum_{k=I}^{N} E\, Y_{kj}\, Y_{ki} = \sum_{k\neq 1} E(Y_{kj}\, Y_{ki}) = \sum_{k=1}^{N} \frac{\binom{n-2}{m-2}}{\binom{n}{m}} + E(Y_{ki})\, E(Y_{kj}) = \text{etc.}\ ]$$

27. Given a lot of size m containing s items of a specified kind. If items are to be drawn *without* replacement until i of the s items have been drawn show that on the average $\frac{i(m+1)}{s-1}$ drawings will be necessary.

(*I.S.I. Calcutta ; M.A., Calcutta*)

28. A box containing k varieties of objects, the number of objects of each variety being the same. These objects are drawn one at a time and put back before the next drawing. Denote by n the smallest number of drawings which produce objects of all varieties find $E(n)$ and $E(n^2)$. Also for large k, find $E(n)$ and Var n.

Sol. Let x_{r+1} denote the number of drawings required to produce any of the (k—r) varieties where r varieties have already been drawn x_{r+1} is a stochastic variate taking the values 1, 2, 3, ..., with respective probabilities.

$$\frac{k-r}{k}, \frac{r}{k}\cdot\frac{k-r}{k}, \left(\frac{r}{k}\right)^2\cdot\frac{k-r}{k}, \left(\frac{r}{k}\right)^3\cdot\frac{k-r}{k}, \ldots\ldots\ldots$$

Hence by definition

$$E(x_{r+1}) = \frac{k-r}{r} \sum_{n=1}^{\infty} n\left(\frac{r}{k}\right)^{n-1} = \frac{k-r}{k}\left(1-\frac{r}{k}\right)^{-2} = \frac{k-r}{k} \cdot \frac{k^2}{(k-r)^2} = \frac{k}{k-r}$$

$$E(x_{r+2}) = \frac{k-r}{k} \Sigma n^2 \left(\frac{r}{k}\right)^{n-1} = \frac{k(k+r)}{(k-r)^2}.$$

Since balls are replaced each time before fresh drawing is made the stochastic variates $x_1, x_2, \cdots\cdots x_k$ are independent.

The last number of drawings required to produce all the k varieties is evidently equal to $n = x_1 + x_2 + \ldots + x_k$.

$$\therefore \quad E(n) = E(x_1 + x_2 + \cdots + x_k) = \sum_{r=0}^{k-1} E(x_{r+1}) = \sum_{r=0}^{k-1} \frac{k}{k-r} = k\left(1 + \tfrac{1}{2} + \cdots + \frac{1}{k}\right)$$

Similarly $\quad E(n^2) = k^2\left(\frac{1}{1^2} + \frac{1}{2^2} + \cdots + \frac{1}{k^2}\right) - k\left(1 + \frac{1}{2} + \cdots + \frac{1}{k}\right)$

$$+ k^2\left(1 + \frac{1}{2} + \cdots + \frac{1}{k}\right)^2$$

For large n, $\quad E(n) = \dfrac{(r-1)k}{k-r+1} - k \log \dfrac{k}{k-r+1}$.

29. A bag contains $2n$ contents, of which half are marked with odd numbers and half with even numbers, the sum of all the numbers being S. A man is to draw counters. If the sum of the numbers drawn is odd, he is to receive that number of rupees, if even, he is to pay that number of rupees. Show that the expectation is $S/n(2n-1)$ rupees.

30. A vessel contains tickets numbered 1 to N. Find the expectation and variance of the largest numbers X drawn in n drawings when random sampling with replacement is used. Also find for large N and fixed k, $E(X)$ and *Var* (X).

[**Hint.** The event $X < k$ means that each of n number drawn is less than or equal to k so that $P(X \le k) = (k/N)^n$.

$$p_k = P(X = k) = P(X \le k) - P(X \le k-1) = \left(\frac{k}{N}\right)^n - \left(\frac{k-1}{N}\right)^n = \frac{k^n - (k-1)^n}{N^n}.$$

$$\therefore \; E(X) = \sum_{k=1}^{N} k p_k = \sum_{k=1}^{N} \frac{k}{N^n}[k^n - (k-1)^n] = \sum_{k=1}^{N} \frac{1}{N^n}[k^{n+1} - k(k-1)^n]$$

$$= \frac{1}{N^n} \sum_{k=1}^{N} [k^{n+1} - (k-1)^{n+1} - (k-1)^n]$$

$$= \frac{1}{N^n}\left[N^{n+1} - \sum_{k=1}^{N} (k-1)^n\right] \sim \frac{1}{N^n}\left[N^{n+1} - \int_0^N t^n\, dt\right] = \frac{nN}{n+1}$$

31. (*a*) With usual notation, prove that $\dfrac{\partial \mu_r'}{\partial \mu_j'} = {}^rC_j \mu'_{r-j}$

(*b*) Deduce that $\mu_2 = k_2, \mu_3 = k_3, \mu_4 = k_4 + 3k_2^{\,2}$, **(*M.A. Madras*)**

[**Hint.** (*a*) Differentiate the following identity with respect to k_j and identify powers of t.

$$\exp\left\{k_1 t + \frac{k_2 t^2}{2!} + \cdots + \frac{k_r t^r}{r!} + \ldots\right\} = 1 + \mu_1' t + \frac{\mu_2' t^2}{2!} + \cdots + \frac{\mu_r' t^r}{r!} + \cdots$$

(*b*) put $j = 1$ in the result of (*a*)].

32. (*a*) Assuming the additive property of cumulative prove that the variance of the sum of a number of independent variates is the sum of their variances.

(*b*) Let $X_1, X_2, \ldots, X_n$ be mutually independent r.v. having cumulants $k_1, k_2, \ldots k_n$ respectively. Prove that the linear combination $X = \sum_1^n a_i X_i$ has cumulants given by $k = \sum_1^n a_i k_i$. Hence deduce the variance of $X = \sum_{i=1}^{n} \frac{a_i X_i}{a}$.

(*c*) Let x_1 and x_2 be two stochastic random variables having variances k and 2 respectively. If the variance of $y = 3x_2 - x_1$ is 25, find k. **[Ans. 7]**

(*d*) Suppose that $\overline{X}$ is a random variable for which $E(X) = 10$ and Var $X = 25$. Find the positive values of a and b such that $Y = aX - b$ has expectation 0 and variance 1. ***(B.Sc. Sardar Patel, 1970)***

[Ans. $a = \frac{1}{5}, b = 2$ **]**

(*e*) let X be a r.v. with m.g.f. $M(t)$ prove that $P(X \geq a) = e^{-ar} E(e^{rx}), r > 0 \leq e^{-at} M(t), 0 < t < h$ and $P(X \leq a) \leq e^{-at} M(t), n < t < h$

(*f*) Show that if $\overline{X}$ is the mean of n independent stochastic variables X, then

$$M_{X=\bar{x}}(t) = \left[M_X\left(\frac{t}{n}\right)\right]^n$$

and if X's are identically distributed independent stochastic variables then

$$\sum_{i=1}^{n} M(t)\ M_{X_i}(t) = [M_{X_i}(t)]^n$$

$$\left[\textbf{Hint.}\ (f)\ M_{\bar{x}}(t) = \frac{M X_i + x2 + \ldots\ x_n}{n}(t) = n\left[M_{x/n}(t)\right] = \left[M_{x/n}(t)\right]^n = \left[M_x\left(\frac{t}{n}\right)\right]^n.\right.$$

33. Show that the variance of the weighted mean $\sum_{i=1}^{N} w_i x_i / \sum_{i=1}^{n} w_i$ of n independent stochastic variates x_i is a minimum when the weights are inversely proportional to the corresponding variances. Also show that this minimum variance is H/n when H is the harmonic mean of the variance $\sigma_i^2, i = 1, 2, \ldots, n$. ***(M.A. Statistics, Delhi)***

[**Hint.** $\sigma^2 =$ variance of the weighted mean $= \dfrac{\sum w_i^2 \sigma_i^2}{(\sum w_i)^2}$

For minimum of σ^2, $\dfrac{d\sigma^2}{dw_i} = 0$, which gives $w_i\sigma_i^2 = \dfrac{\Sigma\, w_i^2\, \sigma_i^2}{\Sigma\, w_i}$ showing that $w_i\sigma_i^2$ is the same for all values of i, *i.e.*, $w_i \propto (1/\sigma_i^2)$. Hence the result .]

34. In a sequence of Bernoullian trials in each of which the probability of success is p, show that

(*a*) (*i*) the expected value of X_1, the number of trials up to and including the first success is $1/p$; and that of x_k, the number of trials up to and including *k-th* success is also $1/p$;

(*ii*) the expected value of $X_1 + X_2 + \cdots + X_n$ of the number of trials up to and including the nth success is $\dfrac{n}{p}$ and variance is $\dfrac{n(1-p)}{p^2}$,

(*b*) if X be the length of run of either successes or failures starting with the first trial, show that

$$E(X) = \frac{q}{p} + \frac{p}{q} \text{ and } \operatorname{var}(X) = \left(\frac{p}{q} - \frac{q}{p}\right)^2 + \left(\frac{p}{q} + \frac{q}{p}\right).$$ **(*B.A. Hons. Stat. Delhi, 1997*)**

[Hint. (*a*) $P(X = r) = P(\underbrace{SS \ldots S}_{r \text{ times}}F) + P(\underbrace{FF \ldots F}_{r \text{ times}}S) = p^r q + q^r p$.

$$\therefore \quad E(X) = \sum_{r=1}^{\infty} rP(x = r) = \text{etc.}; \; E[X(X-1)] = \sum_{r=1}^{\infty} (x)(x-1)P(x = r) = \text{etc.}]$$

35. Let the random variable X assume the value r with the probability law :

$$P(X = r) = q^{r-1}\, p\, ; r = 1, 2, 3, \ldots, q = 1 - p.$$

Find the m.g.f. and hence deduce the mean and variance of X.

Ans. $M_X(t) = \sum_1^{\infty} e^{tr} \,.\, q^{r-1}\, p = pe^t\,(1 - qe^t)^{-1}$; mean $= 1/p$, variance $= q/p^2$.

36. A population of N distinct elements is samples *with replacement.* Prove that (*i*) the expected value and (*ii*) the variance of the size of the sample for getting r distinct elements are respectively given by

$$\left[\frac{1}{N} + \frac{1}{N-1} + \cdots + \frac{1}{N-r+1}\right] \text{ and } N\left[\frac{1}{(N-1)^2} + \frac{2}{(N-2)^2} + \cdots + \frac{r-1}{(N-r+1)^2}\right].$$

[Hint. (*i*) The number, X_k, of drawings up to and including the drawing of a new element equals one plus the number of failure preceding the first success in Bernoullian trials with $p = (N-k)/N$.

$$\therefore \qquad E(X_k) = 1 + q/p = N(N-k)\;;\; q = 1 - p$$

Now $$S_k = 1 + X_1 + \cdots + X_{k-1}$$

(*ii*) Note that the variance of the geometric distribution $q^x p$ is q/p^2 .]

37. A deck of n numbered cards is put into random order so that all arrangements have equal probabilities. If $X_k = 1$ or 0 according as card number k is in its natural place (*i.e.*, *a*, match) or not, prove that

$$P(X_k = 1) = \frac{1}{n},\ P(X_k = 0) = \frac{n-1}{n},\quad E(X_k) = \frac{1}{n},\ Var(X_k) = \frac{n-1}{n^2},$$

$$E(X_h X_k) = \frac{1}{n(n-1)},\quad Cov\ (X_h, X_k) = \frac{1}{n^2\ (n-1)},\ h \neq k$$

Hence deduce that, if $S_n = X_1 + X_2 + \cdots + X_n$ denotes the number of matches, $E(S_n) = 1,\ Var(S_n) = 1$

38. Two similar decks of n distinct cards each are put into random order and are against each other. Prove that the probability of having exactly r matches is given by

$$P_r = \frac{1}{r!} \sum_{k=0}^{n-r} \frac{(-1)^k}{k!},\ r = 0, 1, 2, \ldots\ldots, n.$$

Prove further that the expected number of matches and its variance are equal and are independent of n. **(M.A., Delhi, 1964 ; M.A., Bombay, 1958)**

[**Hint.** $E(x) = \sum_{r=1}^{n} \sum_{k=0}^{n-r} \frac{(-1)^k}{(r-1)!\,k!} = \sum_{r=0}^{n-1} \sum_{k=0}^{n-1-r} \frac{(-1)^k}{r!\,k!} = 1$

Also prove that $E[x(x-1)] = 1 \quad \therefore$ Var $x = 1$.

39. An urn contains a white and b black balls. After a ball is drawn, it is to be returned to the urn if it is white ; but if it is black, it is to be replaced by a white ball from another urn. What is the probability of drawing a white ball after the forgoing operation has been repeated x times ?

$$\left[\textbf{Ans. } 1 - \frac{b}{a+b}\left(1 - \frac{1}{a+b}\right)^x\right]$$

[**Hint.** If M_x denotes the expectation of the number of white balls after x operations, then we have

$$M_{x+1} = M_x \cdot \frac{M_x}{a+b} + (M_x + 1)\left(1 - \frac{M_x}{a+b}\right) = \left(1 - \frac{1}{a+b}\right) M_x + 1$$

The difference equation gives us $M_x = a + b - b\left(1 - \frac{1}{a+b}\right)^x$.]

40. An urn contains n tickets numbered from 1 to n. These tickets are drawn one by one so that a certain number appears in the first place, another number in the second place and so on. Find the mathematical expectation of the number of coincidence (we shall say that there is a coincidence when the number on a ticket corresponds to the place it occupies). [**Ans.** 1]

41. (*a*) Show that if X is non-negative random, variable, then $EX \cdot E\left(\frac{1}{X}\right) > 1$.

Verify this result for a continuous *r.v.* which is such that

$f(x) = x,\ 0 < x < 1\ ; = 2 - x,\ 1 \leq x \leq 2 = 0.$ otherwise.

(*b*) If X, Y are independent continuous random variables with mean $\mu(x)$ and $\mu(y)$ and variances σ_x^2 and σ_y^2 respectively, show that (*i*) $\sigma^2(xy) = \sigma^2(x)\,\sigma^2(y) + \mu^2(x)\,\sigma^2(y) + \mu^2(y)\,\sigma^2(x)$.

(***B.Sc. Gorakhpur, 1992, M.A., Statistics, Delhi*** 1970)

[Hint. $\sigma_{xy}{}^2 = E\,[(x-\mu(x)]\,[y-\mu(y)] + xy + y\mu(x) + x\mu\,(y) - \mu(x)\mu(y)]^2 =$ etc.

(*ii*) $\dfrac{Var\,(xy)}{[E(x)]^2\,[E(y)]^2} = C_x{}^2 C_y{}^2 + C_x{}^2 + C_y{}^2$

where $C_x = \sqrt{Var\ x/E(x)} = \sigma_x/\mu(x)$, $C_y = \sqrt{Var\ y/E(y)} = \sigma_y/\mu(y)$

(***B.Sc. Patna, 1991***)

(*iii*) Let X and Y be two independent r.v. with Var $X > 0$ and Var $Y > 0$. prove that Var XY = Var X. Var Y. if $EX = 0 = EY$.

(***B.A. Hons. Stat. Delhi 1984***)

42. Let X be the number of failure preceding the kth success in an infinite set of Bernoullian with constant probability p of success. Find the probability function $g(x)$ of X and also its mean and variance. Further show that

$$\sum_{x=0}^{n-k} g(x) = \sum_{t=k}^{n} \binom{n}{t} p^t q^{n-1}.$$

(***Bombay, B.Sc.***)

43. A random variable X assumes the values $\lambda_1, \lambda_2, \ldots\ldots$, with probabilities $\mu_1, \mu_2 \ldots\ldots$, show that

$$p_k = \frac{1}{k\,!} \sum_{j=0}^{\infty} \mu_j e^{-\lambda_j} (\lambda_j)^k,\ \lambda_j > 0,\ \sum_{j=0}^{\infty} \mu_j = 1,\ k = 1, \ldots$$

is a probability distribution. Find its generating function and prove that its mean equals $E(X)$ and its variance equals $Var\ (X) + E(X)$.

(***M.Sc. Lucknow, 1967***)

Hint. $\sum_{k=0}^{\infty} p_k = \sum_{j=0}^{\infty} \mu_j\, e^{\lambda_j} \sum_{k=j}^{\infty} (\lambda_j)^k / k1 = \sum_{j=0}^{\infty} \mu_j\, e^{-\lambda_j}\, e^{\lambda j} = 1$

$$\left[\textbf{Ans. } P(s) = \sum_{0}^{\infty} p_k\, s^k = \sum_{j=0}^{\infty} \mu_j\, e^{\lambda_j (s-1)}\right]$$

44. Suppose that shipment involving a varying number of parts arrive each day. If N is the number of items in the shipment the probability distribution of the random variable N is given as :

n :	10	11	12	13	14	15
$P(N = n)$:	·05	·10	10	·20	·35	·20

The probability that any particular part is defective is the same for all parts and equal to 0·10. If X is number of defective parts arriving each day what is the expected value of X ? **[Ans.** 1·33]

[Hint. $E(X) = E(E(X \mid N)] = E(0{\cdot}10N) =$ etc.]

45. In a fairy tale, the fairy grandmother assumes the queen that her baby prince will not die until the following condition is fulfilled :

A scroll is prepared and on the day of the birth and every subsequent birthday a letter of the Greek alphabet, chosen at random, is entered on the scroll. On the day that all the twenty-four letters of the alphabet appear, he will die. Find the expectation of his life.

[It is given that $\log_e 24 = 3{\cdot}17805$ and

$$\lim_{n \to \infty} \left[1 + \frac{1}{2} + \cdots + \frac{1}{n} - \log n \right] = \gamma \text{ (Euler's constant)} = 0{\cdot}57721\ 56649\ldots\ldots j$$

(***M.A. Statistics, Delhi***) [**Ans.** 89·6 years nearly]

46. Let X and Y be two random variables each taking three values –1, 0, 1 and having the joint probability distribution :

Y \ X	– 1	0	1	Total
– 1	0	·1	·1	·2
0	·2	·2	·2	·6
1	0	·1	·1	·2
Total	·2	·4	·4	1·0

(*a*) Find $E(X)$, $E(Y)$, Var X and Var Y. Are X and Y independent ?

(*b*) Given that $Y = 0$, what is the conditional probability distribution of X ?

(*c*) Find Var $(Y \mid X = -1)$. (***B. Sc. Poona ; M.Sc. J. & K.***)

Ans. $E(X) = 0, E(Y) = 0.2, \quad E(XY) = 0 \Rightarrow E(XY) = EXEY \Rightarrow X$ and Y are independent. Var X = ·4, Var Y = ·56,

$P(X = -1 \mid Y = 0) = P(X = - \mid \cap Y = 0)/P(Y = 0) = \cdot 2/\cdot 6 = 1/3$

$P(X = 0 \mid Y = 0) = \frac{1}{3}, P(X = 1 \mid Y = 0) = \frac{1}{3}$

Var $(Y \mid X = -1 = E(Y \mid X = -1)]^2 - [E(Y \mid X = -1)]^2 = 0 - 0^2 = 0.$

47. Two stochastic variables X and Y take the values 1, 2, 3 and their probabilities are as follows :

Y \ X	*1*	2	3	*Total*
1	0·1	0·1	0·1	0·3
2	0·1	0·2	0·1	0·4
3	0·1	0·1	0·1	0·3
Total	0·3	0·4	0·3	1

Find $E(X)$, $E(Y)$, Var X, Var Y, EXY, $E(X + Y)$ and Cov (X, Y). Determine whether X and Y are independent.

(***B.Sc. Poona ; M.Sc. Madras***)

[**Ans.** EX = 2, EY = 2, Var $X = \cdot 6$, Var $Y = \cdot 6$, $E(X + Y) = 4$,]

48. (*a*) A vessel contains *n* different elements. The elements are drawn one at a time with replacement. Show that the expected number of trials to obtain *r* different elements is

$$n\left[\frac{1}{n}+\frac{1}{n-1}+\frac{1}{n-2}+\cdots\cdots+\frac{1}{n-r+1}\right]$$

(*b*) Starting from the origin on OX, a unit step can be taken towards the right with probability p and lowest as left with probability $q = 1 - p$, show that after n independent trials the mean distance from the origin is $n\ (p–q)$.

(*c*) Show that the mean number of trials measured to obtain the consecutive success in a series of independent Bernoullian trials is

$\frac{p+1}{p^2}$ and the variance is $\frac{1+2p-2p^2-p^3}{p^4}$

49. Suppose m persons enter a lift. If the building has n floor above the ground floor, show that the expected number of stops is

$$n\left[1+\left(1-\frac{1}{n}\right)^m\right]$$

if the a prior probability of any one person stopping on the kth floor is $1/n$ for $1 \le k \le n$

50. Show that for a sequence of Bernoullian trials, the expected number of trials until at least m successes and n failures occur is

$$= \sum_{N=m+n}^{\infty} N.\ {}^{N-1}C_{m-1}\, p^m\, q^{N-m} + \sum_{N=m+n}^{\infty} N\ {}^{N-1}\ C_{m-1}\, q^n\, p^{N-m}$$

51. n biased coins, with probability of head for each being p are tossed and all heads are removed. The remaining coins are tossed again and heads are removed. Show that the probability that no coin remain after k successive tosses is $(1-q^k)^n - (1-q^{k-1})^n$

Show that the expected number of tosses until no coin remains is

$$\sum_{r=1}^{n} {}^nC_r(-1)^r\,\frac{1}{q^r-1}.$$

52. The random variable X assumes the values 0, 1, 2......., N–1 each with probability $1/N$. Show that the p.g.f. is $P(s)=\frac{1}{N}\cdot\frac{1-s^N}{1-s}$.

If $s = X_1 + X_2 + \cdots\cdots + X_n$ where $X_1, X_2 \ldots\ldots, X_n$ are independent variables each distribution as X above, show that the *p.g.f.* for S is $Q(s)=\frac{1}{N^n}\left(\frac{1-s^N}{1-s}\right)^n$

Find $P(s = r)$.

53. The number N of animals trapped is a random variable such that $P(N=n)=qp^n, n = 0, 1, 2,\cdots\cdots$ were $0 < p < 1$. If an animal is trapped, the probability of Q not escaping is P. Show that the probability that k do not escape from their trap is $\frac{q}{1-pQ}\left(\frac{pP}{1-pQ}\right)^k$.

54. A group of N men throw their hats into the centre of a room. The hats are mixed up and each man randomly selects one. Find the expected and variance of the number of men that select their own hats.

[**Hint.** Let X denote the number of matches. Then $X = X_1 + X_2 + \cdots + X_N$

where $X_i = \begin{cases} 1 \text{ if the other man selected his own hat} \\ 0 \text{ otherwise} \end{cases}$

$$E(X_i) = P(X_i = 1) = \frac{1}{N}, \text{ for each } i, \quad E(X) = \Sigma E(X_i) = \sum_{i=1}^{N} \frac{1}{N} = \frac{1}{N} \cdot N = 1$$

Hence on the average, exactly one of the men selects his own hat,

$$\text{Var}\, X_i = \frac{1}{N}\left(1 - \frac{1}{N}\right) = \frac{N-1}{N^2}. \text{ Also Cov } (X_i, X_j) = E(X_i X_j) - E(X_i)E(X_j)$$

Now $X_i X_j = \begin{cases} 1 \text{ if the } i \text{ th and } j\text{th men both select their own hats} \\ 0 \text{ otherwise} \end{cases}$

$$E(X_i X_j) = P(X_i = 1, X_j = 1) = P(X_i = 1) P[X_j = 1 \mid X_i = 1] = \frac{1}{N} \cdot \frac{1}{N-1}$$

$$\therefore \text{ Cov}(X_i X_j) = \frac{1}{N(N-1)} - \left(\frac{1}{N}\right)^2 = \frac{1}{N^2(N-1)}$$

$$\text{Var}\, X = \frac{N-1}{N} + 2.\, {}^N C_2 \frac{1}{N^2(N-1)} = \frac{N-1}{N} + \frac{1}{N} = 1. \textbf{ Ans.}$$

55. A freshman dormitory contains 300 white and black students. If room mates were assigned at random, without regard to colour, compute the mean and variance of the number of black students having white room mates.

[**Solution.** Let $X_i = \begin{cases} 1 \text{ if the } i\text{th the black student has a white room mate} \\ 0 \text{ otherwise} \end{cases}$

The number of black students having white room mates $= \sum_{i=1}^{101} X_i$

Now $E(X_i) = P(X_i = 1) = \frac{100}{400}$, $\text{Var}\, X_i = \frac{300}{400} \times \frac{100}{400}$

$$E(X_i X_j) = P(X_i X_j = 1) = P(X_i = 1, X_j = 1)$$

$$= P[X_i = 1 \mid X_j = 1]\, P(X_j = 1) = \frac{299}{398} \times \frac{300}{400}$$

$$E\left(\sum_{i=1}^{101} X_i\right) = \sum_{i=1}^{101} EX_i = 101\left(\frac{3}{4}\right) = 75{\cdot}75$$

$$\text{Var}[X] = \sum_{1}^{101} \text{Var}\, X_i + 2 \sum_{i<j} \text{Cov}\,(X_k, X_j)$$

$$= 101\left(\frac{3}{16}\right) + 2\binom{101}{2}\left[\left(\frac{299}{398}\right)\left(\frac{3}{4}\right) - \left(\frac{3}{4}\right)^2\right] = 27{\cdot}02]$$

56. Suppose that the son of a man of height x inches attains a height that is normally distributed with mean $x + 1$ and variance 4. What is the best prediction of the height at full growth of the son who is 6 feet tall ?

Solution. We have the model Y = X + 1 + e where e is a normal r.v.

independent of X, having mean 0 and variance u, height = 6 feet = 72 inches and Y that of his son

$E(Y \mid X = 72) = E(X + 1 \mid X = 72) = 73 + E(e \mid X = 72) = 73 + E(e) = 73.$

57. A product, sold seasonally, yields a net profit of b dollars for each unit sold and a net loss of l dollar for each unit left unsold when the season ends. The number of units of the product that are ordered at a specific department store during any season a r.v. having probability mass function $p(i), i \geq 0$. If the store must stock this product in advance, determine the number of units the store should stock so as to maximize its expected profit.

58. In a sequence of Bernoullian trials let f_n be the probability that two consecutive successes occur for the first time in trials $n-1$ and n. Show that $f_n = qf_{n-1} + pqf_{n-1}, n \geq 3$, with $f_0 = f_1 = 0$, $f_2 = p^2$. Show that the p.g.f. is

$$P(S) = \frac{p^2 s^2}{1 - sq - pqs^2}.$$

Find the mean and variance.

59. The probability of a person suffering from diabetes is p. Blood sample from $N[= nk]$ persons are pooled into k groups of blood from n persons in each. If the pooled sample is negative, no further test is carried out. If a pooled sample is positive, each of the persons is tested separately. Show that the expected number of tests is

$$N\left(1 - q^n + \frac{1}{n}\right)$$

Show how you would choose n so as minimize the expected number of tests.

60. A newspaper boy buys N newspaper at a price of a each and sells them at a price of b each $(b > a)$. The unsold newspapers are returned at a price c each $(c < a)$. The probability of the boy selling n newspapers is p_n $(n = 0, 1, 2, \ldots\ldots)$. Find his expected profits. How many newspapers he should buy to maximize his expected profits ?

61. Let X be a non-negative random variable with cumulative distribution $F(x) = P(X \leq x)$. Show that

$$E(X) = \int_0^\infty [1 - F(x)]\, dx$$

Conjecture a corresponding expression for $E(X^2)$. ***(I.F.S. 1997)***

[Hint. $EX = \int_0^\infty x f(x)\, dx = \int_0^\infty x\, d F(x) = \int_0^\infty [1 - F(x)]\, dx.$

by integrating by parts.

(*b*) Let X be a non-negative random variable and let

$$X_e = \min(X, C) = X \text{ if } X \leq C, \text{ and } = C \text{ if } X > C$$

where C is a given constant. Find EX_c **Ans.** $E[X_c] = \int_0^\infty (1 - F(x)dx$

62. Suppose we have N chips numbered 1, 2, ..., We take a random sample of size n, without replacement. Let X be the largest number in the random sample. Show that the probability function of X is

$$P(X=k)=\binom{k-1}{n-1}\Big/\binom{N}{n},\ k=n,\ n+1\ldots,N$$

Also $\quad EX=\dfrac{n}{n+1}(N+1)$, $\operatorname{Var}X=\dfrac{n(N-n)(N+1)}{(n+1)^2(n+2)}$

63. Let $C(x)=\sum_{j=1}^{\infty} j^{1+x}$ and X be a random variable with distribution

$$P(X=j)=[C(\alpha)]^{-1}\,\bar{j}^{(1+\alpha)},\ j=1,2,3,\ldots\ldots$$

Show that X has kth moment if and only if $\alpha<k$.

(***B.A. Hons. Statistics Delhi, 1982***)

64. Let X, Y, Z be independent variates distributed as $e^{-m}\,m^x/x\,!$ with $m=a,b,c$ respectively obtain the conditional expectation of X given $X+Y+Z=n$. (***I.C.S. 1985***)

Hint. $P(X=x\mid X+Y+Z=n)=P(X=x,\ X+Y+Z=n)/p(X+Y+Z=n)$

$$=P(X=x)\,P(Y+Z=n-x)/P(X+Y+Z=n)$$

$$=\frac{e^{-a}\,a^x}{x!}\cdot\frac{e^{-(b+c)}\,(b+c)^{n-x}}{(n-x)\,!}\div\frac{e^{-(a+b+c)}\,(a+b+c)}{n\,!}$$

$$={}^nC_x\,p^x\,(1-p)^{n-x},\ p=a/(a+b+c).$$

Conditional expectation $\qquad = np.$

65. A drunk performs a random walk over positions $0,\pm 1,+2,\ldots$ as follows:- He starts at 0. He takes successive one unit steps going to the right with probability p and to the left with probability $1-p$, his steps are independent. Lt X denote his position after n steps. Find the distribution of $(X+n)/2$ and then Ind EX.

[**Hint.** Let Y denote the number of steps to the right. Then the number of steps to the left will be $n-Y$ and hence

$X=$ net position after n steps $=Y-(n-Y)=2Y-n$

$\Rightarrow Y=(X+n)/2$, the number of steps to the right.

Since Y follows binomial distribution with parameters, p, n_c.

so is the distribution of $(X+n)/2$.

is $EX=2np-n=n(2p-1)=n(p-q),\ q=1-p.$]

66. Prove or disprove:

$P(X>Y)=1$ if and only if $E(X)<E(Y)$ (***B.A. Hons. Delhi, 1994***)

Ans. Not true.

Hint. Consider any joint probability distribution of X and Y.

Show that $E(X)>E(Y)$ though $P(X>Y)=P(X=1,Y=0)+$

$P(X=3,Y=0)+P(X=3,Y=2)<1.$

6

Special Discrete Probability Distributions

6·1. Introduction. In chapter 5, we explained what is meant by discrete probability distribution. In this chapter, we shall study some important discrete probability distributions.

6·2. *Binomial Distribution. Suppose that a trial is repeated, so that we have a series of n trials. Let us call the occurrence of an event a 'success' and its non-occurrence a'failure'. Let p be the probability of a success and q be the probability of a failure in a single trial, so that $p+q=1$. We shall assume that the trials are *independent* and the probability p of successes is the same in every trial. The number of successes in n trials may be 0, 1,2, ..., r, ..., n and is obviously a random variate. Let us find the probability that the variate takes a particular value, say r. By the theorem of compound probability, the probability that the first r trials are successes and the remaining $n-r$ are failures, that is, the probability of the sequence of successes and failures being $\underbrace{S\,S.....S}_{r}\ \underbrace{F\,F.....F}_{n-r}$, where S denotes a success and F a failure is $p^r q^{n-r}$

Clearly, this is also the probability for the r successes and $n-r$ failures to occur in *any particular-definite specified order*. But we are interested in any r trials being successes and since r trials can be chosen out of n in $\binom{n}{r}$ mutually exclusive ways, by the theorem of total probability, the chance $P(r)$ of r successes in a series of n independent trials is given by

$$P(x=r)=p\,(r)=\begin{cases}\binom{n}{r}p^r q^{n-r} & =b\,(r\,;n,p),\ \ r=0,1,2\ldots r,n\ldots(6\cdot1)\\ 0. & \text{otherwise},\quad q=1-p\end{cases}$$

Thus the random variate, X *i.e.*, the number of successes, can take

---

* The binomial distribution was discovered early in the eighteenth century by the Swiss mathematician, Jakob Bernoulli (1654–1705). It was published posthumously in 1713.

the non-negative values, 0, 1, 2, ..., r ..., n with corresponding probabilities $q^n, \binom{n}{1} q^{n-1} p, \binom{n}{2} q^{n-2} p^2, ..., \binom{n}{r} q^{n-r} p^r, ..., p^n$.

The probability distribution of the number of successes so obtained is called the binomial probability distribution for the obvious reason that the probabilities are the successive terms in the binomial expansion of $(q+p)^n$.

The sum of the probabilities

$$= \sum_{r=0}^{n} b(r\,; n, p) = \sum_{r=0}^{n} \binom{n}{r} p^r q^{n-r} = (q+p)^n = 1.$$

The binomial distribution contains two independent constants, *viz.*, n and p (or q). They are called *parameters* of the binomial distribution. The distribution function is given by

$$F(x) = 0, x < 0\,; F(x) = \sum_{t=0}^{[x]} \binom{n}{t} p^t q^{n-t}, 0 \le x \le n\,;$$

$$F(x) = 1, x \ge n,$$

where $[x]$ denotes the integral part of x.

If $p = q = \frac{1}{2}$, the binomial distribution is *symmetrical* distribution and when $p \ne q$, it is a skew distribution.

In the figures 6·1 (*a*) and (*b*) are plotted two binomials for

(*a*) $p = \cdot 5, \quad n = 6 \qquad$ (*b*) $p = \cdot 1, \quad n = 20$

Ex. 1. Show that the probability of at least r successes in n Bernoullian trials is

$$= \binom{n}{r} p^r q^{n-r} + \binom{n}{r+1} p^{r+1} q^{n-r-1} + ... + q^n = \sum_{x=r}^{n} \binom{n}{x} p^n q^{n-x}$$

and that of at most r successes is $\sum_{x=0}^{r} \binom{n}{x} p^x q^{n-x}$.

Ex. 2. A and B play a game in which A's chance of winning is 2/3. In a series of 8 games, what is the chance that A will win atleast 6 games ?

Ans. $\dfrac{1024}{2187} = 0{\cdot}46828$

Ex. 3. Let X be a B.D. with parameter p and index n, then

(*i*) $E[(a+bx)^{-1}] = \sum_{i=1}^{n} (a+bi)^{-1} \binom{n}{i} p^i q^{n-i} = \int_0^1 x^{a-1} (q+px^b)^n \, dx, a > 0, b > 0.$

(*ii*) $P(X \le 2) = P(X \ge n-2)$ iff $p = \frac{1}{2}$ **(*B.Sc. Cal. 1989*)**

6·2·1. Binomial Frequency Distribution. Let the n *independent* trials constitute one experiment and let this experiment be repeated N

times. Then we expect r successes to occur $f(r) = Np(r) = N\binom{n}{r}p^r q^{n-r}$ times. This will be called the *expected frequency* of r successes in N experiments and the possible number of successes, together with the expected frequencies will be said to constitute the *binomial (expected) frequency distribution*. In practice, the observed frequencies will differ from the expected frequencies due to chance causes. For N sets, each of n trials, the expected frequencies of 0, 1, 2, ..., n successes are given by the successive terms in the binomial expansion of $N(q+p)^n$, $q+p=1$. [See Fig. 6·1 (*a*) and 6·1 (*b*)]

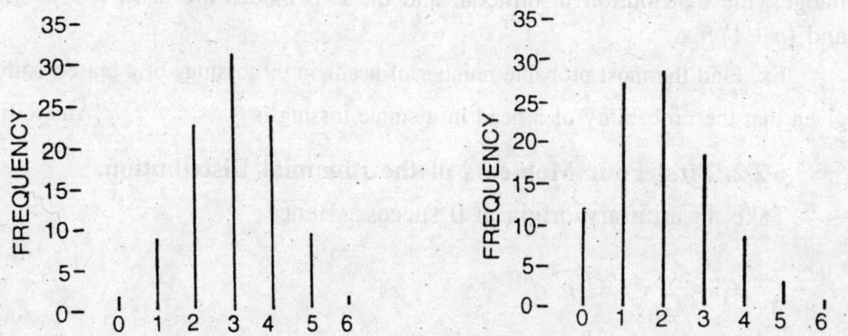

Fig. 6·1 (*a*)(*b*) Graph of Binomial Probability Distribution.

Example 6·1. *Find the most probable numer of successes in a series of n independent trials, the probability of successes in each trial being p.*

(*B.A., Hons. Delhi*)

We are to find the number of successes which has a greater probability than any other.

The chance of r successes will be greater than or equal to that of $r-1$ of $r+1$ successes if

$$b(r-1\,;n,p) \le b(r:n,p) \ge b(r+1\,;n,p)$$

or $$^nC_{r-1}\,p^{r-1}\,q^{n-r+1} \le {}^nC_r\,p^r\,q^{n-r} \ge {}^nC_{r+1}\,p^{r+1}\,q^{n-r+1},\ q = 1-p$$

Dividing each term by the middle term and simplifying, we get

$$\frac{r}{n-r+1}\cdot\frac{q}{p} \le 1 \ge \frac{n-r}{r+1}\cdot\frac{p}{q}.$$

i.e., if $rq \le np - rp + p$ and $rq + q \ge np - rp$

i.e., if $np - q \le r \le np + p$ or $(n+1)p - 1 \le r \le np + p$. ...(6·2)

Case I. $(n+1)p = k$ (an integer) then $P(r)$ increases till $r = k-1$ and then $P(k-1) = P(k)$ and after that it begins to decrease.

Thus if $p = \frac{2}{3}$ and $n = 8$, then $(n+1)p - 1 = 9 \times \frac{2}{3} - 1 = 5$ so that 5 and 6 are the most probable numbers.

Case II. $np = k + f$ where k is an integer and f is a positive proper function, $P(r)$ is maximum when $r = [np + p]$ where $[x]$ means the integral part of x.

Thus if $p = \frac{2}{3}$, the most probable number of successes in 10 trials is $[\frac{20}{2} + \frac{2}{3}] = [\frac{22}{3}]$, that is 7.

The most probable number of successes obviously gives the mode of the Binomial distribution and its value is $[np + p]$. If $(n + 1)\,p - 1$ is an integer, the distribution in bimodal and the two modes are at $(n + 1)\,p - 1$ and $(n + 1)\,p$.

Ex. Find the most probable number of heads in 99 tossings of a biased coin given that the probability of a head in a single tossing is $\frac{3}{5}$. **[Ans.** 60]

6·2·2. First Four Moments of the Binomial Distribution.

Take an arbitrary origin at 0 success. Hence

$$\mu_1' = \sum_{r=0}^{n} r\,.\,{}^nC_r p^r q^{n-r}.$$

$$= 0\,q^n + 1\,.\,{}^nC_1\,q^{n-1}\,p + 2\,.\,{}^nC_2\,q^{n-2}\,p^2 + \ldots + np^n$$

$$= np\,[q^{n-1} + {}^{n-1}C_1\,q^{n-2}\,p + {}^{n-1}C_2\,q^{n-3}\,p^2 + \ldots + p^{n-1}]$$

$$= np\,(q + p)^{n-1} = np. \qquad \ldots(6\cdot3)$$

$$\mu_2' = \sum_{r=0}^{n} r^2\,{}^nC_r p^r q^{n-r} = \sum_{r=0}^{n} [\{r + r(r-1)\}]\,{}^nC_r p^r q^{n-r}]$$

$$= \sum_{r=0}^{n} r\,.\,{}^nC_r p^r q^{n-r} + \sum_{r=0}^{n} r(r-1)\,{}^nC_r p^r q^{n-r}$$

$$= np + n\,(n-1)\,p^2 \sum_{r-2}^{n-r} C_{r-2}^{n-2}\,p^{r-2}\,q^{n-r} = np + n\,(n-1)\,p^2\,(q + p)^{n-2}$$

$$= np\,[1 + (n-1)p] = np\,(q + np) = npq + n^2p^2$$

$$\therefore\ \mu_2 = \mu_2' - \mu_1'^2 = npq \qquad \ldots(6\cdot4)$$

Since $npq < np$, $q < 1$ it follows that for the binomial distribution $(q + p)^n$, mean > *variance.*

Again, $\mu_3' = \sum_{r=0}^{n} r^3\,.\,{}^nC_r p^r q^{n-r}$

Writing r^3 as $r(r-1)(r-2) + 3r\,(r-1) + r$ and proceeding in a similar manner, we have

$$\mu_3' = n(n-1)(n-2)p^3(q+p)^{n-3} + 3n(n-1)p^2(q+p)^{n-2} + np(q+p)^{n-1}$$

$$\mu_3' = n(n-1)(n-2)p^3 + 3n(n-1)p^2 + np$$

$$\mu_3 = \mu_3' - 3\mu_2'\mu_1' + 2\mu_1'^3 = npq(q-p), \text{ on simplification.} \quad ...(6\cdot5)$$

Writing r^4 as $r(r-1)(r-2)(r-3) + 6r(r-1)(r-2) + 7r(r-1) + r$, we get

$$\mu_4' = n(n-1)(n-2)(n-3)p^4 + 6n(n-1)(n-2)p^3 + 7n(n-1)p^2 + np.$$

Easily we find

$$\mu_4 = 3n^2p^2q^2 + npq + (1-6pq) = npq[1+3(n-2)pq]. \quad ...(6\cdot6)$$

Hence $\beta_1 = \dfrac{\mu_3^2}{\mu_2^3} = (npq)^2 \dfrac{(q-p)^2}{(npq)^3} = \dfrac{(1-2p)^2}{npq}, \gamma_1 = \dfrac{1-2p}{\sqrt{npq}}$

$$\beta_2 = \frac{\mu_4}{\mu_2^2} = 3 + \frac{1-6pq}{npq}, \gamma_2 = \frac{1-6pq}{npq}.$$

Note. 1. The skewness is positive for $p < \frac{1}{2}$, negative for $p > \frac{1}{2}$ and zero for $p = \frac{1}{2}$.

2. As n, the number of trials increases indefinitely, $\beta_1 \to 0$ and $\beta_2 \to 3$; both γ_1 and γ_2 tend to zero.

6·2·3. Recurrence Relation for binomial Probabilities. We have

$$b(x; n, p) = f(x) = \frac{n!}{x!(n-x)!} p^x q^{n-x} \quad ...(a)$$

and $$b(x+1); n, p) = f(x+1) = \frac{n!}{(x+1)!(n-x-1)!} p^{x+1} q^{n-x-1} \quad ...(a')$$

Dividing (a') by (a), $$f(x+1) = \frac{n-x}{x+1} \cdot \frac{p}{q} f(x) \quad ...(6\cdot7)$$

(6·7) is called the recurrence relation for probabilities since $f(x+1)$ can be obtained if $f(x)$, n and q are known.

Example 6·2. Fitting Binomial Distribution. *The following data are the number of seeds germinating out of 10 on damp filter paper for 80 sets of seeds. Fit a binomial distribution to these data :*

x :	0	1	2	3	4	5	6	7	8	9	10	Total
f :	6	20	28	12	8	6	0	0	0	0	0	80

Here $n = 10$, $N = 80$ and $\Sigma f_i = 80$

Also the arithmetic mean $= \dfrac{\Sigma f_i x_i}{\Sigma f_i} = \dfrac{174}{80}$ *i.e.*, by (6·3) $np = \dfrac{174}{80}$

$$\therefore \; p = \frac{174}{800} = 0{\cdot}2175, \; q = 1 - p = {\cdot}7825.$$

Hence the binomial distribution to be fitted is $80\,({\cdot}7825 + {\cdot}2175)^{10}$. From (6·7), the successive terms in the expansion of this give the expected or theoretical frequencies of 0, 1, 2, ..., 10 successes. These are approximately.

6·89, 19·14, 23·94, 17·74, 8·63, 2·88, 0·67, 0·1, 0·01, 0·00, 000

Example 6·3. *Prove for the Binomial Distribution the following recurrence relation for central moments :*

$$\mu_{r+1} = pq\left(nr\,\mu_{r-1} + \frac{d\mu_r}{dp}\right).$$

By means of this relation, find μ_2, μ_3 *and* μ_4.

We have $\quad \mu_r = \sum_{j=0}^{n} f_j\,(j - \mu_1')^r = \sum_{j=0}^{n} {}^nC_j\, p^j\, q^{n-j}\,(j - np)^r$

Hence $\quad \dfrac{d\mu_r}{dp} = \Sigma\, {}^nc_j\, jp^{j-1}\, q^{n-j}\,(j - np)^r - \Sigma(n-j)\, {}^nc_j p^j q^{n-j-1}\,(j - np)^r)$

$$- nr\Sigma^n c_j\, p^j q^{n-j}\,(j - np)^{r-1}$$

$$= \sum_{j=0}^{n} \left[\, {}^nc_j\, p^{j-1} q^{n-j-1}\,(j - np)^r\,(jq - np + jp)\right] - nr\mu_{r-1}$$

$\therefore$

$$pq\frac{d\mu_r}{dp} = \sum_{j=0}^{n} {}^rc_j\, p^j q^{n-j}\,(j - np)^{r+1} - pq.nr\,\mu_{r-1} \quad \text{or } \mu_{r+1} = pq\left(\frac{d\mu_r}{dp} + nr\mu_{r-1}\right).$$

Putting r = 1, 2, 3 respectively, we get

$$\mu_2 = pq\left[\frac{d\mu_1}{dp} + n{\cdot}1\,\mu_0\right] = pq\,(0 + n) = npq, \text{ since } \mu_0 = 1$$

$$\mu_3 = pq\left[\frac{d\mu_2}{dp} + n{\cdot}2\mu_1\right] = pq\,[nq - np + 2n\,.\,0] = npq\,(q - p)$$

$$\mu_4 = pq[n(6p^2 - 6p + 1) + 3n\,.\,np\,(1 - p)]$$

$$= npq\,[1 - 6p\,(1 - p) + 3npq] = 3n^2p^2q^2 + npq\,(1 - 6pq).$$

Example 6·4. *A random sample of size n is drawn from the infinite population in which the individuals fall into one of the other of the two categories with probabilities p and 1 – p. The numbers in the sample falling into the two categories are* n_1 *and* n_2 $(n_1 + n_2 = n)$. *Show that the covariance between* n_1 *and* n_2 *is* $-np\,(1 - p)$. *Hence obtain the variance of the difference of the proportions* $n_1/n - n_2/n\,\cdot$

(*a*) Clearly n_1 is a Binomial variate.

$$\because \quad E(n_1) = np, E(n_2) = E(n - n_1) = n - E(n_1) = n - np = nq.$$

$$\therefore \quad Cov(n_1, n) = E(n_1 - np)(n_2 - nq) = E(n_1 - np)(n - n_1 - n + np)$$

$$= - E(n_1 - np)^2 = - Var(n_1) = - npq = - np(1 - p).$$

$$(b)\ Var\left(\frac{n_1}{n} - \frac{n_2}{n}\right) = Var\frac{n_1}{n} + Var\left(\frac{n_2}{n}\right) - 2\,Cov\left(\frac{n_1}{n}, \frac{n_2}{n}\right)$$

$$= \frac{1}{n^2} Var(n_1) + \frac{1}{n^2} Var(n_2) - \frac{2}{n^2} Cov(n_1, n_2)$$

$$= \frac{1}{n^2} \cdot [npq + npq + 2npq] = \frac{4pq}{n}.$$

Ex. If X is a Binomial variate with the probability law

$$P(X = x) = {}^nC_x p^x q^{n-x},\ q = 1 - p,\ x = 0, 1, 2, \ldots, n$$

prove that $Cov\left(\frac{x}{n}, \frac{n-x}{n}\right) = \frac{-pq}{n}$ **(*M.A., Statistics, Delhi*)**

6·2·4. Moment Generating Function and Cumulative Function of the Binomial Distribution

The probability generating function $= \sum_{x=0}^{n} t^x \cdot {}^nC_x p^x q^{n-x} = (q + pt)^n$.

M.G.F. about origin is given by

$$M_{x=0}(t) = E(e^{tx}) = \sum_{x=0}^{n} e^{tx}\, {}^nC_x p^x q^{n-x} = \sum_{x=0}^{n} {}^nC_x q^{n-x} (pe^t)^x$$

$$= (q + pe^t)^n \quad \ldots(6\cdot8)$$

Differentiating $M_0(t)$ once, twice, etc., with respect to t and putting $t = 0$ and using (5·12), we get

$$\mu_1' = np, \qquad \mu_2' = npq + n^2p^2,\ \mu_2 = npq$$

$$\mu_3 = npq(q - p), \qquad \mu_4 = npq[1 + 3(n - 2)pq].$$

Since $M_a(t) = e^{-at} M_0(t)$ we get the *m.g.f.* of the binomial distribution with respect to its mean np.

$$M_{\mu1}'(t) = e^{-npt} (q + pe^t)^n = (qe^{-pt} + pe^{qt})^n. \quad \ldots(6\cdot8')$$

Writing the values of the exponentials and arranging, we get

$$M_{\mu1}'(t) = \left[1 + pq\frac{t^2}{2\,!} + pq(q^2 - p^2)\frac{t^3}{3\,!} + pq(q^3 + p^3)\frac{t^4}{4\,!} + \ldots\right]^n$$

Expanding the right side and equating the coefficients of like powers of t on either side, we deduce that

$$\mu_2 = npq,\ \mu_3 = npq\,(q-p),\ \mu_4 = npq\,[1 + 3(n-2)pq]$$

The cumulative function of the binomial distribution about its mean np is given by $K(t) = \log\,(qe^{-pt} + pe^{-qt})^n$

$$= n \log \left[1 + pq\frac{t^2}{2\,!} + pq\,(q^2 - p^2)\frac{t^3}{3\,!} + pq(q^3 + p^3)\frac{t^3}{4\,!} + \ldots \right]$$

$$= n \left[pq\frac{t^2}{2\,!} + pq(q^2 - q^2)\frac{t^3}{3\,!} + pq(q^3 + p^3)\frac{t^4}{4\,!} + \ldots \right]$$

$$- \frac{n}{2}\left[pq\frac{t^2}{2\,!} + \ldots \right]^2 + \ldots \qquad \ldots(1)$$

Equating the coefficients of $\frac{t^2}{2\,!}, \frac{t^3}{3\,!}, \frac{t^4}{4\,!}$ on either side we get

$$k_2 = npq\ ;\ k_3 = npq\,(q^2 - p^2) = npq(q-p),$$

$$k_4 = npq\,(q^3 + p^3) - 3np^2q^2 = npq[(q+p)^2 - 3pq(q+p) - 3pq]$$

$$= npq(1 - 6pq).$$

Example 6·5. *Show that the factoral moment generating function $w(t)$ of the Binomial Distribution is $(1 + pt)^n$. Hence or otherwise, show that $\mu'_{(r)} = n^{(r)}p^r$.*

(*M.A. Delhi, 1969*)

First Method. $w(t) = E[1+t]^x = \sum_{x=0}^{n} {}^nC_x p^x q^{n-x} \,.\, (1+t)^x$

$$= [q + p(1+t)]^n = (1 + pt)^n \qquad \ldots(6\cdot9)$$

$\therefore\ \mu'_{(r)} = coeff.$ of $t^r/r\,!$ in the expansion of $w(t) = n^{(r)}p^r$.

Second Method. Differentiating $(q+p)^n$ r times partially with respect to p and multiplying by p^r, we have on remembering that

$$n^{(r)}(q+p)^{n-r}p^r = \sum_{j=0}^{n} j^{(r)}\, {}^nC_j\, p^j q^{n-j} = \mu'_{(r)} \quad \text{or} \quad \mu'_{(r)} = n^{(r)}p^r.$$

Thus $\mu'_{(1)} = np$, $\mu'_{(2)} = n(n-1)p^2$, $\mu_{(2)} = \mu'_{(2)} - \mu'^2_{(1)} = npq$.

6·2·5. Truncated Binomial distribution [See Ex. 50, page 216]

6·3. *Poisson Distribution. The binomial distribution assumes a very convenient form in the limiting case when p (or q) is very small and n is large and the average number of successes np is a finite constant m.

* It was first discovered by the French Mathematics S.D. Poisson [1781–1840] in 1837. It is also sometimes known as "The Law of Small Numbers".

In the binomial distribution, the probability of r successes is given by $P(r) = {}^nC_r p^r q^{n-r}$

We write this as $P(r) = \dfrac{n!}{r!(n-r)!}\left(\dfrac{m}{n}\right)^r\left(1-\dfrac{m}{n}\right)^{n-r}$ $\left[\because p = \dfrac{m}{n}\right]$

$$= \frac{m^r}{r!}\left(1-\frac{m}{n}\right)^n \cdot \frac{\left(1-\frac{m}{n}\right)^{-r}}{n^r} \cdot \frac{n!}{(n-r)!}$$

By making n tend to infinity, we have

$$\lim_{n\to\infty} P(r) = \frac{m^r}{r!} e^{-m} \lim_{n\to\infty}\left[\frac{1}{n^r}\cdot\frac{n!}{(n-r)}\right]$$

Using** Stirling's formula for $n!$, *viz*, $n! \sim \sqrt{2\pi}\, e^{-n}\, n^{n+1/2}$ where e is the Napierian base, 2·71828..., we have

$$\lim_{n\to\infty} P(r) = \frac{m^r e^{-m}}{r!} \lim_{n\to\infty} \frac{\sqrt{2\pi}\, e^{-n} n^{n+1/2}}{n^r \sqrt{2\pi}\, e^{-n+r} (n-r)^{n-r+1/2}}$$

$$= \frac{m^r e^{-m}}{r!\, e^r} \lim_{n\to\infty}\left\{\frac{1}{\left(1-\frac{r}{n}\right)^n \left(1-\frac{r}{n}\right)^{-r+1/2}}\right\}$$

$$P(r) = \frac{m^r e^{-m}}{r!}, \quad \text{since } e^r \cdot e^{-r} = 1, r \text{ being finite} \qquad ...(6\cdot10)$$

The chances of 0, 1, 2, ... successes are e^{-m}, me^{-m}, $\dfrac{m^2e^{-m}}{2!}$, ... respectively. In other words, the random variate, the number of successes

---

** Stirling approximation formula which states that the limiting value of the ratio of $n!$ to $\sqrt{2\pi}\, e^{-n}\, n^{n+1/2}$ approaches unity as n approaches infinity, was in fact given by Abraham De moivre [1667-1754] in his book 'Miscellaneous Analysis' in 1930. In the same year James Sterling [1692-1770] gave the following series

$$\log n! \simeq \left(n+\frac{1}{2}\right)\log n - n + \log\sqrt{2\pi} + \frac{B_2}{1\cdot2}\cdot\frac{1}{n} + \frac{B_4}{3\cdot4}\cdot\frac{1}{n^3} + \dots$$

where B's are Bernoulli numbers [defined earlier by James Bernoulli (1654-1705)]. He used it to calculate $\log_{10}$ (1000 !) correctly to ten decimal places, though the form in which we use Stirling's formular, was given by De Moivre, Stirling' series gives a better approximation and that is why the approximation is called Stirling's approximation to n !. By calculation with a pocket calculator we find

	1	2	3	4	5	10
$\dfrac{\sqrt{2\pi}\, e^{-n}\, n^{n+1/2}}{n!}$	·9221	·9591	·9727	·9799	·9835	·9917
	20	30	40	50	60	70
$\dfrac{\sqrt{2\pi}\, e^{-n}\, n^{n+1/2}}{n!}$	·9956	·9912	·9979	9983	·9986	·9988

assumes the values 0, 1, 2, ..., r with the corresponding probabilities $e^{-m}, me^{-m}, \frac{m^2e^{-m}}{2!}, ..., \frac{m^re^{-m}}{r!}, ...$

The probability distribution of the number of successes (with the assumptions mentioned above) is called the Poisson distribution.

The sum of the probabilities $= \sum_{r=0}^{\infty} \frac{m^r e^{-m}}{r!} = e^{-m} \sum_{r=0}^{\infty} \frac{m^r}{r!} = e^{-m} e^m = 1.$

The Poisson Distribution involves only one parameter *viz.*, m.

Fig. 6·2 shows the Poisson distributions from $m = 3$ and $m = 7$.

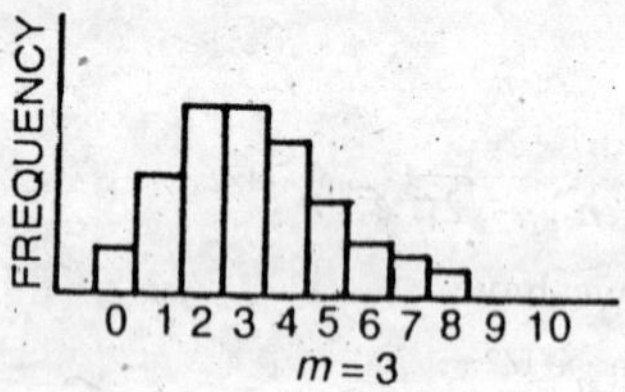

Fig. 6·2(*a*)

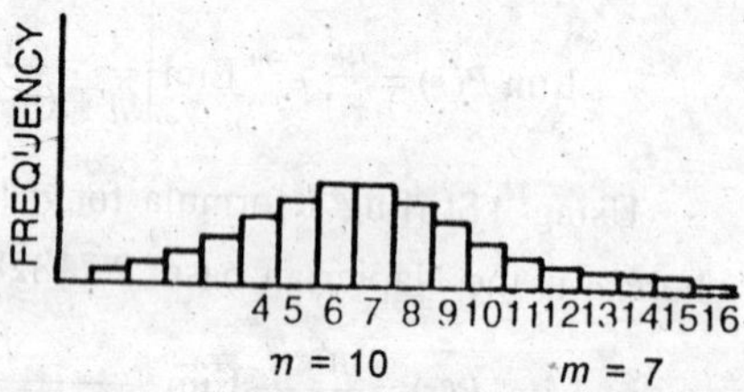

Fig. 6·2(*b*)

The following are some examples of Poisson variates :

(1) The number of typographical errors per page in typed material.

(2) The number of defective screws per box of 100 screws.

(3) The number of bacterial colonies in a given culture per unit area of microscope slab.

(4) The number of deaths in a district in one year by a rare disease.

In each of the above examples, the probability of success is small and n is large.

Ex. Show that $\lim_{n \to \infty} \frac{n!}{(n-r)!} \cdot \frac{1}{n^r} = \text{Lim}_{n \to \infty} \frac{n(n-1) \dots (n-r+1)}{n \,.\, n \dots n \text{ (r times)}}$

$$= \lim_{n \to \infty} \left(1 - \frac{1}{n}\right) \dots \left(1 - \frac{r-1}{n}\right) = 1.$$

Hence show how the result of section 6·3 can be derived without using Stirling formula.

Alternative Derivation of Poisson Distribution.

$$\text{Lim}_{n \to \infty} P(0) = \text{Lim}_{n \to \infty} q^n = \text{Lim}_{n \to \infty} (1-p)^n = \text{Lim}_{n \to \infty} \left(1 - \frac{m}{n}\right)^n = e^{-m}$$

Also $\text{Lim}_{n \to \infty} \frac{P(r+1)}{P(r)} = \text{Lim}_{n \to \infty} \frac{n-r}{r+1} \cdot \frac{p}{q} = \text{Lim}_{n \to \infty} \frac{n-r}{r+1} \cdot \frac{m/n}{1-m/n}$

$$= \lim_{n\to\infty} \frac{n-r}{n} \cdot \frac{1}{1-m/n} \cdot \frac{m}{r+1}$$

$$= \lim_{n\to\infty} \frac{1-r/n}{1-m/n} \cdot \frac{m}{r+1} = \frac{m}{r+1}.$$

$$\therefore \lim P(r) = \lim \frac{P(r)}{P(r-1)} \cdot \lim \frac{P(r-1)}{P(r-2)} \ldots \lim \frac{P(1)}{P(0)} \lim P(0)$$

$$= \frac{m}{r} \cdot \frac{m}{r-1} \ldots \frac{m}{1} \cdot e^{-m} = \frac{e^{-m} m^r}{r\,!}.$$

In this derivation, we have **not** employed Stirling's formula.

Another *Derivation of Poisson Distribution*

The *p.g.f.* of the Binomial distribution is $P(s) = \sum_{k=0}^{n} P(X=k)\, s^k$

$$P(s) = \sum_{k=0}^{n} \binom{n}{k} (ps)^k q^{n-k} = [ps+q]^n,\ q = 1-p$$

Putting $np = m$, we have $P(s) = \left[1 + \frac{m}{n}(s-1)\right]^n$

$$\therefore\ P(s) = \operatorname*{Lt}_{n\to\infty} P(s) = e^{m(s-1)} = e^{-m} \sum_{r=0}^{\infty} \left(\frac{ms}{r}\right)^r$$

$$\therefore\ P(x=r) = \frac{e^{-m} m^r}{r\,!}.$$

Example 6·6. *If X is a random Poisson variate, with parameter m show that* $P(x \ge n) - P(x \ge n+1) = e^{-m} m^n / n\,!$ **(B.A., Delhi, 1971)**

L.H.S. $= [P(x=n) + P(x=n+1) + \ldots] - [P(x=n+1) + P(x=n+2) + \ldots]$

$= P(x=n) = e^{-m} m^n / n\,! =$ R.H.S.

6·3·1. Another Approach to Poisson Distribution. Let the random variate X be the number of calls during time t at a telephone switch board. We assume (*a*) that the probability of a call in time dt is $\lambda\, dt$, (*b*) that the calls are independent.

Let $P_x(t)$ denote the probability of x calls in time t so that $P_x(t+dt)$ is the chance of x calls in time $t + dt$. There are two mutually exclusive possibilities.

(*i*) x calls in time t and no call in time dt.

(*ii*) $x - 1$ calls in time t and 1 call in time dt.

[The possibility of more than one call in time dt is neglected as it would be of order $(dt)^2$].

$$\therefore \qquad P_x(t+dt) = P_x(t)(1-\lambda\, dt) + P_{x-1}(t)\lambda\, dt$$

or
$$\frac{P_x(t+dt) - P_x(t)}{dt} = \lambda\,[P_{x-1}(t) - P_x(t)]$$

Taking limit as $dt \to 0$, we have

$$\frac{dP_x(t)}{dt} = \lambda\,[P_{x-1}(t) - P_x(t)],\ x = 0, 1, 2, \ldots$$

Putting
$$P_x(t) = \frac{(\lambda t)^x}{x\,!} f(t),\ \frac{dP\,(t)}{dt} = \lambda\,[P'_{x-1}(t) - P_x'(t)] \qquad \ldots(1)$$

we get
$$\frac{(\lambda t)^x}{x\,!} f'(t) + f(t)\frac{\lambda^x}{x\,!} x t^{x-1} = \lambda\left[\frac{(\lambda\, t)^{x-1}}{(x-1)\,!} f(t) - \frac{(\lambda\, t)^x}{x\,!} f(t)\right]$$

or
$$\frac{tf'(t)}{x} + f(t) = f(t) - \frac{\lambda t}{x} f(t) \qquad \text{or} \qquad \frac{f'(t)}{f(t)} = -\lambda$$

Integrating $f(t) = e^{\lambda t}$ since from (6·6′) $P_0(t) = f(t)$ so that $f(0) = P_0(0) = 1$.

Hence from (1), $P_x(t) = \dfrac{e^{-\lambda t}\,(\lambda t)^x}{x\,!},\ x = 0, 1, 2, \ldots$

Also
$$\sum_{x=0}^{\infty} P_x(t) = e^{-\lambda t} \sum_{x=0}^{\infty} \frac{(\lambda t)^x}{x\,!} = e^{-\lambda t}\,.\,e^{\lambda t} = 1.$$

Hence the number of calls in a fixed time t is a Poisson variate with parameter λt. For an alternative method of integrating the system of differential equations obtained above, see Ex. 5 below.

The following are further examples of Poisson variates :

(1) The number of cars passing through a certain street in time t.

(2) The number of suicides or deaths by heart attack in time t.

(3) The number or radioactive atoms decaying in time t.

(4) The number of cosmic-ray particles tripping a Geiger-Muller counter in time interval t.

(5) The number of pieces of a certain merchandise sold by a store in time t.

(6) The number of fragments received by a surface area t from a fragmentation bomb.

Ex. 1. Explain how in each of the above examples, the conditions assumed in the derivation of the Poisson distribution are satisfied.

Ex. 2. Give some examples of the occurrence of the Poisson distribution in different fields. ***(M.A. Delhi ; M.A., Patna)***

Ex. 3. If X is a random Poisson variate such that $P(X=1) = P(X=2)$, find $P(X=4)$.
$$\left\{\textbf{Ans. } \frac{2e^{-2}}{3}\right\}$$

Ex. 4. If $\phi(s, t) = \sum_{x=0}^{\infty} P_x(t)\, s^x$,

show that
$$\frac{1}{\phi}\frac{\partial \phi}{\partial r} = \lambda(s-1).$$

Integrate and determine the constant of integration by using $\phi(s, 0) = 1$. Show that $\phi(s, t) = e^{-\lambda t}\, e^{\lambda s t}$. Hence find $P_x(t)$

6·3·2. Mean and Variance of the Poisson Distribution.

Let X be a Poisson variate with probability

$$P(x = r) = m^r e^{-m}/r\,!,\ r = 0, 1, ..., \infty$$

$$\mu_1' = \sum_{r=0}^{\infty} \frac{r\,.\,m^r e^{-m}}{r\,!} = 0\,.\,e^{-m} + 1\,.\,\frac{me^{-m}}{1\,!} + \frac{2 \cdot m^2 e^{-m}}{2\,!} + \ldots$$

$$= me^{-m}\left(1 + \frac{m}{1\,!} + \frac{m^2}{2\,!} + \ldots\right) = me^{-m} e^m = m.$$

$$\mu_2' = \sum_{r=0}^{\infty} \frac{r^2 m^r e^{-m}}{r\,!} = \sum_{0}^{\infty}\left[\{r + r(r-1)\}\frac{m^r e^{-m}}{r\,!}\right]$$

$$= \sum_{0}^{\infty} \frac{r m^r e^{-m}}{r\,!} + m^2 \sum_{0}^{\infty} \frac{m^{r-2} e^{-m}}{(r-2)\,!}$$

$$= m + m^2 \qquad \left(\because \sum_{r=2}^{\infty} \frac{m^{r-2}}{(r-2)\,!} = 1 + m + \frac{m^2}{2\,!} + \ldots = e^m\right)$$

$$\therefore\ \mu_2 = \mu_2' - \mu_1'^2 = m$$

Thus for Poisson distribution, mean = variance. ...(6·10)′

Note. These results can be deduced from the mean and variance of the binomial distribution by putting $p = m/n$ and then letting n tend to infinity. It should be noted that mean and variance of Poisson probability distribution have the same value m.

Ex. 1. Prove that $\mu_3' = m[(m+1)^2 + m]$, $\mu_4' = m[m^3 + 6m^2 + 7m + 1]$. Deduce that $\mu_3 = m$, $\mu_4 = 3m^2 + m$.

Ex. 2. Deduce the first four moments about the mean of the Poisson distribution from those of the binomial distribution.

Ex. 3. Prove that, as m tends to infinity, both γ_1 and γ_2 tend to zero.

Ex. 4. Find $nc_r p^r q^{n-r}$ when $n = 4, p = \frac{1}{4}$; $n = 8, p = \frac{1}{8}$;

$$n = 16, p = \frac{1}{16} \text{ for } n = 0, 1, 2, 3, 4, 5.$$

Find also $\qquad e^{-m}\dfrac{m^r}{r\,!}$ for $m = 1$ for $r = 0, 1, 2, 3, 4, 5$

Show how this illustrates the approach of the binomial to the Poisson distribution.

6·3·3. Moment Generating Function and Cumulative Function of the Poisson Distribution. *M.g.f.* and cumulative function of the Poisson distribution with respect to origin are given by

$$M_{X=0}(t) = E(e^{tx}) = \Sigma e^{tx}\, p(x) = \sum_{x=0}^{\infty} e^{tx} \frac{m^x e^{-m}}{x\,!} = e^{-m} \sum_{x=0}^{\infty} \frac{(me^t)^x}{x\,!}$$

$$= e^{-m} e^{me^t} = e^{m(e^t-1)} = \exp\{m(e^t - 1)\} \qquad \ldots(6\cdot 1')$$

$$K_X(t) = m\,(e^t - 1) = m\left(t + \frac{t^2}{2\,!} + \frac{t^3}{3\,!} + \ldots\right)$$

$$k_r = \text{coeff. of } \frac{t^r}{r\,!} \text{ in } K_X(t) = m \text{ for all } r.$$

Thus all the cumulants of the Poisson distribution are equal to m.

Ex. 1. Find the *m.g.f.* about the mean for the Poisson variate and deduce the values of the first four moments about the mean.

Ex. 2. If x is a Poisson variate with mean m, show that $(x - m)/\sqrt{m}$ is a variate with zero mean and unit variance. Find the *m.g.f.* for this variate and show that it approaches *exp* $(-\frac{1}{2}t^2)$ as m tends to infinity.

6·3·4. Reproductive Property of the Poisson Distribution

Theorem. *If two independent chance variables x_1 and x_2 have Poisson distributions with means m_1 and m_2 respectively, their sum $x_1 + x_2$ is a Poisson variable with mean $m_1 + m_2$.*

Proof. Let $M_1(t)$, $M_2(t)$ be the moment generating functions of x_1 and x_2 and let $M(t)$ be the *m.g.f.* of their sum, then

$$M_1(t) = \exp[m_1(e^t - 1)],\ M_2(t) = \exp[m_2(e^t - 1)]$$

By §5·6·2, the *m.g.f.* of the sum $x_1 + x_2$ is given by

$$M(t) = M_1(t) \times M_2(t) = \exp\,[m_1(e^t - 1)]\,\exp\,[m_2(e^t - 1)]$$

$$= \exp\,[(m_1 + m_2)(e^t - 1)]$$

but this is the *m.g.f.* of a Poisson distribution with mean $m_1 + m_2$. Hence the theorem.

Ex. 1. Show that if X_i $(i = 1, 2, \ldots, n)$ are independent variates having Poisson distribution with means m_i then $\sum_{i=1}^{n} X_i$ is a Poisson variate with mean $\sum_{i=1}^{n} m_i$

Ex. 2. Show that for Poisson distribution with parameter m, $f(r+1) = \frac{m}{(r+1)} f(r)$ and hence $f(1), f(2) \ldots$ can be obtained in terms of $f(0)$.

6·3·5. Mode of the Poisson Distribution. The mode is clearly that value of r for which $e^{-m}m^r/r\,!$ is greater than the term that precedes it and the term that follows it.

$$\therefore \qquad \frac{m^{r-1}\,e^{-m}}{(r-1)\,!} \leq \frac{m^r\,e^{-m}}{r\,!} \geq \frac{m^{r+1}\,e^{-m}}{(r+1)\,!}$$

These give $r \leq m$ and $r+1 \geq m$ *i.e.*, $m-1 \leq r \leq m$.

Thus if m is not an integer, the mode is the integral value between $m-1$ and m. If m is an integer, then the probability of the Poisson variate taking the value $m-1$ or m is the same and hence either can be taken as mode of the Poisson distribution with mean m. The distribution in this case is a bimodal one.

Example 6·7. *Find the probability that at most 5 defective fuses will be found in a box of 200 fuses if experience shows that 2 per cent of such fuses are defective.*

Hence $m = np = 200(\cdot 02) = 4$.

Hence using (6·6), the required probability is given by

$$P(x \leq 5) = \sum_{x=0}^{5} \frac{e^{-4}\,4^x}{x\,!} = e^{-4}\left(1 + 4 + \frac{4^2}{2} + \frac{4^3}{6} + \frac{4^2}{24} + \frac{4^5}{120}\right) = \cdot 785$$

$$[\because e^{-4} = 0{\cdot}0183]$$

Example 6·8. *Fit a Poisson distribution to the following data which gives the number of yeast cells per square for 400 squares.*

No. of cells per square (x)	0	1	2	3	4	5	6	7	8	9	10	Total
No. of squares (f)	103	143	98	42	8	4	2	0	0	0	0	400

The expected or theoretical frequency for r successes is $Ne^{m}m^r/r\,!$, but m is not given in this example. The mean of the Poisson distribution is m $\therefore$ mean m of the distribution of x is estimated by

$$m = \frac{\Sigma fx}{\Sigma f} = \frac{529}{400} = 1{\cdot}32$$

Assuming that the distribution of x is Poisson in the expected frequencies of 0, 1, 2, ..., 10 are the successive values of

$$400\,\frac{e^{-1\cdot 32}\,(1{\cdot}32)^x}{x\,!}$$

Since $e^{-1\cdot 32} = 0{\cdot}2674$, the expected frequencies of 0, 1, ..., 10 cells are 107, 141, 93, 41, 14, 4, 0, 0, 0, 0, 0.

6·3·6. Truncated Poisson Distribution

Suppose Poisson distribution with parameter m is truncated at the origin *i.e.* all the values of $x \le 0$ are discarded. Then

$$f(x) = P(x > 0) \times g(x) \text{ where } P(x = 0) = e^{-m}$$

$$\therefore \quad g(x) = \frac{P(x = m)}{1 - P(x \ge 0)} = \frac{e^{-m}\, m^x / x!}{1 - P(x \ge 0)} = \frac{e^{-m}\, m^x}{(1 - e^{-m})(x!)},\ x = 1, 2, 3, \ldots$$

which is truncated poisson distribution.

$$\text{Mean} = \text{E}X = \sum_{x=1}^{\infty} x g(x) = \sum_{x=1}^{\infty} \frac{x \cdot e^{-m}\, m^x}{(1 - e^{-m})\, x!}$$

$$= \frac{m\, e^{-m}}{(1 - e^{-m})} \sum_{x=1}^{\infty} \frac{m^{x-1}}{(x-1)!} = \frac{m e^{-m}}{1 - e^{-m}} \cdot e^m$$

$$= \frac{m}{1 - e^{-m}} \qquad \left(\because 1 + m + \frac{m^2}{2!} + \ldots = e^m \right)$$

$$\mu_2' = \text{E}x^2 = \Sigma\, x^2 g(x) = \Sigma\, \frac{x^2 \cdot e^{-m}\, m^x}{(1 - e^{-m})\, x!} = \frac{e^{-m}}{1 - e^{-m}} \sum_{x=1}^{\infty} \frac{[x(x-1) + 1]}{x!}\, m^x$$

$$= \frac{e^{-m}}{1 - e^{-m}} \left[m^2 \sum_{x=2}^{\infty} \frac{m^{x-2}}{(x-2)!} + m \sum_{1}^{\infty} \frac{m^{x-1}}{(x-1)!} \right]$$

$$= \frac{e^{-m}}{1 - e^{-m}} \left[m^2 e^m + m e^m \right] = \frac{m(m+1)}{1 - e^{-m}}.$$

$$\mu_2 = \mu_2' - \mu_1'^2 = \frac{m(m+1)}{1 - e^{-m}} - \frac{m^2}{(1 - e^{-m})^2}.$$

6·4. Other Discrete Distributions.

6·4·1. Causal Distribution. Here the variate X takes with certainty one value only, say $x = u$, so that

$P(x = u) = 1$; $P(x - t) = 0$; $t \ne u$. Also $P(x \le t) = 0$, $t < u$; $= 1$. $t \ge u$.

The distribution has one parameter. the distribution obviously corresponds to a causal description met in the exact sciences and this shows that the statistical description of nature includes the casual distribution as a particular case.

6·4·2. Uniform or Rectangular Distribution. Here the variate X takes the values $x_1, x_2, \ldots, x_n$ (and in particular the values 1, 2, ..., n) each with the same probability $1/n$ so that $P(x = x_i) = \frac{1}{n}$, 1, 2, ..., n

$$\text{Also} \qquad \sum_{i=1}^{n} P(x = x_i) = \sum_{1}^{n} \frac{1}{n} = \frac{1}{n} \cdot n = 1.$$

Such a distribution arises in the tossing of a coin or casting of a dice.

6·4·3. Geometric Distribution. In a Bernoullian sequence of trials, probability of success being p, let x be the number of failures preceding the first success then x is a random variate with probability distribution given by

$$P(x) = q^x p, \; x = 0, 1, 2, ..., q = 1 - p.$$

since the first x trials are to be failures and the last trial is to be a success.

Clearly,

$$\sum_{x=0}^{\infty} P(x) = \sum_{x=0}^{\infty} q^x p = p + qp + q^2 p + \ldots = p(1 + q + q^2 + \ldots) = \frac{p}{1-q} = 1.$$

$$M_{x=0}(t) = \sum_{r=0}^{\infty} e^{tx} q^x p = p \sum_{0}^{\infty} (qe^t)^x = \frac{p}{1 - qe^t} = p(1 - qe^t)^{-1}, \; t \le \log\left(\frac{1}{q}\right)$$

$$\mu_1' = \frac{d}{dt} M_{x=0}(t) \Big]_{t=0} = pqe^t (1 - qe^t)^{-2} \Big]_{t=0} = pq(1-q)^{-2} = qp^{-1}.$$

Ex. 1. Show that the mean and variance of the geometric distribution $P(x) = q^x p, x = 0, 1, 2, \ldots$ are respectively qp^{-1}. qp^{-2} and hence mean < variance.

(*B.A. Hons. Delhi*)

Ex. 2. Show that the mode of the distribution $P(x) = (\frac{1}{2})^x, x = 1, 2, 3, \ldots$ is 1.

6·4·4. Negative Binomial Distribution. Let $P(x)$ be the probability that exactly $x + k$ trials will be required to produce k successes in an indefinite series of Bernoullian trials. Clearly the last trial must be a success and its probability is p. Among the other $(x + k - 1)$ trials, there must be $k - 1$ successes and the probability of this is ${}^{x+k-1}c_{k-1} p^{k-1} \cdot q^x$, $q = 1 - p$.

Hence multiplying the two probabilities, we get

$$P(x) = {}^{x+k-1}c_{k-1} p^k q^x, x = 0, 1, 2, ..., \text{ and } k > 0 \qquad ...(6·11)$$

$$\therefore \quad \sum_{0}^{\infty} P(x = k) = p^k \sum_{x=0}^{\infty} {}^{x+k-1}c_{k-1} q^x = p^k (1 - q)^{-k} = 1.$$

Since $\quad {}^{x+k-1}c_{k-1} = \dfrac{(k + x - 1)(k + x - 2) \ldots k}{x!} = (-1)^x \, {}^{-k}C_x$

where $\quad {}^{-k}c_x = \dfrac{(-k)(-k-1) \ldots (-k - x + 1)}{x!} = (-1)^x \, {}^{x+k-1}C_{k-1}$

Thus $P(x)$ can also be written as

$$P(x) = {}^{-k}c_x p^k (-q)^x, x = 0, 1, 2, ... \qquad ...(6·11)$$

This distribution is called *Pascal's distribution.* It contains two parameters p and k. If $k = 1$ it reduces to the geometric distribution and in this sense it is a generalisation of this distribution.

In the above distribution, k is an integer but the distribution remains meaningful even if k is not an integer but $k \geq 0$. In this case the distribution is called the Negative *Binomial Distribution.*

Alternative forms of this distribution are obtained by putting

(*i*) $k = \frac{1}{\beta}, p = \frac{1}{1+\beta\mu}, q = \frac{\beta\mu}{1+\beta\mu}$, so that

$$P(x) = \frac{(1+\beta)(1+2\beta)\ldots[1+(x-1)\beta]}{x!}(1+\beta\mu)^{-\frac{1}{\beta}}\left(\frac{\mu}{1+\beta\mu}\right)^x \quad \ldots(6\cdot12)$$

In this form the distribution with two parameters β and μ is known as *Polya's distribution.*

If $\beta = 1$, then $P(x) = \frac{1}{1+\mu}\left(\frac{\mu}{1+\mu}\right)^x$ ...(6·12)′

which is another form of geometric distribution.

(*ii*) $p = \frac{1}{Q}, q = \frac{P}{Q}$, so that $Q - P = 1$, since $p + q = 1$. Thus

$$P(x) = Q^{-k} \cdot {}^{x+k-1}C_{k-1}\left(\frac{P}{Q}\right)^x \quad \ldots(6\cdot12)''$$

$$= Q^{-k}\ {}^{-k}C_x\left(\frac{-P}{Q}\right)^x, x = 0, 1, 2, \ldots\ldots \quad \ldots(6\cdot12)''$$

which is the general term of the binomial expansion of $(Q-P)^{-k}$.

Let X be a random variable negative binomial law ${}^{-k}c_x\, p^k q^x$. Then the **probability generating function** of X is given by

$$P_x(s) = E(s^x) = \sum_{x=0}^{\infty} e^{sx} P(X=k) = \Sigma\ {}^{-k}c_x\, p^k(-q)^x . e^{sx} = p^k(1-q^s)^{-k}$$

The moment generating function about origin of the negative binomial distribution $(Q-P)^{-k}$ is given by

$$M_{X=0}(t) = \sum_{x=0}^{\infty} Q^{-k} e^{tx}\ {}^{x+k-1}C_{k-1}\left(\frac{P}{Q}\right)^x = \sum_0^{\infty} Q^{-k}\ x+k-1C_{k-1}\left(\frac{Pe^t}{Q}\right)^x$$

$$= (Q - Pe^t)^{-k} \quad \ldots(6\cdot13)$$

$$E(x) = \left[\frac{dM}{dt}\right]_{t=0} = \left[kPe^t(Q-Pe^t)^{-(k+1)}\right]_{t=0} = kP$$

$$E(x^2) = \left[\frac{d^2M}{dt^2}\right]_{t=0} = \left[kPe^t(Q-Pe^t)^{-(k+1)} + k(k+1)P^2e^{2t}(Q-Pe^t)^{-(k+2)}\right]_{t=0}$$

$$= kP + k(k+1)P^2$$

$$\text{Var}(x) = E(x^2) - [E(x)]^2 = kP + kP^2 = kPQ.$$

Cumulative Function of the negative Binomial $(Q-P)^{-k}$

we have $K(t) = \log M_x(t) = \log (Q - Pe^t)^{-k} = -k \log [Q - Pe^t]$.

$$= -k \log [1 + P - Pe^t) = -k \log [1 - P(e^t - 1)]$$

$$= -k \log \left[1 - Pt - \frac{Pt^2}{2!} - \frac{Pt^2}{3!} \right] = k \left[\left(Pt + \frac{Pt^2}{2!} + \frac{Pt^3}{3!} + \frac{Pt^4}{4!} + \cdots \right) \right.$$

$$\left. - \frac{1}{2}\left(Pt + \frac{Pt^2}{2!} + \frac{Pt^3}{3!} + \right)^2 + \frac{1}{3}\left(Pt + \frac{Pt^2}{2!} + \cdots \right)^3 \cdots\cdots \right]$$

Equating the coefficient of $t, t^2/2!, t^3/3!, t^4/4!$ we get

$$k_1 = kP,\ k_2 = kPQ,\ k_3 = kPQ(P+Q),\ k_4 = k\,PQ\,(1 + 6PQ)$$

N. B. 1. The similarity of *m.g.f.* to that of the binomial probability distribution should be noted. It is also to be noted that for the negative binomial $\mu_2 > \mu_1'$ whereas for the ordinary binomial $\mu_2 < \mu_1'$.

2. The negative binomial distribution has been found to occur in many biological situations and can come about as a result of clustering (or contagion) among the successes of an otherwise binomial population *e.g.*, death of insects, number of insect bites per apple. The distribution also arises in inverse sampling from a binomial population and as a weighted average of Poisson distributions.

6·4·4·1. *Poisson distribution as the limiting case of negative Binomial distribution.*

If we let $P \to 0$ and $k \to \infty$ in such a way that limit kP = m (*a* finite constant), then (6·12″, page 196) takes the form

$$P(x) = \lim_{\substack{k \to \infty \\ P \to 0}} (1+P)^{-k} \frac{(k+x-1)....k}{x!} \left(\frac{P}{1+P} \right)^x$$

$$= \lim_{\substack{k \to \infty \\ P \to 0}} \left(1 + \frac{m}{k} \right)^{-k} \frac{1}{x!} \left(1 + \frac{x-1}{k} \right) \cdots\cdots 1.\ m^x \left(1 + \frac{m}{k} \right)^{-x}$$

$$= e^{-m} m^x / x!,$$ which is Poisson distribution.

Ex. Prove the following results for negative binomial distribution $f(x; k, p) = \binom{x+k-1}{k-1} p^k,\ x = 0, 1, 2, \ldots$

(*i*) $f(x+1; k, p) = \dfrac{x+k}{x+1}\, q\, f(x, k, p),\ k > 0.$

(*ii*) *p.q.f.* $= [p\,(1-qs)\,]^{-k}$ as $E(s^x) = \sum_0^\infty \binom{-k}{x} p^k (-qs)^x$.

6·4·5. ***Hyper-geometric Distribution.** Suppose r balls are drawn one at a time, *without replacement*, from a bag containing m white and n black balls. The probability that x will be white out of the r balls so drawn is

$$P(x) = \frac{{}^mC_x \, {}^nC_{r-x}}{{}^{m+n}C_r}, \; x = 0, 1, 2 \ldots\ldots, r, \; r \le m, \; r \le n \qquad \ldots(6\cdot14)$$

where $\sum_{x=0}^{r} P(x) = 1$ since $\sum_{x=0}^{r} {}^mC_x \, {}^nC_{r-x} = {}^{m+n}C_r$

which can be proved by equating the coefficient of y^r in

$$(1+y)^m \, (y+1)^n = (1+y)^{m+n}.$$

The distribution with (6·14) as the probability law is known as Hyper-geometric distribution with three parameters m, n, r.

$$\text{Mean} = E(x) = \sum_{x=0}^{r} xP(x) = \sum_{x=0}^{r} \frac{x \, {}^mC_x \, {}^nC_{r-x}}{{}^{m+n}C_r}$$

$$= \frac{m \sum_{x=1}^{r} {}^{m-1}C_{x-1} \, {}^nC_{r-x}}{{}^{m+n}C_r} = \frac{m}{{}^{m+n}C_r} \sum_{y=0}^{r-1} {}^{m-1}C_y \, {}^nC_{r-y-1};$$

$$x - 1 = y$$

$$\mu_1' = \frac{m}{{}^{m+n}C_r} \times {}^{m-1+n}C_{r-1} = \frac{mr}{m+n}.$$

$$\mu_2' = E(x^2) = \sum_{x=0}^{r} x^2 P(x) = \sum_{x=0}^{r} x(x-1) \, P(x) + \sum_{x=0}^{r} xP(x)$$

Now $\sum_{x=0}^{r} x(x-1) \, P(x)$

$$= \frac{\sum_{x=0}^{r} x\,(x-1) \, {}^mC_x \, {}^nC_{r-x}}{{}^{m+n}C_r} = \frac{m(m-1)}{{}^{m+n}C_r} \sum_{x=2}^{r} {}^{m-2}C_{x-2} \, {}^nC_{r-x}$$

or $$\mu_2' = \frac{m(m-1)}{{}^{n+n}C_r} \sum_{y=0}^{r} {}^{m-2}C_y \, {}^nC_{r-2-y} = \frac{m(m-1)}{{}^{m+n}C_r} \, {}^{m-2+n}C_{r-2}$$

or $$\mu_2' = \frac{m(m-1) \, . \, r\,(r-1)}{(m+n)\,(m+n-1)}.$$

---

* The distribution is so termed since the *m.g.f.* can be expressed in terms of Hyper-geometric Function.

$$\therefore\ \mu_2 = \frac{m(m-1)\,.\,r(r-1)}{(m+n)\,(m+n-1)} + \frac{mr}{m+n} - \left(\frac{mr}{m+n}\right)^2$$

$$= \frac{mnr\,(m+n-r)}{(m+n)^2\,(m+n-1)}, \text{ on simplificatio.}$$

Alternative Form. The Hyper-geometric distribution can also be obtained as follows :–

Suppose a bag contains Np white and Nq black balls $(p+q=1)$. The probability that in drawing balls without replacement a sample of n can be drawn containing r white and $(n-r)$ black balls,

$$P(r) = {}^{Np}C_r\ {}^{Nq}C_{n-r} / {}^{N}C_n \quad \text{or} \quad {}^{n}C_r (Np)^{(n)} (Nq)^{(n-r)} / N^{(n)}$$

where $$x^{(r)} = x(x-1)\,(x-2)\cdots(x-r+1).$$

Consider the probability generating function

$$P(x) = \frac{1}{N^{(n)}} \sum_{r=0}^{n} \{{}^{n}C_r (Np)^{(r)}\,(Nq)^{(n-r)} x^r\} \quad (\text{use } x \text{ instead of } s)$$

$$= \frac{(Nq)^{(n)}}{N^{(n)}} \sum_{r=0}^{n} \left(\frac{(NP)^{(r)} n^{(r)}}{(Nq-n+r)^{(r)}} \cdot \frac{x^r}{r\,!} \right)$$

$$= \frac{(Nq)^{(n)}}{N^{(n)}} F(\alpha, \beta\,;\gamma, x) \qquad \ldots(6{\cdot}15)$$

where $\alpha = -n$, $\beta = -Np$, $\gamma = Nq - n + 1$ and the Hyper-geometric function

$$F(\alpha, \beta\,; \gamma, x) = 1 + \frac{\alpha\beta}{\gamma} \cdot \frac{x}{1\,!} + \frac{\alpha(\alpha+1)\beta(\beta+1)}{\gamma(\gamma+1)}\,\frac{x^2}{2\,!} + \ldots\ldots$$

It can be proved that the function $F(\alpha, \beta\,; \gamma, x)$ satisfies the following differential equation

$$x\,(1-x)\,\frac{d^2 F}{dx^2} + \{\gamma - (\alpha+\beta+1)x\}\,\frac{dF}{dx} - \alpha\beta F = 0$$

Putting $x = e^t$, $\alpha = -n\,.\ \beta = -Np$, $\gamma = Nq - n + 1$, this reduces to

$$(1 - e^{\phi}) \left[\frac{d^2\phi}{dt^2} - (n+Np)\,\frac{d\phi}{dt} - nNp\phi \right] - Nnp\phi + N\frac{d\phi}{dt} = 0$$

But $\phi = \Sigma\,\mu_r'\,t^r/r\,!$ Therefore, equating the constant term, $\mu_1' = np$.

The distribution when transferred about its mean np gives $\phi\,e^{npt}$ for φ and then we have

$$(1 + e^{\theta}) \left[\frac{d^2\varphi}{dt^2} + \frac{d\varphi}{dt}\,\{n(p-q) - Np\} + (N-n)pq^n\,\varphi \right] + \frac{Nd\varphi}{dt} = 0$$

Equating coefficient of $t, t^2, \ldots$, we get

$$\mu_2 = npq\,(N-n)/(N-1)$$

$$\mu_3 = npq(q-p)\,(N-n)\,(N-2n)/[N-1)\,(N-2)].$$

Ex. 1. Deduce the moments of the Binomial distribution.

Ex. 2. Show that as $N \to \infty$, 6·15 approaches the *p.g.f.* for the binomial distribution.

Ex. 3. Show that $\quad {}^{Np}C_r\,{}^{Nq}C_{n-r} \div {}^{N}C_n = {}^{n}C_r\,p^r q^{n-r}$

as $N \to \infty$. Deduce that as $N \to \infty$, the difference in *p.d*, due to drawing balls with or without replacement tends to zero.

Ex. 4. Show that for $f(x\,;N, M, n) = \dfrac{{}^{M}C_x\;{}^{N-M}C_{n-x}}{{}^{N}C_n}$, $x = 0, 1, 2, \ldots, N$

$$\mu_1' = n\frac{M}{N},\quad \mu_2 = \frac{NM\,(N-M)\,(N-n)}{N^2(N-1)}$$

reccurance relation is

$$f(x+1\,;N, M, n) = (n-k)\,(M-k)/(x+1)\,(N-M-n+x+1)\,f(x\colon N, M, n)$$

Ex. 5. Plot the probability functions and the hypergeometric distributions with $M = 8$, $N = 16$, $n = 8$ and that for the binomial distribution for $p = 0{\cdot}5$, $n = 8$.

Ex. 6. Show that for small *p*, large *n* and very large *N*, the hypergeometric distribution can be approximated by the Poisson distribution with mean *np*.

6·4·6. Approach of Hyper-geometric distribution to Binomial and Poisson distributions. In Figure 6·3, we have drawn the probability

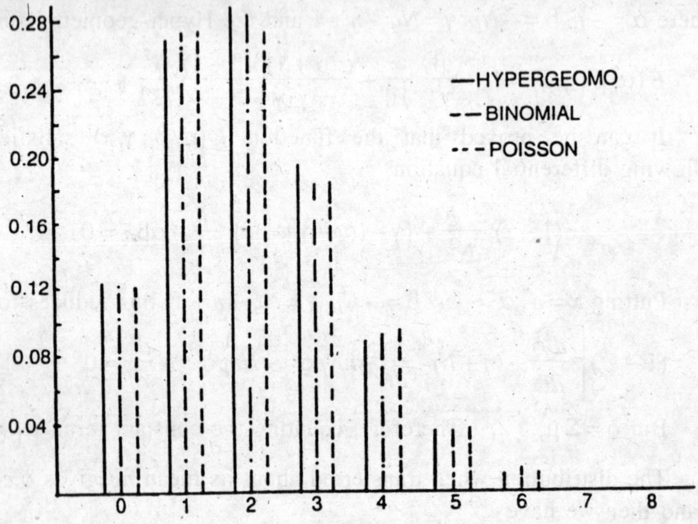

Fig. 6·3

* If $\alpha = 1, \beta = \gamma$, F (1 and x) becomes the geometric series $1 + x + x^2 + \cdots$ Thus hypergeometric series is a generalisation of the geometric series.

functions for the following distributions :

Hyper-geometric for $N = 1000, n = 100, M = 20$

Binomial for $n = 100, p = \cdot 02$; Poisson for $m = \cdot 02$

This figure illustrates that

(*i*) When N is large, Hyper-geometric distribution approaches the Binomial distribution.

(*ii*) When N is large, p is small and np is finite, the Binomial distribution approaches the Poisson distribution.

(*iii*) When N is very large, n is large, p is small and np is finite, the hypergeometric distribution approaches the Poisson distribution.

6·4·7. Factorial Moments of the Hyper-geometric distribution

$$\mu'_{(r)} = \sum_{x=0}^{n} x^{(r)} n_{c_x} (Np)^{(x)} (Nq)^{(n-x)} / N^{(n)} = \sum_{x=r}^{n} x^{(r)}\, {}^n c_x \; (Np)^{(x)} (Nq)^{(n-x)} / N^{(n)}$$

$$= \sum_{y=0}^{n-r} (y+r)^r n_{c_{y+r}} (Np)^{(y+r)} (Nq)^{(n-r-y)} / N^{(n)}, \; y = x - r$$

$$= \frac{n^{(r)} (Np)^{(r)}}{N^{(r)}} \sum_{y=0}^{n-r} n - rc_y \,(Np - r)^{(y)} (Nq)^{n-r-y} / (N-r)^{(n-r)}$$

Now, since $\sum_{x=0}^{n} n_{c_x} (Np)^{(x)} (Nq)^{(n-x)} / N^{(n)} = 1, N^{(k)} (N_r) = N^{(r)} \binom{N-r}{n-r}$

it follows that $\sum_{y=0}^{n-r} n - r_{c_y} (Np - r)^{(y)} (Nq)^{(n-y-r)} / (N-r)^{(n-r)} = 1$

$$\therefore \qquad \mu'_{(r)} = \frac{n^{(r)} (Np)^{(r)}}{N^{(r)}} \quad \therefore \; \mu_{(i)}' = \frac{n.Np}{N} = xp \qquad ...(6{\cdot}16)$$

As $N \to \infty, \mu'_{(r)} \to n^{(r)} p^r$ which is the rth factorial moment of the Binomial distribution. **(*B.Sc. Stat. Hons. Delhi, 1993*)**

Another new way of obtaining this result is based on the relation

$$\frac{d}{dx} F(\alpha, \beta; \gamma; x) = \frac{\alpha\beta}{\gamma} F(\alpha + 1, \beta + 1 \,; \gamma + 1 \,; x) \qquad ...(6{\cdot}17)$$

From (6·15), the p.g.f. of the Hyper-geometric p.d.f. is given by

$$P(x) = \frac{(Nq)^{(n)}}{N^{(n)}} F(\alpha, \beta; \gamma; x), \text{ where } \alpha = -n, \beta = -Np \cdot \gamma = Nq - n + 1.$$

Since $P(1) = 1$, it follows that $F(\alpha, \beta, \gamma; 1) = N^{(n)} / (Nq)^{(n)}$ ...(6·18)

By changing n into $n - 1$, N into $N - 1$ and p into $(Np - 1)/(N - 1)$, we get $F(\alpha + 1, \beta + 1; \gamma + 1; 1) = (N - 1)^{(n-1)} / (Nq)^{(n-1)}$

Similarly we get

$$F(\alpha+r, \beta+r; \gamma+r; 1) = (N-r)^{(n-r)}/(Nq)^{(n-r)}; \; r = 0, 1, 2, \ldots \quad \ldots(6{\cdot}19)$$

The factorial moments are now easily obtained from the p.g.f.

$$\mu'_{(r)} = \left[\frac{d^r P(x)}{dx^r}\right]_{x=1} = \frac{(Nq)^{(n)}}{N^{(n)}}\left[\frac{dr}{dq^r} F(\alpha, \beta; \gamma; x)\right]_{x=1}$$

$$= \frac{(Nq)^{(n)}}{N(n)} \frac{\alpha(\alpha+1)\ldots(\alpha+r-1)\beta(\beta+1)\ldots(\beta+r-1)}{\gamma(\gamma+1)\ldots(\gamma+r-1)}$$

$$\times F(\alpha+r, \beta+r; \gamma+r; 1)$$

$$= \frac{(Nq)^{(n)}}{N^{(n)}} \cdot \frac{n^{(r)}(Np)^{(r)}}{(Nq-n-r)^{(r)}} \cdot \frac{(N-r)^{(n-r)}}{(Nq)^{(n-r)}} = \frac{n^{(r)}(Np)^r}{N^{(r)}}.$$

6·4·8. Another form of Hyper-geometric distribution

The classical hypergeometric distribution is a generalisation of the binomial distribution to the case when there is no replacement in drawing of balls from a vessel and thus should have been called a hyper-binomial distribution. To generalise the classical geometric distribution, we consider the problem of drawing balls one by one and without replacement from a vessel with M white and N-M black balls. We consider the p.d. of the number of black balls drawn before the first white ball is drawn.

P [number of black balls drawn before the first white ball = x]

$$= \frac{N-M}{N} \cdot \frac{N-M-1}{N-1} \cdots \frac{N-M-x+1}{N-x+1} \cdot \frac{M}{N-x},$$

$$x = 0, 1, 2, \ldots N-M \quad \ldots(6{\cdot}19)$$

$$\text{The p.g.f. is } P(s) = \sum_{x=0}^{N-M} \frac{N-M}{N} \cdot \frac{N-M-1}{N-1} \cdots \frac{N-M-x+1}{N-x+1} \cdot \frac{M}{N-x} s^x$$

$$= \frac{M}{N} F[1, -(N-M); -(N-1); s] \quad \ldots(6{\cdot}20)$$

$$\mu'_{(r)} = \left(\frac{d^r}{ds^r} P(s)\right)_{s=1}$$

$$= \frac{M^{(r\,!)}}{N} \frac{(N-M)^{(r)}}{(N-1)^r} F(1+r, -(N-M)+r, -(N-1)+r, 1] \quad \ldots(5{\cdot}21)$$

If $\dfrac{M}{N} = p, \dfrac{N-M}{N} = q$, we get $P(s) = pF[1, -Nq; -(N-1); s]$

$$\mu'_{(r)} = p^{(r\,!)} \frac{(nq)^{(r)}}{(N-1)^{(r)}} F[1+r, -Nq+r; -N+1+r; 1]$$

Ex. 1. Prove that the recurrance relations for.

Ex. 2. Prove that as $N \to \infty$, $P(s) \to \dfrac{p}{1-q^s}$, $\mu'_{(r)} \to \dfrac{r!}{p^r}$

Ex. 3. Prove that $\mu'_1 = \dfrac{N-M}{M+1}$. Find μ_2.

[Hint. $\mu'_1 = \dfrac{M}{N} \cdot \dfrac{N-M}{N-1} F\,[2, -N+M+1; N+2; 1]$

$$= \frac{M}{N}\frac{N-M}{N-1}[(-N+1)\,F\,(1, -N+M+1, -N+1, 1)$$

$$+ NF\,(1, -N+M+1, -N+2; 1)]$$

$$= \frac{M}{N}\frac{N-M}{N-1}\left[(-N+1)\frac{N}{M+1} + \frac{N(N-1)}{M}\right] = \frac{N-M}{M+1}\ \Bigg]$$

6·4·9. Multinomial Distribution. The Binomial distribution can easily be generalized to n repeated independent trials when each trial may result in one of the several outcomes say $E_1, E_2, \ldots, E_k$ with respective probabilities $p_1, p_2, \ldots, p_k$ in each trial where $p_1 + p_2 + \ldots + p_k = 1$.

The probability that *n trials will result in* E_1 occurring n_1 times, E_2 occurring n_2 times, ..., E_k occurring n_k times in a *fixed definite order is* $p_1^{n_1} p_2^{n_2} \ldots p_k^{n_k}$, $\sum_{i=1}^{k} n_i = n$.

But we are interested in events occurring in any order. The number of mutually exclusive ways in which this can happen is

$$\frac{n!}{n_1!\, n_2! \ldots n_k!}$$

Hence the required probability function is

$$P(n_1, n_2, \ldots, n_k) = \frac{n!}{n_1!\, n_2! \ldots n_k!}\, p_1^{n_1} p_2^{n_2} \ldots p_k^{n_k}, \quad \sum_{i=1}^{k} n_i = n \quad \ldots(6{\cdot}22)$$

This distribution is called *multinomial probability distribution*, as the expression is the general term of multinomial expansion of

$$(p_1 + p_2 + \ldots + p_k)^n, \ \sum_{1}^{k} p_i = 1$$

We notice that this is not a univariate distribution but is a multivariate distribution involving k variates $n_1, n_2, \ldots, n_{k-1}, n_k$ but only $(k-1)$ variates are independent since $\sum_{i=1}^{k} n_i = n$. It has application in sampling *with replacement* of which individuals are classified into more than two categories.

The means, variances, covariances can be obtained as follows :

Since $\Sigma \frac{n!}{n_1!\, n_2! \ldots n_k!} p_1^{n_1} p_2^{n_2} \ldots p_k^{n_k} = (p_1 + p_2 + \ldots + p_k)^n =$

$$\therefore \quad p_i \frac{\partial}{\partial p_i} \Sigma \frac{n!}{n_1!\, n_2! \ldots n_k!} p_1^{n_1} p_2^{n_2} \ldots p_k^{nk} = \Sigma\, n_i P(n_1, n_2, \ldots n_k)$$

$$\Rightarrow \quad p_i \frac{\partial}{\partial p_i} (p_1 + p_2 + \ldots + p_k)^n, \qquad = np_i (p_1 + p_2 + \ldots + p_k)^{n-1}.$$

$$p_i \frac{\partial}{\partial p_i} \Sigma n_i P (n_1, n_2, \ldots n_j = \Sigma n_i\, n_j\, P(n_1, n_2 \ldots, n_k)$$

$$= p_i \frac{\partial}{\partial p_i} np_i (p_1 + \ldots + p_k)^{n-1} = n(n-1) p_i p_j (p_1 + \ldots + p_k)^{n-2}$$

Similarly $p_i \frac{\partial}{\partial p_i} \Sigma n_i P (n_1, \ldots, n_k) = \Sigma n_i^2 P (n_1, n_2, \ldots, n_k)$

$$= np_i (p_1 + \ldots + p_k)^{n-1} + n(n-1) p_i^2 (p_1 + \ldots + p_k)^{n-2}$$

These results give $\quad E(n_i) = np_i,\ i = 1, 2, \ldots, k$ ...(1)

$$E(n_i n_j) = n(n-1) p_i p_j,\ i \neq j,\ E(n_i^2) \quad = np_i + n(n-1) p_i^2\ g = i$$

$$\text{Var}(n_i) = E(n_i^2) - [E(n_i)]^2 = np_i (1 - p_i)$$

$\text{Cov}(n_i, n_j) = E(n_i n_j) - E(n_i) E(n_j) = -np_i\, p_j.$ ...(2)

It is note-worthy that the result (1) and (2) can be obtained easily be noticing that each X_i is a binomial variate with p_i as the probability of success of an event in a single trial.

Ex. 1. Show that

$M_0(t) = e^t.$ for Causal distribution.

$M_0(t) = \frac{1}{n} e^t \frac{(e^{nt} - 1)}{e^t - 1}$ for Uniform distribution.

$M_0(t) = \left(\frac{p}{1 - qe^t} \right)^k$ for Pascal distribution.

$M_0(t) = [1 + \beta\mu (1 - e^t)]^{-1/\beta}$ for Polya's distribution.

Deduce *m.g.f.* about the mean in each case. Hence or otherwise. obtain the first four moments about the mean for each distribution.

Ex. 2. Show that the sum of k independent geometric variates with the same parameter a is a Pascal variate. Hence or otherwise prove that the mean and variance of the Pascal distribution are kq/p and kq/p^2 respectively. Show also the sum of a number of independent Pascal variates (with same p) is a Pascal variate.

Ex. 3. The cumulative distribution function (c.d.f.) $F(x)$ of a random variable X is defined by $\ F(x) = P[X \leq x]$ Prove that $\ F(x + 0) - F(x - 0) = P(X = x]$

For the uniform distribution prove that $F(x) = \frac{[x]}{n}$. and draw its graph. [here $[x]$ denotes the integral part of x].

Ex. 4. In question 27 of *Exercise* 5, show that the size of the sample S, can be written as $1 + X_1 + X_2 + \ldots + X_{r-1}$ where X_1 is the number of drawings following the selection of *ith element* and up to and including the drawing of the next new element. Prove that X_1 is a Pascal variate with $p_i = (N - i)/N$. Hence find the mean and variance of S_r.

Ex. 5. In examples 5·7, 5·8 and questions 26, 27 and 28 of Ex. 5 we have found the means and variances of some distribution *without finding the distributions*. What are these distributions ? Show also that the *m.g.f.* for the distribution of the examples is

$$\prod_{i=1}^{n} (q_i + p_i e^t)$$

Hence find the mean and variance for this distribution.

Ex. 6. Show that Poisson's distribution is the limiting form of Polya's distribution as $\beta \to 0$.

Ex. 7. Show that the *m.g.f.* of the multinomial distribution is

$$M_{x_i \ldots x_n}(t_1 \ldots t_n) = E\left[\exp\left(\sum_{i=1}^{k} t_i x_i\right)\right] = p_1 e^{t_1} + p_2 e^{t_2} + \ldots + p_k e^{t_k}]^n, \sum_{1}^{k} p_i = 1$$

and from this deduce by putting $t_2 = -1, t_n = 0$, etc. that the means, variances and covariance are given by

$E(n_i) = np_i \; E(n_i n_j) = n(n-1) p_i p_j, \; Var(n_i) = np_i(1 - p_i), \; Cov(n_i, n_j) = -np_i p_j \; i \neq j$

[M. A., Bombay; M.A. Poona)

Exercises 6

1. (*a*) Six dice are thrown 729 times. How many times do you expect at least three dice to show a five or six ? ***(B. A. , Hons., Delhi)***

(*b*) Take 100 sets of 10 tosses of an unbiased coin. In how many cases do you expect to get 7 heads at least ? [**Ans.** (*a*) 233 (*b*) 17 times]

2. A perfect cubic die is thrown a larger number of times in sets of 8. The occurrence of a 5 or 6 is called a success. In which proportion of the sets would you expect to get 3 successes ? [**Ans.** 27·31 per cent]

3. (*a*) Assuming that half the proportion are consumers of rice, so that the chance of an individual being consumer is $\frac{1}{2}$, and assuming that 100 investigators each took ten individuals to see whether they are consumers, how may investigators would you expect to report that three people or less were consumers ?

(M. A. Eco. Statistics, Delhi; M.A. Punjab)

(*b*) In the above example, if 1,000 investigators each choose 100 individuals how many would you expect to report that more than 60 persons were consumers ? ***(B.A. Lucknow)*** [**Ans.** (*a*) 17; (*b*) 18]

4. An irregular six-faced die is thrown, and the expectation that in ten throws it will give five even numbers is twice the expectation that it will give four even numbers. How many times in 10,000 sets of 10 throws would you expect it to give no even number ? ***(B.A. Hons. Delhi)***

[Ans. 1]

[Hint. ${}^{10}c_5\, p^5\, q^5 = 2 \times {}^{10}c_4\, p^4\, q^6, q = 1 - p.$ find p and then $10{,}000\ q^n$]

5. Eight mice are selected at random and they are divided into two groups of 4 each. Each mouse in group A is given a dose of certain poison a which is expected to kill one in four ; each mouse in group B is given a dose of certain poison b which is expected to kill one in two. Show that nevertheless, there may be fewer deaths in group B than in group A and find the probability of this happening. ***(I.A.S. 1982.)***

[Ans. $\frac{525}{4096} = 0{\cdot}12817$]

6. The following data due to Weldon shows the results of throwing 12 dice 4096 times; a throw of 4, 5, or 6 being called success.

Success	*Frequency*	*Success*	*Frequency*
0	—	7	847
1	7	8	536
2	60	9	257
3	198	10	71
4	430	11	11
5	731	12	–
6	948		
		Total	4096

Find the expected frequencies and compare the actual mean with that of the expected distribution. Calculate also the standard deviation.

(B. Sc., Agra; B.A. Hons., Delhi)

[Ans. Expected frequencies, are : 1, 12, 66, 220, 495, 792, 924, 792, 495, 220, 66, 12, 1, Expected mean = 6, Actual mean = 6·139]

7. Show that if two symmetrical binomial distributions ($p = q = \frac{1}{2}$) of degree n (and of the same number of observations) are so superposed that the *rth* term of the one coincides with the $(r+1)$th term of the other, the distribution formed by adding superposed terms is a symmetrical binomial of degree $(n+1)$.

Use : $\dfrac{1}{2^n}\dbinom{n}{r-1} + \dfrac{1}{2^n}\dbinom{n}{r} = 2 \cdot \dfrac{1}{2^{n+1}}\dbinom{n+1}{r}$

(B. Sc., Delhi, 1976; M. Sc., Agra)

8. The incidence of occupational disease in any industry is such that the workers have a 20% chance of suffering from it. What is the probability that out of six workers 4 or more will catch the disease ? **[Ans.** $\frac{53}{2125} = 0{\cdot}01696$]

9. In a precisions bombing attack there is a 50% chance that any one bomb will strike the target. Two direct hits are required to destroy the target completely How many bombs must be dropped to give a 99% chance or better of completely destroying the target ? ***(M.A. Gujarat, 1992)*** **[Ans.** 11**]**

10. Show that the necessary and sufficient conditions for two given numbers a and b to be respectively the mean and the variance of some binomial distribution are that $a > b > o$ and $a^2/(a-b)$ is an integer. Show further that when these conditions are satisfied the binomial distribution is uniquely determined. ***(M. A. Kanpur, 1979, I.A.S., 1972]***

11. An owner of a small hotel with five rooms is considering buying television sets to rent to room occupants. He expects that about half of his customers would be willing to rent sets, and finally he buys three sets, Assuming 100% occupancy at all times :

(*i*) What fraction of the evenings will there be more requests than TV sets ?

(*ii*) What is the probability that a customer who requests a TV set will receive one ? ***(M.A. Economics, Kanpur, 1973)***

[Ans. (*i*) $P(x > 4) = P(x = 4) + P(x = 5) = {}^5c_4\left(\frac{1}{2}\right)^5 + \left(\frac{1}{2}\right)^5 =$ etc.]

12. (*a*) Five dice are tossed 96 times. Find the expected frequencies when throwing of a 4, 5 or 6 is regarded as a success.

(*b*) Two dice are thrown until a seven is obtained. Find the most probable number of throws and also the expected number of throws.

[Ans. (*a*) 3, 15, 30, 30, 15, 3 (*b*) 1, 6]

13. (*a*) If probability of success is ·01, how many trials are necessary in order that probability of at least one success is $> \frac{1}{2}$? **[Ans.** 69]

(*b*) Bring out the fallacy, if any, in the following statement :

The mean of a binomial distribution is 5 and its s.d. is 3.

(B.A. Hons. Delhi)

[Ans. (*b*) It makes $q = 1{\cdot}8$ which is wrong].

14. (*a*) If X and Y are independent binomial variates with parameters p_1 and p_2 and indices n_1 and n_2 respectively, obtain the probability that (*i*) $X + Y = r$. (*ii*) $|X - Y| = r$.

[Ans. (*i*) $P(X + Y = r) = \Sigma\, P(x = k) \,.\, P(y = r - k) =$ etc.]

(*b*) If X and Y are independent binomial variates $b(5, \frac{1}{2})$ and $b(7, \frac{1}{2})$, find $P\,(X + Y = 3)$. [Use Additive property of binomial variates.]

15. Show that, for the binomial distribution $(q + p)^n$, the factorial moments about the mean are $\mu_{(2)} = npq$, $\mu_{(3)} = -\,2npq\,(p + 1)$.

16. Obtain the characteristic function (C.F.) of the standard binomial variate $\dfrac{X - np}{\sqrt{npq}}$ where X is the number of successes obtained in n independent trials, each with constant probability p of success, $q = 1 - p$. Obtain the limit of this function as n tends to infinity. $\left[\textbf{Ans. } \phi\,(t) = e^{-\frac{1}{2}t^2}\right]$.

17. (*a*) If $P(X \le k) = B(k, n, p)$, show that

$$\frac{d}{dq} B(k, n, p) = n \binom{k}{n-1} p^k q^{n-k-1}, \quad q = 1 - p$$

(*b*) Show that for the binomial distribution $(q+p)^n$, $q = 1 - p$

$$k_{r+1} = pq \frac{dk_r}{dp}, \quad r > 1.$$

Hence deduce that $k_4 = npq(1 - 6pq)$. **(*B.Sc. Stat. Hons. Delhi, 1992*)**

[**Hint** (*b*) $k_r = \left[\frac{d^r}{dt^r} \{ n \log (q + pe^t) \} \right]_{t=0}$ $\quad \therefore \frac{dk_r}{dp} = n \left[\frac{d^r}{dt^r} \left(\frac{-1 + e^t}{q + pe^t} \right) \right]_{t=0}$

Also $k_{r+1} = \left[\frac{d^{r+1}}{dt^{r+1}} n \log (q + pe^t) \right]_{t=0} = np \left[\frac{d^r}{dt^r} \left(\frac{e^t}{q + pe^t} \right) \right]_{t=0}$

Now show that $k_{r+1} - pq \frac{dk_r}{dp}$ simplifies to zero.]

18. (*a*) If $K(t)$ be the cumulative function about the origin of the binomial distribution of size n, show that

$$\frac{d}{dt} K(t) = n \left[1 + e^{-(z+t)} \right]^{-1} \text{ where } z = \log_e \left(\frac{p}{q} \right).$$

(*b*) By expanding the right hand side in powers of t by Taylor's theorem, show that $k_r = n \frac{d^{r-1} p}{dz^{r-1}}$ where k_r is the rth cumulant.

(*c*) Hence or otherwise, obtain the recurrence relation :

$$k_{r+1} = pq \frac{dk_r}{dp}, \quad r > 1.$$

[**Hint.** (*a*) $M_x(t) = (q + pe^t)^n \quad \therefore K(t) = n \log (q + pe^t)$

$$\therefore \frac{d}{dt} K(t) = \frac{n\, pe^t}{q + pe^t} = np(qe^{-t} + p)^{-1} = n \left(1 + \frac{q}{p} e^{-t} \right)^{-1} = \text{etc.}]$$

(*b*) $k_r = \left[\frac{d^{r-1}}{dt^{r-1}} \frac{d}{dt} k(t) \right]_{t=0}$

$$= \left[n \frac{d^{r-1}}{dt^{r-1}} \left(\frac{e^{z+t}}{1 + e^{z+t}} \right) \right]_{t=0}$$

$$= n \frac{d^{n-1}}{dz^{r-1}} \left(\frac{e^{z+t}}{1 + e^{z+t}} \right)_{t=0} \quad \text{as functions is symmetric}$$

= etc.

19. For Bernoulli's distribution $(q+p)^n$, prove the following results :

(*a*) $\sum_{r=0}^{n} {}^nC_r p^r q^{n-r} (r - np) = 0$, (*ii*) $\sum_{r=0}^{n} {}^nC_r p^r q^{n-r} (r - np)^2 = npq$, (*b*) $\mu_r = \left(p\dfrac{\partial}{\partial p}\right)^r (q+p)^n$ when μ_r is the *r*th moment about the origin.

Hence obtain the coefficient β_1 of the Binomial distribution.

[**Hint.** (*a*) Differentiate the following identity twice with respect to u and set $u = 0$.

$$e^{-npu}(pe^u + q)^n = \sum_{r=0}^{n} {}^nC_r p^r q^{n-r} e^{r-npu}).]$$

20. (*a*) Show that in the Binomial distribution $(q+p)^n$, the ratio of the mean deviation about the mean to standard deviation is

$$\frac{2}{\sqrt{npq}}\, {}^{n-1}C_{\mu-1}\, p^{\mu-1} q^{n-\mu}$$

where μ is the greatest integer contained in $np+1$. Find the limits of the above ratio as n tends to infinity.

(*b*) Find the mean deviation about the mean of (*i*) Poisson distribution with parameter *m*, (*ii*) negative binomial distribution.

(*B. A. Hons., Delhi, 1971, 1980*)

[**Sol.** (*a*) By definition, mean deviation, M_d, about the mean is given by

$$M_d = \sum_{r=0}^{n} {}^nC_r p^r q^{n-r} \,|\, r - np \,|.$$

But $\quad 0 = \sum {}^nC_r p^r q^{n-r}(r - np)$. $\therefore$ $M = 2 \sum_{r>np} {}^nC_r p^r q^{n-r}(r-np)$. ...(1)

the summation being for all integers $r > np$.

Let $$F(x, y) = \sum_{r>np} {}^nC_r x^r y^{n-r}$$

$$\therefore \quad \sum_{r>np} {}^nc_r p^r q^{n-r}(r-np) = p\frac{\partial F(p,q)}{\partial p} - npF(p,q)$$

But, by Euler's theorem, $nF(p,q) = p\dfrac{\partial F(p,q)}{\partial p} + q\dfrac{\partial F(p,q)}{\partial q}$

$$\therefore \quad M_d = 2\sum_{r>np} {}^nC_r p^r q^{n-r}(r-np) = 2pq\left(\frac{\partial F}{\partial p} - \frac{\partial F}{\partial q}\right)$$

$$= 2np\, q\, {}^{n-1}C_{\mu-1};\, p^{\mu-1} q^{n-\mu},\ \mu \le np + 1 < \mu + 1.$$

Aliter. We have

$$M_d = 2\sum_{r=0}^{n}\left[\frac{n!}{(n-r)!\,(r-1)!}\cdot p^r q^{n-r+1} - \frac{n!}{r!\,(n-r-1)}p^{r+1}q^{n-r}\right.$$

$$= 2\sum_{r=1}^{n}(v_r - v_{r+1})$$

where $v_r = \dfrac{n!}{(n-r)!\,(r-1)!}p^r q^{n-r+1}$

$\therefore \quad M = 2v_{\mu+1}, \qquad$ since $v_{n+1} = 0$

$$= 2\frac{n!}{\mu!(n-\mu-1)!} p^{\mu+1} q^{n-\mu} = 2npq.\ ^{n-1}C_{\mu} p^{\mu} q^{n-1-\mu}$$

$$(b)\ (i) M = \sum_{0}^{\mu} \frac{e^{-\lambda}\lambda^x}{x!} |x-\lambda| + \sum_{\mu+1}^{\infty} \frac{e^{-\lambda}\lambda^x}{x!}(x-\lambda) = 2e^{-\lambda} \sum_{\mu+1}^{\infty} \left[\frac{\lambda^x}{(x-1)!} - \frac{\lambda^{x+1}}{x!} \right]$$

$$= 2e^{-\lambda} \sum_{\mu+1}^{\infty} \left(v_x - v_{x+1} \right) = 2e^{-\lambda} v_{\mu+1} = 2e^{-\lambda} v_{\mu+1} = 2e^{-\lambda} \frac{\lambda^{\mu+1}}{(\mu+1)!} = \text{etc.}$$

21. Let λ and μ_r denote the mean and central rth moment of a Poisson distribution respectively. Obtain the following recurrence formula :

$$\mu_{r+1} = r\lambda\mu_{r-1} + \lambda \frac{d\mu_r}{d\lambda}, \text{ where } \mu_r = \sum_{j=0}^{\infty} \frac{e^{-\lambda}\lambda^j}{j!}(j-\lambda)^r$$

and hence obtain β_1 and β_2 of the Poisson distribution.

(M. A. Delhi, 1956, 1980; M. A. Eco. Statistics, Delhi, 1958)

[**Ans.** $\beta_1 = 1/\lambda$, $\beta_2 = 3 + 1/\lambda$]

22. From records of 10 Russian army corps kept over 20 years, the following data was obtained showing the number of deaths caused by the kicks of a horse. Determine, *m*, the average number of deaths per-army corps per annum, and calculate the theoretical Poisson frequencies (Data of L. von Borkiewicz) and compare them with the observed frequencies.

Number of deaths per army corps per annum	0	1	2	3	4	Total
Frequency of occurrence	109	65	22	3	1	200

$[e^{-0\cdot61} = 0\cdot5434]$

[**Ans.** $m = 0\cdot61$; 109, 66, 20, 4, 1]

23. In a certain factory turning out razor blades, there is a small chance 1/500 for any blade to be defective. The blades are supplied in packets of 10. Use the Poisson distribution to calculate the approximate number of packets containing no defective, one defective and two defective blades respectively in a consignment of 10,000 packets, given that $e^{-\cdot02} = 0\cdot9802$.

$$\left[\textbf{Ans. } m = 10 \times \frac{1}{500} = \cdot02; \text{ 9802, 196 and 2 packets respectively} \right]$$

24. (*a*) If *X* is a Poissonian variate with mean *m*, what should be the expectation of $e^{-kx}.\ kx$ where *k* is a constant ? ***(B.A. Hons. Delhi, 1980)***

[**Ans.** $m.\ k.\ \exp\{m(e^{-k} - 1) - k\}$]

(*b*) Show that expectation of $(e^{-kx}) = e^{-m(1-e^{-k})}$

(M. A. Delhi; B. A. Hons., Delhi, 1980)

Hence show that, if $\bar{x}$ be the arithmetic means of *n* independent random variables $x_1, x_2, \ldots, x_n$, each having the Poisson distribution with mean *m*, then $e^{-\bar{x}}$ as an estimate of e^{-m} is biased although $\bar{x}$ is an unbiased estimate of m. ***(B. A. Hons., Delhi)***

25. Six coins are tossed 6,400 times. Using the Poisson distribution, find the approximate probability of getting six head x times.

$$\left[\textbf{Ans. } \frac{e^{-100}(100)^x}{x\,!}\right]$$

26. For the Poisson distribution with parameter m, prove that

(*i*) $\beta_1 = \frac{1}{m}$, $\beta_2 = 3 + \frac{1}{m}$ (*ii*) $m\sigma\gamma_1\,\gamma_2 = 1$. (*iii*) $P[X = \text{even}] = \frac{1}{2}(1 + e^{-2m})$

(B.Sc. Stat. Hons. Delhi, 1991)

(*iv*) $\mu'(r) = m^r$, so that $\mu(2) = m, \mu(3) = -2m, \mu(4) = 3m(m + 2)$.

27. A distributor of bean seeds determines from extensive tests that 5 per cent of a large batch of seeds will not germinate. He sells the seeds in packages of 100 and guarantees 20 per cent germination. What is the probability that a given package will violate the guarantee ?

$$\left[\textbf{Ans. } \text{Required probability} = \sum_{r=21}^{100}\left(\frac{e^{-10}10^r}{r\,!}\right) = 1 - \sum_{r=0}^{20}\left(\frac{e^{-10}10^r}{r\,!}\right)\right]$$

28. A manufacturer of cotter pins knows that 5 per cent of his product is defective. If he sells cotter pins in boxes of 100 and guarantees that not more than 10 pins will be defective, what is the approximate probability that a box will fail to meet the guaranteed quality ? $[e^{-5} = \cdot 006738]$

$$\left[\textbf{Ans. } P(x > 10) = \sum_{x=11}^{100}\frac{e^{-5}\cdot 5^x}{x\,!} = 1 - \sum_{x=0}^{10}\left(\frac{e^{-5}\cdot 5^x}{x\,!}\right) = 0{\cdot}013695\right]$$

29. A source of liquid is known to contain bacteria with the mean number of bacteria per cubic centimeter equal to 3. Ten 1-cubic centimeter test-tubes are filled with the liquid. Assuming the Poisson distribution is applicable calculate the probability (*a*) that all 10 test-tubes will show growth, that is, contain at least 1 bacterium each and (*b*) that exactly 7 testtubes will show growth. $[e^{-3} = \cdot 04979]$

[**Ans.** (*a*) $(1 - e^{-3})^{10} = 0{\cdot}60008$ (*b*) ${}^{10}C_7(1 - e^{-3})^7\,(e^{-3})^3 = 0{\cdot}010358$]

30. Suppose the number of telephone calls on an operator received from 9·00 and 9·05 a.m. follow a Poisson distribution with mean 3. Find the probability that (*a*) the operator will receive no calls in that time interval tomorrow; (*b*) in the next 3 days the operator will receive a total of 1 call in that time interval.

[**Ans.** (*a*) $e^{-3} = 0{\cdot}049787$ (*b*) $9e^{-9} = 1{\cdot}1 \times 10^{-3}$]

31. A car hire firm has two cars, which it hires out day by day. The number of demands for a car on each day is distributed as Poisson distribution with mean 1·5. Calculate the proportion of days on which neither car is used and the proportion of days on which some demand is refused. $[e^{-1\cdot 5} = 0{\cdot}2231]$

(B. Sc. Delhi, 1978)

[**Ans.** 0·2231, 0·1913]

32. A large number of observations on a given solution, which contained bacteria, were made taking samples of 1 c.c. each noting down the number of bacteria present in each sample. Assuming the Poisson distribution and given that 10 per cent samples contained no bacteria, find the average number of bateria per c.c. **(*M.A. Statistics, Delhi*)**

[Ans. $\log_e 10$ or 2·3026]

33. Prove that the sum of two independent Poissonian variates with means m_1 and m_2 is also a Poissonian variate with mean $m_1 + m_2$. **(*B.A., Math Hons., Delhi 1996*)**

[Hint. $P(Z=X+Y) = \sum_{x=0}^{z} P(X=x)\,P(Y=z-x) =$ etc.]

34. (*a*) If x and y are independent Poisson variates with mean m_1 and m_2, prove that the probability that $(x-y)$ has the value r is the coefficient of t^r in $\exp(m_1 t + m_2 t^{-1} - m_1 - m_2)$ **(*B. A. Hons., Delhi*)**

(*b*) Does the difference of two independent Poisson variates follows the Poisson distribution ? Given reasons. **(*M. A. Delhi; B. A., Hons. Delhi*)**

[Hint. No, since the *m.g.f.* of the difference is $\exp\,[m_1(e^t - 1) - m_2(e^{-t} - 1)]$ which cannot be put in the form in which Poisson variate has the *m.g.f.*; Further the difference cannot be a Poisson variate is evident from the fact that it may have negative or positive value.]

35. (*a*) If X_1 and X_2 are two independent Poisson variates with means 1 and 2 respectively, show that $P(X_1 + X_2 < 4) = 13e^{-3}$.

(*b*) Let X and Y be two independent Poisson variates. The variance of $X+Y$ is 9 and $P(X=3 \mid X+Y=6) = 5/54$. Find the mean of X.

[Hint. $9 = \text{Var}(x+y) = \text{Var}\,x + \text{Var}\,y = m + m, \;\therefore\; m = 9/2$

Now $\dfrac{5}{54} = P(x=31 \mid y=6-x) = P(x=3 \cap y=6-x)/P(x+y=6)$

$$= \frac{P(x=3, y=3)}{P(x+y=6)} = \frac{P(x=3)P(y=3)}{P(x+y=6)} = \text{etc.]}$$ **(*B. Sc. Kerla, 1970*)**

(*c*) A Poisson variate X is such that $P[X=1] = 2P[X=2]$, find its mean and variance **Ans. 1, 1.**

36. Obtain the characteristic function for the standard Poisson variate $\dfrac{r-m}{\sqrt{m}}$ where r is the number of successes in n independent trials. Obtain the limit of the function as $m \to \infty$. **[Ans.** $e^{-\frac{1}{2}t^2}$ **]**

37. If the probability of exactly r successes is

$$p_r = \frac{1}{r!} \sum_{k=0}^{n-r} \frac{(-1)^k}{k!};\; = 0\ 1.,\ldots.,n.$$

Show that, as $n \to \infty$, this tends to a Poisson probability function.

38. A special die with $(n+1)$ faces is marked in its faces the numbers

$$0, \frac{1}{n}, \frac{2}{n}, \ldots.., \frac{n-1}{n} \text{ and } \frac{n}{n}.$$

The dice is unbiased. For distribution of the random variable corresponding to the number on the uppermost face, find (*a*) the expected value (*b*) the standard deviation (*c*) β_1 and (*d*) β_2. (*I. A. S.*)

$$\left[\textbf{Ans. } \text{Expected value} = \frac{1}{2},\ \text{standard deviation} = \sqrt{\frac{(n+3)}{21\,n}},\ \beta_1 = 0\right]$$

39. If t is positive and real, show that function defined by

$$f(x) = e^{-t}(1 - e^{-t})^{x-1},\ x = 1, 2, 3,$$

can represent a probability density function of the variate X. Find the mean and variance of the distribution. (*M. A. Patna, 1987*)

[**Ans.** mean $= e^t$, $\mu_2 = e^{2t} - (e^t)$]

40. Compound Poisson distribution. For given λ, X is a Poisson variate with mean λ while λ is itself a continuous random variate having the distribution given by

(*i*) $\dfrac{a^v}{\overline{|v}}\, e^{-a\lambda}\, \lambda^{v-1}\, d\lambda,\ (\lambda > 0, a > 0, v > 0).$

(*ii*) exponential distribution with mean $1/c$.

Find the overall distribution of X when the random fluctuation in λ are allowed for.

Calculate the means and variance of the resulting distribution and show that the variance is always greater than the mean.

$$\left[\textbf{Ans. } (i) \text{ Negative binomial distribution, } \mu_1{}' = \frac{vp}{q},\ \text{variance} = \frac{vp}{q}\left(\frac{p}{q} + 1\right)\right]$$

(*ii*) $P(X = k) = c/(c + 1)^{k+1}$.

41. Compound Binomial distribution. Suppose that the probability of an insect laying n eggs is $e^{-\lambda}\lambda^n/n\,!$ and that the probability of an egg developing is p. Assuming natural independence of the eggs, show that the probability of a total of k survivors is given by the Poisson distribution with parameter λp. Hence find the mean and variance of the distribution obtained. (*B.Sc. Poona, 71*)

$$\left[\textbf{Hint. } P(X = k) = \sum_{n=0}^{\infty} \frac{e^{-\lambda}\lambda^n}{n\,!}\cdot {}^nC_k\, p^k (1-p)^{n-k} = \frac{e^{-\lambda}p^k}{k\,!} \sum_{n=0}^{\infty} \frac{\lambda^n (1-p)^{n-k}}{(n-k)\,!}\right.$$

$$\left. = \frac{e^{-\lambda}p^k\lambda^k}{k\,!} \sum_{n=k}^{\infty} \frac{[\lambda(1-p)]^{n-k}}{(n-k)\,!} = \frac{e^{-\lambda}p^k\lambda^k e^{\lambda(1-p)}}{k\,!} = \frac{e^{-\lambda p}(\lambda p)^k}{k\,!}.\right]$$

42. The number of female insects in a give region follows a Poisson distribution with mean m. The number of eggs laid by each insect follows a Poisson distribution with mean λ. Find the probability distribution of the number of eggs in the region and its mean and variance.

(*I.A.S. 1993; B.A., Hons., Calcutta, 1955*)

$$\left[\text{The probability distribution is } \frac{e^{-m}\lambda^x}{x\,!} \sum_{s=0}^{\infty} \frac{(me^{-\lambda})^s s^x}{s\,!}\right.$$

$$\left. m.g.f. = \exp\left[m(\exp\{(\lambda e^{\theta} - 1)\}\right]\right]$$

From this we find mean $= m\lambda$, variance $= m\lambda\,(\lambda + 1)]$.

43. Prove the identify :

$$p^n + {}^nC_1 p^{n-1} q + {}^nC_2 p^{n-2} q^2 + \ldots + {}^nC_k p^{n-k} q^k = \left(\sum_{x=0}^{k} {}^nC_x p^{n-x} q^x \right)$$

$$= \int_0^p x^{n-k-1} (1-x)^k dx + \int_0^1 x^{n-k-1} (1-x)^k dx$$

(***M.A., Statistics, Delhi, 1959***)

[**Hint.** *First Method.* Take derivatives with respect to p.

Second Method. Integrate the numerator of right side by parts repeatedly, taking $(1-x)^k$ as the first function.

Third Method. Use Taylor's theorem in the form

$$f(a+h) = f(a) + hf'(a) + \ldots\ldots + \frac{h^{r-1}}{(r-1)!} f^{r-1}(a) + \int_0^1 \frac{h^r (1-t)^{r-1}}{(r-1)!} f^r(a+th) dt$$

and put $a = q,\ h = p,\ f(a+h) = (q+p)^n$,

$$\therefore \quad (q+p)^n = \sum_{j=0}^{r-1} nc_j p^j q^{r-j} + R_r \text{ where } R_r = \frac{n!}{(n-r)!\,r!} \int_0^p x^{r-1} (1-x)^{n-r} dx$$

or $\quad R_r = \dfrac{B_p(r, n-r+1)}{B(r, n-r+1)}$ and $t = 1 - \dfrac{x}{p}$.]

44. Show that the hpergeometric distribution (§ 6·4·5) tends to Binomial distribution if $m, n \to \infty$ and $\dfrac{m}{m+n} \to p$.

45. The multinomial variables $n_1, n_2, \ldots, n_k$ with $n_1 + n_2 + \ldots + n_k = n$ (constant) are distributed in the form

$$\frac{n!}{n_1!\, n_2! \ldots n_k!} p_1^{n_1} p_2^{n_2} \ldots p_k^{n_k},\ \sum_{i=1}^{k} p_i = 1$$

Show that the mean and variance of $\chi^2 = \sum_{i=0}^{k} \dfrac{(n_i - np_i)^2}{npi}$ are given by

$$(k-1) \text{ and } 2(k-1)\left(1 - \frac{1}{n}\right) + \frac{1}{n}\left(\sum_{i=1}^{k} \frac{1}{p_i} - k^2\right)$$

(***B.A., Hons., & M.A., Madras, 1957***)

[**Hint.** Let $x_1 = \dfrac{(n_i - np_i)}{\sqrt{np_i}}$ so that $\sum_{i=1}^{k} x_i \sqrt{p_i} = 0$ and therefore $\chi^2 = \sum_{1}^{k} x_i^2$.

The joint m.g.f. of the variables $n_1, n_2 \ldots n_k$ is

$$(p_1 e^{t_1} + p_2 e^{t_2} + \ldots + p_k e^{t_k})^n$$

The joint *m.g.f.* of the variables $x_1, x_2, \ldots, x_k$ is

$$\phi(t_1, t_2 \ldots, t_k) = p_1 e^{\frac{t_1}{\sqrt{np_1}}} \cdot p_2 e^{\frac{t_2}{\sqrt{np_2}}} \ldots p_k e^{\frac{t_k}{\sqrt{e^{np_k}}}}$$

Now write down Malaurin expansion of this function and obtain Ex^2.

46. If $r_1, r_2 \ldots, r_k$ be k independent Poisson variates with parameters $\lambda_1, \lambda_2 \ldots, \lambda_k$ prove that (*i*) $r = \sum_{i=1}^{k} r_i$ is a Poisson variate, and (*ii*) the condional distribution $P(r_1, r_2 \ldots, r_k \mid r)$, when the sum $r = (r_1 + r_2 + \ldots + r_k)$ is fixed, is multinomial. **(*B.Sc. Delhi, 1982*)**

[**Hint.** $P(r_1 \ldots, r_k) = e^{-(\lambda_1 + \ldots + \lambda_k)} \dfrac{\lambda_1^{r_1}}{r_1!} \cdot \dfrac{\lambda_2^{r_2}}{r_2!} \cdots \dfrac{\lambda_k^{r_k}}{r_k!}$

$$P(r = r_1 + \ldots + r_k) = \sum_{e}^{k} \lambda_i \frac{\lambda_1^{r_1}}{r_1!} \cdots \frac{\lambda_k{}^{r}k}{r_k!} \div \frac{e^{-\lambda}\lambda^r}{r!}, \lambda = \lambda_1 + \ldots + \lambda_k$$

$$\therefore \quad P(r_1, \ldots, r_k \mid r) = \frac{P(r_1 \ldots r_k)}{P(r)} = \frac{r!}{r_1! \ldots r_k!} \left(\frac{\lambda_1}{\lambda}\right)^{r_1} \cdots \left(\frac{\lambda_k}{\lambda}\right)^{r_k}$$

which is multinomial with probabilities $\lambda_1/\lambda, \ldots, \lambda_k/\lambda$ in the classes.]

47. When n is large and $np_i = \lambda_i\ (i = 1, 2, \ldots, k)$ is moderate, show that the multinomial distribution is approximated by

$$\exp\left[-\sum_{i=1}^{k} \lambda_i\right] \prod_{j=1}^{r} \lambda_j^{k} j / \prod_{j=1}^{r} k_j$$

Prove also that the sum of all the terms is unity.

48. Derive the distribution of the random number N, where N is the smallest number of independent trials required to obtain a given number k of occurrences of an event of which the probability is p in each trial.

Show that as p becomes small the distribution of $2N_p$ tends to that of a χ^2 distribution with $2k$ degrees of freedom.

[**Ans.** The distribution is ${}^{N-1}C_{k-1}\, p^k (1-p)^{N-K}, N = k, k+1, \ldots$]

$$m.g.f. = \left(1 + \frac{e^{-\theta} - 1}{p}\right)^{-k} \quad \therefore \ \mu_1' = \frac{k}{p}, \ \mu_2 = \frac{k}{p^2} + \frac{k}{p}$$

In standard measures *m.g.f.* tends to $(1-\theta)^{-k}$ and hence the result.

49. Show that the probability $P(r+1)$ of at least $(r+1)$ success in $(x+1)$ independent repeated Bernoulli trials with probability p of success at each trial, is equal to

$$(r+1)\binom{n+1}{r+1} \int_0^p x^r (1-x)^{n-r} dx$$

[**Hint.** $P(r+1)$ may be regarded as a function of p for r and n fixed. Verify

$$\frac{d}{dp} P(r+1) = \frac{(n+1)!}{r!\,(n-r)!} p^r q^{n-r}.\Big]$$

50. Find the probability distribution of a binomially distributed random variable (based on n repetitions of an experiment) *truncated* to the right at $X = n$ *i.e.* $X = n$ cannot be observed. Obtain the expected value and the variance of the random variable. **[Truncated Binomiall Distribution]**

[Hint. $P(X = r) = {}^nc_r\, p^r q^{n-r}/(1-p^n),\ r = 0, 1, \ldots, n-1.$

$$Ex = np\,(1 - p^{n-1})/(1-p^n).$$

$$EX^2 = [n(n-1)p^2(1-p^{n-2}) + np(1-p^{n-1})]/(1-p^n).]$$

51. (*a*) If X_1 and X_2 are independent random Poison variates with parameters, λ_1 and λ_2, show that the conditional distribution of X_1 given $X_1 + X_2$ namely $P(X_1 = k \mid X_1 + X_2 = n)$ is binomial with probability $\lambda_1(\lambda_1 + \lambda_2)$.

Generalize this result and prove it. (*B. A. Hons. Delhi, 1984*)

(*b*) If X and Y are independent *r.v.* with some geometric distribution each with parameter p show that (*i*) the conditional distribution of X given that $X + Y = n$ is a uniform distribution. (*ii*) $P[X = Y] = p/(2-p)$.

(*B. Sc. Bombay, 1979 (B.A. Hons. Stat. Delhi, 1984)*

52. For each given p, let X have a binomial distribution with parameters p and N. Suppose p is distributed according to Beta distribution with parameters r and s. Find the resulting distribution of N. When is this distribution uniform ?

[Ans. $P(X = k) = \binom{N}{k} \dfrac{\Gamma r+s\, \Gamma k+s\, \Gamma n-k+s}{\Gamma r\, \Gamma s\, \Gamma N+r+s}$

$P(X = k) = 1/(N+1)$ when $r + s = 1$.

53. Suppose we have N chips marked 1, 2. . . . . , N respectively. We take a random sample of size $rn + 1$ without replacement. Let Y be the median of the random sample. Show that the probability function of Y is

$P(Y = k) = \binom{k-1}{N}\binom{N-1}{N}\binom{N}{2n+1}$ for $k = n+1, n+2, \ldots, N/2$.

Verify $E(Y) = \dfrac{N+1}{2}$ and Var $Y = \dfrac{(N-2n+1)(N+1)}{8n+12}$.

54. Suppose that X has the binomial distribution

$$P(X = x) = \binom{n}{x} p^k (1-p)^{n-k},\ x = 0, 1, 2, \ldots n$$

Let A and B be the events given by

$$A = \{\omega \mid X(\omega) \le 2\}, B = \{\omega \mid X(\omega) \ge n-2\}$$

Show that $P(A) = P(B)$ iff $p = \dfrac{1}{2}$. (*B.A. Hons. Delhi, 1982*)

[Hint. $P(A) = P(B) \Rightarrow q^n + {}^nC_1\, q^{n-1} p + {}^nC_2\, q^{n-2} p^2$

$$= {}^nC_{n-2}\, q^2 p^{n-2} + {}^nC_{n-1}\, qp^{n-1} + p^n$$

$$\Rightarrow (q^n - p^n) + {}^nC_1 pq(q^{n-2} - p^{n-2}) + {}^nC_2 p^2 q^2 (q^{n-4} - p^{n-4}) = 0, q = 1 - p.$$

$\Rightarrow (q - p)$ [factor which can be equal to zero when $q = p$] $= 0$

$\therefore \quad q = p \;\Rightarrow\; 1 - p = p \;\Rightarrow\; 2p = 1 \;\Rightarrow\; p = \frac{1}{2}$.

Conversely, if $q = p$. then clearly $P(A) = P(B)$. Hence the result.]

55. Derive the distribution of the difference of binomial variables with common n and p.

[**Hint.** We want to determine the distribution of Y where $Y = X_1 - X_2$

Now $P[Y = k] = \sum\limits_{r=0}^{k} p[X_1 = r, X_2 = r - k] = \sum\limits_{r=0}^{k} \binom{n}{k}\binom{n}{r-k} p^{2r-k} q^{2n-2r+k}$

$$= q^{2n} \sum_{r=0}^{k} \binom{n}{r}\binom{n}{r-k}\left(\frac{p}{q}\right)^{2r-k}, k = 0, 1, \ldots n]$$

Since $P(Y = -k) = P(Y = k)$, distribution of Y is symmetrical about $Y = 0$.

56. If $X \sim B(n, p)$ and Y has negative binomial distriibution with parameter r and p, prove that $F_x(r-1)\, 1 - F_v(k - r) = P(Y \geq n - 2)$.

(*B.Sc. Stat. Hons. Delhi, 1987*)

$$\left[\textbf{Hint.}\ \text{L.H.S.} = \sum_{k=0}^{r-1}\binom{n}{k} p^k q^{n-k} = p^r q^n \sum_{k=0}^{r-1}\left\{\binom{n}{k}^{-k} q\,(1-q)^{-(rk)}\right\}\right.$$

$$= p^r q^{n-r+1} \sum_{k=0}^{r-1}\left\{\binom{n}{k}\sum_{t=0}^{\infty}\binom{t+r-1-k}{r-1-k} q^{t+r-1k}\right\}, t = z - (r - 1 - 1p)$$

$$\left. = p^r q^{n-r+1} \sum_{k=0}^{r-1}\left\{\binom{n}{k} \sum_{z-k-1-k}^{\infty}\binom{z}{r-1-k} q^z = \text{etc. } z = y - (n - r + 1).\right]\right.$$

57. X is negative binomial variate with p.d.f. given by (6.11) page 195, prove :

$$\mu_{r+1} = q\left[\frac{rk}{p^2}\mu_{r-1} + \frac{d\mu r}{dq}\right], r = 1, 2, \ldots$$ **(*B.Sc. Math. Hons. Pb., 1990*)**

Hint. Show $\mu_r = \sum\limits_{0}^{\infty} (x - \frac{kq}{p})^r . \binom{k+x-1}{x} q^x p^k$ and

then find $\frac{d\mu r}{dq}$, $q = 1 - p$.

7

Univariate Continuous Probability Distributions

7.1. Univariate Continuous Probability Distributions. We have so far dealt with discrete probability distributions *i.e.*, distributions in which the stochastic variate takes either set of values (as in Binomial and Hyper-geometric Distributions) or at most a denumerably infinite set of values (as in Poison and Negative Binomial Distributions). But, when we deal with variates like heights, we find that such variates can taken a non-enumerably infinite set of values, or more precisely, they can take any value in the given interval $a \leq X \leq b$ of the *arithmetic continuum.* Such variates are called, continuous variates, and their probability distributions are accordingly known as *continuous probability distributions*. The concept of a continuous distribution is extremely useful for the following reasons :

(*i*) It is a logical necessity.

(*ii*) It is a theoretical necessity, since in Statistics, we deal not only with distributions occurring in nature but also with theoretical distributions and such distributions are very often continuous distributions.

(*iii*) It is easier mathematically to deal with a continuous distribution through the use of techniques of differentiation and integration than with a discrete distribution with a large number of values for the variate.

(*iv*) Continuous distributions are very good approximations as compared to discrete distributions.

In continuous probability distributions, since we cannot associate a finite non-zero probability with each variate value as the sum of the probabilities will not converge (except possibly when the probability associated with all but denumerable set of points is zero), we find it more useful to associate probabilities with intervals. Thus $f(x)$, a continuous function of x, defines the probability distribution of the stochastic variate X by the relation that the probability of the value of the variate falling in the infinitesimal interval $\left(x - \frac{1}{2}\, dx\right)$ to $\left(x + \frac{1}{2}\, dx\right)$ is expressible in the form $f(x)\,dx$. Symbolically, $P\,(x - \frac{1}{2}\,dx \leq X \leq x + \frac{1}{2}\,dx) = f(x)\,dx$.

Sometimes, we write this as $P\,(x \leq X \leq x + dx) = f_x(x)\,dx = d\,F(x)$.

$f_x(x)$ or simply $f(x)$ is called the *probability density function (p. d. f)* or simply the *density function (d.f.)* and $f(x)\,dx$ is called *probability differential.* The continuous curve $y = f(x)$ is called the *probability density curve* or briefly the *probability curve* and when this is symmetrical, the distribution is said to be *symmetrical probability curve.*

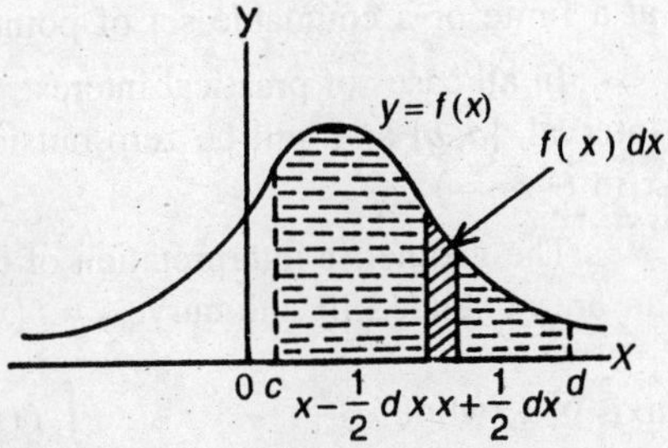

Fig. 7·1

The range of the variate may be finite or infinite, but even when the range is finite, it is convenient to consider the range as infinite, by simply postulating that the density, function is zero, outside the specified range. Thus if X takes values in the closed interval $[a, b]$ or $a \leq X \leq b$ only and its density function is $\phi(x)$, then its distribution is equivalent to the variate with density, function, defined as follows :

$$\left.\begin{aligned} f(x) &= 0, & x &< a \\ &= \phi(x), & a &\leq x \leq b \\ &= 0, & x &> b. \end{aligned}\right\} \qquad \text{...(7·)}$$

The probability function $f(x)$ possesses the following properties :

(*i*) $f(x) \geq 0 \quad \forall x$ *i.e.*, for $-\infty < x < \infty$

(*ii*) $\displaystyle\int_{-\infty}^{+\infty} f(x)\,dx = 1$ ...(7·2)

(*iii*) For any event E, $P(E)$ is given by
$$P(E) = \int_E f(x)\,dx$$

Property (*i*) is obviously necessary since negative probability has no meaning. Property (*ii*) corresponds to the requirement that the probability of an event that is certain to occur should be equal to unity. The probability that the variate X falls in any interval (c, d) is given by

$$P\,(c \leq X \leq d) = \int_c^d f(x)\,dx \qquad \text{...(7·3)}$$

We may note that while we started by requiring $f(x)$ to be a continuous function, the density-function defined in (7·1) can be discontinuous as $x = a$ and $x = b$. Essentially what we require is that in view of (7·2) and (7·3), $f(x)$ should be Riemann integrable*. If $f(x)$ is continuous, it is certainly R-integrable. Even if $f(x)$ is continuous except at a finite or a countable set of points, it is still R-integrable.

In all cases of practical interest, $f(x)$ will be either continuous in the interval, $[a, b]$ and will be zero outside it or it will be continuous for all x in $(-\infty, \infty)$

The geometric interpretation of the integral in (7·3) is that it denotes an area bounded by the curve $y = f(x)$, the ordinates $x = c, x = d$ and the axis of x, while

$$\int_{-\infty}^{+\infty} f(x)\, dx = 1$$

means that total area under the curve $y = f(x)$ where $f(x)$ is the p.d.f. should be unity. Let $F(x) = \int_{-\infty}^{x} f(x)\, dx = P(X \le x)$

then the function $F(x)$ is probability that $X \le x$. Then $F(b) = P(X \le b)$

and $$F(b) - F(a) = \int_{a}^{b} F(x)\, dx = P(a \le X \le b)$$

$F(x)$ is called the *cumulative distribution function* (d.f.). the d.f. $F(x)$ has obviously the following properties :

(*i*) Since $f(x) \ge 0$, $\int_{a}^{x} f(x)\, dx \ge 0$ when $x \ge a$,

In particular $$\int_{-\infty}^{x} f(x)\, dx \ge 0 \text{ or } F(x) \ge 0$$

(*ii*) $F(x)$ is a continuous function since $\int_{a}^{x} f(x)\, dx$ is a continuous of $f(x)$ as an R-integral function so that

$$F(a-0) = F(a) = F(a+0),\quad F(b-0) = F(b) = F(b+0)$$

and $P(a < X < b) = P(a \le X < b) = P(a < X \le b) = P(a \le X \le b)$

(*iii*) Since $f(x) \ge 0$, $F(x)$ is a non- decreasing function of x.

(*iv*) If $F(x)$ is a continuous function, then $F'(x) = f(x)$

* This theory of integrals was given by Riemann in an important paper in 1854.

$$(v)\quad F(-\infty) = \lim_{x \to -\infty} \int_{-\infty}^{x} f(x)\, dx = 0$$

$$(vi)\quad F(\infty) = \lim_{x \to \infty} \int_{-\infty}^{x} f(x)\, dx = \int_{-\infty}^{\infty} f(x)\, dx = 1.$$

Properties (*i*), (*iii*), (*v*) and (*vi*) also hold for probability distribution of discrete random variates, but as we saw earlier, for such distribution, $F(x)$ is discontinuous on the left at those points $x_1, x_2, \ldots, x_n$ where the probability function $f(x)$ is not zero. Moreover $F'(x)$ does not exist at these points.

Also from (7·3) by putting $c = d = k$, it follows that $P(X = k) = 0$, $\forall k$. In other words, it means that in case of continuous *r.v.*, the probability of a.*r.v.x* at any point within the given range is zero.

Example 7·1. *Let $F(x)$ be a distribution function, prove that for any fixed $h \neq 0$, the function $G(x) = \frac{1}{2h} \int_{x-h}^{x+h} F(t)\, dt$ is also a distribution function.*

Since $F(x)$ by hypothesis is a monotonic non- decreasing function, it follows that if $x < x_0$, then $F(X + T) \leq F(x_0 + T)\, dx, h > 0$

$$\therefore \quad \int_{-h}^{+h} F(x + T)\, dx \leq \int_{-h}^{+h} F(x_0 + T)\, dx, h > 0$$

Putting $x + T = t \Rightarrow dx = dt$. Hence $\int_{T-h}^{T+h} F(t)\, dt \leq \int_{T-h}^{T+h} F(t_0)\, dt$

For $x - h \leq t \leq x + h$, $G(x) \leq G(x^0)$; $F(x - h) \leq F(t) \leq F(x + h)$

$F(x - h) \leq G(x) \leq F(x + h)$.

Make $x \to -\infty$ or $+\infty$, we obtain $G(-\infty) = 0$, $G(\infty) = 1$.

$\therefore$ $G(x)$ is continuous every where, Hence $G(x)$ is a distribution function.

Example 7·2. (*a*) *Is the function $f(x)$, defined as follows, a density function ?*

$$
\begin{aligned}
f(x) &= 0 & x &< 2 \\
&= \tfrac{1}{18}(3 + 2x). & 2 &\leq x \leq 4 \\
&= 0, & x &> 4.
\end{aligned}
$$

(*b*) *Find the probability that a variate having this density will fall in the interval* $2 \le X \le 3$.

(*a*) $f(x)$ is clearly ≥ 0 for every x in the given interval and

$$\int_{-\infty}^{\infty} f(x)\, dx = \int_{-\infty}^{2} 0 \,.\, dx + \int_{2}^{4} \frac{1}{18}(3+2x)\, dx + \int_{4}^{\infty} 0 \cdot dx = 1.$$

Hence the function $f(x)$ satisfies the requirements for density function.

(*b*) Required probability $= P(2 \le X \le 3) = \int_{2}^{3} \frac{1}{18} \cdot (3+2x)\, dx = \frac{4}{9}$.

Alternatively, $F(x) = P(x > 2)$

$$= \int_{2}^{x} \frac{1}{18}(3+2x)\, dx = \tfrac{1}{18}(x^2 + 3x - 10). \quad (2 \le x \le 4)$$

$$\therefore \quad P(2 \le x \le 3) = P(x \le 3) - P(x < 2) = F(3) - F(2)$$

$$= \tfrac{1}{18}(9 + 9 - 10) - 0 = \tfrac{4}{9}.$$

Example 7·3. *Find the constant k so that function F (x) defined as follows may be a density function :*

$$f(x) = \frac{1}{k}, \quad a \le x \le b \;;\; = 0, \quad \textit{elsewhere}$$

Find also the cumulative distribution function of the random variable X when k satisfies the requirements for f (x) to be a density function.

(*i*) If $k > 0, f(x) \ge 0$ for every x. Hence for $f(x)$ to be a density function, we must have

$$1 = \int_{a}^{b} f(x)\, dx = \int_{a}^{b} \frac{1}{k}\, dx = \frac{b-a}{k}, \text{ so that } k = b - a.$$

(*ii*) We have $\quad F(x) = 0, \quad\quad x < a$

$$= \int_{a}^{x} \frac{1}{k}\, dx = \frac{x-a}{b-a}, \quad\quad a \le x \le b$$

$$= 1 \quad\quad x > b$$

Its graph is shown in in Fig. 7·2. The density function is shown in Fig. 7·3.

Note. Whenever $f(x)$ is constant throughout its range the variate is said to have a *uniform* or *rectangular probability distribution.*

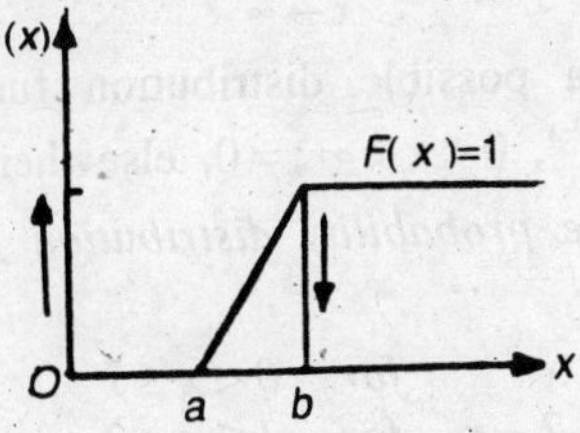

Fig. 7·2. Distribution function for Rectangular distribution.

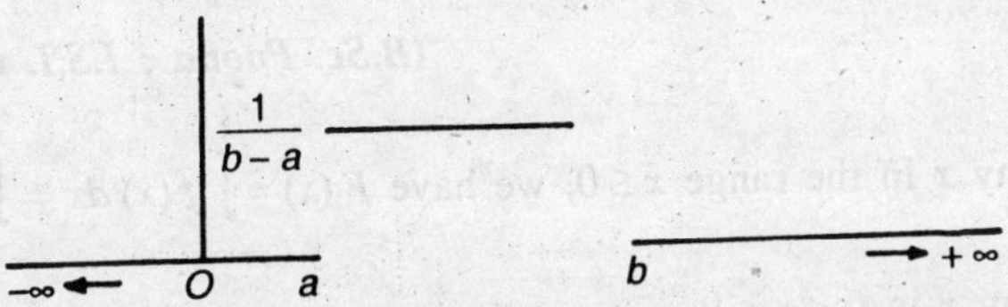

Fig. 7·3. Density function for Rectangular distribution.

Example 7·4.(*a*) *Verify that the following is a distribution function:*

$$F(x) = \begin{cases} 0 & x < -a \\ \dfrac{1}{2}\left(\dfrac{x}{a}+1\right) & -a \le x \le a \\ 1 & x > 1 \end{cases}$$

(*b*) *Show that* $F(x) = 0, \; -\infty < x < 0$

$$= 1 - e^{-x}, 0 \le x < \infty$$

is a possible distribution function and find the density function.

In order that $F(x)$ may be a distribution function, it must satisfy all the five conditions given on page 220. It can be easily verified that the conditions hold good in this case.

Further, $\quad f(x) = \dfrac{d}{dx} F(x) = \begin{cases} \dfrac{1}{2a}, & -a \le x \le a \\ 0, & \text{otherwise} \end{cases}$

Now, $\quad \displaystyle\int_{-\infty}^{\infty} f(x)\,dx = \int_{-\infty}^{-a} 0.dx + \int_{-a}^{a} \frac{1}{2a}\,dx + \int_{a}^{\infty} 0.dx = 1$

Hence $f(x)$ is the probability density function and $F(x)$ is the distribution function.

(*b*) Clearly $F(x)$ is monotonic non- decreasing and continuous every where ; Also $F(-\infty)=0$, $F(\infty)=\lim_{x\to\infty}(1-e^{-x})=1-e^{-\infty}=1-0=1$.

Hence $F(x)$ is a possible distribution function, Differentiating $F(x)$, we get $f(x)=e^{-x}$, $0<x<\infty$; $=0$, elsewhere.

Example 7·5. *The probability distribution function of a random variable X is*

$$f(x)=\begin{cases} x & \text{for} \quad 0<x\le 1 \\ 2-x & \text{for} \quad 1\le x<2 \\ 0 & \text{for} \quad x\ge 2 \end{cases}$$

Compute the cumulative distribution function of X.

(B.Sc. Poona ; I.S.I. Calcutta)

For any x in the range $x\le 0$, we have $F(x)=\int_{-\infty}^{x} f(x)\,dx=\int_{-\infty}^{x} 0.dx=0$

For any x in $0<x<1$,

$$F(x)=\int_{-\infty}^{0} f(x)\,dx+\int_{0}^{x} f(x)\,dx=\int_{-\infty}^{0} 0.dx+\int_{0}^{1} x\,dx=\tfrac{1}{2}x^2$$

For any x in $1<x\le 2$, $F(x)=\int_{-\infty}^{0} f(x)\,dx+\int_{0}^{1} f(x)\,dx+\int_{1}^{x} f(x)\,dx$

$$=\int_{-\infty}^{0} 0\cdot dx+\int_{0}^{1} x\,dx+\int_{1}^{x}(2-x)\,dx=2x-\tfrac{1}{2}x^2-1.$$

For every x in $x\ge 2$,

$$F(x)=\int_{-\infty}^{0} f(x)\,dx+\int_{0}^{1} f(x)\,dx+\int_{1}^{2} f(x)\,dx+\int_{2}^{\infty} f(x)\,dx$$

$$=\int_{-\infty}^{0} 0\cdot dx+\int_{0}^{1} x\,dx+\int_{1}^{2}(2-x)\,dx+\int_{2}^{\infty} 0\cdot dx$$

$$=\tfrac{1}{2}+\left(2-\tfrac{3}{2}\right)=1.$$

Example 7·6. *An angle θ is chosen at random between $-\pi/2$ and $\pi/2$ and a line is drawn through the point P (o, a) at any angle θ with the x-axis. Find the density function for the point x at which this line crosses the x-axis.*

The position X is determined by its distance $OX=x$ from O, this distance being positive or negative according as X is to the right or to the left of the point O. $\angle OPX=\theta$. The relation between θ and x is x

$= a \tan \theta$ or $\theta = \tan^{-1}(x/a)$.

Consequently, the relation between $d\theta$ and dx, obtained by differentiation, is $d\theta = \dfrac{a\,dx}{a^2 + x^2}$.

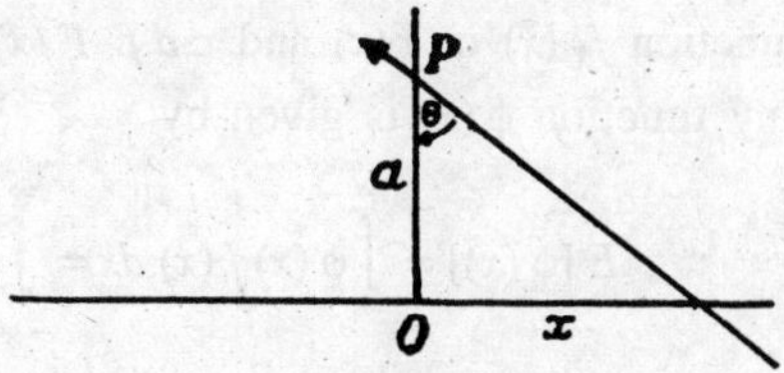

Fig. 7·4

Assuming that the variable θ has uniform distribution of probability in the interval $-\pi/2$ to $+\pi/2$, the probability that θ will lie in the interval θ to $\theta + d\theta$ is $d\theta/\pi$ and the probability that the distance of x from 0 will be contained between x and $x + dx$ is the same.

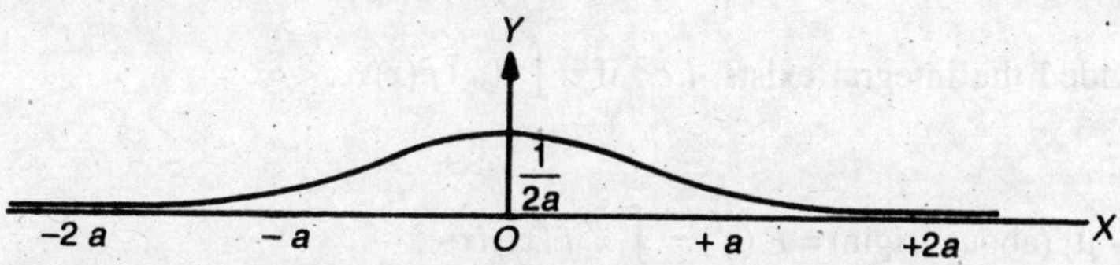

Fig. 7·5. Cauchy's probability curve $f(x) = \dfrac{1}{\pi} \cdot \dfrac{a}{x^2 + a^2}$.

Hence the probability density function for the distribution of the variable X is $\quad f(x) = \dfrac{1}{\pi} \cdot \dfrac{a}{a^2 + x^2}, -\infty < x < +\infty$

Also $\quad \dfrac{1}{\pi} \displaystyle\int_{-\infty}^{+\infty} \dfrac{a\,dx}{a^2 + x^2} = \dfrac{1}{\pi}\left[\tan^{-1}\dfrac{x}{a}\right]_{-a}^{+a} = \dfrac{1}{\pi}\left[\dfrac{\pi}{2} - \left(-\dfrac{\pi}{2}\right)\right] = 1.$

This distribution is called *Cauchy's distribution.* It is named after its discoverer K.L. Cauchy (1789-1857).

Ex. Radio active-particles are being emitted from a point -source P and scintillate against a plane sensitive screen placed at unit distance from the source. If the particles travel in straight lines and all directions are equally likely, show that the frequency (probability) distribution of scintillations along any line on the screen passing through the foot of the perpendicular Q from P *is* given by

$$\frac{1}{\pi} \cdot \frac{dx}{\{1 + (x - a)^2\}}, -\infty < x < +\infty$$

where a is the distance of Q and x that of any point from the origin taken anywhere on the line.

7·2 Definitions of Moments, etc. for Univariate Continuous Probability Distributions. In extending the definition and theorems of discrete probability distributions to continuous probability distributions we shall observe that formally it consists in replacing p_i by $f(x)\,dx$, x_1 by x and summation over i by integration over the permissible range for x.

Let X be a continuous random variable with probability density function $f_X(x)$ or $f(x)$ and *c.d.f.* $F(x)$. The mathematical expectation of any function $\phi(x)$ is given by

$$E[\phi(x)] = \int_{-\infty}^{\infty} \phi(x) f(x)\, dx = \int_{-\infty}^{\infty} \phi(x)\, dF(x), \int_{-\infty}^{+\infty} dF(x) = 1$$

provided this integral *exists* and is called the mean value or expected value of $\phi(x)$. In particular,

$$\text{mean} \quad \bar{x} = \mu_1' = E(x) = \int_{-\infty}^{\infty} x f(x)\, dx = \int_{-\infty}^{\infty} x \,.\, dF(x), \int_{-\infty}^{+\infty} f(x)\, dx = 1;$$

provided the integral exists, *i.e.*, if $\int_{-\infty}^{+\infty} |x| f(x)\, dx < \infty$.

$$\mu_r'(\text{about origin}) = E(x^r) = \int_{-\infty}^{\infty} x^r f(x)\, dx$$

$$\mu_r'(\text{about } x = A) = E[(x-A)^r] = \int_{-\infty}^{+\infty} (x-A)^r f(x)\, dx$$

$$\mu_r = E[(x-\bar{x})^r] = \int_{-\infty}^{+\infty} (x-\bar{x})^r f(x)\, dx \qquad \text{...(7·4)}$$

$$M_{X=0}(t) = E(e^{tx}) = \int_{-\infty}^{\infty} e^{tx} f(x)\, dx \qquad \text{...(7·5)}$$

$$M_u(t) = E\left[e^{t\left(\frac{x-\bar{x}}{\sigma}\right)} \right] = \int_{-\infty}^{\infty} e^{t\left(\frac{x-\bar{x}}{\sigma}\right)} f(x)\, dx$$

where $u = (x-\bar{x})/\sigma = (x - \text{mean}) / \text{standard deviation}$.

or $$M_u(t) = e^{-t\bar{x}/\sigma} \int_{-\infty}^{\infty} e^{tx/\sigma} f(x)\, dx = e^{-t\bar{x}/\sigma} M_{x=0}\left(\frac{t}{\sigma}\right) \qquad \text{...(7·6)}$$

$= e^{-t\bar{x}/\sigma} \cdot$ *m.g.f.* of X with t/σ written for t.

Geometric mean G, is given by

$$\log G = \int_{-\infty}^{\infty} \log x \cdot f(x)\, dx = E(\log x) \qquad \text{...(7·7)}$$

By taking anti-log of both sides of (7·7), we obtain the value of G.

provided the integral exists. Of course, when x is negative, $f(x)$ should be zero for G to be real.

Harmonic mean, H, is defined by

$$\frac{1}{H} = \int_{-\infty}^{\infty} \frac{1}{x} f(x)\, dx = E\left(\frac{1}{x}\right) \qquad \text{...(7·8)}$$

Median 'a' is the value of the variate such that X *is* equally likely to be greater than or less than a, *i.e.* $P(X \le a) = \frac{1}{2} = P(X \ge a)$, $\int_{-\infty}^{+\infty} f(x)\, dx = 1$

or
$$\int_{-\infty}^{a} f(x)\, dx = \tfrac{1}{2} = \int_{a}^{\infty} f(x)\, dx \cdot \qquad \text{...(7·9)}$$

Solving the two equations, (as given by (7·9)), we obtain the value of median 'a' lying within the permissible range of the variate.

The lower and the upper quartiles Q_1 and Q_3 are given by

$$\int_{-\infty}^{Q_1} f(x)\, dx = \tfrac{1}{4} = \int_{Q_3}^{\infty} f(x)\, dx \qquad \text{...(7·10)}$$

Solving these equations, we determine the values of Q_1 and Q_3.

Mode or modal value is that value of the variate for which the probability density is maximum so that if probability density function $f(x)$ is continuous and differentiable, the mode is defined by the root satisfying

$$\begin{Bmatrix} f'(x) = 0 \\ f''(x) < 0 \end{Bmatrix} \qquad \text{...(7·11)}$$

provided that the solution of $f'(x) = 0$ lies within the permissible range of the variate.

Mean deviation from the mean $\bar{x}$ is given by

$$\text{M.D.} = \int_{-\infty}^{\infty} |x - \bar{x}|\, f(x)\, dx \qquad \text{...(7·12)}$$

In terms of the cumulative distribution function or simply distribution functions.

$$\left.\begin{aligned} F(Q_1) &= \tfrac{1}{4} \\ F(Q_2 \text{ or Median}) &= \tfrac{1}{2} \\ F(Q_3) &= \tfrac{3}{4} \\ F(D_r) &= \frac{r}{10} \end{aligned}\right\} \qquad \text{...(7·13)}$$

where D_r is the rth decile.

Mean difference (due to Gini) is defined by

$$\Delta_1 = \int_{-\infty}^{\infty} \int_{-\infty}^{\infty} |x - y| f(x) f(y)\, dx dy \qquad ...(7\cdot 14)$$

Theorems on Expectation

(*a*) If X and Y are continuous random variables, then

$$E(X + Y) = E(X) + E(Y)$$

(*b*) If X and Y are independent continuous random variables, then

$$E(XY) = E(X)\, E(Y)$$

Proof. (*a*) By definition, if $f_{xy}(x, y)$ is joint *p.d.f.* of X, Y then

$$E(X + Y) = \int_{-\infty}^{\infty} \int_{-\infty}^{\infty} (x + y)\, f_{xy}(x, y)\, dx\, dy$$

$$= \int_{-\infty}^{\infty} \int_{-\infty}^{\infty} x f_{xy}(x, y)\, dx dy + \int_{-\infty}^{+\infty} \int_{-\infty}^{+\infty} y f_{xy}(x, y)\, dx\, dy$$

$$= \int_{-\infty}^{\infty} x \left[\int_{-\infty}^{\infty} f_{xy}(x, y)\, dy \right] dx + \int_{-\infty}^{+\infty} y \left[\int_{-\infty}^{+\infty} f_{xy}(x, y)\, dx \right] dy$$

$$= \int_{-\infty}^{\infty} x\, g_x(x)\, dx + \int_{-\infty}^{\infty} y\, h_y(y)\, dy,$$

$= E(X) + E(Y)$, $g_x(x)$ is the marginal distribution of X

(*b*) $$E(XY) = \int_{-\infty}^{-\infty} \int_{-\infty}^{+\infty} xy\, g_x(x)\, h_y(y)\, dx\, dy$$

Since X and Y are independent $\therefore f_{xy}(x, y) = g_x(x) \cdot h_y(y)$, see page 256

$$= \left[\int_{-\infty}^{+\infty} x\, g_x(x)\, dx \right] \left[\int_{-\infty}^{+\infty} y h_y(y)\, dy \right]$$

$$= E(X)\, E(Y).$$

Example 7·7. *If μ_r' exists, then show that μ_s' exists for all $1 \le s \le r$, i.e. if $E(x^r)$ exists, then $E(x^s)$ exists for all $1 \le s \le r$.*

(M.Sc., Nagpur)

We have $E(|x|^s) = \int_{-\infty}^{\infty} |x|^s \, dF(x) = \int_{-1}^{1} |x|^s \, dF(x) + \int_{|x|>1} |x|^s \, dF(x)$

$$\leq \int_{-1}^{1} |x|^s \, dF(x) + \int_{|x|>1} |x|^r \, dF(x)$$

since $|x|^s < |x|^r$ and $|x| > 1$.

Hence $E(|x|^s) \leq \int_{-1}^{1} dF(x) + \int_{|x|>1} |x|^r \, dF(x)$

Thus $E(|x|^s) \leq 1 + E(|x|^r)$. Hence the result follows.

Example 7·8. *Show that the mean diviation from median is less than that measured from any other value.* **(B.A. Hons., Delhi, I.A.S.)**

The mean deviation about an arbitrary point a is given by

$$F(a) = \int_{-\infty}^{\infty} |x - a| f(x) \, dx \quad \text{where} \quad \int_{-\infty}^{\infty} f(x) \, dx = 1 \qquad \text{...(A)}$$

$$\Rightarrow F(a) = \int_{-\infty}^{a} (a - x) f(x) \, dx + \int_{a}^{\infty} (x - a) f(x) \, dx$$

Differentiating with respect to a under the sign of integration*, we have

$$\frac{d}{da} F(a) = \int_{-\infty}^{a} f(x) \, dx - \int_{a}^{\infty} f(x) dx, \text{ and } \frac{d^2}{da^2} F(a) = f(a) + f(a) = 2f(a) > 0,$$

assuming that $f(x)$ is not zero at the median value.

Thus for $F(a)$ to be minimum, we should have $\int_{-\infty}^{a} f(x) \, dx = \int_{a}^{\infty} f(x) \, dx$

Also from (A), $\int_{-\infty}^{a} f(x) \, dx + \int_{a}^{\infty} f(x) \, dx = 1$ Thus $\int_{-\infty}^{a} f(x) \, dx = \frac{1}{2} = \int_{a}^{\infty} f(x) \, dx$.

---

* Theorem if $f(x, \alpha)$ is a continuous function of both variables x and α, *continuous partial derivatives* $\frac{\partial^2 f(x, \alpha)}{\partial x \partial \alpha}, \frac{\partial^2 (f(x, \alpha)}{\partial \alpha \partial x}$ *and a and b* are differntiable functions of α, then

$$\frac{d}{dx}\left[\int_{a}^{b} f(x, \alpha) \, dx)\right] = \int_{a}^{b} \frac{\partial f(x, \alpha)}{\partial \alpha} dx + f(b, \alpha) \frac{db}{d\alpha} - f(a, \alpha) \frac{da}{d\alpha}$$

Example 7·9. *A continuous distribution of a variable X in the range (– 3, 3) is defined by* $f(x) = \frac{1}{16}(3+x)^2, \; -3 \le x \le -1$

$$= \frac{1}{16}(6-2x^2), \; -1 \le x \le 1$$

$$= \frac{1}{16}(3-x)^2, \; 1 \le x \le 3.$$

Verify that the area under the curve is unity. Show that the mean is zero.

(B.Sc. Jiwaji, 1971, B.Sc. Allahabad)

We have $\int_{-3}^{+3} f(x)\,dx = \int_{-3}^{-1} f(x)\,dx + \int_{-1}^{+1} + f(x)dx + \int_{1}^{3} f(x)\,dx$

$$= \int_{-3}^{-1} \frac{(3+x)^2}{16}\,dx + \int_{-1}^{+1} \frac{6-2x^2}{16}\,dx + \int_{1}^{3} \frac{(3-x)^2}{16}\,dx$$

$$= \left| \frac{(3+x)^3}{48} \right|_{-3}^{-1} + \left| \frac{6x - \frac{2x^3}{3}}{16} \right|_{-1}^{+1} + \left| \frac{(3-x)^3}{-48} \right|_{1}^{3} = \frac{1}{48}[\,8+32+8\,] = 1$$

Hence the area under the curve is unity.

Mean $= E(X) = \int_{-3}^{+3} x f(x)\,dx$

$$= \int_{-3}^{-1} \frac{x(3+x)^2}{16}\,dx + \int_{-1}^{1} \frac{x(6-2x^2)}{16} + \int_{1}^{3} \frac{x(3-x)^2}{16}\,dx$$

$$= \frac{1}{16}\left[\int_{-3}^{-1} (9x+6x^2+x^3)\,dx + \int_{-1}^{1} \left(6x - \tfrac{2}{3}x^3\right) dx + \int_{1}^{3} (9x-6x^2+x^3)\,dx \right]$$

$$= \frac{1}{16}\left[\frac{9x^2}{2} + 2x^3 + \frac{x^4}{4} \right]_{-3}^{-1} + \left[3x^2 - \frac{2x^4}{12} \right]_{-1}^{+1} + \left[\frac{9x^2}{2} - 2x^3 + \frac{x^4}{4} \right]_{1}^{3}$$

= 0, on simplification.

Example 7·10. *Show that for the exponential distribution* $dP = y_0\, e^{-x/\sigma}\, dx,\; 0 \le x < \infty,\; \sigma > 0,\; y_0$ *being a constant, the mean and the s.d. are each equal to* σ *and that the interquartile range is* $\sigma \log_e 3$. *Also find* μ_r' *and show that* $\beta_1 = 4$ and $\beta_2 = 9$.

(M.A. Punjab ; I.A.S. ; B.A. Hons., Statistics Delhi 1984)

The constant y_0 is given by

$$y_0 \int_0^{\infty} e^{-\frac{x}{\sigma}}\, dx = 1, \text{ so that } y_0\, \sigma \left[-e^{-x/\sigma} \right]_0^{\infty} = 1 \text{ which gives } y_0 = \sigma^{-1}.$$

$$\mu_r' \text{ (about } x = 0) = \frac{1}{\sigma} \int_0^{\infty} x^r e^{-\frac{x}{\sigma}}\, dx = r\,!\, \sigma^r \text{ by Gamma integral.}$$

$$\therefore\ \mu_1' = \sigma, \mu_2' = 2\sigma^2\ ; \mu_3' = 6\sigma^3, \mu_4' = 24\sigma^4.$$

$$\therefore\ \mu_2 = \mu_2' = \sigma^2, \mu_3 = 2\sigma^3, \mu_4 = 9\sigma^4.$$

Hence $\beta_1 = \dfrac{\mu_3^2}{\mu_2^3} = 4$, $\beta_2 = \dfrac{\mu_4}{\mu_2^2} = 9.$

Also $\displaystyle\int_0^{Q_1} \frac{1}{\sigma} e^{-\frac{x}{\sigma}}\, dx = \frac{1}{4}$ so that $e^{\frac{-Q_1}{\sigma}} = \frac{3}{4}$ ...(A)

and $\displaystyle\int_0^{Q_3} \frac{1}{\sigma} e^{-\frac{x}{\sigma}}\, dx = \frac{2}{4}$ or $e^{\frac{-Q_3}{\sigma}} = \frac{1}{4}$ ...(B)

Dividing (A) by (B), we get $e^{\frac{Q_3 - Q_1}{\sigma}} = 3$ or $Q_3 - Q_1 = \sigma \log_e 3.$

We can obtain from A and B, $Q_1 = -\sigma \log \frac{3}{4}$ and $Q_3 = -\sigma \log \frac{1}{4}$ and $Q_3 - Q_1 = \sigma \log 3$.

Example 7·11. *Show that for the rectangular distribution* $dP = dx,\ 0 \le x \le 1,\ \mu_1' = \frac{1}{2}, \mu_2 = \frac{1}{12}$, mean deviation about mean $= \frac{1}{4}$.

(B.Sc. Mysore, 1971 ; B. Sc.Delhi, 1973 ; I.A.S.)

Here $f(x) = 1,\quad 0 \le x \le 1$

$$\therefore \text{ We have } \mu_1' = \int_0^1 x\, dx = \left[\frac{x^2}{2} \right]_0^1 = \frac{1}{2},\quad \mu_2' = \int_0^1 x^2\, dx = \frac{1}{3}$$

$$\therefore \qquad \mu_2 = \frac{1}{3} - \frac{1}{4} = \frac{1}{12}.$$

Mean deviation (about the mean)

$$= \int_0^1 |\, x - \tfrac{1}{2}\, |\, dx = \int_0^{1/2} (\tfrac{1}{2} - x)\, dx + \int_{1/2}^1 (x - \tfrac{1}{2})\, dx = \left[\frac{x}{2} - \frac{x^2}{2} \right]_0^{1/2} + \left[\frac{x^2}{2} - \frac{x}{2} \right]_{1/2}^1 = \frac{1}{4}.$$

Ex. 1. Show that the mean deviation from the mean is not greater than the standard deviation.

Ex. 2. If (*i*) the origin is taken as the median, (*ii*) $a > 0$ and (*iii*) $\phi(a)$ denotes the mean deviation from the point a, show that

$$\phi(a) - \phi(0) = \int_{-\infty}^{a} (a-x) f(x)\, dx + \int_{a}^{\infty} (x-a) f(x)\, dx - \int_{-\infty}^{0} (-x) f(x)\, dx - \int_{0}^{\infty} x f(x)\, dx$$

$$= 2 \int_{0}^{a} (a-x) f(x)\, dx > 0$$

Similarly, if $a < 0$, show that $F(a) - F(0) > 0$, Now deduce the result of Example 7·8

Ex. 3. If the p.d.f $f(x)$ is discontinuous at $x_1, x_2 \cdots, x_n$ and $x_m < a < x_{m+1}$. show that

$$\phi(a) = \int_{-\infty}^{x_1} (a-x) f(x)\, dx + \int_{x_1}^{x_2} (a-x) f(x)\, dx + \cdots + \int_{x_m}^{a} (a-x) f(x)\, dx$$

$$+ \int_{a}^{x_{m+1}} (x-a) f(x)\, dx + \int_{x_{m+1}}^{x_{m+2}} (x-a) f\, dx + \cdots\cdots + \int_{x_n}^{\infty} (x-a) f(x)\, dx$$

$$\phi'(a) = \int_{-\infty}^{a} f(x)\, dx - \int_{a}^{\infty} f(x) dx, \qquad \phi''(a) = 2f(a)$$

Result also holds good for a grouped frequency distribution.

Ex. 4. For rectangular distribution $dP = dx,\ 1 \le x \le 2$ show that $A > G > H$.

Ex. 5. (*a*) For the triangular distribution

$$dP = \frac{1}{a}\left[1 - \frac{|x-b|}{a}\right] dx,\ |x-b| < a,$$

show that the mean is b and variance is $a^2/6$.

(*b*) For the triangular distribution with the density function

$$f(x) = 1 - |1-x|,\ 0 < x < 2.$$

show that the mean is 1, variance $\frac{1}{6}$ and $\mu_3 = 0$ **(*B.Sc., Shivaji, 1970*)**

Example 7·12. *Show that for a continuous distribution*

$$\Delta_1 = 2 \int_{-\infty}^{\infty} F(x) \{ 1 - F(x) \}\, dx.$$

In particular for the rectangular distribution $dF = dx,\ 0 \le x \le 1$, *show that* $\Delta_1 = \frac{1}{3}.$

$$\Delta_1 = \int_{-\infty}^{\infty} \int_{-\infty}^{\infty} |x-y|\, dF(x)\, dF(y) \text{ where } dF(x) = f(x)\, dx$$

$$= 2 \int_{-\infty}^{\infty} \int_{y=-\infty}^{x} (x - y)\, dF(x)\, dF(y)$$

$$= 2 \int_{-\infty}^{\infty} dF(x) \int_{y=-\infty}^{x} (x - y)\, dF(y) \qquad \text{...(A)}$$

Now $\int_{-\infty}^{x} (x - y)\, dF(y) = [\,(x - y)\, F(y)\,]_{-\infty}^{x} + \int_{-\infty}^{x} F(y)\, dy = \int_{-\infty}^{x} F(y)\, dy$

$\therefore$ from (A); $\Delta_1 = 2 \int_{-\infty}^{\infty} dF(x)\, \phi(x)$ where $\phi(x) = \int_{-\infty}^{x} F(y)\, dy = \int_{-\infty}^{x} F(x)\, dx$

$$\therefore \Delta_1 = 2\,[\,F(x)\,\phi(x)\,]_{-\infty}^{\infty} - 2 \int_{-\infty}^{\infty} F(x)\, F(x)\, dx = 2 \int_{-\infty}^{\infty} F(x)\, \{\, 1 - F(x)\,\}\, dx,$$

since $\quad [\,F(x)\,\phi(x)\,]_{-\infty}^{\infty} = 1 \int_{-\infty}^{\infty} F(x)\, dx - (0)(0) = \int_{-\infty}^{\infty} F(x)\, dx.$

In the particular case, $F(x) = \int_0^x dF = \int_0^x dx = x$

$$\therefore \quad \Delta_1 = 2 \int_0^1 x\,(1 - x)\, dx = 2 \left[\frac{x^2}{2} - \frac{x^3}{3} \right]_0^1 = 2 \cdot \frac{1}{6} = \frac{1}{3}.$$

Example 7·13. *Show that the variance may be defined as half the mean of all possible variate differences, that is to say without reference to deviation from a central value, the mean.*

(*I.A.S.; I.C.A.R., Delhi*)

$$E^2 = \int_{-\infty}^{\infty} \int_{-\infty}^{\infty} (x - y)^2 f(x)\, f(y)\, dx\, dy$$

$$= \int_{-\infty}^{\infty} x^2 f(x)\, dx \int_{-\infty}^{\infty} f(y)\, dy - 2 \int_{-\infty}^{\infty} x f(x)\, dx \int_{-\infty}^{\infty} y f(y)\, dy + \int_{-\infty}^{\infty} f(x)\, dx \int_{-\infty}^{\infty} y^2 f(y)\, dy$$

$$= \mu_2' \,.\, 1 - 2\mu_1' \,.\, \mu_1' + 1 \,.\, \mu_2' = 2\,(\mu_2' - \mu_1'^2) = 2\mu_2 = 2\ (\text{variance}).$$

Hence the result follows.

Ex. 1. Show that, if y and z are independent random values of a variate X, the expected value of $(y - z)^2$ is twice the variance of the distribution of X.

Ex. 2. Show that if mean exists, $\Delta_1 = 4\int_{-\infty}^{+\infty} x\,[F(x) - \frac{1}{2}]\,dF$. where $dF = f(x)\,dx$

In particular, for the rectangular distribution $dF = dx, 0 \le x \le 1$, show that $\Delta_1 = \frac{1}{3}$.

$$[\textbf{Hint. } \Delta_1 = \int_{-\infty}^{\infty}\int_{-\infty}^{\infty} |x - y|\,dF(x)\,dF(y) = 2\int_{-\infty}^{\infty} dF(x)\int_{-\infty}^{x} (x - y)\,dF(y)$$

$$= 2\int_{-\infty}^{\infty} dF(x)\left\{\Big(F(y)(x - y)\Big)_{-\infty}^{x} + \int_{-\infty}^{x} F(y)\,dy\right\} = 2\int_{-\infty}^{\infty} dF(x)\int_{-\infty}^{x} 1\cdot F(y)\,dx$$

$$= 2\int_{-\infty}^{\infty} dF(x)\left\{[yF(y)]_{-\infty}^{x} - \int_{-\infty}^{x} y\,dF(y)\right\}, \text{ integration by parts}$$

$$= 2\int_{-\infty}^{\infty} dF(x)\,[xF(x)]_{-\infty}^{x} - 2\int_{-\infty}^{\infty} dF(x)\int_{-\infty}^{x} y\,dF(y)$$

$$= 2\int_{-\infty}^{\infty} dF(x)\,[xF(x)]_{-\infty}^{x} - 2\int_{-\infty}^{x} y\,dF(y)\int_{y}^{\infty} dF(x)$$

$$= \int_{-\infty}^{\infty} dF(x)\,[xF(x)]_{-\infty}^{x} - 2\int_{-\infty}^{x} y\,dF(y)\int_{y}^{x} dF(x)$$

$$= \int_{-\infty}^{\infty} dF(x)\,[xF(x)]_{-\infty}^{x} - 2\int_{-\infty}^{\infty} x(1 - F(x)\,dF(x) = \text{etc.}$$

7·3. M.G.F. and characteristic functions.

7·3·1. Moment Generating Function. The moment generating function of a continuous, random variable *X*, about origin, having the probability function $f(x)$, is given by

$$M_{x=0}(t) = E(e^{tx}) = \int e^{tx} f(x)\,dx \quad \text{where } t \text{ is the real parameter.}$$

$$= \int \left(1 + tx + \frac{t^2 x^2}{2\,!} + \cdots + \frac{t^r x^r}{r\,!} + \cdots\right) f(x)\,dx$$

$$= \int f(x)\,dx + t\int x f(x)\,dx + \frac{t^2}{2\,!}\int x^2 f(x) + \cdots + t^r/r\,!\int x^r f(x)\,dx + \cdots$$

Then μ_r' about origin = Coeff. of $t^r/r\,!$ in the expansion of $M_x(t)$

$$= \int x^r f(x)\,dx.$$

Also $\left| \dfrac{d^r}{dt^r} M_x(t) \right|_{t=0} = \left| \dfrac{\mu_r'}{r!} \cdot r! + \mu_{r+1}' t + \cdots \right|_{t=0} = \mu_r'$

Thus μ_r' about origin = rth differential coefficient of $M_x(t)$ with $t = 0$. Replacing x by $x - a$, we can obtain the value of μ_r' about $x = a$. We shall illústrate with an example now illustrate with an example that for a continuous probability distribution, all moments exist but *m.g.f.* need not exist.

Example 7·14. *If X has the continuous probability distribution*

$$p(x) = \frac{\theta\, a^\theta}{x^{\theta+1}}\ ;\ x \geq a,\ \theta > 1$$

Show that E (x^r) exists, though m.g.f. does not.

$$E(x^r) = \int x^r\, p(x)\, dx$$

$$= \int_a^\infty x^r \cdot \frac{\theta \cdot a^\theta}{x^{\theta+1}}\, dx = \theta\, a^\theta \int_a^\infty x^{r-\theta-1}\, dx = \theta\, a^\theta \left| \frac{x^{r-\theta}}{r-\theta} \right|_a^\infty$$

$$= \theta\, a^\theta \left| 0 - \frac{a^{r-\theta}}{r-\theta} \right| = \frac{\theta \cdot a^r}{\theta - r}\ ;\ \theta > r$$

$$m.g.f. = E(e^{tx}) = \int_a^\infty \frac{e^{tx} \cdot \theta\, a^\theta}{x^{\theta+1}}\, dx$$

The integral on the right does not exist or is not convergent, since $e^{tx}/x^{\theta+1} \to \infty$ as $x \to \infty$

Example 7·15. *The probability density function of the r.v.X has the following probability law :*

$$p(x) = \frac{1}{2\theta} \exp\left(\frac{-|x-\theta|}{\theta} \right),\ -\infty < x < \infty$$ *Find the m.g.f. of X. Hence find E (X) and var (X).* **(M.A. Eco. Punjab, 1991)**

The *m.g.f.* of X is given by

$$M_x(t) = E(e^{tx}) = \int_{-\infty}^{+\infty} e^{tx} \cdot \frac{1}{2\theta} \exp\left(\frac{-|x-\theta|}{\theta} \right) dx$$

$$= \int_{-\infty}^{\theta} e^{tx} \frac{1}{2\theta} \exp\left(\frac{-(\theta - x)}{\theta} \right) dx + \int_{\theta}^{\infty} e^{tx} \frac{1}{2\theta} \exp\left(\frac{-(x-\theta)}{\theta} \right) dx$$

$$= \int_{-\infty}^{\theta} \frac{e^{tx}}{2\theta} \cdot e^{\left(\frac{-\theta+x}{\theta}\right)} dx + \int_{\theta}^{\infty} \frac{e^{tx}}{2\theta} e^{-\frac{(x-\theta)}{\theta}} dx$$

$$= \frac{e^{-1}}{2\theta} \int_{-\infty}^{\theta} e^{\left(tx+\frac{x}{\theta}\right)} dx + \frac{1}{2\theta} \int_{\theta}^{\infty} e^{tx-\frac{(x-\theta)}{\theta}} dx$$

$$= \frac{e^{-1}}{2\theta} \left| \frac{e^{\left(t+\frac{1}{\theta}\right)x}}{t+\frac{1}{\theta}} \right|_{-\infty}^{\theta} + \frac{e}{2\theta} \left| \frac{e^{\left(t-\frac{1}{\theta}\right)x}}{t-\frac{1}{\theta}} \right|_{\theta}^{\infty}$$

$$= \frac{1}{2}\left(\frac{e^{\theta t}}{\theta t+1} + \frac{e^{\theta t}}{1-\theta t} \right) = \frac{e^{\theta t}}{1-\theta^2 t^2} = e^{\theta t}(1-\theta^2 t^2)^{-1}$$

$$= \left(1+\theta t+\frac{\theta^2 t^2}{2\,!}+\cdots \right)(1+\theta^2 t^2+\cdots) = 1+\theta t+\frac{3\,\theta^2 t^2}{2\,!}+\cdots$$

$\therefore\ \mu_1' =$ Coeff. of t in $M_x(t) = \theta$; $\mu_2' =$ Coeff. of $t^2/2\,!$ in $M_x(t) = 3\theta^2$.

$\therefore\ \mu_2 = \mu_2' - \mu_1'^2 = 3\theta^2 - \theta^2 = 2\theta^2.$

7·3·2. Characteristic Function. The characteristic function of a random variable X for the continuous probability distribution with density function $f(x)$ is defined as

$$\phi_x(t) = E(e^{itx}) = \int_{-\infty}^{\infty} e^{itx} f(x)\, dx = \int_{-\infty}^{\infty} e^{itx}\, dF(x)\,,\ i = \sqrt{-1}$$

and t is a real parameter. $\phi_x(t)$ or $\phi(t)$ has the following *properties* :

(*i*) $\phi(t)$ is continuous in t.

(*ii*) $\phi(t)$ is defined in every finite t interval.

(*iii*) $\phi(0) = 1$.

(*iv*) $\phi(t)$ and $\phi(-t)$ are conjugate quantities.

(*v*) $|\phi(t)| \le \int_{-\infty}^{\infty} |e^{itx}|\, f(x)\, dx = \int_{-\infty}^{\infty} f(x)\, dx = 1,$

Since $|\phi(t)| \le 1$, characteristic function $\phi(t)$ always exists.

These are necessary but not sufficient conditions for $\phi(t)$ to be a characteristic function of some continuous distributions. Thus exp (t^4) satisfies all the above five conditions, yet cannot be a characteristic function since it would mean that the distribution has zero mean and zero variance. In fact no very simple set of sufficient condition for $\phi(t)$ to be a characteristic function are known. One of the simplest set of necessary and sufficient conditions, due to Cramer, is given by the following theorem :

In order that a given bounded and continuous function $\phi(t)$ should be the characteristic function of a distribution it is necessary and suficient that $\phi(0) = 1$ and that the function

$$\phi(z) = \int_0^{\infty} \int_0^{c} \phi(t - z)\, e^{ix(t-b)}\, dt\, dx$$

is real and non-negative for all real x and all $c > 0$.

The *advantage* of the characteristic function over *M.G.F* is that while the integral for the *M.G.F.* may be divergent, integral for the *C.F.* cannot be so in view of property (v) above and, therefore, in proving theorems the *C.F.* is invariably used. The use of the *C.F.* also enables us to solve the converse problem, *i.e.*, given the characteristic function, to find the probability density function.

7·3·3. Fourier's Inversion Theorem.

We state below (without proof) Fourier inversion theorem which enables us to find the distribution function, given the characteristic function.

Let $F(x)$ and $\phi_x(t)$ be the distribution and characteristic function respectively of a random variable X then

$$F(x) - F(0) = \lim_{t \to \infty} \frac{1}{2\pi} \int_{-t}^{t} \frac{1 - e^{-itx}}{it}\, \phi(t)\, dt = P \cdot \frac{1}{2\pi} \int_{-\infty}^{\infty} \frac{1 - e^{-itx}}{it}\, \phi(t)\, dt$$

where P denotes the principal value of the integral.

If $F(x)$ is continuous everywhere and $dF(x) = f(x)\, dx$, we can write

$$f(x) = \frac{1}{2\pi} \int_{-\infty}^{\infty} e^{-itx}\, \phi(t)\, dt \qquad \text{...(7·15)}$$

In the latter case the inversion theorem is equivalent to the Fourier Integral theorem *viz.*

$$f(x) = \frac{1}{2\pi} \int_{-\infty}^{\infty} dt \int_{-\infty}^{\infty} e^{it(y-x)}\, F(y)\, dy$$

and can be easily deduced from the latter by the substitution

$$\phi(t) = \int_{-\infty}^{\infty} e^{ity} f(y)\, dy = \int_{-\infty}^{+\infty} e^{ity}\, F(y) \qquad \text{...(7·16)}$$

The reciprocal relation between the density function $f(x)$ and the characteristic function $\phi(t)$ is obvious on comparing (7·15) and (7·16).

Example 7·16. *Show that the distribution for which the characteristic function is $e^{-|t|}$ has the density function*

$$f(x) = \frac{1}{\pi} \cdot \frac{1}{1+x^2}, \infty < x < \infty.$$

(M.Sc. Rajasthan, 1985, M.A. Delhi ; M.A. Poona

We have $$f(x) = \frac{1}{2\pi} \int_{-\infty}^{\infty} e^{-|t|} e^{-itx} dt$$

$$= \frac{1}{2\pi} \left[\int_{-\infty}^{0} e^{t} . e^{-itx} dx + \int_{0}^{\infty} e^{-t} e^{-itx} dt \right]$$

or $$f(x) = \frac{1}{2\pi} \int_{0}^{\infty} - e^{t} (e^{itx} + e^{-itx}) \, dt = \frac{1}{\pi} \int_{0}^{\infty} e^{-t} \cos tx \, dt$$

Integrating by parts gives

$$f(x) = \frac{1}{\pi} \Big[- e^{-t} \cos tx \Big]_{0}^{\infty} - \frac{x}{\pi} \int_{0}^{\infty} e^{-t} \sin tx \, dt$$

$$= \frac{1}{\pi} + \frac{x}{\pi} \Big[e^{-t} \sin x \Big]_{0}^{\infty} - \frac{x^2}{\pi} \int_{0}^{\infty} e^{-t} \cos tx \, dt$$

$$= \frac{1}{\pi} - x^2 f(x)$$

Thus $$f(x) = \frac{1}{\pi} \cdot \frac{1}{1+x^2}, -\infty < x < \infty.$$

Example 7·17. *Show that if θ is a non-negative function of the random variate X and* $\alpha(t) = \int_{-\infty}^{+\infty} \theta^t dF(x) = \int_{-\infty}^{+\infty} \theta^t f(x) \, dx$

then the density of θ if any, is given by $f(\theta) = \frac{1}{2\pi i} \int_{-i\infty}^{i\infty} \theta^{-t-1} \alpha(t) \, dt$

(M.A., Nagpur ; I.C.A.R. Delhi)

We have $\alpha(t) = \int_{-\infty}^{\infty} \theta^t dF(x) = E(\theta^t) = E(e^{t \log \theta}) = E(e^{tv})$ where $v = \log_e \theta$.

Thus $\alpha(t)$ is the moment generating function of v, if the distribution of v exists. Moreover, if it exists it can be found with the help of the inversion theorem thus :

Characteristic function of $v = \phi(T) = E(e^{iTv}) = \alpha(iT)$

$$\therefore \quad g(v) = \frac{1}{2\pi}\int_{-\infty}^{\infty} e^{-ivT}\,\phi(T)\,dT = \frac{1}{2\pi i}\int_{-i\infty}^{i\infty} e^{-vt}\,\phi(t)\,dt \text{ where } iT = t.$$

Distribution of v is $dF = g(v)\,dv$ so that to find the distribution of θ, we write $dF = g(v)\,\dfrac{dv}{d\theta}\cdot d\theta = \dfrac{1}{\theta}\,g(v)\,d\theta = f(\theta)\,d\theta$ where $f(\theta) = \dfrac{1}{\theta}\,g(v)$.

Hence the frequency function $f(\theta)$, if any, is given by

$$f(\theta) = \frac{1}{\theta}\,g(v) = \frac{1}{2\pi i}\int_{-i\infty}^{i\infty} \frac{1}{\theta}\,e^{-t\log\theta}\,\alpha(t)\,dt = \frac{1}{2\pi i}\int_{-i\infty}^{i\infty} \frac{1}{\theta}\,\theta^{-t}\,\alpha(t)\,dt$$

$$= \frac{1}{2\pi i}\int_{-i\infty}^{+i\infty} \theta^{-t-1}\alpha(t)\,dt.$$

Example 7·18. *If, x, x_2, x_3, x_4 are chance variables whose probability distribution has density function given by*

$$\frac{1}{(2\pi)^2}\,e^{-(x_1^2 + x_2^2 + x_3^2 + x_4^2)/2}$$

show that $Z = x_1x_2 - x_3x_4$ has the density function given by $\frac{1}{2}e^{-|z|}$.

(*M.A. Bombay ; M.A. Delhi*)

We shall first find $\phi_z(t)$ and then the density function by §7·3·3, page 237.

$$\phi_z(t) = \frac{1}{(2\pi)^2}\int_{-\infty}^{+\infty}\int_{-\infty}^{+\infty}\int_{-\infty}^{+\infty}\int_{-\infty}^{+\infty} e^{it(x_1x_2 - x_3x_4)}\;e^{-(1/2)\sum_1^4 x_i^2}\,dx_1\,dx_2\,dx_3\,dx_4$$

$$= \frac{1}{2\pi}\int_{-\infty}^{+\infty}\int_{-\infty}^{+\infty} e^{-(1/2)(x_1^2 - 2it\,x_1x_2 + x_2^2)}\,dx_1dx_2$$

$$\times \frac{1}{2\pi}\int_{-\infty}^{+\infty}\int_{-\infty}^{+\infty} e^{-(1/2)(x_3^2 + 2\,it\,x_3x_4 + x_4^2)}\,dx_3\,dx_4$$

$$= \frac{1}{\sqrt{1+t^2}}\cdot\frac{1}{\sqrt{1+t^2}} = \frac{1}{1+t^2}$$

using $$\frac{1}{\pi}\int_{-\infty}^{+\infty}\int_{-\infty}^{+\infty} e^{-(ax^2 + 2hxy + by^2)}\,dx\,dy = \frac{1}{\sqrt{ab - h^2}}$$

$$\therefore \quad f(z) = \frac{1}{2\pi}\int_{-\infty}^{\infty} \frac{e^{-itz}\,dt}{1+t^2} = \frac{1}{2\pi}\int_{-\infty}^{\infty} \frac{\cos tz\,dt}{1+t^2} = \tfrac{1}{2}\,e^{-|z|}$$, a, b, h are all positive.

Ex. 1. Are $e^{-t^4}, e^{-1/2\,t^2}, |\sin t|, t^{\log t}$ and $\dfrac{1}{1+t^2}$ characteristic functions ? If not, why not ? If so, find the corresponding density functions.

(M.A. Statistics, Delhi ; M.A. Bombay ; M.A. Delhi)

Ex. 2. A variate has moments $\mu_r' = r\,!$. Find the characteristic function and deduce its distribution.

[**Hint.** $\phi(t) = \sum\limits_{r=0}^{\infty} (r\,!) \dfrac{(it)^r}{r\,!} = \sum\limits_{0}^{\infty} (it)^r = \dfrac{1}{1-it}$ [**Ans.** $F(x) = e^{-x}, x \ge 0$]

Ex. 3. (*a*) Find the frequency density of the distribution for which $k_r = (r-1)\,!$

(M.A. Statistics, Delhi)

(*b*) If X is a r.v. for which $k_r = (r-1)!\,pa^{-r}$, $p > 0, a > 0$, find the c.f. of X

(B.A. Hons. Stat. Delhi 1997)

[**Hint.** (*a*) $\log \phi(t) = \sum\limits_{r=1}^{\infty} \dfrac{k_r\,(it)^r}{r\,!} = etc.$] [**Ans.** $f(x) = e^{-x}, x \ge 0$]

Ex. 4. A variate X has moments $\mu_r' = \dfrac{r\,!}{(r/2)\,!}$ when r is even and $\mu_r' = 0$ when r is odd. Deduce the distribution of the variate from its characteristic function.

(M.A. Statistics, Delhi)

Ex. 5. Starting with the theorem of Fourier transform, prove the results of Ex. 7·16, page 238. ***(I.C.A.R. Delhi)***

7·3·4. Theorems on Characteristic Function (c.f.)

1. *If $\phi_x(t)$ is the c.f. of the random variable X, then $e^{itb}\,\phi(at)$ is the c.f. of the random variable $aX + b$, a and b being real constants.*

Proof. We have, by def.,

$$\phi_{aX+b}(t) = E\,[e^{it(aX+b)}] = E\,[e^{it\,b} \cdot e^{itaX}] = e^{itb}\,E\,[e^{it\,ax}] = e^{it\,b}\,\phi_x(at)$$

Cor. $\phi_{ax}(t) = \phi_x(at)$

2. *if ϕ is the c.f. of X, then $\bar{\phi}$ is the c.f. of $-X$ where $\bar{\phi}$ is the conjugate of ϕ.*

Proof. $\phi_{(-x)}(t) = E\,[e^{it(-X)}] = [e^{-itX}] = \phi_x(-t) = \bar{\phi}_x(t)$

3. *If the distribution function of a random variable X is symmetrical about zero i.e. $f(-x) = f(x)$, then $\phi(t)$ is real valued and even function of t.*

Proof. We have $\phi_X(t) = E\,(e^{itX}) = \int_{-\infty}^{+\infty} e^{itx} f(x)\,dx$ Put $x = -y$

$$= \int_{\infty}^{-\infty} e^{-ity} f(-y)\,(-dy)\;;$$

$$= \int_{-\infty}^{+\infty} e^{-ity} f(y)\,dy \quad [\because f(-y) = f(y)]$$

$$= \int_{-\infty}^{+\infty} e^{-itx} f(x)\, dx = \phi_x(-t)$$

Thus $\phi_x(t)$ is an even function of t.

Also $\overline{\phi}(t) = E[e^{-itx}] = E(e^{i(-t)x}] = \phi_x(-t) = \phi_x(t)$

Hence ϕ_x is a real-valued function of t.

4. *The characteristic function of the sum of independent random variables is equal to the product of their individual characteristic functions. In symbols, if* $X_1, X_2, \cdots, X_n$ *are n independent random variables, then*

$$\phi_{X_1+X_2+\cdots\cdots+X_n}(t) = \phi_{X_1}(t)\phi_{X_2}(t)\cdots\phi_{X_n}(t) = \prod_{i=1}^{n} \phi_{X_i}(t)$$

Proof.
$$\phi_{\sum_1^n X_j}(t) = E[e^{it(X_1+X_2+\cdots+X_n)}]$$

$$= E(e^{itX_1} e^{itX_2} \ldots e^{itX_n}) = E[\prod_{j=1}^{n} e^{itX_j}],$$

Since $e^{itX_1}, \ldots, e^{itX_n}$ are independent.

We have
$$\phi_{\sum_1^n X_j}(t) = \prod_{j=1}^{n} E[e^{itx_j}] = \prod_{j=1}^{n} \phi_{X_j}(t)$$

which establishes the result to be proved.

Cor. If $\mathbf{X_j}$ are *identically* distributed, then

$$\phi_{\sum_1^n X_j}(t) = [\phi_x(t)]^n.$$

Remark. We shall now give an *example* to show that the converse of the above theorem does not always hold *i.e.*

$$\phi_{x_1+x_2}(t) = \phi_{x_1}(t)\ \phi_{x_2}(t) \not\Rightarrow X_1 \text{ and } X_2 \text{ are independent.}$$

Illustration. Let X_1 and X_2 be two Cauchy varites each with *p.d.f.*

$$f(x) = \frac{1}{\pi(1+x^2)}, -\infty < x < \infty$$

Then by example 7·16, page 238

$$\phi_{X_1}(t) = e^{-|t|} \text{ and } \phi_{X_2}(t) = e^{-|t|}$$

$$\therefore \quad \phi_{X_1+X_2}(t) = \phi_{2X_1}(t) = e^{-2|t|} = \phi_{X_1}(t) \cdot \phi_{X_2}(t)$$

but X_1 and X_2 are not independent, as $X_1 = X_2 \Rightarrow P(X_1 = X_2) = 1$.

Another example see page 268.

7·3·5. Effect of change of origin and of scale on Characteristic Function.

Let us introduce a new variable U where $U = \dfrac{X-a}{h}$, a and h being real constants. By theorem I, it follows that $\phi_u(t) = e^{-\frac{iat}{h}} \phi_x\left(\dfrac{t}{h}\right)$

In particular, if a = mean = μ, h = s.d = σ, then the c.f. of the standard variable U is given by $\phi_u(t) = e^{-\frac{it\mu}{\sigma}} \phi_x\left(\dfrac{t}{\sigma}\right)$.

Note. It should be noted that a necessary and sufficient condition for the convergence of the sequence $\{F_n(x)\}$ to be a distribution function $F(x)$ is that for every t, the sequence $\{\phi_n(t)\}$ converges to a limit $\phi(t)$ which is continuous for the value $t = 0$. When the condition is satisfied, then limit $\phi(t)$ is identical with the c.f. of $F(x)$.

Example 7·19. *Let $F_n(x)$ be the distribution function defined by*

$$F(x) = 0 \quad \text{for } x \le -n$$
$$= \frac{x+n}{2n} \quad \text{for } -n < x < n;$$
$$= 1 \quad \text{for } x \ge n$$

Is the limit of $F_n(x)$ a distribution function ? If not, why ?

(M.A., Delhi)

$$\text{Now} \quad \phi_n(t) \int_{-n}^{n} e^{itx} f(x)dx = \int_{-n}^{n} e^{itx} \frac{1}{2n} dx = \frac{1}{2n}\left[\frac{e^{itn} - e^{-itn}}{it}\right] = \frac{\sin nt}{nt}$$

$$\therefore \quad \phi(t) = \lim_{n \to \infty} \frac{\sin nt}{nt} = \begin{cases} 1 \text{ if } t = 0 \\ 0 \text{ if } t \ne 0 \end{cases}$$

Thus $\phi(t)$ is discontinuous at $t = 0$. Hence $\phi(t)$ cannot be a distribution function.

7·4. Geometrical Probability. When points are chosen 'at random' in a straight or curved line or in a plane area or in a volume or when lines are drawn 'at random' on a plane area, same lengths or angles or areas are equally likely and their distribution is uniform. Problems of this type can usually be easily solved by geometrical interpretation of the conditions given. We shall illustrate by solving a few examples.

Example 7·20. *Three points are taken at random on the circumference of a circle. Find the probability that they lie on the same semicircle.*

Since we are concerned with relative positions only, we may regard one point as fixed ; call this as A. Selecting one of the other two points

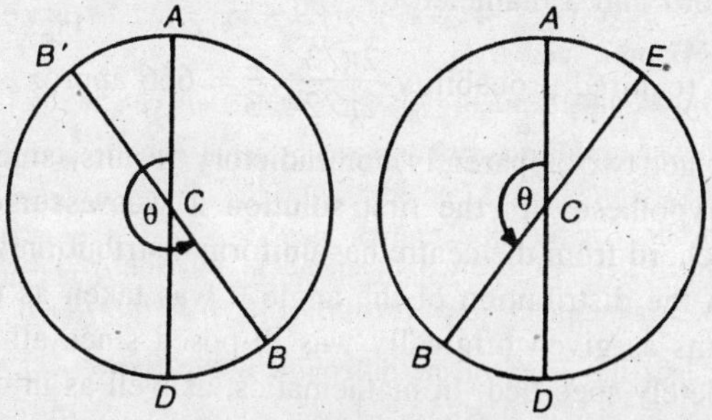

Fig. 7·6

as B, we define the position of B by the are ACB measured anti-clock-wise from CA to CB, C being the centre. Let the third point be D. There are two cases according as $\angle ACB\ (=\theta)$ is (*i*) $< \pi$, or (*ii*) $> \pi$.

Let AD and BE be the diameters. Then the point A, b and D will be on the semi-circle provided that B is not the are DE. The probability for this is :

$$f(\theta) = \frac{2\pi - \theta}{2\pi}, \text{ if } 0 < \theta < \pi \, ;$$

$$\frac{\theta}{2\pi}, \text{ if } \pi < \theta < 2\pi$$

Combining these with the probability that B falls within a small arc of angle $d\theta$ which is $d\theta/2\pi$, we have

the required probability $$= \int_0^{\pi} \frac{2\pi - \theta}{2\pi} \cdot \frac{d\theta}{2\pi} + \int_{\pi}^{2\pi} \frac{\theta}{2\pi} \cdot \frac{d\theta}{2\pi} = \frac{3}{4}.$$

Example 7·21. Bertrand Paradox. *If a chord is taken at random in a circle, what is the chance that its length is not less than the radius of the circle ?*

Sol. 1. If we *assume* that all distance of the chord from the centre of the circle are equally likely, then for all the distances less than $\sqrt{r^2 - \left(\frac{r}{2}\right)^2}$, *i.e.*, $\frac{\sqrt{3}}{2}\, r$. the chord is greater than r *where* r is the radius of the circle.

Hence the required probability

$$=\frac{\sqrt{3}}{2}r/r=\frac{\sqrt{3}}{2}=\cdot 866 \text{ approx.}$$

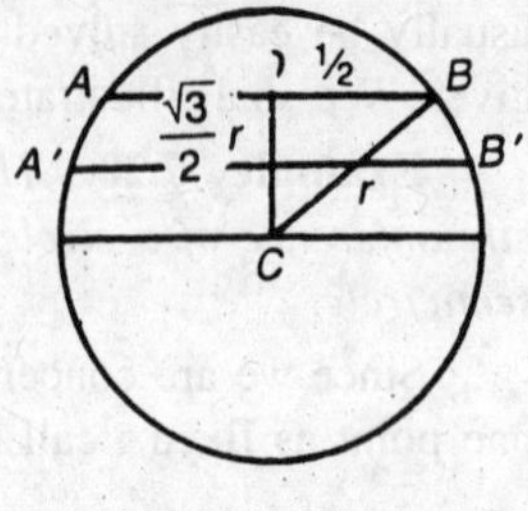

Fig. 7·7

Sol. 2. If we *assume* that all angles between the chord and a diameter are equally likely, then for all values of angle $\theta < \pi/3$, the chord is greater than r where θ is the angle between the chord and a diameter.

Hence the required probability $\frac{2\pi/3}{2\pi/2}=\frac{2}{3}=\cdot 666$ approx.

Note. We get two apparently contradictory results, since we have two different hypotheses. In the first solution it was assumed that the distance of the chord from the centre has uniform distribution while in the second solution the distribution of the angle θ was taken as uniform. In fact the problems as given originally was ill-posed since all hypotheses were not completely specified. In mathematics, as well as in daily life, if all conditions are not fully specified, the answers may not be unique. Thus if one asks how much time it takes to go from Kanpur to Delhi, the answer depends on whether one goes by air, or by train or by bus or by cycle or even on foot.

Example 7·22. *Find the mean distance of two points taken random on the circumference of a circle.*

Let A be any point and AOD be the diameter. Also let B be the second random point and AB make an angle θ with AOD. Then $\angle ABD$ is a right angle and $AB = AD\cos\theta = 2r\cos\theta$, r being the radius of the circle and $-\pi/2 \le \theta \le \pi/2$.

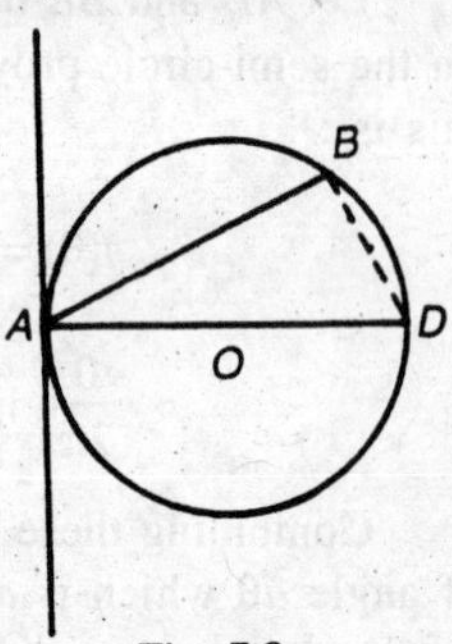

Fig. 7·8

Mean distance of two points

$$=\frac{1}{\pi}\int_{-\pi/2}^{\pi/2} 2r\cos\theta\, d\theta=\frac{4r}{\pi}\cdot$$

Example 7·23. Buffon's Needle Problem. *A smooth table is ruled with parallel lines at distance a apart. A needle of length $l < a$ is dropped on the table. What is the probability that it will cross one of the lines ?*

Take one of the parallel lines for x-axis and any line perpendicular to it for y-axis. The probability that centre of the needles has an ordinate lying between the limits y and $y + dy$ is dy/a ; and the probability that the inclinaton of the needles to OY should be between θ and $\theta + d\theta$ is

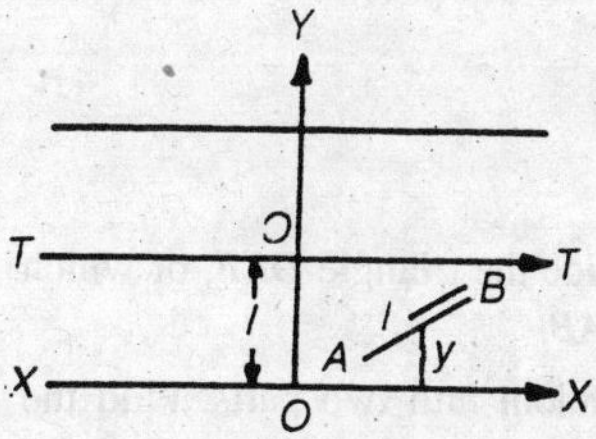

Fig. 7·9

$d\theta/\pi$. Hence the probability that the needles will cross OX is

$$\iint \frac{dy\, d\theta}{a\pi}$$

where the double integral is taken over the range of value of y and θ for which the needle will cross OX. The possible values of y are clearly given by $|y| \le l/2 \cos\theta$ and θ lies in the range $-\pi/2 \le \theta \le \pi/2$.

$$\text{Hence probability} = 2 \int_{-\pi/2}^{\pi/2} \frac{d\theta}{\pi} \int_0^{l/2 \cos\theta} \frac{dy}{a} = \frac{2l}{a\pi}$$

Note. If the needle is thrown a large number of times and if it crosses one of the lines m times, out of n, then

$$\frac{m}{n} \cong \frac{2l}{a\pi} \quad \text{or} \quad \pi \cong \frac{2l}{a}\frac{n}{m}$$

This approximation has been used by many persons to determine the approximate value of π. A few experimental results are :

Experimenter	*Year*	*Number of throws*	*Value of* π
Wolf	1850	5000	3·1596
Smith	1855	3204	3·1553
Fox	1894	1120	3·1415
Lazzaren	1901	3408	3·1415929

This example illustrates the use of probability theory to solve some deterministic problems in mathematics, including evaluation of constants and integrals and solutions of differential and integral equations. This is done on a large scale and the methods used are called *Monte Carlo Methods.*

Ex. 1. A line of given length a is divided into three parts. What is the probability that these form the sides of the triangle ? **[Ans.** $\frac{1}{4}$]

[**Hint.** Let AB be the straight line of length a and let the three parts be of lengths x, y and $a-(x+y)$, so that $x < a$, $y < a$, and $x + y < a$. Thus, the point (x, y) lies inside the triangle OAB where $OA = OB = a$. (Ref. Fig. 7·10)

For these three parts to form a triangle, it is necessary that

$$x + y > a - (x + y)$$

$$x + a - (x + y) > y$$

$$y + a - (x + y) > x.$$

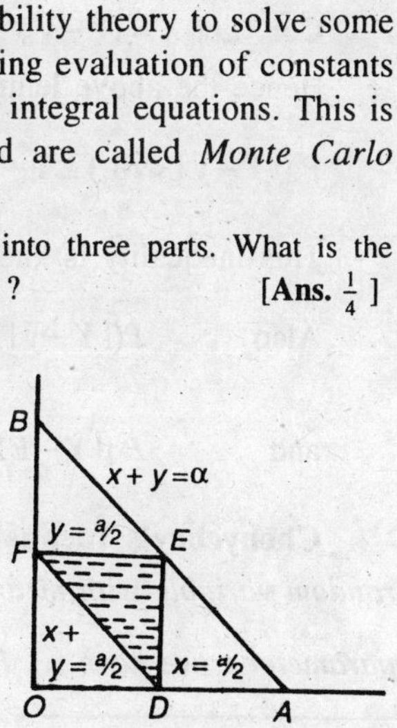

Fig. 7·10

These conditions are equivalent to

$$x+y>\frac{a}{2},\ x<\frac{a}{2},\ y<\frac{a}{2},$$

so that the favourable cases correspond to points inside the triangle *DEF*, of which the area is one-fourth of the area of the triangle *OAB*.

Ex. 2. A st. line of length a is divided at random into two parts. Find the mean value of the rectangle contained by the two parts. **[Ans.** $a^2/6$ **]**

7·5. *Chebychev's Inequality. Weak Law of Large Numbers.

Chebychev's Lemma. *If X is a non-negative stochastic variable with mean $\bar{x}$, then*

$$P(X<\bar{x}t^2)\geq 1-\frac{1}{t^2},\ t \text{ being a real parameter.}$$

Proof. We have $\bar{x}=\int_0^\infty x f(x)\,dx=\int_0^{\bar{x}t^2} x f(x)\,dx+\int_{\bar{x}t^2}^\infty x f(x)\,dx$

or $$\bar{x}\geq\int_{\bar{x}t^2}^\infty x f(x)dx \geq \bar{x}t^2\int_{\bar{x}t^2}^\infty f(x)dx=\bar{x}t^2P(X\geq\bar{x}t^2)$$

$$\therefore\quad \frac{1}{t^2}\geq 1-P(X<\bar{x}t^2)$$

Hence $P(X<\bar{x}t^2)\geq 1-\frac{1}{t^2}\quad \Rightarrow P(|x|\geq a)\leq\frac{1}{a^2}E(X^2)$ for all a.

Cor. Let $X=[Y-\bar{y}]^2$, so that $\bar{x}=E(X)=Var\ Y=\sigma_y{}^2$.

Hence the above lemma reduces to

$$P(|Y-\bar{y}|<t\sigma_y)\geq 1-\frac{1}{t^2}\ \text{ or }\ P(|Y-\bar{y}|\geq t\sigma_y)\leq\frac{1}{t^2}$$

This inequality is known as *Chebychev's Inequality.*

Also $\quad P(|Y-\bar{y}|\geq k)\leq\frac{\text{var } Y}{k^2}$ by putting $t\sigma_y=k$.

and $\quad P[|Y-EY|<k]\geq 1-\frac{\text{Var } Y}{k^2}.$

Chebychev's Inequality. Alternative Proof. *If Y is a continuous random variable with mean $\bar{y}$ and variance σ_y^2 then for any positive real parameter t, we have* $\quad P(|Y-\bar{y}|\geq t\,\sigma_y)\leq\frac{1}{t^2}$

* This inquality was discovered by Chebychev in 1953. It gives an upper limit to the probability.

Proof. Let $f(y)$ be the *p.d.f.* of Y. Then we have

$$\sigma_y^2 = E[Y - \bar{Y}]^2 = \int_{-\infty}^{+\infty} (y - \bar{y})^2 f(y) dy$$

$$= \int_{-\infty}^{\bar{y} - t\sigma_y} (y - \bar{y})^2 f(y) dy + \int_{\bar{y} - t\sigma_y}^{\bar{y} + t\sigma_y} (y - \bar{y})^2 f(y) dy$$

$$+ \int_{\bar{y} + t\sigma_y}^{\infty} (y - \bar{y})^2 f(y)\, dy$$

$$\therefore\ \sigma_y^2 \geq \int_{-\infty}^{\bar{y} - t\sigma_y} (y - \bar{y})^2 f(y) dy + \int_{\bar{y} + t\sigma_y}^{\infty} (y - \bar{y})^2 f(y) dy$$

$$\geq t^2\sigma_y^2 \int_{-\infty}^{\bar{y} - t\sigma_y} f(y) dy + t^2\sigma_y^2 \int_{\bar{y} + t\sigma_y}^{\infty} f(y) dy$$

Since for the first integral $y \leq \bar{y} - t\,\sigma_y \Rightarrow (y - \bar{y})^2 \geq t^2\, \sigma_y^2$ and for the second integral $y \geq \bar{y} + t\sigma_y \Rightarrow (y - \bar{y})^2 \geq t^2\, \sigma^2_y$

Thus $\sigma_y^2 \geq t^2\, \sigma_y^2 [P\,(y - \bar{y} \leq -t\sigma_y) + P(y - \bar{y} \geq t\,\sigma_y)]$

$$= t^2\sigma_y^2\, P[|y - \bar{y}| \geq t\,\sigma_y]$$

$$\Rightarrow\quad P(|\,y - \bar{y}\,| \geq t\,\sigma_y] \leq \frac{1}{t^2}$$

This result can also be stated as $P[\,|\,y - \bar{y}\,| \leq t\sigma_y\,] \geq 1 - 1/t^2$.

Note. This inequality fixes an upper limit to the probability that the value of the variate chosen at random will differ from the mean by more than t times the standard deviation.

Example 7·24. *Use Tchebychev's inequality to find how many times a true coin must be tossed in order that the probability will be at least 0·90 that is X, the proportion of the number of heads will be between 0·4 and 0·6.* **(B. SC. Stat. Hons. Delhi 1988, B. Sc. *Kerala*)**

By Bernoulli's Law of Large Numbers, we have, since coin is fair,

$$P\left[\,\left|\,\frac{X}{n} - p\,\right| \leq t\,\right] > 1 - \frac{1}{4nt^2},\ t > 0, p = \frac{1}{2} = q \qquad ...(1)$$

$$P\left[\,\left|\,\frac{X}{n} - p\,\right| \leq 0{\cdot}1\,\right] = 0{\cdot}9, p = 0{\cdot}5$$

Putting $t = 0{\cdot}1$ in (1), we have

$$P\left[\left|\frac{X}{n}-p\right| \le 0{\cdot}1\right] \ge 1-\frac{1}{4n(0{\cdot}1)^2} = 1-\frac{1}{0{\cdot}04n}$$

Hence $\quad 1-\dfrac{1}{0{\cdot}04n} = 0{\cdot}9 \Rightarrow n = 250$

Thus the number of tosses of the true coin = 250.

Example 7·25. *A random variable X has the density function e^{-x} for $x \ge 0$. Show that Tchebychev's inequality gives*

$$P[|X-1|>2] < \tfrac{1}{4}$$

and show that the actual probability is e^{-3}. **(*B. Sc. Bombay*)**

We have the probability distribution of the random variable X as

$$dP = e^{-x}dx\,,\ x \ge 0.$$

$$E(X) = \int_0^\infty xe^{-x}\,dx = 1\ ; \qquad E(X^2) = \int_0^\infty x^2e^{-x}dx = \Gamma\,3 = 2$$

$\therefore\ \text{Var}\,X = E(X^2) - [E(X)]^2 = 2-1 = 1$

Now $\quad P\left[\left|\dfrac{X-E(X)}{\sigma_x}\right| \ge t\right] \le \dfrac{1}{t^2}$

$\Rightarrow \quad P\,[|X-1 \ge 2] \le \dfrac{1}{4}$ by Tchebychev's inequality.

Now $\quad P\,[|X-1 \ge 2],\ 1-P\,[-2<X-1<2]$

$= 1-P[-1 \le X \le 3]$. Since X cannot be negative, we have

$$= 1-P(0<X<3) = 1-\int_0^3 e^{-x}dx = e^{-3}.$$

Example 7·26. *Does there exist a variate X for which*

$$P\,[\mu-2\,\sigma \le X \le \mu+2\sigma] = 0{\cdot}6\ ?$$

(*B. Sc. Math Hons. Delhi, 1983*)

Chebychev's inequality gives us $P\,[|X-\mu|<2\sigma] \ge 1-\dfrac{1}{2^2} = 1-\dfrac{1}{4} = 0{\cdot}75$

$\Rightarrow \quad P\,[\mu-2\sigma \le X \le \mu+2\sigma] \ge 0{\cdot}75$

Since the lower bound for the given probability is 0·75, there does not exist a *r.v.X* for which the given result holds.

7·5·1 Theorem. Bienayme-Chebychev's Inequality

If g(X) is a non-negative function of a random variable X, then for every k > 0, we have

$$P[g(X) \geq k] \leq \frac{E[g(X)]}{k}.$$

Proof. Let A denote the set of all points satisfying the relation $g(X) \geq k$ and let $\bar{A}$ be the complement of A. Then if $f(x)$ is the probability of $g(X) \geq k$,

$$P[g(X) \geq k] = \int_A f(x)\,dx$$

Again
$$E[g(X)] = \int_A g(x) f(x)\,dx + \int_{\bar{A}} g(x) f(x)\,dx$$

$$\geq k \int_A f(x) dx + \int_{\bar{A}} g(x) f(x)\,dx \quad \left[\because g(x) \geq k \text{ for } A\right]$$

$$\Rightarrow \quad E[g(X)] \geq k \int_A f(x) dx = k\,P[g(x) \geq k]$$

$$\Rightarrow \quad P[g(X) \geq k] \leq \frac{E(g(X)]}{k}$$

Cor. 1. Put $g(x) = (x-\mu)^2$, $k = t\sigma$ and we get

$$P\left[(X-\mu)^2 \geq t^2\sigma^2\right] \geq \frac{E(X-\mu)^2}{t^2\sigma^2} = \frac{\sigma^2}{t^2\sigma^2} = \frac{1}{t^2}$$ which is Tchebycley's inequality.

Cor. 2. Put $g(X) = X$ and we get $P(X \geq k) \leq E(X)/k$ which is **Markoff's inequality.**

Cor. 3. Let $g(X) = |X|$. Then, we have

$$P[|X| \geq k] \leq \frac{E[|X|]}{k}. \qquad [k = \Sigma(y_i - \bar{y})^2]$$

Cor. 4. Let $Y = \dfrac{Y_1 + Y_2 + \ldots + Y_n}{n}$, $\bar{y} = \dfrac{\bar{y}_1 + \bar{y}_2 + \ldots + \bar{y}_n}{n}$ and

$t = \sqrt{\dfrac{B_n}{n^2}} = \omega$ where omega is an arbitrary positive number and

$$B_n = E(u) = E(\Sigma y_1 - \Sigma \bar{y}_i)^2$$

$$= \Sigma E(y_i - \bar{y}_i)^2 = \Sigma \operatorname{Var} y_i,\ P[(Y - \bar{y}) < t\sigma_y] > t - t^2$$

Then from Cor. 4 given above we get

$$P\left[\left|\frac{Y_1 + Y_2 + \ldots + Y_n}{n} - \frac{\bar{y}_1 + \bar{y}_2 + \ldots + \bar{y}_n}{n}\right| < t\sigma_y\right] \leq 1 - \frac{B_n}{n^2\varepsilon^2} \geq 1 - \delta$$

provided $\dfrac{B_n}{n^2\varepsilon^2} < \delta$ *i.e.* $\lim\limits_{n \to \infty} B_n/n^2 \to 0$.

This is known as *Weak Law of Large Numbers* (WLLN). We also write $\overrightarrow{Y_n} \xrightarrow{P} \overline{y}_n$.

It may be stated as follows :

With the probability approaching unity or certainty as near as we please we expect that the arithmetic mean of the values actually assumed by n stochastic variates will differ from the mean of their expectations by less than any given number, however small, provided the number of variates can be taken sufficiently large and provided the condition $B_n / n^2 \to 0$ as $n \to \infty$, is fulfilled.

Remark. 1. The weak law of large numbers (WLLN) holds good if

(*i*) $E(x_i)$ exists for all i

(*ii*) $B_n = \text{Var}\,(x_1 + x_2 + \ldots + x_n)$ exists, and

(*iii*) $\dfrac{B_n}{n^2} \to 0$ as $n \to \infty$.

Remark. 2. It can be proved that if the variables $X_1, X_2, \ldots, X_n$ are *uniformly bounded,* then the condition $\lim\limits_{n \to \infty} \dfrac{B_n}{n^2} = 0$ is necessary as well as sufficient for WLLN to hold.

7·5·2. Khinchin's Theorem. *If X_i's are identically and independently distributed random variables, then the necessary condition for the law of large number to hold is that $E(X_i)$ should exist $\forall$ i.*

The proof of this theorem is beyond the scope of the book.

Example 7·26. *Two good dice whose faces are numbered 1 to 6 are thrown, If X is the sum of the numbers shown up, prove that*

$$P \,|\, | X - 7 | \geq 3 \,| \leq \frac{35}{54}$$

(B. Sc. Karnataka, 1988, B. Sc., Calcutta, 1972)

Values of X and the corresponding probabilities are :

X :	2	3	4	5	6	7	8	9	10	11	12	Total
$P(x)$:	$\frac{1}{36}$	$\frac{2}{36}$	$\frac{3}{36}$	$\frac{4}{36}$	$\frac{5}{36}$	$\frac{6}{36}$	$\frac{5}{36}$	$\frac{4}{36}$	$\frac{3}{36}$	$\frac{2}{36}$	$\frac{1}{36}$	1

Then $EX = 2 \cdot \dfrac{1}{36} + 3 \cdot \dfrac{2}{36} + \ldots + 12 \cdot \dfrac{1}{36} = 7$

$$EX^2 = 2^2 \frac{1}{36} + 3^2 \cdot \frac{2}{36} + \ldots + 12^2 \cdot \frac{1}{36} = \frac{1974}{36} \Rightarrow \text{Var}\, X = \frac{210}{36}$$

By Chebychev's inequality, we have $P[| X - 7 | \geq 3] \leq \dfrac{\text{Var}\, X}{9} = \dfrac{210}{9 \times 36} = \dfrac{35}{54}$

Exact probability of $|X-7|\geq 3 = P[4\leq X\leq 10]$

$$= P[X=4]+P[X=5]+--+P[X=10]$$

$$= 1-\left[\frac{1}{36}+\frac{2}{36}+\frac{2}{36}+\frac{1}{36}\right]=\frac{5}{6}.$$

Example 7·27. *If the variable X_p assumes the value $2^{p-2\log}$ with probability 2^{-p}, $p=1,2,\ldots.$ examine if the law of large number holds.*

We shall apply Khinchin's theorem. Thus we shall see if $E(X_p)$ exists $\forall\, p$. Here the values of the *r.v.* X are

$2^{1-2\log 1}$; $2^{2-2\log 2}$;$2^{3-2\log 3}$, ... (put $p=1,2,\ldots$) :

The value of probability of X taking the value : $\frac{1}{2}$; $\frac{1}{2^2}$, $\frac{1}{2^3}$, ...

$$\therefore \quad E(X_p)=2^{1-2\log 2}\cdot\frac{1}{2}+2^{2-\log 2}\cdot\frac{1}{2^2}+2^{3-\log 3}\cdot\frac{1}{2^3}+\ldots.$$

$$=\sum_{p=1}^{\infty}\frac{1}{2^{2\log p}}=\sum_{p=1}^{\infty}\frac{1}{4^{\log p}}=\Sigma\frac{1}{n^p}.$$

where $n^p=4^{\log p}$ which can be shown to be a convergent series.

Example 7·28. *Examine if the WLLN holds for the sequence $\{X_p\}$ of independent identically distributed random variables (i.i.d.r.v's) with*

$$P\left[X_k=(-1)^{k-1}\cdot k\right]=6/\pi^2k^2,\ k=1,2,3,\ldots;\ p=1,2,\ldots$$

(*B. Sc. Math Hons. 1990*)

We shall apply Khichin's theorem,

$$\text{Now} \quad E(X_k)=\sum_{k=1}^{\infty}(-1)^{k-1}k\cdot\frac{6}{\pi^2k^2}=\frac{6}{\pi^2}\sum_{k=1}^{\infty}\frac{(-1)^{k-1}}{k}$$

$$=\frac{6}{\pi^2}\left[1-\frac{1}{2}+\frac{1}{3}-\frac{1}{4}+\ldots\right]$$

$$=\frac{6}{\pi^2}\log 2. \text{ Hence } E(X) \text{ exists}$$

$\therefore$ WLLN holds for the sequence $\{X_i\}$ of *i.i.d.*

Ex. 1. Use Chebychev's inequality to show that for $n>36$,the probability that in n throws of a fair dice, number of sixes lies between $\frac{1}{6}n-\sqrt{n}$ and $\frac{1}{6}n+\sqrt{n}$ is at least 31/36. **(*B.Sc. Math Hons. Calcutta*)**

Ex. 2. If X is a stochastic variable such that $E(X)=\mu$ and $Var(X)=\sigma^2$ then

$$P(|X-\mu|<\delta)\geq 1-\frac{\sigma^2}{\delta^2} \text{ where } \delta \text{ is any assigned quantity.}$$

[**Hint.** $$\sigma^2 = \int_{-\infty}^{+\infty} (x-\mu)^2 f(x)dx \geq \int_{-\infty}^{\mu-\delta} (x-\mu)^2 f(x)dx + \int_{\mu+\delta}^{\infty} (x-\mu)^2 f(x)dx$$

$$\geq \delta^2 \left[\int_{-\infty}^{\mu-\delta} f(x)\,dx + \int_{\mu+\delta}^{\infty} f(x)dx = \delta^2 P(|x-\mu| > \delta) \right].$$

Ex. 3. For the number of points X on a dice, prove the Chebychev's inequality gives $P[|X - E(X)| > 2{\cdot}5] < {\cdot}47$, while the actual probability is nearly zero.

Ex. 4. For the geometric distribution $p_k = 1/2^k$ $(k = 1, 2, 3\ldots)$, prove that Chebychev's inequality gives $P[|x-2| > 2] < \frac{1}{2}$

while the actual probability is $1/16$

Ex. 5. For the distribution. $dP = y_0 e^{-x} x^k dx, 0 \leq x < \infty \cdot y_0$ being a constant, $k \geq 0$, determine $P[|X - E(X)| > 2\sqrt{Var\,X}]$ and compare this with the value given by Chebychev's inequality.

(***B. Sc., Bombay, 1963***)

Ex. 6. Let x_i assume values i and $-i$, each with probability $\frac{1}{2}$. Show that the law of large number need not hold as B_n/n^2 does not approach zero.

Ex. 7. If x_i can have only two values i^{α} and $-i^{\alpha}$ with equal probabilities show that the law of large number can be applied to $x_1, x_2, \ldots x_n$ if $\alpha < \frac{1}{2}$.

(***I.C.A.R. Delhi***)

Ex. 8. Prove that if $|x_i - E(x_i)| > G$ where G is a constant independent of n, then if the law of large number holds, $B_n/n^2 \to 0$ as $n \to \infty$.

Ex. 9. If x_n takes the values 1 and 0 with corresponding probabilities p_n and $1 - p_n$, examine whether the weak law of large numbers can be applied to the sequence $\{x_n\}$ where x_n's are independent.

Ex. 10. $\{Y_k\}, k = 1, 2, \ldots$ is a sequence of independent random variables each taking the values $-1, 0, 1$. Given that

$$P(X_k = 1) = \frac{1}{k} = P(X_k = -1); \quad P(X_k = 0) = 1 - \frac{2}{k},$$

examine if the law of large numbers holds for this sequence.

Ex. 11. The *r.v.'s* $Y_1, Y_2, \ldots, Y_n$ have equal expectation and finite variance. Is the WLLN applicable to this sequence if all the Covariances are negative.

(***B. Sc. Math. Hons. 1987***)

[**Hint.** $\dfrac{B_n}{n^2} = \dfrac{Var\,(\Sigma x_i)}{n^2} = \dfrac{\Sigma \sigma_i^2 + 2\,\underset{i \neq j}{\Sigma\Sigma}\, \sigma_{ij}}{n^2} < \dfrac{\Sigma \sigma_i^2}{n^2}$ in case σ_{ij}'s are $< 0 \;\forall\; i, j, i \neq j$

Thus $Bn/n^2 \to 0$ as $n \to \infty$. Thus WLLN holds under the given condition.

7·5·3. Bernoulli's Theorem. *Let X be the number of successes in n independent trials with the constant probability p of success. Then given ε and δ, two positive numbers, however small, we can find a number N depending on ε and δ such that* $P\left[\left|\frac{X}{n}-p\right|<\varepsilon\right]$ *will hold is greater than* $1-\delta$*, provided* $n>N$.

Symbolically we have $P\left[\left|\frac{x}{n}-p\right|<\varepsilon\right]>1-\frac{pq}{n\varepsilon^2}$.

Proof. Associate with trials, 1., 2, 3, ... n stochastic variates $X_1, X_2, \ldots, X_n$ as follows :

$$X_i = 1 \quad \text{in case of success in the } i\text{th trial}$$
$$= 0 \quad \text{in case of failure in the } i\text{th trial,}$$

so that $\quad X = X_1 + X_2 + \ldots + X_n$.

Since trials are independent, these variates are independent.

Also $E(X) = p$, $E(X_i^2) = p$, $Var(X_i) = pq$, $Var(X) = npq$, $q = 1-p$

$$E\left(\frac{X}{n}\right) = p,\ Var\left(\frac{X}{n}\right) = \frac{pq}{n} = \sigma^2, \text{ say. Also let } \varepsilon = t\sigma.$$

Applying Chebychev's inequality to the variate m/n, we get

$$P\left(\left|\frac{X}{n}-p\right|<\varepsilon\right) \geq 1-\frac{pq}{n\varepsilon^2}$$

$$\Rightarrow P\left[\left|\frac{x}{n}-p\right| \geq \varepsilon\right] \leq \frac{pq}{n\varepsilon^2} > 1-\delta \text{ when } n > \frac{pq}{\varepsilon^2} = N.$$

Example 7·29. *If you wish to estimate the proportion of engineers and scientists who have studied probability theory and you wish your estimate to be correct within 2% with probability 0·95 or more, how large a sample would you take (i) if you have no idea what the true proportion is, (ii) if you are confident that the true proportion is less than 0·2 ?*

(B. Sc. Hons. Burdwan, 1992)

Let n be the size of the sample. We are given that $\varepsilon = 2\% = \cdot 02$

(*i*) We have $P\left[\left|\frac{x}{n}-p\right|<0{\cdot}02\right] \geq 1-\frac{1}{4n(0{\cdot}02)^2} = 0{\cdot}95$

as $p = q = \frac{1}{2}$ $\quad\therefore\quad 0{\cdot}05 = \frac{1}{4n(0{\cdot}02)^2} \quad \Rightarrow n = 12{,}500$

(*ii*) $p < 0{\cdot}2$, $\quad P\left[\left|\frac{X}{n}-p\right| \leq \varepsilon\right] \geq 1-\frac{p(1-p)}{n\,\varepsilon^2}$

Now $p(1-p) < 0.16 \quad \therefore \; 1 - \dfrac{0.16}{n\varepsilon^2} = 0.95 \; \therefore \; \dfrac{0.16}{n\varepsilon^2} = 0.05 \; \therefore \; n\varepsilon^2 = \dfrac{.16}{.05} = \dfrac{16}{5}$

Hence $n = \dfrac{16}{5} \times \dfrac{1}{(.02)^2} = \dfrac{16}{5} \times 2500 = 8000.$

7·5·4. Stochastic Convergence or Convergence in Probability. A variable t_n is said to *converge stochastically* (or in probability sense) to parameter θ if given ε and δ, two positive numbers however small, we can find a number $N(\varepsilon, \delta)$ such that $P(|t_n - \theta| < \varepsilon) > 1 - \delta$ when $n > N$

i.e.
$$\lim_{n \to \infty} P(|t_n - \theta| \geq \varepsilon) = 0$$

In other words, t_n is said to approach θ in probability sense if we can find a number N such that when n is greater than N we can make a probability t_n being as near to θ as we like as near to unity as we please. It may be noted that when $n > N$, $\left| t_n - \theta \right|$ is not necessarily less than ε though the probability of its being so is very nearly unity. In this sense the concept of stochastic convergence is different from the concept of convergence in Mathematics.

Notation. In symbols, we write $t_n \to \theta$ as $n \to \infty$.

Bernoulli's theorem may now bc stated as follows :

In an infinite series of Bernoullian trials, the relative frequency of success converges stochastically to the value of the probability in each trial as the number of trials increases indefinitely.

The theorem shows how the mathematical and statistical definitions of probability can be reconciled.

Example 7·30. *A symmetric die is thrown 360 times. Determine the lower bound for the probability of getting 50 to 70 ones.*

Let m be the total number of successes. Then

$$m = X_1 + X_2 + \ldots + X_{360} \;\Rightarrow\; E(m) = np = 360 \times \tfrac{1}{6} = 60, \; p = \tfrac{1}{6}$$

$$\text{Var } m = npq = 360 \times \tfrac{1}{6} \times \tfrac{5}{6} = 50$$

Chebychev's inequality gives $P[|m - 60| < t\sqrt{50}] \geq 1 - \dfrac{1}{t^2}$

$$\Rightarrow \qquad P[60 - t\sqrt{50} < m < 60 + t\sqrt{50}] \geq 1 - \frac{1}{t^2}$$

Putting $t = \sqrt{2}$, $P[60 - 10 < m < 60 + 10] \geq 1 - \tfrac{1}{2} \;\Rightarrow\; P[50 < m < 70] \geq \tfrac{1}{2}$.

Ex. 1. If, in Bernoulli's theorem, Q is the probability of the inequality

$|m - np| \geq n\varepsilon$, prove that $Q < \dfrac{pq}{n\varepsilon^2}$,

[**Hint.** In the identity $\sum_{m=0}^{n} T_m(m-np)^2 = npq$ where $T_m = {}^nc_m p^m q^{n-m}$ drop all the terms in which $| m - np |$ and in the remaining terms replace $(m-np)^2$ by $n^2\varepsilon^2$. The resulting inequality is $\sum_{m-np \geq n\epsilon} T_m < \frac{pq}{n\varepsilon^2}$.

Ex. 2. Prove Poisson's theorem : If X represents the number of successes in n independent trials the probability of success at the ith trial being p_i then X/n converges in probability to $\sum_{i=1}^{n} p_i/n$. **(*M.A., Delhi*)**

7·6. Continuous Bivariate Probability Distributions. By analogy it is easy to extend the concept of probability distribution of a single variate to that of two or more variates varying simultaneously. Let X and Y be two stochastic variates and x, y be their particular values. The values x and y can be represented by a point with co-ordinates (x, y) in the $X-Y$ plane. The domain S of all the possible values of X and Y will be represented by a portion (finite or infinite) of the x-y plane with a definite boundary unless its domain coincides with the whole plane. The probability that the point (x, y) will lie in the infinitestimal rectangular region, of area $dxdy$, and defined by $P\left[x - \frac{1}{2}dx \leq X \leq x + \frac{1}{2}dx, y - \frac{1}{2}dy \leq Y \leq y + \frac{1}{2}dy\right]$ is denoted by $f_{XY}(x, y)\, dxdy$, where the function $f_{XY}(x, y)$ is called the probability density function at the point (x, y) of the joint distribution for X and Y. If there is no confusion, we prefer to write $f(x, y)$ instead of $f_{XY}(x,y)$.

The *requirements for the continuous function* $f(x,y)$ to be a joint probability density function for X and Y are :

$$\left.\begin{aligned} &(i)\ f(x,y) \geq 0 \text{ everywhere in the } x-y \text{ plane.} \\ &(ii) \int_{-\infty}^{\infty}\int_{-\infty}^{\infty} f(x, y)dx\,dy = 1 \\ &(iii)\ P[a \leq X \leq b, c \leq Y \leq d] = \int_{x=a}^{x=b}\int_{y=c}^{y=d} f(x, y)dxdy \end{aligned}\right\} \quad ...(7·15)$$

7·6·1. Marginal and Conditional Distributions.

$$\text{If } P\left[y - \tfrac{1}{2}dy \leq Y \leq y + \tfrac{1}{2}dy, -\infty < X < \infty\right] = dy \int_{-\infty}^{\infty} f(x, y)dx$$

$$= h_Y(y)dy. \quad \text{where} \quad h = h_Y(y) = \int_{-\infty}^{+\infty} f_{XY}(x, y)dx$$

then $dP = h(y)dy$ is called the *marginal distribution of Y* and $h(y)$ is called the *marginal density function of Y*.

Similarly, if $P\left[x - \frac{1}{2}dx \le X \le x + \frac{1}{2}dx, -\infty < Y < \infty\right]$

$$= dx \int_{-\infty}^{\infty} f_{XY}(x, y)dy = g_X(x)dx, \text{ where } g(x) = g_X(x) = \int_{-\infty}^{+\infty} f(x, y)dy$$

then $dP = g(x)dx$ is called the *marginal distribution of X and* $g_X(x)$ is called the marginal density function of X.

The conditional density function of Y, given X, for two random variables X, Y is defined as follows : $f(y \mid x) = \dfrac{f(x, y)}{g(x)}$ where x and y are two real numbers.

Similarly, the conditional density function of X, given Y, is

$$f(x \mid y) = \frac{f(x, y)}{h(y)}.$$

provided that $g(x) > 0$ and $h(y) > 0$.

If the conditional density function $f(x \mid y)$ does not depend on Y and if the range of the conditional distribution does not depend on Y, then X is independent of Y in probability sense. It is easily seen that if X and Y are independent in probability sense, then their joint distribution is equal to the product of their marginal distributions, *i.e.*, $f_{XY}(x, y) = g_X(x)h_Y(g)$. The converse of this is also true, *i.e.*, if $f(x,y)$ can be factorised into two functions $g(x)$ and $h(y)$, *i.e.*, $f(x, y) = g(x) \times h(y)$ and *if the range of X and Y do not depend on each other*, then X and Y are independent of each other.

7·6·2. Independent Random Variables. Let X and Y be two random variables jointly distributed with $F_{XY}(x, y)$ as their joint distribution function and $f_{XY}(x, y)$ as their probability density function. Further, let $G_X(x), H_Y(y)$ be their marginal density functions. The random variables X, Y are said to be independent if $F_{XY}(x, y) = G_X(x)H_Y(y)$.

Differentiating partially with respect to x and then with respect to y, we have

$$\frac{\partial^2}{\partial x \partial y} F_{XY}(x,y) = \frac{\partial}{\partial x} G_X(x) \frac{\partial}{\partial y} H_Y(y)$$

$$\Rightarrow \quad f_{XY}(x, y) = g_X(x)\, h_Y(y)$$

$$\Rightarrow \quad f(x,y) = g(x)\, h(y).$$

Theorem. *A necessary and sufficient condition for the independence of two random variables X and Y is that their joint density function should be expressible as the product of a function of X alone and a function of Y alone.*

Proof. The condition is clearly necessary.

To establish the sufficiency, we have to prove

$$F_{XY}(x, y) = G_X(x) H_Y(y)$$

$$\Rightarrow \qquad F_{XY}(\infty, \infty) = G_x(\infty)\, H_Y(\infty) = 1$$

Hence $$F_{XY}(x, y) = \frac{G_X(x)}{G_X(\infty)} \cdot \frac{H_Y(y)}{H_Y(\infty)}$$

Also $G_X(x) = F_{XY}(x, \infty) = \dfrac{G_X(x)}{G_X(\infty)}$ and $H_Y(y) = F_{XY}(\infty, y) = \dfrac{H_Y(y)}{H_Y(\infty)}$

Hence $F_{XY}(x, y) = G_X(x)\, H_Y(y)$ which was to be proved.

Remark 1. If X and Y are independent random variables then $P(X = x_1, Y = y_1) = P(X = x_1)\, P)Y = y_1)$.

2. If X and Y are independent random variables, and

$$A = \{x : a < x < b\},\ B = \{y : c < y < d\}$$

be any two events, then $P(A \cap B) = P(A)\, P(B)$

Proof. We shall assume that X and Y are continuous random variables. $P(A \cap B) = P[(x : a < x < b) \cap (y : c < y < d)]$

$$= \int_a^b \int_c^d f_{XY}(x, y)\, dxdy = \int_a^b \int_c^d g_X(x)\, h_Y(y)\, dxdy,$$

since A and B are independent

$$= \left[\int_a^b g_X(x) dx\right]\left[\int_c^d h_Y(y)\, dy\right] = P(A)P(B)$$

which proves the result.

Def. The mathematical expectation of any function $\phi(x,y)$ is defined by

$$E[\phi(x,y)] = \int_{-\infty}^{\infty} \int_{-\infty}^{\infty} \phi(x,y) f(x, y)\, dx\, dy.$$

In particular, the moment μ_{rs}' about the origin (0, 0) of order r in X and s in Y is defined as $\mu_{rs}' = E(X^r Y^s) = \int_{-\infty}^{\infty} \int_{-\infty}^{\infty} x^r y^s f(x, y) dx\, dy.$ ...(7·17)

Also $$\mu_{rs} = E\left[(x - \bar{x})^r (y - \bar{y})^s\right]$$

$$= \int_{-\infty}^{\infty} \int_{-\infty}^{\infty} (x - \bar{x})^r (y - \bar{y})^s f(x, y)\, dxdy \qquad ...(7·18)$$

$$M(t_1, t_2) = E\left[\exp\{t_1 x + t_2 y\}\right] \quad \text{where } t_1 \text{ and } t_2 \text{ are real.}$$

$$= \int_{-\infty}^{\infty}\int_{-\infty}^{\infty} \exp(t_1 x + t_2 y) f(x, y)\, dxdy. \qquad \text{...(7·19)}$$

When this integral is convergent the exponential may be expanded in powers of t_1 and t_2 and $\mu_{rs}' = \text{coefficient of } \frac{t_1^r t_2^s}{r!\,s!}$ in $M(t_1, t_2)$ ...(7·20)

$$\therefore \quad \mu_{10}' = \int_{-\infty}^{\infty}\int_{-\infty}^{\infty} xf(x, y)\, dxdy = \int_{\infty}^{\infty} xg(x)\, dx.$$

Similarly $$\mu_{01}' = \int_{-\infty}^{\infty}\int_{-\infty}^{\infty} yf(x, y)\, dxdy = \int_{-\infty}^{\infty} yh(y)\, dy$$

$$\mu_{20} = \int_{-\infty}^{\infty}\int_{-\infty}^{\infty} (x - \bar{x})^2 f(x, y\, dxdy = \int_{-\infty}^{\infty} (x - \bar{x})^2 g(x) dx = \sigma_x^2$$

$$\mu_{02} = \int_{-\infty}^{\infty}\int_{-\infty}^{\infty} (y - \bar{y})^2 f(x, y) dxdy = \int_{-\infty}^{\infty} (y - \bar{y})^2 h(y) dy = \sigma_y^2$$

$$\mu_{11} = \int_{-\infty}^{\infty}\int_{-\infty}^{\infty} (x - \bar{x})(y - \bar{y}) f(x, y)\, dx\, dy = \text{Cov}(x, y).$$

The correlation coefficient r_{xy} between X and Y is defined as

$$r_{xy} = \frac{\mu_{11}}{\sqrt{\mu_{20}\mu_{02}}} \qquad \text{...(7·21)}$$

The function $F_{XY}(x, y)$ which is defined by

$$F_{XY}(x, y) = P(-\infty < X \le x, -\infty < Y \le y) = \int_{-\infty}^{x}\int_{-\infty}^{y} f(x, y)\, dx\, dy \quad \text{...(7·22)}$$

is called the *distribution function* of the two dimensional random variables X and Y. Unless there is confusion, we shall prefer to denote $F_{XY}(x, y)$ by $F(x, y)$.

If $F(x, y)$ is continuous, it has the following *properties* :

(*i*) $0 \le F(x, y) \le 1$. (*ii*) $\frac{\partial^2 F(x, y)}{\partial x \partial y} = f(x, y)$.

(*iii*) $F(x, y)$ is a non- decreasing function in two dimensional random variables X and Y.

(iv) $F(x, -\infty) = 0, F(-\infty, y) = 0, F(\infty, \infty) = 1.$

(v) $P(x_1 \le X \le x_2, y_1 \le Y \le y_2)$

$$= P(X \le x_2, Y \le y_2) - P(X \le x_2, Y \le y_1) - P(X \le x_1, Y \le y_2) + P(X \le x_1, Y \le y_1)$$

$$= F(x_2, y_2) - F(x_2, y_1) - F(x_1, y_2) + F(x_1, y_1) \qquad ...(7{\cdot}23)$$

We shall now show by an example that conditions (*i*), (*ii*) and (*iii*) are not *sufficient* to ensure that $F(x, y)$ is a distribution function.

Let $F(x, y) = 0$ for $x + y \le 0$; $= 1$ for $x + y > 0$

i.e., the function $F(x, y)$ takes the value 0 at points on and below the straight $y = -x$ and the value 1 at points above this line. The function $F(x, y)$ so defined satisfies (*i*), (*ii*) (*iii*) and (*iv*).

However, it does not satisfy (*v*) for if we consider a rectangle in the x–y plane whose three vertices lie below and one above the straight line $y = -x$, *e.g.*, the rectangle with vertices (3, 3) ; (3–1) ; (–1, 3) ; (–1, –1), then on applying (7·23), we have

$$P(-1 \le X \le 3, -1 \le Y < 3) = F(3, 3) - F(3, -1) - F(-1, 3) + F(-1, -1)$$
$$= 1 - 1 - 1 + 0 = -1,$$

which is impossible since probability cannot be negative.

Example 7·31. *Show that the conditions that the function*

$$f(x, y) = k \exp\{ Ax^2 + 2Hxy + By^2 \}, -\infty < x, y < \infty$$

may be a density function are (*a*) $A \le 0$, (*b*) $B \le 0$, (*c*) $AB - H^2 > 0$.

Show further that if these conditions are satisfied and the integral of $f(x, y)$ *between* $-\infty$ *and* ∞ *for both variates is unity, then* $k = \sqrt{AB - H^2}/\pi$

Clearly $f(x, y)$ will represent a density function if and only if (iff)

$$\int_{-\infty}^{\infty} \int_{-\infty}^{\infty} k \exp(Ax^2 + 2Hxy + By^2)\, dxdy = 1.$$

We assume that both A and B are not zero, for in that case the integral would not be formed whether it is positive, negative or zero.

Now $Ax^2 + 2Hxy + By^2$ can be written in either of the two forms :

$$A\left[\left(x + \frac{H}{A}y\right)^2 + \frac{AB - H^2}{A^2}y^2\right], A \ne 0$$

$$B\left[\left(y + \frac{H}{B}x\right)^2 + \frac{AB - H^2}{B^2}x^2\right], B \ne 0$$

Hence clearly we should have, for the convergence of the integrals,

$$A \le 0, B \le 0, \text{and} AB - H^2 > 0.$$

Let $A = -a, B = -b, H = h \;\therefore\; AB - H^2 = ab - h^2, a > 0, b > 0$

Also $$k \int_{-\infty}^{\infty} \int_{-\infty}^{\infty} e^{-(ax^2 + 2hxy + by^2)} dxdy = 1$$

or $$k \int_{-\infty}^{\infty} e^{-\frac{(ab-h^2)y^2}{a}} dy \times \int_{-\infty}^{\infty} e^{-\frac{(ax+hy)^2}{a}} dx = 1.$$

or $$k \cdot \sqrt{\frac{\pi a}{ab - h^2}} \times \sqrt{\frac{\pi}{a}} = 1 \qquad \left[\because \int_{-\infty}^{+\infty} e^{-c^2u^2} du = \frac{\sqrt{\pi}}{c} \right]$$

or $$k = \frac{1}{\pi} \sqrt{ab - h^2} = \frac{1}{\pi} \begin{vmatrix} -A & H \\ H & -B \end{vmatrix}^{\frac{1}{2}}.$$

Example 7·32. *If $f(x, y) = 2 - x - y$ $(0 \le x \le 1 \le, 0 \le y \le 1)$; $= 0$ elsewhere, find (a) the marginal probability functions (b) the conditional probability functions (c) μ_{20}' and μ_{02}' and (d) the coefficient of correlation between x and y.* **(M.A. Delhi ; M.A. Statistics, Delhi ; M.Sc. Gauhati)**

We have

(a) $$g(x) = \int_0^1 (2 - x - y)\, dy = \frac{3}{2} - x, \quad h(y) = \int_0^1 (2 - x - y)\, dx = \frac{3}{2} - y.$$

(b) $$f(y \mid x) = \frac{f(x, y)}{g(y)} = \frac{2 - x - y}{\frac{3}{2} - x}, \quad f(x \mid y) = \frac{f(x, y)}{h(y)} = \frac{2 - x - y}{\frac{3}{2} - y},$$

(c) $$\bar{x} = \mu_{10}' = \int_0^1 x \left(\frac{3}{2} - x \right) dx = \frac{5}{12}, \quad \bar{y} = \mu_{01}' = \int_0^1 y \left(\frac{3}{2} - y \right) dy = \frac{5}{12}$$

Hence $$\mu_{20} = \int_0^1 \left(x - \frac{5}{12} \right)^2 \left(\frac{3}{2} - x \right) dx = \frac{11}{144}$$

$$\mu_{02} = \int_0^1 \left(y - \frac{5}{12} \right)^2 \left(\frac{3}{2} - y \right) dy = \frac{11}{144}$$

(d) $$\mu_{11} = \int_0^1 \int_0^1 \left(x - \frac{5}{12} \right) \left(y - \frac{5}{12} \right) (2 - x - y)\, dxdy$$

or $$\mu_{11} = \int_0^1 \left(x - \frac{5}{12}\right)\left\{\int_0^1 \left(y - \frac{5}{12}\right)(2 - x - y)\,dy\right\} dx$$

$$= \int_0^1 \left(x - \frac{5}{12}\right)\left(\frac{1}{24} - \frac{x}{12}\right) dx = \frac{-1}{144}$$

Hence $$r = \frac{\mu_{11}}{\sigma_x \sigma_y} = \frac{\frac{-1}{144}}{\frac{11}{144}} = -\frac{1}{11}.$$

Example 7·33. *Two discrete random variables X and Y have the joint probability density function*

$$p(x, y) = \frac{\lambda^x e^{-\lambda} p^y (1-p)^{x-y}}{y!\,(x-y)!}, \quad y = 0, 1, \ldots\ldots x\,;\ x = 0, 1, 2, \ldots\ldots$$

where λ, p *are constants with* $\lambda > 0$ and $0 < p < 1$. *Find*

(*i*) *the marginal probability density functions of X and Y.*

(*ii*) *The conditional distributions of Y for a given X and of X for a given Y.* (***M.Sc. Gauhati***)

We have $$p_x(x) = \sum_{y=0}^{x} p(x, y) = \sum_{y=0}^{x} \frac{\lambda^x e^{-\lambda} p^y (1-p)^{x-y}}{y!\,(x-y)!}$$

$$= \frac{\lambda^x e^{-\lambda}}{x!} \sum_{y=0}^{x} x! \frac{p^y (1-p)^{x-y}}{y!\,(x-y)!} = \frac{\lambda^x e^{-\lambda}}{x!}(\overline{1-p+p})^x = \frac{\lambda^x e^{-\lambda}}{x!}$$

Similarly $p_Y(y)$ can be obtained.

(*ii*) $$P_{Y|x}(y \mid x) = \frac{p_{xy}(x, y)}{p_x(x)}$$

$$= \frac{[\lambda^x e^{-\lambda} p^y (1-p)^{x-y}]}{y!\,(x-y)!} \Big/ \left(\frac{\lambda^x e^{-\lambda}}{x!}\right) = \binom{x}{y} p^y (1-p)^{x-y}$$

Similarly, conditional probability distribution of X given Y can be obtained.

Ex. 1 . The joint density function of a bivariate distribution is given as follows : $f(x, y) = x + y;\ 0 \le x \le 1,\ 0 \le y \le 1;\ = 0$, otherwise

Determine the marginal distributions and the covariance between x and y.

(***M.A., Punjab***)

[**Ans.** $g(x) = x + \frac{1}{2}\,;\ h(y) = y + \frac{1}{2}$]

Ex. 2. Let X and Y have the joint density function $f(x, y) = 2,\ 0 < x < y < 1$

Find the marginal and conditional probability density functions.

Ans. $g(x) = 2(1-x), 0 < x < 1$ $\quad f(x \mid y) = \dfrac{1}{y}, 0 < y < 1$

$= 0$, elsewhere $\quad = 0$, elsewhere

$h(y) = 2y, 0 < y < 1$ $\quad f\{y \mid x\} = \dfrac{1}{1-x}, 0 < x < 1$

$= 0$, elsewhere $\quad = 0$, elsewhere

Ex. 3. If X and Y are two random variables, having joint density functions

$$f(x, y) = \tfrac{1}{8}(6 - x - y) \quad 0 < x < 2, 2 < y < 4$$
$$= 0, \quad \text{otherwise}$$

Find (*a*) $P(X < 1, Y < 3)$ (*b*) $P(X + Y < 3)$ and (*c*) $P(X < 1 \mid Y < 3)$.

(*B.Sc. Bombay*)

[**Hint.** (*b*) $P(X + Y < 3) = \int_0^1 \int_2^{3-x} \tfrac{1}{8}(6 - x - y)\, dx dy = \text{etc.}$

(*c*) $P(X < 1 \mid Y < 3) = \dfrac{P(X < 1 \cap Y < 3)}{P(Y < 3)}$

where $P(Y < 3) = \int_0^2 \int_2^3 \dfrac{6 - x - y}{8}\, dx dy = \dfrac{5}{8}$. $\left[\textbf{Ans.} \ (a) \dfrac{3}{8} \ (b) \dfrac{5}{24} \ (c) \dfrac{2}{5}\right]$

7·7. Variate Transformations and Distributions.

7·7·1. Univariate Transformations, Distributions

Theorem. *If $x = h(y)$ is an increasing or decreasing function of y and if $f(x)$ is the density function of X then $g(y)$, the density function of Y, is given by*

$$g(y) = f[h(y)] \mid h'(y) \mid$$

Proof. Let the variate X be continuous and have $f(x)$ as its probability function. Suppose $y = H(x)$ is new variable whose probability function is required. We shall assume that the function $H(x)$ is either an increasing or decreasing function throughout the range of X. The relation $y = H(x)$ can be written as $x = H^{-1}(y) = h(y)$, say.

For $H(x)$ an increasing function, the variable y will assume a value less than y_0 where $y_0 = H(x_0)$ if and only if x assumes a value less than x_0; hence the probability that $y < y_0$ must be equal to the probability that $x < x_0$.

Thus we have
$$\int_{-\infty}^{y_0} g(t)\, dt = \int_{-\infty}^{x} f(t)\, dt$$

where $g(y)$ is the density function of y.

Since this result holds for every pair of corresponding values (x_0, y_0) we have, on dropping the subscript,

$$\int_{-\infty}^{y} g(t)\,dt = \int_{-\infty}^{x} f(t)\,dt. \qquad ...(7{\cdot}25)$$

Differentiating both sides of (7·25) with respect to y and using the fundamental theorem of Integral Calculus, we have $g(y) = f[h(y)]h'(y)$

For $H(x)$ a decreasing function we shall obtain similarly

$$g(y) = -f[h(y)]h'(y)$$

Thus whether $H(x)$ is an increasing or decreasing function of x, we get

$$g(y) = f[h(y)]\,|\,h'(y)\,|\,.$$

Example 7·34. (*a*) *Given* $f(x) = e^{-x}, x \geq 0$, *find the probability density function of the variable* $Y = X^2$.

(*b*) *If X has a uniform distribution in* $[0, 1]$ *with* $f(x) = 1$, $0 \leq x \leq 1$; *0, otherwise, find the density function of* $Y = -2 \log X$.

Since $y = x^2$ is an increasing function of x when $x > 0$ the above theorem can be applied.

Hence $\quad h(y) = \sqrt{y} \quad \therefore \quad h'(y) = \dfrac{1}{2\sqrt{y}}$

$$\therefore \qquad g(y) = e^{-\sqrt{y}} \frac{1}{2\sqrt{y}}, y > 0 = 0, \text{ otherwise.}$$

$$G_y(y) = P(Y \leq y) = P(-2 \log X \leq y) = P(\log X \geq -\tfrac{1}{2} y)$$

$$= P(X \geq e^{-\frac{1}{2}y}) = 1 - P(X \leq e^{-y/2}) = 1 - \int_0^{e^{-y/2}} 1\,.\,dx = 1 - e^{-y/2}$$

$$\Rightarrow \qquad g(y) = \frac{d}{dy} G(y) = \tfrac{1}{2} e^{-y/2}, \; 0 \leq y < \infty.$$

Example 7·35. *If X is a random variate having a continuous probability density function and F(x) denotes the probability that X is less than or equal to x then show that F(x) is a random variate having a rectangular distribution in the range (0, 1).*

(*B. Sc. Stat. Hons. 1985, 87, 92*) (*M.A. Delhi*)

We have $\quad P(X \leq x) = F(x) = \displaystyle\int_{-\infty}^{x} f(x)\,dx$

Then $\qquad dP = dF = f(x)\,dx$ and $F'(x) = f(x)$

Let $u = F(x)$ then $dP = f(x)\,dx = f(x)\dfrac{dx}{du}\,du = f(x)\,.\,\dfrac{1}{f(x)}\,du = du, 0 \leq u \leq 1$

Hence u follows a rectangular distribution with unit range.

Note. 1. The example shows that by letting $u = F(x)$ where $F(x)$ is the cumulative distribution of x, any density for a continuous variate x may be transformed to the uniform density $f(u) = 1, 0 \le u \le 1$.

The transformation $u = F(x)$ is called the *probability transformation.*

2. If $Y = 1 - F(x)$ so that $y = \int_x^\infty f(v)\, dv$, then

Y has rectangular distribution with the same range.

Example 7·36. *What transformation will change the variable X to one which will have the uniform distribution over the unit interval if*

$$f(x) = \frac{x-1}{2},\ 1 \le x \le 3,\ = 0, x < 1 \text{ and } x > 3\ ?$$

We have $\quad u = F(x) = \int_1^x \frac{x-1}{2}\, dx = \frac{1}{4}(x-1)^2$

which is the required transformation.

Example 7·37. *Show that whatever be the distribution function F(x) of a r.v. X,* $P[a \le F(x) \le b] = b - a,\ 0 \le (a, b) \le 1$ **(B.Sc. Stat. Hons. Delhi, 1986)**

In Example 7·35, we have shown the $U = F(x)$ is a random variate having a rectangular distribution in the range (0, 1) *i.e.* $dp = F(x)\, dx = 1 \cdot dx, 0 < x < 1$.

Now $f[a \le F(x) \le b] = \int_a^b 1 \cdot dx = b - a.$

Example 7·38. Let $f(x) = 6x(1-x), 0 \le x \le 1$. *Find y as a function of x such that Y has p.d.f.* $g(y) = 3(1 - \sqrt{y}), 0 \le y \le 1$ **(B.A. Stat. Hons. Delhi, 1988)**

Here $F(x) = \int_0^x f(x)\, dx = \int_0^x 6(x - x^2)\, dx = 6 \left| \frac{x^2}{2} - \frac{x^3}{3} \right|_0^x = 3x^2 - 2x^3,$

$$G(y) = \int_0^y g(y)\, dy = 3\int_0^y (1 - y^{\frac{1}{2}})\, dy = 3y - 2y^{3/2}$$

But $\quad F(x) = G(y) \qquad$ since both are $\sim U(0, 1)$

$\therefore$ we find $y = x^2$, which is the required transformation.

Example 7·39. *Let X be uniformly distributed on [–1, 1], find the distribution function and hence the p.d.f. of* $Y = X^2$.

Since the distribution of X is uniform over [–1, 1], $dp = \frac{dx}{2}, -1 < x < 1$

Distribution function $F(x) = P(X \le x) = \int_{-1}^x \frac{dx}{2} = \frac{x+1}{2},$

$\therefore$ Distribution function of Y is $\frac{\sqrt{y}+1}{2} \cdot \frac{1}{2\sqrt{y}} = \frac{1}{4}(1 + y^{-\frac{1}{2}})$

$\therefore$ p.d.f. of $Y = \frac{1}{4} \cdot \frac{-1}{2} y^{-3/2} = \frac{-1}{8} y^{-3/2}$, $0 \le y \le 1$

Example 7·40. *If the p.d.f. of a r v.X. is given by* $f(x) = \frac{1}{2}(x-1)$, $1 \le x \le 3$; $= 0$, *otherwise, determine what interval for* $Y = \int_{-a}^{x} f(x)\,dx$ *will correspond to the interval* $1 \cdot 1 \le x \le 2 \cdot 2$.

$$Y = \int_{-a}^{x} f(x)\,dx = \int_{1}^{x} \frac{1}{2}(x-1)\,dx = \left| \frac{(x-1)^2}{4} \right|_1^x = \frac{1}{4}(x-1)^2,\ 1 \le x \le 3$$

when $x = 1 \cdot 1$, $y = \frac{(1 \cdot 1 - 1)^2}{4} = \frac{(\cdot 1)^2}{4} = \frac{1}{400} = 0 \cdot 0025$

when $x = 2 \cdot 2$, $y = \frac{(2 \cdot 2 - 1)^2}{4} = \frac{1 \cdot 2 \times 1 \cdot 2}{4} = \cdot 36$

Hence $0 \cdot 0025 \le y \le 0 \cdot 36$.

Example 7·41. If $X \sim U\left[\frac{-\pi}{2}, \frac{\pi}{2}\right]$ *distribution, find the p.d.f. of* $Y = \tan X$.

(B.A. Hons. Delhi, 1989)

The p.d.f. of $X = \frac{dx}{\pi}$, $\frac{-\pi}{2} < x < \frac{\pi}{2}$

Now $Y = \tan x \therefore x = \tan^{-1} y \Rightarrow dx = \frac{dy}{1+y^2}$

Also as $\frac{-\pi}{2} < x < \frac{\pi}{2} \Rightarrow -\infty < y < +\infty$

$\therefore$ p.d.f. of Y is $dp = \frac{1}{\pi} \cdot \frac{dy}{1+y^2}$, $-\infty < y < +\infty$.

Ex. 1 If the random variable X is uniformly distributed over $\left(0, \frac{\pi}{2}\right)$, find the distribution of $y = \sin x$. **Ans.** $f(y) = \frac{2}{\pi\sqrt{1-y^2}}$, $0 < y < 1$

Ex. 2 If $u = a + b\,F(x)$, where $F(x)$ is *d.f.*, find the *p.d* function for u.

[Hint. If $F(x) = \int_{c}^{u} f(t)\,dt$ where c is a constant and u is a function of x, then $\frac{d}{dx} F(x) = f(u) \cdot \frac{du}{dx}$.]

7·7·2. Bivariate Transformations. The transformations of the variates X and Y for bivariate distributions follow the ordinary laws for the transformation of differentials.

For example, if $dF = f(x, y)\,dx\,dy$; where $x = x(u, v)$, $y = y(u, v)$, we have $$dF = f[x(u, v), y(u, v)] \mid J \mid du\, dv$$

where J is the Jacobian given by $J = \dfrac{\partial(x, y)}{\partial(u, v)} = \begin{vmatrix} \dfrac{\partial x}{\partial u} & \dfrac{\partial y}{\partial u} \\ \dfrac{\partial x}{\partial v} & \dfrac{\partial y}{\partial v} \end{vmatrix}$

Example 7·42. *If the probability density function f(x, y) of two variates X and Y is given by*

$$f(x, y) = \frac{1}{2\pi\sigma_x\sigma_y \sqrt{1 - p^2}} \cdot \exp\left\{-\frac{1}{2(1 - p^2)}\left(\frac{x^2}{\sigma_x^2} - \frac{2\rho xy}{\sigma_x\sigma_y} + \frac{y^2}{\sigma_y^2}\right)\right\}$$

show that the variates U and V defined as follows are independent :

(*i*) $u = \dfrac{x}{\sigma_x} - \dfrac{\rho y}{\sigma_y},\ v = (1 - \rho^2)^{\frac{1}{2}} \cdot \dfrac{y}{\sigma_y}$.

(*ii*) $u = \dfrac{1}{\sqrt{1 - p^2}}\left(\dfrac{y}{\sigma_y} - \dfrac{\rho x}{\sigma_x}\right),\ v = \dfrac{x}{\sigma_x}$.(**B.A. Hons. Calcutta**)

(*i*) Here $1/J = \begin{vmatrix} \dfrac{1}{\sigma_x} & \dfrac{-\rho}{\sigma_y} \\ 0 & \dfrac{(1 - \rho^2)^{\frac{1}{2}}}{\sigma_y} \end{vmatrix} = \dfrac{(1 - \rho^2)^{1/2}}{\sigma_x\sigma_y}$

and $$\frac{x^2}{\sigma_x^2} - \frac{2\rho xy}{\sigma_x\sigma_y} + \frac{y^2}{\sigma_y^2} = u^2 + v^2.$$

The distribution then becomes

$$dF = \left[\frac{1}{2\pi\sigma_x\sigma_y \sqrt{1 - \rho^2}} \Big/ \frac{\sqrt{1 - \rho^2}}{\sigma_x\sigma_y}\right] \exp\left\{\frac{-1}{2(1 - \rho^2)}(u^2 + v^2)\right\} du dv$$

$$= \left[\frac{1}{\sqrt{2\pi(1 - \rho^2)}} \exp\left\{\frac{-u^2}{2(1 - \rho^2)}\right\} du\right]$$

$$\times \left[\frac{1}{\sqrt{2\pi(1 - \rho^2)}} \exp\left\{\frac{-v^2}{2(1 - \rho^2)}\right\} dv\right]$$

$$= g(u)\, du \times h(v)\, dv,\ -\infty < u < \infty, -\infty < v < \infty.$$

Thus the transformed variates U and V are independent.

(*ii*) Proceed as in (*i*)

Example 7·43. Prove that if the joint density function for the random *variables X and Y is f (x, y), then the density function of the sum Z = X + Y is given by*

$$\int_{-\infty}^{+\infty} f(x, z - x)\, dx$$

In the particular case, when

$$f(x, y) = \frac{1}{4a^2}\left[1 + xy\left(x^2 - y^2\right)\right], |x|, |y| \le a, a > 0$$
$$= 0, \qquad \text{elsewhere}$$

show that the distribution of $Z = X + Y$ is a triangular distribution.

The joint distribution of X and Y is given by

$$dP = f(x, y)\, dx\, dy, -\infty < x, y < \infty.$$

Put $x = x, z - x = y$ then $dP = f(x, z - x)\, dx\, dz$,

Integrate out for x, then the density function of the random variable

Z is given by $\displaystyle\int_{-\infty}^{+\infty} f(x, z - x)\, dx$

Now $$f(x, y) = \frac{1}{4a^2}[1 + xy(x^2 - y^2)]$$

$$\therefore \quad f(x, z - x) = \frac{1}{4a^2}[1 + x(z - x)\{x^2 - (z - x)^2\}]$$
$$= \frac{1}{4a^2}[1 + 3z^2x^2 - 2zx^3 - z^3x]$$

Hence the distribution of Z is given by

$$\theta(z) = \int \frac{1}{4a^2}(1 + 3z^2x^2 - 2zx^3 - z^3x)\, dx. \qquad \text{...(1)}$$

In order to find the limits of integration which depend on z we note that in introducing the variables (x, z) instead of (x, y) ; the square domain $|x| \le a, |y| \le a$ changes into a domain defined by the inequalities

$$|x| \le a, \; x - a \le z \le x + a \quad \text{or } z - a \le x \le z + a.$$

which is a parallelogram bounded by the lines $x = \pm a, x - z = \pm a$

Hence integrating (1) from $-a$ to $z + a$ for $x \le 0$ and $z - a$ to a for $x > 0$.

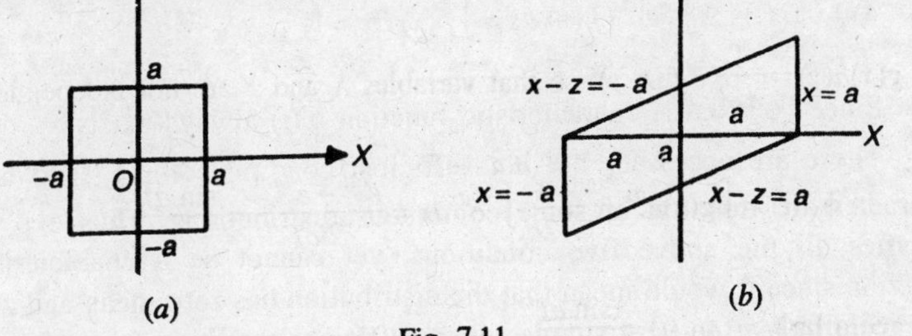

Fig. 7·11

$$\theta(z) = \begin{cases} \displaystyle\int_{-a}^{z+a} \frac{1}{4a^2}(1+3z^2x^2-2zx^3-z^3x)\,dx = \frac{2a+z}{4a^2}, & -2a<z<0 \\ \displaystyle\int_{z-a}^{a} \frac{1}{4a^2}(1+3z^2x^2-2zx^3-z^3x)\,dx = \frac{2a-z}{4a^2}, & 0<z<2a \\ 0, & |z|>2a \end{cases}$$

Thus the distribution of Z is triangular.

Example. 7·44. *Prove the following results for the example 7·43, page 266 where the distribution* $f(x, y) = \frac{1}{4a^2}[1 + xy(x^2 - y^2)]$,

$(|x|, |y| \leq a, a > 0$

$= 0$, elsewhere.

(i) $\phi_x(t) =$ C.F. for $X = \dfrac{\sin at}{at}$;

(ii) $\phi_y(t) =$ C.F. for $Y = \dfrac{\sin at}{at}$;

(iii) $\phi_{x+y}(t) = \phi_z(t) =$ C.F. for $Z = \left(\dfrac{\sin at}{at}\right)^2$.

Hence show that $\phi_z(t) = \phi_x(t) \cdot \phi_y(t)$ *though the variables X and Y are not independent.* **(M.A. Lucknow)**

We have $g(x) = \displaystyle\int_{-a}^{+a} \frac{1}{4a^2}\left[(1 + xy(x^2 - y^2)\right] dy = \frac{1}{4a^2}\left[y + \frac{x^3y^2}{2} - \frac{xy^4}{4}\right]_{-a}^{+a} = \frac{1}{2a}$

$$h(y) = \int_{-a}^{+a} \frac{1}{4a^2}\left[1 + xy(x^2 - y^2)\right] dx = \frac{1}{4a^2}\left[x + \frac{x^4y}{4} - \frac{x^2y^3}{2}\right]_{-a}^{+a} = \frac{1}{2a}$$

$\therefore$ $g(x)h(y) = \dfrac{1}{4a^2} \neq f(x, y)$, so that variables X and Y are not independent.

$$\phi_X(t) = \int g(x)e^{itx}dx = \frac{1}{2a}\int_{-a}^{+a} e^{itx}dx = \frac{e^{iat} - e^{-iat}}{2iat} = \frac{\sin at}{at}$$

Similarly, $\phi_Y(t) = \dfrac{\sin at}{at}$.

Also
$$\phi_Z(t) = \int_{-2a}^{+2a} e^{itz}\,\theta(z)\,dz = \int_{-2a}^{0} \frac{2a+z}{4a^2} e^{itz} dz + \int_{0}^{2a} \frac{2a-z}{4a^2} e^{itz}\,dz$$
$$= \frac{2 - e^{2iat} - e^{-2iat}}{4a^2t^2} = \frac{1}{2a^2t^2}(1 - \cos 2\,at) = \left(\frac{\sin at}{at}\right)^2$$

$$\therefore \qquad \phi_{x+y} = \phi_z(t) = \phi_x(t)\,\phi_y(t).$$

Remark This example illustrates that $\phi_{x+Y}(t) = \phi_x(t) \times \phi_y(t)$ does not imply that X and Y are independent.

7·8. Discrete-cum-continuous probability distributions

Just as we can have a mechanical system with a continuous mass distribution together with a number of point masses, we can have a probability function which can have finite values at an enumerable set of discrete points and also have a probability density function in a certain interval (a, b). Even though such probability distributions are mainly conceptual, the advantage of discussing these is that the unified theory for these includes the theories for discrete and continuous probability distributions as special cases. There are two alternative ways of developing this theory (*i*) through Hermite's unit and Dirac's delta functions* (*ii*) through Riemann-Stieltjes integral.

7·8·1. Unit Function

The unit function $U(x)$ is defined as follows :

$$U(x) = 1 \text{ when } x \geq 0; \qquad = 0 \text{ when } x < 0$$

so that
$$U(x-a) = 1 \text{ when } x \geq a; \qquad = 0 \text{ when } x < a$$

$$U(x-a) - U(x-b) = 1 \text{ when } a \leq x < b$$
$$= 0 \text{ when } x < a \text{ or } x \geq b,$$

so that the probability density function for the uniform continuous distribution can be written as

$$f(x) = \frac{1}{b-a}\,[U(x-a) - U(x-b)]$$

Similarly, the p.d.f. for the exponential distribution function can be

---

* The French mathematician Charles Hermite (1822-1906) devised the unit function for use in electrical engineering and the great English physicist P.A.M. Dirac (1902–) defined his delta function for use in quantum mechanics. We have used here both these functions for giving a unified account of discrete and continuous probability distributions. Though the delta function was used extensively in Physics, it was an improper function. Its use was put on strong foundation by the great French mathematician Lourent Schwarz (1915–) through his theory of Distributions for which he get the Fields Medal, the greatest honour for a mathematician.

written as :

$$f(x) = e^{-x}\, U(x), -\infty < x < \infty$$

In general the p.d.f. for the continuous probability distribution

$$dP = g(x)dx,\ a \le x \le b$$

can be written as $\quad f(x) = g(x)\,[U(x-a) - U(x-b)], -\infty < x < \infty$

If $g(x)$ is continuous in (a, b), then $f(x)$ will be continuous for all x except possibly for a and b and

$$\int_{-\infty}^{\infty} f(x)dx = \int_a^b g(x)\,dx = 1.$$

7·8·2. Dirac's Delta Function

Dirac's delta function $\delta(x)$ is defined to be zero when $x \ne 0$ and has such a large value at $x = 0$, that

$$\int_{-\infty}^{\infty} \delta(x)dx = 1$$

Thus if $\quad f(x) = p_i\delta(x - x_i)\ $ then $\ f(x) = 0$ when $x \ne x_i$

and $$\int_{-\infty}^{\infty} p_i\delta(x - x_i)dx = p_i$$

If $f(x) = \sum_{i=l}^{n} p_i\delta(x - x_i)$ then $\int_{-\infty}^{\infty} f(x)dx = \sum_{i=1}^{n} p_i \int_{-\infty}^{\infty} \delta(x - x_i)dx = \sum_{i=1}^{n} p_i$

7·8·3. Unit-cum-Direct Delta Function

Let $\quad f(x) = \sum_{i=1}^{\infty} p_i\delta(x - x_i) + g(x)[U(x-a) - U(x-b)],$

where $g(x)$ is a continuous function for all x, then

$$f(x) = p_i \text{ when } x = x_i\ (i = 1, 2, 3, ...);\ = g(x) \text{ when } a \le x < b$$

and $$\int_{-\infty}^{\infty} f(x)\,dx = \sum_{i=1}^{\infty} p_i + \int_a^b g(x)dx$$

As an example, we may consider the following :

We throw a coin. If it turns out heads, we choose the point $x = -1$ and if it comes out tails, we choose a number at random between 0 and 1, then $\ f(x) = \frac{1}{2}\,\delta\,(x+1) + \frac{1}{2}[U(x) - U(x-1)]$

Now it is easily seen that $\int_{-\infty}^{x} \delta(x)dx = U(x), \int_{-\infty}^{x} U(x)dx = xU(x)$

so that for the above example

$$F(x) = \int_{-\infty}^{x} f(x)dx = \frac{1}{2}U(x+1) + \frac{1}{2}[xU(x) - (x-1)\,U(x-1)]$$

7·8·4. Riemann-Stieltjes integral

Let $F(x)$ be a function of bounded variation which is always continuous on the left. Thus $F(x)$ can be the distribution function for a probability distribution. Let the interval (a, b) be divided into n subintervals by the set of points $a = x_0, x_1, x_2, \ldots\ldots, x_n = b$ and form the sum

$$I_n = \sum_{i=1}^{n} g(\xi_i)\,[F(x_i) - F(x_i - 1)]$$

where ξ_i is any arbitrary point in (x_{i-1}, x_i). If this sum approaches a unique limit as $n \to \infty$ and the length of each sub-interval tends to zero, we call the limit as the Riemann-Stieltjes integral of $f(x)$ with respect to the distribution function $F(x)$, and write it as

$$I = \int_{a}^{b} g(x)\,dF(x)$$

We also define $\int_{-\infty}^{\infty} g(x)\,dF(x)\,dx = \underset{\substack{a \to -\infty \\ b \to +\infty}}{\text{Lt}} \int_{a}^{b} g(x)\,dF(x)$

Riemann integral is included as a particular case when $F(x) = x$.

If $F(x)$ is the distribution function for a continuous probability density function, we get

$$\int_{-\infty}^{\infty} g(x)\,dF(x) = \int_{-\infty}^{\infty} g(x) f(x)\,dx$$

and if $F(x)$ is the distribution function for a discrete probability distribution, we get

$$\int_{-\infty}^{\infty} g(x)\,d\,F(x) = \sum_{i} g(x_i)\,p(x_i)$$

We can define moments, M.G.F. and C.F. about 'a' for a distribution function $F(x)$ as

$$\int_{-\infty}^{\infty} (x-a)^r dF(x), \quad \int_{-\infty}^{\infty} e^{t(x-a)} dF(x), \quad \int_{-\infty}^{\infty} e^{it(x-a)} dF(x), \; i = \sqrt{-1}$$

and these definitions will hold for the discrete, continuous and discrete-continuous cases.

Note 1. When using Riemann-Stieltjes integral, we have to be cautious about stating whether the limits of the definite integral are included in the range or not

$$\int_a^b g(x)\, dF(x) - \int_{a+0}^{b} g(x) dF(x) = g(a)[F(a+0) - F(a)]$$

$$\int_a^b g(x)\, dF(x) - \int_a^{b-0} g(x)\, dF(x) = g(b)\,[F(b) - F(b-0)]$$

Note 2. We also use the notation $1_{a,\,b}(x)$ to denote a function which has the value unity in the interval (a, b) and is zero elsewhere and $I_A(x, y)$ to denote a function which is unity over the set A and is zero elsewhere. Thus

$F(x) = \frac{1}{a} I_A(x)$ where A is the set of n points *i.e.* is the probability function for the discrete rectangular interval.

$F(x, y) = \frac{1}{ab} I_B(x, y)$ where B is the set of points of the rectangle $0 \le x \le a,\ 0 \le y \le b$ represents the continuous rectangular distribution in two dimensions.

Ex. 1. Let $f(x) = (1 - |x|)\,[U(x+1) - U(x-1)]$, find $F(x)$ and draw the graphs of $f(x)$ and $F(x)$.

Ex. 2. Let $f(x) = \frac{1}{2}\delta(x-2) + \frac{1}{2}\,[U(x) - U(x-1)]$, draw the graphs of $f(x)$ and $F(x)$.

Ex. 3. Give a unified proof for the theorems

$$E(x+y) = E(x) + E(y),\ E(xy) = E(x)\,E(y)$$

applicable for both discrete and continuous probability distributions.

Exercises 7

1. (*a*) Explain the terms (*i*) probability differential, (*ii*) probability density function, and (*iii*) distribution function.

(*B.Sc. Bombay ; B.Sc. Hons, Delhi ; M.A. Delhi*)

(*b*) (*i*) If X is a random variable, then $|E(X)| \le E(|X|)$

(*ii*) If X and Y are two random variables such that $x < y$, then $E(X) \le E(Y)$.

Prove or disprove :

(c) (i) $P[X > Y] = 1 \Rightarrow EX > EY$ (ii) $EX > EY \Rightarrow P[X > Y] > 0$

(iii) If $F_x(Z) > F_y(Z)\ \forall\ Z$, then $EY > EX$ and its converse.

(d) If $g(x) \geq 0$ everywhere then $E[g(x)] > 0$.

(e) Show that whatever the distribution function $F(x)$ of a random variable X $P[a \leq F(a) \leq b] = b - a,\ 0 < a,\ b < 1$.

(f) $P(X^2 \leq k) = P(-\sqrt{k} \leq X \leq +\sqrt{k})$

(g) $P(a \leq X^2 \leq b) = P[(-\sqrt{b} \leq X < -\sqrt{a}) \cup (\sqrt{a} \leq x \leq +\sqrt{b})]$

2. (a) If $f(x) = 2x$ when $0 \leq x \leq 1$ and zero otherwise, find the probability that (i) $x < \frac{1}{2}$ (ii) $\frac{1}{4} < x < \frac{1}{2}$ (iii) $x > 3/4$ given $x < \frac{1}{2}$.

(b) If $f(x) = x \quad 0 \leq x < 1\ ; = 2 - x,\ 1 \leq x < 2$ find $P[\frac{1}{2} \leq x \leq \frac{3}{4}]$.

(c) A random variable X takes values 0, 1, 2,.... with probability proportional to $(x+1)(1/5)^x$. Find the probability that $X \leq 5$,

(B.A. Hons. Delhi)

$$\left[\textbf{Hint. Use } \sum_{x=0}^{\infty} k(x+1)\left(\frac{1}{5}\right)^x = 1\right] \qquad \left[\textbf{Ans. } (a)\ \frac{1}{4}, \frac{3}{16}, \frac{7}{12},\ (b)\ \frac{3}{4}\right]$$

3. If $f(x) = 3x^2$ when $0 \leq x \leq 1$ and zero otherwise, find the number a such that x is equally likely to be greater than or less than a. Find the number b such that the probability that x will exceed b is equal to ·05.

[**Ans.** (i) $a = (\frac{1}{2})^{1/3}$, (ii) $b = (0{\cdot}95)^{1/3}$]

4. (a) If $f(x) = 1,\ 0 \leq x \leq 1$ (a) what is the probability that at least two out of the three sample values will exceed ·6 ? (b) What value of x is such that the probability is $\frac{1}{2}$ that at least two out of 3 sample values will exceed it ?

[**Ans** (a) $=$·352, (b) $x = \frac{1}{2}$]

(b) Let X be a continuous $r.v.$ having the p.d.f.

$$f(x) = 2x^{-3},\ 1 \leq x < \infty$$
$$= 0 \text{ else where}$$

Find the distribution of X. Sketch its graph.

$$\left[\textbf{Hint. } F(x) = \int_{-\infty}^{x} 0.\,dx = 0,\ x < 1\ ;\ F(x) = \int_{1}^{x} \frac{2}{x^3}\,dx = 1 - x^{-2},\ 1 \leq x\right]$$

(c) Let X and Y be independent variate uniformly distributed over the unit internal. Find the distribution function and p.d.f of $Z = X + Y$.

(B.Sc. Math Hons. Delhi 1986)

5. (a) Suppose the life in hours of a certain kind of radio tube has the density

$f(x) = \frac{100}{x^2}$ when $x \geq 100$ and zero when $x > 100$. What is the probability that none of three such tubes in a given radio set will have to be replaced during the first 150 hours of operation ? What is the probability that all three of the original tubes will have been replaced during the first 150 hours? (*M.A. Bombay ; M.A. Punjab ; M.A. Delhi*)

(*b*) A bomb plane carrying three bombs flies directly above a rail road track. If a bomb falls within 40 metres of the track, the track will be sufficiently damaged to disrupt traffic. With a certain bomb sight, the density of points of impact of a bomb is

$$f(x) = \frac{100 + x}{10{,}000} \quad -100 \leq x < 0$$

$$= \frac{100 - x}{10{,}000} \quad 0 \leq \leq 100$$

$$= 0 \quad \text{elsewhere}$$

If all three bombs are used what is the probability that the track will be damaged where x represents the vertical deviation in metres from the aiming point which is the track in this case.

(*B.Sc. Bombay, 1970 ; B.A. Hons. Delhi, 1957*)

(*c*) A country filling station is supplied with petrol once a week. If its weekly volume x of sales in thousands of litres is distributed by $f(x) = 5(1 - x)^1 \quad 0 < x < 1$, what must be the capacity of its tank in order that the probability that its supply will be exhausted in a given week shall be ·01 ?

$\left[\right.$ **Ans.** (*a*) $\frac{1}{27}, \frac{8}{27}$ (*b*) Required Probability $= 1 - \left(\frac{9}{25}\right)^3 = 0{\cdot}9533$

(*c*) Volume $= 1000\,[1 - (\cdot 01)]^{1/5} = 601{\cdot}89$ litres $\left.\right]$

6. The probability of a shot hitting at a point distant r from the centre C of the target follows the law : $p(r) = \frac{\cdot 2}{\pi(1 + r^2)}$.

The target is divided into four regions as follows :

S_1 : Within the circle with centre C and radius $\frac{1}{\sqrt{3}}$

S_2 : Within S_1 but within the concentric circle of radius 1

S_3 : Within S_2 but within the concentric circle of radius $\sqrt{3}$

S_4 : Without S_3.

The scores for hits in the regions S_1, S_2, S_3 and S_4 are 4, 3, 2 and 0 respectively. Let X_1, X_2, X_3 and X_4 be the number of hits registered in the region S_1, S_2, S_3 and S_4 out of a total of N shots. Find

(*i*) the probability distribution of X_1, X_2, X_3, X_4

(*ii*) the expected score after 5 shots, and

(*iii*) the probability that the score exceeds 15.

(*M.Sc. Poona ; B.Sc. Hons. Cal.*)

[**Hint.** Let Y denote the score of a hit. Then

$$(i)\quad P(X = X_1) = P(Y = 4) = \int_0^{1/\sqrt{3}} p(r)dr = \frac{1}{2},\ P(X = X_2) = P(Y = 3) = \frac{1}{3}$$

$$P(X = X_3) = P(Y = 2) = \frac{1}{6},$$

$$P(X = X_4) = P(Y = 0) = 1 - \left(\frac{1}{3} + \frac{1}{6} + \frac{1}{6}\right) = \frac{1}{3}$$

$$E(Y) = 0\,.\,\frac{1}{3} + 2\,.\,\frac{1}{6} + 3\,.\,\frac{1}{6} + 4\,.\,\frac{1}{3} = \frac{13}{6}$$

$$(ii)\quad E(Y_1 + Y_2 + Y_3 + Y_4 + Y_5) = \sum_{i=1}^{5} \frac{13}{6} = \frac{65}{6}$$

$$(iii)\quad \text{Reqd. probability} = P\left(\sum_{i=1}^{5} Y_i > 15\right)$$

$$= P(\Sigma Y_i = 16) + P(\Sigma Y_i = 17) + P(\Sigma Y_i = 18) + P(\Sigma Y_i = 19) + P(\Sigma Y_i = 20) = \text{etc.}$$

7. The probability density function of coded measurements of pitch diameter of a fitting is given by $\quad f(x) = \dfrac{1}{(1+x)^2},\ 0 \le x < \infty$

Find the distribution function of the distribution. Hence obtain (*i*) $P\ (x > 2)$ (*ii*) median and the quartiles of the distribution. Investigate whether the mean and the variance of the distribution exist. **(*M.A. Delhi*)**

8. Find the mode and the median for the distribution $dF = \sin x\ dx,\ 0 \le x \le \pi/2$.

[**Ans.** Modal value $= \pi/2$, Median $= \pi/3$]

9. Find the Arithmetic Mean and the Median for the distribution

$$dF = 2x\ dx,\ 0 \le x \le 1.$$

$$\left[\textbf{Ans. } \text{Arithmetic Mean} = \frac{2}{3},\ \text{Median} = \frac{1}{\sqrt{2}}\right]$$

10. Prove that the Geometric Mean, G of the distribution

$$dF = 6(2 - x)(x - 1)\ dx,\ 1 \le x \le 2$$

is given by $6 \log (16G) = 19$. **(*B.A. Hons. Delhi*)**

11. For the continuous distribution

$$dF = y_0 x(2 - x)\ dx,\ 0 \le x \le 2,\ y_0 \text{ being a constant}$$

find the mean, mean deviation about mean, μ_2', μ_3', μ_4', μ_2, μ_3, μ_4 and also skewness. Also show that for this distribution $\mu_{2n+1} = 0$. Deduce β_2.

(*B.Sc. Delhi, B.A. Punjab ; B.A. Hons. Delhi*)

$$\left[\textbf{Ans. } y_0 = \frac{3}{4},\ \text{Mean} = 1,\ \mu_2' = \frac{6}{5},\ \mu_2 = \frac{1}{5},\ \mu_3' = \frac{3}{5},\ \mu_3 = 0.\right.$$

$$\left.\text{skewness} = 0,\ \text{Mean deviation} = \frac{3}{8}\right]$$

12. (*a*) For the continuous distribution

$$dF = y_0(x - x^2)dx, \; 0 \le x \le 1, \; y_0 \text{ being constant.}$$

find the Arithmetic Mean, the Harmonic Mean, the Mode and the Median.

(B.Sc. Delhi, 1971, B.A. Hons. Delhi ; B.Sc. Hons. Sub Statistics Delhi)

Compute $P(\mu - 2\sigma < x < \mu + 2\sigma)$ **(B.Sc. Sardar Patel)**

(*b*) Find μ_2, μ_3 and μ_4 for the distribution given by

$$dF = \frac{dx}{2a}, \; -a \le x \le a.$$ **(B.Sc. Agra)**

[**Ans.** (*a*) $y_0 = 6, A.M. = 1, H.M. = \frac{1}{3}$, Modal value $= \frac{1}{3}$, Median $= \frac{1}{2}$

(*b*) $\mu_2 = \frac{1}{3}a^2, \mu_3 = 0, \mu_4 = \frac{1}{2}a^4$

(*c*) Prove : $Var\, X = \int_0^\infty 2x[1 - F(x) + F(-x)] - \mu_x^2$

where $F(x)$ is the d.f. of *a r.v. X*.

13. (*a*) Prove that for the distribution

$$dF = \frac{1}{B(m, n)} \frac{x^{m-1}}{(1 + x)^{m+n}} dx, \; 0 \le x < \infty, \; n > 2$$

variance is $\frac{m(m + n - 1)}{(n - 1)^2(n - 2)}$.

Find also the mode and moment of *r*th order about the origin. Show also that the harmonic mean is $(m - 1)/n$. **(B.A. Hons. Delhi)**

(*b*) Calculate the coefficient of variation for the rectangular distribution in (0, *b*), given that the probability law for the rectangular distribution is

$$P(x \le t) = \frac{t}{b}$$ **(B.A. Madras)**

[**Ans.** (*a*) Mode is at $x = \frac{m - 1}{n + 1}, m > 1, \mu_r' = \frac{m(m + 1) \ldots (m + r - 1)}{(n - 1)(n - 2) \ldots (n - r)}, r < n.$

(*b*) $100 \times \frac{1}{\sqrt{3}}$]

14. (*a*) Show that for the continuous distribution

$$dF = \frac{1}{B(m, n)} (1 - x)^{m-1} x^{n-1} dx, \; 0 \le x \le 1$$

Arithmetic mean $= \frac{m}{m + n}$, Harmonic Mean $= \frac{n - 1}{m + n - 1}$, $m > 1$. Verify that *A.M.* > *H.M.* **(M.A. Banaras)**

(*b*) Also show that Variance $= \frac{mn}{(m + n)^2(m + n + 1)}$ **(M.Sc. Agra)** (*c*) Prove that $\log G = \dfrac{1}{B(m, n)} \displaystyle\int_0^1 (1-x)^{m-1} x^{n-1} \log x \, dx$

$$= \frac{1}{B(m, n)} \frac{\partial}{\partial n} B(m, n) = \frac{\partial}{\partial n} [\log \Gamma(n) - \log \Gamma(m+n)]$$

15. Show that for the Gamma distribution

$$dF = \frac{1}{\Gamma(m)} e^{-x} x^{m-1} \, dx, \; 0 < x < \infty, \; m > 0 \; ; \; 0, \text{ else where}$$

mean is m, variance $= m$, harmonic mean $= m - 1$, $m.g.f. = (1-t)^{-m}$, $|t| < 1$.

μ_r' (about origin) $= m(m+1) \ldots (m+r-1)$.

Find the moment coeff. of skewness. **(*I.A.S. 1995*)**

16. (*a*) Express the constants y_0, a and p $(0 < p < 1)$ of the distribution

$$dF = y_0 \left(1 - \frac{x^2}{a^2}\right)^{-p} dx, \; -a < x < a$$

in terms of σ and β_2. **(*B.A. Hons. Delhi*)**

(*b*) Show that for the distribution

$$dF = y_0 \left(1 - \frac{x^2}{a^2}\right)^{-p} dx, \; 0 < p < 1, \; -a \le x \le a, \; y_0 \text{ being a constant}$$

$$\mu'_{r+1} = \frac{ra^2}{r+2-2p} \mu'_{r-1}$$

[Hint. Integrate by parts to get the recurrence relation.]

17. Find y_0, μ_1', μ_2, μ_3 and μ_4 for the distribution

$$dF = y_0 \left(1 + \frac{\alpha}{2}\right)^{\left(\frac{4}{\alpha^2} - 1\right)}, e^{-2x/\alpha} dx, \; \frac{-2}{\alpha} \le x < \infty$$

$$\left[\textbf{Ans. } y_0 = \beta^{2\beta^2 - 1} / e^{\beta^2} \Gamma\beta^2, \text{ where } \beta = \frac{2}{\alpha}, \; \mu_1' = 0, \; \mu_2 = 1, \; \mu_3 = \alpha, \; \mu_4 = \frac{3}{2}\alpha^2 + 3\right]$$

18. The probability law of a continuous random variable x is

$$f(x) = y_0 e^{-b(x-a)}, \; a \le x < \infty,$$

where a, b, y_0 are constants. Show that $y_0 = b = \sigma - 1$ and $a = m - \sigma$ where m and σ are the mean and standard deviation of the distribution. Show also that $\beta_1 = 3$, $\beta_2 = 9$. **(*B.Sc. Hons. Calcutta ; B.A. Hons. Delhi, 1984*)**

19. (*a*) If X is chosen at random in (a, b), what is $Var(X)$? **Ans.** $(b-a)^2/12$

(*b*) If X has negative exponential distribution, $\lambda e^{-\lambda x}$, $x > 0$, with mean 2, find $P(X < 1 | X < 2)$. **(*B.Sc. Hons. Delhi, 1997*)**

[Hint. $2 = \text{mean} = \dfrac{1}{\lambda} \therefore \lambda = 0.5$. Reqd. prob. $= \dfrac{P(X < 1 \cap X < 2)}{P(X < 2)} =$

$$\frac{P(X<1)}{P(X<2)} = \frac{1 - P(X \ge 1)}{1 - P(X \ge 2)} = \frac{1 - e^{-\lambda}}{1 - e^{-2\lambda}} = \frac{e^{\lambda}}{e^{\lambda} + 1}$$

20. A random variate X is known to have the distribution

$$dP = C\left(1 + \frac{x}{a}\right)^{m-1} e^{-\frac{mx}{a}} dx, \; -a \le x < \infty$$

Find the constant C and the first four moments about the mean. Determine the linear relation between β_1 and β_2 of this distribution. Also find the *M.G.F.* about the origin. **(*M.A. Statistics Delhi*)**

$$\left[\textbf{Ans. } C = m^m e^{-m}/a\,\Gamma(m),\ \mu_1' = C,\ \mu_2 = \frac{a^2}{m},\ \mu_3 = \frac{2a^3}{m^2},\ \mu_4 = \frac{3a^4(m+2)}{m^3}\right.$$

$$\left. 2\beta_2 - 3\beta_1 = 6.\ M.G.F. = e^{-at}\left(1 + \frac{at}{m}\right)^{-m}\right]$$

21. Pareto's distribution is defined by the density function

$$f(x) = \frac{\alpha}{\beta}\left(\frac{\beta}{x}\right)^{\alpha+1} \quad \text{for } x > \beta$$
$$= 0, \quad \text{for } x \le \beta.$$

Show that a variable with this distribution has a moment of order r if and only if $\alpha > r$. In particular, if $\alpha > 1$, show that

$$E(x) = \alpha\beta/(\alpha - 1),\ Var\,x = \alpha\beta^2/[(\alpha-1)^2(\alpha-2)].$$

22. Find the position of the mode and the median of the curve

$$f(x) = \frac{abx^{a-1}}{(1 + bx^a)^2},\ b > 0,\ a > 1,\ 0 \le x < \infty$$

$$\left[\textbf{Ans. } \text{Mode} = \left[\frac{a-1}{b(b+1)}\right]^{1/\alpha},\ \text{Median} = \left(\frac{1}{b}\right)^{1/a}\right]$$

23. Can the function $E[(1 - tx)^{-1}]$ be used to generate the moments of a variate X? **[Ans.** Yes]

24. For the distribution $dP = \frac{x}{a^2} e^{-x^2/2a}\, dx,\ 0 \le x < \infty$

find the distance between the quartiles and show that the ratio of this distance to the standard deviation is independent to the parameter a. Find the ratio.

$$\left[\textbf{Ans. } \text{Distance} = a\sqrt{2}[\sqrt{\log 4} - \sqrt{\log 4/3}],\ \text{Ratio} = \sqrt{\log 4}\ \frac{\sqrt{\log \frac{4}{3}}}{\sqrt{1 - \pi/2}}\right]$$

25. (*i*) **Laplace's double exponential distribution** is defined by the density function $f(x) = \frac{1}{2} e^{-|x|},\ -\infty < x < \infty$

Check that $\int_{-\infty}^{\infty} f(x)\, dx = 1$

and find (*a*) arithmetic mean (*b*) standard deviation (*c*) mean deviation about the mean (*d*) characteristic function for the distribution.

(*B.Sc. Hons., Calcutta*)

(*ii*) Let $f(x) = ke^{-\alpha x}(1 - e^{-\alpha x}),\ x \ge 0$. Find k such that $f(x)$ is a *d.f.* and also find corresponding *d.f.* Find $P(X > 1)$.

Ans. (*i*) $k = 2\alpha$ (*ii*) $1 - 2e^{-\alpha x} + e^{2\alpha x}$ (*iii*) $2e^{-\alpha} - e^{-2\alpha}$

(*B.A. Hons. Statistics Delhi, 1989*)

(*iii*) If X has an exponential distribution with mean 2, find $P[X < 1 \mid X < 2]$. **(*B.A. Stat. Hons. Delhi, 1997*)**

[**Ans.** (*a*) mean = 0, (*b*) *s.d.* = $\sqrt{2}$,

(*c*) m.d. about mean $= \frac{1}{2}\int_{-\infty}^{+\infty} |x| e^{-|x|} dx = \int_0^{\infty} |x| e^{-|x|} dx = \int_0^{\infty} xe^{-x} = 1,$

(*d*) $c.f. = \frac{1}{1+t^2}$

26. For the probability density function

$$f(x) = \frac{2(a+x)}{b(a+b)}, -b \le x \le 0$$
$$= \frac{2(a-x)}{(a+b)}, 0 \le x \le a$$

find the mean, variance and median. **(*B.Sc. Bombay*)**

[**Ans.** mean $= \frac{a-b}{3}$, variance $= \frac{1}{18}(a^2 + ab + b)$, median $= a - \sqrt{\frac{1}{2} a(a+b)}$]

27. A continuous cumulative distribution $F(x)$ is defined as follows :

(*a*)
$$F(x) = 0 \qquad x \le 1$$
$$= \frac{1}{16}(x-1)^4 \qquad 1 < x \le 3$$
$$= 1 \qquad x > 3$$

Find the probability density function $f(x)$. Find the mean of X and the median. **[*B.Sc. Bombay*]**

Ans. $f(x) = \frac{1}{4}(x-1)^3, 1 \le x \le 3,$ Mean = 2·65, Median $= 1 + \sqrt[4]{8}$.

$= 0$, otherwise.

(*b*) The *p.d.f.* of a random variable X is

$$f(x) = \begin{cases} 0 & \text{for } x \le -a \\ \dfrac{a+x}{a^2} & \text{for } -a < x \le 0 \\ \dfrac{a-x}{a^2} & \text{for } 0 < x \le a \\ 0 & \text{for } x \ge a \end{cases}$$

Compute cumulative distribution function $F(x)$.

(*I.S.I., B.Sc. Hons. Poona*)

[**Ans.** $F(x) = 0, x \le -a$; $= \frac{ax + \frac{1}{2}x^2 + \frac{1}{2}a^2}{a^2}, -a < x < 0$

$= \frac{\frac{1}{2} + ax - \frac{1}{2}x^2}{a}, 0 < x \le a$; 0 when $x \ge a$.]

(c) A random variable X has a continuous distributions F, where

$$F(x) = 0, \quad \text{if } x < 0$$
$$= kx, \quad \text{if } 0 \le x < 1$$
$$= 1, \quad \text{if } x > 1$$

Determine k and *p.d.f.*

28. Two independent variates X_1 and X_2 are each uniformly distributed within the range $-a$ to $+a$. Show that their sum X has a probability density given by

$$f(x) = \frac{2a + x}{4a^2}, -2a \le x \le 0\ ;\ = \frac{2a - x}{-4a^2}, 0 \le x \le 2a.$$

Find also the p.d.f. of the difference Z.

Verify that the *m.g.f.* calculated from this value of $f(x)$ is equal to

$$\left(\frac{1}{at} \sinh at\right)^2$$ **(M.A. Panjab)**

[**Hint.** The joint *p.d.f.* of X_1 and X_2 is given by $dP(x_1, x_2) = \left(\frac{1}{2a} dx\right)\left(\frac{1}{2a} dy\right)$, $-a < (x_1, x_2) < a$

Let us define new variables X and Z as follows :

$$X = X_1 + X_2, Z = X_1 - X_2$$

$$\Rightarrow x_1 = \frac{x+z}{2}, x_2 = \frac{x-z}{2}, |J| = dx\, dz \begin{vmatrix} \frac{1}{2} & \frac{1}{2} \\ \frac{1}{2} & -\frac{1}{2} \end{vmatrix} = \frac{1}{8a^2} dx\, dz$$

For fixed value of X, $Z = -(x + 2a)$ when $x_1 = -a$, Also when $x_1 = 0$, $x + z = 0$ $\Rightarrow Z = -X$.

$Z = x + 2a$ when $x_2 = -a$, Also, when $x_2 = 0$, $x - z = 0 \Rightarrow Z = X$

$$\therefore \text{ Distribution of } X = g_1(x)\, dx = \int_{Z=-(x+2a)}^{x+2a} \frac{1}{8a^2} dz = \frac{x + 2a}{4a^2}, -2a \le x \le 0$$

Distribution of Z : We break the region into OAB and OAC [See Fig. 7·12)

In region OAB, $h_1(z) = \int_z^{2-z} \frac{1}{2} dx = 1 - z, 0 < z < 1$

In region OAC, $h_2(z) = \int_{-z}^{2+z} \frac{1}{2} dx = 1 + z, -1 < z < 0$]

29. If X and Y are independent rectangular variates on [0, 1] find the distributions of (i) XY (ii) X/Y.

[**Hint.** $f(x, y) = 1, 0 < x, y < 1$

(i) Let $u = XY, v = X \ \therefore\ X = v, Y = u/v, |J| = \begin{vmatrix} 0 & 1 \\ \frac{1}{v} & -\frac{u}{v^2} \end{vmatrix} = \frac{1}{v}$

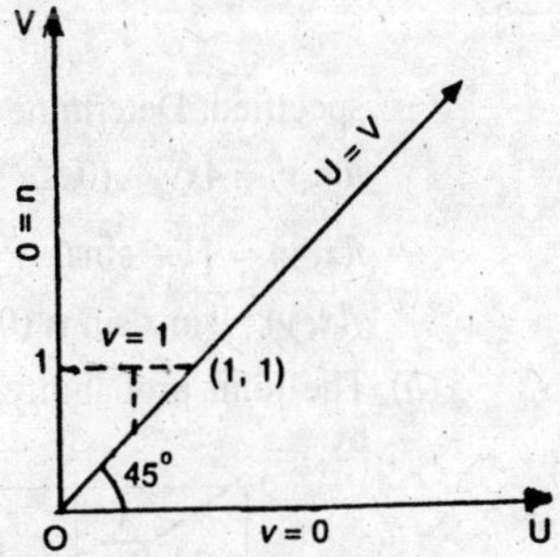

Fig. 7·12 (*a*)

$X = 0 \Rightarrow v = 0, X = 1 \Rightarrow v = 1,$

$Y = 0 \Rightarrow u = 0, Y = 1 \Rightarrow u = v$. Also $v = \dfrac{u}{y} \Rightarrow v \geq u$

$\therefore \quad g(u, 0) = 1 \cdot \dfrac{1}{v}; 0 < u < 1, 0 < v < 1$

$$h(u) = \int_u^1 \frac{1}{v}\, dv = -\log u, 0 < u < 1$$

$\therefore$ Distribution of u is $dP = -\log u\, du, 0 < u < 1$

(*ii*) $X = 1 \Rightarrow uv = 1$ which is a rectangular hyperbola

$Y = 0 \Rightarrow v = 0$ and $Y = 1 \Rightarrow v = 1$

$\therefore \quad g(u, v) = v, 0 < u < \infty, 0 < v < \infty$

In region I, $g_1(u) = \displaystyle\int_0^1 v\, dv = \frac{1}{2}, 0 \leq u \leq 1$

In region II, $g_1(u) = \displaystyle\int_0^{1/u} v\, dv = \frac{1}{2u^2}, 1 < u < \infty$

$$\therefore \quad g_1(u) = \frac{1}{2}, 0 \leq u \leq 1$$

$$= \frac{1}{2u^2}, 1 < u < \infty$$

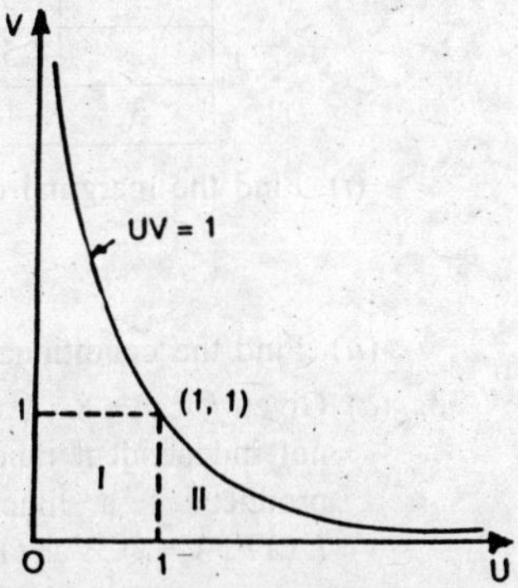

Fig. 7·12 (*b*)

30. If $f(x)$ is the elementary probability law of continuous random variable X in the interval $a \leq x \leq b$, so that

$$f(x) = 0 \text{ for } x < a \text{ and } x > b \text{ and } \int_a^b f(x)\, dx = 1.$$

show that the variable $P = \displaystyle\int_a^x f(x)\, dx$ has a rectangular distribution.

31. Explain the concepts of (*i*) conditional probability distribution, (*ii*) independence of random variables and (*iii*) marginal probability density functions. Given the joint density function

$$f(x, y) = 3x^2y + 3y^2x \qquad 0 \leq x \leq 1, \qquad 0 \leq y \leq 1$$
$$= 0, \qquad \text{otherwise}$$

show that the marginal density functions are $x + \frac{3}{2}x^2, y + \frac{3}{2}y^2$.

Find the conditional probability $P(\frac{1}{2} \leq x \leq \frac{1}{3} \mid \frac{1}{3} \leq y \leq \frac{2}{3})$. $\left[\textbf{Ans. } \dfrac{311}{1033}\right]$

32. (*a*) In each of the following problems a joint density (probability) function $f(x, y)$ is given. It is understood that $f(x, y) = 0$ outside the region specified. Determine whether the variables are dependent or independent.

$f(x, y) = 4xy \quad (0 \le x \le 1, \quad (0 \le y \le 1)$ **[Ans.** Independent]

$f(x, y) = [1 - \sin(x+y)]/4\pi^2, -\pi \le x \le \pi, -\pi \le y \le \pi$ **[Ans.** Dependent]

$f(x, y) = \sin x \sin y, (0 \le x \le \pi/2, 0 \le y \le \pi/2$ **[Ans.** Independent]

(*b*) The joint probability of two discrete random variables X and Y is given by

Y \ X	0	1	2	3
0	1/48	5/48	5/48	1/48
1	5/48	15/48	5/48	0
2	5/48	5/48	–	0
3	1/48	0	0	0

(*i*) Find the marginal densities of X and Y.

$$\left[\textbf{Ans. } (i)\ \frac{1}{4}, \frac{25}{28}, \frac{5}{24}, \frac{1}{48}, (ii)\ \frac{1}{48}\ (iii)\ 1{\cdot}5\right]$$

(*ii*) Find the conditional density of X given $Y = 3$ (*iii*) Find $E(Y \mid X = 0)$

33. (*a*) Given $f(x, y) = 8xy,\ 0 < x < y < 1\ ; = 0$ elsewhere ; show that X and Y are not independent random variables inspite of the fact that $f(x, y)$ is the product of a function of X and a function of Y alone. Find $E(XY \mid X = x)$, $\text{Var}(Y \mid X = x)$ ***(B.Sc. Hons. Math Delhi 1988, 1992)***

(*b*) If X, Y and Z are random variables with *p.d.f.*

$$f(x, y, z) = 1 - \sin 2\pi x \sin 2\pi y \sin 2\pi z,\ 0 \le x, y, z \le 1$$

show that any two of the variables are independent.

(*c*) Let X and Y have joint p.d.f. given by

$$g(x, y) = \begin{cases} \dfrac{e^{-(x+y)} x^3 y^4}{3\,!\,4\,!}, & x > 0, y > 0, = 0, \text{elsewhere} \end{cases}$$

Find the p.d.f. of $U = \dfrac{x}{x+y}$ and identify it.

Ans. (*a*) $E(Y \mid X = x) = \dfrac{2}{3}\left(\dfrac{1 + x + x^2}{1 + x}\right)$, $E(XY \mid X = x) = \dfrac{-2x}{3}\left(\dfrac{1 + x + x^2}{1 + x}\right)$

$$E(Y^2 \mid X = x) = \frac{1 + x^2}{2}.$$

34. (*a*) X and Y have joint probability mass function given by—

X \ Y	–1	1	2	5
–1	a	$3a$	$3a$	a
1	$3a$	$6a$	$3a$	0
5	$4a$	$3a$	0	0

Evalute : P (Y is even), P (XY is odd) and P [Y is even and $X^2 < 2$]

(Delhi Stat. Hons. 1997)

34. (*b*) A random observation on a bivariate population (x, y) can yield one of the following pairs of values with probabilities noted against them.

For each observation pair	Probability	Table probability
(1, 1) ; (2, 1) ; (3, 3) ; (4, 3)	$\frac{1}{20}$	$\frac{4}{20}$
(3, 1) ; (4, 1) ; (1, 2) ; (2, 2) (3, 2) ; (4, 2) ; (1, 3) ; (2, 3)	$\frac{1}{10}$	$\frac{8}{10}$
		= 1

Find the probability that $y = 2$ given $x = 4$ and also the probability that $y = 2$. Examine if the two events $x = 4$ and $y = 2$ are independent.

[**Hint.** $P(y=2) = \frac{4}{10}, P(x=4) = \frac{1}{4}, P(x=4, y=2) = \frac{1}{10}$

$P(y=2, x=4) = \frac{1}{10} \div \frac{1}{4} = \frac{2}{5}$ Hence $P(x=4)P(y=2) = P(x=4, y=2)$].

(*b*) If $F(x, y) = 1$ for $(x + 2y) \geq 1$ and $= 0$ if $x + 2y < 1$, then prove that $F(x, y)$ can not be a joint *p.d.f.*

35. (*a*) If $f(x, y) = x + y \quad 0 \leq x \leq 1, 0 \leq y \leq 1$

$= 0, \quad$ otherwise

find the density function for (*i*) $u = X + Y$ (*ii*) $v = XY$

(*b*) Let X and Y have the joint density function

$$f(x, y) = \frac{3}{2}x, \; 0 < x < 1, -x < y < x; \quad = 0, \text{ otherwise},$$

find the density function of $X - Y$. **(*M.Sc. Bombay*)**

(*c*) The joint *p.d.f.* of X and Y is given by :

$f(x, y) = 3(x + y) \,;\, 0 \leq x \leq 1, 0 \leq y \leq 1, 0 \leq x + y \leq 1$

find (*i*) marginal density of X (*ii*) $P(X + Y < \frac{1}{2})$

(*iii*) $E(Y \mid X = x)$ (*iv*) Cov (X, Y).

[**Ans.** (*a*) (*i*) $h(u) = \begin{cases} u^2 & \text{for } 0 \leq u \leq 1 \\ 2u - u^2 & \text{for } 1 \leq u \leq 2, \text{ 0 otherwise} \end{cases}$

(*ii*) $g(u) = 2 - 2v. \quad 0 < v \leq 1$, 0 otherwise]

(*c*) (*i*) $f(x) = \frac{3}{2}(1 - x^2), 0 \leq x \leq 1$ (*ii*) $\int_0^{0.5} \left[\int_0^{0.5-x} 3(x + y)\, dy \right] dx = \frac{1}{8}$

(*iii*) $\frac{(1 - x)(x + 2)}{3(1 + x)}$ (*iv*) $\frac{1}{10} - \frac{3}{8} \times \frac{3}{8} = \frac{13}{320}$.

36. The random variables X, Y have

$P(x = 0, y = 0) = \frac{2}{9}, \quad P(x = 0, y = 1) = \frac{1}{9}$

$P(x = 1, y = 0) = \frac{1}{9}, \quad P(x = 1, y = 1) = \frac{5}{9}$

Examine whether (*i*) X and Y are identical variables (*ii*) X and Y are independent random variables.

37. (*i*) If $f(x, y) = 1$ when $0 \le x \le 1, 0 \le y \le 1$ and zero otherwise, find the probability that

(*a*) $x < \frac{1}{2}, y < \frac{1}{2}$, (*b*) $x + y < 1$, (*c*) $x + y > 1$, (*d*) $x > 2y$, (*e*) $x > \frac{1}{2}$.

(*f*) $x^2 + y^2 < \frac{1}{4}$, (*g*) $x = y$ (*h*) $x > \frac{1}{2}$, given $y < \frac{1}{2}$, (*i*) $x > y$ given $y > \frac{1}{2}$.

$$\left[\textbf{Ans. } (a)\tfrac{1}{4},\ (b)\tfrac{1}{2},\ (c)\tfrac{1}{2}\ (d)\tfrac{1}{4},\ (e)\tfrac{2}{3},\ (f)\frac{\pi}{16},\ (g)\tfrac{1}{2},\ (h)\tfrac{1}{2}\ (i)\tfrac{1}{4}\right]$$

(*ii*) Given $f(x, y) = 1$ over the square $(0 \le x, y \le 1)$ show that

(*a*) $P(XY > u) = 1 - u + u \log u.$ (*b*) $P(x + y < 0{\cdot}5) = 0{\cdot}125$

(*c*) $P[(X - Y) < 0{\cdot}5] = 0{\cdot}393$ (*d*) $P[\cos \pi y < 0{\cdot}5] = \frac{2}{3}$

$$\left[\textbf{Hint. } (ii)\ P(xy > u) = \int_{x=u}^{1} \int_{y=\frac{u}{x}}^{1} 1\, dx\, dy = \text{etc.}\right.$$

(*iii*) The random variables X and Y are independent and both have the uniform distribution on [0, 1]. Let $Z = |X - Y|$. Prove that for real t,

$$\phi_Z(t) = Z(1 + it - e^{it})/2,\ i = \sqrt{-1}.$$

[**Hint.** (*iii*) $f(x, y) = f_1(x) \times f_2(y) = dx\, dy$

$$\therefore \quad \underset{Z = |X - Y|}{\phi(t)} = E[e^{it|x-y|}] = \int_0^1 \int_0^1 e^{it|x-y|}\, dx\, dy$$

$$= 2\int_0^1 \left(\int_0^x e^{it(x-y)}\, dy \right) dx = \text{etc.}$$

[See Fig. 7·13]

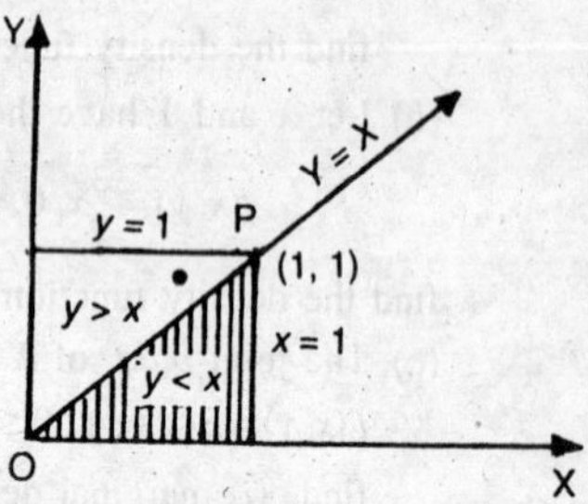

Fig. 7·13 P lies in the region Y > X.

38. (*a*) If $f(x, y) = 2\ ;\ 0 \le x \le 1, 0 \le y \le x$ find the marginal and conditional density functions.

(*b*) If $f(x, y) = xe^{-x(y+1)}$ $(x \ge 0, y \ge 0)$ find the marginal and conditional density functions.

Show also that $E(Y)$ may not exist though $E(XY)$ and $E(Y/X)$ may both exist ? **(*B.A. Stat. Hons. Spl. Course Delhi 1985*)**

$$\left[\textbf{Ans. } (a)\ g(x) = 2x,\ h(y) = 2,\ (b)\ f(x \mid y) = \frac{1}{y}, f(y \mid x) = \frac{1}{x}\ (b)\ g(x) = e^x,\right.$$

$$h(y) = \frac{1}{(y+1)^2}, f_1(x \mid y) = (1 + y)^2 xe^{-x(1+y)}, f_2(y \mid x) = x\, e^{-xy}, x, y < 0$$

$$E(Y) = \int_0^\infty y f(y) = \infty \Rightarrow E(Y) \text{ does not exist.}$$

39. Let X have the density function $f(x) = e^{-x}, x > 0$ and zero for $x > 0$. Find the density function for (*i*) $u = \sqrt{x}$, (*ii*) $u = x^{-1}$.

$$\left[\textbf{Ans. } (i)\ g(u) = 2ue^{-u^2}\ ;\ (ii)\ g(u) = u^{-2}e^{-\frac{1}{u}}\right]$$

40. If X is normally distributed with zero mean and unit variance find the density function for $u = e^x$.

$$\left[\textbf{Ans. } g(u) = \frac{1}{u\sqrt{2\pi}}\, e^{-\frac{1}{2}(\log u)^2},\ u \geq 1\right]$$

$$= 0, \qquad \text{otherwise.}$$

41. (*a*) If $f(x) = 1, 0 \leq x \leq 1$, find the density function of $u = 3x + 1$.

[**Hint.** First find the cumulative distribution of u and then differentiate it.]

(*b*) Show that $F_x(x) + F_y(y) - 1 \leq F_{x,\,y}(x, y) \leq \sqrt{F_x(x)F_y(y)}$ for all x, y

(*c*) If $f(x) = \frac{1}{2}e^{-x/2}, 0 \leq x < \infty$ and zero for $x < 0$, find the marginal density function of $u = \frac{1}{2}(x_1 - x_2)$. [**Ans.** $g(u) = \frac{1}{2}e^{-|u|}, -\infty < u < \infty$]

42. (*a*) If $f(x, y) = e^{-(x+y)} : x > 0, y > 0$ find

(*i*) the probability of the events (*a*) $x < 1$, (*b*) $x > y$ (*c*) $x + y < 1$ (*d*) $x < 2y$

(*ii*) the density function of $u = \frac{1}{2}(x + y)$

and show that $\mu'_{10} = 1 = \mu_{01}', \mu'_{20} = 2 = \mu'_{02}$.

$$\left[\textbf{Ans. } (i)\ 1 - \frac{1}{e}\cdot\frac{1}{2}, 1 - \frac{1}{e}, \frac{2}{3}\ (ii)\ g(u) = 4ue^{-2u}\right]$$

(*b*) Let the random variables X and Y have their density function

$$f(x, y) = \frac{1}{2\pi}\exp\left[-\frac{1}{2}(x^2 - 2xy + 2y^2)\right]$$

show that $\int_{-\infty}^{+\infty}\int_{-\infty}^{+\infty} f(x, y)\,dxdy = 1, \mu_{01} = 0, \mu_{10} = 0, \mu_{11} = 1, \mu_{20} = 2,$

$$\mu_{02} = 1, \rho = 1/\sqrt{2}, g(x) = \int_{-\infty}^{+\infty} f(x, y)\,dy = \frac{1}{2\sqrt{\pi}}\, e^{-\frac{1}{4}x^2}; h(y) = \frac{1}{\sqrt{2\pi}}\, e^{-\frac{1}{2}y^2}$$

$$f(y \mid x) = \frac{1}{\sqrt{\pi}}\, e^{-\left(y - \frac{x}{2}\right)^2}, f(x \mid y) = \frac{1}{\sqrt{2\pi}}\, e^{-\frac{1}{2}(x-y)^2}$$

43. Let X and Y be non-negative continuous chance variables having the joint density

$f(x, y) = 4xy \exp[-(x^2 + y^2)]\,; 0 < x < \infty, 0 < y < \infty.$

find the density function of $U = \sqrt{X^2 + Y^2}$ and $f(x \mid y)$ (*M.Sc. Meerut 1986*)

Ans. $g(u) = 2u^3 \exp[-u^2]\,; 0 < u < \infty, f(x \mid y) = \int_0^{\infty} f(x, y)dy = 2x\,e^{-x^2}, x \geq 0.$

44. Let $f(x, y) = \frac{2}{a^3}, 0 \le x \le y, 0 \le y \le a$, show that

(*i*) $g(x) = \frac{2}{a^2}(a - x)$, (*ii*) $h(y) = \frac{2y}{a^2}$, (*iii*) $f(x \mid y) = \frac{1}{y}$

(*iv*) $\mu_{10}' = \frac{a}{3}$, (*v*) $\mu_{01}' = \frac{2a}{3}$, (*vi*) $r_{xy} = \frac{1}{2}$.

45. (*a*) A point P is taken at random in a line AB of length $2a$, also positions of the point being equally likely. Show that the expected value of the area of the rectangle AP. PB is $2a^2/3$, and that the probability of the area exceeding $\frac{1}{2}a^2$ is $1/\sqrt{2}$. **(*B.Sc. Agra*)**

(*b*) Two points are taken at random on a given straight line of length a. Prove that the probability of their distance exceeding a given length $c (< a)$ is equal to $\left(1 - \frac{c}{a}\right)^2$

46. From a point on the circumference of a circle of a radius a, a chord is drawn in a random direction (*i.e.*, all directions are equally likely. Show that the expected value of the length of the chord is $4a/\pi$ and the variance of the length is $2a^2\left(1 - \frac{8}{\pi^2}\right)$. Also show that the chance is $\frac{1}{3}$ that the length of the chord will exceed the length of the side of an equilateral triangle inscribed in the circle. **(*B.Sc. Agra*)**

47. Two points X_1 and X_2 are taken randomly on a line of finite length l. If X_1 and X_2 are at distances x_1 and x_2 respectively from one given end, find $E[\mid x_1 - x_2 \mid^n]$.

$$\left[\textbf{Ans. } \frac{2l^n}{(n+1)(n+2)}\right]$$

48. (*a*) What is the expected value of the square of the distance between two points P and Q selected at random in a square whose side is of length l ?

(*b*) What is the expected distance between two points selected at random in a circle of radius a ? **[Ans.** (*a*) $l^2/3$ (*b*) $128a/45\pi$]

49. A point is taken at random within the ellipse

$$\frac{x^2}{a^2} + \frac{y^2}{b^2} = 1, (a < b)$$

What is the chance that its distance from the centre exceeds a ?

[Ans. $(b - a)/b$]

50. (*a*) Find the mean value of the ordinates of semi-circle of radius r supposing the ordinates drawn through equidistant points on the circumference.

(*b*) From a point A on the circumference of a circle of radius r two st. lines are drawn to two other points B and C taken at random on the circumference. Find the mean difference in length between the lines AB and AC.

$\left[\textbf{Ans. } (a) \frac{2r}{\pi}\right]$ **(*M.A. Delhi*)**

51. If the roots of the quadratic equation $x^2 - ax + b = 0$ are real and b is positive but otherwise unknown, what are mean values of the roots, assuming that it follows a uniform distribution ? What are the variances of the roots ?

$$\left[\textbf{Ans. } \frac{5a}{6}, \frac{a}{6}, \frac{a^2}{12}, \frac{11}{72}a^2\right]$$

52. Three points are taken at random on the circumference of a circle. Find the probability that they lie in the semi-circle. $\left[\textbf{Ans. } \frac{3}{4}\right]$

[**Hint.** Let P, Q, R, be the random points chosen on the circle with centre O and A a fixed point on it. Let angles made by OP, OQ, OR with OA be X, Y, Z, respectively. Now X, Y, Z are independent each uniformly distributed over $(0, 2\pi)$, $f(x, y, z) = 1/8\pi^3, 0 < x, y, z < 2\pi$; 0, elsewhere.

Consider the case $X < Y < Z$. The event in question may be expressed by the following three inequalities :

$$Z - X < \pi, Z - Y > \pi, Y - X > \pi$$

or (X, Y, Z) lies respectively in the regions

$$0 < x < 2\pi, x < y < x + \pi, y < z < x + \pi ;$$
$$0 < x < \pi, x < y < \pi, y + \pi < z < 2\pi$$
$$0 < x < \pi, x + \pi < y < 2\pi, y < z < 2\pi$$

$\therefore$ Reqd. prob.

$$= \left[\int_0^{2\pi}\int_x^{x+\pi}\int_y^{x+\pi} dxdydz + \int_0^{\pi}\int_x^{\pi}\int_{y+\pi}^{2\pi} dxdydz + \int_0^{\pi}\int_{x+\pi}^{2\pi}\int_y^{2\pi} dxdydz\right] /3\,!$$

53. (*a*) A point is picked at random in a square. What is the probability that it lies inside the inscribed circle which is concentric with the square ?

$\left[\textbf{Ans. } \frac{\pi}{4}\right]$ **(*B.A. Hons. Delhi, 1981*)**

(*b*) A point P is taken at random on the circumference of a circle of radius r. Find the expected value of the distance AP, between P and a given point A on the circumference. **(*M.A. Punjab*)**

$\left[\textbf{Ans. } \frac{4r}{\pi}\right]$

54. (*a*) X and Y are independent variates, each uniformly distributed over the interval (0, 1). Find the probability that the greater of X, Y is less than a fixed number k $(0 < k < 1)$.

[**Hint.** The two dimensional *r.v.* (X, Y) is uniformly distributed over the unit square $R : 0 < x < 1, 0 < y < 1$. The required event is max $(X, Y) < k$. If $Y > X$, then max $(X, Y) = Y$ and the required event means $Y < k$. In other words, (X, Y) lies in the triangular region $R_1 : x > 0, y < k, y > x$.

Similarly, in case $X > Y$, the required event means that (X, Y) lies in the triangular region $R_2 : x < k, y > 0, y < x$.

Now R_1 and R_2 together form the square $R', 0 < x < k, 0 < y < k$. Hence $R = 1$, $R' = k^2$ and the required probability $= R'/R = k^2$.]

(*b*) Two numbers are independently chosen at random between 0 and 1. Find the probability

(*i*) That their product is less than a constant k $(0 < k < 1)$.

(*ii*) These products is less than a fixed constant k

[**Hint.** (*i*) (X, Y) is uniformly distributed over the unit square R, $0 < x < y < 1$. The event in question is $XY < k$ i.e. (X, Y) lies in R'.

$$\therefore \quad R' = k + k \int_k^1 \frac{dx}{x} = k(1 - \log k). \quad \therefore \text{ Reqd. prob.} = R'/R = \text{etc.}$$

(*ii*) Reqd. event is $|X - Y| < k$ or $Y > X - k$ if $X > Y$ and $Y > X + k$ if $Y > X$ i.e. (X, Y) lies in the part R' of R lying between $y = x \pm k$ so that $R' = 1 - (1-k)^2 = k(2-k) \quad \therefore \quad P = R'/R = k(2-k)$.

55. Bertrand Paradox. A chord is drawn at random in a given circle. What is the probability that it is greater than the side of the equilateral triangle inscribed in that circle ? Find the mean and variance of the length of the chord in both cases.

Sol. 1. Assuming that the distance of the chord from the centre has uniform distribution, the required probability $= \frac{1}{2}r/r = \frac{1}{2}$, r being the radius of the circle.

Sol. 2. Assuming that the distribution of the angle between the chord and the tangent at one end of the chord is uniform, the required probability is

$$\left(\frac{2\pi}{3} - \frac{\pi}{3}\right)\Big/\pi = \frac{1}{3}.$$

56. A straight line AB is divided by a point C into two parts $AC = a$, $CB = b$. Points X and Y are taken at random on AC and CB respectively. What is the probability that AX, XY, BY can form a triangle. (***I.S.I. Calcutta***)

$$\left[\textbf{Ans. } \frac{b}{3a}\right]$$

[**Hint.** In order that AX, XY, BY can form a triangle, the following inequalities must be fulfilled :

$x < (a + b - x - y) + y$ or $x < a + b - x$

$y < (a + b - x - y) + x$ or $y < a + b - y$

$a + b - x - y < x + y$ or $\dfrac{a+b}{2} < x + y$

These inequalities are equivalent to

$$x < \frac{a+b}{2}, y < \frac{a+b}{2}, x + y > \frac{a+b}{2}]$$

57. (*a*) A board is ruled with equidistance lines, the width of the strip between two consecutive parallel lines being d. A needle so fine that can be likened to a rectilinear segment of length $l < d$ is thrown on the board.

(*i*) What is the probability that needle will intersect one of the lines ?

(***M.A. Bombay ; M.A. Nagpur***)

(*ii*) What happens when $l < d$?

$$\left[\textbf{Ans. } 2l/d\pi,\ (ii)\ \frac{l}{ad}(1-\cos d) \text{ where } \sin d = \frac{a}{l}\right]$$

(*b*) The radius X of a circle is rectangularly distributed between 1 and 2. Show that the area y of the circle is distributed between π and 4π with probability distribution $f(y)\,dy = \frac{1}{2}\pi^{-\frac{1}{2}}y^{-\frac{1}{2}}\,dy$.

Show that the mean and variance of the area of the circle is $\frac{7\pi}{8}$ and $\frac{34\pi^2}{45}$ respectively.

(*c*) The two equal sides of an isosceles triangle are each of unit length and the angle θ between them is rectangularly distributed between 0 and $\pi/6$. Show that the area of the triangle is described between 0 and 1/4 with probability distribution

$$f(y)dy = \frac{12}{\pi}(1-4y^2)^{-\frac{1}{2}}\,dy.$$

Show that the mean of the area of the triangle is $\frac{3}{2\pi}(2-\sqrt{3})$

58. (*a*) If $f(x)$ and its derivatives vanish at the end points of the range prove by integration by parts that for the distribution $dF = f(x)dx$.

$$\int_{-\infty}^{+\infty} e^{itx} f^r(x)dx = (-1)^r (it)^r\, \psi(t)$$

where $\psi(t)$ is the *C.F.* and $f^r(x) = \frac{d^r}{dx^r} f(x)$.

(*b*) In particular, show that the distribution $dF = \exp\left\{\frac{a}{q\,!}\left(-\frac{d}{dx}\right)^q\right\} f(x)dx$, the qth cumulant exceeds the qth cumulant for the distribution $dF = f(x)dx$ by a and that all other cumulants of the two distributions are equal.

(*M.A . Delhi*)

59. For (*i*) $f(x) = 1, 0 \le x \le 1$

(*ii*) $f(x) = \frac{1}{\sigma\sqrt{2\pi}} e^{-\frac{1}{2}\left(\frac{x-m}{\sigma}\right)^2} \quad -\infty < x < \infty$

determine $P[|x - E(x)| \ge 1{\cdot}5\sigma]$. Compare this value with the value given by Chebychev's inequality.

[**Ans.** (*i*) 0·134 (*ii*) 0·134, Chebychev's inequality gives 0·144]

60. (*a*) If X be a random variable such that $E(X) = 3$ and $E(X^2) = 13$, then using Tchebycheff's lemma, show that

$$P(-2 < X < 8) \ge \frac{21}{25}\cdot$$

(*B.Sc. Hons. Delhi 1990 ; B.Sc. Hons. Gauhati, 1971*)

(*b*) For the discrete variate with density

$$f(x) = \frac{1}{8} I_{(-1)}(x) + \frac{6}{8} I_{(0)}(x) + \frac{1}{8} I_{(1)}(x)$$

evaluate $P[| X - \mu | \geq 2\sigma]$.

[**Hint.** X is a discrete r.v. with values $-1, 0, 1$ with respective probabilities 1/8, 6/8, 1/8. Show that $\mu = 0, \sigma = 1/2$ and apply Chebyshev's inequality.]

$$\left[\textbf{Ans.} \leq \frac{1}{4}\right]$$

Note. For the above r.v. with given μ and σ, Chebyshev's inequality cannot be improved

(*c*) A random variable X takes the values $-1, 1, 3, 5$, with associated probabilities 1/6, 1/6 1/6 and 1/2. Find by direct computation $P[| X - 3 | \geq 1]$. Find an upper bound for the probability by applying Tchebycheff's inequality. ***(B.Sc. Mysore)***

(*d*) Find k if $P[\,| X - E(X)] | < k\sigma_x) \leq \frac{1}{4}$. Let X be a random variable such that $P(X \leq 0) = 0$ and let $\mu = Ex$ exist show that

$$P(X \geq 2\mu) < \frac{1}{2}.$$ ***(I.S.I. 1971)***

(*e*) Two good dice whose faces are numbered 1 to 6 are thrown. If X is the sum of the numbers showing up, prove that $P\{\,| X - 7 | \geq 3\} < \frac{35}{54}$

Compare with the exact probability. ***(B.Sc. Calicut, 1972)***

(*f*) A sample of size n is drawn from a population whose mean is 5 and s.d. 1. Prove that $P\{\,| X - 5 | \leq 0{\cdot}09\} \geq 1 - \frac{10^6}{n}\cdot$ ***(B.Sc. Shivaji, 1971)***

(*g*) Compare the upper bound on the probability $P\{\,| X - E(X) | \geq 2\sqrt{\text{var } X}\}$ obtained from Chebychev's inequality with exact probability if X is uniformly distributed over $(-1, 3)$. ***(B.Sc. Poona, 1971)***

(*h*) Suppose X has a rectangular distribution on $(-1, 1)$. Compute

$$P\left(\left|\frac{X - E(X)}{\sigma_x}\right| \geq 2\right)$$

and compare it with the upper bound given by Tchebycheff's inequality.

(*i*) If X is a $\gamma(n)$ variate, then show that $P(0 < X < 2n) \leq \frac{n-1}{n}$

[**Hint.** $f(x) = \frac{e^{-x} x^{n-1}}{\Gamma n}$, $EX = n$, $\text{Var } X = n$, $\in = n$]

61. (*a*) **Chebychev's inequality.** Show that for any distribution

$$\int_{-\infty}^{+\infty} \left(1 - \frac{x^2}{t^2}\right) dF(x) \leq \int_{-t}^{+t} dF(x)$$

and hence or otherwise deduce that if variance σ^2 of X exists,

$$P(|x - E(x)| \geq k\sigma) \geq \frac{1}{k^2}, k > 0. \qquad \textbf{(I.C.A.R.)}$$

62. If X is a stochastic variable with zero mean and *s.d.* σ, show that

$$P(t) \leq \frac{\sigma^2}{\sigma^2 + t^2} \text{ for } t < 0$$

and
$$1 - P(t) \leq \frac{\sigma^2}{\sigma^2 + t^2} \text{ for } t < 0$$

where $P(t)$ denotes the probability of the inequality $x \leq t$.

(M.S. Statistics Delhi)

[**Hint.** Let $t < 0$ then $\quad -t = \int_{-\infty}^{\infty} (x - t) dF(x) < \int_{x}^{\infty} (x - t) dF(x)$

Apply Schwartz inequality *viz.* $[\int g(x)h(x)dF(x)]^2 \leq \int g^2 \, dF(x) \times \int h^2 \, dF(x)$

so that $t^2 \leq \left[\int_x^{\infty} (x - t) \, dF \, x)\right]^2 \leq \int_x^{\infty} dF(x) \times \int_x^{\infty} (x - t)^2 \, dF(x) = [(1 - P(t)](\sigma^2 + t^2)$

63. If $\gamma_2 = \beta_2 - 3$ then show that it is possible to improve Chebychev inequality in the form, $P\{|x - m| \geq k\sigma\} \leq \dfrac{\gamma_2 + 2}{(k^2 - 1)^2 + \gamma_2 + 2}$,

k is a positive real number where X is r.v. with mean k and s.d. σ.

(M.Sc. Gujarat, M.Sc. Poona)

[**Hint.** From Tchebycheff's inequality $P[|x - m| \geq k\sigma] \leq \dfrac{1}{k^2}$ if $k > 0$

L.H.S. $= P[(X - m)^2 - k^2\sigma^2 \geq 0] = P \dfrac{\{(Y - m)^2 - k^2\sigma^2\}\{\sigma^2(k^2 - 1)\}}{\mu_4 + k^4\sigma^4 - 2k^2\sigma^2} \geq 0$ etc.]

Use $P\{g(x) \leq k\} \leq \dfrac{Eg(x)}{k}$ where $k = 1$.

64. A variate X has uniform distribution over the unit interval. What function of x has the distribution

$$dP = e^{-x} \, dx, 0 \leq x < \infty. \qquad \left[\textbf{Ans. } \log_e\left(\frac{1}{1 - x}\right)\right]$$

65. On the x-axis, $n + 1$ points are taken independently between the origin and $x = 1$, all positions being equally likely. Find the probability that the $(k + 1)^{\text{th}}$ of these points counted from the origin lies in the interval $x - \frac{1}{2}dx$ to $x + \frac{1}{2}dx$. **(B.Sc. Cal. ; M.A. Statistics, Delhi)**

[**Hint.** X is a random variable uniformly distributed in (0 , 1) *i.e.*]

$$f(x) = 1, 0 < x < 1,$$

$$P(0 < X < x) = \int_0^x f(x)dx = \int_0^x 1 \cdot dx = x$$

$$P(X > x) = 1 - x.$$

66. Find the probability function when the characteristic function is given by

(*i*) $\frac{1}{3}\text{Cos } 2t + \frac{2}{3}\text{Cos } t$ (*ii*) $e^{-2it} \cdot e^{-t^2/9}$ (*iii*) $e^{-|t|}$

(*iv*) $\frac{\sin 2t}{2t}$ (*v*) $(1 - t + te^{2it})^{18}$

(*vi*) $\sin^2 t$ (*vii*) $(\cosh^2 t)^{-1}$ **(*B.Sc. Delhi, 1979*)**

67. Show that the distribution

$$dF = \frac{1}{\pi}\left(\frac{1 - \cos x}{x^2}\right)dx, -\infty < x < \infty$$

has the characteristic function

$$\phi(t) = 1 - |t| \quad , \quad |t| \le 1$$
$$= 0 \quad , \quad |t| \le 1$$

[**Hint.** We know that $\int_x^\infty \frac{\sin\alpha x}{x}dx = \frac{\pi}{2}, \alpha < 0$

$$\int_0^\infty \frac{\cos \beta x - \cos \alpha x}{x^2}dx = \frac{\pi}{2}(\alpha - \beta), \alpha < \beta$$

By the double use of identities such as

$\sin\theta \cos m\theta = \frac{1}{2}[\sin(m+1)\theta - \sin(m-1)\theta]$ we get on integrating from

$$f(x) = \frac{2}{\pi}\int_0^\infty \frac{1 - \cos x}{x^2}dx = 1, \alpha, \beta = 0$$

$$\phi(t) = \frac{2}{\pi}\int_0^\infty \frac{1 - \cos x}{x^2} \cdot e^{itx}dx = \frac{2}{\pi}\int_0^\infty \frac{1 - \cos x}{x^2}\cos tx\, dx$$

$$= 1 - |t|, \quad |t| \le 1 ; = 0. \quad |t| > 1.]$$

68. Find the density function $f(x)$ corresponding to the characteristic function defined as follows : $\phi(t) = \begin{cases} 1 + |t|, & |t| < 1 \\ 0 & |t| > 1. \end{cases}$ **(*B.Sc. Hons. Delhi, 1989*)**

[**Hint.** $f(x) = \frac{1}{2\pi}\int_{-\infty}^{+\infty} e^{-itx}\phi(t)dt = \frac{1}{2\pi}\int_{-1}^{0}(1+t)e^{-itx}at + \frac{1}{2\pi}\int_0^1(1-t)e^{-itx}dt$

$$= \frac{1}{2\pi}\int_0^1 (1-t)(e^{itx} + e^{\ itx})] = \frac{1}{\pi}\int_0^1 (1-t)\cos tx\, dt$$

$$= \frac{1}{\pi} \cdot \frac{1 - \cos x}{x^2}, -\infty < x < \infty.]$$

69. Show that the C.F. of the distribution of a discrete random variable X with probability $P(X = 0) = \frac{1}{2}$,

$$P[X = (2k - 1)\pi] = \frac{2}{2(k-1)^2\pi^2}, k = 0, \pm 1, \pm 2 \ldots$$

is $$\phi(t) = \frac{1}{2} + \frac{4}{\pi^2} \sum_{k=1}^{\infty} \frac{\cos (2k-1) t\pi}{(2k-1)^2}.$$

[**Hint** $\phi(t) = \frac{1}{2} + \sum_{k=-\infty}^{\infty} \frac{2}{(2k-1)^2\pi^2} e^{it(2k-1)} \pi = \frac{1}{2} - \frac{4}{\pi^2} \sum_{k=1}^{\infty} \frac{\cos (2k-1)t\pi}{(2k-1)^2}$]

70. Show that in *Q*. 66 $\phi_1(t) = \phi_2(t)$ holds for $|t| \le 1$ and not for $|t| > 1$.

[**Hint.** Developing the function $\phi(t) = |t|$ in the interval $|t| \le 1$ into a Fourier series, we get $\phi(t) = |t| = \frac{1}{2} - \frac{4}{\pi^2} \sum_{k=1}^{\infty} \frac{\cos (2k-1) t\pi}{(2k-1)^2}$.]

71. If $$f_1(t) = \frac{2}{\pi} \int_0^{\infty} \frac{1 - \cos x}{x^2} \cos tx \, dx,$$

$$f_2(t) = \frac{1}{2} + \frac{4}{\pi^2} \sum_0^{\infty} \frac{\cos (2x+1) \pi t}{(2x+1)^2}$$

indicate the form of the distribution of which $f_1(t)$ and $f_2(t)$ are C.F.

(M.A . Bombay)

[**Ans.** Frequency functions are $\frac{1}{\pi} \cdot \frac{1 - \cos x}{x^2}, -\infty < x < \infty$ and $\frac{2}{\pi} \cdot \frac{1}{(2x+1)^2}$ at values $(2x + 1)\pi$ except at $x = 0$ where the value is $\frac{1}{2}$]

72. Show that for the curve

$$y = y_0 \left(1 - \frac{x^2}{a^2}\right)^m, -a \le x \le a, m \ge 1, y_0 \text{ being constant,}$$

the points of inflexion are at distances $\pm \sigma \left\{\frac{\beta_2}{3(\beta_2 - 2)}\right\}$ from the mean.

73. If $\phi(t) = (q + pe^{it})^n$, find the frequency function.

[**Sol.** We have $J_c(x) = \frac{1}{2c} \int_{-c}^{+c} e^{-itx} (q + pe^{it})^n \, dt = \frac{1}{2c} \int_{-c}^{+c} \left\{ e^{-itx} \sum_{j=1}^{n} {}^nc_j \, p^j q^{n-j} e^{itj} \right\} dt$

For values of x other than 1, 2, ... , n, $\underset{c \to \infty}{\text{Lt}} J_c \to 0$.

For $x = j, J_c \to {}^nc_j p^j q^{n-j}$ $\therefore$ Frequency for $x = j$ is ${}^nc_j p^j q^{n-j}$].

74. For what distribution, if any, are the cumulants given by $k_r = m$?

[**Hint.** We have $\psi(t) = \sum_1^{\infty} k_j \cdot \dfrac{(it)^j}{j!} = \sum_1^{\infty} \dfrac{m(it)^j}{j!} = m(e^{it} - 1) \quad \therefore \quad \phi(t) = e^{m(e^{it} - 1)}$

$$\therefore \quad J_c = \frac{1}{2c} \int_{-c}^{+c} e^{-itx} \cdot \exp.\{m(e^{it} - 1)\}\, dt$$

$$= \frac{e^{-m}}{2c} \int_{-c}^{+c} e^{-itx} \left\{ 1 + me^{it} + \frac{m^2}{2!} e^{2it} + \ldots + \frac{m^j}{j!} e^{jit} + \ldots \right\} dt = \text{etc.}\Big]$$

75. A random variable X is distributed in the Poisson's form with parameter λ. If λ is a random variate with a characteristic function (*c.g.*) $\phi(t)$, show that the *c.f.* of X is $\phi\left(\dfrac{e^{it} - 1}{i}\right)$.

(***M.A. Statistics Delhi ; M.A. Poona***)

76. Suppose that $X_1, X_2, \ldots, X_k$ are statistically independent random variables having *p.d.f.*

$$f(x_1, x_2, \ldots, x_k) = \begin{cases} \sum_1^k x_i\,, & x_i \geq 0,\ i = 1, 2, \ldots, k \\ 0, & \text{otherwise} \end{cases}$$

and suppose $L = \sum_{i=1}^{k} x_i$. Find the moments and the *p.d.f.* of L using its characteristic function. (***M.A. Delhi***)

77. (*a*) Examine if the weak law of large numbers is satisfied by the following sequence of independent random variables X_n, $n = 1, 2, \ldots,$

(*i*) $X_n = \begin{cases} \dfrac{1}{n4} \text{ with probability } \dfrac{1}{2} \\ -\dfrac{1}{n4} \text{ with probability } \dfrac{1}{2} \end{cases}$ (***M.Sc. Mysore***)

(*ii*) $X_n = \pm n^{-\frac{1}{4}}$ with probability $\dfrac{1}{2}$ (***B.Sc. Hons. Poona***)

(*iii*) $X_n = \pm\sqrt{2n - 1}$ with probability $\dfrac{1}{2}$.

(*iv*) $X_n = \pm 2^k$ with probability $2^{-(2k+1)}$, $X_k = 0$ with probability $1 - 2^{-2k}$

(***B.Sc. Hons. Math Delhi, 1993, 96***)

(*v*) $X_k = \pm 2^k$ with probability $\dfrac{1}{2}$. (***B.Sc. Hons. Math Delhi, 1997***)

[**Hint.** (*a*) (*i*) $E(X_n) = 0$, $E(X_n^2) = 1$, Var $(X_n) = 1$

(*iv*) Show that $P(X_k = 2^k) + P(X_k = 0) + P(X_k = 2^k) = 1 - \dfrac{1}{2^{2k}} + \dfrac{1}{2^{2k}} = 1$

Also $E(X_k) = 0$, $\text{Var } X_k = 2 \times \dfrac{2^{2k}}{2^{2k+1}} = 1$ is $\Sigma \dfrac{Var\ S_n}{n^n} = 1/n \to 0$ as $n \to \infty$

(v) $EX_k = 0$, $\text{Var } S_n = \sum_1^n \text{var } X_k = \sum_1^n 4^k = \frac{4}{3}(4^n - 1)$, $\lim\limits_{n \to \infty} \dfrac{\text{Var } S_n}{n^2} \to \infty$]

(b) Examine where the weak law of large numbers holds good for the sequence $\{X_n\}$ of independent random variables, X_n having the distribution.

(i) $P\left\{X_n = \dfrac{1}{\sqrt{n}}\right\} = \dfrac{2}{3}$, $P\left\{X_n = -\dfrac{1}{\sqrt{n}}\right\} = \dfrac{1}{3}$. **[Ans. Yes]**

(M.A. Math. Delhi, Old Course, 1974)

(ii) $P\{X_n = +1\} = \dfrac{1 - 2^{-n}}{2} = P\{X_n = -1\}$. ***(B.Sc. Hons. Delhi, 1989)***

(iii) $P\left\{X_n = \dfrac{1}{n}\right\} = p_n$, $P\left\{X_n = 1 + \dfrac{1}{n}\right\} = 1 - p_n$. ***(M.Sc. Poona, 1970)***

Hint. Show that $\lim\limits_{n \to \infty} \dfrac{B_n}{n^2} = 0$ where $B_n = \text{Var}(X_1 + X_2 + \ldots\ldots + X_n)$

78. (a) Let $X_1, X_2, \ldots, X_n$ be n independent uniformly distributed random variates in the interval (0, 1). Show that the probability that the smallest of these be in the interval $(u, u + du)$, the largest lies in the interval $(v, v + dv)$ and the remaining $(n - 2)$ lie in the interval (u, v) is given by

$dP = f(u, v)\, du dv = n(n-1)(v-u)^{n-2}\, du dv\ ;\ 0 \le u \le v \le 1$

$= 0$ elsewhere

(b) Deduce that the density function for u is

$$\int_u^1 n(n-1)(v-u)^{n-2}\, dv = n(1-u)^{n-1}, \quad 0 \le u \le 1$$

(c) Show also that the density function for v is

$$\int_0^v n(n-1)(v-u)^{n-2}\, du = nv^{n-1}, \quad 0 \le v \le 1$$

79. (a) In Ex. 78, if $F(u)$ and $G(v)$ are distribution functions of u and v show that

$F(u) = 1 - (1-u)^n \qquad 0 \le u \le 1$

$G(v) = v^n \qquad 0 \le v \le 1$

(b) Prove that the distribution function for the range $w = v - u$ is

$$H(w) = P(W \le w) = 1 - P(W > w) = w^n + nw^{n-1}(1 - w)0, \le w \le 1$$

80. If n buses have a one hour service each on a route and if buses arrive at random within an hour, show that a passenger arriving will have to wait for more than 6 minutes is $1 - (\cdot 9)^n$.

81. Let $\lambda(t)dt$ be the probability that a piece of equipment will fail in the time-interval $(t, t + dt)$ given that it has functions properly up to time t. If $f(t)\, dt$ represents the probability that the equipment first fails in the interval $(t, t + dt)$, then show that

$$f(t)\, dt = [1 - F(t)]\lambda(t)\, dt \quad \text{or} \quad F'(t) = [1 - F(t)]\lambda(t)$$

Prove that $f(t) = \lambda(t) \exp.\left[-\int_0^t \lambda(D)\, dD\right]$

Show that the exponential and Raleigh p.d.f.'s are special cases when

$$\lambda(t) = \frac{1}{\alpha},\ \lambda(t) = \frac{t}{\beta} \text{ respectively}$$

[This type of problems occur in *Reliability Theory* which is of great engineering importance].

82. (*a*) Let X be a r.v. with *m.g.f.* $M(t), -h < t < h$. Prove that

$$P(X \geq a) \leq e^{-at} M(t), 0 < t < h$$

and that
$$P(X \leq a) \leq e^{-at} M(t), -h < t < 0.$$

(*b*) If $\quad f(x = \frac{1}{2} e^{-|x|}, \quad -\infty < x < \infty,$

show that $F(x) = \frac{1}{2} e^{-x}$for $x < 0$ and $= 1 - \frac{1}{2} e^{-x}$ for $x \geq 0$

Also show that the p.d.f. for Y = X^2 is $g(y) = \frac{1}{2} e^{-\sqrt{y}}\, y^{-\frac{1}{2}}$.

83. For $\quad f(x) = \frac{1}{\sigma\sqrt{2}} \exp.\left[\frac{-\sqrt{2}\,|x - \mu|}{\sigma}\right], -\infty < x < \infty,$

find $F(x)$ and sketch both $f(x)$ and $F(x)$. Find mean and variance of x.

84. (*a*) For
$$f(x) = 1 - |x|, \quad -1 \leq x \leq 1,$$

show that
$$M_X(t) = \frac{e^t + e^{-t} - 2}{t^2}$$

$$\mu_{2n} = \frac{1}{(n+1)(2n+1)}, \mu_{2n+1} = 0\,;\, n = 0, 1, 2, 3, \ldots$$

(*b*) Let
$$f(x, y) = xe^{-x(y+1)}, x > 0, y > 0\,;\, 0, \text{elsewhere.}$$

Find the *m.g.f.* of $Z = XY$

[**Hint.** $M_{XY}(t) = \int_0^\infty \int_0^\infty e^{txy} \cdot xe^{-x(y+1)}\, dx\, dy = \int_0^\infty xe^{-x} \frac{1}{(1-t)^x}\, dx = \frac{1}{1-t}\,;\, t < 1.$

85. For $f(x) = \frac{1}{2} |x| e^{-|x|}, -\infty < x < \infty,$

find the characteristic function and first four moments about the mean.

86. If X is a r.v. with a p.d.f. given by $f(x) = \frac{1}{\alpha^2} xe^{-x/\alpha}, \alpha > 0$

Let $X = x$ be observed and Y be chosen uniformly in the interval (0, 1). If $Z = X + Y$, then For $Z > 0$, show that

$$P\,[Z \leq z] = \left(1 - e^{-\frac{z}{2\alpha}}\right)^2 \text{ and } p(z) = \frac{1}{\alpha}(e^{-z/2\alpha} - e^{-z/a})$$

87. Let $p(x, y) = e^{-(x+y)}$ be the p.d.f. for X, Y. Show that the p.d.f. for $u = \frac{X}{Y}, v = X - Y$ are $f(u) = \frac{1}{(u+1)^2}, g(v) = \frac{1}{2} e^{-|v|}$

88. If $f(x, y, z) = \frac{1}{(2\pi)^{3/2}} \exp\left[-\frac{1}{2\sigma^2}(x^2+y^2+z^2)\right]$

and $x = r\sin\theta\cos\phi$, $y = r\sin\theta\sin\phi$, $z = r\cos\theta$, show that

$$p(r) = \sqrt{\frac{2}{\pi}}\,\frac{1}{\sigma^2}\,r^2\exp\left(-\frac{r^2}{2\sigma^2}\right)$$

89. Show that if n points are chosen at random on the circumference of a circle, then the probability that these points will be at least θ radians apart, $\theta \le \frac{2\pi}{n}$ is $= \left(1 - \frac{n\theta}{2\pi}\right)^{n-1}$

90. Let X_1 and X_2 be independent random variables with uniform distribution over the interval $[\theta - \frac{1}{2}, \theta + \frac{1}{2}]$. Show that $X_1 - X_2$ has a distribution independent of θ and find its density function.

Ans. $f(y) = \begin{cases} 1+y, & -1 \le y \le 0 \\ 1-y, & 0 \le y < 1 \\ 0\;, & |y| > 1 \end{cases}$ $Y = x_1 - x_2$

91. Let $E(X \mid Y=y) = \int_{-\infty}^{+\infty} x f(x \mid y)\,dx$

be the conditional expectation of X given $Y = y$. Show that

$$E(X) = \int_{-\infty}^{+\infty} E_Y(X \mid Y=y)\,f(y)dy = E_y\{E(X \mid Y=y)\}$$

92. X is a positive random variable with distribution function $F(x)$ and density $f(x)$. Show that $P(x \le X \le x + dx \mid x \le X) = g(x)\,dx$ where $g(x) = \frac{f(x)}{1 - F(x)}$

Prove also that $F(x) = 1 - \exp\left(-\int_0^x g(y)\,dy\right)$

Hence deduce that the negative exponential is the only absolutely continuous distribution function with constant $g(x)$. ***(B.A. Hons. Delhi, 1982)***

[Hint. $P(x \le X \le x + dx \mid x \le X) = P(x \le X \le x + dx) \div P(X \le x)$

$$= f(x)\,dx + \int_{-\infty}^{x} f(x)\,dx = \frac{f(x)\,dx}{1 - \int_x^{\infty} f(x)\,dx} = \frac{f(x)}{1 - F(x)}\,dx = g(x)\,dx$$

where $g(x) = \frac{f(x)}{1 - F(x)}$

Also $\int_0^x g(y)\,dx = \int_0^x \frac{f(y)}{1 - F(y)}\,dy = -[\log(1 - F(y))]_0^x$

$$\therefore \quad -\int_0^x g(y)\,dy = \log[1 - F(x)] - \log[1 - F(0)] = \log[1 - F(x)]$$

$$\Rightarrow \quad 1 - F(x) = \exp\{-\int_0^x g(y)\,dy\}$$

$$\Rightarrow \quad F(x) = 1 - \exp\left\{-\int_0^x g(y)\,dy\right\}$$

Let $g(x)$ be constant and $= c$ say. Then

$$F(x) = 1 - \exp\left\{-\int_0^x c\,dy\right\} = 1 - e^{-cx}, x > 0$$

which is the negative exponential density function.

93. For what positive values of C_1 and C_2 is $2^{-1}\{(1 + C_1 s)^{-1} + (1 + C_2 s)^{-1}\}$ the moment generating functiôn of a positive random variable ?

Ans. $C_1 - 2C_1C_2 + C_2 = 0$

94. Compute the mean, variance and *m.g.f.* of each of the probability law, specified by the following distribution functions

(*i*) $f(x) = 0 \quad = 1 - \frac{1}{3}e^{-\frac{x}{3}} - \frac{2}{3}e^{-\left[\frac{x}{3}\right]}$.

(*ii*) $f(x) = 0 \quad = 1 - \frac{1}{2x^2} - \frac{1}{2^{[x]}}$

Ans. (*i*) $m.g.f. = \frac{1}{3} \cdot \frac{1}{1 - 3t} + \frac{2}{3} \cdot \frac{1 - e^{-1}}{1 - e^{3t-1}} e^{3t}$

(*ii*) Mean $\frac{5}{2}$, variance and *m.g.f.* do not exist.

95. If $F(x, y) = 1$ for $x + 2y \geq 1$; $= 0$ for $x + 2y < 1$
then prove that $F(x, y)$ can not be the joint distribution of X and Y.

96. The numbers $X_1, X_2, ..., X_n$ are independently chosen at random in the interval (a, b). Prove that the *p.d.f.* of the *r.v.* $X = \min(X_1, X_2, ..., X_n)$ is given by $f(x) = n(b - x)^{n-1}/(b - a)^n, a < x < b$.

[**Hint.** For $a < x < b, X > x = X_1 < x \ldots X_n > x$

Then $F(X > x) = P(X_1 > x)P(X_2 > x) \ldots P(X_n > x) = (b - x)^{n-1}/(b - a)^n$

$\Rightarrow \quad F(x) \equiv 1 - p(X < x)$ Also $f(x) = F'(x) =$ etc.]

97. Give an example to show that if the *p.d.f.* exists, (*a*) Mean value does not exist (*b*) *MGF* does not exist.

Hint. (*a*) $f(x) = \frac{1}{x^2}, 1 < x < \infty.$ 0 otherwise

$$E(x) = \int_1^\infty x f(x)\,dx = \int_1^\infty \frac{1}{x}\,dx = \lim_{b \to \infty} \int_1^b \frac{1}{x}\,dx$$

$$= \lim_{b \to \infty} (\log b - \log 1) \quad \text{which does not exist.}$$

(*b*) $f(x) = \dfrac{6}{\pi^2 x^2}$; $x = 1, 2 \ldots$ $= 0$, otherwise.

We know that $f(x)$ is *p.d.f.*, since $\dfrac{6}{\pi^2} \sum_1^{\infty} \dfrac{1}{x^2} = \dfrac{6}{\pi^2} \cdot \dfrac{\pi^2}{6} = 1$

$$\text{M.G.F.} = E(e^{tx}) = \sum_1^{\infty} \frac{6e^{tx}}{\pi^2 x^2} = \frac{6}{\pi^2} \sum_1^{\infty} \frac{e^{tx}}{x^2}.$$

This series is divergent, when $t > 0$.

$\therefore$ M.G.F. does not exist, if $|t| > h, h \to 0+$

98. Let $X_1, X_2, \ldots, X_n$ be i.i.d. variates with $f(x) = e^{-(x-1)}, x \geq 1$.

Show that $Y_n \to 1$ in probability where $Y_n = \text{Min } (X_k)$; $1 \leq k \leq n$

(*I.A.S. 1982*)

99. (*a*) Let $\{X_n\}, n = 1, 2, \ldots$ be a sequence of standardised variates and $\text{Corr }(x_m, x_n) = \exp\{-|m-n|\alpha\}, \alpha > 0, m \neq n$. Show that WLLN holds for this sequence.

(*b*) Let $x_1 x_2, \ldots\ldots, x_n$ be jointly normal each with mean 0 and variance 1 and $\text{Cov }(x_i, x_j) = \rho$ if $|j-i| = 1$ and $= 0$, otherwise. Show that WLLN holds for the sequence. **(*I.A.S. 1988*)**

[**Hint.** (*a*) We have $E(X_i) = 0$, $\text{Var }(X_i) = 1$, $\text{Cov }(X_i, X_j) = e^{-|i-j|\alpha}$

$$E(S_n) = E\Big(\sum_{i=1}^{n} X_i\Big) = \sum_{i=1}^{n} E(X_i) = 0$$

$$\sigma^2 = \text{Var }(S_n) = \text{Var }\Big(\sum_{i=1}^{n} x_i\Big) = \sum_{i=1}^{n} \text{Var } X_i + 2 \sum_{i<j_1}^{n} \text{Cov }(X_i, X_j)$$

$$= n + 2(n-1)\, e^{-|m-n|\alpha} > 0 \quad \text{as } \alpha > 0$$

Let $$Y_n = \frac{1}{n} \sum_{i=1}^{n} X_i = \frac{S_n}{n}$$

$$\therefore \quad E\left[\frac{Y_n^2}{1+Y_n^2}\right] = E\left[\frac{S_n^2}{n^2+S_n^2}\right] = \int_{-\infty}^{+\infty} \frac{x^2}{n^2+x^2} \cdot \frac{1}{\sigma\sqrt{2\pi}}\, e^{-\frac{x^2}{2\sigma^2}}\, dx$$

$$= \frac{2\sigma^2}{\sqrt{2\pi}} \int_0^{\infty} \frac{y^2}{n^2+\sigma^2 y^2}\, e^{-y^2/2}\, dy, \; x/\sigma = y$$

$$= \frac{2}{\sqrt{2\pi}} \{n + 2(n-1)\rho\}^{1/2} \int_0^{\infty} \frac{y^2\, e^{-y^2/2}\, dy}{n^2 + \{n + 2(n-1)\rho\}\, y^2}$$

$$\le n + 2(n-1)\frac{\rho}{n^2}\cdot\frac{2}{\sqrt{2\pi}}\int_0^\infty y^2 e^{-y^2/2}\,dy \quad \to 0 \text{ as } n\to\infty \text{ since.}$$

$$\therefore \qquad \lim_{n\to\infty} E\left[\frac{Y_n^2}{1+Y_n^2}\right] \to 0. \text{ Hence WLLN holds for } \{X_n\}.$$

(*b*) **Hint.** Var $S_n = \text{Var}(\Sigma x_1) = n + 2(n-1)\rho,\ Y_n = \dfrac{S_n}{n}$

Show that $E\left(\dfrac{Y_n^2}{1+Y_n^2}\right) \to 0$ as $n\to\infty$.

100. (*a*) Let $X_1, X_2, ..., X_n$ be i.i.d. standard Cauchy variates. Show that the WLLN does not hold for the sequence $\{Y_n/(1+Y_n^2)\}$ where $Y_n = \overline{X}$.

[**Hint.** The necessary and sufficient condition for the WLLN to hold for the sequence $\{X_n\}$ is that $\lim\limits_{n\to\infty} E\left(\dfrac{Y_n^2}{1+Y_n^2}\right) = 0$ where $Y_n = \dfrac{1}{n}\,|\,S_n - E(S_n)\,|$ and $S_n = \sum\limits_{i=1}^{n} X_i = n\overline{X}$. Since X_i's are standard Cauchy i.i.d variates, $\overline{X}$ is also a Cauchy variate. Let $S = \Sigma X_i = n\overline{X} \Rightarrow Y_n = S_n/n$

$$E\left(\frac{Y_n^2}{1+Y_n^2}\right) = E\left(\frac{S_n^2}{n^2+S_n^2}\right) = E\left[\frac{(S_n/n)^2}{1+(S_n/n)^2}\right] = E\left(\frac{x^2}{1+x^2}\right)$$

$$= \frac{1}{\pi}\int_{-\infty}^{+\infty}\frac{x^2}{1+x^2}\cdot\frac{1}{1+x^2}\,dx, \text{ since } f(x) = \frac{1}{\pi}\cdot\frac{1}{1+x^2},\ -\infty < x < +\infty$$

$$= \frac{1}{\pi}\int_{-\infty}^{+\infty}\frac{x^2\,dx}{(1+x^2)^2}. \quad Put\ x = \tan\theta,\ dx = \sec^2\theta\, d\theta$$

$$= \frac{2}{\pi}\int_0^{\pi/2}\sin^2\theta\, d\theta = \frac{1}{\pi}\left|\ \theta - \frac{1}{2}\sin 2\theta\ \right|_0^{\pi/2} = \frac{1}{2} \ne 0$$

$$\therefore \quad \lim_{n\to\infty} E\left(\frac{Y_n^2}{1+Y_n^2}\right) \ne 0. \text{ Hence WLLN does not hold for } \left\{\frac{Y_n}{1+Y_n^2}\right\}.$$

(*b*) Let $X_1, X_2, ..., X_n$ be i.i.d. variables with mean μ and variance σ^2 and $\dfrac{X_1^2 + X_2^2 + \ldots + X_n^2}{n} \to C$ as $n\to\infty$ except for some constant C. Show that

$C = \sigma^2 + \mu^2$ **(*B.Sc. Stat. Hons., Delhi 1989*)**

[**Hint.** Apply Khinchine's theorem]

8

Special Continuous Probability Distributions

8·1. Introduction. In this chapter we shall deal with special continuous distribution.

8·2. *Normal Distribution. This distribution can be regarded as the limiting form of the Binomial distribution when n, the number of trials, is very large and neither p nor q is very small.

Let us introduce a new variate $$Z = \frac{r - np}{\sqrt{npq}} \qquad ...(8\cdot1)$$

where r is the Binomial variate with mean np and standard deviation $\sqrt{npq}$ so that Z is a variate with mean zero and variance unity [see §3·3·2, page 55]. Now as r goes from 0 to n, Z takes the values $-np/\sqrt{npq}$ to $nq/\sqrt{npq}$ and the jump in the value of Z at each stage is $1/\sqrt{npq}$ which decreases as n increases. Also the two limits approach $-\infty$ and $+\infty$ respectively if neither p nor q is zero. Thus the jump is an infinitesimal (*i.e.* it tends to zero as n tends to infinity) and we expect, in the limit, the distribution of Z to be a continuous distribution extending from $-\infty$ to $+\infty$ and having zero mean and unit variance.

To obtain the distribution of Z, we proceed as follows :

In the Binomial distribution, the probability for the variate to take the value r is $\quad P(n, r) = {}^nC_r p^r q^{n-r} = \dfrac{n\,!}{r\,!\,(n-r)\,!} \cdot p^r q^{n-r}.$

To find its limit as n tends to infinity, we apply Stirling's formula to the three factorials.

* Until recently the name of Bernoulli, Laplace and Gauss either separately or in combination have been associated with this distribution. In 1924, however, Karl Pearson announced his discovery that Abrahm De Moivre as early as 1733 had made the discovery of the distribution. During the eighteenth and nineteenth centuries various attempts were made to establish the distribution as the underlying law governing all continuous variates—hence the name 'normal'.

$$P_r = \lim_{n \longrightarrow \infty} P(n, r) = \lim_{n \to \infty} \frac{\sqrt{2}\pi e^{-n} n^{n+\frac{1}{2}} p^r q^{n-r}}{\sqrt{2\pi} e^{-r} r^{r+\frac{1}{2}} \sqrt{2\pi} e^{-(n-r)} (n-r)^{n-r+\frac{1}{2}}}$$

$$= \lim_{n \to \infty} \left(\frac{1}{N} \cdot \frac{1}{\sqrt{2\pi npq}} \right), \text{ where } N = \left(\frac{r}{np} \right)^{r+\frac{1}{2}} \left(\frac{n-r}{nq} \right)^{n-r+\frac{1}{2}}$$

From (8·1) we have $\frac{r}{np} = 1 + z\sqrt{\frac{q}{np}}$ and $\frac{n-r}{nq} = 1 - z\sqrt{\frac{p}{nq}}$.

Therefore, $\log_e N = (np + z\sqrt{npq} + \frac{1}{2}) \log_e \left(1 + z\sqrt{\frac{q}{np}} \right)$

$$+ (nq - z\sqrt{npq} + \tfrac{1}{2}) \log_e \left(1 - z\sqrt{\frac{p}{nq}} \right)$$

Assuming that $| z |$ is less than the smaller of the two quantities $\sqrt{np/q}$ and $\sqrt{nq/p}$, we can expand both the logarithms on the right hand side in uniformly convergent series. Hence we have

$$\log_e N = (np + z\sqrt{npq} + \tfrac{1}{2}) \left[z\sqrt{\frac{q}{np}} - \frac{z^2 q}{2np} + 0\left(n^{-3/2} \right) \right]$$

$$+ (nq - z\sqrt{npq} + \tfrac{1}{2}) \left[-z\sqrt{\frac{p}{nq}} - \frac{z^2 p}{2nq} + 0\left(n^{-3/2} \right) \right]$$

$$= \frac{z}{2\sqrt{n}} \left(\sqrt{\frac{q}{p}} - \sqrt{\frac{p}{q}} \right) + \frac{z^2}{2} - \frac{z^2}{4n} \left(\frac{q}{p} + \frac{p}{q} \right) + 0\left(n^{-3/2} \right)$$

When n tends to infinity, we have $\log_e N \to \frac{1}{2}z^2$ or $N \longrightarrow e^{\frac{1}{2}z^2}$

As r runs through integral values the increments in z are each equal to $(npq)^{-1/2}$ which we denote by dz when n tends to infinity. Thus if dP denotes the probability for the variate z to lie in the interval $z - \frac{1}{2}dz$ to $z + \frac{1}{2}dz$, we have

$$dP = \frac{1}{\sqrt{2\pi}} e^{-\frac{1}{2}z^2} dz, \; -\infty < z < \infty \qquad \text{...(8·2)}$$

This is the required continuous distribution of Z and is called the normal distribution.

Taking $Z = \frac{X-a}{b}$, (8·2) takes the form

$$dP = \frac{1}{b\sqrt{2\pi}} e^{-\frac{1}{2}\left(\frac{x-a}{b} \right)^2} dx, \; -\infty < x < \infty.$$

The mean, μ, of the distribution of X is given by

$$\mu = \frac{1}{b\sqrt{2\pi}} \int_{-\infty}^{+\infty} x\, e^{-\frac{1}{2}\left(\frac{x-a}{b}\right)^2} dx \Rightarrow \mu = \frac{1}{\sqrt{2\pi}} \int_{-\infty}^{+\infty} (a+bz)\, e^{-\frac{1}{2}z^2} dz$$

$$= \frac{a}{\sqrt{2\pi}} \int_{-\infty}^{+\infty} e^{-\frac{1}{2}z^2} dz + \frac{b}{\sqrt{2\pi}} \int_{-\infty}^{+\infty} z\, e^{-\frac{1}{2}z^2} dz$$

$\Rightarrow \quad \mu = a$, since the second integral on the right vanishes.

The variance, σ^2, of the distribution of X is given by

$$\sigma^2 = \int_{-\infty}^{+\infty} (x-\mu)^2 \cdot \frac{1}{b\sqrt{2\pi}}\, e^{-\frac{1}{2}\left(\frac{x-\mu}{b}\right)^2} dx. \quad [\because\ a = \mu]$$

$$= \frac{b^2}{\sqrt{2\pi}} \int_{-\infty}^{+\infty} z^2\, e^{-\frac{1}{2}z^2} dz \qquad \left(Z = \frac{X-\mu}{b} \right)$$

$$= \frac{b^2}{\sqrt{2\pi}} \left[-z.e^{-\frac{1}{2}z^2} \right]_{-\infty}^{+\infty} + \frac{b^2}{\sqrt{2\pi}} \int_{-\infty}^{+\infty} e^{-\frac{1}{2}z^2} dz$$

$$= 0 + \frac{b^2}{\sqrt{2\pi}} \sqrt{2\pi} = b^2.$$

Thus we see that $a = \mu$ and $b^2 = \sigma^2$ so that the distribution of X with mean μ *and standard deviation* σ *is*

$$dP = \frac{1}{\sigma\sqrt{2\pi}}\, e^{-\frac{1}{2}\left(\frac{x-\mu}{\sigma}\right)^2} dx, -\infty < x < \infty\,; -\infty < \mu < \infty,\ \sigma > 0$$

$$= f(x)dx,$$

where $f(x\,;\mu, \sigma) = f(x) = \dfrac{1}{\sigma\sqrt{2\pi}}\, e^{-\frac{1}{2}\left(\frac{x-\mu}{\sigma}\right)^2}$ ...(8·3)

The curve $y = f(x)$ is called the *Normal Probability Curve.*

The continuous variate X which is distributed with probability density $f(x)$ given by (8·3) is called a normal variate with mean μ and variance σ^2 or briefly X is a $N(\mu, \sigma^2)$ variate or $X \sim N(\mu, \sigma^2)$

Ex. 1. Draw a Binomial probability curve with $p = \frac{1}{2}$ and $n = 2, 5, 8, 11$.

Ex. 2. Draw the standardised Binomial probability curve (*i.e.* ${}^nC_r p^r q^{n-r}$. against $z = \dfrac{r - np}{\sqrt{npq}}$ for the above values of n and p.

Ex. 3. Find the mean and the standard deviation of the following probability distribution $\quad f(x) = c \exp\{-\frac{1}{24}(x^2 - 6x + 4)\}$, $-\infty < x < \infty$, c being a constant.

(*M.A. Eco., Statistics Delhi, 1959*)

[**Ans.** mean = 3, *s.d.* = $2\sqrt{3}$]

8·2·1. Important Features of the Normal Curve. The normal probability curve is bell shaped and symmetrical about the ordinate at $x = \mu$.

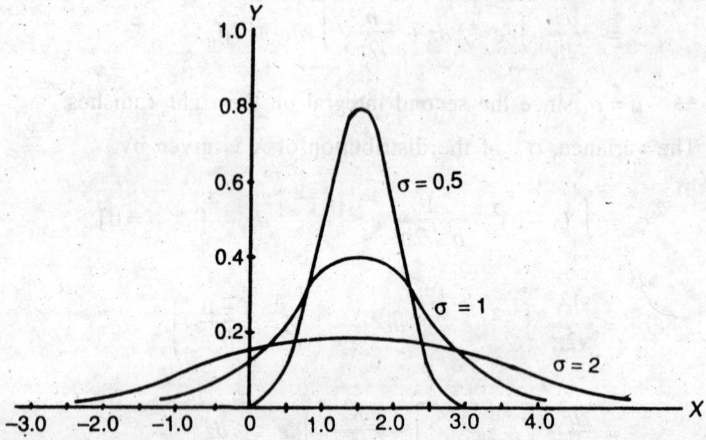

Fig. 8·1. Normal Probability Curves

The ordinate decreases rapidly as x increases numerically. The curve extends to infinity on either side of the mean. The maximum ordinate is at $x = \mu$ and is given by $y_{max.} = 1/\sigma\sqrt{2\pi}$.

Since the ordinate at $x = \mu$ divides the area under the normal curve into two equal parts, the median of the distribution coincides with the mean and the mode. The points of inflexion of the normal curve (obtained by putting $d^2y/dx^2 = 0$ and verifying that at these points of the curve $d^3y/dx^3 \neq 0$) are given by $x = \mu \pm \sigma, y = 1/(\sigma\sqrt{2\pi})$ so that the two points of inflexion of the normal curve are equidistant from the mean. For large values of σ, the normal curve tends to be flatter, while for small values of σ, it has a small peak. x-axis is an asymptate to the normal curve. No portion of the curve lies below the x-axis since normal probability function can not be negative.

8·2·3. Some Properties of the Normal Distribution.

(*a*) *Mean deviation from the mean* μ

$$= \int_{-\infty}^{+\infty} |x - \mu| \cdot \frac{1}{\sigma\sqrt{2\pi}} e^{-\frac{1}{2}\left(\frac{x-\mu}{\sigma}\right)^2} dx$$

$$= \frac{\sigma}{\sqrt{2\pi}} \cdot \int_{-\infty}^{+\infty} |z| e^{-\frac{1}{2}z^2} dz, \quad \left(z = \frac{x-\mu}{\sigma}\right)$$

$$= \frac{\sigma}{\sqrt{2\pi}} \left[\int_{-\infty}^{0} z e^{-\frac{1}{2}z^2} dz + \int_{0}^{\infty} z e^{-\frac{1}{2}z^2} dz \right] = \frac{2\sigma}{\sqrt{2\pi}} \int_{0}^{\infty} z e^{-\frac{1}{2}z^2} dz$$

$$= \sigma \sqrt{\frac{2}{\pi}} = 0{\cdot}7979 \ \sigma \text{ approx.}$$

$$= \frac{4}{5} \sigma \text{ approx.}$$

(*b*) *Moments about the mean* μ

$$\mu_{2n+1} = \int_{-\infty}^{+\infty} (x-\mu)^{2n+1} \cdot \frac{1}{\sigma\sqrt{2\pi}} e^{-\frac{1}{2}\left(\frac{x-\mu}{\sigma}\right)^2} dx$$

$$= \frac{\sigma^{2n+1}}{\sqrt{2\pi}} \int_{-\infty}^{+\infty} z^{2n+1} e^{-\frac{1}{2}z^2} dz, \quad \left(z = \frac{x-\mu}{\sigma} \right)$$

$= 0$, since the integrand is an odd function of z

$\therefore \quad \mu_3 = 0 = \mu_5 = ...$

Thus *all odd order moments about the mean vanish.*

Hence $\quad \beta_1 = \dfrac{\mu_3^3}{\mu_2^3} = 0$;

Even order moments about mean are given by

$$\mu_{2n} = \frac{1}{\sigma\sqrt{2\pi}} \int_{-\infty}^{+\infty} (x-\mu)^{2n} e^{-\frac{1}{2}\left(\frac{x-\mu}{\sigma}\right)^2} dx$$

$$= \frac{\sigma^{2n}}{\sqrt{2\pi}} \int_{-\infty}^{+\infty} z^{2n-1} \cdot z e^{-\frac{1}{2}z^2} dz = \frac{\sigma^{2n}}{\sqrt{2\pi}} \left[z^{2n-1} \left(e^{-\frac{1}{2}z^2} \right) \right]_{-\infty}^{+\infty}$$

$$+ \frac{\sigma^{2n}}{\sqrt{2\pi}} \cdot (2n-1) \int_{-\infty}^{+\infty} z^{2n-2} e^{-\frac{1}{2}z^2} dz$$

$$= (2n-1)\sigma^2 \mu_{2n-2}$$

since the expression in square brackets vanishes at both limits.

$$= (2n-1)(2n-3)\sigma^4 \mu_{2n-4}$$

Applying the recurrence formula successively, we obtain

$$\mu_{2n} = (2n-1)(2n-3).....3{\cdot}1 \cdot \sigma^{2n}. \qquad (\because \ \mu_0 = 1)$$

In particular, $\quad \mu_2 = \sigma^2, \mu_4 = 3\sigma^4$. Hence $\beta_2 = \dfrac{\mu_4}{\mu_2^2} = 3.$

Recurrence Relation for Moments. We shall prove that

$$\mu_{2r+2} = \sigma^2 \mu_{2r} + \sigma^3 \frac{d\mu_{2r}}{d\sigma}. \qquad \textbf{(\textit{B.Sc. Madras, 1989})}$$

We have
$$\mu_{2r} = \frac{1}{\sigma\sqrt{2\pi}} \int_{-\infty}^{\infty} (x-\mu)^{2r} \, e^{-\frac{1}{2}\frac{(x-\mu)^2}{\sigma^2}} \, dx$$

$$\therefore \quad \frac{d\mu_{2r}}{d\sigma} = -\frac{1}{\sigma^2\sqrt{2\pi}} \int_{-\infty}^{\infty} (x-\mu)^{2r} \, e^{-\frac{1}{2}\left(\frac{x-\mu}{\sigma}\right)^2} dx$$

$$+ \frac{1}{\sigma^4\sqrt{2\pi}} \int_{-\infty}^{\infty} (x-\mu)^{2r+2} \, e^{-\frac{1}{2}\left(\frac{x-\mu}{\sigma}\right)^2} dx$$

$$\therefore \quad \sigma^3 \frac{d\mu_{2r}}{d\sigma} = -\sigma^2 \mu_{2r} + \mu_{2r+2}. \quad \text{(multiplying both sides by } \sigma^3\text{)}$$

Hence the result.

Cor. Putting $r = 0$, we get $\mu_2 = \sigma^2 \cdot \mu_0 + \sigma^3 \cdot 0 = \sigma^2$

and $r = 1$,
$$\mu_4 = \sigma^2 \cdot \mu_2 + \sigma^3 \cdot \frac{d\sigma^2}{d\sigma} = \sigma^2 \cdot \sigma^2 + \sigma^3 \cdot 2\sigma = 3\sigma^4.$$

Ex. 1. Find the mean and variance of the normal distribution by differentiating with respect to μ, twice the identity :

$$\frac{1}{\sigma\sqrt{2\pi}} \int_{-\infty}^{\infty} e^{-\frac{1}{2}\left(\frac{x-\mu}{\sigma}\right)^2} dx = 1$$

Find also the variance by differentiating with respect to σ alone.

(*c*) *Moment generating function with respect to origin.*

$$M_{X=0}(t) = E(e^{tX}) = \int_{-\infty}^{+\infty} e^{tx} \cdot \frac{1}{\sigma\sqrt{2\pi}} \, e^{-\frac{1}{2}\left(\frac{x-\mu}{\sigma}\right)^2} dx$$

$$= e^{\mu t} \frac{1}{\sqrt{2\pi}} \int_{-\infty}^{+\infty} e^{-\frac{1}{2}z^2 + t\sigma z} dz, \quad \left(z = \frac{x-\mu}{\sigma} \right)$$

$$= e^{\mu t + \frac{1}{2}t^2\sigma^2} \cdot \frac{1}{\sqrt{2\pi}} \int_{-\infty}^{+\infty} e^{-\frac{1}{2}(z - t\sigma)^2} dz = e^{\mu t + \frac{1}{2}t^2\sigma^2} \qquad \text{....(8·4)}$$

Moment generating function with respect to mean μ.

$$M_{X=\mu}(t) = E[e^{t(x-\mu)}] = e^{-\mu t} E(e^{tx}) = e^{\frac{1}{2}t^2\sigma^2}, \text{ using 8·4} \quad \text{...(8·5)}$$

since $$e^{\frac{1}{2}t^2\sigma^2} = 1 + \frac{(\frac{1}{2}t^2\sigma^2)}{1!} + \frac{(\frac{1}{2}t^2\sigma^2)^2}{2!} + ..., + \frac{(\frac{1}{2}t^2\sigma^2)^n}{n!} + ...$$

we have $\mu_{2n+1} = 0$ as coefficient of $t^{2n+1} = 0, n = 0, 1, 2, ...$

and $\mu_{2n} =$ coefficient of $t^{2n}/(2n)! = \dfrac{(\frac{1}{2}\sigma^2)^n (2n)!}{n!}$

$$= 1{\cdot}3{\cdot}5......(2n-1)\sigma^{2n} \because (2n)! = 1{\cdot}3 ......(2n-1) . 2^n \cdot n!$$

If $Z = (X - \mu)/\sigma$, then $M_Z(t) = e^{\frac{1}{2}t^2}$

(*d*) *Cumulative function*

$$K_x(t) = \log M_{X=0}(t) = \mu t + \tfrac{1}{2}t^2\sigma^2$$

$\therefore \quad k_1 =$ coeff. of $t = \mu$, $k_2 =$ coeff of $t^2/2! = \sigma^2$,

$\therefore \quad k_r =$ coeff. of $t^r/r! = 0, r \geq 3$

so that all the cumulants after the second are equal to zero.

(*e*) *Mode and Median of Normal Distribution*

We have $$f(x) = \frac{1}{\sigma\sqrt{2\pi}} \exp\left\{-\frac{1}{2\sigma^2}\left(x-\mu\right)^2\right\}$$

$$\Rightarrow \quad \log f(x) = c - \frac{1}{2\sigma^2}(x-\mu)^2 \quad \text{where} \quad c = -\log(\sigma\sqrt{2\pi})$$

Differentiating w.r.t. x, we have

$$\therefore \quad \frac{f'(x)}{f(x)} = -\frac{1}{\sigma^2}(x-\mu) \Rightarrow f'(x) = -\frac{x-\mu}{\sigma^2} f(x) \qquad ...(1)$$

and $$f''(x) = -\frac{1}{\sigma^2}[1 . f(x) + (x-\mu)f'(x)] = -\frac{f(x)}{\sigma^2}\left[1 - \left(\frac{x-\mu}{\sigma}\right)^2\right], \text{ using (1)}$$

Hence $f'(x) = 0$ gives $x = \mu$, as $f(x) \neq 0$.

Also $$\left[f''(x)\right]_{x=\mu} = -\frac{1}{\sigma^2}\left[f(x)\right]_{x=\mu} = -\frac{1}{\sigma^2} \cdot \frac{1}{\sigma\sqrt{2\pi}} < 0$$

Hence $x = \mu =$ mean is the mode of the normal distribution.

(*f*) *median of normal distribution.*

If a is the median of the normal distribution, then

$$\int_{-\infty}^{a} f(x)dx = \frac{1}{2} = \int_{a}^{\infty} f(x)dx \qquad ...(8{\cdot}6)$$

Suppose $\mu < a$, then

$$\text{L.H.S.} = \int_{-\infty}^{\mu} \frac{1}{\sigma\sqrt{2\pi}} e^{-\frac{1}{2}\left(\frac{x-\mu}{\sigma}\right)^2} dx + \int_{\mu}^{a} \frac{1}{\sigma\sqrt{2\pi}} e^{-\frac{1}{2}\left(\frac{x-\mu}{\sigma}\right)^2} dx$$

$$= \frac{1}{2} + \int_{\mu}^{a} \frac{1}{\sigma\sqrt{2\pi}} e^{-\frac{1}{2}\left(\frac{x-\mu}{\sigma}\right)^2} dx$$

$$= \frac{1}{2}, \text{ using (8·6)}, \left[\because \int_{-\infty}^{\mu} e^{-\frac{1}{2}\left(\frac{x-\mu}{\sigma}\right)^2} dx = \tfrac{1}{2} \right] \Rightarrow a = \mu$$

since integrand cannot vanish.

Similarly when $\mu > a$, we find $a = \mu$.

Hence the median of the normal distribution is equal to mean.

Thus for the normal distribution,

mean = mode = median.

8·3. A linear combination of independent normal variates.

Theorem. *If $X_i^{'s}$ $(i = 1, 2, \ldots\ldots, n)$ are independent normal variates with means μ_i and variances σ_i^2, then variate $Z = \sum_{i=1}^{n} c_i X_i$ is a normal variate with mean $\sum_{i=1}^{n} c_i\mu_i$ and variance $\sum_{i=1}^{n} c_i^2 \sigma_i^2$ where c_i's are constants.*

Proof. If X is a normal variate with mean μ and variance σ^2 then $E(cX) = cE(X) = c\,\mu$, $Var(cX) = c^2 Var\,X = c^2\sigma^2$ $\therefore$ cX is a normal variate with mean $c\mu$ and variance $c^2\sigma^2$. Hence the *m.g.f.* of the distribution of cX is $exp\ (c\,\mu t + \frac{1}{2} c^2\sigma^2 t^2)$. Since $X_i^{'s}$ are independent variates, so are the variates $c_i X_i$ and hence by § 5·5·2, page 155 the *m.g.f.* of the variate $\Sigma\, c_i X_i$ is

$$\exp\,(t\Sigma\ c_i\,\mu_i + \tfrac{1}{2}t^2\Sigma c_i^2\sigma_i^2).$$

But this is the *m.g.f.* of a normal variate with mean $\Sigma c_i\mu_i$ and variance $\Sigma c_i^2\sigma_i^2$. Hence the theorem follows.

Cor. 1. Let $\mu_i = \mu$, $\sigma_i = \sigma$, $c_i = 1/n$, $i = 1, 2, \ldots, n$ so that

$$z = \sum_{i=1}^{n} c_i x_i = \bar{x},\ Ez = \sum_{i=1}^{n} c_i\mu_i = \sum_{1}^{n} \mu/n = \mu \text{ and Var}\,(z) = \sum_{i=1}^{n} c_i^2\sigma_i^2 = \frac{n\sigma^2}{n^2} = \frac{\sigma^2}{n}.$$

Hence we have the following important result :

If $X_i^{'s}$ $(i = 1, 2) \ldots, n)$ are n independent normal variates with the same mean μ and same variance σ^2, their mean $\bar{x}$ is a normal variate with *mean* μ *and variance* σ^2/n *so that* $\dfrac{\overline{X}-\mu}{\sigma/\sqrt{n}}$ *is a standard normal variate,* $\overline{X} \sim N(\mu, \sigma^2/n)$.

Cor. 2. Putting $c_i = 1 \ \forall \ i$, we get the **additive property** of normal variates. Thus if $X_1 \sim N(\mu_1, \sigma_1^2)$, $X_2 \sim N(\mu_2, \sigma_2^2)$, then $X_1 + X_2 \sim N(\mu_1 + \mu_2, \sigma_1^2 + \sigma_2^2)$

Cor. 3. Putting $c_1 = 1$, $c_2 = -1$, $c_3 = c_4 = \ldots = 0$, we get the following result :

If X_1 and X_2 are independent normal variates with means μ_1, μ_2 and variances σ_1^2, σ_2^2 respectively, then $X_1 - X_2$ will be a normal variate with mean $\mu_1 - \mu_2$ and variance $\sigma_1^2 + \sigma_2^2$, *i.e.* $X_1 - X_2 \sim N(\mu_1 - \mu_2, \sigma_1^2 + \sigma_2^2)$

Example 8·1. *If X, Y are independent normal variates with means 6, 7 and variances 9, 16 respectively, determine* λ *such that*

$$P(2X + Y \leq \lambda) = P(4X - 3Y \geq 4\lambda)$$

(*B.Sc. Stat. Hons. Delhi, 1988*)

We have $E(2X + Y) = 2E(X) + E(Y) = 2 \times 6 + 7 = 19$,

$\text{Var}\,(2X + Y) = 4\ \text{Var}\,X + \text{Var}\,Y = 4 \times 9 + 16 = 52$

Thus if $U = 2X + Y$, then $\dfrac{u - 19}{\sqrt{52}}$ is a standard normal variate.

Also $E\left(\dfrac{4X - 3Y}{4}\right) = \dfrac{4E(X) - 3E(Y)}{4} = \dfrac{3}{4}$. Let $v = \dfrac{4x - 3y}{4}$.

$\text{Var}\left(\dfrac{4X - 3Y}{4}\right) = \dfrac{1}{16}(16\ \text{Var}\,X + 9\ \text{Var}\,Y) = 18$

Thus $\dfrac{v - \frac{3}{4}}{\sqrt{18}} = \dfrac{4v - 3}{12\sqrt{2}}$ is a standard normal variate.

By hypothesis $P(u \leq \lambda) = P(v \geq \lambda)$. we have

$$P\left(Z \leq \frac{\lambda - 19}{\sqrt{52}}\right) = P\left(Z \geq \frac{4\lambda - 3}{12\sqrt{2}}\right)$$

[See cor. §8·3·1 I, page 310]

Hence $-\left(\dfrac{\lambda - 19}{\sqrt{52}}\right) = \dfrac{4\lambda - 3}{12\sqrt{2}}$

Solving for λ, we find $\lambda = \dfrac{114\sqrt{2} + 3\sqrt{13}}{6\sqrt{2} + 4\sqrt{13}}$.

Ex. 1. If X and Y are independent normal variates with common mean and unit variance and if $\quad P(X + 2Y \leq 3) = P(2X - Y \geq 4)$

Obtain the common mean.

[**Hint.** Let $Z=\dfrac{X+2Y}{3}$ and $T=\dfrac{2X-Y}{4}$. Find $E(Z)$, Var Z, $E(T)$ and Var T and proceed as in solved *example* 8·1 .]

Ex. 2. If X_1 and X_2 are two independent normal variates with a common mean zero and variances σ_1^2, σ_2^2 respectively, show that the variates

$$U=X_1+X_2,\ V=-\frac{\sigma_1}{\sigma_2}X_2+\frac{\sigma_2}{\sigma_1}X_1$$

are independent, and that each is normally distributed with mean zero and variance $\sigma_1^2+\sigma_2^2$. [This is another method of proving additive property of normal variates].

[**Hint.** The joint distribution of X_1 and X_2 is $\dfrac{1}{2\pi\sigma_1\sigma_2}\, e^{-\frac{1}{2}\left(\frac{x_1^2}{\sigma_1^2}+\frac{x_2^2}{\sigma_2^2}\right)} dx_1 dx_2$

Also $\dfrac{u^2+v^2}{(\sigma_1^2+\sigma_2^2)}=\dfrac{x_1^2}{\sigma_1^2}+\dfrac{x_2^2}{\sigma_2^2}$ and $\dfrac{\partial(x_1, x_2)}{\partial(u, v)}=\dfrac{\sigma_1\sigma_2}{\sigma_1^2+\sigma_2^2}$.

Hence obtain the joint distribution of U and V.]

Ex. 3. If $(\sum_{i=1}^{n} c_i m_i)^2 = 9\sum_{i=1}^{n}(c_i^2\sigma_i^2)$, find $P(0 \le Y \le 2\sum_{i=1}^{n} c_i m_i)$ when $Y=\sum_{i=1}^{n} c_i X_i$

X_i being a normal variate with mean m_i and variance σ_i^2. **(*B.Sc All., 1988*)**

[**Hint.** We have $\mu^2=9\sigma^2 \Rightarrow \mu=3\sigma, \mu=EY$.

Hence reqd. Prob. $=P(0\le y\le 2\mu) = P(-3\le z\le 3)$] [**Ans.** ·997]

8·3·1. Distribution function ϕ (Z) of standard normal variate. The distribution function $\phi(Z)$ of standard normal variate Z is defined by

$$\phi(z)=P(Z\le z)=\int_{-\infty}^{z} f(u)\,du=\frac{1}{\sqrt{2\pi}}\int_{-\infty}^{z} e^{-\frac{1}{2}u^2}\,du, \text{ where } Z=(X-\mu)/\sigma.$$

Important Properties of ϕ (Z)

I. $\phi(-z)=1-\phi(z)$

Proof. $\phi(-z)=P(Z\le -z)$,

$=P(Z\ge z)$ as the normal curve is symmetrical. ...(1)

$=1-P(Z<z) = 1-\phi(z)$.

Cor. $P(z\le a)=P(z\ge b) \Rightarrow a=-b$

For $P(z\ge b)=P(z\le -b)=P(z=a)$ $\therefore\ -b=a$ or $a=-b$.

II. $P(a\le X\le b)=\phi\left(\dfrac{b-\mu}{\sigma}\right)-\phi\left(\dfrac{a-\mu}{\sigma}\right)$

Proof. $P(a\le X\le b)=P\left(\dfrac{a-\mu}{\sigma}\le Z\le\dfrac{b-\mu}{\sigma}\right),\ Z=\dfrac{X-\mu}{\sigma}$

$$=P\left(Z\le\frac{b-\mu}{\sigma}\right)-P\left(Z\le\frac{a-\mu}{\sigma}\right)=\phi\left(\frac{b-\mu}{\sigma}\right)-\phi\left(\frac{a-\mu}{\sigma}\right)$$

8·3·2. Table for Ordinates. Table 1 on page 325 gives the values of the ordinate $\frac{1}{\sqrt{2\pi}} e^{-x^2/2\sigma^2}$

at interval 0·01, the origin x is *assumed* at the mean. Dividing these values by σ, we get the values of the ordinate y where $y = \frac{1}{\sigma\sqrt{2\pi}} e^{-x^2/2\sigma^2}$

8·3·3. Fitting of the Normal Curve.

Example 8·2. *Two hundred and fifty-five metal rods were cut roughly 6 inches oversize. Finally the length of the oversize amount were measured exactly and grouped with 1-inch interval, there being in all 12 groups $\frac{1}{2}'' - 1\frac{1}{2}''$. $1\frac{1}{2}'' - 2\frac{1}{2}''$,......, $11\frac{1}{2}'' - 12\frac{1}{2}''$. The frequency distribution for the 255 lengths was*

Central value (Length inches) x :	1	2	3	4	5	6	7	8	9	10	11	12
Frequency y :	2	10	19	25	40	44	41	28	25	15	5	1

It is required to fit a normal (or Gaussian) curve to this data.

The equation of the normal curve for N observations is

$$y = \frac{N}{\sigma\sqrt{2\pi}} e^{-\frac{1}{2}\left(\frac{x-\mu}{\sigma}\right)^2}$$

Here $N = 255$. We first require to find the estimates $\bar{x}$ and s^2 for the population mean m and variance σ^2. We find

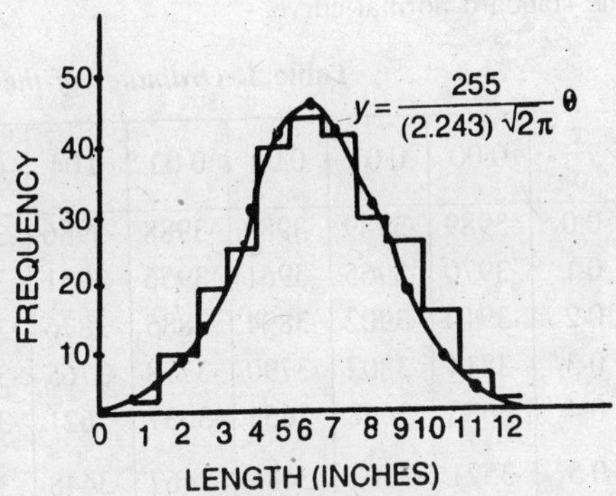

Fig. 8·2

$$\bar{x} = a + \frac{1}{N}\Sigma fu \quad \text{where } u = x - a$$

$$= 6 + \frac{66}{255} = 6{\cdot}259''$$

$$s^2 = \frac{1}{N}\Sigma fu^2 - \left(\frac{\Sigma fu}{N}\right)^2$$

$$= \frac{1300}{255} - (0{\cdot}259)^2 = 5{\cdot}031 \quad \text{or} \quad s = 2{\cdot}243''.$$

Hence $y = \frac{255}{2{\cdot}243\sqrt{2\pi}} e^{-\frac{1}{2}\frac{(x-6\cdot259)^2}{5\cdot031}} = 113{\cdot}68 \frac{1}{\sqrt{2\pi}} e^{-0\cdot99(x-6\cdot259)^2}$.

Consulting the table of ordinates, we have the following table for different values of x :

$x-6{\cdot}259$:	0	±1	±2	±3	±4	±5	±6
y :	45·35	41·06	30·48	11·54	9·25	3·78	1·27

The histogram and the curve had been shown in Fig. 8·2.

Ex. 1. Fit a normal curve to the following data :

Length of line in cms :	8·60	8·59	8·58	8·57	8·56	8·55	8·54	8·53	8·52
Frequency	1	2	3	4	9	10	8	4	1

Ans. Mean = 8·563 cms., S.D. = 0·0175 cm., class-intervals = 1·76.

The equation of the normal curve to be fitted is $y = 9{\cdot}8\, e^{-0{\cdot}163x^2}$

Ex. 2. Draw the probability curve for the standard normal variate and examine how the standard Binomial probability curve of Ex. 2 of section 8·1 approaches it as n increases.

Ex. 3. Draw standardised Poisson probability curve $\left(i.e., \dfrac{e^{-m} m^r}{r!} \right.$ against $z = \dfrac{r-m}{\sqrt{m}}$ for $m =$ 3, 5, 7, 9 and examine how (with increasing m), these approach the standard normal curve.

Table 1. *Ordinates of the Normal Curve*

$\frac{x}{\sigma}$	0·00	0·01	0·02	0·03	0·04	0·05	0·06	0·07	0·08	0·09
0·0	·3989	·3989	·3989	·3988	·3986	·3984	·3982	·3980	·3977	·3973
0·1	·3970	·3965	·3961	·3956	·3951	·3945	·3939	·3932	·3925	·3918
0·2	·3910	·3902	·3894	·3885	·3876	·3867	·3857	·3847	·3836	·3825
0·3	·3814	·3802	·3790	·3778	·3765	·3752	·3739	·3725	·3712	·3697
0·4	·3683	·3668	·3653	·3637	·3621	·3605	·3589	·3572	·3565	·3538
0·5	·3521	·3503	·3485	·3467	·3448	·3429	·3410	·3391	·3372	·3352
0·6	3332	·3312	·3292	·3271	·3251	·3230	·3209	·3187	·3166	·3144
0·7	·3123	·3101	·3079	·3056	·3034	·3011	·2989	·2966	·2943	·2926
0·8	·2897	·2874	·2850	·2827	·2803	·2780	·2756	·2732	·2709	·2685
0·9	·2661	·2637	·2613	·2589	·2565	·2541	·2516	·2492	·2468	·2444
1·0	·2420	·2396	·2371	·2347	·2323	·2299	·2275	·2251	·2227	·2203
1·1	·2179	·2155	·2131	·2107	·2083	·2059	·2036	·2012	·1989	·1965
1·2	·1942	·1919	·1895	·1872	·1849	·1826	·1804	·1781	·1768	·1736
1·3	·1714	·1691	·1669	·1647	·1626	·1604	·1582	·1561	·1539	·1518
1·4	·1497	·1476	·1456	·1435	·1415	·1394	·1374	·1354	·1334	·1315
1·5	·1295	·1276	·1257	·1238	·1219	·1200	·1182	·1163	·1145	·1127
1·6	·1109	·1092	·1074	·1057	·1040	·1023	·1006	·0989	·0973	·0957
1·7	·0940	·0925	·0909	·0893	·0878	·0863	·0848	·0833	·0818	·0804
1·8	·0790	·0775	·0761	·0748	·0734	·0721	·0707	·0694	·0681	·0669
1·9	·0656	·0644	·0632	·0620	·0608	·0596	·0584	·0573	·0562	·0551

$\frac{x}{\sigma}$	0·00	0·01	0·02	0·03	0·04	0·05	0·06	0·07	0·08	0·09
2·0	·0540	·0529	·0519	·0508	·0498	·0488	·0478	·0468	·0459	·0449
2·1	·0440	·0431	·0422	·0413	·0404	·0395	·0387	·0379	·0371	·0363
2·2	·0355	·0347	·0339	·0332	·0325	·0317	·0310	·0303	·0297	·0290
2·3	·0283	·0277	·0270	·0264	·0258	·0252	·0246	·0241	·0235	·0229
2·4	·0224	·0219	·0213	·0208	·0203	·0198	·0194	·0189	·0184	·0180
2·5	·0175	·0171	·0167	·0163	·0158	·0154	·0151	·0147	·0143	·0139
2·6	·0136	·0132	·0129	·0126	·0122	·0119	·0116	·0113	·0110	·0107
2·7	·0104	·0101	·0099	·0096	·0093	·0091	·0088	·0086	·0084	·0081
2·8	·0079	·0077	·0075	·0073	·0071	·0069	·0067	·0065	·0063	·0061
2·9	·0060	·0058	·0056	·0055	·0053	·0051	·0050	·0048	·0047	·0046
3·0	·0044	·0043	·0042	·0040	·0039	·0038	·0037	·0036	·0035	·0034
3·1	·0033	·0032	·0031	·0030	·0029	·0028	·0027	·0026	·0026	·0025
3·2	·0024	·0023	·0022	·0022	·0021	·0020	·0020	·0019	·0018	·0018
3·3	·0017	·0017	·0016	·0016	·0015	·0015	·0014	·0014	·0013	·0013
3·4	·0012	·0012	·0012	·0011	·0011	·0010	·0010	·0010	·0009	·0009
3·5	·0009	·0008	·0008	·0008	·0008	·0007	·0007	·0007	·0007	·0006
3·6	·0006	·0006	·0005	·0005	·0005	·0005	·0005	·0005	·0005	·0004
3·7	·0004	·0004	·0004	·0004	·0004	·0004	·0003	·0003	·0003	·0003
3·8	·0003	·0003	·0003	·0003	·0003	·0002	·0002	·0002	·0002	·0002
3·9	·0002	·0002	·0002	·0002	·0002	·0002	·0002	·0002	·0001	·0001

The table gives the value of $\phi(z) = \dfrac{1}{\sqrt{2\pi}}\, e^{-\frac{1}{2}z^2}$ for z varying from 0·00 to 3·99. It is easily seen that

$$f(x) = \frac{1}{\sqrt{2\pi\,\sigma^2}}\, e^{-\frac{1}{2}\left(\frac{x-m}{\sigma}\right)^2} = \frac{1}{\sigma}\,\phi\left(\frac{x-m}{\sigma}\right)$$

8·4·1. The Normal Probability Integral or Area under the Normal Curve. The probability corresponding to any interval in the range of the variate is given by the area under the curve (8·3) within the interval and the total area under the curve is unity. In particular, the probability P for the interval from the mean μ to the value x is given by the definite integral.

$$P(\mu \le X < z) = \frac{1}{\sigma\sqrt{2\pi}} \int_{\mu}^{z} e^{-\frac{1}{2\sigma^2}(x-\mu)^2}\, dx = \frac{1}{\sqrt{2\pi}} \int_{0}^{z} e^{-\frac{1}{2}z^2}\, dz,\ z = \frac{x-\mu}{\sigma}.$$

The values of P for different values of z at intervals of 0·01 have been tabulated (see Table 8·2). The value of P given by the above definite

Table 8·2. *Area under the Normal Curve*

$\frac{x}{\sigma}$	0·00	0·01	0·02	0·03	0·04	0·05	0·06	0·07	0·08	0·09
0·0	·0000	·0040	·0080	·0120	·0159	·0199	·0239	·0279	·0319	·0359
0·1	·0398	·0438	·0478	·0517	·0557	·0596	·0636	·0675	·0714	·0753
0·2	·0793	·0832	·0871	·0910	·0948	·0987	·1026	·1064	·1103	·1141
0·3	·1179	·1217	·1255	·1293	·1331	·1368	·1406	·1443	·1480	·1517
0·4	·1554	·1591	·1628	·1664	·1700	·1736	·1772	·1808	·1844	·1879
0·5	·1915	·1950	·1985	·2019	·2054	·2088	·2123	·2157	·2190	·2224
0·6	·2257	·2291	·2324	·2357	·2389	·2422	·2454	·2486	·2518	·2549
0·7	·2580	·2611	·2642	·2673	·2704	·2734	·2764	·2794	·2823	·2852
0·8	·2881	·2910	·2939	·2967	·2995	·3023	·3051	·3078	·3106	·3133
0·9	·3159	·3186	·3212	·3238	·3264	·3289	·3315	·3340	·3365	·3389
1·0	·3413	·3438	·3461	·3485	·3508	·3531	·3554	·3577	·3599	·3621
1·1	·3643	·3665	·3686	·3708	·3729	·3749	·3770	·3790	·3910	·3830
1·2	·3849	·3869	·3888	·3907	·3925	·3944	·3962	·3980	·3997	·4015
1·3	·4032	·4049	·4066	·4082	·4099	·4115	·4131	·4147	·4162	·4177
1·4	·4192	·4207	·4222	·4236	·4251	·4265	·4279	·4292	·4306	·4319
1·5	·4332	·4345	·4357	·4370	·4382	·4394	·4406	·4418	·4430	·4441
1·6	·4452	·4463	·4474	·4485	·4495	·4505	·4515	·4525	·4535	·4545
1·7	·4554	·4564	·4573	·4582	·4591	·4599	·4608	·4616	·4625	·4633
1·8	·4641	·4649	·4656	·4664	·4671	·4678	·4686	·4693	·4699	·4706
1·9	·4713	·4719	·4726	·4732	·4738	·4744	·4750	·4756	·4762	·4767
2·0	·4772	·4778	·4783	·4788	·4793	·4798	·4803	·4808	·4812	·4817
2·1	·4821	·4826	·4830	·4834	·4838	·4842	·4846	·4850	·4854	·4857
2·2	·4861	·4865	·4868	·4871	·4875	·4878	·4881	·4884	·4887	·4890
2·3	·4893	·4896	·4898	·4901	·4904	·4906	·4909	·4911	·4913	·4916
2·4	·4918	·4920	·4922	·4925	·4927	·4929	·4931	·4932	·4934	·4936
2·5	·4938	·4940	·4941	·4943	·4945	·4946	·4948	·4949	·4951	·4952
2·6	·4953	·4955	·4956	·4957	·4959	·4960	·4961	·4962	·4963	·4964
2·7	·4965	·4966	·4967	·4968	·4969	·4970	·4971	·4972	·4973	·4974
2·8	·4974	·4975	·4976	·4977	·4977	·4978	·4979	·4980	·4980	·4981
2·9	·4981	·4982	·4983	·4983	·4984	·4984	·4985	·4985	·4986	·4986
3·0	·49865	·4987	·4987	·4988	·4988	·4989	·4989	·4989	·4990	·4990
3·1	·49903	·4991	·4991	·4991	·4992	·4992	·4992	·4992	·4993	·4993

The table gives the value of z varying from 0·0 to 3·19.

$$\phi(z) = P(\ Z \le z\ ) = \frac{1}{\sqrt{2\pi}} \int_0^z e^{-\frac{1}{2}z^2}\, dz$$

integral is called the normal *probability integral*. This name is due to the use of the normal distribution in the theory of sampling.

The probability that a random value of the normal variate will fall within the interval $x = \mu - \sigma$ to $x = \mu + \sigma$ is

$$P[\mu - \sigma < X < \mu + \sigma] = \frac{1}{\sigma\sqrt{2\pi}} \int_{\mu-\sigma}^{\mu+\sigma} e^{-\frac{(x-\mu)^2}{2\sigma^2}} dx$$

or
$$P[-1 < Z < 1] = \frac{1}{\sqrt{2\pi}} \int_{-1}^{+1} e^{-\frac{1}{2}z^2} dz = \frac{2}{\sqrt{2\pi}} \int_{0}^{1} e^{-\frac{1}{2}z^2} dz$$

$$= 0{\cdot}6826, \text{ from the table 8·2 for } z = 1.$$

The probability that a random value of the normal variate will deviate more than σ from the mean is accordingly $(1 - 0{\cdot}6826) = 0{\cdot}3174$ which is less that $\frac{1}{3}$. In other words, the area outside the interval $x = \mu - \sigma$ to $x = \mu + \sigma$ is less than $\frac{1}{3}$. The probability that a random value of the normal variate deviates more than 2σ from the mean is

$$P\left[\left|\frac{x-\mu}{\sigma}\right| \geq 2\right] = 1 - P[\mu - 2\sigma < x < \mu + 2\sigma]$$

$$= 1 - \frac{1}{\sigma\sqrt{2\pi}} \int_{\mu-2\sigma}^{\mu+2\sigma} e^{-\frac{(x-\mu)^2}{2\sigma^2}} dx$$

$$= 1 - P[-2 \leq z < 2] = 1 - 2.\ \frac{1}{\sqrt{2\pi}} \int_{0}^{2} e^{-\frac{1}{2}z^2} dz \text{ where } z = \frac{x-\mu}{\sigma}$$

$$= 1 - 2[0{\cdot}4772] = 0{\cdot}0456 \text{ by using table 8·2 with } z = 2.$$

$\therefore\ P[\,|z| > 2] = 0{\cdot}0456$ which is less than 5%.

Similarly, $P[\,|z| > 3] = 0{\cdot}0027$.

In other words, on an average in 3 cases out of 1000 the deviation will be greater than 3 σ from the mean. From the table 8·2, we find that the deviation from the mean which is exceeded with a probability of 5% is $1{\cdot}96\,\sigma$; and that which is exceeded with a probability of 1% is $2{\cdot}58\,\sigma$.

8·4·2. The *probable error*, λ, or *quartile deviation* is the error such that the probability of an error falling within the limits $\mu - \lambda$ to $\mu + \lambda$ is exactly equal to the chance of an error falling outside these limits, that is, it is an even chance whether a particular error falls within these limits or not.

Hence $$\frac{1}{\sigma\sqrt{2\pi}} \int_{\mu-\lambda}^{\mu+\lambda} e^{-\frac{1}{2}\left(\frac{x-\mu}{\sigma}\right)^2} dx = \tfrac{1}{2}$$

or $$\frac{1}{\sqrt{2\pi}} \int_{0}^{\lambda/\sigma} e^{-\frac{1}{2}z^2} dz = \tfrac{1}{4}. \qquad \left(z = \frac{x-\mu}{\sigma} \right)$$

The table 8·2 gives $\lambda/\sigma = 0{\cdot}6745$ or $\lambda = 0{\cdot}6745\,\sigma = \frac{2}{3}\sigma$ approx.

This gives for the normal distribution :

Lower quartile $= \mu - 0{\cdot}6745\,\sigma$ and upper quartile $= \mu + 0{\cdot}6745\sigma$.

Hence quartile deviation $= \frac{1}{2}(Q_3 - Q_1) \doteq 0{\cdot}6745\ \sigma = (2/3) \times$ (s.d.) approx.

Ex. Using Table 2, show that the deciles of the normal curve are :
$D_1 = \mu - 1{\cdot}282\sigma$, $D_2 = \mu - 0{\cdot}842\sigma$, $D_3 = \mu - 0{\cdot}525\sigma$, $D_4 = \mu - 0{\cdot}253\sigma$,
$D_5 = \mu$, $D_6 = \mu + 0{\cdot}253\sigma$, $D_7 = \mu + 0{\cdot}525\sigma$, $D_8 = \mu + 0{\cdot}842\sigma$, $D_9 = \mu + 1{\cdot}282\sigma$.

8·4·3. *Error Function* The normal probability function is

$$f(x) = \frac{1}{\sigma\sqrt{2\pi}}\ e^{-x^2/2\sigma^2}, \quad -\infty < x < \infty$$

Let $h^2 = 1/2\sigma^2$, then $f(x) = \frac{h}{\sqrt{\pi}}\ e^{-h^2x^2}$

The probability that the value of the r.v. lies in the range $\pm x$ is given by

$$P(-x < X < x) = P = \int_{-x}^{x} f(x)\,dx = \frac{h}{\sqrt{\pi}} \int_{-x}^{x} e^{-h^2x^2} dx$$

$$= \frac{2h}{\sqrt{\pi}} \int_{0}^{x} e^{-h^2x^2} dx = \frac{2}{\sqrt{\pi}} \int_{0}^{x} e^{-h^2x^2} (hdx) = \psi(hx)$$

where $$\psi(y) = \frac{2}{\sqrt{\pi}} \int_{0}^{y} e^{-y^2} dy$$

and is called the error function in Astronomy.

Example 8·3. (*a*) *In a normal distribution whose mean is 12 and s.d. 2, find the probability for the interval (area under the curve) from $x = 9{\cdot}6$ to $x = 13{\cdot}8$.* (*b*) *For a normal distribution with mean 2 and s.d. 3, find a value of the variate such that the probability of the variate from the mean to that value is 0·4115.*

(*a*) $P(9{\cdot}6 \le X \le 13{\cdot}8)$

$$= \frac{1}{2\sqrt{2\pi}} \int_{9\cdot6}^{13\cdot8} e^{-\frac{1}{2}\left(\frac{x-12}{2}\right)^2} dx = \frac{1}{\sqrt{2\pi}} \int_{-1\cdot2}^{0\cdot9} e^{-1/2\, z^2} dz, \quad \left(z = \frac{x-12}{2} \right)$$

$$= \frac{1}{\sqrt{2\pi}} \int_{-1\cdot2}^{0} e^{-\frac{1}{2}z^2} dz + \frac{1}{\sqrt{2\pi}} \int_{0}^{0\cdot9} e^{-\frac{1}{2}z^2} dz$$

$$= \frac{1}{\sqrt{2\pi}} \int_{0}^{1\cdot2} e^{-\frac{1}{2}z^2} dz + \frac{1}{\sqrt{2\pi}} \int_{0}^{0\cdot9} e^{-\frac{1}{2}z^2} dz$$

$= 0{\cdot}3849 + 0{\cdot}3159$ [from table 8·2, page 314] $= 0{\cdot}7008$.

(*b*) We have, $\dfrac{1}{3\sqrt{2\pi}} \displaystyle\int_{2}^{x} e^{-\frac{1}{2}\left(\frac{x-2}{3}\right)^2} dx = 0{\cdot}4115$

or $\quad \dfrac{1}{\sqrt{2\pi}} \displaystyle\int_{0}^{x} e^{-\frac{1}{2}z^2} dz = 0{\cdot}4115 \quad \left(z = \dfrac{x-2}{3} \right).$

From table 8·2, we find that $z = 1{\cdot}35$, and, therefore, $x = 6{\cdot}05$.

Example 8·4. *If skulls are classified as A, B, C according as the length–breadth index is under 75, between 75 and 80, over 80, find approximately (assuming that the distribution is normal) the mean and standard deviation of a series in which A are 58%, B are 38% and C are 4%, being given that if* $f(t) = \dfrac{1}{\sqrt{2\pi}} \displaystyle\int_0^t \exp\left(-\tfrac{1}{2}x^2\right) dx,$

then $f(0{\cdot}20) = 0{\cdot}08$ *and* $f(1{\cdot}75) = 0{\cdot}46$. **(*B.Sc., Delhi 1989*)**

If $t = 0{\cdot}20$, the area of the curve from 0 to t is 0·08, that is the area to the left of the ordinate at t is 0·58. The area corresponds to $x = 75$, *i.e.*, to $t = \dfrac{75-m}{\sigma}$. Hence $\dfrac{75-m}{\sigma} = 0{\cdot}20$ ...(1)

For $t = \dfrac{80-m}{\sigma}$, the area to the right of the ordinate at $x = 80$ is given to be 0·04 and hence the area to the left of this ordinate is 0·96. Thus the area of the curve from 0 to $t = 0{\cdot}96 - 0{\cdot}50 = 0{\cdot}46$ and the corresponding value of t is given as 1·75.

Hence $\dfrac{80-m}{\sigma} = 1{\cdot}75.$ ...(2)

Solving the equations (1) and (2), we get

$m = 74{\cdot}4$ approx. and $\sigma = 3{\cdot}2$ approx.

Example 8·5. *One thousand candidates in an examination were grouped into 3 classes I, II and III in descending order of merit. The numbers in the first two classes were 50 and 350 respectively. The highest and lowest marks in class II were 60 and 50 respectively. Assuming the distribution to be normal, prove that the average mark is approximately 48 and standard deviation is approximately 7.*

You may use the following table :

The area A is measured from the mean zero to any ordinate x.

$\frac{x}{\sigma}$	A	$\frac{x}{\sigma}$	A
0·2	0·079	1·5	0·433
0·3	0·118	1·6	0·445
0·4	0·155	1·7	0·455

(B.A. Aligarh, 1987)

The number of candidates getting III class is 600, that is 0·6 of the total number of candidates. In other words, the area to the left of the total ordinate at $x = 50$ is 0·6, that is, the area from 0 to z is $0·6 - 0·5 = 0·1$.

After interpolating we find that the value of z corresponding to $A = 0·1$ is 0·254. Hence $\dfrac{50 - m}{\sigma} = 0·254$ ...(A)

The number of candidates getting first class is 0·05 of the total number of candidates. In other words, the area to the left of the ordinate at $x = 60$ is 0·05. Hence the area from 0 to z is $0·5 - 0·05 = 0·45$. For this value of A, we find after interpolation, $z = 1·650$

Hence
$$\frac{60 - m}{\sigma} = 1·650 \qquad ...(B)$$

Solving the equations (A) and (B), we get

$$m = 48·2 \text{ approx}, \qquad \sigma = 7·1 \text{ approx.}$$

8·5. Alternative Proof of the Approach of Standard Binomial Probability Distribution to Normal Distribution. D'Moivre-Laplace Theorem.

We know that the mode of the binomial distribution is in the neighbourhood of np and thus for large n, the maximum probability will be of the order $\quad {}^nC_{np}\, p^{np} q^{nq}$, *i.e.*, $\dfrac{n\,!}{np\,!\; nq\,!}\, p^{np} q^{nq}$

By Stirling's approximation to the factorial, this is approx.

$$\frac{e^{-n} n^n \sqrt{2\pi n}\; p^{np} q^{nq}}{(np)^{np}\, e^{-np} \sqrt{2\pi np} \cdot (nq)^{nq}\, e^{-nq} \sqrt{2\pi nq}} \sim \frac{1}{\sqrt{2\pi npq}}$$

and, therefore, tends to zero.

Thus every probability value in the binomial distribution tends to zero, and, therefore, the binomial distribution as such does not tend to any limiting distribution. In § 8·1, it was proved that the distribution of standardised binomial variate $Z \left(= \dfrac{r - np}{\sqrt{npq}} \right)$, tends to the distribution of the standardised normal variate as n tends to infinity. That is

$$P \left\{ a \leq \frac{r - np}{\sqrt{npq}} \leq b \right\} \to \frac{1}{\sqrt{2\pi}} \int_a^b e^{-\frac{1}{2}z^2} \, dz.$$

In view of the importance of the result, we consider an *alternative approach by considering the limiting from of the characteristic function of the standardised binomial variate as n tends to infinity.*

The characteristic function of the binomial variate r is $(q + pe^{it})^n$ and, therefore, using § 5·5, we find that the charactristic function of the standard variate z is

$$\varphi(t) = \left[\exp \left(\frac{-itnp}{\sqrt{npq}} \right) \right] \left(q + pe^{\frac{it}{\sqrt{npq}}} \right)^n$$

$$\therefore \ \log_e \varphi(t) = \frac{-itnp}{\sqrt{npq}} + n \log_e \left(q + pe^{\frac{it}{\sqrt{npq}}} \right)$$

$$= \frac{-itnp}{\sqrt{npq}} + n \log_e \left(1 + \frac{itp}{\sqrt{npq}} - \frac{pt^2}{\sqrt{npq}} + \frac{p\theta t^3}{6(npq)^{3/2}} \right), \quad 0 \leq \theta \leq 1$$

$$= n \left\{ \frac{-pt^2}{2npq} - \frac{1}{2} \frac{(pit)^2}{npq} \right\} + 0(t^2 n^{-\frac{1}{2}}) = -\tfrac{1}{2}t^2 + 0(t^2 n^{-\frac{1}{2}})$$

Thus, for any finite t, $\log_e \varphi(t)$ approaches uniformly to $-\frac{1}{2}t^2$ and hence $\varphi(t) \to e^{-\frac{t^2}{2}}$

But this is the characteristic function of standardised normal distribution.

Thus the standardised binomial probability distribution tends to standardised normal distribution, *viz.*,

$$dp = \frac{1}{\sqrt{2\pi}} \, e^{-\frac{1}{2}z^2} \, dz, \ -\infty < z < \infty.$$

Ex. 1. Show that as m tends to infinity, the probability $\dfrac{e^{-m} m^r}{r!}$ tends to zero for every finite r. Deduce that the Poission distribution as such does not approach any limiting distribution.

Ex. 2. Find the characteristic function of the standard Poission variate

$(r-m)/\sqrt{m}$ and show that it tends to $e^{-\frac{1}{2}t^2}$ as m tends to infinity.

Thus establish that the standardised Poisson distribution converges to normal distribution.

Ex. 3. By putting $x = m + y$ and considering $|y| < m$ show that for large m,

$$\frac{e^{-m}m^x}{x\,!} \sim \frac{1}{\sqrt{2\pi m}} \exp\left\{-\frac{1}{2}\left(\frac{x-m}{\sqrt{m}}\right)^2 + 0(m^{-1/2})\right\}.$$

What conclusion can you draw from this ?

Ex. 4. Find the standardised random variate, for (*i*) discrete rectangular distribution (*ii*) geometric distribution (*iii*) negative binomial distribution (*iv*) hypergeometric distribution. Find its characteristic function in each case and investigate the conditions under which it approaches $e^{-\frac{1}{2}t^2}$

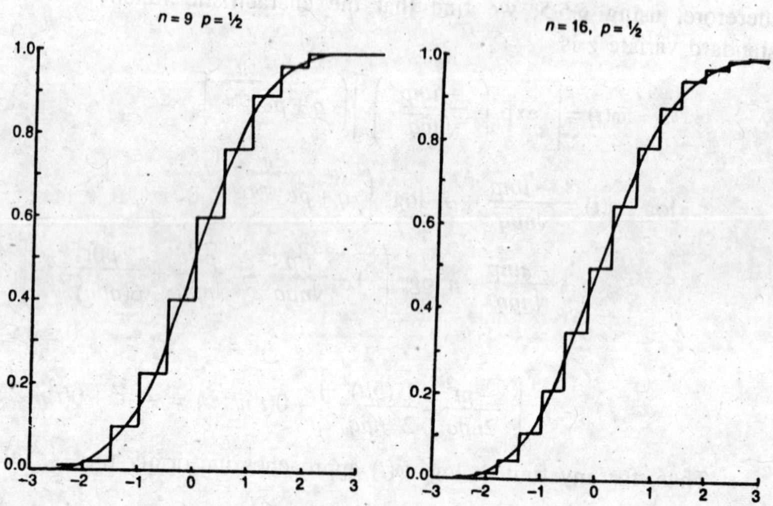

Fig. 8·3 (*a, b*)

Figures 8·3 (*a*), (*b*), (*c*) show how the *c.d.f.* for the standard binomial variate approaches the *c.d.f.* for the standard normal variate as n becomes large and large Figure 8·3 (*d*) illustrates the approach of the Poisson distribution to the normal distribution.

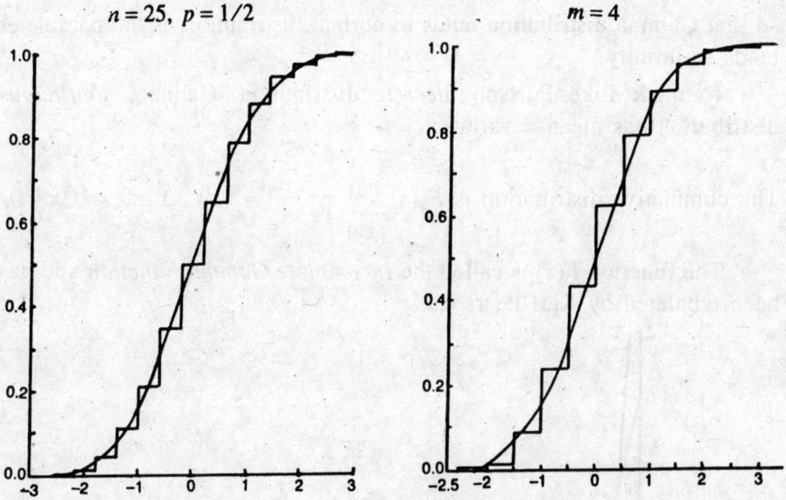

Fig. 8·3 (*c, d*)

8·6. Gamma Distribution. The continuous random variate X which is distributed with probability density

$$f(x) = \frac{m^l}{\Gamma l} x^{l-1} e^{-mx}, x \geq 0, l \geq 0, \ m > 0 \qquad ...(8·6)$$

is called a *Gamma variate with parameters l and m* and its distribution is called the Gamma distribution. We shall denote a Gamma variate with one parameter l by $\gamma(l)$. The function $f(x)$ given by (8·6) represents a probability density, since its integral over the range $(0, \infty)$ is unity.

The *m.g.f.* with respect to origin for gamma distribution with$m = 1$ is

$$M_{x=0}(t) = \int_0^\infty e^{tx} \frac{1}{\Gamma(l)} x^{l-1} e^{-x} \, dx = \frac{1}{\Gamma(l)} \int_0^\infty x^{l-1} \, e^{-x(t-1)} \, dx$$

$$= (1-t)^{-l}, \text{ provided that } |t| < 1$$

Put $\qquad x(1-t) = y \Rightarrow dx = \frac{dy}{1 \ t}$

On differentiating $M_{x=0}(t)$, r times with respect to t and putting $t = 0$, we find $\quad \mu_r' = l(l+1).....(l+r-1)$.

In particular, $\mu_1' = l, \mu_2' = l(l+1)$ and $\mu_2 = l, \ \mu_3 = 2l$ and $\mu_4 = 6l + 3l^2$

Hence $\qquad \beta_1 = \frac{\mu_3^2}{\mu_2^3} = \frac{4}{l}$ and $\beta_2 = \frac{\mu_4}{\mu_2^2} = 3 + \frac{6}{l}.$

Hence as $l \to \infty$, $\beta_1 \to 0$ and $\beta_2 \to 3$

so that Gamma distribution tends to normal distribution as the parameter tends to infinity.

Remark Like Poisson *discrete* distribution, Gamma, *continuous* distribution has mean = variance.

The cumulative distribution is $F_X(x) = \int_0^x \frac{1}{\Gamma l} x^{l-1} e^{-x} dx, \ x \geq 0; = 0, x < 0.$

The function $F(x)$ is called the *Incomplete Gamma Function* and has been tabulated by Karl Pearson.

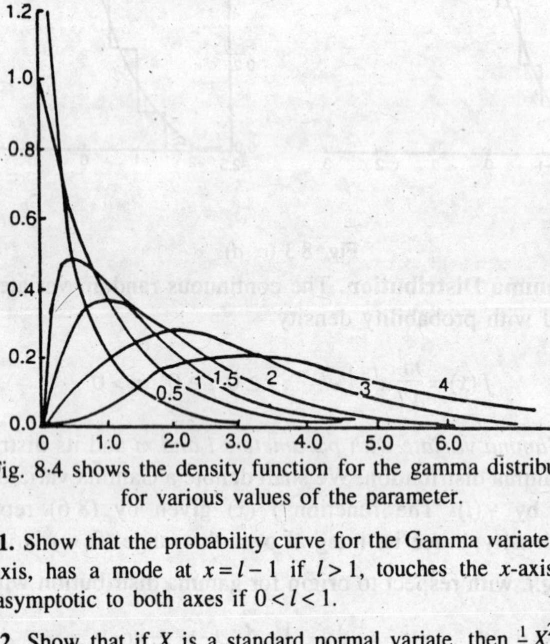

Fig. 8·4 shows the density function for the gamma distribution for various values of the parameter.

Ex. 1. Show that the probability curve for the Gamma variate is asymptotic to the x-axis, has a mode at $x = l - 1$ if $l > 1$, touches the x-axis at origin, if $l > 2$ and asymptotic to both axes if $0 < l < 1$.

Ex. 2. Show that if X is a standard normal variate, then $\frac{1}{2}X^2$ is a Gamma variate with parameter $\frac{1}{2}$.

Ex. 3. Find the *c.f.* for the standard gamma variate and show that it approaches $e^{-\frac{1}{2}t^2}$ as $l \to \infty$

Additive Property of Gamma Variates. *The sum of two independent Gamma variates with parameters l and m is a Gamma variate with parameter $l + m$.*

Let X and Y be two independent Gamma variates with parameters l and m respectively. The *m.g.f.* of their sum is the product of their *m.g.f.* (see § 5·7·2).

Therefore, *m.g.f.* of the sum $= (1-t)^{-l} \times (1-t)^{-m} = (1-t)^{-(l+m)}$.

But this is the *m.g.f.* of a Gamma variate with parameter $l+m$. Hence the result follows.

Note. The result can be generalized to any number of independent Gamma variates.

Ex. 1. Show that if the sum of two independent positive variates is a Gamma variate with parameter $l+m$, and one of them is a Gamma variate with parameter l, then the other is also a Gamma variate with parameter m. [use : §5·7·2]

Ex. 2. Show that if X_i are n independent normal variates with means μ_i and variances σ_i^2, then

$$U = \frac{1}{2}\sum_{i=1}^{n}\frac{(X_i-\mu_i)^2}{\sigma_i^2} \Rightarrow \frac{x-\mu}{\sigma} = \sqrt{2u},\ dx = \frac{\sigma\, du}{\sqrt{2u}}$$

is a Gamma variate with parameter $\frac{1}{2}n$. ***(B.Sc. Meerut, 1993)***

Ex. 3. If X, Y are independent Gamma variates with parameters l and m respectively, then show that

$U = X+Y,\ V = (X-Y)/(X+Y)$ are independent variates

and that U is $\gamma(l+m)$ variate. ***(M.A. Rajasthan, 1984)***

Ex. 4. If X and Y are independent standard normal variates, find the distributions of (*i*) $X+Y$ (*ii*) X^2 (*iii*) X^2+Y^2 (*iv*) X^2/Y^2.

(M.A. Statistics Delhi, 1989)

Ex. 5. If X and Y are independent gamma variates, find the distribution of (*i*) $X+Y$ (*ii*) $X/(X+Y)$ (*ii*) X/Y. ***(B.A. Statistics Hons. Delhi, 1991)***

[**Hint.** Put $x = r\cos^2\theta,\ y = r\sin^2\theta,\ dxdy = 2r\cos\theta\sin\theta\, d\theta\, dr$]

Ex. 6. If X is a gamma variate with parameter show that $P(0 < X < 2l) \le \dfrac{l-1}{l}$.

Ex. 7. Prove that for two parameters gamma distribution (8·6), mean = l/m, variance = l/m^2. Thus variance $\gtreqless$ mean if $m \lesseqgtr l$.

Cumulative Generating Function

The cumulative generating function $K_x(t)$ is given by

$$K_x(t) = \log M_x(t) = \log(1-t)^{-l} = -l\log(1-t),\ |t| < 1$$

$$= l\left(t + \frac{t^2}{2} + \frac{t^3}{3} + \frac{t^4}{4} + \ldots\right)$$

$\therefore$ Mean $= k_1 =$ coeff. of t in $K_x(t) = l$;

$\mu_2 = k_2 =$ coeff. of $t^2/2!$ in $K_x(t) = l$, $k_3 =$ coeff. of $t^3/3!$ in $K_x(t) = 2l$

$k_4 =$ coeff. of $t^4/4!$ in $K_x(t) = 6l$, $\mu_3 = k_3 = 2l$, $\mu_4 = k_4 + 3k_2^2 = 6l + 3l^2$

Hence $\beta_1 = \mu_3^2/\mu_2^3 = 4l^2/l^3 = 4/l$; $\beta_2 = \dfrac{\mu_4}{\mu_2^2} = (6l + 3l^2)/l^2 = 3 + 6/l$

$\therefore \quad \beta_1 \to 0$ and $\beta_2 \to 3$ as $l \to \infty$.

Limiting form of Gamma distribution.

Let us first write down the *m.g.f.* of Z where Z is a standard Gamma variate *i.e.* $Z = (X - \mu)/\sigma = (X - l)/\sqrt{l}$

Now $M_z(t) = e^{-\mu t/\sigma} M_x(t/\sigma) = e^{-\mu t/\sigma}(1 - t/\sigma)^{-l} = e^{-lt/\sqrt{l}}(1 - t/\sqrt{l})^{-l}$

$$\therefore \quad \log M_z(t) = -t\sqrt{l} - l \log\left(1 - \frac{t}{\sqrt{l}}\right)$$

$$= -t\sqrt{l} + l\left[\frac{t}{\sqrt{l}} + \frac{t^2}{2l} + \ldots\right] = \frac{t^2}{2} + 0(l^{-1/2})$$

$$\therefore \log M_z(t) \to \frac{t^2}{2} \text{ as } l \to \infty \quad \Rightarrow \quad M_z(t) \to e^{t^2/2} \text{ as } l \to \infty$$

which is the *m.g.f.* of a standard normal variate. Hence by uniqueness theorem, the Gamma distribution tends to the normal distribution, as the parameter l increases indefinitely.

8·7. Beta Distributions.

8·7·1. Beta Distribution of the First Kind. The continuous random variate X which is distributed with probability density

$$f(x) = \frac{1}{B(l, m)} x^{l-1}(1-x)^{m-1}, \; 0 \le x \le 1, \; l, m > 0 \qquad \ldots(8\cdot7)$$
$$= 0, \text{ elsewhere}$$

is a Beta variate of the first kind with parameters l and m and is briefly referred to as a $\beta_1(l, m)$ variate. the *m.g.f.* with respect to the origin for this distribution does not have a simple form but the moments are easily found directly ;

$$\mu_r' \text{ about origin} = \frac{1}{B(l, m)} \int_0^1 x^{r+l-1}(1-x)^{m-1}\, dx = \frac{B(r+l, m)}{B(l, m)}$$

$$= \frac{l(l+1)\ldots\ldots(l+r-1)}{(l+m)(l+m+1)\ldots\ldots(l+m+r-1)}, \; r = 1, 2, \ldots.$$

In particular, $\mu_1' = \dfrac{l}{l+m}$, $\mu_2' = \dfrac{l(l+1)}{(l+m)(l+m+1)}$, $\mu_2 = \dfrac{lm}{(l+m)^2(l+m+1)}$

Putting $r = 3$ and $r = 4$ we successively obtain μ_3 and μ_4 and hence we can show that $\quad \mu_3 = \dfrac{2lm(m-l)}{(l+m)^3(l+m+1)(l+m+2)}$,

$$\mu_4 = \frac{3lm\{lm(l+m-6) + 2(l+m)^2\}}{(l+m)^4(l+m+1)(l+m+2)(l+m+3)}$$

The cumulative distribution, often called the *Incomplete Beta function, is* $F_X(x) = 0,\ x < 0$

$$f(x) = \begin{cases} 0, & x < 0 \\ \int_0^x \frac{x^{e-1}(1-x)^{m-1}}{B(l, m)}\, dx & 0 \le x \le, \quad l, m > 0 \\ 1, & x > 1 \end{cases}$$

$0 \le x \le 1,\ l, m > 0$

$= 1, \quad x > 1.$

Note. In case $l = 1 = m$,

$$f(x) = \frac{1}{B(1, 1)} \cdot 1 = 1, 0 < x < 1$$

which is the probability function of uniform distri- bution in the range (0, 1).

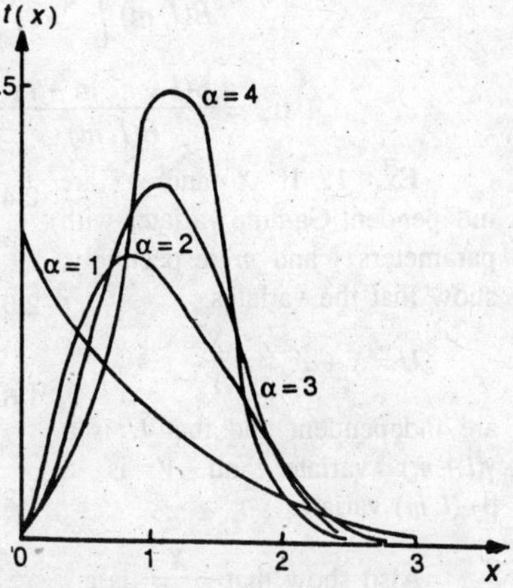

Fig. 8·5. shows the density function for Beta distribution of the first kind for various values of l and m

8·7·2. Beta distribution of the Second Kind. The continuous random variate X which is distributed with probability density

$$f(x) = \frac{1}{B(l, m)} \cdot \frac{x^{l-1}}{(1+x)^{l+m}}, x \ge 0, l, m > 0 \qquad ...(8\cdot8)$$

$$= 0 \qquad , x < 0$$

is a Beta variate of second kind with parameters l and m and may be briefly referred to as $\beta_2(l, m)$ variate. The function given by (8·8) represents a probability density since

$$\int_0^\infty f(x)dx = \int_0^\infty \frac{1}{B(l, m)} \cdot \frac{x^{l-1}}{(1+x)^{l+m}} dx$$

$$= \frac{1}{B(l, m)} \int_0^1 y^{m-1} (1-y)^{l-1} dy,$$

on putting $1 + x = y^{-1},\ dx = dy/y^2$.

$$\therefore \quad \int_0^\infty f(x)dx = 1, \text{ since } B(l, m) = B(m, l)$$

$$\text{Also } \mu_r' \text{ about origin} = \int_0^\infty \frac{1}{B(l, m)} \cdot \frac{x^{l+r-1}}{(1+x)^{l+m}} dx$$

$$= \frac{1}{B(l, m)} \int_0^1 y^{m-r-1} (1-y)^{l+r-1} dy, \quad 1 + x = y^{-1}$$

$$= \frac{1}{B(l, m)} \int_0^1 z^{l+r-1} (1-z)^{m-r-1} \, dz \text{ where } z = 1 - y$$

$$\therefore \qquad \mu_r' = \frac{B(l+r, m-r)}{B(l, m)} = \frac{l(l+1)\ldots(l+r-1)}{(m-1)(m-2)\ldots(m-r)}, r < m.$$

Ex. 1. If X and Y are independent Gamma variates with parameters l and m respectively, show that the variates

$$U = X + Y, \; V = \frac{X}{Y}$$

are independent and that U is a $\gamma(l+m)$ variate and V is a $\beta_2(l, m)$ variate.

Also show that $\frac{X}{X+Y}$ is a $\beta_1(l, m)$ variate.

(B.A. Statistics Hons., Delhi 1991)

Ex. 2. For what values of the parameters does the Beta distribution become uniform distribution over the unit interval ?

Ans. each parameter = 1.

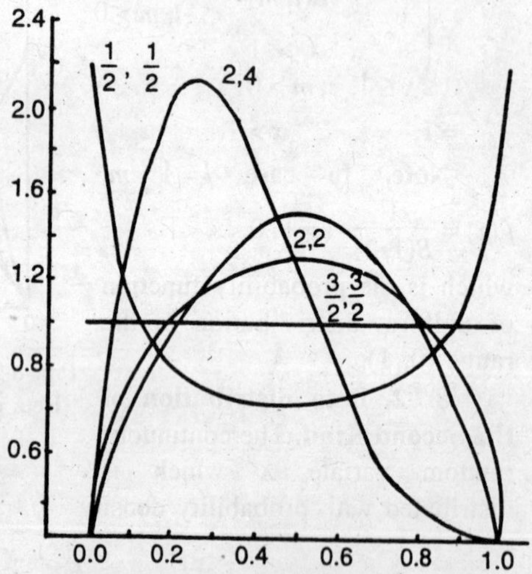

Fig. 8·6 shows the density function for Beta distribution of the second kind for various values of l and m.

8·8. Cauchy Distribution. We have already come across this distribution *viz*,

$$f_X(x) = \frac{1}{\pi} \cdot \frac{1}{1+(x-\mu)^2}, \; -\infty < x < \infty \qquad \ldots(8\cdot9)$$

in Example 7·5. It is called the Cauchy distribution with parameter μ. The distribution is quite useful in statistical theory as providing an example of a distribution for which a number of theorems fail and thus emphasizing that in Statistics also we should be careful to examine the conditions under which theorems hold.

If $\mu = 0$, then (1) is called the Standard Cauchy distribution. If the parameters are λ, μ, then the Cauchy distribution is

$$\frac{\lambda}{\pi\,[\lambda^2 + (x-\mu)^2]}, \; -\infty < x < \infty, \lambda > 0, \mu > 0$$

The density probability curve (8·2) is symmetrical about the point $X = \mu$ and thus $X = \mu$ gives us the median of the distribution. Also the density function is obviously maximum at $X = \mu$ so that this point is also the mode. To find the mean we have,

$$\mu_1' = \frac{1}{\pi}\int_{-\infty}^{\infty} \frac{x\,dx}{1+(x-\mu)^2} = \frac{1}{\pi}\int_{-\infty}^{\infty} \frac{(x-\mu)dx}{1+(x-\mu)^2} + \frac{\mu}{\pi}\int_{-\infty}^{\infty} \frac{dx}{1+(x-\mu)^2}$$

$$= \mu + \frac{1}{\pi}\int_{-\infty}^{\infty} \frac{y\,dy}{1+y^2} \qquad \left[\because \frac{1}{\pi}\int_{-\infty}^{\infty} \frac{dx}{1+(x-\mu)^2} = \frac{1}{\pi}\Big|\tan^{-1}(x-\mu)\Big|_{-\infty}^{+\infty} = 1\right]$$

$$= \mu + \frac{1}{\pi} \underset{\substack{X\to\infty \\ X'\to\infty}}{\text{Lt}} \int_{-X}^{X'} \frac{y\,dy}{1+y^2}$$

If X, X' approach infinity independently, the second integral does not converge to a limit and thus in the most general sense, the mean of the Cauchy distribution does not exist. However, if $X' = X$, the integral vanishes and thus if we take the principal value of the integral, the mean of the Cauchy distribution is at μ. Thus in this sense, the median and mode of the distribution coincide and the distribution is symmetrical.

The variance of the distribution is $\displaystyle \frac{1}{\pi}\int_{-\infty}^{\infty} \frac{(x-\mu)^2}{1+(x-\mu)^2}\,dx$

and the integral does not converge. Thus for Cauchy distribution, the second and higher moments about the mean do not exist.

The distribution of X is $\displaystyle F_X(x) = \frac{1}{\pi}\int_{-\infty}^{x} \frac{dx}{1+(x-\mu)^2} = \frac{1}{\pi}\left[\tan^{-1}(x-\mu) + \frac{\pi}{2}\right]$

The characteristic function of the distribution given by (8·9) about the mean is

$$\phi(t) = E\left[e^{it(x-\mu)}\right] = \int_{-\infty}^{\infty} e^{it(x-\mu)} \cdot \frac{1}{\pi} \cdot \frac{1}{1+(x-\mu)^2}\,dx$$

---

* The proof of this step requires a knowledge of contour integration. By integrating $\dfrac{e^{itz}}{1+z^2}$ round a closed contour consisting of the real axis from $-R$ to R and the semi-circle of radius R in the upper half-plane and then making $R \to \infty$, we get

$$\underset{R\to\infty}{\text{Lt}} \int_{-R}^{R} \frac{e^{itx}}{1+x^2}\,dx = \underset{R\to\infty}{\text{Lt}} \int_{0}^{\pi} \frac{\exp\{it(R\cos\theta + iR\sin\theta)\}}{1+R^2 e^{2i\theta}} Rie^{i\theta}\,d\theta$$

$$= 2\pi i \left[\text{Residue of } \frac{e^{itz}}{1+z^2} \text{ at } z = i\right].$$

$$= \pi e^{-t}$$

or $$\phi(t) = \frac{1}{\pi} \int_{-\infty}^{\infty} \frac{e^{it(x-\mu)}}{1+(x-\mu)^2} dx = \frac{1}{\pi} \int_{-\infty}^{\infty} \frac{e^{ity}}{1+y^2} dy = e^{-|t|*}$$

The density function for the Cauchy distribution is shown in Fig. 8·5.

Ex. 1. Plot the Cauchy curve and the normal curve with zero mean and *s.d.* 2 and compare the two curves.

Ex. 2. By using the characteristic function, prove that if $X_1, X_2 \ldots\ldots X_n$ are independent Cauchy variates with same mean, then $\bar{X}$ is also a Cauchy variate with same mean.

8·9·7 Uniform Distribution A continuous random variable X is said to have a Uniform Distribution over the integral $a \le X \le b$ if its probability density function is constant over the entire range *i.e.*

$f(x) = k, a < x < b \quad$ = otherwise

where k is a constant given by

$$\int_a^b f(x)dx = 1 \Rightarrow k \mid x \mid_a^b = 1 \Rightarrow k = \frac{1}{b-a}$$

Hence $$f(x) = \frac{1}{b-a}, a < x < b$$
$$= 0, \text{ otherwise}$$

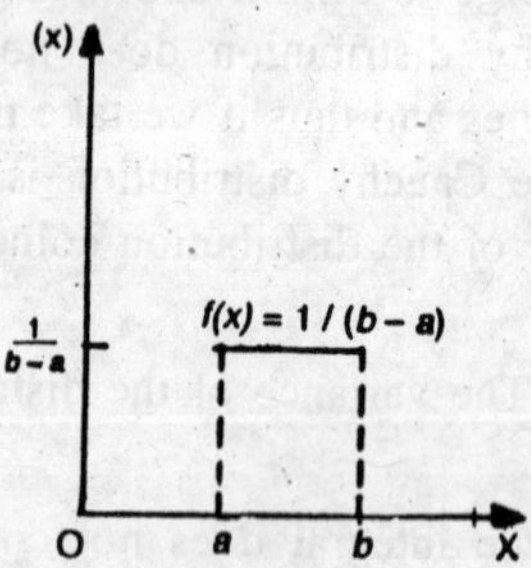

Fig. 8·7. graph of $f(x)$

Since the probability density curve $y = f(x)$ resembles a rectangle on the x-axis and between the ordinates at $x = a$ and $x = b$, the uniform distribution is also known as Rectangular distribution (Fig. 8·9).

(*ii*) *Distribution function* of a uniform variate *X*. The distribution function *F*(*x*) is given by

$$F(x) = \begin{cases} 0, \text{ if} & -\infty < x < a \\ \dfrac{x-a}{b-a}, & a \le x \le b \quad \because \displaystyle\int_a^x f(x)\, dx = \frac{x-a}{b-a} \\ 1 & h < x < \infty \end{cases}$$

Clearly $F(x)$ is not continuous at the two end points $x = a$ and

---

The second integral on the R.H.S. vanishes in limit if $t > 0$.

$$\int_{-\infty}^{\infty} \frac{e^{itx}}{1+x^2} dx = 2\pi i . \frac{e^{-t}}{2i} = \pi e^{-t}.$$

If t is negative, we integrate round the closed contain, which consists of the real axis from $-R$ to R and the semi-circle of radius R in the lower half plane.

$$\because \int_a^x f(x)\,dx = \frac{x-a}{b-a}]]$$

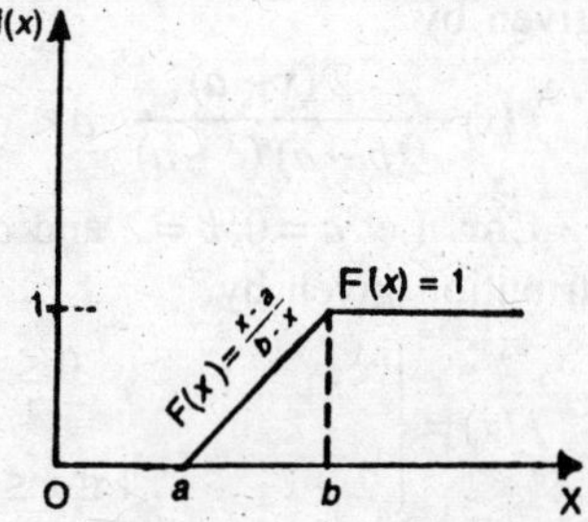

Fig. 8·8 Graph of $f(x)$

Clearly $F(x)$ is not continuous at the two end points $x = a$ and $x = b$, hence it is not differentiable at these points.

The graph of $F(x)$ is shown in the fig. 8·8

Cor. If the range of variation of X is $(-a, a)$, then the *p.d.f.* of X is given by

$$f(x) = \begin{cases} \dfrac{1}{2a}, & -a < x < a \\ 0, & \text{otherwise} \end{cases}$$

(*iii*) *Moments and M.G.F of the Uniform distribution*

We have μ_r' (about origin) $= \int x^r f(x)dx = \int_a^b x^r \cdot \frac{1}{b-a}\,dx = \frac{b^{r+1} - a^{r+1}}{(r+1)(b-a)}$

In particular, putting $r = 1, 2, 3, 4$

$$\text{Mean} = \mu'_1 = \frac{b^2 - a^2}{2(b-a)} = \frac{b+a}{2},$$

$$\mu'_2 = \frac{b^3 - a^3}{3(b-a)} = \frac{b^2 + ba + a^2}{3}, \quad \mu_2 = \mu'_2 - \mu_1'^2 = \frac{1}{12}(b-a)^2, \text{ on simplification}$$

$$\mu'_3 = \frac{b^4 - a^4}{4(b-a)} = \frac{(b^2 + a^2)(b+a)}{4}$$

$$\mu'_4 = \frac{b^5 - a^5}{5(b-a)} = \frac{b^4 + b^3a + b^2a^2 + ba^3 + a^4}{5}$$

$$\underset{X=0}{M}(t) = E\left(e^{tX}\right) = \int_a^b e^{tx} \cdot \frac{1}{b-a}\,dx = \frac{e^{bt} - e^{at}}{(b-a)}.$$

$$= \frac{\left(1 + bt + \frac{b^2t^2}{2!} + \ldots\right) - \left(1 + at + \frac{a^2t^2}{2!} + \ldots\right)}{(b-a)\,t}$$

$$= 1 + \frac{b+a}{2}t + \frac{b^3 - a^3}{3(b-a)} \cdot \frac{t^2}{2} + \frac{b^4 - a^4}{4(b-a)} \cdot \frac{t^3}{3} + \ldots$$

The values of $\mu'_1, \mu'_2, \mu'_3, \mu'_4$ can be written down at once.

8·9·8 Triangular Distribution. *A* continuous *r v. X* is said to have a triangular distribution (Fig 8·11) in the interval $[a, b]$ if the probability law is given by

$$f(x) = \frac{2(x-a)}{(b-a)(c-a)}, \ a < x \leq c \quad = \frac{2(b-x)}{(b-a)(b-c)}, \ c < x < b$$

Cor. Let $a = 0, b = 2$ and $c = 1$, then the *p.d.f.* of *X* is the triangular distribution given by

$$f(x) = \begin{cases} x, & 0 \leq x \leq 1 \\ 2 - x, & 1 \leq x \leq 2 \\ 0, & \text{otherwise} \end{cases}$$

Example 8·6 *Obtain the m.g.f of the above triangular distribution in the range (0, 2). and deduce the values of the mean and variance.*

$$m.g.f. = M_{x=0}(t) = \int e^{tx} f(x)\, dx = \int_0^1 e^{tx} \cdot x\, dx + \int_1^2 e^{tx}(2-x)\, dx$$

$$= \left| \frac{xe^{tx}}{t} \right|_0^1 - \frac{1}{t}\int_0^1 e^{tx}\, dx + \left| (2-x)\frac{e^{tx}}{t} \right|_1^2 + \frac{1}{t}\int_1^2 e^{tx}\, dx$$

$$= \frac{e^t}{t} - \frac{1}{t^2}\left(e^t - 1\right) - \frac{e^t}{t} + \frac{1}{t^2}\left(e^{2t} - e^t\right)$$

$$= \frac{te^t - e^t + 1 - te^t + e^{2t} - e^t}{t^2} = (e^t - 1)^2/t^2$$

$$= \left(t + \frac{t^2}{2!} + \frac{t^3}{3!} + \ldots\right)^2 / t^2 = 1 + t + t^2 + \ldots$$

in $\quad \mu'_1 = 1, \mu'_2 = 2, \mu_2 = \mu'_2 - {\mu'_1}^2 = 1,$

Example 8·7 *If X is uniformly distributed over [1, 2] find Z so that*

$$P\left(X > Z + \mu_x\right) = \frac{1}{6}.$$

The p.d.f. of *X* is given by $\quad f(x) = 1, \ 1 < x < 2$

$$\therefore \quad \mu_x = \int xf(x)\, dx = \int_1^2 xdx = 3/2$$

Now $\quad P(X > Z + \mu_x) = \frac{1}{6} \Rightarrow \int_{z+3/2}^2 dx = \frac{1}{6}$

$$2 - \left(z + \frac{3}{2}\right) = \frac{1}{6} \Rightarrow Z = \frac{1}{3}.$$

Example 8·8 *If* $X \sim U(0,1)$, *prove that for* $b < c$ *fixed,* $Y = (c-b)X + b$ is uniform on $[b, c]$

The p.d.f of X is $dp = \frac{dx}{1}$, $0 < x < 1$

Now $y = (c-b)x + b \Rightarrow dy = (c-b)\,dx$. When $x = 0, y = b$ and $x = 1, y = c$.

$$\therefore \quad y \sim U(b, c). \quad dp = \frac{1}{c-b}\,dy,\ b < y < c$$

8·9·7. Logistic Distribution. A continuous random variate X is said to have a logistic distribution if its distribution is given by

$$F_X(x) = [1 + \exp(-z)]^{-1},\ -\infty < z < \infty \qquad ...(1)$$

where $z = (x - \alpha)/\beta$, α and $\beta\ (> 0)$ being the parameters of the distribution.

$$\text{Since} \quad \left[1 + e^{-z}\right]^{-1} = \frac{1}{1 + e^{-z}} = \frac{e^{\frac{z}{2}}}{e^{z/2} + e^{-z/2}} = \frac{1}{2}\left[1 + \frac{e^{z/2} - e^{-z/2}}{e^{z/2} + e^{-z/2}}\right]$$

$$= \frac{1}{2}\left[1 + \tanh\frac{z}{2}\right]$$

$$\therefore \quad F_X(x) = \frac{1}{2}\left[1 + \tanh\frac{z}{2}\right] = \frac{1}{2}\left[1 + \tanh\frac{1}{2}\left(\frac{x-\alpha}{\beta}\right)\right],$$

The *p.d.f.* $f(x)$ of X is given by

$$f(x) = \frac{d}{dx}F(x) = \frac{1}{\beta}\exp\left(-\frac{x-\alpha}{\beta}\right)\cdot\left[1 + \exp\left(-\frac{x-\alpha}{\beta}\right)\right]^{-2}$$

$$= \frac{1}{\beta}\cdot e^{-z}\left(1 + e^{-z}\right)^{-2}$$

$$\text{Also } f(x) = \frac{1}{4\beta}\operatorname{sech}^2\left(\frac{1}{2}\cdot\frac{x-\alpha}{\beta.}\right) = \frac{1}{4}\operatorname{sech}^2\frac{z}{2},\ -\infty < z < \infty$$

Logarithmic distribution is extensively used in Time Analysis and as a growth function in studies concerning population and demography.

Properties of Standard Logistic Distribution

The *p.d.f.* is given by $f(z) = e^{-z}\left(1 + e^{-z}\right)^{-2}$, $-\infty < z < \infty$ ...(2)

I. *The p.d.f. is symmetrical about the line z = 0*

For $f(-z) = e^{z}\left(1 + e^{z}\right)^{-2}$ [Replace z by $-z$ in (2)]

$$= e^{z}\cdot e^{-2z}\left(e^{-z} + 1\right)^{-2} = e^{-z}\left(1 + e^{-z}\right)^{-2} = f(z)$$

II. *All odd order moments about origin vanish :*

$\mu'_{2r+1} = E\left(z^{2r+1}\right) = \int_{-\infty}^{+\infty} z^{2r+1} f(z)\,dz = 0$, by virtue of properties of

Definitive integral, as $f(z) = f(-z)$.

In particular, $\mu'_1 = 0, \mu'_3 = 0 = \mu_3 \Rightarrow \beta_1 = 0$

III. *Moment Generating Function.* The *m.g.f.* of (2) is given by

$$M_{z=0}(t) = E\left(e^{tz}\right) = \int_{-\infty}^{+\infty} e^{tz} \cdot e^{-z}\left(1+e^{-z}\right)^2 dz$$

$$= \int_1^{\infty} u^2 (u-1)^{-t} du \quad \text{where } u = 1+e^{-z}, du = -e^{-z} dz$$

$$= \int_0^1 v^{-t}(1-v)^t \, dv, v = \frac{u-1}{u} = 1 - \frac{1}{u} \Rightarrow dv = \frac{1}{u^2} du$$

$$= \beta(1-t, 1+t), 1-t > 0, \quad \text{[Properties of the beta function]}$$

$$= \pi t \operatorname{cosec} \pi t, t < 1$$

$$\therefore \quad \mu_2 = \mu'_2 = \frac{d^2}{dt^2} M_z(t) \bigg|_{t=0} = \pi \left[\frac{d^2}{dt^2} \frac{t}{\sin \pi t}\right]_{t=0} = \frac{\pi^2}{3},$$

using $\frac{\sin t}{t} \to 1$ as $t \to 0$

Similarly $\mu_4 = \mu'_4 = \frac{d^4}{dt^4} M_z(t) \bigg|_{t=0} = \frac{7}{15}\pi^4$, on simplification.

$$\therefore \quad \beta_2 = \frac{\mu_4}{\mu_2^2} = \frac{7}{15}\pi^4 \div \left(\frac{\pi^2}{3}\right)^2 = \frac{21}{5} = 4{\cdot}2 > 3, \gamma_2 = \beta_2 - 3 > 0$$

$\therefore$ *p.d.f* is Leptokurtic

IV. *Mean Derivation about mean, which is origin :*

$$\text{M.D. about mean} = \int_{-\infty}^{+\infty} |z| \cdot e^{-z}\left(1+e^{-z}\right)^{-2} dz \quad \left[\text{M.D.} = \int |z| \, g(z) \, dz\right]$$

$$= 2\int_0^{\infty} z e^{-z}\left(1+e^{-z}\right)^{-2} dz \quad [\because g(-z) = g(z)]$$

$$= 2\int_0^{\infty} z e^{z}\left(1+e^{z}\right)^{-2} dz \quad \text{on writing } e^{-z} = 1/e^z$$

$$= 2\left[1 - \frac{1}{2} + \frac{1}{3} - \dots\right] = 2 \log 2, 1 + e^{-z} = u$$

V. *Relation between p.d.g (z) and distribution function G (z).*

From equation (1) $G(z) = \left(1 + e^{-z}\right)^{-1} = \left(1 + \dfrac{1}{e^z}\right)^{-1} = \dfrac{e^z}{1 + e^z}$ ...(3)

$$\Rightarrow 1 - G(z) = \frac{1}{1 + e^z} \qquad ...(4)$$

$$\therefore\ G(z) \cdot [1 - G(z)] = e^z \left(1 + e^z\right)^{-2} = g(z).$$

Thus the relation between $g(z)$ and $G(z)$ is $g(z) = G(z)\,[1 - G(z)]$.

VI. *Determination of Z in terms of* $G(z)$. From (3) and (4), we have

$$G(z)/[1 - G(z)] = e^z \ \Rightarrow Z = \log_e \left[\frac{G(z)}{1 - G(z)}\right].$$

Ex. 1. Show that the mean and variance of the logistic variate X with parameters α and β are given by

$$\text{mean} = \alpha + \beta \cdot 0 = \alpha,$$

$$\text{Var}\, x = \beta^2 \text{ var } y = \beta^2 \cdot \pi^2/3\ ,\ Y = (X - \alpha)/\beta$$

Ex. 2. Obtain the *m.g.f.* of Logistic distribution and hence find its mean and variance **(*B. Sc. Stat. Hons. Delhi, 1993*)**

8·9·1. Exponential Distribution A continuous random variate X is said to have an exponential distribution if the p.d.f. is given by.

$$f(x) = e^x\ ,\ x \ge 0$$
$$= 0, \text{ otherwise}$$

A general form of negative exponential distribution with parameter $\lambda\ (> 0)$ is given by

$$f(x) = \begin{cases} 0 & \text{for } x \le 0 \\ \lambda e^{-\lambda x} & \text{for } x > 0, \end{cases}$$

The distribution function $F(x)$ is given by

$$\underset{X}{F}(x) = \begin{cases} 0 & \text{for } x \le 0 \\ 1 - e^{-\lambda x} & \text{for } x > 0 \end{cases}$$

Note. The exponential distribution plays a key role in the theory of queues and in reliability theory. It is a gamma distribution with parameter unity.

Movement Generating Function of negative exponential distribution $f(x) = \lambda e^{-\lambda x}\ , x > 0, \lambda > 0$

$$M_{X=0}(t) = E\left(e^{tX}\right) = \lambda \int_0^\infty e^{tx} \cdot e^{-\lambda x}\, dx = \frac{\lambda}{\lambda - t}\ ,\ \lambda > t$$

$$= \left(1 - \frac{t}{\lambda}\right)^{-1} = \sum_{r=0}^{\infty} \left(\frac{t}{\lambda}\right)^r$$

$$\therefore \quad \mu'_r = \text{ Coeff. of } t^r/r\,! \text{ in } M_X(t) = \frac{r\,!}{\lambda^r}, r = 1, 2, 3, \ldots$$

$$\therefore \quad \text{Mean} = \mu'_1 = \frac{1}{\lambda}, \mu'_2 = \frac{2!}{\lambda^2};\ \mu_2 = \mu'_2 - \mu_1'^2 = \frac{2}{\lambda^2} - \frac{1}{\lambda^2} = \frac{1}{\lambda^2}.$$

Example 8·9. *If $X_1, X_2, \ldots, X_n$ are independent random variables, X_i having an exponential distribution with parameter θ_i, $i = 1, 2, \ldots, n$ then $Z = \min (X_1, X_2, \ldots, X_n)$ has exponential distribution with parameter $\sum_{i=1}^{n} \theta_i$.* **(*B. Sc. Stat. Hons. Delhi, 1986, 88, 90*)**

$$G_Z = P(Z \le z) = 1 - P(Z > z) = 1 - P[X_i > z,\ i = 1, 2, \ldots n]$$

$$= 1 - \prod_{i=1}^{n} P(X_i > z) = 1 - \prod_{i=1}^{n} [1 - P(X_i \le z)]$$

$$= 1 - \prod_{i=1}^{n} [1 - F_{Xi}(z)] = 1 - \prod_{i=1}^{n} \left[1 - \left(\int_0^{z_i} e^{-\theta_i z} dz \right) \right]$$

$$= 1 - \prod_{i=1}^{n} \left[1 - \left(1 - e^{-\theta_i z} \right) \right] = 1 - \exp \left\{ \left(-\sum_{i=1}^{n} \theta_i \right) z \right\}, z > 0$$

$$= 0, \qquad \text{otherwise}$$

$$\therefore \quad g_Z(z) = \sum_{i=1}^{n} \theta_i \exp \{ (-\Sigma\theta_i)\, z \}, z > 0\ ;\ = 0, \text{ otherwise.}$$

Hence $Z = \min \left(X_1, X_2, \ldots, X_n \right)$ is an exponential variate with parameter $\sum_{i=1}^{n} \theta_i$.

Ex. The p.d.f of a r.v.X has the p.d.f. $f(x) = (2\theta)^{-1} \exp(-|x - \theta|/\theta)$, $-\infty < x < \theta$, find the m.g.f. of X and then determine EX and var X.

(*B.A. Stat. Hons. Delhi 1986*)

8·9·2. Erlang Family of Distributions

A two-parameter generalisation of the exponential distribution is given by the Erlang family of distributions for which

$$f(x) = \frac{(k\,\lambda)^k}{(k-1)\,!} e^{-k\lambda x} x^{k-1}, \quad x \ge 0$$

and k is a positive integer. For $k = 1$, we get the exponential distribution. All the distributions have a common mean at $x = 1/\lambda$. The mode is at $x = 0$, for $k = 1$ and at $x = (k-1)/k\,\lambda$ for other values of k. The variance of the kth member of the family is $1/k\,\lambda^2$. As k increases, the mode moves towards the right to $1/\lambda$ and the variance decreases to zero.

The Erlang family has a close association with the exponential distribution, going beyond the fact that the first Erlang distribution is exponential. If we have k random variables $x_1, x_2, \ldots, x_k$ which are independent and have a common exponential distribution with mean $1/k\,\mu$, then the random variable $X_1 + X_2 + \ldots + X_k$ follows the kth Erlang distribution with parameter μ

If X has Erlang distribution with parameters k and λ, then $k\lambda X$ has a gamma distribution with parameter k.

These distributions play an important role in queueing theory and were developed by A.K. Erlang for use in telephone systems.

8·9·3. Weibull Distribution

The *p.d.f.* of the random variable X is said to be Weibull distribution with two parameters α, β if it is given by

$$f(x\,;\alpha,\beta) = \frac{\alpha x^{\alpha-1}}{\beta}\exp\left\{-\left(\frac{x}{\beta}\right)^{\alpha}\right\}, x > 0, \beta > 0, \alpha > 0 \quad \ldots(1)$$

Here α is referred to as a shape parameter and β *as a scale parameter.*

The distribution function $F(x)$ is given by

$$F_X(x) = \int_0^x f(x)\,dx.$$

$$\Rightarrow \quad F_X(x) = 1 - \exp\left\{-\left(\frac{x}{\beta}\right)^{\alpha}\right\};\, x \ge 0, \beta > 0, \alpha > 0$$

Replace X by $X - \mu$, then $f(x) = \dfrac{\alpha\,(x-\mu)^{\alpha-1}}{\beta}\cdot e^{-\left(\frac{x-\mu}{\beta}\right)^{\alpha}}$

Thus the *r.v.* $Y = (X - \mu)/\beta$ has exponential distribution with p.d.f. $H(y) = e^{-y}, y > 0$

A r.v.x has a Weibull distribution with three parameters $\alpha\,(> 0), \beta\,(> 0)$ and μ if its p.d.f. is given by

$$f_x(x;\alpha,\beta,\mu) = \beta\alpha^{-1}\left(\frac{x-\mu}{\alpha}\right)^{\beta-1}\exp\left[-\left(\frac{x-\mu}{\alpha}\right)^{\beta}\right];\, x > \mu, \beta > 0.$$

This distribution was used by Swedish scientist Weibull in 1951; to describe experimentally observed variation in the fatigue resistance of steel, its elastic limits etc. It has also been used to study the variation of length of service of radio service equipment.

Theoretically, it arises as the limiting distribution, as $n \to \infty$, of the smallest of n independent random variables, with the same distribution *i.e.*

the smallest of n *i.i.d.* independent identically distributed random variables.

The exponential distribution is a special case of the Weibull distribution, if we let $\beta = 1$ in equation (1). For $\alpha = 1$ and $\beta = 1, 2, 3$ we have the following diagram Fig 8·9.

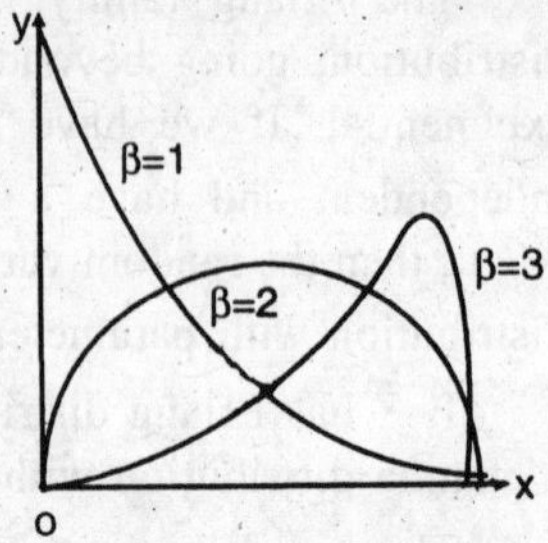

Fig. 8·9.

8·9·4 Double Exponential Distribution

The *p.d.f.* of a Double Exponential Variate with parameters c and α is given by $f(x ; c, \alpha) = \exp(-ce^{-\alpha x})$, $\quad -\infty < x < 0, c > 0, \alpha > 0$

It arises as the distribution for the longest lasting object, out of a large number of apparently identical objects.

8·9·5. Logarithmic Normal or Log-normal Distribution

A non-negative random variable X is said to have a logarithmic normal distribution, if its logarithm has a normal distribution. The *p.d.f.* of the log-normal distribution is given by

$$f(x ; \mu, \sigma) = \frac{1}{x\sigma\sqrt{2\pi}} \exp\left[-\frac{1}{2\sigma^2}(\log x - \mu)^2\right], \ \mu > 0$$

It arises in problems of economics, biology, geology, and reliability theory. In particular, it arises in the study of dimensions of particles under pulverisation.

The expected value of the log-normal distribution is

$$E(X^r) = \int_{-\infty}^{\infty} \frac{x^r}{x} \cdot \frac{1}{\sigma\sqrt{2\pi}} \exp\left\{-\frac{1}{2\sigma^2}(\log x - \mu)^2\right\} dx$$

Put $\quad (\log x - \mu)/\sigma = u \Rightarrow x = e^{\mu + \sigma u}$ and $dx = \sigma e^{\mu + \sigma u} \cdot du$

$$\therefore \quad E(X^r) = \frac{1}{\sqrt{2\pi}} \int_{-\infty}^{+\infty} e^{\mu r + r\sigma u - \frac{1}{2}\mu^2} du$$

$$= \frac{1}{\sqrt{2\pi}} e^{\mu r + \frac{1}{2}\sigma^2 r^2} \int_{-\infty}^{+\infty} e^{-\frac{1}{2}(u - \sigma r)^2} du = \exp\left\{\mu r + \frac{1}{2}\sigma^2 r^2\right\}$$

$$\mu_1' = E(x) = \exp\left\{\mu + \frac{1}{2}\sigma^2\right\}; \ \mu_2' = \exp\{2\mu + 2\sigma^2\}$$

$$\mu_2 = \text{Var } x = E(x^2) - [E(x)]^2 = \exp\left(2\mu + \sigma^2\right)\left[\exp(\sigma^2) - 1\right]$$

$$\mu_3' = \exp\left\{3\mu + \frac{9}{2}\sigma^2\right\}; \ \mu_4' = \exp\{4\mu + 8\sigma^2\}$$

$$\mu_3 = \mu_3' - 3\,\mu_2'\,\mu_1' + 2\mu_1'^3 = \exp\left\{3\mu + \frac{9}{2}\mu^2 - 3\right\}$$

$$\therefore\quad \beta_1 = \mu_3^2/\mu_2^3 = \exp\{6\mu + 9\mu^2 - 6\}/\exp(6\mu + 3\sigma^2)\,[\exp(\sigma^2) - 1]^3.$$

$f(x)$ will be maximum when $u = 0 \Rightarrow x = e^{\mu}$.

Also if a is the median, it is given by

$$= \exp(\mu)\int_{-\sigma}^{a} \frac{1}{\sqrt{2\pi}}\, e^{-\frac{1}{2}u^2 + u\sigma}\, du = \frac{1}{2}$$

From this we can determine a

Example 8·10. *Prove that if X and Y are independent log-normal variates, then XY and X/Y are also log-normal variates.*

Since X is a log-normal variate, it follows that

$$\log X \sim N(\mu_1, \sigma_1^2),\ \text{and}$$

$$\log Y \sim N\left(\mu_2, \sigma_2^2\right)$$

Let $U = XY$ and $V = X/Y$.

Since X and Y are independent variates, log X and log Y are also so.

$$\therefore\quad \log U = \log X + \log Y \sim N\left(\mu_1 + \mu_2, \sigma_1^2 + \sigma_2^2\right)$$

$$\log V = \log X - \log Y \sim N(\mu_1 - \mu_2, \sigma_1^2 + \sigma_2^2)$$

Hence the result follows.

Example 8·11. *Obtain the distribution function for* $Y = \log X$ *when* $X \sim N\left(\mu, \sigma^2\right)$.

$F_X(x) = P(Y \le \log X)$, since $\log X$ is monotonic increasing function.

$$= \frac{1}{\sigma\sqrt{2\pi}}\int_{-\infty}^{\log x} \exp\left\{\frac{-(y-\mu)^2}{2\sigma^2}\right\} dy$$

$$= \frac{1}{\sigma\sqrt{2\pi}}\int_{u=0}^{x} \exp\left\{\frac{-(\log u - \mu)^2}{2\sigma^2}\right\}\cdot\frac{du}{u},\quad y = \log u,\ \log 0 \to -\infty$$

Fig. 8·8 shows the density function for logarithm distribution for various values of σ^2

It can be seen that the distribution is positively skewed about μ

If mean > median > mode if $X_1, X_2, \ldots, X_n$ is a set of independently identically distributed random variables, such that mean of each log X_i is m and its variance is σ^2, then the product $X_1 X_2 \ldots X_n$ is asymptotically distributed according to logarithmic normal distribution and with mean m and variance $n\,\sigma^2$.

8·9·6. Distribution of Mixtures

Let $f_i(x)$ and $F_i(x)$ be the density and distribution functions of X_i,

$$(i = 1, 2, \ldots\ldots, n,) \text{ then } f(x) = \sum_{i=1}^{n} p_i f_i(x), \; F(x) = \sum_{i=1}^{n} p_i F_i(x)$$

where $p_i \geq 0$ and $\sum_{i=1}^{n} p_i = 1$, also gives density and distribution functions which arise in many situations *e.g.* if we have in a box, proportions $p_1, p_2 \ldots, p_n$ of n type of objects whose weights follow normal distributions with means μ_i and variances σ_i^2 $(i = 2, \ldots\ldots, n)$, then the density function for the weight of a random object selected from the box is

$$f(x) = \frac{1}{\sqrt{2\pi}} \sum_{i=1}^{n} \left\{ \frac{p_i}{\sigma_i} \exp \frac{-(x_i - \mu_i)^2}{\sigma_i^2} \right\}$$

Ex. 1. Find the mean, variance and mode for the following distributions : Exponential, Erlang, Weibull, Double Exponential, Logarithmic, Polya.

Ex. 2. Find the moment generating function for an exponential distribution with mean k^{-1}. Deduce the moment generating function of the sum of k independently indentically distributed exponential variables of the type and find the distribution for the sum.

8·10. Pearson's Distribution. *Pearson's type of curves** $y = f(x)$ are obtained as the solution of the differential equation

$$\frac{dy}{dx} = \frac{-(x+a)y}{b_0 + b_1 x + b_2 x^2} \qquad \ldots(8·9)$$

Genesis of Pearson's Curves. Equation (8·9)was obtained by Karl Pearson on account of the following characteristics of a frequency distribution :

1. A frequency distribution generally starts at zero, rises to a single maximum and then falls again, that is, a frequency distribution is generally unimodal. Thus if $y = f(x)$ be the frequency curve, the n

---

* Karl Pearson's first paper dealing with generalized frequency curves appeared in 1895. In this paper, he gave types of frequency curves in addition to normal curve. In two subsequent papers one in 1908 and the other in 1916, he gave seven more types. An excellent account of Pearson's distribution is given in Elderton's book entitled 'Frequency Curves and Correlation'.

$\dfrac{dy}{dx} = 0$ when $x = -a$, say.

2. At the ends of the frequency curve, there is often high contact with the axis of x, *i.e.*,

$$\frac{dy}{dx} = 0 \text{ when } y = 0.$$

3. Generally first four moments are sufficient to determine the distribution. The conditions (1) and (2) give $\dfrac{dy}{dx} = -\dfrac{y(x+a)}{F(x)}$ where $F(x)$ is an arbitrary function of x not vanishing at $x = -a$. Expanding $F(x)$ by the Maclaurin's series, we get

$$F(x) = b_0 + b_1 x + b_2 x^2 + b_3 x^3 + \ldots$$

Empirically we shall retain only 3 terms in the expansion of $F(x)$ so that

$$F(x) = b_0 + b_1 x + b_2 x^2.$$

Then
$$\frac{dy}{dx} = \frac{-(x+a)y}{b_0 + b_1 x + b_2 x^2}.$$

From (8·9), we get

$$(b_0 x^n + b_1 x^{n+1} + b_2 x^{n+2}) \frac{dy}{dx} \cdot dx = -y(x^{n+1} + ax^n)\, dx$$

Integrating by parts over the total range of variate x,

$$\left[\left(b_0 x^n + b_1 x^{n+1} + b_2 x^{n+2}\right) y\right]_{-\infty}^{+\infty} - \int_{-\infty}^{+\infty} \left[\{nb_0 x^{n-1} + (n+1)\, b_1 x^n + (n+2) b_2 x^{n+1}\}\, ydx\right]$$

$$= -\int_{-\infty}^{+\infty} y\,(x^{n+1} + ax^n)\, dx, \text{ provided integrals exist.}$$

Assume that the first bracket on left side vanishes at either limit we have
$$nb_0 \mu'_{n-1} + (n+1) b_1 \mu'_n + (n+2) b_2 \mu'_{n+1} = (\mu'_{n+1} + a\mu'_n)$$

Measure x from the mean of the curve and put $n = 0, 1, 2, 3$ we get

$$b_1 = a,\ b_0 + 3b_2\mu_2 = \mu_2,\ 3b_1\mu_2 + 4b_2\mu_3 = \mu_3 + a\mu_2,$$

$$3b_0\mu_2 + 4b_1\mu_3 + 5b_2\mu_4 = \mu_4 + a\mu_3$$

Solving these, we get $b_0 = \dfrac{\sigma^2(4\beta_2 - 3\beta_1)}{2\,(5\beta_2 - 6\beta_1 - 9)}$

$$b_1 = \frac{\sigma\sqrt{\beta_1}(\beta_2+3)}{2(5\beta_2-6\beta_1-9)}$$

$$b_2 = \frac{2\beta_2-3\beta_1-6}{2(5\beta_2-6\beta_1-9)} \qquad K' = \frac{\beta_1(\beta_2-3)^2}{4(4\beta_2-3\beta_1)(2\beta_2-3\beta_1-6)}$$

where $\mu_2=\sigma^2$, $\beta_1 = \dfrac{{\mu_3}^2}{{\mu_2}^3}$, $\beta_2 = \dfrac{\mu_4}{{\mu_2}^2}$.

The solution of the differential equation (8·9) depends on the nature of the roots of the equation $b_0+b_1x+b_2x^2=0$. The discriminant of this equation is ${b_1}^2-4b_0b_2$. We now define a quantity $k={b_1}^2/4b_0b_2$, the values of which, it will appear, will determine the nature of various particular distributions. Also discriminant $=4b_0b_2(k-1)$.

8·10·1. *Roots of $b_0+b_1x+b_2x^2=0$ are real and of opposite signs so that $k<0$, since product of roots* $=\dfrac{b_0}{b_2}=\dfrac{{b_0}^2}{b_0b_2}$ and $b_0b_2<0$.

Shifting the origin to the mode $x=-a$, we have

$$\frac{1}{y}\frac{dy}{dx}=\frac{x}{B_0+B_1x+B_2x^2}=\frac{-x}{B_2(x-\alpha_1)(x-\alpha_2)}$$

or $\dfrac{1}{y}\dfrac{dy}{dx}\equiv\dfrac{m_1}{x-\alpha_1}+\dfrac{m_2}{x-\alpha_2}$ where $m_1\alpha_2+m_2\alpha_1=0$ and $m_1+m_2=-1/b_2$.

$\therefore\quad \log y = m_1\log(x-\alpha_1)+m_2\log(x-\alpha_2)$

or $$y=y_0\left(1+\frac{x}{a_1}\right)^{m_1}\left(1-\frac{x}{a_2}\right)^{m_2},\ -a_1\le x\le\alpha_2, \qquad \textit{Type I}$$

where $\dfrac{m_1}{a_1}=\dfrac{-m_2}{a_2}$ or $\dfrac{-m_1}{a_1}=\dfrac{m_2}{a_2}$ where a_1 and a_2 have same sign.

8·10·2. *Roots of $b_0+b_1x+b_2x^2=0$ are real and of same sign so that $k>1$.*

Let the roots be α_1 and α_2 so that $\alpha_1=-a_1$, $\alpha_2=-a_2$, a_1 and $a_2>0$. $\therefore\quad \dfrac{m_1}{a_1}=\dfrac{-m_2}{a_2}$ and the equation of the curve is

$$y=y_0\left(1+\frac{x}{a_1}\right)^{m_1}\left(1+\frac{x}{a_2}\right)^{-m_2} \qquad \textit{Type VI}$$

This equation can be written as $y=y_0x^{-m_2}(x+a_1)^{m_1},\ -a_1\le x<\infty$

8·10·3. *Roots of $b_0+b_1x+b_2x^2=0$ are imaginary, so that $0<k<1$.*

We have $\dfrac{1}{y}\dfrac{dy}{dx} = \dfrac{-(x+a)}{b_0 + b_1 x + b_2 x^2}$

$$= \frac{-x}{b_2 x^2 + b_1 x + b_0} \text{ on shifting origin to } x = -a$$

or $$\frac{1}{y}\,dy = \frac{-x}{b_2[(x+c)^2 + d^2]}\,dx = -\frac{(x+c) - c}{b_2[(x+c)^2 + d^2]}\,dx$$

$$\therefore \quad \log y = \text{const.} - \frac{1}{2b_2}\log\left[(x+c)^2 + d^2\right] + \frac{c}{b_2 d}\tan^{-1}\left(\frac{x+c}{d}\right)$$

or $$y = y_0\left[(x+c)^2 + d^2\right]^{-\frac{1}{2b_2}} \times \exp\left[\frac{c}{b_2 d}\tan^{-1}\left(\frac{x+c}{d}\right)\right]$$

$$= y_0\left(1 + \frac{x^2}{a^2}\right)^{-m} e^{-\nu\tan^{-1}(x/a)},\ \nu, m > 0, -\infty < x < \infty \qquad \textit{Type IV}$$

8·10·4. *One root of $b_0 + b_1 x + b_2 x^2 = 0$ is infinite.*

$b_2 = 0, b_1 \neq 0 \quad \therefore k \to \pm\infty.$ We have $\dfrac{1}{y}\dfrac{dy}{dx} = -\dfrac{(x+a)}{b_0 + b_1 x}$.

Shifting the origin, we get $\dfrac{1}{y}\,dy = -\dfrac{x\,dx}{b_1(x+c)} = \dfrac{a-1}{b_1} + \dfrac{c}{b_1(x+c)}$

$$\therefore \quad \log y = \text{const.} - \frac{x}{b_1} + \frac{c}{b_1}\log(x+c)$$

or $$y = y_0\left(1 + \frac{x}{c}\right)^p e^{-\frac{px}{c}}, -c \le x < \infty \qquad \textit{Type III}$$

where $p = c/b$. It can be transformed to Gamma form

$y = c\,e^{-y}y^b,\ 0 < y < \infty.$

8·10·5. Both *roots of $b_0 + b_1 x + b_2 x^2 = 0$ are infinite i.e., $b_2 = 0 = b_1$.*

We have $\dfrac{1}{y}\,dy = -\dfrac{x+a}{b_0}\,dx$

$$\therefore \quad \log y = \text{const.} - \frac{1}{2b_0}(x+a)^2$$

or $$y = y_0\,e^{-\frac{1}{2b_0}(x+a)^2}, -\infty < x < \infty. \textit{ Normal curve.} \qquad \textit{Type VII}$$

8·10·6 *Roots of $b_0 + b_1 x + b_2 x^2 = 0$ are equal so that $k = 1$.*

We have $\dfrac{1}{y}\dfrac{dy}{dx} = -\dfrac{x+a}{b_2(x+d)^2}$

$$= -\frac{x}{b_2(x+c)^2} \text{ by a proper choice of origin.}$$

$$= -\frac{1}{b_2}\left[\frac{1}{x+c} - \frac{c}{(x+c)^2}\right]$$

$$\therefore \quad \log y = \text{const.} - \frac{1}{b_2}(x+c) + \frac{c}{b_2}(x+c)^{-1}$$

$$\therefore \quad y = y_0 (x+c)^{-p} e^{\frac{-q}{x+c}}$$

$$= y_0 x^{-p} e^{-\frac{q}{x}}, \; 0 \le x < \infty, p, q > 0 \qquad \textit{Type V}$$

8·10·7 *Roots of $b_0 + b_1 x + b_2 x^2 = 0$ are equal but of opposite signs so that $k = 0$.*

We have $b_1 = 0$ and b_0 and b_2 are of opposite signs. $\therefore \; b_0/b_2 > 0$

$$\therefore \quad \frac{1}{y}\frac{dy}{dx} = -\frac{x}{b_2\left(x^2 + \frac{b_0}{b_2}\right)} = -\frac{x}{b_2(x^2+c^2)}$$

$$\therefore \quad \log\frac{y}{y_0} = -\frac{1}{2b_2}\log(x^2+c^2) \quad \text{or} \quad y = y_0(x^2+c^2)^{-\frac{1}{2b_2}}$$

$$\text{or} \qquad y = y_0\left(1 + \frac{x^2}{c^2}\right)^{-m}, \; -c \le x < c \qquad \textit{Type II}$$

Cor. If $m = 0$, then y= constant and the distribution is rectangular.

Ex. 1. Obtain the following distributions with p.d.f. given by

Type VIII. $b_0 = 0, b_1 > 0, \; f(x) = \frac{1-m}{a}\left(1 + \frac{x}{a}\right)^{-m}, \; 0 < m < 1, -a \le x \le 0.$

Type IX. $b_0 = 0, b_1 < 0, \; f(x) = \frac{1+m}{a}\left(1 + \frac{x}{a}\right)^{m}, -a \le x \le 0.$

Type X. $b_0 = b_2 = 0, \; f(x) = \frac{1}{\sigma} e^{-x/\sigma}, \; 0 \le x < \infty, \; \sigma > 0.$

Type XI. $b_0 = b_1 = 0, k > 1, \; f(x) = b^{m-1}(m-1)x^{m-1}, b \le x < \infty.$

Ex. 2. Show that the following distributions can all be transformed into the Pearson Type I, *viz.* $dP = y_0(1-x)^{p-1} x^{q-1}\, dx, 0 \le x \le 1, p, q > 0$

(i) $dP = r_0(1-r^2)^{\frac{1}{2}(n-4)} \; dr, -1 \le r \le 1$ [Put $r^2 = x$]

(ii) $dP = t_0\left(1 + \frac{t^2}{\nu}\right)^{-\frac{1}{2}(\nu+1)} \; dt. -\infty \le t \le \infty$ [Put $t = \sqrt{\nu}\tan\theta$,] (*iii*) $dP = z_0 e^{\nu_1 z}\left(\nu_1 e^{2z} + \nu_2\right)^{-\frac{\nu_1+\nu_2}{2}} dz, -\infty < z < \infty$ [Put $\nu_1 e^{2z} = \nu_2 \tan^2 \theta$]

Ex. 3. Derive normal and rectangular distributions as particular cases of Pearson's system of frequency curves. ***(M.A., Delhi 1954, 59)***

Type of Pearson's Curves for different values of k

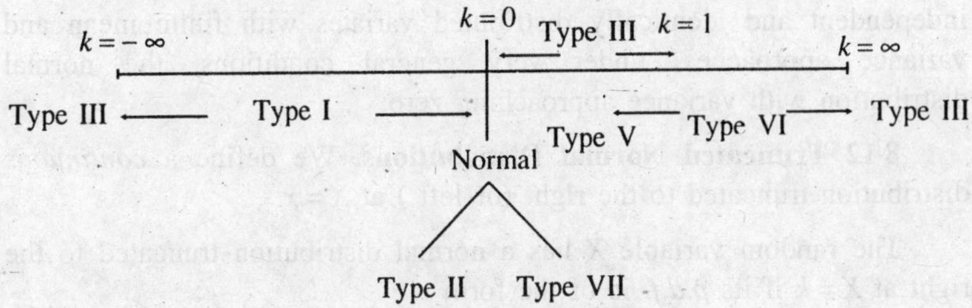

8·11. Significance of the normal distribution

(*i*) We have already established in sections 8·2 and 8·5 the Laplace-De Moivre theorem that the distribution of the standardised binomial variate approaches the distribution of the standardised normal variate as $n \to \infty$, provided p is not very small.

(*ii*) In Ex. 2. section 8·5, we gave a method for showing that the standardised Poisson variate approaches $N(0, 1)$ as $m \to \infty$.

(*iii*) It can be similarly proved that the standardised variate in the following cases approaches $N(0, 1)$:

(*a*) hypergeometric distribution when N and n approach infinity,

(*b*) gamma distribution when the parameter tends to infinity,

(*c*) beta distribution of both kinds when parameters tend to infinity.

*(*iv*) The mathematician John Herschel proved the following result : If shots are fired at a target taken as origin and x, y be the random vector variate denoting the errors of a shot relative to two orthogonal axes through the origin and if (*a*) the marginal density functions of X and Y are continuous (*b*) the probability density at (x, y) depends only on the radial distance $r(=\sqrt{x^2+y^2})$ (*c*) the errors in x, y *directions* are independent, then the probability density of displacement Z in any direction is the normal density.

*(*v*) The English mathematician Maxwell arrived at the normal distribution for the velocity of molecules of a gas under the following assumptions : (*a*) the component of velocities u, v, w in three orthogonal directions are independently distributed (*b*) the marginal distributions of u, v, w are same (*c*) the phase space is isotropic, that is, the density of molecules with given velocity components is a function of total velocity and not of the direction.

(*vi*) **Huygen** based his proof of the normal law of errors on the assumptions (*a*) an error is the sum of a large number of infinitesimal erros, of different magnitudes, due to different causes (*b*) different components of error are independent (*c*) each component of error has an equal chance of being positive or negative.

(*vii*) **According to the General Limit Theorem**, the mean of *n* independent and identically distributed variates with finite mean and variance approaches, under very general conditions, the normal distribution with variance approaching zero.

8·12 Truncated Normal Distributions. We define a *continuous* distribution truncated to the right (or left) at $X = x$.

The random variable X has a normal distribution truncated to the right at $X = k$ if its *p.d.f.* is of the form

$$f(x) = C \cdot \frac{1}{\sigma\sqrt{2\pi}} \exp\left[-\frac{1}{2}\left(\frac{x-\mu}{\sigma}\right)^2\right] \text{ if } x \leq k, \sigma > 0 \qquad \text{...(8·10)}$$

$$= 0 \text{ if } x > k,$$

where C is determined by the condition

$$\int_{-\infty} f(x)\, dx = 1 \quad \Rightarrow C = \frac{1}{\phi\left(\dfrac{k-\mu}{\sigma}\right)} = \frac{1}{P(Z \leq k)},$$

where X is $N(\mu, \sigma^2)$, ϕ is the *d.f.* of Z which is$N(0, 1)$,

Similarly if X has a normal distribution truncated to the left at $X = k$, we have

$$f(x) = \frac{C}{\sigma\sqrt{2\pi}} \exp\left[-\frac{1}{2}\left(\frac{x-\mu}{\sigma}\right)^2\right] \text{ if } x \geq k \qquad \text{...(8·11)}$$

$$= 0 \text{ if } x < k$$

where C is given by $\int^{+\infty} f(x)\, dx = 1 \Rightarrow \quad C = 1 - \phi\left(\dfrac{k-u}{\sigma}\right).$

Example. 8·11. *Determine the mean of the general normal distribution truncated to the right of* $X = k$. *as given by* (8·10).

We have $E(X) = \int_{-\infty}^{+\infty} x\, f(x)\, dx$

$$= \frac{1}{\phi\left[\frac{k-\mu}{\sigma}\right]} \int_{-\infty}^{k} x \cdot \frac{1}{\sigma\sqrt{2\pi}} e^{-\frac{1}{2}\left(\frac{x-\mu}{\sigma}\right)^2} dx$$

$$= \frac{1}{\phi\left[\frac{k-\mu}{\sigma}\right]} \cdot \frac{1}{\sigma\sqrt{2\pi}} \int_{-\infty}^{(k-\mu)/\sigma} (s\sigma+\mu)\, e^{-\frac{1}{2}s^2} ds, \left(s = \frac{x-\mu}{\sigma} \right)$$

$$= \frac{1}{\phi\left[\frac{k-\mu}{\sigma}\right]} \left[\mu\, \phi\left(\frac{k-\mu}{\sigma}\right) + \sigma \cdot \frac{1}{\sqrt{2\pi}} \int_{-\infty}^{(k-\mu)/\sigma} s \cdot e^{-\frac{1}{2}s^2} ds \right]$$

$$= \mu + \frac{\sigma}{\phi\left[\frac{k-\mu}{\sigma}\right]} \frac{1}{\sqrt{2\pi}} \left[e^{-\frac{1}{2}s^2} (-1) \right]_{-\infty}^{(k-\mu)/\sigma}$$

$$= \mu - \frac{\sigma}{\phi\left[\frac{k-\mu}{\sigma}\right]} \cdot \frac{1}{\sqrt{2\pi}} \exp\left[-\frac{1}{2}\left(\frac{k-\mu}{\sigma}\right)^2 \right]$$

Ex. Determine k for the above example, if $\mu = 10, \sigma = 1$ and $E(X) = 9{\cdot}5$. **Ans.** $k = 10{\cdot}52$

Example. 8·12. ***Truncated Cauchy Distribution.*** *Obtain the mean and variance of a standard Cauchy distribution truncated at both ends, with relevant range of variation as* $-\beta, \beta$ ***(M. Sc. Stat. Delhi, 1987)***

Let $f(x)$ and $g(x)$ be the *p.d.f.* of standard and truncated Cauchy distribution. Then

$$f(x) = g(x) \times P(-\beta < x < \beta)$$

$$\therefore \quad g(x) = \frac{f(x)}{P(-\beta < x < \beta)} = \frac{\frac{1}{\pi} \cdot \frac{1}{1+x^2}}{\int_{-\beta}^{\beta} \frac{1}{\pi} \cdot \frac{1}{1+x^2} dx} = \frac{1}{1+x^2} \cdot \frac{1}{2\tan^{-1}\beta}$$

$$\text{Mean} \quad = \int_{-\beta}^{\beta} x\, g(x) = \frac{1}{2\tan^{-1}\beta} \int_{-\beta}^{\beta} \cdot \frac{x}{1+x^2} dx = 0$$

as integrand $= x/(1+x^2)$ and is an odd function of x

$$\mu_2' = \int_{-\beta}^{\beta} x^2 g(x)\, dx = \frac{2}{\tan^{-1}\beta} \int_{-\beta}^{\beta} \frac{x^2 \cdot dx}{1+x^2} = \frac{1}{\tan^{-1}\beta} \left| x - \tan^{-1} x \right|_{-\beta}^{\beta}$$

$$= (\beta - \tan^{-1}\beta)/\tan^{-1}\beta. \quad \therefore \ \text{Var}\, x = \mu_2' - \mu_1'^2 = (\beta - \tan^{-1}\beta)/\tan^{-1}\beta.$$

Exercises 8

1. (*a*) Mention the conditions under which B.D. tends to Normal Distribution.

(*b*) Deduce the first four moments about the mean of the normal distribution from those of (*i*) the Binomial distribution (*ii*) the Poisson distribution. **(*B. A. Hons, Delhi, 1950*)**

(*c*) For the general normal distribution $N(\mu, \sigma^2)$ prove that

$$\mu'_{r+2} = 2\mu\mu'_{r+1} + (\sigma^2 - \mu^2)\mu_r' + \frac{\sigma^3 d\mu_r'}{d\sigma}, \text{ where } \mu_r' = E(x^r).$$

(*B.A. Pass & Hons. Delhi 1971*)

$$\left[\textbf{Hint. } \frac{d\mu'_r}{d\sigma} = \int_{-\infty}^{+\infty} \frac{x^r}{\sigma^2\sqrt{2\pi}} \exp\left\{-\frac{(x-\mu)^2}{2\sigma^2}\right\} dx + \int_{-\infty}^{+\infty} \frac{x^r (x-\mu)^2}{\sigma^4\sqrt{2\pi}} \exp\left\{-\frac{(x-\mu)^2}{2\sigma^2}\right\} dx\right.$$

$$\left. = -\frac{\mu'_r}{\sigma} + \frac{\mu'_{r+2}}{\sigma^3} - \frac{2\mu\mu'_{r+1}}{\sigma^3} + \frac{\mu^2\mu'_r}{\sigma^3} = \text{etc.}\right]$$

2. (a) For a certain normal distribution, the first moment about 10 is 40 and that the fourth moment about 50 is 48. What is the arithmetic mean and standard deviation of the distribution ? **(*B.Sc. Alla., 1990*)**

(*b*) If X is a $N(50, 10)$ find $P(Y \le 3137)$ where $Y = X^2 + 1$.

$$\left[\frac{1}{\sqrt{2\pi}} \int_0^{0\cdot6} e^{-x^2/2}\, dx = 0\cdot2258\right]$$

(*B.Sc. Hons. Delhi, 1990*)

[**Ans.** (*a*) mean = 50 , s.d. = 2 (*b*) 0·7258]

[**Hint.** (*b*) $P(X^2 + 1 \le 3137) = P(-56 < X < 56)$

3. (*a*) If two normal universes A and B have the same total frequency but the standard deviation of universe A is k times that of the universe B, show that the maximum frequency of universe A is $1/k$ that of universe B.

(*b*) Let $X \sim N(\mu, \sigma^2)$, calculate

(*i*) $P(2X + 3 \le t)$ (*ii*) $P(|X| > t)$. **(*B.E., Poona, 1991*)**

4. (*a*) If X is normally distributed with zero mean and variance unity, what is the expectation and variance of (*i*) X^2 (*ii*) $e^{aX}, a > 0$?

(*b*) If X is a standard normal random variate, show that

$$P[a < X^2 < b] = 2P[-\sqrt{b} < X < -\sqrt{a}], a > 0$$

(*c*) If $X \sim N(\mu, 9^2)$ and $Y \sim N(\mu, 12^2)$ are independent and if

$$P(X + 2Y \le 3) = P(2X - Y \ge 4), \text{ determine } \mu.$$

(*B. Sc. Hons Math Calcutta 1989*)

[**Hint.** $X + 2Y \sim N\left(3\mu, 9^2 + 4 \times 12^2\right)$ and $2X - Y \sim N\left(\mu, 4 \times 9^2 + 12^2\right)$

i.e. $X + 2Y \sim N(3\mu, 657)$ and $2X + Y \sim N(\mu, 387)$]

See Ex. 8·1, page 309.

(*d*) If $X \sim N(0, 1)$ and $Y \sim N(0, 1)$, prove that

(*i*) $\text{Var}(\sin X) > \text{Var}(\cos X)$ (*ii*) $E|X - Y| \le \sqrt{8/\pi}$

[**Hint.** (*i*) $X \sim N(0, 1)$ $\therefore$ $\phi_X(t) = e^{-t^2/2} \Rightarrow E\left[e^{itX}\right] = e^{-t^2/2}, i = \sqrt{-1}$

$\Rightarrow E[\cos tX + i \sin tX] = e^{-t^2/2}$ $\therefore$ $E(\cos tX) = e^{-t^2/2}, E(\sin tX) = 0$

Putting $t = 1$ and 2, we get $E(\cos X) = e^{-1/2}, E(\cos 2X) = e^{-2}$

$$E(\sin X) = 0 = E(\sin 2x)$$

Now $\text{Var}(\cos X) = E(\cos^2 X) - (E \cos X)^2 = E\left[\dfrac{1 + \cos 2X}{2}\right] - e^{-1}$

$$= \frac{1}{2} + \frac{1}{2}\left(e^{-2}\right) - e^{-1} = \frac{1}{2}\left(e^{-2} - 2e^{-1} + 1\right) = \frac{1}{2}\left(1 - e^{-1}\right)^2 \simeq 0.99$$

$\text{Var}(\sin X) = E[\sin^2 X] - [E(\sin X)]^2 = E\left[\dfrac{1 - \cos 2X}{2}\right] - 0$

$= \dfrac{1}{2}\left(1 - e^{-2}\right) \simeq 0.43$ $\therefore$ Hence the result .

(*ii*) We know that $|X - Y| \le |X| + |Y| \Rightarrow E|X - Y| \le E|X| + E|Y|$

$\therefore$ $E|X| = \sqrt{2/\pi} = E|y|$ $X - Y \sim N(0, 2)$

(*e*) If $X \sim N(0, 1)$, find the *p.d.f.* of $|X|$ and $E(|x|)$.

(*B. Sc. Math Hons. Delhi, 1989*)

Let $Y = |X|$, then $G_Y(y) = P((Y \le y) = P(|X| < y)$

$$= P(-y < X \le y) = P(X \le y) - P(X \le -y)$$

$$= F_X(y) - F_X(-y)$$

Differentiating *w.r.t.* *y*, we get

$g_Y(y) = f_X(y) + f_x(-y) = 2f_X(y)$ as $f(-y) = f(y) = \sqrt{2/\pi}\, e^{-y^2/2}, y \ge 0$

5. (*a*) **Log-Normal distribution.** A variate *X* has the density function

$$f(x) = \frac{1}{x\sqrt{2\pi}}\, e^{-\frac{1}{2}(\log x)^2}, x > 0$$

is said to have a logarithmico-normal distribution. Find the expectation, mode and variance of *X*. Also obtain the coefficient of skewness.

Show that $e^a X^b$ has a log-normal *p.d.f.* $f(x; a + 2\mu, b\sigma)$

[Ans. $\sqrt{e}, 1/e, e^2 - e, (1 - e^{-3/2})/\sqrt{e} - 1$].

(*b*) *Y* is a standard normal variate and $X = \log_e Y$. Find the distribution of *X* and show that $\text{Var}\, X = e^e - e$. **(*B.Sc. Sardar Patel, 1982*)**

6. The random variate *X* has mean *m* and standard deviation *s*. If $Y = \log_e X$ is normally distributed with mean *M* and standard deviation S, prove that

(*i*) $m = e^{M + \frac{1}{2}S^2}$. (*ii*) $1 + \dfrac{s^2}{m^2} = e^{S^2}$

7. X is a normal variate with Mean $\bar{x}$ and *s.d.* σ. Find the mean and variance of variate Y defined as follows :

$$Y = \frac{1}{2}(X - \bar{x})^2.$$

[**Ans.** mean $= \frac{1}{2}$, variance $= \frac{1}{2}$]

8. Let X and Y be independent variates and let each be normally distributed with common mean zero and unit variance. Find the density function for

(*i*) $U = \sqrt{X^2 + Y^2}$, (*ii*) $V = X^2 + Y^2$, (*iii*) $W = \dfrac{X+Y}{\sqrt{|X-Y|^2}}$

$$\left[\textbf{Ans. } (i)\, f(u) = u e^{-\frac{u^2}{2}} \quad (ii)\, f(v) = \frac{1}{2} e^{-\frac{v}{2}}, v \geq 0\right]$$

9. (*a*) If $f(x)$ is the density function for $X \sim N(0, \sigma^2)$ show that

$$\int_{-\infty}^{\infty} \{f(x)\}^2 dx = \frac{1}{2\sigma\sqrt{\pi}}$$

(*b*) If a normal distribution is grouped in intervals of total frequency N and S is the sum of squares of the frequencies, an estimate of *s.d.* σ is given by $\dfrac{N^2}{2S\sqrt{\pi}}$ (***B.Sc. Gujarat, 1992***)

Hint. $S = \int_{-\sigma}^{+\infty} \{Nf(x)\}^2 dx = \dfrac{N^2}{e\sigma\sqrt{\pi}}$.

10. A variate X is normally distributed with mean m and variance σ^2. Show that the mean of the conditional distribution of X, given $a < X < b$, is

$$m + \frac{n(a) - n(b)}{N(b) - N(a)} \cdot \sigma^2 \quad \text{where} \quad N(t) = \frac{1}{\sigma\sqrt{2\pi}} \int_{-\infty}^{t} e^{-\frac{1}{2\sigma^2}(x-m)^2} dx$$

and
$$n(t) = \frac{1}{\sigma\sqrt{2\pi}} e^{-\frac{1}{2\sigma^2}(t-m)^2}$$

[**Hint.** Mean of the conditional distribution of X is

$$\int_a^b x f(x)\, dx \div \int_a^b f(x)\, dx \text{ where } f(x) = \frac{1}{\sigma\sqrt{2\pi}} e^{-\frac{1}{2\sigma^2}(x-m)^2} \Bigg]$$

11. (*a*) X is a *r.v.* with *p.d.f.* $f(x)$ given by

$$f(x) = \frac{1}{\sigma\sqrt{2\pi}} e^{-\frac{1}{2\sigma^2}(x-m)^2}, -\infty < x < \infty$$

is truncated at the point $X = x$ and the values $X > x$ are discarded. Find the mean and variance of that part for which $X \leq x$

(*b*) An establishment uses 1,000 bulbs which are kept burning approximately for 4 hours every day. Past experience of costs incurred in replacing burnt out bulbs has shown that it is profitable to replace all the 1,000 bulbs, whether some of them are burnt out or not, in 4 months (about 120 days). Bulbs burning out during the period are not replaced till the end and the establishment was prepared to suffer the inconvenience so caused. Assuming that life of a bulb is nearly normally distributed with a mean of 450 hours and standard deviation of 30 hours, find the expected number of hours for which a bulb is dead at any point during four months period.

[it is known that the area under the normal curve beyond a distance of one standard deviation from the mean is equal to 0·159. You need not simplify any expression which needs the use of tables not supplied to you]. ***(I.A.S. 1947)***

$$\text{[Ans. (a) If } M_0 = \int_x^{\infty} f(x)dx, \text{ mean} = m + \frac{\sigma}{M_0\sqrt{2\pi}} e^{-\frac{1}{2}\left(\frac{X-M}{\sigma}\right)^2}$$

$$\text{Variance} = \sigma^2 + \frac{\sigma^2}{M_0\sqrt{2\pi}}\left(\frac{X-m}{\sigma}\right)\exp\left[-\tfrac{1}{2}\left(\frac{X-m}{\sigma}\right)^2\right] - \frac{\sigma}{M_0{}^2 2\pi}\exp\left[-\left(\frac{X-m}{\sigma}\right)^2\right]$$

$$(b)\ 480 - \left[450 + \frac{30}{450\sqrt{2\pi}} e^{-\frac{1}{2}}\right] - \frac{1}{15\sqrt{2\pi}} e^{-\frac{1}{2}}$$

12. X is a normal variate with mean 1 and standard deviation 3, find the probability that

(*i*) $3{\cdot}43 \le X \le 6{\cdot}19$ (*ii*) $-1{\cdot}43 \le X \le 6{\cdot}19$ [**Ans.** (*i*) ·1672 (*ii*) ·7492]

13. If X is normally distributed with mean 11 and *s.d.* 1·5, find the number x_0 such that (*a*) $P[X > x_0] = {\cdot}3$ and (*b*) $P[X > x_0] = {\cdot}09$.

[**Ans.** (a) 11·7875 (*b*) 13·01]

14. Prove that, for normal distribution, the quartile deviation, mean deviation and the standard deviation are approximately in the ratio 10 : 12 : 15.

(B.Sc. Dibrugarh, 1993)

15. Let X be a chance variate having a normal distribution with mean 30 and *s.d.* 5. What is the probability that (*i*) $26 \le X \le 40$ (*ii*) $|X - 30| > 5$?

[**Ans.** (*i*) ·7654 (*ii*) ·3174]

16. In certain examination test 2,000 students appeared in Statistics. The average marks obtained were 50% and the standard deviation was 5%. How many students do you expect to obtain more than 60% marks supposing the marks to be distributed normally ? [**Ans.** 46]

17. A normal distribution has mean 13 and *s.d.* 2, find the area under the curve

(*a*) from x = 10 to $x = 13{\cdot}5$. (*b*) from $x = 11{\cdot}4$ to $x = 14{\cdot}2$

(*c*) from $x = 6$ to $x = 18$. [**Ans.** (*a*) 0·6147 ; (*b*) 0·4822 ; (*c*) 0·9973]

18. (*a*) If $\log_e X$ is normally distributed with mean 1 and variance 4 find $P(\frac{1}{2} < X < 2)$, given $\log_e 2 = 0{\cdot}693$.

(b) If $\log_{10}X$ is normally distributed with mean 4 and variance 4, find the probability of $1.202 < X < 83180000$,
[Given $\log_{10} 1202 = 3.08$, $\log_{10} 8318 = 3.92$]

(B.Sc., Statistics, Bombay 1981)

$$\text{[Ans. (a) } \int_{.153}^{8.41} \frac{1}{\sqrt{2\pi}} e^{-\frac{1}{2}y^2} dy. \text{ (b) } .95$$

(c) If $\log Z \sim N(7, 3)$ and $\log Y \sim N(3, 1)$, and X, Y and independent, show that $P(1.202 < X/Y < 83180000) = .95$.

(d) Prove that if the independent random variables X and Y have the probability densities $\frac{h}{\sqrt{\pi}} e^{-h^2x^2}$ and $\frac{k}{\sqrt{\pi}} e^{-k^2y^2}$ then the random variate $U = X + Y$ has the p.d.f $f(u)$ given by **(I.F.S. 1997)**

$$f(u) = \frac{l}{\sqrt{\pi}} e^{-l^2u^2} \quad \text{where } l^{-2} = h^{-2} + k^{-2}, -\infty < (x, y) < \infty.$$

[**Hint.** $X \sim N\left(0, \frac{1}{2h^2}\right)$, $Y \sim N\left(0, \frac{1}{2k^2}\right)$

$\therefore$ $E(U) = E(X + Y) =$ etc. and $\text{Var } U = \text{Var } X + \text{Var } Y$]

19. In firing a target assume that the horizontal distance of a shot from the centre is normally distributed with *s.d.* 2 feet.

In 200 shots, how many would be expected to miss the target, if it is 10 feet wide and sufficiently high ?

[**Hint.** Using table 8·2, the area outside the range $x = -2.5\,\sigma$ to $x = 2.5\,\sigma$ is 0·012 which is about $1\frac{1}{4}\%$ of the whole. Therefore, the number of shots expected to miss the target $= 200 \times \frac{5}{4} \times \frac{1}{100} = \frac{5}{2}$, that is 3.]

20. Steel rods are manufactured to be 3 inches in diameter, but they are acceptable if they are inside the limits of 2.99 inches and 3·01 inches. It is observed that about 5% are rejected over-size and 5% are rejected under-size. Assuming that the diameters are normally distributed, find the standard deviation of the distribution. Hence calculate what proportion of rejects would be, if the permissible limits were widened to 2·983 inches and 3·015 inches.

You may use the following table. The area A is measured from the mean zero to any ordinate x.

$\frac{x}{\sigma}$	A	$\frac{x}{\sigma}$	A
1·64	·4995	2·5	·4938
1·65	·4505		

$$\left[\textbf{Hint. } (a) \frac{3 - 2.99}{\sigma} = 1.645\right]$$

[**Ans.** (a) $S.D. = 0.806$ inch approximately ; (b) 1·24%]

21. Five thousand candidates appeared in a certain examination paper, carrying a maximum of 100 marks. It was found that the marks were normally distributed, with mean 39·5 and standard deviation 12·5. Determine approximately the number of candidates who secured a first class for which a minimum of 60 marks is necessary. You may use the table given below. [x denotes the deviation from the mean].

The proportion A of the whole area of the normal curve lying to the left of the ordinate at the deviation x/σ.

$\frac{x}{\sigma}$	1·5	1·6	1·7	1·8
A	·9932	·9452	·9554	·9641

$$\left[\textbf{Hint. } \frac{x}{\sigma}=\frac{50-39.5}{12.5}=1.64\right]$$ **[Ans. 253]**

22. Assume the mean height of soldiers to be 68·22 inches with variance of 10·8 (in.)^2. How many soldiers in a regiment of 1,000 would you expect to be over 6 feet tall, given that the area under the standard normal curve between $x = 0$ and 0·35 is 0·1368; and between $x = 0$ and $x = 1·15$ is 0·3746. (6 feet = 72 inches)

[Ans. 125]

23. The marks obtained by candidates in Statistics in a certain examination are found to be normally distributed. If 12·5% of the candidates obtain 60% or more marks, 39% obtain less than 30 marks ; Find the mean number of marks obtained by the candidates. You may use the table given below.

The proportion A of the whole area of the normal curve lying to the left of the ordinate at the deviation x/σ, [x is the distance from the mean].

x/σ	0·27	0·28	0·29	1·14	1·15	1·16
A	·6064	·6102	·6141	·8727	·8749	·8770

[Ans. 36 approx.]

$$\left[\textbf{Hint. } \frac{60-m}{\sigma}=1.15,\ \frac{m-30}{\sigma}=0.28,\text{ Dividing we get } m=\frac{5130}{143}=36\text{ approx.}\right]$$

24. The distribution of the yields of one thousand beds, taken at random, is found to be normal with mean 20·5 seers. The yield of 160 beds is 17·75 seers or less. Find out the standard deviation of each bed. Find also how many beds have got the yields between 13·9 seers and 27·1 seers.

[Ans. *s.d.* = 2·75 seers, 984 beds]

25. In a sample of 1000 cases, the mean of certain test is 14 and *s.d.* 1·5. Assuming the normality of the distribution, find (*i*) how many candidates score between 12 and 15 ? (*ii*) How many score above 18 ? (*iii*) How many score below 8 ? (*iv*) What is the probability that a candidate selected at random will score above 15 ?

[Ans. (*i*) 444 ; (*ii*) 55 ; (*iii*) 8 ; (*iv*) 0·345 *i.e.*, 34·5%]

26. The scores made by candidates in a certain test are normally distributed with mean 500 and *s.d.* 100. What per cent of the candidates receive scores (*i*) greater than 700 ; (*ii*) less than 400 ; (*iii*) between 400 and 600 ; (*iv*) which differ from 500 by more than 150 marks ; (*v*) if a candidate gets a score of 680, what per cent of the candidates have higher score than he ?

[**Ans.** (*i*) 2·28 ; (*ii*) 15·87 ; (*iii*) 68·26 ; (*iv*) 13·36 ; (*v*) 3·509]

27. The height measurements of 300 individuals are arranged in ascending order and it is observed that 120th measurement is 64·50 inches and the 210th measurement is 67·35 inches. Assuming that the heights form a normal population, estimate the height of 50th and 270th individuals.

You may use the following table which gives the probability p of standardized normal variate x falling outside the range $(-x, x)$.

p	·1	·2	·3	·4	·5	·6	·7	·8	·9
x	1·6449	1·2816	1·0364	0·8416	0·6745	0·6244	0·5244	0·3853	0·1255

[**Hint.** Solving the equations $\dfrac{m-64{\cdot}50}{\sigma}=0{\cdot}2533, \dfrac{67{\cdot}35-m}{\sigma}=0{\cdot}5234,$

we get $m=65{\cdot}43, \sigma=3{\cdot}66$].

[**Ans.** 62·36 inches, 70·11 inches]

28. In a normal distribution, 31% of the items are under 45 and 8% are over 64. Find the mean and *s.d.* of the distribution.

[**Ans.** Mean = 50, S.D. = 10]

29. In a certain examination, the percentage of passes and distinctions were 45 and 9 respectively. Estimate the average marks obtained by the candidates, the minimum pass and distinction marks being 40 and 75 respectively. (Assume the distribution of marks to be normal). [**Ans.** 36·4]

Also determine what would have been the minimum qualifying marks for admission to a re-examination of the failed candidates had it been desired that the best 25% of them should be given another opportunity of being examined ?

[**Ans.** 291]

30. In any examination, it is laid down that a student passes if he secures 30 per cent or more marks. He is placed in the first, second or third division according as he secures 60 per cent or more marks, between 45 per cent and 60 per cent marks and between 30 and 45 per cent respectively. He gets a distinction in case he secures 80 per cent or more marks. It is noted from the results that 10 per cent of the students fail in the examination whereas 5 per cent of them obtained distinction. Calculate average marks, *s.d.* and the percentage of students placed in the second division. (Assume normal distribution of marks.)

[**Ans.** average marks = 52, *s.d.* = 17·12, required percentage = 33]

[**Hint.** $\dfrac{m-30}{\sigma}=1{\cdot}38, \dfrac{80-m}{\sigma}=1{\cdot}647.$

31. If $X_1, X_2, \ldots\ldots, X_n$ be independent normally distributed chance variables each with mean m and variance σ^2, prove that $Y_n=\dfrac{1}{n}\sum_{i=1}^{n} X_i$ is normally distributed with mean m and variance σ^2/n. Hence show that for any $\varepsilon > 0$

$$\underset{n \to \infty}{\text{Lt}} \; P(|Y_n - m| > \varepsilon) = 0$$

State whether the assumption of normality of X_i is necessary for its validity.

32. Show that the mean value of the positive square root of a $\beta(l, m)$ variate is $\overline{|l + \frac{1}{2}}\,\overline{|l + m} / \overline{|l + m + \frac{1}{2}}$ and that of a $\gamma(l)$ variate is $\Gamma(l + \frac{1}{2}) / \Gamma(l)$).

(***B.Sc. Stat. Hons. 1986***)

33. If the probability that a deviation lies between $-x$ and $+x$ is given by

$$\phi(hx) = \frac{2h}{\sqrt{\pi}} \int_0^x e^{-h^2x^2} dx \quad \text{where} \quad \phi(y) = \frac{2}{\sqrt{\pi}} \int_0^y e^{-y^2} dy,$$

show that (*i*) $\phi(y) = 1 - \dfrac{e^{-y^2}}{y\sqrt{\pi}} \left[1 - \dfrac{1}{2y^2} + \dfrac{1 \cdot 3}{(2y^2)^2} - \dfrac{1 \cdot 3 \cdot 5}{(2y^2)^3} + \cdots\cdots \right]$

(*ii*) $\phi(y) = \dfrac{2y}{\sqrt{\pi}} e^{-y^2} \left[1 + \dfrac{1}{3} \cdot (2y^2) + \dfrac{1}{3 \cdot 5} \cdot (2y^2)^2 + \dfrac{1}{3 \cdot 5 \cdot 7} (2y^2)^3 + \cdots \right]$

Sol. (*i*) $\sqrt{\pi} - \sqrt{\pi}\,\phi(y) = 2 \int_0^\infty e^{-y^2} dy - 2 \int_0^y e^{-y^2} dy = 2 \int_y^\infty e^{-y^2} dy$

$$= \int_t^\infty e^{-t} \cdot t^{-\frac{1}{2}} dt \qquad \text{where } y^2 = t$$

$$= \left[-e^{-t} t^{-\frac{1}{2}} \right]_t^\infty - \tfrac{1}{2} \int_t^\infty e^{-t} \cdot t^{-3/2} dt = e^{-t} t^{-\frac{1}{2}} - \frac{1}{2} \int_t^\infty e^{-t} t^{-3/2} dt$$

Successive integration by parts gives

$$\sqrt{\pi} - \sqrt{\pi}\phi(t) = e^{-t} t^{-1/2} - e^{-t} t^{-3/2} + \frac{1 \cdot 3}{2 \cdot 2} e^{-t} t^{-5/2} \;\ldots\ldots$$

$$\text{Hence } \phi(y) = 1 - \frac{e^{-y^2}}{y\sqrt{\pi}} \left[1 - \frac{1}{2y^2} + \frac{1 \cdot 3}{(2y^2)^2} - \cdots \right]$$

(*ii*) Let $H(y) = \int_0^y e^{-y^2} dy = e^{-y^2} t$, so that $\phi(y) = \dfrac{2}{\sqrt{\pi}} \cdot e^{-y^2} t$

Differentiating under the integral sign with respect to t, we get

$$e^{-y^2} = e^{-y^2} \frac{dt}{dy} - 2yt\, e^{-y^2} \quad \text{or} \quad \frac{dt}{dy} - 2yt = 1. \qquad \ldots(1)$$

As $H(y)$ is a function of y, $e^{-y^2} t$ is also a function of y

Hence let $\quad t = y + ay^3 + by^5 + \ldots\ldots \qquad \ldots(2)$

From (1) and (2) we get $(1 + 3ay^2 + 5by^4 + \cdots\cdots) - 2y(y + ay^2 + \cdots\cdots) - 1 = 0.$

Equating coefficients of different powers of y on the two sides we have

$$3a - 2 = 0, 5b - 2a = 0, 7c - 2b = 0$$

$$\therefore \quad t = y + \tfrac{2}{3}y^3 + \frac{2^2}{3 \cdot 5}y^5 + \cdots\cdots \text{ and hence } \phi(y) = \text{etc.}$$

34. If $F(x) = \dfrac{1}{\sqrt{2\pi}} \displaystyle\int_{-\infty}^{x} e^{-\frac{1}{2}x^2}\, dx$, show that

(*i*) $F(x) = \frac{1}{2} + \dfrac{1}{\sqrt{2\pi}}\left[x - \dfrac{x^3}{2 \cdot 3} + \dfrac{1}{2\,!}\dfrac{x^5}{2 \cdot 5} - \cdots\cdots \right]$ when x is small,

(*ii*) $F(x) = 1 - \dfrac{e^{-\frac{1}{2}x^2}}{x\sqrt{\pi}}\left[1 - \dfrac{1}{x^2} + \dfrac{3}{x^4} - \dfrac{3 \cdot 5}{x^6} + \cdots\cdots \right]$ when x is large.

35. The conditional probability that the random variate X should lie within the range dx for a given σ is given by

$$\frac{1}{\sigma\sqrt{2\pi}} \exp\left\{ -\frac{1}{2}\left(\frac{x - m}{\sigma}\right)^2 \right\} dx, -\infty < x < \infty$$

while the probability of σ itself lying within the range $d\sigma$ is

$$\frac{1}{{\sigma_0}^2}\, \sigma \exp\left\{ -\frac{1}{2}\frac{\sigma^2}{{\sigma_0}^2} \right\} d\,\sigma, 0 < \sigma < \infty$$

where σ_0 is a constant. Show that the unconditional (*i.e.*, marginal) distribution of X has the following probability.

$$\frac{1}{2{\sigma_0}^2}\, e^{-|x - m|/\sigma_0}, -\infty < x < \infty$$

36. Fit a normal curve to the following data :

x :	10–12	12–14	14–16	16–18	18–20	20–22	22–24	24–26	26–28	Total
f :	4	30	106	206	272	219	120	37	6	1000

[**Hint.** $\bar{x} = 13 \cdot 14, \sigma = 2 \times 1 \cdot 445 = 2 \cdot 890$]

37. Discuss the differential equation for the Pearsonian system of curves and find for what values of the constants we get the following χ^2 distribution having n degrees of freedom:

$$dF = \text{const.}\, e^{-\frac{1}{2}\chi^2}\,(\chi^2)^{n/2 - 1}\, d(\chi^2), 0 < \chi^2 < \infty.$$

38. Show that, when the constant are suitably selected, the solution of $\dfrac{dy}{dx} = \dfrac{-y(x + a)}{b + cx}$ leads to the χ^2 distribution with $2k$ degrees of freedom where $c^2k = c(a + c) - b$.

$$\left[\textbf{Hint. } \frac{-1}{y}\, dy = \left[\frac{1}{c} + \frac{\left(a - \frac{b}{c}\right)}{cx + d} \right] dx \quad \therefore \quad y = y_0 e^{-x/c}\left(1 + \frac{x}{b}\right)^{2k - 1} \right]$$

39. (*a*) Show that for a Pearson distribution $\dfrac{df(x)}{dx} = \dfrac{(x+a)f(x)}{b_0 + b_1 x + b_2 x^2}$, the characteristic function ϕ obeys the relation

$$b_2\theta \frac{d^2\phi}{d\theta^2} + (1 + 2b_2 + b_1\,\theta)\frac{d\phi}{d\theta} + (a + b_1 + b_0\theta)\phi = 0 \quad \text{where} \quad \theta = i\,t.$$

(*b*) Hence deduce the recurrence relation for moments.

[**Hint.** (*b*) Differentiate the given relation *n* times with respect to θ then put $\theta = 0$ remembering that $\left(\dfrac{d^n\phi}{d\theta^n}\right)_{\theta=0} = \mu'_n$. Now transfer the origin to the mean] **[Ans.** $[(n+2)b_2 + 1]\mu_{n-1} + [(n+1)b_1 + a]\mu_n + nb_0\mu_{n-1} = 0]$

40. Obtain Pearson Type III curve from the basic differential equation of the Pearsonian System of curves and prove that for this curve $2\beta_2 = 3(\beta_1 + 2)$.

41. Show that for the Pearson system of frequency curves given by the differential equation. $\dfrac{1}{y}\dfrac{dy}{dx} = \dfrac{m - x}{a + bx + cx^2}$

the moments of the standardised variate $z = (x - \mu)/\sigma$ satisfy the relation

$$(1 - 3c)\,r\mu_{r-1} - mr\mu_r + [c(r+2) - 1]\,\mu_{r+1} = 0$$

Show that if the first four moments of a frequency distribution exist, then $\beta_2 - \beta_1 - 1 \geq 0$.

42. (*a*) If $X_1, X_2, \ldots\ldots, X_n$ are *n* independent gamma variates each with parameter *l*, find the *m.g.f.* for $(X_1 + X_2 + \cdots\cdots + X_n)/n$.

(*b*) Find β_1 and β_2 for (*i*) gamma distribution (*ii*) Beta distribution of the first kind (*iii*) Beta distribution of the second kind. Under what conditions do they approach 0 and 3 respectively ?

(*c*) The daily consumption of milk in a city in excess of 20,000 gallons, is approx distributed as a Gamma variate with the parameters $\nu = 2$ and $\lambda = 1/10{,}000$. The city has a daily stock of 30,000 gallons. What is the probability that the stock is insufficient on a particular day ?

[**Hint.** Let the *r.v.* *X* denote daily consumption of milk (in gallons) in a city, then the *r.v.* $Y = X - 20{,}000$ has a γ (2, 1/10,000).

$$p = P(X > 30{,}000) = P(Y > 10{,}000) = \int^{\infty} 10{,}000 f(y)\,dy \text{ where}$$

$f(y) = \dfrac{1}{(10{,}000)^2\sqrt{2}}\; y^{2-1}\, e^{-y/10{,}000}$; Put $z = y/10{,}000$ and show that $p = 2/e$.]

43. Show that if *X* and *Y* are independent random variables and $\phi_1(t)$, $\phi_2(t)$ and $\phi(t)$ are the characteristic functions of *X*, *Y* and $aX + bY$ respectively, *a*, *b*, being constants then $\phi(t) = \phi_1(at)\,\phi_2(bt)$

Hence prove that a linear function of *n* independent normal variables is distributed normally.

44. If X_1 and X_2 are independent normal variates with means m_1 and m_2 and variances σ_1^2 and σ_2^2 respectively, find the relation between α, β, γ and δ so that

$$P\{c_1x_1 + c_2x_2 < \alpha\} = \gamma \quad \text{and} \quad P\{c_1x_1 + c_2x_2 < \beta\} = \delta$$

45. If X and Y are independent normal variates possessing a common mean μ such that

$$P(2X + 4Y \le 10) + P(3X + Y \le 9) = 1$$
$$P(2X - 4Y \le 6) + P(-3X + Y > 1) = 1$$

determine the values of μ and the ratio of the variances of X and Y.

Hint. $2X + 4Y \sim N(6\mu, 4\sigma_1^2 + 16\sigma_2^2)$, etc.

Let $4\sigma_1^2 + 16\sigma_2^2 = \alpha^2$ and $9\sigma_1^2 + \sigma_2^2 = \beta^2$

$$\therefore \frac{10 - 6\mu}{\alpha} = -\left(\frac{9 - 4\mu}{\beta}\right). \text{ Similarly } \frac{6 + 2\mu}{\alpha} = \frac{1 + 2\mu}{\beta}.$$

[**Ans.** $\mu = 2$, or $-1{\cdot}6$, $\lambda = \frac{64}{87}$.]

46. Calculate β_1 and β_2 of the following distribution. Fit an appropriate Pearsonian curve.

(*a*) The following frequency distribution reports the scores obtained by students in a certain test examination.

Score (Central Value)	Frequency	Score (*Central Value*)	Frequency
15	7	95	80
25	9	105	80
35	11	115	65
45	14	125	40
55	19	135	18
65	35	145	6
75	57	155	3
85	64		

(*b*)

Class interval	0–2	2–4	4–6	6–8	8–10	10–12	12–14	14–16	16–18
Frequency	4	23	51	69	59	57	35	14	18

47. On the *x*-axis, $(n + 1)$ points are taken independently between origin and $x = 1$, all positions being equally. Show that the probability that the $(k + 1$th of these points, counted from the origin lies in the interval $x \pm \frac{1}{2} dx$ is

$$^nC_k (n + 1)\, x^k (1 - x)^{n-k}\, dx$$

Deduce the distribution of the (*i*) smallest (*ii*) the longest observation (*iii*) the median observation when $n = 4m + 3$.

48. (*a*) If X is uniformly distributed with mean 1 and variance 4/3, find $P(X < 0)$.

(***B.A. Stat. Hons., Delhi, 1989***)

[**Hint.** Solve for a and b : $\dfrac{b+a}{2} = 1$ and $\dfrac{(b-a)^2}{12} = \dfrac{4}{3}$ to get $a = -1, b = 3$

$$P(X < 0) = \int_{-1}^{0} f(x)dx = \int_{-1}^{0} \frac{1}{3-(-1)}\,dx = \frac{1}{4}\,\Big|\, x \Big|_{-1}^{0} = \frac{1}{4}.\Big]$$

(*b*) Find the mean deviation about mean of the rectangular distribution.

[**Hint.** M.D. $= \dfrac{1}{b-a}\displaystyle\int_a^b \left| x - \frac{b+a}{2} \right| dx$; Put $x - \dfrac{b+a}{2} = y$ **Ans.** $\dfrac{1}{4}(b-a)$

49. If X_1, X_2 are independent rectangular variates over (0, 1), show that the distribution of

(*i*) $u = \dfrac{X_1}{X_2}$ is $f(u) = \dfrac{1}{2}, 0 \le u \le 1, = \dfrac{1}{2u}, 1 \le u < \infty$

(*ii*) $v = X_1 X_2$ is $g(v) = -\log v, 0 < v < 1$

Find also the distributions of $X_1 + X_2$ and $X_1 - X_2$.

50. If $X_1, X_2, \ldots\ldots, X_n$ are n independent rectangular variates over [0, 1] show that the density function for $Y = X_1 X_2 \ldots\ldots X_n$ is

$$f(y) = \frac{(-\log y)^{n-1}}{(n-1)}, 0 < y < 1 \;; = 0, \text{ otherwise.}$$

Find the density function of $X_1 + X_2 + \cdots\cdots + X_n$

[**Hint.** Let $Y_1 = X_1, Y_2 = X_1X_2, \ldots\ldots, Y_n = X_1 X_2 \ldots\ldots X_n$

$\Rightarrow x_1 = y_1, x_2 = y_2/y_1, \ldots\ldots, x_n = y_n/y_{n-1}$

$$p.d. \text{ of } y_n = \int_{-\infty}^{+\infty}\int_{-\infty}^{\infty} \cdots\cdots \int_{-\infty}^{\infty} f\left(y_1, \frac{y_2}{y_1}, \ldots, \frac{y_n}{y_{n-1}}\right) \cdot \frac{dy_1\, dy_2 \cdots\cdots dy_{n-1}}{y_1\, y_2 \cdots\cdots y_{n-1}}$$

$$= \int\int \cdots\cdots \int (y_1\, y_2 \cdots\cdots y_{n-1})^{-1}\, dy_1\, dy_2 \cdots\cdots dy_{n-1},$$

$$y_2 < y_1 < 1, \frac{4}{3} < y_2 < 1, y_n < y_{n-1} < 1.$$

51. Let $X \sim N(\mu, \sigma^2)$. If $\sigma^2 = \mu^2, \mu > 0$ express $P(X < -\mu \mid X < \mu)$ in terms of distribution function of $N(0,1)$.

(*B.Sc. Math. Hons. Delhi, 1988, B.Sc. Stat. Hons. Delhi, 1993*)

[**Hint.** Regd. prob. $= P(X < -\mu \cap X < \mu) / P(X < \mu) = P(X < -\mu) \div P(X < \mu)$

$$= P(Z < -2) \div P(Z < 0) = P(z > 2) \div \tfrac{1}{2} = 2\,[1 - \phi(2)\,].$$

52. (*a*) Let X and Y be independent variates with the common Cauchy density

$$\frac{1}{\pi} \cdot \frac{1}{1+x^2}, -\infty < x < +\infty$$

Prove that $u = XY$ has the density $f(u) = \frac{2}{\pi^2}\left\{\frac{\log|u|}{u^2-1}\right\}, -\infty < u < +\infty$

(M.A. Statistics Delhi, 1991)

(*b*) Let $X \sim \gamma(l, a)$ and $Y \sim \gamma(l, b)$ be independent *r.v.*, show that.

$$E\left(\frac{X}{X+Y}\right) = \frac{E(X)}{E(X+Y)}.$$

[**Hint.** (*a*) Put $u = xy,\ v = \frac{x}{y},\ dxdy = \frac{1}{u}\,dudv$ in the joint distribution of X and Y]

(*b*) We know that $U = X/(X+Y)$ and $V = X+Y$ are independent

Hence $E(UV) = E(U)\,E(V) \Rightarrow E(X) = E\left(\frac{X}{X+Y}\right).\,E(X+Y) \quad \therefore$ etc.

53. Show that the mean and variance of Weibull distribution

$P_X(x) = \alpha\, x^{\alpha-1} \exp(-x^{\alpha})\ ;\quad x > 0, \alpha > 0$ are given by

$EX = \Gamma(1/\alpha + 1)$, and $\text{Var } X = \{\Gamma(2/\alpha + 1) - \{\Gamma(1/\alpha + 1)\}\}^2$.

54. X and Y are independent random variables each exponentially distributed with the same parameter θ. Find the *p.d.f.* for (*i*) $X/(X+Y)$ (*ii*) $X - Y$

(B.Sc. Hons. Stat. Delhi, 1985, 88, 89)

[**Hint.** (*i*) Joint *p.d.f.* of X and Y is, $(\theta e^{-x\theta} dx)(\theta e^{-\theta y} dy) = \theta^2 e^{-\theta(x+y)}\,dxdy$

Let $Z = X/(X+Y),\ U = X+Y$ so that $X = ZU,\ Y = U - Z,\ dx\,dy =$ etc. $= u$

(*ii*) Show that $\phi_{X-Y}(t) = \frac{\theta^2}{(1-it\theta)(1+it\theta)} = \frac{\theta^2}{1+\theta^2 t^2}$ which is the *c.f.*

where X has Laplace's distribution $g(u) = \frac{1}{2\theta}\, e^{-\theta|u|}, -\infty < u < \infty$.]

55. A random variable X has a uniform distribution in [0, 1] *i.e.* $f(x) = 1$. Find the *p.d.f.* of $-2\log_2 X$ and identify its distribution.

(B.A. Hons. Statistics, Delhi, 1984)

[**Hint.** $G(y) =$ distribution function G of Y

$$= P(Y \le y) = P(-2\log X \le y) = P\left(\log_X \ge \frac{-y}{2}\right)$$

$$= P(X \ge e^{-y/2}) = 1 - P(X \le e^{-y/2}) = 1 - \int_0^{e^{-y/2}} \frac{1}{1-0}\,dx = 1 - e^{-y/2}$$

$\therefore \quad g(y) = p.d.f$ of $y = \frac{d}{dy}G(Y) = \frac{1}{2}e^{-y/2}, 0 < y < \infty$ which is chi-square distribution with 2 *d.f.*

The range of Y are obtained by putting the ranges of X *viz.* [0, 1] in $Y = -2\log_2 X$.

56. Show that the *p.d.f.* of a random variable X defined over $[-1, 1]$ satisfies the differential equation $\quad \frac{df}{dx} = cx\,(1-x^2)^{-1}$, c being constant.

57. If X is a $N(\mu, \sigma^2)$, then *p.d.f.* of $Y = X^2$ is

$$\frac{1}{\sigma\sqrt{2\pi}} \exp\left(-\frac{\mu^2}{2\sigma^2}\right) \exp\left(\frac{y}{2\sigma^2}\right) y^{-\frac{1}{2}} \cosh\left(\frac{\mu\sqrt{y}}{\sigma^2}\right), \qquad 0 < y < \infty$$

[**Hint.** Divide the whole real line, the range of u into the mutually exclusive parts $(-\infty, 0]$ and $[0, \infty)$.

p.d.f. of $\quad y = f(-\sqrt{y}) \cdot \frac{1}{2\sqrt{y}} + f(\sqrt{y}) \cdot \frac{1}{2\sqrt{y}} = \text{etc. } 0 < y < \infty$]

58. Show that for any $x > 0$, $\; 1 - \Phi(x) \leq \frac{1}{x}\phi(x)$

where $\quad \Phi(x) = \int_{-\infty}^{x} \phi(t)\,dt$ and $\phi(t) = \frac{1}{\sqrt{2\pi}} e^{-\frac{1}{2}t^2}$

59. If X, Y are independent standard normal variates, (*a*) find the mean value of the greater of $|X|$ and $|Y|$. (*b*) find the distribution of X/Y.

(*B.Sc. Stat. Hons. Delhi 1990*)

[**Hint.** Clearly, $f(x, y) = \frac{1}{2\pi} \exp.\left[-\frac{1}{2}(x^2 + y^2)\right], -\infty < x, y < \infty$

Ans. (*b*) $\frac{1}{\pi} \cdot \frac{1}{1 + u^2}, -\infty < u < \infty$

where $u = x/y$, $v = y$ $\therefore$ $x = uv$, $y = v$

(*a*) $E\,[\text{Max}\,(|x|, |y|)] = \frac{1}{2\pi} \cdot \int_{-\infty}^{\infty}\int_{-\infty}^{\infty} \max\,(|x|, |y|)\, f(x,y)\, dx\,dy$

$$= \frac{1}{2\pi} \int_{|x|>|y|} |x|\, e^{-\frac{1}{2}(x^2+y^2)}\, dx\,dy + \frac{1}{2\pi} \int_{|y|>|x|} |y|\, e^{-\frac{1}{2}(x^2+y^2)}\, dx\,dy$$

$$= \frac{1}{\pi} \int\!\!\int_{|x|>|y|} |x|\, e^{-\frac{1}{2}(x^2+y^2)}\, dx\,dy \qquad \text{by symmetry}$$

$$= \frac{2}{\pi} \int_{-\infty}^{+\infty} e^{-\frac{1}{2}y^2}\, dy \int_{|y|}^{\infty} x\, e^{-x^2/2}\, dx, \text{ by symmetry } = \frac{2}{\pi} \int_{-\infty}^{+\infty} e^{-y^2/2}\, dy = \frac{2}{\pi}.$$

60. Let X_i $(i = 1, 2, \ldots, n)$ be *i.i.d.* random variables. Then show that $\min\,(X_1, X_2, \ldots X_n)$ has a Weibul distribution, if the common distribution of X_i's is a Weibul distribution.

[**Hint.** $Y = \min\left(X_1, X_2, .., X_n\right) \Rightarrow P\,(Y > y) = P\left[\min\left(X_1, X_2, \ldots X_n\right) > y\right]$

$$= P\left[\bigcap_{i=1}^{n} X_i > y\right] = \prod_{i=1}^{n} P(x_i > y)$$

$$= \left[P(x_i > y)\right]^n, \text{ since } x\text{'s are } i.i.d. \ r.v.$$

$$\text{Now } P(x_i > y) = \int_y^{\infty} \alpha\beta^{-1}\left(\frac{x-\mu}{\beta}\right)^{-1} \exp\left[-\left(\frac{x-\mu}{\beta}\right)^{\alpha}\right] dx = \text{etc.}\Bigg]$$

61. (*a*) If X has exponential distribution with mean 2, find $P(X < 1 \mid X < 2)$

Ans. $\left(1 - e^{-\theta}\right) / \left(1 - e^{-2\theta}\right)$ where $\theta = 1/2$. (***B. A. Hons Stat. Delhi 1989***)

$$\Bigg[\textbf{Hint. } f(x) = \theta e^{-\theta x}, x > 0; 0 \text{ else where, } M_x(t) = \left(1 - \frac{t}{\theta}\right)^{-1} = \sum_{r=0}^{\infty} \left(\frac{t}{\theta}\right)^r$$

$$\mu_1' = \frac{1}{\theta} = 2 \text{ (given)} \Rightarrow \theta = 1/2.$$

(*b*) Suppose that during rainy season, on a tropical island, the length of shower has an exponential distribution with parameter $\theta = 2$, time being measured in minutes. What is the probalility that a shower will last more than three minutes ? If a shower has already lasted for 2 minutes, what is the probability that it will last for atleast one more minute ? (***B. Sc. Statistic Madras 1988***)

$$\Bigg[\textbf{Hint. } f(x) = 2e^{-2x}, x > 0. \quad P(x \geq 3) = \int_3^{\infty} 2e^{-2x}\, dx = \left| -e^{-2x} \right|_3^{\infty} = e^{-6}.$$

$$P[x \geq 3 \mid x \geq 2] = \frac{P(x > 3)}{P(x > 2)} = \frac{e^{-6}}{e^{-4}} = e^{-2}.\Bigg]$$

62. (*a*) If $X \sim \text{Expo}(\lambda)$ with $P(X \leq 1) = P(X > 1)$, find Var X

(***B. Sc. Hons Maths. Delhi, 1989***)

(*b*) Show that $Y = -(1/\lambda)\log F(x)$ is Expo (λ).

(***B.A. Hons. Stat. Delhi, 1985***)

$$\Bigg[\textbf{Hint. } (a) \text{ We have } P(x \leq 1) + P(x > 1) = 1 \quad \therefore \quad P(x \leq 1) = 1/2$$

$$f(x) = \lambda e^{-\lambda x}, \text{ Var } x = 1/\lambda^2$$

(*b*) We know that $Z = F(x) \sim U(0,1) \quad \therefore \quad f(Z) = 1, 0 < Z < 1$

$$\text{M.G.F.} = E\left[e^{ty}\right] = E\left[e^{t\left(-\frac{1}{\lambda}\log F(x)\right)}\right]$$

$$= E\left[e^{t\log (F(x))^{-1/\lambda}}\right] = E\left[e^{\log (F(x))^{-t/\lambda}}\right]$$

$$= E\left[\{F(x)\}^{-t/\lambda}\right] = E\left[Z^{-t/\lambda}\right]$$

$$= \int_0^1 z^{-t/\lambda}\, dz = \left|\frac{z^{-t/\lambda+1}}{-t/\lambda+1}\right|_0^1 = \frac{1}{1-t/\lambda} = \left(1 - \frac{t}{\lambda}\right)^{-1}$$

which is the *m.g.f.* of an expo. variate with parameter λ. Hence by uniqueness theorem, it follows that Y is Expo (λ).]

63. Let X and Y be independent and identically distributed each with probability function $\theta e^{-\theta x}$, $0 < x < \infty, \theta > 0$. Find the *p.d.f.* of

(*i*) $X - Y$ (*ii*) $|X - Y|$

64. Consider a truncated standard normal distribution, truncated at both ends with relevant range of variates (A, B). Obtain the *p.d.f.* and determine the mean, and variance of the truncated distribution. ***(M. Sc. Statistics, Delhi, 1988)***

$$\left[\textbf{Hint. } g(z) = \frac{\phi(z)}{P(A \le Z \le B)} = \frac{\phi(z)}{\phi(B) - \phi(A)} = \frac{(1/\sqrt{2\pi})\, e^{-\frac{1}{2}z^2}}{k}, \; k = \phi(B) - \phi(A)\right.$$

$$\text{Mean} = \int_A^B z g(z) dz = \frac{1}{k\sqrt{2\pi}} \int_A^B z e^{-\frac{1}{2}z^2} dz = \text{etc.}$$

$$\text{Mode} = \begin{cases} B & \text{if } A < 0, B < 0 \\ 0 & \text{if } A < 0, B > 0 \\ A & \text{if } A > 0, B > 0 \end{cases}; \quad \left.\text{Variance} = \int_A^B z^2 g(z) - \mu^2 = \text{etc.}\right]$$

65. A random variable X has Gamma distribution with parameters r and λ where r is positive integer. Show that $F_x(x) = 1 - \sum_{j=0}^{r-1} \frac{e^{-\lambda x} (\lambda x)^2}{j!}$

(B.Sc. Hons. Math. Delhi 1993)

$$\textbf{[Hint. } P(X = x) = P\left[\sum_{i=1}^{x+r-1} Z_i = r \cap Z_{x+r} = 1\right]$$

$$= P\left[\sum_{i=1}^{x+r-1} Z_i = r\right] P\left[z_{x+r} = 1\right]$$

$$= \binom{x+r-1}{r-1} p^{r-1} q^x p = \binom{x+r-1}{r-1} p^r q^x$$

$\therefore$ $Y = \sum_{i=1}^{n} z_i$ is a Binomial variate with parameters n and 1

Here $n = x + r - 1$ are independent.

$\sum_{i=j}$ and Z_{x+r} are independent.

$$(b)\; F_X(x) = \int_0^x \frac{\lambda^x}{r-1!} e^{-\lambda x} x^{r-1} dx = \frac{1}{(r-1)!} \int_0^b e^{-t} t^{r-1} dt, \; t = \lambda x$$

Integrate by part and continue the process till we get

$$F_X(x) \frac{e^{-t} t^{r-2}}{(r-1)!} - \frac{e^{-t} t^{r-2}}{(r-2)!} \;\ldots\ldots\ldots - \frac{-e^t t}{t!} + \left\{-e^{-t}\right\}_0^t$$

$$= 1 - \left\{e^{-t} + e^{-t}\frac{t}{1!} + \ldots\ldots\ldots + \frac{e^{-1} t^{r-1}}{(r-1)!}\right\} = 1 - \sum_{j=0}^{r-1} e^{-\lambda x} \frac{(\lambda x)^j}{j!}.$$

9

Principle of Least Squares, Fitting of Curves & Orthogonal Polynomials

9·1. ***The Principle of Least Squares.** *Def.* Suppose we have a number of independent linear equations in n unknowns $x_1, x_2, \ldots\ldots, x_n$,

$$\left.\begin{array}{l} a_{11}x_1 + a_{12}x_2 + \cdots + a_{1n}x_n = b_1 \\ a_{21}x_1 + a_{22}x_2 + \cdots + a_{2n}x_n = b_2 \\ \ldots\ldots\ldots\ldots\ldots\ldots\ldots\ldots\ldots\ldots \\ a_{m1}x_1 + a_{m2}x_2 + \cdots + a_{mn}x_n = b_m \end{array}\right\} \text{ or } \sum_{j=1}^{n} a_{ij}\, x_j = b_i,\ i = 1, 2, \ldots, m \quad \ldots(9{\cdot}1)$$

where a's and b's are constants. If $m = n$, we can, in general,** find a unique set of values satisfying the given system of equations. However, if $m > n$ that is if the number of equations is greater than the number of unknowns, no such solution exists. We, therefore, try to find those values of $x_1, x_2, \ldots, x_n$ which will satisfy the given system of equations as nearly as possible.

The principle of least squares asserts that these values are those which make S a minimum where

$$S = \sum_{i=1}^{m} (a_{i1}x_1 + a_{i2}x_2 + \cdots + a_{in}x_n - b_i)^2 = \sum_{i=1}^{m} E_i^2,$$

where $\quad E_i = a_{i1}x_1 + a_{i2}x_2 + \cdots\cdots + a_{in}x_n - b_i$

* The problem of solving m observation equations for unknowns, numbering less than m seems to have engaged the attention of astronomers since about 1750. The principle of least squares as means of accomplishing this solution approximately, appears to have been used originally by Gauss in 1795 but it was named and first published in 1805 by Legendre.

** The condition for a unique solution is that the rank of the $m \times n$ matrix of coefficients viz., $A = (a_{ij})$ should be equal to the rank of the augmented matrix (A, b) which is an $m \times (n+1)$ matrix obtained by adding to the n columns of A one more column of b's] and this common rank is n. If this common rank is $>n$, then we get no solution.

We shall call these values the *best* or *most plausible values* in the least square sense.

By a well-known theorem of differential calculus, the extreme values of the function $f \equiv f(x_1, x_2 \cdots\cdots, x_n)$ are given by $\frac{\partial f}{\partial x_1} = \frac{\partial f}{\partial x_2} = \cdots = \frac{\partial f}{\partial x_n} = 0$, provided that the partial derivatives exist.

Applying this theorem, S will have a maximum or minimum, for those values of the unknowns x_j $(j = 1, 2, \ldots, n)$ which satisfy the following n equations :

$$\frac{\partial S}{\partial x_j} = 0, j = 1, 2, \ldots, n \text{ i.e. } \sum_{i=1}^{m} a_{i1} E_i = 0, \sum_{i=1}^{m} a_{i2} E_i = 0, \ldots, \sum_{i=1}^{m} a_{in} E_i = 0 \quad \ldots(9 \cdot 2)$$

These n equations are called the *normal equations* corresponding to the equations (9·2). They can be solved like simultaneous equations for the n unknowns $x_1, x_2, \ldots, x_n$. If the common rank of the matrix of coefficients and the augmented matrix is n, we can get a unique solution for $x_1, x_2, \ldots, x_n$.

Note. To see the values which we obtain from the normal equations will make S a minimum, we calculate the second order partial derivatives

$$\frac{\partial^2 f}{\partial x_i^2} = 2 \sum_{i=1}^{m} a_{ij} \quad \frac{\partial^2 f}{\partial x_j \partial x_l} = 2 \sum_{i=1}^{m} a_{ij} a_{il} \, ; j, l = 1, 2, \ldots, n, j \neq l.$$

These values make $\frac{\partial^2 f}{\partial x_1^2} y_1^2 + \cdots + \frac{\partial^2 f}{\partial x_n^2} y_n^2 + 2 \frac{\partial^2 f}{\partial x_1 \partial x_2} y_1 y_2 + \cdots \ldots(9 \cdot 3)$

positive definite and hence S is a minimum.

Alternatively. If $x_{10}, x_{20}, \ldots, x_{n0}$ are the solutions of the normal equations (9·2), we can write S as

$$S = \sum_{i=1}^{m} \Big[(a_{i1} x_{10} + a_{i2} x_{20} + \cdots + a_{in} x_{n0} - b_i) + \{a_{i1}(x_1 - x_{10}) + \cdots + a_{in}(x_n - x_{n0})\} \Big]^2$$

$$= \sum_{i=1}^{m} (a_{i1} x_{10} + a_{i2} x_{20} + \cdots + a_{in} x_{n0} - b_i)^2 + [2(x_1 - x_{10}) \Sigma a_{i1} E_{i0} + 2(x_2 - x_{20}) \Sigma a_{i2} E_{10} + \cdots + 2(x_n - x_{n0}) \Sigma a_{in} E_{i0}] + \sum_{i=1}^{m} [a_{i1}(x_1 - x_{10}) + \cdots + a_{in}(x_n - x_{n0})]^2$$

Since $x_{10}, x_{20} \ldots\ldots, x_{n0}$ are the solutions of the normal equations, the second group of terms vanishes. $\therefore\ S \geq \sum_{i=1}^{m} (a_{i1}x_{10} + \cdots\cdots + a_{in}x_{n0} - b_i)^2$.

Hence S is a minimum when the values are obtained by solving the normal equations.

Ex. Verify that (9·3) gives a positive definite quadratic form for n = 2, 3.

Example 9·1. *Form normal equations and hence find the most plausible values of x and y from the following equations :*

$$x + y = 3{\cdot}01,\ 2x - y = 0{\cdot}03,\ x + 3y = 7{\cdot}03,\ 3x + y = 4{\cdot}97.$$

Here $S = (x + y - 3{\cdot}01)^2 + (2x - y - 0{\cdot}03)^2 + (x + 3y - 7{\cdot}03)^2 + (3x + y - 4{\cdot}97)^2$

$= f(x, y)$

The normal equations are $\dfrac{\partial S}{\partial x} = 0$ and $\dfrac{\partial S}{\partial y} = 0$

i.e., $(x + y - 3{\cdot}01) + 2(2x - y - 0{\cdot}03) + (x + 3y - 7{\cdot}03) + 3(3x + y - 4{\cdot}97) = 0$

and $(x + y - 3{\cdot}01) - (2x - y - 0{\cdot}03) + 3(x + 3y - 7{\cdot}03) + (3x + y - 4{\cdot}97) = 0$

i.e., $15x + 5y = 25$ and $5x + 12y = 29{\cdot}01$.

Solving these equations, we get $x = 0{\cdot}9997$, $y = 2{\cdot}0009$ approx.

Example 9·2. *Three independent measurements on each of the angles A, B, C of a triangle are as follows :*

	A	B	C
	39·5	60·3	80·1
	39·3	62·2	80·3
	39·6	60·1	80·4
Total	118·4	182·6	240·8

Obtain the best estimate of the three angles taking into account the relation that sum of the angles is equal to 180°.

Let the three observations on A be denoted by x_1, x_2, x_3, on B by y_1, y_2, y_3, and on C by z_1, z_2, z_3 respectively. Let the best estimate of A, B and C be θ_1, θ_2 and $180 - \theta_1 - \theta_2$. Then by the principle of least squares, we want E to be minimum where

$$E = \sum_{i=1}^{3} (x_i - \theta_1)^2 + \sum_{i=1}^{3} (y_i - \theta_2)^2 + \sum_{i=1}^{3} (z_i - 180 + \theta_1 + \theta_2)^2$$

Thus the equations to determine the best values of θ_1, θ_2 are given by

$$0 = \frac{\partial E}{\partial \theta_1} = -\Sigma(x_i - \theta_1) + \Sigma(z_i - 180 + \theta_1 + \theta_2),$$

and

$$0 = \frac{\partial E}{\partial \theta_2} = -\Sigma(y_i - \theta_2) + \Sigma(z_i - 180 + \theta_i + \theta_2).$$

These give $\Sigma x + 3\theta_1 + \Sigma z - 540 + 3\theta_i + 3\theta_2 = 0$

or $$6\theta_1 + 3\theta_2 - 417{\cdot}6 = 0 \qquad ...(1)$$

and $$\Sigma x + 3\theta_2 + \Sigma z - 540 + 3\theta_1 + 3\theta_2 = 0$$

or $$3\theta_1 + 6\theta_1 - 481{\cdot}8 = 0 \qquad ...(2)$$

Solving (1) and (2) for θ_1 and θ_2, we get

$$\hat{\theta}_1 = 39{\cdot}27,\ \hat{\theta}_2 + 60{\cdot}66,\ \hat{\theta}_3 = 180 - \hat{\theta}_1 - \hat{\theta}_2 = 80{\cdot}07$$

9·2. Fitting of Curves. The fitting of a curve to a given data is very important both from the point of view of theoretical and practical Statistics. In theoretical Statistics, we shall find that the study of regression and correlation (whether total, partial or multiple) can be regarded as fitting of linear curves (st. lines, planes, hyper-planes) to the given bivariate or multivariate frequency or probability distributions. In practical Statistics, it enables us to get a close functional relation between x and y. The relation is, in general, expressed by a polynomial, but other types of relationship, algebraic, exponential or logarithmic can also be fitted by using the principle of least squares.

9·2·1. Fitting of Polynomials or Parabolic Curves. Let it be required to fit the curve whose equation is

$$Y = a_0 + a_1 x + a_2 x^2 + \cdots\cdots + a_p x^p, \qquad ...(9{\cdot}4)$$

the coefficients $a_0, a_1 \cdots\cdots, a_p$, being constants and $a_p \neq 0$, to a number n $(n > p + 1)$ of given points or values (x_1, y_1) ; (x_2, y_2)......; (x_n, y_n) of the two variables so that we have to determine the $(p + 1)$ constants a_0, a_1,........, a_p. The curve whose equation is (9·4), is called the *parabola of degree p* of best fit. Substituting x_i for x in (9·4), we get

$$Y_i = a_0 + a_1 x_i + a_2 x_i^2 + \ldots. + a_p x_i^p \ (i = 1, 2, \ldots.., n).$$

The quantity Y_i is generally called the *expected value* of y corresponding to $x = x_i$, contrasted with the value y_i which is called the *observed value* of y corresponding to the same value x_i of x. These two values are, in general, different as the observed points do not necessarily lie on a parabolic curve. The difference $y_i - Y_i$ is called the residual or *error of estimate* corresponding to $x = x_i$ and, according to the *principle of least squares*, we choose the constants $a_0, a_1, \ldots\ldots., a_p$ so that sum of squares of residuals is minimum.

$$\text{Now } S = \sum_{i=1}^{n} (y_i - Y_i)^2 = \sum_{i=1}^{n} (y_i - a_0 - a_1 x_i - \cdots - a_p x_i^p)^2. \qquad ...(9{\cdot}5)$$

Equating to zero the partial derivatives of (9·5) with respect to $a_1, a_2, \ldots, a_p$, we get $(p + 1)$ normal equations. On simplification, these equations reduce to

$$\left.\begin{aligned}\Sigma y &= na_0 + a_1\Sigma x + \dots\dots + a_p\Sigma x^p\\ \Sigma xy &= a_0\Sigma x + a_1\Sigma x^2 + \dots\dots + a_p\Sigma x^{p+1}\\ \Sigma x^2 y &= a_0\Sigma x^2 + a_1\Sigma x^3 + \dots\dots + a_p\Sigma x^{p+2}\\ &\dots\dots\dots\dots\dots\dots\dots\dots\\ \Sigma x^p y &= a_0\Sigma x^p + a_1\Sigma x^{p+1} + \dots\dots + a_p\Sigma x^{2p}\end{aligned}\right\} \quad \dots(9{\cdot}6)$$

where Σ everywhere stands for $\sum\limits_{i=1}^{n}$.

These equations are called the *normal equations* for fitting the parabola of degree p. They are $(p + 1)$ in number and when solved as simultaneous equations give the values of the $(p + 1)$ unknowns $a_0, a_1, \dots\dots, a_p$.

In particular, when $p = 1$, the normal equations for fitting the straight line $Y = a_0 + a_1 x$ of best fit to the given set of points (x_i, y_i), $i = 1, 2, \dots, n$ are

$$\left.\begin{aligned}\Sigma y &= a_0 n + a_1\Sigma x\\ \Sigma xy &= a_0\Sigma x + a_1\Sigma x^2\end{aligned}\right\} \quad \dots(9{\cdot}7)$$

Thus the best fitting linear function for the points (x_i, y_i), $i = 1, 2, \dots\dots, n$ may be expressed in the form

$$\begin{vmatrix} x & y & 1\\ \Sigma x & \Sigma y & n\\ \Sigma x^2 & \Sigma xy & \Sigma x\end{vmatrix} = 0.$$

When $p = 2$, the normal equations for fitting the second degree parabola $Y = a_0 + a_1 x + a_2 x^2$ of best fit to the set of given points (x_i, y_i) are

$$\left.\begin{aligned}\Sigma y &= a_0 n + a_1\Sigma x + a_2\Sigma x^2\\ \Sigma xy &= a_0\Sigma x + a_1\Sigma x^2 + a_2\Sigma x^3\\ \Sigma x^2 y &= a_0\Sigma x^2 + a_1\Sigma x^3 + a_2\Sigma x^4\end{aligned}\right\} \quad \dots$$

Note. *Change of origin.* When the values x_i $(i = 1, 2, \dots\dots, n)$ correspond to equal increment h and n is *odd* say $= 2m + 1$, the equations (9·6) can be simplified by taking the origin of x at the middle of the value and the common increment h as unit of measurement. Thus if a be the middle value, $u = (x - a)/h$ takes the values $-m, \dots\dots, -1, 0, 1, \dots\dots, m$. Hence, we get $\Sigma u = 0 = \Sigma u^3 = \dots\dots$

If n is *even*, say $2m$ we take the origin of x at the mean of the middle pair of values and $h/2$ as the new unit *i.e.*

$$u = \frac{x - (\text{mean of two suitable or middle terms})}{\frac{1}{2}h}$$

The values of x then become $-(2m-1), -(2m-3)$,........, $-3, -1, 1$ $3, ..., (2m-3), (2m-1)$, and again $\Sigma u = 0 = \Sigma u^3 = ...$

Example 9·3. *The weights of a calf taken at weekly intervals are given below. Fit a straight line using the method of least squares, and calculate the average rate of growth per week.*

Age (x)	*1*	*2*	*3*	*4*	*5*	*6*	*7*	*8*	*9*	*10*
Weight (y)	*52·5*	*58·7*	*65·0*	*70·2*	*75·4*	*81·1*	*87·2*	*95·5*	*102·2*	*106·4*

Since n is even, we take the origin of x at the mean of the middle pair of values, and $\frac{1}{2}$ as the new unit that is we take $u = \dfrac{x - 5{\cdot}5}{\frac{1}{2}} = 2x - 11$. The values of u, then are $-9, -7, -5, -3, -1, 1, 3, 5, 7, 9$

We find that $\Sigma u = 0, \Sigma u^2 = 330, \Sigma y = 796{\cdot}2$ and $\Sigma uy = 1016{\cdot}8$

The normal equations for fitting the straight line $y = a + bu$

are $$\Sigma y = na + b\Sigma u, \; \Sigma uy = a\Sigma u + b\Sigma u^2.$$

On substituting the values of Σy, etc., we have

$$796{\cdot}2 = 10a, \;\; 1016{\cdot}8 = 330b \; ; \;\; \therefore \;\; a = 79{\cdot}62, \; b = 3{\cdot}08 \;\; \text{approx.}$$

The regression equation is, therefore,

$$y = 79{\cdot}62 + 3{\cdot}08u = 79{\cdot}62 + 3{\cdot}08\,(2x - 11) \quad \text{or} \quad y = 45{\cdot}74 + 6{\cdot}16x$$

The average rate of growth per week is 6·16 units.

Example 9·4. *Fit a second degree parabola to the following data, taking x as the independent variable.*

x :	*1*	*2*	*3*	*4*	*5*	*6*	*7*	*8*	*9,*
y :	*2*	*6*	*7*	*8*	*10*	*11*	*11*	*10*	*0*

Introduce two new variables u and v connected by the relations $u = x - 5, v = y - 7$. Since n is odd and the values of x are equidistant, we take the origin for the x series at the middle value 5. This makes $\Sigma u = 0 = \Sigma u^3$. Let us fit the curve $v = a_0 + a_1 u + a_2 u^2$.

The calculations of $\Sigma uv, \Sigma u^2 v$, etc., are shown in the following table :

u	v	uv	u^2	u^2v	u^3	u^4

	–4	–5	20	16	–80		256
	–3	–1	3	9	–9		81
	–2	0	0	4	0		16
	–1	1	–1	1	1		1
	0	3	3	0	0		0
	1	4	4	1	4		1
	2	4	8	4	16		16
	3	3	9	9	27		81
	4	2	8	16	32		256
Total	0	11	54	60	–9	0	708

Substituting the values of Σu, etc., in the normal equations (9·8), we get

$$11 = 9a_0 + 0 + 60a_2\,;\ 54 = 0 + 60a_1 + 0\,;\ -9 = 60a_0 + 0 + 708a_2.$$

Solving these simultaneous equations we obtain

$$a_0 = 3{\cdot}0,\ a_1 = 0{\cdot}85,\ a_2 = -\,0{\cdot}27.$$

Hence the equation of the curve is $v = 3 + 0{\cdot}85u - {\cdot}027u^2$

or $\quad y - 7 = 3 + 0{\cdot}85(x-5) - 0{\cdot}27(x-5)^2$ or $y = -\,1 + 3{\cdot}55x - 0{\cdot}27x^2$

Ex. Show that if the pair (x_i, y_i), $i = 1, 2, \ldots\ldots, n$ occurs with the frequency f_i where $\Sigma f_i = N$, the normal equations, for fitting the parabola of the second degree $y = a_0 + a_1x + a_2x^2$ are

$$\Sigma f_i\, y_i = N\, a_0 + a_1\, \Sigma f_i\, x_i + a_2\, \Sigma f_i\, x_i^2,$$

$$\Sigma f_i\, x_i\, y_i = a_0\, \Sigma f_i\, x_i + a_1\, \Sigma f_i\, x_i^2 + a_2\, \Sigma f_i\, x_i^3,$$

$$\Sigma f_i\, x_i^2\, y_i = a_0\, \Sigma f_i\, x_i^2 + a_1\, \Sigma f_i\, x_i^3 + a_2\, \Sigma f_i\, x_i^4.$$

9·3. Fitting of the Curves of the types $y = ab^x$ and $y = ax^b$.

(Curves reducible to st. lines.) Whenever it is required to fit curves of exponential types to the data, the procedure is to take common logarithms of both sides of the equation and then form normal equations to find the *best* values of the constants. For example, to fit the curve $y = ab^x$ we proceed as follows :

We have $\log y = A + Bx$ where $A = \log a$, $B = \log b \Rightarrow Y = A + Bx$

we have $\quad S = \Sigma(\log y - y)^2 = \Sigma\,(\log y - A - Bx)^2 = f(A, B)$

The normal equations are

$$0 = \frac{\partial S}{\partial A} \Rightarrow \Sigma \log y = nA + B\Sigma x\,;\ 0 = \frac{\partial S}{\partial B} \Rightarrow \Sigma x \log y = A\Sigma x + B\Sigma x^2$$

In the second case $y = ax^b$, we have $\log y = \log a + b \log x = A + bX$

In this case it may be useful to plot the data on a log— paper.

The normal equations can be solved for A and b and then by taking the antilogs, a and b can be found out.

Example 9·5. *Fit the exponential curve $y = ae^{bx}$ to the following data, e being Napierian base, 2·71828 :*

x :	0	2	4
y :	5·012	10	31·62

(I.A.S. 1993)

The equation to the curve can be written as $\log y = \log_{10} a + (b \log e)x$

or $\qquad Y = A + Bx$ where $Y = \log_{10} y, A = \log_{10} a, B = b \log_{10} e$

The normal equations are $nA + B\Sigma x = \Sigma Y$; $A\Sigma x + B\Sigma x^2 = \Sigma xY$

The calculations of $\Sigma x, \Sigma Y, \Sigma x^2$ and ΣxY are shown in the following table :

	x	y	$Y(=\log y)$	x^2	xY
	0	5·012	0·7	0	0
	2	10·00	1·0	4	2
	4	31·62	1·5	16	6
Total	6		3·2	20	8

Thus we have $3A + 6B = 3·2$ and $6A + 20B = 8$

Solving, we get $A = 0·666, \quad B = 0·2$; $\therefore a = 4·642, \ b = 0·46$ approx

Hence equation of the curve fitted to the data is $y = 4·642e^{0·46x}$.

9·4. Fitting of Logistic curve. $y = \dfrac{a}{1 + be^{cx}}, c < 0$

This is an S-shaped curve, As $x \to \infty, y \to a$,

Thus a can be estimated from the data since it is the value of y when x is reasonably large.

The given equations can be written as

$$\frac{1}{y} = \frac{1}{a} + \frac{b}{a} e^{cx}$$

$$\frac{1}{y} - \frac{1}{a} = \frac{b}{a}(e^c)^x \Rightarrow Y = AB^x$$

where $Y = \dfrac{1}{y} - \dfrac{1}{a}, A = \dfrac{b}{a}, B = e^c$

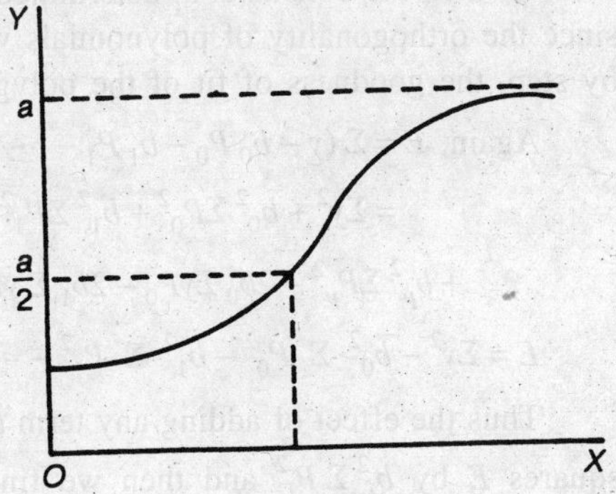

Fig. 9·1

Knowing x, we can plot $\log y$ against x and use the method of least squares to estimate A and B and therefore to estimate b and c.

Such curves are widely used in the study of growth of populations.

9·5. Fitting of orthogonal Polynomials to the polynomial of degree p.

Two polynomials $P_1(x)$ and $P_2(x)$ are said to be orthogonal to each other, if $\Sigma P_1(x) P_2(x) = 0$,

where x is a discontinuous variable, summation is carried over the given set of values of x ;

$\int_a^b P_1(x) P_2(x) = 0$, if x is a continuous variable in the range from a to b.

The *pth* degree polynomial equation (9·4) can be rewritten as

$$Y = b_0 P_0 + b_1 P_1 + b_2 P_2 + \cdots\cdots + b_p P_p \qquad \ldots(9·8)$$

where P's are polynomials in x, P_i being the polynomial in x of degree i ($i = 0, 1, \ldots, p$). Now P's shall be determined by applying the condition of orthogonality of P_i and P_j *i.e.* $\sum_x P_i P_j = 0, i \neq j$. ...(9·9)

the summation is carried over the observed values of x.

The normal equations for *estimating* the constant $b^{'s}$ are obtained by minimizing the error of estimation

viz. $\quad E = \Sigma (y - Y)^2 = \Sigma(y - b_0 P_0 - b_1 P_1 \cdots - b_p P_p)^2$

$$\therefore \ \Sigma P_i (y - b_0 P_0 - b_1 P_1 \cdots - b_i P_i - \cdots - b_p P_p) = 0$$

$$\Rightarrow \Sigma P_i y - b_i \Sigma P_i^2 = 0, \Rightarrow b_i = \Sigma y P_i / \Sigma P_i^2, \ i = 0, 1, 2, \ldots. p \quad \ldots(9·10)$$

Thus b_i is determined only when P_i is known.

However, if we want to fit a polynomial of order $p + 1$, instead of order p, then the constants b_i determined by (9·10) need not be changed, since the orthogonality of polynomials will enable us to determine, step by step, the goodness of fit of the polygomial curve (9·8).

Again, $E = \Sigma (y - b_0 P_0 - b_1 P_1 - \cdots\cdots b_p P_p)^2$

$$= \Sigma y^2 + b_0^2 \Sigma P_0^2 + b_1^2 \Sigma P_1^2 + \cdots$$

$$+ b_p^2 \Sigma P_p^2 - 2b_0 \Sigma y P_0 - 2b_1 \Sigma y P_1 - \cdots - 2b_p \Sigma y P_p, \quad \text{using (9·9)}$$

$$E = \Sigma y^2 - b_0^2 \Sigma P_0^2 - b_1^2 \Sigma P_1^2 - \cdots\cdots - b_p^2 \Sigma P_p^2 \quad \text{on using (9·10)}$$

Thus the effect of adding any term $b_i P_i$ is to reduce the error sum of squares E by $b_i^2 \Sigma P_i^2$ and then we find the effect of this term on E separately. In case, the addition of any term $b_i P_i$ does not reduce E significantly, then we conclude that the addition of any term is not desired.

9·5·1. To determine the orthogonal polynomial P_p (for fitting a polynomial curve of degree p)

Consider $P_p = \sum_{i=0}^{p} c_{pi} x^i = c_{p0} + c_{p1} x + ... + c_{pp} x^p$, ...(9·11)

where the constants $c_{p0}, c_{p1}, ..., c_{pp}$ are $(p + 1)$ in number

In (9·8), the number of unknown constants in all the polynomials upto and including that of pth order

$$= 1 + 2 + 3 + ... + (p+1) = \frac{1}{2}(p+1)(p+2)$$

Also the total number of orthogonality conditions

$$= {}^{p+1}C_2 = \frac{1}{2}(p+1)p$$

Thus the number of *arbitrary* constants c's

$$= \frac{1}{2}(p+1)(p+2) - \frac{1}{2}(p+1)p = p+1.$$

We shall now assign unity, the arbitrary value to c_{ii} (coeff. of x^i in P_i)

$\Rightarrow \qquad c_{ii} = 1, \quad (i = 0, 1, 2, ..., p)$ ...(9·12)

In particular, $\qquad c_{00} = p_0 = 1$ ...(9·13)

$$\sum_x P_p P_i = 0, \; i < p \; (i = 0, 1, 2, ..., p-1)$$

when $\quad i = 0, \; \Sigma P_p P_0 = 0 \Rightarrow \Sigma P_p = 0 \qquad (\because P_0 = 1)$ ...(9·14)

$i = 1, \; \Sigma P_p P_1 = 0 \Rightarrow \Sigma P_p (x + A) = 0 \Rightarrow \Sigma x P_p = 0$, on using (9.14),...(9·15)

$$i = 2, \; \Sigma P_p P_2 = 0 \Rightarrow \Sigma P_p (x^2 + A_1 x + A_2) = 0$$

$$\Rightarrow \Sigma x^2 P_p = 0, \quad \text{on using (9·15) and (9·14)}$$

Proceeding in this way we have $\Sigma P_p x^r = 0, r = 0, 1, 2, .. \; p-1$

$$\Rightarrow \sum_x \left(\sum_{i=0}^{p} c_{pi} x^i \right) x^r = 0 \Rightarrow \sum_{i=0}^{p} c_{pi} x^{i+r} = 0 \Rightarrow \sum_{i=0}^{p} c_{pi} \mu_{i+r} = 0, r = 0, 1, ..., p-1.$$

assuming x to be measured from the mean. Putting $r = 0, 1, 2, ...,$ successively, we obtain the following equations for determining c_{pi}'s.

$$c_{p0}\mu_0 + c_{p1}\mu_1 + \cdots + c_{pi}\mu_i + ... + c_{pp}\mu_p = 0$$

$$c_{p0}\mu_1 + c_{p1}\mu_2 + ... + c_{pi}\mu_{i+1} + \cdots + c_{pp}\mu_{p+1} = 0$$

$$c_{p0}\mu_{p-1} + c_{p1}\mu_p + \; + c_{pi}\mu_{i+p-1} + ... + c_{pp}\mu_{2p-1} = 0$$

Applying Cramer's Rule, we get, noting that $c_{pp} = 1$,

$$c_{pi} = \frac{\begin{vmatrix} \mu_0 & \mu_1 & \cdots\cdots & \mu_p & \cdots\cdots & \mu_{p-1} \\ \mu_1 & \mu_2 & \cdots\cdots & \mu_{n+1} & \cdots\cdots & \mu_p \\ \mu_{p-1} & \mu_p & \cdots\cdots & \mu_{2p-1} & \cdots\cdots & \mu_{2p-2} \end{vmatrix} l}{\begin{vmatrix} \mu_0 & \mu_1 & \cdots\cdots & \mu_i & \cdots\cdots & \mu_{p-1} \\ \mu_1 & \mu_2 & \cdots\cdots & \mu_{i+1} & \cdots\cdots & \mu_p \\ \vdots & \vdots & & \vdots & & \vdots \\ \mu_{p-1} & \mu_p & & \mu_{i+p-1} & & \mu_{2p-2} \end{vmatrix}} = \frac{\Delta^{(b)}_{pi}}{\Delta^{(p-1)}} \quad \text{...(9·16)}$$

where $\Delta^{(b)}{}_{pi}$ = minor of the element in the last row and $(i + 1)$th column in $\Delta^{(p)}$. Thus from (9·16) on substituting the values of c_{pi}, $i = 0, 1, 2, ...,p)$ we get

$$P_p = \frac{1}{\Delta^{p-1}}\left[\Delta^{(p)} p_0 + x\, \Delta^{(p)}_{p1} + ... + x^p\, \Delta^{(p)}\, pp \right]$$

$$= \frac{1}{\Delta^{(p-1)}} \begin{vmatrix} \mu_0 & \mu_1 & \mu_2 & \cdots & \mu_p \\ \mu_1 & \mu_2 & \mu_3 & \cdots\cdots\cdots & \mu_{p+1} \\ \cdots & \cdots & \cdots & \cdots\cdots\cdots & \cdots \\ \mu_{p-1} & \mu_p & \mu_{p+1} & & \mu_{2p-1} \\ 1 & x & x^2 & & x^p \end{vmatrix} \quad \text{...(9·16)}$$

This result enables us to determine all $P_p^{'s}(x)$.

In order to get the values of P_0, P_1, P_2, etc. we assume that, x is a standardized variate *i.e.* $E(x) = 0$, Var $x = 1$ *i.e.* $\mu = \mu_1{}' = 0, \mu_2 = 1$.

Hence $P_0 = 1$

$$P_1(x) = \begin{vmatrix} \mu_0 & \mu_1 \\ 1 & x \end{vmatrix} / \mu_0 = x \quad P_2(x) = \begin{vmatrix} \mu_0 & \mu_1 & \mu_2 \\ \mu_1 & \mu_2 & \mu_3 \\ 1 & x & x^2 \end{vmatrix} / \begin{vmatrix} \mu_0 & \mu_1 \\ \mu_1 & \mu_2 \end{vmatrix}$$

$$= \begin{vmatrix} 1 & 0 & 1 \\ 0 & 1 & \mu_3 \\ 1 & x & x^2 \end{vmatrix} \div \begin{vmatrix} 1 & 0 \\ 0 & 1 \end{vmatrix} = x^2 - x\,\mu_3 - 1 = x^2 - 1.$$

Generally, if we assume that x is a standard normal variate, such that $\mu_3 = \mu_5 = ... = 0$, and $\mu_{2n} = (2n - 1)\,\mu_{2n-2}$

$$\therefore \quad \mu_4 = 3\mu_2 = 3, \mu_6 = 5\mu_4 = 15, \cdots$$

$$P_3(x) = \begin{vmatrix} \mu_0 & \mu_1 & \mu_2 & \mu_3 \\ \mu_1 & \mu_2 & \mu_3 & \mu_4 \\ \mu_2 & \mu_3 & \mu_4 & \mu_5 \\ 1 & x & x^2 & x^3 \end{vmatrix} \div \begin{vmatrix} \mu_0 & \mu_1 & \mu_2 \\ \mu_1 & \mu_2 & \mu_3 \\ \mu_2 & \mu_3 & \mu_4 \end{vmatrix}$$

$$= \begin{vmatrix} 1 & 0 & 1 & 0 \\ 0 & 1 & 0 & 3 \\ 1 & 0 & 3 & 0 \\ 1 & x & x^2 & x^3 \end{vmatrix} \div \begin{vmatrix} 1 & 0 & 1 \\ 0 & 1 & 0 \\ 1 & 0 & 3 \end{vmatrix} = x^3 - 3x$$

$$P_4(x) = \begin{vmatrix} \mu_0 & \mu_1 & \mu_2 & \mu_3 & \mu_4 \\ \mu_1 & \mu_2 & \mu_3 & \mu_4 & \mu_5 \\ \mu_2 & \mu_3 & \mu_4 & \mu_5 & \mu_6 \\ \mu_3 & \mu_4 & \mu_5 & \mu_6 & \mu_7 \\ 1 & x & x^2 & x^3 & x^4 \end{vmatrix} \div \begin{vmatrix} \mu_0 & \mu_1 & \mu_2 & \mu_3 \\ \mu_1 & \mu_2 & \mu_3 & \mu_4 \\ \mu_2 & \mu_3 & \mu_4 & \mu_5 \\ \mu_3 & \mu_4 & \mu_5 & \mu_6 \end{vmatrix}$$

$$= \begin{vmatrix} 1 & 0 & 1 & 0 & 3 \\ 0 & 1 & 0 & 3 & 0 \\ 1 & 0 & 3 & 0 & 15 \\ 0 & 3 & 0 & 15 & 0 \\ 1 & x & x^2 & x^3 & x^4 \end{vmatrix} \div \begin{vmatrix} 1 & 0 & 1 & 0 \\ 0 & 1 & 0 & 3 \\ 1 & 0 & 3 & 0 \\ 0 & 3 & 0 & 15 \end{vmatrix} = x^4 - 6x^2 + 3$$

Note. Take $w(x) = \dfrac{1}{\sqrt{2\pi}} e^{-\frac{1}{2}x^2}, -\infty < x < \infty$, we can show that

$$\int_{-\infty}^{+\infty} P_i(x), P_j(x) w(x) = 0, i \neq j.$$

where $P_i(x), P_j(x)$ are called Hermite Polynomials.

9·5·2. To get the simplified values of b_i's as given by (9·10), page 370.

We have $b_p = \Sigma y Pp / \Sigma Pp^2$

Now $\quad \Sigma P_p^2 = \sum_x P_p \,.\, P_p = \sum_x P_p \,[c_{p0} + c_{p1}x + \ldots + c_{pp}x^p]$

$$= \sum_x P_p x^p. \quad \because \sum_x P_p x^r = 0, r = 0, 1, \ldots, p-1, c_{pp} = 1$$

$$= \sum_x \left(\sum_{i=0}^{p} c_{ij} x^i \right) x^p = \sum_{i=0}^{p} \left(c_{ij} \Sigma x^{p+i} \right)$$

$$= N \sum_{i=0}^{p} c^{ij} \mu_{p+i} = N \sum_{i=0}^{p} \frac{\Delta^{(p)} pi}{\Delta^{(p-1)}} \mu_{p+i}$$

$$= \frac{N\Delta^{(p)}}{\Delta^{(p-1)}},$$

where $\Delta^{(p)}i$ is defined in (9·16).

$$\text{Also} \quad \Sigma y P_p = N \sum_{i=0}^{p} \frac{\Delta^{(p)}}{\Delta^{(p-1)}} = \frac{N\Delta^{(p)}}{\Delta^{(p-1)}}$$

$$b_0 = \frac{\Sigma y p_0}{\Sigma P_0^{\,2}} = \frac{\Sigma y}{N},\ b_i = \frac{\Sigma y P_i}{\Sigma P_i^{\,2}},\ i = 1, 2, ..., p$$

in case (9·8) is the orthogonal polynomial fitted to the data.

Cor. If (*i*) N, the number of observations is odd, then we take $y = (x_i - a)/h$

(*ii*) N, the number of observations is even, then we take $y = (x_i - b)/b/2$.

Exercises 9

1. (*a*) Write a short note on 'Method of Least Squares'.

   (*b*) Given a set of m linear equations involving n unknowns ($m > n$), show how the most probable values of the unknowns are determined.

2. Find the most plausible values of x, y, and z from the following equations :—

   (*a*) $x - y + 2z = 3,\ 3x + 2y - 5z = 5,\ 4x + y + 4z = 21,\ -x + 3y + 3z = 14,$

   (*b*) $x + 2y + z = 1, 2x + y + z = 4,\ -x + y + 2z = 4,\ 4x + 2y - 5z = -7.$

   [**Ans.** (*a*) $x = 2{\cdot}47,\ y = 3{\cdot}55,\ z = 1{\cdot}92$

   (*b*) $x = 1{\cdot}16,\ y = -0{\cdot}76,\ z = 2{\cdot}80$]

3. Fit a straight line to the following points :—

x :	0	1	2	3	4
y :	1	1·8	3·3	4·5	6·3

   [**Ans.** $y = 0{\cdot}72 + 1{\cdot}33x$]

4. Fit a parabola of second degree to the following data :

x :	1·0	1·5	2·0	2·5	3·0	3·5	4·0
y :	1·1	1·3	1·6	2·6	2·7	3·4	4·1

   [**Hint.** Take 2·5 as the working origin for x series.]

   [**Ans.** $y = 1{\cdot}04 - 0{\cdot}20x + 0{\cdot}24x^2$]

5. The profit of a certain company in the xth year of its life are given by :

x :	1	2	3	4	5
y :	1250	1400	1650	1950	2300

   Taking $u = x - 3$ and $v = (y - 1650)/50$ show that the parabola of the second degree of v on u is $v + 0{\cdot}086 = 5{\cdot}30u + 0{\cdot}64u^2$ and deduce that the parabola of second degree of y on x is $y = 1140 + 72x + 32{\cdot}15x^2$.

6. Fit a parabola of the second degree to the following data taking x as the independent variable :

x :	0	1	2	3	4
y :	1	1·8	1·3	2·5	6·3

Find out the difference between the actual value of y and the estimated value of y obtained from the fitted curve when $x = 2$.

[**Ans.** $y = 1{\cdot}48 + 1{\cdot}13\,(x-2) + 0{\cdot}55(x-2)^2$, Difference $= -\,{\cdot}18$]

7. The pressure (p) and volume (v) of a gas are known to be related by an equation of the form $pv^k =$ constant.

Indicate how you would proceed to find the value of k by fitting a straight line to the set of observed values of p and v, taking p to be the independent variable.

8. (*a*) Derive the least square equations for fitting a curve of the type $y = ax^2 + bx^{-1}$ to a set of n points.

Fit the above curve to the following data :

x :	1	2	3	4
y :	–1·51	0·99	3·88	7·66

(*b*) It is known that the readings for x and y given below should follow a law of the form $y = ax + bx^{-1}$ when a and b are constants.

x :	1	2	3	4	5	6	7	8
y :	5·43	6·28	8·23	10·32	12·63	14·86	17·27	19·51

Use the method of least squares to find the best values of a and b.

[**Hint.** (a) The normal equations for determining the best values of a and b are $\quad \Sigma y x^2 = a\,\Sigma x^4 + b\,\Sigma x\,,\ \Sigma \dfrac{y}{x} = a\,\Sigma x + b\,\Sigma \dfrac{1}{x^2}\cdot$

Ans. $159{\cdot}930 = 354a + 10b$, $2{\cdot}1933 = 10a + 1{\cdot}4238b$, $a = 0{\cdot}509$, $b = -\,2{\cdot}04$

9. Write down normal equations for fitting $y = ae^{bx}$ to a set of points. Fit the curve $y = ae^{bx}$ to the following data :

x :	1	2	3	4	5	6	7	8
y :	15·3	20·5	27·4	36·6	49·1	65·6	87·8	117·6

[**Ans.** $y = 11{\cdot}58\, e^{0{\cdot}2898\, x}$]

10. Use the method of least squares to determine a and b in the formula $y = ax + bx^2$ for the following observations :

x :	1	2	3	4	5
y :	1·8	5·1	8·9	14·1	19·8

Estimate the value of y, when $x = 2$.

[**Ans.** $a = 1{\cdot}521$, $b = 0{\cdot}49$; $y = 5{\cdot}002$ when $x = 2$]

11. For a given bivariate distributions, find the straight line for which the sum of squares of the normal deviations is a minimum.

[**Hint.** Let the line be $x \cos\alpha + y \sin\alpha - p = 0$. Then

$$S = \Sigma f_i (x_i \cos\alpha + y_i \sin\alpha - p)^2$$

Normal equations are $\frac{\partial S}{\partial \alpha} = 0$ and $\frac{\partial S}{\partial p} = 0$. These determine p and α. The first of these equations shows that the line passes through the mean of the distribution. Taking the mean as origin, the second equation gives α where $\tan 2\alpha = 2\mu_{11}/(\sigma_x^2 - \sigma_y^2)]$

12. Show that the parabola of best fit of degree 2 for the $(2n + 1)$ points

$$(x_1, y_1) ; (x_2, y_2) ...... (x_{2n+1}, y_{2n+1})$$

where the values of x are in A.P. with common difference unity and $\bar{x} = 0$, can be expressed in the form **(B.A. Delhi, 1984)**

$$\begin{vmatrix} y & x^2 & x & 1 \\ \Sigma y_i & \frac{n(n+1)(2n+1)}{3} & 0 & 2n+1 \\ \Sigma x_i y_i & 0 & \frac{n(n+1)(2n+1)}{3} & 0 \\ \Sigma x_i^2 y_i & \frac{n^2(n+1)^2}{2} & 0 & \frac{n(n+1)(2n+1)}{3} \end{vmatrix} = 0$$

13. Show that for the modified exponential curve $y = a + be^{vx} = a + bq^x$, if y_0, y_1, y_2 are the ordinates for $x - a$, x and $x + a$, then estimated value of a is given by $a = (y_0 y_2 + y_1^2)/(y_2 - 2y_1 + y_0)$

Determine the estimates for b and q.

14. Discuss how you would fit the given data to the following :

(*i*) Gompertz curve $y = a(b)^{q^x}$ or $\log y = \log a + q^x \log b$

(*ii*) Makhem curve $y = k\, s^x (b)^{q^x}$ or $\log y = \log k + x \log s + q^x \log b$.

15. Fit the straight lines $y = a + bx$ to the following data, using orthogonal polynomials :

x :	0	1	2	3	4
y :	1	1·6	2·9	3·4	5·1

16. A physicist wants to approxinate the following data:

x :	0.0	0.5	1.0	2.0
$f(x)$:	0.00	0.57	1.46	5.05

using a function a $e^{bx} + c$. Assume that $b \approx 1$, compute the values of a and c that gives the best least squares approximation assuming $b = 1$.

[**Hint.** Take $b = 1$ and determaine a and c such that

$$\sum_{i=1}^{4} [f(x_i) - a e^{x_i} - c]^2 = \text{minimun}$$

The normal equations are : $\Sigma f(x_i) - a \sum_{i=1}^{4} e^{x_i} - 4c = 0$,

$$\Sigma f(x_i) e^{x_i} - a \sum_{i=1}^{4} e^{x_i} - c \sum_{i=1}^{4} e^{x_i} = 0]$$

10

Correlation and Regression

10·1. Bivariate Distribution. Correlation Coefficient. In earlier chapters, we have been concerned with univariate distributions only. We shall now discuss bivariate distributions and, in particular, study the simultaneous variation of two variables, say X and Y. The variables may, for example, be heights and weights of students in a class, ages of husbands and wives, or records of rainfall and yields of crops. The relationship between the variables may be of any type, but the relationship which is easiest to handle and is of great interest in Statistics, is of the linear type.

Just as $E(X-EX)^2$ gives us a measure for the variation in X and $E(Y-EY)^2$ gives us a similar measure for the variation in Y, we may expect $E(X-EX)(Y-EY)$ to give us a measure for the simultaneous variations of X and Y. But this will depend on the units used for X and Y. To make it independent of these units, we should divide it by a quantity having the same dimension as $E(X-EX)(Y-EY)$. Keeping this in view, Karl Pearson defined in 1890 the measure r_{XY} defined by

$$r_{XY}=\frac{E[X-EX)(Y-EY]}{\sqrt{E(X-EX)^2}\sqrt{E(Y-EY)^2}}=\frac{{}^*\text{Cov}(X,Y)}{\sqrt{\text{var}(X)\,.\,\text{Var}(y)}}=\frac{\mu_{11}}{\sigma_X\sigma_Y} \quad \text{...(10·1)}$$

$$=\frac{\Sigma xy-\dfrac{(\Sigma x)(\Sigma y)}{n}}{\sqrt{\Sigma x^2-\dfrac{(\Sigma x)^2}{n}}\times\sqrt{\Sigma y^2-\dfrac{(\Sigma y)^2}{n}}},$$

the summation extending over i from 1 to n.

r_{XY} is called the *product moment correlation coefficient* or simply the *correlation coefficient*. By symmetry, it is clear that $r_{XY}=r_{YX}$. Since

---

* $Cov(x,y)=\frac{1}{n}(x_i-\bar{x})(y_i-\bar{y})=\frac{1}{n}\Sigma(x_i y_i-\bar{x}y_i-x_i\bar{y}+\bar{x}\bar{y})$

$=\frac{1}{n}\Sigma x_i y_i-\bar{x}\bar{y}=\frac{1}{n}\left[\Sigma x_i y_i-\frac{\Sigma(x_i)(\Sigma y_i)}{n}\right]$

σ_X and σ_Y are essentially positive, the correlation coefficient between X and Y has the *same sign as covariance* μ_{11} of the two variables.

Karl Pearson's coefficient of correlation is based on the *assumptions :*

1. There is a linear relationship between the two variables. *i.e.* if the paired observations of both the variables are plotted on a scatter diagram, the plotted points will form a straight line.

2. The variations in the two variables follow the normal law.

Example 10·1. *Two independent variates X_1 and X_2 have means 5 and 10 and variances 4 and 9 respectively. Obtain the correlation coefficient between* $y_1 = 3x_1 + 4x_2$ and $y_2 = 3x_1 - x_2$.

We have $E(y_1) = 3\,E(x_1) + 4E(x_2)$, $E(y_2) = 3E(x_1) - E(x_2)$

Hence $y_1 - E(y_1) = 3[x_1 - E(x_1)] + 4\,[x_2 - E(x_2)]$, $\bar{y}_1 = E(y_1)$

and $y_2 - E(y_2) = 3[x_1 - E(x_1)] - [x_2 - E(x_2)]$

$$\therefore \quad E[y_1 - \bar{y}_1][y_2 - \bar{y}_2]$$

$$= 9E[(x_1 - \bar{x}_1)^2] - 4E[(x_2 - \bar{x}_2)^2] + 9E[(x_1 - \bar{x}_1)(x_2 - \bar{x}_2)] \quad ...(1)$$

$$= 9\,Var(x_1) - 4\,Var(x_2) = 9 \times 4 - 4 \times 9 = 0$$

The last term on the right of (1) vanishes since x_1 and x_2 are independent variates.

$$\text{Hence } r_{Y_1Y_2} \text{ with } r(Y_1, Y_2) = \frac{E(y_1 - \bar{y}_1)(y_2 - \bar{y}_2)}{\sigma_{y_1}\sigma_{y_2}} = \frac{0}{\sigma_{y_1}\sigma_{y_2}} = \frac{0}{\sigma_{y_1}\sigma_{y_2}} = 0$$

Note. It is obvious that the answer is independent of the values which the means $E(x_1)$ and $E(x_2)$ take.

10·1·1. Effect of Change of Origin and Scale of reference on correlation coefficient Let U and V be two new variables defined by

$$U = \frac{X-a}{h}, V = \frac{Y-b}{k}$$

so that $X = a + hU$; $EX = a + hE(U)$ $Y = b + kV$; $EY = b + k\,EV$

Substituting in (10·1) the expression for r_{XY} we get

$$r_{XY} = \frac{E(X-EX)(Y-EY)}{\sqrt{E(X-EX)^2}\sqrt{E(Y-EY)^2}} = \frac{hkE(U-EU)(V-EV)}{\sqrt{h^2k^2}\sqrt{E(U-EU)^2}\sqrt{E(V-E(V))^2}}$$

$$= \frac{hk}{\sqrt{h^2k^2}} r_{UV}$$

so that if h and k are both positive (they are positive if we change the units of measurement only), we have $r_{xy} = r_{uv}$ ...(10·2),

Thus r_{UV} is independent of *a, b* (corresponding to change of origin) and *h, k* (corresponding to change of scale).

Thus the value of the correlation coefficient is independent of the origin of reference and the unit of measurement. In other words, *r* has no dimensions and is just a real number. This property is quite useful in numerical evaluation of *r* since due to this property, we can choose any convenient origin and scale.

Ex. If *h* and *k* are of opposite sign, show that $r_{uv} = -r_{xy}$.

Example 10·2 *If* $X \sim N(0, 1)$, *find Corr* (X, Y) *where* $Y = a + bx + cx^2$. **(B.Sc. Hons. Delhi, 1985)**

$$X \sim N(0, 1), \; \therefore \; E(x) = 0, = \text{Var}\, x = 1, Ex^2 = 1, \; Ex^3 = 0, Ex^4 = 3.$$

$$E(Y) = E(a + bx + cx^2) = a + b.0 + c.1 = a + c, \sigma_y^{-2} = E[\,Y - (a + c)\,]^2$$

$$= E(Y^2) - (a + c)^2 . = b^2 + 2c^2 - ac$$

$$\text{Cor}\,(X, Y) = E[(X - EX)(Y - E(Y)] = E[x(bx + c(x^2 - 1)] = 6.$$

$$= E(Y^2) - (a + c)^2 . b^2 + 2c^2 - ac$$

$$\therefore \quad r(X, Y) = \text{Cov}\,(X, Y) / \sigma_x \sigma_y = \frac{b}{\sqrt{b^2 + 2c^2 - ac}}.$$

Example 10·3 *If X and Y are random variables and a, b, c, d are any numbers such that* $a \neq 1$; $c \neq 0$, *then*

$$r(ax + b, cy + d) = \frac{ac}{|\,ac\,|} r(x, y).$$

$$r(ax + b, cy + d) = \frac{\text{Cov}\,(ax + b, cy + d)}{[\,\text{Var}\,(ax + b)\,\text{Var}\,(cy + d)\,]^{1/2}} = \frac{ac\,\text{Cov}\,(x, y)}{\sqrt{a^2\,\text{Var}\,x \,.\, c^2\,\text{Var}\,y}}$$

$$= \frac{ac}{\sqrt{a^2 c^2}} \cdot \frac{\text{Cov}\,(x_i\, y)}{\sqrt{\text{Var}\,x\,\text{Var}\,y}} = \frac{ac}{|\,ac\,|} r(x, y)$$

Case I. If $ac > 0$ which will be the case when both *a* and *c* are of same sign, then $ac/|\,ac\,| = +1$. $\therefore$ $r(u, v) = r(x, y)$, $u = ax + b$, $v = cy + d$

Case II. If $ac < 0$ *i.e* when *a* and *c* are of opposite signs, then $ac/|\,ac\,| = -1$. $\therefore$ $r(u, v) = -r(x, y)$

10·1·2. Theorem. *If X and Y are independent random variables, they are uncorrelated.*

Since *X* and *Y* are independent, we have

$$\text{Cov}\,(X, Y) = E(X - EX)(Y - EY) = E(X - EX)\,E(Y - EY)$$

$$= [E(X) - E(X)]\,[E(Y) - E(Y)\,] = 0$$

$\therefore$ $r = 0$. Hence the theorem.

The converse of this theorem is not necessarily true,

i.e., Variables may be un-correlated but dependent.

For example, if X is a variate with a constant density function, say

$$f(x) = \tfrac{1}{2}, \quad -1 \le x \le 1 \quad \text{and if} \quad Y = X^2,$$

Then $E(X) = \int_{-1}^{1} x \cdot f(x)dx = \int_{-1}^{+1} \tfrac{1}{2}\, x dx = 0.$ so that $E(X)\, E(Y) = 0.E(Y) = 0$

Furthermore, $E\,(XY) = E(X^3) = \int_{-1}^{+1} x^3 \,.\, \tfrac{1}{2}\, dx = 0$

Hence $\quad \text{Cov}\,(X, Y) = E\,(X - EX)\,(Y - EY) = E(XY) - E(X)\,E(Y)$

$$= 0 - 0 = 0.$$

Thus X and Y are un-correlated. However, for each value of X, there is only one possible value for Y ; and for each value of Y, there are only two possible values of X. Therefore, X and Y are far from being independent.

Note. For other examples of this type, see Ex. 3, page 412.

10·1·3. Limits for the Coefficient of Correlation. We shall show that

$$|\, r\,| \le 1, \text{ that is, } -1 \le r \le 1.$$

Lemma. (Schwarz's Inequality). *For any stochastic variates X and Y,*

$$[E(XY)]^2 \le E(X^2)\, E(Y^2)\,.$$

Proof of the Lemma. For every real constant a, the variate $(aX - Y)^2$ is always non-negative and therefore has a non-negative expectation.

That is $E(aX - Y)^2 \ge 0$

or $\quad a^2 E(X^2) - 2aE(XY) + E(Y^2) \ge 0$ for every real a.

In particular, this holds for $a = \dfrac{E\,(XY)}{E(X^2)}$.

Substituting this value of a, we have

$$\frac{[E(XY)]^2}{E(X^2)} - 2\,\frac{[E(XY)]^2}{E(X^2)} + E(Y^2) \ge 0$$

The result now follows, if we multiply by $E(X^2)$ and transpose.

Proof of theorem. If X', Y' denote deviations of the variates X and Y from their respective means, we have by the Lemma,

$$[\,E(X'\,Y')]^2 \le E(X'^2)\, E(Y'^2)$$

Dividing both sides by $E(X'^2)\, E(Y'^2)$, we get $r_{xy}{}^2 \le 1$ or $-1 \le r_{xy} \le 1$.

Note. For other proofs of this important result, see Ex. 1, page 411 and § 10·4,.

10·2. Bivariate Frequency Distribution.

Let the variable value pair (x_i, y_i) occur with frequency f_i, $(i = 1, 2, ..., n)$, $\sum_{i=1}^{n} f_i = N$.

Then for a bivariate frequency distribution we define

$$r_{XY} = \frac{\frac{1}{N}\Sigma f_i (x_i - \bar{x})(y_i - \bar{y})}{\sqrt{\frac{1}{N}\Sigma f_i (x_i - \bar{x})^2}\ \sqrt{\frac{1}{N}\Sigma f_i (y_i - \bar{y})^2}}$$

$$= \frac{\Sigma f_i x_i y_i - (1/N)(\Sigma f_i x_i)(\Sigma f_i y_i)}{\sqrt{\Sigma f_i x_i^2 - (1/N)(\Sigma f_i x_i)^2}\ \sqrt{\Sigma f_i y_i^2 - (1/N)(\Sigma f_i y_i)^2}} \quad ...(10·3)$$

The points (x_i, y_i) can be represented in *xy*-plane to give what is called a *scatter* or dot *diagram*. This diagram is useful in giving an intuitive idea of the relationship between *X* and *Y* but for a quantitatively precise measure, we use the measure *r*.

Shapes of Scatter diagrams for various values of r

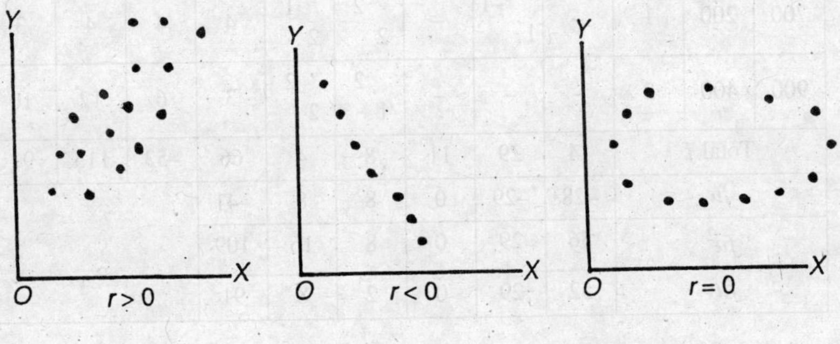

Fig. 10·1

Example 10·4. *Calculate from the data reproduced pertaining to 66 selected villages in Meerut district, the value or r, between 'total cultivable area' and 'the area under wheat'.*

Total Cultivable Area (*in Bighas*)

Area under wheat (in Bighas)	0–500	500 –1000	1000 –1500	1500 –2000	2000 –2500	Total
0–200	12	6	–	–	–	18
200–400	2	18	4	2	1	27
400–600	–	4	7	3	–	14
600–800	–	1	–	2	1	4
800–1000	–	–	–	1	2	3
Total	14	29	11	8	4	66

Let us introduce two new variables u and v given by the relations

$$u = \frac{x - 1250}{500}, \; v = \frac{y - 500}{200}.$$

x : 250 750 1250 1750 2250

y ;			−1000	−500	0	+500	+1000	Total			
	u / v		−2	−1	0	+1	+2	f	fv	fv^2	fuv
100	−400	−2	12 (48)	6 (12)	–	–	–	18	−36	72	60
300	−200	−1	2 (4)	18 (18)	4 (0)	2 (−2)	−2	27	−27	27	18
500	0	0	–	4 (0)	7 (0)	3 (0)	–	14	0	0	0
700	200	1	–	1 (−1)	–	2 (2)	2 (1)	4	4	4	3
900	400	2	–	–	–	1 (2)	2 (2)	3	6	12	10
Total f			14	29	11	8	4	66	−53	115	91
fu			−28	−29	0	8	8	−41			
fu^2			56	29	0	8	16	109			
fuv			52	29	0	2	8	91			

We have $\Sigma fuv = 12 \times (-2) \times (-2) + 6 \times (-1) \times (-2) + \cdots\cdots$

$$= 48 + 12 + \cdots\cdots = 91$$

Now $$r_{xy} = r_{uv} = \frac{\Sigma fuv - \frac{1}{n} (\Sigma fu)(\Sigma fv)}{\sqrt{\Sigma fu^2 - \frac{1}{n} (\Sigma fu)^2} \; \sqrt{\Sigma fv^2 - \frac{1}{n} (\Sigma fv)^2}}$$

$$r_{xy} = \frac{91 - \frac{1}{66}(-41)(-53)}{\sqrt{109 - \frac{(-41)^2}{66}}\sqrt{115 - \frac{(-53)^2}{66}}}$$

$$= \frac{91 \times 66 - 41 \times 53}{\sqrt{109 \times 66 - (-41)^2}\sqrt{115 \times 66 - (-53)^2}}$$

$$= \frac{6006 - 2173}{\sqrt{7194 - 1681}\sqrt{7590 - 2809}} = + \cdot 749.$$

Thus the coefficient of correlation between 'total cultivable area' and the 'area under wheat' is + ·749.

10·3. Regression and its analysis. The literal meaning of regression is 'moving backward' or 'the return to the mean value' or 'stepping back towards the average'. Sir Francis Galton used this term in the study of heredity. According to him with a correlation coefficient of 0·8 between heights of fathers and children if the average height of a certain set of fathers is x cm. above the general average, average height of children shall be ·8x cm. above the general average. Thus, there is a tendency to move towards the average. The term 'regression; in Statistics is used without making any reference to biometry.

Regression analysis is 'mathematical' measure of the average relationship between two or more variables, in terms of the original units of the data'.

In regression analysis, there are two types of variables. The variable, the value of which can be predicted, is called *predicted* or *dependent* variable and the variable which is used for prediction is called *independent* or *predictor* variable.

10·3·1. Curvilinear Regression, Linear Regression. In case there exists association or relationship between two variables X and Y, the dots of the scatter diagram will be more or less concentrated round a curve, which may be called the *curve* of *regression* and the relationship is said to be expressed by means of *curvilinear regression*. In the particular case, when the curve is a straight line, (AB, in Fig 10·2, 10·3) it is called the *line of regression* and the regression is said to be linear. More precisely, the line of regression is the straight line, which gives the 'best' fit in the least square sense to the given frequency or probability distribution.

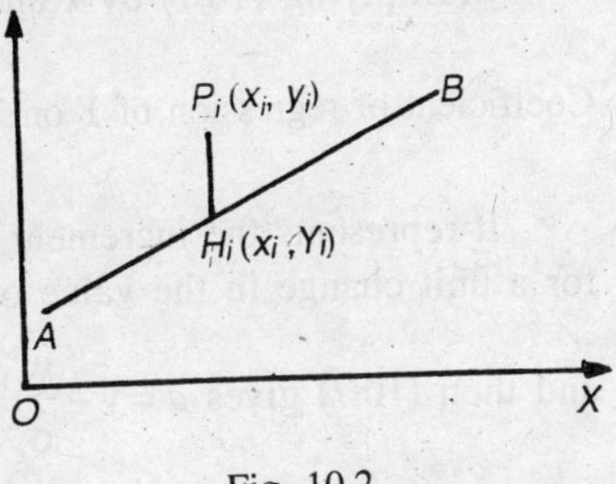

Fig. 10·2

If the straight line is so chosen that the sum of squares of deviations parallel to the axis of y is minimized, it is called the line of regression of y on x and it gives the *best estimate of y for any given value of x*. If, on the other hand, the sum of squares of the deviations parallel to the x-axis is minimized, the resulting straight line is the line of regression of x on y and it gives the best estimate of x for any given value of y. The two lines of regression would obviously, in general, be different.

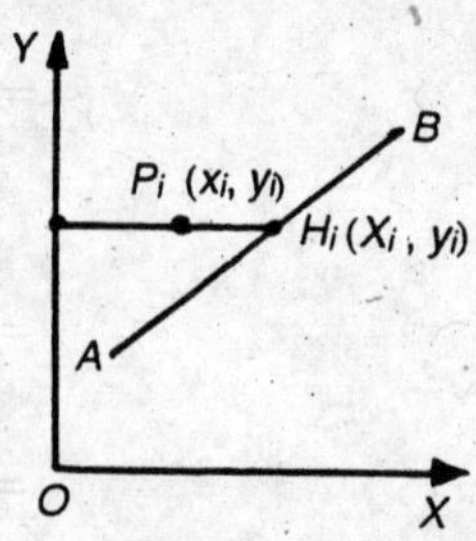

Fig. 10·3

10·3·2. Equations to the lines of regression. Regression coefficients.

Let $Y = a + bx$ be the equation of the line of regression of Y on X, so that corresponding to $x = x_i$, the expected value of Y is $Y_i\ (= a + bx_i)$ and S, the sum of the squares of deviations of observed value y_i from the expected value Y_i is given by $S = \sum_{i=1}^{n} f_i (y_i - Y_i)^2 = \sum_{i=1}^{n} f_i (y_i - a - bx_i)^2$.

According to the principle of least squares, we have to estimate a and b so that S is minimum. The normal equations are : $\frac{\partial S}{\partial a} = 0, \frac{\partial S}{\partial b} = 0,$

that is, $$\Sigma f_i (y_i - a - bx_i) = 0 \qquad ...(10·4)$$

and $$\Sigma f_i x_i (y_i - a - bx_i) = 0 \qquad ...(10·5)$$

Since $\Sigma f_i = N, \Sigma f_i x_i = N\bar{x}, \Sigma f_i y_i = N\bar{y}, \Sigma f_i x_i^2 = N(\sigma_x^2 + \bar{x}^2),$

$$\Sigma f_i y_i^2 = N(\sigma_y^2 + \bar{y}^2), \Sigma f_i x_i y_i = N(\mu_{11} + \bar{x}\,\bar{y}), \qquad ...(10·6)$$

the equations (10·4) and (10·5) reduce respectively to

$$\bar{y} = a + b\bar{x} \quad ...(10·7) \qquad \text{and} \qquad \mu_{11} + \bar{x}\,\bar{y} = a\bar{x} + b(\bar{x}^2 + \sigma_x^2) \quad ...(10·8)$$

Multiplying (10·7) by $\bar{x}$ and subtracting from (10·8), we get

Coefficient of regression of Y on $X = b_{yx} = \frac{\mu_{11}}{\sigma_x^2} = \frac{r\sigma_x\sigma_y}{\sigma_x^2} = \frac{r\sigma_y}{\sigma_x}$, $\sigma_x \neq 0$...(10·9)

It represents the increment in the value of the dependent variable Y for a unit change in the value of the independent variable X.

and then (10·7) gives $a = \bar{y} - \frac{\mu_{11}}{\sigma_x^2}\bar{x}$,

so that the equation to the line of regression of y on x is

$$y-\bar{y}=\frac{\mu_{11}}{\sigma_x^2}(x-\bar{x}) \qquad ...(10\cdot10)$$

Cor. The slope of the line of regression of y on x is $\frac{\mu_{11}}{\sigma_x^2}$ or $r\frac{\sigma_y}{\sigma_x}$ as $\mu_{11}=r\sigma_x\sigma_y$ from (10·1), page 376.

Similarly, the equation to the line of regression of x on y is

$$x-\bar{x}=\frac{\mu_{11}}{\sigma_y^2}(y-\bar{y}) \qquad ...(10\cdot11)$$

The slope of the line of regression of x on y is $\frac{\sigma_y^2}{\mu_{11}}$ or $\frac{\sigma_y}{r\,\sigma_x}$.

Notes 1. It is obvious from (10·7) that the straight line $Y=a+bx$ passed through the mean point $(\bar{x}, \bar{y})$. Similarly $X=c+dy$ passes through the mean $(\bar{x}, \bar{y})$. Thus both lines of regression pass through the mean $(\bar{x}, \bar{y})$ of the bivariate distribution.

2. The coefficients μ_{11}/σ_x^2 and μ_{11}/σ_y^2 are called the *regression coefficients* of y on x, and of x on y respectively and give *estimate* of changes of y and x corresponding to unit changes in x and y respectively. They are usually denoted by b_{yx} and b_{xy} respectively, and unlike the correlation coefficient, these are not symmetrical in x and y.

Properties of Regression Coefficients.

I. *The correlation coefficient is the geometric mean between the regression coefficients.*

Proof. $$b_{yx}=\frac{\mu_{11}}{\sigma_x^2}=\frac{r\,\sigma_x\,\sigma_y}{\sigma_x^2}=\frac{r\,\sigma_y}{\sigma_x} \qquad ...(10\cdot12)$$

Similarly $$b_{xy}=\frac{r\,\sigma_x}{\sigma_y} \qquad ...(10\cdot13)$$

Hence $$b_{yx}\times b_{xy}=r^2 \Rightarrow r=\pm\sqrt{b_{yx}\times b_{xy}}\,.$$

II. *The sign of r is the same as that of any regression coefficient.*

Proof. Since σ_x, σ_y *are always positive*, from (10·12), (10·13) *and* (10·14) b_{yx}, b_{xy}, μ_{11} *and r have the same sign.* In other words, if any one of b_{yx}, b_{xy}, μ_{11} *and r is negative (positive), then the other three will also be negative (positive).*

III. *One of the two regression coefficient must be numerically less than unity and the other must be numerically greater than unity or both will be unity or both will be numerically less than unity. Both coefficients cannot be greater than unity.*

Proof. Since $b_{yx} \cdot b_{xy} = r^2 \leq 1$ or $b_{yx} \leq \frac{1}{b_{yx}}$, the result follows.

IV. *The regression coefficients are independent of the change of origin, but not of scale.*

Proof. Let $U = \frac{X-a}{h}$ and $V = \frac{Y-b}{k}$. Then $\sigma_x = h\sigma_u, \sigma_y = k\sigma_v$.

$$\therefore \qquad b_{yx} = \frac{r\sigma_y}{\sigma_x} = r \cdot \frac{k\sigma_v}{h\sigma_u} = \frac{k}{h}\left(\frac{r\sigma_v}{\sigma_u}\right) = \frac{k}{h} b_{vu}$$

Similarly $\quad b_{xy} = \frac{h}{k} b_{uv}$

so that b_{yx} and b_{xy} are independent of a and b but not of h and k.

V. *Arithmetic mean of the regression coefficients is greater than the correlation coefficient, provided the correlation coefficient is positive.*

Proof. $\quad \frac{1}{2}(b_{yx} + b_{xy}) = \frac{1}{2}\left(\frac{r\,\sigma_y}{\sigma_x} + \frac{r\,\sigma_x}{\sigma_y}\right) \geq r$

$$\Rightarrow \qquad \sigma_y^2 + \sigma_x^2 \geq 2\sigma_x\,\sigma_y,\; r > 0$$

$$\Rightarrow \qquad (\sigma_y - \sigma_x)^2 \geq 0 \text{ which is always true.}$$

Note. b_{yx} and b_{xy} *depend on the units in which X and Y are measured*. Thus if X denotes heights in cms. and Y denotes weights in kgms., then b_{yx} and b_{xy} will be measures in kgm/cm and in cm/kgm respectively.

If $\quad b_{yx} = 2$ kgm/cm $= \cdot 2$ kgm/mm $= 2000$ gm/cm $= 200$ gm/mm

and $\quad b_{xy} = \frac{1}{4}$ cm/kgm $= 2\cdot5$ mm/kgm $= \cdot00025$ cm/gm $= \cdot0025$ mm/gm

it means that the expected change in weight for a unit cm. change in height is 2 kgm and the expected change in height for a unit kgm change in weight is $\frac{1}{4}$ cm.

Remarks :

1. It may be noted that in the above example whatever be the units used, r^2 is always half $\left(=\frac{1}{2}\right)$ and the numerical measure of the coefficients are never both greater than unity.

2. If X and Y are marks obtained in two papers or are heights of husbands and wives, measured in the same units of length or are weights of fathers and eldest sons, measured in the same units of time, then b_{yx} and b_{xy} can be dimensionless. It is only in this case that Note makes sense. If

X and *Y* are heights and weights, we can never talk of the arithmetic mean of the two regression coefficients.

3. In our example, if we use b_{yx}, a change of 1 cm in height will correspond to change of 2 kgm in weight, where if we use *bxy*, change of 1 cm in height will correspond to a change of 4 kgm in weight. Thus apparent discrepancy arises because we are using two regression coefficients for two different purposes. The discrepancy will not arise if $r = \pm 1$.

Cor. When do two lines of Regression reduce to one; In §10·3·2, we hare seen that there are two lines of regression, one of *Y* on *X* and the other of *X* on *Y*, the first one gives the estimate of *Y* when *x* is known, while the other gives the estimate of *X* when *y* is known. In the first one, *Y* is the dependent variable and *X* the independent variable, while in the other, *Y* is the independent variable and *X* the dependent variable. Since the assumptions for deriving the equations of two lines of regression are different, the two equations are not reversible. In other words the one equation gives the estimate of *Y* while the other gives the estimate of *X*.

However, if $r = \pm 1$, the two lines of regression viz.

$$Y - \bar{y} = \pm \frac{\sigma_y}{\sigma_x}(x - \bar{x}); \quad \text{and} \quad X - \bar{x} = \pm \frac{\sigma_x}{\sigma_y}(y - \bar{y})$$

can be reduced to the same form viz. $\dfrac{Y - \bar{y}}{\sigma_y} = \pm \dfrac{X - \bar{x}}{\sigma_x}$.

Thus when there is perfect correlation coefficient, positive or negative, the two lines of regression coincide. In other words, if $r = \pm 1$, then two lines coincide, either in the same or opposite direction(s).

Example 10·5. *The following marks have been obtained by a class of students in Statistics (out of 100).*

Paper I : 45 55 56 58 60 65 68 70 75 80 85

Paper II : 56 50 48 60 62 64 65 70 74 82 90

Compute the coefficient of correlation for the above data. Find also the equations of the lines of regression.

Calculation of the coefficient of correlation between marks obtained by the class of students in Statistics paper 1 and Statistics paper II.

Paper I			Paper II			
Marks out of 100	Deviation from assumed average 65	Square of deviation	Marks out of 100	Deviation from assumed average 70	Square of deviation	Product of deviations
x	u	u^2	y	v	v^2	uv
45	–20	400	56	–14	196	280
55	–10	100	50	–20	400	200
56	–9	81	48	–22	484	198
58	–7	49	60	–10	100	70
60	–5	25	62	–8	64	40
65	0	0	64	–6	36	0
68	3	9	65	–5	25	–15
70	5	25	70	0	0	0
75	10	100	74	4	16	40
80	15	225	82	12	144	180
85	20	400	90	20	400	400
$n = 11$	$\Sigma u = +2$	$\Sigma u^2 = 1414$	$n = 11$	$\Sigma v = -49$	$\Sigma v^2 = 1865$	$\Sigma uv = 1393$

Substituting the above values in Pearson's formula, we get

$$r = \frac{1393 - \dfrac{2 \times -49}{11}}{\sqrt{1414 - \dfrac{(2)^2}{11}}\ \sqrt{1865 - \dfrac{(49)^2}{11}}.}$$

$$= \frac{1393 \times 11 + 2 \times 49\ .}{\sqrt{1414 \times 11 + 2^2}\ \sqrt{1865 \times 11 - (-40)^2}}$$

$$= \frac{15323 + 98}{\sqrt{15554 - 4}\ \sqrt{20515 - 2401}} = \frac{15421}{\sqrt{15550 \times 18114}} = \cdot 918$$

Regression coefficient of x on $y = b_{xy} = r\dfrac{\sigma_x}{\sigma_y} = r\dfrac{\sigma_u}{\sigma_v}$

$$= \frac{\Sigma\, uv - \dfrac{(\Sigma u)\ (\Sigma v\ )}{n}}{\Sigma\, v^2 - \dfrac{(\Sigma\ v\ )^2}{n}} = \frac{1393 - \dfrac{2 \times 49}{11}}{1865 - \dfrac{(-49)^2}{11}}$$

$$= \frac{1393 + 8{\cdot}9}{1865 - 218{\cdot}3} = \frac{1401{\cdot}9}{1646{\cdot}7} = {\cdot}85$$

Regression coefficient of y on $x = b_{yx} = r\frac{\sigma_y}{\sigma_x} = r\frac{\sigma_v}{\sigma_u}$

$$= \frac{\Sigma uv - \frac{(\Sigma u)(\Sigma v)}{n}}{\Sigma u^2 - \frac{(\Sigma u)^2}{n}} = \frac{1393 - \frac{2 \times -49}{11}}{1414 - \frac{(2)^2}{11}}$$

$$= \frac{1393 + 8{\cdot}9}{1414 - 0{\cdot}4} = \frac{1401{\cdot}9}{1413{\cdot}6} = 0{\cdot}99$$

Arithmetic average of the marks of the 1st paper.

$$\bar{x} = \text{assumed average} + \frac{\Sigma u}{n} = 65 + \tfrac{2}{11} = 65{\cdot}2 \text{ marks.}$$

Arithmetic average of the marks of the 2nd paper.

$$\bar{y} = \text{assumed average} + \frac{\Sigma v}{n} = 70 + \frac{-49}{11} = 65{\cdot}55 \text{ marks.}$$

Substituting the above values in the regression equation of x on y; we get $(x - 65{\cdot}2) = {\cdot}85\ (y - 65{\cdot}55)$, $(x - \bar{x}) = b_{xy}(y - \bar{y})$

or $\qquad x = {\cdot}85y - 55{\cdot}72 + 65{\cdot}2 = {\cdot}85y + 9{\cdot}48$

Similarly, the equation of the line of regression of y on x is

$$y - 65{\cdot}55 = 0{\cdot}99\,(x - 65{\cdot}2) \quad \text{or} \quad y = {\cdot}99x + 1.$$

Example 10·6. *Show that θ, the acute angle between two lines of regression, is given by*

$$\tan\theta = \left(\frac{1 - r^2}{r}\right)\left(\frac{\sigma_x\sigma_y}{\sigma_x^{\ 2} + \sigma_y^{\ 2}}\right).$$

Interpret the cases when r = 0, r = ± 1.

From (10·11) and (10·11) we find that the slopes of the two lines of regression are respectively $\frac{r\sigma_y}{\sigma_x}$ and $\frac{\sigma_y}{r\sigma_x}$. $\left[\tan\theta = \frac{m_1 \sim m_2}{1 + m_1 m_2}\right]$

$$\text{Hence } \tan\theta = \frac{\frac{r\,\sigma_y}{\sigma_x} \sim \frac{\sigma_y}{r\sigma_x}}{1 + \frac{r\sigma_y}{\sigma_x}\cdot\frac{\sigma_y}{r\sigma_x}}$$

$$= \frac{1 - r^2}{r}\cdot\frac{\sigma_x\sigma_y}{\sigma_x^{\ 2} + \sigma_y^{\ 2}},$$

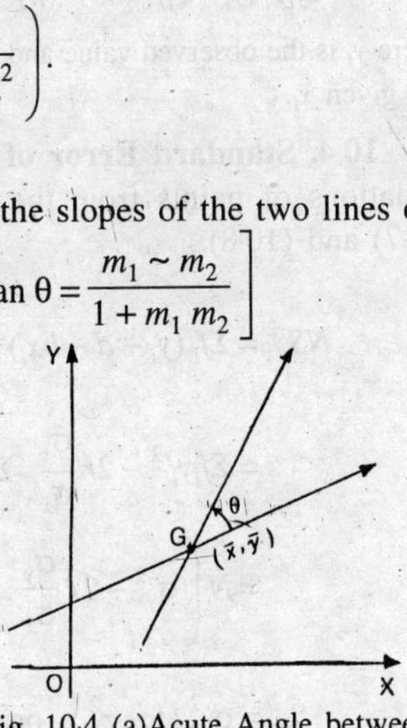

This equation gives the values of θ *i.e.*

Fig. 10·4 (a)Acute Angle between regression

$$\theta = \tan^{-1}\left[\frac{1-r^2}{r}\cdot\frac{\sigma_x\sigma_y}{\sigma_x^2+\sigma_y^2}\right].$$

When $r = \pm 1$, $\tan\theta = 0$, so that $\theta = 0$ or π.

Hence the two lines of regression coincide (in same or opposite direction) and there is a perfect correlation between the two variables x and y.

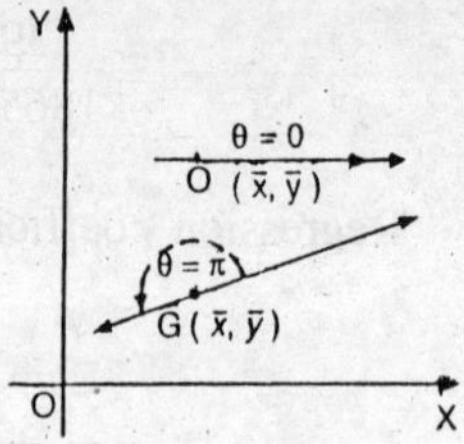

Fig. 10·4 (*b*)

When $r = 0$, the two lines of regression are $x = \bar{x}$ and $y = \bar{y}$ and are perpendicular to each other. The estimated value of y (or x) is the same for all values of x (or y).

Cor. The smaller is the angle between the two regression lines, the greater is the degree of correlation between the variables and vice versa.

Ex. 1. Obtain (10·11) from the *first principles.*

Ex. 2. Construct three examples in which $r = +1, 0, -1$ respectively.

Ex. 3. For a continuous probability distribution with density function $f(x, y)$, choose a and b so as to minimize $\int\int (y-a-bx)^2 f(x, y)\, dxdy$.

Deduce the equations of the two lines of regression.

Ex. 4. Obtain the equations of the lines of regression for a discrete probability distribution.

Ex. 5. If the origin is taken at the mean point $(\bar{x}, \bar{y})$, show that (10·7) and (10·8) give $a = 0, b = \mu_{11}/\sigma_x^2$. Hence deduce (10·10).

Ex. 6. If $r = 0$, show that the two lines of regression are parallel to the axes.

Ex. 7. Prove the following results :

(*i*) $\Sigma f_i Y_i (y_i - Y_i) = 0$ (*ii*) $\Sigma f_i (y_i - Y_i)(Y_i - \bar{y}) = 0$ (*iii*) $\sigma_y = |r|\sigma_y$.

where y_i is the observed value and Y_i is the corresponding least square estimate for any given x_i.

10·4. Standard Error of Estimate. The sum of the squares of the deviations of points from the line of regression of y on x is by using (10·7) and (10·8).

$$NS_y^2 = \Sigma f_i (y_i - a - bx_i)^2 = \Sigma f_i\left(y_i' - \frac{r\sigma_y}{\sigma_x}x_i'\right)^2 \qquad \left[\because a = 0, b = \frac{r\sigma_y}{\sigma_y}\right]$$

$$= \Sigma f_i y_i'^2 - 2r\frac{\sigma_y}{\sigma_x}\cdot\Sigma f_i x_i' y_i' + \frac{r^2\sigma_y^2}{\sigma_x^2}\cdot\Sigma f_i x_i'^2$$

$$= N\left(\sigma_y^2 - 2r\frac{\sigma_y}{\sigma_x}\cdot r\sigma_x\sigma_y + r^2\sigma_y^2\right)$$

or $$S_y^2 = \sigma_y^2(1-r^2) \text{ or } S_y = \sigma_y(1-r^2)^{\frac{1}{2}} \qquad ...(10·14)$$

This value of S_y is sometimes called the root-mean square error of estimate of y or more briefly the standard error of estimate of y. Similarly, the standard error of estimate of x given by

$$S_x = \sigma_x (1 - r^2)^{1/2} \quad \ldots(10\cdot14)$$

Since the sum of squares of deviations is non-negative, it follows that $\quad r^2 < 1 \quad$ or $\quad -1 \le r \le 1 \quad$ or $\quad |r| \le 1$.

If $r = 1$ or -1, the sum of squares of deviations from either line of regression is zero. Consequently each deviation is zero, and all the points lie on both the lines of regression. These two lines, then coincide and there is a linear functional relation between the variates x and y, giving perfect correlation. As r^2 comes nearer to unity, S_x^2 and S_y^2 approach zero, so that the points are closer to the lines of regression, which are nearer to coincidence. Thus *the departure of the value of* r^2 *from unity is a measure of departure of the relationship between two variables from linearity.*

Ex. 1. For continuous probability distribution, prove that

$$\iint \left[(y - \bar{y}) - \frac{\mu_{11}}{\sigma_x^2} (x - \bar{x}) \right]^2 f(x, y)\, dxdy = \sigma_y^2 (1 - r^2).$$

Ex. 2. Find the correlation coefficient, r, given

x :	1	2	3	4
y :	1	4	9	16

Explain why r^2 is not unity though there is a functional relation $y = x^2$ between x and y. **[Ans.** $r = +\cdot 98$]

Example 10·7. *Establish the formula* : $nr\sigma_x\sigma_y = n_1 r_1 \sigma_{x_1} \sigma_{y_1} + n_2 r_2 \sigma_{x_2} \sigma_{y_2} + n_1 d_{x_1} d_{y_1} + n_2 d_{x_2} d_{y_2}$ *where* n_1, n_2 *and* n *are respectively the sizes of the first, second and combined samples,* $(\bar{x}_1, \bar{y}_1)$, $(\bar{x}_2, \bar{y}_2)$, $(\bar{x}, \bar{y})$ *their means,* r_1, r_2 *and* r *their coefficients of correlation,* $(\sigma_{x_1}, \sigma_{y_1})$, $(\sigma_{x_2}, \sigma_{y_2})$, (σ_x, σ_y) *their standard deviations and* $d_{x_1} = \bar{x}_1 - \bar{x}$, $d_{y_1} = \bar{y}_1 - \bar{y}$, $d_{x_2} = \bar{x}_2 - \bar{x}$, $d_{y_2} = \bar{y}_2 - \bar{y}$.

By Equation 3·5, we have $\bar{x} = \dfrac{n_1 \bar{x}_1 + n_2 \bar{x}_2}{n_2 + n_2}, \bar{y} = \dfrac{n_1 \bar{y}_1 + n_2 \bar{y}_2}{n_1 + n_2}$

$$n\,\sigma_x^2 = n_1 (\sigma_{x_1}^2 + d_{x_1}^2) + n_2 (\sigma_{x_2}^2 + d_{x_2}^2)$$

$$n\,\sigma_y^2 = n_2 (\sigma_{y_1}^2 + d_{y_1}^2) + n_2 (\sigma_2^2 + dy_2^2)$$

$$r_1 = \frac{\text{Cov}(x_1, y_1)}{n_1 \sigma_{x_1} \sigma_{y_1}}, r_2 = \frac{\text{Cov}(x_2, y_2)}{n_2 \sigma_{x_2} \sigma_{y_2}}$$

For the pooled sample, we have $r = \dfrac{\text{Cov}(x, y)}{n\,\sigma_x \sigma_y}$

$$\therefore\ n\, r\, \sigma_x \sigma_y = \sum_1^n (x-\bar{x})(y-\bar{y}) = \sum_1^{n_1} (x-\bar{x})(y-\bar{y}) + \sum_{n_1+1}^{n} (x-\bar{x})(y-\bar{y}) \quad ...(1)$$

$$\text{Now } \sum_1^{n_1} (x-\bar{x})(y-\bar{y}) = \sum_1^{n_1} [(x-\bar{x}_1)+(\bar{x}_1-\bar{x})]\,[(y-\bar{y}_1)+(\bar{y}_1-\bar{y})]$$

$$= \sum_1^{n_1} (x-\bar{x}_1)(y-\bar{y}_1) + n_1(\bar{x}_1-\bar{x})(\bar{y}_1-\bar{y}) = n_1 r_1 \sigma_{x1}\sigma_{y1} + n_1 d_{x1} d_{y1}$$

$$\text{Similarly } \sum_{=1}^{n_2} (x-\bar{x})(y-\bar{y}) = n_2 r_2 \sigma_{x2}\sigma_{y2} + n_2 d_{x2} d_{y2}.$$

Hence on substituting these values in (1) we get the required result.

10·5. Regression Curves.

For a continuous bivariate distribution, the conditional mean is defined by

$$E(Y \mid X=x) = \int_{-\infty}^{+\infty} y f(x, y)\, dy \Big/ \int_{-\infty}^{+\infty} f(x, y)\, dy = \int_{-\infty}^{\infty} y f(x, y)\, dy / f_X(x)$$

and is called the regression function of Y on X. The curve

$$y = E(Y \mid X=x) \quad ...(10{\cdot}15)$$

is called the regression curve of Y on X, or the regression curve for the mean of Y.

Similarly we define the regression function of X on Y as

$$E(X \mid Y=y) = \int_{-\infty}^{+\infty} x f(x, y)\, dx / f_Y(y) \text{ and the curve } x = E(X \mid Y=y) \quad ...(10{\cdot}15')$$

as the regression curve of X on Y.

Thus the two regression curves are given by equation (10·15) and (10·15′) respectively. We **note** that if one of the regression curve is a straight line, so that the regression in linear, it does not follow that the other is also linear.

Remarks 1. For a discontinuous bivariate distribution, the regression curve of Y on X will be a disconnected set of points.

2. If regression curve of Y on X is linear *i.e.* $E(Y \mid X) = a + bx$, then a and b can be proved to be given by

$$a = \bar{y} - \frac{\mu_{11}}{\sigma_x^2}\bar{x} \text{ and } b = \frac{\mu_{11}}{\sigma_x^2} = \frac{r\sigma_y}{\sigma_x} \text{ and thus } E(Y \mid X) = \bar{Y} + \frac{r\sigma_y}{\sigma_x}(X - \bar{X}).$$

Example 10·8. *For the joint p.d.f.*

$$f(x, y) = \frac{1}{2} x^3 \exp\,[-x\,(y+1)],\ y \geq 0,\ x \geq 0\ ;\ = 0,\ \text{otherwise}$$

find : (*i*) *Regression curves for the means* (*ii*) η^2_{xy} *and* η^2_{yx}

(B.A. Hons. Stat. Delhi, 1987)

(*i*) Regression function of Y on X is given by

$$y = E(Y \mid X = x) = \int_0^\infty y f(y \mid x)\, dy$$

where
$$f(y \mid x) = f(x, y)/f_1(x) = \frac{\frac{1}{2} x^3 \exp\,[-x\,(y+1)]}{\int_0^\infty \frac{1}{2} x^3 \exp\,[-x\,(y+1)]\, dy}$$

$$= \frac{\exp\,[-x(y+1)]}{\frac{-1}{x} \Big| \exp\,[-x\,(y+1)\,] \Big|_{y=0}^{\infty}} = \frac{-x \exp\,[-x\,(y+1)]}{\exp\,(-x)} = -x\,e^{-xy}$$

$$\therefore \qquad y = -\int_0^\infty xy\,e^{-xy} dy = -x \left[\frac{\left(\dfrac{y e^{-xy}}{-2} \right)_0^\infty}{-x} - \int_0^\infty \frac{e^{-xy}}{-x}\, dy \right]$$

$$= -\int_0^\infty e^{-xy}\, dy = \frac{1}{x}\cdot$$

Thus $xy = 1$ is the regression curve rectangular hyperbola of y on x.

Regression of X on Y : $m_x(x_1) = E\,(X \mid Y = y) = \int_0^\infty x f(x \mid y)\, dx$

where
$$f(x \mid y) = \frac{f(x, y)}{f_Y(y)} = \frac{\frac{1}{2} x^3 \exp\,[-x\,(y+1)]}{\int_0^\infty \frac{1}{2} x^3\ \exp\,[-x\,(y+1)]\, dx}$$

$$f_y(y) = \int_0^\infty f(x, y)\, dx = \int_0^\infty \tfrac{1}{2} x^3\, e^{-x(y+1)}\, dx$$

$$= \tfrac{1}{2} \left| \frac{x^3\, e^{-x(y+1)}}{-(y+1)} \right|_0^\infty - \frac{3}{2} \int_0^\infty x^2\ \frac{e^{-x(y+1)}}{-(y+1)}\, dx = \int_0^\infty x^2\, e^{-x(y+1)} dx.$$

$$= \frac{3}{2(y+a1)} \cdot \frac{\Gamma 3}{(y+1)^3}, \quad \because \int_0^\infty x^{n-1} e^{-ax} dx = \Gamma n / a^n.$$

$$= 3/(y+1)^4.$$

Thus $x(x+1)^4 = 3$ is the regression curve of X on Y.

$$\sigma^2_{yx} = E\left(y - m_y x\right)^2 = E\left(y - \frac{1}{x}\right)^2$$

$$= \tfrac{1}{2} \int_0^\infty \int_0^\infty \left(y - \frac{1}{x}\right)^2 \cdot x^3 e^{-x(y+1)} \, dx\, dy$$

$$= \tfrac{1}{2} \left[\int_0^\infty x^3 \left(\int_0^\infty y^2 e^{-x(y+1)} a dy \right) dx - 2 \int_0^\infty x^z \left(\int_0^\infty e^{-x(y+1)} dy \right) ds \right.$$

$$\left. + \int_0^\infty x\, a \left(\int_0^\infty y\, e^{-x(y+1)} a\, dy \right) dx \right.$$

$$= \frac{1}{2} \cdot \qquad \left(\because \int_0^\infty a y a^{-1} e^{-ay} dy = T_n / a^n \right)$$

Now $\sigma^2_{yx} = \sigma^2_y \left(1 - \eta^2_{\cdot yx}\right) \Rightarrow \frac{1}{2} = \frac{3}{4}\left(1 - \eta^2_{yx}\right)$:

$$\therefore \quad 1 - \eta^2_{yx} = \frac{2}{3} \Rightarrow \eta^2_{yx} = \frac{1}{3} \text{ so that } \eta_{yx} = \frac{1}{\sqrt{3}} \cdot$$

Similarly η_{xy} can be obtained.

10·6. Correlation of Ranks. If we have a group of individuals ranked according to different characters, it is natural to enquire whether the ranks can be made to give us the measure of degree of relation between the two characters e.g., beauty, intelligence. Suppose we have n individuals, whose ranks, according to character A are $x_1, x_2 ...., x_n$, and according to character B are $y_1, y_2, ..., y_n$ where x's and y's are merely different permutations of n numbers from 1 to n. Assuming that no two individuals are bracketed or tied in either classification, both x and y take all integral values from 1 to n. *i.e.* $x, y = 1, 2, \cdots, n$.

$$\therefore \quad \sum_{i=1}^n x_i = \sum_1^n n = \frac{n(n+1)}{2} = \sum_{i=1}^n y_i \Rightarrow E(X) = \frac{n+1}{2} = E(Y)$$

Hence $\bar{x} = \bar{y} = \frac{1}{2} n(n+1)$ and $\sum_{i=1}^n (x_i - \bar{x})^2 = \frac{1}{12}(n^2 - n) = \sum_{i=1}^n (y_i - \bar{y})^2$,

$$Var\ X = \frac{n^2 - 1}{12}\ Var\ Y$$

If d_i stands for the difference in ranks of the ith individual, we have

$$d_i = x_i - y_i = (x_i - \bar{x}) - (y_i - \bar{y}) = x_i' - y_i', \quad [\because \bar{x} = \bar{y}]$$

where x_i', y_i' are the deviations of x_i and y_i from the means $\bar{x}$ and $\bar{y}$.

$$\therefore \quad \sum_{i=1}^{n} d_i^2 = \Sigma x_i'^2 + \Sigma y_i'^2 - 2\Sigma x_i' y_i' \quad \Rightarrow \quad \Sigma x_i' y_i' = \tfrac{1}{2}\left[\frac{n^3 - n}{6} - \sum_{i=1}^{n} d_i^2\right]$$

The coefficient of correlation between x and y is given by

$$r = \frac{\Sigma x_i' y_i'}{\sqrt{\Sigma x_i'^2}\,\sqrt{\Sigma y_i'^2}} = \frac{\tfrac{1}{2}\left[\dfrac{n^3 - n}{6} - \sum_{1}^{n} d_i^2\right]}{\tfrac{1}{12}(n^3 - n)} = 1 - \frac{6 \sum_{i=1}^{n} d_i^2}{n(n^2 - 1)} \qquad \ldots(10{\cdot}16)$$

This formula for rank correlation coefficient is due to E Spearman.

Tied Ranks. Suppose m individuals with ranks $k+1$, $k+2$, ..., $k+m$ are tied with mean $= \dfrac{1}{m}[(k+1) + \ldots + (k+m)]$ $= k + \dfrac{1}{2}(m+1)$.

Thus each of the m individuals is assigned

The rank $= k + \dfrac{1}{2}(m+1)$

New $V =$ variance of m untied ranks

$= (k+1)^2 + (k+2)^2 + (k+m)^2 - [k + \dfrac{1}{2}(m+1)]^2$

$= mk^2 + m(m+1)k + \dfrac{1}{6} m(m+1)(2m+1)[(k + \dfrac{1}{2}(m+1)]^2$

$V' =$ variance of m tied ranks

$= m[k + \dfrac{1}{2}(m+1)]^2 - [k + \dfrac{1}{2}(m+1)]^2$

$= mk^2 + m(m+1)k + \dfrac{1}{4} m(m+1)^2 - [k + \dfrac{1}{2}(m+1)]^2$

$\therefore$ Adjustment $= V - V' = \dfrac{1}{12} m(m+1)[2(2m+1) - 3(m+1) - \dfrac{1}{2} m(m^3 - 1).]$

Example 10·9. *The rankings of ten students in two subjects A and B are as follows :*

A :	*3*	*5*	*8*	*4*	*7*	*10*	*2*	*1*	*6*	*9*
B :	*6*	*4*	*9*	*8*	*1*	*2*	*3*	*10*	*5*	*7*

What is the coefficient of rank correlation ? **(M.A., Delhi, 1951)**

The differences d_i are : $-3, 1, -1, -4, 6, 8, -1, -9, 1, 2$.

$\Sigma d_i = 0, \Sigma d_i^2 = 214$

$$\therefore \quad r = 1 - \frac{6 \times 214}{10(100-1)} = -\frac{147}{495} = -0{\cdot}3 \text{ approx.}$$

Ex. 1. If $d_i = 0$ for all i, show that $r = 1$ and conversely if $r = 1$ then $d_i = 0$ for all i.

Ex. 2. If $x_i + y_i = n + 1$, show that $r = -1$ and conversely. Explain the result.

[**Hint.** $x_i - y_i = d_i, x_i + y_i = n + 1 \quad \therefore \quad d_i = 2x_i - (n+1),$

$$\therefore \quad \Sigma d_i^2 = 4\Sigma x_i^2 - 4(n+1)\ \Sigma x_i + n(n+1)^2 = \tfrac{1}{2} n\,(n^2 - 1)].$$

Ex. 3. Limits for rank correlation. Show that the maximum value of $\Sigma\, d_i^2$ is $\frac{1}{2}\left(\dfrac{n^3 - n}{n}\right)$ and is always even. Hence or otherwise show that the rank correlation coefficient lies between —1 and +1.

Ex. 4. Show that if X, Y be identically distributed with common probability mass function $P(X = k) = \dfrac{1}{N}$. for $k = 1, 2, \ldots\ldots, N$, $N > 1$, then $\rho_{x,y}$ the correlation coeff. between X and Y, is given by $1 - \dfrac{6E(X - Y)^2}{N^2 - 1}$. ***(B.Sc. Math Hons. Delhi, 1994)***

Hint. R.H.S. $= 1 - \dfrac{6}{N^2 - 1}\,[E\,(x - \mu_x) - (y - \mu_y)]^2$

$$= 1 - \frac{6}{N^2 - 1}\ [Var\,X + Var\,Y - 2\rho\,\sqrt{Var\,X.Var\,y}]$$

10·7. Correlation Ratio. So far we have assumed that there is a single observed value y_i, corresponding to any given value x_i, but in general there may be more than one such value of y. Let the value y_{ij} occur with frequency f_{ij}, then the definitions given earlier can be extended to get

$$\sum_i \sum_j f_{ij} = \sum_i n_i \bar{y}_i = N\bar{y} \quad \text{where} \quad \bar{y}_i = \frac{\sum_j f_{ij}\, y_{ij}}{\sum_j f_{ij}} = \frac{\sum_j f_{ij}\, y_{ij}}{n_i},$$

$$\sum_i \sum_j f_{ij} = N, \ \sum_i n_i = N, \quad \text{and} \quad \bar{y} = \frac{\sum_i \sum_j f_{ij}\, y_{ij}}{\sum_i \sum_j f_{ij}} = \frac{\sum_i n_i \bar{y}_i}{\sum_i n_i}$$

so that $\bar{y}_i$ is the mean of the ith array, consisting of all y's corresponding to x_i and $\bar{y}$ is the general mean, that is the weighted mean of the array means, the weights being the array frequencies. Let (x_i, y_{ij}) the typical pair of values in the ith array, be represented by the point P_{ij} and the mean $(x_i, \bar{y}_i)$ of the array by the point G_i. (see Fig. 10·5).

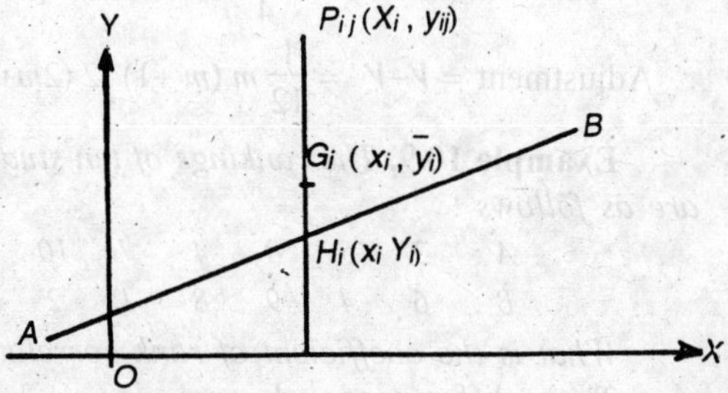

Fig. 10·5

Let $y = a + bx$ be the equation of the line of regression, then by the principle of least squares, a and b are determined by minimizing

$$S = \sum_i \sum_j f_{ij}\,(y_{ij} - a - bx_i)^2$$

The normal equations are $\sum_i \sum_j f_{ij}(y_{ij} - a - bx_i) = 0$

and $\sum_i \sum_j f_{ij}\, x_i (y_{ij} - a - bx_i) = 0$

These equations give for the equation of the line of regression

$$y - \bar{y} = \frac{\mu_{11}}{\sigma_x^2}(x - \bar{x}),$$

where $\mu_{11} = \frac{1}{N} \sum_i \sum_j f_{ij}(y_{ij} - \bar{y})(x_i - \bar{x}) = \frac{1}{N} \sum_i n_i(y_i - \bar{y})(x_i - \bar{x})$

$$\sigma_x^2 = \frac{1}{N} \sum_i \sum_j f_{ij}(x_i - \bar{x})^2 = \frac{1}{N} \sum_i n_i (x_i - \bar{x})^2$$

$$\sigma_y^2 = \frac{1}{N} \sum_i \sum_j f_{ij}(y_{ij} - \bar{y})^2 = \frac{1}{N} \sum_i n_i (y_i - \bar{y})^2.$$

Thus we get the same equations for the lines of regression and in the same way, we get $\sum_i \sum_j f_{ij}(y_{ij} - Y_i)^2 = N\sigma_y^2(1 - r^2)$.

On analogy with the relation (10·14), we define correlation ratio η_{yx} of y on x by the relation

$$NS_y'^2 = \sum_i \sum_j (y_{ij} - \bar{y}_i)^2 = N\sigma_y^2(1 - \eta_{yx}^2) \qquad (10·17)$$

where $S_y'^2$ is the mean square of deviations from the means of arrays.

From (10·17), it follows that $0 \le \eta_{yx}^2 \le 1$ (10·18)

Also since the sum of squares of deviations in any array is least when they are measured from the mean of the array, we have

$S_y'^2 \le S_y^2$. This gives $\eta_{yx}^2 \ge r_{yx}^2$ ...(10·19)

Combining (10·18) and (10·19) we have $0 \le r_{yx}^2 \le \eta_{yx}^2 \le 1$. ...(10·20)

Now $\sum_i \sum_j f_{ij}(y_{ij} - Y_i)^2 = \sum_i \sum_j f_{ij}(y_{ij} - \bar{y}_i)^2 + \sum_i n_i(\bar{y}_i - Y_i)^2$

or $N\sigma_y^2(1 - r^2) = N\sigma_y^2(1 - \eta_{yx}^2) + \sum n_i(\bar{y}_i - Y_i)^2$

we deduce that

(*i*) $r^2 = 1 \Rightarrow y_{ij} = Y_i \Rightarrow$ all observed points lie on a st. line. $\Rightarrow$ relation between y and x is linear.

(*ii*) $\eta_{yx}^2 = 1 \Rightarrow y_{ij} = \bar{y}_i \Rightarrow$ all observed values in an array coincide with its mean *i.e.* there is a single value of y corresponding to any given $x \Rightarrow$ there is a functional relationship between y and x which of course need not be linear.

(*iii*) $\eta_{yx}^2 = r^2 \Rightarrow \bar{y}_i \Rightarrow$ array means lie on a st. line

(*iv*) From (10·20), we deduce that $r^2 = 1 \Rightarrow \eta_{yx}^2 = 1$; $\eta_{yx}^2 = 0 \Rightarrow r^2 = 0$

it follows that

(*a*) $r^2 = 1 \Rightarrow$ relationship between y and x is linear,

(*b*) $\eta^2_{yx} = 1 \Rightarrow$ there is a functional relationship between y and x,

(*c*) $\eta^2_{yx} = 1,\ r^2 \neq 1 \Rightarrow$ there is a non-linear relationship between y and x,

(*d*) If $r^2 = 1$, η^2_{yx} has to be unity and the relationship between y and x has to be linear.

(*e*) If $\eta_{yx} = 0$, then r has to be zero, there is no relationship between y and x or we get the same value of y for all values of x. This can also be deduced from the results of section 10·7.

The scatter diagrams in various cases are shown in Fig. 10·6 :

The next case is of great importance. In this case, array means lie on a straight line which is the line of regression of y an x and the variance

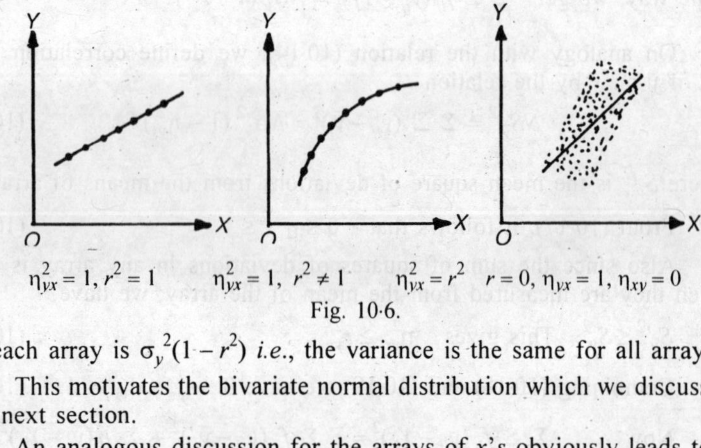

$\eta^2_{yx} = 1,\ r^2 = 1$ $\qquad$ $\eta^2_{yx} = 1,\ r^2 \neq 1$ $\qquad$ $\eta^2_{yx} = r^2$ $\quad$ $r = 0,\ \eta_{yx} = 1,\ \eta_{xy} = 0$

Fig. 10·6.

of each array is $\sigma_y^2(1 - r^2)$ *i.e.*, the variance is the same for all arrays.

This motivates the bivariate normal distribution which we discuss in the next section.

An analogous discussion for the arrays of x's obviously leads to

$$\eta^2_{xy} = 1 - \frac{S_x'^2}{\sigma_x^2} \qquad \text{...(10·21)}$$

giving η_{xy}, the correlation ratio of x on y.

For some purposes, it is convenient to express the correlation ratio in a form involving the standard deviation of the means of arrays. The result is proved below.

10·7·1. Theorem. *The correlation ratio of y on x is the ratio of the standard deviation of the weighted means of the arrays of y's weighted by the corresponding array frequencies) to the standard deviation of all y's of the distribution.*

Proof. We have $N\sigma_y^2 = \sum_i \sum_j f_{ij}(y_{ij} - \bar{y})^2$. $\qquad$ ...(10·22)

Writing $y_{ij} - \bar{y}$ as $(y_{ij} - \bar{y}_i) + (\bar{y}_i - \bar{y})$ and observing that

$$\sum_j f_{ij}(y_{ij}-\bar{y}_i) = (\sum_j f_{ij}y_{ij} - \bar{y}_i \sum_j f_{ij}) = n_i\bar{y}_i - \bar{y}_i n_i = 0$$

(10·22) can be expressed in the form

$$N\sigma_y^2 = \sum_i \sum_j f_{ij}(y_{ij}-\bar{y}_i)^2 + \sum_i \sum_j f_{ij}(\bar{y}_i - \bar{y})^2 + 2\sum_i (\bar{y}_i - \bar{y}) \sum_j f_{ij}(y_{ij}-\bar{y}_i)$$

$$= \sum_i \sum_j f_{ij}(y_{ij}-\bar{y}_i)^2 + \sum_i n_i(\bar{y}_i-\bar{y})^2 = NS_y'^2 + N\sigma_{my}^2$$

where $N\sigma_{my}^2 = \sum_i n_i(\bar{y}_i - \bar{y})^2$

so that σ_{my} is the s.d. of the means of arrays of y's, each mean being weighted by the frequency n_i of that array.

Therefore, using (10·17), $\eta_{yx}^2 = \dfrac{\sigma_y^2 - S_y'^2}{\sigma_y^2} = \dfrac{\sigma_{my}^2}{\sigma_y^2}$ ...(10·23)

Hence the theorem is established.

Example 10·10. *Calculate* η_{xy} *from the following correlation table :*

y \ *x*	*u* \ *v*	*10*	*15*	*20*	*25*	*Total f*	*Mean of the rows*
		–2	*–1*	*0*	*1*		
7	*—2*	*3*	*2*	—	—	*5*	*12*
9	*—1*	—	*1*	*4*	*6*	*11*	*22·3*
11	*0*	—	*3*	*4*	*2*	*9*	*19·4*
13	*1*	*2*	*1*	*5*	—	*8*	*16·9*
15	*2*	—	*6*	—	*1*	*7*	*16·4*
Total f		*5*	*13*	*13*	*9*	*40*	
Mean of the columns		*9·4*	*12·2*	*11·2*	*10·1*		

We calculate $\bar{x}$, σ_x and σ_{xy}.

$$\bar{x} = a + \frac{h\Sigma uf_x}{\Sigma f_x} \text{ where } u = \frac{x-20}{5} \text{ or } \bar{x} = 20 + \frac{(-14)}{40} \times 5 = 18{\cdot}25$$

$$\sigma_x = h\sqrt{\frac{\Sigma u^2 f_x}{\Sigma f_x} - \left(\frac{\Sigma uf_x}{\Sigma f_x}\right)^2}$$

$$= 5 \times \sqrt{\frac{42}{40} - \left(\frac{-14}{40}\right)^2} = 5 \times {\cdot}968 = 4{\cdot}84.$$

The table given below shows the calculation of σ_{my}.

Mean of the rows	$f\bar{x}$	u' (= *Deviation from 19*)	$f\bar{x} \times u'$	$f\bar{x} \times u'^2$
12	5	—7	—35	245·00
22·3	11	3·3	36·3	119·79
19·4	9	0·4	3·6	1·44
16·9	8	− 2·1	− 16·8	35·28
16·4	7	− 2·0	− 18·2	47·32
Total	40		− 30·1	448·83

$$\bar{x} = 19 + \frac{(-30 \cdot 1)}{40} = 18 \cdot 25.$$

This is same as already obtained.

$$\sigma_{my}^2 = \frac{448 \cdot 83}{40} - \left(-\frac{30 \cdot 1}{40}\right)^2 = 10 \cdot 6583 \quad \therefore \quad \sigma_{my} = 3 \cdot 26$$

Hence $\quad \eta_{xy} = \dfrac{\sigma_{my}}{\sigma_x} = \dfrac{3 \cdot 26}{4 \cdot 84} = 0 \cdot 68.$

Ex. 1. Show that η_{xy} calculated for the distribution given in table of the above example is 0·37.

Ex. 2. Show that

$$\sum_i n_i(\bar{y}_i - y)^2 = \sum_i \left(\frac{T_i^2}{n_i}\right) - \frac{T^2}{N}$$

where T_i is the sum of the y's in the ith vertical array and T is the sum of the y's for the whole distribution.

[**Hint.** $n_i \; \bar{y}_i = \sum_j f_{ij} \; y_{ij} = T_i, \sum_i n_i \bar{y}_i = \sum_i T_i = N\bar{y} = T.$]

Ex. 3. Show that η_{xy} can be calculated from the formula

$$\eta^2_{yx} = \left(\sum_i \frac{T_i^2}{n_i} - \frac{T^2}{N}\right) \Big/ N\sigma_y^2$$

show that the value of η_{xy}^2 calculated from this formula is unaffected by the change of origin and unit. State the corresponding formula and result for η_{xy}.

Ex. 4. Apply the above formula to calculate η_{yx} and η_{xy} for the data of Example 10·10.

Ex. 5. Prove from first principles that

(*i*) $1 \geq \eta_{xy}^2 \geq r^2$

(*ii*) $\eta_{xy}=\dfrac{\sigma_{xm}}{\sigma_x}$

Ex. 6. Prove the relations :

(*i*) $\sum\limits_i n_i\,(\bar{y}_i - Y_i) = N\sigma_y^2\,(\eta_{yx}^2 - r^2)$

(*ii*) $\sum\limits_i n_i\,(\bar{y}_i - \bar{y})^2 = \sum\limits_i n_i(\bar{y}_i - Y_i)^2 + \sum\limits_i n_i\,(Y_i - \bar{y})^2.$

Ex. 7. Prove the above theorem for a continuous probability distribution.

10·7·2. Regression Curves. For a continuous distribution, the conditional mean $E(Y \mid X = x)$ is called the regression function of Y on X. The graph of this function is called the regression curve of Y on X or for the mean of Y on X. Similarly the regression curve of X on Y is defined and the regression function of X on Y is $E\,(X \mid Y = y)$. It should be clearly noted that any one or both of the regressions may be linear. In other words, if one of the regressiona curves is linear, it can not be concluded that the other is also linear.

Theorem. Let $\bar{X}, \bar{Y}, \sigma_x^2, \sigma_y^2$ and r be the means, variances and correlations between X and Y. If the regression of Y on X is linear, then

$$E\,(Y \mid x) = \bar{y} + r\frac{\sigma_y}{\sigma_x}\,(X - \bar{x})$$

and similar result follows in case regression of X on Y is linear.

Proof. Let the regression equation of Y on X be

$$E(Y \mid x) = a + bx \qquad \text{...(1)}$$

But $$E(Y \mid x) = \int_{-\infty}^{+\infty} y f(y \mid x)\,dx = \int_{-\infty}^{+\infty} \frac{y f(x, y)}{f_X(x)}\,dx = \frac{1}{f_X(x)}\int_{-\infty}^{+\infty} y f(x, y)\,dx \qquad \text{...(2)}$$

where $f(x, y) = f(y \mid x) \times f_X(x)$

On multiplying both sides by $f_x(x)$ it follows that on integrating both sides w.r.t. x,

$$\int_{-\infty}^{+\infty}\int_{-\infty}^{+\infty} y f(x, y)\,dy\,dx = a\int_{-\infty}^{+\infty} f_x(x)\,dx + b\int_{-\infty}^{+\infty} x f_x(x)\,dx$$

$$\Rightarrow \quad \int_{-\infty}^{+\infty} y\left[\int_{-\infty}^{+\infty} f(x, y)\,dx\right]dy = a + b\,E(X)$$

$$\Rightarrow \quad \int_{-\infty}^{+\infty} y f_Y(y)\,dy = a + b\,E(X) \Rightarrow E(Y) = a + b\,E(X) \qquad \text{...(3)}$$

On multiplying both sides of (2) by $x f_x(x)$ and integrating w.r.t. x.

$$\int_{-\infty}^{+\infty} \int_{-\infty}^{+\infty} xy\, f(x, y)\, dy\, dx = a \int_{-\infty}^{+\infty} x f_x(x)\, dx + b \int_{-\infty}^{+\infty} x^2 f_x(x) dx$$

$$\Rightarrow E(XY) = aE(X) + b\, E(X^2) \quad \Rightarrow \quad \mu_{11} + \overline{x}\overline{y} = a\overline{x} + b\,(\sigma_x^2 + \overline{x}^2) \quad \text{...(4)}$$

Taking $EX = 0 = EY$, we find from (4), $b = \dfrac{\mu_{11}}{\sigma_x^2}$ and then from (3),

$$a = \overline{Y} - \frac{\mu_{11}}{\sigma_x^2} \overline{X}$$

Hence on substituting the values of a and b in (1),

$$E(Y \mid x) = \overline{Y} + \frac{\mu_{11}}{\sigma_x^2} (x - \overline{x})$$

Similarly, $$E(X \mid y) = \overline{X} - \frac{\mu_{11}}{\sigma_y^2} (y - \overline{y})$$

Example 10·11. *Given* $f(x, y) = \begin{cases} 1 & \text{if } |y| < x,\ 0 < x < 1 \\ 0, & \textit{otherwise} \end{cases}$

show that the regression of Y on X is linear but regression of X on Y is non-linear.

The marginal p.d.f. of $X = \displaystyle\int_{|y| < x} f(x, y)\, dy = \int_{-x}^{+x} 1 \cdot dy = 2x,\ 0 < x < 1$

The marginal p.d.f. of $Y = f_2(y) = \displaystyle\int_{x = |y|}^{1} 1 \cdot dx = 1 - |y|,\ -1 < y < 1.$

$\therefore$ Conditional density function of $X = f_1(x \mid y) = f(x, y) / f_2(y)$

$$= \frac{1}{1 - |y|},\ -1 \le y \le 1,\ 0 < x < 1$$

$$= \begin{cases} \dfrac{1}{1 + y}, & -1 < y < 0,\ 0 < x < 1 \\ \dfrac{1}{1 - y}, & 0 < y < 1,\ 0 < x < 1 \end{cases}$$

Also $$f_2(y \mid x) = \frac{f(x, y)}{f_1(x)} = \frac{1}{2x},\ 0 < x < 1,\ |y| < x$$

$$E(Y \mid X = x) = \int_{-x}^{x} y f_2(y \mid x)\, dy = \int_{-x}^{x} \frac{y}{2x}\, dy = \frac{1}{4x} \int_{-x}^{x} y\, dy = 0$$

$\therefore$ Regression Curve of Y on X is $y = 0$ which is a straight line.

Also $E(X \mid Y=y) = \int x f_1(x \mid y)\, dx$

$$= \begin{cases} \displaystyle\int_0^1 x\left(\frac{1}{1-y}\right) dx = \frac{1}{2(1-y)}, & 0<y<1 \\ \displaystyle\int_0^1 x\left(\frac{1}{1+y}\right) dx = \frac{1}{2(1+y)}, & -1<y<0 \end{cases}$$

Thus regression curve of X on $Y = E(X \mid Y=y)$ is not linear.

Example 10·12. *Let the joint density function of X and Y be given by $f(x, y) = 8xy$, $0<x<y<1$; $=0$ elsewhere*

Find (i) $E \mid Y \mid X=x)$ (ii) $E[XY \mid X=x]$ (iii) $Var[Y \mid X=x]$

(B.Sc. Math Hons. Delhi 1988)

$$\text{We have } f(y \mid x) = \frac{f(x, y)}{f_x(x)} = \frac{f(x, y)}{\int_x^1 8xy\, dy} = \frac{8xy}{8x\left[\frac{y^2}{2}\right]_x^1} = \frac{2y}{1-x^2}$$

$$\text{Now } E(Y \mid X=x) = \int_x^1 y f(y \mid x) dy = \int_x^1 \frac{y \cdot 2y}{1-x^2}\, dy = \frac{2}{3(1-x^2)} \left| y^3 \right|_x^1$$

$$= \frac{2}{3} \frac{(1+x+x^2)}{1+x}.$$

$$E(XY \mid x) = \int_x^1 x y f(y \mid x) dy = x \int_x^1 y \cdot \frac{2y}{1-x^2}\, dy = x \cdot \frac{2}{3}\left(\frac{1+x+x^2}{1+x}\right)$$

$$E(Y^2 \mid x) = \int_x^1 y^2 f(y \mid x) dy = \int_x^1 y^2 \cdot \frac{2y}{1-x^2} dy = \frac{1}{2} \frac{[1-x^4]}{1-x^2} = \frac{1+x^2}{2}$$

$$Var(Y \mid x) = E(y^2 \mid x) - [E(y \mid x)]^2 = \frac{1+x^2}{2} - \left[\frac{2}{3} \cdot \frac{1+x+x^2}{1+x}\right]^2.$$

10·8. Bivariate Normal Distribution. The bivariate normal distribution is a generalization of the normal distribution for a single variate. It is derived briefly as follows under the three *assumptions*.

Assume first that one of the variates, say X, is distributed normally with mean μ_1 and *s.d.* σ_x so that the probability that the value of X taken at random will lie in the interval dx is

$$g(x)dx = \frac{1}{\sigma_x \sqrt{2\pi}} \exp\left\{\frac{-(x-\mu_1)^2}{2\sigma_x^2}\right\} dx$$

Assume next that each array of Y is homoscedastic corresponding to an assigned value of X and has a normal distribution with the standard deviation of an array given by $\sigma_y \sqrt{(1-\rho^2)}$; independent of x, ρ is the coefficient of correlation between the variates.

Finally, assume that the regression of y on x is linear. Since the mean of each array is on the line of regression $Y = \mu_2 + \dfrac{\rho\sigma_y}{\sigma_x}(x - \mu_1)$, the probability that a value of y taken at random from an assigned array of y will be in the interval dy is

$$f_y(y \mid x) dx = \frac{1}{\sigma_y \sqrt{2\pi(1-\rho^2)}} \cdot \exp\left[\frac{-1}{2\sigma_y^2(1-\rho^2)}\left\{(y-\mu_2) - \frac{\rho\sigma_y}{\sigma_x}\left(x-\mu_1\right)\right\}^2\right] dy$$

By the theorem of compound probability, the joint probability that x will fall in the interval dx and y in the interval dy is

$$g(x)\, h(y \mid x)\, dxdy \Rightarrow dP(x, y) = f(x, y)\, dx\, dy, -\infty < x, y < \infty,$$

where
$$f(x, y) = g(x)\, h(y \mid x) = \frac{1}{2\pi\sigma_x\sigma_y\sqrt{1-\rho^2}} \cdot \exp\left[-\frac{1}{2(1-\rho^2)} \cdot \left\{\frac{(x-\mu_1)^2}{\sigma_x^2} - \frac{2\rho(x-\mu_1)(y-\mu_2)}{\sigma_x\sigma_y} + \frac{(y-\mu_2)^2}{\sigma_y^2}\right\}\right] \quad ...(10{\cdot}23)$$

Thus the joint p.d.f. of X and Y is $f(x, y)$ where $f(x, y)$ is

$$\text{BVN}(\mu_1, \mu_2, \sigma_x, \sigma_y, \rho).$$

This distribution is called the bivariate normal distribution (see figure 10·7 below) and the variates are said to be normally correlated. The surface in three dimensions with equation $Z = f(X, Y)$ is called the normal correlation surface, $f(x, y)$ is given by (10·23)

The function $f(x, y)$ actually represents a density function and is proved below by showing that its integral over the whole plane is unity

i.e., $$\int_{-\infty}^{+\infty}\int_{-\infty}^{+\infty} f(x, y)\, dxdy = 1 \quad ...(10{\cdot}24)$$

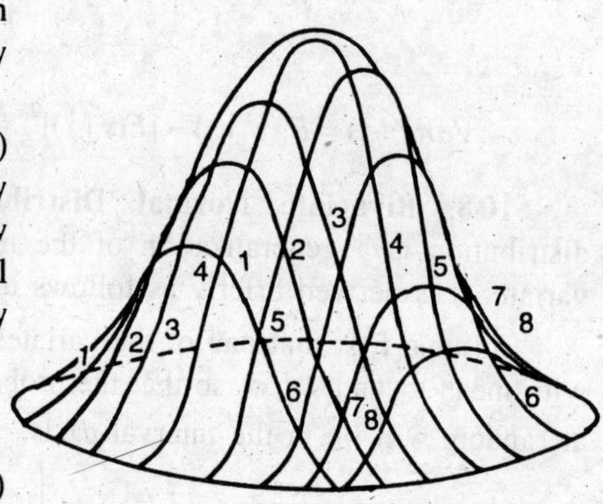

Fig. 10·7. Bivariate Normal Surface.

Put $u = \dfrac{x-\mu_1}{\sigma_x}$, $v = \dfrac{y-\mu_2}{\sigma_y}$ and then L.H.S. of (10·24) becomes

$$\int_{-\infty}^{+\infty}\int_{-\infty}^{+\infty} \frac{1}{2\pi\sqrt{1-\rho^2}} \exp\left[-\frac{1}{2(1-\rho^2)}(u^2 - 2\rho uv + v^2)\right] du dv \quad ...(10·25)$$

$$= \int_{-\infty}^{+\infty}\int_{-\infty}^{+\infty} \frac{1}{2\pi\sqrt{1-\rho^2}} \exp\left[-\frac{1}{2(1-\rho^2)}\{(u-\rho v)^2 + (1-\rho^2)v^2\}\right] du dv$$

Put $$w = \frac{u-\rho v}{\sqrt{1-\rho^2}},\ v = v$$

and the integral may be written as the product of two simple integrals

$$\int_{-\infty}^{+\infty} \frac{1}{\sqrt{2\pi}} e^{-\frac{w^2}{2}} dw \times \int_{-\infty}^{+\infty} \frac{1}{\sqrt{2\pi}} e^{-\frac{v^2}{2}} dv$$

both of which are equal to unity. Thus equation (10·24) is verified.

Note. (10·24) can alternatively be proved by using

$$\int_{-\infty}^{+\infty}\int_{-\infty}^{+\infty} \exp\left[-(ax^2 + 2hxy + by^2)\right] dx\, dy = \pi / \sqrt{ab - h^2}.$$

where $$a > 0,\ b > 0,\ ab - h^2 > 0.$$

10·8·1. Properties of the Normal Correlation Distribution.

I. *The m.g.f. of the BVN* $(\mu_1, \mu_2, \sigma_x, \sigma_y, \rho)$ *distribution about the mean* (μ_1, μ_2) is given by

$$M_{XY}(t_1, t_2) = E\left[\exp\{t_1(x-\mu_1) + t_2(y-\mu_2)\}\right]$$

$$= \int_{-\infty}^{+\infty}\int_{-\infty}^{+\infty} \exp\left[t_1(x-\mu_1) + t_2(y-\mu_2)\right] f(x, y)\, dx\, dy$$

Making the substitutions

$$\xi = x - \mu_1 - \sigma_x^2 t_1 - \rho\sigma_x\sigma_y t_2;\quad \eta = y - \mu_2 - \rho\sigma_x\sigma_y t_1 - \sigma_y^2 t_2$$

we find that

$$M_{XY}(t_1, t_2) = \exp\left[\tfrac{1}{2}(t_1^2\sigma_x^2 + 2t_1 t_2 \sigma_x \sigma_y \rho + t_2^2\sigma_y^2)\right] \times \frac{1}{2\pi\sigma_x\sigma_y\sqrt{1-\rho^2}}$$

$$\times \int_{-\infty}^{+\infty}\int_{-\infty}^{+\infty} \exp\left[-\frac{1}{2(1-\rho^2)}\left(\frac{\xi^2}{\sigma_x^2} - \frac{2\rho\xi\eta}{\sigma_x\sigma_y} + \frac{\eta^2}{\sigma_y^2}\right)\right] d\xi\, d\eta$$

$$= \exp\left[\tfrac{1}{2}(t_1^2\sigma_x^2 + 2\, t_1 t_2 \sigma_x \sigma_y \rho + t_2^2 \sigma_y^2)\right] \quad ...(10·26)$$

Now μ_{rs} = coefficient of $\dfrac{t_1^r t_2^s}{r!\, s!}$ in the expansion of $M_{XY}(t_1, t_2)$.

we find that

$$\mu_{20} = \sigma_x^2,\ \mu_{11} = \rho\sigma_x\sigma_y,\ \mu_{02} = \sigma_y^2,\ \mu_{30} = \mu_{21} = \mu_{12} = \mu_{03} = 0$$

$$\mu_{40} = 3\,\sigma_x^4,\ \mu_{31} = 3\,\rho\,\sigma_x^2\sigma_y, \mu_{04} = 3\,\sigma_y^4$$

$$\mu_{22} = (1 + 2\sigma^2)\sigma_x^2\sigma_y^2;\ \mu_{13} = 3\,\rho\,\sigma_x\,\sigma_y^3.$$

Cor. If $\rho = 0$, then

$M_{XY}(t_1, t_2) = \exp[\tfrac{1}{2}(t_1^2\,\sigma_x^2 + t_2^2\sigma_y^2) = M_X(t_1) \times M_Y(t_2)$

$\Rightarrow$ X and Y are independent,

Conversely, if X and Y are independent, then $\rho = 0$

II. *The marginal density of* X is given by

$$f_X(x) = g(x) = \int_{-\infty}^{+\infty} f(x, y)\, dy \qquad \text{Put } v = \frac{y - \mu_2}{\sigma_y}$$

$$= \int_{-\infty}^{+\infty} \frac{1}{2\pi\sigma_x\sqrt{1-\rho^2}} \exp\left[-\frac{1}{2}\frac{(x-\mu_1)^2}{\sigma_x^2} - \frac{1}{2(1-\rho^2)}\left(v - \frac{\rho(x-\mu_1)}{\sigma_x}\right)^2\right] dv$$

Put $w = v - \rho\left(\dfrac{x - \mu_1}{\sigma_x}\right)\Big/\sqrt{1-\rho^2}$

$$\therefore\ f_X(x) = g(x) = \frac{1}{\sigma_x\sqrt{2\pi}}\, e^{-\frac{1}{2}\frac{(x-\mu_1)^2}{\sigma_x^2}}$$

Similarly, marginal density of Y is given by

$$f_Y(y) = h(y) = \frac{1}{\sigma_y\sqrt{2\pi}}\, e^{-\frac{1}{2}\frac{(y-\mu_2)^2}{\sigma_y^2}}$$

Hence $X \sim N(\mu_1, \sigma_x^2)$ and $Y \sim N(\mu_2, \sigma_y^2)$.

III. *The conditional density of X for a given value of Y is*

$$f(x \mid y) = \frac{f(x, y)}{h(y)}$$

$$= \frac{1}{\sigma_x\sqrt{2\pi(1-\rho^2)}} \times \exp\left[\frac{-1}{2\sigma_x^2(1-\rho^2)}\left\{(x - \mu_1) - \frac{\rho\sigma_x}{\sigma_y}(y - \mu_2)\right\}^2\right]$$

which is a univariate normal density with mean

$$E(X \mid Y) = \mu_1 - \frac{\rho\,\sigma_x}{\sigma_y}(y - \mu_2)$$

and with variance Var $(X \mid Y)$ given by $\sigma_x^2 (1-\rho^2)$. Interchanging x and y, μ_1 and μ_2 and σ_x and σ_y throughout, we get the conditional distribution of Y for a given value of X.

IV. The exponent in the bivariate normal distribution can be written as

$$-\frac{1}{2(1-\rho^2)}\left[\left(\frac{x-\mu_1}{\sigma_x}-\frac{\rho(y-\mu_2)}{\sigma_y}\right)^2+\frac{(y-\mu_2)^2}{\sigma_y^2}(1-\rho^2)\right] \quad ...(10{\cdot}27)$$

or $$-\frac{1}{2(1-\rho^2)}\left[\left(\frac{y-\mu_2}{\sigma_y}-\frac{\rho(x-\mu_1)}{\sigma_x}\right)^2+\frac{(x-\mu_1)^2}{\sigma_x^2}(1-\rho^2)\right] \quad ...(10{\cdot}27)'$$

From (10·27), $$E(X \mid Y=y)=\mu_1-\frac{\rho\sigma_x}{\sigma_y}(y-\mu_2)$$

and $$Var(X \mid Y=y)=\sigma_x^2(1-\rho^2)$$

The conditional mean of either variate, given the values of the other, is a linear function of that value and that its conditional variance is constant, independent of that value. Thus the regression of x on y (or of y on x) is *linear* and *homoscedastic* (Greek word meaning 'equal scattering').

Theorem *(X, Y) possesses a bivariate normal distribution iff every linear combination of X and Y is a normal variate.*

Proof. *Necessity* : Let (X, Y) be a $BVN(\mu_1, \mu_2, \sigma_1^2, \sigma_2^2, \rho)$. we shall prove that $aX+bY$, $(a, b) \neq 0$ is a normal variate.

$$M_{(X,Y)}(t_1, t_2)=E(e^{t_1 X+t_2 Y})$$

$$=\exp\left[t_1\mu_1+t_2\mu_2+\tfrac{1}{2}(t_1^2\sigma_1^2+2\rho\, t_1 t_2\sigma_1\sigma_2+t_2^2\sigma_2^2)\right]$$

$$M_Z(t)=E[e^{tZ}]=E[e^{t(aX+bY)}]=E[e^{(at)X+(bt)Y}]$$

$$=\exp\left[t(a\mu_1+b\mu_2)+\frac{t^2}{2}(a^2\sigma_1^2+2\rho\, a b\sigma_1\sigma_2+b^2\sigma_2^2)\right]$$

which is the *m.g.f.* of a normal variate with parameters

$$\mu=a\mu_1+b\mu_2,\ \sigma^2=a^2\sigma_1^2+2\rho\, a b\sigma_1\sigma_2+b^2\sigma_2^2 \quad ...(10{\cdot}27)$$

Hence by uniqueness theorem, we have $L=aX+bY \sim N(\mu, \sigma^2)$ where μ and σ^2 are defined in (10·27).

Sufficiency : Let $Z=aX+bY$, $(a, b) \neq 0$ be a normal variate. we shall prove that $(X, Y) \sim BVN(\mu_x, \mu_y, \sigma_x^2, \sigma_y^2, \rho)$.

Proof. $\mu=E(Z)=E(aX+bY)=a\mu_x+b\mu_y$

$$\sigma^2 = Var(Z) = Var(aX + bY) = a^2\sigma_x^{\ 2} + 2ab\,\rho\,\sigma_x\,\sigma_y + b^2\,\sigma_y^{\ 2}$$

$$\text{Now} \quad M_Z(t) = \exp\left[\ t\,\mu + \frac{t^2}{2}\sigma^2\ \right]$$

$$= \exp\left[\ t\,(a\,\mu_x + b\,\mu_y) + \frac{t^2}{2}(a^2\sigma_x^{\ 2} + 2ab\,\rho\sigma_x\,\sigma_y + b^2\sigma_y^{\ 2})\ \right]$$

$$= \exp\left[\, t_1\,\mu_x + t_2\,\mu_y + \tfrac{1}{2}(t_1^{\ 2}\,\sigma_x^{\ 2} + 2\,\rho\,t_1\,t_2\,\sigma_x\,\sigma_y + t_2^{\ 2}\,\sigma_y^{\ 2}\,\right]$$

where $t_1 = at$ and $t_2 = bt$.

But the R.H.S. is the m.g.f. of BVN distribution.

Hence by uniqueness theorem, it follows that

$$(X, Y) \sim BVN\,(\mu_x, \mu_y, \sigma_x^{\ 2}, \sigma_y^{\ 2}, \rho).$$

Example 10·13. *Let X and Y have bivariate normal distribution with parameters* $\mu_x = 5$, $\mu_y = 10$, $\sigma_x^{\ 2} = 1$, $\sigma_y^{\ 2} = 25$ *and* $Corr\,(X, Y) = \rho$

If $\rho > 0$, *find* ρ *when* $P(4 < Y < 16 \mid X = 5 = 0{\cdot}954$

(*B.Sc. Math Hons. Delhi 1993*)

Since $(X, Y) \sim BVN\,(5, 10, 1, 25, \rho)$, it follows that the conditional distribution of Y given $X = x$ is normal.

$$\therefore \quad Y \mid X = x \sim N\left[\ \mu_y + \frac{\rho\sigma_y}{\sigma_x}(x - \mu_x),\ \sigma_y^{\ 2}\,(1 - \rho^2)\ \right]$$

$$\Rightarrow \quad Y \mid X = 5 \sim N\left[\ 10 + \rho\frac{5}{1}(5 - 5),\ 25(1 - \rho^2\ \right] = N\,[10, 25(1 - \rho^2)]$$

We want to determine P such that $P(4 < Y < 16 \mid X = 5) = 0{\cdot}954$

$$\text{Now} \quad Z = \frac{Y - \mu}{\sigma} = \frac{Y - 10}{5\sqrt{1 - \rho^2}} \sim N(0, 1)$$

$$\therefore\ P\left(\frac{4 - 10}{\sigma} < Z < \frac{16 - 10}{\sigma}\right) = 0{\cdot}954$$

$$\Rightarrow P\left(\frac{-6}{\sigma} < Z < \frac{6}{\sigma}\right) = 0{\cdot}954 = P\ (-2 \le Z \le 2) \text{ from normal table}$$

$$\therefore\ \frac{6}{\sigma} = 2\ \therefore\ \sigma = 3. \text{ Also } \sigma^2 = 5(1 - \rho^2) \text{ gives } \rho^2 = \frac{16}{25} \Rightarrow \rho = \frac{4}{5} = 0.8$$

Ex. 1. (*a*) Write down the parameters dentsity function of bivariate normal distribution in the variates X and Y, and show that

(*i*) X is distributed normally with variance $\sigma_x^{\ 2}$.

(*ii*) Y is distributed normally with variance $\sigma_y^{\ 2}$.

(*iii*) For any given X, Y is distributed normally about mean $(\rho\sigma_y/\sigma_x)x$ with variance $\sigma_y^{\ 2}\,(1 - \rho^2)$.

(*iv*) For any given X, Y is distributed normally about mean $(\rho\sigma_x/\sigma_y)$ with variance $\sigma_x^2(1-\rho^2)$.

(*v*) Both the regressions of Y and X and of X on Y are linear.

(*vi*) The variances of all the y-arrays are same (*i.e.*, regression of Y on X is *homoscedastic*).

(*b*) Find the parameters of the simultaneous materialization of the inequalities $X > E(X)$ and $Y > E(Y)$.

(*c*) If $\rho = 0$ in Ex.10·34 Page 421 $P(X+Y \le 16) = \Phi(1/\sqrt{26})$

[**Hint.** $X + Y \sim N\left(\mu_x + \mu_2, \sigma_x^2 + \sigma_y^2\right) \sim N(15, 26)$

[**Hint.** (*b*) $P(x > 0 \cap y > 0) = \dfrac{1}{2\pi\sqrt{1-\rho^2}} \displaystyle\int_0^\infty \int_0^\infty e^{-(ax^2 + 2hxy + by^2)} dxdy$

$$= \frac{1}{2\sqrt{ab-h^2}} \cot^{-1} \frac{h}{\sqrt{ab-h^2}},$$

or Put $x = r\cos\theta, y = r\sin\theta, dxdy = rdrd\theta$

$a > 0, b > 0, h > 0, ab - h^2 > 0$] **Ans.** $\dfrac{1}{4} + \dfrac{\sin^{-1}\rho}{2\pi}$

Ex. 2. By differentiating $\displaystyle\int_{-\infty}^{+\infty} dP = 1$ (as given by (10·23) with respect to ρ, show that $E(X-\mu_1)(Y-\mu_2) = \rho\sigma_x\sigma_{xy}$. Find also the mean and variance of X and Y by the same method.

Ex. 3. Obtain the regression curve for the mean of Y for a given X and the marginal distributions of X for the distribution

(*i*) $f(x, y) = \dfrac{9(1+x+y)}{2(1+x)^4(1+y)^4}, \; 0 \le x, y < \infty.$

(*ii*) $f(x, y) = \dfrac{y}{(1+x)^4}) \exp\left(\dfrac{-y}{1+x}\right); \; x, y > 0$

Ans. (*i*) $y = \dfrac{x+3}{2x+3}, f_1(x) = \dfrac{3}{4}\dfrac{2x+3}{(1+x)^4}$

(*ii*) $y = \dfrac{1}{(1+x)^2} \displaystyle\int_0^\infty y^2 e^{\frac{-y}{1+x}} dx = 2(1+x). \; f_1(x) = \dfrac{1}{(1+x)^2}, x \ge 0$

Ex. 4. Prove that the curve of regression of Y on X for the distribution

(*i*) $f(x, y) = \begin{cases} 2\,(x-y), & 0 < x < 1, \; 0 < y < 1 \\ 0, & \text{elsewhere} \end{cases}$

(*ii*) $f(x, y) = xe^{-x(y+1)}, \; x \ge 0, y \ge 0$

is a hyperbola.

Ex. 5. If $(X, Y) \sim BVN(\mu_1, \mu_2, \sigma_1^2, \sigma_2^2, \rho)$, find $r(e^x, e^y)$

[**Hint.** $\mu'_{rs} = E(U^r V^s) = E[e^{rX+sY}] = \exp[r_1\mu_1 + s_1\mu_2 + \frac{1}{2}(r^2\sigma_1^2 + 2\rho rs + s^2\sigma_2^2)$

thus $E(u) =$ etc.

Ans. $\rho(u, v) = (e^{\rho\sigma_1\sigma_2} - 1)/[(e^{\sigma_1^2} - 1)(e^{\sigma_2^2} - 1)]^{1/2}$

Ex. 6. For the bivariate normal distribution given by

$$Z = \text{const. exp}\left[\frac{-1}{2(1-\rho^2)}(x^2 - 2\rho xy + y^2)\right], -\infty < x, y < \infty$$

i.e. $(x, y) \sim BNV(0, 0, 1, 1, \rho)$,

(*a*) Prove that

(*i*) $\dfrac{\partial^2 z}{\partial x \partial y} = \dfrac{\partial z}{\partial \rho}$ (*ii*) $M(t_1, t_2) = \exp[\frac{1}{2}(t_1^2 + 2\rho t_1 t_2 + t_2^2)]$

(*b*) Obtain the values of $\dfrac{\partial^2 M}{\partial t_1 \partial t_2}$, $t_1 \dfrac{\partial M}{\partial t_1}$ and $t_2 \dfrac{\partial M}{\partial t_2}$.

(*c*) Show that $\dfrac{\partial^2 M}{\partial t_1 \partial t_2} - \rho\left\{t_1 \dfrac{\partial M}{\partial t_1} + t_2 \dfrac{\partial M}{\partial t_2}\right\} = \rho(1-\rho^2)t_1 t_2 M.$

Get a relation between these quantities and M and identifying coefficient show that $\mu_{rs} = (r + s - 1)\mu_{r-1, s-1} + (r-1)(s-1)(1-\rho^2)\mu_{r-2, s-2}$.

Hence or otherwise show that $\mu_{rs} = 0$ if $r + s$ is odd ; $\mu_{21} = 3\rho, \mu_{22} = 1 + 2\rho^2$.

(*d*) independence implies zero correlation and zero correlation implies independence. ***(B.A. Hons. Statistics, 1989)***

10·9. Intra-class Correlation Coefficient. Sometimes specially in biological situations, it is desired to study the correlation between some characteristic of members of one or more families. Thus we may be interested in the correlation of heights of brothers or weights of sisters or lengths of leaves in trees. The correlation coefficients between heights of tallest and shortest brothers or between the weights of eldest and youngest sisters would not serve our purpose, as here we are interested in correlations in which all the members of a family have the same status.

Let there be h families containing $k_1, k_2, \ldots\ldots, k_h$ members and let x_{ij} $[i = 1, 2, \ldots\ldots, h, j = 1, 2, \ldots\ldots, k_i]$ denote the value of the jth member of the ith family.

The product-moment correlation coefficient obtained from the

$$N = \sum_{i=1}^{h} k_i(k_j - 1)$$

pairs x_{ij}, x_{il} $[i = 1, 2, \ldots\ldots, h; j, l = 1, 2 \ldots\ldots, k_i ; j \neq l]$ is called the *intra-class correlation coefficient* and the table giving all these variate values is called *intra-class correlation table*. There is an obvious symmetry in the table, since the value of each member of the ith family occur

$(k_i - 1)$ times as a value of the x-variate and the same number of times as value of the y-variate. Therefore,

$$\bar{x} = \bar{y} = \frac{\sum_i \sum_j (k_i - 1) x_{ij}}{\sum_i k_i(k_i - 1)} = \frac{1}{N}\left[\sum_i (k_i - 1) \sum_j x_{kj}\right]$$

Similarly, $\sigma_x^2 = \sigma_y^2 = \frac{1}{N} \sum_i (k_i - 1) \sum_j (x_{ij} - \bar{x})^2,$

and $\quad \text{Cov}(x, y) = \frac{1}{N} \sum_i \sum_j \sum_l{}' (x_{ij} - \bar{x})(x_{il} - \bar{x}),\ i \neq l$

$$= \frac{1}{N} \sum_i \sum_j (x_{ij} - \bar{x}) \sum_l{}' (x_{il} - \bar{x})$$

Σ' indicating that the summation over l has to exclude the value j, so that

$$\sum_l{}' (x_{il} - \bar{x}) = \sum_l (x_{il} - \bar{x}) - (x_{ij} - \bar{x}) = k_i(\bar{x}_i - \bar{x}) - (x_{ij} - \bar{x}),$$

where $\bar{x}_i$ denotes the mean of the ith class *i.e.*, $\bar{x}_i = \sum_{ij} x_{ij}/k_i$

$$\therefore \quad \text{Cov}(x, y) = \frac{1}{N}\left[\sum_i k_i(\bar{x}_i - \bar{x}) \sum_j (x_{ij} - \bar{x}) - \sum_i \sum_j (x_{ij} - \bar{x})^2\right]$$

$$= \frac{1}{N}\left[\sum_i k_i^2 (\bar{x}_i - \bar{x})^2 - \sum_i \sum_j (x_{ij} - \bar{x})^2\right]$$

$$\therefore \quad \rho_{xy} = \rho = \frac{\sum_i k_i^2 (\bar{x}_i - \bar{x})^2 - \sum_i \sum_j (x_{ij} - \bar{x})^2}{\sum_i (k_i - 1) \sum_j (x_{ij} - \bar{x})^2}$$

Case of equal number of members in each family.

Let $k_1 = k_2 = \ldots = k_h = k$, then $\rho = \dfrac{k^2 \sum (\bar{x}_i - \bar{x})^2 - \sum\sum (x_{ij} - \bar{x})^2}{(k-1) \sum\sum (x_{ij} - \bar{x})^2}$

Let σ^2 denote the variance of x and σ_m^2 the variance of means of families (about the mean $\bar{x}$), so that

$$\sigma^2 = \frac{1}{hk} \sum_i \sum_j (x_{ij} - \bar{x})^2 \ ; \ \sigma_m^2 = \frac{1}{h} \sum_i (\bar{x}_i - \bar{x})^2$$

$$\therefore \quad \rho = \frac{1}{k-1}\left(\frac{k\sigma_m^2}{\sigma^2} - 1\right) = \frac{k\sigma_m^2 - \sigma^2}{(k-1)\sigma^2} \qquad \ldots(10{\cdot}28)$$

Notes. 1. For distinction, the ordinary correlation coefficient *e.g.*, the correlation between the heights of eldest and youngest brothers would be called an *inter-class correlation coefficient.*

2. Since $\sigma_m^2 \geq 0$, the intra-class correlation coefficient cannot be less than $-1/(k-1)$, though it may attain the value $+1$. This is thus a skew coefficient in the sense that a negative value for it has not the same significance as an equivalent positive value. In fact if k is very large, ρ is almost always positive.

3. ρ will attain the value $-1/(k-1)$ if $\sigma_m^2 = 0$ *i.e.* if there is no variation between family means. In particular, if there is only one family of N members, the intra-class correlation coefficient has the value $-1/(N-1)$. In §12·12, we shall deduce a similar result for sampling from a finite population.

4. ρ is obviously independent of the origin and the scale.

Ex. 1. In five families of 3, the heights of brothers are in inches 69, 70, 71 ; 70, 71, 72 ; 71, 72, 73 ; 72, 73, 74 ; 73, 74, 75.

Find the intra-class coefficient of correlation.

Ex. 2. From the intra-class correlation table for the above heights, calculate intra-class coefficient without using the formula.

Ex. 3. In Ex. 1, find the correlation coefficient between the heights of the tallest and shortest brothers and compare with the value of ρ.

Exercises 10

1. Show that, if x', y' are the deviations of the variables X, Y from their means,

$$r = 1 - \frac{1}{2N}\sum_i f_i \left(\frac{x_i'}{\sigma_x} - \frac{y_i'}{\sigma_y}\right)^2 \text{ and } r = -1 + \frac{1}{2N}\sum_i f_i \left(\frac{x_i'}{\sigma_x} + \frac{y_i'}{\sigma_y}\right)^2, \sum_i f_i = N$$

Deduce that $-1 \leq r \leq 1$.

2. Let X and Y be continuous chance variables having the joint density $f(x, y)$. Define r, the coefficient of correlation between X and Y, in terms of $f(x, y)$. Prove that if X and Y are independent, then $r = 0$. Is the converse also true ?

3. (*a*) Give examples to illustrate that two uncorrelated variables may not be independent.

(*b*) Let X have the marginal density $f_1(x) = 1, -\frac{1}{2} < x < 1$, and let the conditional density of Y be

$$f(y \mid x) = 1, x < y < x+1, -\frac{1}{2} < x < 0 ;$$
$$= 1, -x < y < 1-x, 0 < x < \frac{1}{2} ;$$
$$= 0, \text{ otherwise}$$

Show that $E(x) = 0$, $E(xy) = 0$ and hence deduce that the correlation between x and y is zero.

[**Hint.** (*a*)

x :	–3	–2	–1	1	2	3
y :	9	4	1	1	4	9.

If $P[X=i, Y=j] = p_{ij}$ such that $p_{1,0} = p_{1,0} = \frac{3}{16}, p_{0,0} = \frac{1}{8}, p_{-1,1} = p_{1,1} = \frac{1}{16}, p_{0,1} = \frac{3}{8}$.

Show that X and Y are uncorrelated but not independent.

If $p_{1,1} = p_{-1,1} = 1/32, p_{-1,-1} = p_{1,-1} = p_{1,0}\, p_{0,1} = \frac{3}{32}, p_{-1,0} = p_{0,-1} = \frac{-5}{32}; p_{0,0} = 8/32.$

Show that X and Y are not independent though X^2 and Y^2 are independent.

Another Example : Let X be a normal variate with zero mean and unit variate and let $Y = X^2$, then

$\text{Cov}(x, y) = E(x - \bar{x})(y - \bar{y}) = E(x - 0)(x^2 - \bar{y}) = E(x^3) - \bar{y}\, E(x) = 0 - \bar{y} \,.\, 0 = 0]$.

4. (*a*) Calculate the coefficient of correlation and obtain the lines of regression for the following data :

x :	1	2	3	4	5	6	7	8	9
y :	9	8	10	12	11	13	14	16	15

Obtain an estimate of y which should correspond on the average to $x = 6{\cdot}2$.

[**Ans.** $r = {\cdot}95$, the lines of regression are $y - 12 = {\cdot}95\,(x - 5)$ and $x - 5 = 95(y - 12)$. Required estimate is 13·41]

[**Hint.** Take $u = x - 5, v = y - 12$, then $\Sigma u = 0 = \Sigma \upsilon$

$\Sigma u^2 = 60, \Sigma v^2 = 60, \Sigma uv = 57. \;\therefore\; r =$ etc.]

(*b*) Calculate thee coefficient of correlation between the values of x and y from the following table :

x :	78	89	97	69	59	79	61	61
y :	125	137	156	112	107	136	123	108

(You may use 69 as working mean for x and 112 as that for y)

[**Ans.** $r = +{\cdot}957$]

5. Calculate the coefficient of correlation from the following table giving the ages of 100 husbands and their wives in years.

Age of husbands

Age of wives	20–30	30–40	40–50	50–60	60–70	Total
15–25	5	9	3	...	...	17
25–35	...	10	25	2	...	37
35–45	...	1	12	2	...	15
45–55	...	...	4	16	5	25
55–65	...	...	...	4	2	6
Total	5	20	44	24	7	100

[**Ans.** $r = {\cdot}796$]

6. The variables X and Y are correlated by the equation $ax + by + c = 0$. Show that the correlation between them is –1 if the signs of a and b are alike, and +1 if they are different.

[**Hint.** $ax + by + c = 0 \;\;\therefore\;\; a.E(x) + bE(y) + c = 0.$

Hence $a[x - E(x)] = -b[y - E(y)]$]

7. (*a*) If σ_x and σ_y are the standard deviations of two uncorrelated statistical variates x and y, prove that the standard deviation of $ax + by$ is

$$\sqrt{a^2\sigma_x^2 + b^2\sigma_y^2}.$$

(*b*) Show that if a and b are constants, and r is the correlation between x and y, then the correlation between ax and by is equal to r if the signs of a and b are alike and to $-r$ if they are different. Also show that, if the constants a, b, c are positive, correlation between $ax + by$ and cy is equal to $$(ar\sigma_x + b\sigma_y)/\sqrt{a^2\sigma_x^2 + b^2\sigma_y^2 + 2abr\sigma_x\,\sigma_y}.$$

8. (*a*) Let x_1, x_2 and x_3 be un-correlated variables, each having the same standard deviation. Obtain the coefficient of correlation between $(x_1 + x_2)$ and $(x_2 + x_3)$.

(*b*) If X and Y are standardised variates, with correlation coefficient ρ, prove that $E\,[\max\,(X^2, Y^2)] \le 1 + \sqrt{1-\rho^2}$

[**Hint.** $E\,[\max\,(X^2, Y^2)] = E\,[\frac{1}{2}\,|\,X^2 - Y^2\,| + \frac{1}{2}\,(X^2 + Y^2)]$

$E\,X = E\,Y = 0,\ E\,(X^2) = E\,(Y^2) = 1,\ E\,(XY) = \rho,\ E\,(X \pm Y)^2 = 2 \pm 2\rho = 2(1 \pm \rho)$

Use $E\,[\,|\,X - Y\,|\,.\,|\,X + Y\,|\,]^2 \le E\,(X - Y)^2\,.\,E\,(X + Y)^2$

$\Rightarrow\ E\,[\,|\,X^2 - Y^2]^2 \le 4\,(1 - \rho^2)\ \Rightarrow\ E\,[\frac{1}{2}\,|\,X^2 - Y^2\,|\,] \le \sqrt{1 - \rho^2}$

and $E\,[\frac{1}{2}\,(X^2 + Y^2] = \frac{1}{2}\,[1 + 1] = 1$ $\quad\therefore$ etc.] [**Ans.** (*a*) $r = \frac{1}{2}$]

9. If $u = ax + by$ and $v = bx - ay$ where x and y are measured from their respective means, and r is the coefficient of correlation between x and y and if u and v are uncorrelated, show that

(*i*) $\sigma_u\sigma_v = (a^2 + b^2)\sigma_x\sigma_y\,(1 - r^2)^{\frac{1}{2}}$

(*ii*) $ab(\sigma_x^2 - \sigma_y^2) = r\sigma_x\sigma_y(a^2 - b^2)$

(*iii*) $\sigma_u^2 + \sigma_v^2 = (a^2 + b^2)(\sigma_x^2 + \sigma_y^2)$

$$\left[\textbf{Hint.}\ (i)\ \text{Use the result.} \begin{bmatrix} Var\,u & Cov(u, v) \\ Cov(u, v) & Var\,v \end{bmatrix} = \begin{bmatrix} a & b \\ b & -a \end{bmatrix} \begin{bmatrix} Var\,x & Cov(x, y) \\ Cov(x, y) & Var\,y \end{bmatrix} \times \begin{bmatrix} a & b \\ b & -a \end{bmatrix}\right]$$

10. (*a*) If x and y are independent random variables, show that

$$r^2_{x+y\,:\,x-y} = r^2_{x\,:\,x+y} - r^2_{y\,:\,x+y}$$

where $r_{x+y\,:\,x-y}$ denotes the coefficient of correlation between $x + y$ and $x - y$.

(*b*) If X has Laplace distribution $p\,(x) = \frac{1}{2}\,\lambda \exp\,[-\,\lambda\,|\,x\,|\,],\ -\infty < x < +\infty$ and $Y = a + bx + cx^2$, show that $\rho(x, y) = \lambda\,b\,\sqrt{\lambda^2 b^2 + 10c^2}$

(**B.A. Stat. Hons. Delhi, 1989**)

[**Hint.** $E(\,X^{2k+1}) = 0,\ E\,(X_{2k}) = (2k\,!)/\lambda^{2k}$]

11. (*a*) Show that the coefficient of correlation r between two variables x and y is given by $r = (\sigma_x^2 + \sigma_y^2 - \sigma_{x-y}^2)/\sigma_x \sigma_y$

where σ_x^2, σ_y^2 and σ_{x-y}^2 are the variances of x, y and $x - y$ respectively.

(*B.Sc. Cal. 1992*) (*B.A., Madras, 1947*)

[**Hint.** $\sigma_{x-y}^2 = \text{Var}(x - y) = \sigma_x^2 + \sigma_y^2 - 2\,r\,\sigma_x \sigma_y$]

(*b*) X_1, X_2, X_3 are three variables each with variance σ^2 and the correlation coefficient between any two of them is r. If $\overline{X} = \frac{1}{3}(X_1 + X_2 + X_3)$, show that $\text{Var}\,\overline{X} = \frac{2}{3}(1 + 2r)$. Deduce that $r \geq -\frac{1}{2}$.

12. (*a*) The variates x and y have zero means, the same variance σ^2 and zero correlation. Show that the variates

$$u = (x \cos \alpha + y \sin \alpha) \text{ and } v = (x \sin \alpha - y \cos \alpha)$$

have the same variance σ^2 and zero correlation.

Further show that $\text{var}\, v = 0$ if $\tan \alpha = \rho\sigma_y/\sigma_x$ where $\text{var}\, x \neq \text{var}\, y$.

(*B.Sc. Hons. Stat., 1989*)

(*b*) If the variance σ_x^2 and σ_y^2 of x and y are different, show that the correlation coefficient between the two variates u and v where $u = x \cos \alpha + y \sin \alpha$ and $v = x \sin \alpha - y \cos \alpha$ is given by

$$r_{uv} = \frac{\sigma_x^2 + \sigma_y^2}{[(\sigma_x^2 - \sigma_y^2)^2 + 4\sigma_x^2\sigma_y^2 \operatorname{cosec}^2 2\alpha)]^{\frac{1}{2}}}$$

Deduce that $$-\frac{\sigma_x^2 \sim \sigma_y^2}{\sigma_x^2 + \sigma_y^2} \leq r_{uv} \leq \frac{\sigma_x^2 \sim \sigma_y^2}{\sigma_x^2 + \sigma_y^2}.$$

(*c*) Prove that $1 \geq \eta_{yx}^2 > \rho^2$ where ρ is the coefficient of correlation between x and y. How are the values of η_{yx} interpreted ? Bring out the difference in the interpretation of the values of η_{yx} and those of ρ.

13. x_1 and x_2 are two variates with variances σ_1^2 and σ_2^2 respectively and r is the correlation between them. Determine the value of the constant k such that $u = x_1 + kx_2$ and $v = x_1 + \frac{\sigma_1}{\sigma_2} x_2$ are un-correlated.

(*B.A. Hons. Statistics Delhi, 1984*) [**Ans.** $k = -\sigma_1/\sigma_2$]

14. (*a*) If x_i ($i = 1, 2, ..., n$) are the values of the variable x and w_i are the corresponding weights, then show that the weighted arithmetic mean $\overline{x}_w \left(= \sum_{i=1}^{n} w_i x_i / \sum_{i=1}^{n} w_i \right)$ is greater or less than the unweighted mean $\overline{x} \left(= \sum_{i=1}^{n} x_i / n \right)$ according as the weights and their values of the variable are positively or negatively correlated.

(b) If X and Y are standardized random variables, and

$$r(aX+bY, bX+aY) = (1+2ab)/(a^2+b^2)$$

find $r(X, Y)$. **(B.A. Hons. Stat. Delhi, 1997)***

[**Ans.** $(a^2+b^2)/[(a^2-b^2)^2-2ab]$

15. (i) If two random variates X and Y assume only two values each, and Cov $(X, Y) = 0$, show that X and Y are independent.

(ii) Carefully examine each of the following statements (a) to (f) and say whether the statement is true or false adding each case a few lines to substantiate your conclusion.

If the correlation coefficient between the random variables X and Y be negative, then it means that

(a) The correlation coefficient between $-X$ and $-Y$ is positive.

(b) $E(XY) < E(X)E(Y)$.

(c) the regression coefficient of X on Y and that of Y on X are both negative.

(d) The regression of Y on X is linear and the line is downward sloping.

(e) Y^2 and X^2 are dependent random variables.

(f) If correlation between X and Y is zero, then correlation between X^2 and Y^2 is also zero.

16. (a) Find the most likely price in Bombay (x) corresponding to the price of Rs. 70 at Calcutta (y) from the following data :

Average price at Calcutta = 65 ; Average price at Bombay = 67

S.D. at Calcutta = 2·5 ; S.D. at Bombay = 3·5

Coefficient of correlation is + ·8 between the two prices of the commodity in two towns.

(b) Two random variables have the least square regression lines with equations $3x+2y-26=0$ and $6x+y-31=0$. Find the mean values and the correlation coefficient between x and y.

(c) In a partially destroyed laboratory record of an analysis of correlation data, the following results only are legible : Variance of $x = 9$.

Regression equations : $8x-10y+66=0$; $40x-18y=214$.

What were (a) the mean values of x and y, (b) the standard deviation of y, and (c) the coefficient of correlation between x and y ?

[**Ans.** (a) The equation to the line of regression of x on y is

$x = 1{\cdot}12y - 5{\cdot}8$, $x_{70} ==$ Rs. 72·6

(b) $\bar{x}=4, \bar{y}=7, r=-\cdot5$ (c) $\bar{x}=13, \bar{y}=17, \sigma_y=4, r_{xy}=+\cdot6$]

17. (a) A computer while calculating the correlation coefficient between two variates x and y from 25 pairs of observations obtained the following constants :

$$n=25,\ \Sigma x=125,\ \Sigma x^2=650,\ \Sigma y=100,\ \Sigma y^2=460,\ \Sigma xy=50.$$

It was, however, later discovered at the time of checking that he had copied down two pairs as $\begin{array}{c|c} x & y \\ \hline 6 & 14 \\ 8 & 6 \end{array}$ while the correct values were $\begin{array}{c|c} x & y \\ \hline 8 & 12 \\ 6 & 8 \end{array}$

Obtain the correct value of the correlation coefficient.

(*b*) In two sets of variables *x* and *y* with observations each, the following data were observed : $\bar{x} = 10, \sigma_x = 3, \bar{y} = 6, \sigma_y = 2, r = \cdot 3$

But on subsequent verification, it was found that one value of *x* (= 10) and one value of *y* (= 6) were found inaccurate and hence weeded out. With the remaining 49 pairs of values, how is the original value of *r* affected. **[Ans.** (*a*) $\frac{2}{3}$ (*b*) $\bar{x} = 13. \bar{y} = 17, \sigma_y = 4, r_{xy} = +\cdot 6$]

(*c*) If x_1, x_2 and x_3 are three variates with means $\bar{x}_1, \bar{x}_2$ and $\bar{x}_3$ respectively and each variate occurs the same number of times, show that $r_{12} + r_{23} + r_{31} > -\frac{3}{2}$ where r_{12} is the coefficient of correlation between x_1 and x_2.

$$\left[\textbf{Hint. } Let\ A = \frac{1}{n}\Sigma\left(\frac{x_1 - \bar{x}_1}{\sigma_1} + \frac{x_2 - \bar{x}_2}{\sigma_2} + \frac{x_3 - \bar{x}_3}{\sigma_3}\right)^2\right.$$

$$\left. = \frac{\sigma_1^2}{\sigma_1^2} + \frac{\sigma_2^2}{\sigma_2^2} + \frac{\sigma_3^2}{\sigma_3^2} + \frac{2}{\sigma_1\sigma_2} r_{12}\,\sigma_1\,\sigma_2 + r_{23} + r_{31} = 3 + 2\,(r_{12} + r_{23} + r_{31})\right].$$

(*d*) For random variables *X* and *Y* with the same mean, the two regression equations are $Y = aX + b$ and $X = \alpha Y + \beta$. Show that

$$\frac{b}{\beta} = \frac{1 - a}{1 - \alpha}.$$ ***(B.A. Hons. Statistics, Delhi, 1989, 96)***

[**Hint.** $\bar{Y} = a\,\bar{x} + b$ and $\bar{X} = \alpha\,\bar{y} + \beta$. Also $\bar{X} = \bar{Y}$. Equate the values of $\bar{X}$].

18. (*a*) If θ is the acute angle between the two regression lines with coefficient of correlation *r*, then $\tan\theta = \frac{1 - r^2}{r} \cdot \frac{2\sigma_x\sigma_y}{\sigma_x^2 + \sigma_y^2}$ and hence deduce that $\sin\theta \le 1 - r^2$. ***(B.Sc. Hons. Delhi, 1984)***

(*b*) If the tangent of the angle between the lines of regression of *y* on *x* is 0·6 and the s.d. of *y* is twice the s.d. of *x*, find the value of the correlation between the two variables. **[Ans.** $r = \frac{1}{2}$]

19. (*a*) Two variables *X* and *Y* are known to be related to each other by the relation $Y = X/(aX + b)$. How is the theory of linear regression to be employed to estimate the constants *a* and *b* from a set of *n* pairs of observations $(x_i, y_i), i = 1, 2, ..., n$.

$$\left[\textbf{Hint. } \frac{1}{Y} = a + \frac{b}{X} \Rightarrow \upsilon = a + bu \text{ where } \frac{1}{X} = u, \frac{1}{Y} = v.\right]$$

(*b*) Explain why there are two lines of regression. Write down two regression equations that may be associated with the following pairs of values :

x :	152	114	138	158	144	153	141	117	136	154
y :	198	300	414	594	676	549	320	483	481	659

(*c*) Obtain an equation of the line of regression of yield of rice on water from the data given in the following table which shows the amount of water applied in inches and the yield of rice in tons per acre in an experimental farm. Estimate the most probable yeild of rice of 40 inches of water.

Water x	12	18	24	30	36	42	48
Yield y	5·27	5·68	6·25	7·21	8·02	8·71	8·42

[**Ans.** (*b*) $x = 0{\cdot}03y + 12{\cdot}693$, $y = 3{\cdot}575x - 34{\cdot}67$.

(*c*) $y = 0{\cdot}10x + 4{\cdot}08$, 808 tons.]

20. If x_1, x_2 and x_3 are three variates measured from their respective means as origin and of equal variances, find the coefficient of correlation between $x_1 + x_3$ and $x_2 + x_3$ in terms of r_{12}, r_{13} and r_{23} and show that it is equal to

(*i*) $\dfrac{r_{12}+1}{2}$ if $r_{13} = r_{23} = 0$ or (*ii*) $\dfrac{r_{12}+3}{4}$ if $r_{13} = r_{23} = 1$

21. Two judges in a beauty contest rank the ten competitors in the following order :

6	4	3	1	2	7	9	8	10	5
4	1	6	7	5	8	10	9	3	2

Do the two judges appear to agree in their standard ?

[**Ans.** Rank correlation = 0·224]

This indicates some sort of concordance between the standards of two judges.

22. The ranks of the same 16 students in two subjects A and B where as follows. Two numbers within brackets denote the ranks of the students in A and B respectively.

(1, 1), (2, 10), (3, 3), (4, 4), (5, 5), (6, 7), (7, 2), (8, 6), (9, 8), (10, 11), (11, 15), (12, 9), (13, 14), (14, 12) (15, 16), (16, 13).

Calculate the rank correlation for proficiencies of this group in subjects A and B.

[**Ans.** 0·8]

23. (*a*) If the variables $x_1, x_2, ..., x_{2n}$ all have the same variance σ^2 and the correlation coefficients between x_i and x_j $(i \neq j)$ has the same value, show that the correlation between $\sum_1^n x_i$ and $\sum_{n+1}^{2n} x_j$ is given by $n\rho/\{1 + (n-1)\rho\}$.

(*b*) The independent random variables x and y are defined by

$f(x) = 4ax \quad 0 \le x \le r$; $= 0$ elsewhere ; $f(y) = 4by \quad 0 \le y \le s$; $= 0$ elsewhere.

Find the coefficient of correlation between u and v in terms of a and b where $u = x + y$ and $v = x - y$. $\left[\textbf{Ans.}\ (b)\ r_{uv} = \dfrac{b-a}{b+a}\right]$

24. The n variates y_i satisfy the regression relations

$$y_i = \alpha + \beta(x_i - \bar{x}) + e_i, i = 1, 2, ..., n$$

where α, β are population parameters, the x's are fixed and the e_i are independently distributed with mean zero and variance σ^2. Prove from the first principles that

(*i*) the least square estimate b is an unbiased estimate of β with variance $\sigma^2/\Sigma(x_i - \bar{x})^2$.

(*ii*) If $\eta_x = \alpha + \beta[x - \bar{x}]$ and $y_x = a + b(x - \bar{x})$, denote the population and sample regression lines, show that y_x is an unbiased estimate of η_x and calculate its variance.

The variate $u = \Sigma m_i y_i$ where m's are fixed numbers. If u is to be an unbiased estimate of β for all possible values of α and β, find the restrictions that must be satisfied by the m's. By choosing the m's so as to minimize the variance of u subject to the restrictions, prove that the smallest variance is reached where $u = b$.

$$\left[\textbf{Ans. } Var\,(y_x) = \frac{\sigma^2}{n} + \frac{\sigma^2(x - \bar{x})^2}{\Sigma(x - \bar{x})^2}\right.$$

Restrictions to be satisfied by m's are $\Sigma m_i = 0, \Sigma m_i x_i = 1$]

25. By minimizing $\Sigma f_i\,(x_i \cos\alpha + y_i \sin\alpha - p)^2$ for variations in α and p, show that there are two straight lines passing through the mean of the distribution for which the sum of squares of normal deviations has an extreme value. Prove also that their slopes are given by $\tan 2\alpha = \dfrac{2\mu_{11}}{\sigma_x^2 - \sigma_y^2}$

26. (*a*) The variates x and y are normally correlated and u, v are defined by $u = x\cos\alpha + y\sin\alpha$, $v = y\cos a - x\sin\alpha$.

Show that u, v will be uncorrelated if $\tan 2\alpha = \dfrac{2\rho\sigma_x\sigma_y}{\sigma_x^2 - \sigma_y^2} = \dfrac{2\mu_{11}}{\sigma_x^2 - \sigma_y^2}$

(*b*) Show further that, if in this case σ_u^2, σ_v^2 are their variances, then

$$\sigma_u\sigma_v = \sigma_x\sigma_y\sqrt{1 - \rho^2}, \sigma_u^2 + \sigma_v^2 = \sigma_x^2 + \sigma_y^2.$$

27. If x and y are standard normal variates, with coefficient of correlation ρ, show that

(*i*) regression of y on x is linear,

(*ii*) $x + y$ and $x - y$ are independently distributed,

(*iii*) $Q = \dfrac{x^2 - 2\rho xy + y^2}{(1 - \rho^2)}$ is distributed like a chi-square, *i.e.*, as that of the sum of squares of standard normal variates.

[**Hint.** (*ii*) Let $u = x + y$ and $v = (x - y) \Rightarrow x = (u + v)/2, y = (u - v)/2$

$$\therefore\ dx\,dy = \frac{1}{2}\,du\,dv \quad \therefore\ dP = c_1 \exp\left(\frac{-u^2}{4(1+\rho)}\right)du.\ c_2 \exp\left(\frac{-v^2}{4(1-\rho)}\right)dv$$

(*iii*) $$M_Q(t) = \int_{-\infty}^{+\infty} \int_{-\infty}^{+\infty} e^{tQ} \times \frac{1}{2\pi\sqrt{1-\rho^2}} \exp\left[\frac{-1}{2(1-\rho^2)}(x^2 - 2\rho xy + y^2)\right] dxdy$$

$$= \frac{1}{2\pi\sqrt{1-\rho^2}} \int_{-\infty}^{+\infty} \int_{-\infty}^{+\infty} \exp\left(tQ - \frac{Q}{2}\right) dxdy$$

$$= \frac{1}{2\pi\sqrt{1-\rho^2}} \int_{-\infty}^{+\infty} \int_{-\infty}^{+\infty} \exp\left[\frac{-Q}{2}(1-2t)\right] dxdy$$

Put $\sqrt{(1-2t)x} = u$ and $\sqrt{(1-2t)y} = v$, $Q = \dfrac{1}{1-\rho^2}\left[\dfrac{u^2 - 2\rho uv + v^2}{1-2t}\right]$

Show that $M_Q(t) = (1-2t)^{-1}$. Hence etc.]

28. Show that , if x and y are independent normal variates with zero means, the points of inflexion of the curve of intersection of the normal correlation surface by planes through the z-axis, lie on the elliptic cylinder $\dfrac{x^2}{\sigma_x^{\ 2}} + \dfrac{y^2}{\sigma_y^{\ 2}} = 1.$

29. The bivariate random variables x_1, x_2 have a bivariate normal distribution with means 50 and 75 and s.d. 6 and 12 with a correlation coefficient 0·55. Find the following probabilities :

(*i*) $P(65 \le x_1 \le 75)$ (*ii*) $P(71 \le x_2 < 80/x_1) = 55)$ and (*iii*) $P(|x_1 - x_2| \ge 15)$

30. From a standard bivariate normal population a sample of n values $(x_1, y_1) ; (x_2, y_2), ..., (x_n, y_n)$ is drawn. Show that the distribution of $Z_1 = \dfrac{1}{n}\Sigma x^2$ and $Z_2 = \dfrac{1}{n}\Sigma y^2$

has $$m.g.f. = \text{const} \times \left\{\left(1 - \frac{2t_1}{n}\right)\left(1 - \frac{2t_2}{n}\right) - \frac{4\rho t_1 t_2}{n^2}\right\}^{-\frac{n}{2}}$$

[Use : $$\int_{-\infty}^{+\infty}\int_{-\infty}^{+\infty} \exp[-(ax^2 + 2hxy + by^2)]\, dxdy = \pi/\sqrt{ab - h^2}$$

and *m.g.f.* of $\dfrac{1}{n}x^2$ and $\dfrac{1}{n}y^2$ is $M(t_1, t_2) = E\left(\dfrac{t_1 x^2}{n} + \dfrac{t_2 y^2}{n}\right) = $ etc.]

31. Show that the curves along which the density function is constant for a bivariate normal distribution are the ellipses

$$\frac{x^2}{\sigma_x^{\ 2}} - \frac{2\rho\, xy}{\sigma_x \sigma_y} + \frac{y^2}{\sigma_y^{\ 2}} = \lambda^2.$$

For the ellipse, show that

(*i*) the line of regression of y on x is conjugate to y-axis.

(*ii*) the line of regression of x on y is conjugate to x-axis.

(*iii*) the principal axes are inclined to the x-axis at an angle θ where

$$\tan 2\theta = \frac{2\rho\sigma_x\sigma_y}{\sigma_x^2 - \sigma_y^2}.$$

32. In a bivariate normal population, the standard deviation of an x-array is k times that of x variate as a whole. Show that the correlation is $\sqrt{1-k^2}$.

33. Let X and Y be independent standard normal variates. Obtain the *m.g.f.* of XY. **(*M.Sc. Gauhati, 1992*)**

[**Hint.** $M_{XY}(t) = E[e^{txy}] = \int_{-\infty}^{+\infty}\int_{-\infty}^{\infty} e^{txy} \cdot \frac{1}{2\pi} e^{-x^2/2} \cdot e^{-y^2/2}\, dxdy$

$$= \frac{1}{2\pi}\int_{-\infty}^{+\infty}\int_{-\infty}^{+\infty} \exp\left[\frac{-1}{2(1-t^2)}\left\{\frac{x^2}{1/(1-t^2)} - \frac{-2txy}{\frac{1}{\sqrt{1-t^2}}\cdot\frac{1}{\sqrt{1-t^2}}} + \frac{y^2}{1/(1-t^2)}\right\}\right] dxdy$$

Compare with *m.g.f.* of $BVN(0, 0, \sigma_1^2, \sigma_2^2, \rho)$,

$$\sigma_1^2 = \sigma_2^2 = 1/(1-t^2), \rho = t, M_{XY}(t) = (1-t^2)^{1/2}, -1 < t < -1.$$

34. For the bivariate normal distribution

$$(X, Y) \sim BVN\left(5, 9, 3^2, 4^2, \frac{4}{5}\right), \text{ find}$$

(*i*) $P(5 \le X \le 9)$ (*ii*) $P(7 \le Y \le 8 \mid X = 4)$ and (*iii*) $P(|X - Y| \ge 2)$.

35. Let (X, Y) follow the bivariate normal distribution. Show that the conditional distribution given $X + Y =$ Constant $= c$ does not exist.

[**Hint.** Show that $P(X + Y = C) = 0$].

36. If $(X, Y) \sim BVN(0, 0, 1, 1, \rho)$, show that $E[\max(X, Y)] = \sqrt{\frac{1-\rho}{\pi}}$

[**Hint.** $\max(X, Y) = \frac{1}{2}(X + Y) + \frac{1}{2}|X - Y|$ and $Z = X - Y \sim N[0, 2(1-\rho)]$]

37. If x and y follow the bivariate normal law with zero means and variances σ_1^2 and σ_2^2 respectively and correlation coefficient ρ, show that

(*i*) $u = \frac{x}{\sigma_1} + \frac{y}{\sigma_2}$ and $v = \frac{x}{\sigma_1} - \frac{y}{\sigma_2}$

are independently distributed with zero means and variances $2(1+\rho)$ and $2(1-\rho)$. **(*B.Sc. Poona, 1971*)**

(*ii*) $E(x^2y^2) = \sigma_1^2\sigma_2^2(2\rho^2 + 1)$

Hint. (*i*) $x = \frac{\sigma_1}{2}(u + v), y = \frac{\sigma_2}{2}(u - v), dxdy = \frac{\sigma_1\sigma_2}{2} dudv$

(*ii*) $E(x_1^2 x_2^2) =$ coeff. of $t_1^2 t_1^2$ in $M(t_1, t_2)$]

38. Show that X_1 and X_2 are standard normal variates with correlation coefficient ρ between them, then the correlation coefficient between X_1^2 and X_2^2 is ρ^2. **(B.A. Hons. Delhi, 1984)**

[**Hint.** $E[(X_1^2X_2^2) = 2\rho^2 + 1, E X_1^4 = 3 = E X_2^4, E X_1^2 = 1 = E X_2^2$

$$\rho_{x_1^2, x_2^2} = \frac{2\rho^2 + 1 - 1}{\sqrt{(3-1)(3-1)}} = \rho^2 \Big]$$

39. A bivariate distribution in two discrete variables, X and Y is defined by probability generating function

$$\phi(u, v) = \text{Exp}\,[a(u-1) + b(v-1) + c(u-1)(v-1)$$

Simultaneous probability of $x = r, y = s$ where r and s are integers being the coefficient of $u^r v^s$. Find the correlation coefficient of x and y.

$$\left[\textbf{Hint. } E(x) = \left(\frac{\partial\phi}{\partial u}\right)_{u=v=1}, \ \text{var}\, x = \left(\frac{\partial^2\phi}{\partial u^2} + \frac{\partial\phi}{\partial u}\right)_{u=v=1}\right.$$

$$\left.\text{cov}\,(x, y) = \left[\frac{\partial^2\phi}{\partial u\partial v} - \frac{\partial\phi}{\partial u}\cdot\frac{\partial\phi}{\partial v}\right]_{u=v=1}\right]$$

40. For any a, b ; show that if $E\{u(a-X)u(b-Y)\} = E\{u(a-X)\}\,E\{u(b-Y)\}$ then X and Y are independent where

$$u(t) = 1 \text{ if } t \geq 0, \quad 0 \text{ if } t < 0$$ **(B.A. Math. Hons., Delhi 1981)**

41. Show that if a is any constant such that $|\,a\,| < 1$, then

$$f(x, y) = 1 + a(1-2x)(1-2y), 0 < x, y < 1$$

is a bivariate density function with equal marginal densities. Find also the correlation coefficient ρ. **(B.A. Hons. Delhi, 1982)**

$$\left[\textbf{Hint. } \int_0^1\!\!\int_0^1 f(x, y)dxdy = \int_{y=0}^{1} \{x + a(1-2y)(x - x^2)\}_0^1\, dy = \int_0^1 (1+0)dy = 1\ \forall a.\right]$$

42. Show that $f(x, y) = \frac{1}{4}[1 + xy(x^2 - y^2)], |\,x\,| < 1, |\,y\,| < 1$

is the density function of a bivariate random variable (X, Y). Obtain the marginal density functions of X and Y.

43. Let $F(x, y) = \begin{cases} 0, \text{ if } x < 0, y < 0 \text{ or } x + y < 1 \\ 1, \text{ elsewhere} \end{cases}$

Show that $F(x, y)$ is not a distribution function.

[**Hint.** $P[\frac{1}{2} < X, < 1, \frac{1}{2} < Y < 1] = F(1, 1) - F(\frac{1}{2}, 1) - F(1, \frac{1}{2}) + F(0, 1)$

$$= 1 - 1 - 1 + 0 = -1.]$$

44. The joint *p.d.f.* of two random variables X, Y is given by

(*a*) $f(x, y) = (6 - x - y)/8, 0 < x < 2, 2 < y < 4$

(*b*) $f(x, y) = k(1 - x - y)$ inside the triangle formed by the axes and the line $x + y = 1$ and zero elsewhere. Find the value of k. Find if X and Y are independent. **(I.A.S. 1990)**

Ans. (*b*) $k = 6$, dependent as $f(x, y) \neq f_X(x) f_Y(y)$

45. If $f(x, y) = 3x^2 - 8xy + 6y^2, 0 < x, y < 1$, find $f_x(x \mid y)$ and $f_y(y \mid x)$ and show that X and Y are independent. Find the regression curves for the means.

$$\textbf{Ans. } y = \frac{9x^2 - 16x + 9}{6(3x^2 - 4x + 2)}, \quad x = \frac{36y^2 - 32y + 9}{12(6y^2 - 4y + 1)}$$

46. A random point (X, Y) is uniformly distributed over a circular region $x^2 + y^2 < a^2$, Find the marginal distributions of X and Y and the conditional distribution of Y assuming $X = x, |x| < a$.

$$\textbf{Ans. } \frac{2}{\pi a^2}\sqrt{a^2 - x^2}, -a < x < a, \frac{2\sqrt{a^2 - y^2}}{\pi a^2}, -a < y < a$$

$$\tfrac{1}{2}\sqrt{a^2 - x^2}, -\sqrt{a^2 - x^2} < y < \sqrt{a^2 - x^2}$$

47. If $f(x, y) = x + y$ is the joint *p.d.f.* of X and Y, find the distribution of $X + Y$.

[**Hint.** $U = X + Y, V = X; u = x + y, v = x \Rightarrow x = v, y = u - v, dx\, dy = du\, dv$

Now u ranges from 0 to 2 and v from 0 to 1.

$$\text{Distribution of } u = \int_{-\infty}^{+\infty} f(x, y)\, dy = \int_{-\infty}^{+\infty} f(v, u - v)\, dv = \int_{-\infty}^{+\infty} u\, dv \text{ for } 0 < v < 1,$$

$= 0$, otherwise

$$\text{Hence } f(u) = u\int_0^u dv = u^2, 0 < u < 1, = u\int_{u-1}^{1} dv = u(2 - u), 1 < u < 2$$]

48. If (X, Y) has the normal distribution in two dimensions with zero means, unit variances and correlation coefficient ρ, then prove that expectation of the greater of X and Y is $\sqrt{(1 - \rho)/\pi}$.

$$\textbf{[Hint. } E[\max(X, Y)] = \int_{-\infty}^{+\infty}\int_{-\infty}^{+\infty} \max(x, y) f(x, y)\, dx\, dy$$

$$= \frac{1}{2\pi\sqrt{1 - \rho}} \cdot 2 \int_{-\infty}^{+\infty} x \exp\left\{-\frac{1}{2(1 - \rho^2)}(x^2 - 2\rho xy + y^2)\, dx\right\} = etc.$$

49. The joint probability density function of two variates X and Y is given by

$$f(x, y) = (6 - x - y)/8, 0 < x < 2, 2 < y < 4, = 0 \text{ elsewhere}$$

Calculate (*i*) $P(X < 1, Y < 3)$ (*ii*) $P(< X + Y < 3$ (*iii*) $P(X < 1 \mid Y = 3$ (*iv*) $P(X < 1 \mid Y < 3)$ (*v*) Var X.

$$\textbf{Ans. } \frac{3}{8}, \frac{5}{24}, \frac{5}{8}, \frac{3}{5}, \frac{11}{36}.$$

50. Show that the regression equation of Y on X for the distribution $f(x, y) = \frac{4}{5}(x + 3y)\, e^{-x-2y}, x, y \geq 0$ is $\frac{x + 3}{2x + 3}$.

(*M.Sc. Sardar Patel, 1992*)

11

Multiple and Partial Correlations

11·1. Introduction. So far we have considered correlation between two variables only. But, often it is necessary to find correlation between three or more variables. For example, the statures of men are influenced by those of all their ancestors, and the yield of grain is affected by the amounts of different fertilizers used.

Whenever we are interested in the combined influence of a group of variables, upon a variable, not included in the group, our study is that of *multiple regression* and *multiple correlation.*

In a multi-variate population, the different variables may be mutually correlated and this correlation will, in general, be influenced by the other variables of the population. To study the relationship between any two variables, we have two methods. Firstly, we may consider only those members of the observed data, in which the other members have specified values. Secondly, we may eliminate mathematically the effect of other variables on the two variables under study. The first method has the disadvantage that it limits the size of the data and also the results of this will be applicable to only those data in which the other variables. have assigned values. In the second method, it may not be possible to eliminate the entire influence of the other variables, but we can easily eliminate the linear effects. The correlation between two variables, when the linear effect of the other variables in them has been eliminated from both, is called *partial correlation.*

11·2. Distribution of three variables. The theory of partial and multiple correlations was developed by Karl Pearson (1896) for three variables and then generalized by G. Udny Yule (1897). In what follows, we shall use Yule's notation. Also for the sake of simplicity, we shall study the distribution of three variables only, though the *arguments will apply to the case of p variables also.*

Suppose we are given N sets of corresponding values of three variables. Let these variables be *measured from their respective means* and the quantities so obtained be denoted by x_1, x_2 and x_3. The regression equation of x_1 say, on x_2 and x_3 may be written in the form

$$X_1 = a + b_{12\cdot3}\, x_2 + b_{13\cdot2}\, x_3 \qquad \text{...(11·1)}$$

where the constants a and b's are such as to give, on the average, the 'best' estimate of x_1 corresponding to any assigned values of x_2 and x_3. Thus we are to find a and b's, such that the sum of squares of the residuals *viz.*

$$S = \Sigma\,(x_1 - X_1)^2 = \Sigma\,(x_1 - a - b_{12\cdot3}\, x_2 - b_{13\cdot2}\, x_3)^2 = \Sigma\, x_{1\cdot23}^2$$

where $x_{1\cdot23} = x_1 - a - b_{12\cdot3}\, x_2 - b_{13\cdot2}\, x_3$,

so that $S = \Sigma\, x_{1\cdot23}^2$ is a minimum, and the summation is taken over all sets of values of x_2 and x_3.

The three normal equations for determining a and two b's are

$$\left.\begin{aligned} 0 = \frac{\partial S}{\partial a} &\Rightarrow \Sigma\,(x_1 - a - b_{12\cdot3}\, x_2 - b_{13\cdot2}\, x_3) = 0 \\ 0 = \frac{\partial S}{\partial b_{12\cdot3}} &= \Sigma\, x_2\,(x_1 - a - b_{12\cdot3}\, x_2 - b_{13\cdot2}\, x_3) = 0 \\ 0 = \frac{\partial S}{\partial b_{13\cdot2}} &= \Sigma x_3\,(x_1 - a - b_{12\cdot3}\, x_2 - b_{13\cdot2}\, x_3) = 0 \end{aligned}\right\}$$

$$\left.\begin{aligned} \Sigma\, x_{1\cdot23} &= 0 \\ \Sigma\, x_2\, x_{1\cdot23} &= 0 \\ \Sigma\, x_3\, x_{1\cdot23} &= 0 \end{aligned}\right\} \qquad \text{...(11·2)}$$

The first of these equations gives $a = 0$ and the last two equations may be written in the form

$$\Sigma\, x_1\, x_2 - b_{12\cdot3}\, \Sigma\, x_2^2 - b_{13\cdot2}\, \Sigma\, x_2\, x_3 = 0,$$

$$\Sigma\, x_1\, x_3 - b_{12\cdot3}\, \Sigma\, x_2\, x_3 - b_{13\cdot2}\, \Sigma\, x_3^2 = 0$$

Expressing the summation in terms of standard deviations and correlation coefficients, we have

$$N r_{12}\, \sigma_1 \sigma_2 = N b_{12\cdot3}\, \sigma_2^2 + N b_{13\cdot2}\, r_{23}\, \sigma_2 \sigma_3$$

$$N r_{13}\, \sigma_1 \sigma_3 = N b_{12\cdot3}\, r_{23}\, \sigma_2 \sigma_3 + N b_{13\cdot2}\, \sigma_3^2$$

or

$$\left.\begin{aligned} r_{12}\, \sigma_1 &= b_{12\cdot3}\, \sigma_2 + b_{13\cdot2}\, r_{23}\, \sigma_3 \\ r_{13}\, \sigma_1 &= b_{12\cdot3}\, r_{23}\, \sigma_2 + b_{13\cdot2}\, \sigma_3 \end{aligned}\right\} \qquad \text{...(11·3)}$$

where r_{ij} is the coefficient of correlation between x_i and x_j for the N given pairs of values and σ_j is the standard deviation of x_j for the N values.

11·3. Determination of Regression Coefficients. Equation to Regression Plane Residuals.

Solving equations (11·3) for $b_{12\cdot3}$ and $b_{13\cdot2}$, we obtain

$$b_{12\cdot3} = \begin{vmatrix} r_{12}'\, \sigma_1 & r_{23}\, \sigma_3 \\ r_{13}\, \sigma_1 & \sigma_3 \end{vmatrix} \div \begin{vmatrix} \sigma_2 & r_{23}\, \sigma_3 \\ r_{23}\, \sigma_2 & \sigma_3 \end{vmatrix}$$

or
$$b_{12\cdot3}=\frac{-\dfrac{\sigma_1}{\sigma_2}\begin{vmatrix} r_{12} & r_{23}\\ r_{13} & 1\end{vmatrix}}{\begin{vmatrix} 1 & r_{23}\\ r_{23} & 1\end{vmatrix}}=-\frac{\sigma_1}{\sigma_2}\frac{\Delta_{12}}{\Delta_{11}} \qquad ...(11\cdot4)$$

and
$$b_{13\cdot2}=\frac{-\dfrac{\sigma_1}{\sigma_3}\begin{vmatrix} 1 & r_{12}\\ r_{23} & r_{13}\end{vmatrix}}{\begin{vmatrix} 1 & r_{23}\\ r_{23} & 1\end{vmatrix}}=-\frac{\sigma_1}{\sigma_3}\frac{\Delta_{13}}{\Delta_{11}} \qquad (11\cdot5)$$

where Δ_{ij} is the co-factor of the element in the ith row and jth column in the determinant

$$\Delta=\begin{vmatrix} 1 & r_{12} & r_{13}\\ r_{21} & 1 & r_{23}\\ r_{31} & r_{32} & 1\end{vmatrix}$$

Hence on substituting the values of $b_{12\cdot3}$ and $b_{13\cdot2}$ the equation to the regression plane of x_1 on x_2 and x_3 is

$$X_1=\left[-\frac{\sigma_1}{\sigma_2}\frac{\Delta_{12}}{\Delta_{11}}\right]x_2+\left[-\frac{\sigma_1}{\sigma_3}\frac{\Delta_{13}}{\Delta_{11}}\right]x_3 \qquad ...(11\cdot6)$$

Note. The quantity $x_{1\cdot23}=x_1-b_{12\cdot3}\,x_2-b_{13\cdot2}\,x_3$ is called the *residual* of second order. The order of the residual is the number of subscripts after the point. The quantities *b's* are called *partial regression coefficients.* The first subscript attached to the *b's* is the subscript of the letter on the left (the dependent variable), the second subscript is that of the x to which it is attached. These are called *primary* subscripts. The subscripts separated from the primary subscripts by a point are called the *secondary* subscripts.

A standard deviation denoted by a symbol with p secondary subscripts will be termed as *standard deviation of the pth order*, the standard deviations σ_1, σ_2, etc., being regarded as of order zero ; the standard deviations $\sigma_{1\cdot23}, \sigma_{2\cdot13}$ etc. of the second order and so on.

Ex. 1. Show that, if the origin is not necessarily at the mean, the equation of the plane of regression is

$$\begin{vmatrix} \dfrac{X_1-\bar{x}_1}{\sigma_1} & \dfrac{x_2-\bar{x}_2}{\sigma_2} & \dfrac{x_3-\bar{x}_3}{\sigma_3}\\ r_{21} & 1 & r_{23}\\ r_{31} & r_{32} & 1\end{vmatrix}=0$$

or
$$\frac{X_1-\bar{x}_1}{\sigma_1}\Delta_{11}+\frac{x_2-\bar{x}_2}{\sigma_2}\Delta_{12}+\frac{x_3-\bar{x}_3}{\sigma_3}\Delta_{13}=0$$

Ex. 2. Write the equations of the planes of regression of X_2 on x_1 and x_3, and of X_3 on x_1 and x_2.

Ex. 3. For p variates, show that the equation of the regression hyperplane of X_i on the rest of the $(p-1)$ variables is

$$\frac{x_1 - \bar{x}_1}{\sigma_1} \Delta_{i1} + \frac{x_2 - \bar{x}_2}{\sigma_2} \Delta_{i2} + \cdots + \frac{x_i - \bar{x}_i}{\sigma_i} \Delta_{ii} + \cdots + \frac{x_p - \bar{x}_p}{\sigma_p} \Delta_{ip} = 0$$

11·4. Properties of the Residuals.

I. Let the equation to the plane of regression of x_1 on x_2 and x_3 be

$$X_1 = b_{12 \cdot 3}\, x_2 + b_{13 \cdot 2}\, x_3$$

Then the normal equations for determining the b's are

$$\Sigma x_2\, x_{1 \cdot 23} = 0 = \Sigma x_3\, x_{1 \cdot 23}$$

Similarly, the equations for determining the b's are

$$\Sigma x_1\, x_{2 \cdot 13} = 0 = \Sigma x_3\, x_{2 \cdot 13}$$

$$\Sigma x_2\, x_{3 \cdot 12} = 0 = \Sigma x_1\, x_{3 \cdot 12}$$

These normal equations express the following result :

The sum of the products of corresponding values of a variable and a residual is zero (or the covariance of the variable and the residual is zero), provided the subscript of the variable occurs among the secondary subscripts of the residual.

II. Writing $x_{1 \cdot 2} = x_1 - b_{12} x_2$, we get by virtue of normal equations

$$\Sigma x_{1 \cdot 23}\, x_{1 \cdot 2} = \Sigma x_{1 \cdot 23}\, (x_1 - b_{12}\, x_2) = \Sigma x_{1 \cdot 23}\, x_1$$

and $$\Sigma x_{1 \cdot 23}\, x_{1 \cdot 23} = \Sigma x_{1 \cdot 23}\, (x_1 - b_{12 \cdot 3}\, x_2 - b_{13 \cdot 2}\, x_3) = \Sigma x_{1 \cdot 23}\, x_1.$$

Thus the sum of the products (or covariance) of two residuals is unaltered by omitting from one residual any or all of the secondary subscripts which are common to both.

III. By virtue of normal equations, we have

$$\Sigma x_{3 \cdot 2}\, x_{1 \cdot 23} = \Sigma (x_3 - b_{32} x_2)\, x_{1 \cdot 23} = 0$$

Similarly, $$\Sigma x_{2 \cdot 3}\, x_{1 \cdot 23} = 0$$

Thus it follows that the *sum of the products (or covariance) of two residuals is zero, provided all the subscripts of one residual occur among the secondary subscripts of the second residual.*

Ex. 1. Prove that (*i*) $\Sigma x_2\, x_{1 \cdot 23} = 0$, (*ii*) $\Sigma x_3\, x_{1 \cdot 23} = 0$

Ex. 2. For the distribution of p variates, prove the following results :

(*i*) $\Sigma x_k\, x_{1 \cdot 23 \ldots p} = 0, \; k = 2, 3, \cdots, p.$

(*ii*) $\Sigma x_{1 \cdot 34 \cdots p}\, x_{2 \cdot 3 \cdots p} = \Sigma x_{1 \cdot 34 \cdots p}\, x_2 = \Sigma x_1\, x_{2 \cdot 34 \cdots p}$

Ex. 3. Prove that

(*i*) $\Sigma x_{2\cdot 34 \cdots p}\, x_{1\cdot 23 \cdots p} = 0$

(*ii*) $\Sigma x_{2\cdot 34 \cdots p} (x_1 - b_{12\cdot 4 \cdots p}\, x_2 -) = 0$

Deduce that $b_{12\cdot 34 \cdots p} = (\Sigma x_{2\cdot 34 \cdots p} x_{1\cdot 34 \cdots p}) / \Sigma (x_{2\cdot 34 \cdots p})^2$

so that if q stands for the group of suffixes $3, 4, \cdots, p$

$$b_{12\cdot q} = \frac{\text{cov}(x_{1\cdot q}, x_{2\cdot q})}{\text{var}(x_{2\cdot q})}.$$

Similarly, show that

$$b_{31\cdot q} = \frac{\text{cov}(x_{1\cdot q}, x_{3\cdot q})}{\text{var}(x_{1\cdot q})}.$$

11·5 Variance of a Residual. We shall now obtain formula for the variance $\sigma_{1\cdot 23}^2$ of the residual $x_{1\cdot 23}$ (the deviation of the observed value x_1 from its corresponding computed value on the regression plane) in terms of σ_1^2 and the correlation coefficients.

We have

$$\begin{aligned} N\sigma_{1\cdot 23}^2 &= \Sigma x_{1\cdot 23}^2 = \Sigma x_1 x_{1\cdot 23} \\ &= \Sigma x_1 (x_1 - b_{12\cdot 3} x_2 - b_{13\cdot 2} x_3) \\ &= N\sigma_1^2 - N b_{12\cdot 3}\, \sigma_1 \sigma_2\, r_{12} - N b_{13\cdot 2}\, \sigma_1\, \sigma_2\, r_{13} \end{aligned}$$

or $$\sigma_1 \left(1 - \frac{\sigma_{1\cdot 23}^2}{\sigma_1^2} \right) = b_{12\cdot 3}\, \sigma_2\, r_{12} + b_{13\cdot 2} \sigma_3\, r_{13}$$

Eliminating $b_{12\cdot 3}$ and $b_{13\cdot 2}$ between this equation and equations (11·3), we have

$$\begin{vmatrix} 1 - \dfrac{\sigma_{1\cdot 23}^2}{\sigma_1^2} & r_{12} & r_{13} \\ r_{21} & 1 & r_{23} \\ r_{31} & r_{32} & 1 \end{vmatrix} = 0$$

or $\Delta - \dfrac{\sigma_{1\cdot 23}^2}{\sigma_1^2} \Delta_{11} = 0.$ Hence $$\sigma_{1\cdot 23}^2 = \sigma_1^2 \cdot \frac{\Delta}{\Delta_{11}} \qquad \text{...(11·7)}$$

Thus the variance of a residual of order 2 is expressible in terms of variance of zero order and correlations of zero order.

Similarly, show that the standard error of estimating x_1 from assigned terms of values of x_2 and x_3 is given by

$$\sigma_1\left(\frac{\Delta}{\Delta_{11}}\right)^{1/2}$$

where Δ_{11} is the cofactor of r_{11} in

$$\Delta=\begin{vmatrix} r_{11} & r_{12} & r_{13} \\ r_{21} & r_{22} & r_{23} \\ r_{31} & r_{32} & r_{33} \end{vmatrix}$$

Ex. Show that, for the distribution of p variates,

$$\sigma_{1\cdot 23\cdot p}^2=\left|\frac{\Delta}{\Delta_{11}}\right|\sigma_1^2$$

where

$$\Delta=\begin{vmatrix} 1 & r_{12} & r_{13}\cdots\cdots r_{1p} \\ r_{21} & 1 & r_{23}\cdots\cdots r_{2p} \\ \cdots & \cdots & \cdots \quad \cdots \\ r_{p1} & r_{p2} & r_{p3}\cdots\cdots 1 \end{vmatrix}$$

and Δ_{11} is the minor of the element in the first row and first column of Δ.

11·6. Multiple Correlation Coefficient. Consider the regression equation of x_1 on x_2 and x_3 *viz.*,

$$X_1=b_{12\cdot 3}\,x_2+b_{13\cdot 2}\,x_3=x_1-x_{1\cdot 23}$$

Next the correlation between x_1 (observed value of the variable and X_1 (expected value of the variable) is given by

$$R_{1\,(23)}=\frac{\Sigma\, x_1X_1}{\sqrt{(\Sigma\, x_1^2)\,(\Sigma\, X_1^2)}}$$

We have $\Sigma\, x_1\, X_1=\Sigma\,\{\, x_1\,(x_1-x_{1\cdot 23}\,)\}=\Sigma\, x_1^2-\Sigma\, x_1x_{1\cdot 23}$

$$=\Sigma\, x_1^2-\Sigma\, x_{1\cdot 23}^2 \qquad \text{by property II}$$

$$=N\,(\sigma_1^2-\sigma_{1\cdot 23}^2)$$

Also $\Sigma\, X_1^2=\Sigma\,(x_1-x_{1\cdot 23})^2=\Sigma\, x_1^2-2\,\Sigma\, x_1\, x_{1\cdot 23}+\Sigma\, x_1x_{1\cdot 23}$

$$=\Sigma\, x_1^2-\Sigma\, x_{1\cdot 23}^2 \qquad [\because\ \Sigma\, x_1\, x_{1\cdot 32}+\Sigma\, x_{1\cdot 32}^2=x_{1\cdot 23}^2]$$

$$=N\,(\sigma_1^2-\sigma_{1\cdot 23}^2\,)$$

$$\therefore\quad R_{1\,(23)}=\frac{\sigma_1^2-\sigma_{1\cdot 23}^2}{\sigma_1\sqrt{\sigma_1^2-\sigma_{1\cdot 23}^2}}=\left(1-\frac{\sigma_{1\cdot 23}^2}{\sigma_1^2}\right)^{\frac{1}{2}} \qquad ...(11\cdot 7')$$

or
$$1-R_{1\,(23)}^2=\frac{\sigma_{1\cdot 23}^2}{\sigma_1^2}=\frac{\Delta}{\Delta_{11}} \quad\Rightarrow R_{1(23)}^2=(\sigma_1^2-\sigma_{1\cdot 23}^2\,)/\sigma_1^2 \qquad ...(11\cdot 8)$$

From (11·7), $\sigma_1^2\ge\sigma_{1\cdot 23}^2$ and therefore from (11·7′), $R_{1\,(23)}\ge 0$.

11·6·1 Properties of Multiple Correlation Coefficient.

1. $R_{1(23)}$ is necessarily positive or zero *i.e.* non-negative. Thus

$$0 \le R_{1(23)} \le 1$$

2. If $R_{1(23)} = 1$, then from (11·8) we have $\sigma^2_{1\cdot 23} = 0$, *i.e.*, all the residuals $x_{1\cdot 23}$ are zero, the observed and expected values of x_1 coincide, and thus the observed x_1 is a linear function of x_2 and x_3.

3. It measures the closeness of the association between the observed values and the expected values of a variable obtained from the multiple linear regression of that variable on other variables.

4. $R^2 \ge R_0^2$ where R = multiple correlation coefficient of x_1 on $x_2, x_3, \cdots, x_p$ and R_0 = correlation of x_1 with any arbitrary linear function of $x_2, x_3, \cdots, x_p$. [See Q 25 (*i*) Page 438]

Ex. 1. Establish the following alternative forms for $R_{1(23\cdots p)}$, the multiple correlation coefficient between x_1 and $x_2, x_3 \cdots\cdots, x_p$:

(*i*) $R^2_{1(23\cdots p)} = 1 - (\sigma^2_{1\cdot 23\cdots p}/\sigma_1^2)$

(*ii*) $1 - R^2_{1(23\cdots p)} = \dfrac{\Delta}{\Delta_{11}}$

where Δ and Δ_{11} have the usual meanings.

(*iii*) $R^2_{1(23)} = b_{12\cdot 3}\, r_{12} \dfrac{\sigma_2}{\sigma_1} + b_{13\cdot 2}\, r_{13} \dfrac{\sigma_3}{\sigma_1}$ ***(B.Sc. Sardar Patel, 1991)***

Ex. 2. For a trivariate population prove that the coefficient of multiple correlation $R_{1(23)}$ satisfies the following relation :

$$R^2_{1(23)} = \frac{r_{12}^2 + r_{13}^2 - 2r_{12}r_{23}r_{31}}{1 - r_{23}^2} \ge r_{12}^2.$$

11·7. Partial Correlation Coefficient. Let us find the partial correlation coefficient $r_{12\cdot 3}$ which may be defined as the correlation coefficient between $x_{1\cdot 3}$ and $x_{2\cdot 3}$ *i.e.*, between x_1 and x_2 after the linear effect of the third variable x_3 has been eliminated from both x_1 and x_2.

We have, by property III,

$$\begin{aligned} 0 = \Sigma\, x_{2\cdot 3}\, x_{1\cdot 23} &= \Sigma\, x_{2\cdot 3}\,(x_1 - b_{12\cdot 3}\, x_2 - b_{13\cdot 2} x_3) \\ &= \Sigma\, x_1 x_{2\cdot 3} - b_{12\cdot 3}\, \Sigma\, x^2_{2\cdot 3} \\ &= \Sigma\, x_{1\cdot 3}\, x_{2\cdot 3} - b_{12\cdot 3}\, \Sigma\, x^2_{2\cdot 3} \end{aligned}$$

This gives $b_{12\cdot 3} = \dfrac{\Sigma\, x_{1\cdot 3}\, x_{2\cdot 3}}{\Sigma\, x^2_{2\cdot 3}}$ ...(11·9)

This shows that $b_{12\cdot 3}$ is the coefficient of regression of $x_{1\cdot 3}$ on $x_{2\cdot 3}$.

Similarly, $b_{21\cdot3}$ is the coefficient of regression of $x_{2\cdot3}$ on $x_{1\cdot3}$ so that $r_{12\cdot3}$, the coefficient of correlation between $x_{1\cdot3}$ and $x_{2\cdot3}$ is given by

$$r_{12\cdot3}^2 = b_{12\cdot3} b_{21\cdot3} = \frac{\Delta_{12}}{\Delta_{11}} \cdot \frac{\Delta_{21}}{\Delta_{22}} = \frac{\Delta_{12}^2}{\Delta_{11}\Delta_{22}}$$

Since $r_{12\cdot3}$ has the same sign as $b_{12\cdot3}$, which is the sign of $-\Delta_{12}$ by (11·4), we have

$$r_{12\cdot3} = -\frac{\Delta_{12}}{\sqrt{\Delta_{11}\,\Delta_{22}}} = \frac{r_{12} - r_{13}\, r_{23}}{\sqrt{(1 - r_{13}^2)\,(1 - r_{23}^2)}} \qquad \text{...(11·10)}$$

Note. Formula for $r_{23\cdot1}$ and $r_{31\cdot2}$ can be obtained similarly. We observe that the partial correlation coefficients of the first order can be expressed in terms of the total correlation coefficients.

Cor. Since $r_{1\cdot23}^2 \le 1$., we have $(r_{12} - r_{13}\, r_{23})^2 \le (1 - r_{13}^2)\,(1 - r_{23}^2)$

that is $$r_{12}^2 + r_{13}^2 + r_{23}^2 - 2r_{12}\, r_{13}\, r_{23} \le 1$$

If we take r_{12}, r_{13} as known, this gives as limits for r_{23} :

$$r_{12}\, r_{13} \pm \sqrt{1 - r_{12}^2 - r_{13}^2 + r_{12}^2\, r_{13}^2} \qquad \text{...(11·11)}$$

Note. Writing down similar equations for $b_{lm\cdot qp}$ and $b_{ml\cdot qp}$ multiplying and then taking square root, we can show that

$$r_{lm\cdot qp} = \frac{r_{lm\cdot q} - r_{lp\cdot q}\, r_{mp\cdot q}}{\sqrt{(1 - r_{lp\cdot q}^2)\,(1 - r_{mp\cdot q}^2)}} \qquad \text{...(11·12)}$$

l, m, q, p are unequal.

given the correlation coefficient in terms of those of lower order.

Ex. 1. Defining $r_{12\cdot q}$, the product moment coefficient of correlation between $x_{1\cdot q}$ and $x_{2\cdot q}$ as

$$r_{12\cdot q} = (b_{12\cdot q}\, b_{21\cdot q})^{1/2}$$

and using Ex. 3, page 428, show that

$$r_{21\cdot q} = \frac{\text{cov}\,(x_{1\cdot q}, x_{2\cdot q})}{[\,\text{var}\,(x_{1\cdot q})\ \text{var}\,(x_{2\cdot q})]^{1/2}}.$$

Ex. 2. Prove that for a trivariate distribution,

(*i*) $b_{12\cdot3} = \dfrac{b_{12} - b_{13}\, b_{32}}{1 - b_{23}\, b_{32}}$

(*ii*) $r_{12\cdot3}\dfrac{\sigma_{1\cdot3}}{\sigma_{2\cdot3}} = \dfrac{\sigma_1}{\sigma_2}\left(\dfrac{r_{12} - r_{13}\, r_{23}}{1 - r_{23}^2}\right).$

Ex. 11·1. *If* $r_{12} = +0{\cdot}80$, $r_{13} = -0{\cdot}40$, $r_{23} = -0{\cdot}56$, *find the values of* $r_{12\cdot3}$, $r_{13\cdot2}$ and $r_{23\cdot1}$.

We have $r_{12\cdot3} = \dfrac{r_{12} - r_{13}\,r_{23}}{\sqrt{(1 - r_{13}^2)\,(1 - r_{23}^2)}}$

$$= \frac{0{\cdot}80 - (-0{\cdot}40)\,(-0{\cdot}56)}{\sqrt{\{1 - (0{\cdot}40)^2)\}\,\{1 - (0{\cdot}56)^2\}}}$$

$$= 0{\cdot}759$$

Similarly, $r_{13\cdot2} \dfrac{r_{13} - r_{12}\,r_{32}}{\sqrt{(1 - r_{12}^2)\,(1 - r_{32}^2)}}$

$$= 0.097$$

and $r_{23\cdot1} = -0{\cdot}436.$

Ex. 1. if $r_{12} = 0{\cdot}86, r_{13} = 0{\cdot}65, r_{23} = 0{\cdot}72$, show that $r_{12\cdot3} = 0{\cdot}743$.

Ex. 2. Show that the values $r_{12} = {\cdot}6, r_{13} = {\cdot}4, r_{23} = -{\cdot}7$ are inconsistent.

Ex. 11·2. *If the correlation coefficient of zero order in a set of p variables were equal to* ρ, *show that*

(*a*) *every partial correlation of the sth order is* $\rho/(1 + s\rho)$; *and*

(*I.A.S. 1993*)

(*b*) *the coefficients of multiple correlation R of a variable with the other (p – 1) variables is given by*

$$1 - R^2 = (1 - \rho)\cdot\frac{[1 + (p - 1)\rho]}{1 + (p - 2)\rho}$$

(*a*) We are given that

$$r_{ij} = \rho,\ i, j = 1, 2, \cdots, p,\ i \neq j$$

We have $r_{lm\cdot h} = \dfrac{r_{lm} - r_{lh}\,r_{mh}}{\sqrt{(1 - r_{mh}^2)\,(1 - r_{lh}^2)}}\ l, m, h = 1, 2, \cdots, p, l \neq m \neq h$

$$= \frac{\rho - \rho\cdot\rho}{\sqrt{(1 - \rho^2)\,(1 - \rho^2)}} = \frac{\rho}{1 + \rho}$$

Thus every partial correlation of the 1st order is $\rho/(1 + \rho)$.

$$r_{lm\cdot hk} = \frac{r_{lm\cdot k} - r_{lh\cdot k}\,r_{mh\cdot k}}{\sqrt{(1 - r_{lh\cdot k}^2)\,(1 - r_{mh\cdot k}^2)}}$$

$$= \frac{\dfrac{\rho}{1 + \rho} - \dfrac{\rho}{1 + \rho}\cdot\dfrac{\rho}{1 + \rho}}{\sqrt{\left[1 - \left(\dfrac{1}{1 + \rho}\right)^2\right]\left[1 - \left(\dfrac{\rho}{1 + \rho}\right)^2\right]}} = \frac{\rho}{1 + 2\rho}$$

that is, every partial correlation coefficient of the second order is $\rho/(1 + 2\rho)$.

Now assuming that every correlation coefficient of order s is $\rho/(1+s\rho)$, the partial correlation coefficient of order $s+1$ would be

$$=\frac{\frac{\rho}{1+sp}-\frac{\rho}{1+sp}\cdot\frac{\rho}{1+sp}}{\sqrt{1-\left(\frac{\rho}{1+sp}\right)^2}\sqrt{1-\left(\frac{\rho}{1+sp}\right)^2}}=\frac{\rho}{1+(s+1)\rho}.$$

Thus the result is proved by *Induction.*

(*b*) Since all the arguments hold good for a distribution of p variables, we have

$$1-R^2=\frac{\Delta}{\Delta_{11}}$$

where Δ is the determinant which has its j th element of the ith row equal to r_{ij} $(i, j = 1, 2, \ldots\ldots, p)$, $r_{ij}=1$, $i=j$ and Δ_{11} is the determinant obtained by omitting the first row and the first column of Δ.

Since all the correlation coefficients of zero order are equal to ρ, we have

$$\Delta=\begin{vmatrix} 1 & \rho\ldots\ldots\ldots\ldots\rho \\ \rho & 1\ldots\ldots\ldots\ldots\rho \\ \ldots\ldots & \ldots\ldots\ldots\ldots \\ \rho & \rho\ldots\ldots\ldots\ldots 1 \end{vmatrix}=[1+(p-1)\rho]\,(1-\rho)^{p-1}$$

and then clearly $\Delta_{11}=[1+(p-2)\,\rho]\,(1-\rho)^{p-2}$.

Hence $\quad 1-R^2=(1-\rho)\dfrac{[\,1+(p-1)\,\rho\,]}{1+(p-2)\,\rho}.$

[For another proof of this result, see Ex. 11, page 434]

Exercises 11

1. Prove the identity :

$b_{12\cdot3}\, b_{23\cdot1}\, b_{31\cdot2}=r_{12\cdot3}\, r_{23\cdot1}\, r_{31\cdot2}$

(B.Sc. Gujarat, 1992, B.A. Hons., Delhi, 1983)

[**Hint.** $b_{12\cdot3}=\dfrac{\Sigma x_{1\cdot3}\, x_{2\cdot3}}{\Sigma x_{2\cdot3}^2}=\dfrac{r_{12\cdot3}\,\sigma_{1\cdot3}}{\sigma_{2\cdot3}}$. Similarly for $b_{23\cdot1}$ and $b_{31\cdot2}$]

2. Show that $\sigma_{1\cdot23}\,\sigma_{2\cdot13}/r_{12\cdot3}=-\Delta_{11}\,\sigma_1\,\sigma_2/\Delta_{12}$.

3. Is it possible to get the following from a set of experimental data ?

$r_{12}=0{\cdot}6, r_{23}=0{\cdot}8, r_{31}=-0{\cdot}5$. [**Ans.** No]

4. If the relation $ax_1+bx_2+cx_3=0$ holds for all sets of values of x_1, x_3 and x_2, what must be the partial correlations ?

[**Ans.** $r_{12\cdot3}=r_{31\cdot2}=r_{23\cdot1}=-1$]

5. Show that if $x_3 = ax_1 + bx_2$, the three partial correlations are numerically equal to unity, $r_{13\cdot2}$ having the sign of a, $r_{23\cdot1}$ the sign of b and $r_{12\cdot3}$ the opposite sign of a/b.

6. X, Y, Z are independent random variables with the same variance. If

$$X_1 = \frac{1}{\sqrt{2}}(X - Z);$$

$$X_2 = \frac{1}{\sqrt{3}}(X + Y + Z),$$

$$X_3 = \frac{1}{\sqrt{6}}(X + 2Y + Z)$$

show that X_1, X_2, X_3 have equal variances. Calculate $r_{12\cdot3}$ and $R_{1(23)}$.

7. Show that correlation coefficient between the residuals $x_{1\cdot23}$ and $x_{2\cdot13}$ is equal and opposite to that between $x_{1\cdot3}$ and $x_{2\cdot3}$. ***(B.Sc., Punjab, 1991)***

[**Hint.** Coeff. of correlation between $x_{2\cdot13}$ and $x_{1\cdot23}$

$$= (1/N)\,\Sigma\, x_{2\cdot13}\, x_{1\cdot23}/\sigma_{2\cdot13}\,\sigma_{1\cdot23} = (1/N)\,\Sigma\, x_{2\cdot13}\,(x_1 - b_{12\cdot3}\,x_2 - b_{13\cdot2}\,x_3)/\sigma_{2\cdot13}\,\sigma_{1\cdot23}$$

$$= [\,0 - b_{12\cdot3}\,\Sigma\, x_{2\cdot13}\, x_2 - 0\,] \div [\,(1/N)\,\sigma_{2\cdot13}\,\sigma_{1\cdot23}\,] = etc.]$$

8. (*a*) Show that for given r_{12} and r_{13}, r_{23} must lie in the range

$$r_{12}\, r_{13} \pm (1 - r_{12}^2 - r_{13}^2 + r_{12}^2\, r_{13}^2)^{1/2}$$

(*b*) If $r_{12} = k$, $r_{13} = -k$, show that r_{23} will lie between -1 and $1 - 2k^2$.

[**Hint.** $r_{12\cdot3}^2 \le 1$ gives on simplification $r_{13}^2 + 2k^2\, r_{13} + (2k^2 - 1) \le 0$.]

9. Show that $R_{1\,(23)} \ge r_{12}$. What is the relation between $R_{1\,(23)}$ and r, the coefficient of correlation between two variables ?

[**Hint.** $N\sigma_{1\cdot23}^2 = \Sigma\, x_{1\cdot23}^2 = \Sigma\, x_{1\cdot23}\,(x_1 - b_{12\cdot3}\, x_2 - b_{13\cdot2}\, x_3\,)$.

$$= \Sigma\, x_{1\cdot2}^2 - b_{13\cdot2}\,\Sigma\, x_{1\cdot2}\, x_{3\cdot2} \qquad \text{using Property II}$$

$$= N\sigma_{1\cdot2}^2\,(1 - b_{13\cdot2}\, b_{31\cdot2})$$

$$\therefore \quad \sigma_{1\cdot23}^2 = \sigma_1^2\,(1 - r_{12}^2\,)\,(1 - r_{13\cdot2}^2\,).$$

From equation (11·8) $\sigma_{1\cdot23}^2 = \sigma_1^2\,[1 - R_{1\,(23)}^2]$

$\therefore\ 1 - R_{1(23)}^2 = (1 - r_{12}^2\,)\,(1 - r_{13\cdot2}^2\,) \le 1 - r_{12}^2$. Hence etc.]

10. Establish the following results :

(*i*) $\sigma_{1\cdot23\cdots p}^2 = \sigma_1^2\,(1 - r_{12}^2\,)\,(1 - r_{13\cdot2}^2) \cdots (1 - r_{1p\cdot23\cdots\overline{p-1}}^2)$

(*ii*) $1 - R_{1\,(23\cdots p\,)}^2 = (1 - r_{12}^2)\,(1 - r_{13\cdot2}^2\,) \cdots (1 - r_{1p\cdot23\cdots\overline{p-1}}^2\,)$

(M.Sc. Stat. Delhi, 1987)

Hence show that if the multiple correlation coefficient $R_{1(23\cdots p)}$ is equal to zero, x_1 is uncorrelated with any of the other variables.

11. If all the correlation coefficients of zero order in a set of p variables were equal to ρ, then (*i*) every partial correlation of the *s*th order is $\rho/(1+s\rho)$,

and (*ii*) $1-R^2_{1(23\cdots p)}=\dfrac{(1-\rho)\,[1+(p-1)\,\rho]}{1+(p-2)\,\rho}$.

[**Hint.** (*i*) In Ex. 10 (*ii*) above, put $r_{12}=\rho$, $r_{13\cdot 2}=\dfrac{\rho}{1+\rho}$, etc. and simplify].

12. Show that $1-r^2_{1p\cdot 23\cdots \overline{p-1}}=[1-R^2_{1(23\cdots p)}]\,/[1-R^2_{1\,(23\cdots p-1)}]$

which shows how a partial correlation coefficient of order $p-2$ may be expressed in terms of multiple correlation coefficient of order $p-1$ and $p-2$.

13. (*a*) Show that for *p* variates, there are ${}^pC_2$ correlation coefficients of order zero and $({}^{p-2}C_2)\,({}^pC_2)$ of order *s*. Show further that there are

(*i*) $({}^pC_2)\,2^{p-2}$ correlation coefficients and

(*ii*) $({}^pC_2)\,2^{p-1}$ regression coefficients.

(*b*) Show that if r_{12} is zero, $r_{12\cdot 3}$ will not be zero unless one at least of r_{13} and r_{23} is zero.

14. In a *p*-variate distribution, all the total (order zero) correlation coefficients are equal to $\rho_0\neq 0$. Let ρ_k denote the partial correlation coefficient of order *k* and R_k be the multiple correlation coefficient of one variate on *k* other variates. Prove that

(*i*) $\rho_0\geq -\dfrac{1}{p-1}$ (*ii*) $\rho_k-\rho_{k-1}=-\rho_{k-1}\,\rho_k$ and

(*iii*) $R_k^2=\dfrac{k\rho_0^2}{1+(k-1)\,\rho_0}$ (*M. Sc. Stat. Delhi, 1987*)

[**Hint.** In Ex. 11·2 we have proved that

$$\rho_k=\frac{\rho_0}{1+k\,\rho_0}$$

$$\therefore\qquad |\,\rho_{p-2}\,|=\left|\frac{\rho_0}{1+(p-2)\,\rho_0}\right|$$

But $|\,\rho_{p-2}\,|\leq 1\quad\therefore\ -1\leq\dfrac{\rho_0}{1+(p-2)\,\rho_0}$ which gives $\rho_0\geq\dfrac{-1}{(p-1)}$

Also from Ex. 11·2, $1-R_k^2=(1-\rho_0)\left[\dfrac{1+k\,\rho_0}{1+(k-1)\,\rho_0}\right]$

$$\Rightarrow R_k^2=1-\frac{(1-\rho_0)\,(1+k\,\rho_0)}{1+(k-1)\,\rho_0}=\frac{k\,\rho_0^2}{1+(k-1)\,\rho_0}$$

15. Prove the formulae :

(*a*) $b_{12}=\dfrac{b_{12\cdot 3}+b_{13\cdot 2}\;b_{21\cdot 3}\;b_{32\cdot 1}}{1-b_{13\cdot 2}\;b_{23\cdot 1}}$

(*b*) $r_{12}=\dfrac{r_{12\cdot 3}-r_{13\cdot 2}\,r_{23\cdot 1}}{\sqrt{(1-r^2_{13\cdot 2})\,(1-r^2_{23\cdot 1})}}$

16. If x_1, x_2 and x_3 are three variables measured from their respective means as origin and if e_1 is the expected value of x_1 for given values of x_2 and x_3 from the linear regression of x_2 and x_3, prove that

$\text{Cov}(x_1, e_1) = \text{Var}(e_1) = \text{Var}(x_1) = \text{Var}(x_1 - e_1).$

17. Prove that the necessary and sufficient condition for the coincidence of the three regression planes for a trivariate distribution is

$$r_{12}^2 + r_{13}^2 + r_{23}^2 - 2r_{12}\, r_{13}\, r_{23} = 1.$$

18. If $r_{23} = 1$, prove that

(i) $r_{12}^2 = r_{13}^2$.

(ii) $\sigma_{1\cdot 23}^2 = \sigma_1^2 (1 - r_{12}^2)$.

19. (a) If $r_{23} = 0$, prove that

(i) $R_{1(23)}^2 = r_{12}^2 + r_{13}^2$.

(ii) $\sigma_{1\cdot 23}^2 = \sigma_1^2 (1 - r_{12}^2 - r_{13}^2)$.

(b) If $R_{1\cdot 23} = 0$, prove that $r_{12} = r_{13} = 0$.

[**Hint.** $1 - R_{1(23)}^2 = (1 - r_{12}^2)(1 - r_{13\cdot 2}^2)$]

20. If $x_1 = y_1 + y_2, x_2 = y_2 + y_3, x_3 = y_3 + y_1$.

where y_1, y_2, y_3 are uncorrelated variables and each of which has zero mean and unit *s.d.*, find the multiple correlation coefficient between x_1 and two variables x_2 and x_3.

Ans. $R_{1\cdot 23} = \left(1 - \dfrac{\Delta}{\Delta_1}\right)^{1/2} = \dfrac{1}{\sqrt{3}}$

21. $x_0, x_1, \ldots\ldots, x_n$ are $(n + 1)$ variates. Obtain a linear function of $x_1, x_2, \ldots\ldots, x_n$ which has maximum correlation with x_0. Show that the correlation R of x_0 with the linear function is given by

$$R = \left(1 - \frac{\Delta}{\Delta_{11}}\right)^{1/2}$$

where Δ, Δ_{11} have usual significance.

22. For a trivariate distribution, show that

$$1 + 2\, r_{12}\, r_{13}\, r_{23} \geq r_{12}^2 + r_{23}^2 + r_{13}^2$$

where r_{12} is correlation coefficient between the random variables x_1 and x_2, etc.

23. Five hundred students were examined in three subjects I, II and III, each subject carrying 100 marks. A student getting 120 or more, but less than 150 marks was put in pass class, a student getting 150 or more, but less than 180 was put in second class and a student getting 180 or more marks was put in first class. The following results were obtained :

	I	*II*	*III*
Mean	35·8	52·4	48·8
S.D.	4·2	5·3	6·1
Corr. Coeff.	$r_{12} = 0{\cdot}6$	$r_{13} = 0{\cdot}7$	$r_{23} = 0{\cdot}8$

(*i*) Find the number of students in each of the three classes.

(*ii*) Estimate the number of students with total marks lying between 180 and 190.

(*iii*) Calculate the proportion of students who failed.

(*iv*) If r_{23} is not known, obtain the limits within which it may lie when r_{12} and r_{13} have the given values.

Solution. Let Z denote the total marks of the students in the three subjects and X_1, X_2, X_3 the total marks of the students in subjects, I, II and III respectively.

$$Z = X_1 + X_2 + X_3$$

$$\therefore \quad EZ = EX_1 + EX_2 + EX_3 = 35{\cdot}8 + 52{\cdot}4 + 48{\cdot}8 = 137$$

$$\text{Var } Z = \text{Var } X_1 + \text{Var } X_2 + \text{Var } X_3 + 2 \text{ Cov } (X_1, X_2) + 2\text{Cov } (X_2, X_3) + 2\text{Cov } (X_3, X_1)$$

$$= 17{\cdot}64 + 28{\cdot}09 + 37{\cdot}21 + 26{\cdot}712 + 35{\cdot}868 + 51{\cdot}728$$

$$= 197{\cdot}248$$

$$\therefore \quad \sigma_z = 14{\cdot}045 \quad \therefore \quad t = \frac{Z - E(Z)}{\sigma_Z} \sim N(0, 1)$$

Z	t	$p = \int_{-\infty}^{t} p(t)\, dt$	Class	Area under the curve in class (A)	Frequency 500 × (A)
120	1·21050	0·11314	120—150	0·70937	354·685
150	0·92567	0·82251	150—180	0·17636	88·195
180	3·06180	0·99890	180—190	0·00102	0·510
190	3·77400	0·99992	190—240	0·88678	443·390
240	7·33410	1·00000	240—	0·00000	0·000

(*i*) The number of students in the first, second and third classes are approx, 355, 88 and 0.

(*ii*) Total number of students with total marks between 120—190 is 443.

(*iii*) Probability that student gets more than 240 marks is zero.

(*iv*) The corr. coeff. between marks in subject I and II of the students, who secured equal marks in subjects III is $r_{12{\cdot}3}$.

Hence $$r_{12{\cdot}3} = \frac{r_{12} - r_{23}\, r_{13}}{\sqrt{(1 - r_{23}^2)(1 - r_{13}^2)}} = \text{etc.}$$

(*v*) By (11·11), we can find the limits for r_{23}.

24. In a certain examination, there are three papers I, II and III, each of 100 marks. The distribution of marks in each paper is normal with the following means, s.d. and intercorrelation.

	I	*II*	*III*
Mean	53	43	48
S.D.	11	8	9

$$r_{12} = 0{\cdot}8,\ r_{13} = 0{\cdot}3,\ r_{23} = 0{\cdot}4$$

The first class, second class and pass class marks, are on the total 60%, 50% and 40%. Find the percentage of students who will pass in each class.

25. Prove ;

(*i*) $R^2 \geq R_0^2$ where R = multiple correlation coefficient of x_1 on $x_2, x_3, \cdots x_p$ and R_0 = correlation of x_1 with any arbitrary linear function of $x_2, x_3, \ldots\ldots, x_p$.

(*ii*) $R^2 \geq R_1^2$ where R_1 = multiple correlation of x_1 with $x_2, x_3 \ldots\ldots, x_k$, $k < p$.

(*iii*) $1 - R^2 = \prod_{j=2}^{p} \left(1 - r^2_{ij\cdot 23 \cdots j-1}\right)$

26. Define $r_{1(2\cdot 34 \ldots\ldots p)}$ as the correlation between x_1 and the residual of x_2 removing the regression on $x_3, x_4, \ldots\ldots, x_p$, show that

(*i*) $r^2_{1\,(2\cdot 34 \ldots\ldots p)} \leq r^2_{12\cdot(34 \ldots\ldots p)}$

(*ii*) $r^2_{1\,(2\cdot 34 \ldots\ldots p)} = r^2_{1p} + r^2_{1\,(p-1.p)} + \cdots\cdots + r^2_{1\,(2\cdot 34 \ldots p)}$

(M.Sc., Banaras, 1989)

27. If X_1, X_2 follow bivariate normal law with zero means and variances σ_1^2, σ_2^2 respectively and ρ is the correlation, show that rth semi-invariant K_r of X_1, X_2 is given by

$$K_r(X_1, X_2) = \tfrac{1}{2}(r-1)!\,\sigma_1^2\,\sigma_2^2\,\{(\rho-1)^r + (\rho+1)^r\}.$$

(M.Sc., Banaras, 1988)

28. X_1, X_2 and X_3 are three correlated variables where

$$S_1 = \sqrt{\frac{1}{3}\Sigma(X_1 - \overline{X}_1)^2} = 1,\ S_2 = \sqrt{\frac{1}{3}\Sigma(X_2 - \overline{X}_2)^2} = 1{\cdot}3,$$

$$S_3 = \sqrt{\frac{1}{3}\Sigma(X_3 - \overline{X}_3)^2} = 1{\cdot}9.$$

and $r_{12} = 0{\cdot}370$, $r_{13} = -0{\cdot}641$ and $r_{23} = -0{\cdot}736$. Compute $r_{13\cdot 2}$. If $X_4 = X_1 + X_2$, obtain $r_{42\cdot 3}$ and $r_{43\cdot 2}$.

29. Show that (*i*) $\sigma_1 \geq \sigma_{1\cdot 2} \geq \sigma_{1\cdot 23} \geq \ldots\ldots \geq \sigma_{1\cdot 23 \ldots n}$ **(M.Sc. Stat. Delhi, 1987)**

(*ii*) $R_{1\cdot 2} \leq R_{1\cdot 23} \leq \ldots\ldots \leq R_{1\cdot 23 \ldots\ldots n}$ **(M.Sc. Stat. Delhi, 1989)**

30. In a p-variate distribution, all the total (order zero) correlation coefficient are equal to $\rho_0 \neq 0$. Let ρ_k denote the parttial corelation coefficient of order k and

R_k be the multiple correlation coefficientt of one variable on k, other variables. Prove that

(*i*) $\rho_0 \geq -\dfrac{1}{p-1}$ (*ii*) $\rho_k - \rho_{k-1} = -\rho_k \rho_{k-1}$ and

(*iii*) $R_k^2 = \dfrac{k\rho_0^2}{1+(k-1)\rho_0}$ **(*M.Sc. Stat. Delhi, 1989*)**

[**Hint.** We have earlier proved that $\rho_k = \dfrac{\rho_0}{1+k\rho_0}$

$$\rho_{k-2} = \frac{\rho_0}{1+(p-2)\rho_0}$$

Since $\quad |\rho_{p-2}| \leq | \Rightarrow -| \leq \rho_{p-2} \leq |$

Hence $-| \leq \rho_{p-2} \leq \dfrac{\rho_0}{1+(p-2)\rho_0} \Rightarrow \rho_0 \geq \dfrac{-1}{p-1}$

(*iii*) Take $\rho = \rho_0$ and $k = p-1$ so that

$$1 - R_k^2 = (1-\rho_0)\left[\frac{1+k\rho_0}{1+(k-1)\rho_0}\right] \quad \therefore \text{ etc.}]$$

31. Let x_1 and x_2 be r.v. with means 0, variances 1 and correlation coeff. ρ. Show that

$$E[\max(x_1^2, x_2^2)] \leq 1 + \sqrt{1-\rho^2}.$$

32. If $X_1 = Y_1 + Y_2, X_2 = Y_2 + Y_3$ and $X_3 = Y_3 + Y_1$, where $Y_1, Y_2\ Y_3$ are uncorrelated variates with mean zero and variance unity, find the multiple correlation of X_1 on Y_2 and X_3 and partial correlation of X_1 and X_2.

12

Theory of Sampling

12·1. Sampling. Before we consider sampling theory, we first define the term population. The word '*population*' or *universe*' in Statistics is used to refer to any collection of individuals or of their attributes or of results of operations, which can be numerically specified. Thus, we may speak of the populations of weights, heights, prices of wheat, mileages of automobile tyres, etc.

A population containing a finite number of individuals or members is called a *finite population*. Such, for instance, is the population of ages of twenty boys in a class. A population with infinite number of members is known as *infinite population*. The population of pressures at various points in the atmosphere is an example of this type of population. We call the population of concrete individuals an *existent population*, while the collection of all possible ways in which an event can materialize, as the *hypothetical population*. The population of heads and tails, obtained by tossing a coin an infinite number of times, is a hypothetical population. A part or small section selected from the population is called a *sample* and the process of such selection is called *sampling*. Sampling is resorted to when either it is impossible to enumerate the whole population or when it is too costly to enumerate in terms of time and money or when the uncertainty inherent in sampling is more then compensated by the possibilities of errors in complete enumeration. To serve a useful purpose, sampling should be unbiased or representative.

The aim of the theory of sampling is to get as much information as possible, ideally the whole of the information, about the population from which the sample has been drawn. In particular, given the form of the parent population, we would like to *estimate* the parameters of the population or specify the limits within which the population parameters are expected to lie, with a specified degree of confidence. It is, however, to be clearly understood that the logic of the theory of sampling is the logic of induction, that is we pass from particular (*i.e.*, sample) to general (*i.e.*, population) and hence all results will have to be expressed in terms of probability.

The fundamental assumption underlying most of the theory of sampling is *random sampling*, which consists in selecting the individuals from the populations, in such a way that each individual of the population has the same chance of being selected. The selection of random samples is very much facilitated by the use of *table of random numbers* due to Tippet, Kendall, Smith, Fisher and Yates or Rand Corporation (U.S.A.)

Tippet's Random Numbers. The random number table, as given by L.H.C. Tippet, consists of 10,400 four-digit numbers, so that there are 41,600 digits. These numbers were taken from the census report in U.K. Any number taken either from row or column or diagonal of these random number table, which is random in character, may be taken to constitute the sample.

Types of Sampling. Some of the important types of sampling are :—

(*i*) Purposive sampling.

(*ii*) Random sampling.

(*iii*) Simple sampling.

(*iv*) Stratified sampling.

12·2. Sampling of Attributes. The sampling of attributes may be regarded as the drawing of samples from a population whose members possess attribute *A* or not – *A*. For example, in sampling from a population of persons, we may have persons who are smokers and non-smokers. The choosing or drawing of an individual in sampling may be called an 'event' or 'trial' and the possession of the specified attribute *A* by the individual selected a 'success'.

By *simple sampling* we mean random sampling in which each event has the same probability p of success and the probability of an event is independent of the success or failure or events in the preceding trials. Thus, simple sampling is a special case of random sampling, in which trials are independent and probability of success is constant. For example, counting the number of success in the throwing of a dice or tossing of a coin is a case of simple sampling, since the probability of getting heads with a coin is unaffected by the previous trials and remains constant irrespective of the number of trials made provided the coin remains unbiased. Following is the illustration of sampling which is random but not simple.

Suppose in an urn, there are 2 white balls and 3 black balls. Now the probability of drawing a white ball is $\frac{2}{5}$. Suppose, at the first trial we draw a white ball. The probability of getting a white ball at the second trial, when the ball is not replaced, is $\frac{1}{4}$, which is clearly not the same as the probability in the first trial. Thus the sampling, though random, is not simple. However, if the ball drawn at the first trial was put back in the urn before the next trial, success or failure having been noted, the random sampling would become simple sampling. The two cases are sometimes distinguished as *sampling without replacement* and *sampling with replacement*.

It should be clear that random sampling from an infinite population would always be *simple*, since drawing of one individual cannot effect the probability distribution but when sampling from a finite population, the sampling may or may not be simple according as the members drawn are replaced or not.

12·2·1. Mean and Standard Deviation in Simple Sampling of Attributes. Suppose we take a simple sample of n members. This is clearly identical with that of a series of n independent trials, with constant probability p of success. The probabilities of 0, 1, 2, ..., n successes are the terms in the binomial expansion of $(q+p)^n$, $q = 1 - p$. The distribution so obtained is called *sampling distribution* of the number of successes in the sample. The expected value, or the mean value, of the member of successes is, therefore, np and the standard deviation [also called the *standard error* (s.e.)] of the number of successes is $\sqrt{npq}$. By the *proportion of successes* in a sample, we mean the number of successes divided by the number of members in the sample. The mean and the standard error of the proportion of successes can be obtained by dividing the corresponding results for the number of successes by n. Thus

mean of the proportion of successes = p

and e = *s.e* of the proportion of successes = $\sqrt{pq/n}$

Hence *precision* (reciprocal of *s.e.*) of the proportion of successes $= \sqrt{n/pq}$, which is proportional to $\sqrt{n}$, since p and (therefore) q are constants. In other words, *the precision of the proportion of the successes varies as the square root of the number of members in the simple sample.*

12·3. Large Samples. Testing the Significance for a single Proportion. Suppose a large number n of independent Bernoullian trials is performed and x successes are obtained. We wish to test the hypothesis that the probability of success in each trial is p. Assuming the hypothesis to be correct, the mean and variance of the sampling distribution of the number of successes are np and npq respectively, $q = 1 - p$.

It was shown in §8·1 that, for large n,

$$z = \frac{x - np}{\sqrt{npq}}$$

is distributed as a standard normal variate. Hence from table 8·2, we have

$$P(|z| > 3) = 0{\cdot}0027.$$

Therefore, in random sampling such a result is extremely unlikely. We conclude that the truth of the hypothesis is itself very improbable and we say that the difference between the observed number of successes x and the expected number of successes np is highly *significant*.

If $|z| < 1{\cdot}96$, we say that the data is consistent with the hypothesis. If $|z| > 1{\cdot}96$, we say that the difference or deviation is significant at 5% *level of significance* and if $|z| > 2{\cdot}58$, the difference is said to be significant at 1% level of significance.

Cor 1. Let the observed proportion of successes in a sample of n observation be p, then $E\left(\frac{x}{n}\right) = p$, $\text{Var}\left(\frac{x}{n}\right) = \frac{1}{n^2}\text{Var}\, x = \frac{pq}{n}$.

Thus
$$z = \frac{\frac{x}{n} \sim p}{\sqrt{\frac{pq}{n}}} \sim N(0, 1).$$

Cor. 2. The probability for the proportion in the population is

$$p \pm 3\sqrt{\frac{pq}{n}}.$$

Notes. 1. The above argument holds if instead of the number of successes we use the proportion of successes.

2. The above test is valid for *large samples* only, since the binomial distribution may not be very near the normal for small samples. For small samples, binomial distribution is to be used.

3. It may be remembered that while the above test may furnish evidence against the hypothesis, it cannot prove the hypothesis to be correct. It can, at the most, prove no evidence against it.

4. The hypothesis tested above is a typical example of a *statistical hypothesis by which we understood some assumption about the density function of a random variate*. In the above case, the variate is the number of successes and it is not being tested that it follows the binomial distribution. In fact, this is assumed. The statistical hypothesis here is that the population parameter, the probability of success in a single trial, has the value p.

5. *By the testing of a statistical hypothesis is meant a procedure for deciding whether to accept or reject the hypothesis*. The procedure usually consists in assuming or accepting the hypothesis as correct and then calculating the probability of getting the observed or *more extreme* sample. If this probability is less than a certain pre-assigned small value, the hypothesis is rejected, since samples with small probabilities should be rare and we assume that a rare event has not happened.

6. The probability level below which we reject the hypothesis is called the *level of significance*. Usually, two levels are used, *viz*., the 5% and 1%, but in any particular problem the level would depend on the risk, rejecting the hypothesis when it is true, would entail.

7. The tests are usually called *tests of significance* since here we test whether the difference between the sample values and the population values or between the values given by two samples are so large that they *signify* evidence against the hypothesis, or these differences are small enough to be accounted for as due to fluctuations of sampling, *i.e.*, they may be regarded as due only to the fact that we are dealing with a sample and not with the whole population.

8. According to Prof. Fisher, the hypothesis, a definite statement about the population parameter, tested for possible rejection under the assumption that it is true, is usually called the *null hypothesis* and is usually denoted by H_0. To test whether there is a difference between two populations, we shall assume that there is no difference. Similarly, to test whether there is relationship between two variates, we assume there is no relationship. The reason is that a hypothesis can be rejected but cannot be proved. Rejection of no difference will mean a difference, while rejection of no relationship will imply a relationship.

9. A region corresponding to a statistic in the sample space which leads to the rejection of H, is called a *critical region* or *region of rejection*.

Example 12·1. *A coin is tossed 400 times and it turns up head 216 times. Discuss whether the coin may be an unbiased one.*

On the assumption of an unbiased coin, the expected number of heads in 400 tosses $= 400 \times \frac{1}{2} = 200$.

Deviation of the actual number of heads from expected is therefore, $216 - 200 = 16$.

Now $e = s.e.$ of the number of heads $= \sqrt{400 \times \frac{1}{2} \times \frac{1}{2}} = 10$.

∴ Deviation is only 1·6 times the *s.e.* and hence it is likely to appear as a result of fluctuations of simple sampling. We conclude that the coin may be an unbiased one.

Example 12·2. *A die is thrown 9,000 times and a throw of 3 or 4 is reckoned as a success. Suppose that 3,240 throws of a 3 or 4 have been made out. Do the data indicate an unbiased die ?*

Suppose the die is unbiased. Then the chance of success, that is of throwing a 3 or 4 with one die in one throw $= \frac{2}{6} = \frac{1}{3} = p$.

∴ The expected number of successes $= 9000 \times \frac{1}{3} = 3000$

$e =$ standard deviation of simple sampling

$$\Rightarrow \qquad e = \sqrt{npq} = \sqrt{9000 \times \tfrac{1}{3} \times \tfrac{2}{3}} = 44{\cdot}72.$$

The deviation 3240 – 3000, *i.e.*, 240 is nearly 5·4 *e*. Therefore, it is most improbable that it is due to fluctuation of sampling, that is, the die is almost certainly biased and that $p \neq \frac{1}{3}$.

The estimate of *p* obtained from the given sample is

$$\frac{3240}{9000} = 0{\cdot}36.$$

$$\text{Now } s.e. \text{ of the proportion of success} = \sqrt{\frac{0{\cdot}36 \times 0{\cdot}64}{9000}} = 0{\cdot}005$$

Hence the true value of *p* almost certainly lies within the range $0{\cdot}36 \pm 3 \times {\cdot}005$, *i.e.*, it lies between 0·345 and 0·375.

Example 12·3. *In a locality of 18,000 families, a sample of 840 families was selected. Of these 840 families, 206 families were found to have a monthly income of Rs. 50 or less. It is desired to estimate how many out of the 18,000 families have a monthly income of Rs. 50 or less. Within what limits would you place your estimate ?*

p = proportion of families having monthly income of Rs. 50 or less

$$= \frac{206}{840} = {\cdot}245 \text{ so that } q = {\cdot}755.$$

Assume that the condition of this problem will give a simple sample.

S.E. of the proportion of families having monthly income of Rs. 50 or less $= \sqrt{\frac{pq}{n}} = \sqrt{\frac{{\cdot}245 \times {\cdot}755}{840}} = 0{\cdot}015 = 1{\cdot}5\%.$

Hence taking ·245 to be the estimate of families having a monthly income of Rs. 50 or less the limits are 24·5 per cent $\pm (3 \times 1{\cdot}5)$ per cent, *i.e.* 20 per cent and 29 per cent, *i.e.*, between 3,600 and 5,220 families.

Ex. 1. Are the following statistical hypotheses ?

(*i*) There is life on planet Mars ;

(*ii*) The orbit of a newly discovered comet is a parabola ;

(*iii*) A new instrument can take measurements with less than $\frac{1}{2}\%$ error.

Ex. 2. A coin is thrown 10 times. Assuming it to be unbiased find

(*i*) the probability of 2 successes or less ;

(*ii*) the probability of the number of successes differing from 5 by more than 3.

At 5% level of significance, what number of successes will enables you to reject the hypothesis of an unbiased coin ?

12·4. Testing of Significance for Difference of Proportions. Two large simple samples of n_1 and n_2 members respectively are taken from two populations. The samples give proportions of attribute *A*'s as p_1 and

p_2 respectively. We are to find :

(a) *Is the difference, $p_1 \sim p_2$ due to fluctuations of simple sampling significant, the two populations being similar in respect of the proportion of A's therein ?*

(b) *If the difference in population proportions is real, could it be hidden in sampling, i.e., could it vanish or change sign owing to the fluctuations of sampling in samples taken from the two populations ?*

(a) On the hypothesis that the populations are similar as regards the given attribute, we can combine the two samples to give an estimate of the common value of the proportion of A's in the populations. This estimate is given by

$$p = \frac{n_1 p_1 + n_2 p_2}{n_1 + n_2} \qquad \text{...(12·1)}$$

If e_1 and e_2 be the standard errors in the proportions of successes in independent samples, the estimate of the standard error, *i.e.*, of proportion of successes in the parent universe is obtained from

$$e^2 = e_1^{\,2} + e_2^{\,2} = \frac{pq}{n_1} + \frac{pq}{n_2}$$

or $$e^2 = pq\left(\frac{1}{n_1} + \frac{1}{n_2}\right). \qquad \text{...(12·2)}$$

On the hypothesis that the two populations are identical,

$$E(p_1 - p_2) = E(p_1) - E(p_2) = 0.$$

If n_1 and n_2 be large then

$$z = \frac{p_1 \sim p_2}{e} \qquad \text{...(12·3)}$$

can be treated as a standard normal variate. If the difference $p_1 \sim p_2$ is less than $2e$, it may be due to fluctuations of sampling, but if it is greater than $3e$, there is likely to be real difference in population proportions but if the difference lies between $2e$ and $3e$, then it is significant at 5% level of significance.

Example 12·4. *A machine puts out 16 imperfect articles in a sample of 500. After machine is overhauled, it puts out 3 imperfect articles in a batch of 100. Has the machine been improved ?*

Here $p_1 = \frac{16}{500} = \cdot 032$, $p_2 = \frac{3}{100} = \cdot 03$, so that $p_1 - p_2 = \cdot 002$.

On the assumption that the machine has not been improved, we have, on combining together as our estimate of proportions,

$$p = \frac{16+3}{500+100} = \frac{19}{600}$$

$$\therefore \quad e = \sqrt{pq\left(\frac{1}{n_1} + \frac{1}{n_2}\right)} = \sqrt{\frac{19}{600} \times \frac{581}{600}\left(\frac{1}{500} + \frac{1}{100}\right)} = 0{\cdot}0192$$

Since the difference in proportions, *viz.*, 002 is less than $2e$, the difference is not significant and we conclude that the machine has not been improved.

(*b*) Suppose that the two samples of large sizes n, and n_2 have been drawn from two populations in which the proportions are respectively p_1 and p_2 where $p_2 < p_1$. The question arises :

Is it likely that the proportions x_1/n_1 and x_2/n_2 in the samples will be such that $x_1/n_1 - x_2/n_2 \leq 0$? In other words :

Is the real difference between the populations likely to be hidden in simple sampling ?

$$\text{Since } E\left(\frac{x_1}{n_1}\right) = p_1, E\left(\frac{x_2}{n_2}\right) = p_2,$$

$$\text{Var}\left(\frac{x_1}{n_1}\right) = \frac{p_1 q_1}{n_1}, \text{ var}\left(\frac{x_2}{n_2}\right) = \frac{p_2 q_2}{n_2}$$

$$\therefore \quad z = \frac{\left(\frac{x_1}{n_1} - p_1\right) - \left(\frac{x_2}{n_2} - p_2\right)}{\sqrt{\frac{p_1 q_1}{n_1} + \frac{p_2 q_2}{n_2}}} = \frac{\left(\frac{x_1}{n_1} - \frac{x_2}{n_2}\right) - p_1 + p_0}{e} \quad \text{...(12·4)}$$

can be treated for large n_1, n_2 as an asymptotically standard normal variate.

We require $$P\left[\frac{x_1}{n_1} - \frac{x_2}{n_2} < 0\right]$$

or $$P\left[ze + p_1 - p_2 \leq 0\right]$$

where $$e = \sqrt{\frac{p_1 q_1}{n_1} + \frac{p_2 q_2}{n_2}}$$

i.e., we require $$P\left[z \leq -\left(\frac{p_1 - p_2}{e}\right)\right].$$

If $p_1 - p_2 > 1{\cdot}6{\cdot}45e$, then from table 2, this probability is less than 0·5, in other words the probability of the difference being hidden is less than 5% and we say that it is unlikely that the difference will be hidden in sampling.

Example 12·5. *In two large populations, there are 30% and 25% respectively of fair haired people. Is this difference likely to be hidden in samples of 1,200 and 900 respectively from the two populations ?*

Here $p_1 = 0{\cdot}30, p_2 = 0{\cdot}25$ so that $p_1 - p_2 = 0{\cdot}05$

$$e^2 = \frac{(\cdot 3)(\cdot 7)}{1200} + \frac{(\cdot 25)(\cdot 75)}{900} = \frac{138}{3{,}60{,}000},$$

so that $$e = 0{\cdot}0195.$$

The difference $p_1 - p_2$ is about $2\frac{1}{2}$ times the standard error.

Hence it is unlikely that the real difference will be hidden.

Ex. 1. If $p_1 - p_2 > 2{\cdot}327e$, prove that the difference is unlikely to be hidden at 1% level of significance.

Ex. 2. What is level of significance of $p_1 - p_2 > 1{\cdot}06\,e$?

12·5. Sampling of values of a Variable. We now consider the sampling of values of a variable such as, height, weight, etc. Each individual or member of the population provides a value of the variable and the population is a frequency distribution of the variable. From the population, we can draw a random sample of n members and this is the same as choosing n values of the variable from those of the distribution.

Parameters and Statistics. The statistical constants of the population, *viz.*, the mean, the variance, etc. are known as the parameters. The statistical concepts of the sample, computed from the members or observations of the sample, to determine or estimate the parameters of the population from which the sample has been drawn, are known as statistics.

12·5·1. Sampling Distribution. Any function of the observed sample values calculated to estimate some parameters of the population is called a *statistic*. We can draw Nc_n different samples of size n from a finite population of size N and an infinite member of samples of size n from an infinite population. Each sample gives its own values for the statistic under consideration. The assemblage or aggregate of all these values of the statistics, together with their relative frequencies or the probabilities with which they occur, constitute the sampling distribution of the corresponding statistic. Thus we can talk of the sampling distribution of the sample mean $\bar{x}$, of the sampling variance s^2, etc.

Let the density function for the sampling distribution of the statistic x be $f(x)$, then the probability that any sample chosen at random from all possible sample will give a value of X between x_1 and x_2 is given by

$$P(x_1 < X < x_2) = \int_{x_1}^{x_2} f(x)\,dx \div \int_{-\infty}^{\infty} f(x)\,dx$$

$$= \int_{x_1}^{x_2} f(x)\,dx \qquad \left[\because \int_{-\infty}^{\infty} f(x)\,dx = 1.\right]$$

In particular, the probability that the sample will give a value of x greater than x_0 is given by

$$P(x > x_0) = \int_{x_0}^{\infty} f(x)\,dx \div \int_{-\infty}^{\infty} f(x)\,dx = \int_{x_0}^{\infty} f(x)\,dx$$

Ex. From a rectangular population with 6 members 1, 2, 3, 4, 5, 6 samples of sizes 3 are drawn. Find the sampling distribution of the mean and variance and deduce the probability that

(*i*) $\bar{x} > 4$ (*ii*) $\left|\bar{x} - \frac{7}{2}\right| > 1$ (*iii*) $s^2 > 1$ (*iv*) $1 < s^2 < 2$

where $$s^2 = \frac{1}{n} \cdot \sum_{i=1}^{n} (x_i - \bar{x})^2.$$

12·6. Sampling Distribution of the Mean.

Central Limit Theorem. We shall find the distribution of the mean of random samples of given size drawn from a specified population.

Case 1. *Parent population is normal.*

Let the population from which the random sample is drawn be a normal population with mean μ and variance σ^2. By the definition of a random sample, variate values $x_1, x_2, \ldots, x_n$ of the sample are independent and from §8·2 it follows that the *sample mean is distributed normally about mean* μ *and variance* σ^2/n. The result shows how the *precision* of a sample mean increases as the sample size increases.

Case II. *Parent population is non-normal.*

Linderberg-Levy Central Limit Theorem. *If the variable X has a non-normal distribution with mean* μ *and finite standard deviation* σ *and if the moment generating function exists, then the variate*

$$Z = \frac{(\bar{x} - \mu)\sqrt{n}}{\sigma}$$

has a distribution that approaches the standard normal distribution as n tends to infinity.

Proof. $M_z(t) = M_{\frac{\bar{x}-u}{\sigma/\sqrt{n}}}(t)$

$$M_X(t) = e^{-\frac{\mu\sqrt{n}}{\sigma}t} M_{\bar{x}}\left(\frac{\sqrt{n}}{\sigma}t\right)$$

$$= e^{-\frac{\mu\sqrt{n}}{\sigma}t}\left[M_x\left(\frac{t}{\sigma\sqrt{n}}\right)\right]^n$$

$$\therefore \quad \log_e M_X(t) = -\frac{\mu\sqrt{n}}{\sigma}t + n\log M_e\left(\frac{t}{\sigma\sqrt{n}}\right)$$

$$= -\frac{\mu\sqrt{n}}{\sigma}t + n\log\left[1 + \mu_1'\frac{t}{\sigma\sqrt{n}} + \mu_2'\frac{t^2}{2n\sigma^2} + \cdots\right]$$

$$= -\frac{\mu\sqrt{n}}{\sigma}t + n\left[\left(\mu_1'\frac{t}{\sigma\sqrt{n}} + \mu_2'\frac{t^2}{2n\sigma^2} + \ldots\ldots\right) - \frac{1}{2}\left(\frac{\mu_1' t}{\sigma\sqrt{n}} + \ldots\right)^2 + \ldots\ldots\right]$$

$$= \frac{t^2}{2} + 0\left(\frac{1}{\sqrt{n}}\right) \qquad [\because \mu_1' = \mu]$$

$\therefore$ As n tends to infinity, we have

$$\log_e M_z(t) \to \frac{t^2}{2} \quad \text{or} \quad M_z(t) \to \exp\left(\frac{t^2}{2}\right).$$

Now exp $(\frac{1}{2}t^2)$ is the *m.g.f.* of a standard normal variate.

Hence the theorem is established.

Note. 1. This theorem, known as the *Linderberg-Levy Central Limit theorem* (CLT), can be proved to hold for all populations with *finite* variance. The existence of a moment generating function is not necessary. The condition that the variance is finite is, however, not very restrictive since most of the distributions have a finite range and, therefore, a finite variance or an infinite range but a finite variance. The degree of approximation to normal distribution depends on (*i*) the degree of departure of the parent population from the normal and (*ii*) the sample size n. *For a parent normal population, the distribution of the sample mean is always exactly normal i.e., it is normal for all n.*

Note 2. *Case of Equal Components.* If the random variables $X_1, X_2, ..., X_n$ have the same distribution with mean μ and variance σ^2, then

$$S_n = X_1 + X_2 + \ldots\ldots + X_n.$$

has a finite mean $n\mu$ and variance $n\sigma^2$.

Proof. Let $f(x)$ be the common density function of X_i.

$$\text{Now} \lim_{n \to \infty} \frac{1}{n\sigma^2} \sum_{k=1}^{n} \int_{|x-\mu_\alpha| \geq \gamma\sqrt{n}\,\sigma} (x-\mu_k)^2 f(x)\, dx$$

$$= \frac{1}{\sigma^2} \lim_{n \to \infty} \int_{|x-\mu| \geq \gamma\sqrt{n}\,\sigma} (x-\mu)^2 f(x)\, dx = 0$$

Since $\sigma^2 = \int_{-\infty}^{+\infty} (x-\mu)^2 f(x)\, dx$ exists, it follows that R.H.S. vanishes for any fixed $\gamma > 0$. Hence the proof.

We note $(s_n - n\mu)/\sigma\sqrt{n} = \dfrac{(\bar{x}-\mu)}{\sigma/\sqrt{n}}$ and hence the *statement* $\bar{X}$ is asymptotically normal with mean μ and variance σ^2/n.

Case of failure of Cauchy Distribution. If each random variable of the sequence $X_1, X_2, ...$ has a Cauchy distribution, having parameters (μ, λ), then $\bar{X}$ has also a Cauchy distribution with the same parameters and thus is not asymptotically normal. The central limit theorem fails as Cauchy distribution does not posses finite mean and variance.

Note 3. If Central limit theorem (CLT), holds then the weak Law of Large Number (WLLN) holds. The converse of this result is not true *i.e.* the law of large number holds even for sequence of random variables, for which the variance of their common distribution does not exist.

Note 4. CLT and WLLN both hold for a sequence $\{x_n\}$ of independent identically distributed (i.i.d) r.v. with finite mean μ and variance σ^2.

Ex. 1. Verify the theorem for the following distributions :—

(*i*) $f(x) = p^x (1-p)^{1-x}$ $[x = 0, 1]$

(*ii*) $f(x) = \dfrac{e^{-m} m^x}{x!}$ $[x = 0, 1, 2, ...]$

(*iii*) $f(x) = 1$ $[0 \leq x \leq 1]$

(*iv*) $f(x) = \dfrac{1}{\Gamma(l)} e^{-x} x^{l-1}, l > 0,$ $[x \geq 0]$

(*v*) $f(x) = e^{-x}$ $[x \geq 0]$

Ex. 2. Prove that the moment generating function for a Cauchy distribution is the same as that of the parent population. Explain the failure of the central limit theorem in this case.

Ex. 3. Show that the *r*th cumulant for the sample mean is k_r/n^{r-1} where k_r is *r*th cumulant for the parent population.

Deduce that for the sampling distribution of the sample mean,

(*i*) mean is same as the population mean,

(*ii*) variance is σ^2/n where σ^2 is the variance of the population,

(*iii*) third moment about mean is μ_3/n^2,

(*iv*) skewness is $1/\sqrt{n}$ of the skewness of the population,

(*v*) excess of kurtosis is $1/n$ of the excess of kurtosis for the population.

Ex. 4. Let $\{X_k\}, k \geq 1$ be a sequence of i.i.d. variates with

$$f(x) = \frac{1}{2} e^{-|x|}, -\infty < x < \infty$$

Find the constants a_n and b_n such that

$$\{|x_1| + |x_2| + \cdots + |x_n| - a_n\}/b_n \xrightarrow{p} N(0, 1)$$ ***(I.A.S. 1982)***

Hint. to Ex. 5, page 542.

Let $X_1, X_2, ..., X_n$ be i.i.d. P (1). Then

$S_n = X_1 + X_{2,} + - +X_n \sim P(X) \Rightarrow Y_n = S_n$

$$\therefore P\left[a \leq \frac{Y_n - n}{\sqrt{n}} \leq b\right] = P\left[a \leq \frac{S_n - n}{\sqrt{n}} \leq b\right] \to \bar{\phi}(b) - \bar{\phi}(a) \text{ as } n \to \infty$$

Taking in partiular, $a \to -\infty, b = 0$ **then** $P\left(\frac{Y_n - n}{\sqrt{n}}\right) \leq 0 \Rightarrow P(Y_n \leq n)$

& R.H.S. $= \bar{\phi}(0) - \bar{\phi}(-\infty) = \frac{1}{2}$

We can prove that $P[S_n \leq E(S_n)] \to \frac{1}{2}$ as $n \to \infty$.

Ex. 5. Using *C L T*, If Y_n is distributed as $p(n)$, then

$$\lim_{n \to \infty} P\left[a \leq \frac{Y_n - n}{\sqrt{n}} < b\right] = \phi(b) - \phi(a) \text{ show that}$$

$$\lim_{n \to \infty} P(Y_n \leq n) = \lim_{x \to \infty}\left[e^{-n} \sum_{k=0}^{n} \frac{n^k}{k!}\right] = \frac{1}{2} = \lim_{x \to \infty} \int_0^n \frac{e^{-t} t^{n-1}}{(n-1)!} dt$$ ***(I.A.S. 1984)***

Example 12·6. *A sample of 400 male students is found to have a mean height of 67·47". Can it be reasonably regarded as a sample from a large population, with mean height 67·39"* and s.d. *1·30"* ?

$\sigma_{\bar{x}}$, the standard error (*s.e.*) of the mean of the sample of 400 from such population is $1·30/\sqrt{400} = 0·065$ inch. The deviation of the mean of the sample from that of the population is ·80 inch and this is less that twice the *s.e.* of the mean and is, therefore, not significant. We conclude that the given sample might have been drawn from the population specified.

12·7. Confidence or Fiducial Limits for Unknown Mean.

Let the population from which a random sample of size n is drawn, have mean μ and *s.d.* σ. If σ be known, but not μ, there is a range of possible values of μ, for which the observed mean $\bar{x}$ of the sample is not significant at any specified level of probability. The relative deviation of $\bar{x}$ from μ is the ratio of $(\bar{x} \sim \mu)$ to the standard error of $\bar{x}$ *i.e.*,

$(\bar{x} \sim \mu) \sqrt{n}/\sigma$. If $\bar{x}$ is not significant at 5% level of probability, this relative deviation must be less than that which (in a normal distribution) is exceeded numerically with a probability of 5%. Such a relative deviation is 1·96.

Hence the observed mean will not be significant at 5% level of probability if
$$\left| \left(\frac{\bar{x} \sim \mu}{\sigma} \right) \sqrt{n} \right| < 1{\cdot}96$$

i.e., if
$$\bar{x} - 1{\cdot}96\, \sigma/\sqrt{n} < \mu < \bar{x} + 1{\cdot}96\, \sigma/\sqrt{n} \qquad \text{...(12·5)}$$

The values $\bar{x} \pm 1{\cdot}96\sigma/\sqrt{n}$ are called 95% *Fiducial Limits* or *Confidence Limits*, for the mean of the population corresponding to the given sample. They give the limits for μ in order that the observed sample mean $\bar{x}$ should not be significant at the prescribed level of probability.

Note. For normal population with known variance, the above test of significance is valid for all sample size.

Example 12·7. *Suppose that distribution of the statures of men is a normal distribution with s.d. 2·8 inches. One hundred male students in a large university are measured and their average height is found to be 68·52 inches. Determine the 98 per cent confidence limits for the mean height of the men of the university.*

We have $\sigma_{\bar{x}} = \dfrac{2{\cdot}48}{\sqrt{100}} = 0{\cdot}248$ inch.

The 93 per cent fiducial limits for the mean of the population are $\bar{x} \pm 2{\cdot}33\, \sigma/\sqrt{n}$, *i.e.*, 68·52 inches $\mp$ ·578 inch *i.e.*, 67·942 inches and 69·098 inches. Thus at 2% level of confidence the confidence limits are 67·942 and 69·098 inches.

Ex. 1. If p is the observed proportion of success in n independent Bernoullian trials, prove that 95% fiducial limits for the population proportion p' are, for the large samples, given by
$$p \pm 1{\cdot}96 \sqrt{\frac{p - p^2}{n}}$$

Ex. 2. Prove in the above exercise that the 99% fiducial limits are the roots of the quadratic $\quad (p - p^2)(2{\cdot}58)^2 = n(p' - p)^2 \quad$ (***M.A. Patna 1956***)

Ex. 3. An unbiased coin is thrown n times. It is desired that the relative frequency of the appearance of heads should lie between ·49 and ·51. Find the smallest value of n that will ensure this result with 90% confidence. [**Ans.** 6745]

Ex. 4. p_1, p_2 are the observed proportions of successes in two independent sets of trials of large sizes n_1, n_2. Prove that the 95% fiducial limits for the difference $p_1' - p_2'$ of the proportions of successes in the population are.
$$p_1 - p_2 \pm 1{\cdot}96 \sqrt{\frac{p_1 q_1}{n_1} + \frac{p_2 q_2}{n_2}}\ ; \quad q_1 = 1 - p_1 \text{ and } q_2 = 1 - p_2.$$

12·8. Testing the Significance of the Difference between the Means of two Large Samples. Suppose two random samples of n_1 and n_2 members respectively have been drawn from the *same* population of *s.d.* σ. We would like to know if the difference $\bar{x}_1 \sim \bar{x}_2$ of their means is due to fluctuations of sampling. On the assumption that the samples are independent and drawn from the same population, the standard error, e, of the difference of their means is given by

$$e^2 = e_1^2 + e_2^2$$

where e_1 and e_2 are the *s.e.* of the means of the two samples and are $\sigma/\sqrt{n_1}$ and $\sigma/\sqrt{n_2}$ respectively so that

$$e = \sigma\left(\frac{1}{n_1} + \frac{1}{n_2}\right)^{\frac{1}{2}} \qquad ...(12{\cdot}6)$$

If n_1 and n_2 be sufficiently large then $\bar{x}_1 \sim \bar{x}_2$ is asymptotically normally distributed with mean zero and *s.d.e.* Consequently, if the difference $\bar{x}_1 \sim \bar{x}_2$ exceeds $3e$ the difference can hardly be accounted for by the fluctuations of sampling and our assumption is unlikely to be correct while if the difference exceeds $2e$, it is regarded as significant at the 5% level of probability.

If two independent samples of n_1 and n_2 members respectively be drawn from different populations with variances σ_1^2 and σ_1^2 respectively we can examine whether the two populations from which samples have been drawn differ in mean apart from the difference in dispersion. Since the samples are independent the *s.e.*, e, of the difference of their means is given by

$$e^2 = \frac{\sigma_1^2}{n_1} + \frac{\sigma_2^2}{n_2} \qquad ...(12{\cdot}7)$$

Assuming that n_1 and n_2 are large and the two populations have the same mean, the difference of the means of the samples will be normally distributed with mean zero and *s.d.e.* given by (12·7). If the difference of the means of the samples exceeds $3e$, it can hardly be accounted for on the basis of fluctuations of sampling and our assumption that the two populations have the same mean is almost certainly wrong.

Note 1. In the above discussion, we have assumed that the population variances σ_1^2, σ_2^2 are known. In practice this is hardly the case and accordingly in the expressions for e, these have to be replaced by their estimates obtained for the samples, *viz.*, by the sample variances s_1^2, s_2^2 respectively

$$\text{where } s_1^2 = \frac{1}{n_1}\sum_{i=1}^{n_1}(x_i - \bar{x}_1)^2.$$

2. The above tests are valid for only large samples for two reasons :

(*i*) the parent populations may not be normal, though we are assuming that they do not depart strikingly from it. In particular, we assume that the populations of finite variances. For populations like Cauchy's where the variance is not finite, the tests would break down completely even for 'infinitely large samples'.

(*ii*) The population variances are not known and have to be replaced by their estimates.

3. For *normal* populations with *known* variances, the above tests are *valid for all sample sizes.*

4. If the hypothesis to be tested is that the population means are μ and μ', we can carry out the test of significance as above, but in this case

$$z = \frac{(\bar{x}_1 - \bar{x}_2) \sim (\mu - \mu')}{\sqrt{\dfrac{s_1^2}{n_1} + \dfrac{s_2^2}{n_2}}} \qquad ...(12{\cdot}8)$$

will be asymptotically a standard normal variate for large n_1, n_2.

Example 12·8. *If 60 new entrants in a given university are found to have a mean height of 68·60 inches and 50 seniors a mean height of 69·51 inches ; is the evidence conclusive that the mean height of the seniors is greater than that of the new entrants ? Assume the standard deviation of heights to be 2·48 inches.*

Since the two samples are independent and come from the same population, the *s.e., e,* of the difference of the two mean heights is

$$e = \sigma\sqrt{\frac{1}{n_1} + \frac{1}{n_2}} = 2{\cdot}48\sqrt{\frac{1}{60} + \frac{1}{50}} = 0{\cdot}47 \text{ inch}$$

The observed difference of the two mean heights is 0·91 inch which is less than $2e$ and so could be due to fluctuations of sampling.

Hence it cannot be said that the mean height of the seniors is necessarily greater than that of the new entrants.

Example 12·9. *A random sample of 100 farms in a certain year gives an average yield of barley of 2,000 lbs. per acre with a s.d. of 192 lbs. A random sample of 100 farms in the following year gives an average yield of 2,100 lbs. per acre with a s.d. of 224 lbs. Show that the data are inconsistent with the hypothesis that the average yield in the country as a whole were same in the two years.*

The *s.e.* of the average yields of the two samples are $192/\sqrt{1000}$ and $224/\sqrt{1000}$ respectively.

Since the two samples are independent, the *s.e.* of the difference of their mean yields is given by

$$e = \sqrt{\frac{(192)^2}{1000} + \frac{(224)^2}{1000}} = \sqrt{87{\cdot}04} = 9{\cdot}33 \text{ per acre.}$$

The observed difference between the means of the samples is 100 lbs. per acre which is about 10 times its *s.e.* and so cannot be due to fluctuations of sampling. Hence the data are inconsistent with the assumption that the average yield in the country as a whole were same in the two years.

Ex. 1. If $\bar{x}_1, \bar{x}_2$ are the sample means of two independent samples of sizes n_1, n_2 from normal populations with known variances σ_1^2, σ_2^2, prove that the 98% fiducial limits for the difference between population means are

$$\bar{x}_1 - \bar{x}_2 \pm 2{\cdot}325 \sqrt{\frac{\sigma_1^2}{n_1} + \frac{\sigma_2^2}{n_2}}.$$

Ex. 2. How will the above limits be effected if

(*i*) the populations are not known to be normal,

(*ii*) the variances are not known,

(*iii*) confidence coefficient is 95% ?

12·8·2. Testing the Significance of Difference between Standard Deviations of two Large Samples. Let two random samples of large sizes n_1 and n_1 having standard deviations s_1 and s_2 be drawn from two normal populations with standard deviations σ_1 and σ_2 respectively. We would like to know if the difference between s_1 and s_2, *i.e.*, $s_1 - s_2$ is due to fluctuations of sampling.

We set up the null hypothesis that the samples have been drawn from normal populations having the same standard deviation *i.e.*, $\sigma_1 = \sigma_2 = \sigma$. Under the null hypothesis, the statistic.

$$z = \frac{s_1 \sim s_2}{\sqrt{\text{Var}(s_1 - s_2)}} \sim N(0, 1).$$

since the samples are of large sizes

and $\quad E(s_1 - s_2) = E(s_1) - E(s_2) = \sigma - \sigma = 0$

Now $\text{Var}(s_1 - s_2) = \text{Var}\, s_1 + \text{Var}\, s_2 = \dfrac{\sigma_1^2}{2n_1} + \dfrac{\sigma_2^2}{2n_2}$,

since samples are independent.

$$\therefore \qquad z = \frac{s_1 \sim s_2}{\sqrt{\dfrac{\sigma_1^2}{2n_1} + \dfrac{\sigma_2^2}{2n_2}}} \sim N(0, 1)$$

If the value of $z > 2$ but < 3, then the deviation is significant at 5% level.

Example 12·10. Random samples drawn from two countries gave the following data relating to the heights of adult males.

	Country A	*Country B*
Mean height in inches	*67·42*	*67·25*
Standard deviation	*2·58*	*2·50*
Number in samples	*1000*	*1200*

Is the difference between the standard deviations significant ?

We set up the null hypothesis that there is no significant difference between sample standard deviations. Under the null hypothesis,

$$z = \frac{s_1 - s_2}{\sqrt{\dfrac{s_1^2}{2n_1} + \dfrac{s_2^2}{2n_2}}} \sim N(0, 1)$$

Now $$z = \frac{2{\cdot}58 - 2{\cdot}50}{\sqrt{\dfrac{(2{\cdot}58)^2}{2 \times 1000} + \dfrac{(2{\cdot}50)^2}{2 \times 1200}}} = 1{\cdot}03, \text{ on simplification}$$

Since $z < 1{\cdot}96$, we conclude that data do not provide any evidence against the null hypothesis at 5% level.

Hence the sample standard deviations do not differ significantly.

STANDARD ERRORS OF STATISTICS

12·9·1. Introduction. We have seen earlier that for large samples, the distribution of most of the statistics tends to normality. Thus, if to estimate a population parameter we construct from the sample values the values, of the parameter in the sub-population, composed by the sample, its distribution will, in general, tend to normality about the true population parameter as mean and with a sampling variance which will in general be of order $1/n$ for large samples. The positive square root of the sampling variance is called the *standard error* of the estimate and for large samples, it is a valid measure of precision for such statistics of which the distribution tends to normality. Now we obtain the standard errors of important statistics.

12·9·2. Standard Error of Sample Moments about the arbitrary values. It is the usual convention to denote the parameters of the population by Greek letters and the corresponding statistics of the sample by Roman letters. Thus, if μ'_r and μ_r denote respectively the rth moment of the population about the arbitrary value and about the mean, then m'_r and m_r denote the corresponding moments of the sample. If n be the size of the sample, we have

$$m_r' = \frac{1}{n} \sum_{i=1}^{n} x_i^r \,, \; m_r = \frac{1}{n} \sum_{i=1}^{n} (x_i - m_1')^r \qquad ...(12{\cdot}9)$$

$$\therefore \quad E(m_r') = E\left(\frac{1}{n} \Sigma x_i^r\right) = \frac{1}{n} \Sigma_1^n E(x_i^r) = E(x^r) = \mu_r'$$

$$\text{Var}(m_r') = E(m_r' - \mu_r')^2 = E\left(\frac{1}{n} \Sigma x_i^r - \mu_r'\right)^2$$

$$= \frac{1}{n^2} E\,[(\Sigma\, x_i^{\,r})^2 - 2n\,\mu_r'\,(\Sigma x_i^{\,r}) + n^2\,\mu_r'^2\,]$$

$$= \frac{1}{n^2} E\,[(\Sigma x_i^{\,r})^2 - \mu_r^{\,2}\,] = \frac{1}{n^2} E\,[\,\Sigma\, x_i^{2r} + \sum_{i \neq j} x_i^{\,r}\, x_j^{\,r}\,] - \mu_r'^2$$

$$= \frac{1}{n}\,[n\,E\,(\,x_i^{2r}\,) + n\,(n-1)\,E(\,x_i^{\,r}\,)\,E\,(\,x_j^{\,r}\,)\,] - \mu_r'^2\,,$$

since $x_i'^s$ are independent

$$= \frac{1}{n^2}\,[\,n\,\mu_{2r}' + n\,(n-1)\,\mu_r'^2\,] - \mu_r'^2$$

$$\therefore \quad \text{Var}\,(m_r'\,) = \frac{1}{n}\,(\,\mu_{2r}' - \mu_r'^2\,)\,. \qquad \text{...(12·10)}$$

$$\text{Cov}\left(m_r', m_s'\right) = E\left[\left(m_r' - \mu_s'\right)\left(m_z - \mu_s'\right)\right]$$

$$= E\left[\left(\frac{1}{n}\Sigma\, x_i^{\,r} - \mu_r'\right)\left(\frac{1}{n}\Sigma\, x_i^{\,s} - \mu_s'\right)\right]$$

$$= \frac{1}{n} E\left(\Sigma\, x_i^{\,r+s}\right) + \frac{1}{n^2} E\left(\sum_{i \neq j} x_i^{\,r} x_j^{\,s}\right) - \frac{1}{n}\,\mu_r'\,E\left(\Sigma x_i^{\,s}\right)$$

$$- \frac{1}{n}\,\mu_s' E\left(\Sigma x_i^{\,r}\right) + E\left(\Sigma \mu_r'\,\mu_s'\right)$$

$$= \frac{1}{n^2}\cdot \mu'_{r+s} + \frac{1}{n^2}\,n\,(n-1)\mu_r'\,\mu_s' - \frac{1}{n}\,\mu_r' n\,\mu_s'$$

$$- \frac{1}{n}\cdot \mu_s' n\,\mu_r' + \mu_r'\,\mu_s'$$

$$= \frac{1}{n}\left(\mu'_{r+s} - \mu_r'\,\mu_s'\right). \qquad \text{...(12·11)}$$

$$E\,(m_2) = E\left[\frac{1}{n}\Sigma\left(x_i - m_1'\right)^2\right] = \frac{1}{n} E\left[\,\Sigma\, x_i^{\,2} - 2\,m_1'\,\Sigma\, x_i + n m_1'^2\,\right]$$

$$= \frac{1}{n} E\left[\,\Sigma x_i^{\,2} - \frac{2}{n}\left(\Sigma x_j\right)\left(\Sigma x_i\right) + \frac{1}{n}\left(\Sigma x_i\right)^2\right]$$

$$= \frac{1}{n} E\left[\,\Sigma\, x_i^{\,2} - \frac{1}{n}\,\Sigma\, x_i^2 - \frac{2}{n}\sum_{i \neq j} x_i x_j\right]$$

$$= \mu_2' - \frac{1}{n}\,\mu_2' - \frac{n(n-1)}{n^2}\,\mu_1'^2 = \frac{n-1}{n}\left(\mu'_2 - \mu_1'^2\right)$$

$$= \left(1 - \frac{1}{n}\right)\mu_2 \qquad \text{...(12·12)}$$

$$\text{Var}\left(m_2\right)=E\left(m_2-\frac{n-1}{n}\,\mu_2\right)^2$$

$$=E\left[m_2^{\,2}-2\,\frac{(n-1)}{n}\,m_2\,\mu_2+\left(\frac{n-1}{n}\right)^2\mu_2^{\,2}\right]$$

$$=E\left(m_2^{\,2}\right)-\left(\frac{n-1}{n}\right)^2\mu_2^{\,2} \qquad \text{using } (12\cdot12)$$

Now $$m_2^{\,2}=\left[\frac{1}{n}\Sigma\,x_i^{\,2}-\frac{1}{n^2}\left(\Sigma\,x_i\right)^2\right]^2$$

$$=\frac{1}{n^2}\left(\Sigma\,x_i^{\,2}\right)^2-\frac{2}{n^3}\left(\Sigma\,x_i^{\,2}\right)\left(\Sigma\,x_i\right)^2+\frac{1}{n^4}\left(\Sigma\,x_i\right)^4$$

$$=\frac{1}{n^2}\left(\Sigma\,x_i^{\,4}+2\sum_{i\neq j}x_i^{\,2}\,x_j^{\,2}\right)-\frac{2}{n^2}\left(\Sigma\,x_i^2\right)\times$$

$$\times\left(\Sigma x_i^{\,2}+2\sum_{i\neq j}x_i\,x_j\right)+\frac{1}{n^4}\left(\Sigma\,x_i^{\,2}+2\sum_{i\neq j}x_i\,x_j\right)\left(\Sigma\,x_i^{\,2}+2\sum_{i\neq j}x_i\,x_j\right)$$

$$=\frac{1}{n^2}\left(\Sigma\,x_i^{\,4}+2\,\Sigma\,x_i^{\,2}\,x_j^{\,2}\right)-\frac{2}{n^2}\left(\Sigma\,x_i^{\,4}+2\sum_{i\neq j}x_i^{\,2}\,x_j^{\,2}\right.$$

$$\left.+\,2\sum_{i\neq j}x_i^{\,3}x_j+2\sum_{i\neq j\neq k}x_i^{\,2}\,x_i\,x_k\right)+\frac{1}{n^4}\left(\Sigma\,x_i^{\,4}+6\sum_{i\neq j}x_i^{\,2}\,x_j^{\,2}\right)$$

$$\therefore\;E\left(m_2^{\,2}\right)=\frac{1}{n}\,\mu_4+\frac{n-1}{n}\,\mu_2^{\,2}-\frac{2}{n^2}\,\mu_4-\frac{2(n-1)}{n^2}\cdot\mu_2^{\,2}+\frac{1}{n^3}\cdot\mu_4$$

$$+\frac{3(n-1)}{n^3}\,\mu_2^{\,2},\ \text{taking } \mu_1{}'=0$$

$$=\left(\frac{1}{n}-\frac{2}{n^2}+\frac{1}{n^3}\right)\mu_4+\frac{n-1}{n}\left(1-\frac{2}{n}+\frac{3}{n^2}\right)\mu_2^{\,2}$$

$$=\frac{(n-1)^2}{n^3}\,\mu_4+\frac{(n-1)\,(n+1)(n-3)}{n^3}\,\mu_2^{\,2}$$

$$\therefore\;\text{Var}\left(m_2\right)=\frac{(n-1)^2}{n^3}\,\mu_4+\frac{(n-1)(n+1)(n-3)}{n^3}\,\mu_2^{\,2}-\frac{(n-1)^2}{n^2}\,\mu_2^{\,2}$$

$$=\frac{(n-1)^2}{n^3}\,\mu_4-\frac{n-1}{n^3}\left(\,\overline{n-1}-2\,\right)\mu_2^{\,2}$$

$$=\frac{(n-1)^2}{n^3}\left(\mu_4-\mu_2^{\,2}\right)+\frac{2\,(n-1)}{n^3}\,\mu_2^{\,2}$$

Hence $\text{Var}\left(m_2\right)=\frac{1}{n}(\mu_4-\mu_2^2)$ to order $1/n$, and that for normal parent

$$\text{Var}\left(m_2\right)=\frac{3\sigma^4-\sigma^4}{n}=\frac{2\sigma^4}{n}$$

Cor. $$\text{Var}\left(s^2\right)=\text{Var}\left(\frac{n}{n-1}m_2\right)=\frac{n^2}{(n-1)^2}\text{Var}\left(m_2\right)$$

$$=\frac{1}{\left(\frac{1}{n}-1\right)^2}\text{Var}\, m_2=\text{Var}\, m_2 \text{ to order } \frac{1}{n}$$

$$=\frac{1}{n}\left(\mu_4-\mu_2^2\right)+\frac{2}{n(n-1)}\mu_2^2$$

$$=\frac{1}{n}\left(\mu_4-\mu_2^2\right)+\frac{2}{n-1}\mu_2^2 \text{ to order } 1/n.$$

$$=\frac{2\sigma^4}{(n-1)} \qquad \text{[for normal parent].}$$

12·10. Theorem. *If $y_i{}^s$, $i=1, 2, \ldots, k$ are k statistics calculated from a sample of size n and if each y_i, is distributed about mean ξ, with* variance Var (y_i) of order *$1/n$, then*

$$Var(f)\equiv\sum_{i=1}^{k}\left(\frac{\partial f}{\partial \xi_i}\right)^2 Var(y_i)+\sum_{i\neq j}\left(\frac{\partial f}{\partial \xi_i}\right)\left(\frac{\partial f}{\partial \xi_j}\right)cov\left(y_i, y_j\right),$$

where $f\equiv f(y_1, y_2, \ldots, y_k)$ is any function of y's and possesses partial derivatives.

Proof. By Taylor's theorem, we have

$$f(y_1, y_2, \ldots, y_k)=f(\xi_1, \xi_2, \ldots, \xi_k)+\Sigma\left[(y_i-\xi_i)\frac{\partial f}{\partial \xi_i}\right]$$

$$+\frac{1}{2!}\left[\Sigma(y_i-\xi_i)^2\frac{\partial^2 f}{\partial \xi_i^2}+\sum_{i\neq j}(y_i-\xi_i)(y_j-\xi_j)\frac{\partial^2 f}{\partial \xi_i\,\partial \xi_j}\right]+\ldots$$

since Var (y_i) is of order $1/n$, we have $(y_i-\xi_i)\propto 1/\sqrt{n}$. Hence neglecting terms of order higher than $1/n$, we have

$$f(y_1, y_2, \ldots, y_k)=f(\xi_1, \xi_2, \ldots, \xi_k)+\Sigma\left[(y_i-\xi_i)\frac{\partial f}{\partial \xi_i}\right]$$

$$\therefore\ \text{Var}(f)=E\left[f(y_1, y_2, \ldots, y_k)-f(\xi_1, \xi_2, \ldots \xi_k)\right]^2=E\left[\Sigma(y_i-\xi_i)\frac{\partial f}{\partial \xi_i}\right]^2$$

$$= E\left[\Sigma\left\{\left(y_i - \xi_i\right)^2 \left(\frac{\partial f}{\partial \xi_i}\right)^2\right\} + \right.$$

$$\left. + \sum_{i \neq j}\left\{(y_i - \xi_i)(y_j - \xi_j)\frac{\partial f}{\partial \xi_i}\frac{\partial f}{\partial \xi_j}\right\}\right]$$

$$\Rightarrow \quad Var(f) = \left[\left(\frac{\partial f}{\partial \xi_i}\right)^2 \text{Var}(y_i)\right] + \sum_{i \neq j}\left[\frac{\partial f}{\partial \xi_i}\frac{\partial f}{\partial \xi_j}\text{Cov}(y_i, y_j)\right]$$

Cor. 1. If y_i's are independent, then Var $(f) = (\frac{\partial f}{\partial \xi_i})^2$ Var y_i

2. $f(y) \sim N\left[\xi, \left(\frac{\partial f}{\partial \xi}\right)^2 \text{Var}(y)\right].$

Note. If $f_1 = f_1(y_1, y_2 ..., y_k)$ and $f_2 = f_2(y_1, y_2, ..., y_k)$ be two functions of y's, then $f_1(y_1, y_2, ... y_k) = f_1(\xi_1, \xi_2 .... \xi_k) + \Sigma\left[(y_i - \xi_i)\frac{\partial f_1}{\partial \xi_i}\right]$

$$f_2(y_1, y_2, ..., y_k) = f_2(\xi_1, \xi_2, ..., \xi_k) + \Sigma\left[(y_i - \xi_i)\frac{\partial f_2}{\partial \xi_i}\right]$$

$$\therefore \quad \text{Cov}(f_1, f_2) = E[f_1(y_1, y_2, ... y_k) - f_1(\xi_1, \xi_2, ..., \xi_k)].$$

$$[f_2(y_1, y_2, ..., y_k)] - f_2(\xi_1, \xi_2, ..., \xi_k)]$$

$$= E\left[\Sigma\left\{\left(y_i - \xi_i\right)\frac{\partial f_1}{\partial \xi_i}\right\}\right]\left[\Sigma\left\{\left(y_i - \xi_i\right)\frac{\partial f_2}{\partial \xi_i}\right\}\right].$$

$$= \sum_{i \neq j}^{k}\left[\frac{\partial f_1}{\partial \xi_i}\frac{\partial f_2}{\partial \xi_i}\text{var}(y_i)\right] + \sum_{i \neq j}\left[\frac{\partial f_1}{\partial \xi_i}\frac{\partial f_2}{\partial \xi_i}\text{Cov}(y_i, y_j)\right]$$

Cor. *Standard Error of Sample Moments about Mean.*

$$m_r = \frac{1}{n}\Sigma(x_i - m_1')^r$$

$$= m_r' - {}^rC_1 m_{r-1}' m_1' + ...(-1)^i \, {}^rC_i m_{r-i}' m'^i_1 - .... + (-1)^r m'^r_1$$

$$= f(m_r', m_{r-1}', ..., m_1')$$

$$= f(\mu_r, \mu_{r-1}, ..., \mu_1) + \sum_{i=1}^{r}\left[(m_i' - \mu_i)\frac{\partial f}{\partial \mu_i}\right]$$

since m_r, is distributed about the mean μ_r, with variance $\propto 1/n$.

$$= f(\mu_r, \mu_{r-1}, ..., \mu_1) + (m_1' - \mu_1) + (m_1' - \mu_1)^r \, {}^rc_1 \, \mu_{r-1}$$

Cor. Variance of the central moment based on a large sample of size n.

$$\therefore \quad \text{Var}(m_r) = E[m_r - f\{\mu_r, \mu_{r-1}, ..., \mu_1\}]^2$$

$$= E[(m_r' - \mu_r) - r(m_1' - \mu_1)\mu_{r-1}]^2$$

$$= (m_r' - \mu_1)^2 + r^2\mu^2_{r-1}(m_1' - \mu_1)^2 - 2r\mu_{r-1}(m_1' - \mu_1) \times$$

$$(m_r' - \mu_r')]$$

$$= \text{Var}(m_r') + r^2 \mu^2_{r-1} \text{Var}(m_1') - 2r\mu_{r-1} \text{Cov}(m_1', m_r')$$

$$= \frac{1}{n}(\mu_{2r} - \mu_r^2) + r^2 \mu^2_{r-1}(\mu_2 - \mu_1^2)$$

$$- 2r\mu_{r-1}(\mu_{r+1} + \mu_r \mu_1)]$$

on using (12·8) and $\mu_1' = 0$ ...(12·13)

or $\text{Var}(m_r) = \frac{1}{n}(\mu_{2r} - \mu_r^2 + r^2\mu_2 \mu_{r-1}^2 - 2r\mu_{r-1}\mu_{r+1})$, Approximately

(I.A.S., 1979)

$$\text{Cov}(m_r, m_s) = E[\{m_r - f(\mu_r, ..., \mu_1)\}\{m_s - f(\mu_s, ...\mu_1)\}]$$

$$= E[\{(m_r' - \mu_r) - r(m_1' - \mu_1)\mu_{r-1}\}\{(m_s - \mu_s)$$

$$- s(m_1' - \mu_1)\mu_{s-1}\}]$$

$$= E[(m_r' - \mu_r)(m_s' - \mu_s) - r\mu_{r-1}(m_1' - \mu_1)(m_s' - \mu_s)$$

$$- s\mu_{s-1}(\mu_1')(\mu_s' - \mu_r) + rs\mu_{r-1}\mu_{s-1}(m_1' - \mu_1)^2]$$

$$= \text{Cov}(m_r', m_s') - r\mu_{r-1}\text{Cov}(m_s', m_1')$$

$$- s\mu_{s-1}\text{Cov}(m_r', m_1') + rs\mu_{r-1}\mu_{s-1}\text{Var}(m_1')$$

$$= \frac{1}{n}[\mu_{r+s} - \mu_r\mu_s - r\mu_{r-1}(\mu_{s+1} - \mu_s\mu_1)$$

$$- s\mu_{s-1}(\mu_{r+1} - \mu_r\mu_1) + rs\mu_{r-1}\mu_{s-1}(\mu_2 - \mu_1^{2\prime})]$$

$$= \frac{1}{n}(\mu_{r+s} - \mu_r\mu_s - r\mu_{r-1}\mu_{s+1} - s\mu_{s-1}\mu_{r+1}$$

$$+ rs\mu_{r-1}\mu_{s-1}\mu_2) \quad ...(12·14)$$

Covariance between the Sample Mean and any Sample Mean moment

$$\text{Cov}(m_r, m_s') = E[\{m_r - f(\mu_r, ..., \mu_1)\}\{m_s' - \mu_s\}]$$

$$= E[\{m_r' - \mu_r) - r(m_1' - \mu_1)\mu_{r-1}\}\{m_s' - \mu_s\}]$$

$$= \text{Cov}(m_r', m_s') - r\mu_{r-1}\text{Cov}(m_1', m_s')$$

$$= \frac{1}{n}(\mu_{r+s} - \mu_r\mu_s - r\mu_{r-1}\mu_{s+1}) \quad [\because \mu_1' = 0] ...(12·15)$$

Example 12·11. *Prove that in samples of size n from a symmetrical population, the first product-moment between the mean and any moment of even order vanishes to order 1/n.*

To order $1/n$. we have from (12·13)

$$\text{Cov}(m_1', m_r) = \text{Cov}(m_r, m_1') = \frac{1}{n}(\mu_{r+1} - r\mu_2\mu_{r-1})$$

Now if r is even, μ_{r+1} and μ_{r-1} are moments of odd order and, therefore, vanish for a symmetrical population.

$$\text{Cov}\,(m_1', m_r) = 0$$

Example 12·12 *Prove that in samples of size n from a symmetrical population, the variance of the even moments of samples about sample mean is the same as the variance of the corresponding even moments about the population mean.*

For a symmetrical distribution, μ_{r+1} and μ_{r-1} vanish and therefore, from (12·11) we have

$$\text{Var}\,(m_{2r}) = \frac{1}{n}(\mu_{4r} - \mu_{2r}^2) = \text{Var}\,(m_{2r}') \text{ from (12·10)}$$

Example 12·13 *Find the sampling variance of the following statistics : (i) Standard deviation (ii) Coefficient of variance (iii)* β_1.

(*i*) $m_2 = S^2 \quad \therefore\ \delta m_2 = 2S\delta S$

Hence squaring and taking mean values, we have

$$\text{Var}\, m_2 = 4\mu_2 \,\text{Var}\, S$$

or
$$\text{Var}\, S = \frac{1}{4\mu_2}\text{Var}\, m_2 = \frac{1}{4\mu_2}\cdot\frac{1}{n}(\mu_4 - \mu_2^2)$$

$$= \frac{\sigma^2}{2n} \quad \text{for a normal population}$$

Hence for a normal population, standard error of standard deviation is $\sigma/\sqrt{2n}$. ...(12·15)

(*ii*) The coefficient of variation *V*, is given by

$$V = \frac{100\sqrt{m_2}}{m_1'}$$

Taking logarithmic differentiation, we have

$$\frac{dV}{V} = \frac{dm_2}{2m_2} - \frac{dm_1'}{m_1'}$$

Hence, squaring and taking mean values, we get

$$\frac{\text{Var}\, V}{V^2} = \frac{1}{4\,m_2^2}\text{Var}\, m_2 - \frac{1}{m_2\, m_1'}\text{Cov}\,(m_2, m_1') + \frac{1}{m_1'^2}\text{Var}\, m_1'$$

$$= \frac{1}{n}\left(\frac{\mu_4 - \mu_2^2}{4\,m_2^2} - \frac{\mu_3}{m_2\, m_1'} + \frac{\mu_2}{\mu_1'^2}\right)$$

Taking $m_r = \mu_r$, we have

$$\text{Var } V = \frac{V^2}{n}\left(\frac{\mu_4 - {\mu_2}^2}{4\,{m_2}^2} - \frac{\mu_3}{\mu_2\,\mu_1'} + \frac{\mu_2}{\mu_1'^2}\right)$$

For a normal population, this reduces to

$$\text{Var } V = \frac{V^2}{n}\left(\frac{1}{2} + \frac{\mu_2}{\mu_1'^2}\right) = \frac{V^2}{2\,n}\left(1 + \frac{2V^2}{\mu_1'^2}\right)$$

$$\text{Var } V = \frac{V^2}{2n} \text{ approx.}$$

(*iii*) Let b_1 be the sample estimate of the coefficient β_1, then

$$\beta_1 = \frac{{\mu_3}^2}{{\mu_2}^3} = {\mu_3}^2 \times \frac{1}{{\mu_2}^3}$$

$$\therefore \quad d\beta_1 = \frac{2\,\mu_3}{{\mu_2}^3}\,d\mu_3 - \frac{3{\mu_3}^2}{{\mu_2}^4}\,d\mu_2$$

$$\therefore \text{ Var}(\beta_1) = \left(\frac{2\,\mu_3}{{\mu_2}^3}\right)^2 \text{Var}\,\mu_3 + \left(\frac{3\,{\mu_3}^2}{{\mu_2}^4}\right)^2 \text{Var}\,\mu_2 - 2\left(\frac{6\,{\mu_3}^3}{{\mu_2}^7}\right)\text{Cov}(\mu_3, \mu_2)$$

$$\therefore \text{ Var}(b_1) = \left(\frac{2\mu_3}{{\mu_2}^3}\right)^2 \cdot \frac{\mu_6 - {\mu_3}^2 + 9{\mu_2}^3 - 6\,\mu_2\,\mu_4}{n} + \left(\frac{3\,{\mu_3}^2}{{\mu_2}^4}\right) \cdot \frac{\mu_4 - {\mu_2}^2 + 6\,\mu_2{\mu_1}^3 - 4\,\mu_1\mu_2}{n}$$
$$+ 2\left(\frac{-6\,{\mu_3}^3}{{\mu_2}^7}\right) \cdot \frac{\mu_5 - \mu_3\,\mu_2 - 2\,\mu_1\,\mu_2 - 3\,\mu_2\,\mu_3 - 6\,\mu_0\,\mu_1\,\mu_2}{n}$$
$$= 4\beta_1 \cdot \left(\frac{\beta_4 - \beta_1 + 9 - 6\beta_2}{n}\right) + 2{\beta_1}^2\left(\frac{\beta_2 - 1}{n}\right) - 12\beta_1\left(\frac{\beta_3 - 4\beta_1}{n}\right)$$

where $\quad \beta_{2n+1} = \mu_3\,\mu_{2n+3}/{\mu_2}^{n+3},\ \beta_{2n} = \mu_{2n+2}/{\mu_2}^{n+1}$ and $\mu_1 = 0,$

Hence $\quad \text{Var}(\beta_1) = \beta_1(4\beta_4 - 24\beta_2 + 36 + 9\beta_1\beta_2 - 12\beta_3 + 35\beta_1)/n$

$$\text{Var}(b_1) = 6\beta_1/n \text{ for a normal parent.}$$

Ex. 1. Show that for a symmetrical distribution :

$$\text{Cov}(m_{2r}, m_{2s}) = \text{Cov}(m_{2r}', m_{2s}').$$

Ex. 2. Prove the following results :

(*i*) $\text{Var}(m_3) = \frac{1}{n}(\mu_6 - {\mu_3}^2 - 6\,\mu_4\,\mu_2 + 9{\mu_2}^3)$

(*ii*) $\text{Var}\left(m_4\right)=\frac{1}{n}\left(\mu_8-\mu_4^2-8\,\mu_5\,\mu_3+16\,\mu_2\,\mu_3^2\right)$

(*iii*) $\text{Var}\left(k_4\right)=\text{Var}\,\mu_4-12\mu_2\,\text{Cov}\left(\mu_1,\mu_2\right)+36\mu_2^2\,\text{Var}\left(\mu_2\right)$

$$=\frac{1}{n}\left(\mu_8-12\,\mu_6\,\mu_2-8\,\mu_5\,\mu_3-\mu_4^2+48\,\mu_4\,\mu_2^2+64\,\mu^2{}_3\,\mu_2-36\,\mu^4{}_2\right)$$

(*iv*) $\text{Var}\left(b_2\right)=\frac{1}{n}\left(\beta_6-4\,\beta_4\,\beta_2+4\,\beta_2^3-\beta_3^2+16\,\beta_4\,\beta_1-8\,\beta_3+16\,\beta_1\right)$

where $\qquad \beta_{2n+1}=\mu_3\,\mu_{2n+3}/\mu_2^{n+3},\ \beta_{2n}=\mu_{2n+2}/\mu_2^{n+1}.$

Deduce that, for a normal parent, $\text{Var}\left(m_3\right)=6\sigma^6/n$, $\text{Var}\left(m^4\right)=96\sigma^8/n$, $\text{Var}\left(k_4\right)=26\sigma^8/n.$

Ex. 3. Given a distribution of values x_i, $i=1, 2, \ldots, n$, show that

(*i*) $E\left|x_i-\mu_1'\right|=\delta$ $\qquad$ (*ii*) $E\left|x_i-\mu_1'\right|^2=\sigma^2$

where μ_1', δ and σ^2 are the parent mean deviation and variance respectively.

Hence show that, if d is the sample mean deviation,

$$\text{Var}\,(d)=\frac{1}{n^2}\left(\sigma^2-\delta^2\right)\text{ approx.}$$

Deduce that, for a normal population,

$$\text{Var}\,(d)=\frac{\sigma^2}{n}\left(1-\frac{2}{\pi}\right)\text{approx.}$$

[Hint. $\qquad d_\alpha=\frac{1}{n}\sum_{i=1}^{n}|x_i-\alpha|$

where dx = sample mean deviation measured from the fixed point α

Now $\qquad E\left(d_\alpha\right)=E\,|x_i-\alpha|=\delta_\alpha.$

where δx is corresponding population mean deviation.

$$\text{Var}\,d_\alpha=\frac{1}{n}\text{Var}\,(|x-\alpha|),$$

since the n sample observations are independent.

$$=\frac{1}{n}\{E\,(|x-\alpha|^2)-E^2\,(|x-\alpha|)\}$$

$$=\frac{1}{n}\left\{\sigma^2+\left(\alpha-\mu_1'\right)^2-\delta_\alpha^2\right\}$$

Let $\alpha=\mu_1'$, then $\quad \text{Var}\,d_{\mu_1'}=\frac{1}{n}\left[\,(\sigma^2-\delta\mu_1'^2)\,\right]$

For a normal parent, $\quad d_{\mu_1'}=\sigma\sqrt{\frac{2}{\pi}}$

Ex. 4. If $\bar{x}=\frac{1}{n}\sum_{i=1}^{n}x_i$ and $m_2=\frac{1}{n}\sum_{i=1}^{n}\left(x_i-\bar{x}\right)^2$ are the sample mean and the

sample variance, show that

$$\text{Cov}\left(\bar{x}, m_2\right) = \frac{n-1}{n^2}\mu_3,$$

where μ_3 is the third central moment of the parent population.

Show that if the population is symmetrical, $\bar{x}$ and m_2 are un-correlated.

Ex. 5. (*a*) Show that

$$E\left(m_r'\right) = \mu_r', \ \text{Var}\left(m_r'\right) = \frac{1}{n}\left(\mu_{2r}' - \mu_r'^2\right)$$

where m_r' is the rth sample moment about origin for a random sample $x_1, x_2, \ldots, x_n$ from a population. ***(M.A. Madras, 1980)***

(*b*) Find the sampling variance (or standard error) of the coefficient of variance for samples of size n from a normal population with variance σ^2.

How will you use these variances to test the significance of m_r' and the coefficient of variation ?

(*c*) Show that $\frac{1}{n-1} \cdot \Sigma(x - \bar{x})^2$ is an unbiased estimate of the population variance and that the sampling variance of the estimate is given by

$$\frac{1}{n}\left(\mu_4 - \frac{n-3}{n-1}\mu_2^2\right).$$

[**Hint.** (*c*) $S^2 = \frac{1}{n}\Sigma(x_i - m)^2 - (\bar{x} - m)^2$,

$$S^4 = \frac{1}{n^2}\Sigma(x_i - m)^2(x_j - m)^2 - \frac{2}{n^2}\Sigma(x_i - m)^2(x_j - m) + (\bar{x} - m)^4$$

$$\therefore \quad E\left(S^4\right) = (n-1)^2\mu_4/n^3 + (n-1)(n^2 - 2n + 3)\sigma^4/n^3$$

$$\therefore \quad \sigma^2(S^2) = E(S^4) - [E(S^2)]^2 = \text{etc.}$$]

(*d*) Obtain the formula for the variance of a function of a statistic in large samples, when the variance of the statistic is known. Hence calculate the variance of b_1 where b_1, is the sample estimate of β_1.

Ex. 6. A random sample of N individuals is drawn from an infinite population and out of this sample, a sub-sample of n is randomly drawn. Then, mean of a character x, in the sample is $\bar{X}$ and in the sub-sample is $\bar{x}$. Show that if the process of sampling and sub-sampling is repeated again and again, the standard error of $d = \bar{x} - \bar{X}$ is

$$\sigma_d = \frac{\sigma}{\sqrt{n}}\sqrt{1 - \frac{n}{N}}$$

where σ is the standard deviation of x in population.

If a sample of $4n$ be drawn, how large a sub-sample do you consider must be taken to make it practically certain that $\bar{x}$ will differ from $\bar{X}$ by less than $\sigma/5$?

Ex. 7. M_r is the rth central moment in a sample of size n. If n is large, show that the covariance between M_r and M_s is

$$[\mu_{r+s} - s\,\mu_{r-1}\,\mu_{s+1} - r\,\mu_{s-1}\mu_{r+1} + r\,s\,\mu_{r-1}\,\mu_{s-1}\,\mu_2]/n$$

μ_r being the rth central moment in the population.

Show that in large samples from symmetrical distribution, the correlation between two consecutive central moments is zero.

12·11. Sampling from a Finite Population (without replacement). Let the values $y_1, y_2, ..., y_N$, constitute a finite population with mean μ and variance σ^2, so that

$$\mu = \frac{1}{N}\left(y_1 + y_2 + ... + y_N \right)$$

$$\sigma^2 = \frac{1}{N}[(y_1 - \mu)^2 + (y_2 - \mu)^2 + ... + (y_N - \mu)^2]$$

$$= \frac{1}{N}[y_1^{\ 2} + y_1^{\ 2} + ... + y_N^{\ 2}] - \mu^2$$

Let $x_1, x_2, ..., x_n$ be a random sample drawn without replacement from the population and let $\bar{x}$ be the sample mean ; then

$$\bar{x} = \frac{1}{n}\left(x_1 + x_2 + ... + x_n \right)$$

and $$E(\bar{x}) = \frac{1}{n}[E(x_1) + E(x_2) + ... + E(x_n)]$$

Now for a fixed value of i, the variate x_i takes values $y_1, y_2, ..., y_N$ each with probability $1/N$ so that

$$E\left(x_i \right) = \frac{1}{N}\left(y_1 + y_2 + ... + y_N \right) = \mu, \ (i = 1, 2, ..., n)$$

$$\therefore \quad E(\bar{x}) = \frac{1}{n}(\mu + \mu + ... + \mu) = \mu. \qquad ...(12·16)$$

Thus *the expected value of the sample mean in random sampling is the population mean* μ, *irrespective of sample size.*

The variance $\sigma_{\bar{x}}^2$ of the sample mean $\bar{x}$ is

$$\sigma_{\bar{x}}^{\ 2} = E[\bar{x} - E(\bar{x})]^2 = E(\bar{x} - \mu)^2$$

$$= \frac{1}{n^2} E[(x_1 - \mu) + (x_2 - \mu) + ... + (x_n - \mu)]^2$$

$$\sigma_{\bar{x}}^{\ 2} = \frac{1}{n^2}[\sum_{i=1}^{n} E(x_j - \mu)^2 + 2 \sum_{\substack{j=1 \\ i \neq j}}^{n} \sum_{i=1}^{n} (x_i - \mu)(x_j - \mu)]$$

Since $x_1, x_2, ..., x_n$ is a random sample, the n expected values in the first sum are the same, and for the same reason, the $n(n-1)$ expected values in the second sum are identical. Therefore

$$\sigma_{\bar{x}}^2 = \frac{1}{n^2}[\,nE(x_1-\mu)^2 + n(n-1)E(x_1-\mu)(x_2-\mu)\,]$$

$$= \frac{1}{n}E(x_1-\mu)^2 + \frac{n-1}{n}E(x_1-\mu)(x_2-\mu) \quad ...(12\cdot15)$$

Now $\quad E(x_1-\mu)^2 = \frac{1}{N}[(y_1-\mu)^2 + (y_2-\mu)^2 + \ldots + (y_N-\mu)^2] = \sigma^2$

and *the covariance between any two members of the sample is*

$$Cov(x_i, x_j) = E[(x_i-\mu)(x_j-\mu)]$$

$$= \frac{1}{N(N-1)} \sum_{i=1}^{N} \sum_{\substack{j=1 \\ j\neq i}}^{N} (y_i-\mu)(y_j-\mu)$$

$$= \frac{1}{N(N-1)} \sum_{i=1}^{N} (y_i-\mu)[(y_1-\mu)+(y_2-\mu) + \ldots + (y_N-\mu) - (y_i-\mu)]$$

$$= \frac{1}{N(N-1)}\left[\sum_{i=1}^{N}(y_i-\mu)\{0-(y_i-\mu)\}\right]$$

$$= \frac{-1}{N(N-1)}\sum_{i=1}^{N}(y_i-\mu)^2 = -\frac{N\sigma^2}{N(N-1)}$$

$$= -\frac{\sigma^2}{N-1}$$

$$\therefore \quad \operatorname{Var}\bar{x} = \frac{1}{n}\sigma^2 + \frac{1}{n}(n-1)\left(-\frac{\sigma^2}{N-1}\right) = \frac{\sigma^2}{n}\left(1-\frac{n-1}{N-1}\right)$$

$$= \frac{\sigma^2}{n}\cdot\frac{N-n}{N-1}. \quad ...(12\cdot17)$$

Thus *the variance of the sample mean for random sampling from finite population is*

$$\frac{\sigma^2}{n}\cdot\frac{N-n}{N-1}.$$

In the same way, we can find μ_3, μ_4 for the distribution of sample mean $\bar{x}$. The method is similar, though obviously the algebra is a little more complicated.

Ex. 1. Show that covariance between any two members tends to zero as N tends to infinity. Explain this.

Ex. 2. Deduce the $\sigma_{\bar{x}}^2$ for a random sample from (*i*) a finite population when each member is replaced before the next drawing (*ii*) an infinite population, is, in each case, σ^2/n.

Ex. 3. Show that, without any loss of generality, the population mean can be assumed to be zero. Rewrite the above proof for Var $\bar{x}$ with the simplification.

Ex. 4. A box contains N tickets bearing numbers $y_1, y_2, ..., y_N$, n tickets are drawn at random. If s denotes the sum of the numbers drawn, prove that

$$E(s) = n\,m, \ \text{Var}(s) = n\,\sigma^2 \cdot \frac{N-n}{N-1}.$$

where m is the mean of the numbers appearing on all tickets in the box.

Ex. 5. Show that the correlation coefficient between any two members of the sample is $\frac{1}{N-1}$ and explain why it is negative.

Ex. 6. (*a*) Verify formulae (12·7) and (12·8) for the set 2, 4, 6 taking samples of two.

(*b*) Let the population consist of 6 values 1, 2, 3, 4, 5, 6;

(*i*) write the means of all the 15($= {}^6c_2$) possible samples of size 2;

(*ii*) find the mean of all these 15 means and verify that it is the same as the population mean *viz.,* 3·5;

(*iii*) find the variance of these 15 values and verify that it satisfies the formula deduced above ;

(*iv*) find $\mu_2(\bar{x})$ for these 15 values and verify that it is satisfies the formula

$$\mu_3(\bar{x}) = \frac{(N-n)(N-2n)}{n^2(N-1)(N-2)}\,\mu_3$$

where μ_3 is the population value;

(*v*) find s^2 for each sample and show that the average value of the 15 values of s^2 is $\frac{n-1}{n}\,\mu_2$.

Ex. 7. Show that the sampling variance of the proportion of males in a random sample of n people drawn from a population of N is given by

$$\frac{N-n}{N-1}, \ \frac{p(1-p)}{n},$$

where p is the proportion of males in the population. Obtain also an unbiased estimate of this sampling variance in terms of sample estimates.

[**Hint.** If each male is assigned a number unity and each female a number zero, then $\sigma^2 = p(1-p)$.]

Ex. 8. From the formula for the sampling variances of the sample mean, deduce an expression for the variance of the sample proportion in sampling from a binomial population of size N.

Ex. 9. Deduce the result, of the two exercises, from the expression for the variance of the hypergeometric distribution.

Ex. 10. From a finite population of size N and variance

$$\sigma^2 = \frac{1}{N-1} \sum_{i=1}^{N} \left(x_i - \bar{x}\right)^2,$$

a random sample of n is drawn without replacement. Show that the arithmetic mean and variance given by

$$\frac{1}{n}\sum_{1}^{n} x_i \text{ and } \frac{1}{n-1}\sum_{1}^{n}\left(x_i-\bar{x}\right)^2,$$

are unbiased estimates of the corresponding population values.

Ex. 11. Show that if N is large the error in writing $\sigma_{\bar{x}} = \sigma/\sqrt{n}$ is approx. $50(n-1)/N$ per cent of the value of $\sigma_{\bar{x}}$.

Exercises 12

1. (*a*) Explain the terms :

(*i*) Confidence limits, (*ii*) Critical region,

(*iii*) Null hypothesis, (*iv*) Statistic and parameter.

(*b*) Suggest a possible source of bias in the following :

(*i*) The mean income per family in a certain town is sought to be estimated by sampling from motor owners.

(*ii*) Readers of newspapers are sampled by printing in it an invitation to them to send up their observations on some typical events.

2. (*a*) A coin is tossed 10,000 times and it turns up head 5,195 times. Is it reasonable to think that the coin is unbiased ?

(*b*) A coin is tossed 400 times and it turns up head 216 times. Discuss whether the coin may be unbiased one.

[**Ans.** (*a*) No, the coin is biased.
(*b*) The coin may be regarded as an unbiased one]

3. Certain crosses of the peas gave 5,321 yellow and 1,804 green seeds. The expectation is 25 per cent green seeds on Mendelian hypothesis. Is the divergence significant or might have occurred as due to fluctuations of simple sampling ?

[**Hint.** Total number of seeds in 7,125 and the expected green seeds = 1781. The deviation = 1804 − 1781 = 23 which is about 0·6 *s.d.* and is less than 2 *s.d.* and hence it might have occurred as due to fluctuations of simple sampling.]

4. (*a*) A random sample of 500 pineapples was taken from a large consignment and 65 ware found to be bad. Show that *S.E.* of the proportion of bad one's in a sample of this size is 0·015 and deduce that the percentage of bad pineapples in the consignment almost certainly lies between 8·5 and 17·5.

(*b*) 400 oranges are taken from a large consignment and 50 are found to be bad. Estimate the percentage of bad oranges in the consignment and assign the limits within which the percentage lies.

[**Hint.** (*a*) S.E. of the proportion of bad pineapples

$$= \sqrt{\frac{65}{500} \times \frac{435}{500} \times \frac{1}{500}} = 0{\cdot}015 \text{ or } 1{\cdot}5\%$$

Limits of the bad pineapples in the consignment would be

$$\left(\frac{65}{500}\times 100\right)\pm 3\times 1{\cdot}5 = \text{etc.}$$

(*b*) Percentage of bad oranges = 12·5.

Probable limits are 7·7% and 17·4%.]

5. (*a*) In a simple sample of 600 men from a certain large city, 450 are found to be smokers. In one of 900 from another large city, 450 are smokers. Do the data indicate that the cities are significantly different with respect to prevalence of smoking among men ?

[**Ans.** The hypothesis that the populations are similar is almost certainly wrong.]

(*b*) A sample of 900 days in taken from meteorological records of a certain district and 100 of them are found to be foggy. What are the probable limits to the percentage of foggy days in the district ?

[**Ans.** Limits are : $\left(\frac{1}{9}\times 100\right)\pm 3\times 1{\cdot}05$ per cent *i.e.,* 8% and 14·25%]

(*c*) In a year there are 956 births in a town *A*, of which 52·5% were males while in towns *A* and *B* combined, this proportion in a total of 1,406 births was 0·496. Is there any significant difference in the proportion of male births in the two towns ? **(*I.F.S. 1997*)**

$$\Bigg[\textbf{Hint. } 0{\cdot}496 = \frac{n_1 p_1 + n_2 p_2}{n_1 + n_2} = \frac{956\times{\cdot}525 + 450\times p_2}{1406}$$

$$\text{This gives } p_2 = 0{\cdot}432,\ z = \frac{p_1 \sim p_2}{\sqrt{pq\left(\frac{1}{n_1}+\frac{1}{n_2}\right)}}\Bigg]$$

[**Ans.** There is significant difference in the proportion of male births in the towns *A* and *B*.]

6. In a random sample of 500 persons from town *A*, 200 are found to be consumers of rice. In a sample of 400 from town *B*, 200 are found to be consumers of rice. Discuss the question whether the data reveal a significant difference between *A* and *B* so far as the proportion of rice consumers is concerned. [**Ans.** Yes, the deviation is significant.]

7. One thousand articles from a factory are examined and found to be 3 per cent defective. Fifteen hundred similar articles from a second factory are found to have only 2 per cent defective. Can it reasonably be concluded that the products of the first factory are inferior to the second ?

[**Ans.** It cannot be concluded that the products of the first factory are inferior to the second, since the deviation which is 1 per cent is about 1·6 times the S.E. of the proportion of defectives, which is

$$\left[2{\cdot}4\times 97{\cdot}6\left(\frac{1}{1000}+\frac{1}{1000}\right)\right]^{1/2} \quad \textit{i.e.,} 0{\cdot}625 \text{ per cent approx.}]$$

8. The subject under investigation is the measure of dependence of Tamil on words of Sanskrit origin. One newspaper article, reporting the proceeding of the Constituent Assembly, contained 2,025 words of which 729 were declared by a literary critic to be of Sanskrit origin. A second article by the same author, describing atomic research, contained 1,600 words, of which 640 were declared by the same critic to be of Sanskrit origin. Assuming that the simple sampling conditions hold, estimate the limits for the proportion of Sanskrit words in the writer's vocabulary and examine whether there is any significant difference in the dependence of the writer on words of Sanskrit origin in writing on these two subjects.

[**Hint.** The estimate of percentage of Sanskrit terms in the writer's vocabulary in both the articles taken together is given by

$$p = \frac{(729 + 640) \times 100}{2025 + 1600} = \frac{1369}{3625} \times 100 = 37{\cdot}77$$

S.E. of the difference between the percentages

$$\sqrt{37{\cdot}77 \times 62{\cdot}23 \left(\frac{1}{2025} + \frac{1}{1600} \right)} = 1{\cdot}16 \text{ per cent}$$

$\therefore$ Observed difference is about 3·4 times the S.E. and so could not have arisen through fluctuations of sampling.

$$\text{Limits are } \frac{1369}{3625} \text{ per cent} \pm \sqrt{\frac{1369}{3625} \times \frac{2256}{3625} \times \frac{1}{3625}} \text{ per cent}$$

i.e., 35·34 and 40·20 per cent approx.

9. A sample of 900 members is found to have a mean of 3·4 cms. Can it be reasonably regarded as a simple sample from a large population with mean 3·2 cms and S.D. 2·3 cms. ?

[**Ans.** No. The deviation 0·2 cm. is 2·6 times the S.E. of the mean which is ·077 cm.]

10. The mean of a certain normal distribution is equal to the standard error of the mean of samples of 100 from that distribution. Find the probability that the mean of a sample of 25 from the distribution will be negative.

$$\left[\textbf{Ans. } \frac{1}{\sqrt{2\pi}} \int_{1/2}^{\infty} e^{-\frac{1}{2}t^2} . dt \text{ or } 0{\cdot}3085 \right]$$

11. A normal population has mean 0·1 and S.D. 2·1. Find the probability that the mean of a simple sample of 900 members will be negative. [**Ans.** 0·0764]

[**Hint.** $z = \dfrac{\bar{x} - 0{\cdot}1}{\left(\dfrac{2{\cdot}1}{30}\right)}$ $\therefore$ $\bar{x} = 0{\cdot}1 + {\cdot}07z$ $\therefore$ $\bar{x} < 0$ gives $z < -\dfrac{10}{77}$

$\therefore$ $P(\bar{x} < 0) = P(z > 1{\cdot}43) = 0{\cdot}5 - P(0 < z < 1{\cdot}43) =$ etc.]

12. The sampling distribution of a statistic is a normal population with mean 9 units and S.D. 2 units. What is the probability that the sample will give a value of the statistic greater than 12 units ? [**Ans.** 0·0668]

13. The means of simple samples of 1,000 and 2,000 are 67·5 and 68·8 inches respectively. Can the samples be regarded as drawn from the same population of S.D. 2·5 inches ?

[**Ans.** No. the difference 0·5 is more than 5 times the S.E. of the difference which is 0·097 inch nearly.]

14. A research worker wishes to estimate the mean of a population using a sample sufficiently large that the probability will be 0·95 that the sample mean will not differ from the true mean by more than 25 per cent of the standard deviation. How large a sample should be taken ?

(***B.Sc., Agra, 1981***)

[**Ans.** 62 or more]

15. The guaranteed average life of a certain type of electric light bulb is 1,000 hours with a standard deviation of 125 hours. It is decided to sample the output so as to ensure that 90 per cent of the bulbs do not fall short of the guaranteed average by more than 2·5 per cent. What must be minimum size of the sample ? **[Ans. 41]**

[Hint. We wish $\bar{x} > (100 - 2{\cdot}5)$ of $1000 = 975$.

$$z = \frac{\bar{x} - \mu}{\sigma/\sqrt{n}} > \frac{975 - 1000}{125/\sqrt{n}} \text{ or } z = \frac{-\sqrt{n}}{5}$$

We have $P\left(z > \frac{-\sqrt{n}}{5}\right) = 0{\cdot}90$ or $P(0 < z < \sqrt{n}/5) = 0{\cdot}90$

$\therefore \quad \sqrt{n}/5 = 1{\cdot}28 \quad$ or $n = 41$ approx.]

16. Given that on the average 4% of insured men of age 65 die within a year, and that 60 of a particular group of 1,000 such men died within a year. Can this group be regarded as *a* representative sample ?

[**Ans.** No, since the deviation of a proportion of deaths *i.e.* (·06 – ·04) or ·02 is more than three times the S.E. of the proportion which is

$$\sqrt{\frac{\cdot 04 \times \cdot 96}{1000}} \text{ } i.e. \cdot 0062.]$$

17. A random sample of 1,000 men from Northern India gives their mean wage to be Rs. 2·50 per day with a S.d. of Rs. 1·50. A sample of 1,500 men from Southern India gives a mean wage of Rs. 2·70 per day with a S.D. of Rs. 2. Discuss whether the mean rate of wages varies as between the two regions.

[**Ans.** No. The difference of the means is less than three times the S.E. of the difference]

18. A simple sample of heights 6,400 Englishmen has a mean of 67·85 inches and a S.D. of 2·56 inches while a simple sample of heights of 1,600 Australians has a mean of 68·55 inches and a S.D. of 2·52 inches. Do the data indicate that the Australians are on the average taller than Englishmen ?

[**Ans.** Australian are, on the average, taller than Englishmen. The difference of the means *viz.*, 0·70 in, is ten times the S.E. which is

$$\sqrt{\frac{(2{\cdot}56)^2}{6400} + \frac{(2{\cdot}52)^2}{1600}} \text{ } i.e., 0{\cdot}07 \text{ inch.}]$$

19. A potential buyer of light bulbs bought 30 bulbs of each of 2 brands. Upon testing these bulbs, he found that brand *A* had a mean life of 1, 282 hours with a S.D. of 80 hours, brand *B* had a mean life of 1,208 hours with a S.D. of 94 hours. Can the buyer be quite certain that the two brands differ in quality ?

[**Ans.** Brand *A* is to be preferred. The difference of the means, *i.e.*, 74 hours is greater than 3 times its S.E. which is

$$\sqrt{\frac{(80)^2}{50} + \frac{(94)^2}{50}} = 17{\cdot}5 \text{ hours.}$$

20. The mean yield of two sets of plots and their variability are as given. Examine :

(*i*) whether the difference in the mean yields of the two sets of plots is significant. and

(*ii*) whether the difference in the variability in yields is significant.

	Set of 40 plots	Set of 60 plots
Mean yields per plot	1258 lb.	1243 lb.
S.D. per plot	34	28

[**Ans.** (*i*) significant at 5% level, (*ii*) Not significant at 5% level.]

21. Two populations have their means equal, but S.D. of one is twice the other. Show that in the samples of size 2,000 from each drawn under simple sampling conditions, the difference of means will, in all probability, not exceed $0 \cdot 15\sigma$ where σ is the smaller S.D. What is the probability that the difference will exceed half this amount ?

[**Hint.** $z = \dfrac{|\bar{x}_1 - \bar{x}_2|}{\sqrt{\dfrac{\sigma^2}{n_1} + \dfrac{(2\sigma)^2}{n_2}}} = \dfrac{|\bar{x}_1 - \bar{x}_2|}{0 \cdot 05\sigma}$, $n_1 = 2000 = n_2$.

We should have $|z| < 3$. Hence etc.

We want $P\left[\,\left|\,\bar{x}_1 - \bar{x}_2\,\right| > \dfrac{1}{2} \times 0 \cdot 15\,\sigma\right] = P\left[|z| > \dfrac{3}{2}\right.$

$= 1 - P\left[\,|z| \le \dfrac{3}{2}\right] =$ etc.]

22. A sample of n objects taken from an infinite population in which the proportion $p = (1 - q)$ of objects possess a certain attribute A. Prove that $\bar{p}\,\bar{q}/(n-1)$ is an unbiased estimator of pq/n where $\bar{p}\,(= 1 - \bar{q})$ is the proportion of objects possessing attribute A in the sample.

Instead of taking a sample in the above manner, the objects of the sample are selected at random in succession until a fixed number m of objects possessing A is obtained in the sample. Find the probability distribution of the size n of the sample and prove that $(m-1)/(n-1)$ is an unbiased estimate of p, *but* m/n is not an unbiased estimate of p.

23. A bag contains N balls numbered 1 to N. Let X denote the largest number in n drawings when random sampling with replacement is used, each drawing consisting of one ball and noting the number on the ball. Find the probability distribution of X and show that the expectation of X is

$$\left[N^{n+1} - \sum_{k=1}^{N} (k-1)^n\right] / N^n$$

and show that an approximation for the latter for large N is $nN/(n+1)$.

24. Suppose X takes values 0 and 1 with probability $(1-p)$ and p respectively. If n independent observations $x_1, x_2, \ldots, x_n$ are taken and r denotes the number of observations which are as large as x_1, find the probability distribution of r.

[**Ans.** Reqd. distribution of r is $p \times {}^{n-1}c_{r-1} p^{r-1} q^{n-r} + q \times {}^{n-1}c_{r-1} q^{r-1} p^{n-r}$]

25. It is known that 40% of a population die within a year, 30 persons are vaccinated of which only 5 die in a year. The inventor of the vaccine claims that the vaccine has reduced the mortality by 20%. Examine the following :

(*a*) Justify his claim.

(*b*) Discuss whether the vaccine is effective at all.

(*c*) Find the least value of n such that if $n/6$ die after vaccination, his claim is definitely accepted.

[**Hint.** (*a*) $z = \dfrac{0\cdot32 \sim 0\cdot96}{\sqrt{\dfrac{0\cdot32 \times 0\cdot68}{36}}} > 1\cdot96$. Hence the claim is rejected.

(*b*) $z = \dfrac{\cdot4 - \cdot166}{\sqrt{\dfrac{\cdot4 \times \cdot6}{36}}} > 1\cdot96$ Hence the vaccine is effective.

(*c*) n is the least integer given by $\dfrac{\cdot32 - \dfrac{n}{6}}{\sqrt{\dfrac{\cdot32 \times \cdot68}{n}}} > 1\cdot96$.]

26. Experience has shown that 40% of the young of a certain breed of animals die of a certain disease during the first year of their life. The discoverer of a new vaccine claims that its use at least have mortality and in support of his claim cites an experiment in which out of 40 youngs treated with the vaccine only 8 died. Examine how far his claim is justified

27. If for one half of n events, the chance of success is p and the chance of failure is q, whilst for the other half, the chance of successes is q and the chance of failure is p. Show that the standard deviation of the number of successes is the same as if the chance of successes were p in all cases but the mean of the number of successes is $\frac{1}{2}n$ and not np.

[**Hint.** If x_1 and x_2 denote the number of successes in the first and the second half of the n events respectively, then we have.

$$E(n_1) = \tfrac{1}{2}np,\ E(n_2) = \tfrac{1}{2}nq,\ \text{Var}(n_1) = \tfrac{1}{2}npq,\ \text{Var}(n_2) = \tfrac{1}{2}npq$$

$$\therefore \quad E(n_1 + n_2) = E(n_1) + E(n_2) = n(p+q)/2 = \tfrac{1}{2}n,\ p + q = 1.$$

and $$\text{Var}(n_1 + n_2) = \text{Var}\, n_1 + \text{Var}\, n_2 = \tfrac{1}{2}npq + \tfrac{1}{2}npq = npq;]$$

28. (*a*) In samples of size n from an infinite population find the covariance between the arithmetic mean and variance.

(*b*) Show that $E\left(x_i x_j\right) = \frac{1}{2}E\left(x_i + x_j\right)^2 - \frac{1}{2}Ex_i^2 - \frac{1}{2}Ex_j^2 = n(n-1)p_i p_j$

29. The moment statistic m_r is defined as $\dfrac{1}{n}\Sigma(x - \bar{x})^r$. Show that in normal samples $\quad \text{Cov}\left(m_r, m_n\right) = \dfrac{\sqrt{3(n-1)}}{\sqrt{4n^2 - 9n + 6}}$

30. Show that the coefficient of correlation between the mean of a sample of n individuals and its square *s.d.* is given by $\dfrac{\sqrt{\beta_1}(n_1 - 1)}{\sqrt{\beta_2(n-1) - n + 3}}$.

31. Let $X_i^{'s}$ be i.i.d. r.v. with $E(X_i) = \mu$, $\text{Var}\, X_i = \sigma^2$, $0 < \sigma^2 < \infty$ and $E(X_i - \mu) = \sigma + 1$, find $\lim\limits_{n \to \infty} P\left[\sigma^2 - \dfrac{1}{\sqrt{n}} \le \dfrac{1}{n}\Sigma(X_i - \mu)^2 \le \sigma^2 + \dfrac{1}{\sqrt{n}}\right]$

(B.A. Stat. Hons. Delhi, 1986)

[Hint. Reqd. limit $= \lim_{n\to\infty} P[-1 \le s_n/\sqrt{n} \le 1]$ where $s_n^2 = \sum_1^n [(X_i - \mu)^2 - \sigma^2]$

Also $E(s_n) = \sum_{i=1}^n [E(X_i - \mu)^2 - \sigma^2] = \sum_{i=1}^n [\sigma^2 - \sigma^2] = 0,$

$$\text{Var}\, s_n = \sum_{i=1}^n \text{Var}[(X_i - \mu)^2 - \sigma^2] = \sum_{i=1}^n \text{Var}(X_i - \mu)^2$$

$$= \sum_{i=1}^n E(X_i - \mu)^4 - \sum_{i=1}^n [E(X_i - \mu)^2] = \sum_1^n [\sigma^4 + 1 - \sigma^4] = n$$

$\therefore \quad [s_n - E(s_n)]/\sqrt{\text{Var}\, s_n} \Rightarrow s_n/\sqrt{n} \xrightarrow{L} N(0, 1)$ as $n \to \infty$

$\therefore \quad$ Reqd. limit $= P(-1 \le z \le 1) = 0{\cdot}6826.]$

32. Let X_i, $i = 1, 2, \ldots$ be i.i.d. Poisson variates with mean 2. Use CLT to estimate $P(0 \le S_n \le 60)$.

Where $S_n = x_1 + x_2 + - + x_{60}$

Hint. Since x_i's are i.i.d poisson variates eah with mean $\lambda = 2$ and variate = 2.

So that $E(S_n) = 60 \times 2 = 120$, Var $S_n = Var(x_1 + \ldots + x_{60}) = 60\ \lambda = 120.$

$\therefore$ by Lindeberg-Levy theorem, $P(90 \le S_n \le 150)$

$$= P\left(\frac{100-120}{\sqrt{120}} \le Z \le \frac{150-120}{\sqrt{120}}\right)$$

$$= P\left(-20/\sqrt{120} \le Z \le \frac{30}{\sqrt{120}}\right) = P\left(0 \le Z \le \frac{20}{\sqrt{120}}\right)$$

$$+ P\left(0 \le Z \le \frac{30}{\sqrt{120}}\right) = \text{etc. where } z \sim N(0, 1)$$

13

Exact Sampling Distributions

13·1. Introduction. The sampling distributions play an important role in statistical inference. We construct, from the observations, different statistics to estimate the population parameters. The sampling distributions of these statistics will, in general (but not always, see Distribution Free Methods, Chapter 18), depend on the form and the parameters of the parent population. The probability of the observed value of the statistics then permits the making of statements about the parameters and hence we are able to draw conclusions about the parent population. The sampling distributions are thus fundamental to the whole subject and we shall obtain the sampling distributions of important *statistics*, when the population from which the sample is drawn, is *specified*. We shall *assume* that the sampling is simple and random. Thus, if the parent distribution is $dF(x)$, the simultaneous distribution of n values $x_1, x_2, \ldots\ldots, x_n$ is

$$dF(x_1)\, dF(x_2), \ldots\ldots dF(x_n)$$

where $dF(x) = f(x)\, dx, f(x)$ being the *p.d.f.* of x,

and U is the statistic $\quad U = U(x_1, x_2, \ldots\ldots, x_n)$.

The distribution function $F(u)$ of u is given by

$$F(u) = \int \int \ldots \int d\,F(x_1)\, dF(x_2) \ldots, dF(x_n) \qquad \ldots(13\cdot1)$$

the integration being taken over the domain of the x's, such that $u(x_1, x_1 \ldots\ldots, x_n) \leq u_0$ where u_0 is the fixed value of U.

The evaluation of these integrals is generally a complicated task and either of the following four techniques may be used in simplifying the task.

(*a*) *Geometrical Method* : It involves the use of geometrical terminology *i.e.*, the geometry of Euclidean Hyper-space.

(*b*) *Analytical Method* : The evaluation of integrals is done by the changes of variates or transformation of variates.

(*c*) *Method of Characteristic Functions.*

(*d*) *Method of Induction.*

In this chapter, we shall use only the first method ; the other three methods will be explained in the next chapter.

If $x_1, x_2, \ldots\ldots, x_n$ is a random sample, drawn from a standard normal population, we shall first derive the sampling distributions of

$$(i)\ \chi^2 = x_1^2 + x_2^2 + \cdots + x_n^2 = \sum_{i=1}^{n} x_i^2,$$

$$(ii)\ \bar{x} = \frac{1}{n}(x_1 + x_2 + \ldots + x_n) = \frac{1}{n} \sum_{i=1}^{n} x_i, \text{ and}$$

$$(iii)\ s^2 = \frac{1}{n}\left[(x_1 - \bar{x})^2 + (x_2 - \bar{x})^2 + \cdots + (x_n - \bar{x})^2 \right] = \frac{1}{n} \sum_{i=1}^{n} (x_i - \bar{x})^2$$

From the sampling distribution of χ^2, $\bar{x}$ and s^2 and the distributions of sum and ratio of two independent χ^2 variates we shall deduce the exact sampling distributions of the statistics 't', 'F' and 'z'. The distributions of these three statistics together with those of χ^2, $\bar{x}$ and s^2 constitute the most important exact sampling distributions of Statistics.

13·2. Distribution of the sum of squares of independent standard normal variates. [The Chi-Square Distribution]

The following geometrical proof is due to Prof. R.A. Fisher.

Let $X_1, X_2, \ldots\ldots, X_n$ be n independent normal variates, each of which is distributed normally with mean zero and variance unity. We require the sampling distribution of

$$\chi^2 = X_1^2 + X_2^2 + \cdots\cdots + X_n^2 = \sum_{i=1}^{n} X_i^2$$

The probability that random value of the variable X_i will fall in the interval dx_i is

$$\frac{1}{\sqrt{2\pi}} e^{-\frac{1}{2}x_i^2}\, dx_i$$

Since x_i's are independent variates, the probability that the values of the variates will fall simultaneously in the respective intervals dx_i, $i = 1, 2, \ldots\ldots, n$ is

---

* The distribution was discovered by Helmert in 1875 and rediscovered independently in 1900 by Karl Pearson, who devised by means of it the χ^2 test of goodness of fit, which we shall discuss in Chapter 16.

$$dP = \frac{1}{(2\pi)^{n/2}} e^{-\frac{1}{2}\chi^2} dx_1\, dx_2 \ldots dx_n \qquad \ldots(13{\cdot}2)$$

Let us represent the sample by the point E, with co-ordinates $(x_1, x_2, \ldots\ldots, x_n)$ in Euclidean hyper-space of n dimensions. Then dP is the probability that E will fall in the element of volume $dx_1\, dx_2 \ldots dx_n$.

To obtain the distribution of χ^2 or χ, we find out how the volume element $\prod_{i=1}^{n} dx_i$ transforms, when χ lies between χ and $(\chi + d\chi)$.

Now, when χ is constant, $\sum_{i=1}^{n} x_i^2 = \chi^2$ represents a hyper-sphere, with radius χ and centre at the origin. As $\sqrt{\Sigma x_i^2}$ lies between χ and $(\chi + d\chi)$ the sample point E lies between two concentric hyperspheres of radii χ and $(\chi + d\chi)$ and the volume of the annulus between the two concentric hyperspheres is proportional to $d(\chi^n)$ *i.e.*, to $\chi^{n-1}\, d\chi$. Hence the probability that the value of χ from the random sample will fall in the interval $d\chi$ is, from (13·2), proportional to $e^{-\frac{1}{2}\chi^2} \chi^{n-1}\, d\chi$ *i.e.*, proportional to $e^{-\frac{1}{2}\chi^2} (\frac{1}{2}\chi^2)^{\frac{n-2}{2}} d(\frac{1}{2}\chi^2)$.

Since χ^2 must lie between 0 and ∞ and the integral of the probability density function over the whole range must be unity, the constant factor is therefore found out to be $1/\Gamma(n/2)$

Hence the probability that χ^2 lies in interval $d\chi^2$ is given by

$$dP = \frac{1}{\Gamma(n/2)} e^{-\frac{1}{2}\chi^2} (\tfrac{1}{2}\chi^2)^{\frac{n-2}{2}} d(\chi^2)$$

or
$$dP = \frac{1}{2^{n/2}\Gamma(n/2)} \cdot e^{-\frac{1}{2}\chi^2} (\chi^2)^{\frac{n-2}{2}} d(\chi^2),\ 0 < \chi^2 < \infty \ldots(13{\cdot}3)$$

The distribution (13·3) is called the χ^2 distribution, with n *degrees of freedom* (*d.f.*). The number of independent variates is called the number of degrees of freedom. (The variate having the distribution (13·3) is called the χ^2, chi-square, variate with *n.d.f.*)

Cor. 1. If X_i's are n independent normal variates with means μ_i and *s.d.* σ_i $(i = 1, 2, \ldots\ldots, n)$, then $Z_i = (X_i - \mu_i)/\sigma_i$ is a standard normal variate and all Z_i's are independent variates.

$$\therefore \quad \sum_{i=1}^{n} \left(\frac{X_i - \mu_i}{\sigma_i} \right)^2 = \sum_{i=1}^{n} Z_i^2$$

is a χ^2 variate with *n.d.f.*

Cor. 2. Since for random sampling from a normal population with mean μ and *s.d.* $\sigma, \bar{x}$ is distributed normally about mean μ with *s.d.* $\sigma/\sqrt{n}$ *i.e.* $Z = (\bar{x} - \mu) / \dfrac{\sigma}{\sqrt{n}}$ is a standard normal variate. From §8·2, Cor. 1, it follows that $\left[\dfrac{\sqrt{n}\,(\bar{x} - \mu)}{\sigma} \right]^2$ is a χ^2 variate with 1 *d.f.*

Cor. 3. Let $X_1, X_2, \ldots\ldots, X_n$ be n standard normal variates, but instead of being independent, let these be subject to p *independent linear constraints* of the type

$$a_{i1}\, x_1 + a_{i2}\, x_2 + \ldots\ldots + a_{in}\, x_n = b_i, \; (i = 1, 2, \ldots\ldots, p).$$

This means that in Euclidean hyper-space, the sample point is restricted to lie upon p hyperplanes. The first hyperplane cuts the hypersphere of constant density in a hypersphere of one lower dimension, the second hyperplane cuts this hypersphere in hypersphere of one dimension further less and so on. The result of p intersections will be a hypersphere of p lower dimensions and the distributions of χ^2 subject to p independent linear constraints, would be obtained by replacing n by $\nu = (n - p)$ in (13·3), so that

$$dP = \frac{1}{2^{\nu/2}\, \Gamma(\nu/2)} \cdot e^{-\frac{1}{2}\chi^2} \, (\chi^2)^{\frac{1}{2}(\nu - 2)} \, d(\chi^2), \; 0 < \chi^2 < \infty \quad \ldots(13{\cdot}4)$$

Thus we observe that ______

Each independent linear constraint reduces the number of degrees of freedom by one.

13·2·1. Chief features of the Chi-square Probability Curve. The equation to the Chi-square probability curve is

$$y = \frac{1}{2^{n/2}\Gamma(n/2)} \, e^{-\frac{1}{2}\chi^2} \, (\chi^2)^{(n-2)/2} \quad \ldots(13{\cdot}5)$$

In Fig. 13·1, the curve has been drawn for $\nu = 1, 2 \ldots, 6$, where $\nu = n - 1$.

The probability that value of χ^2 from a random sample will not exceed a fixed value χ_0^2 is obtained by integrating (13·5) with respect to χ^2 from 0 to χ_0^2. Similarly, if P be the probability that the value of χ^2 from a random sample will exceed χ_0^2, we have

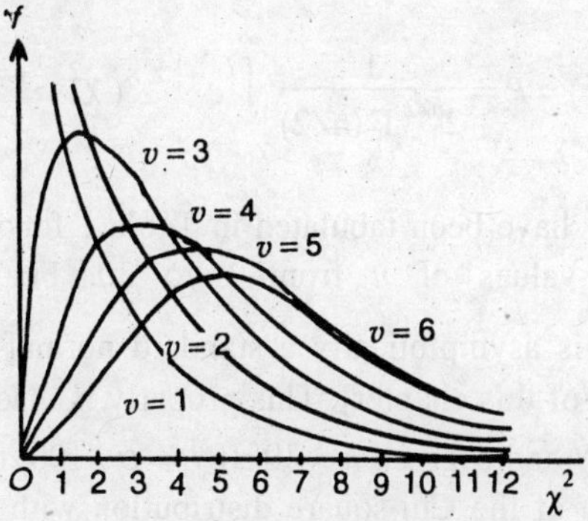

Fig. 13·1. χ^2 – Probability Curve.

Table 13·1. Values of χ^2 with probability P of being exceed in random sampling; v = number of degrees of freedom

v \ P	0.99	0.95	0.50	0.30	0.20	0.10	0.05	0.01
1	0·0002	0·004	0·46	0·07	0·64	2·71	3·84	6·64
2	0·020	0·103	1·39	2·41	3·22	4·00	5·99	9·21
3	0·115	0·35	2·17	3·66	4·64	6·25	7·82	11·34
4	0·30	0·71	3·86	4·88	5·99	7·78	9·49	13·28
5	0·55	1·14	4·35	6·06	7·29	9·24	11·07	15·09
6	0·87	1·64	5·35	7·23	8·56	10·64	12·59	16·81
7	1·24	2·17	6·35	8·38	9·80	12·02	14·07	18·48
8	1·65	2·73	7·34	9·52	11·03	13·36	15·51	20·09
9	2·09	3·32	8·34	10·66	12·24	14·68	16·92	21·67
10	2·56	3·94	9·34	11·78	13·44	15·99	18·31	23·21
11	3·05	4·58	10·34	12·90	14·63	17·28	19·68	24·72
12	3·57	5·23	11·34	14·01	15·81	18·55	21·03	26·22
13	4·11	5·89	12·34	15·12	16·98	19·81	22·36	27·69
14	4·66	6·57	13·34	16·22	18·15	21·06	23·68	29·14
15	5·23	7·26	14·34	17·32	19·31	22·31	25·00	30·58
16	5·81	7·96	15·34	18·42	20·46	23·54	26·30	32·00
17	6·41	8·67	16·34	19·51	21·62	24·77	27·59	33·41
18	7·02	9·39	17·34	20·60	22·76	25·99	28·87	34·80
19	7·63	10·12	18·34	21·69	23·90	27·20	30·14	36·19
20	8·26	10·85	19·34	22·78	25·04	28·41	31·41	37·57
21	8·90	11·59	20·34	23·86	26·17	29·62	32·67	38·93
22	9·54	12·34	21·34	24·94	27·30	30·81	33·92	40·29
23	10·20	13·09	22·34	26·02	28·43	32·01	35·17	41·64
24	10·86	13·85	23·34	27·10	29·55	33·20	36·42	42·98
25	11·52	14·61	24·34	28·17	30·68	34·38	37·65	44·31
26	12·20	15·38	25·34	29·25	31·80	35·56	38·88	45·64
27	12·88	16·15	26·34	30·32	32·91	36·74	40·11	46·96
28	13·56	16·93	27·34	31·39	34·03	37·92	41·34	48·28
29	14·26	17·71	28·34	32·46	35·14	39·09	42·56	49·59
30	14·95	18·49	29·34	33·53	36·25	40·26	43·77	50·89

Reproduced by permission of the author, Professor R.A. Fisher, from his book on *Statistical Methods for Research Workers.*

$$P(\chi^2 \geq \chi_0^2) = P = \frac{1}{2^{n/2}\,\Gamma(n/2)} \int_{\chi_0^2}^{\infty} e^{-\frac{1}{2}\chi^2} (\chi^2)^{\frac{1}{2}n-1}\, d(\chi^2).$$

The value of χ_0^2 have been tabulated in Table 1 for various fixed values of P and for values of n from 1 to 30. For values of $n > 30$, $\sqrt{2\chi^2} - \sqrt{2n-1}$ is asymptotically a standard normal variate. [See Ex. 4 and 7 at the end of this chapter]. This property is made use of for finding χ_0^2 for a given P and a given $n > 30$. If $P = \alpha/100$, then χ^2 is called the α per cent value of the Chi-square distribution with *n.d.f.*

13·2·2. Some Properties of the Chi-square Distribution.

I. The moment generating function of χ^2 distribution with respect to origin is :

$$M_{\chi^2}(t) = E[e^{t\chi^2}] = \int_0^{\infty} e^{t\chi^2} f(\chi^2)\, d\chi^2$$

$$= \frac{1}{2^{n/2}\,\Gamma(n/2)} \int_0^{\infty} e^{-\frac{1}{2}(1-2t)\chi^2} \cdot (\chi^2)^{\frac{n}{2}-1}\, d\chi^2$$

$$= \int_0^{\infty} e^{t\chi^2} \frac{1}{2^{n/2}\Gamma(n/2)}\, e^{-\frac{1}{2}\chi^2} (\chi^2)^{\frac{1}{2}n-1}\, d(\chi^2)$$

$$= \frac{1}{(n/2)\,\Gamma(n/2)} \int_0^{\infty} e^{\frac{1}{2}\chi^2(1-2t)} (\chi^2)^{\frac{n}{2}-1}\, d\chi^2 = \frac{2^{n/2}\,\Gamma(n/2)}{2^{n/2}\,\Gamma(n/2)\,(1-2t)^{n/2}}$$

$$= (1-2t)^{-\frac{n}{2}}, \; |2t| < 1 \qquad \text{...(13·6)}$$

$$= 1 + \frac{n}{2}(2t) + \frac{\left(\frac{n}{2}\right)\left(\frac{n}{2}+1\right)}{2!}(2t)^2 + \cdots\cdots$$

$$+ \frac{\frac{n}{2}\left(\frac{n}{2}+1\right)\cdots\cdots\left(\frac{n}{2}+r-1\right)}{r!}(2t)^r + \cdots\cdots$$

$\mu_r' =$ Coeff. of $t^r/r!$ in the expansion of $M_{\chi^2}(t)$,

$$\therefore \; \mu_r' = 2^r \cdot \frac{n}{2}\left(\frac{n}{2}+1\right)\left(\frac{n}{2}+2\right)\cdots\cdots\left(\frac{n}{2}+r-1\right),$$

$$\mu'_r = n(n+2)(n+4)\cdots\cdots(n+2r-2) = 2^r\,\Gamma\left(\frac{n}{2}+r\right)\Big/\Gamma(n/2) \text{ if } n \text{ is even.}$$

Consequently, the mean value of χ^2 is n and variance of χ^2 is $n(n+2)-n^2=2n$ and thus $(\chi^2-n)/\sqrt{2n}$ is a standard variate.

Example 13·1. *For Chi-square distribution with n.d.f., establish the following recurrence relation for central moments*

$$\mu_{r+1}=2r(\mu_r+n\,\mu_{r-1}),\ r\geq 1. \quad \textbf{(\textit{B.Sc. Stat. Hons. Delhi, 1991})}$$

Hence find β_1 and β_2.

m.g.f. about the mean, n, of the χ^2 distribution with *n.d.f.* is

$$M(t)=E\,[e^{(\chi^2-n)t}]=E\,(e^{-nt}\,.\,e^{\chi^2.t})$$
$$\chi^2=n$$

$$M(t)=e^{-nt}(1-2t)^{-\frac{n}{2}} \qquad [\because E\exp(\chi^2 t)=(1-2t)^{-n/2}]$$

Taking logarithmic differentiation with respect to t, we have

$$\frac{M'(t)}{M(t)}=-n+\frac{n}{2}\cdot\frac{2}{1-2t}=\frac{2nt}{1-2t}$$

$$\Rightarrow (1-2t)\,M'(t)=2nt\,M(t)$$

Differentiating r times *w.r.t.* t by Leibnitz's theorem, we have

$$(1-2t)\,M^{r+1}(t)+(r)(-2)\,M^r(t)=2\,n\,t\,M^r(t)+2nr\,M^{r-1}(t) \qquad ...(1)$$

Remembering that $\mu_r=\left[\dfrac{d^r}{dt^r}M(t)\right]_{t=0}=M^r(0)$

Putting $t=0$ in (1)

and equating coefficient of $t^{r+1}/(r+1)!$ on both sides, we have

$$\mu_{r+1}-2r\,\mu_r=2n_r\,\mu_{r-1} \ \Rightarrow\ \mu_{r+1}=2r\,(\mu_r+n\,\mu_{r-1}).$$

Putting $r=1,2,3$ we obtain

$$\mu_2=2n;\quad \mu_3=4\,(\mu_2+n\mu_1)=8n$$

$$\mu_4=6\,(\mu_3+n\,\mu_2)=48n+12n^2$$

and hence $\beta_1=\mu_3^2/\mu_2^3=8/n$

and $\beta_2=\mu_4/\mu_2^2=12/n+3.$

II. Cumulant Generating Function of Chi-square distribution

$$M_{\chi^2(t)}=(1-2t)^{-\frac{n}{2}},$$

$$K(t)=\log M_{\chi^2}(t)=\frac{-n}{2}\log(1-2t)$$

$k_r=$ Coeff. of $t^r/r!$ in the expansion of $K(t)$

$$\therefore \quad k_r = n \, . \, 2^{r-1} (r-1)\,!$$

$$\mu_1' = k_1 = n, \ \mu_2 = k_2 = 2n$$

$$\mu_3 = k_3 = 8\,n, \ \mu_4 = k_4 + 3k_2^2 = 48n + 12n^2$$

Cor. $\beta_1 = \dfrac{\mu_3^2}{\mu_2^3} = \dfrac{8}{n}$ and $\beta_2 = \dfrac{\mu_4}{\mu_2^2} = \dfrac{12}{n} + 3.$

As $n \to \infty$, $\beta_1 \to 0$ and $\beta_2 \to 3$.

Hence *for large degrees of freedom, chi-square distribution tends to normal distribution.*

III. Additive Property of Independent Chi-square Variates.

If χ_1^2 and χ_2^2 *are independent* χ^2 *variates with* n_1 *and* n_2 *d.f. then* $\chi_1^2 + \chi_2^2$ *is a* χ^2 *variate with* $(n_1 + n_2)$ *d.f.*

Proof. *m.g.f.* of $(\chi_1^2 + \chi_2^2)$

$$= m.g.f. \text{ of } \chi_1^2 \times m.g.f. \text{ of } \chi_2^2 \ (\S 5{\cdot}6{\cdot}2)$$

$$= (1 - 2\,t)^{-n_1/2} \times (1 - 2t)^{-n_2/2} = (1 - 2t)^{-(n_1 + n_2)/2}$$

But this is the *m.g.f.* of a chi-square variate with $(n_1 + n_2)$ *d.f.* Hence by uniqueness theorem, the result follows.

Aliter. for another proof, see §13·4.

The Converse of the additive property. If X and Y are independent non-negative variates, such that $(X + Y)$ has χ^2 – distribution with $(n_1 + n_2)$ d.f., then if X is a chi-square variate with n_1 *d.f.*, then Y is a chi-square variate with n_2 d.f.

Proof. Sine X and Y are independent variates,

$$M_{X+Y}(t) = M_X(t) \times M_Y(t)$$

$$\Rightarrow \quad (1 - 2t)^{-\frac{1}{2}(n_1 + n_2)} = (1 - 2t)^{\frac{-n_1}{2}} \times M_Y(t)$$

$\therefore \quad M_Y(t) = (1 - 2t)^{\frac{-n_2}{2}}$ which is *m.g.f.* of a χ^2-variate with n_2 *d.f.*

Hence the result follows from the uniqueness theorem.

IV. Chi Square Distribution Tends to Normal Distribution as n Tends to Infinity.

Proof. By § 5·5·1, we have

$$M(t) = e^{-\frac{nt}{\sqrt{2n}}} M_0\left(\frac{t}{\sqrt{2n}}\right) = e^{-\frac{nt}{\sqrt{2n}}}\left(1 - \frac{2t}{\sqrt{2n}}\right)^{-\frac{n}{2}}$$

$$\therefore \quad \log_e M(t) = \frac{-nt}{\sqrt{2n}} - \frac{n}{2}\log_e\left(1 - \frac{2t}{\sqrt{2n}}\right)$$

$$\Rightarrow \quad \log_e M(t) = -\frac{nt}{\sqrt{2n}} + \frac{n}{2}\left[\frac{2t}{\sqrt{2n}} + \frac{1}{2}\left(\frac{2t}{\sqrt{2n}}\right)^2 + \cdots\right.$$

$$\Rightarrow \quad \log_e M(t) = \tfrac{1}{2}t^2 + 0\left(\frac{1}{n}\right)$$

Further, as $n \to \infty$, $\log_e M(t) \to \frac{1}{2}t^2$

$\Rightarrow \quad M(t) \to e^{\frac{1}{2}t^2}$ which is the *m.g.f.* of a standard normal variate.

Hence by uniqueness theorem, the result follows.

Ex. 1. (*a*) Show that if χ is a standard normal variate, then $\frac{1}{2}\chi^2$ is a Gamma variate with parameter $\frac{1}{2}$.

(*b*) Use Ex. 1 (*a*) above and the additive property of Gamma variates to show that $\frac{1}{2}\chi^2$ is a Gamma variate with parameter $n/2$. Hence deduce the χ^2 distribution.

Ex. 2. From the table 13·1 find χ_0^2, given

$$P[\chi^2 \geq \chi_0^2] = \cdot 5 \text{ for } \nu = 1, 4, 7, 10.$$

Ex. 3. If the sum of two independent positive variates is a χ^2 variate with $(m+n)$ *d.f.* and if one of them is a χ^2 variate with *m.d.f.* show that the other is a χ^2 variate with *n.d.f.*

Ex. 4. Let x_1, x_2 be independent random Chi-square variates each having 2 *d.f.* then the *m.g.f.* of the variate Y defined by $Y = \frac{1}{2}(x_1 - x_2)$ is $(1-t^2)^{-1}$, $-1 < t < 1$. Deduce that the probability density function of Y is

$$g(y) = \tfrac{1}{2}e^{-|y|}, -\infty < y < \infty$$

Ex. 5. A linear transformation $\xi_i\left(= \sum_{j=1}^{n} c_{ij}x_i\right)$, $(i = 1, 2, \ldots\ldots\ldots\ldots, n)$ is said to be orthogonal if

$$\sum_{i=1}^{n} c_{ij}c_{ik} = 1 \text{ if } j \neq k$$

$$= 0 \text{ if } j = k.$$

Show that for such a transformation

(*i*) $\sum_{i=1}^{n} \xi_i^2 = \sum_{i=1}^{n} x_i^2.$

(*ii*) If $x_1, x_2, \ldots\ldots\ldots, x_n$ are independent standard normal variates, so are $\xi_1, \xi_2, \ldots, \xi_n$.

Ex. 6. Show that the mode of the χ^2-distribution with n *d.f.* is given by $\chi^2 = n - 2$ and skewness $= \sqrt{2/n}$ and for $n \geq 1$, χ-distribution is positively skewed..

(*M.A. Delhi*)

[**Hint.** Skewness = (mean – mode)/s.d.]

Ex. 7. Derive the sampling distribution of $\chi^2 = \sum_{i=1}^{n} \chi_i^2$ by using the integral

$$\underset{\Sigma\chi_i^2 = \mu}{\int_{-\infty}^{+\infty} \cdots\cdots\cdots \int} f(\Sigma\chi_i^2)\, d\chi_i^2 = \tfrac{1}{2} c(n)\, u^{\frac{n-2}{n}} f(u) \text{ where } c(n) = (2\pi)^{\frac{1}{2}n} / \sqrt{n}.$$

13·3. Simultaneous Sampling Distribution of the Mean and Variance in Random Samples drawn from a Normal Population. The following geometrical proof is due to Prof. Fisher.

Let $x_1, x_2, \ldots\ldots, x_n$ be the n values of the random sample drawn from a normal population with mean μ and variance σ^2. Let $\bar{x}$ and s^2 be the mean and variance of the sample. Then we require the simultaneous distribution of $\bar{x}$ and s^2.

First step. Joint distribution of the values $x_1, x_2, \ldots\ldots, x_n$.

Since the values $x_1, x_2, \ldots\ldots\ldots, x_n$ are independent, the probability that these n values of the sample will fall in the respective intervals $dx_1, dx_2, \ldots\ldots, dx_n$ is, by the theorem of compound probability,

$$dp = \left(\frac{1}{\sigma\sqrt{2\pi}}\right)^n \exp.\left[\frac{-1}{2\sigma^2} \sum_{i=1}^{n} (x_i - \mu)^2\right] dx_1 dx_2 \ldots\ldots, dx_n. \qquad \ldots(13\cdot7)$$

Second step. Transformation of the density function

We have $\sum_{i=1}^{n} (x_i - \mu)^2 = \sum_{i=1}^{n} [(x_i - \bar{x}) + (\bar{x} - \mu)]^2$

$$= \sum_{i=1}^{n} (x_i - \bar{x})^2 + n(\bar{x} - \mu)^2,$$

the sum of the products vanishes since $\sum_{i=1}^{n} (x_i - \bar{x}) = 0$.

Thus
$$\sum_{i=1}^{n} (x_i - \mu)^2 = ns^2 + n(\bar{x} - \mu)^2. \qquad \ldots(13\cdot8)$$

Third step. Transformation of the volume element.

Let us represent the sample by a point P with co-ordinates $(x_1, x_2, \ldots\ldots\ldots, x_n)$ in Euclidean hyperspace of n dimensions (Fig. 13·2). Let OE be the unit vector in the n dimensional space. The direction cosines (*d.c.*)

of the unit vector are proportional to 1, 1, ..., 1 so that the co-ordinates of any point on this line are $(\alpha, \alpha, ..., \alpha)$.

Let PM be the perpendicular from P on OE. If M be $(\alpha, \alpha, ..., \alpha)$ then since (the *d.c.*'s. of PM are proportional to $x_1 - \alpha$, $x_2 - \alpha, ..., x_n - \alpha$ and that of OM are proportional to $\alpha, \alpha, ..., \alpha$ and these are at right angles, we have

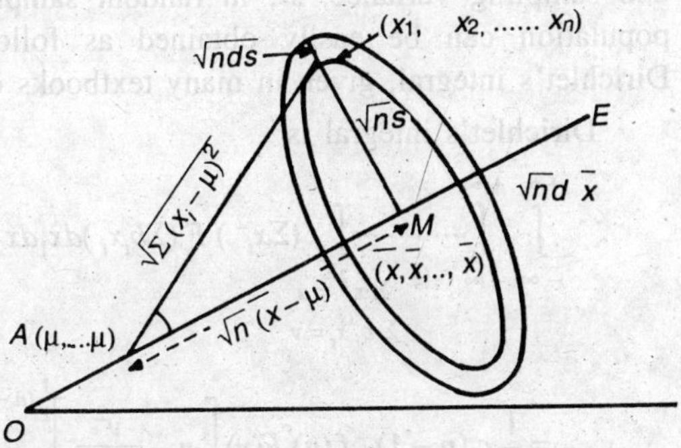

Fig. 13·2

$$\alpha(x_1 - \alpha) + \alpha(x_2 - \alpha) + \cdots\cdots + \alpha(x_n - \alpha) = 0.$$

Since $\alpha \neq 0$, this gives $\alpha = \dfrac{x_1 + x_2 + \cdots + x_n}{n} = \bar{x}$ so that the co-ordinates of M are $(\bar{x}, \bar{x}, ..., \bar{x})$.

Then $OM^2 = \bar{x}^2 + \bar{x}^2 + ... + \bar{x}^2 = n\bar{x}^2$, *i.e.*, $OM = \bar{x}\sqrt{n}$;

and $PM^2 = (x_1 - \bar{x})^2 + (x_2 - \bar{x})^2 + ... + (x_n - \bar{x})^2 = ns^2$ *i.e.*, $PM = s\sqrt{n}$.

The probability density factor in (13·7) has been expressed in terms of $\bar{x}$ and s. We are now to find out the transformation of the volume element $dx_1, dx_2 .... dx_n$ as $\bar{x}$ and s are allowed to take values between the limits $(\bar{x}, \bar{x} + d\bar{x})$ and $(s, s, + ds)$ respectively. When $\bar{x}$ and s are kept constant, OM and PM become constant so that the sample point P moves on the surface of a hypersphere of radius $s\sqrt{n}$ and centre at M in the $(n-1)$ dimensional space which is absolutely orthogonal to the line OE. The spherical shell in which P moves has a thickness $d(PM)$ *i.e.*, $\sqrt{n}.\, ds$ and suffers a displacement of length $d(OM)$ *i.e.*, $\sqrt{n}\, d\bar{x}$ along the unit vector vector OE. Therefore, the volume element described by P, as $\bar{x}$ lies in $(\bar{x}, \bar{x} + d\bar{x})$ and s in $(s, s + ds)$, is proportional to $(s\sqrt{n})^{n-2} . \sqrt{n}\, ds. \sqrt{n}\, d\bar{x}$, that is volume element $= \text{const.} s^{n-2}\, ds\, d\bar{x}$.

Fourth Step. The required probability, therefore, is

$$dp = \text{const.}\, e^{-\frac{n}{2\sigma^2}(\bar{x}-\mu)^2}\, e^{-\frac{ns^2}{2\sigma^2}}\, s^{n-2}\, ds d\bar{x}.$$

Since dp is expressible in the form $f_1(\bar{x})\, d\bar{x} . f_2(s) ds$

we conclude that *the distributions of $\bar{x}$ and s are independent.*

Alternative Proof. The sampling distributions of sample mean, $\bar{x}$, and sampling variance, s^2, in random samples drawn from a normal population can be easily obtained as follows, using the following Dirichlet's integral, given in many textbooks on advanced calculus.

Dirichlet's integral is

$$\underset{\substack{\Sigma x_i^2 = u \\ \Sigma b_i x_i = v}}{\int_{-\infty}^{+\infty}\int_{-\infty}^{+\infty} \cdots\cdots\cdots \int} f(\Sigma x_i^2)\, F(\Sigma b_i x_i)\, dx_1 dx_2 \ldots\ldots dx_n$$

$$= \frac{1}{2} c\,(n-1)\,.\, f(u)\, F(v) \left[n - \frac{v^2}{\Sigma b_i^2} \right]^{\frac{1}{2}(n-3)} \left(\Sigma b_i^2 \right)^{-\frac{1}{2}}$$

where $c(n) = 2(\pi)^{\frac{1}{2}} / \Gamma(\frac{1}{2}n)$

To obtain the distributions of $\bar{x}$ and s^2, we have

$$\left(\frac{1}{2\pi} \right)^{\frac{1}{2}n} \underset{\substack{\Sigma x_i^2 = u \\ \frac{1}{n}\Sigma x_i = \bar{x}}}{\int_{-\infty}^{+\infty} \cdots\cdots\cdots \int} e^{-\frac{1}{2}\Sigma x_i^2}\, dx_1 dx_2 \ldots dx_n$$

$$= \left(\frac{1}{2\pi} \right)^{\frac{1}{2}n} \frac{1}{2} c(n-1)\,.\, e^{-\frac{1}{2}u} \left(u - n\bar{x}^2 \right)^{\frac{1}{2}(n-3)} n^{\frac{1}{2}}\, du\, d\bar{x}$$

which is the joint distribution of u and $\bar{x}$.

Put $u = ns^2 + n\bar{x}^2$, so that the joint distribution of ns^2 and $\bar{x}$ is

$$\left(\frac{1}{2\pi} \right)^{\frac{1}{2}n} \cdot \frac{1}{2} c\,(n-1).\, e^{-\frac{1}{2}ns^2} \left(ns^2 \right)^{\frac{1}{2}(n-3)} n^{\frac{1}{2}}\, e^{-\frac{1}{2}n\bar{x}^2}\, ds^2 d\bar{x}$$

Thus the distribution of the sample mean $\bar{x}$ is

$$dp_1 = c_1 e^{-\frac{n}{2}\bar{x}^2}\, d\bar{x}, \quad -\infty < \bar{x} < \infty$$

and the distribution of the sample variance s^2 is

$$dp_2 = c_2 e^{-\frac{n}{2}s^2} (s^2)^{\frac{1}{2}(n-3)}\, ds^2, 0 < s^2 < \infty.$$

Cor 1. The distribution of sample mean $\bar{x}$ is

$$dp_1 = c_1 e^{-\frac{n}{2\sigma^2}(\bar{x}-\mu)^2}\, d\bar{x}, \quad -\infty < \bar{x} < \infty \qquad \ldots(13{\cdot}8)$$

where the constant c_1 is given by

$$c_1 \int_{-\infty}^{\infty} e^{-\frac{n}{2\sigma^2}(\bar{x}-\mu)^2} \, d\bar{x} = 1, \text{ so that } c_1 = \frac{1}{\sigma}\sqrt{\frac{n}{2\pi}} .$$

Thus $\bar{x}$ is distributed normally with mean μ and variance σ^2/n. In other words, $(\bar{x}-\mu)\sqrt{n}/\sigma$ is a standard normal variate and hence $n(\bar{x}-\mu)^2/\sigma^2$ is a χ^2 variate with 1 d.f.

Cor. 2. The distribution of s^2 is

$$dp_2 = c_2 e^{-\frac{ns^2}{2\sigma^2}} (s^2)^{\frac{n-3}{2}} \, ds^2, \; 0 < s^2 < \infty \qquad \text{...(13·9)}$$

where $\int_0^{\infty} c_2 e^{-\frac{ns^2}{2\sigma^2}} (s^2)^{\frac{1}{2}(n-3)} \, ds^2 = 1$ so that $c_2 = (n/2\sigma^2)^{(n-1)/2} \div \Gamma\left(\frac{n-1}{2}\right)$

Writing (13·9) as

$$dp_2 = \frac{1}{2^{\frac{1}{2}(n-1)} \Gamma\left(\frac{n-1}{2}\right)} e^{-\frac{ns^2}{2\sigma^2}} \left(\frac{ns^2}{\sigma}\right)^{\frac{n-3}{2}} d\left(\frac{ns^2}{\sigma^2}\right) \qquad \text{...(13·9)}$$

we conclude that ns^2/σ^2 is a χ^2 *variate with (n–1) d.f.*

Further, we observe that $\frac{1}{2} ns^2/\sigma^2$ *is a Gamma variate with parameter* $\frac{1}{2}(n-1)$.

i.e., $\frac{1}{2} n s^2/\sigma^2$ is $\gamma\{\frac{1}{2}(n-1)\}$.

$$\text{Also } E(s^2) = \frac{1}{\Gamma\left(\frac{n-1}{2}\right)} \left(\frac{n}{2\sigma^2}\right)^{\frac{n-1}{2}} \int_0^{\infty} s^2 \, e^{-\frac{ns^2}{2\sigma^2}} (s^2)^{\frac{n-3}{2}} \, ds^2$$

$$= \frac{1}{\Gamma\left(\frac{n-1}{2}\right)} \left(\frac{n}{2\sigma^2}\right)^{\frac{n-1}{2}} \Gamma\left(\frac{n+1}{2}\right) \left(\frac{2\sigma^2}{n}\right)^{\frac{n+1}{2}}$$

$$\Rightarrow \quad E(s^2) = \frac{n-1}{n}\sigma^2. \qquad \therefore \; E\left(\frac{ns^2}{n-1}\right) = \sigma^2.$$

In other words, $\dfrac{ns^2}{n-1} = \left[\dfrac{1}{n-1} \sum_{i=1}^{n} (x_i - \bar{x})^2 = S^2\right]$ *is an unbiased estimate of σ^2 and is based on (n – 1) d.f.* This is expected, since the n variates $x_i - \bar{x}$ are not all independent, but are connected by the linear relation $\sum_{i=1}^{n} (x_i - \bar{x}) = 0$ *i.e.* $(x_1 - \bar{x}) + (x_2 - \bar{x}) + \cdots + (x_n - \bar{x}) = 0.$

[**Definition.** A statistic $t = t(x_1, x_2, \ldots\ldots, x_n)$ is said to be an unbiased estimate of the parameter θ if $E(t) = \theta$.]

Note. The three main results of this section hold for normal population only and in a sense characterise the normal population. It can be proved that for no other distribution—

(*i*) the sample mean and sample variance are independently distributed.

(*ii*) the sample mean is exactly normally distributed. or

(*iii*) the sum of squares of deviations from either the population of sample mean is exactly distributed by the Chi-square distribution.

Ex. 1. In Fig. 13·2, (page 487) using $PA^2 = PM^2 + MA^2$, deduce that

$$\sum_{i=1}^{n} (x_i - \mu)^2 = ns^2 + n(\bar{x} - \mu)^2$$

Ex. 2. Deduce the mean, variance and characteristic function of the Chi-square distribution from the mean, variance and characteristic function of a Gamma variate.

Ex. 3. Show that $\frac{1}{\sigma^2} \sum_{i=1}^{n} (x_i - \mu)^2$ is distributed as χ^2 with *n.d.f.* Explain why $\frac{1}{\sigma^2} \sum_{i=1}^{n} (x_i - \bar{x})^2$ is distributed as χ^2 with $(n-1)$ *d.f.*

Ex. 4. Write down the density function for $S^2 = \frac{1}{n-1} \sum_{i=1}^{n} (x_i - \bar{x})^2$ and show that it is maximum when $S^2 = \sigma^2$.

Ex. 5. Prove the following result for a random sample $x_1, x_2, \ldots\ldots, x_n$ drawn from a population with finite variance σ^2 :

$$E\left[\frac{1}{n} \sum_{i=1}^{n} (x_i - \bar{x})^2\right] = \frac{n-1}{n} \sigma^2$$

[**Hint.** $E s^2 = \frac{1}{n} E[\Sigma(x_i - \bar{x})^2] = \frac{1}{n}[\Sigma E(x_i - \mu)^2 - nE(\bar{x} - \mu)^2]$

$$= \frac{1}{n}\left[n\sigma^2 - \frac{n\sigma^2}{n}\right] = \text{etc.}$$

Ex. 6. Show that *m.g.f.* of the distribution of s^2 is $\left(1 - \frac{2t\sigma^2}{n}\right)^{-\frac{n-1}{2}}$

Hence $\mu_1' = \frac{n-1}{n}\sigma^2, \mu'_2 = \frac{(n^2-1)}{n^2}\sigma^4$ and $Var\,(s^2) = 2\frac{(n-1)^4}{n}\sigma^4$.

Deduce for large sample, $Var\,(s^2) = \frac{2\sigma^4}{n}$.

Ex. 7. For what value of k, will $E\,(ks^2 - \sigma^2)$ be minimum ?

13·4. Distribution of Sum and Ratio of two independent Chi-square Variates. *If χ_1^2 and χ_2^2 are independent χ^2 variates with n_1 and n_2 degrees of freedom respectively, then*

(a) $\chi^2 = \chi_1^2 + \chi_2^2$ *is a* χ^2 *variate with* $(n_1 + n_2)$ *d.f. and*

(b) $T^2 = \frac{\chi_1^2}{\chi_2^2}$ *is a* $\beta_2\left(\frac{n_1}{2}, \frac{n_2}{2}\right)$ *variate.*

Proof. The joint probability that a random value χ_1^2 will fall in the interval $d\,(\chi_1^2)$ and, at the same time, a random value of χ_2^2 will fall in the interval $d\,(\chi_2^2)$ is, by the theorem of compound probability,

$$dP = \frac{1}{2^{(n_1+n_2)/2}\,\lceil\overline{n_1/2}\,\lceil\overline{n_2/2}}\, e^{-\frac{1}{2}(\chi_1^2+\chi_2^2)}\,(\chi_1^2)^{\frac{1}{2}(n_1-2)}\,(\chi_2^2)^{\frac{1}{2}(n_2-2)}\, d\,(\chi_1^2)\,d\,(\chi_2^2) \qquad ...(13\cdot9)$$

In other words, the probability that χ_1 and χ_2 (the positive square roots of χ_1^2 and χ_2^2) will fall simultaneously in the intervals $d\chi_1$ and $d\chi_2$ respectively is

$$dP = \frac{4}{2^{(n_1+n_2)/2}\,\left\lceil\overline{\frac{n_1}{2}}\right.\left\lceil\overline{\frac{n_2}{2}}\right.}\, e^{-\frac{1}{2}(\chi_1^2+\chi_2^2)}\,\chi_1^{\,n_1-1}\,\chi_2^{\,n_2-1}\,d\chi_1 d\chi_2$$

Let us introduce the new variates χ and θ given by

$$\chi_1 = \chi\sin\theta,\ \chi_2 = \chi\cos\theta \quad i.e.,\ \chi = \sqrt{\chi_1^2+\chi_2^2},\ \theta = \tan^{-1}\left(\frac{\chi_1}{\chi_2}\right)$$

so that $$d\chi_1\,d\chi_2 = \left|\frac{\partial\,(\chi_1, \chi_2)}{\partial\,(\chi, \theta)}\right| d\chi\,d\theta = \chi d\chi\,d\theta,$$

and as χ_1^2 and χ_2^2 range from 0 to ∞, χ^2 ranges from 0 to ∞ and θ from 0 to $\pi/2$.

Hence the probability that simultaneously χ will fall in the interval $d\chi$ and θ in the interval $d\theta$ is

$$dp = \frac{4}{2^{\frac{1}{2}(n_1+n_2)}\,\Gamma\left(\frac{n_1}{2}\right)\Gamma\left(\frac{n_2}{2}\right)}\, e^{-\frac{1}{2}\chi^2} (\chi \sin\theta)^{n_1-1} (\chi\cos\theta)^{n_2-1} \chi\, d\chi\, d\theta$$

$$= \frac{1}{2^{\frac{n_1+n_2}{2}}\,\Gamma\left(\frac{n_1+n_2}{2}\right)}\, e^{-\frac{1}{2}\chi^2} (\chi^2)^{\frac{1}{2}(n_1+n_2-2)}\, d(\chi^2)$$

$$\times \frac{2}{B\left(\frac{n_1}{2}, \frac{n_2}{2}\right)} (\sin\theta)^{n_1-1} (\cos\theta)^{n_2-1}\, d\theta.$$

Thus, it follows that

(*i*) the variates χ^2 and θ are independently distributed ;

(*ii*) $\chi^2 = \chi_1^2 + \chi_2^2$ is distributed like χ^2 with $(n_1 + n_2)$ degrees of freedom.

The variate θ is distributed as

$$dp_1 = \frac{2}{B\left(\frac{n_1}{2}, \frac{n_2}{2}\right)} \sin^{n_1-1}\theta \cos^{n_2-1}\theta\, d\theta,\ 0 \le \theta \le \pi/2 \qquad \text{...(13·9')}$$

Now $\theta = \tan^{-1}\left(\frac{\chi_1}{\chi_2}\right) = \tan^{-1} T$, so that $d\theta = \frac{dT}{1+T^2}$.

$\therefore\ \sin\theta = \frac{T}{\sqrt{1+T^2}},\ \cos\theta = \frac{1}{\sqrt{1+T^2}}.$

As θ ranges from 0 to $\pi/2$, T ranges from 0 to ∞ and the variate T is distributed as

$$dp_2 = \frac{2}{B\left(\frac{n_1}{2}, \frac{n_2}{2}\right)} \cdot \frac{T^{n_1-1}}{(1+T^2)^{\frac{1}{2}(n_1+n_2)}}\, dT$$

$$= \frac{1}{B\left(\frac{n_1}{2}, \frac{n_2}{2}\right)} \frac{(T^2)^{\frac{n_1}{2}-1}}{(1+T^2)^{\frac{1}{2}(n_1+n_2)}}\, d(T^2),\ 0 < T^2 < \infty \qquad \text{...(13·10)}$$

Hence T^2 is a $\beta_2\left(\frac{n_1}{2}, \frac{n_2}{2}\right)$ variate.

Example 13·2. *Show that if χ_1^2 and χ_2^2 are independently distributed as χ^2 with n_1 and n_2 d.f. respectively, then*

$$U = \frac{\chi_1^2}{\chi_1^2 + \chi_2^2} \text{ is a } \beta_1\left(\frac{n_1}{2}, \frac{n_2}{2}\right) \text{ variate.}$$

First Method. $U = \dfrac{\chi_1^2}{\chi_1^2 + \chi_2^2} = 1 \div \left(1 + \dfrac{\chi_2^2}{\chi_2^2}\right) = \dfrac{1}{1 + 1/T^2}$

so that $1/T^2 = \dfrac{1}{U} - 1 = \dfrac{1-U}{U}$, $T^2 = \dfrac{U}{1-U}$ and $d\,(T^2) = \dfrac{du}{(1-u)^2}$.

The range of T^2 is from 0 to ∞ and of U is from 0 to 1. The probability differential of T^2 is (13·10). Consequently, the probability differential of U is

$$dp = \frac{1}{B\left(\frac{n_1}{2}, \frac{n_2}{2}\right)} \left(\frac{u}{1-u}\right)^{\frac{n_1}{2}-1} (1-u)^{\frac{n_1+n_2}{2}} \frac{du}{(1-u)^2}$$

$$= \frac{u^{\frac{n_1}{2}-1} (1-u)^{\frac{n_2}{2}-1}}{B\left(\frac{n_1}{2}, \frac{n_2}{2}\right)} du, \; 0 \le u \le 1.$$

Hence u is a $\beta_1\left(\dfrac{n_1}{2}, \dfrac{n_2}{2}\right)$ variate.

Second Method. We require the distribution of

$$u = \frac{{\chi_1}^2}{{\chi_1}^2 + {\chi_2}^2} = \sin^2 \theta.$$

Now $\theta = \sin^{-1} \sqrt{u}$, where $\chi_1 = \chi \sin \theta$ and by (13·9″) distribution of θ is

$$dp = \frac{1}{B\left(\frac{n_1}{2}, \frac{n_2}{2}\right)} (\sin \theta)^{n_1 - 1} (\cos \theta)^{n_2 - 1} \, d\theta$$

We easily find that the probability differential of U is

$$dp = \frac{1}{B\left(\frac{n_1}{2}, \frac{n_2}{2}\right)} u^{n_1/2-1} (1-u)^{n_2/2-1} \, du, \; 0 < u < 1$$

so that u is a $\beta_1\left(\dfrac{n_1}{2}, \dfrac{n_2}{2}\right)$ variate.

Example 13·3. *Show that if X and Y are independent normal variates with means m_1, m_2 and variances ${\sigma_1}^2, {\sigma_2}^2$ respectively, then quotient*

$$Z = \frac{X - m_1}{Y - m_2}$$

follows Cauchy's distribution

$$dp = \frac{\sigma_1 \sigma_2}{\pi\,(\sigma_1^2 + \sigma_2^2 z^2)}\,dz, \; -\infty < z < \infty.$$

We have $\dfrac{\sigma_2^2 z^2}{\sigma_1^2} = \left(\dfrac{x - m_1}{\sigma_1}\right)^2 \Big/ \left(\dfrac{y - m_2}{\sigma_2}\right)^2 = \dfrac{u}{v}$

where, by Cor. 1 page 480, $u = \left(\dfrac{x - m_1}{\sigma_1}\right)^2$ is a χ^2 variate with 1 *d.f.* and so is v and hence $\sigma_2^2 z^2/\sigma_1^2$ is the quotient of two independent χ^2 variates. Therefore, by § 13·4, page 491 $\sigma_2^2 z^2/\sigma_1^2$ is a $\beta_2\left(\frac{1}{2}, \frac{1}{2}\right)$ variate with probability differential

$$dp = \frac{\left(\dfrac{\sigma_2^2}{\sigma_1^2} z^2\right)^{\frac{1}{2} - 1} d\left(\dfrac{\sigma_2^2}{\sigma_1^2} z^2\right)}{B\left(\frac{1}{2}, \frac{1}{2}\right)\left(1 + \dfrac{\sigma_2^2}{\sigma_1^2} z^2\right)} = \frac{\sigma_1 \sigma_2}{\pi\,(\sigma_1^2 + \sigma_2^2 z^2)\sqrt{z^2}}\,d\,(z^2\,).$$

The range of z^2 is from 0 to ∞ and of z is from $-\infty$ to $+\infty$. The probability differential of z is

$$dp = \frac{\sigma_1 \sigma_2}{\pi\,(\sigma_1^2 + \sigma_2^2 z^2)}\,dz, \; -\infty < z < \infty,$$

the factor 2 disappearing since the integral from $-\infty$ to ∞ must be equal to unity.

Ex. 1. Putting $\dfrac{\chi_1^2}{\chi_2^2} = y, \chi_1^2 + \chi_2^2 = z$ in (13·9′) and using

$$J\,(\chi_1^2, \chi_2^2) = \frac{z}{(1 + y)^2},$$

show that the joint probability density of y and z is given by

$$\frac{y^{\frac{n_1 - 1}{2}}}{2\,\left|\overline{\dfrac{n_1}{2}}\right.\,\left|\overline{\dfrac{n_2}{2}}\right.\,(1 + y)^{\frac{1}{2}(n_1 + n_2)}}\left(\frac{z}{2}\right)^{\frac{(n_1 + n_2 - 2)}{2}} e^{-\frac{z}{2}}, \; 0 < z < \infty, \; 0 < y < \infty$$

Hence show that the density function of y is

$$\frac{1}{B\left(\frac{n_1}{2}, \frac{n_2}{2}\right)} \cdot \frac{y^{\frac{n_1-2}{2}}}{(1+y)^{\frac{1}{2}(n_1+n_2)}}.$$

and that of z is $e^{-\frac{z}{2}} z^{\frac{n_2+n_2-2}{2}} \Big/ 2^{\frac{1}{2}(n_1+n_2)} \left|\frac{1}{2}(n_1+n_2)\right.$.

Ex. 2. Deduce from Ex. 1, the distribution of the statistic F defined by

$$F = \frac{n_2}{n_1} \cdot \frac{\chi_1^2}{\chi_2^2}.$$

[**Hint.** Put $y = \frac{n_1}{n_2} F$ in the distribution of y.]

13·5. *Student's *t* Distribution. Let $x_1, x_2, \ldots, x_n$ be the members of a random sample drawn from a normal population with mean μ and variance σ^2.

Then we define the statistic t as

$$t = \frac{(\bar{x} - \mu)\sqrt{n}}{S} \Rightarrow t^2 = \frac{(\bar{x} - \mu)^2 (n-1)}{s^2} \qquad \ldots(13\cdot11)$$

where $S^2 = \frac{1}{n-1} \sum_{i=1}^{n} (x_i - \bar{x})^2$ *i.e.*, $(n-1)S^2 = \sum_{i=1}^{n} (x_i - \bar{x})^2 = ns^2$,

Hence $\frac{t^2}{v} = \frac{(\bar{x} - \mu)^2 \div \frac{\sigma^2}{n}}{ns^2/\sigma^2}$ where $v = n - 1$.

We proved in § 13·3 Cor 2, page 489 that $n(\bar{x} - \mu)^2/\sigma^2$ is distributed as a Chi-square variate with 1 *d.f.* and that ns^2/σ^2 is a Chi-square variate with $(n-1)$ *d.f.*

---

* The distribution of the statistic t was first found by W.S. Gosset, who wrote under the *nom de plume* of Student'. His paper entitled 'the probable error of the mean', published in 1908, ushered in the modern era of Statistics. Student has found the distribution of $(\bar{x} - \mu)/S$. The correct definition of t and its distribution were rigorously established by Fisher in 1926.

The t distribution is independent of population parameters μ and σ^2. The definition of t is independent of σ^2 and, therefore, it will not be necessary to know σ^2 before using this distribution. This is the chief merit of the t distribution. Before its discovery, it was usual to replace the unknown σ by S in

$$t = \frac{\bar{x} - \mu}{\sigma/\sqrt{n}} \text{ to get } t = \frac{\bar{x} - \mu}{S/\sqrt{n}}$$

and still to regard t as normally distributed. While for large samples, the distribution of t is asymptotically normal, it is far from normal in small samples. The discovery of its distribution supplied an exact distribution applicable to small samples and in fact to samples of all sizes.

Thus t^2/ν is the ratio of two independent χ^2 variates, distributed with 1 and ν *d.f.* respectively and, therefore, by § 13·4, t^2/ν is a $\beta_2\,[\frac{1}{2}, \nu/2]$ variate and its distribution is

$$dp = \frac{\left(\dfrac{t^2}{\nu}\right)^{-\frac{1}{2}} d\left(\dfrac{t^2}{\nu}\right)}{B\left(\dfrac{1}{2}, \dfrac{\nu}{2}\right)\left(1+\dfrac{t^2}{\nu}\right)^{\frac{1}{2}(\nu+1)}}, \; 0 < t^2 < \infty$$

or
$$dP = \frac{dt}{\sqrt{\nu}.\,B\left(\dfrac{1}{2}, \dfrac{\nu}{2}\right)\left(1+\dfrac{t^2}{\nu}\right)^{\frac{1}{2}(\nu+1)}}, \; -\infty < t < \infty \qquad \text{...(13·12)}$$

the factor 2 disappearing since the integral $-\infty$ to ∞ must be equal to unity. This distribution is known as the t distribution with ν *d.f.*

Alternative Proof. We can deduce the t distribution from the distributions of $\bar{x}$ and s as follows :

The probability differential for $\bar{x}$ is

$$dp = \frac{\sqrt{n}}{\sigma\sqrt{2\pi}}\, e^{-\frac{n}{2\sigma^2}(\bar{x}-\mu)^2}\, d\bar{x}$$

We require the probability differential of $t = \dfrac{(\bar{x}-\mu)\sqrt{n}}{S} = \dfrac{(\bar{x}-\mu)\sqrt{\nu}}{s}$

where $\nu S^2 = \sum\limits_{i=1} (x_i - \bar{x})^2 = ns^2, \nu = n-1, \; d\bar{x} = s\,dt/\sqrt{\nu}.$

Hence, for a fixed value of s in the interval ds, the probability that t will fall in the interval dt is

$$dp = \frac{\sqrt{n}}{\sigma\sqrt{\nu}\sqrt{2\pi}}\, e^{-\frac{ns^2t^2}{2\nu\sigma^2}}\, s\, dt.$$

The probability that a random value of s will fall in the interval ds is

$$dp' = 2\left(\frac{n}{2\sigma^2}\right)^{\frac{\nu}{2}} \frac{e^{-\frac{ns^2}{2\sigma^2}}, s^{\nu-2}}{\left|\overline{\dfrac{\nu}{2}}\right.}\, d\,s, \; 0 < s < \infty$$

The probability that simultaneously t will lie in the interval dt and s in the interval ds is the product $dp.\,dp'$. Integrating this product for s from 0 to ∞ we get the probability that t will lie in the interval dt as

$$dP = \frac{2\sqrt{n}}{\sigma\sqrt{2\pi}\,\left|\frac{\nu}{2}\right.} \cdot \frac{n^{\nu/2}}{(2\sigma^2)^{\nu/2}}\, dt \times \int_0^\infty e^{-\frac{ns^2}{2\sigma^2}\left(1+\frac{t^2}{\nu}\right)} s^{\nu-2} ds$$

$$= \frac{dt}{\sqrt{\nu}\, B\left(\frac{1}{2}, \frac{\nu}{2}\right)\left(1+\frac{t^2}{\nu}\right)^{\frac{1}{2}(\nu+1)}}.$$

Ex. 1. Show that, in Fig. 13·2, $t = \sqrt{n-1}\cot\phi$, so that, t is constant over the cone obtained by rotating by AP about the unit vector OE (page 487).

Show that the distribution of ϕ is given by

$$dF = \text{Const.} \sin^{n-2}\phi\, d\phi,\ 0 \le \phi \le \pi,$$

Hence deduce the distribution of t.

Ex. 2. Show that in both the cases discussed above, t can be regarded as the ratio of a statistic distributed normally with zero means to an unbiased estimate of the variance of that statistic in random sampling.

Ex. 3. If $I_x(p, q)$ represents the incomplete Beta function defined by

$$I_x(p, q) = \frac{1}{B(p, q)} \int_0^x t^{p-1}(1-t)^{q-1}\, dt,\ p > 0,\ q > 0$$

show that the distribution $P(t)$, given by (13·12), is given by

$$P(t) = 1 - \tfrac{1}{2} I_x\left(\frac{n}{2}, \frac{1}{2}\right), \text{ where } x = 1\Big/\left(1 + \frac{t^2}{\nu}\right).$$

[**Hint.** $P(t) = 1 - \int_t^\infty f(u)\, du = \text{etc.}$] ***(M.Sc. Stat. Delhi, 1990)***

13·5·1. Chief Features of the t Probability Curve.

The equation to the t probability curve is

$$y = \frac{1}{\sqrt{\nu}\, B\left(\frac{1}{2}, \frac{\nu}{2}\right)} \cdot \frac{1}{\left(1+\frac{t^2}{\nu}\right)^{\frac{1}{2}(\nu+1)}} \quad ...(1)$$

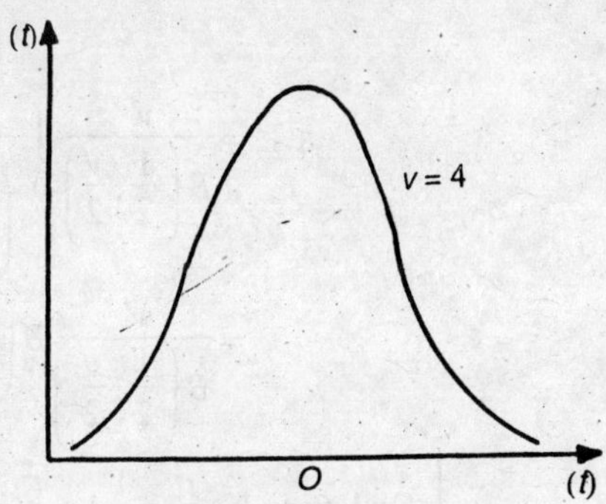

Fig. 13·3. t Probability Curve.

The curve is symmetrical about the line $t = 0$, since only even powers of t appear in the equation of the curve. Further since $\dfrac{1}{1+t^2/\nu}$ decreases rapidly as $|t|$ increases, the t-curve will tail off to zero on either side of the origin. O, (Fig. 13·3). In other words, the curve is asymptotic to the t-axis at each end. The curve has a maximum ordinate at $t = 0$.

The probability that the value of t from a random sample will be between two fixed values t_1 and t_2 is given by integrating y with respect to t from t_1 to t_2. Similarly, if P be the probability that the value of $|t|$ from a random will exceed t_0 then we have

$$P = 2\int_{t_0}^{\infty} \frac{1}{\sqrt{\nu}B\left(\frac{1}{2}, \frac{\nu}{2}\right)} \cdot \frac{dt}{\left(1+\frac{t^2}{\nu}\right)^{\frac{1}{2}(\nu+1)}}$$

Values of t_0 have been tabulated Table 13·2, page 500, for various fixed values of P for values of ν from 1 to 60. For values of $\nu > 60$, the fact t is asymptotically a standard normal variate can be used.

13·5·2. Some Properties of t Distribution. It can be easily proved :

I. *All the moments of odd order about the origin vanish.*

The moment of order $2r$ about the origin (which is also the mean) :

$$\mu_{2r}' = 2\int_0^{\infty} \frac{t^{2r}}{\sqrt{\nu}B\left(\frac{1}{2}, \frac{\nu}{2}\right)} \cdot \frac{dt}{\left(1+\frac{t^2}{\nu}\right)^{\frac{1}{2}(\nu+1)}}$$

$$= \int_0^{\infty} \frac{t^{2r}\left(\frac{t^2}{\nu}\right)^{-\frac{1}{2}} d\left(\frac{t^2}{\nu}\right)}{B\left(\frac{1}{2}, \frac{\nu}{2}\right)\left(1+\frac{t^2}{\nu}\right)^{\frac{1}{2}(\nu+1)}}$$

$$\mu_{2r}' = \frac{\nu^r}{B\left(\frac{1}{2}, \frac{\nu}{2}\right)} \int_0^{\infty} \frac{\left(\frac{t^2}{\nu}\right)^{r+\frac{1}{2}-1}}{\left(1+\frac{t^2}{\nu}\right)^{\frac{1}{2}(\nu+1)}} d\left(\frac{t^2}{\nu}\right)$$

$$= \frac{\nu^r}{B\left(\frac{1}{2}, \frac{\nu}{2}\right)} B\left(r+\frac{1}{2}, \frac{\nu}{2}-r\right), \quad r < \frac{\nu}{2}.$$

$$\left[\text{On letting } 1+\frac{t^2}{\nu} = \frac{1}{y} \text{ i.e. } \frac{t^2}{\nu} = \frac{1-y}{y}\right]$$

Hence $$\mu_{2r} = \frac{(2r-1)(2r-3)\ldots\ldots 1}{(\nu-2)(\nu-4)\ldots\ldots(\nu-2r)}\nu^r \qquad \ldots(2)$$

In particular, we have $\mu_2 = \dfrac{1}{\nu - 2} \cdot \nu = \dfrac{\nu}{\nu - 2} > 1$

$$\mu_4 = \frac{3 \cdot 1}{(\nu - 2)(\nu - 4)} \cdot \nu^2 = \frac{3\nu^2}{(\nu - 2)(\nu - 4)}$$

$$\therefore \qquad \beta_2 = \frac{\mu_4}{\mu_2^{\,2}} = 3\left(\frac{\nu - 2}{\nu - 4}\right).$$

As $\nu \to \infty$, $\beta_2 = \lim\limits_{\nu \to \infty} 3 \left(\dfrac{1 - \dfrac{2}{\nu}}{1 - \dfrac{4}{\nu}}\right) = 3.$

Hence for large degrees of freedom, *t*-distribution tends to normal distribution.

II. *Recurrence Relation for the moments of the t-distribution*

We have $\quad \mu_{2r} = \dfrac{(2r-1)(2r-3)......1}{(\nu-2)(\nu-4)....(\nu-2r)} \nu^r$, from (2), page 498

and, $\therefore$, $\quad \mu_{2r-2} = \dfrac{(2r-3)(2r-5).......1}{(\nu-2)(\nu-4)\dots(\nu-2r+2)} \nu^{r-1}$

Hence $\quad \mu_{2r} = \dfrac{\nu(2r-1)}{\nu - 2r} \cdot \mu_{2r-2}$

which is the required relation. **(*B. Sc. Delhi Stat. Hon., 1992*)**

Ex. 1. Show that, as $\nu \to \infty$

$$\frac{1}{\sqrt{\pi \nu}} \frac{\Gamma\left(\frac{\nu+1}{2}\right)}{\Gamma\left(\frac{\nu}{2}\right)\left(1 + \frac{t^2}{\nu}\right)^{\left(\frac{1}{2}\nu + 1\right)}} \to \frac{1}{\sqrt{2\pi}} e^{-\frac{1}{2}t^2}$$

Hence deduce that *t* distribution approaches normality as the number of degrees of freedom increases indefinitely.

[**Hint.** Use $x\,! = \sqrt{2\pi}\, e^{-x} x^{x + \frac{1}{2}}$]

Ex. 2. Show that standardised *t* variate is $t\sqrt{(\nu - 2)/\nu}$ and that its distribution also approaches normality as the number of degrees of freedom increases.

Ex. 3. From the table, find t_0 such that

$$P[-t_0 \leq t \leq t_0] = \cdot 95$$

for $\nu = 5, 10, 15, 20, 25$ and 30 and show that this illustrates the approach of the distribution to normality.

Ex. 4. Show that for $\nu = 1$ the Student's *t*-distribution reduces to Cauchy distribution.

Table 13·2.

Values of mod. t with a probability of being exceeded in random sampling

ν = number of degrees of freedom

u \ P	0·50	0·10	0·05	0·02	0·01
1	1·000	6·34	12·71	31·82	63·66
2	0·816	2·92	4·30	6·96	9·92
3	0·765	2·35	3·18	4·54	5·84
4	0·741	2·13	2·78	3·75	4·60
5	0·727	2·02	2·57	3·36	4·03
6	0·718	1·94	2·45	3·14	3·71
7	0·711	1·90	2·36	3·00	3·50
8	0·706	1·86	2·31	2·90	3·36
9	0·703	1·83	2·26	2·82	3·25
10	0·700	1·81	2·23	2·76	3·17
11	0·697	1·80	2·20	2·72	3·11
12	0·695	1·78	2·18	2·68	3·06
13	0·694	1·77	2·16	2·65	3·01
14	0·692	1·76	2·14	2·62	2·98
15	0·691	1·75	2·13	2·60	2·95
16	0·690	1·75	2·12	2·58	2·92
17	0·689	1·74	2·11	2·57	2·90
18	0·688	1·73	2·10	2·56	2·88
19	0·688	1·73	2·09	2·24	2·86
20	0·687	1·72	2·09	2·53	2·84
21	0·686	1·72	2·08	2·52	2·83
22	0·686	1·72	2·07	2·51	2·82
23	0·685	1·71	2·07	2·50	2·81
24	0·685	1·71	2·06	2·49	2·80
25	0·684	1·71	2·06	2·48	2·79
26	0·684	1·71	2·06	2·48	2·78
27	0·684	1·70	2·05	2·47	2·77
28	0·683	1·70	2·05	2·47	2·76
29	0·683	1·70	2·04	2·46	2·76
30	0·683	1·70	2·04	2·46	2·75
35	0·682	1·69	2·03	2·44	2·72
40	0·681	1·68	2·02	2·42	2·71
45	0·680	1·68	2·02	2·41	2·69
50	0·679	1·68	2·01	2·40	2·68
60	0·678	1·67	2·00	2·39	2·66
∞	0·674	1·64	1·96	2·33	2·58

Reproduced by permission of the author, Professor R.A. Fisher.
from his book on *Statistical Methods for Research Worker*.

13·6. Distribution of *t* for Comparison of two Sample Means.

Let x_{1i} $(i = 1, 2, \ldots\ldots, n_1)$ and x_{2j} $(j = 1, 2, \ldots\ldots, n_2)$ be two independent random samples drawn from two normal populations with mean

μ and μ' respectively but with *same* variance σ^2. Let $\bar{x}_1, \bar{x}_2$ be sample means and let

$$S_1^2 = \frac{1}{n_1 - 1} \sum_{i=1}^{n_1} (x_{1i} - \bar{x}_1)^2 \text{ and } S_2^2 = \frac{1}{n_2 - 1} \sum_{j=1}^{n_2} (x_{2j} - \bar{x}_2)^2$$

Then we define the statistic t by the equation

$$t = \frac{(\bar{x}_1 - \bar{x}_2) - (\mu - \mu')}{\sqrt{S^2 \left(\frac{1}{n_1} + \frac{1}{n_2} \right)}} \quad \ldots(13 \cdot 13)$$

where

$$S^2 = \frac{(n_1 - 1)S_1^2 + (n_2 - 1)S_2^2}{n_1 + n_2 - 2}$$

$$= \frac{\sum_{i=1}^{n_1} (x_{1i} - \bar{x}_1)^2 + \sum_{j=1}^{n_2} (x_{2j} - \bar{x}_2)^2}{n_1 + n_2 - 2} \quad \ldots(13 \cdot 14)$$

and is called the *pooled estimate* of the population variance.

Hence if $\nu = n_1 + n_2 - 2$ we have, from (13·13) and (13·14)

$$\frac{t^2}{\nu} = \frac{[(\bar{x}_1 - \mu) - (\bar{x}_2 - \mu')]^2 / \sigma^2 \left(\frac{1}{n_1} + \frac{1}{n_2} \right)}{[(n_1 - 1)S_1^2 + (n_2 - 1)S_2^2]/\sigma^2} \quad \ldots(13 \cdot 15)$$

since samples have been drawn from normal populations with means μ and μ' respectively and same variance σ^2. $\bar{x}_1 - \mu$ and $\bar{x}_2 - \mu'$ are distributed normally about mean zero with variances σ^2/n_1 and σ^2/n_2 respectively. Further, since the samples are independent, the difference $(\bar{x}_1 - \mu) - (\bar{x}_2 - \mu')$ is distributed normally about zero mean with variance $\sigma^2 \left(\frac{1}{n_1} + \frac{1}{n_2} \right)$. Hence the numerator of second member of (13·15) is a χ^2 variate with 1 *d.f.* We have already proved in § 13·3, Cor. 2, page 489 that $(n_1 - 1) S_1^2/\sigma^2$ and $(n_2 - 1) S_2^2/\sigma_2^2$ are chi-square variates with $(n_1 - 1)$ and $(n_2 - 1)$ *d.f.* respectively and, therefore, by virtue of the additive property of χ^2, the sum

$$\frac{1}{\sigma^2} [(n_1 - 1) S_1^2 + (n_2 - 1) S_2^2]$$

is a χ^2 variate with $\nu \; (= n_1 + n_2 - 2)$ *d.f.* Also

$$E(S^2) = E\left[\frac{(n_1 - 1)S_1^2 + (n_2 - 1)S_2^2}{n_1 + n_2 - 2} \right]$$

$$E(S^2) = \frac{(n_1-1)\sigma^2 + (n_2-1)\sigma^2}{n_1+n_2-2} = \sigma^2 \qquad \text{...(13·16)}$$

Thus the second member of (13·15) is the ratio of two independent χ^2 variates with 1 and ν *d.f.* respectively. Therefore, by § 13·4, $\frac{t^2}{\nu}$ is $\beta_2\left(\frac{1}{2}, \frac{\nu}{2}\right)$ variate so that t defined by (13·13) conforms to the t distribution with ν *d.f.*

Ex. 1. Can we use $S'^2 = \left[\sum_{i=1}^{n_1} (x_{1i} - \bar{x})^2 + \sum_{j=1}^{n_2} (x_{2j} - \bar{x})^2\right] \Big/ (n_1 + n_2 - 1)$

instead of S^2 where

$$\bar{x}' = \frac{n_1\bar{x}_1 + n_2\bar{x}_2}{n_1 + n_2} ? \qquad \textbf{[Ans. No]}$$

Ex. 2. For three independent samples of sizes 5, 6, 7, the sample variances are 16, 25, 36. Find an unbiased estimate of σ^2 by pooling the three samples.

13·7. *Snedecor's *F* Distribution. *Distribution of the ratio of independent estimates of the population variance.*

Let x_{1i} $(i = 1, 2, \ldots, n_1)$ and x_{2j} $(j = 1, 2, \ldots, n_2)$ be the values of two independent random samples drawn from the same normal population with variance σ^2. Let $\bar{x}_1$ and $\bar{x}_2$ be the sample means and let

$$S_1^2 = \frac{1}{n_1 - 1}\sum_{i=1}^{n_1} (x_{1i} - \bar{x}_1)^2 \text{ and } S_2^2 = \frac{1}{n_2 - 1}\sum_{j=1}^{n_2} (x_{2j} - \bar{x}_2)^2.$$

Then we define the statistic F by the relation

$$F = \frac{S_1^2}{S_2^2}, \quad (S_1^2 > S_2^2) \qquad \text{...(13·17)}$$

Hence $\frac{\nu_1 F}{\nu_2} = \frac{(n_1-1)S_1^2/\sigma^2}{(n_2-1)S_2^2/\sigma^2}$, where $\nu_1 = n_1 - 1$, $\nu_2 = n_2 - 1$.

The numerator and denominator of the second member are independent χ^2 variates with ν_1 and ν_2 *d.f.* respectively. Hence $\nu_1 F/\nu_2$ is a $\beta_2(\nu_1/2, \nu_2/2)$ variate so that the probability that a random value of F will fall in the interval dF is

---

* Prof. R.A. Fisher had found the distribution of $z = \frac{1}{2}\log_e F$ and published (in 1925), tables giving the values of z that will be exceeded with probability 0·05 and 0·01 respectively corresponding to specified values of *d.f.* From these, Prof. G.W. Snedecor (1934) prepared a table for the variance ratio F and called it 'F' in honour of Fisher.

$$dp = \frac{\nu_1^{\frac{\nu_1}{2}} \nu_2^{\frac{\nu_2}{2}} F^{\frac{\nu_1 - 2}{2}} dF}{B\left(\frac{\nu_1}{2}, \frac{\nu_2}{2}\right)(\nu_1 F + \nu_2)^{\frac{1}{2}(\nu_1 + \nu_2)}}, \; 0 < F < \infty \qquad \text{...(13·18)}$$

This distribution is called the distribution of the variance – ratio F with ν_1 and ν_2 *d.f.*

Ex. 1. If $\nu_1 = \nu_2 = n - 1$, show that the distribution of $F' = 1/F$ is the same as the distribution of F with $n - 1$ and $n - 1$ *d.f.* and that the curve $F' = 1/F$ has to be traced from the right to the left. Explain the practical utility of this.

Ex. 2. If two independent samples are drawn from normal populations with variances σ_1^2 and σ_2^2 respectively show that the distribution of $F = S_1^2/S_2^2$ is

$$dp = \frac{\sigma_1^{\nu_1} \sigma_2^{\nu_2} \nu_1^{\frac{\nu_1}{2}} \nu_2^{\frac{\nu_2}{2}}}{B\left(\frac{\nu_1}{2} \cdot \frac{\nu_2}{2}\right)} \cdot \frac{F^{\frac{\nu_1 - 2}{2}}}{\left(\frac{\nu_1}{\sigma_1^2} F + \frac{\nu_2}{\sigma_2^2}\right)^{\frac{1}{2}(\nu_1 + \nu_2)}} dF, 0 < F < \infty.$$

Ex. 3. Prove that $u = \dfrac{\frac{\nu_1}{\nu_2} F}{1 + \frac{\nu_1}{\nu_2} F}$ is a $\beta_1\left(\frac{\nu_1}{2} - 1, \frac{\nu_2}{2} - 1\right)$ variate. Hence show that the tables of incomplete Beta distribution can be used to tabulate the distribution function of F.

13·7·1. Important Properties of the F – Distribution and the F – Curve.

I. The rth moment about the origin of the F distribution is given by

$$\mu_r' = E(F^r) = \frac{\left(\frac{\nu_1}{\nu_2}\right)^{\frac{\nu_1}{2}}}{B\left(\frac{\nu_1}{2}, \frac{\nu_2}{2}\right)} \int_0^\infty F^{r + \frac{\nu_1}{2} - 1} \left(1 + \frac{\nu_1}{\nu_2} F\right)^{-\frac{\nu_1 + \nu_2}{2}} dF$$

$$= \frac{\left(\frac{\nu_1}{\nu_2}\right)^{-r}}{B\left(\frac{\nu_1}{2}, \frac{\nu_2}{2}\right)} \int_0^1 y^{\frac{\nu_2}{2} - r - 1} (1 - y)^{r + \frac{\nu_1}{2} - 1} dy,$$

where $1 + \frac{\nu_1}{\nu_2} F = \frac{1}{y}$ and $dF = -\frac{\nu_2}{\nu_1} \cdot \frac{dy}{y^2}$.

$$\therefore \qquad \mu_r' = \frac{\left(\dfrac{\nu_1}{\nu_2}\right)^{-r}}{B\left(\dfrac{\nu_1}{2}, \dfrac{\nu_2}{2}\right)} \cdot B\left(\frac{\nu_1}{2} + r, \frac{\nu_2}{2} - r\right), \quad r < \frac{\nu_2}{2}$$

or
$$\mu_r' = \frac{\Gamma\left(\dfrac{\nu_1}{2} + r\right)\Gamma\left(\dfrac{\nu_2}{2} - r\right)}{\Gamma\left(\dfrac{\nu_1}{2}\right)\Gamma\left(\dfrac{\nu_2}{2}\right)}\left(\frac{\nu_2}{\nu_1}\right)^r, \quad r < \frac{\nu_2}{2}.$$

In particular,
$$\mu_1' = \frac{\Gamma\left(\dfrac{\nu_1}{2} + 1\right)\Gamma\left(\dfrac{\nu_2}{2} - 1\right)}{\Gamma\left(\dfrac{\nu_1}{2}\right)\Gamma\left(\dfrac{\nu_2}{2}\right)}\left(\frac{\nu_2}{\nu_1}\right) = \frac{\nu_2}{\nu_2 - 2},$$

which is *independent* of ν_2 and is always greater than unity.

$$\mu_2' = \frac{\Gamma\left(\dfrac{\nu_1}{2} + 2\right)\Gamma\left(\dfrac{\nu_2}{2} - 2\right)}{\Gamma\left(\dfrac{\nu_1}{2}\right)\Gamma\left(\dfrac{\nu_2}{2}\right)}\left(\frac{\nu_2}{\nu_1}\right)^2 = \frac{(\nu_1 + 2)\nu_2^2}{(\nu_2 - 2)(\nu_2 - 4)\nu_1}.$$

$$\therefore \qquad \mu_2 = \mu_2' - \mu_1'^2 = \frac{2\nu_2^2(\nu_1 + \nu_2 - 2)}{\nu_1(\nu_2 - 2)^2(\nu_2 - 4)} \sim 2\left(\frac{1}{\nu_1} + \frac{1}{\nu_2}\right).$$

The distribution of F is independent of the population variance σ^2 and depends on ν_1 and ν_2 only. The F curve is J shaped, if $\nu_2 \le 2$ and bell shaped for $\nu_1 > 2$. For $\nu_1 > 4$ the shape of the F curve is as shown in Fig. 13·4.

Fig. 13·4. F Probability Curve.

II. Mode of F-distribution. We have from (13·18), page 503,

$$\log f(F) = c + \left[\frac{\nu_1}{2} - 1\right]\log F - \left(\frac{\nu_1 + \nu_2}{2}\right)\log\left(1 + \frac{\nu_1}{\nu_2}F\right)$$

$$\text{Now} \quad 0 = \frac{\partial}{\partial F}[\log f(F)] = \left(\frac{\nu_1}{2} - 1\right) \cdot \frac{1}{F} - \frac{\nu_1 + \nu_2}{2} \cdot \frac{1}{1 + \dfrac{\nu_1}{\nu_2}F} \cdot \frac{\nu_1}{\nu_2}$$

$$\Rightarrow \qquad F = \frac{\nu_2(\nu_1 - 2)}{\nu_1(\nu_2 + 2)} \qquad \text{...(1)}$$

It can be easily shown that $\frac{\partial^2}{\partial F^2}[\log f(F)] < 0$ for the value of F given by (1). Hence the mode of the F-distribution is

$$\frac{\nu_2(\nu_1 - 2)}{\nu_1(\nu_2 + 2)} \quad \text{or} \quad \frac{\nu_2}{\nu_2 + 2} \cdot \frac{\nu_1 - 1}{\nu_1}.$$

which is less than unity.

Ex. 1. For $\nu_1 = \nu_2 = n - 1$, prove that (*i*) A.M.=H.M. $= \frac{n-1}{n-3}$, (*ii*) median is at F=1, (*iii*) $Q_1 Q_3 = 1$, (*iv*) $D_1 D_3 = 1$.

Ex. 2. Find Pearson's measure of skewness *viz.*, $\frac{\text{mean} - \text{mode}}{\text{standard deviation}}$ for the F distribution and show that it is always positively skew.

Also show that for $\nu_1 = \nu_2 = n - 1$, $\frac{\text{mean} - \text{mode}}{\text{mean} - \text{median}} \to 3$ as $n \to \infty$.

Ex. 3. If $F(\nu_1, \nu_2)$ represents an F variate with ν_1, ν_2 *d.f.* prove that $F(\nu_2, \nu_1)$ is distributed as $1/F(\nu_1, \nu_2)$ variate. Deduce that if $c > 0$, then

$$P\left[F(\nu_1, \nu_2) \geq c\right] = P\left[F(\nu_2, \nu_1) \leq \frac{1}{c}\right]$$

Show how you can make use of this property to deduce the values of F which will be exceeded with probability ·95 and ·99 from the table values of F exceeded with probability ·05 and ·01.

Ex. 4. Show how that probability points of $F(\nu_2, \nu_1)$ can be obtained from those of $F(\nu_1, \nu_2)$.

Ex. 5. If F has a F distribution with $n_1 = 5$, $n_2 = 10$, find a and b so that $P(F \leq a) = \cdot 05$ and $P(F \leq b) = \cdot 95$ and accordingly $P(a < F < b) = \cdot 90$.

[**Hint.** $P(F \leq a) = P\left(\frac{1}{F} \geq \frac{1}{a}\right) = 1 - P\left(\frac{1}{F} \leq \frac{1}{a}\right)$ $\left[\because F' = \frac{1}{F}\right]$

$= 1 - P[F(n_2, n_1) \leq \frac{1}{a}] \Rightarrow P[F(n_2, n_1) \geq \frac{1}{a}] = \cdot 05$]

[**Ans.** $a = \frac{1}{4 \cdot 74}$, $b = 3 \cdot 33$]

Ex. 6. Prove that if X has the F-distribution with (m, n) *d.f.* and Y has the F-distribution with (n, m) *d.f.*, then for every $a > 0$

$$P\left(x \leq a\right) + P\left(y \leq \frac{1}{a}\right) = 1.$$

Example 13·4. (*a*) *Let* χ_1^2, χ_2^2 *be independently distributed variates each having a Chi-square distribution with* n_1 and n_2 *d.f. Derive the* *distribution of* $F=\dfrac{\chi_1^2/n_1}{\chi_2^2/n_2}$.

(*b*) *If* s_1^2, s_2^2 *are the sample variances based on n independent observations each from two normal populations with variances* σ_1^2, σ_2^2. *Calculate 90% confidence interval for* σ_1^2/σ_2^2, *in terms of c where the probability of F exceeding c based on n − 1, n − 1 d.f. is 0·5.*

(*a*) The joint distribution of χ_1^2 and χ_2^2 is

$$\frac{1}{2^{\frac{n_1}{2}}\Gamma\left(\frac{n_1}{2}\right)} e^{-\frac{1}{2}\chi_1^2} (\chi_1^2)^{\frac{n_1}{2}-1} d\chi_1^2 \times \frac{1}{2^{\frac{n_2}{2}}\Gamma\left(\frac{n_2}{2}\right)} e^{-\frac{1}{2}\chi_2^2} (\chi_2^2)^{\frac{n_2}{2}-1} d\chi_2^2$$

Let us make the change of variates

$$F=\frac{\chi_1^2/n_1}{\chi_2^2/n_2},\ v=\chi_2^2,\ 0\le F<\infty, 0\le v<\infty$$

Then $\chi_1^2=\dfrac{vn_1}{n_2}F,\ \chi_2^2=v.$

The Jacobian of transformation is

$$J=\begin{vmatrix} \dfrac{n_1 v}{n_2} & \dfrac{n_1}{n_1}F \\ 0 & 1 \end{vmatrix}=\frac{n_1}{n_2}v$$

The distribution of the transformed variates F and v is

$$\frac{\left(\dfrac{n_1}{n_2}\right)^{\frac{n_1}{2}}}{2^{\frac{1}{2}(n_1+n_2)}\Gamma\left(\frac{n_1}{2}\right)\Gamma\left(\frac{n_2}{2}\right)} F^{\frac{n_1}{2}-1} v^{\frac{n_1+n_2}{2}-1} e^{-\frac{1}{2}\left(\frac{n_1}{n_2}F+1\right)v}\, d F dv$$

Integrating out for variable v throughout the whole range, we get the distribution of F as

$$dP=\frac{\left(\dfrac{n_1}{n_2}\right)^{\frac{n_1}{2}}}{\Gamma\left(\frac{n_1}{2}\right)\Gamma\left(\frac{n_2}{2}\right)} F^{\frac{n_1}{2}-1}\left(1+\frac{n_1}{n_2}F\right)^{-\frac{n_1+n_2}{2}} d F$$

$$\times \int_0^\infty \left[\left(1 + \frac{n_1}{n_2} F \right) \left(\frac{v}{2} \right) \right]^{\frac{n_1+n_2}{2} - 1} e^{-\frac{v}{2}\left(1 + \frac{n_1}{n_2} F\right)} dv$$

$$= \frac{\Gamma\left(\frac{n_1+n_2}{2}\right)}{\Gamma\left(\frac{n_1}{2}\right)\Gamma\left(\frac{n_2}{2}\right)} \left(\frac{n_1}{n_2} \right)^{\frac{n_1}{2}} F^{\frac{n_1}{2} - 1} \left(1 + \frac{n_1}{n_2} F \right)^{-\frac{n_1+n_2}{2}} dF, 0 < F < \infty$$

This is F distribution with n_1 and n_2 *d.f.*

(*b*) We have $F = \dfrac{s_1^2/\sigma_1^2}{s_2^2/\sigma_2^2} \times \dfrac{n-1}{n-1}$. $\quad (\because \; n_1 = n_2 = n)$

$$\therefore \quad \frac{1}{F} = \frac{s_2^2/\sigma_2^2}{s_1^2/\sigma_1^2} \times \frac{n-1}{n-1}.$$

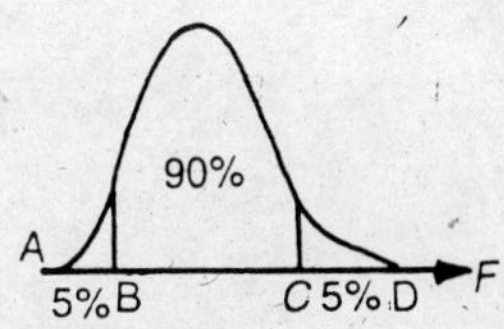

Fig. 13·5

By Ex. 1, page 503, the distribution of F and $1/F$ are same, each is based on $n-1$, $n-1$ *d.f.*

Now $P(F > c) = \cdot 05$ (given) ...(1)

$\therefore \; P\left(\dfrac{1}{F} < \dfrac{1}{c} \right) = 0 \cdot 5$ ...(2) or $P\left(F < \dfrac{1}{c} \right) = \cdot 05$ ...(3)

Combining (1) and (3), we have $P\left[\dfrac{1}{c} \le F \le c \right] = \cdot 90$

or $\quad P\left[\dfrac{1}{c} \le \dfrac{s_1^2/\sigma_1^2}{s_2^2/\sigma_2^2} \le c \right] = \cdot 90$ or $P\left[\dfrac{1}{c} \dfrac{s_2^2}{s_1^2} \le \dfrac{\sigma_2^2}{\sigma_1^2} \le c \dfrac{s_1^2}{s_2^2} \right] = \cdot 90$

Similarly, $P\left[\dfrac{1}{c} \dfrac{s_1^2}{s_2^2} \le \dfrac{\sigma_1^2}{\sigma_2^2} \le c \dfrac{s_1^2}{s_2^2} \right] = \cdot 90$

$\therefore \; P\left[c \dfrac{s_1^2}{s_2^2} \ge \dfrac{\sigma_1^2}{\sigma_2^2} \ge \dfrac{1}{c} \dfrac{s_1^2}{s_2^2} \right] = \cdot 90$ or $P\left[\dfrac{1}{c} \dfrac{s_1^2}{s_2^2} \le \dfrac{\sigma_1^2}{\sigma_2^2} \le c \dfrac{s_1^2}{s_2^2} \right] = \cdot 90$

Hence 90% confidence limits for $\dfrac{\sigma_1^2}{\sigma_2^2}$ are $\dfrac{1}{c} \dfrac{s_1^2}{s_2^2}$ and $c \dfrac{s_1^2}{s_2^2}$.

Example 13·5. (*i*) *Show that m.g.f. of* $Y = \log_e \chi^2$ *where* χ^2 *follows Chi-square distribution with n d.f. is given by*

$$M_Y(t) = 2^t \,\Gamma\left(\frac{n}{2}+t\right) \Big/ \Gamma\left(\frac{n}{2}\right)$$

(*ii*) *If* χ_1^2 and χ_2^2 *are independent values of the above* χ^2*-statistics and* $F = \chi_1^2/\chi_2^2$, *deduce for positive integer k,*

$$E(F^k) = \Gamma\left(\frac{n}{2}+k\right) \Gamma\left(\frac{n}{2}-k\right) \Big/ \left(\Gamma\left(\frac{n}{2}\right)\right)^2$$ **(B.A. Hons. Delhi, 1986)**

(*i*) $$M_Y(t) = E[e^{tY}] = E\,[e^{t\log\chi^2}] = E\,[(\chi^{2t})]$$

$$= \int_0^\infty \frac{1}{2^{\frac{n}{2}}\Gamma\frac{n}{2}} e^{\frac{1}{2}-1} (\chi^2)^{-\frac{n}{2}\chi^2} \cdot (\chi^2)^t \, d\chi^2$$

$$= \int_0^\infty \frac{1}{2^{n/2}\Gamma\frac{n}{2}} e^{-\frac{1}{2}\chi^2} (\chi^2)^{\frac{n}{2}-1} \, d\chi^2$$

$$= \frac{2^{\frac{n}{2}+t}\,\Gamma\left(\frac{n}{2}+t\right)}{2^{\frac{n}{2}}\Gamma\frac{n}{2}} = 2^t \,\Gamma\left(\frac{n}{2}+t\right) \Big/ \Gamma\left(\frac{n}{2}\right)$$

(*ii*) Now $$E(F^k) = E\left[\left(\frac{\chi_1^2}{\chi_2^2}\right)^k\right] = E\left[\exp\left\{\log\left(\frac{\chi_1^2}{\chi_2^2}\right)^k\right\}\right]$$

$$= E\left[e^{k\log\chi_1^2 - k\log\chi_2^2}\right] = E\left(e^{k.\log\chi_1^2}\, e^{-k\log\chi_2^2}\right)$$

$$= M_{\log\chi_1^2}(k).\ M_{\log\chi_2^2}(-k)$$ as χ_1^2 and χ_2^2 are independent

$$= \frac{2^k\,\Gamma\left(\frac{n}{2}+k\right)}{\Gamma\left(\frac{n}{2}\right)} \cdot 2^{-k} \frac{\Gamma\left(\frac{n}{2}-k\right)}{\Gamma\left(\frac{n}{2}\right)}$$

$$E(F^k) = \frac{\left|\overline{\frac{n}{2}+k}\right. \left|\overline{\frac{n}{2}-k}\right.}{\left(\left|\overline{\frac{n}{2}}\right.\right)^2}$$

Ex. Let x_i, s_i^2, n_i $(i = 1, 2, ..., k)$ be means, variances and sizes of sample drawn from k normal populations with means μ_t $(i = 1, 2, ..., k)$ and same variance σ^2, show that

(*i*) $$\frac{\sum_1^k n_i(\bar{x}_i - \mu_i)^2}{\sum_1^k (n_i - 1)s_i^2} \cdot \frac{n_1 + n_2 + \ldots + n_k - k}{k}$$

is distributed as F with k and $(n_1 + n_2 + \ldots + n_k - k)$ *d.f.*

(*ii*) $$\frac{\sum_1^k n_i(\bar{x}_i - \bar{x})^2}{\sum_1^k (n_i - 1)s_i^2} \cdot \frac{n_1 + n_2 + ... + n_k - k}{k - 1}$$

is distributed as F with $k - 1, n - k$ *d.f.* where $n = \sum_{i=1}^{k} n_i$.

13·8. Fisher's *z* distribution. Making the change of variable viz.

$$z = \tfrac{1}{2} \log_e F \quad \text{or} \quad F = e^{2z} \qquad ...(13·19)$$

in the F distribution, we get

$$dP = \frac{2\, \nu_1^{\frac{\nu_1}{2}}\, \nu_2^{\frac{\nu_2}{2}}\, e^{\nu_1 z}\, dz}{B\left(\frac{\nu_1}{2}, \frac{\nu_2}{2}\right)\left(\nu_1 e^{2z} + \nu_2\right)^{\frac{1}{2}(\nu_1 + \nu_2)}}, -\infty < z < \infty \qquad ...(13·20)$$

This distribution is known as Fisher's z distribution and is more nearly symmetrical than the distribution of F. It is symmetrical function of n_1 and n_2. Table 13·3 has been prepared showing, *variance ratio* ν_1/ν_2 and 1% '*points*' of F.

for various values of ν_1 and ν_2, the values of z that will be exceeded in simple sampling with probabilities 0·05 and 0·01 respectively.

Table 13·3

The variance ratio, 5 *and* 1% *'points' of F*

ν_1 is the number of degrees of freedom for the greater estimate of variance, and ν_2 for the smaller

ν_2 \ ν_1	1	2	3	4	5	6	8	12	24	∞
2	18·51 98·49	19·00 99·00	19·16 99·17	19·25 99·25	19·30 99·30	19·32 99·33	19·37 99·36	19·41 99·42	19·45 99·46	19·50 99·50
3	10·13 34·12	9·55 30·82	9·28 29·46	9·12 28·71	9·01 28·24	8·94 27·91	8·84 27·49	8·74 27·05	8·64 26·60	8·63 20·12
4	7·71 21·20	6·94 18·00	6·59 16·69	6·39 15·98	6·26 15·52	6·16 15·21	6·04 14·80	5·91 14·37	5·77 13·93	5·63 13·46
5	6·61 16·26	5·79 13·27	5·41 12·06	5·19 11·39	5·05 10·97	4·95 10·67	4·82 10·27	4·68 9·89	4·53 9·47	4·36 9·02
6	5·99 13·74	5·14 10·92	4·76 9·78	4·53 9·15	4·39 8·75	4·28 8·47	4·15 8·10	4·00 7·72	3·84 7·31	3·67 6·88
7	5·59 12·25	4·74 9·55	4·35 8·45	4·12 7·85	3·97 7·46	3·87 7·19	3·73 6·84	3·57 6·47	3·41 6·07	3·23 5·65
8	5·32 11·26	4·46 8·65	4·07 7·59	3·84 7·01	3·69 6·63	3·58 6·37	3·44 6·03	3·28 5·67	3·12 5·28	2·93 4·86
9	5·12 10·56	4·26 8·02	3·86 6·99	3·63 6·42	3·48 6·06	3·37 5·80	3·23 5·47	3·07 5·11	2·90 4·73	2·71 4·31
10	4·96 10·04	4·10 7·56	3·71 6·55	3·48 5·99	3·33 5·64	3·22 5·39	3·07 5·06	2·91 4·71	2·74 4·33	2·54 3·91
12	4·75 9·33	3·88 6·93	3·49 5·95	3·26 5·41	3·11 5·06	3·00 4·82	2·85 4·50	2·69 4·16	2·50 3·78	2·30 3·36
14	4·60 8·86	3·74 6·51	3·34 5·56	3·11 5·03	2·96 4·69	2·85 4·46	2·70 4·14	2·53 3·80	2·35 3·43	2·13 3·00
16	4·49 8·53	3·63 6·23	3·24 5·25	3·01 4·77	2·85 4·44	2·74 4·20	2·59 3·89	2·42 3·55	2·24 3·18	2·01 2·75
18	4·41 8·28	3·55 6·01	3·16 6·09	2·93 4·58	2·77 4·25	2·66 4·01	2·51 3·71	2·34 3·37	2·15 3·01	1·92 2·57
20	4·35 8·10	3·49 5·85	3·10 4·94	2·87 4·43	2·71 4·10	2·60 3·87	2·45 3·56	2·28 3·23	2·08 2·86	1·84 2·42
25	4·24 7·77	3·38 5·57	2·99 4·68	2·76 4·18	2·60 3·86	2·49 3·63	2·34 3·32	2·16 2·99	1·96 2·62	1·71 2·17
30	4·17 7·56	3·32 5·39	2·92 4·51	2·69 4·02	2·53 3·70	2·42 3·47	2·27 3·17	2·09 2·84	1·89 2·47	1·62 2·01
40	4·08 7·31	3·23 5·18	2·84 4·31	2·61 3·83	2·45 3·51	2·34 3·29	2·18 2·99	2·00 2·66	1·79 2·29	1·51 1·81
60	4·00 7·08	3·15 4·98	2·76 4·13	2·52 3·65	2·37 3·34	2·25 3·12	2·10 2·82	1·92 2·50	1·70 2·12	1·39 1·60
80	3·96 6·96	3·11 1·88	2·72 4·04	2·49 3·56	2·33 3·25	2·21 3·04	2·06 2·74	1·88 2·41	1·65 2·03	1·32 1·49

Extracted from G.W. Snedecor's *Statistical Methods*, pp. 184—87, by courtesy of the author and the Iowa Collegiate press

Characteristic function of the z-distribution.

$$\phi(t) = E(e^{itz}) = \int_{-\infty}^{\infty} \frac{2\, \nu_1^{\frac{\nu_1}{2}}\, \nu_2^{\frac{\nu_2}{2}}\, e^{\nu_1 z}\, e^{itz}}{B\left(\frac{\nu_1}{2}, \frac{\nu_2}{2}\right)\left(\nu_1 e^{2z} + \nu_2\right)^{\frac{1}{2}(\nu_1+\nu_2)}}\, dz$$

Let $z = \frac{1}{2}\log_e F \Rightarrow e^{2z} = F$ and $dz = \frac{1}{2}\frac{dF}{F}$

$$\phi(t) = \int_0^{\infty} \frac{\nu_1^{\frac{\nu_1}{2}}\, \nu_2^{\frac{\nu_2}{2}}\, F^{\frac{\nu_1}{2}+\frac{\theta}{2}-1}}{B\left(\frac{\nu_1}{2}, \frac{\nu_2}{2}\right)\left(\nu_1 F + \nu_2\right)} \text{ where } \theta = it$$

$$= \left(\frac{\nu_2}{\nu_1}\right)^2 \frac{1}{B\left(\frac{\nu_1}{2}, \frac{\nu_2}{2}\right)} B\left(\frac{\nu_1+\theta}{2}, \frac{\nu_2-\theta}{2}\right)$$

We find $\mu_1 = k_1 = \frac{1}{2}\left(\frac{1}{\nu_2} - \frac{1}{\nu_2}\right)$,

$$\mu_2 = k_2 = \frac{1}{2}\left(\frac{1}{\nu_2} + \frac{1}{\nu_1} + \frac{1}{\nu_1^2} + \frac{1}{\nu_2^2}\right)$$

$$= \left[\frac{1}{2}\left(\frac{1}{\nu_2} + \frac{1}{\nu_1}\right)\right]$$

for large ν_1 and ν_2:

Ex. 1. Writing $x = \dfrac{\nu_1 e^z}{\nu_1 e^z + \nu_2}$, show that (13·20) reduces to Pearson Type 1

$$dP = \frac{1}{B\left(\frac{\nu_1}{2}, \frac{\nu_2}{2}\right)} x^{(\nu_1-2)/2} \quad (1-x)^{(\nu_2-2)/2}\, dx,\ 0 \le x \le 1.$$

Ex. 2. Show that the characteristic function of z is

$$\phi(t) = \left(\frac{\nu_2}{\nu_1}\right)^{\frac{\theta}{2}} \left|\overline{\frac{\nu_2-\theta}{2}}\right. \left|\overline{\frac{\nu_1-\theta}{2}}\right. \Big/ \left|\overline{\frac{\nu_1}{2}}\right. \left|\overline{\frac{\nu_2}{2}}\right. \text{ where } \theta = it.$$

Taking logarithm and using Stirling expansion

$$\log \left|\overline{1+x}\right. = \tfrac{1}{2}\log 2\pi + \left(x + \tfrac{1}{2}\right)\log x - x + \frac{1}{12x} - \ldots$$

show that for large ν_1 and ν_2

$$\log \phi(t) = \frac{\theta}{2}\left(\frac{1}{\nu_2} + \frac{1}{\nu_1}\right) + \frac{\theta^2}{4}\left(\frac{1}{\nu_1} + \frac{1}{\nu_2}\right)\ldots.$$

Hence, deduce that for large ν_1 and ν_2, z is distributed approximately normally with mean $= \frac{1}{2}\left(\frac{1}{\nu_2}+\frac{1}{\nu_1}\right)$ and variance $= \frac{1}{2}\left(\frac{1}{\nu_2}+\frac{1}{\nu_1}\right)$. **(M.A. Bombay)**

Ex. 3. It is required to test that the mean of a normal population is zero. A random sample drawn from the population gives the values $x_1, x_2, \ldots\ldots, x_n$. Show that the t test for acceptance of the hypothesis reduces to

$$\left(\sum_{i=1}^{n} x_i\right) \leq \frac{n\, t_\alpha^2}{t_\alpha^2+(n-1)}\left(\sum_{i=1}^{n} x_i^2\right)$$

13·9. Inter-relations between t, χ^2 and F Distributions.

13·9·1. Theorem. *The square of t variate with n.d.f. is distributed as F with 1 and n.d.f.*

Putting $F=t^2$, $n_1=1$, $n_2=n$ in the distribution of F with n_1, n_2 *d.f.*. page 503 we get

$$dP = \frac{\left(\frac{1}{n}\right)^{1/2} (t^2)^{\frac{1-2}{2}}\, d(t^2)}{B\left[\frac{1}{2}, \frac{n}{2}\right]\left[1+\frac{t^2}{n}\right]^{\left(\frac{1}{2}+\frac{1}{2}n\right)}}$$

$$= \frac{1}{\sqrt{n}} \cdot \frac{1}{B\left[\frac{1}{2}, \frac{n}{2}\right]} \cdot \frac{dt}{\left[1+\frac{t^2}{n}\right]^{\frac{1}{2}(n+1)}}$$

which is t distribution with *n.d.f.*

13·9·2. Theorem. *As n_2 tends to infinity, n_1 F tends to be distributed as a χ^2 variate with n_1 d.f.*

The distribution of F with n_1, n_2, *d.f.* can be written as

$$dP = \frac{n_1^{\frac{n_1}{2}} F^{\frac{n_1}{2}-1}\, dF}{\left|\overline{\frac{n_1}{2}}\right.\left[1+\frac{n_1}{n_2}F\right]^{\frac{n_1+n_2}{2}}} \times \frac{\left|\overline{\frac{n_1+n_2}{2}}\right.}{n_2^{\frac{n_1}{2}}\left|\overline{\frac{n_2}{2}}\right.}$$

Now using Stirling's formula, we get

$$\frac{\left|\overline{\frac{n_1+n_2}{2}}\right.}{n_2^{\frac{n_1}{2}}\left|\overline{\frac{n_2}{2}}\right.} = \frac{e^{-\frac{n_1+n_2}{2}-1}\left[\frac{n_1+n_2}{2}-1\right]^{\frac{n_1+n_2}{2}-\frac{1}{2}}\sqrt{2\pi}}{n_2^{\frac{n_1}{2}}\, e^{-\frac{n_2}{2}-1}\left[\frac{n_2}{2}-1\right]^{\frac{n_2}{2}-\frac{1}{2}}\sqrt{2\pi}}$$

$$= \frac{e^{-\frac{n_1}{2}} \left[\frac{n_1 + n_2 - 2}{2} \right]^{\frac{n_1 + n_2 - 1}{2}}}{n_2^{\frac{n_1}{2}} \left[\frac{n_2 - 2}{2} \right]^{\frac{n_2 - 1}{2}}}$$

$$\sim \frac{e^{-\frac{n_1}{2}} \left[1 + \frac{n_1 - 2}{n_2} \right]^{\frac{n_1 - 1}{2}} \left[1 + \frac{n_1 - 2}{n_2} \right]^{\frac{n_2}{2}}}{n_2^{\frac{n_1}{2}} \left(1 - \frac{2}{n_2} \right)^{\frac{n_2 - 1}{2}}}$$

$$\therefore \qquad \frac{\left\lceil \frac{n_1 + n_2}{2} \right.}{n_2^{\frac{n_1}{2}} \left\lceil \frac{n_2}{2} \right.} \to \frac{e^{-\frac{n_1}{2}} e^{\frac{n_1}{2}}}{2^{\frac{n_1}{2}}} \text{ or } 2^{-n_1/2} \text{ as } n_2 \to \infty$$

Thus, when n_2 tends to infinity, we get

$$dP = \frac{n_1^{\frac{n_1}{2}} F^{\frac{n_1}{2} - 1} dF}{2^{\frac{n_1}{2}} \left\lceil \frac{n_1}{2} \right. e^{\frac{n_1}{2} F}}$$

Putting $F = \frac{\chi^2}{n_1}$, we have

$$dP = \frac{n_1^{n_1/2} \left(\frac{\chi^2}{n_1} \right)^{\frac{n_1}{2} - 1} d\left[\frac{\chi^2}{n_1} \right]}{2^{n_1/2} \left\lceil \frac{n_1}{2} \right. e^{\frac{1}{2}\chi^2}}$$

or $$dP = \frac{1}{2^{n_1/2} \left\lceil \frac{1}{2} n_1 \right.} e^{-\frac{1}{2}\chi^2} (\chi^2)^{\frac{1}{2}(n_1 - 2)} d(\chi^2), \; 0 < \chi^2 < \infty$$

which is χ^2 distribution based on n_1 *d.f.*

Note. In Example 13·4, we deduced the F distribution from χ^2 distribution. In the present example, we have deduced the χ^2 distribution as a limiting form of the F distribution.

Ex. Show how the F distribution is related to the t distribution and the Chi-square distribution.

13·10. Non-Central distributions. In § 13·2, 13·5 and 13·7 we have derived the central—distributions of the central statistics χ^2, t and F. We shall, in this section, derive the non-central distributions of non-central statistics χ^2, t and F.

13·10·1. Non-Central Chi-square distributions Let $X_1, X_2, ..., X_n$ be independent normal variates with means μ_i and variances σ_i^2, $i = 1, 2, ..., n$.

We wish to derive the distribution of non-central χ^2

$$\chi^2 = \sum_{i=1}^{n} x_i^2/\sigma_i^2.$$

with n degrees of freedom and non-central parameter $\psi^2 = \sum_{i=1}^{n} \mu_i^2/\sigma_i^2$.

Proof. Let us transform the variables x_i's to the variables y's by

$$y_i = \left(\frac{\mu_1}{\sigma_1} \cdot \frac{x_1}{\sigma_1} + \frac{\mu_2}{\sigma_2} \cdot \frac{x_2}{\sigma_2} + ... + \frac{\mu_n}{\sigma_n} \cdot \frac{x_n}{\sigma_n} \right) \Big/ \sqrt{\sum_1^n \frac{\mu_i^2}{\sigma_i^2}}$$

$$= \sum_{i=1}^{n} \left(\frac{\mu_i}{\sigma_i} \cdot \frac{x_i}{\sigma_i} \right) \Big/ \sqrt{\sum_{i=1}^{n} \frac{\mu_i^2}{\sigma_i^2}},$$

$$y_j = C_{j1} \cdot \frac{x_1}{\sigma_1} + C_{j2} \cdot \frac{x_2}{\sigma_2} \cdot + ... + C_{jn} \cdot \frac{x_n}{\sigma_n} = \sum_{i=1} C_{ji} \cdot \frac{x_i}{\sigma_i},\ j = 2, 3, ..., n$$

where the coefficients C_{ji} satisfy the relations

$$\sum_{i=1}^{n} C^2_{ji} = 1, \qquad [i.e., \text{ vectors } (C_j' = (C_{j1}, C_{j2}, ..., C_n) \text{ are unitary}]$$

$$\sum_j C_{ji} C_{ji}' = 0 = \sum_i C_{ji}\, C_{ji}',\ i' \neq i\ [\ i.e. \text{ vectors } C_j \text{ and } C_j', \textit{are} \text{ orthogonal}]$$

It can be proved that the variates y_i's are independent normal variates with

$$E(y_1) = \sum_{i=1}^{n} \frac{\mu_i}{\sigma_i} E\left(\frac{x_i}{\sigma_i} \right) \Big/ \sqrt{\sum_{i=1}^{n} \frac{\mu_i^2}{\sigma_i^2}} = \sqrt{\sum \frac{\mu_i^2}{\sigma_i^2}} = \psi$$

$$E(y_j) = \sum_{i=1}^{n} C_{ji}\, E\left(\frac{x_i}{\sigma_i} \right) = \sum_{i=1}^{n} C_{ij} \frac{\mu_i}{\sigma_i} = 0,\ j = 2, 3, ..., n$$

Since the vectors C_j' are mutually orthogonal to the unit vector

$$\left(\frac{\mu_1}{\sigma_1}, \frac{\mu_2}{\sigma_2}, \ldots, \frac{\mu_n}{\sigma_n}\right) \Big/ \sqrt{\sum_1^n \frac{\mu_i^2}{\sigma_i^2}}$$

Also $\quad \text{Var } y_1 = \sum_{i=1}^{n} \frac{\mu_i^2}{\sigma_i^2} \cdot \frac{1}{\sigma_i^2} \text{ Var } x_i \Big/ \sum_{i=1}^{n} \frac{\mu_i^2}{\sigma_i^2} = 1.$

$$\text{Var } y_j = \sum_{i=1}^{n} C^2_{ji} \cdot \frac{1}{\sigma_i^2} \text{ Var } x_i = \sum_{i=1}^{n} C^2_{ji} = 1.$$

Thus the variable y_1 is $N(\psi, 1)$ and y_j is $N(0, 1)$, $j = 2, 3, \ldots, n$.

Now $\quad \chi^2 = \sum_{i=1}^{n} \frac{x_i^2}{\sigma_i^2} = \sum_{i=1}^{n} y_i^2 = y_1^2 + \sum_{j=2}^{n} y_j^2 = \chi_1^2 + \chi_2^2$

where χ_1^2 has the distribution

$$\frac{1}{\sqrt{2\pi}} e^{-(\chi_1^2 + \psi^2)/2} \cosh(\psi \chi_1) \cdot (\chi_1^2)^{-\frac{1}{2}} \, d\chi_1^2$$

$$\Rightarrow \frac{1}{\sqrt{2\pi}} \exp\left[-\tfrac{1}{2}(\chi_1^2 + \psi^2)\right] \sum_{k=0}^{\infty} \frac{(\psi \chi_1)^{2k}}{2^k} \cdot (\chi_1^2)^{-1/2} \, d\chi_1^2$$

$$\Rightarrow \frac{1}{2} \sum_{k=0}^{\infty} \exp\left[-\tfrac{1}{2}\psi^2\right] \frac{(\frac{1}{2}\psi^2)^k}{k\,!} \cdot \frac{\exp\left[-\frac{1}{2}\chi_1^2\right] (\frac{1}{2}\chi_1^2)^{k-\frac{1}{2}}}{\Gamma(k + \frac{1}{2})} \, d\chi_1^2$$

and the distribution of χ_2^2 is

$$\frac{1}{2^{\frac{n-1}{2}} \Gamma\left(\frac{n-1}{2}\right)} \exp\left[-\tfrac{1}{2}\chi_2^2\right] (\chi_2^2)^{\frac{n-3}{2}} \, d\chi_2^2$$

In the joint distribution of χ_1^2 and χ_2^2, let us make the transformations

$$\chi_1 = \chi \cos\theta, \; \chi_2 = \chi \sin\theta, \; 0 < \chi < \infty, \; 0 < \theta < \pi/2$$

We find that the *p.d.f.* of χ^2 $(= \chi_1^2 + \chi_2^2)$ is

$$\exp\left[-\tfrac{1}{2}\psi^2\right] \exp\left[-\tfrac{1}{2}\chi^2\right] \sum_{k=0}^{\infty} \frac{(\frac{1}{2}\psi^2)^k}{k\,!} \cdot \frac{(\chi^2)^{k + \frac{(n-2)}{2}}}{2^{\frac{n}{2}+k} \Gamma\left(\frac{n}{2} + k\right)}$$

Remark 1. The *p.d.f.* of χ^2 is the weighted average of central χ^2 *p.d.f.*'s with $(2k + n)$ *d.f.*, the weights being the Poisson probabilities $(\psi^2/2)^k \exp(-\frac{1}{2}\psi^2)/k\,!$, $k = 0, 1, \ldots, n$.

2. Let $\psi = 0$ and $k = 0$, then the distribution of χ^2 is the central χ^2-distribution with *n.d.f.*

Moment Generating Function of Non-Central χ^2-distribution. For simplicity, we shall assume $X_i \sim N(\mu, 1)$.

$$\therefore M_{x^2}(t) = E[e^{tx^2}] = \frac{1}{\sqrt{2\pi}} \int_{-\infty}^{+\infty} e^{tx^2} e^{-(1/2)(x-\mu)^2} dx$$

$$= \frac{1}{\sqrt{2\pi}} \int_{-\infty}^{+\infty} \exp\left[-\{(\tfrac{1}{2}-t)x^2 - \mu x + \tfrac{1}{2}\mu^2\}\right] dx$$

$$= \frac{1}{\sqrt{2\pi}} \int_{-\infty}^{+\infty} \exp\left[-\left(\frac{1}{2}-t\right)\left\{\left(x-\frac{\mu}{1-2t}\right)^2 + \frac{\mu^2}{1-2t} - \frac{\mu^2}{(1-2t)^2}\right\}\right] dx$$

$$= \frac{1}{\sqrt{2\pi}} \exp\left(\frac{t\mu^2}{1-2t}\right) \int_{-\infty}^{+\infty} \exp\left[-\left(\frac{1-2t}{2}\right)\left(x-\frac{\mu}{1-2t}\right)^2\right] dx$$

$$= \frac{1}{\sqrt{2\pi}} \exp\left(\frac{t\mu^2}{1-2t}\right) \int_{-\infty}^{+\infty} \exp\left(\frac{1}{2}Z^2\right) \frac{dZ}{\sqrt{1-2t}},\quad Z = \left(x - \frac{\mu}{1-2t}\right) \times \sqrt{1-2t}$$

$$= (1-2t)^{-\frac{1}{2}} \exp\left(\frac{t\mu^2}{1-2t}\right),\quad t < \frac{1}{2}.$$

Thus if X_i's $(i = 1, 2, \ldots, n)$ are *independent* $N(\mu_i, 1)$, and

$$X'^2 = \sum_{i=1}^{n} X_i^2, \text{ then } M_{x'^2}(t) = M_{\sum_1^n X_i^2}(t) = \prod_{i=1}^{n} M_{x_i^2}(t)$$

$$= \prod_{i=1}^{n} \left[(1-2t)^{-\frac{1}{2}} \exp\left(\frac{t\mu_i^2}{1-2t}\right)\right]$$

$$= (1-2t)^{-\frac{n}{2}} \exp\left(\frac{t}{1-2t}\right) \prod_{i=1}^{n} \exp \mu_i^2$$

$$= (1-2t)^{-\frac{n}{2}} \exp\left(\frac{t\mu}{1-2t}\right);\ t < \frac{1}{2},$$

where $\psi = \sum_{1}^{n} \mu_i^2$ is the non–centrality parameter.

Cor. 1. Additive or Reproductive property of Non-central

χ^2**-variates.** It can be easily proved that

If Y_i $(i = 1, 2, \ldots\ldots, k)$ are independent non-central χ^2-variates with n_i, *d.f.* and non-central parameter ψ_i, then $\sum_{i=1}^{k} y_i$ is also a non-central χ^2-Variate with $\sum_{i=1}^{k} n_i$ *d.f.* and non-centrality parameter $\psi = \sum_{i=1}^{k} \psi_i$.

2. $k_r = r$th cumulant = coeff. of $t^r/r\,!$ in $\log M_{\chi'^2}(t)$ *i.e.*

$$-(n/2)\log(1-2t) + 2t\psi(1-2t)^{-1}$$

$$K_{\chi'^2}(t) = (n+2\psi)\,t + (n+4\psi)\,t^2 + \ldots\ldots + \left|\frac{2^{r-1}}{r}\cdot n + 2\psi\, 2^{r-1}\right| t^r + \cdots$$

$$\therefore \qquad k_r = \text{coeff. of } t^r/r! \text{ in } K_{\chi'^2}(t) = r\,!\left(\frac{n}{r} + 2\psi\right)2^{r-1}$$

$$k_r = 2^{r-1}\ (r-1)\,!\ (n + 2\psi r)$$

and $\quad k_{r-1} = 2^{r-2}\ (r-2)\,!\ [n + 2\psi(r-1)]$

$$\therefore\ \frac{d}{d\psi}k_{r-1} = 2^{r-2}\ (r-2)\,!\ 2(r-1) = 2^{r-1}\ (r-1)\,! = \frac{k_r}{n+2\psi r}$$

$$\Rightarrow\ k_r = (n+2\psi r)\frac{d}{d\psi}(k_{r-1}).$$

13·10·2. Non-central t-distribution

Theorem. *If X is distributed as* $N(\mu, \sigma^2)$ *and Y is an independent* χ^2*-variate with n.d.f., then*

$$t = \frac{X/\sigma}{\sqrt{Y}/\sqrt{n}}$$

has a non-central t-distribution with n.d.f. and with non-centrality parameter μ/σ.

Proof. Since X and Y are independent, their joint *p.d.f.* is

$$\frac{1}{\sigma\sqrt{2\pi}\, 2^{\frac{n}{2}} \left|\frac{n}{2}\right.} e^{-\frac{1}{2}y}\, y^{\frac{1}{2}n-1} \cdot e^{-\frac{1}{2}\frac{(x-\mu)^2}{\sigma^2}}$$

$$\Rightarrow \frac{1}{\sigma\sqrt{2\pi}} \cdot \frac{1}{2^{\frac{n}{2}}\Gamma\left(\frac{n}{2}\right)} e^{-\frac{1}{2}\frac{(\mu^2+x^2+y)}{\sigma^2}}\, y^{\frac{n}{2}-1} \sum_{j=0}^{\infty}\left(\frac{\mu}{\sigma}\right)^j\left(\frac{X^j}{j\,!}\right)$$

$$\Rightarrow \frac{1}{\sqrt{2\pi}} \cdot \frac{1}{2^{\frac{n}{2}-1}\Gamma\left(\frac{n}{2}\right)} \cdot e^{-\frac{\mu^2}{2\sigma^2}} \cdot e^{-\frac{1}{2}y} \cdot y^{\frac{n}{2}-1} \cdot e^{-\frac{1}{2}z^2}$$

$$\times \sum_{j=0}^{\infty} \left(\frac{\mu}{\sigma}\right)^j \frac{z^j}{j\,!}; \text{ where } z = x/\sigma$$

Now make the transformation $(y, z) \to (t, u)$ such that

$$t = \frac{\sqrt{n}\,Z}{\sqrt{Y}} \text{ and } u = \sqrt{Y} \Rightarrow Y = u^2 \text{ and } Z = tu/\sqrt{n}$$

$$\text{Jacobian of transformation } = J\begin{pmatrix} y & z \\ t & u \end{pmatrix} = \begin{vmatrix} 0 & 2u \\ \frac{u}{\sqrt{n}} & \frac{t}{\sqrt{n}} \end{vmatrix} = \frac{2u^2}{\sqrt{n}}$$

Hence the joint *p.d.f.* of t and u is

$$\frac{1}{2^{\frac{n-1}{2}}\Gamma\left(\frac{n}{2}\right)\Gamma\left(\frac{1}{2}\right)} e^{-\frac{\mu^2}{2\sigma^2}} \cdot \sum_{j=0}^{\infty} \frac{1}{j\,!}\left(\frac{\mu}{\sigma}\right)^j$$

$$\times \exp\left[-\frac{u^2}{2}\left(1+\frac{t^2}{n}\right)\right]\left(\frac{t\,u}{\sqrt{n}}\right)^j u^{n-1}\,|\,J\,|$$

$$= \frac{1}{2^{\frac{(n-1)}{2}}\Gamma\left(\frac{n}{2}\right)\Gamma\left(\frac{1}{2}\right)} e^{-\frac{\mu^2}{2\sigma^2}} \cdot \sum_{j=0}^{\infty} \frac{1}{j\,!}\,\frac{(\mu/\sigma)^j}{(j+1)/2}$$

$$\times \exp\left[-\frac{u^2}{2}\left(1+\frac{t^2}{n}\right)\right] t^j\, u^{n+j}$$

Integrating this w.r.t. u between the limits 0 and ∞, we get the *p.d.f.* of non-central t with *n.d.f.* as

$$\frac{1}{\sqrt{\pi}\,\Gamma\left(\frac{n}{2}\right)} \exp\left[-\frac{\mu^2}{2\sigma^2}\right] \times$$

$$\times \sum_{j=0}^{\infty} \frac{1}{j\,!}\left(\frac{\mu}{\sigma}\right)^j \cdot n^{\frac{j}{2}} \frac{\Gamma\left(\frac{n+j+1}{2}\right)}{n^{(j+1)/2}} \cdot \frac{1}{\left(1+\frac{t^2}{n}\right)^{(n+j+1)/2}}\, dt, \; -\infty < t < \infty$$

Cor. If $\mu = 0$, then *p.d.f.* of central t with *n.d.f.* is

$$\frac{1}{\sqrt{\pi}\,\Gamma\left(\frac{n}{2}\right)} \frac{\Gamma\left(\frac{n+1}{2}\right)}{\sqrt{n}} \cdot \frac{1}{\left(1+\frac{t^2}{n}\right)^{(n+1)/2}}, \quad -\infty < t < \infty$$

$$= \frac{1}{\sqrt{n}\; B\left(\frac{1}{2}, \frac{n}{2}\right)} \cdot \left[1+\frac{t^2}{n}\right]^{-(n+1)/2}, \quad -\infty < t < \infty.$$

13·10·3. Non-Central F Distribution. Let χ_1^2 be a non-central χ^2 with *d.f.* n_1 and non-centrality parameter ψ^2 and let χ_2^2, independent of χ_1^2, be a central χ^2 with *d.f.* $= n_2$. Then the ratio

$$\frac{\chi_1^2/n_1}{\chi_2^2/n_2}$$

is known as a non-central F statistic with (n_1, n_2) *d.f.* and with non-centrality parameter ψ^2,

Now the joint *p.d.f.* of χ_1^2 and χ_2^2 is

$$\exp\left[-\tfrac{1}{2}(\chi_1^2+\chi_2^2)\right]$$

$$\times \sum_{j=0}^{\infty} \frac{(\psi^2/2)^j}{j\,!} \times \frac{(\chi_1^2)^{\frac{n_1}{2}-1+j}\;(\chi_2^2)^{\frac{n_2}{2}-1}}{2^{\frac{1}{2}(n_1+n_2)+j}\;\Gamma\left(\frac{n_1}{2}+j\right)\Gamma\left(\frac{n_2}{2}\right)}$$

Making the transformation $(\chi_1^2, \chi_2^2) \to (F, v)$, such that

$$\left.\begin{aligned} F &= \frac{\chi_1^2/n_1}{\chi_2^2/n_2} \\ \text{and}\quad v &= \chi_2^2 \end{aligned}\right\} \Rightarrow \begin{aligned} \chi_1^2 &= \frac{n_1}{n_2}\,F\,v \\ \text{and}\quad \chi_2^2 &= v \end{aligned}$$

the joint *p.d.f.* of F and v can be obtained.

Integrating with respect to v between 0 and ∞ we find the *p.d.f.* of F is

$$\frac{n_1}{n_2}\exp\left(-\frac{1}{2}\psi^2\right)\sum_{j=0}^{\infty}\left(\frac{\frac{1}{2}\psi^2}{j\,!}\right)^j \cdot$$

$$\times \frac{\Gamma\left(\frac{n_1+n_2}{2}+j\right)}{\Gamma\left(\frac{n_1}{2}+j\right)\Gamma\left(\frac{n_2}{2}\right)} \cdot \frac{\left(\frac{n_1}{n_2}F\right)^{\frac{n_1}{2}-1+j}}{\left(1+\frac{n_1}{n_2}F\right)^{\frac{n_1+n_2}{2}+j}}, 0 < F < \infty$$

Cor. For $\psi = 0$, we get the central F-distribution.

Exercises 13

1. Show that for the χ^2 distribution with *n.d.f.*, the moments about the mean obey the relation $\mu_{j+1} = 2j(\mu_j + n\mu_{j-1}), j \geq 1$. Obtain also the values of β_1, β_2. **(*B.Sc. Hons. Delhi, 1984*)**

2. (*a*) Show that the mode of χ^2 distribution is $n-2$ and that of the χ distribution is given by $\chi^2 = n-1$.

(*b*) Also show that for χ^2 distribution with *n.d.f.*

$$\gamma_1 = \sqrt{\frac{8}{n}}, \quad \gamma_2 = \frac{12}{n} \text{ for large } n.$$

3. Show that, for large n, $\sqrt{2\chi^2}$ where χ^2 is a chi-square variate with *n.d.f.* is normally distributed about mean $\sqrt{2n-1}$ with variance unity.

First Proof. Using Stirling's formula for $n!$ we find, after some reduction moments of the χ-distribution about origin are given by

$$\mu_1' = \sqrt{n}\left(1 - \frac{1}{4n} + \frac{1}{32n^2} - \frac{17}{192n^2} + \ldots\ldots\right)$$

whence $$\mu_1'^2 = n\left(1 - \frac{1}{2n} + \frac{1}{8n^2} - \frac{37}{192n^2} + \ldots\ldots\right)$$

$$\left.\begin{aligned} \mu_2' &= n \\ \mu_3' &= (n+1)\mu_1' \\ \mu_4' &= (n+2)n \end{aligned}\right] \quad \text{and} \quad \begin{aligned} \mu_2 &= \frac{1}{2} - \frac{1}{8n} + \ldots\ldots \\ \mu_3 &= \frac{1}{4\sqrt{n}} + \ldots\ldots \\ \mu_4 &= \frac{3}{4} + \frac{31}{48n} + \ldots\ldots \end{aligned}$$

Hence for the moments of $\sqrt{2\chi^2}$ we have

$$\mu_2 = 1 - \frac{1}{4n}, \quad \gamma_1 = \frac{1}{\sqrt{2n}} + \ldots\ldots; \quad \gamma_2 = \frac{3}{12n} + \ldots\ldots$$

Comparing with the results of Ex. 2 (*b*) we conclude that $\sqrt{2\chi^4}$ tends to normality with considerably greater rapidity than χ^2.

Moreover, the expression for μ_1' of χ is equal to $\sqrt{n-\frac{1}{2}}$ to order

$n^{-3/2}$ and hence $\sqrt{2\chi^2}$ is distributed about mean $\sqrt{2n-1}$ to the order with unit variance to order n^{-1}.

Second Proof. $\sqrt{2\chi^2} \le \sqrt{2n-1} + z$ is equvalant to $\chi^2 \le (n - \frac{1}{2}) + z\sqrt{2n-1}$ $+ \frac{1}{2}z^2$ *i.e.* $\dfrac{\chi^2 - 1}{\sqrt{n}} < z + \dfrac{z^2}{2\sqrt{2n}}$

As $n \to \infty$, this expression is asymptotic to $n + z\sqrt{2n}$, and therefore

$$P\left[\frac{\chi^2 - n}{\sqrt{2n}} \le z\right] \sim P\left[\sqrt{2\chi^2} \le \sqrt{2n-1} + z\right]$$

and we know by the central limit theorem that as the number n of degrees of freedom increases the distribution of $(\chi^2 - n)/\sqrt{2n}$ tends to normal.

4. (*a*) Show that for 2 *d.f.* the probability P of a value of χ^2 greater than χ_0^2 is $e^{-\frac{1}{2}\chi_0^2}$ and hence that $\chi_0^2 = 2 \log_e (1/P)$. Deduce the value of χ_0^2 when $P = 0.5$.

(*b*) The variates $x_1, x_2, \ldots\ldots, x_n$ are independently distributed in the rectangular form $\qquad dF = dx, \quad 0 \le x \le 1.$

Show that $-2 \log (x_1 \, x_2 \ldots\ldots x_n)$ is distributed as χ^2 with $2n.d.f.$

(B.A. *Hons. Statistics, 1989*)

(*c*) Let X_1, X_2 be *r.v.* following the density law $f(x) = e^{-x}, x > 0$. Show that $Z = X_1/X_2$ has a F-distribution.

[**Hint.** (*b*) Show that $y = -2 \log x$ or $x = e^{-\frac{1}{2}y}$ is distributed as χ^2 variate with 2 *d.f.* and then use additive property of χ^2-distribution]

5. (*a*) Let X_1, X_2 be *r.v.* following the density law $f(x) = e^{-x}$, $0 < x < \infty$. Show that $Z = X_1/X_2$ has a F distribution.

(*b*) Integrating by parts, show that if r is a positive integer,

$$\int_\beta^\infty \frac{1}{r!} e^{-t} t^r \, dt = \frac{1}{r!} \beta^r e^{-\beta} + \int_\beta^\infty \frac{1}{(r-1)!} e^{-t} t^{r-1} \, dt$$

and by continuing the process, show that the value of the intergal is

$$e^{-\beta} \left(1 + \beta + \frac{\beta^2}{2!} + \ldots\ldots + \frac{\beta^r}{r!}\right)$$

Hence show that the probability, in random sampling that the value of χ^2 with ν (even) *d.f.* will exceed χ_0^2 is obtained from the expression by putting

$$\beta = \tfrac{1}{2}\chi^2 \text{ and } r = \tfrac{1}{2}(\nu - 2).$$

Hence or otherwise show that the values of $P(\chi^2 \ge \chi_0^2)$ for a given value of χ^2 can be obtained from tables of the Poisson exponential limit.

[**Hint.** Repeated application of (1) of 5(*b*) above gives us

$$\frac{1}{r!}\int_{\beta}^{\infty} e^{-t}\, t^r\, dt = e^{-\beta}\left[1+\beta+\frac{\beta^2}{r}+\ldots\ldots+\frac{\beta^{r-1}}{(r-1)!}+\frac{\beta^r}{r!}\right]$$

Putting $\beta = \chi^2/2$ and $r = \dfrac{\nu-2}{2} = \dfrac{\nu}{2} - 1$ so that since r is an integer, ν is an even integer, $t = \dfrac{\chi^2}{2}$ we get

$$P = \frac{1}{2^{\frac{\nu}{2}-1}\,\overline{\left|\frac{\nu}{2}\right.}}\int_{\lambda}^{\infty} e^{-\chi^2/2}\,\chi^{\nu-1}\,d\chi = e^{-\chi^2/2}\left[1+\frac{\chi^2}{2}+\frac{\chi^4}{2\cdot 4}+\ldots\ldots+\frac{\lambda^{\nu-2}}{2\cdot 4\cdots(\nu-2)}\right]$$

(*c*) For χ^2 distribution with $(2v+2)$ *d.f.* prove that

$$P(\chi^2 \geq \chi_0^2) = \frac{1}{v!}\int_{\lambda}^{\infty} e^{-y}\,y^v\,dv = e^{-\lambda}\sum_{0}^{v}\frac{\lambda^r}{r!}$$

where $\lambda = \frac{1}{2}\chi_0^2$. Explain the uses of the result.

6. (*a*) If X is a normal variate with zero mean and *s.d.* unity, obtain the distribution of $t = \dfrac{x\sqrt{n}}{\chi} = \dfrac{x}{\sqrt{\dfrac{1}{n}\sum\limits_{1}^{n} x_i^2}}$

Find also the mean and variance of the distribution.

(*b*) Let U be a normal variate with zero mean and unit variance and V^2 have a χ^2 distribution with *n.d.f.* and be distributed independently of U. Derive the distribution of $\dfrac{u\sqrt{n}}{v}$.

(*c*) Show that the mean deviation about origin of t distribution with *n.d.f.* is

$$\sqrt{n}\,\overline{\left|\frac{n-1}{2}\right.}\Big/\sqrt{\pi}\,\overline{\left|\frac{n}{2}\right.}$$

(*d*) Let $x_1, x_1, \ldots\ldots, x_n$ be independent observations from a normal universe with mean μ and variance σ^2 and let x' be one more observation, independent of previous one. Show that

$$\frac{x'-\bar{x}}{s}\left[\frac{n(n-1)}{n+1}\right]^{1/2}$$

has Student t-distribution with $(n-1)$ degrees of freedom where $s^2 = \sum\limits_{i=1}^{n}(x_i - \bar{x})^2$. **(*B.Sc. Stat. Hons., Delhi 1989*)**

7. (*a*) Let $x_1, x_2, \ldots\ldots, x_n$ be a random sample from $N(\mu, \sigma^2)$ and $\bar{x}$ and S^2 respectively be the sample mean and sample variance. Let

$X_{n+1} \sim N(\mu, \sigma^2)$ and assume that $X_1, X_2, \ldots, X_n, X_{n+1}$ are independent. Obtain the sampling distribution of

$$U = \frac{(X_{n+1} - \overline{X})}{S} \cdot \sqrt{\frac{n}{n+1}}.$$

(*b*) Express the constants y_0, a and m of the distribution

$$f(x) = y_0 \left(1 - \frac{x^2}{a^2}\right)^m, \quad -a \le x \le a$$

in terms of its μ_2 and β_2. Show that if x is related to a variable t by the following equation $x = \dfrac{at}{[2(m+1) + t^2]^{\frac{1}{2}}}$

then t is distributed as Student's t-distribution with $2(m+1)$ degree freedom. Use the transformation to calculate the probability that $t \ge 2$ when the degree of freedom are (*i*) 2 and (*ii*) 4. **(*M.Sc. Madras, 1991*)**

$\left[\textbf{Ans. } y_0 = 1/a\, B(\tfrac{1}{2}, m+1), a^2 = (2m+3)\,\mu_2, m = (9 - 5\beta_2)/2\,(\beta_2 - 3), \cdot 092, \cdot 5 - \dfrac{5\sqrt{2}}{16}\right]$

8. If x_1, x_2, x_3, x_4 are four independent standard normal variates, obtain the sampling distribution of the following statistic :

(*a*) $u = \dfrac{x_1^2 + x_2^2}{x_1^2 + x_2^2 + x_3^2}$ (*b*) $\dfrac{x_1 + x_2 + x_3}{[\frac{1}{2}(x_1 - x_3)^2 + \frac{1}{6}(x_1 - 2x_2 + x_3)^2 + x_4^2]^{\frac{1}{2}}}$

(*c*) $w = \dfrac{x_1^2 + x_2^2}{x_3^2 + x_4^2}$ **(*B.Sc. Bombay 1981*)**

(your may assume the additive property of χ^2)

[**Hint.**(*b*)$x_1 + x_2 + x_3$ is normally distributed about zero mean and with variance 3.

Hence $\left(\dfrac{x_1 + x_2 + x_3}{\sqrt{3}}\right)^2$ is distributed like χ^2 with 1 *d.f.* Similarly, each of $\left(\dfrac{x_1 - x_3}{\sqrt{2}}\right)^2$, $\left(\dfrac{x_1 - 2x_2 + x_3}{\sqrt{6}}\right)^2$, $(x_4)^2$ is distributed independently (to be proved) like χ^2 with 1 *d.f.* and hence their sum is distributed like χ^2 with 3 *d.f.* Hence v_2 is the ratio of two variates independently distributed like χ^2 with 1 and 3 *d.f.* respectively. Now apply § 13·4]

9. If x_1, x_2, x_3 obey the joint distribution

$$(x_1\ x_2\ x_3)^{-\frac{1}{2}} f(x_1 + x_2 + x_3)\, dx_1\, dx_2\, dx_3$$

in the range $(0, \infty)$, obtain the sampling distribution of (*i*) $u = x_1 + x_2 + x_3$. (*ii*) $v = \dfrac{x_2}{x_1}$, (*iii*) $w = \dfrac{x_3}{x_1 + x_2}$ **(M.A., Calcutta, 1982)**

[Put $U = x_1 + x_2 + x_3$, $UV = x_2 + x_3$, $UVW = x_3$, $dx_1\, dx_2\, dx_3 = U^2 V\, dU dV d\, W$]

10. Show that the probability density of sample variance s^2 is maximum when $s^2 = \sigma^2$ as sample size $n \to \infty$ but the probability density of s is not maximum when $s = \sigma$ as $n \to \infty$. Show that s^2 satisfies the equation

$$(n-1)\, s^2 = \sum_{2}^{n} [k(x_k - \bar{x}_k)^2/(k-1)],\ \bar{x}_k = (x_1 + \ldots.. + x_k)/k.$$

11. If the r.v. x_1 and x_2 are independent and follows χ^2-distribution with nd.f, show that $\sqrt{n}\,(x_1 - x_2)/2\sqrt{x_1 x_2}$ follows t-distribution with n.d.f. independently of $x_1 + x_2$.

[**Hint.** Substitute for x_1 and x_2, from $u = \dfrac{\sqrt{n}\,(x_1 - x_2)}{2\sqrt{x_1 x_2}}$ and $v = x_1 + x_2$.

in $f(x_1, x_2) = \dfrac{1}{2^n\, (\overline{|n/2})^2}\, e^{-\frac{1}{2}(x_1 + x_2)}\, x_1^{n/2-1}\, x_2^{n/2-1}\, dx_1\, dx_2$.

12. (*a*) If u, v posses independent χ^2 distributions with n_1 and n_2 *d.f.* respectively, then $F = \dfrac{u/n_1}{v/n_2}$ has the F distribution with n_1 and n_2 *d.f.*

(*b*) Let X_i be a random sample from $N(0, 1)$. Find the distributions of

(*i*) $\frac{1}{2}(\bar{x}_k + \bar{x}_{n-k})$ (*ii*) $\bar{x}_k^2 + (n-k)\,\bar{x}_{n-k}^2$. **(B.A. Stat. Hons. Delhi, 1989)**

Clearly $\vec{x_k} \sim N\left(0, \dfrac{1}{k}\right)$ and $\bar{x}_{n-k} \sim N\left(0, \dfrac{1}{n-k}\right)$ where $\bar{x}_{n-k} = \dfrac{1}{n-k}\sum_{k+1}^{n} X_i$.

Also $\bar{x}_k$ and $\bar{x}_{n-k}$ are independent Hence

$$\frac{1}{2}(\bar{x}_k + \bar{x}_{n-k}) \sim N\left(0, \frac{1}{4k} + \frac{1}{4\,(n-k)}\right) \sim N\left(0, \frac{n}{4k\,(n-k)}\right)$$

(*ii*) $\bar{x}_k/k \sim N(0, 1)$ and $\bar{x}_{n-l}/\sqrt{1/(n-k)} \sim N(0, 1)$

$$\therefore \quad k\,\bar{x}_k^{\,2} \sim X^2_{(1)} \text{ and } (n-k)\,\bar{x}^2_{n-k} \sim X^2_{(2)}.$$

13. If X is distributed as F with 2 and n *d.f*, show that

$$p = P(x \geq k) = \left(1 + \frac{2F}{n}\right)^{\frac{-n}{2}}$$

14. (*a*) Show that the distribution of the statistic $F = \dfrac{s_1^{\,2}\,\sigma_2^{\,2}}{s_2^{\,2}\,\sigma_1^{\,2}}$ where $\sigma_1^{\,2}$, $s_2^{\,2}$ are the independent estimates of $\sigma_1^{\,2}$ and $\sigma_2^{\,2}$ respectively based on n_1, n_2 degrees of freedom depends only on n_1 and n_2.

(*b*) Show also that the mean of the F distribution does not involve n_1.

(*c*) When $n_1 = 2$, show that the significance level of F corresponding to a significance probability p is

$$F = \frac{n_2}{2}\left(p^{\frac{-2}{n_2}} - 1\right)$$

(*d*) Find the ·95 probability level of F for 2 and 4 degrees of freedom by direct integration of the distribution function.

15. (*a*) Let x_1, x_2 be a random sample from a distribution having probability density function $f(x) = e^{-x}, 0 < x < \infty$ and 0 elsewhere. Show that

(*i*) $z = x_1/x_2$ has an F distribution ;

(*ii*) $x_1 + x_2$ and x_1/x_2 are independently distributed.

(*b*) Show that the distribution of the quotient of two independent standard normal variates follows Cauchy's distribution.

16. Show that the mean of Fisher's z-distribution is the value of

$$\tfrac{1}{2}\log\left(\frac{n_2}{n_1}\right) - \sum_{r=0}^{\infty}\left(\frac{1}{n_1+t+2r} - \frac{1}{n_2-t+2r}\right)$$

when t is put equal to zero and variance is the differential coefficient of this expression with respect to t put equal to *zero*.

17. (*a*) If s_1^2 and s_2^2 are the variance in two independent samples of the same size taken from common normal population, determine the distribution of s_1^2/s_2^2.

[**Hint.** It can be shown that *m.g.f.* of s^2 in samples of n from a normal population with zero mean and variance σ^2 is

$$M(t) = \left(1 - \frac{2t\sigma^2}{n}\right)^{-\frac{1}{2}(n-1)}$$

$$\therefore \quad E[e^{t(s_1^2 + s_2)^2}] = [M(t)]^2 = \left(1 - \frac{2t\sigma^2}{n}\right)^{-(n-1)}$$

Hence the probability differential for $s_1^2 + s_2^2$ is

$$dp = \frac{\left(\frac{n}{2\sigma^2}\right)^{n-1}}{n-1}\exp\left\{-n\frac{(s_1^2+s_2^2)}{2\sigma^2}\right\} \times (s_1^2+s_2^2)^{n-2}\, d(s_1^2+s_2^2)]$$

(*b*) Given that $P[F(10, 12) > 2{\cdot}753] = 0{\cdot}05$ and $P[F(1, 12) > 4{\cdot}7143]$, find

(*i*) $P\left[F(12, 10) > \frac{1}{2{\cdot}753}\right]$ (*ii*) $P[-\sqrt{4{\cdot}747} < t\ (12) < \sqrt{4{\cdot}774}] = 0{\cdot}05$

18. Show that if χ_1^2, χ_2^2 are two independent χ^2 variates with n_1, n_2 *d.f.* and if t follows Student's distribution, then the statistics

$$\frac{\chi_1^2}{\chi_1^2+\chi_2^2},\ \chi_1^2,\ \frac{1}{\chi_1^2},\ \frac{\chi_1^2}{\chi_2^2},\ t$$

follow respectively Pearson Types I, III, V, VI and VII.

19. x_i $(i = 1, 2, ..., n)$ are independently and normally distributed with zero mean and common variance σ^2. Let

$$\xi_i = \sum_{j=1}^{n} c_{ij}\, x_i \qquad i = 1, 2, ..., p < n$$

and $\sum_j c_{ij}\, c_{i'j} = \delta_{ij}$

where δ_{ij}' is Kronecker delta.

Show that $\left[\sum_{i=1}^{n} x_i^2 - \sum_{i=1}^{p} \xi_i^2\right] \Big/ \sigma^2$

is distributed as χ^2 with $(n-p)$ degrees of freedom. (***R.A. Fisher***).

Sol. Since we are given that the coefficients c_{ij} satisfy the relations :

$$\left.\begin{aligned} \sum_i c^2_{ij} &= 1 = \sum_i c^2_{ij}{}' \\ \sum_i c_{ij}\, c_{ij}' &= 0 = \sum_j c_{ji} c_{ji}' \end{aligned}\right] (i' \neq i) \qquad ...(1)$$

we immediately have $\sum_{i=1}^{n} \xi_i^2 = \sum_{i=1}^{n} x_i^2$

so that $\left[\sum_{i=1}^{n} x_i^2 - \sum_{i=1}^{p} \xi_i^2\right] \Big/ \sigma^2 = \sum_{i=p+1}^{n} \left(\frac{\xi_i^2}{\sigma^2}\right).$

Also like x_i, ξ_i are normally distributed with zero mean and common variance σ^2. For the probability that simultaneously the n values of the independent variates x_i will fall in the respective intervals dx_i. is by the theorem of compound probability,

$$dp = (2\pi)^{\frac{n}{2}}\, e^{-\frac{1}{2\sigma^2}\Sigma x_i^2}\, dx_1 dx_2 ... dx_n = (2\pi)^{\frac{n}{2}}\, e^{-\frac{1}{2}\Sigma \frac{\xi_i^2}{\sigma^2}}\, dx_1 dx_2 \ldots dx_n$$

But $dx_1 dx_2 \ldots dx_n = J\left(\begin{matrix} x_1, x_2, \ldots x_n \\ \xi_1, \xi_2, \ldots \xi_n \end{matrix}\right) d\xi_1 .\ d\xi_2 \,.....\, d\xi_n$

$$= |\ c_{ij}\ |\ d\xi_1 d\xi_2 \,.....\, d\xi_n = d\xi_1 d\xi_2 \,.....\, d\xi_n$$

since $|\ c_{ij}^2\ | = 1$ by virtue of the condition (1).

Hence the probability that ξ_i's will fall simultaneously in their respective intervals $d\xi_i$ is

$$dp = \frac{1}{\sqrt{2\pi}} e^{-\frac{1}{2}\xi_1^2/\sigma^2} d\xi_1 \cdot \frac{1}{\sqrt{2\pi}} e^{-\frac{1}{2}\xi_2^2/\sigma^2} d\xi_2 .... \frac{1}{\sqrt{2\pi}} e^{-\frac{1}{2}\xi_n^2/\sigma^2} d\xi_n$$

It follows that ξ_i's *are independently and normally distributed about zero mean with same variance* σ^2.

Hence $\sum_{i=p+1}^{n} \frac{\xi_i^2}{\sigma^2}$ is the sum of the squares of $(n-p)$ independently and normally distributed variates and therefore by §13·2 cor. 1 it is distributed like χ^2 with $(n-p)$ *d.f.*

20. Use Exercise 19 to show that x_i are object to p linear constraints the distribution of the sum of the squares is that of χ^2 with $(n-p)$ *d.f.*

21. $x_1, x_2, ..., x_n$ is a sample of n independent observations from a normal population with mean μ and variance σ^2, show by means of an orthogonal transformation.

$$y_i = \frac{x_1}{\sqrt{n}} + \frac{x_2}{\sqrt{n}} + ... + \frac{x_n}{\sqrt{n}} = \bar{x}\sqrt{n}$$

and $$y_i = c_{i1}x_1 + c_{i2}x_2 + ... + c_{in}\, x_n, \quad i = 1, 2, ..., n$$

or otherwise, that the distribution of $u = \sqrt{n}\ (\bar{x} - \mu)$ and $v = \sum_{1}^{n} (x_i - \bar{x})^2$ are independent and derive their distributions. Hence obtain the mean and variance of $w = \frac{1}{n-1} \sum_{1}^{n} (x_i - \bar{x})^2$

22. If $\bar{x}$ and s^2 are the mean and the variance of a random sample of n members from an infinite population, derive the formulae for :

(*i*) Var $(\bar{x})$. (*ii*) Var (s^2), for large n (*iii*) Var (s), for large n.

If $a = \bar{x} + ks$ and $b = \bar{x} - ks$ where $k = 2(\beta_2 - 1)^{-\frac{1}{2}}$, show that, in large samples, a and b are uncorrelated.

23. If $X_i (i = 1, ..., n)$ are independent random variables each distributed normally with mean zero and variance one, show that ΣX_i and $\Sigma(X_i - \bar{X})^2$ are independent. Find the mean and variance of $(1/n)\, \Sigma\, (x_i - \bar{x})^2$ and now show that they approach 1 and 2/n as n increases.

24. Show that the distribution of the sum Y of three independent observations from the distribution

$$f(x) = \tfrac{1}{2}\, e^{-|x|}, -\infty < x < \infty$$

is given by $\frac{1}{16} (3 + 3\,|y| + y^2)\, e^{-|y|}\, dy$

[**Hint.** Distribution of Y is obtained by integrating $e^{-ity} \cdot (1 + t^2)^{-3}$ as contour integral, the characteristic function being $1/(1+t^2)$].

25. Show that the distribution of the standard deviation of samples of two from the rectangular population $dF = dx,\ 0 \le x \le 1$, is given by

$$f(s) = 4(1-2s),\ 0 \le s \le \tfrac{1}{2}.$$

[**Hint.** $dF = dx_1\, dx_2,\ s = \dfrac{x_1 - x_2}{2},\ u = \dfrac{x_1 + x_2}{2},\ dx_1\, dx_2 = 2\, ds du$. Integrate for u from s to $1-s$ and double the result.]

26. The variables X and Y with zero means and standard deviations σ_1 and σ_2 are normally correlated with coefficient ρ. Show that u and v, defined by $u = \dfrac{x}{\sigma_1} + \dfrac{y}{\sigma_2}$; $v = \dfrac{x}{\sigma_1} - \dfrac{y}{\sigma_2}$ are independent normal variates with variances $2(1+\rho)$ and $2(1-\rho)$ respectively ; and that, if R is the correlation between u and v in random samples of n pairs from the bivariate population, R^2 is a $\beta_1(\frac{1}{2}, \frac{1}{2}n - 1)$ variate.

$$\left[\textbf{Hint.}\ \text{Since}\ \frac{-1}{2(1-\rho^2)}\left[\frac{x^2}{\sigma_1^2} - \frac{2\rho xy}{\sigma_1\sigma_2} + \frac{y^2}{\sigma_2^2}\right]\right.$$

$$= -\frac{1}{2}\left[\frac{1}{2(1+\rho)}\left\{\frac{x}{\sigma_1} + \frac{y}{\sigma_2}\right\}^2 + \frac{1}{2(1-\rho)}\left\{\frac{x}{\sigma_1} - \frac{y}{\sigma_2}\right\}^2\right]$$

We conclude that $u = \dfrac{x}{\sigma_1} + \dfrac{y}{\sigma_2}$ and $v = \dfrac{x}{\sigma_1} - \dfrac{y}{\sigma_2}$ are independent normal variates with variances $2(1+\rho),\ 2(1-\rho)$, respectively. Hence etc.]

27. Derive the distribution of r and show that when $n = 5$ and if $P(|r| \ge c) = \alpha$, c is a root of the equation $c\sqrt{1-c^2} + \sin^{-1} c + \frac{1}{2}\pi(\alpha - 1) = 0$.

$$\left[\textbf{Hint.}\ P(|r| \ge c) = 1 - P(-c \le r \le c) = 1 - 2P(0 \le r \le c)\right.$$

$$\left. = 1 - 2\frac{\Gamma c}{\Gamma(1/2)\,\Gamma(1/2)}\int_0^c (1-r^2)^{1/2}\, dr = \text{etc.}\right]$$

28. (*a*) X is a Poisson variate with parameter λ and χ^2 is a Chi-square variate with $2k$ *d.f.* Prove that for all positive integer k, $P(x < k-1) = P(\chi^2 \ge 2\lambda)$.

[B.Sc. Stat. Hons. 1996]

(*b*) Prove that if $n_1 = n_2$, median of F-distributions is at $F = 1$ and $Q_1 Q_3 = 1$

[B.Sc. Stat. Hons. Delhi, 1989]

[**Hint.** (*a*) Since $n_1 = n_2 = n$, say, we have

$$P[F(n, n) \le M] = \frac{1}{2} \Rightarrow P\left[\frac{1}{F(n, n)} \ge \frac{1}{M}\right] = \frac{1}{2}$$

$$\Rightarrow\ P[F(n, n) \ge \frac{1}{M}] = \frac{1}{2}\ \Bigg]$$

$$\therefore \quad P\left[F(n, n) \geq \frac{1}{M}\right] = \frac{1}{2}$$

$$\therefore \quad P\left[F(n, n) \leq \frac{1}{M}\right] = 1 - \frac{1}{2} = \frac{1}{2}$$

$$\therefore \quad M = \frac{1}{M} \Rightarrow M^2 = 1. \text{ Hence } M = 1$$

Also $P[F(n, n) \leq Q_1) = 0{\cdot}25$ and $P[F(n, n) \geq Q_3) = 0{\cdot}25$

$$\Rightarrow P\left[\frac{1}{F(n, n)} \leq \frac{1}{Q_3}\right] = 0{\cdot}25$$

$$\Rightarrow Q_1 = \frac{1}{Q_3} \Rightarrow Q_3 Q_1 = 1$$

[**Hint.** (*a*) $P(\chi^2 > 2\lambda) = \dfrac{1}{2^k \Gamma k} \displaystyle\int_{2\lambda}^{\infty} e^{-\frac{1}{2}\chi^2} (\chi^2)^{k-1} d\chi^2$

$$= \frac{1}{(k-1)!} \int_{\lambda}^{\infty} e^{-y} y^{k-1} dy = etc.$$]

29. If X and Y are independent *r.v.'s* having rectangular distribution in the interval (0, 1), show that $U = \sqrt{-2 \log X} \cos 2\pi Y$ and $V = \sqrt{-2 \log x} \sin 2\pi Y$ are independently distributed as $N(0, 1)$.

[**Hint.** $u^2 + v^2 = -2 \log x$]

30. If X and Y are independent *r.v.*, having a common *p.d.f.* find the *p.d.f.* of $X - Y$.

$$f(x) = \begin{cases} e^{-x}, & x \geq 0 \\ 0, & x < 0 \end{cases}$$

[**Hint.** Put $U = X - Y$, $V = X + Y$, $dxdy$ = etc.]

31. If X and Y are independent *r.v.*, with common Cauchy *p.d.f.*

$$\frac{1}{\pi} \cdot \frac{1}{1 + x^2}, \quad -\infty < x < \infty$$

prove that $Z = XY$ has the *p.d.f.* $f(z) = \dfrac{2}{\pi^2} \left\{ \dfrac{\log z}{z^2 - 1} \right\}$

[**Hint.** $W = X/Y$, $\therefore X^2 = ZW$, $Y^2 = Z/W$, $dXdY = J \begin{vmatrix} X & Y \\ Z & w \end{vmatrix} dzdw$]

32. (*a*) If $X_1, X_2, \ldots, X_n$ are independenty and normally distributed with same mean μ but with different variances $\sigma_1^2, \sigma_2^2, \ldots, \sigma_n^2$ and assume that

$$U = \Sigma\left(\frac{X_i}{\sigma_i^2}\right) \Big/ \Sigma\left(\frac{1}{\sigma_i^2}\right) \text{ and } V = \Sigma(X_i - \mu)^2/\sigma^2$$

are independenty distributed, show that U is normal and V has chi-square distribution with $(n - 1)$ *d.f.* **(*B.A. Hons., Statistics Delhi, 1985*)** (*b*) $X_1, X_2, \ldots\ldots, X_n$ is a set of n mutually independent variates, each normal with mean μ and variance σ^2. Then prove that the sample mean $\bar{X}$ and sample variance S^2 are independent variates.

[**Hint.** $S^2 = \dfrac{1}{n} \Sigma (X_i - \mu)^2 - (\bar{X} - \mu)^2$

$$\therefore \quad \chi^2 = \frac{nS^2}{\sigma^2} = \Sigma \left(\frac{X_i - \mu}{\sigma} \right)^2 - \left\{ \frac{\sqrt{n}\,(\bar{X} - \mu)}{\sigma} \right\}^2$$

Now $\dfrac{X_1 - \mu}{\sigma}, \dfrac{X_2 - \mu}{\sigma}, \ldots\ldots, \dfrac{X_n - \mu}{\sigma}$ are n mutually independent standard normal variates,

and $\dfrac{\sqrt{n}\,(\bar{X} - \mu)}{\sigma} = \dfrac{1}{\sqrt{n}} \left(\dfrac{X_1 - \mu}{\sigma} \right) + \dfrac{1}{\sqrt{n}} \left(\dfrac{X_2 - \mu}{\sigma} \right) + \ldots\ldots + \dfrac{1}{\sqrt{n}} \left(\dfrac{X_n - \mu}{\sigma} \right)$

is a linear combination of them, such that $\left| \dfrac{1}{\sqrt{n}} \right|^2 + \left(\dfrac{1}{\sqrt{n}} \right)^2 + \ldots\ldots + \left(\dfrac{1}{\sqrt{n}} \right)^2 = 1$

Thus χ^2 is χ^2-distributed with $v = n - 1$ *d.f.* and χ^2 is independent of $U = \sqrt{n}\ (\bar{X} - \mu)/\sigma$, so that $\bar{X}$ and S^2 are independent.

33. $X_1, X_2, \ldots X_n$ are independent random variables, each having the same negative exponential distribution. Prove that $X_1 + \frac{1}{2} X_2 + \ldots. + \frac{1}{n} X_n$ and $\max (X_1 \ldots.., X_n)$ have the same distribution.

34. Show that X is stochastically independent of $X_1 - \bar{X}, X_2 - \bar{X}, \ldots\ldots, X_n - \bar{X}$. Deduce that X is stochostically independent of S^2 where $nS^2 = \sum_{i=1}^{n} (X_i - \bar{X})^2$ and X_i's are $N(\mu, \sigma^2)$.

(*B.A. Hons. Delhi, 1988 ; B.A. Hons. Statistics Delhi, 1989*)

Sol. Let $M(t, t_1, \ldots..t_n)$ be the moment-generating function of $\bar{X}$, $X_1 - \bar{X}, \ldots., X_n - \bar{X}$. Then

$$M(t, t_1, \ldots\ldots t_n) = \int_{-\infty}^{+\infty} \ldots. \int_{-\infty}^{+\infty} \left(\frac{1}{\sigma \sqrt{2\pi}} \right)^n \exp\ [t\,x + t_1(x_1 - \bar{x}) + \ldots.$$

$$+ t_n(x_n - \bar{x}) - \frac{1}{2\sigma^2} \sum_{1}^{n} (x_i - \mu)^2]\ dx_1 \ldots. dx_n$$

Consider the integral on x_i viz.

$$\int_{-\infty}^{+\infty} \frac{1}{\sigma_i \sqrt{2\pi}} \exp \left[\{ t + nt_1 - (t_1 + \ldots.. + t_n) \} \frac{x_i}{n} - \frac{1}{2\sigma^2} (x_i - \mu)^2 \right] dx_i$$

This integral exists for all real values of $t, t_1, \ldots\ldots, t_n$ and it may be interpreted as the *m.g.f.* of 0 distribution which is $N(\mu, \sigma^2)$ with t replaced by

$$\frac{1}{n}[t+nt_i-\Sigma t_j]=\frac{1}{n}[t+n(t_1-\bar{t})] \text{ where } \bar{t}=\frac{1}{n}\sum_1^n t_i$$

Hence the given integral is equal to

$$\exp\left[\sum_{i=1}^{n}\left\{\frac{\mu}{n}t+\mu(t_i+t)+\frac{\sigma^2}{2n^2}(t+n(t_1+\bar{t})^2\right\}\right]$$

Since $\sum_1^n (t_1-\bar{t})=0$, we have $M(t, t_1, ..., t_n)$

$$=\exp\left[\mu t+\frac{\sigma^2}{2n}t^2+\frac{\sigma^2}{2}\Sigma(t_i-\bar{t})^2\right]$$

$$=\exp\left[\mu t+\frac{\sigma^2 t^2}{2n}\right]\exp\left[\frac{\sigma^2}{2}\sum_{i=1}^{n}(t_i-\bar{t})^2\right]$$

$$=M_{\bar{x}}(t)\times M_{\Sigma(x_i-\bar{x})}(t)$$

Thus the *r.V.* $\bar{X}$ and the *r.v.*'s $(X_1-\bar{X}, X_2-\bar{X}, ..., X_n-\bar{X}]$ are stochastically independent.

Hence $\bar{X}$ and nS^2 are stochastically independent.]

Aliter. $M(t_1, t_2)=E[e^{t_1\bar{x}}\, e^{t_2(x-\bar{x})}]=E[e^{t_2 x}\cdot e^{(t_1-t_2)\bar{x}}]$

35. Let X, Y, and Z be independent random variables, each having normal distribution $N(0, \sigma^2)$, then prove that

$$S=\sqrt{X^2+Y^2+Z^2} \quad \text{has a } p.d.f.$$

$$g(s)=\sqrt{\frac{2}{\pi}}\cdot\frac{s^2}{\sigma^3}e^{-s^2/2\sigma^2}, s\geq 0.$$

36. Suppose that the volume V of a box with a square base is chosen randomly from the chi-squares distribution with 2 *d.f.* In addition, let the length S of a side of base be closen randomly and independently of V according to the following procedure.

$S=|T|$ where T is chosen from the normal distribution with mean zero and *s.d.* one. What is the probability that the height H of the box exceeds 400 ?

[Hint. $V=HS^2$

T is distributed as $N(0, 1)$, then $S^2=T^2$ is distributed as a chi-square with 1 *d.f.* Then $H=\dfrac{V}{S^2}$ implies that $H/2=\dfrac{V/2}{S^2}$

Now $\frac{1}{2}H$ has an distribution with 2 *d.f.* in the numerater and 1 *d.f.* in the denominator.

Thus $P(H>400)=P(\frac{1}{2}H>200)=1-P(\frac{1}{2}H\leq 200)$

$$=0{\cdot}05, \text{ using } F \text{ tables, } S=F_{2,1}$$

37. Prove that $\overline{X} = \frac{1}{n} \sum_{1}^{n} X_i$ and $X_i - \overline{X} = i = 1, 2, ..., n$ are independently distributed where $X_1, X_2, ..., X_n$ is a random sample from a normal population with mean μ and variance σ^2.

Sol. The joint *m.g.f.* of $\overline{X}$ and $(X_i - \overline{X})$ is given by

$$M(t_1, t_2) = E\left[e^{t_1 \overline{X} + t_2 (X_i - \overline{X})} \right] = E\left[e^{(t_1 - t_2)\overline{X} + t_2 X_i} \right]$$

$$= E\left[\exp\left\{ \frac{t_1 - t_2}{n} \Sigma X_i + t_2 X_i \right\} \right]$$

$$= E\left[\exp\left\{ \left(\frac{t_1 - t_2}{n} + t_2 \right) X_i + \frac{t_1 - t_2}{n} \sum_{\substack{j=1 \\ j \neq i}}^{n} X_j \right\} \right]$$

$$= E\left[\exp\left\{ \left(\frac{t_1 - t_2}{n} + t_2 \right) X_i \right\} \right] E\left[\exp\left(\frac{t_1 - t_2}{n} \sum_{\substack{j=1 \\ j \neq i}}^{n} (X_j) \right) \right] \quad ...(i)$$

Now $U = \sum_{\substack{j=1 \\ j \neq i}}^{n} X_j$ being the sum of $(n-1)$ *i·i·d* $N(\mu, \sigma^2)$ variates is a

$N\left\{ (n-1)\mu, (n-1)\sigma^2 \right\}$ variate

$$\therefore \quad M_u(t) = \exp\left[t \cdot (n-1)\mu + t^2 \cdot (n-1)\sigma^2/2 \right] \quad ...(ii)$$

Hence $E\left[\exp\left\{ \frac{t_1 - t_2}{n} \sum_{\substack{j=1 \\ j \neq i}}^{n} X_j \right\} \right] = E\left[\exp\left(\frac{t_1 - t_2}{n} ; U_j \right) \right] = M_u \frac{(t_1 - t_2)}{n}$

$$= \exp\left[\left(\frac{t_1 - t_2}{n} \right)(n-1)\mu + \left(\frac{t_1 - t_2}{n} \right)^2 (n-1)\frac{\sigma^2}{2} \right]. \text{ on using } (i), (ii)$$

and $E\left[\exp\left\{ \left(\frac{t_1 - t_2}{n} + t_2 \right) X_i \right\} \right] = M_{x_i}\left(\frac{t_1 - t_2}{n} + t_2 \right)$

$$= \exp\left[\left(\frac{t_1 - t_2}{n} + t_2 \right)\mu + \left(\frac{t_1 - t_2}{n} + t_2 \right)^2 \frac{\sigma^2}{2} \right] \quad ...(iii)$$

Substituting from (*iv*) and (*v*) in (*ii*), we get

$$M(t_1, t_2) = \exp\left[\left\{ \left(\frac{t_1 - t_2}{n} \right)(n-1) + \left(\frac{t_1 - t_2}{n} + t_2 \right) \right\} \mu \right]$$

$$\times \exp\left[\left\{\left(\frac{t_1 - t_2}{n}\right)^2 \cdot (n-1) + \left(\frac{t_1 - t_2}{n} + t_2\right)^2\right\}\frac{\sigma^2}{2}\right]$$

$$= \exp\left[t_1 \mu + \tfrac{1}{2} t_1^2 \frac{\sigma^2}{n}\right] \times \exp\left[\tfrac{1}{2} t_2^2 \left(\frac{n-1}{n} \sigma^2\right)\right]$$

$$= M(t_1) \times M(t_2)$$

Thus $\overline{X}$ and $X_i - \overline{X}$ are independently distributed.

and $\overline{X} \sim N\left(\mu, \dfrac{\sigma^2}{n}\right)$ and $X_i - \overline{X} \sim N\left(0, \dfrac{n-1}{n}\sigma^2\right)$

38. Let x_1 and x_2 the two independent normal variates $N(\mu, \sigma^2)$. Obtain the distribution of

$$Z = \frac{x_1 + x_2 - 2\mu}{\sqrt{|x_1 - x_2|^2}}$$

[See Ex. 13·3]

14

Exact Sampling Distributions (*Contd.*)

14·1. Introduction. In the previous chapter, we have obtained the sampling distributions of the statistics $\bar{x}, s^2, \chi^2, t, z, F, \chi_1^2 + \chi_2^2$ and χ_1^2/χ_2^2 for random samples from normal population, by using the geometrical method. In the present chapter, we shall illustrate the use of direct analytic method, by deriving the χ^2 distribution, by its use. We next shall discuss the use of the method of characteristic function to derive the sampling distributions specially that of the sample mean and illustrate its use by deriving the distributions of the sample means from Binomial, Poisson, Normal and Cauchy distributions. The use of the method of mathematical induction is discussed in § 14·4. In § 14·5 we show how to derive the distribution of the sample range and in the next section we derive the distributions of the various quantiles.

Up to § 14·6, we consider samples from uni-variate universes only. In § 14·7, we consider random samples from a bi-variate normal population and derive the distributions of the sample means, variances and in particular the important distribution of the sample correlation coefficient.

14·2. Analytic Method. We shall illustrate this method by finding the χ^2 distribution *viz.*, sampling distribution of the sum of squares of n independent variates, each of which is distributed normally, with zero mean and unit variance.

The distribution of the variate X_i is $\frac{1}{\sqrt{2\pi}} \exp\left(-\frac{1}{2} x_i^2\right) dx_i,\ i = 1, 2, ..., n$

Hence the joint distribution of n independent variates X_i is

$$dF = (2\pi)^{-\frac{1}{2}n} \exp\left[-\frac{1}{2}\left(x_1^2 + x_2^2 + ... + x_n^2\right)\right] dx_1 dx_2 .... dx_n,$$

The distribution function of

$$Z = X_1^2 + X_2^2 + ... + X_n^2 \qquad ...(14·1)$$

is given by

$$F(x) = \int\int\int \ldots\ldots \int (2\pi)^{-\frac{1}{2}n} \exp\left[-\frac{1}{2}\sum_{i=1}^{n} x_i^2 \right] dx_1\, dx_2 \ldots dx_n$$

where the domain of x's is conditioned by

$$x_1^2 + x_2^2 + \ldots + x_n^2 \le z.$$

Let us change the n variates x_i's to $z, \theta_1, \theta_2, \ldots, \theta_{n-1}$

where $x_1 = \sqrt{z} \cos\theta_1 \cos\theta_2 \ldots \cos\theta_{n-1}$,

$x_2 = \sqrt{z} \cos\theta_1 \ldots \cos\theta_{n-2} \sin\theta_{n-1}$,

....

....

$x_n = \sqrt{z} \sin\theta_1$

and hence $x_1^2 + x_2^2 + \ldots + x_n^2 = z$.

It can be easily seen that

$$\frac{\partial (x_1, x_2, \ldots\ldots x_n)}{\partial (z, \theta_1, \theta_2 \ldots\ldots \theta_{n-1})} = z^{\frac{n-2}{2}} f(\theta_1, \theta_2, \ldots, \theta_{n-1})$$

where $f(\theta_1, \theta_2, \ldots\ldots, \theta_{n-1})$ is a function of θ's only. Hence

$$F(z) = (2\pi)^{-\frac{1}{2}n} \int\int \ldots\ldots \int z^{\frac{1}{2}(n-2)} e^{-\frac{z}{2}} f(\theta_1, \theta_2, \ldots\ldots, \theta_{n-1})\, d\theta_1 \ldots d\theta_{n-1}\, dz \qquad \ldots(14\cdot2)$$

in which z varies from 0 to z and θ's vary from 0 to π or 2π.

The right hand side of (14·2) is the product of n integrals of which those containing $\theta_1, \theta_2, \ldots\ldots, \theta_{n-1}$ are constants. Thus we get

$$F(z) = k \int_0^z z^{\frac{n-2}{2}} e^{-\frac{z}{2}} dz,$$

where k is a constant given by

$$1 = k \int_0^\infty z^{\frac{n-2}{2}} e^{-\frac{1}{2}z} dz = k\, 2^{\frac{n}{2}} \Gamma(n/2)$$

Thus the required distribution of z is

$$dF = 2^{-\frac{1}{2}n} \left[\Gamma\left(\tfrac{1}{2}n\right) \right]^{-1} e^{-\frac{1}{2}z} z^{\frac{1}{2}(n-2)}\, dz, \quad 0 < z < \infty.$$

Ex. 1. Obtain from first principles the sampling distribution of

$$Z = X_1^2 + X_2^2 + X_3^2$$

where X's are each distributed independently and normally with unit variance and zero mean.

Ex. 2. Derive the χ^2 distribution from that of the sum of the squares of n standard normal variates subject to linear restriction $\sum_{i=1}^{n} a_i x_i = 0$ by the substitution $x_i = u_i \sqrt{\chi}$ subject to $\sum_{i=1}^{n} u_1^2 = 1$.

14·3. The Method of Characteristic Functions. We shall make use of the following properties of characteristic functions :

If $x_1, x_2, \ldots, x_n$ is a random sample of size n, so that, $x_1, x_2, \ldots, x_n$ are independent and identically distributed and $\bar{x}$ be the sample mean, then

$$(i)\quad \phi_{x_1 + x_2 + \ldots + x_n}(t) = \phi_{x_1}(t) \times \ldots \times \phi_{x_n}(t) = [\phi_x(t)]^n$$

$$(ii)\quad \phi_{\bar{x}}(t) = \phi_{x_1}\left(\frac{t}{n}\right)\phi_{x_2}\left(\frac{t}{n}\right) \ldots \phi_{x_n}\left[\frac{t}{n}\right] = \left[\phi_x\left(\frac{t}{n}\right)\right]^n$$

We illustrate the method by solving a few examples.

14·3·1. Distribution of the Mean of the Poisson Distribution. The characteristic function (*c.f.*) of the Poisson distribution with parameter λ is

$$\exp\left[\lambda\left(e^{it} - 1\right)\right]$$

Therefore, by the property (*ii*) of *c.f., the c.f.* of the mean is

$$\left[\exp \lambda\left\{e^{\frac{i.t}{n}} - 1\right\}\right]^n = \exp\left[n\lambda\left\{e^{i.t/n} - 1\right\}\right]$$

Hence the distribution of the mean is the Poisson distribution with parameter $n\lambda$, the interval being $1/n$, *instead* of unity.

14·3·2. Distribution of the Mean of the Normal Population. The *c.f.* of the normal distribution with mean μ and *s.d.* σ is

$$\exp\left[-\tfrac{1}{2}t^2\sigma^2 + it\mu\right]$$

The *c.f.* of the distribution of the mean of n values is then

$$\exp\left\{n\left(-\frac{1}{2}\frac{t^2\sigma^2}{n^2} + \frac{it\mu}{n}\right)\right\} = \exp\left\{-\frac{1}{2}\frac{t^2\sigma^2}{n} + it\mu\right\}$$

This is the *c.f.* of the normal distribution with mean μ and *s.d.* $\sigma/\sqrt{n}$, which is, therefore, the distribution required.

Ex. 1. Show that the distribution of the mean for the Binomial distribution $(q+p)^r$ is $(q+p)^{rn}$, the interval being $1/n$ instead of unity.

Ex. 2. By using characteristic function, obtain the sampling distribution of the mean in samples of size n from Cauchy's population.

$$dF = \frac{dx}{\pi(1+x^2)}, \quad -\infty < x < \infty. \qquad \left[\textbf{Ans. } \frac{1}{\pi} \cdot \frac{d\bar{x}}{1+\bar{x}^2}\right]$$

$$\left[\textbf{Hint. use} \int_{-\infty}^{\infty} \frac{e^{itx} dx}{\pi(1+x^2)} = e^{-|t|} \right]$$

Ex. 3. Derive by the method of characteristic functions the sampling distribution of the sum of the squares of n independent standard normal variates.

14·4. The Method of Induction. If $X_1, X_2, \ldots, X_n$ are n independent variates, each having the same distribution, the general form of the distribution of their sum may be suggested, when the results for two or three variates have been worked out and then the validity may be tested by induction. We, therefore, give the method of obtaining the distribution of the sum of two independent variates, whose individual distributions are known.

Let X_1 and X_2 be two independent variates with density functions $f_1(x_1)$ and $f_2(x_2)$ respectively then the distribution of $Z = X_1 + X_2$ is given by

$$F(z) = \int\int f_1(x_1) f_2(x_2)\, dx_1 dx_2.$$

the range of integration being that for which $x_1 + x_2 \le z$.

$$\therefore \qquad F(z) = \int_{-\infty}^{\infty} \left[\int_{-\infty}^{z-x_2} F_1(x_1) \right] f_2(x_2) dx_2$$

where $\int_{-\infty}^{x_1} f_1(x_1)\, dx = F_1(x_1)$

Hence
$$F(z) = \int_{-\infty}^{\infty} F_1(z - x_2) f_2(x_2) dx_2$$

Differentiating with respect to z, we get

$$f(z) = \int_{-\infty}^{\infty} f_1(z - x_2) f_2(x_2) dx_2$$

This is the density function of Z.

Example 14·1. *Derive (by the method of Induction) the sampling distribution of the mean of a sample of n values $x_1, x_2, \ldots, x_n$ from Cauchy's distribution*

$$dF = \frac{dx}{\pi(1+x^2)}, \quad -\infty < x < \infty.$$

We first obtain the distribution of the sum

$$z = x_1 + x_2 + \ldots + x_n.$$

The distribution of the sum $z = x_1 + x_2$ is

$$\int_{-\infty}^{\infty} \pi^{-2}\left[\frac{1}{1+(z-x_2)^2}\right]\left(\frac{1}{1+x_2^2}\right)dx_2 = \frac{2}{\pi\,(2^2+x^2)}$$

This suggests the general form $\dfrac{n}{\pi\,(n^2+x^2)}$ for the probability function of n variates.

If this is true, then the probability function of $(n+1)$ variates is

$$\int_{-\infty}^{\infty} \frac{n}{\pi^2}\left[\frac{1}{n^2+(z-x)^2}\right]\left(\frac{1}{1+x^2}\right)dx$$

$$= \frac{n}{\pi}\left(\frac{n+1}{n}\right)\Big/ [(n+1)^2+z^2]$$

$$= \frac{1}{\pi}\cdot\frac{n+1}{[(n+1)^2+z^2]}$$

The result holds for $n = 1, 2$ and is, therefore, true in general.

But $\bar{x} = z/n$ so that $dz = n\,d\bar{x}$. Hence the distribution of $\bar{x}$ is

$$dP = \frac{1}{\pi}\cdot\frac{d\bar{x}}{1+\bar{x}^2},\ -\infty < \bar{x} < \infty$$

Ex. Use the method of induction to obtain the sampling distribution of the sum of square of n independent variates each of which is normally distributed with zero mean and standard deviation.

14·5. Distribution of the Range. In industrial quality control work, the range is a useful measure of dispersion because of its ease in computation. Also for small samples from a normal population, it can be shown, that in a certain sense, the range is as efficient for estimating σ as is the sample standard deviation. We, therefore, are interested in obtaining the sampling distribution of the range w. Since the range of sample of values is the difference between the highest and lowest values, we shall first obtain the joint distribution of these values.

Case 1. *Infinite range of population.* Consider a random sample of n values drawn from a continuous population, whose density function is $f(x)$. Let these sample values be arranged in order of increasing magnitude, and denote the ordered set by $x_1, x_2, ..., x_n$.

Let the infinite range be divided into five intervals $(-\infty, u)$, $(u, u+du)$, $(u+du, v)$, $(v, v+dv)$, $(v+dv, \infty)$ where u and $v\,(>u)$ are any two values of x. Then the probability that a values of x will fall in the interval $(u, u+du)$ is $f(u)du$.

$-\infty \leftarrow \quad u \quad v \quad u+du \quad v+dv \rightarrow \infty$

Similarly, the probability that a value of x will fall in the interval $(v, v+dv)$ is $f(v)dv$ and the probability that a value of x will fall in the interval $(u+du, v)$ is

$$\int_{u+du}^{v} f(x)\,dx = F(v) - F(u+du) \quad \text{where} \int f(x)\,dx = F(x) + \text{constant.}$$

Consequently, the probability that in the drawing of n values, no value of x will fall in the first interval, one in the second interval du, $(n-2)$ values in the third interval, one in the fourth interval of length dv, and no value in the fifth interval is,

$$dP = \frac{n\,!}{(n-2)}\,[f(u)du][F(v) - F(u+du)]^{n-2}[f(v)dv]$$

$$= n(n-1)\,f(u)\,f(v)\,[F(v) - F(u)]^{n-2}\,du\,dv,$$

neglecting infinitesimals of order higher than second in the R.H.S.

or $$dP = n(n-1)\,f(u)\,f(v)\left[\int_u^v f(x)\,dx\right]^{n-2} du\,dv \qquad \ldots(14{\cdot}2)$$

Thus if u and v denote the lowest and the highest values respectively, (14·2) is the joint distribution of u and v.

To obtain the distribution of the range w, let $w = v - u$ or $v = u + w$, then $J\begin{pmatrix} u, & w \\ u, & v \end{pmatrix} = 1$ so that the probability that u will fall in the interval du and w in the interval dw is

$$dP' = n\,(n-1)\,f(u)\,f(u+w)\left[\int_u^{u+w} f(x)\,dx\right]^{n-2} du\,dw$$

Hence the probability that w will fall in the interval dw, irrespective of the value of u, is

$$dP = n\,(n-1)\left[\int_{-\infty}^{\infty}\left\{f(u)\,f(u+w)\left(\int_u^{u+w} f(x)dx\right)^{n-2} du\right\} dw\right. \qquad \ldots(14{\cdot}3)$$

This is the required probability differential of w.

(14·3) can be written as follows :

$$dP = n\,(n-1)\,dw\int_{-\infty}^{+\infty}\left\{F(u+w) - F(u)\right\}^{n-2} . f(u+w)\,dF(u).$$

$$E(w) = n(n-1) \int_{-\infty}^{+\infty} dF_n \int_{-\infty}^{x_n} (F_n - F_1)^{n-2}(x_n - x_1)\, dF_1$$

$$= \int_{-\infty}^{+\infty} dx_n \int_{-\infty}^{x_n} \frac{d^2}{dx_1 dx_n} (F_n - F_1)^n (x_n - x_1)\, dx_1$$

On integration by parts, under the integral sign, we have

$$E(w) = \int_{-\infty}^{\infty} dx_n \left\{ \left[-(x_n - x_1) \frac{d}{dx_n} F_n^{\ n} \right]_{x_1 = -\infty} + \int_{-\infty}^{x_n} \frac{d}{dx_n} (F_n - F_1)^n dx_1 \right\}$$

Since $(F_n - F_1)^n = 0$ at the upper limit of integration, the order of integration and differentiation may be interchanged in this last integral so that

$$E(w) = \int_{-\infty}^{+\infty} \left[(x_n - x_1) \frac{d}{dx} F_1^{\ n} \right]_{x_1 = -\infty} dx_n - \int_{-\infty}^{+\infty} (1 - F_1)^n dx_1$$

$$= \int_{-\infty}^{+\infty} \left\{ \frac{d}{dx_n} [(x_n - x_1) F_n^{\ n}] - F_n^{\ n} \right\}_{x_1 = -\infty} - \int_{-\infty}^{+\infty} (1 - F_1)^n dx_1$$

$$= \Big[(x_n - x_1) F_n^{\ n}) \Big]_{-\infty, +\infty}^{-\infty, \infty} - \int_{-\infty}^{+\infty} F_n^{\ n}\, dx_n - \int_{-\infty}^{+\infty} (1 - F_1)^n dx_1$$

$$= \int_{-\infty}^{+\infty} \{1 - F^n - (1 - F)^n\} dx, \qquad \text{where } dF = f(x)dx.$$

This result is due to L.H.C. Tippett (1925).

Case II. *Finite range of the population.* Let the density function be defined in the finite interval (a, b). Proceeding as in **Case I,** it can be shown that the result (14·2) holds good. Since $u = v - w$ for w fixed, u cannot exceed the value $b - w$. Hence the probability that w will fall in the interval dw, irrespective of the value of u, is

$$dp = n(n - !) \left[\int_a^{b-w} \left\{ f(u) f(u + w) \left(\int_u^{u+w} f(x) dx \right)^{n-2} \right\} du \right] dw \qquad \ldots(14{\cdot}4)$$

This is the required probability differential of w.

The expected value of *w is* $E(w) = \int_a^{b-a} w\, h(w)\, dw$

where $h(w)$ is the coefficient of dw in (14·4).

Example 14·2. *Suppose a random sample of n observations is taken from a population with the density function* $f(x) = e^{-x}$, $x \geq 0$. *Find the density function of the range w.*

Also show that the mean value of w is

$$1 + \tfrac{1}{2} + \tfrac{1}{3} + \ldots\ldots + \frac{1}{n-1}.$$

Here $\displaystyle\int_u^{u+w} f(x)\, dx = e^{-u} - e^{-u-w} = e^{-u}(1 - e^{-w})$

$$\therefore \phi(w) = n(n-1)\int_0^\infty e^{-u} e^{-(u+w)} e^{-(n-2)u} (1 - e^{-w})^{n-2}\, du, \text{ using (14·3)}$$

$$= (n-1)e^{-w}(1 - e^{-w})^{n-2}$$

$$\therefore E(w) = (n-1)\int_0^\infty w e^{-w}(1 - e^{-w})^{n-2}\, dw \quad \ldots(1)$$

$$= (n-1)\int_0^1 [-\log(1-z)]\, z^{n-2}\, dx, \text{putting } 1 - e^{-w} = z$$

$$= (n-1)\int_0^1 \left(z + \frac{z^2}{2} + \frac{z^3}{3} + \ldots \right) z^{n-2}\, dz$$

[The series is uniformly convergent in (0, 1)]

$$= (n-1)\left[\frac{1}{n} + \frac{1}{2}\cdot\frac{1}{n+1} + \frac{1}{3}\cdot\frac{1}{n+2} + \ldots \right]$$

$$= \left[\frac{1}{1} - \frac{1}{n} + \frac{1}{2} - \frac{1}{n+1} + \frac{1}{3} - \frac{1}{n+2} + \ldots \right]$$

$$= 1 + \frac{1}{2} + \frac{1}{3} + \ldots + \frac{1}{n-1}.$$

Ex. 1. In (1) Ex. 14·2 expand $(1 - e^{-w})^{n-2}$ by binomial theorem and integrate term by term, using the result $\displaystyle\int_0^\infty w e^{-rw}\, dw = \frac{1}{r^2}$.

Deduce the algebraic identity :

$$\frac{{}^{n-1}c_2}{1} - \frac{{}^{n-1}c_3}{2} + \frac{{}^{n-1}c_4}{3} - \cdots + \frac{(-1)^{n-1}\,{}^{n-1}c_{n-1}}{n-1} = 1 + \frac{1}{2} + \frac{1}{3} + \ldots + \frac{1}{n-1}.$$

Ex. 2. Suppose that samples of size 4 are taken from the population with distribution $f(x) = e^{-x}, x \geq 0$. Determine the limits w_1 and w_2 for w such that

$$P(w < w_1) = \cdot 025 \text{ and } P(w > w_2) = \cdot 025$$

Ex. 3. For a sample of size n from a population with the uniform distribution $f(x) = b^{-1}, 0 \leq x \leq b$, show that wb^{-1} is a $\beta_1(n-1, 1)$ variate. Obtain the mean and variance of the distribution of the statistic $w' = wb^{-1}$.

[**Ans.** $E(w') = (n-1)/(n+1)$]

Ex. 4. Show that the probability that, in a sample of n random numbers drawn from the numbers between 0 and 1, the range will exceed 0·5 is $1 - \left(\dfrac{n+1}{2n}\right)$. Show also that 8 is the least value of n for which the probability exceeds 0·95.

[Here $f(x) = 1$, $P(w > 0 \cdot 5)] = \int_{1/2}^{1} n(n-1)w^{n-2}(1-w)dw =$ etc. Solve by trial for small value of n]

Ex. 5. Find how many random numbers between 0 and 1 would not be drawn in order that the probability will exceed 0·95 that the range will exceed 0·9.

[n is the least integer given by $\int_{0 \cdot 9}^{1} n(n-1)\,w^{n-2}(1-w)\,dw > \cdot 95$]

14·6. Distribution of Quantiles. Consider a random sample of size n drawn from the continuous population whose density function is $f(x)$. The probability that one value of x will fall in the interval $(-\infty, u)$ is

$$\int_{-\infty}^{u} f(x)\,dx = F(u) = F, \text{ say,}$$

and the probability that one value of x will fall in the interval $(u + du, \infty)$ is $\int_{u+du}^{\infty} f(x)\,dx = 1 - \int_{-\infty}^{u+du} f(x)\,dx = 1 - F(u + du).$

Consequently, the probability that out of a sample of $n, l-1$ values will fall below the value u, one in the infinitesimal interval $(u, u + du)$ and the remaining $(n - l)$ fall above u, is

$$dp = \frac{n!\,F^{l-1} f(u)du\,[1 - F(u + du)]^{n-l}}{(l-1)!\,(n-l)!}$$

$$= \frac{n!}{(l-1)!\,(n-l)!} F^{l-1}(1-F)^{n-l}\,dF \qquad \ldots(14 \cdot 5)$$

where $dF = f(u)$ and infinitesimals of order higher than the first have been neglected on the right hand side.

This is the distribution of u, the number of the sample to which a proportion l/n of the members fall *i.e.*, of the lth quantile.

Example 14·3. *Show that in odd samples of n from a rectangular population of unit range, the sampling mean and variance of the distribution of the median, is given exactly by* $\frac{1}{2}$ *and* $\frac{1}{4}(n+2)$ *respectively.*

Let $n = 2m + 1$, then the median is $(m+1)th$ member of the sample so that in this case $l = m + 1$.

Hence the distribution of the median is

$$dP = \frac{(2m+1)!}{(m!)^2} F^m (1-F)^m \, dF$$

where $\quad dF = du, \quad 0 \le u \le 1$

and $$F = \int_0^u f(x)\,dx = \int_0^u 1\,dx = u.$$

Therefore $$dP = \frac{(2m+1)!}{(m!)^2} u^m (1-u)^m \, du, \; 0 \le u \le 1$$

$$= \frac{1}{B(m+1, m+1)} u^m (1-u)^m \, du, \; 0 \le u \le 1.$$

$$E(u) = \frac{1}{B(m+1, m+1)} \int_0^1 u^{m+1} (1-u)^m \, du$$

$$= \frac{B(m+2, m+1)}{B(m+1, m+1)} = \tfrac{1}{2}.$$

Similarly, $$E(u^2) = \frac{B(m+3, m+1)}{B(m+1, m+1)} = \frac{m+2}{2(2m+3)}$$

Therefore $$Var(u) = E(u)^2 - [E(u)]^2 = \frac{1}{4(2m+3)} = \frac{1}{4(n+2)}.$$

Example 14·4 *Show that for large samples the distribution of the median tends to normality.*

Let a be the population median, then $F(a) = \int_{-\infty}^{a} f(x)\,dx = \frac{1}{2}$

Also let $f(a) = y_0$. Then, by Talyor's theorem,

$$F(u) = F(a) + (u-a) F'(a) + \frac{(u-a)^2}{2!} F''(a) + \dots$$

$$= \tfrac{1}{2} + (u-a)\, y_0 + \frac{(u-a)^2}{2\,!} f\,(a) + \ldots$$

Differentiating,

$$f(u) = y_0 + (u-a) f'(a) + \ldots$$

For a symmetrical distribution $f'(a) = 0$ and for a moderately asymmetrical distribution $f'(a)$ will be small. Let

$$Z = \frac{(2m+1)\,!}{(m\,!)^2} F^m (1-F)^m f(u) \qquad \text{where } n = 2m+1.$$

For large m, Stirling's formula for factorials,

$$Z \cong \frac{\sqrt{2\pi} e^{-(2m+1)} (2m+1)^{2m+3/2}}{(\sqrt{2\pi}\, e^{-m} m^{m+1/2})^2} [\tfrac{1}{2} + (u-a) y_0 + \ldots]^m$$

$$\times [\tfrac{1}{2} - (u-a) y_0 + \ldots]^m y_0$$

$$\cong \frac{e^{-1}}{\sqrt{2\pi}} 2^{2m+3/2} e^1 \sqrt{m}\, (\tfrac{1}{4})^m [1 - 4(u-a) y_0^2 \ldots.]^m\, y_0$$

$$= \frac{1}{\sqrt{\pi}} \left(\frac{1}{2\sqrt{2my_0}} \right)^{-1} e^{-4m(u-a)^2 y_0^2}, \quad \log_{m \to \infty} \left(1 + \frac{1}{2m} \right)^{2m} = e$$

$$\cong \frac{1}{\sqrt{\pi}} \cdot \frac{1}{\sigma} e^{-\frac{(u-a)^2}{2\sigma^2}} \quad \text{where } \sigma^2 = \frac{1}{8\, m y_0^2} \sim \frac{1}{4\, n y_0^2}$$

Thus for large sample size n, the median is normally distributed about the population median at mean and variance approximately equal to $1/(4ny_0^2)$.

Ex. 1. For large samples from a normal distribution with variance σ^2 show that the variance of the sample median is $\pi\,\sigma^2/2n$ and for Cauchy distribution its value is $\pi^2/4\,n$.

Ex. 2. Obtain the large sample distribution of the median in samples from any population, and deduce the form it takes when the population is

$$dP = \theta e^{-\theta x} dx, \; 0 \le x < \infty, \; \theta > 0.$$

Ex. 3. Show that the distribution of the largest number x_n, of the sample drawn from the population

$$d\,P = \frac{2x}{2\theta + 1} dx, \; 0 \le x \le \theta + 1,$$

is

$$\frac{2nx_n}{(2\theta+1)^n} \cdot (x_n^2 - \theta^2)^{n-1}\, dx_n.$$

Ex. 4. Show that the distribution of the *m*th value from the top is

$$\frac{n\,!}{(n-m)\,!\,(m-1)\,!}\,F^{n-m}(1-F)^{m-1}\,d\,F$$

when the sample of size *n* is drawn from a population with a continuous density function $f(x)=1, 0 \le x \le 1$.

In samples from the rectangular distribution $dF = dx, 0 \le x \le 1$

[**Hint.** $\therefore$ $F(x) = x$, the distribution of the *m*th value from top is

$$d\,P = \text{const.}\; x^{n-m}(1-x)^{m-1}\,dx,\; 0 \le x \le 1.]$$

Ex. 5. Let x_1, x_2, x_3, x_4, x_5 be an order statistic random sample of size 5 from a distribution having probability density function $f(x)=e^{-x}, 0<x<\infty$, zero elsewhere. Find *p.d.f.* of the median x_3 of the random sample.

$$\left[\textbf{Ans. } g(x_3)=\frac{5\,!}{2\,!\,2\,!}\,e^{-3x_3}(1-e^{-x_3})^2,\; 0<x_3<\infty, g(x_3)=0, \text{elsewhere}\right]$$

14·7. *Distribution of the Correlation Coefficient. The sampling distribution of the correlation coefficient, ρ, for random samples of size *n* from a correlated bivariate normal population will be derived in four steps.

First Step. *Geometrical interpretation or r.*

Let $x_1, y_1; x_2, y_2; \ldots; x_n, y_n$ be a random sample of size *n* drawn from a correlated bivariate normal population

$$d\,P=\frac{1}{2\pi\sigma_1\sigma_2\sqrt{1-\rho^2}}\exp\left[-\frac{1}{2(1-\rho^2)}\left\{\frac{(x-m_1)^2}{\sigma_1^{\,2}}-\frac{2\rho\,(x-m_1)(y-m_2)}{\sigma_1\sigma_2}+\frac{(y-m_2)^2}{\sigma_2^{\,2}}\right\}\right]dxdy \qquad \ldots(14\cdot6)$$

Let us take a sample space of $2n$ dimensions and in this space let $OX_1, OX_2, \ldots, OX_n$ and $OY_1, OY_2, \ldots, OY_n$ be rectangular axes, *O* being the origin. Then these axes form two mutually orthogonal spaces, each of *n* dimensions, and the sample may be represented by the point *P* with co-ordinates $(x_1, x_2 \ldots, x_n\,;\, y_1, y_2, \ldots, y_n)$

Let $P_1\,(x_1, x_2, \ldots, x_n)$ and $P_2(y_1, y_2, \ldots, y_n)$ be the projections of *P* on the *x*-space and *y*-space respectively.

Further, let OE_1, OE_2 be the equiangular lines in the *x*-space and *y*-space respectively and let M_1, M_2 be the feet of the perpendiculars from P_1 and P_2 on these. Then, as in § 13·3

* The distribution of the correlation coefficient, *r, in* random sample of *n* pairs from a bivariate normal popmation in which the correlation is ρ was given by prof. R.A. Fisher in 1915.

$$OM_1 = \bar{x}\sqrt{n}, P_1M_1 = s_1\sqrt{n}, OM_2 = \bar{y}\sqrt{n}, P_2M_2 = s_2\sqrt{n},$$

where the variances s_1^2, s_2^2 of x and y in the sample space are given by

$$s_1^2 = \frac{1}{n} \cdot \sum_{i=1}^{n} (x_i - \bar{x})^2, s_2^2 = \frac{1}{n} \sum_{i=1}^{n} (y_i - \bar{y})^2,$$

and the correlation, r; in the sample is given by

$$r = \frac{\sum_{i=1}^{n} (x_i - \bar{x})(y_i - \bar{y})}{n\, s_1\, s_2}$$

If we rotate the y-space round O so that it coincides with the x-space *i.e.*, the axis OY_i falls along OX_i falls along $OX_i (i = 1, 2, ..., n)$ then P_2 will occupy a new position P'_2 with coordinates $y_1, y_2 ..... y_n$ in the x-space and OE_2 will coincide with OE_1

Let M'_2 be the foot of the perpendicular from P_2' on OE_1' then $OM'_2 = \bar{y}\sqrt{n}$ and $P_2M_2' = s_2\sqrt{n}$, M_2 and M_2' are $(\bar{x}, x_1 - \bar{x})$ and $(\bar{y}, y_1 - \bar{y})$ respectively.

If θ be the angle between P_1M_1 and $P_2' M_2'$, we have

$$\cos\theta = \frac{\Sigma(x_i - \bar{x})(y_i - \bar{y})}{\sqrt{\Sigma(x_i - \bar{x})^2 \Sigma(y_i - \bar{y})^2}} = r.$$

Second Step. *Transformation of the density function.*

$$\text{Now } \sum_{i=1}^{n} (x_i - m_1)^2 = \sum_{i=1}^{n} (x_i - \bar{x})^2 + n(\bar{x} - m_1)^2 = n[s_1^2 + (\bar{x} - m_1)^2]$$

$$\text{Similarly, } \sum_{i=1}^{n} (y_i - m_2)^2 = n[s_2^2 + (\bar{y} - m_2)^2] \text{ and}$$

$$\sum_i (x_i - m_1)(y_i - m_2) = \sum_i (x_i - \bar{x} + \bar{x} - m_1)(y_i - \bar{y} + \bar{y} - m_2)$$

$$= \Sigma(x_i - \bar{x})(y_i - \bar{y}) + n(\bar{x} - m_1)(\bar{y} - m_2)$$

$$= n[rs_1s_2 + (\bar{x} - m_1)(\bar{y} - m_2)]$$

Therefore, the density factor in (14·6) reduces to

$$\text{const.} \exp\left[\frac{-n}{2(1-\rho^2)}\left\{\frac{(\bar{x} - m_1)^2}{\sigma_1^2} - \frac{2\rho(\bar{x} - m_1)(\bar{y} - m_2)}{\sigma_1\sigma_2} + \frac{(\bar{y} - m_2)^2}{\sigma_2^2}\right\}\right]$$

$$\times \exp\left[\frac{-n}{2(1-\rho^2)}\left(\frac{s_1^2}{\sigma_1^2} - \frac{2\rho r s_1 s_2}{\sigma_1\sigma_2} + \frac{s_2^2}{\sigma_2^2}\right)\right]$$

Third Step. *Transformation of the volume element.*

We will now find out how the volume element changes when $\bar{x}, \bar{y}, s_1, s_2$, and r (or θ) are allowed to vary within infinitesimal ranges. For the purpose let us first keep $\bar{x}$ and s_1 fixed, then P_1 will move on the surface of an $(n-1)$ dimensional hypersphere with centre at M_1 and radius $s_1 \sqrt{n}$. Hence, if $\bar{x}$ lies between $\bar{x}$ and $\bar{x}+d\bar{x}$, s_1 between s_1 and s_1+ds_1, the element of volume described by P_1 is constant $\times s_1^{n-2} ds d\bar{x}$. Now keeping $\bar{x}$ and s_1 fixed, if we allow $s_2, \bar{y}$ and θ to lie in the intervals (s_2, s_2+ds_2), $(\bar{y}, \bar{y}+d\bar{y})$, $(\theta, \theta+d\theta)$, then the volume element described by P_2' in the x-space and hence P_2 in the y-space is const. $\times (s_2 \sin\theta)^{n-3} ds_2\, d\bar{y} \cdot s_2\, d\theta$ *i.e.,* const. $s_2^{n-2} \sin^{n-3}\theta\, ds_2\, d\bar{y}\, d\theta$.

Now since P_1 and P_2 are the projections of P on the mutually orthogonal spaces, the volume element described by P as $\bar{x}, \bar{y}, s_1, s_2\ \theta$ vary between their infinitesimal ranges is simply the product of these two volume elements and is thus

$$\text{Const. } s_1^{n-2} s_2^{n-2} \sin^{n-2}\theta\, ds_1\, ds_2\, d\theta\, d\bar{x}\, d\bar{y}$$

i.e.,

$$\text{Const. } s_1^{n-2}\, s_2^{n-2} (1-r^2)^{\frac{1}{2}(n-4)}\, ds_1\, ds_2\, dr\, d\bar{x}\, d\bar{y}$$

Hence the joint sampling distribution of $\bar{x}, \bar{y}, s_1, s_2$ and r *is*

$$C_1 \exp\left[\frac{-n}{2(1-p^2)}\left\{\frac{(\bar{x}-m_1)^2}{\sigma_1^2} - \frac{2\rho(\bar{x}-m_1)(\bar{y}-m_2)}{\sigma_1\sigma_2} + \frac{(y-m_2)^2}{\sigma_2^2}\right\}\right] d\bar{x}\, d\bar{y}$$

$$\times C_2 \exp\left[\frac{-n}{2(1-\rho^2)}\left(\frac{s_1^2}{\sigma_1^2} - \frac{2\rho r_1 s_1 s_2}{\sigma_1\sigma_2} + \frac{s_2^2}{\sigma^2}\right)\right] s_1^{n-2}\, s_2^{n-2} (1-r^2)^{\frac{n-2}{2}}\, ds_1\, ds_2\, dr.$$

This shows that *joint distribution of* $\bar{x}$ *and* $\bar{y}$ *and the joint distribution of* s_1, s_2 *are mutually independent and hence in normal samples, the distribution of means is independent of that of variances and covariances.*

Fourth Step. In order to obtain the sampling distribution of r, we are to integrate out the second factor of the above expression for s_1 and s_2 from 0 to ∞. We shall consider two cases.

Case I. $\rho \neq 0$

Let us set $\xi = \dfrac{s_1 s_2}{\sigma_1 \sigma_2}$, $e^z = \dfrac{s_1 \sigma_2}{s_2 \sigma_1}$

then $\dfrac{s_1^2}{\sigma_1^2} - \dfrac{2\rho r s_1 s_2}{\sigma_1\sigma_2} + \dfrac{s_2^2}{\sigma_2^2} = \xi e^z - 2\rho\xi r + \xi e^{z} = 2\xi(\cosh z - \rho r)$

Also Jacobian of transformation

$$= \frac{\partial(\xi, z)}{\partial(s_1, s_2)} = \begin{vmatrix} \dfrac{s_2}{\sigma_1\sigma_2} & \dfrac{s_1}{\sigma_1\sigma_2} \\ \dfrac{1}{s_1} & \dfrac{-1}{s_2} \end{vmatrix} = \frac{-2}{\sigma_1\sigma_2} = \text{const.}$$

Hence the joint distribution of r, z, and ξ is

$$\text{const. } \exp\left[\frac{-n\xi}{1-\rho^2}(\cosh z - \rho r)\right] \times \xi^{n-2}(1-r^2)^{\frac{n-4}{2}}\, d\xi\, dz\, dr$$

Thus limits for ξ and z are from 0 to ∞. Integrating out for ξ, we have joint distribution of z and r as

$$\text{const.} \frac{(1-\rho^2)^{\frac{n-1}{2}}}{(\cosh z - \rho r)^{n-1}}(1-r^2)^{\frac{n-4}{2}}\, dz\, dr$$

Hence the distribution of r is

$$\text{const.} (1-r^2)^{\frac{n-4}{2}}\, dr \cdot \int_0^\infty \frac{dz}{(\cosh z - \rho r)^{n-1}}$$

Putting $-\rho r = \cos\theta$, evaluating the constant, distribution of r can be expressed in the form.

$$^* dP = \frac{(1-\rho^2)^{\frac{n-1}{2}}}{\pi\,\Gamma(n-2)}(1-r^2)^{\frac{n-4}{2}} \frac{d^{n-2}}{d(r\rho)^{n-2}}\left[\frac{\cos^{-1}(-\rho r)}{\sqrt{1-\rho^2 r^2}}\right] dr.$$

Case II. $\rho = 0$ (*i.e.*, sample is drawn from an uncorrelated bivariate normal population). In this case the second factor reduces to

$$\text{Cons. } \exp\left[-\frac{n}{2}\left(\frac{s_1^2}{\sigma_1^2} + \frac{s_2^2}{\sigma_2^2}\right)\right] s_1^{n-2} s_2^{n-2} (1-r^2)^{\frac{1}{2}(n-4)}\quad ds_1\, ds_2\, dr$$

* Prof. Harold Hotteling (1953) made an extensive study of the distribution of r. He has obtained the following form of the distribution of r :

$$\frac{n-2\,\overline{|n-2}}{\sqrt{2\pi}\,\overline{\left|n-\frac{1}{2}\right.}}(1-\rho^2)^{\frac{n-1}{2}}(1-r^2)^{\frac{n-4}{2}}(1-\rho r)^{-n+\frac{2}{3}} F\left(\frac{1}{2}, \frac{1}{2}, n-\frac{1}{2}, \frac{1+\rho r}{2}\right)$$

where
$$F(a, b, c, x) = \sum_{i=0}^{\infty} \overline{\left|\frac{a+j}{a}\right.} \cdot \overline{\left|\frac{b+j}{b}\right.} \cdot \overline{\left|\frac{c}{c+j}\right.} \cdot \frac{x^j}{j!}$$

is a hypergeometric function. Hotteling discusses methods of integrating the density function and calculates moments of r.

This shows that the distribution of r is independent of the distributions of s_1 and s_2 and is, therefore, given by

$$dp = c\,(1-r^2)^{\frac{n-4}{2}}\,dr, \quad -1 \le r \le 1 \qquad \ldots(14\cdot7)$$

where $$1 = \int dp = c\int_0^1 (r^2)^{-\frac{1}{2}}(1-r^2)^{\frac{n-4}{2}}\,d(r^2),$$

i.e., $$c = \frac{1}{B\left(\dfrac{1}{2}, \dfrac{n-2}{2}\right)}.$$

Hence the distribution of r when $\rho = 0$ is

$$dP = \frac{1}{B\left(\dfrac{1}{2}, \dfrac{n-2}{2}\right)}\cdot(1-r^2)^{\frac{n-4}{2}}\,dr, \quad -1 \le r \le 1.$$

14·7·1. Alternative Method of Derivation of the Distribution of r when $\rho = 0$. In the case of an uncorrelated population *i.e., when* $\rho = 0$, the distribution of the correlation coefficient, r, can be obtained easily as follows :

Let the n values $y_i (i = 1, 2, \ldots, n)$ be subjected to an orthogonal transformation yielding n independent variates $\eta_1, \eta_2, \ldots, \eta_n$ so that

$$\sum_{i=1}^{n} \eta_i^2 = \sum_{i=1}^{n} y_i^2.$$

Choosing $\eta_1 = \dfrac{1}{\sqrt{n}}\cdot\sum_{i=1}^{n} y_i = \sqrt{n}\,\bar{y}$

(the sum of the squares of the coefficients of y_i 's is unity), we get

$$\sum_{i=1}^{n}\eta_i^2 = \sum_{i=1}^{n} y_i^2 = \sum_{i=1}^{n}(y_i - \bar{y} + \bar{y})^2 = \sum_{i=1}^{n}(y_i - \bar{y})^2) + n\bar{y}^2 = ns_2^2 + \eta_1^2$$

or $$ns_2^2 = \sum_{i=2}^{n}\eta_i^2$$

Also $$\sqrt{n}\,rs_2 = \frac{\Sigma(x_i - \bar{x})(y_i - \bar{y})}{\sqrt{n}\,s_1} = \frac{\Sigma(x_i - \bar{x})y_i}{\sqrt{n}\,s_1}$$

Since the sum of the squares of the coefficient of y_i is unity, we may take

$$\eta_2 = \frac{\Sigma(x_i - \bar{x})y_i}{\sqrt{n}\,s_1} = \sqrt{n}\,r\,s_2$$

Hence $\sum_{i=3}^{n} \eta_i^2 = n(1-r^2)s_2^2.$

Since η_i/σ^2 are independent standard normal variates we conclude that $\frac{nr^2s_2^2}{\sigma_2^2}$ and $\frac{n(1-r^2)s_2^2}{\sigma_2^2}$ are distributed independently like χ^2 with 1 and $(n - 2)$ *d.f.* respectively.

$$\text{Now } r^2 = \frac{r^2 s_2^2}{s_2^2} = \frac{nr^2s_2^2/\sigma_2^2}{nr^2s_2^2/\sigma_2^2 + n(1-r^2)s_2^2/\sigma_2^2} = \frac{\chi_1^2}{\chi_1^2+\chi_2^2}$$

where χ_1^2 and χ_2^2 are distributed like χ^2 with 1 and $(n - 2)$ *d.f.* respectively and therefore, by Ex. 13·1, page 483, r^2 is a $\beta_1\left[\frac{1}{2}, \frac{n-2}{2}\right]$ variate.

Hence the distribution of r^2 is

$$dp = \frac{(r^2)^{\frac{1}{2}-1}(1-r^2)^{\frac{n-4}{2}}}{B\left[\frac{1}{2}, \frac{n-2}{2}\right]} d(r^2), \; 0 \le r^2 \le 1$$

Thus the distribution of r is

$$dp = \frac{1}{B\left(\frac{1}{2}, \frac{n-2}{2}\right)}(1-r^2)^{\frac{n-4}{2}} dr, \; -1 \le r \le 1.$$

Ex. 1. Show that $\bar{x}, \bar{y}$ are distributed in a bivariate normal distribution with means m_1, m_2; variances σ_1^2/n_1, σ_2^2/n_2 and correlation coefficient ρ.

Ex. 2. Show that, if $\rho = 0$, $\bar{x}$ and $\bar{y}$ are independently distributed and so are s_1^2 and s_2^2.

Ex. 3 . Show that when $\rho = 0$, $t = \frac{r \times \sqrt{n-2}}{\sqrt{1-r^2}}$ is distributed as t with $(n-2)$ *d.f.*

Ex. 4. Using $$\int_0^\infty \frac{dz}{\cosh z + \cos\theta} = \frac{\theta}{\sin\theta}$$

deduce that $$\int_0^\infty \frac{d^{n-2}}{d(\cos\theta)^{n-2}}\left(\frac{\theta}{\sin\theta}\right) = \int_0^\infty \frac{dz}{(\cosh z + \cos\theta)^{n-1}}.$$

Ex. 5. Show that in samples of size 4 from a bivariate formula population for which $\rho = 0$, r is distributed as a rectangular variate.

Ex. 6. Show that if $\rho = 0$, the $2m$th moment about origin of the distribution of r, is $\mu'_{2m} = \left\lceil \frac{n-1}{2} \right. \left\lceil m + \frac{1}{2} \right. \Big/ \sqrt{\pi} \left\lceil \frac{n-1}{2} + m. \right.$

14·8 Fisher's z Transformation. The ordinates and the density function of the correlation coefficient are not expressible in terms of simple mathematical functions and have been tabulated by F.N. David (1938).

The distribution or r is far from normal and its probability curve is very skew in the neighbourhood of $\rho = \pm 1$ even for large n. We cannot, therefore, use the standard error. Fisher (1921), however, showed that for the variate $z - \xi$ where

$$z = \tfrac{1}{2} \log_e \left[\frac{1+r}{1-r} \right] = \tanh^{-1} r \text{ where } r = \tanh z = \frac{e^{2z} - 1}{e^{2z} + 1}.$$

$$\xi = \tfrac{1}{2} \log_e \left[\frac{1+\rho}{1-\rho} \right] = \tanh^{-1} \rho \text{ where } \rho = \tanh \xi$$

We have,

$$\mu_1' = \frac{\rho}{n-1} + \frac{5\rho + \rho^3}{4(n-1)^2}, \gamma_1 = \frac{\rho}{(n-1)^{3/2}} \left[\rho^2 - \frac{9}{16} \right] + \ldots$$

$$\mu_2 = \frac{1}{n-1} + \frac{4 - \rho^2}{(n-1)^2} + \ldots\ldots, \gamma_2 = \frac{32 - 3\rho^4}{(16\,(n-1)} + \ldots.$$

so that $z - \xi$ is approximately normally distributed with mean zero and variance which is approximately $1/(n-3)$. For $n > 50$, this approximation is quite good, but can be used even for smaller values of n for many practical purposes.

14·8·1. Fisher's transformation, when $\rho = 0$. In this case, the distribution of z by putting $r = \tanh z$ *i.e.* $z = \frac{1}{2} \log_e \left(\frac{1+r}{1-r} \right)$ in (14·7), is

$$d\,p = \frac{\text{sech}^{n-2} z\, dz}{B\left[\frac{1}{2}, \frac{n-2}{2} \right]} \propto \exp\left[-\tfrac{1}{2} (n-2)\, z^2 \right] dz \text{ approximately.}$$

Thus the distribution of z is approximately normal with variance $1/(n-2)$. Fisher has shown, however, that a better approximation to the variance is $1/(n-3)$.

14·9. Distributions of regression coefficients in normal samples.

Putting $b_2 = \dfrac{r\,s_2}{s_1}$ in the second factor of (14·7), we have the joint distribution of s_1, s_2 and b_2.

$$d\,F = \text{const.}\exp\left[\frac{-n}{2(1-\rho^2)}\left\{\frac{s_1^{\,2}}{\sigma_1^{\,2}} - \frac{2\rho\, s_1 s_2 b_2}{\sigma_1\sigma_2} + \frac{s_2^{\,2}}{\sigma_2^{\,2}}\right\}\right] s_1^{\,n-1} s_2^{\,n-2}$$

$$\times\left(1 - \frac{s_1^{\,2} b_2^{\,2}}{s_2^{\,2}}\right)^{\frac{1}{2}(n-4)} ds_1\, ds_2\, db_2$$

To obtain the distribution of s_1 and b_2, we intergrate *w.r.t.* s_2

$$\therefore \quad dF = \text{const.}\exp\left[\frac{-n}{2(1-p^2)}\left\{\frac{s_1^{\,2}}{\sigma_1^{\,2}} - \frac{2\rho s_1 s_2 b_2}{\sigma_1\sigma_2} + \frac{s_1^{\,2} b_2^{\,2}}{\sigma_2^{\,2}}\right\}\right] \times s_1\, ds_1\, db_2$$

Integrating *w.r.* to s_1, we obtain the distribution of b_2

$$dF = \text{const.}\left(1 - \frac{2\rho\sigma_1}{\sigma_2} b_2 + \frac{\sigma_1^{\,2}}{\sigma_2^{\,2}} b_2^{\,2}\right)^{-\frac{1}{2}n} db_2$$

$$= \frac{\Gamma(n/2)\sigma_2^{\,n-1}(1-\rho^2)^{\frac{1}{2}(n-1)}\, db_2}{\sqrt{\pi}\,\Gamma\left(\frac{n-1}{2}\right)\sigma_1^{\,n-1}\cdot\left\{\frac{\sigma_2^{\,2}}{\sigma_1^{\,2}}(1-\rho^2) + \left(b_2 - \frac{\rho\sigma_2}{\sigma_1}\right)^2\right\}^{\frac{1}{2}n}}$$

Th distribution of b_2 is symmetrical about the point $\beta_2 = \rho\sigma_2/\sigma_1$ the population regression coefficient. It tends to normality fairly rapidly and the use of the standard error for regression is, therefore, valid for lower values of n than in the case of the correlation coefficient.

Putting $u = \dfrac{(b_2 - \beta_2)\, s_1}{(s_2^{\,2} - b_1^{\,2} s_1^{\,2})}$, we can show that the distribution of u is

$$dF = \text{const.}\frac{du}{(1+u^2)^{1/2}(n-1)}$$

It follows that $\quad t = u\sqrt{n-2} = \dfrac{(b_2 - \beta_2)s_1\sqrt{n-2}}{(s_2^{\,2} - b_2^{\,2} s_1^{\,2})}$

is distributed in Student's t form with $(n-2)$ *d.f.*

Exercises 14

1. A transformation rectangular to polar coordinates is

$$y_1 = u\sin\theta_1,\; y_2 = u\cos\theta_1\sin\theta_2,$$

....

$$y_{n-1} = u\cos\theta_1\cos\theta_2\ldots\cos\theta_{n-2}\sin\theta_{n-1},$$

$$y_n = u\cos\theta_1\cos\theta_2\ldots\cos\theta_{n-2}\cos\theta_{n-1},$$

where $\quad -\frac{\pi}{2} < \theta_i \le \frac{\pi}{2}$, $i = 1, 2, \ldots\ldots, n-2$ and $-\pi < \theta_{n-1} \le \pi$

Prove that (*a*) $u^2 = \sum\limits_1^n y_i^2$

(*b*) Jacobian of transformation is $u^{n-1} \cos^{n-2}\theta_1 \cos^{n-3}\theta_2 \ldots\ldots \cos\theta_{n-2}$.

Deduce that the surface of a sphere of unit radius in n dimension is

$$(2\pi)^{n/2} / \overline{|n/2}$$

2. Derive the distribution of section 14·2 by using the transformations

$$x_1^2 + x_2^2 + \ldots\ldots + x_n^2 = u_1$$
$$x_1^2 + x_2^2 + \ldots\ldots + x_{n-1}^2 = u_1 u_2$$
$$x_1^2 + x_2^2 + \ldots\ldots + x_{n-2}^2 = u_1 u_2 u_3$$
$$\ldots\ldots\ldots\ldots\ldots\ldots\ldots\ldots\ldots\ldots\ldots\ldots$$
$$x_1^2 \qquad = u_1 u_2 u_3 \ldots u_n$$

and integrating over $u_2, u_3, \ldots\ldots, u_n$ from 0 to 1 and over u_1 from 0 to z.

3. If a variable possesses a continuous density function and if z denotes the proportion of the population that lies between the extreme values of a random sample of size n from the population, then the density function of z is given by

$$f(z) = n(n-1)z^{n-2}(1-z).$$

4. Show that the distribution of the range in samples of n from rectangular population $dF = \frac{1}{2}\,dx, -1 \le x \le 1$ is given by

$$\phi(w) = \frac{n(n-1)}{2n}\, w^{n-2}(2-w). \qquad \textbf{(\textit{M.A., Bombay})}$$

5. Show that the distribution of range in samples of n from the population

$$dF = 2(1-x)dx, \quad 0 \le x \le 1$$

is $\quad \phi(w) = \dfrac{n(n-1)w^{n-2}}{2}\left[\dfrac{2w^{n+1}}{n^2-1} + \dfrac{(2-w)^{n+1}}{n+1} - \dfrac{w^2(2-w)^{n-1}}{n-1}\right]$

6. Show that the distribution of the range w in samples of three from a normal population of unit variance is given by $\phi(w) = \dfrac{6}{\pi\sqrt{2}}\, e^{-\frac{1}{4}w^2} \displaystyle\int_0^{w^2/\sqrt{6}} e^{-\frac{1}{2}z^2}\, dz.$

7. If two independent variates X and Y have the density functions $f_1(x)$ and $f_2(y)$ respectively, show that the density function for the ratio $V = X/Y$ where V is non-negative is given by $\quad f(v) = \displaystyle\int_{-\infty}^{\infty} y f_1(vy) f_2(y)\, dy.$

Use this formula to derive the distribution of t and F.

8. Show that in samples of n from the distribution $dF = e^{-x}dx,\ 0 \le x < \infty$ the variance of the smallest sample value is $1/n^2$

9. Use the method of characteristic function to show that the distribution of the sample mean $\bar{x}$ from Pearson Type III population

$$dp = \frac{1}{\Gamma l}\, e^{-\frac{x}{a}} \left(\frac{x}{a}\right)^{l-1} \frac{dx}{a},\ a > 0, x \ge 0$$

is given by $$dp = \frac{1}{\Gamma(nl)}\, e^{-\frac{n\bar{x}}{a}} \left(\frac{n\bar{x}}{a}\right)^{nl-1} \cdot \frac{n}{a}\, d\bar{x},$$

Hence or otherwise obtain Var $(\bar{x})$.

10. If X, Y, Z have the joint density

$$f(x, y, z) = \frac{6}{(1 + x + y + z)^2},\ x, y, z > 0.$$

find the distribution of $U = X + Y + Z$.

$$\left[\textbf{Ans. } dP = \frac{3u^2}{(1+u)^4}\, du\right]$$

[**Hint.** Put $u = x + y + z,\ uv = y + z, uvw = z$ and use that $\left| J\left(\frac{x, y, z}{u, v, w}\right)\right| = u^2 v$]

11. (*a*) Use the method of Induction to derive the χ^2 distribution.

(*b*) By using characteristic function obtain the distribution of the mean of normal and rectangular distributions the rectangular distribution having the range 0 to 1.

12. Obtain the characteristic functions of the sampling distribution of the mean in samples of size n from the rectangular population $dF = dx, 0 \le x \le 1$

Show that $$\beta_1 = 0,\ \beta_2 = 3 - \frac{6}{5n}$$

13 (*a*) Show that the characteristic function for $y = \log x$ is

$$\phi(t) = \int_{-\infty}^{\infty} e^{itx} \log x\, f(x)\, dx = \int_{-\infty}^{\infty} x^{it} f(x)\, dx$$

and the characteristic function of log G where G is geometric mean of a random sample of size n *is* $\left[\phi\left(\frac{t}{n}\right)\right]^n$. Hence deduce that the density function for G, by inversion theorem, is

$$f(y) = \frac{1}{2\pi} \int_{-\infty}^{\infty} e^{-it \log G} \left[\phi\left(\frac{t}{n}\right)\right]^n dt.$$

(*b*) If $x_1, x_2, \ldots.. x_n$ is a random sample of size n drawn from the rectangular population $dF = dx, 0 \le x \le 1$

show that the density function of (*i*) $X = x_1, x_2, \ldots\ldots, x_n$ and

(*ii*) the geometric mean $Y = (x_1\, x_2, \ldots\ldots, x_n)^{1/n}$ of the sample is represented by

$$(i)\ f(x) = \frac{1}{(n-1)},\ [-\log x]^{n-1},\ (ii)\ f(y) = \frac{n^n y^{n-1}}{\lceil n}\ (-\log y)^{n-1},\ 0 \le y \le 1.$$

14. Write a brief explanatory note on transformation of variates. Show that if X is an exponential variate, then e^{-x} is a rectangular variate.

Also prove that the sum of a number of independent identical exponential variates is a Chi-square variate and deduce the additive property of Chi-square variates.

15. A frequency function $f(x)$ has derivative $f'(x)$ and a single mode at $x = m$ such that $f'(x) > a$ or $< a$ according as $x < m$ or $> m$. Show that for a given β the range from u to $u + v$ determined by $\beta = \int_u^{u+v} f(x)\, dx$ is shortest when $f(u) = f(u + v),\ u < m < u + v.$

16. Let $X_i\ (i = 1, 2, \ldots\ldots, n)$ be n identically independently distributed random variables having Weiball distribution with three parameters. Show that the variable

$$Y = \min (X_1, X_2, \ldots, X_n)$$

has also Weibull distribution and identity its parameters.

(*B.A. Hons. Statistics, Delhi, 1984*)

17. Show that if X the *p.d.f.* given by

$$f(x, \theta) = \frac{1}{\theta},\ 0 < x \le \theta,\ 0 < \theta < \infty$$

$$= 0, \text{ elsewhere}$$

then $E\,[\max.(x_i)] = \dfrac{n}{n+1}\,\theta.$

[**Hint.** If X_n is the max. x_i, then its distribution is given by

$$n\left(\int_0^x \frac{1}{\theta}\, dx\right)^{n-1} \cdot \frac{1}{\theta} \cdot dx = n\left(\frac{x}{\theta}\right)^{n-1} \cdot \frac{1}{\theta}\, dx$$

$$\therefore \quad E\,[\max.(x_i)] = \frac{n}{\theta^n}\int_0^\theta x^n\, dx = \frac{n}{\theta^n}\left(\frac{x^{n+1}}{n+1}\right)_0^\theta = \frac{n}{n+1}\,\theta.\ ]$$

18. If X, Y are independent *r.v.*, whose *p.d.f.* are given by

$$f_x(x) = \frac{1}{\pi\sqrt{1-x^2}},\ -1 < x < 1,\ f_y(y) = 2y\, e^{-y^2},\ 0 < y < \infty,$$

prove that the *r.v.* XY has a normal distribution.

[**Hint.** Let $U = xy,\ V = x$, so that $u = xy,\ v = x,\ x = v,\ y = u/v,\ -1 < v < 1,$

$-1 < u < 1, \dfrac{\partial (u, v)}{\partial (x, y)} = -v < 0$ everywhere. Hence $f_u(u) = \dfrac{2u}{\pi} \int_0^1 \dfrac{e^{-u^2/v^2}}{v^2\sqrt{1-v}}\, dv$

$= \dfrac{1}{\sqrt{\pi}} e^{-u^2}$. For $u < 0$, $f_u(u)$ has the same expression.]

19. The numbers $X_1, X_2, \ldots, X_n$ are independently chosen at random in the interval (a, b), Prove that the probability density function of the *r.v.* $X = \min (X_1, X_2, \ldots, X_n)$ is given by $f(x) = \dfrac{n (b-x)^{n-1}}{(b-a)^n}, a < x < b$.

[**Hint.** For $a < x < b, (X > x) = (x_1 > x, x_2 > x, \ldots, x_n > n)$

$\because \quad P(X > x) = P(x_1 > x) \ldots . P(x_n > x) = (b-x)^n/(b-a)^n$

or $F(x) = 1 - (b-x)^n/(b-a)^n \quad \therefore f(x) = F'(x) =$ given result.]

20. Show that for a random sample of size 2 from $N(0, \sigma^2)$, $E(x_{(1)}) = -\sigma/\sqrt{\pi}$

(*M. Sc. State Delhi, 1982, 88*)

[**Hint.** For $n = 2, f_1(x) = \dfrac{1}{\beta(1, 2)} [1 - F(x)] f(x)$ where $f(x) = \dfrac{1}{\sigma\sqrt{2\pi}} e^{-x^2/2\sigma^2}$.]

21. Let $X_1 \; X_2 \ldots X_n$ be *i, i, d* with *p. d. t.*

$$F(x) = \begin{cases} \dfrac{1}{\sigma} e^{-(x-\theta)}, & \text{if } x > \theta \\ 0, & \text{if } x < \theta \end{cases}$$

Show that $X_{(1)}, \; X_{(2)}, \; -X_{(1)}, \; X_{(3)} - X_{(2)} \ldots X_{(n)} - X_{(n-j)}$ are independent. **(*B.Sc. State Hons. Delhi, (1996*)**

15

Tests of Significance Based on t, f and z Distributions

15·1. Introduction. In Chapter 12, the following tests of significance were discussed :

(*a*) In a *large* number n of independent Bernoullian trials, x successes are obtained, to test the hypothesis that the constant probability of success in each trial is p. [§ 12·3].

(*b*) Two independent sets of n_1, n_2 (*large*) independent Bernoullian trials are performed and x_1, x_2 successes respectively are obtained, to test the hypothesis that the constant probability of success is the same for both sets. [§ 12·4]

(*c*) Given a random sample $x_1, x_2, \ldots\ldots, x_n$ of *large size* n from a *normal* population with *known* variance σ^2, to test the hypothesis that the population mean is μ. [§ 12·7]

(*d*) Given a random sample $x_1, x_2, \ldots\ldots, x_n$ of *large size* n from a population (which is not necessarily normal) with a finite (but unknown) variance, to test the hypothesis that the population mean is μ, [§ 12·8].

(*e*) Given two independent random samples of *any size* n_1, n_2 from *normal* populations with *known* variances σ_1^2, σ_2^2 to test the hypothesis that the means of the populations are the same. [§ 12·8]

(*f*) Given two independent random samples of large sizes n_1, n_2 from two populations (not necessarily normal) with unknown but finite variance, to test the hypothesis that the population means are the same. [§ 12·8].

It will be noted that tests in cases (*c*) and (*e*) were applicable to samples of any size, since we were able to find the *exact* sampling distributions of the sample mean and of the difference between the sample means. On the other hand, the other four tests were applicable to large samples only, and were based on the fact that in each case, we could find statistics which were normally distributed for large samples. The asymptotically standard normal statistics were :

$$z_1 = \frac{x - np}{\sqrt{npq}} = \frac{\dfrac{x}{n} - p}{\sqrt{\dfrac{pq}{n}}} \quad \text{for case } (a);$$

$$z_2 = \frac{\dfrac{x_1}{n_1} - \dfrac{x_2}{n_2}}{\sqrt{\bar{p}\,\bar{q}\left(\dfrac{1}{n_1} + \dfrac{1}{n_2}\right)}} \quad \text{for case } (b)$$

where $\bar{p} = \dfrac{x_1 + x_2}{n_1 + n_2}, \bar{q} = 1 - \bar{p}$;

$$z_3 = \frac{(\bar{x} - \mu)}{s/\sqrt{n}} \quad \text{for case } (d)$$

where $\bar{x} = \dfrac{1}{n} \sum\limits_{i=1}^{n} x_i, s^2 = \dfrac{1}{n} \sum\limits_{i=1}^{n} (x_i - \bar{x})^2$

$$z_4 = \frac{\bar{x}_1 - \bar{x}_2}{s \times \sqrt{\dfrac{1}{n_1} + \dfrac{1}{n_2}}} \quad \text{for case } (f)$$

where $s^2 = \left\{ \sum\limits_{i=1}^{n_1} (x_{1i} - \bar{x}_1)^2 + \sum\limits_{j=1}^{n_2} (x_{2j} - \bar{x}_2)^2 \right\} / (n_1 + n_2)$

To develop tests of significance applicable to all sample sizes, we have to use the exact sampling distributions of those or other suitable statistics. We have accordingly in Chapters 13 and 14 obtained the exact sampling distribution of t (due to Student) and of F (due to Fisher) and of other statistics like z, s^2, χ^2, etc., to carry out different tests of significance for samples of all sizes. We discuss tests of significance based on exact sampling distribution of t, F and r in the present chapter. Tests based on χ^2 distribution will be discussed in Chapter 16.

For all the tests developed below, it is *assumed* that the parent population is normal. Also while the tests in the next two sections are applicable to all sample sizes, they are, in practice, applied to samples of small or moderate sizes only, since for large sample the tests based on the normal approximation are easier and quicker to apply and give fairly satisfactory results.

15·2. Tests of significance based on t distribution. We consider five tests of significance based on the t distribution.

15·2·1. Testing the significance of the mean of a random sample from a normal population. Given a random sample $x_1, x_1, \ldots\ldots, x_n$ from a normal population, we have to test the hypothesis, that the mean of the population is μ. For this purpose, we assume the hypothesis to be correct and then see whether the data is consistent with the hypothesis or whether it provides some evidence against the hypothesis. To carry out the test, we calculate the statistic

$$t = \frac{(\bar{x} - \mu)\sqrt{n}}{S} \qquad \ldots(15\cdot1)$$

where $\bar{x} = \frac{1}{n} \sum_{i=1}^{n} x_i, \; S^2 = \frac{1}{n-1} \sum_{i=1}^{n} (x_i - \bar{x})^2.$

t-distribution has been derived in § 13·5 and a table based on it is given on Table 13·2 page 500. For different values of the degrees of freedom $v\,(= n - 1)$ the table gives the values of $t_{0\cdot05}$ and $t_{0\cdot01}$, which are defined by the relations $P[\,|t| \geq t_{0\cdot05}\,] = \cdot05$; $P[\,|t| \geq t_{0\cdot01}\,] = \cdot01$

If our calculated value of $|t|$ exceeds $t_{0\cdot05}$, we say that the difference between $\bar{x}$ and μ is significant at 5% level of significance ; if it exceeds $t_{0\cdot01}$, the difference is said to be significant at 1% level or just highly significant. If $|t| < t_{0\cdot05}$, we say that the data is consistent with the hypothesis that μ is the mean of the population.

Example 15·1. *Show that 95% fiducial limits for the mean μ of the population are* $\bar{x} \mp St_{0\cdot05}/\sqrt{n}$. *Deduce that for a random sample of 16 values with mean* 41·5 *inches and the sum of the squares of the deviations from the mean 135 (inches)*^{2} *and drawn from a normal population 95% fiducial limits for the mean of population are 39·9 and 43·1 inches.*

We have $P\left[\left|\frac{(\bar{x} - \mu)\sqrt{n}}{S}\right| \leq t_{0\cdot05}\right] = \cdot95$

or $P\left[\bar{x} - \frac{St_{0\cdot05}}{\sqrt{n}} \leq \mu \leq \bar{x} + \frac{St_{0\cdot05}}{\sqrt{}}\right] = \cdot95$

so that we can say with a confidence coefficient ·95 that the confidence interval $\bar{x} \mp \frac{S}{\sqrt{n}} t_{0\cdot05}$ contains the population mean μ. The limits of this confidence interval give us the 95% confidence or fiducial limits for μ.

In the particular case,

$$n = 16, v = n - 1 = 15, S = \sqrt{\frac{1}{15} \times 135} = 3$$

Also from the table, $t_{0\cdot05} = 2\cdot13 \;\therefore\; \frac{S}{\sqrt{n}} t_{0\cdot05} = \frac{3}{4} \times 2\cdot13 = 1\cdot6$ approx.

Therefore, the required fiducial limits are $41{\cdot}5 \mp 1{\cdot}6$, *i.e.*, 39·6 and 41·3 inches.

Example 15·2. *Ten individuals are chosen at random from a normal population and the heights are found to be in inches 63, 63, 66, 67, 68, 69, 70, 71 and 71. In the light of these data, discuss the suggestion that the mean height in the population is 66 inches.*

Here $\bar{x} =$ sample mean $= 67{\cdot}8$ inches, $n = 10$.

$$S^2 = \frac{1}{n-1}\left[\sum_{i=1}^{n}(x_i - \bar{x})^2\right] = \frac{1}{n-1}\left[\sum_{i=1}^{n} d_i'^2 - \frac{(\Sigma d_i')^2}{n}\right]$$

where d_i' is the deviation from the assumed mean, say 67·8.

$$\therefore\ S = \sqrt{\frac{1}{9}(8 - 6{\cdot}4)^2} = 3{\cdot}011 \text{ inches} \ ;\ t = \frac{67{\cdot}8 - 66}{3{\cdot}011/\sqrt{10}} = 1{\cdot}89 \text{ approx.}$$

Also ν = number of degrees of freedom $= n - 1 = 9$.

For 9 *d.f.*, we get from the table, $t_{0{\cdot}05} = 2{\cdot}26$.

As the calculated value of $|\,t\,|$ is less than $t_{0{\cdot}05}$, the difference between $\bar{x}$ and μ may be due to fluctuations of random sampling. In other words, the data does not provide any significant evidence against the hypothesis that the population mean is 66″

Ex. 1. Obtain the 99% fiducial limits in Example 15·1.

Ex. 2. Obtain the 95% fiducial limits in Example 15·2.

Ex. 3. For a given sample show that the length of the 'symmetrical' confidence interval increases with the confidence coefficient. Comment on this result.

Ex. 4. Show that the length of the evidence interval $\frac{2S}{\sqrt{n}}\, t_{0{\cdot}05}$ need not necessarily decrease with n. Show that the expected value of its square in random sampling is $\frac{4\sigma^2}{n}(t_{0{\cdot}05})^2$ and definitely decreases with n.

15·2·2. Testing the significance of difference between two sample means. Here we consider three tests of significance.

(*a*) Given two independent random samples $x_{11}, x_{12}, \ldots\ldots x_{1n_1}$ and $x_{21}, x_{22} \ldots . x_{2n_2}$, with means $\bar{x}_1$ and $\bar{x}_2$ and standard deviations S_1 and S_2 from normal populations with the *same variance*, we have to test the hypothesis that the population means are the same. In other words, since a normal distribution is completely specified by its mean and variance, we have to test the hypothesis that the two independent samples come from the *same* normal population.

To carry out the test, we calculate the statistic

$$t = \frac{\bar{x}_1 \sim \bar{x}_2}{S\sqrt{\dfrac{1}{n_1}+\dfrac{1}{n_2}}} \qquad \ldots(15 \cdot 2)$$

where $\bar{x}_1 = \dfrac{1}{n_1} \sum_{i=1}^{n_1} x_{1i}$, $\bar{x}_2 = \dfrac{1}{n_2} \sum_{j=1}^{n_2} x_{2j}$

and $$S^2 = \frac{1}{n_1+n_2-2}\left[(n_1-1)S_1^{\,2} + (n_2-1)\,S_2^{\,2}\right]$$

$$= \frac{1}{n_1+n_2-2}\left[\sum_{i=1}^{n_1}(x_{1i}-\bar{x}_1)^2 + \sum_{j=1}^{n_2}(x_{2j}-\bar{x}_2)^2\right]$$

It has been shown in § 13·6. that this statistic follows the t-distribution with n_1+n_2-2 degrees of freedom. If the calculated value of $|t|$ be $> t_{0 \cdot 05}$ $(t_{0 \cdot 01})$, the difference between the sample means is said to be significant at 5% (1%) level of significance ; otherwise the data are said to be consistent with the hypothesis.

Note. If the *size of the two samples is the same*, say equal to n, and the data are paired [see (*b*), page 461] when we calculate the statistic t defined by

$$t = \frac{\bar{d}-0}{\left(\dfrac{S}{\sqrt{n}}\right)} \qquad \ldots(15 \cdot 3)$$

where $S^2 = \dfrac{1}{n-1} \sum_{i=1}^{n} (d_i - \bar{d})^2,$

d_i = difference between the ith members of the samples and $\bar{d}$ = mean of the difference $= \Sigma d_i / n$, and the number of *d.f.* on which t is based $= n-1$. For non-paired data, we continue to use (15·2).

Example 15·3. For a random sample of 10 pigs fed on diet A, the increases in weight in pounds in a certain period were

10, 6, 16, 17, 13, 12, 8, 14, 15, 9 lbs.

For another random sample of 12 pigs, fed on diet B, the increases in the same period were

7, 13, 22, 15, 12, 14, 18, 8, 21, 23, 10, 17 lbs

Test whether diets A and B differ significantly as regard the effect on increases in weight (or test whether the mean increase in the two samples are significantly different). You may use the fact that 5% value of t for 20 degrees of freedom is 2·09.

We first calculate the mean and standard deviations for the two samples.

Mean gain in weight on diet $A = \bar{x}_1 = \dfrac{\Sigma x_{1i}}{n_1} = \dfrac{120}{10} = 12$ pounds

Mean gain in weight on diet $B = \bar{x}_2 = \dfrac{\Sigma x_{2j}}{n_2} = \dfrac{180}{12} = 15$ pounds

The difference in the sample means = 3 pounds.

$$\text{Again } S = \sqrt{[\sum_i (x_{1i} - \bar{x}_1)^2 + \sum_j (x_{2j} - \bar{x}_2)^2] / (n_1 + n_2 - 2)}$$

$$= \sqrt{\frac{120 + 314}{10 + 12 - 2}} = \sqrt{\frac{434}{20}} = \sqrt{21 \cdot 1} = 4 \cdot 65.$$

Diet A			*Diet B*		
Gain in weight x_{1i}	Deviation from mean $(x_{1i} - \bar{x}_1)$	*Square of deviation* $(x_{1i} - \bar{x}_1)^2$	*Gain in weight* x_{2j}	*Deviation from mean* $(x_{2j} - \bar{x}_2)$	*Square of deviation* $(x_{2j} - \bar{x}_2)^2$
10	–2	4	7	–8	64
6	–6	36	13	–2	4
16	+4	16	22	+7	49
17	+5	25	15	0	0
13	+1	1	12	–3	9
12	0	0	14	–1	1
8	–4	16	18	+3	9
14	+2	4	8	–7	49
15	+3	9	21	+6	36
9	–3	9	23	+8	64
			10	–5	25
			17	+2	4
120	0	120	180	0	314

Substituting these values in the formula (15·2) we get

$$|t| = \frac{3}{4 \cdot 65 \sqrt{\frac{1}{10} + \frac{1}{12}}} = 1 \cdot 6.$$

The number of degrees of freedom $= \nu = n_1 + n_2 - 2 = 20.$

For 20 *d.f.* we are given that $t_{0 \cdot 05} = 2 \cdot 09.$

The calculated value of $|t|$ is less that $t_{0 \cdot 04}$ and hence the difference between the sample means is not significant.

(*b*) In the above example, the difference in the increases in weight are regarded as due to differences in diets *A* and *B*, apart from the random differences due to sampling but it is possible that the pigs in the two samples may differ in their ages, sexes, initial weights, earlier nutritional standards etc:, and part of the differences in increases in weights may be due to these non-random factors. To get a more precise test the data are sometimes paired. Thus in the above example instead of taking 22 pigs at random, we may choose 11 pairs, each pair being as homogeneous as possible *e.g.*, a pair may be taken from a litter. Similarly, in the case of dieting experiments on children we may use twins. If we take n pairs, the variate to be considered is the difference between the values for a pair and the hypothesis to be tested is that the sample of these differences comes from a normal population with mean zero so that t-test with $(n-1)$ *d.f.* would be required.

Example 15·4. *The yields of two types 'Type 17' and 'Type 51' of grains in pounds per acre in 6 replications are given below. What comments would you make on the differences in the mean yields ? You may assume that if there be 5 degrees of freedom and $P = 0{\cdot}2$, t is approximately 1·476.*

Replication	*Yield in pounds* *Type 17*	*Yield in pounds* *Type 51*
1	*20·50*	*24·86*
2	*24·60*	*26·39*
3	*23·06*	*28·19*
4	*29·98*	*30·75*
5	*30·37*	*29·97*
6	*23·83*	*22·04*

We calculate the mean and standard deviation of the differences in the yields of the two types of grains.

Replication	*Yield in pounds* *Type 17*	*Yield in pounds* *Type 51*	*Difference* d	*Deviation from mean* $d-\bar{d}$	*Square of deviations* $(d-\bar{d})^2$
1	20·50	24·86	4·36	+ 2·717	7·38
2	24·60	26·39	1·79	+ 0·147	0·02
3	23·06	28·19	5·13	+ 3·487	12·15
4	29·98	30·75	0·77	− 0·873	0·76
5	30·37	29·97	− 0·40	− 2·043	4·17
6	23·83	22·04	− 1·79	− 3·433	11·79
			9·86		36·27

$$\bar{d} = \text{mean difference of the yields} = \frac{\Sigma d}{n} = \frac{9{\cdot}86}{6} = 1{\cdot}643 \text{ pounds}$$

$$S = S.D. \text{ of difference of yields } = \sqrt{\frac{\Sigma(d-\bar{d})^2}{n-1}}$$

$$= \sqrt{\frac{36\cdot7}{5}} = 2\cdot69 \text{ pounds.}$$

We set up the null hypothesis that the difference in types has no effect on yields, *i.e.*, the population mean of the difference is zero, then

$$t = \frac{1\cdot643-0}{2\cdot59}\sqrt{6} = 1\cdot489.$$

This value of t is less than $t_{0\cdot05}$ for 5 *d.f.* and, therefore, the difference is not significant at 5% level, but is significant at 20% level.

(*c*) Given two independent random samples from normal populations with the same variance, we have to test the hypothesis that the population means are μ_1 and μ_2 respectively. For this case

$$t = \frac{(\bar{x}_1-\bar{x}_2)\sim(\mu_1-\mu_2)}{S\sqrt{\frac{1}{n_1}+\frac{1}{n_2}}} \qquad \text{...(15·4)}$$

with (n_1+n_2-2) *d.f.* and the test of significance is carried out as in case (*a*).

Note about assumptions for the t test.

For testing the significance of a sample mean, t test has a greater range of validity and precision than the normal approximation test of § 12·6, but this is achieved by placing an important restriction *viz.*, that the parent population must be normal. If the parent population is not normal, the distribution of the statistic $\sqrt{n}\,(\bar{x}-\mu)/S$ may be quite different from the t-distribution since, in that cases the distribution of $\bar{x}$ and s^2 will not be independent and also the distribution will depend on parameters giving the departure of the parent distribution from normality.

For testing the significance of the difference between two sample means, we make an additional assumption that the variances of the two populations are the same. Before applying the test it may be desirable to test this assumption by applying the F test as discussed in § 15·2·1. If the two variances are different, t-test is no longer valid and we can apply* Behren '*d*' test based on fiducial intervals. We put

$$d = \frac{\bar{x}_1-\bar{x}_2}{\sqrt{S_1^2+S_2^2}}$$

then we carry out the test of significance by using the tables of Sukhatme

* See Statistical Inference by H.C. Saxena .

and Fisher for various values of n_1, n_2 and S_1^2 / S_2^2, the values of d corresponding to various probability levels.

Ex. 1. Show that the 95% fiducial limits for the difference of means of the populations are $(\bar{x}_1 - \bar{x}_2) \pm \bar{x}\, S t_{0.05} \sqrt{\frac{1}{n_1} + \frac{1}{n_2}}$ where $t_{0.05}$ is the 5% value of t for $(n_1 + n_2 - 2)$ *d.f.* Deduce that if zero lies outside this range for $\mu - \mu'$ the difference between the sample means is significance at 5% level.

Ex. 2. Eleven school boys were given a test in Drawing. They were given a month's further tuition and second test of equal difficulty was held at the end of it. Do the marks give evidence that students have benefited by the extra coaching ?

Roll No.	1	2	3	4	5	6.	7	8	9	10	11
Marks of Ist test	23	20	19	21	18	20	18	17	23	16	19
Marks of 2nd test	24	19	22	18	20	22	20	20	23	20	17

[**Hint.** $\bar{d} = 1$, $S = 2{\cdot}24$, $t = 1{\cdot}482$, $v = 10$, $t_{0.05} = 2{\cdot}228$, difference is not significant. Hence the students do not seem to gain by extra-coaching].

15·2·3. Testing the significance of an observed correlation coefficient. Given a random sample $(x_1, y_1), (x_2, y_2)........, (x_n, y_n)$, from a bivariate normal population we have to test the hypothesis that the correction coefficient of the population is zero.

If the hypothesis is true, it was shown in § 14·7·1 that the statistic

$$t = \frac{r\sqrt{n-2}}{\sqrt{1-r^2}} \qquad ...(15{\cdot}5)$$

is a t variate with $(n-2)$ *d.f.* If our calculated value of $|t|$ exceeds $t_{0.05}$ for $(n-2)$ *d.f.* we say that the value of r is significant at 5% level of significance. If $|t| < t_{0.05}$ the sample is consistent with the hypothesis of an uncorrelated population.

Example 15·5. *A random sample of 15 from a normal universe gives a correlation coefficient of* $-\,0{\cdot}5$. *Is this significant of the existence of correlation in the population ?*

Here $\qquad n = 15, \qquad r = -\,0{\cdot}5,$

$$|t| = \frac{0{\cdot}5\sqrt{15-2}}{\sqrt{1-(-\,0{\cdot}5)^2}} = 2{\cdot}07.$$

Also for 13 *d.f.*, $t_{0.05} = 2{\cdot}16.$

$\therefore$ Sample correlation coefficient is not significant, to warrant the existence of a correlation in the population.

Example 15·6. *Find the least value of r in a sample of 18 pairs from a bivariate normal population which is significant at 5% level.*

We have $$t=\frac{r\sqrt{18-2}}{\sqrt{1-r^2}}=\frac{4r}{\sqrt{1-r^2}}$$

Also for 16 $d.f.$, $t_{0.05}=2{\cdot}06$.

$\therefore$ For significance at 5% level, we should have

$$\left|\frac{4r}{\sqrt{1-r^2}}\right|>2{\cdot}06$$

This gives $|r|>{\cdot}47$ approximately.

Thus ·47 is the least value of $|r|$ required.

Ex. 1. Is a value of $r=0{\cdot}5$ significant if obtained from a random sample of 11 pairs of values from a normal population ?

Ex. 2. Verify the following table giving the least value of $|r|$ that are significant at 5% level for sample of size n.

n	6	10	15	20	27	42	52	82	102
$\vert r \vert$	·811	·032	·514	·444	·381	·384	·273	·217	·195

15·2·4. Testing the significance of an observed partial correlation coefficient. Fisher has shown that the sampling distribution of a partial correlation coefficient of order k (*i.e.*, with k secondary subscripts) is of the same form as that of correlation coefficient from a bivariate normal population with the sample size n reduced by k. In particular, if we are given a random sample from a multivariate normal distribution and we have to test the hypothesis that a particular partial correlation coefficient of order k in the population is zero, we make use of the fact that

$$t=\frac{r}{\sqrt{1-r^2}}\sqrt{n-k-2} \qquad \text{...(15·6)}$$

is a t variate with $(n-k-2)$ $d.f.$

Example 15·7. *Show that in a sample of size 20 from a normal population, a correlation coefficient $r_{12\cdot34}=0{\cdot}5$ is significant at 5% level.*

Here $r=0{\cdot}5$, $n=20$, $k=2$.

$$\therefore \quad t=\frac{0{\cdot}5}{\sqrt{1{\cdot}25}}\sqrt{16}=\frac{2}{\sqrt{{\cdot}75}}=\frac{4\sqrt{3}}{{\cdot}3}=\frac{6{\cdot}928}{3}=2{\cdot}31 \text{ approx.}$$

As for 16 $d.f.$, $t_{0.05}=2{\cdot}12$.

Also $|t|<t_{0.05}=2{\cdot}12$.

Also $|t|<t_{0.06}$, the value of $r_{12\cdot34}$ is significant at 5% level.

Ex. Are the following values of r significant at 5% level ?

(*i*) $r_{12\cdot345}=0{\cdot}40$ for $n=21$? **Ans. No.**

(*ii*) $r_{12\cdot345}=0{\cdot}457$ for $n=32$? **Ans. Yes.**

15·2·5. Testing the significance of an observed regression coefficient. We are given a random sample $(x_1, y_1), (x_2, y_2), \ldots\ldots, (x_n, y_n)$

from a bivariate normal population. Let the regression equation of y on x obtained from the sample be

$$y - \bar{y} = b(x - \bar{x})$$

so that the estimated value of y corresponding to a given x_i is

$$Y_i = \bar{y} + b(x_i - \bar{x}).$$

We have to test the hypothesis that the regression coefficient of y on x in the population is β.

If the hypothesis is correct, we have shown in § 14·2 that the statistic

$$t = (b - \beta)[(n-2)\sum_i (x_i - \bar{x})^2 / \sum_i (y_i - Y_i)^2]^{\frac{1}{2}} \quad ...(15·7)$$

conforms to t distribution with $(n-2)$ *d.f.*

Example 15·8. *For a sample of size 30 where x takes the values 1, 2, 3, ..., 30 it is found that* $\Sigma(x_i - \bar{x})(y_i - \bar{y}) = 599·62$ *and* $\Sigma(y_i - \bar{y})^2 = 1020·6$. *Test the significance of regression coefficient of y on x.*

Here $$\Sigma(x_i - \bar{x})^2 = \frac{n(n^2-1)}{12} = \frac{30(899)}{12} = 2247·5$$

$$b = \frac{\Sigma(x_i - \bar{x})(y_i - \bar{y})}{\Sigma(x_i - \bar{x})^2} = \frac{599·62}{2247·5} = 0·2668.$$

Also $$\Sigma(y_i - Y_i)^2 = \Sigma(y_i - \bar{x})^2 - b^2\Sigma(x_i - \bar{x})^2$$
$$= 1020·6 - 159·97 = 860·59$$

Also the hypothesis gives $\beta = 0$

Substituting in (15·7)

$$t = 2·28 \text{ with } \nu = 28$$

This value of t is significant at 5% level but not at 1% level.

Ex. Show that 95% fiducial limits for the regression coefficient are

$$b \pm t_{0·05}[(n-2)\Sigma(x_i - \bar{x})^2/\Sigma(y_i - Y_i)^2]^{-\frac{1}{2}}$$

where $t_{0·05}$ is based on $n-2$ *d.f.*

15·2·6. Testing the significant of a rank correlation coefficient. For samples of moderate size, it can be shown that if r is the rank correlation coefficient from a sample of size n, and the hypothesis that the population rank correlation coefficient is zero is correct then

$$t = r\left(\frac{n-2}{1-r^2}\right)^{1/2}$$

conforms to the t-distribution with $(n-2)$ *d.f.*

This is otherwise obvious since the rank correlation coefficient can be regarded as the correlation between two variates.

Ex. Test the significance of the following rank correlation coefficient at 5% level.

(*a*) $r = 0{\cdot}139, n = 10,$ [$t = {\cdot}397$, not significant]
(*b*) $r = 0{\cdot}618, n = 20,$ [$t = 3{\cdot}34$, highly significant]

15·3. Tests based on the *F* distribution. We consider five tests of significance based on the F distribution.

15·3·1. Testing the significance of the ratio of two independent estimates of the population variance. Given two independent random samples from normal populations, we have to test the hypothesis that the population variances are the same. If the hypothesis is correct, it has been shown in § 13·7, that the statistic S_1^2/S_2^2, where

$$S_1^2 = \frac{1}{n_1 - 1} \sum_{i=1}^{n_1} (x_{1i} - \bar{x}_1)^2, \; S_2^2 = \frac{1}{n_2 - 1} \sum_{j=1}^{n_2} (x_{2j} - \bar{x}_2)^2$$

and $S_1^2 > S_2^2$ conforms to the F distribution with $n_1 - 1, n_2 - 1$ *d.f.* For carrying out the test of significance, therefore, we calculate

$$F = \frac{S_1^2}{S_2^2} \quad (S_1^2 > S_2^2) \qquad \text{...(15·8)}$$

If the calculated value exceeds $F_{0{\cdot}05}$ for $n_1 - 1, n_2 - 1$ *d.f.* we say that ratio is significant at 5% level. However, if $F < F_{0{\cdot}05}$ we say that the same could have come from two normal populations with the same variance.

Example 15·9. *Two samples of sizes 9 and 8 give the sum of squares of deviations from their respective means equal to 160 inches square and 91 inches square respectively. Can they be regarded as drawn from the same normal population ?*

Here $$S_1^2 = \frac{160}{8} = 20, \; S_2^2 = \frac{91}{7} = 13$$

$$\therefore \quad F = \frac{20}{13} = 1{\cdot}54 \text{ approx.}$$

Also $F_{0{\cdot}05}$ for 8 and 7 *d.f.* is 3·37 and, therefore, the calculated values of F is not at all significant. Hence the two samples can be regarded as drawn from two normal populations with the same variance. If the populations are to be the same their means should also be the same and whether this is so can be tested by the *t*-test.

Ex. 1. In two groups of ten children each the increases in weight due to two different diets in the same period were in pounds.

8,	5,	7,	8,	3,	2,	7,	6,	5,	7,
3,	7,	5,	6,	5,	4,	4,	5,	3,	6,

Find whether the variances are significantly different.

[**Ans.** Not significant]

Ex. 2. Prove that $\dfrac{S_1^2}{S_2^2} \cdot \dfrac{1}{F_{0\cdot25}}, \dfrac{S_1^2}{S_2^2} \cdot \dfrac{1}{F_{0\cdot975}}$

are 95% fiducial limits for the ratio σ_1^2/σ_2^2 of the population variances σ_1^2 and σ_2^2 respectively, when the samples are known to have been drawn from populations.

15·3·2. Testing the significance of correlation ratio. Let (x_i, y_{ij}) $(i = 1, 2, ..., h\,; j = 1, 2, ..., n_i)$ denote a random sample of size $N\,(= \Sigma n_i)$ from a bivariate normal population and let η by the correlation ratio of y on x defined by

$$\sum_i \sum_j (y_{ij} - \bar{y}_i)^2 = (1 - \eta^2) \,.\, \sum_i \sum_j (y_{ij} - \bar{y})^2$$

To calculate the significance of η, it can be shown that the ratio

$$F = \frac{\eta^2}{1 - \eta^2} \times \frac{N - h}{h - 1} \qquad ...(15\cdot9)$$

conforms to the F distribution with $v_1 = h - 1, v_2 = (N - h)$ *d.f.*

Example 15·10. *A random sample of 100 pairs of values from a normal populations grouped in 10 arrays of y's gives a correlation ratio* $\eta_{yx} = 0{\cdot}3$. *It it significant of association between the variates ?*

Here $n = 100, h = 10, \eta = 0{\cdot}3$.

$$F = \frac{0{\cdot}09}{1 - {\cdot}09} \times \frac{90}{9} = \frac{0{\cdot}90}{0{\cdot}91} = \frac{90}{91}, v_1 = 9, v_2 = 90$$

It is obvious that this value is not significant even at 5% level.

Ex. The significance of correlation ratio of Example 10·6.

15·3·3. Testing for the significance of a regression function. The appropriate statistic for testing whether the sample indicates any degree of association as represented by the regression equation is

$$F = \frac{r^2 (n - 2)}{1 - r^2} \qquad ...(15\cdot10)$$

which is a F variate with 1 and $n - 2$ *d.f.* This is equivalent to testing the significance of r and it is obvious that F is the square of the statistic t so that the two tests are equivalent.

Ex. Solve examples 15·5 and 15·6 by the use of the F table and show that these give the same results as obtained there.

15·3·4. Testing for non-linearity of regression. For a sample of size N arranged in h arrays from a normal sample the appropriate statistic for testing the hypothesis of linearity of regression is

$$F = \frac{\eta^2 - r^2}{1 - \eta^2} \times \frac{N - h}{h - 2} \qquad ...(15\cdot11)$$

with $(h{-}2, N{-}h.)$ *d.f.*

Example 15·11. *A random sample of 150 pairs from a bivariate normal population when grouped in 15 arrays of y's gives $r = 0{\cdot}4$, $\eta = 0{\cdot}5$. Are the results consistent with the assumption of linearity of regression ?*

Here $N = 150, h = 15, r = 0{\cdot}4, \eta = 0{\cdot}5$

$$F = \frac{\cdot 25 - \cdot 16}{1 - \cdot 25} \times \frac{135}{13} = 1{\cdot}25,\ \nu_1 = 13,\ \nu_2 = 135$$

which is not significant at 5% level. Hence the results are consistent with the hypothesis of linearity of regression.

Ex. From 823 observations arranged in 9 arrays.

$$\Sigma\, n_i\,(y_i - \bar{y}\,)^2 = 396,\ \Sigma\,\Sigma\,(y_{ij} - \bar{y}_i\,) = 3832.$$

Is the regression linear ? **[Ans.** Not significant]

15·3·5. Testing the significance of an observed multiple correlation. Let R be the multiple correlation coefficient between a variate and p other variates in a random sample of size N from a $(p + 1)$ variate normal distribution. Fisher has found the distribution of R and he has shown that if the population multiple correlation coefficient is zero, then the statistic

$$F = \frac{R^2}{1 - R^2} \cdot \frac{N - p - 1}{p} \qquad ...(15{\cdot}12)$$

conforms to the F distribution with $p, N - p - 1$ *d.f.*

Ex. 1. For a sample of 30 sets of values from a normal population $R_{2(31)}$ is found to be 0·5. Show that this is significant of correlation in the population between x_2 and x_1, x_3.$(F = 4{\cdot}5,\ \nu_1 = 2,\ \nu_2 = 27)$

Ex. 2. From a sample of size 20, $R_{1(23)}$ is found to be ·5433. Show that this is significant at 1% level.

15·4. Further applications of the *F* distribution.

15·4·1. Testing homogeneity of means. Analysis of variance. The t test enables us to test the significance of the difference between two sample means but if we have a number of means and we have to test the hypothesis that means are homogeneous or that they do not differ significantly, then the technique known as *Analysis of Variance* developed by Fisher is useful.

Consider a set of N values of the variate x arranged into h rows and k columns so that $N = hk$. Let x_{ij} represent the value of the member belonging to ith row and jth columns, $\bar{x}_j$ the mean of the jth column and $\bar{x}_i$, the mean of the ith row. The data is then arranged as follows :

	Ist col.	2nd col.	.....	jth col.	.....	kth col.	Row totals	Row means
1st row	x_{11}	x_{12}	....	x_{1j}	....	x_{1k}	$T_{1\cdot}$	$\bar{x}_{1\cdot}$
2nd row	x_{21}	x_{22}	....	x_{2j}	....	x_{2k}	$T_{2\cdot}$	$\bar{x}_{2\cdot}$
⋮	⋮	⋮		⋮		⋮	⋮	⋮
ith row	x_{i1}	x_{i2}	....	x_{ij}	....	x_{ik}	$T_{i\cdot}$	$\bar{x}_{i\cdot}$
⋮	⋮	⋮		⋮		⋮	⋮	⋮
hth row	x_{h1}	x_{h2}	....	x_{hj}	....	x_{hk}	$T_{h\cdot}$	$\bar{x}_{h\cdot}$
Column totals	$T_{\cdot 1}$	$T_{\cdot 2}$	.....	$T_{\cdot j}$	.....	$T_{\cdot k}$	T	$\bar{x}_{\cdot\cdot} = \frac{T}{N}$
Column means	$\bar{x}_{\cdot 1}$	$\bar{x}_{\cdot 2}$	.....	$\bar{x}_{\cdot j}$	.....	$\bar{x}_{\cdot k}$	$\bar{x}_{\cdot\cdot}$	

We regard the data as classified into h classes, each row corresponding to a class according to some factors of classification. Thus each row may represent the price of commodity obtained from k shops in a district and the different rows may correspond to different districts or the data may represent marks obtained by h groups of students, each group consisting of k students.

The various values x_{ij} differ among themselves and this variability may be due to :

(*i*) variation within classes or groups :

(*ii*) variation from class to class.

The object of the analysis of variance would be to break up the total variation into components due to each of the factors and then compare them by the F test.

If we assume that each group is homogeneous within itself, then the variation within itself is due to chance or random causes. If the data is homogeneous with respect to the factor of classification, *i.e.*, if there is no effect due to this factor then the variation between classes would also be due to random causes and the two variations would be of the same order.

15·4·2. Analysis of Variance for one-way classification.

Let $$x_{ij} = \mu_i + e_{ij}$$

where x_{ij} stands for the jth observation in the ith set, μ_i is the effect of fertilizer number i and e_{ij} is the chance error which is due to chance variation in the jth observation of the ith set. We shall prefer a general model for our experiment *i.e.* we shall take $\mu_i = \mu + \alpha_i$ where μ is the general effect and α_i is equal to the deviation. Thus we have

$$x_{ij} = \mu + \alpha_i + e_{ij},$$

wher $i = 1, 2, ..., h$; $j = 1, 2, ..., k$, with

Assumptions

1. e_{ij},*s* are independently and identically normally distributed with mean zero and variance σ^2 *i.e.* e_{ij}'*s* are $N(0, \sigma^2)$.

2. α' *s* are constants with the restriction $\sum_{i=1}^{h} \alpha_i = 0$

3. μ is constant

Null hypothesis. The null hypothesis to be set up is

$$H_0 : \mu_1 = \mu_2 = ... = \mu_i = ... = \mu_k.$$

or equivalently, $H_0 : \alpha_1 = \alpha_2 = ... = \alpha_i = ... = \alpha_k$.

Principle of Least Squares will be helpful in estimating the parameters μ and α_i, $i = 1, 2,...., h$. Hence if

$$L = \sum_{i=1}^{h}\sum_{j=1}^{k} e_{ij}^2 = \sum_{i=1}^{h}\sum_{j=1}^{k} (x_{ij} - \mu - \alpha_i)^2 = \sum_{i=1}^{h}\sum_{j=1}^{k} (x_{ij} - \mu_i)^2) \quad ...(1)$$

The normal equations for determining the μ_i'*s* are

$$\frac{\partial L}{\partial \mu_i} = 0 \quad \Rightarrow \sum_{j=1}^{k} (x_{ij} - \mu_i) = 0$$

$$\Rightarrow \sum_{j=1}^{k} x_{ij} - \sum_{j=1}^{k} \mu_i = 0 \quad i.e., T_i. = k\mu_i \quad \text{where } T_i. \sum_{j=1}^{k} x_{ij}.$$

Thus the estimated value of μ_i is obtained by $\hat{\mu}_i = \frac{1}{k} T_i.$

Also $\hat{\alpha}_p - \hat{\alpha}_q = (\hat{\mu}_p - \mu) - (\hat{\mu}_q - \mu) = \hat{\mu}_p - \hat{\mu}_q = \frac{1}{k}(T_{p.} - T_{q.})$

$\therefore$ Minimum value of $L = \sum_{i,j} (x_{ij} - \hat{\mu}_i)^2$

$$= \sum_{i,j} \left(x_{ij} - \frac{T_{i.}}{k} \right)^2 = \sum_{i,j} x_{ij}^2 - \frac{1}{k} \sum_{i=1}^{h} T_{i.}^2$$

Hence
$$S^2 = \sum_{i=1}^{h} \sum_{j=1}^{k} \left(x_{ij} - \frac{T_{i.}}{k} \right)^2$$

is called the measure of variation due to variation within set.

If we set up the hypothesis that the difference among the observed value may be attributed to chance variation, then the model (15·4·2) reduces to

$$x_{ij} = \mu + e_{ij} \;;$$

$$\therefore \qquad L_0 = \sum_{i,j} (x_{ij} - \mu)^2$$

and, therefore, the normal equation for estimating μ, is

$$0 = \frac{\partial L_0}{\partial \mu} = \sum_{i,j} (x_{ij} - \mu)$$

$$\therefore \qquad \hat{\mu} = \frac{1}{hk} \sum_{i,j} x_{ij} = \frac{1}{hk} \overline{x}_{..}$$

$$\therefore \text{ Under } H_0 : S_0^2 = \sum_{i,j} \left(x_{ij} - \frac{\overline{x}_{..}}{hk} \right)^2$$

where S_0^2 is called measure of the total variation in the data.

$\therefore$ $S_0^2 - S^2$ is called the measure of the variation due to the hypothesis H_0 and is given by

$$S_0^2 - S^2 = \sum_{i,j} \left(x_{ij} - \frac{\overline{x}_{..}}{hk} \right)^2 - \sum_{i,j} \left(x_{ij} - \frac{\overline{x}_{i.}}{k} \right)^2$$

$$= \sum_i \left(\frac{\overline{x}_{i.}^2}{k} \right) - \frac{\overline{x}_{..}^2}{hk}$$

But $\qquad S_0^2 = S + (S_0^2 - S^2)$

i.e. Total variation = Within set variation + Between set variation

Again, $\qquad E\left[\dfrac{S_0^2 - S^2}{h-1} \right] = \sigma^2, \; E\left[\dfrac{S^2}{hk-h} \right] = \sigma^2.$

Hence $(S_0^2 - S^2)/(h-1)$ and $S^2/(hk-h)$ are unbiased estimates of σ^2.

Also $(S_0^2 - S^2)/\sigma^2$ is distributed as χ^2 with $(h-1)$ *d.f. i.e.* χ^2_{h-1} and S^2/σ^2 is distributed as χ^2 with $h(k-1)$ *d.f. i.e.* $\chi^2_{h(k-1)}$.

Also χ^2_{h-1} and $\chi^2_{h(k-1)}$ are independent χ^2 variates.

Hence $\qquad \dfrac{(S_0^2 - S^2)/(h-1)}{S^2/h(k-1)} = F_{h-1,\, h(k-1)}$

In view of (15·13), it is obvious that the three unbiased estimates of σ^2 are *not independent.* However, in view of the independence of the distribution of sample mean and variance in random sampling from a normal distribution established in § 13·3, it can be shown that the second and third estimates are independent and their ratio can be treated by the F test

$$\text{where } F = \frac{\Sigma k(\bar{x}_{i.} - \bar{x}_{..})^2}{\Sigma\Sigma(x_{ij} - \bar{x}_{i.})^2} \times \frac{N-h}{h-1} \text{ with } v_1 = h-1, v_2 = N-h.$$

The above results are expressed conveniently in the *analysis of variance table given on the below.*

Source of variation	*D.F.*	*sum of squares*	*Mean square*
Between class means	$h-1$	$\sum_i k(\bar{x}_{i.} - \bar{x}_{..})^2$	$\sum_i k(\bar{x}_{i.} - \bar{x}_{..})^2/(h-1)$
Within classes	$N-h$	$\sum_i\sum_j (x_{ij} - \bar{x}_{i.})^2$	$\sum_i\sum_j (x_{ij} - \bar{x}_i)^2/(N-h)$
Total	$N-1$	$\sum_i\sum_j (x_{ij} - \bar{x}_{..})^2$,	

Notes 1. It will be noted that the mean squares are obtained by dividing the S.S. by the corresponding *d.f.* For the practical calculation of S.S. we may note first that all the S.S. are independent of change of origin and any convenient origin may, therefore, be chosen.

2. Secondly, all the S.S. can be expressed in terms of row totals $T_{i.}$, column total $T_{.j}$ and grand total T.

$$\text{Thus } \sum_i\sum_j (x_{ij} - \bar{x}_{..})^2 = \sum_i\sum_f x_{ij}^2 - 2\bar{x}_{..}\sum_i\sum_j x_{ij} + N\bar{x}_{..}^2 = \sum_i\sum_j x_{ij}^2 - \frac{T^2}{N}, \quad ...(2)$$

$$\text{and } \sum_i k(\bar{x}_{i.} - \bar{x}_{..})^2 = k\sum_i \left(\frac{T_{i.}}{k} - \frac{T}{hk}\right)^2 = \sum_i \frac{T_{i.}^2}{k} - \frac{T^2}{N}, \quad ...(3)$$

Now $\Sigma\Sigma(x_{ij} - \bar{x}_{i.})^2$ can be obtained by subtracting (3) from (2) on making use of (15·13). The term T^2/N is usually called the *correction factor* on account of measurement being taken from the mean.

Example 15·12. *10 varieties of wheat are grown in 3 plots each and following yields, in maunds, per acre, obtained:*

Plots/Variety	*1*	2	3	4	5	6	7	8	9	10	*Total*
I	7	7	14	11	9	6	9	8	12	9	
II	8	9	13	10	9	7	13	13	11	11	
III	7	6	16	11	12	5	12	11	11	11	
Total	22	22	43	32	30	18	34	32	34	31	298

Test the significance of the differences between variety yields.

Since the S.S. are unaffected by change of origin, we subtract 10 maunds from each yield.

Now $h = 10, k = 3, T_{.1} = -8, T_{.2} = -8, \ldots\ldots, T = -2, N = 30$

$$\Sigma\Sigma x_{ij}^2 = (-3)^2 + (-3)^2 + (-3)^2 + (-3)^2 \ldots\ldots = 204$$

$$\frac{1}{3}\sum_j T_{.j}^{\ 2} = \tfrac{1}{3}[(-8)^2 + (-8)^2 + (13)^2 + \ldots\ldots + (1)^2] = 160{\cdot}67$$

$$\text{Total S.S.} = \sum_i\sum_j x_{ij}^2 - \frac{T^2}{N} = 204 - \frac{4}{30} = 203{\cdot}87.$$

$$\text{S.S. between classes} = \tfrac{1}{3}\sum_j T_{.j}^2 - \frac{T^2}{N} = 160{\cdot}67 - \frac{4}{30} = 160{\cdot}54$$

S.S. between classes $= 203{\cdot}87 - 160{\cdot}54 = 43{\cdot}33$

Thus we get

Analysis of variance Table

Source of variation	*D.F.*	S.S.	*M.S.*
Between varieties	9	160·54	17·84
Within varieties	20	43·33	2·17
Total	29	203·87	

$$F = \frac{17{\cdot}84}{2{\cdot}17} = 8{\cdot}22,\ \nu_1 = 9,\ \nu_2 = 20$$

This is highly significant at 1% level.

15·4·3. Analysis of Variance for two way classification. Suppose the data are divided according to both rows and columns. Thus if we study the effects of two types of treatments, say fertilizers and varieties, then the observation corresponding to the *i*th treatment of one type and the *j*th treatment of the other type may be denoted by x_{ij}.

Consider the model $x_{ij} = \mu + \alpha_i + \beta_j + e_{ij}$, $i = 1, 2, \ldots, h\,;\ j = 1, 2, \ldots\ldots, k$

where α_i is the effect due to the *i*th fertilizer, β_j is the effect due to the *j*th variety and e_{ij} is the chance or random error and are $N(0, \sigma^2)$ *i.e.* $E(e_{ij}) = 0$, Var $(e_{ij}) = \sigma^2$. With the restrictions

$$\sum_{i=1}^{h} \alpha_i = 0,\quad \sum_{j=1}^{k} \beta_j = 0$$

By the principle of least squares,

$$P = \sum_{i=1}^{h}\sum_{j=1}^{k} e_{ij}^2 = \sum_{i=1}^{h}\sum_{j=1}^{k} (x_{ij} - \mu - \alpha_i - \beta_j)^2$$

$$= \sum_{i=1}^{h} \sum_{j=1}^{k} (x_{ij} - \mu_i - \beta_j)^2 \quad ...(1)$$

Normal equations for estimation of the constants μ_i and β_j are:

$$0 = \frac{\partial P}{\partial \mu_i} = -2 \sum_{j=1}^{k} (x_{ij} - \mu_i - \beta_j) \quad ...(2)$$

and

$$0 = \frac{\partial P}{\partial \beta_j} = -2 \sum_{i=1}^{h} (x_{ij} - \mu_i - \beta_j) \quad ...(3)$$

Equation (2) gives $\hat{\mu}_i = \frac{1}{k} \sum_{j=1}^{k} x_{ij} = \frac{1}{k} \bar{x}_{i.}$

and (3) gives $\hat{\beta}_j = \frac{1}{h} \sum_{i=1}^{h} x_{ij} = \frac{1}{h} \bar{x}_{.j}$

Also $\hat{\alpha}_p - \hat{\alpha}_q = \hat{\mu}_p - \hat{\mu}_q = \frac{\bar{x}_{.p}}{k} - \frac{\bar{x}_{.q}}{k}, \; \hat{\beta}_p - \hat{\beta}_q = \frac{\bar{x}_{.p}}{h} - \frac{\bar{x}_{.q}}{h}$

$$\therefore \sum_{i=1}^{h} \sum_{j=1}^{k} (x_{ij} - \bar{x}_{..})^2 = \sum_{ij} \left(x_{ij} - \frac{\bar{x}_{i.}}{k} - \frac{\bar{x}_{.j}}{h} + \frac{\bar{x}_{..}}{hk} \right)^2$$

$$= k \sum_{i=1} (\bar{x}_{i.} - \bar{x}_{..})^2 + h \sum_{j=1} (\bar{x}_{.j} - \bar{x}_{..})^2 + \sum_{i=1}^{h} \sum_{j=1}^{k} (x_{ij} - \bar{x}_{i.} - \bar{x}_{.j} + \bar{x}_{..})^2$$

As in the last sub-section, on the basis of normality of distribution of the variates and the null hypothesis of the homogeneity of both the row and column factors of classification the above four sums on being divided respectively by $hk - 1, h - 1, k - 1$ and $(h - 1)(k - 1)$ give unbiased estimates of the population variance σ^2 and by using the theorem due to Cochran it can be shown that estimates given by S.S. the between row 'means' and also those given 'between column means' are independent of the estimate given by the last error, *i.e.*, S.S. The analysis of variance table is as follows.

Analysis of Variance Table

Source of variation	*D.F.*	*S.S.*	*M.S.*	*F*
Between rows	$h-1$	$\sum_i k(\bar{x}_{i.} - \bar{x}_{..})^2$	S.S./D.F.	
Between columns	$k-1$	$\sum_j h(\bar{x}_{.j} - \bar{x}_{..})^2$		
Error	$(h-1)(k-1)$	$\sum_i \sum_j (x_{ij} - \bar{x}_{i.} - \bar{x}_{.j} + \bar{x}_{..})^2$		
Total	$hk-1$	$\sum_i \sum_j (x_{ij} - \bar{x}_{..})^2$		

Example 15·13. *Yields of 4 varieties of wheat in 3 blocks are given below :*

Varieties ↓ \ *Blocks* →	*1*	*2*	*3*
I	*10*	*9*	*8*
II	*7*	*7*	*6*
III	*8*	*5*	*4*
IV	*5*	*4*	*4*

(M.A., Punjab)

Is the difference between varieties significant ?

Subtracting 5 from each element, we get

Varieties	*Blocks*			*Total*
I	5	4	3	12
II	2	2	1	5
III	3	0	– 1	2
IV	0	– 1	– 1	– 2
Total	10	5	2	17

$\Sigma \Sigma x_{ij}^2 = 25 + 16 + 9 + 4 + 4 + 1 + 9 + 0 + 1 + 0 + 1 + 1 = 71,$

$\frac{1}{3} T_{i\cdot}^2 = \frac{1}{3} [\, 12^2 + 5^2 + 2^2 + (-2)^2 \,] = \frac{1}{3} (177) = 59,$

$\frac{1}{4} T_{\cdot j}^2 = \frac{1}{4} [\, 10^2 + 5^2 + 2^2 \,] \qquad = \frac{1}{4} (129) = 32{\cdot}25. \; \frac{T^2}{N} = \frac{(17)^2}{12} = 24{\cdot}1$

Source of variation	*D.F.*	*S.S.*	*M.S.*	*F*
Varieties	3	59 – 24·1 = 34·9	11·63	18·1
Blocks	2	32·25 – 24·1 = 8·15	4·07	6·4
Error	6	3·85	0·64	—
Total	11	71–24·1 = 46·9	—	—

The difference between varieties is highly significant.

The difference between blocks is significant at 5% level but not at 1% level.

Ex. 1. Prove the following :

(*i*) $\sum_i \sum_j (\bar{x}_{i\cdot} - \bar{x}_{\cdot\cdot})(x_{ij} - \bar{x}_{i\cdot} - \bar{x}_{\cdot j} + \bar{x}_{\cdot\cdot}) = 0$

(*ii*) $\sum_j \sum_i (\bar{x}_{..} - \bar{x}_{.j})(x_{ij} - \bar{x}_{i.} - \bar{x}{\cdot}j + \bar{x}_{..}) = 0$

Ex. 2. For the data of example 15·12, establish the following table:

Due to	*D.F.*	*S.S.*	*M.S.*	*F*
Between rows	2	8·27	4·13	2·12
Between columns	9	160·54	17·84	—
Error	18	35·06	1·96	—
Total	29	203·87	—	—

15·5. Tests of significance based on Fisher's z transformation. Fisher's z transformation

$$z = \tfrac{1}{2}\log_e\left(\frac{1+r}{1-r}\right) = 1{\cdot}1513\log_{10}\left(\frac{1+r}{1-r}\right)$$

have been discussed in § 14·8. Fisher has shown that for samples of a large size n, z is distributed asymptotically normally about mean ξ, where

$$\xi = \tfrac{1}{2}\log_e\left(\frac{1+\rho}{1-\rho}\right) = 1{\cdot}1513\log_{10}\left(\frac{1+\rho}{1+\rho}\right)$$

with variance $1/(n-3)$. This transformation forms the basis of the tests of significance discussed below, but will be obvious that unlike the tests of the previous two sections the present tests will hold for large samples only.

15·5·1. To test whether r differs significantly from ρ. Given a random sample $(x_1, y_1), \ldots\ldots, (x_n, y_n)$ from a bivariate normal population, we have to test the hypothesis that the population correlation coefficient is ρ. To do this, we calculate z and ξ by applying Fisher's transformation and then calculate the asymptotically standard normal variate

$$\frac{z-\xi}{1/\sqrt{n-3}} \qquad \ldots(15{\cdot}15)$$

If the absolute value of this statistic exceeds 1·96 the difference is significant at 5% level.

Example 15·14. *A correlation coefficient of 0·7 is discovered in a sample of 28 pairs. Apply z transformation to find out if this differs significantly* (*a*) *from 0* (*b*) from *0·5*.

(*a*) $\rho = 0, \qquad \xi = 0$

Also $\quad z = 1{\cdot}1513\log_{10}\left(\frac{1+{\cdot}7}{1-{\cdot}7}\right) = 0{\cdot}87$

$\therefore \quad t = \dfrac{0{\cdot}87 - 0}{1/\sqrt{28-3}} = 4{\cdot}35 > 1{\cdot}96$

The hypothesis of zero correlation is discredited and we conclude that there is a correlation in the population.

(*b*) $\rho = 0{\cdot}5, \xi = 1{\cdot}1513 \log_{10}\left(\dfrac{1+0{\cdot}5}{1-0{\cdot}5}\right) = 0{\cdot}55,\ t = \dfrac{0{\cdot}87 - 0{\cdot}50}{1/\sqrt{28-3}} = 0{\cdot}6 < 1{\cdot}96$

The data is consistent with the hypothesis that $\rho = 0{\cdot}5$

Ex. 1. Show that the 95% fiducial limits for ζ are $z \pm 1{\cdot}96\sqrt{n-3}$.

Ex. 2. Prove that the 95% fiducial limits for ρ in the above example are 0·45 and 0·85.

15·5·2. To test the significance of the difference between two independent correlation coefficients. We have to test the hypothesis that the two samples of sizes n_1, n_2 giving correlation coefficients r_1, r_2 may be regarded as drawn from the sample population or from two populations with the same correlation coefficient. If the hypothesis is correct, the statistic

$$t = \frac{z_1 - z_2}{\sqrt{\dfrac{1}{n_1-3} + \dfrac{1}{n_2-3}}} \qquad \text{...(15·16)}$$

where $z_1 = \frac{1}{2}\log_e\left(\dfrac{1+r_1}{1-r_1}\right), z_2 = \frac{1}{2}\log_e\left(\dfrac{1+r_2}{1-r_2}\right)$

is asymptotically a standard normal variate. If absolute value of this statistic is greater than 1·96, the difference with be significant at 5% level.

Example 15·15. *The correlation coefficients between temperatures of rice and breakage percentage calculated from two samples of 12 and 16 are ·8912 and ·8482 respectively. Do the two estimates differ significantly ?*

Here $n_1 = 12, n_2 = 16, r_1 = {\cdot}8912, r_2 = {\cdot}8482$

$\therefore \quad z_1 = 1{\cdot}4276, z_2 = 1{\cdot}2496$

Now $$\frac{z_1 - z_2}{\sqrt{\dfrac{1}{n_1-3} + \dfrac{1}{n_2-3}}} = \frac{1{\cdot}4276 - 1{\cdot}2496}{\sqrt{0{\cdot}1880}} = 0{\cdot}41$$

which is not at all significant at 5% level.

Ex. Two samples of sizes 23, 28 give correlation coefficients 0·5 and 0·8 respectively. Show that these are not significantly different.

15·5·3. Combination of estimates of a correlation coefficient. Correlation coefficients $r_1, r_2, \ldots\ldots, r_k$ are obtained from samples of sizes $n_1, n_2, \ldots\ldots, n_k$ from the same bivariate normal population with correlation coefficient ρ, then $z_1, z_2, \ldots, z_k$ are independent estimates of the same quantity $\zeta = \frac{1}{2}\log_e\left(\dfrac{1+\rho}{1-\rho}\right)$. To get a pooled estimate for ζ, we take weighted average of $z_1, z_2, \ldots, z_k$ and to get an estimate with the minimum variance we use weights inversely proportional to the variance (*cf.* Ex 34. page 590).

The estimate is accordingly $\bar{z} = \dfrac{\Sigma\,(n_i - 3)\, z_i}{\Sigma\,(n_i - 3)}$

and from this we can get an estimate for ρ from the relation $\rho = \tanh \bar{z}$. Then since the variates z_i are approximately normally distributed about ξ with variance $1/(n_i - 3)$, the sum $\Sigma\,(n_i - 3)\,(z_i - \bar{z})^2$ is distributed approximately as χ^2 with $(k-1)$ *d.f.* The significance can be ascertained from the table of χ^2. Not significance shows that the estimate r_i of correlation in the population may be regarded as homogeneous.

Example 15·16. *Combine the two estimates of r in Example 15·15 to get a pooled estimate.*

$$\bar{z} = \frac{(n_1 - 3)\, z_1 + (n_2 - 3)\, z_2}{(n_1 - 3) + (n_2 - 3)} = \frac{12{\cdot}8484 + 16{\cdot}2448}{9 + 13} = \frac{29{\cdot}0932}{22} = 1{\cdot}3224.$$

The corresponding estimate of ρ is 0·8674.

Ex. 1. Independent sample of 21, 30, 39, 26, 35 pair of values give correlation coefficients 0·39, 0·61, 0·43, 0·54, 0·48. Prove that the pooled estimate for ρ is 0·5 nearly and the value of χ^2 is 1·4 approx.

Ex. 2. Two independent samples of sizes 20 and 25 give correlation coefficients ·60 and ·80. Show that these are not significantly different. Show also that the weighted average value of $r = {\cdot}7267$.

Ex. 3. Show that in exercise of § 15·4·2 the 95% fiducial limits for r_1 from the given value of r_2 are 0·47 and 0·93.

15·5·4. Tests of significance for partial correlation coefficients. The results of the last three sections hold equally for partial correlation coefficients, only in each case the sample number should be reduced by k, the order of the partial correlation coefficients indicated by the number of secondary subscripts.

Ex. From independent sample of 30 and 20 sets of values, partial correlation of order three are found to be ·4 and ·6 respectively. Examine

(*i*) whether the first sample could have come from a population with the corresponding correlation coefficient equal to ·6 ;

(*ii*) whether the two samples could have come from the same normal population and if this is true find the pooled estimate.

Exercises 15

1. Obtain tests of significance for testing the following hypothesis :

(*a*) σ_0 is the hypothetical value of σ in a normal population $N(\mu, \sigma^2)$ with known μ

(*b*) σ_0 is the hypothetical value of σ in a normal population $N(\mu, \sigma^2)$ with unknown μ.

(*c*) $\mu_1 = \mu_2$ in two normal populations $N(\mu_1, \sigma_1^2)$ and $N(\mu_2, \sigma_2^2)$ where μ_1 and μ_2 are unknown but σ_1^2 and σ_2^2 are known.

[**Hint** (*a*) $\Sigma \dfrac{(x_i - \mu)^2}{\sigma_0^2}$ is distributed as χ^2 with *n.d.f.*

(*b*) $\Sigma \dfrac{(x_i - \bar{x})^2}{\sigma_0^2}$ is distributed as χ^2 with $(n-1)$ *d.f.*

(*c*) $\dfrac{(\bar{x}_1 - \bar{x}_2) - (\mu_1 - \mu_2)}{\sqrt{\dfrac{\sigma_1^2}{n_1} + \dfrac{\sigma_2^2}{n_2}}}$ is a standard normal variate.]

2. Find Student's *t* for the following variate value in a sample of eight,

$$-4, -2, -2, 0, 2, 2, 3, 3$$

taking the mean of the universe to be zero **(*M.Sc., Agra*)**

[**Ans.** = 0·29 nearly]

3. A machine which produce mica insulating washers for use in electric devices is said to turn out washers having a thinkness of 10 mils (1 mil = 0·001 inch). A sample of 10 washers has an average thickness of 9·52 mils with a standard deviation of 0·60 mil. Find out the significance of such a deviation.

[**Ans.** $t = 2{\cdot}528$, Deviation is significant]

4. A certain stimulus administered to each of 12 patients resulted in the following increases of blood pressures :

$$5, 2, 8, -1, 3, 0, 6, -2, 1, 5, 0, 4.$$

Can it be concluded that the stimulus will be, in general, accompained by an increase in blood pressure ? **(*B. Sc. Delhi, 1989*)**

[**Ans.** $t = 2{\cdot}347$. The value is significant and therefore it can be concluded that stimulus will be in general accompanied by blood pressure.]

5. A sample of 11 rats from a central population had an average blood viscosity of 3·92 with a *s.d.* of 0·61. On the basis of this sample, establish 95% limits of μ, the mean blood viscosity of control population.,

[**Ans.** Limits are $3{\cdot}92 \pm \dfrac{2{\cdot}228\,(0{\cdot}61)}{\sqrt{11}}$ *i.e.* 3·51 and 4·33]

6. Suppose X is a continuous chance variate such that all values of X in the interval $(0, \theta)$ are equally likely where θ is unknown. Suppose a sample of 20

values of X is drawn and the sample mean $\bar{x}$ has the value 3·2. What are the 90% confidence limits of θ ?

$$\left[dF = \frac{dx}{\theta}, 0 \leq x \leq \theta : \mu_1' = \tfrac{1}{2}\theta, \mu_2' = \frac{1}{12} \cdot \theta^2 \Rightarrow \sigma = \frac{\theta}{\sqrt{12}} \right.$$

From table, t for 11 $d.f.$ at 1% level = 1·73.

$$\frac{\bar{x} - \theta/2}{\dfrac{\theta}{\sqrt{12} \cdot \sqrt{n}}} = 1 \cdot 73. \text{ Since } \bar{x} = 3 \cdot 2 \therefore \theta = etc.]$$

7. The sleep of 10 patients was measured for the effects of the soporific drugs referred to in the following table at Drug A and Drug B. From the data given below show that there is a significant difference between the effects of two drugs, on the assumption that difference between the effects of two drugs, on the assumption that different random samples of patients were used to test the two drugs A and B.

Additional hours of sleep gained by use of soporific drugs.

Patient	*Drug A*	*Drug B*
1	+0·7	1·9
2	−1·6	0·8
3	−0·2	1·1
4	−1·2	0·1
5	−0·1	0·1
6	−3·4	4·4
7	3·7	5·5
8	0·8	1·6
9	0	4·9
10	2·0	3·6

[**Hint.** This is an example of paired samples. Using the formula (15·3)

$$t = \frac{1 \cdot 6 \times 3 \cdot 1}{1 \cdot 23} = 3 \cdot 78$$

This value is significant and hence there is a significant difference between the effect of the two drugs which is in favour of the second].

8. The following table shows the mean number of bacterial colonies per plate obtained by four slightly different methods from soil samples taken at 4 P.M. and 8 P.M. repectively.

	Method A	*Method B*	*Method C*	*Method D*
4 P.M.	29·75	27·50	30·25	27·80
8 P.M.	39·20	40·60	36·20	42·40

Are there significantly more bacteria at 8 P.M. than at 4 P.M. ?

[**Ans.** Treating as paired samples, $t = 5 \cdot 560, n = 3$. There are more bacteria at 8 P.M. than at 4 P.M.]

9. In a certain experiment to compare two types of pig-foods A and B, the following results of increase in weights were observed in pigs :

Pig Number		1	2	3	4	5	6	7	8	Total
Increase in weight in lbs.	*Food A*	49	53	51	52	47	50	52	53	407
	Food B	52	55	52	53	50	54	54	53	423

Assuming that the two samples of pigs are independent, can it be concluded that food B is better than food A. Examine the case when the same set of eight pigs were used in both the foods ?

[**Ans.** Food B is better than Food A]

10. A farmer grows crops on two fields A and B. On A he puts Rs. 10 worth of manure per acre and on B Rs. 20 worth. The net returns per acre, exclusive of the cost of manure, on the two fields are :

Year A	*Yield A, Rs. per acre*	*Yield B, Rs. per acre*
1	34	36
2	28	33
3	42	48
4	37	38
5	44	50

Other things being equal, discuss the question whether it is likely to pay the farmer to continue the more expensive dressing ?

[**Ans.** $\bar{d}$ = Rs. 4, S = Rs. 2·345, $t = 3{\cdot}814$, $\nu = 4$. It is likely to pay the farmer to continue the more expensive dressing].

11. The heights of six randomly chosen sailors are in inches 63, 65, 68, 69, 71, and 72. Those of the randomly chosen soldiers are 61, 62, 65, 69, 69, 70, 71, 72, 73.

Discuss in the light that the data throw that the soldiers are on the average, taller than the sailors.

[**Ans.** The soldiers on the average are not likely to be taller than the sailors].

12. Eight pots growing three wheat plants each were exposed to a high tension discharge while nine similar pots were enclosed in an earthen wire case. The number of tillers in each pot were as follows :—

Caged	17	26	18	25	27	28	26	23	17
Electrified	16	16	22	16	21	18	15	20	

Discuss whether electrification exercises any real effect on tillering.

[**Ans.** The electrification does exert some effect on tillering]

13. Mitchell conducted a paired feeding experiment with pigs on the relative value of limestone and bone meal for bone development. The results are :

Ash content in percentages of scapulas of pairs of pigs fed on limestone and bone meal.

Pair	*Limestone*	*Bone meal*
1	49·2	51·5
2	53·3	54·9
3	50·6	52·2
4	52·0	53·3
5	46·8	51·6
6	50·5	54·1
7	32·1	54·2
8	53·0	54·3

Determine the significance of the difference between the means in two ways (1) By assuming that the values are paired and (2) by assuming that the values are not paired.

[**Ans.** It can be reasonably concluded that the difference between meals is significant]

14. The means of two random samples of sizes 9 and 7 respectively, are 196·40 and 198·82 respectively. The sum of the squares of the deviations from the means are 26·94 and 18·73 respectively. Can the samples be considered to have been drawn from the same normal population, it being given that the value of t for *d.f.* at 5% level of significance is 2·145 and at 1% level of significance is 2·977 ?

[**Ans.** The sample cannot be considered to have been drawn from the same normal population]

15. Two horses A and B were tested according to the time (in seconds) to run a particular track with the following results :

Horse A	28	30	32	33	33	29	and	34
Horse B	29	30	30	24	27	and	29	

Analyse, as completely and in as many ways as you can, the above data and report whether or not you can discriminate between the two horses, and if so, with what degree of confidence. You may use the following extracts from satistical tables. Probability that t will exceed the tabular entry is absolute.

Degrees of freedom	*Percentage*		*Probability*	
	5	2·50	0·05	0·1
11	2·20	2·59	3·50	4·44
12	2·18	2·56	3·43	4·32
13	2·16	2·53	3·37	4·22
18	2·10	2·44	3·20	3·92

[**Hint.** $\bar{x}_1 = \dfrac{219}{7} = 31 \cdot 3$ approx., $\bar{x}_2 = \dfrac{169}{6} = 28 \cdot 1$ approx.

Taking 32 and 19 an the working means for series A and B respectively, we get $S^2 = \dfrac{31 \cdot 43 + 26 \cdot 86}{11} = 2 \cdot 5$ nearly. Hence $t = \dfrac{3 \cdot 2}{\sqrt{5 \cdot 3}\sqrt{\dfrac{1}{6} + \dfrac{1}{7}}} = 2 \cdot 5$ approx.

As 5% level of significance the value of t for 11 *d.f.* $= 2 \cdot 20$. Hence we can discriminate between the two horses. We can discriminate between the horses only with 97·5 degree of confidence. F test can also be applied].

16. A set of 15 observations gives mean $= 68{\cdot}57$, standard deviation $= 2{\cdot}40$, another set of 7 observations gives mean $= 65{\cdot}14$, standard deviation $= 27{\cdot}70$. Use the t test to find whether the two sets of data were drawn from populations with the same mean it being assumed that the standard deviations in the two populations were (1) equal (2) unequal. You may base your answer on the following extracts from t table. Significance limits of t probability of a value numerically greater that t.

Degrees of freedom	*Percentage*		*Probability*
	0·5	·02	·01
6	2·447	3·143	3·707
14	2·145	2·624	2·977
20	2·086	2·528	2·845

[**Hint.** : $t = \dfrac{68{\cdot}57 - 65{\cdot}14}{\sqrt{\dfrac{(2{\cdot}40)^2}{15} + \dfrac{(2{\cdot}70))^2}{7}}} = 2{\cdot}336$]

17. Two random samples drawn from two normal populations are :

Sample I : 20, 16, 26, 27, 23, 22, 18, 24, 25, and 19

Sample II : 27, 33, 42, 42, 35, 32, 34, 38, 28, 41, 43, 30 and 37.

Obtain the estimates of the variances of the population and test whether the two populations have the same variance.

[**Hint.** $\bar{x}_1 = 32$, $\bar{x}_2 = 35$; $\sum_{i=1}^{10} (x_{1i} - \bar{x}_1)^2 = 1{\cdot}30$, $\sum_{j=1}^{12} (x_{8j} - \bar{x}_2)^2 = 3{\cdot}14$

The estimates for the population variances furnished by the samples are 120/9 and 314/11 corresponding to 9 and 11 *d.f.* respectively. The second estimate is larger, so that $F = \dfrac{314/11}{120/9} = 2{\cdot}14$.

This value of F is much below the 5% point as indicated by the F table . It is, therefore, not at all significant, so that the samples may very well be regarded as drawn from the same population.]

18. Show how you would use Student's t-test and Fisher's z-test to decide whether the two sets of observations

	17,	27,	18,	25,	27,	29,	27,	23,	17
and	16,	16,	20,	16,	20,	17,	15,	21	

indicate samples drawn from same universe.

$$\left[\textbf{Hint. } t = 3{\cdot}05, \quad z = \tfrac{1}{2} \log_e \frac{S_1^2}{S_2^2} = \tfrac{1}{2} \log_e \frac{23}{5{\cdot}4107} = 0{\cdot}724. \right]$$

19. Test whether correlation is significant if $r = {\cdot}6$ and $n = 38$.

[**Ans.** Correlation is significant]

20. In a random sample of 28 pairs of values from a bivariate normal population the correlation was found to be 0·7. Is this value consistent with the assumption that the correlation in the population is 0·5 ? Find the 95% fiducial limits for ρ, giving that $r = 0{\cdot}7, z = 0{\cdot}87$ and for $z = 0{\cdot}5, \xi = 0{\cdot}55$.

[**Ans.** $t = \dfrac{z - \xi}{\sqrt{\dfrac{1}{n-3}}} = \dfrac{0{\cdot}87 - 0{\cdot}55}{\sqrt{\dfrac{1}{25}}} = 1{\cdot}6 < 2.$

Hence value is not significant so that the correlation in the population might very well be 0·5, 95% fiducial limits are 0·45 and 0·85 approx.]

21. What is the probability that a correlation coefficient of 0·75 or less, can arise in a sample of 30 from a normal population in which the true correlation is +0·9 ?

[**Hint.** $t = \dfrac{z - \xi}{\sqrt{\dfrac{1}{n-3}}} =$ etc. $= 2{\cdot}6$] **[Ans.** ·048]

22. The first of two samples consists of 23 pairs, and gives a correlation of 0·5 while the second of 28 pairs has a correlation of 0·8. Are these values significantly different ?

[**Hint.** Hypothesis is that they are drawn from the normal population

$$z_1 = 1{\cdot}1513 \log_{10}\left(\frac{1+{\cdot}5}{1-{\cdot}5}\right) = 0{\cdot}55,\ z_2 = 1{\cdot}1513 \log_{10}\left(\frac{1+{\cdot}8}{1-{\cdot}8}\right) = 1{\cdot}10$$

$$n_1 = 23, n_2 = 28, t = \frac{z_1 - z_2}{\sqrt{\dfrac{1}{n_1-3} + \dfrac{1}{n_2-3}}} = \frac{0{\cdot}55}{\sqrt{\dfrac{1}{20} + \dfrac{1}{25}}} = \frac{0{\cdot}55}{0{\cdot}3} = 1{\cdot}83 < 2$$

and is not quite significant at 5% level. The hypothesis is not discredited.]

23. Four breeds of cattle B_1, B_2, B_3, B_4 where fed on three different rations R_1, R_2, R_3. Gain in weight in pounds over a given period were recorded.

	B_1	B_2	B_3	B_4
R_1	46·5	62	41	45
R_2	47·5	41·5	22	31·5
R_3	50	40	25·5	28·5

Is there a significant difference : (*a*) between breeds ; (*b*) between rations ?

[**Ans.** No. Significant difference between breeds or between rations]

24. 12 pictures submitted in a competition were ranked by two judges with results as shown in the table given below :

Table

Picture	A	B	C	D	E	F	G	H	I	J	K	L
Rank assigned by Ist judge	5	9	6	7	1	3	4	12	2	11	10	8
Rank assigned by 2nd judge	8	8	9	11	3	1	2	10	4	12	7	6

Calculate ρ, the rank correlation coefficient. Is there a lack of independence in these ranking ? Assume that on the hyothesis of independence of two sets on n rankings $t = \rho\left(\dfrac{n-2}{1-\rho^2}\right)^{1/2}$ follows the t—distribution with $n-2$ degrees of freedom.

Table of t distribution

Degrees of freedom	10	11	12
Values of t significant at 5% level of probability	2·23	2·20	2·18

[Ans. $\rho = \cdot 79, t = 4{\cdot}089$ approx. Yes, ranking is significant]

25. (*a*) Find out the approximate probability that a correlation coefficient of 0·75 or less can arise in a sample of 18 from a normal population in which the true coefficient is +0·9.

(*b*) Test on the 5% level the signifiance of the following :

(*i*) A multiple correlation coefficient of 0·65 of y as 4 variates, x_1, x_2, x_3 and x_4 based on a sample of size 25.

(*ii*) A partial correlation $-0{\cdot}45$ between x_1 and x_2 (eliminating effect of x_3 and x_4).

(*iii*) The correlation coefficient between mathematical aptitude and linguistic aptitude for a group of 42 boys is 0·72 and for a group of 275 girls is 0·75. Is the difference significant ?

(*c*) In a sample of 25 sets of values from a normal population $r_{1.234}$ was found out to be 0·4. Find whether there is significance of correlation in the population between x_1 and the variables x_2, x_3, x_4.

(M.A., Statistics, Delhi, 1959)

26. In a feeding experiment on swins, three rations R_1, R_2, R_3 were tried. The animals put into three classes of three each according to littre in lbs. in a certain period.

	Class I	*Class II*	*Class III*	Total
R_1	4	16	10	30
R_2	14	18	19	51
R_3	3	14	7	24
Total	21	48	36	105

Analyse the data and state your conclusion. Has division into classes proved effective ? (5 per cent value of F for $n_1 = 2, n_2 = 4$ is 6·94 and for $n_1 = 2, n_2 = 6$ is 5·14)

[Sol. Total sum of squares $= (221 + 776 + 510) - \dfrac{(105)^2}{9} = 282,$

$$\frac{T^2}{N} = \frac{(105)^2}{9} = 1225$$

$$\text{Between class means} = \frac{(21)^2 + (48)^2 + (36)^2}{3} - 1225 = 122$$

$$\text{Between ration means} = \frac{(30)^2 + (51)^2 + (24)^2}{3} - 1225 = 134$$

$$\text{Within sum of squares} = 282 - (122 + 134) = 26.$$

Analysis of Variance Table

Sources of variance	*S.S.*	*D.F.*	*M.S.*	*Variance ratio*
Between classes	122	2	61·0	$F_1 = \frac{61}{6 \cdot 5} = 9 \cdot 38$
Between rations	134	2	67·0	$F_2 = \frac{67}{6 \cdot 5} = 10 \cdot 30$
Errors	26	4	6·5	
Total	282	8		

If we do not consider variation between rations, then the table is

S.V.	S.S.	D.F.	M.S.	V.R.
Between classes	122	2	61·0	$F = 2 \cdot 5$
Error	160	6	26·7	
Total	382	8		

F_1, F_2 are significant at 5% level ; F is not significant at 5% level.

28. The varieties A, B, C of a crop are tested in a randomized block design with four replications, the layout being given in the diagram appended. The plot yields in pounds are also indicated therein. Analyse the experimental yield and state your conclusions.

A	C	A	B
6	5	8	9
C	A	B	C
8	4	6	9
B	B	C	A
7	6	10	6

(5 per cent value of F for $n_1 = 2, n_2 = 9$ is 1·236).

[**Ans.** The difference in varieties of the given crop do not have any significant effect on the yield].

29. (*a*) Explain the principles : randomization. replication and local control used in field experiments.

Plots	Varieties			
	a	*b*	*c*	*d*
1	200	230	250	300
2	190	270	300	270
3	240	150	145	180

Ans. TABLE of A.V.

Source of variation	*S.S.*	*D.F.*	*M.S.*
Within varieties	29184	8	3648
Between varieties	2829	2	1414·5
Total	32013	10	

The variance ratio is not significant at 5% level.

30. A certain company had four salesmen *A, B, C* and *D* each of whom was sent for a week into three types of area ; Country area *K*, outskirts of a city *O* and shopping centre of a city *S*. The sales in pound per week are shown below :—

District	Salesman			
	A	B	C	D
K	30	70	30	30
O	80	50	40	70
S	100	60	80	80

Carry out an analysis of variance and interpret the results stating the limitation under which your results are valid.

31. To test the significance of the variation of the retail price of a certain commodity among four large cities *A, B, C,* and *D,* seven shops were chosen at random in each city, and the prices observed in pence were as follows :

A :	82,	79,	73,	69,	69,	63,	61.
B :	84,	82,	80,	79,	76,	68,	62.
C :	88,	84,	80,	68,	68,	62,	66.
D :	79,	77,	76,	74,	72,	68,	64.

Do the data indicate that the prices for the four cities are significantly different ?

Ans.

Source of variation	*D.F.*	*S.S.*	*M.S.*	*F*
Between cities	3	94·96	31·65	10
Within cities	24	1446·00	60·25	—
Total	27	1540·96	—	—

The value of *F* is not significant and the data is consistent with the hypothesis that the prices are the same in the four cities.

32. The following table gives the weights of 12 individuals drawn from four different sources situated in 3 different districts. Carry out an analysis of variance.

Sources	Districts			Shift the origin to 64			Total
	A	*B*	*C*				
S_1	61	62	63	–3	–2	–1	–6
S_2	63	64	63	–1	0	–1	–2

S_3	66	67	68	2	3	4	9
S_4	68	69	69	4	5	5	14
[**Hint.**				2	6	7	15

Total S.S. $= [\, 3^2 + 2^2 + \cdots + 5^2 - \frac{[13]^2}{12} = 111 - 18\cdot 75 = 92\cdot 25$

Between S.S. (Rows) $= \frac{6^2 + 2^2 + 9^2 + 14^2}{3} - 18\cdot 75 = 86\cdot 92$

Between S.S. (columns) $= \frac{2^2 + 6^2 + 7^2}{3} - 18\cdot 75 = 3\cdot 5$

Error (within S.S.) = Total S.S. - B.S.S. Rows - B.S.S. Column = 1·83

Source of variation	D.F.	S.S.	M.S.	V.R.
B.S.S. Rows	3	86·92	28·97	$\frac{28\cdot 97}{3\cdot 05} > F(3, 6)$ 1%
B.S.S. Column	2	3·5	1·75	
W.S.S.	6	18·3	3·05	$\frac{1\cdot 75}{3\cdot 05} > F(3, 6)$ 5%
Total	11	92·25		

33. Carry out the analysis of the following Latin Square design :

50 A	70 B	70 C	80 D	90 E
70 B	90 C	80 D	80 E	50 A
60 C	50 D	90 E	80 A	90 B
50 D	60 E	80 A	50 B	70 C
80 E	90 A	50 B	70 C	60 D

(B.A. Vikram)

34. In § 15·5 3, prove that if w_i is the weight of z_i and $\Sigma_i w_i z_i / \Sigma_i w_i$ then $E(z) = \xi$. Show that Var z is minimum for variation in w_i if $1/\text{Var}\, z_i = (n_i - 3)$ and the minimum variance of z is $1/\Sigma_i (n_i - 3)$.

16

Tests of Significance Based on Chi-square Distribution

16·1. Introduction. The chi-square distribution has a number of applications in statistics. We shall now consider these.

16·2. χ^2–test as a Test of Goodness of Fit. An important problem of statistical inference, is to test the hypothesis that the given data has been obtained by random sampling from a specified population, with definite values for its parameters. The data etc., usually given, can be arranged in the form of a frequency distribution, wherein, we are given the observed frequencies for the various 'classes' or 'cells'. The corresponding theoretical frequencies are obtained from the knowledge of the population and our problem is to test the compatibility of observed and theoretical frequencies or to determine, whether the deviations of the observed frequencies from the theoretical frequencies, are small enough to be regarded, as due to fluctuations of random sampling or whether they indicated that the data could not have possibly come from a population giving rise to theoretical frequencies.

We have already come across a problem of this type in Chapter 12, when for sampling from a binomial population, we had two classes–success and failure. In n independent trials the observed frequencies were x and $n–x$ respectively and the expected frequencies were np and nq, $q = 1 - p$ respectively. The deviation of the observed frequencies from theoretical frequencies were thus

$$x - np \text{ and } (n - x) - nq = np - x$$

respectively. The statistic used for testing the hypothesis was

$$z = \frac{x - np}{\sqrt{npq}}$$

so that we can write

$$\chi^2 = \frac{(x - np)^2}{npq} = \frac{q\,(x - np)^2 + p\,(n - x - nq)^2}{npq}$$

$$= \frac{(x-np)^2}{np} + \frac{(n-x-nq)^2}{nq} \qquad ...(16\cdot1)$$

For large samples, z was found to be an asymptotic standard normal variate so that z^2 is a χ^2 variate with 1 *d.f.*

If o_1, o_2 represent the observed frequencies in the two classes and e_1, e_2 the expected frequencies in these classes then (16·1) can be written as

$$z^2 = \frac{(o_1-e_1)^2}{e_1} + \frac{(o_2-e_2)^2}{e_2} \qquad ...(16\cdot2)$$

(16·2) suggests a very natural generalization to the case where we have k classes, instead of the two above, with $o_1, o_2, \ldots\ldots, o_k$ as the observed frequencies and $e_1, e_2, \ldots\ldots, e_k$ as the expected frequencies (with $\sum\limits_i o_i = \sum\limits_i e_i$). We form the statistic

$$\chi^2 = \frac{(o_1-e_1)^2}{e_1} + \frac{(o_2-e_2)^2}{e^2} + \ldots\ldots + \frac{(o_k-e_k)^2}{e_k} \qquad ...(16\cdot3)$$

and on the basis of the distribution of z^2, we expect that for large samples χ^2 will be distributed as a χ^2 variate with $(k-1)$ *d.f.* We proceed to establish this result in the next sub-section. Just as the distribution of z^2 was obtained from the binomial distribution, the distribution of χ^2 will be obtained from a consideration of the multinominal distribution.

16·2·1. Theorem. We shall prove that

$$\chi^2 = \sum_{i=1}^{k} \frac{(n_i - np_i)^2}{np_i}$$

has approximately χ^2 – distribution with $(k-1)$ *d.f.* where n_1 and np_i are the observed and expected frequencies of the ith class and $\Sigma n_i = n$. We consider a large sample of n members divided into k classes or cells and suppose that the members are distributed at random in these classes. If the probabilities for different cells are $p_1, p_2, \ldots\ldots, p_k$ then the probability P of there being o_1 members in the first cell, o_2 in the second,....and o_k in the kth cell is the term $p_1^{o_1} p_2^{o_2} \ldots p_k^{o_k}$ in the multinomial expansion of $(p_1 + p_2 + \cdots\cdots + p_k)^n$ so that

$$P = \frac{n!}{o_1!\, o_2! \ldots o_k!} p_1^{o_1} p_2^{o_2} \ldots\ldots p_k^{o_k}$$

where $\sum_{i=1}^{k} o_i = n, \ \sum_{i=1}^{k} p_i = 1.$

If the sample is sufficiently large, so that $o_1, o_2, \ldots\ldots, o_k$ are not small, then using Stirling's approximation for factorials, we have

$$P \sim p_1^{o_1} \, p_2^{o_2} \ldots\ldots p_k^{o_k} \, . \, \frac{\sqrt{2\pi} e^{-n} \, n^{n+1/2}}{\prod_{i=1}^{k} \left[\sqrt{2\pi} \, e^{-o_i} (o_i)^{o_i + \frac{1}{2}} \right]}$$

i.e. $$P \sim C \left(\frac{np_1}{o_1} \right)^{o_1 + \frac{1}{2}} \left(\frac{np_2}{o_2} \right)^{o_2 + \frac{1}{2}} \ldots\ldots\ldots \left(\frac{np_k}{o_k} \right)^{o_k + \frac{1}{2}} \qquad \ldots(16{\cdot}4)$$

which C is a constant, independent of $o_1, o_2, \ldots\ldots\ldots\ldots, o_k$. Now the expected frequency for the ith class is $e_i = np_i$ *i.e.* $E(o_i) = np_i$.

Let $\xi_i = \dfrac{o_i - e_i}{\sqrt{e_i}} = \dfrac{o_i - np_i}{\sqrt{np_i}} \Rightarrow o_i = e_i + \xi_i \sqrt{e_i}$

so that $\chi^2 = \sum_{i=1}^{k} \dfrac{(o_i - e_i)^2}{e_i} = \sum_{i=1}^{k} \xi_i^2 .$

Taking logarithms of both sides of (16·4), we have

$$\log_e \left(\frac{P}{C} \right) \sim \sum_i \left(o_i + \tfrac{1}{2}\right) \log_e \left(\frac{e_i}{o_i} \right) \sim -\sum_i \left[\left(e_i + \tfrac{1}{2} + \xi_i \sqrt{e_i}\,\right) \log_e \left(1 + \frac{\xi_i}{\sqrt{e_i}} \right) \right].$$

If e_i is large, ξ_i will be small compared to $\sqrt{e_i}$ and to first order we have on expanding the logarithms on the right

$$\log_e \left(\frac{P}{C} \right) \sim -\sum_i \left(e_i + \tfrac{1}{2} + \xi_i \sqrt{e_i}\,\right) \left(\frac{\xi_i}{\sqrt{e_i}} - \frac{1}{2} \frac{\xi_i^2}{e_i} + \ldots\ldots \right)$$

$$= -\tfrac{1}{2} \sum_{i=1}^{k} \xi_i^2 - \sum_{i=1}^{k} \xi \sqrt{e_i} - o \left(\frac{1}{\sqrt{e_i}} \right), \qquad \ldots(16{\cdot}5)$$

where $o \left(\dfrac{1}{\sqrt{e_i}} \right)$ denotes quantities of the order of $1/\sqrt{e_i}$, or smaller.

Now $\sum_{i=1}^{k} \xi_i \sqrt{e_i} = \sum_{i=1}^{k} o_i - \sum_{i=1}^{k} e_i = n - n = 0,$

Hence to order $1/\sqrt{e_i}$ we have

$$P \sim C \exp \left\{ -\tfrac{1}{2} \sum_{i=1}^{k} \xi_i^2 \right\} \qquad \ldots(16{\cdot}6)$$

which shows that the probability varies as that of the sum of squares of k normal variates of unit variance which are subject to the linear constraint $\Sigma\xi_i \sqrt{e_i} = 0$ but otherwise independent.

It thus appears that χ^2 as defined in (16·3) conforms to the χ^2 distribution with $(k-1)$ *d.f.* and if we write

$$\chi^2 = \sum_{i=1}^{k} \frac{(o_i - e_i)^2}{e_i} \qquad \text{...(16·7)}$$

the variate follows the chi-square distribution

$$dP = \frac{1}{2^{\frac{1}{2}(k-1)} \Gamma\left(\frac{k-1}{2}\right)} e^{-\frac{1}{2}\chi^2} (\chi^2)^{\frac{1}{2}(k-1)} d(\chi^2) \qquad \text{...(16·8)}$$

16·2·2. Remarks on the above distribution.

1. In chapter 13, we proved that the sum of squares of n standard normal variates is distributed as a χ^2 variate with $n-p$ *d.f.* where p denotes the number of independent linear constraints on the variates. In the above case χ^2 is the sum of the squares of k variates with one linear constraint imposed by the fact that we require the total of the observed frequencies to be the same as the total of the expected frequencies. In our further discussion, we shall find it convenient to impose further linear constraints and the number of degrees of freedom for ξ^2 as defined in (16·7) would be correspondingly reduced. Thus if m of the parameters of population are not specified, but have to be estimated from the data itself so that the observed and expected frequencies have not only to agree in the totals but also in m other statistics like mean, variance, etc., the number of degrees of freedom will be $k-1-m$, provided the constraints imposed by the process are linear.

2. In order that χ^2 test for *compatibility of observed and theoretical frequencies of 'goodness of fit'* may be applicable, it is clear from the proof of the above distribution and further discussion of the same by Fisher that following **conditions for the validity of χ^2-test** must be satisfied :

(*i*) *The members of the sample should be independent.*

(*ii*) *Constraints on the cell frequencies, if any, should be linear.*

(*iii*) *N should be reasonably large, say, greater than 50.*

(*iv*) *The theoretical cell frequencies should not be less than 5.*

In alternative case, re-grouping of this frequency with the preceding frequency (*frequencies*) *should be made to make the total frequency greater than 5.*

(*v*) *In case parameter values have to be estimated from the data the*

> *method of estimation should be such as to minimize χ^2. This is equivalent in large sample to choosing a maximum likelihood estimator.**

Both experience and theoretical considerations show that the χ^2 approximation will be satisfactory provided all $e_i \geq 5$ and $k \geq 5$. If k is less than 5 it is best to have e_i somewhat larger than 5. If some of the cell frequencies do not exceed 5, they may be combined with other preceding cell frequencies so that the condition is satisfied.

3. If $\chi^2 = 0$, all the observed and the expected frequencies coincide. On the other hand, if observed and expected frequencies differ greatly, χ^2 will be large. Thus χ^2 affords a measure of the correspondence between theory and observations. We may note, however, that in the expression for χ^2 all the deviations of observed from theoretical frequencies are squared so that their signs are ignored. If the signs show some marked peculiarities it is desirable to supplement the χ^2 test with other tests that may be available.

4. The χ^2 table gives for any number of degrees of freedom $\chi^2_{0.05}$ and $\chi^2_{0.01}$ which will be exceeded in random sampling with probabilities 0·05 and 0·01 respectively. If the observed value of χ^2 exceeds $\chi^2_{0.05}$, we say that the difference between observed and theoretical frequencies is significant at 5% level.

Alternatively, we may determine the probability P of getting a value of χ^2 equal to or larger than the actual observed value. If this is small, we are justified in suspecting a significant divergence between theory and observation. If the probability is not small it does not prove the hypothesis, that is, it shows that the data are consistent with the hypothesis.

5. Not only small values of P lead us to suspect the hypothesis but a value of P very near to unity may also lead to a similar result, for a large value of P will imply in general a small value of χ^2 showing too good an agreement between theory and observation to be true. Thus if $\chi^2 = 0$, $P = 1$ but getting such a remarkable agreement can only be a rare event.

6. Since the density function of χ^2 depends only on ν, the number of degrees of freedom and not on any population parameter, the χ^2 test is considered sometimes as a non-parametric test, though it is often applied

* See § 17·4·3

to parametric problems. The χ^2 distribution is concerned with the values of e_i but not with the form of the parent distributions. This property of χ^2 distribution explains why χ^2 distribution is found useful in such a large variety of cases.

7. An **alternative expression for** χ^2 as defined in (16·7) is

$$\chi^2 = \sum_{i=1}^{k} \frac{o_i^2}{e_i} - 2\sum_{i=1}^{k} o_i + \sum_{i=1}^{k} e_i \qquad \ldots(16\cdot 9)$$

$$= \sum_{i=1}^{k} \left[\frac{o_i^2}{e_i}\right] - N$$

where $\sum_{i=1}^{k} o_i = \sum_{i=1}^{k} e_i = N.$

Note. If the expected frequencies are not integers, (16·7) may be easier to use than (16·9).

Example 16·1. *12 dice were thrown 4096 times and a throw of 6 was reckoned as a success, the observed frequencies were as given below :*

Number of successes	0	1	2	3	4	5	6	7 & over	total
Frequencies	447	1145	1181	796	380	115	24	8	4096

Find the value of χ^2 *on the hypothesis that the dice were unbiased and hence show that the data are consistent with the hypothesis so far as the* χ^2 *test is concerned.*

Here $q = \frac{5}{6}, p = \frac{1}{6}$. On the hypothesis of unbiased dice the theoretical frequencies in 4096 throws are the terms in the binomial expansion of $4096\left(\frac{5}{6}+\frac{1}{6}\right)^{12}$ and are given below.

Number of successes	0	1	2	3	4	5	6	7 & over	total
Theoretical	459	1102	1212	808	364	116	27	8	4096

$$\therefore \chi^2 = \frac{(447-459)^2}{459} + \frac{(1145-1102)^2}{1102} + \cdots + \frac{(24-27)^2}{27} + \frac{(8-8)^2}{8} = 5\cdot 811$$

The number of classes is 8 but the total of the observed and theoretical frequencies agree. Therefore, ν = number of degrees of freedom = 8–1 = 7. From χ^2 table, $\chi^2_{0\cdot 05}$ for 7 *d.f.* = 14·07. The calculated value of χ^2 is, therefore, not significant and thus the observed frequency distribution is consistent with the hypothesis.

Example 16·2. *May the data in Example 6, page 611 be regarded as those of a random sampling from the Poisson distribution ?*

The parameter m of the Poisson distribution is not specified by the hypothesis and since the mean of Poisson distribution is m, an estimate of the population parameter is given by the sample mean *viz.*, by

$$\frac{1}{1096}[0\times364+1\times376+2\times218+3\times89+4\times33+5\times13+6\times2+7\times1]$$

$$=\frac{1}{1096}[1295]=1{\cdot}18.$$

The theoretical frequencies, with the value of the parameter, are given by

$$1096\times\frac{e^{-1\cdot18}(1{\cdot}18)^r}{r\,!},\ r=0,1,2,......,7\ ;\ e^{-1\cdot18}=\cdot3075$$

No. of deaths per day	0	1	2	3	4	5	6	7	Total
Observed frequency	364	376	218	89	33	13	2	1	1096
Expected frequency	336·5	397·1	234·6	92·5	27·3	6·5	1·3	0·2	1096

Since the expected frequencies in the last two classes are less than five, we group together the last class with two preceding classes to make totals greater than 5, then

The adjusted distribution is:

No. of deaths per day	0	1	2	3	4	5 & over	Total
Observed frequency	364	376	218	89	33	16	1096
Expected frequency	336·5	397·1	234·6	92·5	27·3	8	1096.

$$\chi^2=\frac{(364-336{\cdot}5)^2}{336{\cdot}5}+\frac{(376-397{\cdot}1)^2}{397{\cdot}1}+.....+\frac{(16-8)^2}{8}=9{\cdot}99.$$

Here the number of the classes is 6 (since the last three classes have been grouped) but observed and the expected frequencies agree both in their totals as well as their means so that

$\nu=$ number of degrees of freedom $=6-1-1=4$.

From the table, the value of χ^2 for 4 *d.f.* at 5% level is 9·49. The calculated value is, therefore, significant at 5% level, but is barely significant there.

Ex. 1. A die was thrown 60 times and the following frequency distribution was observed :

Face	1	2	3	4	5	6
Observed frequency	15	6	4	7	11	17

Can you regard the die as honest? ($\chi^2=13{\cdot}6$, $\nu=5$. significant at 5% level.)

16·3. χ^2 – Test as a Test of Independence. A special and important class of classes where compatibility of observed and expected frequencies has to be tested is testing the independence of two attributes when for convenience the data is arranged in two-way tables usually called *contingency tables.*

16·3·1. Contingency table. Let the data be classified into s classes $A_1, A_2, \ldots, A_s$ according to attribute A and into t classes B_1, B_2, ..., B_t according to attribute B. Let O_{ij} denote the observed frequency of the cell belonging to both the classes A_i and B_j [$i = 1, 2, \ldots, s$; $j = 1, 2, \ldots, t$.] Such a classification where two attribution are divided into more than two classes each is called by the name manifold classification.

Let (A_i) and (B_j) denote the totals of all the frequencies belonging to classes A_i and B_j respectively, then the data can be set into a $s \times t$ contingency table of s rows and t columns as follows :

A↓ Classes B→	B_1	B_2............B_j.............B_{t-1}	B_t	Total
A_1	O_{11}	O_{12}...........O_{1j}............O_{1t-1}	O_{1t}	(A_1)
A_2	O_{21}	O_{22}...........O_{2j}............O_{2t-1}	O_{2t}	(A_2)
$\vdots$		....		$\vdots$
A_i	O_{i-11}	O_{i-12}.........O_{ij}..........O_{it-1}	$O_{i.t}$	(A_i)
$\vdots$		....		$\vdots$
A_{s-1}	O_{s-11}	$O_{s-1,2}$.......$O_{s-i,j}$.......$O_{s-1,t-1}$	O_{s-1t}	(A_{s-1})
A_s	O_{s1}	O_{s2}...........O_{sj}...........$O_{s\ t-1}$	$O_{s,t}$	(A_s)
Total	(B_1)	(B_2).........(B_j)...........(B_{t-1})	(B_t)	N

16·3·2. Calculation of χ^2 to test the hypothesis of independence. To test whether there is any relationship between the attributes A and B, we set up the null hypothesis of *no relationship or of independence* between A and B, and calculate the expected frequencies in all the cells on this hypothesis. Since we wish to test independence only and not any hypothesis as to the total number belonging to different classes the expected frequencies are calculated by keeping the class totals fixed. This is equivalent to imposing linear constraints so that the class totals of the observed and expected frequencies must agree.

Let e_{ij} denote the expected frequency in the cell common to A_i and B_j. Then since A and B are independent, in any class the proportions of members belonging to $A_1, A_2, \ldots, A_i, \ldots, A_s$ should be same and equal to the proportions in the total. Thus

$$\frac{e_{1j}}{(A_1)}=\frac{e_{2j}}{(A_2)}=\cdots=\frac{e_{ij}}{(A_i)}=\cdots=\frac{e_{sj}}{(A_s)}=\frac{\Sigma e_{ij}}{\Sigma(A_i)}=\frac{(B_j)}{N}$$

so that
$$e_{ij}=\frac{(A_i)\times(B_j)}{N} \qquad \text{...(16·10)}$$

Thus on the basis of independence the expected frequency in any cell is equal to the product of the class totals of the two classes to which the cell belongs divided by the total number of observations.

Hence
$$\chi^2=\sum_{j=1}^{t}\sum_{i=1}^{s}\left(\frac{(O_{ij}-e_{ij})^2}{e_{ij}}\right)=\sum_{j=1}^{t}\sum_{i=1}^{s}\left[\frac{\left(O_{ij}-\frac{(A_i)\times(B_j)}{N}\right)^2}{\frac{(A_i)\times(B_j)}{N}}\right] \qquad \text{...(16·11)}$$

16·3·3. Number of degrees of freedom associated with a $s \times t$ contingency table. There are in all $s \times t$ $(= st)$ cells but since the marginal totals are fixed there are $(s+t)$ constraints. These constraints are, however, not independent since the sum of the border column frequencies must be equal to that of the border row frequencies and thus there are only $(s+t-1)$ independent linear constraints. Hence v, the number of degrees of freedom, associated with a $s \times t$ contingency table, is

$$v = st-(s+t-1)=(s-1)(t-1).$$

This is the number on the cells whose frequencies can be arbitrarily assigned. This is obvious from the above table where the frequencies of $(s-1)(t-1)$ cells enclosed by the dotted line can be fixed arbitrarily and when the frequencies of the remaining $(s+t-1)$ cells will be determined from the class totals.

Example 16·3. *In a locality 100 persons were randomly selected and asked about their educational achievements. The results are given as follows :*

		Education			
		Middle	*High School*	*College*	*Totals*
Sex	*Male*	10	15	25	50
	Female	25	10	15	50
	Total	35	25	40	100

Can you say that education depends on sex ?

We set up the null hypothesis that there is no association between education and sex. On this hypothesis the expected frequencies are tabulated as follows, using (16·10) :

	Middle	*High School*	*College*	*Total*
Male	17·5	12·5	20	50
Female	17·5	12·5	20	50
Total	35·0	25·0	40	100

$$\chi^2 = \frac{(10-17\cdot5)^2}{17\cdot5} + \frac{(15-12\cdot5)^2}{12\cdot5} + \frac{(25-20)^2}{20} + \frac{(25-17\cdot5)^2}{17\cdot5}$$

$$+ \frac{(10-12\cdot5)^2}{12\cdot5} + \frac{(15-20)^2}{20} = 9\cdot93 \text{ approximately.}$$

v = the number of degrees of freedom $= (s-1)(t-1) = (3-1)(2-1) = 2.$

From the table for 2 *d.f.*, $\chi^2_{0\cdot05} = 5\cdot99$ and $\chi^2_{0\cdot01} = 9\cdot21$.

Thus the calculated value of χ^2 is significant and the hypothesis that the education does not depend on sex is discredited.

Example 16·4. *Show that in a* 2×2 *contingency table, wherein the frequencies are* $\frac{a}{c} \Big| \frac{b}{d}$, χ^2 *calculated from independent frequencies is*

$$\chi^2 = \frac{(a+b+c+d)(ad-bc)^2}{(a+b)(c+d)(b+d)(a+c)} \quad ...(16\cdot12)$$

Classes	A	$\bar{A}$	Totals
B	a	b	$a+b$
$\bar{B}$	c	d	$c+d$
Totals	$a+c$	$b+d$	$a+b+c+d=N$

Since the marginal totals are fixed the probability for a member to belong to class A is $(a+c)/N$ and is constant.

Further the attributes being independent, the probability for it to belong to both classes A and B is $\left(\frac{a+b}{N}\right)\left(\frac{a+c}{N}\right)$, and therefore, the expected frequency of this class, denoted by $E(a)$, is given by

$$E(a) = N \times \frac{a+b}{N} \times \frac{a+c}{N} = \frac{(a+b)(a+c)}{N}$$

Similarly,

$$E(b) = \frac{(a+b)(b+d)}{N}, \quad E(c) = \frac{(c+d)(a+c)}{N}$$

$$E(d) = \frac{(c+d)(b+d)}{N}.$$

By definition,

$$\chi^2 = \frac{[a-E(a)]^2}{E(a)} + \frac{[b-E(b)]^2}{E(b)} + \frac{[c-E(c)]^2}{E(c)} + \frac{[d-E(d)]^2}{E(d)}$$

$$\chi^2 = \frac{\left[a - \dfrac{(a+b)(a+c)}{a+b+c+d}\right]^2}{\dfrac{(a+b)(a+c)}{a+b+c+d}} + \text{ three similar terms.}$$

$$= \frac{(ad-bc)^2}{(a+b+c+d)}\left[\left\{\frac{1}{(a+b)(a+c)} + \frac{1}{(a+b)(b+d)}\right\} + \left\{\frac{1}{(c+d)(a+c)} + \frac{1}{(c+d)(b+d)}\right\}\right]$$

$$= \frac{(ad-bc)^2}{(a+b+c+d)}\left[\frac{a+b+c+d}{(a+b)(a+c)(b+d)} + \frac{a+b+c+d}{(d+c)(b+d)(c+d)}\right]$$

$$= (ad-bc)^2\left[\frac{c+d+a+b}{(a+b)(a+c)(b+d)(d+c)}\right]$$

$$= \frac{(ad-bc)^2(a+b+c+d)}{(a+b)(a+c)(b+d)(c+d)}.$$

Ex. 1. Show that in § 16·3·3 the linear constraints on expected frequencies are

$$\sum_{j=1}^{t} e_{ij} = (A_j),\quad \sum_{i=1}^{s} e_{ij} = (B_j),$$

$$\sum_{i=1}^{s}\sum_{j=1}^{t} e_{ij} = \sum_{i=1}^{s}(A_i) = \sum_{j=1}^{t}\sum_{i=1}^{s} e_{ij} = \sum_{j=1}^{t}(B_j) = N.$$

Show that only $s+t-2$ constraints are independent.

Ex. 2. Prove that the marginal totals are held fixed,

$$P\ (a \text{ member belonging to class } A_i) = \frac{(A_i)}{N}.$$

$$P\ (a \text{ member belonging to class } B_j) = \frac{(B_j)}{N}.$$

Hence on the basis of independence of A and B, prove that

$$e_{ij} = \frac{(A_i)\times(B_j)}{N}.$$

Ex. 3. Deduce the value of χ^2 for a 2×2 table from that of a $2\times n$ table.

Ex. 4. Prove that for a $s\times t$ table,

(*i*) χ^2 is less than or equal to $N(s-1)$ or $N(t-1)$ whichever is less.

(*ii*) $E(\chi^2) = N(r-1)(s-1)/(N-1)$

(*iii*) $\max(C) = \left[\left(\dfrac{s-1}{s}\right)\right]^{\frac{1}{2}}$ $r=s$ where C is the coefficient of contingency.

[Hint. Using the facts that $e_{ij} = \frac{(A_i) \times (B_j)}{N}$, $O_{ij} \le A_i$, $O_{ij} \le B_j$

we have $\chi^2 = \Sigma\Sigma \frac{(O_{ij} - e_{ij})^2}{e_{ij}} = \Sigma\Sigma \frac{O_{ij}^2}{e_{ij}} - N = N\Sigma\Sigma\left[\frac{O_{ij}}{(A_i)} \times \frac{O_{ij}}{(B_j)}\right] - N$

$\le N\left\{\sum_{i=1}^{s}\sum_{j=1}^{t}\left(\frac{O_{ij}}{(A_i)}\right) - 1\right\} = N\left\{\sum_{i=1}^{s}\left(\frac{A_i}{A_i}\right) - 1\right\}$

$\therefore \chi^2 \le N(s-1)$. Similarly, $\chi^2 \le N(t-1)$]

16·4. Further consideration of 2 × 2 Contingency Table.

a	b
c	d

16·4·1. *Yates Correction for Continuity. The distribution of χ^2 as obtained in chapter 13 is a continuous distribution while the distribution of frequencies is by its very nature discontinuous. The continuous χ^2 distribution may be regarded as the limit to which the true discontinuous distribution tends as the sample size increases. One of the reasons why we postulated earlier that the expected cell frequencies should not be less than 5 was to avoid irregularities due to this discontinuity.

In a 2×2 table there is only one *d.f.* and the frequency of only one cell can be assigned arbitrarily and if the frequency of a cell is small, say 3, the sum of the probabilities for 3 or more extreme (*i.e.*, 2, 1, 0) observations in the discontinuous case can be approximated by the tail area of the continuous χ^2 curve corresponding to $3\frac{1}{2}$ frequency. F. Yates, in 1934, therefore, suggested the following adjustment to the 2×2 contingency table.

Calculate ad and add $\frac{1}{2}$ to both factors of the smaller and subtract $\frac{1}{2}$ from both factors of the larger. This does not change the marginal totals. The correction is known as *Yates correction for continuity.* If N $(= a + b + c + d)$ is large, the use of Yates correction will make very little difference in the value of χ^2. [see Ex. 16·5, page 602]

Example 16·5. *Two batches, each of 13 experimental animals, are inoculated and the other not inoculated, were exposed to the infection of a disease. The following frequencies of dead and surviving animals were noted in the two cases. Can the inoculation be regarded as effective against the disease ?*

	Dead	*Survived*	*Total*
Inoculated	2	10	12
Not Inoculated	8	4	12
Total	10	14	24

After applying Yates correction, the adjusted table is :

	Dead	*Survived*	*Total*
Inoculated	2·5	9·5	12
Not Inoculated	7·5	4·5	12
Total	10	14	24

Applying formula of Example 16·4, page 600 or otherwise

$$\chi^2 = \frac{\left(\frac{3}{2} \times \frac{9}{2} - \frac{19}{2} \times \frac{17}{2}\right)^2 \times 24}{12 \times 12 \times 10 \times 14} = 4{\cdot}286$$

For 1 *d.f.* the value of χ^2 at 5% level is the square of the standard normal variate for 5% *i.e.* of 1·96. Since 4·286 is greater than $(19{\cdot}6)^2$ 3·84, χ^2 is significant at 5% level. Hence at this level, the inoculation may be said to be effective against the disease.

16·4·2. Fisher Exact Method for 2×2 contingency table. The testing of independence by χ^2 test in a 2×2 table is an approximation which is useful for the comparative simplicity of the calculations. The exact method is somewhat more laborious though necessary in case of doubt and valuable as displaying the true nature of the inference which the method of χ^2 is designed to draw.

2×2 table

a	b	$a+b$
c	d	$c+d$
$a+c$	$b+d$	$a+b+c+d=n$

The number of ways in which the body of this table can be completed is

$$\frac{n!}{a!\,b!\,c!\,d}, n = a+b+c+d$$

and assuming independence of the two attributes the total number of ways in which marginal totals can be obtained is

$$\frac{n!}{(a+c)\,(b+d)!} \times \frac{n!}{(a+b)\,(c+d)!}$$

Consequently the probability of the distribution of the table is

$$\frac{(a+c)\,(b+d)!\,(a+b)\,(c+d)!}{a!\,b!\,c!\,d!\,(a+b+c+d)!}$$

and from here we can find the probabilities for $a = 0, 1, 2, \ldots\ldots$

Example 16·6. *For the data of Example 16·5 compare the result of the exact treatment with that obtained from the χ^2 approximation.*

The probability for the number of dead-inoculated to be, or less corresponds to $a = 0, 1$ and

$$= \frac{12!\,12!\,10!\,14!}{24!\,0!\,12!\,10!\,2!} + \frac{12!\,12!\,10!\,14!}{24!\,1!\,11!\,9!\,3!} + \frac{12!\,12!\,10!\,14!}{24!\,21!\,10!\,8!\,4}$$

$$= \frac{12\,!\,14\,!}{24\,!\,2\,!}\left[1+\frac{12\times 10}{1\times 3}+\frac{12\times 11\times 10\times 9}{1\times 2\times 3\times 4}\right] = \cdot 018$$

Without correction $\chi^2 = 6{\cdot}172$ and the corresponding standard normal variate is 2·48 giving $P = \cdot 007$ with correction $\chi^2 = 4{\cdot}286$ corresponding $P = \cdot 02$ nearly. Thus Yates correction gives a great improvement. It may be noted here that since in the exact test we have found the probabilities of a being, or less, to find the probability for a given χ^2 we use either one tail of the normal curve or half the probability obtained from the χ^2 table.

Ex. 1. (*a*) If p is the probability of A and $q\ (= 1 - p)$ that of $\bar{A}$, prove that if the two attributes are independent, the joint probability of a and c belonging to A out of $a + b + c + d$ respectively is

$$\frac{(a+b)\,!\,(c+d)\,!}{a\,!\,b\,!\,c\,!\,d\,!}\,p^{a+c}\,q^{b+d}$$

(*b*) Summing over possible samples with same marginal totals show that

$$\Sigma\,\frac{1}{a\,!\,b\,!\,c\,!\,d\,!} = \frac{n\,!}{(a+b)\,!\,(c+d)\,!\,(a+c)\,!\,(b+d)\,!}$$

(*c*) Deduce the result of Ex. 16·4·2.

Ex. 2. For Hellman's data $\begin{array}{c|c} 4 & 16 \\ 1 & 21 \end{array}$ show that the exact test gives $P = \cdot 1435$, χ^2 uncorrected gives $P = 0{\cdot}0612$ and χ^2 after Yates correction gives $P = 0{\cdot}1427$.

Ex. 3. For Langes data $\begin{array}{c|c} 10 & 8 \\ 2 & 15 \end{array}$, show that [*P*] exact $= \cdot 00046$

[*P*] χ^2 uncorrected $= \cdot 00015$ [*P*] χ^2 corrected $= \cdot 00061$

Ex. 4. Show that for a 2×2 table the value of χ^2 after applying Yates correction for a continuity is

$$\frac{\left(ad - bc - \frac{N}{2}\right)^2 N}{(a+b)\,(a+c)\,(b+d)\,(c+d)} \quad \text{or} \quad \frac{\left(ad - bc + \frac{N}{2}\right)^2 N}{(a+b)\,(a+c)\,(b+d)\,(c+d)}$$

according as $ad - bc \geq 0$ or ≤ 0. **(*M. Sc., Marathwada, 1991*)**

$$\left[\textbf{Hint. } \chi^2 = N\left[\left(a \mp \frac{1}{2}\right)\left(d \mp \frac{1}{2}\right) - \left(b \pm \frac{1}{2}\right)\left(c \pm \frac{1}{2}\right)\right]^2 \Big/ (a+b)\,(a+c)\,(b+d)\,(c+d)\,.\right.$$

$$\text{Numerator} = N\left[(ad - bc) \mp \frac{N}{2}\right]^2 = N\left[\,|\,ad - bc\,| - \frac{N}{2}\right]^2 = \text{etc.}$$

16·5. Other Tests of Significance Based on χ^2 distribution.

16·5·1. Test for variance of a normal population. Given a random sample $x_1, x_2, \ldots\ldots, x_n$ from a normal population, we have to test the hypothesis that the variance of the population is σ^2.

If the hypothesis is correct, the statistic

$$\chi^2 = \sum_{i=1}^{n} \frac{(x_i - \bar{x})^2}{\sigma^2} = \frac{ns^2}{\sigma^2}$$

is distributed as a χ^2 variate with $(n-1)$ *d.f.* If the calculated value of χ^2 is less than $\chi^2_{0.05}$ for $(n-1)$ *d.f.*, the data will be consistent with the hypothesis.

Example 16·7. *Explain how you will find confidence intervals for the variance of a normal population.*

From χ^2 distribution with $n-1$ *d.f.* we can find two numbers a and b corresponding to any given confidence coefficient α, such that

$$P[a < \chi^2 < b] = \int_a^b \frac{1}{2^{\frac{1}{2}(n-1)} \Gamma\left(\frac{n-1}{2}\right)} e^{-\frac{1}{2}\chi^2} (\chi^2)^{\frac{1}{2}(n-3)} d(\chi^2) = \alpha,$$

so that $$P\left[\frac{\Sigma (x_i - \bar{x})^2}{b} < \sigma^2 < \frac{\Sigma (x_i - \bar{x})^2}{a}\right] = \alpha \qquad \left[\because \chi^2 = \frac{\Sigma (x_i - \bar{x})^2}{\sigma^2}\right]$$

giving a confidence interval for σ^2 of length $\Sigma (x_i - \bar{x})^2 \left(\frac{1}{a} - \frac{1}{b}\right)$.

In practice we may choose a and b so that the two tail areas are equal. Thus for a 95 per cent confidence interval, $a = \chi^2_{.975}$, $b = \chi^2_{.025}$.

Ex. A random sample of 15 gives an unbiased estimate of sample variance equal to 1·05 cm². Find a 90 per cent confidence interval for σ^2.

16·5·2. Combination of probabilities from independent tests of significance. Sometimes, we have a number of independent tests of significance, giving different probabilities, some of which may give significant and others non- significant results. Our problem is to combine or pool various probabilities, to get a single probability, from which to decide on the significance of the aggregate of these tests. For 2 *d.f.*, the *p.d.f.* of χ^2 is given by $f(\chi^2) = \frac{1}{2} e^{-\chi^2/2}, 0 < \chi^2 < \infty$.

(*Delhi B.Sc. Stat., Hons., 1990*)

$\therefore$ For 2 *d.f.*, $p_1 = P[\chi^2 \geq \chi_1^2] = \frac{1}{2} \int_{\chi_1^2}^{\infty} e^{-\frac{1}{2}\chi^2} d(\chi^2) = e^{-\frac{1}{2}\chi_1^2}$,

so that $$\chi_1^2 = -2 \log_e p_1 = 2 \log_e (1/p_1).$$

Similarly for the other probabilities $p_2, p_3, \ldots\ldots, p_k$ we have

$$\chi_2^2 = -2 \log_e p_2$$

$$\ldots\ldots\ldots\ldots\ldots\ldots\ldots$$

$$\chi_k^2 = -2 \log_e p_k$$

so that $\chi^2 = \chi_1^2 + \chi_2^2 + \ldots\ldots + \chi_k^2 = -2 \log_e (p_1 p_2 \cdots p_k)$.

But the sum of independent χ^2 variates is again a χ^2 variate.

$$\therefore \quad 2 \log_e \left(\frac{1}{p_1 p_2 \cdots p_k} \right) = 2{\cdot}3026 \times \log_{10} \left(\frac{1}{p_1 p_2 \cdots p_k} \right)^2$$

is a χ^2 variate with $2k$ *d.f.* and if this value exceeds $\chi^2_{0{\cdot}05}$ for $2k$ *d.f.*, the combined result will be said to be significant at 5% level.

Example 16·8. *Three independent tests of significance give the probabilities* ·15, ·26, ·11 ; *test whether the aggregate of these tests is significant.*

We have

P	$-2 \log_e P$	$D.F.$
·15	3·79424	2
·26	2·69414	2
·11	4·41454	2
	10·90292	6

The 5% and 1% values of χ^2 for *d.f* are 12·592 and 10·645 respectively. Therefore, the probability of the combined test is slightly less than 0·10.

Ex. 1. Four tests of significance have yielded probabilities ·145, ·263, ·087 and 0·75. Test whether the aggregate of these four tests should be regarded as significant.

Ex. 2. The variables $x_1, x_2, \ldots\ldots, x_n$ are independently distributed in the rectangular form $df = dx, 0 \le x \le t$.

Then if $P = X_1 X_2 \ldots X_n$, show that $-2 \log_e P$ has χ^2-distribution with $2n$ degrees of freedom.

[**Hint.** $-2 \log P = -2 \log (X_1, X_2 \ldots X_n) = Y_1 + Y_2 + \ldots + Y_n$

where $Y_i = -2 \log X_n \Rightarrow X_n = \exp\left(-\frac{1}{2} Y_n\right)$

The *p.d.f.* of Y_i is given by $g(y_i) = f(x_i) \left| \frac{dx_i}{dy_i} \right| = \frac{1}{2} \exp\left(\frac{-y_i}{2}\right)$

which is the χ^2-distribution with 2 *d.f.*

16·5·3. Testing homogeneity of several independent estimates of population variances. Bartlett Test.

Let k independent samples give unbiased estimates $S_1^2, S_2^2, \ldots\ldots, S_k^2$ of the population variance based on $v_1, v_2, \ldots\ldots, v_k$ *d.f.*, then on the hypothesis that the samples are drawn from the same population —

$$S^2 = \frac{\Sigma v_i S_i^2}{\Sigma v_i} = \frac{\sum_{i=1}^{k} v_i S_i^2}{v}, \; v = \sum_{i=1}^{k} v_i, \; S_i^2 = \frac{1}{v_i - 1} \sum_{j=1}^{n_i} (x_{ij} - \bar{x}_i)^2 \quad \ldots(1)$$

is also an unbiased estimate of the population variance σ^2, since

$$E(S^2) = \frac{1}{v} \sum_{i=1}^{k} v_i [E(S_i^2)] = \frac{1}{v} \sum_{i=1}^{k} v_i \sigma^2 = \sigma^2.$$

Bartlett has shown that on the hypothesis that the various samples come from the same population, the statistic

$$\sum_{i=1}^{n} \left(v_i \log_e \frac{S^2}{S_i^2} \right) \Bigg/ \left[1 + \frac{1}{3(k-1)} \left\{ \left(\sum_{i=1}^{k} \frac{1}{v_i} \right) - \frac{1}{v} \right\} \right] \quad \ldots(2)$$

is distributed approximately as χ^2 with $(k-1)$ *d.f.*

The value of χ^2, calculated from the data, would thus enable us to test the homogeneity of the several estimates.

It is clear, however, that if the largest and the smallest variance of a set do not differ significantly, as tested by the F test, then those lying between these cannot differ significantly either, and the whole set can be reasonably regarded as coming from a single population. In these circumstances, there is no need to employ the Bartlett test.

Ex. Three independent samples give the following results :—

Size	*Observations*
5	34, 40, 47, 60, 84
9	40, 59, 60, 67, 86, 92, 95, 93, 108
3	46, 93, 100

Prove that $S_1^2 = 324, S_2^2 = 512{\cdot}25, S_3^2 = 862{\cdot}833, S^2 = 528{\cdot}475, \chi^2 = 0{\cdot}391$, $v = 2$. Deduce that the apparent difference in variance is not significant.

16·5·4. Testing homogeneity of several independent estimates of an ordinary or partial correlation coefficient.

Let $r_1, r_2, \ldots\ldots\ldots, r_k$ be k independent estimates of the correlation coefficient. We have to test whether these can be regarded as having come from the same bivariate population.

Let $z_i, i = 1, 2, \ldots\ldots, k$ and $\bar{z}$ be defined by

$$z_i = \frac{1}{2}\log_e\left(\frac{1+r_i}{1-r_i}\right); \text{ and } \bar{z} = \frac{\sum_{i=1}^{k}(n_i-3)z_i}{\sum_{i=1}^{k}(n_i-3)} \text{ respectively.}$$

Then, since for large samples $z_1, z_2, \ldots\ldots, z_k$ are all distributed normally about the same mean, the sum $\sum_{i=1}^{k}(n_i-3)(z_i-\bar{z})^2$ is distributed approximately as χ^2 with $(k-1)$ *d.f.*, one *d.f.* is lost as z is determined from the data and the test of independence can be carried out accordingly.

For partial coefficient of order k, n_i will be replaced by $n_i - k$.

Example 16·9. *The correlation coefficient between daily ration of green grass and rate of growing calves on the basis of observations taken on 10, 14, 16, 20, 25 and 28 cows at six farms were found to be 0·318, 0·106, 0·253, 0·340, 0·116 and 0·112. Can these be considered homogeneous ? If so, estimate the common correlation coefficient.*

We set up the hypothesis that the given values are homogeneous or the samples are drawn from equally correlated population.

We have
$$z_i = \frac{1}{2}\log_e\left(\frac{1+r_i}{1-r_i}\right) = \frac{1}{2}\log_e\left(\frac{1+0{\cdot}318}{1-0{\cdot}318}\right)$$
$$= \frac{1}{2}\log_e 10 \times \log_{10}\left(\frac{1{\cdot}318}{{\cdot}682}\right)$$
$$= 1{\cdot}1513\,(\log 1{\cdot}318 - \log {\cdot}682)$$
$$= {\cdot}3294, \text{ on simplification}$$

Similarly
$$z_2 = {\cdot}1063,\ z_3 = {\cdot}2586,\ z_4 = {\cdot}354;\ z_5 = {\cdot}1165,\ z_6 = {\cdot}1125$$

or
$$\bar{z} = \frac{\Sigma(n_i-3)\,z_i}{\Sigma(n_i-3)} = 0{\cdot}1919, \text{ on simplification}$$

Also
$$\chi^2_{6-1} = \Sigma(n_i-3)(z_i-\bar{z})^2 = 1{\cdot}008, \text{ on simplification}$$

But tabulated χ^2_5 at 5% level of significance = 11·07

Hence the null hypothesis appears to be correct.

In other words, the sample correlation coefficients are homogeneous.

Again $\bar{z} = \frac{1}{2}\log_e\left(\frac{1+\rho}{1-\rho}\right)$ where $\rho = \frac{e^{2\bar{z}}-1}{e^{2\bar{z}}+1} = \frac{1{\cdot}468-1}{1{\cdot}468+1} = 0{\cdot}1894$

Hence the pooled estimate population correlation coefficient is 0·1894.

Ex. From examples of 24, 29, 33, 38, 42 a partial correlation coefficient of order 3 is found to be 0·38, 0·54, 0·60, 0·50, 0·42.

Test their homogeneity and find the pooled estimate.

16·6. Association of Attributes and Contingency

16·6·1. Measures of association for contingency tables.

For measuring the degree of association between two attributes *A* and *B*, the data may be arranged in a 2×2 table.

	B's	not – *B's* or β's	Total
A's	$(AB) = a$	$(A\beta) = b$	$a + b$
not – *A's* or α's	$(\alpha B) = c$	$(\alpha\beta) = d$	$c + d$
Total	$a + c$	$b + d$	$a + b + c + d = N$

The three meausres of association that have to be proposed are :

$$Q = \frac{ad - bc}{ad + bc} \quad [\textit{Yule's coefficient of association}]$$

$$Y = \frac{1 - \sqrt{\dfrac{bc}{ad}}}{1 + \sqrt{\dfrac{bc}{ad}}} \quad [\textit{Yule's coefficient of colligation}]$$

$$V = \frac{ad - bc}{[(a+b)(a+c)(b+d)(c+d)]^{1/2}}$$

For a $s \times t$ contingency table, we may use the *square contingency* χ^2 *or the mean-square contingency* χ^2/N usually denoted by ϕ^2. Karl Pearson proposed another measure

$$C = \sqrt{\frac{\chi^2}{N + \chi^2}} = \sqrt{\frac{\phi^2}{1 + \phi^2}} \qquad \text{...(16·13)}$$

called Pearson's *coefficient of contingency.*

Tchuprow proposed the coefficient

$$T = \left[\frac{\chi^2}{N(s-1)(t-1)}\right]^{1/2} \qquad \text{...(16·14)}$$

for the same purpose.

The value of *C* lies between 0 and 1.

Ex. 1. In a 2×2 table the association is called positive or negative according as

$$\delta = a - \frac{(a+b)(a+c)}{N} > \text{or} < 0$$

A	a b
not A	c d

Show that Q, Y, V are positive or negative, according as the association is positive or negative.

Ex. 2. Show that Q, Y, V all vanish when the two attributes are independent.

Ex. 3. Show that in the table

Q and Y become $+1$ if $bc = 0$ in which case there is *complete association* and these become -1 if $ad = 0$ in which case there is *complete disassociation.*

Ex. 4. Show that $V = 1$ if $b = 0, d = 0$ (*i.e.*, if all A's are B's and *vice versa*) and $V = -1$ if $a = 0, d = 0$ (*i.e., if* no A's are B's and B's are A's). When $V = 1$ the association is said to be *absolute.*

Ex. 5. Prove that $Q = \dfrac{2Y}{1 + Y^2}$. Hence or otherwise show that the coefficient of association is greater in absolute value than the coefficient of the colligation except when both are zero or unity in absolute value.

Ex. 6. (*a*) Prove that χ^2 vanishes if and only if the two attributes are independent. (*b*) For a 2×2 table, prove that $V^2 = \chi^2/N$.

Ex. 7. Prove that if $s < t$, then $T \le \left(\dfrac{s-1}{t-1}\right)^{\frac{1}{4}}$ and hence that T can be unity only if $s = t$.

Exercises 16

1. (*a*) Define χ^2. What hypothesis does it test ? What are the conditions for the application of this test ?

(*b*) 200 digits were chosen at random from a set of tables. The frequencies of the digits were :

Digits	0	1	2	3	4	5	6	7	8	9
Frequencies	18	19	23	21	16	25	22	20	21	15

Use the χ^2 test to assess the correctness of the hypothesis that the digits were distributed in equal numbers in the table from which these numbers were taken. ***(B.Sc. Delhi, 1992)***

[**Ans.** $\chi^2 = 4{\cdot}3$, $v = 9$. Hypothesis appears to be correct)

2. The following table gives the number of aircraft accidents that occurred during the various days of the week. Find whether the accidents are uniformly distributed over the week.

Days	Sun.	Mon.	Tue.	Wed.	Thu.	Fri.	Sat.	Total
No. of accidents	14	16	8	12	11	9	14	84

[**Ans.** Accidents seem to be uniformly distributed over all days of the week.]

3. Four coins were tossed at a time and this operation is repeated 160 times. It is found that 4 beads occur 6 times, 3 heads occur 43 times, 2 heads occur

69 times, one head occurs 34 times. Discuss whether the coins may be regarded as unbiased.

[**Ans.** Coins are unbiased]

4. Five dice were thrown 96 times and the number of times 4, 5 or 6 were thrown are given below

No. of dice throwing 4, 5 or 6	5	4	3	2	1	0
Frequency	7	19	35	24	8	3

Calculate the value of χ^2 up to two places of decimals.

[The expected frequencies are given by the various terms in the expansion of $\left(\frac{1}{2}+\frac{1}{2}\right)^5$. Combining the first two classes and the last two classes, respectively for 4 d.f.,

$$\chi^2 = \frac{(26-18)^2}{18} + \frac{(35-30)^2}{30} + \frac{(24-30)^2}{30} + \frac{(11-8)^2}{18} = 8{\cdot}31.]$$

5. In a biochemical experiment 9 insects were put each into 100 jars and were subjected to fumigant. After 3 hours, the number of living insects in each jar was counted. The distribution is shown below. Can this be fitted by a Binomial distribution ?

No. of insects alive	x	0	1	2	3	4	5	6	7	8	9	Total
No. of jars	f	3	8	11	15	16	14	12	11	9	1	100

[**Hint.** If every insect had an equal chance p of surviving then since the mean np of the population has to be estimated from the mean of the distribution.

$$np = \frac{\Sigma f_i x_i}{\Sigma f_i} = \frac{439}{100}$$

$\therefore\ p = 0{\cdot}2195$ and the theoretical frequencies obtained from $100\,(0{\cdot}7805 + 0{\cdot}2195)^{20}$ are 0·7, 4·0, 10·6, 17·8 21·3, 19·2, 13·5, 7·6, 3·5, 1·3, 0·4, 0·1. Combining the first three and the last three classes $\chi^2 = 11{\cdot}36,\ \nu = = 5 - 14.$]

6. The following data show suicides of women in 8 German States during 14 years :

State per year	0	1	2	3	4	5	6	7	Total
Observed frequency	364	376	218	89	33	13	2	1	1096

Fit a Possion distribution to the data and show that the fit is not satisfactory.

7. Show that for a $2 \times n$ table

$$\chi^2 = \sum_{r=1}^{n} \left[N_1 N_2 \frac{\left(\frac{a_{1r}}{N_1} - \frac{a_{2r}}{N_2}\right)^2}{a_{1r} + a_{2r}} \right]_r$$

where a_{1r} and a_{2r} are the frequencies in the rth column and N_1 and N_2 are the border sums of the two rows. ***(B.Sc. Allahabad, 1992)***

[**Hint.** $E(a_{1r}) = N_1 (a_{1r} + a_{2r})/(N_1 + N_2), E(a_{2r}) = E(a_{1r} + a_{2r})/(N_1 + N_2)$.]

8. Show that for entries in the $2 \times r$ contingency table

							Totals
	a_1	a_2	......	a_i	......	a_r	a
	b_1	b_2	......	b_i	......	b_r	b
Totals	n_1	n_2	......	n_i	......	n_r	n

the value of χ^2 for testing homogeneity is $\chi^2 = \sum_{i=1}^{r} w_i (p_i - p)^2$

where $p_i = \dfrac{a_i}{n_i}, p = \dfrac{a}{n}, w_i = \dfrac{n_i}{pq}, q = \dfrac{b}{n}, q_i = 1 - p_i, n = a + b,$

$q = 1 - p, q_i - q = -(p_i - p)$ **(*I.C.A.R., Delhi*)**

$$[\textbf{Hint. } \chi^2 = \sum_{i=1}^{r} \left[\frac{\left(a_i - \dfrac{n_i a}{n} \right)^2}{\left(\dfrac{n_i a}{n} \right)} \right] + \sum_{i=1}^{r} \left[\frac{\left(b_i - \dfrac{n_i b}{n} \right)^2}{\left(\dfrac{n_i b}{n} \right)} \right]$$

$$= \sum_{i=1}^{r} \left[n_i \frac{(p_i - p)^2}{p} \right] + \sum_{i=1}^{r} \left[\left(\frac{n_i (q_i - q)^2}{q} \right) \right] = \text{etc.}]$$

9. (*a*) Prove **Brandt and Snedecor's formula** for a $2 \times r$ contingency table with a, a' as any pair of observed frequencies and n, n' the corresponding totals.

2×2 Table	Total
$a_1, a_2, \ldots\ldots a_i \ldots\ldots a_r$	n
$a_1' \; a_2' \ldots\ldots a_i' \ldots\ldots a_r'$	n'

$$\chi^2 = \frac{1}{\bar{p}\,\bar{q}} \left[\sum_{i=1}^{r} (a_i p_i) - n\bar{p} \right]$$

where $\bar{p} = \dfrac{n}{n + n'}, \; p = \dfrac{a_i}{a_i + a_i'}, \; \bar{q} = 1 - \bar{p}, q = 1 - p.$

(*b*) Use this to prove that for the following 2×3 table, $\chi^2 = 1{\cdot}51$

				Total
	42	13	33	88
	20	8	25	53
Total	62	21	58	141

$$[\textbf{ Hint } (a)\; \chi^2 = \sum_{i-1}^{r} \left\{ a_i - \frac{n \times (a_i + a_i')}{n + n'} \right\}^2 + \sum_{i=1}^{r} \left\{ a_i' - \frac{n' \times (a_i + a_i')}{n + n'} \right\}^2$$

$$= \left(\frac{n'a_i - a'_i}{n + n'} \right)^2 + \left(\frac{na'_i - n'a_i}{n + n'} \right)^2$$

$$= \sum_{i=1}^{r} \left[\frac{(a_i + a_i')}{n} \left\{ \frac{a_i}{a_i + a_i'} - \frac{n}{n + n'} \right\} \right]^2 \qquad \because \; q_i - \bar{q} = -(p_i - \bar{p})$$

$$= \sum_{i=1}^{r} \left[\frac{(a_i + a_i')}{\bar{p}} (p_i - \bar{p})^2 + \frac{(a_i + a_i')}{\bar{q}} (q_i - \bar{q})^2 \right] = \frac{1}{\bar{p}\,\bar{q}} \Sigma n_i (p_i^2 + \bar{p}^2 - 2p_i\bar{p})$$

$$= \frac{1}{\bar{p}\,\bar{q}} \left[\sum_1^r n_i p_i^2 + \bar{p}^2 \sum_1^r n_i - 2\bar{p} \sum_1^r n_i p_i \right]$$

$$= \frac{1}{\bar{p}\,\bar{q}} \left[\sum_1^r n_i p_i^2 + \bar{p}^2 n - 2\bar{p} \cdot n\bar{p} \right] = \frac{1}{\bar{p}\,\bar{q}} \left[\sum_1^r n_i p_i^2 - n\bar{p}^2 \right] = \text{etc.}$$

10. The following table occurs in a memoir of Karl Pearson :

		Eye colour in sons	
		No light	light
Eye colour in fathers	No light	230	148
	Light	151	471

Test whether the colour of the son's eyes is associated with that of the father's. ($\chi^2 = 3{\cdot}34$, $v = 1$).

[**Ans.** The colour of son's eye appears to be associated with that of the father's].

11. In an experiment on immunization of cattle from tuberculosis the following results were obtained :

	Affected	Unaffected
Inoculated	12	26
Not Inoculated	16	6

Examine the effect of vaccine in controlling susceptibility to tuberculosis. (*I.A.S.*)

[**Ans.** The vaccine is effective in controlling susceptibility to tuberculosis.]

12. In an experiment on the immunization of goats from anthrax, the following results were obtained. Derive your inference on the efficiency of the vaccine :

	Died	Survived	Total
Inoculated	2	10	12
Not Inoculated	6	6	12
Total	8	16	24

[**Ans.** $\chi^2 = 3$ without Yates correction,
$\chi^2 = 1{\cdot}7$ with Yates correction, both are significant].

13. The normal rate of infection for a certain disease in cattle is known to be 50%. In an experiment with seven animals injected with a new vaccine, it was found that none of the animals caught infection. Can the evidence be regarded as conclusive at 1% level of significance to prove the value of the new vaccine ?

[**Ans.** The evidence may be regarded as conclusive.]

14. An operation's research group studying the libraries of a certain university found that the number (x) of books withdrawn by individual users during any one visit to the science library was distributed with probability function

$$f(x) = (0{\cdot}4)\,(0{\cdot}6)^x, \quad x = 0, 1, 2.$$

Test this hypothesis for goodness of fit using the following data, which represents an independent random sample of 100 observations.

Value of x	0	1	2	3	4 or more
Frequency	37	26	17	5	15

15. In experiments on pea breeding, Mendel obtained the following frequencies of seeds : 315 round and yellow, 101 wrinkled and yellow, 108 round and green, 32 wrinkled and green. Total 556. Theory predicts that the frequencies should be in the ratio 9 : 3 : 3 : 1. Find χ^2 and examine correspondence between theory and experiment.

[**Ans.** The theoretical frequencies are approx. 313, 104, 104, 35 and $\chi^2 = 0{\cdot}51, \nu = 1$. There appears to be correspondence between theory and experiment.]

16. (*a*) A certain hypothesis was tested by two similar experiments which gave $\chi^2 = 14{\cdot}7$ for $\nu = 9$ and $\chi^2 = 14{\cdot}9$ for $\nu = 11$. Show that the two experiments combined gave less reason for confidence in the hypothesis than either experiment alone.

(*b*) A certain hypothesis is tested by the similar experiments. These gave $\chi^2 = 1{\cdot}19$ for $\nu = 6$, $\chi^2 = 14{\cdot}2$ for $\nu = 8$ and $\chi^2 = 18{\cdot}3$ for $\nu = 11$. Show that the three experiments together provide more justification for rejecting the hypothesis than any one experiment alone.

17. According to the genetic model the proportions in three groups should be $p^2 : 2pq : q^2$ where $p + q = 1$. Are the data cosistent with the sample 9, 51, 45 ?

18. According to the genetic model the proportions of individuals having the four blood types should be related by $O = q^2$, $A = p^2 + 2pq$; $B = r^2 + 2qr$, $AB = 2pr$ with $p + q + r = 1$. Given the sample $O = 176, A = 182$; $B = 60, AB = 17$, find if it is consistent with the model. [**Ans.** Consistent]

19. Show that the expected value of χ^2 for a $s \times t$ table is $\dfrac{n}{n-1}(s-1)(t-1)$.

[**Hint.** Let p_{ij} denote the probability that a randomly chosen individual belongs to the ith row and jth column of the table where n_{ij} is the number of observed individuals in the ith row and jth column. The hypothesis of independence in then equivalent to the hypothesis that there exist $s + t$ constraint $p_{i\cdot}$ and $p_{\cdot j}$ such that

$$p_{ij} = p_{i\cdot}\, p_{\cdot j}\,, \quad \sum_i p_{i\cdot} = \sum_j p_{\cdot j} = 1$$

We have
$$\chi^2 = \sum_{i,j} \frac{(n_{ij} - n\, p_{i\cdot}\, p_{\cdot j})^2}{n p_{i\cdot}\, p_{\cdot j}}$$

The estimate of $p_{i\cdot}$ and $p_{\cdot j}$ can be shown to be $n_{i\cdot}/n$ and $n_{\cdot j}/n$ respectively.

Thus
$$\chi^2 = n \sum_{i,j} \frac{\left(n_{ij} - \dfrac{n_{i\cdot} n_{\cdot j}}{n} \right)^2}{n_{i\cdot} n_{\cdot j}} = n \left(\sum_{i,j} \frac{n_{ij}^2}{n_{i\cdot} n_{\cdot j}} - 1 \right)$$

Since we have here st groups and $s + t - 2$ parameters determined from the sample, the limiting distribution of χ^2 has $st - (s + t - 2) - 1 = (s - 1)\, t - 1)$ *d.f.* of frequency. Assuming that the independence hypothesis is true, we have

$$E\,(\chi^2) = \frac{n}{n-1}\,(s-1)\,(t-1).]$$

20. $x_{ij}\ (i = 1, 2, \ldots, m,\ j = 1, 2, \ldots, n)$ are the cell frequencies in a $m \times n$ contingency table. The marginal totals of the rows and columns are denoted by $r_i\ (i = 1, 2, \ldots, m)$ and $c_j\ (j = 1, 2, \ldots, n)$ and the grand total by N. Show that the means, variances, and covariances of the cell-frequencies are given by

$$E\,(x_{ij}) = \frac{r_i c_j}{N},$$

$$\text{Var}\,(x_{ij}) = r_i\, c_j\,(N - r_i)\,(N - c_j)/N\,(N-1)$$

$$\text{Cov}\,(x_i, x_j) = r_i\, c_j\, r_k\, c_1/N\,(N-1),\ i = k, j \neq l$$

$$= r_i\,(N - r_i)\, c_j\, c_i\,/N^2\,(N-1),\ i = k, j = l.$$

21. (*a*) What is meant by a class frequency of (*i*) first order (*ii*) third order. How would you express a class frequency of first order in terms of class frequencies of third. order. **(*B.Sc., Agra Univ.*)**

(*b*) What do you understand by consistency of given data ? How do you check it ? **(*B.Sc., Allahabad*)**

(*c*) What do you mean by independence of attributes ? Give a criteria of independence for attributes A and B. **(*B.Sc., Nagpur*)**

(*d*) When are two attributes said to be positively associated and negatively associated ? Also define complete association and disassociation of two attributes. **(*B.Sc., Bombay; B.Sc., Nagpur*)**

(*e*) What is association of attributes ? Write a note on the strength of association and how it is measured ? **(*B.Sc., Madras*)**

(*f*) Define Yule's coefficient of association and the coefficient of colligation. Obtain the relation between the two. **(*B.Sc., Shivaji*)**

22. (*a*) Given that $(A) = (\alpha) = (B) = (\beta) = N/2$ show that $(A\,B) = (\alpha, \beta)$, $(A\beta) = (\alpha B)$. **(*B.Sc., Agra*)**

(*b*) Given that $(A) = (\alpha) = (B) = (\beta) = (C) = (\gamma) = N/2$ and also that (ABC) $= (\alpha\beta\gamma)$ show that

$$2\,(ABC) = (AB) + (AC) + (BC) - \frac{N}{2}$$ **(B.Sc., Hons., Delhi)**

23. Show that if

$$\frac{(A)}{N} = x, \frac{(B)}{N} = 2x, \frac{(C)}{N} = 3x \text{ and } \frac{(AB)}{N} = \frac{(BC)}{N} = \frac{(CA)}{N} = y$$

neither the values of x nor of y can exceed 1/4.

24. (*a*) Given that $(A) = (B) = (C) = N/2$ and 80 per cent of A's are B's, 97 per cent of A's are B's, find the limits to the percentage of B's that are C's.

[**Ans.** 55% and 95%]

25. If $1000 = N = 1\frac{2}{3}(A) = 2\,(B) = 2\frac{1}{2}(C) = 5\,(AB)$ and $(AC) = (BC)$ what should be the minimum value of (BC). [**Ans.** 150]

26. Among the adult population of a certain town 50 per cent are males, 60 per cent wage-earners and 50 per cent are 45 years of age or over, 10 per cent of the males are not wage earners and 40 per cent of the males are under 45. Make the best possible inference about the limits within which the percentage of persons (male or female) of 45 years or over are wage earners.

[**Hint**. $N = 100, (A) = 50, (B) = 60, (C) = 50$]

[**Ans**. $25 \le (BC) \le 45$]

27. In an university examination, 65% of the candidates passed in English, 90% passed in the second language and 60% passed in the optional subjects. Find how many at least should have passed the whole examination.

[**Ans.** 15%]

28. Show that if $\delta = (AB) - (\alpha\beta)$, then with usual notations, prove that

(*i*) $[(A) - (\alpha)]\,[(B) - (\beta)] + 2N\delta = (AB)^2 + (\alpha\beta)^2 - (A\beta)^2 - (\alpha B)^2$.

(*ii*) $\delta = \dfrac{(B)(\beta)}{n}\left\{\dfrac{(AB}{(B)} - \dfrac{(A\beta)}{(\beta)}.\right\} = \dfrac{(A)\,(B)}{N}\left\{\dfrac{(AB)}{(A)} - \dfrac{(\alpha\beta)}{(\alpha)}\right\}$

29. χ is a Poisson variate with parameter λ and χ^2 is a chi-square variate with $2k$ *d.f.* Prove that for all positive integer k,

$$P(\chi < k - 1) = P(\chi^2 > 2\lambda)$$

[**Hint.** $P(\chi^2 > 2\lambda) = \dfrac{1}{2^k \,|\, \underline{k}} \displaystyle\int_{2\lambda}^{\infty} e^{-(1/2)\chi^2} (\chi^2)^{k-1}\, d\chi^2 = \dfrac{1}{(k-1)\,!} \int_{\lambda}^{\infty} e^{-y} y^{k-1}\, dy$

$$= \frac{1}{(k-1)\,!}\left[\lambda^{k-1} e^{-\lambda} + (k-1)\int_{\lambda} e^{-y} y^{k-2}\, dy\right]$$

Repeated integration by parts gives the required result.]

30. In a 2 × 3 Contingency table if $N = x + y + z$, $N' = x' + y' + z'$ and $N = N'$, prove that

$$\chi^2 = \frac{(x - x')^2}{x + x'} + \frac{(y - y')^2}{y + y'} + \frac{(z - z')^2}{z + z'}$$

(*B.A. Hons. Math. Delhi, 1983, B.Sc., Poona, 1990*)

[**Hint.**

				Total
	x	y	z	N
	x'	y'	z'	N
Total	$x + x'$	$y + y'$	$z + z'$	$2N$

$$\chi^2 = \frac{\left[x - N\dfrac{(x + x')}{2N}\right]^2}{(x + x)} + \frac{\left[y - N\dfrac{(y + y')}{2\,N}\right]^2}{y + y'} + \frac{\left[z - N\dfrac{(z + z')}{2\,N}\right]^2}{z + z'} + \frac{\left[x' - N\dfrac{(x + x')}{2N}\right]^2}{x + x'} + \ldots$$

$$= \frac{(x - x')^2}{x + x'} + \frac{(y - y')^2}{y + y'} + \frac{(z - z')^2}{z + z'}.\ ]$$

31. In the past the standerd deiration of weights of certain 40.0 gram packages filled by a machine was 0.25 gram. A random sample of 20 packages has shown that a standard deviation of 0.32 gram. Is the apparent increase in variability significant at a 0.01 level of significant? (Give that $X^2_{0.99}$ for 19 d.f. = 36.2) **(*I.F.S. 1997*)**

[**Hint.** $X^2 = \sum_{i=1}^{n} \dfrac{(x_1 - \bar{x})^2}{\sigma_0^2} = \dfrac{1}{\sigma_0^2}\ [\sum_{1}^{n} x_i^2 - \dfrac{(\Sigma x_i)^2}{n}\ ] = ns^2/\sigma_0^{\,2} = \dfrac{20\,(.32)^2}{(.25)^2} = 32.8$

where, $H_0 : \sigma = .25$ ounces $H_1 : \sigma \geq .25$ onces & variability has increased. Use a one-tailed test, we reject H_0 at a .01 level of significance]

17

*Statistical Theory of Point Estimation

17·1. Introduction : Given a random sample $x_1, x_2, \ldots, x_n$ of size n from a population, with probability density function (*p.d.f*) of known form $f(x, \theta)$ but with unknown parameter θ, our problem is to find an estimate for θ in terms of the sample values. In general, we can construct a number of statistical functions of $x_1, x_2, \ldots, x_n$, which may claim to estimate θ. For example, in order to estimate the population mean, we may use as estimator the sample mean or sample median or the sample geometric mean or the sample harmonic mean. Thus, it is essential to discriminate between the various statistics, so as to arrive at the *best* estimator; the word 'best' being used in Fisherian sense.

Statistical Inference
(Testing of Hypothesis)
|
Estimation

Point Estimation | Interval Estimation

17·2. Fisher's Criteria for the Best Estimator. According to Fisher, who founded the theory of estimation in 1930, an estimator is said to be the *best,* if it is (*i*) unbiased (*ii*) consistent (*iii*) efficient and (*iv*) sufficient.

17·2·1. An estimator $t_n = t(x_1, x_2, \ldots, x_n)$, drawn from a sample of size n is said to be an ***unbiased*** estimator of a population parameter θ if

$$E(t_n) = \theta.$$

If a random sample ($x_1, x_2, \ldots, x_n$) of size n is drawn from a Normal population, with mean μ and Standard deviation σ, then

* Fora detailed discussion, see 'Statistical Inference' and for more solved problems, see 'Problems in Statistics' by H.C. Saxena, published by S. Chand.

$$E(\bar{x}) = \mu \text{ and } E(S^2) = \sigma^2$$

where $$S^2 = \frac{1}{n-1} \sum_1^n (x_i - \bar{x})^2.$$

Thus $\bar{x}$ and S^2 are unbiased estimates of the population μ mean and population variance σ^2 respectively. It can be proved that $E(s^2) \neq \sigma^2$, where $s^2 = \frac{1}{n} \sum_{i=1}^{n} (x_i - \bar{x})^2$. In other words, sample variance is a biased estimate of population variance, bias being $E(s^2) - \sigma^2$. The quantity $E(t_n) - \theta$ is called the bias of the estimate θ, the estimate is said to be positively or negatively biased, according as the bias is positive or negative. The bias is positive if $E(t_n) - \theta > 0$ and negative if $E(t_n) - \theta < 0$.

Example 17·1. *$x_1, x_2, \ldots, x_n$ are random observations on a Bernoulli variable X, taking the value 1 with probability θ and the value 0 with probability $1 - \theta$. Show that $T(T-1)/[n(n-1)]$ is an unbiased estimate of θ^2, where $T = \sum_{i=1}^{n} x_i$.* **(B.Sc. Gujarat)**

We have $E(x_i) = 1 \cdot \theta + 0 \cdot (1 - \theta) = \theta, \forall i = 1, 2, \ldots, n$

$$E(x_i^2) = 1^2 \cdot \theta + 0^2 \cdot (1 - \theta) = \theta$$

$$\therefore \quad \text{Var } x_i = E(x_i^2) - [E(x_i)]^2 = \theta - \theta^2 = \theta(1 - \theta)$$

Now $$E(T) = E\left(\sum_1^n x_i\right) = \sum_1^n E(x_i) = n\theta \qquad \ldots(1)$$

$$\text{Var } T = E(\Sigma x_i)^2 = \Sigma\theta(1 - \theta) = n\theta(1 - \theta)$$

$$\therefore \quad E(T^2) = \text{Var}(T) + \{E(T)\}^2 = n\theta(1 - \theta) + n^2\theta^2 = n\theta(1 - \theta + n\theta) \qquad \ldots(2)$$

Hence $$E\left[\frac{T(T-1)}{n(n-1)}\right] = \frac{1}{n(n-1)}(ET^2 - ET) = \theta^2, \text{ using (1) \& (2).}$$

Hence the result.

Example 17·2. *Let X be distributed in the Poisson form with parameter θ. Show that the only unbiased estimator of $\exp[-(k+1)\theta], k > 0$ is $T(x) = (-k)^x$ so that*

$T(x) > 0$ if x is even

and $T(x) < 0$ if x is odd. **(B.Sc. Stat. Hons. Delhi, 1988, 1993)**

Now $$E\{T(x)\} = E[(-k)^x] = \sum_{x=0}^{\infty} (-k)^x \cdot e^{-\theta}\, \theta^x / x!$$

$$= e^{-\theta} \sum_{x=0}^{\infty} \frac{(-k\theta)^x}{x!} = e^{-\theta} \cdot e^{-k\theta} = e^{-(1+k)\theta}.$$

$\therefore\ T(x) = (-k)^x$ is an unbiased estimate of $\exp[-(1+k)\theta]$, $k > 0$.

17·2·2. Consistent Estimator. An estimator $t_n = t(x_1, x_2, \ldots, x_n)$ computed from a random sample of n values, is said to be a consistent estimator for a population parameter θ, if it converges in probability to θ as n tends to infinity, *i.e.*, if given two positive numbers ε and η, however small, we can find a positive integer N, depending on ε and η, such that

$$P[\,|t_n - \theta| < \varepsilon] > 1 - \eta, \text{ for all } n > N.$$

$$i.e.\ P\{\lim_{n\to\infty} t_n = \theta\} = 1 \quad \Rightarrow \quad t_n \xrightarrow[inf]{P} \theta \text{ as } n \to \infty.$$

Note 1. The property of consistency ensures that the difference between t_n and θ would become smaller and smaller in probability sense as n increases indefinitely. In other words, it would give increasing accuracy with the increasing of the size of the sample.

2. The property of consistency is a limiting property, *i.e.*, it refers to the behaviour of t_n as the sample size increases. Moreover, if there exists one consistent statistic t_n, we can construct an infinite number of consistent statistics since each of statistics $\dfrac{an+b}{an+c}\,t_n$, $\dfrac{an^2+bn+c}{an^2+dn+e}\,t_n$, etc., converges in probability to θ.

3. A consistent estimator is unbiased in the limit but an unbiased estimator may or may not be a consistent estimator.

Example 17·3. *Show that in random sampling from a normal population, the sample mean is a consistent estimator for the population mean.*

Let $x_1, x_2, \ldots, x_n$ be the sample drawn from the normal population with mean μ and finite variance σ^2. We have proved in § 8·2, § 12·6, § 13·2 that the statistic $z = \dfrac{(\bar{x}-\mu)\sqrt{n}}{\sigma}$

is a standard normal variate.

$$\therefore\ P[\,|\bar{x}-\mu| < \varepsilon] = [P\ |z| < \varepsilon\sqrt{n}/\sigma] = \int_{-\varepsilon\sqrt{}/\sigma}^{+\varepsilon\sqrt{n}/\sigma} \frac{1}{\sqrt{2\pi}} \cdot e^{-\frac{1}{2}z^2}\, dz$$

Hence, given any number η, however small, we can choose n, so that area under the standard normal curve between $-\varepsilon\sqrt{n}/\sigma$ and $+\varepsilon\sqrt{n}/\sigma$ becomes greater then $1-\eta$.

Example 17·4. *Show that for random sampling from Cauchy population with density function*

$$f(x, \mu) = \frac{1}{\pi} \cdot \frac{1}{1 + (x - \mu)^2}, -\infty < x < \infty,$$

the sample mean is not a consistent estimator for the population mean but the sample median is a consistent estimator.

Let $\bar{x}$ be the mean of the sample $x_1, x_2, \ldots\ldots, x_n$ drawn from Cauchy population. We have proved in § 14·3 and § 14·4 that sample mean $\bar{x}$ is distributed as single observation x and that its distribution is

$$dp = \frac{1}{\pi} \cdot \frac{1}{1 + (\bar{x} - \mu)^2} d\bar{x}, -\infty < \bar{x} < \infty$$

$$\therefore \; P[\,|\bar{x} - \mu| < \varepsilon\,] = P[\,|x - \mu| < \varepsilon\,] = P[\,|z| < \varepsilon\,], z = x - \mu$$

$$= \frac{1}{\pi} \int_{-\varepsilon}^{+\varepsilon} \frac{1}{1 + z^2} dz = \frac{2}{\pi} \tan^{-1} \varepsilon.$$

Since this does not involve $n, P[\,|\bar{x} - \mu| < \varepsilon]$ cannot always be greater than $1 - \eta$, however large n may be. Hence sample mean is not a consistent estimator. In fact for estimating μ, a sample of 1000 observations has as much accuracy as a sample of 1 (one) observation only and thus by using the sample mean for estimating μ we would be wasting 999 observations.

In Ex. 14·4, we have shown that for large samples the statistic $z = (\hat{x} - \mu) \sqrt{4n/\pi}$ is a standard normal variate where $\hat{x}$ (is read as x cap or x caret) is the sample median and population mean is same as population median. Hence, for large samples,

$$P[\,|\hat{x} - \mu| < \varepsilon\,] = P\left(|z| < \frac{2\varepsilon\sqrt{n}}{\pi}\right) = \int_{-2\varepsilon\sqrt{n}/\pi}^{+2\varepsilon\sqrt{n}/\pi} \frac{1}{\sqrt{2\pi}} e^{-1/2 z^2} dz$$

By making n large, this can be made as near to unity as we please. Hence sample median is a consistent estimator for the population mean.

Ex. 1. Show that the sample variance is a consistent estimator for the population variance of a normal distribution.

[**Hint.** We have earlier proved that for a large sample, $E(s^2)$ is asymptotically equal to σ^2 and Var (s^2) is equal to $2\sigma^4/n$.]

Ex. 2. Show that a statistic t_n whose mean value differs from θ by order $1/n^2$, whose variance is of order $1/n$ and which tends to normality as n increases, is a consistent estimator for θ.

Ex. 3. Show that the sample median $\hat{x}$ is a consistent estimator for the population mean of a normal distribution.

[**Hint.** For a large sample, Var $\hat{x} = \frac{\sigma}{\sqrt{2n}}$, $\hat{x}$ is a standard normal variate].

Ex. 4. Show that, for a random sampling from a normal population, in estimating for the mean, the estimator x_1 (= first observation of the sample) is unbiased but not consistent.

Ex. 5. If T is an unbiased estimator of θ, show that T^2 is a biased estimator of θ^2 but (*ii*) If T is a consistent estimator of θ then T^2 is also a consistent estimator of θ^2. ***(B.A. Hons. Statistics, Delhi, 1989)***

[**Hint.** (*i*) $0 \neq \text{Var}\, T = E\,(T-\theta)^2 = E\,(T^2) - 2\,\theta\, E\,(T) + \theta^2 = E\,(T^2) - \theta^2$.

Hence $E\,(T^2) \neq \theta^2$]

Ex. 6. (*a*) A statistic t_n for which $E\,(t_n) = \theta_n \to 0$ and Var $(t_n) \to 0$ as $n \to \infty$ is consistent. ***(M.A., Bombay)***

$$P\,[\,|\,t_n - \theta\,| > t\,] \leq \frac{E\,(t_n - \theta)^2}{t^2} = \frac{1}{t^2}\,[\,\text{Var}\; t_n + E\,(t_n - \theta)^2\,) \to 0$$

(*b*) t_n an unbiased estimate of θ_n with variance δ_n and $\theta_n \to \theta$, $\delta_n \to 0$ as $n \to \infty$. Show that t_n is a consistent estimator of θ.

(B.A. Hons, Statistics, Delhi, 1989 ; M.A. Statistics, Delhi)

Hence obtain consistent estimators for

(*i*) mean of the normal distribution,

(*ii*) variance of the normal distribution when mean is known

(B.A. Hons. Statistics, Delhi, 1989)

Ex. 7. Show that the mean $\bar{x}$ of a random sample of size n from the distribution $f(x, \theta) = \theta^{-1}\, e^{-x/\theta}$, $0 < x < \infty$, $0 < \theta < \infty$, zero elsewhere, is an unbiased statistic for θ and has variance θ^2/n.

Ex. 8. Let $x_1, x_2, \ldots\ldots\ldots, x_n$ be a random sample from a normal distribution (*i*) with mean zero and variance σ^2 (*ii*) with mean μ and variance 1. Show that $\sum_{1}^{n} x_i^2/n$ is an unbiased statistic for (*i*) σ^2 and has variance $2\sigma^4/n$ (*ii*) $\mu^2 + 1$.

(M.A. Stat. Delhi, 1980)

Ex. 9. (*i*) $x_1, x_2, \ldots\ldots, x_n$ are random observations on a Bernoulli variable x taking the value 1 with probability θ and the value 0 with probability $1 - \theta$. Show that $T\,(T-1)/n\,(n-1)$ is an unbiased estimator of θ^2 where $T = \sum_{i=1}^{n} x_i$.

(B.A. Hons. Statistics, Delhi, 1989 ; B.Sc., Bombay, 1980 ; B.Sc., Gujarat)

(*ii*) If $x_1, x_2, \ldots\ldots, x_n$ are random observations on a random variable X taking the value 1 with probability p and the value 0 with probability $(1-p)$, show that $\frac{\Sigma x_i}{n}\left(1 - \frac{\Sigma x_i}{n}\right)$ is a consistent estimator of $p\,(1-p)$

(B.A. Hons. Stat. Delhi, 1988)

[**Hint.** $T = \sum_{i=1}^{n} x_i \sim B(n, p)$;

$$E(T) = np,\ \text{Var}(T) = npq,\ q = 1 - p;\quad E(\bar{x}) = E\left(\frac{t}{n}\right) = \frac{1}{n} \cdot np = p$$

$$\text{Var}(\bar{x}) = \text{Var}\left(\frac{t}{n}\right) = \frac{1}{n^2} \cdot npq = \frac{pq}{n} \to 0 \text{ as } n \to \infty$$

Since $E(\bar{x}) \to p$ and $Var(\bar{x}) \to 0$ as $n \to \infty$

$\therefore\ \bar{x}$ is a consistent estimator of p.

Also $\dfrac{\Sigma x_i}{n}\left(1 - \dfrac{\Sigma x_i}{n}\right) = \bar{x}(1 - \bar{x})$ is continuous, as it is a polynomial in $\bar{x}$ and hance by the invariance property of consistent estimators, it follows that $\bar{x}(1 - \bar{x})$ is a consistent estimator of $p(1 - p)$.

[**Hint.** (*i*) $E(x_i) = 1 \cdot \theta + 0 \cdot (1 - \theta) = \theta$, $\text{Var}\, x_i = \theta(1 - \theta)$, $E(T) = E\left(\sum_{i=1}^{n} x_i\right) = n\theta$,

$\text{Var}(T) = \text{Var}\left(\sum_{i=1}^{n} x_i\right) = n\theta(1 - \theta) \Rightarrow ET^2 = n\theta(1 - \theta) + (n\theta)$

Now $E[T(T - 1)] = E(T^2 - T) = E(T^2) - E(T) = etc.$]

Ex. 10. Let $y_1 < y_2 < y_3$ be the order statistics of a random sample of size 3 from the uniform distribution having density function

$$f(x, \theta) = \frac{1}{\theta},\ 0 < x < \theta,\ 0 < \theta < \infty,$$
$$= 0, \text{ elsewhere}$$

Show that $4y_1$, $2y_2$ and $(4/3)\, y_3$ are unbiased statistics for θ. Find the variance of each of these unbiased statistics. [**Ans.** $3\theta^2/5, \theta^2/5, \theta^2/15$]

Ex. 11. Let $y_1 < y_2 < y_3 < y_4 < y_5$ be the order statistics of *a r.v.* of size 5 from the uniform distribution having *p.d.f.* $f(x; \theta) = \dfrac{1}{\theta},\ 0 < x < \theta, 0 < \theta < \infty$ *show that* $2y^3$ *is an unbiased estimator of* θ. *Find the conditional expectation* $[E\{2y_3 \mid y_5\}$. **(*B.A. Hons. Statistics, Delhi, 1989*)**

Invariance Property of Consistent Estimators

Theorem. *If* T_n *is a consistent estimator of* $\gamma(\theta)$ *and* $\psi\{\gamma(\theta)\}$ *is a continuous function of* $\gamma(\theta)$, *then* $\psi(T_n)$ *is a consistent estimator of* $\psi\{\gamma(\theta)\}$.

Proof. By hyp. T_n is a consistent estimator of $\psi(\theta)$, $\therefore$ for every $\varepsilon > 0, \eta > 0$, a positive integer $n > N(\varepsilon, \eta)$ such that

$$P[\,|T_n - \gamma(\theta)| < \varepsilon\,] > 1 - \eta,\ \forall\, n \geq N$$

Further since $\psi(\gamma(\theta))$ is a continuous function of $\gamma(\theta)$, it follows that $|\psi(T_n) - \psi(\gamma(\theta)| < \varepsilon'$ whenever $|T_n - \gamma(\theta)| < \varepsilon$ where ε' is very small, corresponding to ε.

If A and B are two events, then $A \Rightarrow B$ implies $A \leq B$

i.e. $P(B) \geq P(A)$

$\therefore P[\,|\,\psi(T_n) - \psi(\gamma(\theta))\,| < \varepsilon_1\,] \geq [\,|\,T_n - \gamma(\theta)\,| < \varepsilon\,]$

$\geq 1 - \eta \;\forall\; n > N$

Hence $\psi(T_n)$ is a consistent estimator of $\psi\{\gamma(\theta)\}$.

17·2·3. Efficient Estimator. Let for large samples two consistent estimators t_n and t'_n be both distributed asymptotically normally about the true value of the parameter θ and with variance σ^2/n and σ'^2/n respectively. This will usually be the case in virtue of the central limit theorem. then t_n is said to be more efficient estimator than t'_n if $\sigma^2 < \sigma'^2$. However, if var $(t_n) <$ var (t_n') for all n, then t_n will be said to be more efficient than t'_n for all n. The estimator with the smaller variance will be grouped more closely round the true value and on the average, will deviate less from the true value than the estimator with large variance and thus it may be reasonably regarded as more efficient than the other.

If we can find a consistent estimator t_n whose variance is less than that of all other consistent estimators for all n, then t_n will be said to be the most efficient and the *efficiency E,* or any other estimator is defined as the ratio of the variance of the most efficient estimator to the variance of the given estimator. The efficiency of a statistic represents, in large samples, the fraction of the relevant information available in the sample which is utilized by the statistic in question.

Example 17·5. *Show that, in estimating for the mean for random sampling from a normal populating sample mean is more efficient than the sample median. Discuss also the populations of the two estimators for Cauchy population.*

For a normal population with finite variance σ^2, the sample mean and sample variance are consistent estimators for the population mean and variance respectively.

For any n, Var (mean) $= \dfrac{\sigma^2}{n}$

and for large n, Var (median) $= \dfrac{\pi\sigma^2}{2n}$.

Since Var (mean) is smaller then Var (median), the mean is more efficient than the median for large n. For small n, by actual calculations, it can be proved that Var (mean) is smaller than Var (median). Also the efficiency of the median relative to the mean (which will be proved later to be most efficient) is $2/\pi$, *i.e.*, 0·637.

For Cauchy population, a direct comparison is not possible since the mean is not even a consistent estimator. However,

$$\text{Var (median)} = \pi^2/4n$$

and obviously in this case, median is a better estimator than mean.

Example 17·6. *If two most efficient estimators t_n and t'_n estimating the same parameter θ, tend in the limit to be distributed in the bivariate normal form*

$$dp = \frac{1}{2\pi v_n \sqrt{1-\rho^2}} \exp\left[-\frac{1}{2v_n(1-\rho^2)} \times \right.$$

$$\left. \left\{ (t_n-\theta)^2 - 2\rho\,(t_n-\theta)\,(t_n'-\theta) + (t_n'-\theta)^2 \right\} \right] dt_n\, dt_n'$$

then show that $\rho = 1$ *where* $\text{Var}\,(t_n) = v_n = \text{Var}\,(t_n')$, $\text{Cov}\,(t_n, t_n) = \rho v_n$.

Let $u_n = \frac{1}{2}(t_n + t_n')$ ...(1), $u_n' = \frac{1}{2}(t_n - t_n')$ ...(2)

then clearly u_n is consistent estimator for θ and $E(u_n') = 0$.

$$\text{Var } u_n = \text{Var}\left[\tfrac{1}{2}(t_n + t_n')\right] = \tfrac{1}{4}(\text{Var } t_n + \text{Var } t_n' + 2\rho \text{ Var } t_n) = \frac{1+\rho}{2} v_n$$

Since t_n and t_n' are most efficient, it follows from (1) that

$$\text{Var } u_n \geq \text{Var } t_n$$

or $\qquad v_n(1+\rho)/2 \geq v_n \quad$ or $\quad 1+\rho \geq 2 \quad$ or $\quad \rho \geq 1.$

But ρ cannot be greater than 1, it follows that $\rho = 1$, then

$$\text{Var } u_n = v_n$$

Thus, in the limit, all most efficient estimators tend to equivalence.

Ex. 1. Given that in samples from normal population with *s.d.* σ,

$$\text{Var (mean deviation)} = \frac{2}{\pi}\cdot\frac{n-1}{n^2}\left(\frac{\pi}{2} + \sqrt{n(n-2)} - n + \sin^{-1}\left(\frac{1}{n+1}\right)\right)$$

$\approx \dfrac{\sigma^2}{n}\left(1 - \dfrac{2}{n}\right)$ and Var (standard deviation) $= \dfrac{\sigma^2}{n}$, compare their relative efficiencies, n is the size of the sample. [$E = \dfrac{1}{\pi - 2} = 0{\cdot}876$ and use of the mean deviation with 100 observations is equivalent to using the *s.d.* with 876 observations yet the mean deviation may be preferred on account of its ease in calculation].

Ex. 2. (*a*) Show that for samples of size *n* from normal population with mean μ and variance σ^2, the statistic $\hat{\mu} = \dfrac{1}{n+1}\sum_{i=1}^{n} x_i$ is most efficient for estimation μ, though it is biased. **(*B.S.c.Bombay*)**

$$E(\hat{\mu}) = \left(\frac{1}{n+1}\sum_1^n x_i\right) = \frac{n\mu}{n+1},\ \text{Var}\,\bar{x} = \sigma^2/n,\ \text{Var}\,\hat{\mu} = \frac{n}{(n+1)^2}\cdot\sigma^2.$$

(*b*) If $x_1, \ldots\ldots\ldots, x_n$ is a random sample from a $N(\mu, 1)$ population, show that $\sum^{n} x_i^2$ is an unbiased estimator of $\mu^2 + 1$. **(*B.Sc. Bombay 1980*)**

Ex. 3. If t is a most efficient estimator and t' a less efficient estimator with efficiency E and if the correlation coefficient between t and t' is ρ, show by considering estimator t'' defined by

$$(1 + E - 2\rho\sqrt{E})\, t'' = (1 - \rho\sqrt{E})\, t + (E - \rho\sqrt{E})\, t', \text{ prove that } \rho = \sqrt{E}.$$

[**Hint.** If t and t' approach θ in probability sense so does t''. Moreover,

$$(1 + E - 2\rho\sqrt{E})^2 \text{Var}(t'') = (1 - \rho\sqrt{E})^2 \text{Var}(t) + (E - \rho\sqrt{E})^2 \text{Var } t' + 2(1 - \rho\sqrt{E})(E - \rho\sqrt{E}) \text{Cov}(t, t').$$

Let $\text{Var } t = v$, then $\text{Var } t' = \dfrac{v}{E}$, $\text{Cov}(t, t') = \rho\sqrt{v}\sqrt{\dfrac{v}{E}} = \dfrac{\rho v}{\sqrt{E}}$.

$$\therefore \text{Var } t'' = v\,[(1 - \rho\sqrt{E})^2 + E(\sqrt{E} - \rho)^2 + 2(1 - \rho\sqrt{E})(\sqrt{E} - \rho)\rho]/(1 + E - 2\rho\sqrt{E})^2$$

$$= \frac{v(1 - \rho^2)}{(1 - \rho^2) + (\rho - \sqrt{E})^2} \le v \qquad \ldots(1)$$

But $\qquad \text{Var } t'' \ge v \quad \therefore \text{Var}(t'') = v, \quad \therefore$ from (1), $\rho = \sqrt{E}$.]

Ex. 4. Deduce from Exercise 3, the result of example 17·4.

Ex. 5. Show that in large sample the correlation between mean and median is ·7979.

Ex. 6. Prove that if $t'' = \frac{1}{2}(t + t')$, where t and t' are both most efficient with variance v, then $\text{Var } t'' = v\dfrac{(1 + \rho)}{2}$ Hence deduce $\rho = 1$.

Ex. 7. Prove the result of Ex. 3, by using the method of Example 17·4.

Ex. 8. If the sample mean necessarily efficient estimator of the population mean for every population ?

Ex. 9. Deduce from Exercise 3 , page 626 that $\text{Cov}(t, t - t') = 0$.

[**Hint.** $\text{Cov}(t, t - t') = -\text{Cov}(t, t') + \text{Var } t$.].]

Ex. 10. Find the maximum likelihood estimator of θ based on a sample of size n from a rectangular population in the range θ to 2θ. Show that if ξ, η are the maximum and minimum of the observations the estimate $(2\xi + \eta)/5$ is more efficient than maximum likelihood estimate.

Ex. 11. Suppose $X_1, X_2 \ldots, X_n$ are sample values independently drawn from population with mean m and variance σ^2. Consider the estimates

$$Y_n = \frac{1}{n-1} \sum_{i=1}^{n} X_i, \quad Z_n = \frac{1}{n^2} \sum_{i=1}^{n} iX_i$$

Discuss whether Y_n and Z_n are unbiased and consistent. What is the efficiency of y_n over z_n?

Minimum Variance Unbiased (MVU) Estimators. If a statistic $T \equiv T(x_1, x_2, \ldots, x_n)$ based on sample of size n is such that

(*i*) T is unbiased for $\gamma(\theta)$ for all $\theta \in \Theta$ and (*ii*) it has the smallest variance among the class of all unbiased estimators of $\gamma(\theta)$, *i.e.,* Var θ $(T) \in$ Var_θ (T) for all $\theta \varepsilon \theta$, T' is any other unbiased estimator then T is called the minimum variance unbiased estimator of $\gamma(\theta)$.

Ex. Define MVU estimator. Show that it is unique.

(B.Sc. Stat. Hons. 1985)

17·2·4. Sufficient Estimator. An estimator t_n is said to be sufficient for estimating a population parameter θ, if it contains all the information in the samples about the parameter. The criterion can be made more precise in terms of the concept of *likelihood function.* If $f(x, \theta)$ is the density function for a population, then the likelihood function $L(x_1, x_2, ..., x_n; \theta)$ for a random sample $x_1, x_2, ..., x_n$ is defined by

$$L(x_1, x_2, ..., x_n ; \theta) = \prod_{i=1}^{n} f(x_i, \theta) = f(x_1, \theta) f(x_2, \theta), ... f(x_n, \theta).$$

Now if it is possible to write

$$L(x_1, x_2, ..., x_n ; \theta) = L_1(t_n, \theta) L_2(x_1, x_2, ..., x_n)$$

where the second function on the right hand side does not contain θ then t_n will be said to be sufficient for θ. In this form it is obvious that any other function t_n' cannot provide any additional information about θ ; for given t_n, the conditional distribution of t_n' cannot involve θ and, therefore, if t_n is known, t_n' cannot give us more information about θ. Thus the statistic t_n is said to be sufficient for the parameter θ if with respect to any other statistic t_n', the joint probability density $P(t_n, t_n')$ of t_n, t_n' of the form

$$P(t_n, t_n') = P_1(t_n, \theta) P_2(t_n' \mid t_n)$$

where $P_2(t_n' \mid t_n)$ is the relative probability density of t_n' give t_n and is independent of θ.

If sufficient estimator exists, we need not use any other, for it is most desirable, but unfortunately sufficient estimators are the exceptions rather than the rule and, in practice, we have to be contented with less satisfactory estimators.

17·2·4·1. Factorization theorem. (Without Proof) Let $(x_1, x_2, ..., x_n)$ be the random sample from a distribution with *p.d.f.* $f(x_i, \theta)$. Then the necessary and sufficient condition that the statistic $y_1 = u_1(x_1, x_2, ..., x_n)$ is sufficient for θ if there exist two non-negative functions L_1 and L_2 such that

$$L(x_1, x_2, ..., x_n ; \theta) = L_1[u_1(x_1, x_2, ..., x_n ; \theta] L_2(x_1, x_2, ..., x_n)$$

Example 17·7. *Let* $(x_1, x_2, ..., x_n)$ *denote a random sample from a distribution with p.d.f.* $f(x, \theta) = \theta x^{\theta - 1}, 0 < x < 1, \theta > 0 ; = 0,$ *elsewhere.*

Prove that the product $u_1(x_1, x_2, ..., x_n) = x_1 x_2 ... x_n$ *is a sufficient statistic for* θ. **(B.Sc. Stat. Hons. Delhi 1988)**

The joint *p.d.f.* of $x_1, x_2, ..., x_n$ is

$$\theta^n (x_1, x_2 ..... x_n)^{\theta - 1}, 0 < x_i < 1, i = 1, 2, ..., n$$

which can be written as $[\theta^n (x_1 x_2 ... x_n)^{\theta}] \left(\dfrac{1}{x_1 x_2 ... x_n} \right)$

Hence by factorization theorem, $x_1 x_2 ... x_n$ is a sufficient statistic for θ.

Note. The domain of *p.d.f.* should *not* depend on θ.

Example 17·8. *Show that, if,* σ^2 *is known in random sampling from normal population,* $\bar{x}$ *is a sufficient estimator for* μ *; but if* μ *is known,* $s^2 = \dfrac{1}{n} \sum_{i=1}^{n} (x_i - \bar{x})^2$ *is not a sufficient estimator for* σ^2. *Suggest a sufficient estimator for* σ^2.

For the normal population

$$f(x, \mu) = \frac{1}{\sigma \sqrt{2\pi}} \exp \left[-\tfrac{1}{2} \frac{(x - \mu)^2}{\sigma^2} \right]$$

$$\therefore L(x_1, x_2, ..., x_n, \mu) = \left(\frac{1}{\sigma \sqrt{2\pi}} \right)^n \exp \left[-\tfrac{1}{2} \sum_{i=1}^{n} \frac{(x_i - \mu)^2}{\sigma^2} \right] \quad ...(i)$$

Here $(x_i - \mu)^2 = [(x_i - \bar{x}) + (\bar{x} - \mu)]^2 = ns^2 + n(\bar{x} - \mu)^2$

$$\therefore \quad = \left(\frac{1}{\sigma \sqrt{2\pi}} \right)^n \exp \left[-\frac{n}{2\sigma^2} \{ (\bar{x} - \mu)^2 + s^2 \} \right]$$

$$= \exp \left[-\frac{n}{2\sigma^2} (\bar{x} - \mu)^2 \right] \times \left\{ \left(\frac{1}{\sigma \sqrt{2\pi}} \right)^n \exp \left[-\frac{ns^2}{2\sigma^2} \right] \right\}$$

Thus L has been broken up into two factors, the first factor contains only $\bar{x}$ (estimator) and μ (parameter) and the second factor does not contain μ. Therefore, knowing σ^2, $\bar{x}$ is sufficient for estimating μ.

On the other hand if μ is known, then while the second factor involves only s^2 and σ^2 the first factor also contains σ^2 and hence σ^2 is not sufficient for σ^2.

However, if we define.

$$S^2 = \frac{1}{n} \sum_{i=1}^{n} (x_i - \mu)^2$$

Here instead of $\bar{x}$ we are taking μ.

then from (1), $L = \left(\frac{1}{\sqrt{2\pi\sigma^2}}\right)^n e^{\frac{-nS^2}{2\sigma^2}}$

and S^2 is obviously sufficient for σ^2. In fact nS^2/σ^2 is distributed as σ^2 with n *d.f.* This example, of course, only illustrates that if μ is sample mean, when we can get a better estimator S^2 by taking deviations from the known population mean.

Example 17·9. *Let $x_1, x_2, ..., x_n$ be a random sample from a uniform population on* $[0,\theta]$, *Find a sufficient estimator for* θ.

We have $L = \prod_{i=1}^{n} f(x, \theta) = \frac{1}{\theta^n}, 0 < x_i < \theta$

If $t = \max(x_1, x_2, ..., x_n) = x_{(n)}$ say, then *p.d.f.* of T is given by

$$g(t, \theta) = n\left[F\left(x_{(n)}\right)\right]^{n-1} \cdot f\left(x_n\right) \qquad \text{[See §14·6 ....]}$$

$$= n\left[\int_0^x f(x, \theta)\, dx\right]^{n-1} \qquad f\left(x_n\right) = n\left[\frac{x_n}{\theta}\right]^{n-1} \frac{1}{\theta} = \frac{n}{\theta^n}\left[x_{(n)}\right]^{n-1}$$

$$\therefore \qquad L = \frac{n}{\theta^n}\left[x_{(n)}\right]^{n-1}$$

$$= g(t,\theta)\, h(x_1, x_2, ..., x_n)$$

Hence by Fisher Neymarn Criterion, the statistic $t = x_{(n)}$ is sufficient estimator for θ.

Example 17·10. *Let $x_1, x_2, ... , x_n$ be a random sample from Cauchy population* $f(x, \theta) = \frac{1}{\pi} \cdot \frac{1}{1 + (x - \theta)^2}, -\infty < x < \infty, -\infty < \theta < \infty$. *Does there exist a sufficient statistic for* θ ?

$$L = \prod_{i=1}^{n} f\left(x_i, \theta\right) = \frac{1}{\pi^n} \cdot \prod_{i=1}^{n}\left[\frac{1}{1 + (x_i - \theta)^2}\right]$$

Now L cannot be factorized, such that one factor does not contain θ. Hence there does not exist a sufficient estimator for θ.

Example 17·11. *Show that a sufficient estimator is most efficient if a most efficient estimator exists.*

We assume that sufficient estimate t_n and any other estimator t_n' tend to joint normality for large n in the form

$$dP = \frac{1}{2\pi \sqrt{v_n v_n'} \sqrt{1-\rho^2}} \exp\left[-\frac{1}{2(1-\rho^2)} \times \left\{ \frac{(t_n-\theta)^2}{v_n} - \frac{2\rho (t_n-\theta)(t_n'-\theta)}{\sqrt{v_n v_n'}} + \frac{(t_n'-\theta)^2}{v_n'} \right\} \right] dt_n dt_n'. \quad ...(1)$$

Since t_n is sufficient, the distribution of t_n', given t_n cannot contain θ. The distribution of t_n is $dP_1 = \frac{1}{\sqrt{2\pi v_n}} \exp\left[-\frac{1}{2} \frac{(t_n-\theta)^2}{v_n} \right] dt_n$ ...(2)

and, therefore, conditional distribution of t_n', given t_n, is obtained by dividing (1) by (2). On simplification, we find that

$$\therefore\ dP_2 = \frac{1}{\sqrt{2\pi v_n} \sqrt{1-\rho^2}} \exp\left[\frac{-1}{2(1-\rho^2)} \left\{ \frac{\rho (t_n-\theta)}{\sqrt{v_n}} - \frac{t'_n-\theta}{\sqrt{v'_n}} \right\}^2 \right] dt_n'.$$

Since p.d. does not depend on θ *i.e.*, coefficient of θ is zero.

$$\therefore\ \frac{\rho}{\sqrt{v_n}} = \frac{1}{\sqrt{v_n'}} \text{ or } \rho = \sqrt{\frac{v_n}{v_n'}}. \quad \text{But } \rho \le 1 \ \therefore\ v_n \le v_n'.$$

$\therefore$ Variancc of the sufficient estimator is less than or equal to that of any other estimator—atleast of those estimators which tend to be distributed normally with the sufficient estimator. Accordingly in this class of estimators the sufficient is most efficient.

Ex. 1. Show that if t_n is a sufficient estimator for θ, then $(\partial/\partial\theta)(\log L)$ is a function of t_n and θ only.

Ex. 2. Show that in estimating for the parameter μ in the Poisson distribution $f(x, \mu) = \frac{e^{-\mu} \mu^x}{x!}$, $\bar{x}$ is a sufficient estimator for μ.

Ex. 3. Show that for Pearson type III distribution

$$dp = -\frac{e^{-x/\theta} x^{p-1}}{\theta^p \Gamma p} dx, x \ge 0, p \ge 0.$$

where p is known, $\bar{x}/p$ is a sufficient estimator for θ

Ex. 4. Let x_1 and x_2 denote a random sample from the normal distribution with mean $\theta, -\infty < \theta < \infty$, and s.d. unity. Show that $y_1 = x_1 + x_2$ is a sufficient statistic for θ.

[**Hint.** put $y_2 = x_2 \ \therefore\ x_1 = y_1 - y_2, x_2 = y_2, dx_1 dx_2 = dy_1 dy_2$]

Ex. 5. Examine whether it is possible to obtain a sufficient estimator for the parameter θ of the distribution

$f(x, \theta) = \theta (\sin x)^{\theta - 1} \cos x, 0 \le x \le \pi/2.$

[**Ans.** $t_1 = \sin x_1 \sin x_2 \ldots \sin x_n$ is a sufficient statistic for θ]

Ex. 6. Show that if m is known, the estimator

$$S = \left[\frac{1}{n} \Sigma (x - m)^2 \right]^{1/2}$$

is sufficient for σ in samples of n from a normal population with mean m and *s.d.* σ.

Ex. 7. If t_1 and t_2 are two unbiased estimates of a parameter with variances σ_1^2, σ_2^2 and correlation ρ, what is the best unbiased linear combination of t_1 and t_2 and what is the variance of such a compound ?

(*B.A. Stat. Hons., Delhi, 1990*)

[**Hint.** Let $t = l_1 t_1 + l_2 t_2$ where l_1 and l_2 are constants.

Then $E(t) = l_1\theta + l_2\theta = \theta$ (given),

$$\therefore \qquad l_1 + l_2 = 1 \qquad ...(1)$$

Also $Var(t) = Var(l_1 t_1 + l_2 t_2) = l_1^2 \sigma_1^2 + 2\rho l_1 l_2 \sigma_1 \sigma_2 + l_2^2 \sigma_2^2$, Cov $(t_1, t_2) = \rho$ ...(2)

We want to minimize (2) subject to the condition (1)

$\therefore$ We have $\qquad 1 + \dfrac{dl_2}{dl_1} = 0 \Rightarrow \dfrac{dl_2}{dl_1} = -1.$

$$2l_1 \sigma_1^2 + 2\rho\sigma_1\sigma_2 \left(l_2 + l_1 \frac{dl_2}{dl_1} \right) + 2\sigma_2^2 l_2 \frac{dl_2}{dl_1} = 0$$

or $$l_1 \sigma_1^2 + \rho\sigma_1\sigma_2 l_2 - \rho\sigma_1\sigma_2 l_1 - l_2 \sigma_2^2 = 0$$

or $$l_1 (\sigma_1^2 - \rho\sigma_1\sigma_2) = l_2 (\sigma_2^2 - \rho\sigma_1\sigma_2)$$

which gives the ratio between l_1 and l_2. Also we can find l_1 and l_2 such that $l_1 + l_2 = 1$.

Ex. 8. Let t_1 be an unbiased estimator with minimum variance σ^2 and t_2 be any other unbiased estimator with variance σ^2/e. What is the correlation between t_1 and t_2 ?

[**Hint.** From Ex. 7 above we have the best linear compound t_1 and t_2 with coefficient l_1 and l_2 satisfying the conditions

$$l_1 (e - \rho\sqrt{e}) = l_2 (1 - \rho\sqrt{e}) \qquad ...(8)$$

Now use the condition that the variance of the linear compound is not less than σ^2 or since t_1 cannot be improved upon, $l_2 = 0$ $\therefore$ from (3), $e = \rho\sqrt{e}$ or $\rho = \sqrt{e}$.]

Ex. 9. If t_1 and t_2 are two unbiased statistics having the same variance then the correlation is $\geq 2e - 1$ where e is the ratio of the variance of the best estimate to the common variance of t_1 and t_2. **(*B.A. Hons. Statistics Delhi, 1989*)**

[**Hint.** Let $t = \frac{1}{2}(t_1 + t_2)$ and use that $V(t) \geq$ the least variance.]

***Ex. 10.** If e_1, e_2 are the efficiencies of two unbiased estimators A and B and ρ is coefficient of correlation between A and B, then

$$\sqrt{e_1 e_2} - \sqrt{(1-e_1)(1-e_2)} < \rho < \sqrt{e_1 e_2} + \sqrt{(1-e_1)(1-e_2)}.$$

(B.Sc. Stat. Hons. Delhi, 1986, 1988, 1993)

Solution. [Due to Prof. V.D. Thawani]. Let Var $A = V/e_1$, Var $B = V/e_2$ where V is the minimum variance.

Let
$$C = \lambda A + \mu B$$
where λ and μ are constants such that $\lambda A + \mu B = 1$

As A and B are unbiased so is C.

$$\therefore \text{ Var } C = \left(\frac{\lambda^2}{e_1} + \frac{\mu^2}{e_2} + \frac{2\lambda\mu\rho}{\sqrt{e_1 e_2}}\right)$$

But
$$\text{Var } C \geq V$$

$$\therefore \quad \frac{\lambda^2}{e_1} + \frac{\mu^2}{e_2} + \frac{2\lambda\mu\rho}{\sqrt{e_1 e_2}} \geq 1 = \lambda^2 + \mu^2 + 2\lambda\mu,$$

or
$$\lambda^2\left(\frac{1}{e_1} - 1\right) + 2\lambda\mu\left(\frac{\rho}{\sqrt{e_1 e_2}} - 1\right) + \mu^2\left(\frac{1}{e_2} - 1\right) \geq 0$$

$$\left(\frac{\rho}{\sqrt{e_1 e_2}} - 1\right)^2 - \left(\frac{1}{e_1} - 1\right)\left(\frac{1}{e_2} - 1\right) \geq 0$$

[discriminant of a quadratic expression]

$$\left[\text{The conditions } \frac{1}{e_1} - 1 > 0 \text{ and } \frac{1}{e_2} - 1 > 0 \text{ are satisfied.}\right]$$

$$\therefore \quad \frac{\rho^2}{e_1 e_2} - \frac{2\rho}{\sqrt{e_1 e_2}} + \left(\frac{1}{e_1} + \frac{1}{e_2} - \frac{1}{e_1 e_2}\right) \leq 0;$$

$$\rho^2 - 2\sqrt{e_1 e_2}\,\rho + (e_1 + e_2 - 1) \leq 0$$

$\therefore$ ρ lies between the roots of the quadratic equation

$$x^2 - 2\sqrt{e_1 e_2}\, x + (e_1 + e_2 - 1) = 0$$

The roots are $\sqrt{e_1 e_2} - \sqrt{e_1 e_2 - (e_1 + e_2 + 1)}$

$$\therefore \sqrt{e_1 e_2} - \sqrt{(1-e_1)(1-e_2)} < \rho < \sqrt{e_1 e_2} + \sqrt{(1-e_1)(1-e_2)}.$$

Cor. 1. If $e_1 = 1, \rho = \sqrt{e_2}$

Cor. 2 If $e_1 = e_2 = e, e - (1-e) < \rho < e + 1 - e$ or $2e - 1 < \rho < 1$

i.e. $\rho > (2e - 1)$ which is due to Prof. C.R. Rao.

***Aliter.** See Problems in Statistics by H.C. Saxena.

We state below (without Proof) the **invariance property of sufficient estimator.**

If t is a sufficient estimator for the parameter θ and if $\psi(t)$ is one to one function of t, then $\psi(t)$ is sufficient for $\psi(\theta)$.

Rao-Blackwell Theorem

Theorem. Let X and Y denote random variables such that $EY = \mu$ and var $Y = \sigma_y^2$. Let $E(Y \mid x) = \varphi(x)$, then we shall prove that

(*i*) $E[\varphi(X)] = \mu$

and (*ii*) $\text{Var}(\varphi(x)) \leq \text{Var}(y)$

Proof. Let $f_{XY}(x, y), f_1(x), f_2(y)$ be the joint *p.d.f.* of X and Y the two marginal *p.d.f* also let $h(y \mid x)$ be the conditional *p.d.f.* of Y, given $X = x$.

Now $$E(Y \mid x) = \int_{-\infty}^{+\infty} y\, h(y \mid x)\, dx$$

But $$f(x, y) = h(y \mid x) f_1(x)$$

$\therefore$ $$L(Y \mid x) = \int_{-\infty}^{+\infty} y \frac{f(x, y)}{f_1(x)} dy$$

$\Rightarrow$ $$\varphi(x) = \frac{1}{f_1(x)} \int_{-\infty}^{+\infty} y f(x, y)\, dy.$$

Further $$E(Y \mid x) = \int_{-\infty}^{+\infty} \varphi(x) f_1(x)\, dx.$$

$$= \int_{-\infty}^{+\infty} \left[\int_{-\infty}^{+\infty} y f(x, y)\, dx \right] dx$$

$$= \int_{-\infty}^{+\infty} y \left[\int_{-\infty}^{+\infty} f(x, y)\, dx \right] dy$$

$$= \mu = \int_{-\infty}^{+\infty} y f_2(y)\, dy$$

Next $$\sigma_y^2 = E[(Y - \mu)^2]$$
$$= E[\{Y - \varphi(x)\} + \{\varphi(x) - \mu\}]^2]$$
$$= E[Y - \varphi(x)]^2 + E[\varphi(x) - \mu]^2, \qquad \text{...(1)}$$

the cross-product term vanishes since

$$E[(Y - \varphi(X))(\varphi(X) - \mu)]$$

$$= \int_{-\infty}^{+\infty} \int_{-\infty}^{+\infty} \{y - \varphi(x)\}\{\varphi(x) - \mu\} f(x, y)\, dx\, dy$$

$$= \int_{-\infty}^{+\infty} \{\varphi(x) - \mu\} \left[\int_{-\infty}^{+\infty} \{y - \phi(x)\}\ h(y \mid x)\, dy \right] \cdot f_1(x)\, dx$$

$$= \int_{-\infty}^{+\infty} \{\varphi(x) - \mu\} \cdot 0 \, dx = 0.$$

Again $\sigma^2_{\varphi(x)} = E[\varphi(X) - \mu]^2$

$$= E[\{Y - \mu\} - \{Y - \varphi(X)\}]^2$$

$$= E[Y - \mu]^2 + E[Y - \phi(X)]^2$$

$$\Rightarrow \quad \sigma_y^{\,2} \geq \sigma^2_{\varphi(x)}.$$

Remark. The above theorem enables us to obtain unbiased estimators with minimum variance with the help of sufficient statistic.

Example 17·12.. *Let X and Y have a bivariate normal distribution with means* μ_1 *and* μ_2 *and variances* σ_1^2 *and* σ_2^2 *and correlation coefficient* ρ. *Show that the variance of* $Z = Y \mid X$ *is* $\rho^2\sigma^2$.

Now $E(Y \mid x) = \varphi(x) = \mu_2 + \rho \dfrac{\sigma_2}{\sigma_1}(X - \mu_1)$.

Then $\varphi(X) = \mu_2 + \dfrac{\rho\sigma_2}{\sigma_1}(X - \mu_1)$ and $E[\varphi(X)] = \mu_2$.

Also $\operatorname{Var} z = \sigma^2 = E[\phi(x) - \mu_2]^2 = E\left[\dfrac{\rho\,\sigma_2}{\sigma_1}(x - \mu_1)\right]^2 = \rho^2\,\sigma^2$

Complete family of distributions. Def. Let $f(x\,;\theta)$, $\theta \,\varepsilon\, \Omega\}$ denote, a family of probability density functions. Let $u(x) = 0$ be a continuous function of x. Then if $E[u(x)] = 0$ for every θ implies $u(x) = 0$ at each point x at which at least one member of the family of *p.d.f.* is positive, then the family of *p.d.f.* is called a complete family.

Example. 17·13. *Let* $f(x\,;\theta) = 1/\theta$, $0 < x < \theta$, *Zero else where. Let* $E[u(x)] = 0$ *for every* $\theta > 0$. *We shall prove that this implies* $u(x) = 0$ *for every* $x > 0$, *at which at least one member of p.d.f. is positive*

Now $E[u(x)] = \displaystyle\int_{-\infty}^{+\infty} u(x) \cdot f(x\,;\theta)\,dx = \int_0^{\theta} u(x) \cdot \frac{1}{\theta}\,dx = 0$ for $\theta > 0$

$$\Rightarrow \int_0^{\theta} u(x)\,dx = 0 \quad \text{for } \theta > 0$$

Differentiating w.r.t. x, we have

$u(\theta) = 0$ for $\theta > 0$

i.e. $u(x) = 0$ for $x > 0$, Hence the result.

Example. 17·14. *Show that the family $f(x, \theta)$ of p.d.f. where $f(x, \theta) = \theta^x (1-\theta)^{1-x}$, $x = 0, 1$; Zero else where, is complete.*

$$\text{Now } E[u(x)] = \sum_x u(x) f(x, \theta) = \sum_{x=0}^{1} u(x)\, \theta^x (1-\theta)^{1-x}$$

$$= u(0) \cdot (1-\theta) + u(1) \cdot \theta.$$

$$= \theta [u(1) - u(0)] + u(0), \quad 0 < \theta < 1$$

$$= a\theta + b.$$

where a, b are constants.

Now if $E(u(x)] = 0$, then $a\theta + b = 0$ for all θ, $0 < \theta < 1$

Hence we have $a = 0, b = 0$

i.e. $\quad u(1) = u(0)$ and $u(0) = 0$

$\therefore \quad u(0) = 0, u(1) = 0.$

Ex. 1. Prove that the family $\{b(x, \theta), 0 < \theta < 1\}$ is complete.

Ex. 2. Show that the family $f(x, \theta), 0 < \theta < \infty\}$ is not complete where

(*i*) $f(x, \theta) = 1/2\theta, -\theta < x < \theta$; Zero elsewhere

(*ii*) $f(x, \theta) = (1/\sqrt{2\pi})\, e^{-\frac{x^2}{2\theta}} \quad -\infty < x < +\infty$

Ex. 3. Show that the family $f(x, \theta)$ with $f(x, \theta) = {}^2C_x\, \theta^x (1-\theta)^{2-x}, x = 0, 1, 2$ is complete. **(*B.Sc. Stat. Hons, Delhi, 1993*)**

17·3. Method of Maximum Likelihood. We have discussed above the four criteria of unbiased, consistency, efficiency and sufficiency which according to Fisher, optimum or best estimators should satisfy, but these criteria by themselves do not give any method for getting these estimators. A number of methods for estimation have accordingly been proposed, the most important of which is the method of maximum likelihood due to Fisher.

According to the principle of maximum likelihood, we should choose the estimator which makes the likelihood function a maximum, *i.e.*, t_n will be the maximum likelihood estimator if

$$L_1(x_1, x_2, \ldots\ldots, x_n, t_n) > L(x_1, x_2, \ldots., x_n, t_n')$$

for any other estimator t_n'. If L is a differentiable function of θ, then $t_n(x_1, x_2, \ldots\ldots\ldots, x_n)$ is the solution (if any) of

$$\frac{\partial L}{\partial \theta} = 0 \quad \text{with } \frac{\partial^2 L}{\partial \theta^2} < 0.$$

Since L is positive, the first equation is equivalent to $\dfrac{\partial \log L}{\partial \theta} = 0$, a form which is more convenient in practice.

The principle of maximum likelihood can also be supplied to the case when the population is bivariate or multivariate and also for the simultaneous estimation of a number of parameters.

Note. It is a convention to denote the maximum likelihood estimator of parameter by placing ^ (caret) over the parameter. Thus the maximum likelihood estimator of θ is denoted by $\hat{\theta}$.

Example 17·15. *For random sampling from a normal population, find the maximum likelihood estimators for*

(*a*) *the population mean, when the population variance is known.*

(*b*) *the population variance, when the population mean is known,*

(*c*) *the simultaneous estimation of both the population mean and variance.*

Here the likelihood function is

$$L = \left(\frac{1}{\sqrt{2\pi\sigma^2}}\right)^n \cdot e^{-\frac{1}{2\sigma^2}\sum_{i=1}^{n}(x_i-\mu)^2}$$

$$= \left(\frac{1}{2\pi\sigma^2}\right)^{\frac{n}{2}} \exp\left[\frac{-n}{2\sigma^2}\left\{(\bar{x}-\mu)^2+s^2\right\}\right]$$

$$\therefore \quad \log L = \text{Const.} - \frac{n}{2}\log\sigma^2 - \frac{n}{2\sigma^2}[(\bar{x}-\mu)^2+s^2]$$

(*a*) If σ^2 is known, then in order to estimate for μ, we have

$\dfrac{\partial \log L}{\partial \mu} = 0$ which gives $\bar{x} - \mu = 0$ Hence $\hat{\mu} = \bar{x}$.

(*b*) If μ is known, then to estimate for σ^2, we have

$$0 = \frac{\partial}{\partial\sigma^2}\log L = -\frac{n}{2}\cdot\frac{1}{\sigma^2} + \frac{1}{2}\frac{n}{\sigma^4}[(\bar{x}-\mu)^2+s^2]$$

$$\hat{\sigma}^2 = \frac{1}{n}\cdot\sum_{i=1}^{n}(x_i-\mu)^2$$

(*c*) For simultaneous estimation of μ and σ^2

$$\left.\begin{aligned} 0 &= \frac{\partial}{\partial\mu}\log L = \frac{n}{\sigma^2}(\bar{x}-\mu), \\ \text{and}\quad 0 &= \frac{\partial}{\partial\sigma^2}\log L = -\frac{n}{2}\cdot\frac{1}{\sigma^2} + \frac{1}{2}\frac{n}{\sigma^4}[(\bar{x}-\mu)^2+s^2] \end{aligned}\right\}$$

$$\left.\begin{aligned} \therefore \quad \hat{\mu} &= \bar{x} \\ \text{and} \quad \hat{\sigma}^2 &= \sum_{i=1}^{n}(x_i-\bar{x})^2 \end{aligned}\right\}$$

Example 17·16. *A random sample* $x_1, x_2, \ldots\ldots\ldots, x_n$ *is drawn from the* *exponential population with density function*

$$f(x\,;\alpha,\beta)=y_0\,e^{-\beta(x-\alpha)},\alpha\le x<\infty,\ \beta>0.$$

y_0 being a constant, obtain the maximum likelihood estimators for α and β. **(B.A. Hons. Statistics, Delhi, 1980)**

First we evaluate y_0 in terms of α and β.

We have $\quad y_0\int_{\alpha}^{\infty}e^{-\beta(x-\alpha)}\,dx=1$

This gives $\quad y_0=\beta.$

$$L=\prod_{i=1}^{n}f(x_i\,;\alpha,\beta)=\beta^n\exp\left[-\beta\sum_{i=1}^{n}(x_i-\alpha)\right] \quad ...(17\cdot1)$$

Now $\quad \log L=n\log\beta-n\beta\,(\bar{x}-\alpha)$

$$\therefore\quad 0=\frac{\partial}{\partial\alpha}\log L=n\beta \quad ...(17\cdot1\,a)$$

and $$0=\frac{\partial}{\partial\beta}\log L=\frac{n}{\beta}-n\,(\bar{x}-\alpha)\quad\therefore\quad\beta=\frac{1}{z-\alpha}\cdot \quad ...(17\cdot1\,b)$$

The first equation gives $\beta=0$ which is inadmissible as $\beta>0$ and with this the second equation would fail to determine a finite α. In this case the method of differentiation clearly fails. Our object, however, is to choose α and β so that L is maximum. From (17·1) it is obvious that whatever $\beta>0$), L would be maximum when α takes its largest possible value but since from definition, the population range extends α to ∞, it is obvious that α must be less than or equal to every member of the sample. Thus the greatest value which α can take, consistent with the sample, is the least value of the sample, *i.e.*, x_1 when the sample values are arranged in ascending order. Thus $\hat{\alpha}=x_1$, and then from (17·16), $\hat{\beta}=1/(\bar{x}-x_1)$.

Ex. 1. Show that for the rectangular population with $f(x,\theta)=1/\theta$, $(0\le x\le\theta)$, $0<\theta<\infty$, $=0$, elsewhere, the maximum likelihood estimator for θ is the largest member of the sample, *i.e.*, can be made no larger than $\dfrac{1}{[\max.\,(x_i)]}$ and the unique *m.l.e.* $\hat{\theta}$ of θ is the nth order statistic max. (x_i). Also $E\,[\max.\,(x_i)]$ $=n\,\theta/(n+1)\ne\theta$ *i.e. m.l.e. is biased.* Thus property of unbiasedness is not in general a property of a *m.l.e.*

Ex. 2. Show that for the rectangular population with *p.d.f.* $f(x)=\dfrac{1}{\beta-\alpha}$, $0<\alpha\le x\le\beta)$ the maximum likelihood estimators x_a and x_b for α and β they from a sufficient set of statistics for α and β and their joint distribution is $n\,(n-1)\,(\beta-\alpha)^{-n}\,(x_a-x_b)^{n-2}\,dx_a\,dx_b.$ **(B.A. Hons. Statistics, Delhi, 1989)**

Ex. 3. Show that for the binomial population with density function

$f(x, p) = {}^nc_x p^x q^{n-x}, x = 0, 1, 2, ..., n$ the maximum likelihood estimator for p is x/n and Var $\hat{p} = pq/n$. Show it is also sufficient.

Ex. 4. Find the maximum likelihood estimator for θ for $f(x) = \theta x^{\theta}$, $0 < x < 1$?

[**Ans.** $\hat{\theta} = -n/\log(x_1 x_2 \ldots x_n)$]

Ex. 5. What is the maximum likelihood estimator for α in the density $f(x) = (\alpha + 1)x^{\alpha}, 0 < x < 1$?

[**Ans.** $\hat{\alpha} = -1 - [n/\log(x_1 x_2 \ldots x_n)]$]

Ex. 6. Find the estimate for α in samples of size 2 when the density function is

$$f(x, \alpha) = \frac{2}{\alpha^2}(\alpha - x), 0 < x < \alpha$$

Ex. 7. Let y be a random variable which depends on the parameter x where x is not a random variable. We assume that the random variable y has normal distribution with mean $ax + b$ and *s.d.* σ for every value x, where a, b and $\sigma > 0$ are unknown constants. Let n pairs of observations (x_k, y_k) be given where $y_1, y_2, ..., y_n$ are independent. Determine the constants a, b and σ^2.

[**Hint.** $f(y; a, b, \sigma) = \dfrac{1}{\sigma\sqrt{2\pi}} \exp\left[-\dfrac{(y - ax - b)^2}{2\sigma^2}\right]$

$$\therefore \quad L = \left(\frac{1}{\sigma\sqrt{2\pi}}\right)^n \exp\left[\frac{-1}{2\sigma^2} \sum_{k=1}^{n} (y_k - ax_k - b)^2\right]$$

Equate the partial derivatives to zero.]

[**Ans.** $a = \dfrac{rs_2}{s_1}, b = \bar{y} - \dfrac{rs_2}{s_1}\bar{x}, \sigma^2 = \dfrac{1}{n}\sum_{k=1}^{n}(y_k - ax_k - b)^2 = s_2^2(1 - r^2)$]

Ex. 7. Independent samples of sizes n_1 and n_2 are taken two normal populations with equal mean μ and variances respectively equal to $\lambda\sigma^2, \sigma^2$. Find the *m.l.e.* of μ and show that its large sample variance is

$$\text{Var}(\hat{\mu}) = \sigma^2 / \left(\frac{n_1}{\lambda} + n_2\right)$$

Hence show that the unbiased estimator $t = (n_1\bar{x}_1 + n_2\bar{x}_2)/(n_1 + n_2)$ has efficiency $\lambda(n_1 + n_2)^2/(n_1\lambda + n_2)(n_1 + n_2\lambda)]$ which attains the value 1 if and only if $\lambda = 1$

Solution. Let $x_{1i}, i = 1, 2, ..., n_1$ and $x_{2j}, j = 1, 2, ,......, n_2$ be two independent samples, from the given normal population respectively. Since the samples are independent,

$$L = \text{Likelihood of } (n_1 + n_2) \text{ sample observations}$$

$$= \frac{1}{(2\pi\lambda\sigma^2)^{\frac{1}{2}n_2}} \exp\left[\frac{-1}{2\lambda\sigma^2}\sum_{i=1}^{n_1}(x_{1i} - \mu)^2\right]$$

$$\times \frac{1}{(2\pi\sigma^2)^{\frac{1}{2}n_2}} \exp\left[\frac{-1}{2\sigma^2}\sum_{j=1}^{n_2}(x_{2j} - \mu)^2\right]$$

$$\Rightarrow \qquad \log L = \text{const.} - \frac{1}{2\lambda\sigma^2} \sum_{i=1}^{n_1} (x_{1i} - \mu)^2 - \frac{1}{2\sigma^2} \sum_{j=1}^{n_2} (x_{2j} - \mu)^2$$

Now $\dfrac{\partial}{\partial \mu} \log L = 0$

gives $\hat{\mu} = (n_1 \bar{x}_1 + n_2 \bar{x}_2) \Big/ \left(\dfrac{n_1}{\lambda} + n_2 \right)$

Now $\text{Var}\, \hat{\mu} = \dfrac{1}{\left(\dfrac{n_1}{\lambda} + n_2 \right)^2} \left[\dfrac{n_1^2}{\lambda^2} \text{Var}\, \bar{x}_1 + n_2^2 \text{Var}\, \bar{x}_2 \right] = \text{etc.}$

Also $E(t) = \dfrac{[n_1 E(\bar{x}_1) + n_2 E(\bar{x}_2)]}{n_1 + n_2} = \dfrac{n_1 \mu + n_2 \mu}{n_1 + n_2} = \mu.$

Hence t is an unbiased estimate of μ.

Also $\text{Var}\, t = \dfrac{1}{(n_1 + n_2)^2} [n_1^2 \text{Var}\, \bar{x}_1 + n_2^2 \text{Var}\, \bar{x}\,] = \text{etc.}$

Also efficiency $= \dfrac{\text{Var}\, \hat{\mu}}{\text{Var}\, t} = \dfrac{\lambda (n_1 + n_2)^2}{(n_1 \lambda + n_2)(n_1 + n_2 \lambda)}$

Now efficiency = 1 gives $\lambda (n_1 + n_2)^2 = (n_1 \lambda + n_2)(n_1 + n_2 \lambda)$.

By comparing the coeff. of $n_1 n_2$, we get

$$2\lambda = 1 + \lambda^2 \Rightarrow \lambda = 1.$$

17·3·1. Cramer-Rao Inequality. *Let X be a continuous r.v. with p.d.f. f(x; θ) and likelihood function L. Let t be an unbiased estimator of some function of θ, say τ (θ). then*

$$Var\, t \geq \{ \tau'(\theta) \}^2 / E\left[\left(\frac{\partial}{\partial \theta} \log L \right)^2 \right], \quad \tau'(\theta) = \frac{d}{d\theta} T(\theta)$$

Proof. By hyp., $\tau(\theta) = E(t) = \int t\, L\, dx$

where $L = \prod_{i=1}^{n} f(x_i, \theta)$ and

$$\int f(x_1, \theta) . f(x_2, \theta), \ldots\ldots f(x_n, \theta)\; dx_1, dx_2, \ldots\ldots, dx_n = \int L\, dx$$

$$\therefore \qquad \tau(\theta) = \int t \left(\frac{\partial}{\partial \theta} \log L \right) L\, dx$$

$$= \int \{ t - T(\theta) \} \left\{ \frac{\partial}{\partial \theta} \log L \right\} L\, dx$$

Hence using Cauchy-Schwarz inequality, we have

$$\{\tau'(\theta)\}^2 \leq \int \{t - \tau(\theta)\}^2 L\,dx \cdot \int \left(\frac{\partial}{\partial\theta}\log L\right)^2 L\,dx.$$

or $$Var\,t = E\,[t - \tau(\theta)]^2 / [\tau(\theta)]^2 \geq E\left(\frac{\partial}{\partial\theta}\log L\right)^2$$

or $$Var\,t \geq -[\tau'(\theta)]^2 / E\left(\frac{\partial^2}{\partial\theta^2}\log L\right)$$

Def. Var t is called the maximum variance bound (*MVB*) for estimation of $\tau(\theta)$ and the estimator attaining this bound is known as *MVB* estimator.

Cor. 1. If $t = 0$, then $\tau(\theta) = \theta \quad \therefore \tau'(\theta) = 1$

$$\therefore \qquad \text{Var}\,t \geq 1/E\left(\frac{\partial}{\partial\theta}\log L\right)^2.$$

2. If $\frac{\partial}{\partial\theta}\log L = K\,[t, \tau(\theta)]$ where K is independent of x but may depend on θ, then Cauchy-Schwartz inequality becomes an equality.

Example. 17·17 *Given the p.d.f.* $f(x;\theta) = \frac{1}{\pi}\cdot\frac{1}{1+(x-\theta)^2}$, $-\infty < x < \infty,\ -\infty < \theta < \infty$; *show that the Cramer-Rao bound of variance of an unbiased estimators of* θ *is* $2/n$, *where n is the size of the random sample from this distribution.*

$$f = f(x, \theta) = \frac{1}{\pi}\cdot\frac{1}{1+(x-\theta)^2} \Rightarrow \log f = -\log\pi - \log\{1 + (x-\theta)^2\}$$

$$\frac{\partial \log f}{\partial\theta} = \frac{2\,(x-\theta)}{1+(x-\theta)^2}$$

$$E\left(\frac{\partial}{\partial\theta}\log f\right)^2 = \int_{-\infty}^{+\infty} \frac{4\,(x-\theta)^2}{[1+(x-\theta)^2]^2}\cdot\frac{1}{\pi}\cdot\frac{1}{1+(x-\theta)^2}\,dx$$

$$= \frac{1}{\pi}\int_{-\infty}^{+\infty} \frac{4\,(x-\theta)^2}{[1+(x-\theta)^2]^3}\,dx$$

$$= \frac{2}{\pi}\int_0^{\pi/2} \frac{4\tan^2\varphi}{sec^6\varphi}\,sec^2\varphi\,d\varphi \text{ where } \tan\varphi = x - \theta$$

$$= \frac{2}{\pi}\int_0^{\pi/2} 4\sin^2\varphi\cos^2\varphi = \frac{8}{\pi}\int_0^{\pi/2} (\cos^2\varphi - \cos^4\varphi)\,d\theta$$

$$= \frac{8}{\pi}\left[\frac{1}{2}\cdot\frac{\pi}{2} - \frac{3\cdot1}{4\cdot2}\,\frac{\pi}{2}\right] = \frac{1}{2}, \text{ on simplifications.}$$

$\therefore$ Cramer-Rao lower bound

$$= 1/n\,E\left(\frac{\partial}{\partial\theta}\log f\right)^2 = \frac{1}{n\left(\frac{1}{2}\right)} = \frac{2}{n}.$$

Ex. 1. Let $x_1, x_2, \ldots\ldots, x_n$ be a random sample drawn from $N(\theta, \sigma^2)$ where σ^2 is known. Find the Cramer-Rao lower bound for θ. (**Ans.** σ^2/n)

Ex. 2. Let $x_1, x_2, \ldots\ldots, x_n$ be a random sample of size n from $f(x, \theta) = \frac{1}{\theta} e^{-x/\theta}$, $0 < x < \infty, 0 < \theta < \infty$. Obtain Cramer-Rao lower bound unbiased estimator of θ. (**Ans.** θ^2/n) (***B.A. Hons. Statistics, Delhi, 1989***)

17·3·2. Independence tests in contingency tables. Let p_{ij} be the probability that an element selected at random from the universe belongs to the ith class with respect to the attribute A and jth class with respect to the attribute B (Refer § 16· 2 · 1) and let $p_{i\cdot}$ and $p_{\cdot j}$ be their respective boundary probabilities, then according to the hypothesis that the attributes A and B are independent in the universe whence the observed sample has been made.

$$p_{ij} = p_{i\cdot}\,p_{\cdot j}$$

for every pair (i, j), where

$$\sum_{i=1}^{s} p_{i\cdot} = \sum_{j=1}^{t} p_{\cdot j} = 1. \qquad \ldots(17\cdot2)$$

The hypothesis H_0 formulated in this manner does not define the values of $s + t$ unknown parameters, *i.e.*, the values $p_{\cdot j}$ and $p_{i\cdot}$ Since we can determine two of these parameters from (17·2), there remains $s + t - 2$ unknown parameters. To verify the hypothesis H_0 by applying a χ^2 – test we would determine the unknown parameters from the sample by means of maximum likelihood principle. We have

$$L = \prod_{i=1}^{s}\prod_{j=1}^{t} p_{ij}^{o_{ij}} = \prod_{i=1}^{s}\prod_{j=1}^{t} (p_{i\cdot}p_{\cdot j})^{o_{ij}} = \prod_{i=1}^{s}\prod_{j=1}^{t} p_{i\cdot}^{o_{ij}}\, p_{\cdot j}^{o_{ij}}$$

$$= \left[\prod_{i=1}^{s} p_{i\cdot}^{o_{i\cdot}}\right]\left[\prod_{j=1}^{t} p_{\cdot j}^{o_{\cdot j}}\right]$$

$$= \left\{1 - \sum_{i=1}^{s-1} p_{i\cdot}\right\}^{o_{s\cdot}}\left\{1 - \sum_{j=1}^{t-1} p_{\cdot j}\right\}^{o_{\cdot t}} \sum_{i=1}^{s-1} p_{i\cdot}^{o_{i\cdot}} \prod_{j=1}^{t-1} p_{\cdot j}^{o_{\cdot j}}$$

since from (17·2), $p_{s\cdot} = 1 - \sum_{i=1}^{s-1} p_{j\cdot}, p_{\cdot t} = 1 - \sum_{j=1}^{t-1} p_{\cdot j}.$

and $\log L = O_{s\cdot} \log\left(1 - \sum_{i=1}^{s-1} p_{i\cdot}\right) + O_{\cdot t} \log\left(1 - \sum_{j=1}^{t-1} p_{\cdot j}\right) + \sum_{i=1}^{s-1} O_{i\cdot} \log p_{i\cdot}$

$$+ \sum_{j=1}^{t-1} O_{\cdot j} \log p_{\cdot j}$$

$$\therefore \frac{\partial}{\partial p_{i\cdot}} \log L = -\frac{O_{s\cdot}}{s-1} + \frac{O_{i\cdot}}{p_{i\cdot}} = \frac{O_{i\cdot}}{p_{i\cdot}} - \frac{O_{s\cdot}}{p_{s\cdot}} = 0, i = 1, 2, \ldots, s-1 \; ; \; 1 = \sum_{i=1}^{s} p_{i\cdot}$$

$$\frac{\partial}{\partial p_{\cdot j}} \log L = -\frac{O_{\cdot i}}{s-1} + \frac{O_{\cdot j}}{p_{\cdot j}} = \frac{O_{\cdot j}}{p_{\cdot j}} - \frac{O_{\cdot t}}{p_{\cdot t}} = 0, j = 1, 2, \ldots, t-1 \; ; 1 = \sum_{j=1}^{t} p_{\cdot j}$$

$$\therefore \quad p_{i\cdot} = \frac{O_{i\cdot}}{C} \text{ where } C = \frac{O_{s\cdot}}{p_{s\cdot}}, i = 1, 2, \ldots, s$$

and $\quad p_{\cdot j} = \dfrac{O_{\cdot j}}{D}$ where $D = \dfrac{O_{\cdot t}}{p_{\cdot j}}, j = 1, 2, \ldots, t$

Hence $\displaystyle\sum_{i=1}^{s} p_{i\cdot} = \frac{\sum_{i=1}^{s} O_{i\cdot}}{C} = \frac{N}{C} = 1$

and $\quad \displaystyle\sum_{j=1}^{t} p_{\cdot j} = \frac{\sum_{j=1}^{t} O_{\cdot j}}{D} = \frac{N}{D} = 1$

Thus $C = D = N \; ; \therefore \; p_{i\cdot} = \dfrac{O_{i\cdot}}{N} \; ; p_{\cdot j} = \dfrac{O_{\cdot j}}{N}$

$$\therefore \qquad \chi^2 = \sum_{i=1}^{s} \sum_{j=1}^{t} \frac{\left(O_{ij} - \dfrac{O_{i\cdot} O_{\cdot j}}{N} \right)^2}{O_{i\cdot} O_{\cdot j}}$$

Since we have determined $(s + t - 2)$ parameters from the sampling,it follows that Pearson statistics χ^2 has approximately a χ^2 distribution with $st - (s + t - 2) - 1 = (s - 1)(t - 1)$ *d.f.* Therefore, we can apply a χ^2 test to verify hypothesis H_0 that the variables A and B are independent.

17·3·3. Properties of Maximum Likelihood Estimators. We state below, without proof, some of the important properties of maximum likelihood estimators, which give a strong *a positeriori* justification for the use of this method, whenever it can be applied.

(*a*) Maximum likelihood estimators are, in general, not un-biased.

(*b*) *Maximum likelihood estimators are consistent.* More precisely, if (*i*) the density function $f(x, \theta)$, is continuous in x throughout its range and if

(*ii*) $f(x,\theta)$ is continuous and monotonic in θ in some θ interval containing the true value θ_0 and for all x, in some x-interval, then the maximum likelihood estimator, say $\hat{\theta}$, is consistent.

(*c*) *The distribution of maximum likelihood estimator tends to normality for large samples.* More precisely, if

(*i*) $f(x,\theta)$ is continuous in x throughout its range, and if

(*ii*) in a θ-interval containing the true value θ_0, $\dfrac{\partial f}{\partial \theta}$ is continuous in θ for every x, $x^2 \dfrac{\partial f}{\partial \theta}$ approaches continuous function of θ as x tends to infinity, and $\dfrac{\partial f}{\partial \theta}$ does not vanish in some interval, then for large n, the maximum likelihood estimator θ will tend to be normally distributed with variance given by

$$\frac{1}{\text{var}\,\hat{\theta}} = \int_{-\infty}^{\infty} \frac{1}{f}\left(\frac{\partial f}{\partial \theta}\right)^2 dx = n\left(\int_{-\infty}^{\infty} \frac{\partial}{\partial \theta}\log f\right)^2 \cdot dx$$

$$= nE\left(\frac{\partial}{\partial \theta}\log f\right)^2, \qquad \text{...(17·2 a)}$$

where θ is to be put equal to θ_0 on the right.

If the range is independent of f, or if f and $\partial f/\partial \theta$ vanish at the extremity of the range, which depends on θ, we have an alternative form for variance, *viz.*,

$$\frac{1}{\text{var}\,\hat{\theta}} = -\,nE\left(\frac{\partial^2}{\partial \theta^2}\log f\right) \qquad \text{...(17·2}b\text{)}$$

(*d*) *Maximum likelihood estimators are most efficient, i.e.*, in the class of estimators, which for large n tend to be normally distributed about population parameters as mean, the variance of the maximum likelihood estimators will be less than or equal to that of any other estimator, *i.e.*, if t is any other such estimator

$$\text{var}(t) \geq \text{var}(\hat{\theta}).$$

(*e*) *Maximum likelihood estimators are sufficient, if sufficient estimators exist.* In other words, if a sufficient estimator exists, it is a function of the maximum likelihood estimator.

(*f*) *Maximum likelihood estimators have the invariance property i.e.*, if θ is a maximum likelihood estimator for θ, then $f(\hat{\theta})$ will be a maximum likelihood estimator for $f(\theta)$.

Example 17·18. *Find the efficiency of the median as an estimator for the population mean of the Cauchy distribution.*

We have seen (Ex. 14·4, page 543) that for large samples,

$$\text{var (median)} = \pi^2/4n.$$

To find the efficiency, we have to find the variance of the most efficient estimator, which would be maximum likelihood estimator, if it exists for Cauchy distribution

$$L = \left(\frac{1}{\pi}\right)^n \prod_{i=1}^{n} \left[\frac{1}{1+(x_i-\mu)^2}\right]$$

$$\therefore \qquad -\frac{\partial}{\partial\mu} \log L = \sum_{i=1}^{n} \frac{2\,(x_i-\mu}{1+(x_i-\mu)^2} = 0.$$

It is not easy to solve this equation for μ but the variance of μ can be easily found by using (17·2 *b*).

$$\frac{1}{\text{var}\,(\hat{\mu})} = -nE\left(\frac{\partial^2}{\partial\mu^2}\log f\right) = -\frac{n}{\pi}\int_{-\infty}^{\infty} \frac{2(x-\mu)^2-2}{[1+(x-\mu)^2]^3}\,dx = \frac{n}{2}\cdot$$

Hence $\qquad \text{var}\,(\hat{\mu}) = 2/n$

$\therefore \qquad$ Efficiency of the median $= 8/\pi^2 \cong 0{\cdot}8$.

Ex. 1. Show that in estimating λ for a Poisson distribution with parameter λ.

$$\text{var}\,(\hat{\lambda}) = \lambda/n.$$

Ex. 2. Show that in estimating μ for normal population with known variance σ^2.

$$\text{var}\,(\hat{x}) = \sigma^2/n\,\cdot$$

Ex. 3. Show that if a most efficient estimator *t* of a parameter θ exists, the maximum likelihood equation will have a unique solution equal to *t*. Also prove that if a sufficient estimator exists the maximum likelihood solution will be a function of the sufficient estimator.

Ex. 4. Deduce [17·2 (*b*), from 17·2 (*a*) page 543], using integration by parts.

17·4. Other Methods of Estimation. In order to estimate a population parameter, the theory of probability which is purely deductive has to be supplemented by some axioms to make inductive inference possible. That the estimator should make likelihood function a maximum is one such axiom. We study below five other axioms giving five other methods of estimation. *A priori* which axiom is more plausible can be a matter of opinion, but any method of estimation will be judged by the properties of the estimators it gives rise to.

17·4·1. Method of Moments. This method, due to Karl Pearson, is the oldest and is freely used in fitting distributions, specially of the

Pearson type. According to it, to choose m parameters of a population, we equate, in general, the first m moments of the sample values to the first m moments of the population and solve from these m equations for the m estimators. Fisher, however, showed that these estimators are, in general, inefficient, *i.e.*, they do not have minimum variance in the class of estimators distributed normally for large n. For normal population, the method of moments gives sample mean and variance as estimators for population mean and variance and these estimators are efficient, but from this point of view the normal distribution is an exception and for all other Pearson distributions, moment- estimators are not efficient.

17·4·2. Method of Minimum Variance. The axiom for this method is that we choose those estimators which are unbiased and have minimum variance. Thus, if t_n is the estimator for θ, we should have $E(t_n) = \theta$, Var (t_n) should be minimum.

In other words, if $L(x_1, x_2, ..., x_n, \theta') \equiv L$,

$$\int L dx = 1.\ dx_1 dx_2 \dots dx_n = dx,$$

$$\int \dots \iint t_n L(x_1, x_2, \dots x_n, \theta')\, dx_1\, dx_2 \dots dx_n = \theta \Rightarrow \int t_n L dx = \theta.$$

and $$\int \dots \iint (t_n - \theta)^2 L(x_1, x_2, ..., x_n, \theta)\, dx_1\, dx_2 \dots dx_n$$

or $$\int (t_n - \theta)^2 L dx = \text{minimum}.$$

By using calculus of variations, it can be showed that if there exists an estimator t_n satisfying these conditions, we can express

$$\frac{\partial}{\partial \theta} \log L = \frac{t_n - \theta}{\lambda}, \qquad ...(1)$$

where λ may depend on θ, but not on $x_1, x_2, ..., x_n$. In fact λ would be the variance of t_n.

We have $\int L dx = 1 \Rightarrow \int \frac{\partial L}{\partial \theta} dx = 0 \Rightarrow \int \left(\frac{1}{L} \frac{\partial L}{\partial \theta}\right)(L\, dx) = 0$

$$\Rightarrow \int \left(\frac{\partial}{\partial \theta} \log L\right) \cdot L\, dx = 0 \Rightarrow E\left(\frac{\partial}{\partial \theta} \log L\right) = 0 \qquad ...(2)$$

Also $$E\left(\frac{\partial^2}{\partial \theta^2} \log L\right) = \int \left(\frac{\partial^2}{\partial \theta^2} \log L\right) L\, dx$$

$$= \frac{\partial}{\partial \theta}\left[\left(\frac{\partial}{\partial \theta} \log L\right) L\, dx - \int \left(\frac{\partial}{\partial \theta} \log L\right)^2 L\, dx\right]$$

$$= \frac{\partial}{\partial \theta}\left[E\left(\frac{\partial}{\partial \theta} \log L\right) - E\left(\frac{\partial}{\partial \theta} \log L\right)^2\right]$$

$$= \theta - E\left(\frac{\partial}{\partial\theta}\log L\right)^2, \text{ using (2)}$$

$$\therefore \quad E\left(\frac{\partial}{\partial\theta}\log L\right)^2 = -E\left(\frac{\partial^2}{\partial\theta^2}\log L\right)$$

Ex. 1. For the normal population, show that $\bar{x}$ is the minimum variance estimator and its variance is σ^2/n.

Ex. 2. For the Cauchy population, show that a minimum variance estimator does not exist.

Hint : Cauchy population is

$$dF(x, \theta) = \frac{1}{\pi} \cdot \frac{1}{1+(x-\theta)^2}\, dx, -\infty < x < \infty$$

$$= f(x, \theta)\, dx$$

$$\therefore \quad L = \prod_{i=1}^{n} f(x_i, \theta) = \frac{1}{\pi^n} \prod_{i=1}^{n} \left[\frac{1}{1+(x_i-\theta)^2}\right]$$

$$\Rightarrow \quad \log L = -n \log \pi - \sum_{i=1}^{n} \log\left[1+(x_i-\theta)^2\right]$$

$$\therefore \quad \frac{\partial}{\partial\theta}\log L = 2 \sum_{i=1}^{n} \frac{x_i-\theta}{1+(x_i-\theta)^2}$$

Thus $\frac{\partial}{\partial\theta}\log L$ cannot be expressed in the form (1). Hence the result.

Ex. 3. Show that for a minimum variance estimator t_n,

$$\frac{1}{\text{var}(t_n)} = -nE\left(\frac{\partial^2}{\partial\theta^2}\log f\right)$$

holds for all n.

Ex. 4. For the type III population,

$$\text{const. } e^{\frac{-x}{\theta}} x^{p-1}\, dx, 0 < x < \infty, p > 1.$$

where p is known, show that the minimum variance estimator for θ is $\bar{x}/p$ and its variance is θ^2/np.

Ex. 5. (*a*) The random variate X follows the frequency distribution $qp^{x-1}, x = 1, 2, \ldots ; q = 1-p$. A sample of n values $x_1, x_2, \ldots, x_n$ is drawn from this distribution. Find the maximum likelihood estimate of p and calculate its variance. Discuss whether this estimate is unbiased when $n = 1$.

$$\left[\textbf{Ans. } \hat{p} = 1 - \frac{1}{x}\right]$$

(*b*) Let $x_1, x_2, \ldots, x_n$ be a random sample from *B.D.* with parameter θ,

$0 < \theta < 1$. Obtain a sufficient statistic for θ and show that it is complete. Hence obtain *MVE* of θ. **(*B.A. Hons. Statistics, Delhi, 1989*)**

17·4·3. Method of Minimum Chi-square. The axiom of this method is that we choose estimators for the population parameters in such a way as to minimise

$$\chi^2 = \sum_i \frac{(o_i - e_i)^2}{e_i}$$

where e_i's are expected frequencies calculated from this population. In practice a modified version of this where we minimize

$$\chi'^2 = \sum_i \frac{(o_i - e_i)^2}{o_i}$$

is used.

Now the likelihood function, for getting o_1 members in the first cell, o_2 in the second cell ... and o_r in the rth cell, is obtained by using the multinomial distribution

$$L = \frac{n\,!}{o_1\,!\,o_2\,!\,\ldots\,o_r\,!}\left(\frac{e_1}{o_1}\right)^{o_1}\left(\frac{e_2}{o_2}\right)^{o_2}\ldots\left(\frac{e_r}{o_r}\right)^{o_r}$$

$$\therefore \quad \log L = \text{const.} + \sum_{i=1}^{r}\left[o_i \log\left(\frac{e_i}{o_i}\right)\right]$$

For large samples we may put

$$\Rightarrow \qquad e_i = o_i + a_i n^{\frac{1}{2}},$$

where a_i is finite and, therefore, small compared with o_i ;

$$|\,a_i n^{\frac{1}{2}}\,| < o_i \text{ and } \sum_i a_i = 0$$

$$\therefore \quad \log L = C + \sum_{i=1}^{r} o_i\left[\log\left(1 + \frac{a_i n^{1/2}}{o_i}\right)\right], \text{ where } C \text{ is a constant}$$

$$= C - \tfrac{1}{2}\Sigma\frac{na_i^2}{o_i} + O(n^{-\frac{1}{2}})$$

$$= C - \tfrac{1}{2}\Sigma\chi'^2 + O(n^{-\frac{1}{2}}) \qquad \text{...(17·3)}$$

Again

$$\chi^2 - \chi'^2 = \Sigma(o_i - e_i)^2\left(\frac{1}{e_i} - \frac{1}{o_i}\right)$$

$$= \Sigma \frac{(e_i - o_i)^2}{o_i} \left[\left(1 + \frac{a_i n^{1/2}}{o_i} \right)^{-1} - 1 \right]$$

$$= O(n^{-\frac{1}{2}}) \qquad \text{...(17·4)}$$

From (17·3) and (17·4), we see that for large samples, up to order $n^{-\frac{1}{2}}$, the estimators obtained by minimising χ'^2 or χ^2 are equivalent to the estimators obtained by maximizing the likelihood function.

17·4·4. Method of Inverse Probability. The axiom of this method is that for a given '*a priori*' density function $g(\theta)$ of the parameter θ, we should choose our estimator θ so that the '*a posteriori*' probability function for θ as obtained from the given random sample is a maximum. Now according to Baye's theorem.

"*A Posteriori*" probability $P \propto$ likelihood function $\times$ '*a priori*' density function of θ

so that $$\frac{\partial}{\partial \theta} \log P \propto \frac{\partial}{\partial \theta} \log L + \frac{\partial}{\partial \theta} \log g(\theta)$$

Since on the R.H.S., the first factor has ordinarily a variance of order n, while the second term is independent of n, it appears that, for large n, the inverse probability estimators will approach the maximum likelihood estimators, whatever be the density function $g(\theta)$

According to the relative frequency definition of probability, θ cannot have a density function (except the trivial one, *viz.*, that $P(\theta = \theta_0) = 1, P(\theta \neq \theta_0) = 0$ where θ_0 is the true value). According to Baye's postulate $g(\theta)$ is constant for a finite range and according to others there is a definite interpretation of $g(\theta)$ as denoting '*a priori*' measure of belief. The method has been subject to a great deal of controversy.

17·4·5. Method of Least Squares. The application of this method in the estimation of the parameters $b_0, b_1, ..., b_p$ for fitting a curve $Y = b_0 + b_1 x + ... + b_p x^p$ has already been discussed in chapter 9.

In the more general case when we have n equations

$$\phi(x_i, \theta_0, \theta_1, ..., \theta_p) = 0 \; ; i = 1, 2, ..., n$$

to solve for $(p+1)$ parameters $\theta_0, \theta_1, ..., \theta_p$ and $n > p+1$, we apply the method of least squares and choose the parameters so as to minimize the sum of squares of residuals, *i.e.*, choose θ's so as to satisfy the $(p+1)$ equations

$$\frac{\partial}{\partial \theta_j} \Sigma\{\phi(x_i, \theta_0, \theta_1, ..., \theta_p)\}^2 = 0, j = 0, 1, 2, ..., p$$

In the case when the residuals are distributed normally with variance σ^2, the logarithm of likelihood function is given by

$$\log L = \text{const.} - n \log \sigma - \frac{1}{2\sigma^2} - \sum_{i=1}^{p} \phi^2 (x_i, \theta_0, \theta_1, ..., \theta_p).$$

This is clearly examined by minimizing the sum of squares of residuals so that in this case the method of least squares will yield the same result of the method of maximum likelihood.

(For examples, see problems in Statistics by H.C. Saxena, published by S. Chand & Co.)

17·5. Interval Estimation. In the last three sections we have discussed the criteria which optimum '*point estimators*' should satisfy and the method by which such estimators may be obtained. A single estimator, however good it may be, is not expected to coincide with the population parameter and the distribution of errors is determined by the sampling distribution of the estimator. It is, therefore, desirable to give for any population parameter an interval in which the population parameter may be expected to lie, with a specified degree of confidence. This is the problem of 'interval estimation' and the intervals for the parameters are called *confidence intervals*.

We have already determined confidence intervals for a number of cases.

Confidence Interval for the Mean of any Population For Large Samples

Consider the estimation of the population mean from a random sample $x_1, x_2, ..., x_n$ drawn from a normal population with known variance σ^2. Since $\bar{x}$, in this case, is distributed normally about mean μ with variance σ^2/n, we have

$$P\left[-1{\cdot}96 \le \frac{\bar{x} - \mu}{\sigma/\sqrt{n}} \le 1{\cdot}96\right] = {\cdot}95 \qquad ...(17{\cdot}5)$$

or

$$P\left[\bar{x} - 1{\cdot}96 \frac{\sigma}{\sqrt{n}} \le \mu \le \bar{x} + 1{\cdot}96 \frac{\sigma}{\sqrt{n}}\right] = {\cdot}95 \qquad ...(17{\cdot}6)$$

and the interval $(\bar{x} - 1{\cdot}96\, \sigma/\sqrt{n}, \bar{x} + 1{\cdot}96\sigma/\sqrt{n})$ is called the 95% confidence interval for the population mean μ. What does this statement mean ?

We note first that it does not mean that the probability for μ to lie in this interval is ·95, for this would imply that μ is a random variate, which is not the case, μ is a fixed, though unknown, constant and either the specified interval contains μ or it does not contain μ. If it contains μ, the probability is 1, if it does not contain μ, the probability is zero.

Though μ is not a random variate the interval $\bar{x} \pm 1{\cdot}96\, \sigma/\sqrt{n}$ is a random interval, since its limits depend on chance and would be different for different samples. If we take a large number of random samples, we shall get a corresponding large number of random intervals and equation (17·6) is interpreted to mean that on an average, 95% of these intervals are expected to include the true population parameter μ.

To make it clear that in (17·6) μ is not to be interpreted as if μ is a random variate, we write it as

$$P_F\left[\bar{x} - 1{\cdot}96 \frac{\sigma}{\sqrt{n}} \leq \mu \leq \bar{x} + 1{\cdot}96 \frac{\sigma}{\sqrt{n}}\right] = {\cdot}95 \quad \ldots(17{\cdot}7)$$

and call it the *fiducial probability* and this gives the probability even before drawing the sample that the resulting interval will contain the true parameter value. Another alternative way of writing this is

$$P\,[\mu\ \varepsilon\ \delta_n] = {\cdot}95$$

where δ_n is the interval $\bar{x} - 1{\cdot}96\, \sigma/\sqrt{n}, \bar{x} + 1{\cdot}96\, \sigma/\sqrt{n}$ and 'ε' is read and 'is contained in' or is 'covered by'.

It may be noted that even for the same confidence ·95, say we can have any number of confidence intervals. Thus from the normal distribution table, we find

$$P_F\left[\bar{x} - 1{\cdot}68 \frac{\sigma}{\sqrt{n}} \leq \mu \leq \bar{x} + 2{\cdot}70 \frac{\sigma}{\sqrt{n}}\right] = {\cdot}95$$

$$P_F\left[\bar{x} - 1{\cdot}86 \frac{\sigma}{\sqrt{n}} \leq \mu \leq \bar{x} + 2{\cdot}08 \frac{\sigma}{\sqrt{n}}\right] = {\cdot}95$$

But these intervals are of different lengths and for obvious reasons we prefer, for a given confidence cofficient, the intervals of shortest length.

Ex. 1. Prove :

$$P_F\left(\bar{x} - \frac{2\sigma}{\sqrt{n}} \leq \mu \leq \bar{x} + \frac{2\sigma}{\sqrt{n}}\right) = P_F\left(\bar{x} - \frac{2{\cdot}326}{\sqrt{n}} \leq \mu \leq \bar{x} + \frac{1{\cdot}806\sigma}{\sqrt{n}}\right) = {\cdot}9543$$

Ex. 2. Find at least four confidence intervals for μ with confidence coefficient ·99.

Ex. 3. Show that for a symmetrical unimodal distribution, the confidence intervals of shortest length is central.

17·5·1. 95 per cent confidence intervals for special cases. In earlier chapters we have found confidence intervals for a number of population parameters. In each case our success was due to our *finding a suitable* *function of sample and parameter whose distribution was independent of the population parameters.* We summarise the earlier results on the next page and then proceed to find confidence interval in a few more cases. For definiteness we give confidence limits corresponding to a confidence coefficient 95.

Example 17·19. *Show that for the rectangular distribution*

$$dP = \frac{dx}{\theta}, \; 0 \leq x \leq \theta$$

with confidence cofficient α, *confidence limits for* θ *are* R *and* R/ψ_α, *where* R *is the sample range and* ψ_α is given by

$$\psi_\alpha^{\,n-1}\,[n - (n-1)\,\psi_\alpha] = 1 - \alpha.$$

The distribution of R has been obtained in § 14·5, as

$$dP = n(n-1)\theta^{-n}\,R^{n-2}\,(\theta - R)\,dR, \quad 0 \leq R \leq \theta.$$

Therefore the distribution of $\psi = R/\theta$

is
$$dP = n(n-1)\,\psi^{n-2}\,(1-\psi)\,d\psi, \quad 0 \leq \psi \leq 1.$$

Thus we have found ψ, a function of the sample and parameter, whose distribution is independent of the parameter θ. Now let ψ_n be defined by

$$\int_{\psi_\alpha}^{I} n\,(n-1)\,\psi^{n-2}\,(1-\psi)\,d\psi = \alpha$$

i.e.,
$$\psi_\alpha^{\,n-1}\,[n - (n-1)\,\psi_\alpha\,] = 1 - \alpha$$

Now
$$P_F\,[\psi_\alpha \leq \psi \leq 1] = \alpha$$

or
$$P_F\left[\psi_\alpha \leq \frac{R}{\theta} \leq 1\right] = \alpha$$

or
$$P_F\left[R \leq \theta \leq \frac{R}{\psi_\alpha}\right] = \alpha$$

so that the confidence limits for θ are R and R/ψ_α.

Example 17·20. *Obtain the 95% confidence region for mean and variance of normal distribution.*

For obtaining the confidence limits for μ and σ^2 separately, we have used respectively the t and χ^2 distributions, but for the two together we cannot combine these since t and χ^2 are not independently distributed ; but in §13·3 we saw that the mean and variance of a random sample from a normal distribution are independently distributed. Accordingly we find the three numbers a, b_1, b_2 such that

	Population or Populations	Parameter	Confidence limits	Ref.
1.	Normal with known variance σ^2	μ	$\bar{x} \mp 1{\cdot}96 \dfrac{\sigma}{\sqrt{n}}$	§12·7
2.	Normal with unknown variance	μ	$\bar{x} \mp \dfrac{t_{0\cdot05}}{\sqrt{n}} \sqrt{\sum_{i=1}^{n} \dfrac{(x_i-\bar{x})^2}{n-1}}$	§15·2
3.	Two normal with unknown variances σ_1^2, σ_2^2	$\mu_1-\mu_2$	$\bar{x}_1 - x_2 \mp 1{\cdot}96 \sqrt{\dfrac{\sigma_1^2}{n_1}+\dfrac{\sigma_2^2}{n_2}}$	§12·8
4.	Two normal with unknown but same variances	$\mu_1-\mu_2$	$\bar{x}_1 - x_2 \mp t_{0\cdot05} S \sqrt{\dfrac{1}{n_1}+\dfrac{1}{n_2}}$	§15·2
5.	Bivariate normal	ρ	$\tanh\left[\tanh^{-1} r_1 \mp 1{\cdot}96 \sqrt{\dfrac{1}{n-3}}\right]$	§15·5
6.	Bivariate normal	$\rho_1-\rho_2$	$\tanh\left[\tanh^{-1} r_1 - \tanh^{-1} r_2 \mp 1{\cdot}96 \sqrt{\dfrac{1}{n_1-3}+\dfrac{1}{n_2-3}}\right]$	§15·5
7.	Normal	σ^2	$\dfrac{\Sigma(x_i-\bar{x})^2}{\chi^2_{0\cdot025}}, \dfrac{\Sigma(x_i-\bar{x})^2}{\chi^2_{0\cdot975}}$	§16·4
8.	Binomial	p	$\dfrac{x}{n} \mp 1{\cdot}96 \sqrt{\dfrac{\dfrac{x}{n}\left(1-\dfrac{x}{n}\right)}{n}}$	§12·3
9.	Binomial	p_1-p_2	$\dfrac{x_1}{n_1} - \dfrac{x_2}{n_2} \mp 1{\cdot}96 \sqrt{\dfrac{\dfrac{x_1}{n}\left(1-\dfrac{x_1}{n_1}\right)}{n_1} + \dfrac{\dfrac{x_2}{n_2}\left(1-\dfrac{x_2}{n_2}\right)}{n_2}}$	§12·4
10.	Normal	σ_1^2/σ_2^2	$\dfrac{S_1^2}{S_2^2} \cdot \dfrac{1}{F_{0\cdot025}}, \dfrac{S_1^2}{S_2^2} \cdot \dfrac{1}{F_{0\cdot975}}$	§15·3
11.	Normal		$\beta b \mp t_{0\cdot05}\{[n-2)\sum_i (x_i-\bar{x})]^2/\sum_i (y_i-Y_i)^2\}^{-\frac{1}{2}}$	§15·2

$$P\left[-a \le \frac{\bar{x}-\mu}{(\sigma/\sqrt{n})} \le a\right] = 0{\cdot}95$$

and $$P\left[b_1 \le \frac{\Sigma(x_i-\bar{x})^2}{\sigma^2} \le b_2\right] = 95$$

so that $$P\left[-a \le \frac{\bar{x}-\mu}{\sigma/\sqrt{n}} \le a\,;\, b_1 \le \frac{\Sigma(x_i-\bar{x})^2}{\sigma^2}\right] = {\cdot}95$$

because of the independence of the two distributions.

Thus the 95% confidence region is the area bounded by the straight lines

$$\sigma^2 = \frac{\Sigma(x_i-\bar{x})^2}{b_1}\,;$$

$$\sigma^2 = \frac{\Sigma(x_i-\bar{x})^2}{b_2}$$

and the parabola

$$(\mu-\bar{x})^2 = \frac{a^2\sigma^2}{n}$$

in the parametric (μ, σ^2) space.

Example 17·21. *Given one observation from a population with p.d.f.*

$$f(x, \theta) = \frac{2}{\theta^2}(\theta - x),\ 0 \le x \le \theta$$

Obtain $100(1-\alpha)\%$ *confidence interval for* θ.

(B.Sc. Stat. Hons. Delhi, 1991)

Let us find the p.d.f. of $u = x/\theta$. $\therefore\ dx = \theta\, du$

Now $$g(u) = f(x, \theta)\left|\frac{dx}{du}\right| = \frac{2}{\theta^2}(\theta - x)\,.\,\theta\, du$$

$$= 2(1-u),\ 0 \le u \le 1$$

Now we choose two quantities a and b such that

$$P[a \le u \le b] = 1 - \alpha$$

and $$P[u < a] = P[u > b] = \frac{\alpha}{2}$$

$$\therefore \quad \int_0^a 2(1-u)\,du = \frac{\alpha}{2} \quad \text{and} \quad \int_b^1 2(1-u)\,du = \frac{\alpha}{2}$$

$$\therefore \quad a^2 - 2a + \frac{\alpha}{2} = 0 \quad \text{and} \quad b^2 - 2b + \left(1 - \frac{\alpha}{2}\right) = 0 \qquad \ldots(1)$$

$$\therefore \quad P\left[a \le \frac{x}{\theta} \le b\right] = 1 - \alpha \Rightarrow P\left[\frac{x}{b} \le \theta \le \frac{x}{a}\right] = 1 - \alpha$$

Hence the required interval for θ is $\left(\frac{x}{b}, \frac{x}{a}\right)$ where a and b are given by (1).

Ex. 1. The largest observation L of a sample n from a rectangular distribution with density function $f(x) = \frac{1}{\theta}$, $(0 \le x \le \theta)$ has the distribution

$$dP = n\left(\frac{L}{\theta}\right)^{n-1} \frac{dL}{\theta}, \qquad 0 \le L \le \theta$$

so that the distribution of $v = L\theta^{-1}$ is

$$dP = n\, v^{n-1}\, dv, \qquad \theta \le v \le 1$$

and is independent of θ. Hence deduce that the confidence limits for confidence coefficient α are

$$L, L/(1-\alpha)^{\frac{1}{n}}.$$

Find also the shortest confidence interval.

Ex. 2. Show that the length and the variance of the length of the t confidence interval approaches zero with increasing sample size.

Ex. 3. Show that the length of the confidence interval for σ^2 from normal population with variance σ^2 from a normal population with variance σ^2 approaches zero with sample size.

Ex. 4. Compute two different 95% confidence intervals for the range of a rectangular distribution, given the sample 1·5, 2·4, 3·8, 4·2, 2·8, given that the lower limit of the range is zero.

Ex. 5. A random sample of two members from a rectangular population with range θ is $9, -1$. Obtain 81% confidence limits for θ from the sample.

Ex. 6. Show that the 95% confidence limits for the parameter α in a random sample of size one from the population $f(x, \theta) = \frac{2}{\theta^2}(\theta - x), 0 < x < \theta$ are

$$\frac{2x}{1-\sqrt{0{\cdot}25}}, \frac{2x}{1-\sqrt{{\cdot}975}}.$$

$$\left[\textbf{Hint. } \frac{\theta - x}{\theta^2} \text{ is distributed independently of } \theta.\right]$$

17·5·2. Confidence Intervals for Large Samples. We know that, under very general conditions, the maximum likelihood estimator $\hat{\theta}$ is distributed normally about mean θ with variance

$$\sigma^2(\hat{\theta}) = \frac{-1}{nE\left[\frac{\partial^2}{\partial\theta^2}\log f(x, \theta)\right]}$$

$\therefore$ For large samples

$$P\left[-1{\cdot}96 \le \frac{\hat{\theta}-\theta}{(\sigma\,\hat{\theta})} \le 1{\cdot}96\right] = {\cdot}95$$

For large samples, $\sigma(\theta)$ may be replaced by $\sigma(\hat{\theta})$ so that

$$P_F\,[\hat{\theta} - 1{\cdot}96\sigma(\hat{\theta}) \le \theta \le \hat{\theta} + 1{\cdot}96\,\sigma\,(\hat{\theta})] = {\cdot}95$$

Similarly the confidence interval for any confidence coefficient α can be obtained.

Example 17·22. *Show that for the distribution*

$$dP = \theta\, e^{-x\theta}\, dx, \quad 0 \le x < \infty$$

central confidence limits for large sample with $\alpha = 0{\cdot}95$ *are given by*

$$\theta = \frac{1 \pm \dfrac{1{\cdot}96}{\sqrt{n}}}{\bar{x}}$$ ***(B.Sc. Stat. Hon. Delhi, 1990)***

Here $\qquad L = \theta^n\, e^{-(x_1 + x_2 + \dots + x_n)\theta}$

$\therefore \qquad \log L = n \log \theta - n\bar{x}\theta$

Hence $\quad \dfrac{\partial}{\partial\theta} \log L = \dfrac{n}{\theta} - n\bar{x}.$

$\therefore$ maximum likelihood estimator of θ is $\hat{\theta} = 1/\bar{x}$

$$\text{Also Var}\,\hat{\theta} = \frac{1}{nE\left[\dfrac{\partial^2}{\partial\theta^2}\log f(x,\theta)\right]} = -\frac{1}{nE\left[-\dfrac{1}{\theta^2}\right]} = \frac{\theta^2}{n}.$$

$$P\left[-1{\cdot}96 \le \left(\frac{\dfrac{1}{\bar{x}} - \theta}{\dfrac{\theta}{\sqrt{n}}}\right) \le 1{\cdot}96\right] = {\cdot}95$$

or $$P\left[1 - \frac{1{\cdot}96}{\sqrt{n}} \le \frac{1}{\bar{x}\theta} \le 1 + \frac{1{\cdot}96}{\sqrt{n}}\right] = {\cdot}95$$

or $$P_F\left\{\frac{1}{\bar{x}\left(1 + \dfrac{1{\cdot}96}{\sqrt{n}}\right)} \le \theta \le \frac{1}{\bar{x}\left(1 - \dfrac{1{\cdot}96}{\sqrt{n}}\right)}\right\} = {\cdot}95$$

Note. The length of a confidence interval depends strongly on what estimator is used. Thus in estimating for μ from a normal population, the estimator x_i, where x_i is the smallest or largest or any other member of the sample given an interval $\sqrt{n}$ times as long as obtained by using the sample mean from sample of size n. In general large sample confidence intervals and regions based on maximum likelihood estimators will be smaller, on an average, than intervals and region determined by other estimators of the parameters.

17·5·3. Approximate Confidence Interval for binomial population parameter *p* for large samples

We know that $\dfrac{X-np}{\sqrt{np(1-p)}} \sim N(0,1)$ for large n.

Hence $$P\left(-u_\in < \frac{X-np}{\sqrt{np(1-p)}} \le u_\in\right) = \int_{-u_\in}^{u_\in} \frac{e^{-\frac{1}{2}x^2}}{\sqrt{2\pi}}\,dx = 1-\in \qquad \text{...(1)}$$

where the points $\pm u_\in$ are given by

$$\int_{-u_\in}^{u_\in} \frac{1}{\sqrt{2\pi}} e^{-\frac{1}{2}x^2}\,dx = 1-\in \quad \text{or} \quad \int_{u_\in}^{\infty} \frac{1}{\sqrt{2\pi}} e^{\frac{-1}{2}x^2}\,dx = \frac{1}{2}\in \qquad \text{...(2)}$$

Re-writing (1) as

$$P(A < p < B) = 1 - \in$$

we find that A, B are the roots of the quadratic equation in p viz.

$$(x-np)^2 = u_\in^2 \,.\, np\,(1-p)$$

$$\Rightarrow n(n+u_\in^2)\,p^2 - np\,(2v + u_\in^2) + v^2 = 0$$

where v is the value of the $r.v. X$.

$$\therefore \qquad A, B = \frac{n\,(2v+u_\in^2) \pm u_\in \sqrt{4\pi v(n-v) + n^2 u_\in}}{2\,n\,(n+u_\in^2)}$$

$$= \frac{v}{n} \pm u_\in \sqrt{\frac{v\,(n-v)}{n^3}} \text{ up to } 0\left(\frac{1}{\sqrt{n}}\right).$$

Thus the approximate confidence interval of p is

$$\left(\frac{v}{n} - u_\in \sqrt{\frac{v\,(n-v)}{n^3}},\ \frac{v}{n} + u_\in \sqrt{\frac{v\,(n-v)}{n^3}}\right)$$

where $u_\in$ is given by (2).

Ex. 1. Show that approximate confidence limits for large samples for the parameter μ of a Poisson population having confidence coefficient $(1-\in)$ are given by the roots of the quadratic equation in μ.

$$n\,(\bar{x}-\mu)^2 = u_\in^2 \mu$$

which, to the order of $n^{-\frac{1}{2}}$ are approximately $\bar{x} \mp u_\in \sqrt{\bar{x}/n}$.

[**Hint.** For large x, $\bar{x} \sim N(\mu, \mu/n)$ or $(\bar{x}-\mu)\sqrt{n/\mu} \sim N(0,1)$]

Ex. 2. Let v_1, v_2 be independent samples of unit size from two binomial populations (n_1, p_1) and (n_2, p_2) respectively, where n_1, n_2 are known and large. Prove that confidence limits for $p_1 - p_2$ having confidence coefficient $(1 - \in)$ are approximately

$$\frac{v_1}{n_1} - \frac{v_2}{n_2} \pm u_{\in} \sqrt{\frac{v_1(n_1 - v_1)^2}{n_1^3} + \frac{v_2(n_2 - v_2)^2}{n_2^3}}$$

where $u_{\in}$ is given by (2).

[**Hint.** $X_1 \sim N(n_1 p_1, n_1 p_1 q_1)$ for large n_1, $q_1 = 1 - p_1$. Also $\hat{p}_1 = \frac{v_1}{n_1}$ where v_1 is the observed value of x_1.

$\therefore \quad X_1 \sim N(n_1 p_1, v_1(n_1 - v_1)/n_1$. Similar results for X_2.

$\therefore \quad [(X_1/n_1 - X_2/n_2) - (p_1 - p_2)]/[v_1(n_1 - v_1)/n_1^3 + v_2(n_2 - v_2)/n_2^3)^{1/2}$ is $N(0, 1)$.

Exercises 17

1. If in n trials of an event with probability x there are x successes, show that a maximum likelihood estimator of p is x/n. Find the sampling variance and show that it is sufficient.

2. Show that the distribution

$$dP = \tfrac{1}{2}\exp[-|x - \theta|]\, dx, \quad -\infty < x < \infty$$

has a likelihood function for a sample of n which is maximum at the mediun if n is odd and between the $\frac{n}{2}$ th and $\left(\frac{n}{2} + 1\right)$ th members if n is even.

[**Hint.** Use §3·2·3.]

3. Show that the most general form of distribution for which the maximum likelihood estimator is

(*i*) the arithmetic mean of the sample is

$$f(x, \theta) = k \exp\left[(x - \theta)\frac{\partial \psi}{\partial \theta} + \psi(\theta) + \phi(x)\right]$$

(*ii*) the geometric mean of the sample is

$$f(x, \theta) = (x)^{\theta \frac{\partial \psi}{\partial \theta}} \,.\, \exp[\psi(\theta) + \varphi(x)].$$

(*iii*) the harmonic mean of the sample is

$$f(x, \theta) = \exp\left[\frac{1}{x}\left\{\frac{\partial \psi}{\partial \theta} . - \psi(\theta)\right\} - \frac{\partial \psi}{\partial \theta} + \varphi(x)\right]$$

4. Find the maximum-likelihood estimator for ρ in samples of n from

$$dP \frac{1}{2\pi(1 - \rho^2)^{1/2}} \exp\left\{-\frac{1}{2(1 - \rho^2)}(x^2 - 2\rho xy + y^2)\right\} dxdy$$

and show that it is not product moment correlation coefficient. Prove also that

$$\text{Var}(\hat{p}) = \frac{(1-\rho^2)}{n(1+\rho^2)}.$$

5. Show that the maximum likelihood estimators for the five parameters form the bivariate normal population

$$dP = \frac{1}{2\pi\sigma_1\sigma_2(1-\rho^2)^{\frac{1}{2}}} \exp\left[-\frac{1}{2(1-\rho^2)}\left\{\left(\frac{x-\alpha}{\sigma_1}\right)^2 - \frac{2\rho(x-\alpha)(y-\beta)}{\sigma_1\sigma_2} + \left(\frac{y-\beta}{\sigma}\right)^2\right\}\right] dxdy, \quad -\infty \le x, y \le \infty$$

are $\hat{\alpha} = \hat{x}, \hat{\beta} = \hat{y},$

$$\hat{\sigma}_1{}^2 = \frac{1}{n}\Sigma(x-\bar{x})^2, \hat{\rho}\,\hat{\sigma}_1\,\hat{\sigma}_2 = \frac{1}{n}\Sigma(x_i-\bar{x})(y_i-\bar{y}), \hat{\sigma}_2{}^2 = \frac{1}{n}\Sigma(y_i-\bar{y})^2.$$

6. (*a*) Show that if a most efficient estimator *A* and a less efficient estimator *B* tend to joint normality for large sample *B* – *A* tend to zero correlation with *A*.

Show that the error in *B* may be regarded as composed for large samples of two parts which are independent, the error in *A* and the error in *B* – *A*. Interpret the first as the sampling error, necessarily inherent in the problem of estimation, the second as error due to inefficiency of the estimator.

(*b*) $x_1, x_2, ..., x_n$ are random observations on a Bernoulli variable *X* taking the value 1 with probability θ and the value 0 with probability $1-\theta$, show that

$$T(T-1)/n(n-1)$$

is an unbiased estimate of θ^2 where $T = \sum_{i=1}^{n} x_i$.

[Hint. (*b*) $E(T) = E\left(\sum_{1}^{n} x_i\right) = \sum_{i=1}^{n} E(x_i) = \sum_{i=1}^{n} [1\cdot\theta + 0\cdot(1-\theta)] = n\theta$

and $E(T^2) = Var(T) + [E(T)]^2$

where $\text{Var}\, T = \text{Var}\, \Sigma x_i = \Sigma\, \text{Var}\, x_i = \sum_{1}^{n} [E(x_i^2) - \{E(x_i)\}^2] = \sum_{i=1}^{n} (\theta - \theta^2)$

$= n\,\theta\,(1-\theta)]$

7. Find the maximum likelihood estimator and variance for β for the Gamma distribution. $f(x) = \frac{1}{\beta^\alpha\, \Gamma\alpha} e^{-\frac{x}{\beta}} x^\alpha, 0 \le x < \infty$, *x* being known.

[Ans. $\hat{\beta} = \bar{x}/a$, $\text{Var}(\hat{\beta}) = \beta^2/n\alpha$]

8. Find the maximum likelihood estimator for *p* for

$$f(x) = {}^{2}_{x}c\; p^x q^{2-x}; x = 1, 2, q = 1-p.$$

9. A sample of size n is drawn from each of four normal populations all of which have same variance σ^2. The means of the four populations are $a+b+c, a+b-c, a-b+c, a-b-c$. What are the maximum likelihood estimators for a, b, c and σ^2?

[**Hint.** The sample observations may be denoted by x_{ij}, $i = 1, 2, 3, 4$ and $j = 1, 2, 3, ..., n$.

$$L = \frac{1}{(2\sigma^2\pi)} \exp\left[-\frac{1}{2\sigma^2}\{\Sigma(x_{ij} - \overline{a+b+c})^2 + \Sigma(x_{ij} - \overline{a+b-c})^2 + \Sigma(\ )^2 + \Sigma(\ )^2\right]'$$

$$\therefore \quad \frac{\partial}{\partial \alpha} \log L = 0 \text{ gives } \hat{\alpha} = \frac{1}{4}\Sigma\Sigma x_{ij} = \bar{x}, \text{ etc.}]$$

10. Prove that a sufficient condition for t_n to be a consistent statistic for estimating θ is that $E(t_n) \to \theta$ and $\text{Var}(t_n) \to 0$ as $n \to \infty$

11. Develop a method for finding the 95% confidence interval for the parameter of (*i*) a Binomial distribution (*ii*) a Poisson distribution.

12. Show that for the rectangular distribution $f(x) = 1/\theta, (0 \le x \le \theta)$ the confidence limits for samples x_1, x_2 of two are

$$\frac{x_1 + x_2}{1 + \sqrt{1 + \alpha}}, \frac{x_1 + x_2}{1 + \sqrt{1 - \alpha}}.$$

13. For the type III distribution

$$dP = \frac{e^{-\frac{x}{\theta}} x^{p-1}}{\Gamma p\, \theta^p} dx, 0 \le x < \infty, p > 1$$

where p is known, find the exact sampling distribution of the maximum likelihood estimator θ and show that $np\hat{\theta}/\theta$ is distributed independently of θ, hence find the confidence interval for θ with fiducial probability α.

14. Show that for the distribution

$$dp = \frac{1}{2(1 - e^{-\theta})} e^{-|x|} dx, -\theta \le x \le 0,$$

the maximum likelihood estimate of θ is the largest member among the members $|x_1| \ldots |x_n|$.

15. For the double Poisson distribution

$$P(x = i) = \tfrac{1}{2}\frac{e^{-m_1} m_1^{\,i}}{i\,!} + \tfrac{1}{2}\frac{e^{-m_2} m_2^{\,i}}{i\,!}, i = 0, 1, 2, \ldots$$

show that the moment method estimates for m_1, m_2 are

$\mu_2' \pm \sqrt{\mu_2' - \mu_1' - \mu_1'^2}$. **(*B.Sc. Stat. Hons. Delhi, 1993*)**

[**Hint.** $\mu_1' = \Sigma x p(x) = \frac{1}{2} m_1 + \frac{1}{2} m_2$, $\mu_2' = \frac{1}{2}[(m_1^2 + m_1) + m_2^2 + m_2)] =$ etc.]

16. According to a certain genetic model the proportions of insects in three cases should be $q^2 : 2pq : p^2$ where $p + q = 1$. If n_1, n_2, n_3 are the observed frequencies in the classes, prove that $L = q^{2n_1}(2pq)^{n_2} p^{2n_3}$

and that the maximum likelihood estimator for q is

$$\hat{q} = \frac{2n_1 + n_2}{2(n_1 + n_2 + n_3)}.$$

17. For the multinomial distribution, the observed frequencies in the k classes are $n_1, n_2, ..., n_k$. By maximizing the likelihood function

$$L = \frac{n!}{n_1!\, n_2! \,..., n_k!}\, p_1^{n_1} p_2^{n_2} \dots p_k^{n_k}$$

subject to $\sum_1^n . p_i = 1, \; \sum_{i=1}^k n_i = n,$

show that the maximum likelihood estimators are

$$\hat{p}_1 = \frac{n_1}{n}, \hat{p}_2 = \frac{n_2}{n}, ..., \hat{p}_k = \frac{n_k}{n}$$

[**Hint.** $\log L = \Sigma\, n_i \log p_i +$ terms independent of p_i's $= \Sigma \frac{n_i}{p_i} dp_i = 0, \Sigma dp_i = 0$

$\therefore$ If λ is Lagrange's multiplier, then we have

$$\frac{n_i}{p_i} + \lambda = \text{ for all } i$$

$$\therefore \quad \frac{n_1}{p_1} = \frac{n_2}{p_2} = \dots = \frac{n_k}{p_k} = \frac{\Sigma n_i}{\Sigma p_i} = n \qquad \therefore\; p_i = n_i/n.$$

Tests of Hypotheses $\lambda^{-1} = \left(\frac{n_1}{np_{01}}\right)^{n_1} \left(\frac{n_2}{np_{02}}\right)^{n_2} \dots \left(\frac{n_k}{np_{0k}}\right)$

$$\therefore \quad -\log \lambda = \Sigma\, n_i \log \frac{n_p}{np_{0i}} = \Sigma \left[(n_i - np_{0i}) \log \left(1 + \frac{n_k - np_i k}{np_{0i}}\right)\right.$$

$+\, n\, p_{0i} \log \left(1 + \frac{v_i - np_{0i}}{n\, p_{0i}}\right)$ under H_0 : each $n_i\, n - v_{0i}$ is a small quantity for large

$$\therefore \quad -\log \lambda \simeq \Sigma \left[\frac{(n_i - np_{0i})^2}{np_{oi}} + n\, p_{0i} \left\{ \frac{n_i - np_{0i}}{np_{0i}} - \frac{(n_i - np_{0i})^2}{2n^2 p_{0i}^2} \right\} \right]$$

Since $\Sigma\, (v_i - np_{0i}) = 0 \quad \therefore \; -2 \log \lambda \simeq \Sigma \frac{(n_i - np_{0i})^2}{np_{0i}} = \lambda^2$

$$\therefore \quad \lambda^2 = \Sigma \frac{(n_i - np_{0i})^2}{np_{0i}} \Bigg]$$

18. If $x_1, x_2, ..., x_n$ is a random sample from a population with mean θ, prove that

$$a_1 x_1 + a_2 x_2 + \dots + a_n x_n$$

will be an unbiased estimate for θ if $a_1 + a_2 + \dots + a_n = 1$. Show also that the variance of an unbiased linear estimate will be least if that estimate is sample mean.

[**Hint.** Var $T = \sigma^2(a_1^2 + a_2^2 + \dots + a_n^2)$ $\therefore a_1 da_1 + a_2 da_2 + \dots + a_n da_n = 0$

$\therefore \quad a_1 = a_2 = \dots = a_n = p$, say $= \dfrac{1}{n}$ $\quad \therefore$ Var $T = \dfrac{\sigma^2}{n}$.].

19. Prove that for a random sample for a population with finite variance σ^2.

$$E(s^2) = E\left[\frac{1}{n}\Sigma(x_i - \bar{x})^2\right] = E\left[\frac{1}{n}\Sigma(x_i - \mu)^2 - \frac{1}{n}(\bar{x} - \mu)^2\right]$$

$$= \frac{1}{n}\sum_{i=1}^{n}\sigma^2 - \frac{1}{n}\sigma^2 = \sigma^2\left(1 - \frac{1}{n}\right).$$

Deduce that $\dfrac{n-1}{n}s^2$ is an unbiased estimator for σ^2.

20. If T be an unbiased estimate of the parameter θ, prove that in large samples

$$Var_{F(T)} = Var_T\left\{\frac{d}{d\theta}F(\theta)\right\}^2$$

where $F(T)$ is any continuous function of T, $Var\,T$ and $Var\,F(T)$ are the variances of T and $F(T)$ respectively. Also show that in large samples if θ be the expectation of T, then $F(\theta)$ is the expected value of $F(T)$. Show how the above result can be used in transforming variates so that its variance may be free from parameter value.

[**Hint.** In large samples, it is assumed that T is distributed normally about mean θ with variance $T \propto O\left(\dfrac{1}{\sqrt{n}}\right)$ and, therefore, $[T - \theta] \propto O\left(\dfrac{1}{\sqrt{n}}\right)$, n being the size of sample. Assuming the derivatives of $F(T)$ are continuous at $T = \theta$, we have by expanding $F(T)$ by Taylor's theorem at $T = \theta$,

$$F(T) = F(\theta) + (T - \theta)F'\theta) + O\left(\frac{1}{n}\right)\cdot \text{ Suppose that } F'(\theta) \neq 0.$$

$$\therefore \quad T - \theta = \frac{F(T) - F(\theta)}{F'(\theta)} \quad \text{and } dF(T) = F'(T)\,dT$$

$$\therefore \quad dT = \frac{dF(T)}{F'(T)} = \frac{dF(T)}{F'(\theta) + (T - \theta)F''(\theta)}$$

$$dt = \frac{dF(T)}{F'(\theta)} \times \left[1 + (T - \theta)\frac{F''(\theta)}{F(\theta)}\right]^{-1}$$

$$= \frac{dF(T)}{F'(\theta)}, \quad \text{neglecting terms higher than } \frac{1}{\sqrt{n}}.$$

The distribution of T is

$$dF = \frac{1}{\sqrt{2\pi\,Var(T)}} \cdot e^{-\frac{1}{2}\frac{(T-\theta)^2}{Var\,T}}$$

$$dT = \frac{1}{\sqrt{2\pi\,Var\,T}}\,e^{-\frac{1}{2}\left[\frac{F(T) - F(\theta)}{F'(\theta)}\right]^2}\,\frac{1}{Var\,T} \times \frac{dF(T)}{F(\theta)}$$

Thus $F(T)$ is distributed normally about the mean $F(\theta)$ with variance $[F(\theta)]^2 \, Var\, T$.

Finally, let $Var_{F(T)} = \text{const.} = c^2$, then we see $\dfrac{dF(\theta)}{d\theta} = \dfrac{c^2}{Var\,(T)}$

and, therefore, $F(\theta) = c \displaystyle\int \frac{d\theta}{f(\theta)}$ where $Var\,(T) = [f(\theta)]^2$.

The Square Root Transformation for the Poisson variate.

If X is a Poisson variate, then $E(X) = \mu$, $Var\,(X) = \mu$.

The transformation is obtained by

$$F(\mu) = c \int \frac{d\mu}{\sqrt{\mu}} \sqrt{\mu}, \quad \text{by choosing } c \text{ suitably.}$$

Fisher's angular or sin^{-1} transformation of the Binomial Proportion.

The binomial proportion r/n has the mean value p and variance $p(1-p)/n$.

Transformation is obtained by $F(x) = c \displaystyle\int \frac{\sqrt{n}}{\sqrt{p(1-p)}} dp = \sin^{-1} \sqrt{p}$.

by choosing c suitably and $Var\left(\sin^{-1} \sqrt{\dfrac{r}{n}} \right) = \dfrac{1}{4n}$ approx.

(*b*) Show that if variance of X is approx. proportional to $(1-\alpha^2)^2$, then a suitable transformation for producing a variate with approx. constant variance is

$$Y = \frac{1}{X} \cdot \log_e \left(\frac{1+X}{1-X} \right).$$

21. Show that for Kapteyn's class of distributions

$$dp = \frac{1}{\sigma\sqrt{2\pi}} \exp\left[-\frac{G(x) - \mu}{2\sigma^2} \right]^2 \left| \frac{dG'(x)}{dx} \right| dx$$

(*i*) If σ is known, the variance of the maximum likelihood estimator for μ is σ^2/n.

(*ii*) If μ is known, the variance of the maximum likelihood estimator for σ^2 is $2\sigma^4/n$.

(*iii*) If μ is known, the variance of the maximum likelihood estimator for σ is $\sigma^2/2n$.

22. Show that for the binomial and Poisson distributions the maximum likelihood estimator and moment estimators are identical.

23. For variable X following the law

$$f(x) = \frac{1}{2\sigma\sqrt{2\pi}} \left[e^{\frac{-(x-a)^2}{2\sigma^2}} + e^{\frac{-(x-b)^2}{2\sigma^2}} \right]$$

calculate the first four moments about the arithmetic mean.

Hence show that in terms of the sample moments, the best estimates of a, b and σ are

$$a = m_1 + \left[\frac{3-\beta_2}{2} m_2{}^2\right]^{\frac{1}{4}}, b = m_1 - \left[\frac{3-\beta_2}{2} m_2{}^2\right]^{\frac{1}{4}}, \sigma^2 = m_1\left[1 - \sqrt{\frac{3-\beta_2}{2}}\right]$$

where $m_1 = \dfrac{a+b}{2}$.

$$\left[\textbf{Ans. } \mu_1' = m_1, \mu_2 = \sigma_2{}^2 + \left(\frac{a-b}{2}\right)^2, \mu_3 = 0, \mu_4 = 3\sigma^4 + \frac{3}{2}(a-b)^2\sigma^2 + \frac{(a-b)^4}{16}, \beta_1 = 0\right]$$

24. If X is a standard normal variate $N(0, 1)$, obtain the distribution of Z where $Z = e^{a+bx}$, $b < 1$, a and b being constants.

Prove that its rth moment is given by $\mu'_r = \exp\left\{ra + \dfrac{r^2b^2}{2}\right\}$.

Hence obtain, using the method of fitting moments, the estimates of the parameters a and b written as $\hat{a}$ and $\hat{b}$ in terms of m_1' and m_2', the moments of the sample.

25. If e_1 and e_2 are the efficiencies of two unbiased estimates and ρ is the coefficient of correlation between them, show that

$$\sqrt{e_1e_2} - \sqrt{(1-e_1)(1-e_2)} < \rho < \sqrt{e_1e_2} + \sqrt{(1-e_1)(1-e_2)}.$$

26. If T is the most efficient estimator and if T' is an unbiased estimator with efficiency e, obtain the efficiency of $lT + (1-l)\,T'$, where l is a constant.

(*B.Sc. Hons. Stat. Delhi, 1987*) [**Ans.** $\sqrt{e}$]

27. A variate X is such that its logarithm has a normal distribution with zero mean and unit variance. Show the efficiency of the sample mean $\bar{x}$ in estimating the true mean is $1/(e-1)$.

[**Hint.** The distribution of X is $g(x)dx$ where $g(x) = \dfrac{1}{\sqrt{2\pi}} e^{-\frac{1}{2}(\log x)^2} \cdot \dfrac{1}{x}$ and the expected value of X is $\int xg(x)\,dx$

$$= \frac{1}{\sqrt{2\pi}} \int_{-\infty}^{+\infty} e^{-\frac{1}{2}y^2} e^y \, dx \text{ when } \log x = y$$

$$= \sqrt{e_1}.$$

Similarly, $\mu_2' = e^2$ so that $\mu_2 = e(e-1)$.]

28. If T is the most efficient estimate of θ and T' an estimate of efficiency η, the variance of $T - T'$ is $\left(\dfrac{1}{\eta} - 1\right)$ Var (T).

29. Estimate by the method of moments the parameters a, ξ, and σ in the distribution $p(x) = \dfrac{1}{\sqrt{2\pi}[\sigma(x-a)]} \cdot e^{-\frac{1}{2\sigma^2}[\log(x-a)-\xi]^2}$, $x \geq a$.

30. Find the maximum likelihood estimators of the parameters α and λ (λ being large) of the distribution. $f(x, \alpha, \lambda) = \frac{1}{\Gamma\lambda}\left(\frac{\lambda}{\alpha}\right)^{\lambda} e^{-\frac{\lambda x}{\alpha}} e^{\lambda-1}, 0 \le x \le \infty$

assume that for large values of λ,

$$\psi(\lambda) = \frac{d}{d\lambda} \log \Gamma\lambda = \log \lambda - \frac{1}{2\lambda},\ \psi'(\lambda) = \frac{1}{\lambda} + \frac{1}{2\lambda^2}.$$

$$\left[\textbf{Ans. } \bar{x},\ 1/\Sigma\left(2 \log x - \frac{1}{n} \log x\right)\right].$$

31. State Rao-Blackwell theorem. Hence or otherwise obtain the maximum variance unbiased estimator of $e^{-\theta}$ where θ is the parameter of the Poisson distribution from which a sample of size n is available. ***(M.Sc., Delhi 1980)***

32. Define a sufficient statistic. If $x_1, x_2, ..., x_n$ is a random sample from the population with the discrete probability function

$$p(x, \theta, \xi) = (1 - \xi)\, \xi^{x-\theta}, x = \theta, \theta + 1, ..., 0 < \xi < 1,$$

find a sufficient statistic for θ, given ξ. ***(M.Sc. Bombay, 1980)***

33. If $x_1, x_2, ..., x_n$ is a random sample from any distribution having finite kth moment $E(x^k)$. show that $\frac{1}{n} \sum_{i=1}^{n} x_i^k$ is a consistent estimator for $E(x^k)$.

34. If $\hat{\theta}$ is an efficient estimator for θ and θ' is any other unbiased estimator for θ and if σ^2 and $k\sigma^2, k > 1$ are the variances of $\hat{\theta}$ and θ', show that the correlation coefficient between $\hat{\theta}$ and θ' equals $1/\sqrt{k}$.

35. If $x_1, x_2, ..., x_n$ is a random sample from the distribution having p.d.f. $\lambda^{k-1} x^k e^{\lambda x}/\Gamma(k+1), x > 0, \lambda > 0, k$ is constant. Show that m.l.e. for λ is $(k + 1)/x$. Show also that this estimator is biased but consistent.

36. A random sample x_1, x_2, x_3, x_4, x_5 of size 5 is drawn from a normal population with unknown mean μ. Consider the following estimator to estimate μ.

(*i*) $t_1 = x_1 + x_2 + x_3 + x_4 + x_5$ (*ii*) $t_2 = \frac{x_1 + x_2}{2} + x_3$ (*iii*) $t_3 = \frac{2x_1 + x_2 + \lambda x_3}{3}$

where λ is such that t_3 is an unbiased estimator. Find λ. Are t_1 and t_2 unbiased ? Which estimator t_1, t_2, t_3 is best ? ***(B.Sc. Bombay, 1981)***

37. Let $X_1, ..., X_n$ be independent *r.v.* each having Poisson distribution $P_\theta(x) = P_\theta(X = x) = e^{-\theta}\theta^x/x\,!$. Find the lower bound for the variance of an unbiased estimate $t(x_1, x_2, ..., x_n)$ of θ.

$$\left[\textbf{Hint. } \frac{\partial}{\partial\theta} \log p_\theta(x) = \frac{x}{\theta} - 1 = \frac{x - \theta}{\theta}.\right.$$

$$\text{Also } E\left[\frac{\partial}{\partial\theta} \log p_\theta\right]^2 = E\left[\frac{x - \theta}{\theta}\right]^2 = \frac{\theta}{\theta^2} = \frac{1}{\theta}.$$

$$\therefore \quad \sigma_E^2(\theta) \geq \frac{1}{n\, E\left[\frac{\partial}{\partial\theta}\log P_\theta(x)\right]^2} = \frac{\theta}{n}$$

Now $\overline{X} = \frac{1}{n}\sum_1^n x_i$ is an unbiased estimate of θ and has variance θ^2/n.]

38. Let Y_1, Y_2, Y_3, Y_4 be un-correlated observations with common variance σ^2 and expectations as under

$$E(Y_1) = \theta_1 + \theta_2 + \theta_3 = E(Y_2)\ ;\ E(Y_3) = \theta_1 - \theta_2 = E(Y_4)$$

Show that $\theta_1 + \theta_2 + \theta_3$ is estimable. Hence obtain the Best linear unbiased estimator of this parameteric function and its variance.

[**Hint.** Let $E(w_1y_1 + w_2y_2 + w_3y_3 + w_4y_4) = \theta_1 + \theta_2 + \theta_3$

so that $(w_1 + w_2)(\theta_1 + \theta_2 + \theta_3) + (w_3 + w_4)(\theta_1 - \theta_2). = \theta_1 + \theta_2 + \theta_1$

with $w_1 + w_2 + w_3 + w_4 = 1,\ w_1 + w_2 - w_3 - w_4 = 1,\ w_1 + w_3 = 1\ ;\ w_2 + w_4 = 0$

and hence $\theta_1 + \theta_2 + \theta_3$ is estimable.

Also $V[w_1y_1 + w_2y_2 + w_3y_3 + w_4y_4]$ is minimum

such that, $w_1 + w_2 + w_3 + w_4 = 1$, etc.]

39. If $X_1, X_2, \ldots\ldots, X_n$ is a random sample of size n from $N(\mu, \sigma^2)$ where μ is known and if $\quad T = \frac{1}{n}\sum_{i=1}^{n} |\, X_i - \mu \,|$

examine if T is unbiased for σ. If not, obtain an unbiased estimator of σ.

(B.Sc. Stat. Hons. Delhi, 1987)

Ans. $E(T) = \sigma\sqrt{\frac{2}{\pi}} \neq \sigma$ $\therefore$ No., $\sqrt{\frac{\pi}{2}}\, T$.

40. X is a uniform *r.v.* with range $[0, \theta] \cdot x_1, x_2, \ldots\ldots, x_n$ are independent observations on X.

Let $\hat{\theta}_1 = \frac{2}{\theta}(x_1 + x_2 + \ldots\ldots + x_n)$, $\hat{\theta}_2 = \frac{n+1}{n}\max(x_1, x_2, \ldots\ldots, x_n)$.

Show that $\hat{\theta}_1$ and $\hat{\theta}_2$ are unbiased for θ.

41. $(x_1, x_2, \ldots\ldots, x_n)$ is a random sample from the distribution with *p.d.f.* $f(x\,;\theta) = \theta\, x^{\theta - 1},\ 0 < x < 1,\ \theta > 1$

If $Z = -\sum_{i=1}^{n}\log x_i$, show that $(n-1)/Z$ is an unbiased estimator for θ and its efficiency is $(n-2)/n$.

[**Hint.** find $U = -\log X$ and $Y = -\sum_{i=1}^{n}\log x_i$.]

42. If $x_1, x_2, ..., x_n$ is a random sample from the distribution

$$f(x, \theta) = \theta^x (1-\theta)^x \; ; \; x = 0, 1 \; ; 0 < \theta < 1$$
$$= 0 \; ; \text{ elsewhere}$$

show that $Y_1 = x_1 + x_2 + ... + x_n$ is a sufficient statistic for θ.

(B.Sc. Madras, 1987)

43. Let (x_1, x_2) be a random sample from a Poisson distribution with parameter θ. Prove that $t = 2X_1 + 3X_2$ is not sufficient for θ.

44. Show that there exists a parameter function $\psi(\theta)$ in the case of the following distribution.

$$f(x, \theta) = (1-\theta)\,\theta^{x}, \; x = 0, 1, 2, ..., 0 < \theta < 1$$

such that there exists an MVB unbiased estimator T of $\psi(\theta)$. obtain $\psi(\theta)$, T and $V(T)$ **(M.Sc. Agra, 1988)**

45. Let $x_1, x_2, ..., x_n$ be a random sample from the rectangular distribution

$$f(x, \theta) = 1/\theta, \; 0 \le x \le \theta$$

Find MVU estimators of θ and $3\theta + 5$

(B.Sc. Stat. Hons. Delhi, 1987, 1993)

46. Let $(x_1, x_2, ..., x_n)$ be random sample from poisson distribution with parameter θ. Find an MVU estimator for $e^{-\theta}\theta^4/4!$

47. Let $(x_1, x_2, ..., x_n)$ be a random sample from Bernoulli population with parameter $\theta, 0 < \theta < 1$. Obtain a sufficient statistic for θ and show that it is complete. Hence obtain. MVU estimator for θ.

48. A *r.v.* X takes the values 0, 1, 2 with respective probabilities

$$\frac{\theta}{4N} + \frac{1}{2}\left(1 - \frac{\theta}{N}\right), \frac{\theta}{2N} + \frac{\alpha}{2}\left(1 - \frac{\theta}{N}\right) \text{ and } \frac{\theta}{4N} + \frac{1-\alpha}{2}\left(1 - \frac{\theta}{N}\right)$$

where N is a known number and α, θ are unknown parameters. If 75 independent observations on X yielded the values 0, 1, 2 with frequencies 27, 38, 10 respectively, respectively

θ and α by the method of moments. **(B.Sc. Stat. Hons. Delhi, 1988)**

[**Hint.** Show that $\mu_1' = EX = \dfrac{\theta}{N} + \left(1 - \dfrac{\theta}{N}\right)\left(1 - \dfrac{\alpha}{2}\right) = 1 - \dfrac{\alpha}{2}\left(1 - \dfrac{\theta}{N}\right)$

$$\mu_2' = E(X^2) = \frac{3\theta}{2N} + \left(1 - \frac{\theta}{N}\right)\left(2 - \frac{3\alpha}{2}\right) = 2 - \frac{\theta}{2N} - \frac{3}{2}\alpha\left(1 - \frac{\theta}{N}\right)$$

Again $\quad \mu_1' = \dfrac{1}{N}\,\Sigma fx = \dfrac{58}{75}, \mu_2' = \dfrac{1}{N}\,\Sigma fx^2 = \dfrac{78}{75}$

Equate the corresponding values and obtain

$$\hat{\theta} = \frac{42}{75}N, \; \hat{\alpha} = \frac{34}{33}.\Bigg]$$

49. In sampling from a power series distribution with *p.d.f.*

$$f(x, \theta) = a_x\,\theta^x/\psi(\theta) \; ; x = 0, 1, 2, 3$$

where a_x may be zero for some x, show that *MLE* of θ is a root of the equation $\bar{x} = \theta\, \psi'(\theta)/\psi(\theta) = \mu(\theta)$

where $\mu(\theta) = E(x)$. **(*B. Sc. Stat. Hons., Delhi 1989*)**

$$\left[\textbf{Hint. } L = \prod_{i=1}^{n} f(x_i, \theta) = \prod_{i=1}^{n} \left[\frac{a_{x_i}\theta^{x_i}}{\psi(\theta)}\right] = \left[\prod_{i=1}^{n} a_{x_i}\right] \frac{\theta^{\Sigma x_i}}{[\psi(\theta)]^n}\right.$$

$$\Rightarrow \log L = \sum_{i=1}^{n} \log a_{x_i} + \log\theta \sum_{i=1}^{n} x_i - n \log \psi(\theta)$$

$$\frac{\partial}{\partial\theta} \log L = 0 \Rightarrow \frac{\Sigma x_i}{\theta} - \frac{n\,\psi'(\theta)}{\psi(\theta)} = 0 \Rightarrow \bar{X} = \frac{\Sigma x_i}{n} = \frac{\theta\,\psi'(\theta)}{\psi(\theta)} = \mu(\theta)$$

Also show $E(x) = \sum_{x=0}^{\infty} \cdot \left[x\left\{\frac{a_x\theta^x}{\psi(\theta)}\right\}\right]$ and $\sum_{0}^{\infty} f(x, \theta) = 1 \to \sum_{0}^{\infty} a_x \theta^x = \psi(\theta)$

Also find $\psi'(\theta)$.$\Big]$

50. Let *r.v.* X and Y have the joint *p.d.f.*

$$f(x, y) = \frac{2}{\theta^2} \exp\left[-\frac{(x+y)}{\theta}\right], \quad 0 < x < y < \infty, = 0 \text{ elsewhere.}$$

Show that $E(Y \mid x) = x + \theta$'

obtain $E(X + \theta)$ and compare Var $(X + \theta)$ with Var (Y).

(*B. Sc. Stat. Hons., Delhi, 1992*)

51. The *r.v.* X takes the values 0 and 1 with respective probabilities $1 - \theta$ and θ. Show that if $\eta = x_1 + x_2 + \cdots + x_n$ where x_i's are independent observations X, then $\eta(n - \eta)/n(n-1)$ is an unbiased estimate of $\theta(1 - \theta)$.

52. (*a*) Verify that there exists an M.V.B. estimator for the parameter θ of the distribution

$$f(n, \theta) = e^{-\theta}\theta^x/x!, x = 0, 1, 2, \ldots$$

and hence obtain the value of M.V.B. **(*M.Sc., Maratharada, 1993*)**

(*b*) Show the there exists a parameter function $\psi(\theta)$ in case of $f(x, \theta) = (1 - \theta)\theta^x$, $x = 0, 1, 2, \ldots, 0 < \theta < 1$. Such the there exists an M.V.B. unbiased estimator T of $\psi(\theta)$. Obtain $\psi(\theta)$, T and $V(T)$.

(*M.Sc., Agra, 1988*)

53. Estimate α and β by the method of moments for the distribution

$$f(x;\ \alpha_1 \beta) = \beta^{\alpha} x^{\alpha - 1} e^{-\beta x}/\Gamma\alpha, \quad 0 \le x < \infty$$

(*B.Sc. Stat. Hons. Delhi, 1983, 1987*)

18

*Testing of Hypotheses, Sequential Analysis, Distribution-Free Methods, Statistical Decision Theory

18·1. Introduction. As remarked in Chapter 1, the present book is concerned mainly with problems of *Statistical Inference* and in the preceding seventeen chapters, we have given an account of this subject as developed by Karl Pearson and R.A. Fisher, till about 1930. During the last two decades, a number of important developments, associated with the names of J. Neyman, E.S. Pearson, Abraham Wald, S.S. Wilks and others, have taken place in this subject. It is not possible, within the scope of an introductory course, to cover these completely. Nevertheless, it is desirable that a student of Statistics should be aware of the developments. With this object, we give in this chapter some of the more elementary and basic results of these developments. Unlike the preceding chapter, however, proofs of a number of results have been omitted.

Statistical Inference may be regarded as consisting of—

(*i*) *Estimation of Parameters.*

(*ii*) *Testing of Hypothesis.*

We have discussed the first aspect in chapter 17. We shall discuss the *theory of testing hypotheses*, as developed by Neyman and Pearson, in section 2 of the present chapter. The testing of a hypothesis may be based on a sample, whose size may be fixed in advance or the size itself may be determined during the course of the experiment, by criteria, which depend on the observations, as they arise. The second aspect, known as *Sequential analysis*, was developed by Abraham Wald and would be studied in section 3. Again, our tests may require a specification of the form of the parent population or they may require no assumption about this form. The latter tests are called *non-parameteric test or distribution-free tests* and we shall discuss these in section 4. The

* For a detailed study of these topics, *see* the book 'Statistical Inference' by H.C. Saxena.

statistical decision theory will be discussed in section 5.

18·2. Testing of Hypothesis.

18·2·1. Simple and Composite Hypotheses. Any statement about the population, from which a given random sample $x_1, x_2, \ldots\ldots, x_n$ may have been drawn, is called a *Statistical Hypothesis.* If it determines the population completely, the hypothesis is called a *Simple Hypothesis*, otherwise it is called a *Composite Hypothesis*. Thus, in sampling from a bivariate normal population, characterised by five parameters, *viz.*, the two means, the two variances and a correlation coefficient, any hypothesis specifying one, two, three or four of the parameters is composite. The hypothesis would be simple, only if all the five parameters are specified in addition to the normality of the population.

A hypothesis which specifies the population completely, except for v parameters is said to be a *composite hypothesis with v degrees of freedom.* Such a hypothesis may be regarded as an aggregate of (or a composite of) simple hypotheses, *e.g.*, the hypothesis that a population is normal with mean μ is composed of the aggregate, for all possible values of σ^2, of the simple hypotheses that the population is normal with mean μ and variance σ^2.

Ex. 1. "There is life on planet Mars." Is it a statistical hypothesis ?

Ex. 2. Give some examples of statistical hypotheses with 0, 1, 2, 3, 4 degrees of freedom.

Ex. 3. How many parameters are there in a multivariate normal distribution with p variates ? State some composite hypotheses about a population and explain how these can be regarded as aggregates of simple hypotheses.

[**Ans.** p means, p variances and $\dfrac{p(p-1)}{2}$ covariances]

18·2·2. Two Types of Errors. Critical Region. Power Function of the test Hypotheses. Operating Characteristic Function. A test of a statistical hypothesis H_0 is a rule for determining a region w_0 in the n-dimensional Euclidean sample space, which is such that the hypothesis H_0 will be rejected, if the sample point falls within w_0 and it will be accepted if the sample point lies outside this region. The region w_0 so defined is called the *critical region* or *region of rejection.* The rest of the sample space may be called the *region of acceptance.* (Fig. 18·1).

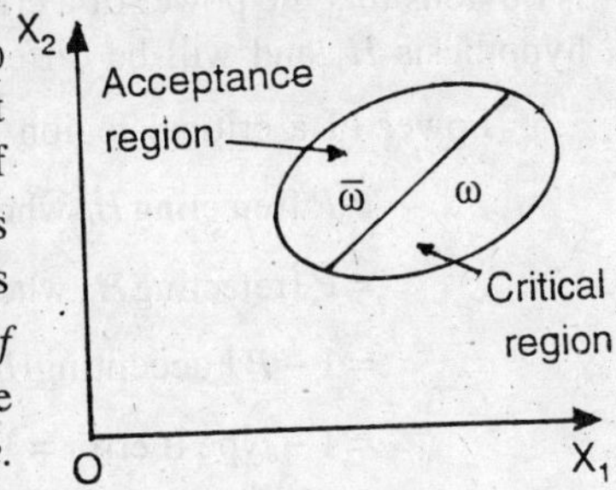

Fig. 18·1 Acceptance and critical regions

In testing a statistical hypothesis,

we can make two possible types of error :

(*i*) we may reject the hypothesis when we ought to accept it *i.e.*, when it is true, or

(*ii*) we may accept the hypothesis when we ought to reject it, *i.e.* when it is false.

We present the above ideas in the following tabular form :

Test	*Decision drawn from Sample about Population*	
	Reject H_0	*Accept* H_1
H_0 True	Wrong (Type I Error)	Correct
H_0 False (H_1 True)	Correct	Wrong (Type II Error)

These are known as *errors of the first* and *second kind* respectively and we shall denote their probabilities by α and β respectively. We would like to carry out test, *i.e.,* choose our critical region so as to minimize both types of errors simultaneously, but this is, in general, not possible for a sample of fixed size. In fact, decreasing one type of error may very likely increase the other type. Thus, by deciding to always accept the hypothesis, we can reduce type I error to zero, but in that case β would have its largest value. In practice, we keep type I error fixed at a specified value and then, out of these critical regions all of which give this type I error, we choose that region which minimizes the type II error. The type I error, which is the same for all these regions, is sometimes called *size* of these critical regions.

$$\alpha = \text{ probability of type I error } i.e. \int L_0\, dx = \alpha$$

$$= \text{probability of rejectin when } H_0 \text{ is true.}$$

The *power of a critical region* (or of a test) is defined as the probability of the sample point falling in this critical region when the given hypothesis H_0 is false and some other hypothesis, say H_1 is true. It is obvious that the power of a critical region will depend on the alternative hypothesis H_1 and will be different for different hypotheses H_1. Thus

Power of a critical region for testing H_0 against H_1

$$= P\,[\text{rejecting } H_0 \text{ when } H_1 \text{ is true}]$$

$$= P\,[\text{rejecting } H_0 \text{ when } H_0 \text{ is false}]$$

$$= 1 - P\,[\text{accepting} H_0 \text{ when} H_0 \text{ is false}]$$

$$= 1 - \text{type II error} = 1 - \beta$$

where $\beta =$ probability of type II error

$$= \text{probability of accepting } H_0 \text{ when } H_0 \text{ is false } i.e. \int L\, dx = \beta$$

$$L_0 = \prod_{i=1}^{n} f(x_i, \theta_0).$$

The power of a critical region refers to its 'power of detecting the falsehood of H_0 when H_0 is false.

In some fields of application it is more common to use the *operating characteristic function* (*OC* function) in place of the power function. The *OC* function of a test is the probability of accepting H_0 when some other hypothesis H_1 is true. The *OC* function is a function of H_1 and can be plotted as such. It is obvious that for any H_1, the *OC* function is 'one minus the power function'.

For a given type I error, we choose our critical region so as to minimize the type II error or minimize the operating characteristic function or maximize the power function. If there exists a particular region w_0 with greater power than any other region given same type I error, we say it is the *best critical region* and the test based on it will be called the *most powerful test.*

Ex.1. "A judge can acquit a guilty person or convict an innocent person. An examiner may pass a bad student or may fail a good student." Discuss the relevance of the concept of the two types of errors in these two cases. Give other examples also.

Ex. 2. Show that the *OC* function for testing H_1 against H_0 is the power function for testing H_0 against H_1.

Ex. 3. For testing the hypothesis that a random sample comes from a normal population with mean zero, when its variance is known to be unity, a critical region of size ·05 is given by $|x| \geq 1{\cdot}96/\sqrt{n}$. Find its power for the hypothesis $\mu = -3, -2, -1, 0, 1, 2, 3$ and plot the power function as a function for μ.

Ex. 4. Explain the following terms :

(*i*) Errors of first and second kind

(*ii*) Best critical region (B.C.R.)

(*iii*) Power function (*iv*) *O.C.* curve

(*v*) Simple and Composite hypotheses. (*vi*) most powerful test.

Ex. 5. Give a critical region of size ·05 for testing the hypothesis that a random sample comes from a normal population with variance σ_0^2 and determine its power function as a function of σ^2. If the test is based on tails of 'χ^2' distribution with equal areas show that the power function is less for $\frac{1}{2}\sigma_0^2$ than for σ_0^2.

Ex. 6. If $x \geq 1$ is the critical region for testing $H_0 : \theta = 2$ against the alternative $H_1 : \theta = 1$, on the basis of single observation from the population $f(x, \theta) = \theta e^{-\theta x}, \; x \geq 0$

Obtain the values of type I and type II errors.

(B.Sc. Allahabad, 1993, B.Sc. Stat. Hons. Delhi, 1988)

$$\alpha = \text{ size of type I error } = P(x \in W \mid H_1) = P[x \geq 1 \mid \theta = 2]$$

$$= \int_1^\infty 2e^{-2x}\, dx = e^{-2} = 1/e^2 \qquad \because f(x, 2) = 2e^{-2x}$$

$$\beta = \text{ size of type II error } = P(x \in \overline{W} \mid H_1) = P(x < 1 \mid \theta = 1)$$

$$= \int_0^1 e^{-x}\, dx = 1 - e^{-1} = \frac{e-1}{e} \because f(x, 1) = e^{-x},$$

Ex. 7. Let $X \sim N(\mu, 4)$, μ unknown. To test $H_0 : \mu = -1$ against $H_1 : \mu = 1$ based on a sample of size 10 from this population, we use the critical region $x_1 + 2x_2 + \cdot + 10x_{10} \geq 0$. What is its size and power of the test ?

Critical Region $W = \{\ x = x_1 + 2x_2 + ... + 10x_{10} \geq 0\ \}$

Now since x_i's are *i.i.d.*

$$\therefore X = x_1 + 2x_2 + ... + 10\, x_{10} \sim N\left(\overline{1 + 2 + ... + 10}\, \mu, (1^2 + 2^2 + ... + 10^2)\, \sigma^2\right)$$

$$\simeq N(55\,\mu, 385\,\sigma^2) \simeq N(55\mu, 1540) \text{ as } \sigma = 2$$

$\therefore\ \alpha =$ size of the critical region

$$= P(x \in W \mid H_0) = P(x \geq 0 \mid H_0)$$

Now H_0 ; $\mu = -1 \therefore x \sim N(-55, 1540)$

$$\text{Now } Z = \frac{X - E(X)}{\sqrt{Var\,X}} = \frac{X + 55}{\sqrt{1540}} = \frac{0 + 55}{39{\cdot}3} = 1{\cdot}4 \text{ approx}$$

$$\alpha = P(Z \geq 1{\cdot}4) = {\cdot}5 - P(0 \leq Z \leq 1{\cdot}4) = {\cdot}5 - {\cdot}4192 = {\cdot}0808$$

Power of the test : $1 - \beta = P(X \in W \mid H_1) = P(X \geq 0 \mid H_1)$

$$\text{Now } Z = \frac{0 - 55}{\sqrt{1540}} = -1{\cdot}4$$

$$\therefore 1 - \beta = P(Z \geq -1{\cdot}4) = {\cdot}5 + P(-1{\cdot}4 \leq z \leq 0)$$

$$= {\cdot}5 + {\cdot}4192 = {\cdot}9192.$$

18·2·3. Testing a Simple Hypothesis H_0 against an Alternative Simple Hypothesis H_1. Neyman–Pearson Theorem. Let $f(x, \theta)$ be the density function of the population from which a random sample $x_1, x_2 ..., x_n$; n being fixed, is drawn. Let the simple hypothesis be H_0 : $\theta = \theta_0$ and the alternative simple hypothesis be $H_1 : \theta = \theta_1$, the likelihood functions corresponding to H_0 and H_1 are denoted by L_0 and L_1 where

$$L_0 = \prod_{i=1}^{n} f(x_i, \theta_0), \qquad L_1 = \prod_{i=1}^{n} f(x_i, \theta_1) \qquad ...(18{\cdot}1)$$

Then the *best critical region* (BCR) W of α for testing $H_0 : \theta_0$ against H_1 ; $\theta = \theta_1$ which minimizes the type II error, is given by

$$L_1 \geq k\, L_0 \qquad ...(18{\cdot}2)$$

where the constant k is determined from the condition that

$$P\ [(x_1, x_2, \ldots\ldots, x_n) \in w\ ;\ H_0] = \alpha \quad \Rightarrow \int_w L_0\, dx = \alpha. \qquad \ldots(18{\cdot}3)$$

This theorem is known as Neyman-Pearson theorem.

Proof. Let c be another critical region of size α

i.e. $\displaystyle\int_c L_0\, dx = \alpha.$

Now $\displaystyle\int_w L_1\, dx - \int_c L_1\, dx = \int_{w \cap c} L_1\, dx + \int_{w \cap \overset{\times}{c}} L_1\, dx - \int_{c \cap w} L_1\, dx - \int_{c \cap \overset{\times}{w}} L_1\, dx$

where $\overset{\times}{w}$ is the complement of the set w.

$$= \int_{w \cap \overset{\times}{c}} L_1\, dx - \int_{c \cap \overset{\times}{w}} L_1\, dx$$

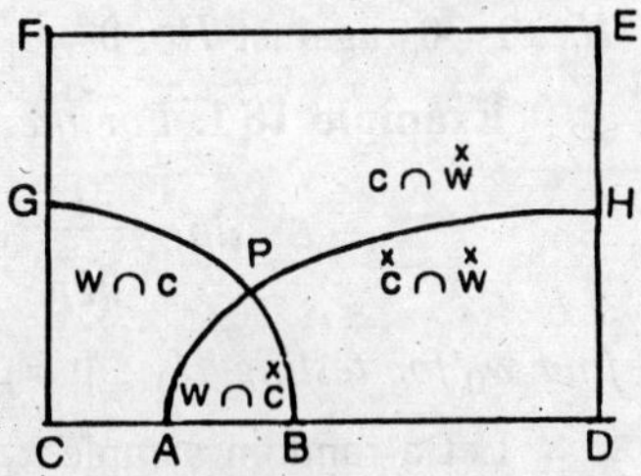

Fig. 18·2

Area		
Area	OAPG is	$W \cap C$
	ABP	$W \cap C^x$
	BDHP	$C^x \cap W^x$
	P1+EF	$C \cap W^x$

By hypothesis, we have $L_1 \geq k\ L_0$ at each point of w and consequently at each point of $w \cap \overset{\times}{c}$.

$$\therefore \qquad \int_{w \cap \overset{\times}{c}} L_1\, dx \geq k \int_{w \cap c} L_0\, dx$$

Further, at each point of $\overset{\times}{w}$, $L_1 \leq k\, L_0$ and hence at each point of $c \cap \overset{\times}{w}$.

$$\int_{c \cap \overset{\times}{w}} L_1\, dx \leq k \int_{c \cap \overset{\times}{w}} L_0\, dx$$

Hence by subtraction,

$$\int_{w \cap \overset{\times}{c}} L_1\, dx - \int_{c \cap \overset{\times}{w}} L_1\, dx\, dx \geq k \left[\int_{w \cap \overset{\times}{c}} L_0\, dx - \int_{c \cap \overset{\times}{w}} L_0\, dx \right]$$

$$= k \left[\int_{w \cap \overset{\times}{c}} L_0\, dx + \int_{w \cap c} L_0\, dx - \int_{c \cap w} L_0\, dx - \int_{c \cap \overset{\times}{w}} L_0\, dx \right]$$

$$\therefore \quad \int_{w \cap \bar{c}} L_1\, dx - \int_{c \cap \bar{w}} L_1\, dx \geq k \left[\int_{w} L_0\, dx - \int_{c} L_0\, dx \right]$$

$$\Rightarrow \quad \int_{w} L_1\, dx - \int_{c} L_1\, dx \geq k \left[\alpha - \alpha \right] = 0 \text{ using (18·3)}$$

$$\Rightarrow \quad \int_{w} L_1\, dx \geq \int_{c} L_1\, dx$$

Hence the result,

Remarks. This theorem enables us to find BCR for testing $H_0 : \theta = \theta_0$ against $H_1 : \theta = \theta_1\ (\neq \theta_2)$

Example 18·1. *For the normal population*

$$dp = \frac{1}{\sqrt{2\pi}}\, e^{-\frac{1}{2}(x-\mu)^2}\, dx, \; -\infty < x < \infty$$

find w_0 for testing $H_0 : \mu = \mu_0$ against $H_1 : \mu = \mu_1$.

Let a random sample $x_1, x_2 \ldots\ldots, x_n$, be drawn from the population.

$$\therefore \quad L_1 = \left(\frac{1}{\sqrt{2\pi}} \right)^n \exp \left[-\tfrac{1}{2} \sum_{i=1}^{n} (x_i - \bar{x})^2 - \tfrac{1}{2} n (\bar{x} - \mu_1)^2 \right]$$

$$L_0 = \left(\frac{1}{\sqrt{2\pi}} \right)^n \exp \left[-\tfrac{1}{2} \sum_{i=1}^{n} (x_i - \bar{x})^2 - \tfrac{1}{2} n (\bar{x} - \mu_0)^2 \right]$$

(18·1) then gives

$$n\,(2\bar{x} - \mu_0 - \mu_1)(\mu_1 - \mu_0) \geq \log_e k.$$

If $\mu_1 > \mu_0$, the critical region is given by $\bar{x} \geq A$, and

If $\mu_1 < \mu_0$, the critical region is given by $\bar{x} \leq B$,

where (18·3) gives $A = \mu_0 + 1{\cdot}645/\sqrt{n},\ B = \mu_0 - 1{\cdot}645/\sqrt{n}$.

Ex. 1. Show that if random sample of size n is taken from the population.

$$dP = \frac{1}{\sigma\sqrt{2\pi}} \exp\left(-\frac{x^2}{2\sigma^2} \right) dx, \; -\infty < x < \infty$$

the critical region for testing the hypothesis H_0: $\sigma = \sigma_0$ against H_1: $\sigma = \sigma_1$ are given by

$$\sum_{i=1}^{n} x_i^2 \leq A \; ; \qquad \text{if } \sigma_1 > \sigma_0 \; ;$$

$$\sum_{i=1}^{n} x_i^2 \leq B \, , \qquad \text{if } \sigma_1 < \sigma_0 .$$

and determine A, B by using the χ^2 distribution.

Ex. 2. Suppose a random sample of n taken from Poisson population $e^{-\lambda}\lambda^x/x!$ Given the most powerful critical region of size α test for testing the hypothesis $H_0 : \lambda = \lambda_0$ against $H_1 : \lambda = \lambda_1, \lambda_1 > \lambda_0$. ***(M.Sc. Madras)***

How can you use the above result to find a confidence interval for λ?

Write an expression for the power function of the test for the hypothesis $\lambda = \lambda_0$ against $\lambda > \lambda_1$. ***(M.A. Statistics, Delhi)***

18·2·4. Testing a Simple Hypothesis against a Composite Hypothesis. A composite hypothesis may be regarded as an aggregate of simple hypothesis and as above, we may determine the best critical region for each of these hypothesis. In general this will vary from one simple alternative hypothesis to another. If there exists a region which is best for all the alternative hypothesis, it is called the *uniformly most powerful region* and the test based on it is the *uniformly most powerful test* (*U.M.P.* test). Unfortunately, as the above example shows, a *U.M.P.* test does not, in general, exist but such a test will, in general, exist if the alternative hypothesis belongs to a suitably restricted class. In general for testing $\theta = \theta_0$ a *U.M.P.* test exists if the alternative hypothesis are restricted to either $\theta > \theta_0$ or $\theta < \theta_0$.

Example 18·2. *A random sample $x_1, x_2, \ldots\ldots, x_n$ is drawn from a normal population N (0, 1). Show that there is no uniformly most powerful test of the simple hypothesis $H_0 : \theta = \theta_0$ where θ_0 is a fixed positive number against the alternative Composite hypothesis $H_1 : \theta = \theta_1 \neq \theta_0$.*

We have $\quad L_0 \leq k L_1$

where definition $\quad L_0 = \left(\dfrac{1}{2\pi}\right)^{n/2} \exp\left[-\dfrac{1}{2}\sum_{i=1}^{n}(x_i - \theta_0)^2\right]$

On simplification, we have

$$\exp\left\{-(\theta' - \theta_0)\right\}\sum_{1}^{n} x_i + \frac{n}{2}(\theta'^2 - \theta_0^2))\} \leq k$$

$$\Rightarrow \quad (\theta' - \theta_0)\sum_{1}^{n} x_i \geq \frac{n}{2}(\theta'^2 - \theta_0^2) - \log k$$

$$\Rightarrow \quad \sum_{1}^{n} x_i \geq \frac{n}{2}(\theta' + \theta_0) - \frac{\log k}{\theta' - \theta_0} \quad \text{if } \theta' > \theta_0$$

and
$$\sum_{1}^{n} x_i \leq \frac{n}{2}(\theta' + \theta_0) - \frac{\log k}{\theta' - \theta_0} \quad \text{if } \theta' < \theta_0$$

Thus there will not be one best critical region for testing $H_0 : \theta = \theta_0$ against the alternative simple hypothesis $H_1 : \theta < \theta_0$ and the alternative simple hypothesis $\theta > \theta_0$. For testing $H_0 : \theta = \theta_a$, θ_0 being fixed positive number against $H_1 : \theta < \theta_0$, the UMPT is given by $\sum_1^n x_i^2 \le C$.

Ex. 1. Let a random sample of 10 be drawn from the distribution with *p.d.f.* $f(x ; \theta) = \theta^x (1-\theta)^{1-x}$, $x = 0, 1$; Zero elsewhere. We test the simple hypothesis $H_0 : \theta = \frac{1}{4}$ against the alternative composite hypothesis $H_1 : \theta < \frac{1}{4}$ and reject $H_0 : \theta = \frac{1}{4}$ iff the observed values $x_1, x_2, \ldots, x_{10}$ are such that $\sum_1^{10} x_i \le 1$. Find the power function. $K(\theta)$, $0 < \theta \le \frac{1}{4}$ of this test **Ans.** $(1-\theta)^9 (1+9\theta)$

2. Show that if random sample $x_1, x_2 \ldots, x_n$ is taken from the normal distribution $N(\theta, 0)$, then find the BCR for testing the simple hypothesis $H_0 : \theta = 0$ against the alternative simple hypothesis $H_1 : \theta = 1$. If $n = 25$, $\alpha = 0{\cdot}05$, find C_1 and the power of the test.

[**Hint.** BCR $w = \{ (x_1, x_2, \ldots, x_n), \sum_{i=1}^n x_i \ge C$ or $\bar{x} \ge C_1 \}$

$$C_1 = 1{\cdot}645 \sqrt{25} = 0{\cdot}329, \text{ Power of the test } = \int_{0{\cdot}329}^{\infty} \frac{1}{\sqrt{2\pi}} \frac{1}{\sqrt{\frac{1}{5}}} e^{-\frac{1}{2}\frac{(\bar{x}-1)^2}{\frac{1}{2}}}$$

$= 0{\cdot}999$, when H_1 is true.

3. Let $x_1, x_2, \ldots, x_{10}$ be a random sample from a normal distribution $N(0, \sigma^2)$. Find BCR of size $\alpha = 0{\cdot}05$ for testing $H_0 : \sigma^2 = 1$ against $H_1 : \sigma^2 = 2$.

Ans. $\sum_1^{10} x_i^2 \ge 18{\cdot}3$

4. Examine whether a BCR exists for testing to the hypothesis $H_0 : \theta = \theta_0$ against the alternative hypothesis $H_1 : \theta = \theta_1 > \theta_0$ for the parameter θ of the distribution $f(x, \theta) = \dfrac{1+\theta}{(x+\theta)^2}$, $1 \le x < \infty$. ***(B.Sc. Bangalore, 1993)***

$$\left[\textbf{Hint. } \prod_{i=1}^n f(x_i, \theta) = \frac{(1+\theta)^n}{\prod_{i=1}^n (x_i+\theta)^2}. \text{ use } L_1 \ge kL. \right]$$

Ans. No BCR exists.

18·2·5. Likelihood Ratio Test. Let the density function $f(x, \theta_1, \theta_2, \ldots, \theta_k)$ have k parameters and let the composite hypothesis H_0 to be tested specify $(k-r)$ of these, then the critical region for testing H_0 is given by

$$0 < \lambda < A \quad \ldots(18{\cdot}4)$$

where $$\lambda = \frac{\text{max. likelihood for variation of unspecified parameters}}{\text{max. likelihood for variation of all parameters}} \quad \ldots(18{\cdot}5)$$

λ is a function of the random sample values $x_1, x_2, \ldots, x_n$ and the constant A is to be determined from the sampling distribution of λ when H_0 is true, so that

$$P(0 < \lambda < A) = a. \qquad \ldots(18\cdot6)$$

For large samples, it can be shown that when H_0 is true, $-2 \log \lambda$ is asymptotically distributed as χ^2 with ν degrees of freedom.

Example 18·3. *A random sample $x_1, x_2, \ldots, x_n$ is drawn from a normal population $N(\mu, \sigma^2)$. Test the null hypothesis $H_0 : \mu = \mu_0$.*

Likelihood function

$$= \left(\frac{1}{2\pi\sigma^2} \right)^{\frac{n}{2}} \exp\left[-\frac{1}{2\sigma^2} \sum_{i=1}^{n} (x_i - \bar{x})^2 - \frac{n}{2\sigma^2} (\bar{x} - \mu)^2 \right]$$

Its maximum value when μ and σ are allowed to vary

$$= e^{-\frac{n}{2}} \left\{ \frac{n}{2\pi \sum_{i=1}^{n} (x_i - \bar{x})^2} \right\}^{\frac{n}{2}}$$

Its maximum value when $\mu = \mu_0$ and σ alone is allowed to vary

$$= e^{-\frac{n}{2}} \left\{ \frac{n}{2\pi \sum_{i=1}^{n} (x_i - \mu_0)^2} \right\}^{\frac{n}{2}}$$

$$\therefore \quad \lambda = \left[\frac{\sum_{i=1}^{n} (x_i - \bar{x})^2}{\sum_{i=1}^{n} (x_i - \mu_0)^2} \right]^{\frac{n}{2}} = \left[\frac{1}{1 + \frac{t^2}{n-1}} \right]^{\frac{n}{2}}$$

where
$$t = \frac{(\bar{x} - \mu_0)\sqrt{n(n-1)}}{\left[\sum_{i=1}^{n} (x_i - \bar{x})^2 \right]^{\frac{1}{2}}}$$

Also λ is a monotonic function of t^2, so that the interval $0 < \lambda < A$ is equivalent to the critical region $t^2 > B$. For $\alpha = 0{\cdot}05$, the critical region is $t_{0\cdot05}$ where $t_{0\cdot05}$ is the 5% value of t for $(n-1)$ degrees of freedom. Thus the present method leads to the usual Student's t test for testing the significance of a mean.

Ex. 1. Explain the terms : (*i*) U.P.M.T. test (*ii*) likelihood ratio test.

Ex. 2. Develop the likelihood ratio test for testing the significance of the difference between two sample means assumed to come from normal population with same variance and show that it leads to the t test of section 15·2·2.

Ex. 3. Test the null hypothesis that the variance of a normal population has a specified value σ_0^2 on the basis of a random sample of size n.

Ex. 4. Given random samples from each of two normal populations with means and variances (μ_1, σ_1^2) and (μ_2, σ_2^2), show that for testing the hypothesis

$$H_0 : \sigma_1^2 = \sigma_2^2. \qquad \lambda = \frac{\left(\dfrac{n_1 - 1}{n_2 - 1} F\right)^{n_1/2}}{\left(1 + \dfrac{n_1 - 1}{n_2 - 1} F\right)^{\frac{n_2 + n_1}{2}}}$$

Ex. 5. Given k independent random samples of sizes n_1, n_2 ....., n_k from k normal populations, find λ for testing the hypothesis

$$H_0 : \sigma_1^2 = \sigma_2^2 = \ldots = \sigma_k^2.$$

18·2·6. Bias in Statistical Tests. Let $f(x, \theta)$ be the population density function and let it be desired to test the hypothesis $\theta = \theta_0$ against the other possible values of θ. For a given critical region of size α, we can find the power function for different values of θ and plot it. It happens sometimes that the power function does not have a minimum at $\theta = \theta_0$ [*cf.* Ex. 5, page 692], but at some other value of θ, say θ_1. This would imply that we would have a greater probability of accepting the hypothesis when it is false than of rejecting it when it is true. We say that in this case the critical region is biased and we try to get unbiased critical region, *i.e.*, those for which the power function has the minimum value at $\theta = \theta_0$. For this purpose Neyman and Pearson defined unbiased critical regions of several types A, A_1, B, C, etc., and proved their existence.

18·3. Sequential Analysis.

18·3·1. Sequential Likelihood Ratio Test. In the Neyman-Pearson Theory of testing hypotheses discussed above, the sample size n and the type 1 error α are kept fixed and the test is so designed as to minimize the type II error β. In sequential test, on the other hand, α and β are fixed in advance and sample size n is a random variable which is determined by the *sequence* of observations as they arise. The sequential tests, lead on an average, to a saving of as many as 50 per cent observation as compared with fixed size samples for the same control of the errors.

The *sequential likelihood ratio* test for testing $H_0 : \theta = \theta_0$ against alternative $H_1 : \theta = \theta_1$ employes the ratios

$$\lambda_m = \prod_{i=1}^{m} \frac{f(x_i, \theta_1)}{f(x_i, \theta_0)} \quad (m = 1, 2, 3.....) \qquad ...(18{\cdot}7)$$

and two numbers A and B, with $A > 1$ and $B < 1$. We go on taking the observations and after the mth observation, we take one of the following three directions :—

(*i*) If $\lambda_m \le B$, we accept H_0. (*ii*) If $\lambda_m \ge A$, we accept H_1.

(*iii*) If $B < \lambda_m < A$, we continue the observations.

Thus the observations continue so long as λ_m lies between A and B and terminate as soon as either limit is crossed. *It can be shown that the probability is one that the process will terminate whatever be the distribution of n.*

To determine the numbers A and B, we compute the errors α and β.

$$\alpha = P\,[H_0 \text{ is rejected when } H_1 \text{ is true }]$$

$$= P\,[\lambda_m \text{ exceeds } A \text{ before it becomes less than } B\,]$$

$$= P\,(\lambda_1 \ge A) + P(B < \lambda_1 < A,\ \lambda_2 \ge A\,) + \ldots \qquad ...(18{\cdot}8)$$

$$\beta = P\,(\,\lambda_m \le B \text{ before } \lambda_m \text{ exceeds } A\,)$$

$$= P\,(\lambda_1 \le B\,) + P(\,B < \lambda_1 < A, \lambda_2 \le B\,) + ... \qquad ...(18{\cdot}9)$$

The determination of A and B from the equations would be extremely difficult. Fortunately, this is not necessray as very simple and accurate approximations are available. These are

$$A \cong \frac{1-\beta}{\alpha},\ B \cong \frac{\beta}{1-\alpha} \qquad ...(18{\cdot}10)$$

These approximations, independent of the population distribution, make the task of carrying out a sequential test amazingly simple, since no sampling theory is needed.

18·3·2. Power Function and Average Sample Size. The power function $P(\theta)$ is the probability of rejecting $H_0 : \theta = \theta_0$ when θ is the true value and this can be obtained by adding the probabilities that H_0 will be rejected at each observation when θ is the true value so that

$$P(\theta) = P(\lambda_1 > A) + P(B < \lambda_1 < A, \lambda_2 > A), + ... \qquad ...(18{\cdot}11)$$

The exact computation would be tedious but again a very good approximation is available, *viz.*,

$$P(\theta) \cong \frac{1 - B^h}{A^h - B^h}, \qquad ...(18{\cdot}12)$$

where h is the non-zero number for which

$$\int_{-\infty}^{\infty} \left[\frac{f(x, \theta_1)}{f(x, \theta_c)} \right]^h f(x, \theta)\, dx = 1. \qquad ...(18\cdot13)$$

It can be shown that such a non-zero number h exists.

The probability $p(n)$ that the size of the sample would be n can be calculated for all n, *e.g.*,

$$p(1) = P(\lambda_1 < B) + P(\lambda_1 > A)$$

$$p(2) = P(B < \lambda_1 < A, \lambda_2 < B) + P(B < \lambda_1 < A, \lambda_2 > A)$$

and so on. The average sample size $E(n)$ can then be computed. A good approximation is given by

$$E(n) \cong \frac{P(\theta) \log A + [1 - P(\theta)] \log B}{E[\log f(x, \theta_1) - \log f(x, \theta_0)]} \qquad ...(18\cdot14)$$

where $P(\theta)$, A and B are determined from (18·10), (18·12) and (18·13).

$E(n)$ regarded as a function of θ is called the *average sample number function* or simply the *ASN* function. The *OC* function is easily found from (18·12) to be

$$OC \text{ function} \cong \frac{A^h - 1}{A^k - B^h} \qquad ...(18\cdot15)$$

The *ASN* function enables us to compare sequential tests with fixed size example tests.

Ex. 1. Write two more terms on the right of each of the equation (18·8), (18·9) and (18·11).

Ex. 2. Write expressions for $p(3)$, $p(4)$ and $p(5)$.

Ex. 3. Show that $P(\theta_0) = \alpha$, $P(\theta_1) = 1 - \beta$.

Ex. 4. For testing the null hypothesis that the mean of a normal population with known variance is μ_0 against the alternative that the mean is μ_1 $(\mu_1 > \mu_0)$, show that

$$h = \frac{\mu_1 + \mu_0 - 2\mu}{\mu_1 - \mu_0}$$

and find $P(0)$ and $E(n)$

Ex. 5. In the above exercise, show that $E(n) \cong \theta$, if

$$\mu_0 = \mu = 0, \mu_1 = \alpha = \cdot01, \beta = \cdot01$$

Ex. 6. Show that (18·13) is satisfied by $h = 0$.

18·4. Distribution Free Methods.

18·4·1. Comparison with Parametric Case. Most of the tests which we have considered in earlier chapters have been based on the assumption that the parent population is normal. Even if the parent population is not normal, we can often find transformations to reduce it to the normal form.

In practice, however, our knowledge about the parent population may not be sufficient to enable us to find such a transformation and in such case we need tests which do not depend on any assumption about the form of the population. We shall consider some of these non-parametric or distribution-free tests in this section. It is obvious, however, that the collection of these results cannot be so comprehensive as in the normal case, but these tests are not only capable of wider application, but also are simpler to apply and do not need complicated sampling theory. Distribution free methods are based on *order statistics* or *ordered samples, i.e.,* we suppose that the sample $x_1, x_2, \ldots, x_n$ is ordered, so that, the observations in it are in ascending order of magnitude. Also, while in the parametric case the measures of location and dispersion, which are most commonly used are the mean and the standard deviation respectively, which do not depend on order in the sample ; in the non-parametric case, we prefer to use the median, quartiles, interquartile range, etc., for which an order sample is desirable.

18·4·2. Point Estimation and Confidence Intervals. The population median ν is estimated by the sample median $\hat{x}$, which however is not an unbiased estimate, though the bias is not serious and tends to zero as n tends to infinity. Similarly, to estimate the population quartiles or deciles, we use the corresponding sample quartiles or deciles as estimators.

To construct a confidence interval for ν, the population median, we use the fact that the probability of an observation, falling to the left or right of ν, is one-half in either case. The probability that x_r, rth order statistic, exceeds ν is given by

$$P(x_r > \nu) = \sum_{i=0}^{r-1} ({}^nC_i)\left(\tfrac{1}{2}\right)^n.$$

Similarly, $$P(x_s < \nu) = \sum_{i=0}^{n} ({}^nC_i)\left(\tfrac{1}{2}\right)^n$$

and $$P(x_r < \nu < x_s) = \sum_{i=r}^{s-1} ({}^nC_i)\left(\tfrac{1}{2}\right)^n \quad \ldots(18\cdot16)$$

(x_r, x_s) gives the confidence interval for the confidence coefficient given by the R.H.S. of (18·16). It is obvious, however, that we cannot construct a confidence interval for every confidence coefficient as the R.H.S. of (18·16) can take only a certain set of values for different values of r and s. We can use linear interpolation, but usually we restrict ourselves to the confidence levels available with the simple order statistics. In fact the paucity of confidence levels for small samples constitutes one of drawbacks of distribution free methods.

For moderate sample sizes, we can compute the R.H.S. of (18·16) either directly or by use of tables of incomplete beta functions. For large samples we can use the result that the number of successes will be asymptotically normally distributed with mean $n/2$ and variance $n/4$.

Ex. 1. Obtain estimates for Q_1, Q_2, D_2, D_7 from an ordered sample of size 15.

Ex. 2. For a sample of size 6, prove that

$$P(x_1 < v < x_6) \cong \cdot 97 \; ; \; P(x_2 < v < x_5) \cong \cdot 78 \; ; \; P(x_1 < Q_1 < x_4) \cong \cdot 78$$

Ex. 3. For an ordered sample of size 100 gives 99% confidence interval for v.

Ex. 4. For an ordered sample of size n, list the various confidence levels available for estimating the upper quartile.

18·4·3. The Sign Test. Let $(x_1, x_2 ..., x_n)$ and $(y_1, y_2, ..., y_n)$, be two random samples of same size n, from two populations and let the sample values be paired by pairing the ith member of one with the ith member of the other and consider the signs of the differences $x_i - y_i$ $(i = 1, 2, ..., n)$. If the two populations are continuous and *identical* and probability of a difference being positive or negative is one half, so we can test this null hypothesis by treating the number of positive signs as a binomial variable with mean $n/2$ and variance $n/4$.

In case of 'ties' $x_i = y_i$, we can either ignore them or decide to allot them positive or negative signs by tossing a coin or assign half of them positive signs and the other half negative signs. Usually the first choice of ignoring them is preferred and in that case the conditional probability of the sign being positive, given that the difference is non-zero, is one-half.

When applicable, a t test is generally more efficient than a sign test, but the sign test is much easier to apply and is applicable even when t-test is not applicable *viz*, when the parent population is not necessarily normal.

Ex. For 20 paired observations without ties what number of positive difference would you consider significant at 5% level.

18·4·4. The Run Test. We have to test the null hypothesis that two ordered random samples $x_1, x_2, ..., x_n$, and $y_1, y_2, ..., y_n$, come from the same population. The two sets of observations are combined and arranged in order of magnitude as, say,

$$x_1, y_1, y_2, x_2, x_3, x_4, y_3..... \qquad ...(18\cdot17)$$

and in this new arrangement we find the total number of *runs*, where a run is defined as a sequence of letters the same kind bounded by letters of the other kind. Thus (18·17) starts with a run of one x, followed by a run of two y's which is succeeded by a run of three x's and so on.

It is obvious that if the two samples are from the same population, x's and y's will be well-mixed and r, the number of runs, would be large. If the two populations are so widely separated, that they do not overlap, the number of runs would be only two. If the two populations have the same mean or median, but the first population has a larger variance, there will be long runs of x's at the end. Thus, any difference in mean or variance tends to reduce r. The run test for testing the null hypothesis that the two populations are identical, accordingly consists in counting the number of runs in the combined ordered sample and rejecting a null hypothesis is $r \le r_0$, where r_0 is a number to be determined from the distribution of runs and depends on n_1, n_2 and the level of significance but not on the form of the distribution of the parent population.

The probability of exactly r runs is given by

$$p\,(r) = \begin{cases} 2.\ \dfrac{{}^{n_1-1}C_{k-1} \times {}^{n_2-1}C_{k-1}}{{}^{n_1+n_2}C_{n_1}}; & r = 2k = \text{even} \\[2ex] \dfrac{{}^{n_1-1}C_k \times {}^{n_2-1}C_{k-1} + {}^{n_1-1}C_{k-1} \times {}^{n_2-1}C_k}{{}^{n_1+n_2}C_{n_1}}; & r = 2k+1 = \text{odd} \end{cases} \tag{18·18}$$

The probability distribution of R, the total number of runs of $n = n_1 + n_2$ objects, n_1 of type 1 and n_2 of type 2, in a random sample is

$$p(r) = f_n(r) = \begin{cases} 2\dbinom{n_1-1}{\frac{r}{2}-1}\dbinom{n_2-1}{\frac{r}{2}-1} \div \dbinom{n_1+n_2}{n_1}, & r = 2k \text{ i.e. if } r \text{ is even} \\[2ex] \left[\dbinom{n_1-1}{\frac{r-1}{2}}\dbinom{n_2-1}{\frac{r-3}{2}} + \dbinom{n_1-1}{\frac{r-3}{2}}\dbinom{n_2-1}{\frac{r-1}{2}}\right] \div \dbinom{n_1+n_2}{n_1} & \\ \qquad \text{if } r = 2k+1 \text{ and is odd} & \end{cases}$$

for $r = 2, 3, \ldots\ldots, n_1 + n_2$

Proof. For r even, there must be the same number of runs of both types. Thus the only values of r_1 and r_2 are $r_1 = r_2 = r/2$ and (2·1) is summed over this pair. If $r_1 = r_2 \pm 1$, r is odd. In this case, (2·1) is assumed over the two pairs of values

$$r_1 = \frac{(r-1)}{2} \text{ and } r_2 = \frac{(r+1)}{2}, \quad r_1 = \frac{(r+1)}{2} \text{ and } r_2 = \frac{(r-1)}{2},$$

obtaining the given result.

$$E(R) = \sum_{i=1}^{n_1} 4i \binom{n_1-1}{i-1}\binom{n_2-1}{i-1} + \sum_{i=1}^{n_1} (2i+1)\binom{n_1-1}{i}\binom{n_2-1}{i-1}$$

$$+ \sum_{i=1}^{n_1} (2i+1)\binom{n_1+1}{i-1}\binom{n_2-1}{i}$$

To evaluate these three sums, the following lemmas are useful.

Lemma 1. $\sum_{r=0}^{c}\binom{m}{r}\binom{n}{r}=\binom{m+n}{m}$ where $c=\min(m, n)$.

Proof. $(1+x)^{m+n}=(1+x)^m(1+x)^n$ for all x,

$$\sum_{i=0}^{m+n}\binom{m+n}{i}x^i=\sum_{j=0}^{n}\binom{m}{j}x^j\sum_{k=0}^{n}\binom{n}{k}x^k$$

Assume without loss of generality that $c=m$ and equate the coeffi. of x^m on both sides of this equation, we have

$$\binom{m+n}{n}=\sum_{r=0}^{m}\binom{m}{m-r}\binom{n}{r}.$$

Lemma 2. $\sum_{r=0}^{c}\binom{m}{r}\binom{n}{r+1}=\binom{m+n}{m+1}$

where $c=\min(m, n-1)$

[See also page 147 of Math. Statistics by S. Wilks]

Proof. Proof follows, as in Lemma 1, on equating the coeffs. of x^{m+1}.

However it is difficult to find $E(R)$ and var R and is left as an exercise for the Reader. A simpler approach however can be made with the help of *Indicator Function*. Consider R as the sum of indicator variables as follows for $n=n_1+n_2$. Let

$$R=1+I_2+I_3+.....+I_n$$

where in an ordered sequence of the two types of symbols, we define

$$I_k=\begin{cases}1 & \text{if the } k\text{th element} \neq (k-1) \text{ st element}\\ 0, & \text{otherwise}\end{cases}$$

Then I_k is a Bernoullian random variable with parameter

$$p={}^{n_1}C_1\,.\,{}^{n_2}C_1/{}^{n_1+n_2}C_2=n_1n_2\Big/\binom{n}{2},\ n_1+n_2=n.$$

so that $E\left(I_k\right)=E\left(I^2_k\right)=\dfrac{2\,n_1\,n_2}{n\,(n-1)}$

$$E(R)=1+\sum_{k=2}^{n}E\left(I_k\right)=1+(n-1)\cdot\frac{2\,n_1\,n_2}{n(n-1)}=1+\frac{2n_1,\,n_2}{n}$$

$$\therefore\quad \lim_{n\to\infty}E\left(\frac{R}{n}\right)=2\,\lambda\,(1-\lambda),\ \lambda=n_1/n.$$

$$\text{Var}\, R = \text{Var}\left(\sum_{k=2}^{n} I_k\right) = (n-1)\,\text{Var}\, I_k + \sum_{2 \le j \ne k \le n}\sum \text{Cov}(I_j, I_k)$$

$$= (n-1)\, E(I_k^2) + \sum_{2 \le j \ne k \le n}\sum E(I_j, I_k) - (n-1)^2 [E(I_k)]^2$$

To evaluate the $(n-1)(n-1)$ joint moments of the type

$$E(I_j, I_k) \text{ for } j \ne k,$$

the subscript choice can be classified as follows :

1. For the $2(n-2)$, selections where $j = k-1$ or $j = k+1$,

$$E(I_j, I_k) = \frac{n_1 n_2 (n_1 - 1) + n_2 n_1 (n_2 - 1)}{n(n-1)(n-2)} = \frac{n_1 n_2}{n(n-1)}$$

2. For the remaining $(n-1)(n-2) - 2(n-2) = (n-2)(n-3)$ selections of $j \ne k$

$$E(I_j, I_k) = \frac{4 n_1 n_2 (n_1 - 1)(n_2 - 1)}{n(n-1)(n-2)(n-3)}$$

$$\therefore \quad \text{Var}\, R = \frac{2n_1 n_2}{n} + \frac{2(n-2) n_1 n_2}{n(n-1)} + \frac{4 n_1 n_2 (n_1 - 1)(n_2 - 1)}{n(n-1)} - \frac{4 n_1^2 n_2^2}{n^2}$$

$$= \frac{2 n_1 n_2 (2 n_1 n_2 - n_1 - n_2)}{(n_1 + n_2)^2 (n_1 + n_2 - 1)}$$

$$\therefore \quad \lim_{r=\infty} \text{Var}\left(\frac{R}{n^{1/2}}\right) = 4\lambda^2 (1-\lambda)^2$$

To test the null hypothesis at probability level P, we find r_0 from

$$\sum_{r=0}^{r_0} p(r) = P \qquad \text{...(18·19)}$$

as nearly as possible and reject H_0 if $r \le r_0$.

For large n_1, n_2 the distribution of r is asmptotically normal with

$$E(r) = \hat{r} = \frac{2 n_1 n_2}{n_1 + n_2} + 1 \qquad \text{...(18·20)}$$

$$\text{Var}\ r = \sigma_r^2 = \frac{2 n_1 n_2 (2 n_1 n_2 - n_1 - n_1)}{(n_1 + n_2)^2 (n_1 + n_2 - 1)} \qquad \text{...(18·21)}$$

and we reject the null hypothesis if $r < \hat{r} - 1{\cdot}645\sigma_r$. The above approximation can be used if both n_1, n_2 exceed 10.

That run test can also be used for *testing the randomness of a given sample, i.e.,* to test whether the sample observations are independent and identically distributed or in other words to test whether the phenomenon yielding the data is in the *statistical control.* Let $x_1, x_2, \ldots, x_n$ denote the observations in the order obtained. For each observation we note whether it is above or below the sample median. We write a if it is above and b if it is below the sample median, obtaining a sequence of a's and b's such as, say,

$$aa \;\; bb \;\; ab \;\; aa \;\; bb\ldots \qquad \ldots(18\cdot22)$$

If the sample is *actually* a random sample, the number of runs above and below the sample median is a random variable with

$$E(r) = \frac{n+2}{2}, \quad \text{Var }(r) = \frac{n}{4}\left(1 - \frac{1}{n-1}\right) \qquad \ldots(18\cdot23)$$

The formulae are exact only if n is even and population is continuous. For large n, say above 25, the asymptotic normality of the distribution of r can be used.

Another test for randomness can be obtained by writing + or – between each pair of successive observations, depending on whether the second observation of the pair is larger or smaller than the first and counting the number of runs of positive or negative signs. Under the assumption of randomness, the distribution of the number of runs is asymptotically normal with

$$E(r) = \tfrac{1}{3}\,(2n-1) \text{ and Var }(r) = \tfrac{1}{90}\,(16n-29) \qquad \ldots(18\cdot24)$$

Ex. 1. For $\mu_1 = 6, n_2 = 8$, give the probability distribution of r and find r_0 for 5 per cent level of significance.

Ex. 2. If in a sample of size 50, there are 20 runs above and below the median : is the sample random ?

Ex. 3. Thirty observations given below are obtained :

14, 25, 2, 40, 50, 60, 58, 39, 70, 15, 59, 18, 18, 1, 73, 2.

27, 24, 44, 65, 35, 85, 65, 16, 33, 47, 84, 38, 53, 35, 33, 9.

Test their randomness by considering the sequence of positive and negative signs.

18·4·5. The Median Test. The run test discussed above is sensitive to both differences in shape and location of the two distributions. If we are interested in differences in location only, the median test is to be preferred as it is very little sensitive to differences in shapes.

Let $x_1, x_2, \ldots, x_{n_1}$ and $y_1, y_2, \ldots, y_{n_2}$ be two ordered random samples from the two populations $f(x)$ and $f(y)$ respectively and let $z_1, z_2, \ldots, z_{n_1+n_2}$ be the combined ordered sample z_a as its median. The null hypothesis is that the two populations have the same median. Under this null hypothesis, the probability that m_1 of the r's and m_2 of the y's are less than z_a is given by

$$^{n_1}C_{m_1} \cdot {}^{n_2}C_{m_2} \Big/ {}^{n_1+n_2}C_{m_1+m_2} \qquad \ldots(18\cdot25)$$

where we take a to $b\,(n_1+n_2)/2$ if n_1+n_2 is even and $(n_1+n_2+1)/2$ if n_1+n_2 is odd. If the null hypothesis is true, m_1 has a hypergeometric distribution with mean $n_1/2$ and variance $(n_1n_2)/4\,(n-1)$. For small n_1, n_2, the computation can be done numerically, while for large values of n_1, n_2 we can use the asymptotic normal distribution. We can also arrange observations in a 2×2 contingency table and use χ^2 test.

Ex. The random samples of size 20 each are drawn and 5 of the one and 15 of the other are found to be below the overall median. Prove that the probability of seven or less observations being below the overall mediam in 0·56. Find the probability of six or less and discuss the significance of the observed result.

18·4·6. Wilcoxon - Mann Whitney U Test. Let two ordered samples be combined and ordered in order of magnitude as follows, say

$$x_1, y_1, x_2, x_3, y_2, y_3 \ldots. \qquad \ldots(18{\cdot}26)$$

For each y we count the number of '*inversions*' *i.e.*, the number of x's that precede that y in the sequence. Thus for y_1, y_2, y_3 in the above sequence, there are 1, 3, 3 inversions. Alternatively, for each pair of observations x_i, y_j, let

$$z_{ij} = \begin{cases} 1 & \text{if } x_i < y_j \\ 0 & \text{if } x_i > y_j \end{cases} \qquad \ldots(18{\cdot}27)$$

Then the total number of inversions would be

$$U = \sum_{j=1}^{n_2} \sum_{i=1}^{n_1} z_{ij} \qquad \ldots(18{\cdot}28)$$

Under the null hypothesis z_{ij} is a Bernoullian variate with $p = \frac{1}{2}$, and

$$E\,(u) = \frac{n_1n_2}{2}, \ \text{Var}\, u = \frac{n_1n_2}{2}\,(n_1+n_2+1) \qquad \ldots(18{\cdot}29)$$

Also n is asymptotically normal with these parameters.

Example. Test the hypothesis of no difference between the ages of male and female employees of a certain company using the **Mann - Whitney U test** for the sample data. Use the 0·10 level of significance.

Males	3·1	2·5	3·8	3·3	4·2	4·0	4·4	2·6	4·3	3·5
Females	44	30	34	47	35	32	35	47	48	34

Sol.

Rank	Age	Employee	Rank	Age	Employee	Rank of Males	Rank of Females
1	25		10	35		4	
2	26		12	38		1	
3	30		13	40		12	
4	31		14	42		6	
5	32		15	43		14	
6	33		16·5	44		13	
7.5	34		17·5	44		16·5	
7.5	34		18·5	47		2	
10	35		18·5	47		15	
10	35		20	48		10	
						93·5	

Here $n_1 = 10$, $n_2 = 10$ (Sample sizes need not be same)

Sum of ranks, $R_1 = 93{\cdot}5$

$$\text{U-statistic. } U = n_1 n_2 + \frac{n_1(n_1+1)}{2} - R_1$$

$$= 10 \times 10 - \frac{10 \times 11}{2} - 93{\cdot}5 = 61{\cdot}5$$

$H_0 = n_1 + n_2$ observations came from identical population.

$$\text{Mean of U statistics} = E(U) = \frac{n_1 n_2}{2} = \mu_u = \frac{10 \times 10}{2} = 50$$

$$\text{S.E. of U statistics} = \sigma_u = \sqrt{\frac{n_1 n_2 (n_1 + n_2 + 1)}{12}}$$

$$= \sqrt{\frac{10 \times 10 \times 21}{12}} = = \sqrt{\frac{100 \times 7}{4}} = \sqrt{175} = 13.2$$

Testing the hypothesis

$H_0 \Rightarrow \mu_1 = \mu_2$ *i.e.* there is no difference between the two population and hence they have the same mean.

$H_1 \Rightarrow \mu_1 \neq \mu_2$ *i.e.* there is difference between the two population *i.e.* they have different means.

α = Level of significance for testing the high H_0 and H_1

Limits of the acceptance region

Lower limit $= \mu_u - 1{\cdot}64\sigma_u = 50 - 1{\cdot}64 \times 13{\cdot}2$

Upper limit $= \mu_u + 1{\cdot}64\,\sigma_u = 28{\cdot}32 = 50 + 1{\cdot}64 \times 13{\cdot}2 = 71{\cdot}66$

Since 61·5 lies within the limits of acceptance, we accept the null hypothesis H_0 and hence the mean ages are not significantly different.

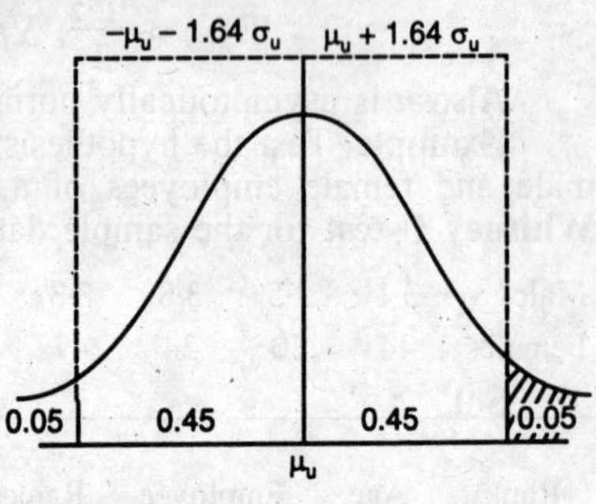

Ex. 1. The rank of any *y* (or *x*) is its position in the combined ordered sample. Let $R_y, R_x; U_y, U_x$ be the sum of ranks and inversion respectively of *y*'s and *x*'s. Show that

$$R_y = U_y + \frac{n_2(n_2+1)}{2} \qquad R_x = U_x + \frac{n_1(n_1+1)}{2}$$

$$E(R_y) = \frac{n_1 n_2 + n_1(n_1+1)}{2}, \ \text{Var}\, R_x = \text{Var}\, U_x.$$

Ex 2. Two sets of data are given below. Test the hypothesis that there has been no change in the centre of population at 5% level.

(1) – ·1, 3·6, – 7, 3, 1, 2, – 8, 3, 2, 1, 8, 1 – 4, 2, 1, 2, 3, 5, 6, – 1

(2) 3, 1, 6, – 1, 1, 2, 3, 4, – 4, 1, 0, 6, 2, 0, – 5, 1, 1, 2.

18·4·7. Kolmorgov Smirnov Test. The test for testing the goodness of fit between observed and expected frequencies uses the statistic

$$D_n = l.u.b. \mid F_n(x) - F(x) \mid \qquad ...(18·30)$$

where $F(x)$ is the specific population distribution function being tested and $F_n(x)$ is the sample distribution function given by

$$F_n(x) = \sum_{x_i \le x} \frac{f_i}{n} \qquad ...(18·31)$$

and *l.u.b.* refer to the least upper bound of the absolute difference between $F_n(x)$ and $F(x)$ *for all x.* Both the large sample and small sample distributions of D_n are independent of $F(x)$.

For samples of moderate size, some of the 5% significant values are:

n	1	3	5	7	9	11	13	15	20	50	100
5% value	·915	·708	·563	·486	·432	·391	·361	·338	·204	·19	·14

For large n the asymptotic formulae are :

Significance level	·20	·10	·05	·10
Formulae	$\frac{1·07}{\sqrt{n}}$	$\frac{1·22}{\sqrt{n}}$	$\frac{1·36}{\sqrt{n}}$	$\frac{1.63}{\sqrt{n}}$

For testing the significance of the difference between two samples of sizes n_1, n_2. *Kolmorgov-Smirnov* test uses the statistic

$$D = l.u.b. \mid F_{n_1}(x) - F_{n_2}(x) \mid \qquad ...(18·32)$$

For small values of n_1, n_2 tables are available. For large values of n_1, n_2 we reject the null hypothesis of the identity of two populations if D exceeds $a \times \sqrt{\frac{n_1 + n_2}{n_1 n_2}}$, where a is 1·36 at 5% level and is 1·63 at 1% level.

Example. At the 0·05 level of significance, can we conclude that the following data came from the Poisson distribution with $\lambda = 3$

Number of arrivals per day:	0	1	2	3	4	5	6 or more	total
Number of states:	6	18	03	24	11	2	9	100

Sol. We set up the null hypothesis H_0 that the data has been taken from the Poisson distribution with mean $\lambda = 3$.

$H_1 \Rightarrow$ Alternative hypothesis that the data has not been taken from the Poisson distribution, $\alpha = 0.05$

Number of arrivals x	Observed fr.	ob. cum. fr.	F_0 ob. relative cum. fr.	F_e-expected relative fr. $\lambda = 3$	cum. fr.	$\lvert F_0 - f_e \rvert$
0	6	6	0·06	0·0498	0·0498	0·002
1	18	24	0·24	0·1494	0·1992	0·001
2	30	54	0·54	0·2240	0·4232	0·117
3	24	78	0·78	0·2240	0·6472	0·133
4	11	89	0·89	0·1680	0·8152	0·075
5	2	91	0·91	0·1008	0·9060	0·004
6 or more	9	100	1·00	0·0504	0·9564	0·000
	100					

$n = 100 \cdot \alpha$ = significance level = 0·05.

From last column, $D_n = 0{\cdot}133$ for $x = 3$

$$\text{Calculated } D_n = \frac{1{\cdot}63}{\sqrt{n}} = \frac{1{\cdot}63}{10} = 0{\cdot}163$$

The upper tail critical value =

Hence we reject H_0 and hence Poisson distribution with mean $\lambda = 3$ is unlikely to be correct. The table value for the chosen significance level is less than the calculated value of D_n to reject the null hypothesis.

Ex. Using the 0·05 significance level, determine whether the binomial distribution with probability 0·45 fits the following data:

Number of sales per week	0	1	2	3	4	5	6	7
Frequency of the number of sales	25	32	61	47	39	21	18	12

Ans. $D_n = 0{\cdot}1463$, upper-tail critical value = 0·0852. Reject H_0 and therefore the binomial distribution is unlikely to be correct

18·5. Statistical Decision Theory. Statistics may be considered as the science of decision making in the presence of uncertainty. This testing of a hypothesis H_0 against a hypothesis H_1 may be regarded as a decision between two actions (*i*) accepting H_0 or (*ii*) accepting H_1. In estimation, out of a finite or infinite set of possible values of the parameter, we have to choose a subset. This again involves decision in the face of uncertainty. The problems of design of experiments is again one of decision, since here we have to decide, in the face of uncertainty, as to how many and what sort of observations should be made during the course of experimentation.

Abraham Wald was the first to give a comprehensive 'statistical decision theory', which includes estimation, testing of hypothesis and design of experiments, as particular cases. The theory has had tremendous impact on the development of statistics as well as on fields like economics, military science, business, industry and operations research, for in all these fields, we are concerned with decisions in the presence of uncertainty.

An Example. Consider the problem of determining the proportions of defectives in a lot. We take a sample and on the basis of the sample we may do the following : (*i*) we may like to estimate the proportion of defectives in the population. This is the problem of estimation ; (*ii*) we may like to test the hypothesis that the proportion of defectives in the population is p. This is the problem of testing of hypothesis ; (*iii*) we may decide between the following actions : (*a*) rejection of the lot ; (*b*) acceptance of the lot ; (*c*) selling it at reduced price. This will be a problem in the general statistical decision theory. When we have to make a choice out of a finite number of possible decisions, we call it a multiple classification problem.

18·5·1. Basic Concepts. Each possible sample $x = (x_1, x_2......x_n)$ may be regarded as a point in a Euclidean space. The set of all such possible points is called the *outcome space X*.

If there are m parameters $\omega_1, \omega_2,......., \omega_m$, we may regard $\omega = (\omega_1, \omega_2......, \omega_m)$ as a point in m dimensional Euclidean space. The set of all such possible points is called *parameter space Q*.

Corresponding to each ω in Ω, there is a probability distribution P_ω. The set of all such probability distributions will be denoted by P.

The set A of the possible actions which the statistics can take in the light of the observations is called the action space A (Fig. 18·1).

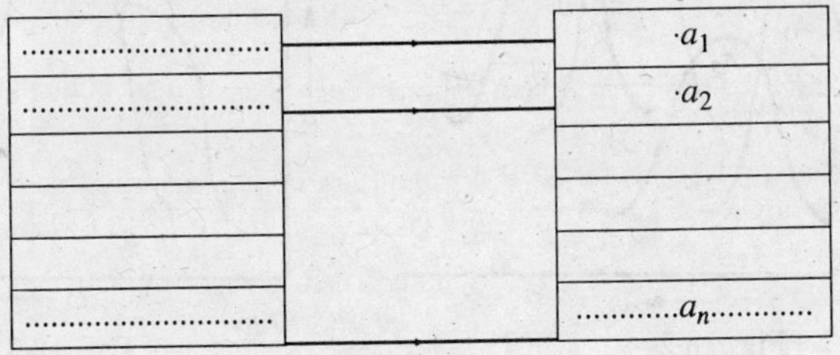

Fig. 18·1

The statistician partitions the outcome space X into a number of disjoint subjects. If the observed sample is in the first subset, the statistician takes the first action, if it lies in the second, he chooses the second action and so on. The outcome space X may, however, be divided into a countably infinite or even non-countably infinite number of subsets, depending on the number of possible actions available. In every case, however, we define a mapping of every possible outcome uniquely to some action. This mapping of function is called *decision function*. A decision function d is thus a rule which assigns to every outcome x a unique action so that $a = d(x)$

Given an outcome x, we choose an action as determined by the decision function. Since there is uncertainly, there can be some loss, positive and negative, which depends on a and the unknown parameter ω. This *loss function* is denoted by $L(\omega, a) = L[\omega, d(x)]$

For given (ω, d) the loss function depends on the outcome x and is thus a random variable. Its expected value, *i.e.*, its average over all possible outcome is called the *risk function* and is denoted by

$$p(\omega, d) = E[L\{\omega, d(x)\}]$$

The risk function depends on the unknown ω and the decision function d.

18·5·2. Minimax Decision Rule. Our object in statistical decision theory is to determine the decision function d. For this, we want to take such a decision, as will minimise risk. But the risk function depends also upon the unknown ω. We now adopt the principle of minimizing the minimum risk. For a fixed d that we may choose, $\rho(w, d)$ depends upon ω. We find that the value of ω in Ω for which this is maximum. Call this value ω_d. Then $\rho(\omega_d, d)$ is a function of d only and for each d, it gives the maximum risk. The minimax principle asserts that we choose such a decision rule d which minimises $\rho(\omega_d, d)$.

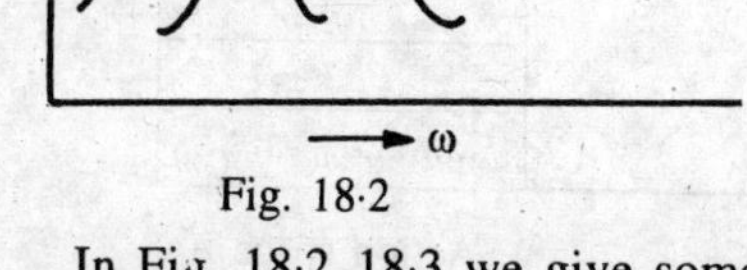

Fig. 18·2

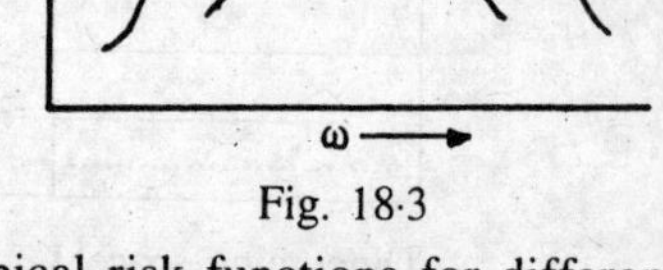

Fig. 18·3

In Fig. 18·2, 18·3 we give some typical risk functions for different decision rules, A_1, A_2, A_3, A_4 give the maximum risks for different decision rules. We choose decision rule d_2 since it gives the minimum of the maximum risks.

18·5·3. Bayes Decision Rule. In the maximum decision rule we give importance to one point *viz.*, the maximum point in the decision curve.

This may not always be desirable *e.g.*, in the Fig. 18·3, according to minimax rule we will prefer d_2 to d_1, since its maximum is smaller than that of d_1, but it is obvious that except for a narrow range of ω, d_1 always gives a smaller risk and as such should be preferred.

This suggests that instead of the maximum of the risk curve we may try to find some measure of average risk for each *d*. The average of risk function ρ (*omega, d*) can be found only if the a *priorie* distribution, of ω is known. For a given a *priorie* distribution we find the expected value of $p(w, d)$ and then we choose the *d* so as to minimize this average. Such a decision rule is called Bayes decision rule with respect to the given a *priorie* distribution for ω.

In many statistical situations, the a *priorie* distribution cannot be postulated and even if it can be postulated, it is not known to the experimenter. As such, the practical value of Bayes decision rule is limited. It is, however, important in the theory of decision functions. In fact, it appears that a good decision must be a Bayes decision or the limit of a sequence of Bayes decisions.

18·5·4. An Example. We use the above principle in everyday life also. An engineer, when he designs a dam, has to take decisions in the face of uncertainty, represented by weather conditions. For any set of decision rules he makes, he considers the design, which covers the maximum risk due to earthquakes and large amounts of rainfall in the catchment area. He then chooses a design out of these which minimizes the cost.

18·5·5. Games against Nature. A decision rule is quite often regarded as a game against nature. There are two players *viz.*, the statistician and Nature. The statistician can choose *d* and Nature can choose ω. For any rule statistician chooses, Nature chooses ω, so as to maximize $p(\omega, d)$. The statistician therefore chooses *d* so as to get the minimum of these maxima.

There is, however, no reason why Nature should be perverse (some persons tried to prove the perversity of Nature by throwing a buttered piece of bread a large number of times. They found that in a majority of cases, the buttered face came down !). Mathematically, however it is easier to determine the consequences of this complete opposition of Nature than those of partial opposition.

18·5·6. Admissible Decision Functions and Complete Classes of Decision Functions. A decision function *d* is said to be *admissible*, if there exists on other decision function d^+ which is uniformly better than *d i.e.*, if there exists no decision function d^+ satisfying the following two conditions

(*i*) $\rho(\omega, d^+) \leq \rho(\omega, d)$ for all ω in Ω

(*ii*) $\rho(\omega, d^+) < \rho(\omega, d)$ for at least one ω in Ω

A class C of decision function d is said to be *complete* if for any d not in C, we can find a d^+ in C such that d^+ is uniformly better than d. A complete class is said to be a minimal *complete class* if no proper sub-class of it is complete. It can be shown that

(*i*) if a minimal complete class exists, it must be equal to the class of all admissible functions.

(*ii*) if the class of admissible decision functions is complete, it must be minimal.

The problem of finding minimal complete classes of decision functions is important in statistics. The principles of sufficiency and invariance have been used for this purpose, but we shall not go into these. It is sufficient to say that there are many unsolved mathematical problems in this area.

Exercises 18

1. Write an essay on 'Statistical Inference' explaining clearly the basic concepts of point and interval estimation, testing of hypotheses, sequential distribution-free methods and statistical decision theory.

2. A sample of size n is drawn from each of k normal populations with the same variance. Obtain the likelihood ratio criteria for testing the hypothesis that the means are all zero. Show that the criterion is a function of ratio which has the F distribution.

3. Describe the likelihood ratio test for testing whether the correlation of a bivariate normal population is zero.

4. Two random samples of same size n are drawn from two bivariate normal populations with means m_1, m_2, standard deviations σ_1, σ_2 and correlation coefficients ρ_1, ρ_2. Show that for testing the hypothesis $\rho_1 = \rho_2$, the likelihood ratio is given by

$$\lambda = [\text{sech}\, \tfrac{1}{2}(z_1 - z_2)]^{\frac{1}{2}n}$$

where $r_1 = \tanh z_1$, $r_2 = \tanh z_2$, are the same correlation coefficients.

5. In testing between the values μ_0 and μ_1 for the mean of normal population, show that the probabilities of both types of errors can be made arbitrarily small by taking a sufficiently large sample.

6. How would you make a two-sided test of $\mu = \mu_0$ for normal population with known variance ? Is this a uniformly most powerful test ?

7. It is desired to test the hypothesis $\lambda_t = 1$ against $\lambda_t = 2$ assuming that the population is Poisson type with parameters λ_t. Using the mean as the basis of the test, determine the critical region and sample size corresponding to $\alpha = \beta = \cdot 01$. **(*M.A. Delhi*)**

8. Show that the one-side sequential test for the mean of a normal population with known variance σ^2 may be performed by plotting the lines

$$y = a \log A + cn, \; y = b \log B + cn$$

where $a = \dfrac{\sigma^2 \log A}{\mu_1 - \mu_0}$, $b = \dfrac{\sigma^2 \log B}{\mu_1 - \mu_0}$, $c = \dfrac{\mu_0 - \mu_1}{2}$

in $x - y$ plane concluding the test when point $\left(n, \sum_{i=1}^{n} y_i\right)$ crosses one of the lines.

9. Show that the power function and the average sample size of the above test are given by

$$P(\mu) \cong \frac{1 - e^{2(c-\mu)b/\sigma^2}}{e^{2(c-a)/\sigma^2} - e^{2(c-\mu)b/\sigma^2}};$$

$$E(\mu) \cong \frac{b + P(\mu)(b - a)}{\mu - c}.$$

10. (*a*) Develop likelihood ratio for testing binomial probability $p = p_0$ against $p = p_1$. Prove that if T_m represents the number of successes in first m trials,

$$m = \frac{p_1^{S_m} q_1^{m - S_m}}{p_0^{S_m} q_0^{m - S_m}}.$$

Discuss the test for $p_0 = 0{\cdot}5, p_1 = 0{\cdot}7, \alpha = 0{\cdot}5, \beta = {\cdot}10.$

(*b*) Prove that h is given by

$$p\left(\frac{p_1}{p_0}\right)^h + (1 - p)\left(\frac{1 - p_1}{1 - p_0}\right)^h = 1$$

and the operating characteristic curve is obtained by eliminating h between the above equation and

$$L(p) = \frac{A^h - 1}{A^h - B^h}.$$

(*c*) Prove that for the above test

$$E(n) \cong \frac{[1 - L(p)] \log A + L(p) \log B}{p \log \dfrac{p_1}{n} + (1 - p) \log [(1 - p_1)/(1 - p_0)]}$$

11. Show that the expected value of larger of a sample of two observations from a normal population with mean μ and variance σ^2 is $\mu + \sigma/\sqrt{\pi}$.

12. If x, y, is an observation from a bivariate normal population with zero mean, variances unity and correlation ρ, show that the expected value of the larger of x and y is $\sqrt{(1 - \rho)/\pi}$.

$$\left[\textbf{Hint. } E[\max(x, y)] = \int_{-\infty}^{+\infty} \int_{-\infty}^{+\infty} \max(x, y) f(x, y)\, dx\, dy = \frac{1}{2\pi\sqrt{1 - \rho^2}}, \right.$$

$$2\int_{-\infty}^{+\infty} dy \int_{-\infty}^{+\infty} x \exp\left[-\frac{(x^2 - 2\rho xy + y^2)}{2(1-\rho^2)}\right] dx = \text{etc.}\Bigg]$$

13. Derive the L.R.T. based on n independent observations for the hypothesis that $\theta = 1$, where θ is the parameter occurring in the distribution

$$\text{Const. } e^{-\theta|x|} x^2\, dx, x > 0$$

and obtain the distribution of the test criterion.

14. Draw the sequential OC curve for the binomial distribution to test p_0 against p_1.

$$p_0 = 0{\cdot}01, \quad p_1 = 0{\cdot}15, \quad \alpha = 0{\cdot}01, \quad \beta = 0{\cdot}02$$

obtain from the curve the probability of accepting a lot which is 20% defective.

15. Random samples of size 10 are taken from $dF = \theta^{-1}\, e^{-x/\theta}\, dx, x \geq 0, \theta > 0$ obtain U.M.P. test for testing $H_0 : \theta = 4$ taking $0 \cdot 05$ as the level of significance. Draw the power curve of the test.

16. Obtain the SPRT for the hypothesis $m = m_0$ against the alternative hypothesis $m = m\,(> m_0)$ where m is the mean of a Poisson population. Derive the OC and ASN functions of the test. What will be the OC and ASN curve of the test if there is a population which is normally distributed with *s.d.* 3 and H_0 : $m = 5$ against $H_1 : m = 6$ and the probabilities of both types of error are 0·05.

17. A sequential sampling plan is to be adopted for acceptance of lots of very large size. The probability of rejection of a lot with 20% defective is to be kept at 0.05 and the probabilities of rejection of a lot with 8% defectives is to be kept at 0·1. Give the graphical and tabular procedure for carrying out the above plan.

18. From each lot of very large size, 100 articles are taken at random by inspector A. He passes the lot unless there are more than 2 defectives amongst the 100 he has chosen. Inspector B follows the same procedure with another sample of 100 articles. If the two inspectors agree the lot is accepted or rejected as the case may be. If they disagree a further sample of size 200 is taken. The lot is finally accepted unless there are more than 4 defectives amongst the 200 chosen. Obtain the OC and ASN curves of the plan.

19. Discuss testing of hypothesis point and interval estimation as a special case of the general decision problem.

20. What is the relation of admissible decision functions to U.M.P. tests ?

21. A decision d_0 is said to be uniformly best if there is decision which is uniformly better than d_0. Do uniformly best decisions exist in the case of one-sided tests hypotheses ?

22. If D is a complete class of decisions and C is any other class, show that DUC is also complete.

23. Show that a complete class of decisions always exists.

24. (*a*) Explain the terms : (*i*) Critical region and Acceptance region. (*ii*) Type

I and II errors (*iii*) Null hypothesis and Alternation hypothesis.

(*b*) Let X have a p.d.f. of the form $f(x;\theta)=\theta x^{\theta-1}$, $\theta>0, 0\le x\le 1$; zero elsewhere where $\theta\in(\theta;\theta=1,2)$. To test the simple hypothesis $H_0:\theta=1$ against $H_1:\theta=2$, use a random sample X_1, X_2 of size $n=2$ and define the critical region to be $C=\{(x_1,x_2)\,;\,3/(4x_1)\le x_2\}$. Find the power function of the test. **(M.A., Delhi, 1980)**

$$\left[\textbf{Ans. } \frac{1}{4}+\frac{3}{4}\log\frac{3}{4}\,;\,\frac{7}{16}+\frac{9}{8}\log\frac{3}{4}\right]$$

[**Hint.** When H_0 is true *i.e.* $\theta=1, L=dx_1 . dx_2$. Integrate for x_2 from $3/4x_1$ to 1 and for x_1 from 3/4 to 1.]

25. Let X have a binomial distribution with parameter $n=10$ and $p\,\varepsilon\{p:p=\frac{1}{4},\frac{1}{2}\}$. The simple hypothesis $H_0:p=\frac{1}{2}$ is rejected and the alternative simple hyp. $H_1:p=\frac{1}{4}$ is accepted if the observed value of x_1, a random sample of size 1, is less than or equal to 3. Find the power function of the test. **(M.A., Delhi)**

[**Hint.** When H_0 is true, B.D. is $\left\{\frac{1}{2}+\frac{1}{2}\right\}^{10}$. find $P(X\le 3)$.]

$$\left[\textbf{Ans. } (i)\ \frac{11}{64},\ (ii)\ \frac{3^n}{4^n}\right]$$

26. Let X have a *p.d.f.* of the form $f(x;\theta)=\theta^{-1}e^{-x/\theta}$; $0<x<\infty$; zero elsewhere where $\theta\,\varepsilon\,(\theta:=2,4)$. Test $H_0=2$ against the alternative sample hyp. $H_1:\theta=4$ by using a random sample of size $n=2$ defining the critical region to be $0=\{(x_1,x_2)\,;\,9{\cdot}5\le x_1+x_2<\infty\}$. Find the significance level of the test. **(M.A. Delhi)**

[**Hint.** $P(x_1,x_2)=1-\int_0^{9\cdot5}\int_0^{9\cdot5-x_2}\frac{1}{4}e^{-(x_1+x_2)/2}\,dx_1dx_2=0{\cdot}05\approx;\theta=2$

When $\theta=4, P[(x_1,x_2)\,\varepsilon\,C]=0{\cdot}31$ approx.]

27. If $x\ge 1$ is the critical region for testing $H_0:\theta=2$ against the alternative $\theta=1$, on the basis of the single observation from the population

$$f(x;\theta)=\theta\exp(-\theta x),\ 0\le x<\infty$$

Obtain the values of type I and type II errors

(B.Sc. Stat. Hons. Delhi 1988) (B.Sc. Allahabad 1993)

[**Hint.** $W=\{x:x\ge1\}$, $H_0:\theta=2, H_1=\theta=1$

$$\alpha=P[x\ge1\mid\theta=2]=2\int_1^{\infty}e^{-2x}\,dx=e^{-2}=1/e^2$$

$$\beta=P[x<1\mid\theta=1=\int_0^1 e^{-x}\,dx=(1-e^{-1})=(e-1)/e\,.]$$

28. For the distribution $dF = B\,e^{(x-r)}\,dx,\ x \ge r$, 0 when $\alpha < r$, show that for a hyp. H_0 that $\beta = \beta_0,\ r = r_0$ with $H_1 : \beta = \beta_1,\ r = r_1$ BCR is given by

$$\bar{x} \le \frac{1}{\beta_1 - \beta_0}\left\{r_1\,\beta_1 - r_0\,\beta_0 - \frac{1}{x}\log k + \log\frac{\beta_1}{\beta_2}\right\}$$

provided that the admissible hyp. is restricted by the conditions $r_1 \le r_0,\ \beta_1 \ge \beta_0$.

$$\left[\textbf{Hint. } L_1 = \beta_1^n \exp\left[-\beta_1 \sum_{i=0}^{n}(x_i - r_1)\right],\ \underset{\beta = \beta_0}{L_0 = (L)},\quad \text{use } \frac{L_1}{L_0} \ge k\right.$$

$$\left.\Rightarrow \quad \left(\frac{\beta_1}{\beta_0}\right)\exp\left[-\beta_1 n\,(\bar{x} - r_1) + \beta_0 n\,(\bar{x} - r_0)\right] \ge k.\right]$$

29. Examine if a *BCR* exists for testing $H_0 : \theta = a_0$ against $H_1 : \theta = \theta_1\ (> \theta_2)$ for parameter θ of the distribution $f(x, \theta) = (1 + \theta)/(x + \theta)^2,\ 1 \le x < \infty$

(*B.Sc. Bangalore*)

$$\left[\textbf{Hint. } (1 + \theta_1)^n \prod_{i=1}^{n}\frac{1}{(x_i + \theta_1)^2} \ge k\,(1 + \theta_0)^n \prod_{i=1}^{n}\frac{1}{(x_i + \theta_0)^2}.\ \text{Take log of both sides.}\right]$$

30. Given a random sample $x_1, x_2, \ldots, x_n$ from the distribution with *p.d.f.* $f(x, \theta) = \theta\,e^{-\theta x},\ x > 0$, show that there exists no UMP test for testing

$$H_\partial : \theta = \theta_\partial \text{ against } H_1 : \theta \ne \theta_\partial.$$

(*B.Sc. Stat. Hons Delhi, 1988*)

[**Hint.** Using $L \ge kL_0 > BCR$ gives $\Sigma\,x_i\dfrac{k_1}{\theta_2 - \theta} = \lambda_1$ if $\theta_1 > \theta_0$ and $\Sigma\,x_i \ge \dfrac{k_1}{\theta_\partial - \theta} = \lambda_2$ where λ_1, λ_2 are determined by $\rho\,[\,\Sigma\,x_i \le \lambda_1 \mid H_0\,] = \alpha$ and $\rho\,[\,\Sigma\,x_i \ge \lambda_2 \mid H_0\,] = \alpha$.

19

Elements of Stochastic Processes

19·1. Introduction. Basic concepts and illustrations. So far, in this book, we have been concerned with the 'static' aspects of statistical theory. In the present chapter, we shall be concerned with the 'dynamic' aspect or with 'statistics of change'. A stochastic process is a process which has some random or stochastic element involved in its structure.

Mathematically, a stochastic process is a set of random variables $\{x_t\}$ or $\{x(t)\}$ depending on some real parameter like the time t. For examples we may consider the following :

(*i*) **A queuing system.** Here the number of persons joining the queue in a time-interval t, the number of persons being serviced in the time-interval t, the number of persons in the queue at time t, are all random variables, depending on the parameter t.

(*ii*) **Turbulent fluid flow.** Here the fluid velocity components u, v and w are random variables depending on the space co-ordinates x, y, z and the time t.

(*iii*) **A random walk model.** A particle moves in a straight line in steps of unit length. At each stage, it can move one step to the right with probability p or one step to the left with probability q, where $(p+q)=1$. If it starts from the origin, its position after n moves is a random variable, depending on the discrete parameter n.

(*iv*) **An epidemic process.** Here the number of persons infected with a disease, the number of persons susceptible to the disease, the number of persons recovered and immune from the disease, are all the discrete random variables, depending on the time parameter t.

(*v*) **A communication process.** Here the amplitude of the signals to be transmitted, and amplitude of the noise or disturbance produced in the channel, are both random variables, depending on time.

We shall come across many more examples of stochastic processes in this chapter.

A stochastic process may be classified according as

(*i*) the random variable is discrete or continuous

(*ii*) the parameter is discrete or continuous.

If the joint distribution of x_{t1} and x_{t2} depends on only $t_1 - t_2$, the process is called *stationary*, otherwise it may be termed *evolutionary*. A stochastic process is called *Gaussian* or *normal*, if the joint distribution of any number of variables x_{t_1}, x_{t_2},.... is a normal distribution.

Ex. 1. Give some more examples of stochastic processes.

Ex. 2. Show that $x_i = A\cos\lambda_t + B\sin\lambda_t$ where A and B are independent normal variates with zero means and unit variances, is a normal stochastic process. Show also it is a stationary process.

Ex. 3. Show that

$$x_i = R\cos(wt + \theta) = R\cos\theta\cos wt - R\sin\theta\sin wt = A\cos wt + B\sin wt$$

is a stochastic process.

19·2. Gamblers Ruin Problem. A gambler with capital z plays against an adversary with capital $(a - z)$. The game is played in stages. At each stage, the gambler can win one unit with probability p and lose one unit with probability q, where $(p + q) = 1$. The capital of the gambler, after n stages, is a random variate, depending on n. Even without finding the distribution of the variate, we can study some important characteristics of this process.

19·2·1. Probability of ruin. Let q_z denote the probability of ultimate ruin of the gambler. After the first trial, the gambler's capital is $(z + 1)$ if he wins (with probability p) or it is $(z - 1)$ if he loses (with probability q). The probability of his ruin, after one trial, will be q_{z+1} and q_{z-1} respectively, so that

$$q_z = p\,q_{z+1} + q\,p_{z-1}\,,\quad 1 < z < a - 1 \qquad ...(19·1)$$

It is also easily seen that

$$q_1 = p\,q_2 + q, \qquad z = 1 \qquad ...(19·2)$$

$$q_{z-1} = q\,q_{a-2}, \qquad z = a - 1 \qquad ...(19·3)$$

All these three equations are included in

$$q_z = p\,q_{z+1} + q\,q_{z-1}\,.\quad 1 \le z \le a - 1. \qquad ...(19·4)$$

If we use the convention

$$q_0 = 1,\ q_a = 0 \qquad ...(19·5)$$

To solve (19·4), we try the solution $q_z = p^z$. Substitution in (19·4) gives

$$p\,\rho^2 - \rho + q = 0. \qquad ...(19·6)$$

which has roots $\rho = 1,\ \rho = \dfrac{q}{p}$

Thus, if $p \neq \frac{1}{2}$, (19·4) is a linear difference equation, its general solution is

$$q_z = A + B\,(q/p)^z. \qquad \text{...(19·7)}$$

Using (19·5) to find A and B, we get the final solution

$$q_z = \frac{\left(\frac{q}{p}\right)^a - \left(\frac{q}{p}\right)^z}{\left(\frac{q}{p}\right)^a - 1} = \frac{\left(\frac{q}{p}\right)^z\left[\left(\frac{q}{p}\right)^{a-z} - 1\right]}{\left(\frac{q}{p}\right)^a - 1} \qquad \text{...(19·8)}$$

The probability p_z of the opponent's ruin can be obtained from this by writing $a - z$ for z and interchanging p and q so that

$$p_z = \frac{\left(\frac{p}{q}\right)^{c-z}\left[\left(\frac{p}{q}\right)^{2} - 1\right]}{\left(\frac{p}{q}\right)^a - 1} = \frac{1 - \left(\frac{p}{q}\right)^{-z}}{1 - \left(\frac{q}{p}\right)^a} = \frac{\left(\frac{q}{p}\right)^z - 1}{\left(\frac{q}{p}\right)^a - 1} \qquad \text{...(19·9)}$$

From (19·8) and (19·9) $p_z + q_z = 1$, ...(19·10)

so that the game has to terminate.

If $p = q = \frac{1}{2}$, (19·4) becomes

$$2\,q_z = q_{z+1} + q_{z-1}, \qquad \text{...(19·11)}$$

of which the solution, subject to (19·5), is easily found to be

$$q_z = 1 - \frac{z}{a} = \frac{a-z}{a} \qquad \text{...(19·12)}$$

In this case writing $a - z$ for z, we get

$$p_z = \frac{z}{a}, \qquad \text{...(19·13)}$$

so that (19·10) is still true.

19·2·2. Expected Duration of the game. Let d_z denote the duration expected of the game. If the first trial leads to a win for the gambler, the expected total duration would be $(1 + d_{z+1})$ and if it leads to a loss, the duration would be $(1 + d_{z-1})$, so that we get the difference equation

$$\begin{aligned} d_z &= p\,(1 + d_{z+1}) + q\,(1 + d_{z-1}) \\ &= 1 + p\,d_{z+1} + q\,d_{z-1}. \end{aligned} \qquad \text{...(19·14)}$$

It is easily verified that this equation is true for $z = 1$ and $z = a - 1$ if also we take

$$d_0 = 0,\ d_a = 0. \qquad \text{...(19·15)}$$

The equation (19·14) is non-homogeneous and its homogeneous part is the same as (19·4) when q_z is replaced by d_z. We get general solution by adding to (19·7) a particular solution λz which is seen to be a particular solution of (19·4) if

$$\lambda z = 1 + p\,\lambda\,(z+1) + q\,\lambda\,(z-1) \quad \text{or } \lambda = 1/(q-p) \qquad ...(19{\cdot}16)$$

The general solution of (19·4) is therefore

$$d_z = A + B\left(\frac{q}{p}\right)^z + \frac{1}{q-p}\,z \qquad ...(19{\cdot}17)$$

Using (19·15), we get A and B, so that finally

$$d_z = \frac{z}{q-p} - \frac{d}{q-p}\cdot\frac{1-\left(\dfrac{q}{p}\right)^z}{1-\left(\dfrac{q}{p}\right)^a} \qquad ...(19{\cdot}18)$$

If $p = q = \frac{1}{2}$, we try the particular solution $\mu\, z$, so that

$$\mu\, z^2 = 1 + \mu\,\tfrac{1}{2}\,(z+1)^2 + \mu\,\tfrac{1}{2}\,(z-1)^2 \text{ giving } \mu = 1 \qquad ...(19{\cdot}19)$$

The general solution is then $d_z = A + Bz - z^2$. ...(19·20)

Using (19·15), we get $d_z = z(\,a - z\,)$ ...(19·21)

19·2·3. Gambler's Ruin with Correlation. In this case the probabilities of the gambler's winning or losing at a trial are p_1, q_1 if he wins in the earlier trial and these are q_2, p_2 if he loses in the earlier trial. If x, y denote his respective gains in the two trials and if the score of +1 has the same probability as a score of –1 in the first game, then

$x = 1, \quad y = 1,$ with probability $\frac{1}{2}\,p$

$x = 1, \quad y = -1,$ with probability $\frac{1}{2}\,q$

$x = -1, \quad y = -1,$ with probability $\frac{1}{2}\,p$

$x = -1, \quad y = 1,$ with probability $\frac{1}{2}\,q$

giving $\Sigma\, p\, x = 0,\ \Sigma py = \frac{1}{2}\,(p_1 + q_2 - p_2 - q_1),\ \Sigma px^2 = 1$;

$\Sigma\, p\, y^2 = 1,\ \Sigma pxy = p_1 - q_2$

so that the correlation coefficient is given by

$$R_{xy} = \frac{p_1 - q_2}{\sqrt{\frac{1}{4}\,(p_1 + q_2 - p_2 + q_1)^2 - \frac{1}{4}\,(p_1 + q_2 - p_2 + q_1)^2}}$$

or

$$R_{xy} = \frac{p_1 - q_2}{\sqrt{(\,p_1 + q_2\,)\,(\,p_2 + q_1\,)}} \qquad ...(19{\cdot}22)$$

If consecutive trials are independent, $R_{xy} = 0$, otherwise R_{xy} will influence the probability of ruin q_z, the expected duration of the game d_z and all other characteristics of the stochastic process.

Ex. 1. If $p=\frac{1}{2}, q=\frac{1}{2}, z=500, a=1000$, prove that $d_z=250{,}000$

Ex. 2. If $p=\frac{1}{3}, q=\frac{1}{2}, z=1\ a=1000$, prove that $d_z=999$.

Ex. 3. Consider the gamblers ruin problem, when the chances of winning and losing one unit are p and q respectively and the chance of a tie or draw is r, $(p+q+r=1)$. Show that the probability of ruin is

$$\frac{\left(\frac{q}{p}\right)^a-\left(\frac{q}{p}\right)^z}{\left(\frac{q}{p}\right)^{z-1}} \qquad \text{...(19·23)}$$

Ex. 4. Show that in the above example, the expected duration of the game is

$$\frac{1}{(q-p)(1-r)}\left[z-a\cdot\frac{\left\{\frac{q}{p}\right\}^z-1}{\left\{\frac{q}{p}\right\}^{z-1}}\right] \qquad \text{...(19·24)}$$

Ex. 5. Show that if the stakes are doubled while the initial capitals remain unchanged, then the probability of ruin decreases for the player whose probability of success is less than $\frac{1}{2}$.

19·3. Random walk models. The random walk model described in section 19·1 is called *symmetric*, if $p=q=\frac{1}{2}$. If $p>\frac{1}{2}$, we say there is a *drift to the right*, if $q>\frac{1}{2}$, we say there is *drift to the left*. If the particle can go to any point on the line, then walk is said to be *un-restricted.*

The gamblers ruin problem is also represented by a random walk model, but it is restricted random walk, since gambler's capital is restricted to point between 0 and a. In fact, the point 0 and a are *absorbing barriers*, since when the gambler's capital becomes 0 or a, the walk ceases.

We can also consider random walks, with reflecting or elastic barriers. If the particle, when it reaches $x=1$, can move to the right with probability p and can remain as 1 (*i.e.*, it can go to 0 and be reflected from there) with probability p and can move to the left with probability q. A random walk with both reflecting barriers can continue indefinitely. A barrier $x=0$ will be called *elastic*, if it can either absorb with probability δq or reflect with probability $(1-\delta q)$.

In *two dimensional random walk*, the particle moves in unit steps in one of the four directions, parallel to x and y-axes. In *three-dimensional random walk*, the particle moves in unit steps in one of the six directions, parallel to the three axes. In symmetrical random walk all the probabilities are equal.

There is an interesting theorem due to Polya which we state without proof. The thereom reads. "In the symmetric random variable in one and two dimensions, there is probability one that the particle will sooner or later return to its initial position. In three dimensions, however, this probability is only about 0·35. The theorem is interesting inasmuch as the theorem shows that results, which hold in one or two dimension, need not necessarily hold in three or higher dimensions.

Ex. 1. Give interpretation of gamblers ruin problem with reflecting and elastic barrier.

Ex. 2. In the unrestricted random walk problem, find the probability that the particle is at the position n after N steps. Find also its expected position after N-steps.

19·4. Markov chains. We consider a system which can be in any one of a finite numbers of states $E_1, E_2, \ldots\ldots E_n$. If at any time, the system is in state E_i, the probability of it being in the state E_j at the next trial is p_{ij}. We assume that probability of the system being in a given state at the next trial depends just on its state at present and not upon the states it may have been in earlier times. This property of a process is known as the *Markov property*.

It is convenient to consider the matrix*

$$P = \begin{bmatrix} p_{11} & p_{12} & \cdots & \cdots & p_{1n} \\ p_{21} & p_{22} & \cdots & \cdots & p_{2n} \\ \cdots & \cdots & \cdots & \cdots & \cdots \\ \cdots & \cdots & \cdots & \cdots & \cdots \\ p_{n1} & p_{n2} & \cdots & \cdots & p_{nn} \end{bmatrix} \quad \ldots(19{\cdot}25)$$

This matrix of transition probability p_{ij} is called the *transition probability matrix*. Its elements are non-negative. Also

$$\sum_{j=1}^{n} p_{ij} = 1, \quad \ldots(19{\cdot}26)$$

since it is certain that the system must go from the ith state Ei one of the mutually exclusive states $E_1, E_2, \ldots\ldots E_n$. The sum of the elements of every row is unity. Such a matrix is called a *stochastic matrix*.

The probability of system going from ith to jth stage in two steps is given by,

---

* See 'Matrices' by H.C. Saxena, published by Ram Prasad & Sons, Agra .

$$\sum_{k=1}^{n} p_{ik}\, p_{kj} = P_{ij} \qquad \text{...(19·27)}$$

It is easily seen that $[P_{ij}] = P^2$, so that proceeding in the same way, i–jth element of P^n gives the probability of going from E_i to E_j in n steps.

If $[\pi_1, \pi_2, ..., \pi_n]$ gives the probabilities of the system being initially in the states $E_1, E_2, ..., E_n$, then the probability of its being in the state E_j after one step is

$$\sum_{i=1}^{n} \pi_i\, p_{ij},$$

but these are elements of matrix ΠP where Π is the row matrix

$$\Pi = (\pi_1, \pi_2 ...., \pi_n) \qquad \text{...(19·28)}$$

Similarly, the probability of the system being, in the jth stage after n transition is given by jth element of the $m \times n$ matrix ΠP^n.

Consider the transition probability matrix :

$$P = \begin{array}{c} \\ E_1 \\ E_2 \\ E_3 \\ E_4 \\ E_5 \end{array} \begin{array}{c} \begin{array}{ccccc} E_1 & E_2 & E_3 & E_4 & E_5 \end{array} \\ \begin{bmatrix} 1 & 0 & 0 & 0 & 0 \\ \frac{1}{3} & 0 & \frac{2}{3} & 0 & 0 \\ 0 & \frac{1}{3} & 0 & \frac{2}{3} & 0 \\ 0 & 0 & \frac{1}{3} & 0 & \frac{2}{3} \\ 0 & 0 & 0 & 0 & 1 \end{bmatrix} \end{array}$$

Its graph is shown below :

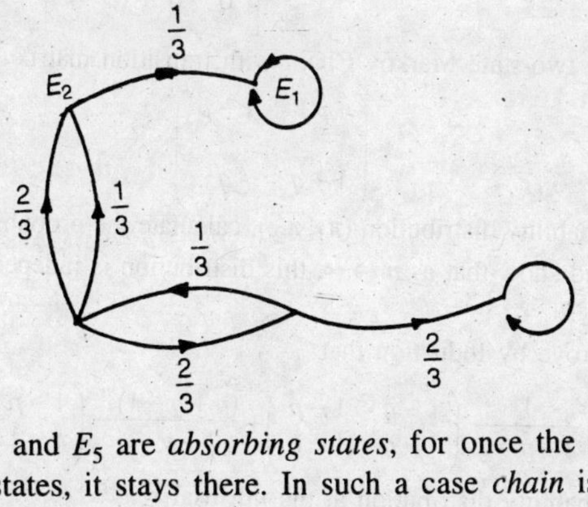

States E_1 and E_5 are *absorbing states*, for once the system reaches either of the states, it stays there. In such a case *chain* is said to be an *absorbing chain.*

A chain in which every two states can communicate with each other and in which the system cannot be left is called an *ergodic chain*. An important special case of an ergodic chain is the *regular chain* in which for some n, P^n has no zero elements.

The following theorems are stated without proof.

Theorem 1. For an ergodic chain

$$\lim_{n \to \infty} \frac{P + P^2 + ... + P^n}{n} = A \qquad ...(19.29)$$

where A is a matrix with each row the same vector $\alpha = (a_1, a_2, ..., a_n)$.

The vector α has all components positive and is such that $\alpha P = \alpha$.

Theorem 2. For a regular chain,

$$\lim_{n \to \infty} P^n = A \qquad ...(19.30)$$

In this case the probabilities from E_i to E_j in n steps (where n is large) is essentially independent of the initial system.

Ex. 1. Write transition probability matrices for

(*i*) gamblers ruin problems

(*ii*) random walk with reflecting barriers

(*iii*) random walk with elastic barriers.

Ex. 2. Draw graphs for the Markov chain with the following transition probability matrices.

(*a*) (*i*) $\begin{bmatrix} 0 & \frac{1}{2} & \frac{1}{2} \\ \frac{1}{2} & 0 & \frac{1}{2} \\ \frac{1}{2} & \frac{1}{2} & 0 \end{bmatrix}$ (*ii*) $\begin{bmatrix} 0 & 0 & 0 & 1 \\ 0 & 0 & 0 & 1 \\ \frac{1}{2} & \frac{1}{2} & 0 & 0 \\ 0 & 0 & 1 & 0 \end{bmatrix}$

(*b*) For a two-state Markov Chain with transition matrix

$$\begin{pmatrix} p & 1-p \\ 1-q & q \end{pmatrix}$$

and initial probability distribution (π_1, π_2), calculate the probability distribution at the nth trial and show that as $n \to \infty$, this distribution is independent of the initial distribution.

[**Hint.** Prove by Induction that

$$P^n = \frac{1}{2-p-q}\begin{pmatrix} 1-q & 1-p \\ 1-q & 1-p \end{pmatrix} + \frac{(p+q-1)^n}{2-p-q}\begin{pmatrix} 1-p & p-1 \\ q-1 & 1-q \end{pmatrix}$$

The probability distribution at the nth trial $= P^{n-1}$

$$= \frac{1}{2-p-q}(1-q, 1-p) + \frac{(p+q-1)^{n-1}}{2-p-q}.$$

$$[\pi_1(1-p)+\pi_2(q-1)+\pi_1(p-1)+\pi_2(1-q)].$$

$$\Rightarrow \frac{1}{2-p-q}(1-q, 1-p) \text{ as } n \to \infty \text{ since } |p+q-1|<1.$$

Ex. 3. In each of the above cases, find P^2 and discuss the asymptotic behaviour of P^n.

Ex. 4. For the Markov chain with transition probabilities matrix

$$P=\begin{bmatrix} p & q \\ q & p \end{bmatrix}$$

and the initial distribution $[\alpha, 1-\alpha]$, find the probabilities of the spetem being in the two states after (*i*) two steps (*ii*) three steps (*iii*) n steps (*iv*) infinite number of steps.

19·5. Poisson Process. A Poisson process is characterised by any of the following three equivalent properties :–

(*a*) The total number of arrivals during any given time interval is independent of the number of arrivals that have already occurred and the probability of an arrival in time dt is $\lambda dt + 0(dt)^2$ where λ is constant. The probability of more than one arrival in time dt is of the order of $(dt)^2$ and can be neglected. Here λ represents the mean arrival rate, *i.e.*, number of units of the incoming traffic per unit time and $1/\lambda$ is the mean length of time interval between consecutive arrivals.

(*b*) If n is the stochastic variate giving the number of arrivals in some time interval of length T, then n obeys a Poisson distribution with parameter λT *i.e.*

$$P\ (n \text{ arrials in time } T) = \frac{e^{-\lambda T}(\lambda T)^n}{n\,!}$$

(*c*) The distribution of time intervals between consecutive arrivals is exponential, *i.e.*, the density function for the time interval between consecutive arrivals is

$$a(T) = \lambda e^{-\lambda T}$$

To show that there these three properties are equivalent, we show that $(a) \Rightarrow (b) \Rightarrow (c) \Rightarrow (a)$.

We have already shown that $(a) \Rightarrow (b)$ in Chapter 6.

To show that $(b) \Rightarrow (c)$, we note that

P[time interval between consecutive arrivals lies between T and $T+dT] = P$ [no arrival during time interval T] $\times P$[one arrival during time interval dt

$$= e^{-\lambda T}\lambda dT$$

P [time interval between consecutive arrivals $\geq T$]

$$\int_t^{\infty} \lambda e^{-\lambda T}\, dT = e^{-\lambda T}.$$

$\therefore$ P [no arrival taking place in time T] $= 1 - e^{-\lambda T}$

If T is small, P [one or more arrivals in time dT]

$$= 1 - e^{-\lambda dT} = \lambda dT + 0(\,dT\,)^2$$

Ex. 1. If the random variates X_t and Y_t represent the independent stochastic processes parameters λ and μ, prove that the process defined by $S_t = X_t + Y_t$ is also a Poisson process with parameter $\lambda + \mu$.

Ex. 2. Let X_t be a random variate representing a Poisson process with parameter λ, so that

$$E(X_t) = \lambda t, \qquad \text{Var}\,(X_t) = \lambda t, \qquad E\,[X_t(X_{t+\tau} - X_t)] = 0$$

$$E[X_t X_{t+\tau}\,] = \lambda t(\lambda t + \tau t + 1)$$

Deduce that the correlation coefficient between X_t and $X_{t+\tau}$ is

$$\left[\frac{t}{t+\tau}\right]^{\frac{1}{2}}$$

19·6. The simple birth process. Consider a population of organisms multiplying in accordance with the following rules :

(*a*) the sub-populations generated by two co-existing individuals develop in complete independence of one another.

(*b*) an individual existing at time t has a chance $\lambda dt + 0(dt)^2$ of multiplying by binary fission during the following time-interval of length dt.

(*c*) the 'birth-rate' λ is the same for all individuals in the population at time t.

The 'birth' in (*b*) will usually be interpreted in the sense that at each birth just one new member is added to the population, but mathematically it is not possible to distinguish between this and an alternative interpretation in which a parent dies at birth and is replaced by two new members.

Let N_0 be the number of individuals at time $t = 0$. For the present we take $N_0 = 1$. Also let $P_n\,(t)$ be the probability that the population size $N\,(t)$ be the probability that the population size $N\,(t)$ has the value n at time t, then from the given rules it follows that

$$P_n(t + dt) = P_n(t)\,[1 - n\lambda dt] + P_{n-1}\,(t)\,[(n-1)\lambda dt] + 0(dt)^2,\ (n > 1).$$

Proceeding to the limit as $dt \to 0$, we get

$$P'_n(t) = -\,n\lambda P_n(t) + (n-1)\lambda P_{n-1}(t),\ (n > 1) \qquad \text{...(19·31)}$$

Proceeding in the same way, we get

$$P'_1(t) = -\lambda P_1(t) \qquad ...(19{\cdot}32)$$

so that (19·31) also holds for $n = 1$.

Consider the probability generating function,

$$\phi(z, t) = \sum_{n=1}^{\infty} P_n(t)\, z^n \qquad ...(19{\cdot}33)$$

so that

$$\frac{\partial \phi}{\partial t} = \sum_{n=1}^{\infty} P_n'(t)\, z^n \; ; \; \frac{\partial \phi}{\partial z} = \sum_{n=1}^{\infty} n P_n(t) z^{n-1} \qquad ...(19{\cdot}34)$$

Using (19·31) and (19·32) we get from (19·34)

$$\frac{\partial \phi}{\partial t} = \lambda z(z-1) \frac{\partial \phi}{\partial z} \qquad ...(19{\cdot}35)$$

Let
$$\frac{dz}{z(z-1)} = du$$

On integration, $e^u = A\left(1 - \dfrac{1}{z}\right)$, A being an arbitrary constant. ...(19·36)

then (19·35) becomes

$$\frac{\partial \phi}{\partial t} - \lambda \frac{\partial \phi}{du} = 0, \qquad ...(19{\cdot}37)$$

so that

$$\phi(z, t) = f(u + \lambda t) = \psi(e^{u + \lambda t}) = \phi\,[(1 - 1/z)(e^{\lambda t})] \qquad ...(19{\cdot}38)$$

but
$$\phi(z, 0) = \sum_{n=1}^{\infty} P_n(0)\, z^n = z \qquad ...(19{\cdot}39)$$

since there is only one individual at time $t = 0$.

From (19·38) and (19·39)

$$\phi\left(1 - \frac{1}{z}\right) = z$$

or
$$\phi(z) = \frac{1}{1-z},$$

Writing $\left(1 - \dfrac{1}{z}\right) e^{-\lambda t}$ for z, we get

or
$$\phi\left[\left(1 - \frac{1}{z}\right) e^{-\lambda t}\right] = \frac{1}{1 - \left(1 - \dfrac{1}{z}\right) e^{\lambda t}} = z\, e^{-\lambda t}$$

so that
$$\phi(z, t) = z e^{-\lambda t} \left[1 + z\left(1 - e^{-\lambda t}\right)\right]^{-1} \qquad ...(19{\cdot}40)$$

Expanding in powers of z we get the probability distribution for the population size at any time t, so that

$$P_n(t) = e^{-\lambda t}(1-e^{-\lambda t})^{n-1}, (n \geq 1) \qquad ...(19.41)$$

Remarks. (*i*) The above probability distribution is of the geometrical form and the common ratio tends to unity as t tends to infinity.

(*ii*) The mean or the expected number of individuals at time t is given by using (19·40),

$$E[N(t)] = \overline{N}(t) = \sum_{1}^{\infty} nP_n(t) = \left[\frac{\partial \phi}{\partial z}\right]_{z=1} = e^{\lambda t} \qquad ...(19.42)$$

(*iii*) The variance of the number of individuals at time t is given by

$$\text{Var}[N(t)] = E[N(t) - \overline{N(t)}]^2 = \left.\frac{\partial^2 \phi}{\partial z^2}\right]_{z=1} = \overline{N}(\overline{N}-1)$$

$$= e^{\lambda t}(e^{\lambda t} - 1) \qquad ...(19.43)$$

(*iv*) In the more general case when the number of individuals in the population at time $t = 0$ is N_0, the probability generating function (*p.g.f.*) in (19·40) will have to be raised to power N_0, since the *p.g.f.* for the sum of n independent stochastic variates is the product of the n *p.g.f.* Also since the mean and variance of n independent stochastic variables are the sum of the means and variances respectively, the means and variances in (19·42) and (19·43) will have to be multiplied by N_0. Thus if there are N_0 individuals in the population at $t = 0$ then

$$\phi(z, t) = \sum_{n=1}^{N_0} e^{-N_0 \lambda t}[1 + z(1 - e^{-\lambda t})]^{-N_0} \qquad ...(19.44)$$

$$P_n(t) = C e^{-N_0 \lambda t}(1 - e^{-\lambda t})^{n - N_0} \text{ for } n \geq N_0 \qquad ...(19.45)$$

$$= 0 \text{ for } n < N_0 \qquad ...(19.46)$$

$$\overline{N}(t) = N_0 e^{\lambda t}$$

$$\text{Var}[\overline{N}(t)] = N_0 e^{\lambda t}(N_0 e^{\lambda t} - 1) \qquad ...(19.47)$$

Alternatively (19·44) could be derived as follows :

In this case, (19·39) would become

$$\phi(z, 0) = \sum_{n=1}^{\infty} z^n P_n(0) = z^{N_0} \qquad ...(19.48)$$

$$\phi\left(1 - \frac{1}{z}\right) = z^{N_0} \quad \text{or} \quad \phi(z) = \left(1 - \frac{1}{z}\right)^{N_0} \qquad ...(19.49)$$

$$\phi\left[\left(1-\frac{1}{z}\right)e^{\lambda t}\right]=\frac{N_0}{\left[1-\left(1-\frac{1}{z}\right)e^{\lambda t}\right]}$$

$$\phi(z, t) = z^{N_0} e^{-N_0\lambda t}\,[1 + z(1 - e^{-\lambda t})]^{-N_0} \qquad \text{...(19·50)}$$

which is same as (19·44).

Ex. 1. Solve (19·32) subject to $P_1(0) = 1$. Then solve (19·31) for $n = 2, 3, 4...$ in succession and verify (19·41).

Ex. 2. Solve (19·31) and (19·32) subject to the condition $P_n(0) = 1$ if $n = N_0$ and $P_n(0) = 0$ if $n \neq N_0$ and verify (19·45) and (19·46).

Ex. 3. Starting with (19·45) and (19·46) verify (19·44), (19·47) and (19·48).

Ex. 4. Show that the large t, variance (N_0T) is asymptotically equal to $N_0\, e^{2\lambda t}$.

Ex. 5. A population of initial N_0 grows deterministically in such a fashion that at any time its rate of increase is λ times its size. Show that its size at time t is the same as the average size in the pure birth stochastic case.

19·7. The simple birth and death process. To the conditions (*a*), (*b*), (*c*), given in the last section, we add the following conditions :

(*d*) an individual existing at time t has a chance $\mu\, dt + 0\,(dt)$ of dying in the following time interval dt.

(*e*) the 'death rate' μ is the same for all individuals at time t.

From conditions (*a*) to (*e*), we get

$$\begin{aligned} P_n(t + dt) &= P_n(t)[1 - n\lambda dt - n\mu dt] \\ &+ P_{n-1}(t)[(n-1)\lambda\, dt]\,[1 - (n-1)\mu dt] \\ &+ P_{n+1}(t)](n+1)\mu\, dt]\,[1 - (n+1)\lambda\, dt] + 0\,(dt) \end{aligned}$$

Proceeding to the limit as $dt \to 0$, we get

$$P'_n(t) = -nP_n(t)(\lambda + \mu) + (n-1)\lambda P_{n-1}(t) + (n+1)\mu\, P_{n+1}(t),$$

so that $$\Sigma z^n P'_n(t) = -(\lambda + \mu)\,\Sigma n z^n P_n(t) + \lambda\Sigma(n-1)z^n P_{n-1}(t) + \mu\Sigma(n+1)z^n P_{n+1}(t) \qquad \text{...(19·51)}$$

If $$\phi(z, t) = \Sigma\, z^n P_n(t) \qquad \text{...(19·52)}$$

denotes the probability generating function, then (19·51) gives

$$\frac{\partial\phi}{\partial t} = -(\lambda + \mu)\,z\frac{\partial\phi}{\partial z} + \lambda z^2\frac{\partial\phi}{\partial z} + \mu\frac{\partial\phi}{\partial z}$$

or $$\frac{\partial\phi}{\partial t} = (\lambda z - \mu)(z-1)\frac{\partial\phi}{\partial z} \qquad \text{...(19·53)}$$

Let $$\frac{dz}{(\lambda z - \mu)(z-1)} = du$$

or
$$\frac{\lambda z-\mu}{z-1}=e^{(\mu-\lambda)u} \qquad \text{...(19·54)}$$

$$\phi(z,t)=\phi\left[\frac{\lambda z-\mu}{z-1}e^{(\mu-\lambda)t}\right] \qquad \text{...(19·55)}$$

But $\phi(z,0)=z$

so that
$$\phi(z)=\frac{\mu-z}{\lambda-z}$$

$$\phi(z,t)=\frac{\mu-\dfrac{\lambda z-\mu}{z-1}e^{(\mu-\lambda)t}}{\lambda-\dfrac{\lambda z-\mu}{z-1}e^{(\mu-\lambda)t}}$$

$$=\frac{\mu-\lambda-\lambda/\eta}{\lambda}+\frac{1}{\eta t}\cdot\frac{(1-\xi)(1-\eta)}{1-z\eta} \qquad \text{...(19·56)}$$

where
$$\frac{\xi}{\mu}=\frac{\eta}{\lambda}=\frac{e^{(\lambda-\mu)t}-1}{\lambda e^{(\lambda-\mu)t}-\mu} \qquad \text{...(19·57)}$$

Expanding (19·56) in powers of t, we get

$$P_n(t)=(1-\xi)(1-\eta)(\eta)^{n-1}\ (\eta>1) \qquad \text{...(19·58)}$$

In the present case, it is possible that at time $t=0$, there is no individual in the population. To find its probability we have

$$P_0(t)=1-\sum_{n=1}^{\infty}P_n(t)=1-\sum_{n=1}^{\infty}(1-\xi)(1-\eta)(\eta)^{n-1}$$
$$=1-(1-\xi)=\xi \qquad \text{...(19·59)}$$

The *same result* can be seen from (19·56) and (19·57).

Since from here

$$\phi(1,t)=\frac{\mu}{\lambda}+1-\frac{1}{\eta}+\frac{1}{\eta}-\frac{\xi}{\eta}-(1-\xi)$$

and the constant term in $\phi(z,t)$

$$=\frac{\mu}{\lambda}+1-\frac{1}{\eta}+(1-\xi)\left(\frac{1}{\eta}-1\right) \qquad \text{...(19·60)}$$

This can also be seen from *first principle*, since

$$P_0(t+dt)=P_0(t)(1-0\lambda\,dt-0\mu\,dt)+P_1(t)\,\mu\,dt+0\,(dt)$$

so that

$$P'_0(t)=\mu P_1(t)=\mu(1-\xi)(1-\eta),\ \text{using (19·58)} \qquad \text{...(19·61)}$$

or
$$\frac{dP_0(t)}{dt}=\frac{\mu(\lambda-\mu)^2e^{(\mu-\lambda)t}}{[\lambda-\mu e^{(\mu-\lambda)t}]^2}$$

Integrating and remembering that $P_0(0) = 0$, we get

$$P_0(t) = \frac{\mu e^{(\lambda-\mu)t} - 1}{\lambda e^{(\lambda-\mu)t} - \mu} = \xi \qquad \text{...(19·62)}$$

Remarks : The probabilities at a given time t form a geometric series with modified first term (corresponding to no individual). The common ratio is

$$\eta = \lambda \frac{e^{(\lambda-\mu)t} - 1}{\lambda e^{(\lambda-\mu)t} - \mu} = \lambda \frac{1 - e^{-(\lambda-\mu)t}}{\lambda - \mu e^{(\lambda-\mu)t}} \qquad \text{...(19·63)}$$

so the ratio $\to 1$ if $\lambda > \mu$, it approaches λ/μ if $\lambda < \mu$.

If $\lambda = \mu$.

$$\xi = \eta = \frac{\lambda t}{1 + \lambda t} = \frac{1}{(\lambda t)^{-1} + 1} \qquad \text{...(19·64)}$$

and again the ratio approaches unity.

(*iii*) $$E[N(t)] = \overline{N}(t) = \left.\frac{\partial \phi}{\partial z}\right|_{z=1} = e^{(\lambda-\mu)t} \qquad \text{...(19·65)}$$

$$\text{Var}\,[N(t)] = \left.\frac{\partial^2 \phi}{\partial z^2}\right|_{z=1} = \overline{N}\,(\overline{N} - 1)$$

$$= \frac{\lambda+\mu}{\lambda-\mu}\, e^{(\lambda-\mu)t}\left\{ e^{(\lambda-\mu)t} - 1 \right\} \qquad \text{...(19·66)}$$

If there are N_0 individuals in the population at time $t = 0$, both mean and variance have to be multiplied by N_0.

When $\lambda = \mu, \overline{N} = 1, Var\,[N(t)] = 2\lambda t.$ ...(19·67)

Thus if birth and death rates are equal, the average number of individuals at any time t will be the same as that the time $t = 0$, but the variations in numbers will increase with both N_0 and t,

(*iii*) We have, for all N_0

$$\left.\begin{aligned} \underset{t\to\infty}{Lt}\; P_0(t) &= 1, && \text{if } \lambda \le \mu \\ &= \left(\frac{\mu}{\lambda}\right)^{N_0} && \text{if } \lambda > \mu \end{aligned}\right\} \qquad \text{...(19·68)}$$

so that the probability of extinction is certain when $t \to \infty$ not only when birth rate is less than death rate, but even when birth and death rates are equal. When birth rate is greater than the death rate, there is still a definite probability of extinction which decreases as N_0 increases.

Ex. 1. For the simple death process, prove that

(*i*) $\phi(z, t) = \{-e^{-\mu t}(1-z)\}^{N_0}$

(*ii*) $P_n(t) = {}^{N_0}C_n e^{-n\mu t}(1-e^{-\mu t})^{N_0-n} \quad 0 \le n \le N_0$

(*iii*) $P_0(t) \to 1$ as $t \to \infty$

(*iv*) mean $= N_0 e^{-\mu t}$

(*v*) Variance $= N_0 e^{-\mu t}(1-e^{-\mu t})$.

Ex. 2. For the birth and death process, prove that

$$P_n(t) = \sum_{j=1}^{\min(N_0, n)} {}^{N_0}C_j \, {}^{N_0+n-j-1}C_{N_0-1} \, \xi^{N_0-j} \eta^{n-j} (1-\xi-\eta^{\varphi}) \, ; \; P_0(t) = \xi^{N_0}$$

Ex. 3. For the case $\lambda = \mu$, show that

$$\phi = (z, t) = \left\{ \frac{1-(\lambda t-1)(z-1)}{1-\lambda t(z-1)} \right\}^{N_0}$$

Ex. 4. For the case $\lambda = \mu$, $N_0 = 1$, show that

$$P_n(t) = \frac{(\lambda t)^{n-1}}{(1+\lambda t)^{n+1}}, n \ge 1 \; P_n(t) = \frac{\lambda t}{1+\lambda t}; \; n = 0$$

Find mean and variance $N(t)$.

Ex. 5. Prove that $P_0(t) = \left[\dfrac{\mu e^{(\lambda-\mu)t} - 1}{\mu e^{(\lambda-\mu)t} - \mu} \right]^{N_0}$ when $\lambda \neq \mu$

$$= \left(\frac{\lambda t}{\lambda t + 1} \right)^{N_0} \quad \text{when } \lambda = \mu$$

Ex. 6. Show that when $\lambda = \mu$, the mean remains constant, yet the chance of extinction is unity. Explain this.

Ex. 7. Find the differential equation satisfied by the cumulant function $K(z, t)$.

[**Hint.** Put $\quad K(z, t) = k_1 z + k_2 \dfrac{z^2}{2!} + \cdots$

and equate the various coefficients of z. Solve the resulting equating for k_1, k_2].

Ex. 8. By substituting $P_n(t) = \phi(t)[\psi(t)^{n-1}] \, P_0(t) = 1 - \sum_{n=1}^{\infty} P_n(t)$ shows that $\phi(t), \psi(t)$ can be independent of n. Find $\mu_1(t)$ and $\mu_2(t)$ from these expressions.

Ex. 9. From (19·65) obtain differential equations for $\mu_1(t)$ and $\mu_2(t)$ also solve them.

Ex. 10. Compare the expression $\mu_1(t)$ and $\mu_2(t)$ obtained in questions 8 and 9 to get $\phi(t)$ and $\lambda(t)$. Verify the results of question 4.

19·8. Birth, Death and Immigration Processes.

One interesting variant of the simple birth and death process is obtained if we add the condition :

(*f*) in an infinitesimal time interval dt, there is a chance $k\,dt + 0\,(dt)$ that a single member will be added to the population by immigration from the outside would.

The characteristic feature of immigration effect is that it acts at an expected rate which is independent of the population size.

The generating function for (λ, μ, k) process satisfies the differential equation

$$\frac{\partial \phi}{\partial t} = (\lambda z - \mu)(z-1)\frac{\partial \phi}{dz} + k(z-1)\phi \qquad ...(19\cdot 69)$$

Also $\phi(z, 0) = 1$, initial population is zero ...(19·70)

On solving (19·69), subject to (19·70), we get

$$\phi(z, t) = \left[\frac{\lambda - \mu}{\lambda e^{(\lambda-\mu)^t} - \mu}\right]^{\frac{k}{\lambda}} \left[1 - z\frac{\lambda\{(\lambda e^{(\lambda-\mu)t} - 1\}}{\lambda e^{(\lambda-\mu)t} - \mu}\right]^{\frac{-k}{\mu}} \text{ when } \lambda \neq \mu \qquad ...(19\cdot 71)$$

and $$\phi(z, t) = (1 + \lambda t)^{\frac{k}{\lambda}} \left(1 - \frac{\lambda t z}{1 + \lambda t}\right\}^{\frac{-k}{\mu}} \quad \text{when } \lambda = \mu. \qquad ...(19\cdot 72)$$

The mean number of individuals is

$$\bar{n} = \frac{k}{\lambda - \mu}[e^{(\lambda-\mu)t} - 1] \quad \text{when } \lambda \neq \mu \qquad ...(19\cdot 73)$$

$$= kt \quad \text{when } \lambda = \mu \qquad ...(19\cdot 74)$$

Remarks : (*i*) When $\lambda < \mu$ and $t \to \infty$, the limiting generating function is

$$\phi(z) = \left(1 - \frac{\lambda}{\mu}\right)^{\frac{-k}{\lambda}} \left(1 - \frac{\lambda z}{u}\right)^{\frac{-k}{\lambda}} \qquad ...(19\cdot 75)$$

and in the limit the mean population size is

$$\bar{n} = \frac{k}{\mu - \lambda}\cdot \qquad ...(19\cdot 76)$$

The is related to the stable distribution of population which immigration can just maintain against the excess of μ over λ.

(*ii*) When $\mu = 0$, *i.e.*, when there are only births and immigration,

$$\phi(z, t) = e^{kt}[1 - z(1 - e^{-\lambda t})]^{\frac{-k}{\lambda}} \qquad ...(19\cdot 77)$$

which is the *p.g.f.* for the negative binomial distribution.

(*iii*) when $\lambda = 0$ *i.e.*, when there is immigration and emigration the latter depending on the size of the population

$$\phi(z) = \exp\left[\frac{k}{\mu}(1 - e^{-\mu t})(z-1)\right] \qquad ...(19{\cdot}78)$$

which is the *p.g.f.* of Poisson process.

When $t \to \infty$, it gives

$$\phi(z) = \exp\left[\frac{k}{\mu}(z-1)\right] \qquad ...(19{\cdot}79)$$

(*iv*) When $\lambda = 0, \mu = 0$ *i.e.*, when there is only immigration

$$\phi(z, t) = \exp[kt(z-1)] = e^{-kt} \sum_{n=0}^{\infty} \frac{(kt)^n}{n!} \qquad ...(19{\cdot}80)$$

$$P_n(t) = e^{-kt}\frac{(kt)^n}{n!} \qquad ...(19{\cdot}81)$$

which again gives probability distribution for a Poisson distribution with parameter kt.

Ex. 1. Show that for the case when the initial population size is N_0.

$$\phi(z, t) = \frac{(\lambda - \mu)^{k/\lambda}\{\mu e^{(\lambda-\mu)t} - 1) - (\mu e^{(\lambda-\mu)t} - 1)^z\} V_0}{\left\{(\lambda e^{(\lambda-\mu)t} - \mu) - \lambda(e^{\lambda-\mu)t} - \lambda)^z\right\}^{N_0 + \frac{k}{\gamma}}}$$

Deduce (19·71).

Ex. 2. Discuss the limiting case when $k \to 0$.

Ex. 3. Obtain $\overline{N}(t)$ from (19·77), (19·7), (19·79) and (19·80).

Ex. 4. Discuss the model when immigration is independent of the population size and emigration is proportional to the rate of population size. Interpret this model to discuss the colloidal particle suspension or the movements in spermatozoa.

Ex. 5. Find $\overline{N}(t)$ from problems 1.

19·9. Generalised Birth and Death Process. In the present section, we postulate that the probability of going from state E_n to state E_{n+1} in time dt is $\lambda_n dt + 0(dt)$ and the probability of going state E_n to E_{n-1} is $\mu_n dt + 0(dt)$, then following the method earlier sections, we get

$$P_n(t + dt) = P_n(t)[1 - (\lambda_n + \mu_n)dt] + P_{n+1}(t)\mu_{n+1}dt + P_{n-1}(t)\lambda_{n-1}dt + 0(dt)$$

so that

$$P_n'(t) = -(\lambda_n + \mu_n)P_n(t) + \lambda_{n-1}P_{n-1}(t) + \mu_{n+1}P_{n+1}(t), n \geq 1 \qquad ...(19{\cdot}82)$$

$$P_0'(t) = -\lambda_0 P_0(t) + \mu_1 P_1(t) \qquad ...(19{\cdot}83)$$

If at time zero, the system is in a state E_r, the initial conditions are :

$$P_r(0) = 1 \; ; P_n(0) = 0 \text{ for } n \neq r \qquad ...(19{\cdot}84)$$

Thus to determine $P_n(t)$ for all n, we have got an infinite system of simultaneous differential equations. For a pure birth or a pure birth and death process, there is a recurrence relation of the type for which the uniqueness of solution can be easily seen. For the present case, the result is :

"For arbitrarily prescribed coefficients, $\lambda_n \geq 0, \mu_n \geq 0$, there always exists a positive solution $P_n(t)$ such that $\Sigma P_n(t) \leq 1$. If the coefficients are bounded or increase sufficiently slowly, this solution is unique and satisfies the regularity condition $\Sigma P_n(t) = 1$. It is possible to choose the coefficients so that $\Sigma P_n(t) < 1$ and there exist in finitely many solutions."

Ex. 1. Obtain the values of λ_n, μ_n for pure birth process, pure death process, simply birth and death process and birth, death and immigration process.

Ex. 2. In the logistic process for which $\lambda_n = \alpha(N_1 - n), \mu_n = \beta(N_2 - n)$, $N_1 > N_2$, show that the *p.g.f.* satisfies the differential equation.

$$\frac{\partial \phi}{\partial t} = \beta(1-z)\frac{\partial}{\partial z}\left(z\frac{\partial \phi}{\partial z} - N_1 \phi\right) - \alpha(1-z)z\frac{\partial}{\partial z}\left(N_1 \phi - z\frac{\partial \phi}{\partial z}\right)$$

Ex. 3. Discuss the following cases :

(*i*) $\lambda_n = \lambda, \mu = n_1$.

(*ii*) $\lambda_n = (m-n)\lambda, \mu_0, \mu_n = \mu, n \neq 0$

(*iii*) $\lambda_n = (a-n)\lambda, \mu_n = \mu, 0 \leq n \leq a$

19·10. The Simplest Trunking Problem. Infinitely Many Channels. Suppose that infinitely many trunk lines or channels are available and that the probability or a call coming in time dt is $\lambda dt + 0(dt)$ and of a call ending in this time is $\mu dt + 0(dt)$. This corresponds to immigration and death process. The corresponding differential equation are :

$$P'_n(t) = -(\lambda + n\mu)P_n(t) + \lambda P_{n-1}(t) + (n+1)\mu P_{n+1}(t), (n \geq 1) \quad ...(19·85)$$

$$P_0'(t) = -\lambda P_0(t) + \mu P_1(t) \quad ...(19·86)$$

The differential equation for the probability generating function $\phi(z, t)$ is

$$\frac{\partial \phi}{\partial t} = (1-z)\left[-\lambda\phi + \mu\frac{\partial \phi}{\partial z}\right] \quad ...(19·87)$$

of which the solution is

$$\phi = e^{-\lambda(1-z)}(1-e^{-\mu t})/\mu\,[1-(1-z)e^{-\mu t}] \quad ...(19·88)$$

where i is the number of calls at time $t = 0$.

For $i = 0$, we get $\phi = e^{-\lambda(1-z)}(1-e^{-t})/\mu$

$$\phi = e^{-m(1-z)} \quad ...(19·89)$$

where
$$m = \frac{\lambda}{\mu}(1-e^{-\mu t}) \quad ...(19·90)$$

which gives a Poisson distribution with parameter $\frac{\lambda}{\mu}(1-e^{-\mu t})$

As $t \to \infty$, we get a Poisson distribution with parameter λ/μ.

We also consider the equilibrium state when $P_n(t)$ is independent of t so that $P_n'(t) = 0$ and $P_n(t) = p_n$ (*say*)

then the basic equations become $\lambda p_0 = \mu p_1$ ...(19·91)

$$(\lambda + n\mu)\, p_n = \lambda p_{n-1} + (n+1)\mu p_{n+1} \qquad \text{...(19·92)}$$

so that

$$\begin{aligned}\lambda p_n - (n+1)\,\mu p_{n+1} &= \lambda p_{n-1} - n\mu p_n \\ &= \lambda p_{n-2} - (n-1)\mu p_{n-1} \\ &\ldots\ldots\ldots\ldots\ldots \\ &= \lambda p_0 - \mu p_1 = 0.\end{aligned}$$

Hence $$\frac{p_{n+1}}{p_n} = \frac{\lambda}{\mu} \cdot \frac{1}{n+1} \qquad \text{...(19·93)}$$

or $$p_n = p_0 \frac{\left(\frac{\lambda}{\mu}\right)^n}{n\,!} \qquad \text{...(19·94)}$$

But $$\sum_{n=0}^{\infty} p_n = p_0 \sum_{n=0}^{\infty} \frac{(\lambda/\mu)^n}{n\,!} = p_0 e^{\lambda/\mu} = 1 \qquad \text{...(19·95)}$$

Hence $$p_n = e^{-\lambda/\mu} \frac{(\lambda/\mu)^n}{n\,!} \qquad \text{...(19·96)}$$

Thus the limiting distribution is a Poisson distribution with parameter λ/μ and is independent of the initial state.

The mean number of calls in time t is giving by

$$\bar{n} = \left.\frac{\partial \phi}{\partial z}\right|_{z=1} = \frac{\lambda}{\mu}(1-e^{-\mu t}) + \lambda e^{-\mu t} \qquad \text{...(19·97)}$$

At $t \to \infty$, then mean tends to λ/μ.

Ex. 1. Prove equation 19·88.

Ex. 2. Find $N(t)$ from (19·88) and discuss its variation with it.

Ex. 3. Find Var $(N(t))$ from (19·88) and deduce the result for $i = 0$.

Ex. 4. What are λ_n and μ_n for this problem.

19·11. Waiting lines for a Finite Number of Channels. We now assume that the number of channels is finite and equal to k. If all channels are busy, each new call joins a waiting line and waits until a channel is free.

We say that the system is in a state E_n if there are exactly n persons either being serviced or in the waiting line. Such a line exists only when, $n > k$ and there are $n - k$ persons in it.

As long as at least one channel is free, we are in exactly the same situation as before. However if $n > k$, only k conversations are going on and $\mu_n = k\mu$ for $n \geq k$. Accordingly, the differential equations are :

$$P_0'(t) = -\lambda P_0(t) + \mu P_1(t) \qquad \text{...(19·98)}$$

$$P_n'(t) = -(\lambda + \mu n) P_n(t) + \lambda P_{n-1}(t) + (n+1)\mu P_{n+1}(t), (n < k) \qquad \text{...(19·99)}$$

$$P_n'(t) = -(\lambda + \mu k) P_n(t) + \lambda P_{n-1}(t) + k\mu P_{n+1}(t), (n \geq k) \qquad \text{...(19·100)}$$

The equilibrium equations are : $\lambda p_0 = \mu p_1$ ...(19·101)

$$(\lambda + n\mu)p_n = \lambda p_{n-1} + (n+1)\mu p_{n+1}, (n < k) \qquad \text{...(19·102)}$$

$$(\lambda + n\mu)p_n = p\lambda_{n-1} + k\mu p_{n+1}, (n \geq k) \qquad \text{...(19·103)}$$

By recursion, we find again that

$$p_n = p_0 \frac{(\lambda/\mu)^n}{\lfloor n}, n \leq k, \qquad \text{...(10·104)}$$

$$p_n = p_0 \frac{(\lambda/\mu)^n}{(\lfloor k)\, k^{n-k}}, n \leq k, \qquad \text{...(19·105)}$$

The series $\Sigma \frac{p_n}{p_0}$ converges if $\frac{\lambda}{\mu} < k$.

If $\lambda/\mu > k$, a limiting distribution $[p_k]$ does exist and $p_n = 0$ for all n which means that gradually the waiting line grows over all bounds. On the other hand if $\lambda/\mu < k$,

$$p_0 \left[1 + \frac{\lambda}{\mu} + \text{........} + \frac{(\lambda/\mu)^k}{\lfloor k} + \frac{k^k}{\lfloor k} \left(\frac{(\lambda/\mu k)^{k+1}}{1 - \lambda/\mu k} \right) \right] \qquad \text{...(19·106)}$$

and we can find p_0.

From the explicit expression for $P_n(t)$, it can be shown that p_n thus obtained really represents the limiting distribution of $P_n(t)$.

The following table gives the limiting probabilities for the case of three channels ($k = 3$) and $\lambda/\mu = 2$.

n	0	1	2	3	4	5	6	7
Lines busy	0	1	2	3	3	3	3	3
Persons waiting	0	0	0	0	1	2	3	4
P_n	0·11111	0·2222	0·222	0·1481	0·988	0·0658	0·0439	0·0293

Ex. 1. What are λ_n and μ_n for this problem ?

Ex. 2. Prepare a table for limiting probability for the case of the channels and $\lambda/n = 1$

19·12. Servicing of Machines with only one Serviceman. We consider m automatic machines which call for service only in the event of a breakdown. $\lambda dt + 0\,(dt)$ is the probability of calling for service in time dt if it is working at time t and $\mu\, dt + 0\,(dt)$ is the probability of its reverting back to work if it is being serviced at time t. We say that the system is in a state E_n if n machines are not working.

A transition $E_n \rightarrow E_{n+1}$ is caused by breakdown of one of the working machines, while transition $E_n \rightarrow E_{n-1}$ is caused by the return to work of a machine being serviced.

Thus we have a birth and death process with coefficients :

$$\lambda_n = (m - \lambda)\lambda,\ \mu_0 = 0,\ \mu_n = \mu,\ \lambda_0 = m\lambda,\ (n \neq 0),$$

so that the basic differential equations become

$$P_0'(t) = -m\lambda P_0(t) + \mu P_1(t) \qquad \text{...(19·107)}$$

$$P_n'(t) = -[(m-n)\lambda + \mu]P_n(t) + (m-n+1)\lambda P_{n-1}(t) + \mu P_{n+1}(t)$$
$$(n = 1, 2, \ldots\ldots., m-1) \qquad \text{...(19·108)}$$

$$P_m'(t) = -\mu P_m(t) + \lambda P_{m-1}(t). \qquad \text{...(19·109}$$

These give a finite number of differential equation which can be solved by the usual methods. For the equilibrium state

$$m\lambda p_0 = \mu p_1 \qquad \text{...(19·110)}$$

$$[(m-n)\lambda + \mu]p_n = (m-n+1)\lambda p_{n-1} + \mu p_{n+1} \qquad \text{...(19·111)}$$

$$\mu p_m = \lambda p_{m-1} \qquad \text{...(19·112)}$$

Equation (19·111) gives

$$(m-n)\lambda p_n - \mu p_{n-1} = (m-n+1)\lambda p_{n+1} - \mu p_n$$
$$= m\lambda p_0 - \mu p_1 = 0 \qquad \text{...(19·113)}$$

so that we have easily the recurrence relation

$$(m-n)\lambda p_n = \mu p_{n+1}. \qquad \text{...(10·114)}$$

Substituting successively $n = m-1, m-2, \ldots\ldots, 1, 0$, we get

$$\lambda p_{m-1} = \mu p_m,\ 2\lambda p_{m-2} = \mu p_{m-1}, \ldots\ldots, m\lambda p_0 = \mu p_1$$

so that
$$p_{m-k} = \frac{1}{k}\left(\frac{\mu}{\lambda}\right)^k p_m \qquad \text{...(19·115)}$$

The constant p_m is to be so evaluated that the probabilities add up to unity. This gives

$$p_m = \left\{1 + \frac{\mu}{\lambda} + \cdots + \frac{1}{\lfloor m}\left(\frac{\mu}{\lambda}\right)^m\right\}^{-1} \qquad \text{...(19·116)}$$

The formula is known as Erlang's loss formula.

Now $$p_0 = \frac{1}{\lfloor m}\left(\frac{\mu}{\lambda}\right)^m p_m \qquad \text{...(19·117)}$$

which gives the probability of a repairman being idle. The expected number of machines in the waiting line is

$$w = \sum_{k=1}^{m} (k-1)p_k = \sum_{k=1}^{m} kp_k - (1-p_0) \qquad \text{...(19·118)}$$

Since $\sum_{n=0}^{m-1} (m-n)\lambda p_n = \sum_{n=0}^{m-1} \mu p_{n+1}$

$$m\lambda(1-p_m) - \lambda(w+1-p_0-mp_m) = \mu(1-p_0)$$

or $$m\lambda - \lambda w - \lambda(1-p_0) = \mu(1-p_0)$$

or $$w = m - \frac{\lambda+\mu}{\lambda}(1-p_0). \qquad \text{...(19·119)}$$

The following table gives the probabilities for the case

$$\frac{\lambda}{\mu} = 0{\cdot}1,\ m = 6.$$

n	*machines in waiting line*	P_n
0	0	0·4845
1	0	0·2907
2	1	0·1454
3	2	0·0582
4	3	0·0175
5	4	0·0035
6	5	0·0003

Ex. 1. What are λ_n and μ_n for the problem ?

Ex. 2. Prepare a table for limiting probability of the case. $\lambda/\mu = 0{\cdot}2$, $m = 6$ and compare with the table given above.

19·13. Servicing of Machines : Several Repairmen. Let m machines be now serviced by r repairers ($r < m$). Thus for $n \le r$, the state E_n means that n machines are being serviced and $r-n$ repairmen are idle. For $n > r$, the situation is that r machines are being serviced and $n-r$ are in the waiting line. The set-up now is

$$\lambda_0 = m\lambda,\ \mu_0 = 0 \qquad \text{...(19·120)}$$

$$\lambda_n = (m-n)\lambda,\ \mu_n = n\mu\ (1 \le n < r) \qquad \text{...(19·121)}$$

$$\lambda_n = (m-n)\lambda,\ \mu_n = n\mu\ (r \le n < m) \qquad \text{...(19·122)}$$

For the limiting probability p_n, we have

$$m\lambda p_0 = \mu p_1 \qquad ...(19\cdot123)$$

$$[(m-n)\lambda + n\mu]\, p_n = (m-n+1)\lambda p_{n-1} + (n+1)\mu p_{n-1}\ (1 \le n < r) \qquad ...(19\cdot124)$$

$$[(m-n)\lambda + r\mu]\, p_n = (m-n+1)\lambda p_{n-1} + r\mu p_{n+1}\ (1 \le n \le m) \qquad ...(19\cdot125)$$

From the first equation, we get the ratio of p_1/p_0. From the second and third, we get by induction

$$(n+1)\lambda p_{n+1} = (m-n)\lambda p_m\ \ (n < r) \qquad ...(19\cdot126)$$

$$r\mu p_{n+1} = (m-n)\lambda p_n\ \ (n \ge r) \qquad ...(19\cdot127)$$

These equations permit calculation successively of p_n/p_0. Finally p_0 follows from the condition $\Sigma p_k = 1$.

The following table gives the probability p_n, for the case $\lambda/\mu = 0{\cdot}1$, $m = 20$, $r = 3$.

n	*Machines being serviced*	*machines waiting*	*idle repairmen*	p_n
0	0	0	3	·13625
1	1	0	2	·27250
2	2	0	1	·25888
3	3	0	0	·15533
4	3	1	0	·08802
5	3	2	0	·04695
6	3	3	0	·02347
7	3	4	0	·01095
8	3	5	0	·00475
9	3	6	0	·00190
10	3	7	0	·00070
11	3	8	0	·00023
12	3	9	0	·00007

We define coefficients of loss for machine by

$$\frac{w}{m} = \frac{\text{average number of machines in waiting}}{\text{number of repairmen}}$$

and coefficient of loss for repairmen by

$$\frac{\rho}{r} = \frac{\text{average number of repairmen idle}}{\text{number of repairmen}}$$

Then we have the following comparison :

	I	II
Number of machines	6	20
Number of repairmen	1	3

Machines per repairmen	6	6·2/3
Coefficient of loss for repairmen	0·4845	0·4042
Coefficient of loss for machines	0·0549	0·01695

The comparison show that for $\lambda/\mu = 0{\cdot}1$, the second process of 3 men for 20 machines is more efficient than of three men for 6 machines.

Ex. 1. Prepare a table for $\lambda/\mu = 0{\cdot}2$, $m = 20$, $r = 3$ and compare with the table in this section.

Ex. 2. Find the coefficient of loss for repairmen and machines for the above example.

19·14. A Power supply Problem. A welder working independently draws current from some circuit. If at time t, a welder is using current, the probability that he ceases using it at time $t + dt$ is $\mu\, dt + 0\,(dt)$. If at time t he does not require current, the probability that he calls for current in next time interval dt is $\lambda dt + 0\,(dt)$. We say that the system is in a state E_n if n welders are using current. Thus we have only a finite number of states $E_0, E_1, ..., E_a$.

If the system is in a state E_n, $a - n$ welders are not using current and

$$\lambda_n = (a - n)\lambda \,;\, \mu_n = n\mu_n \quad 0 \le n \le a, \qquad ...(19{\cdot}128)$$

The basic differential equations become

$$P_0'(t) = -a\lambda\, P_0(t) - \mu P_1(t) \qquad (19{\cdot}129)$$

$$P_0'(t) = -[n\mu + (a-n)\lambda]\, P_n(t) - (n+1)\mu P_{n+1}(t)$$
$$+ (a - n + 1)\lambda P_{n-1}(t); \; (1 \le n \le a - 1) \qquad ...(19{\cdot}130)$$

$$P_a'(t) = -a\mu\, \lambda P_a(t) + \lambda P_{a-1}(t) \qquad ...(19{\cdot}131)$$

It can be easily verified that the limiting probabilities are given by the binomial distribution.

$$p_n = C_n^a \left(\frac{\lambda}{\lambda + \mu}\right)^n \left(\frac{\mu}{\lambda + \mu}\right)^{a-n} \qquad ...(19{\cdot}132)$$

Ex. 1. Find the average number of welders using current in the steady case.

Ex. 2. Find the variance of the number of welders using current in the steady case.

Ex. 3. Find the variance in the number of welders not using current in the steady case.

19·15. Waiting Line for a Single Channel. In this case from (19·104) and (19·105), we get

$$p_n = p_0\,(\lambda/\mu)^n, \; n \; 0 \qquad ...(19{\cdot}133)$$

and since $$\sum_{n=1}^{\infty} p_n = p_0 \sum_{n=0}^{\infty} (\lambda/\mu)^n = \frac{p_0}{1 - \lambda/\mu} = 1$$

we have

$$p_0 = 1 - \lambda/\mu \qquad \text{...(19·134)}$$

and

$$p_n = \left(1 - \frac{\lambda}{\mu}\right)\left(\frac{\lambda}{\mu}\right)^n, \quad n \geq 0 \qquad \text{...(19·135)}$$

Equations 19·134) and (19·135) are valid only if the mean service rate μ exceeds the mean arrival rate λ.

Let ω denote the waiting time of an arrival, before being taken into service, then

$$f(\omega)\, d\omega = P(\omega < \text{waiting time} < \omega + d\omega)$$

$= \sum_{n=1}^{\infty} P$ (n units in system just before arrival $n-1$ services in time interval ω and one service in time $d\omega$)

$$= \sum_{n=1}^{\infty} \left(1 - \frac{\lambda}{\mu}\right)\left(\frac{\lambda}{\mu}\right)^n \frac{e^{-\omega\mu}\,(\mu w)^{n-1}}{(n-1)\,!}\,\mu d\omega$$

$$-\lambda\left(\lambda - \frac{\lambda}{\mu}\right) e^{-\mu\omega}\, d\omega \sum_{n=1}^{\infty} \left(\frac{\lambda}{\mu}\right)^{n-1} (\mu\omega)^{n-1}$$

$$-\lambda\left(1 - \frac{\lambda}{\mu}\right) e^{(\lambda-\mu)\omega}\, d\omega$$

Now

$$\int_0^{\infty} f(\omega)\, d\omega = \lambda/\mu, \qquad \text{...(19·136)}$$

the reason for this not being equal to unity is that there is a non-zero probability for $\omega = 0$, *viz.*

$$P(\omega = 0) = p_0 = 1 - \lambda/\mu \qquad \text{...(19·137)}$$

Thus the probability distribution for ω is partly discrete (at $w = 0$) and partly continuous (for $\omega > 0$).

The conditional density function for waiting time, given that a person has to wait is obtained by dividing $f(w)$ by $p(w)$ so that

$$f(w \mid w > 0) = (\mu - \lambda) e^{(\lambda-\mu)\omega} \qquad \text{...(19·138)}$$

and that

$$\int_0^{\infty} f(w \mid w > 0) dw = 1. \qquad \text{...(19·139)}$$

Let v denote the random variable denoting the total time an arrival spends in the system so that

$$v = \text{waiting time} + \text{service time} = w + s \qquad \text{...(19·140)}$$

The distribution function w is given by (19·136) and 19·141) and the function s is given by

$$dP = \mu e^{-\mu s}\, dx \qquad ...(19\cdot141)$$

then
$$\int \varphi(v)\, dv = \int\int \lambda\left(1-\frac{\lambda}{\mu}\right) e^{-(\lambda-\mu)w}\, d\omega.\, \mu e^{\mu-2}\, ds$$

$$+\int_0^v \left(1-\frac{\lambda}{\mu}\right) \mu e^{-\mu v}\, dv \qquad ...(19\cdot142)$$

Putting $w+s=v, w=w$ in the first integral on the *R.H.S.* and integrating for w from 0 to v,

$$\varphi(v) = (e^{\lambda v}-1)\, e^{-\mu v}(\mu-\lambda)+(\lambda-\mu)e^{-\mu v}.$$

$$= (\mu-\lambda)e^{(\mu-\lambda)v}. \qquad ...(19\cdot143)$$

Average length of the queue $= E(n)$

$$= \sum_{n=2}^{\infty}(n-1)p_n = \sum_{n=2}^{\infty}(n-1)\left(1-\frac{\lambda}{\mu}\right)\left(\frac{\lambda}{\mu}\right)^n$$

$$= \left(1-\frac{\lambda}{\mu}\right)\left(\frac{\lambda}{\mu}\right)\sum_{n=1}^{\infty} n\left(\frac{\lambda}{\mu}\right)^n \qquad ...(19\cdot144)$$

$$= \frac{\lambda^2}{\mu(\lambda\mu-\lambda)}. \qquad ...(19\cdot145)$$

Average length or units in the system $= E(n) = \sum_{n=0}^{\infty} np_n$

$$\sum_{n=1}^{\infty} n\left(\frac{\lambda}{\mu}\right)^n\left(1-\frac{\lambda}{\mu}\right) = \left(1-\frac{\lambda}{\mu}\right)\frac{\frac{\lambda}{\mu}}{1-\frac{\lambda}{\mu}} = \frac{\lambda}{\lambda-\mu}.$$

Average length of non-empty queues $= E(m \mid m>0)$

$$= \sum_{n=1}^{\infty}(n-1)p_n = \sum_{n=2}^{\infty}\frac{(n-1)p_n}{1-p_0-p_1} = \frac{\mu}{\mu-\lambda}, \text{ using (19·145)} \qquad ...(19\cdot146)$$

Average waiting of an arrival

$$E(w) = \lambda\left(1-\frac{\lambda}{\mu}\right)\int_0^{\infty} we^{(\lambda-\mu)w}\, dw = \frac{\lambda}{\mu(\mu-\lambda)} \qquad ...(19\cdot147)$$

Average waiting time of arrival who waits

$$= E(w \mid w>0) = \int_0^{\infty} w(\mu-\lambda)\, e^{(\lambda-\mu)w}\, dw$$

$$= \frac{\mu - \lambda}{(\mu - \lambda)^2} = \frac{1}{(\mu - \lambda)} \qquad \text{...(19·148)}$$

Average time an arrival spends in the system

$$= E(v) = \int_0^\infty v(\mu - \lambda)\, e^{(\lambda - \mu)v}\, dv = \frac{1}{\mu - \lambda}, \text{ using}(19·143) \qquad \text{...(19·149)}$$

Ex. 1. Show that the variance of the number of persons in the queue is

$$\frac{\rho}{(1-\rho)^2} \text{ where } \rho = \frac{\lambda}{\mu}.$$

Ex. 2. Discuss the variation of all the quantities found in this section, (*i*) when λ increases (*ii*) when λ decreases (*iii*) when μ increases (*iv*) when μ decreases.

Ex. 3. Discuss the particular case $\lambda = \mu$.

19·16. Queues with several Servicing Stations. In this case the probability that an arrival has to wait is just the probabilities that there are at least k units in the system at any time. Thus the required probability is given by

$$P[n \geq k] = \sum_{n=k}^{\infty} p_n = p_0 \sum_{n=k}^{\infty} \frac{(\lambda/\mu)^n}{k!\, k^{n-k}} k$$

$$= \frac{p_0}{k!} k^k \sum_{n=k}^{\infty} \left(\frac{\lambda}{\mu k}\right)^n = \frac{p_0}{k!} k^k \frac{(\lambda/\mu k)^k}{1 - \lambda/\mu k}$$

$$= \frac{\mu(\lambda/\mu)^k}{(k-1)!\,(k\mu - \lambda)} p_0 \qquad \text{...(19·150)}$$

$$\text{Average queue length} = \sum_{n=k+1}^{\infty} (n-k) p_n = \sum_{n=k+1}^{\infty} \frac{(n-k) p_0 (\lambda/\mu)^n}{k!\, k^{n-k}}$$

$$= \sum_{m=1}^{\infty} m p_0 \frac{(\lambda/\mu)^{m+k}}{k!\, k^m} = \frac{p_0}{k!}\left(\frac{\lambda}{\mu}\right)^k \sum_{m=1}^{\infty} m \left(\frac{\lambda}{\mu k}\right)^m$$

$$= \frac{p_0}{(k-1)!} \frac{(\lambda/\mu)^k \lambda\mu}{(\mu k - \lambda)^2}. \qquad \text{...(19·151)}$$

Average number of units in the system

$$E(n) = \sum_{n=1}^{\infty} n p_n = \sum_{n=1}^{\infty} n p_0 \frac{(\lambda/\mu)^n}{n!} + \sum_{n=k+1}^{\infty} n p_0 \frac{(\lambda/\mu)^n}{k!\, k^{n-k}}$$

$$= p_0 \frac{\lambda}{\mu} \sum_{n=1}^{\infty} \frac{(\lambda/\mu)^{n-1}}{(n-1)!} + \frac{p_0}{k!} k^k \sum_{n=k+1}^{\infty} n\,(\lambda/\mu)^n$$

$$= p_0 \frac{\lambda}{\mu}\left[\frac{1}{p_0} - \frac{(\lambda/\mu)^k}{k} - \frac{k^k}{k\,!} \cdot \frac{(\lambda/\mu)^{k-1}}{1-\lambda/\mu k}\right]$$

$$+ p_0 \frac{k^k}{k\,!} \frac{(k+1)(\lambda/\mu k)^{k+1} - k(\lambda/\mu k)^{k+1}}{(1-\lambda/\mu k)^2} \qquad ...(19{\cdot}152)$$

$$= \frac{\lambda}{\mu} + p_0 \frac{\lambda\mu (\lambda/\mu)^k}{(k-1)\,!\,(k\mu - \lambda)^2}.$$

Average waiting time of an arrival

$$E(w) = \frac{\mu (\lambda/\mu)^k}{(k-1)\,!\,(k\mu-\lambda)^2} p_0 \qquad ...(19{\cdot}153)$$

Average time an arrival spends in the system

$$E(v) = \frac{\mu(\lambda/\mu)^k}{(k-1)\,!\,(k\mu-\lambda)^2} p_0 + \frac{1}{\mu} \qquad ...(19{\cdot}154)$$

Ex. 1. The expected number of waiting individual in the queue is

$$\sum_{n=k}^{\infty} (n-k)p_n = p_k . \frac{\rho}{(1-\rho)^2}$$

Ex. 2. The probability that all the k channels are occupied is

$$\sum_{n=k}^{\infty} p_n = p_k \cdot \frac{1}{1-\rho}$$

Ex. 3. The probability that there is someone waiting is

$$\sum_{n=k+1}^{\infty} p_n = p_k \cdot \frac{\rho}{1-\rho} \cdot$$

Ex. 4. The mean waiting time in the queue for all arrivals

$$\frac{1}{\mu} \sum_{n=k}^{\infty} (n-k)p_n = p_k \cdot \frac{1}{kp(1-\rho)^2}$$

19·17. Arrival and Service Time Distributions. Arrivals or inputs into a queuing system may

(*i*) occur at intervals of constant length T. In this case the cumulative distribution function $A(T)$ of times between arrivals with density function $a(T)$ to be given by

$$A(T) = 0 \text{ for } T < P, \; A(T) = 1 \text{ for } T \geq P \qquad ...(19{\cdot}155)$$

(*ii*) follow the Poisson distribution discussed in section 19.

(*iii*) follow the Erlangian frequency distribution of time between arrival for which the density function is

$$f_k(T) = \frac{e^{-\lambda k T} (\lambda k)^k T^{k-1}}{k\,!} \qquad ...(19{\cdot}156)$$

When $k = 1$, the Erlangian distribution reduces to the exponential distribution and thus corresponds to a Poisson input. When k tends to infinity

$$\frac{e^{-\lambda kT}(\lambda k)^k T^{k-1}}{k\,!} \sim \frac{e^{-\lambda kT}(\lambda k)^k T^{k-1}}{\sqrt{2\pi}\, e^{-(k-1)}(k-1)^{(k-1/2)}}$$

$$\sim \frac{1}{\sqrt{2\pi}} e^{-\lambda kT} k^{1/2} \lambda\, e^k (\lambda T)^{k-1} \sim \frac{1}{\sqrt{2\pi}} e^{-k(\lambda T)\lambda - 1} \lambda k^{1/2} (\lambda T)^{k-1}$$

and thus the density function tends to zero except for $\lambda T = 1$, so that the constant input distribution is obtained as a limiting case, as $k \to \infty$, of the Erlangian distribution.

If a Poisson input is filtered in such a way that only every kth item is admitted into the system, the resulting density function of arrival time is $f_k(T)$.

It thus appears that a regular input gives rise to a constant input distribution of arrival time, a Poisson input to an exponential distribution of arrival time and a filtered Poisson input to an Erlangian type distribution of arrival times.

The **Erlang family of distribution** has many important **properties.** The mean of the kth distribution is

$$\int_0^\infty T \frac{e^{-\lambda kT}}{(k-1)\,!} (\lambda k)^k T^{k-1}\, dT$$

$$= \int_0^\infty \frac{e^{-x} x^k\, dx}{\lambda\,[(k-1)\,!]} = \frac{1}{\lambda}, \text{ independent of } k,\ \lambda hT = x \qquad ...(19\cdot157)$$

Thus all members have the same mean $= 1/\lambda$.

The mode of (19·150) is given by

$$-\lambda k + \frac{k-1}{T} = 0, \quad \therefore\ T = \frac{k-1}{\lambda k}. \qquad ...(19\cdot158)$$

The variance of the kth member of the family is

$$\int_0^\infty T^2 \frac{e^{-\lambda kT}(\lambda k)^k T^{k-1}\, dT}{(k-1)\,!} - \frac{1}{\lambda^2} = \frac{(k+1)}{\lambda^2 k^2} - \frac{1}{\lambda^2} = \frac{1}{\lambda^2 k} \qquad ...(19\cdot159)$$

As k increases the mode move to the right towards $1/\lambda$ and the variance decreases towards zero. For $k \to \infty$, the mode is at $1/\lambda$ and the variance is zero ; this again suggests that the limiting case gives the constant input case.

From (19·161) it appears that λkT is a Gamma variate with parameter k, so that λT is a Gamma variate with paramerter unity *i.e.,* it has the exponential distribution. From the additive property of independent Gamma variates, it follows that if $X_1, X_2, ..., X_k$ are k independent random variates each having a common exponential distribution with mean $1/k\lambda$, then $X_1 + X_2 + ... + X_k$ follows the kth Erlangian distribution with parameter λ.

Thus if the time interval between consecutive arrival follows a Poisson distribution, then time interval between an arrival and the $(k-1)$th arrival will follow the kth Erlangian distribution.

Similarly a servicing station in which a unit goes through k independent places each exponential with mean time $T/\lambda k$ will have service time obeying kth Erlangian distribution with parameter λ.

Thus we find Erlang distributions have physical interpretations, but even if this physical interpretation is not there, the formulae may give an adequate fit to the observed distribution of service times or of arrival times.

Ex. 1. Establish all the results of this section.

Ex. 2. Deduce the mean, mode and variance of constant input and Poisson's distributions from those of Erlangian distribution.

19·18. Queuing Processes. Most of the stochastic processes we have considered in this chapter are included under the general heading 'queuing processes'. Any queuing process is characterised by the following three characteristics :

(*i*) **arrival stochastic process.** The intervals between consecutive arrivals are all supposed to be independently and identically distributed. The common distribution is called the arrival distribution.

(*ii*) **service stochastic process.** The duration of various services are again supposed to be independently and identically distributed. The common distribution is called the service distribution.

(*iii*) **the queue discipline.** In most cases the is, 'first come, first served'. In some cases it may be 'last come, first served' *e.g.*, when things are taken out of a large godown when items which arrive last are taken out first. The most difficult queue discipline to handle is that of 'random selection' or of 'might is right'.

Given these three characteristics, it is usually required to find

(*i*) **system size** (or queue length) **distribution,** *i.e.*, the distribution of the number of individuals in the system (those in queue + those in service) or in the queue at any time.

(*ii*) **waiting time distribution,** *i.e.*, the probable distribution of the waiting time of each customer.

(*iii*) **busy period distribution,** *i.e.,* the distribution of the the time for the system to go into state E_0 for the firstt time from its initial state E_1. (E_n refers to the state when there are n persons in the system). It is found useful to denote a queue system as follows : arrival time distribution | service time distribution | number of services. The exponential distribution is denoted by M, the Erlangian distribution by E_k, the general input distribution by GI and the general service distribution by G. Thus $M \mid M \mid 1$ system denotes exponential arrival distribution, exponential service distribution and one server. This is the most important queuing system which has been investigated most extensively and which we have discussed in section 19·15. The system $M \mid M \mid k$ has been investigated in sections 19·11 and 19·16. In Section 19·9 we gave a general families of queuing systems of which those in sections 19·6, 19·7, 19·8, 19·10 19·11, 19·12, 19·13, 19·14, 19·15, 19·16 are particular cases. Some other interesting particular cases are :

(*i*) queuing system $M \mid M \mid 1$ with finite waiting space.

In that case the system can accommodate only N units so that a unit which arrives when the system is in state E_n is not allowed to join so that

$$\lambda_n = \lambda,\ (n < N)\ ;\ \lambda_n = 0,\ (n \geq N)\ ;\ \mu_n = \mu \qquad \text{...(19·160)}$$

(*ii*) **queues with discouragement**

$$\text{Let} \quad \left.\begin{aligned} \lambda_n &= \frac{\lambda}{n+1} \qquad (n = 0, 1, 2, \ldots) \\ \mu_n &= \mu \qquad (n = 1, 2, 3, \ldots) \end{aligned}\right] \qquad \text{...(19·161)}$$

As n increases λ_n decreases, so that units are being discouraged to join the queue. Instead of $\lambda/(n+1)$ we can use $\lambda f(n)$ where $f(n)$ is a monotonic decreasing function of n.

(*iii*) **Calls on telephone exchange with k lines**

We consider a telephone exchange with k *lines with no facility of waiting. Then*

$$\left.\begin{aligned} \lambda_n &= \lambda \qquad n = 1, 2, \ldots, k-1 \\ &= 0 \qquad n = k, k+1, \ldots \\ \mu_n &= n\mu \qquad n = 1, 2, \ldots \end{aligned}\right] \qquad \text{...(19·162)}$$

(*iv*) **queues with balking**

Let n/N be the probability that a unit arriving when the system is in state E_n , *i.e.,* when there are n persons in the system, decides to balk, *i.e.,* not to join the queue so that $1-(n/N)$ is the probability of his joining the queue. Thus

$$\left.\begin{aligned}\lambda_n &= \lambda\left(1-\frac{n}{N}\right); \quad n=0,1,2,\ldots\ldots,N\\ &= 0, \qquad\qquad n>N\end{aligned}\right] \qquad \ldots(19\cdot163)$$

$$\mu_n = \mu$$

(*v*) **queues with reneging**

Here a unit joins the queue, but after some time it gets fed up and leaves the queue. When the system is in state E_n, any one of the units can renege so that in this case

$$\left.\begin{aligned}\lambda_n &= \lambda \qquad & n&=0,1,2,\ldots\ldots\\ \mu_n &= \mu+(n-1)\alpha & n&=1,2,\ldots\ldots\end{aligned}\right] \qquad \ldots(19\cdot164)$$

(*vi*) **queues with balking and reneging**

In this case

$$\left.\begin{aligned}\lambda_n &= \lambda\left(1-\frac{n}{N}\right); \qquad & n&=0,1,2,\ldots.,N\\ &= 0\,; & n&>N\\ \mu_n &= \mu+(n-1)\alpha & n&=1,2\ldots.\end{aligned}\right] \qquad \ldots(19\cdot165)$$

Ex. 1. Write the system of difference-differential equation for each the above six cases.

Ex. 2. Find the system length distribution for the steady state for the following cases :

(*i*) $\mu_n = \mu$, $\qquad \mu_n = \mu + c_n$

(*ii*) $\mu = \mu + d_n$, $\qquad \mu = \mu + c_n$

Exercises 19

1. A system has initially i units. Units arrive at regular time intervals of length a and are serviced at regular time intervals of length b. Discuss the development of the queue when $a<b, a=b, a>b$.
2. In section 19·5, show that (*a*) ⇒ (*b*) by using the generating function method used in section 19·6.
3. Show that when service times are exponentially distributed, the probability, that a unit which is undergoing service, will finits its services after time t from now, is independent of how long it has been in service till now.
4. For the polya process, *i.e.*, for the non-stationary pure birth process with

$$\lambda_n = \frac{1}{t} + an$$

show that the solution $P_o(0) = 1$ is $P_0(t) = (1 + at)^{\frac{1}{\alpha}}$

$$P_n(t) = \frac{t^a (1 + at)^{-n-1/a} (1 + a)(1 + 2a) \ldots [1 + (n-1)a]}{n!}$$

Show from the differential equation or otherwise that the mean and variance are t and $t(1 + at)$ respectively.

5. For the Polya process for which $\lambda_n = \dfrac{1 + \mu n}{1 + \lambda \mu t}$

show that $P_n(t) = \dfrac{(\lambda t)^n}{n!} (1 + \lambda \mu t)^{-n - \frac{1}{\mu}} \prod\limits_{j-1}^{n=1} (1 + j\mu)$

and that the mean and variances are λt, $\lambda t (1 + \lambda \mu t)$ respectively.

6. Assuming a trial solution of (19·53) of the form $P_n(t) = A(t) B(t) B^n(t)$, prove that the solution with $P_1(0) = 1$

$P_n(t) = \mu B(t) B^n(t) = (1 - \lambda B(t))(1 - (\lambda B(t))^{n-1}$ with $B(t) = \dfrac{1 - e^{-(\lambda-\mu)t}}{\mu \lambda e^{(\lambda-\mu)t}}$

7. Let $\mu_1(t) = \sum\limits_n P_n(t)$, $\mu_2(t) = \sum\limits_n n^2 P_n(t)$ deduce for (19·58) or otherwise that

$$\frac{d}{dt} \mu_1(t) = (\lambda - \mu) \mu_1(t) \, ; \frac{d}{dt} \mu_2(t) = 2(\lambda - \mu) \mu_2(t) + (\lambda + \mu) \mu_1(t).$$

8. Find $\mu_1(t)$ and $\mu_2(t)$ In terms of $A(t)$ and $B(t)$ of Ex. 6 and find the differential equations for $A(t)$ and $B(t)$ from Ex. 7.

9. From the system defined by (19·132)–(19·134) show that the generating function satisfies the partial differential equation $(\mu + \lambda z) \dfrac{\partial \phi}{\partial z} = a\lambda\phi$ with the solution $\phi(z, t) = \left\{\dfrac{\mu + \lambda z}{\mu + \lambda}\right\}^a$

10. In equation (19·58) let λ and μ be function of t. Write the subsidiary equation. Show that $\phi =$ constant is one integral. By making the substitution $z = 1 + 1/y$ show that the second integral is

$$\frac{e^{\rho(t)}}{z-1} - \int_0^t \lambda(t) e^{\rho(t)} dt - const \text{ where } P(t) = \int_0^t \{\mu(t) - \lambda(t)\} dt.$$

Deduce $\phi(z, t) = c + z^4$.

11. Discuss the particular case when

$$\lambda(t) = a + bt, \mu(t) = c + dt.$$

12. Suppose a particle performa unrestricted symmetrical random walk in two dimensions starting from the origin. Let r_n be the distance of the moving particle from the origin after the nth step. Prove that $E(r_n^2) = n$.

13. For the queue with discouragement, prove that the queue length is a Poisson distribution with parameter λ/μ.

14. Show that the following processes are identically distributed

(i) $X(t) = A\cos\omega t + B\sin\omega t$. (ii) $X(t) = R\cos(\omega t + \theta)$.

when A and B are independent normally distributed random variables with means 0 and variances σ^2, θ is uniformly distributed on 0 to 2π and R is Rayleigh distribution with probability density function

$$\frac{r}{\sigma^2} e^{-\frac{1}{2}\frac{r^2}{\sigma^2}} ; r > 0$$

15. Let $X_1(t)$ and $X_2(t)$ be the independent Poisson processes with mean rates λ_1, λ_2 respectively. For example $X_1(t)$ may represent number of waiting taxis and $X_2(t)$ represents the number of waiting passengers at a taxi stand. Show that if $X(t) = X_1(t) - X_2(t)$, then $X(t)$ is not a Poisson processes. Also find $P[Y(t) = X(s) = k]$ for $t > s \geq 0$ and $k = 0 \pm 1, \pm 2, \ldots$

[Such stochastic processes are useful in studying double ended queuing systems.]

16. For any Poisson process $X(t)$, show that for $s < t$.

$$P\{N(S)\} = k\{N(t)\} = n\} = {}^nC_k \left\{\frac{s}{t}\right\}^k \left\{1 - \frac{s}{t}\right\}^{n-k}$$

17. For the stochastic process (i) $X(t) = A + Bt + Ct^2$ where A, B, C are independent random variables each with mean 1 and variance 1 (ii) $X(t) = A\cos wt + B\sin wt$, in which w is positive constant and A and B are uncorrelated random variables with 0 mean and variance σ^2, find

(a) $E\{X(t)\}$ (b) $\text{cov}\{X(s), X(t)\}$.

18. Particles arrive in group. For $k = 1, 2, \ldots$, suppose tat the event that k particles arrive simultaneously occurs in accordance with a Poisson process $Nk(t)$ at a mean rate μ_k per unit time Assume. independence of that process. Let $N(t)$ be the total number of particles that arrive in interval (0, t(). Show that

(i) $N(t) = N_1 + 2N_2(t) + .. + kN_k(t) + \ldots$

(ii) $E[N(t)] = \mu_1 t$, where $\mu_1 = \sum_{k=1}^{\infty} k\lambda_k$

(iii) $\text{Var}[N(t)] = \mu_2 t$, where $\mu_2 = \sum_{i=1}^{\infty} k^2\lambda_k$

20

Order Statistics

20.0 Let the random variates be X_1, X_2 .... X_n arranged in non-decreasing order of magnitude and written in symbol form.

$$\text{as } X_{(1)} \leq X_{(2)} \leq .... \leq X_{(n)} \qquad .... (1)$$

We regard $X_{(i)}$ as the i-th orderstatistic, (i = 1, 2.... , n).

Note 1. X_1, X_2 ,....> X_n need note be dependent variates but the X_i's are necessarily dependent because of the inequality relation (1) among them.

2. Rewriting the independent ordered variables $X_{(1)}$ $X_{(2)}$ ..... $X_{(n)}$ as

$Y_1 \leq Y_2 \leq ..... \leq Y_n$, then

$Y_r = X_{(r)}$ = rth smallest of X_1, X_2, ... , X_n

$Y_1 = X_{(1)}$ = the smallest of X_1, X_2, ... , X_n

$Y_n = X_{(n)}$ = the largest of X_1, X^1, ... X_n

We shall now deal with the properties and applications of these ordered randon variates and of functions involving them. The examples of order statistics are:

(*i*) The extremes $X_{(n)}$ and $X_{(1)}$, which arise in the statistical study of floods and droughts, as well as in problems of breaking strength and fatigue failure.

(*ii*) The range $R = X_{(n)} - X_{(1)}$, which is well-known quick estimator of σ and is widely accepted in the field of quality control.

(*iii*) The extreme deviate $X_{(n)} - \overline{X}$, which is a basic tool in the detection of outliers, large value of $(X_{(n)} - \overline{X})/\sigma$ indicates, the presence of an excessively large observation.

(*iv*) The studentized range R/S_r where S_r is a root mean squre estimator of σ based on r degrces of freedom when the sampling is done from the normal $N(\mu, \sigma^2)$ population, which is used when the out are not confined to one direction.

(*v*) The linear functions of order statistics which can be used to estimate local and scale parameters by using Gauss-Markov theorem of least squares.

20.1. Distribution of Single Order Statistic

Let us assume that $X_1, X_2, \ldots, X_n$ are independent variated each with cumulative distribution function (cdf) $F(x)$. Let $G_r(x)$ $r = 1, 2, \ldots, n$) be the cumulative distribution, function of $X_{(r)}$ which is given by for any x.

$$G_r(x) = P(X_{(r)} \le X)$$

$$= P \text{ (atleast } r \text{ of the } n, X\text{'s are less than or equal to } x)$$

$$= \sum_{i=r}^{n} P \text{ [Exactly } j \text{ of the } n\text{' } X\text{'s are less than or equal to } x)$$

$$= \sum_{j=r}^{n} \binom{n}{j} F^j(x)\,(1 - F(x)^{n-j},$$

$$= \left[B(r, n-r+1) \right]^{-1} \int_0^{F(x)} u^{r-1}\,(1-u)^{(n-r+1)-1}\,du,\ u = F(x).$$

$$= I_{F(x)}(r, n-r+1), \qquad \ldots (20.1)$$

which is the Incomplete Beta function.

Case I. X is Continuous, the p.d.f. of $X_{(r)}$ is given by

$$g_r(x) = \frac{d\,G_r(x)}{dx} = [B(r, n-r+1)]^{-1} \frac{d}{dx} \int_0^{F(x)} u^{r-1}\,(1-u)^{n-r}\,du$$

$$= \frac{n!}{(r-1)!\,(n-r)!} F^{r-1}(x)\,[1 - F(x)]^{n-r} f(x) \qquad \ldots(20.2)$$

Case II. X is discrete, $x = 0, 1, 2, \ldots\ldots$, the p.m.f. of $X_{(r)}$ is given by

$$P(X_{(r)} = x) = P(X_{(r)} \le x) - P(X_{(r)} \le x - 1)$$

$$= [B(r, n-r+1)]^{-1} \int_{F(x-1)}^{F(x)} u^{r-1}\,(1-u)^{n-r}\,du$$

$$= I_{F(x)}(r, n-r+1) - I_{F(x-1)}(r, n-r+1) \qquad \ldots (20.3)$$

Alternate Derivation

Let us divide the real line into three partitions:

$(-\infty, x)$, $[x, x+dx)$, $[x+dx, \infty]$.

Consider the event $x < X_{(r)} \le x + dx$ which may be realized as follows:

$X_i \le x$ for $r - 1$ of the X_i, $x < X_i \le x + dx$ for one X_i, and $X_i > x + dx$ for the remaining $n - r$ of the X_i the number of ways in which n observations can be so divided into three partitions is.

$$\frac{n!}{(r-1)!(n-r)!} = \frac{1}{B(r, n-r+1)}$$

and each such way has probability

$F^{r-1}(x)[F(x+dx) - F(x)] 1 - F(x+dx)]^{n-r}$

Treating dx as small, we have

$$P(x < X_{(r)} \le x + dx) = \frac{1}{B(r, n-r+1)} F^{r-1}(x) f(x)\, dx$$

$$\times [1 - F(x + dx\ 1]^{n-r} + O(dx^2) \quad \text{.... (20.4)}$$

where $O(dx^2)$ means terms of order $(dx)^2$ and includes the probability of realizations of $x < X_{(r)} \le x + dx$ in which more than one X_i is in $(x, x+dx)$. Dividing both sides by dx and letting $dx \to 0$, we have

$$\lim_{dx \to 0} \frac{P(x < X_{(r)} \le x + dx)}{dx} = \frac{1}{B,(r_1\ n-r+1)} F^{r-1}(x) f(x)$$

$$\times \lim_{dx \to 0} [1 - F(x+dx)]^{n-r} + \lim_{dx \to 0} \frac{0(dx^2)}{dx}$$

$$g_r(x) = \frac{1}{B(r_1\ n-r+1)} F^{r-1}(x)[1 - F(x1)]^{n-r} f(x) \quad \text{... (20.5)}$$

which is the required p.d.f. of single order statistic $X_{(r)}$.

Ex. If $\underset{x}{I}(r, s) = B_x(r, s)\ \ B(r, s)$ .... (1), prove that $I_x(r, s) = 1 - I_{1-x}(s, r)$,

Where $B_x(r, s)$ is the incomplete beta function given by (1).

Hint. $\underset{x}{I}\, x(r, s)\, B(r, s) = \underset{x}{B}(r, s) \int_0^x x^{r-1}(1-x)^{s-1}\, dx$

$$= \int_0^1 x^{r-1}(1-x)^{s-1}\, dx - \int_x^1 x^{r-1}(1-x)^{s-1}\, dx$$

$$= B(r, s) - \int_x^1 x^{r-1}(1-x)^{s-1}\, dx$$

Now put $1 - x = y$ so that the R.H.S. integral $= \int_{1-x}^{0} (1-y)^{r-1} y^{s-1}(-dy) = \underset{1-x}{B}(s, r) = \underset{1-x}{I}(s\ r)\, B(s, r)$.

20.2. Joint Distribution of two or more order Statistics

For continuous X, the joint distribution of $X_{(r)}$ and $X_{(s)}$ $(1 \le r \le s \le n)$ is denoted by $g_{rs}(x, y)$, $(x < y)$. The derivation of an expression corresponding to (20.5) is straight forward by noting that the compound event $x < X_{(r)} \le x + dx$, $y < X_{(s)} \le y + dy$ is realized by the configuration meaning that $r - 1$ of the X_r are less than x, one is in $(x, x + dx]$, $s - r - 1$ are in $(x + dx, y]$, one in $(y, y + dy]$ and $x - s$ are in $(y + dy)\ \infty$. It follows that for $x < y$,

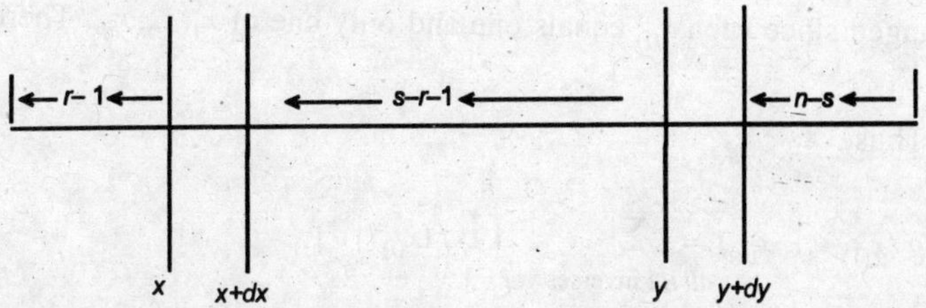

$$g_{rs}(x, y) = \frac{n!}{(r-1)!\,(s-r-1)!\,(n-s)!} F^{r-1}(x)\,[F(y) - F(x)]^{s-r-1}$$

$$\times\ [1 - F(y)]^{n-s} f(x) f(y) \quad \ldots (20.6)$$

Generalizing the result (20.6), the joint p.d.f. of $X_{(r_1)}, X_{(r_2)}, \ldots, X_{(r_k)}$

$(1 \le r_1 < r_2 < \ldots, r_k \le n;\ 1 \le k \le n)$ is

$$P_k(x_{(r_1)}, x_{(r_2)}, \ldots, x_{(r_k)}) = \frac{n!}{\prod_{i=1}^{k+1} (r_i - r_{i-1} - 1)!}$$

$$\prod_{i=1}^{k+1} [F(x_{(r_i)}) - F(x_{(r_i - 1)})]^{r_i - r_{i-1} - 1} \cdot \prod_{i=1}^{k} f(x_{(r_i)}) \text{ for } x_{(r_1)} < \ldots < x_{(r_k)} \quad \ldots (20.7)$$

where $r_0 = 0$, $r_{k+1} = n + 1$ and $x_{(r_0)} = -\infty, x_{(r_k + 1)} = +\infty$

For discrete X, the p.m.f. of $X_{(r)}$ and $X_{(s)}$ is given by

$$P(X_{(r)} = x, X_{(s)} = y) = C_{rs} \int_{F(x-1)}^{F(x)} \int_{F(y-1)}^{F(y)} w^{r-1} (u - v)^{s-r-1} (1 - v)^{n-s}\, dv dw \quad \ldots (20.8)$$

where $C_{rs} = \dfrac{n!}{(r-1)!\,(s-r-1)!\,(n-s)!}$

which is an expression corresponding to (20.3).

Theorem. The joint pdf of $X_{(1)}, X_{(2)}, \ldots, X_{(n)}$ is given by

$$g(x_{(1)}, x_{(2)}, \ldots, x_{(n)}) = \begin{cases} n! \prod_{i=1}^{n} f(x_{(i)}, x_{(1)} \le x_{(2)} \le \ldots \le x_{(n)} (x) \\ 0, \text{ otherwise}. \end{cases}$$

Proof. The transformation from $(X_1, X_2, \dots X_n)$ to $(X_{(1)}, X_{(2)}, \dots, X_{(n)})$ is not one-to-one. In fact there is a total of $n!$ possible arrangements of $x_1, x_2, \dots, x_n$ in increasing order of magnitude.

Thus there are $n!$ inverses to the transformation. The Jacobian of this transformation is the determinant of an $n \times n$ identity matrix with row rearranged since each $x_{(i)}$ equals one and only one of $x_1, \dots, x_n$. Therefore $J = \pm 1$

Thus

$$g(x_{(1)}, \dots x_{(n)}) = \sum_{\text{all } n! \text{ inverses}} \prod_{i=1}^{n} f(x_{(i)}) \, |J|$$

$$= \begin{cases} n!\, f(x_{(1)}), f(x_{(2)}) \dots f(x_{(n)}), & \text{if } x_{(1)} \le x_{(2)} \le \dots \le x_{(n)} \\ 0, & \text{otherwise} \end{cases}$$

Example. *Let $X_1, X_2, \dots, X_n$ be independent identically distributed random variables with common p.d.f.*

$$f(x) = \begin{cases} 1, & \text{if, } 0 < x < 1 \\ 0, & \text{otherwise} \end{cases}$$

The joint p.d.f. of $X_{(1)}, \dots, X_{(n)}$ is

$$g(x_{(1)}, \dots, x_{(n)}) = \begin{cases} n! & \text{if } 0 < x_{(1)} < x_{(2)} < \dots < x_{(n)} < 1 \\ 0, & \text{otherwise} \end{cases}$$

20.3. Distribution of the Range; $x_{(n)} - x_{(1)}$

The joint p.d.f. of $X_{(1)}$ and $X_{(n)}$ is given by

$$g(x_{(1)}, x_{(n)}) = n(n-1)\,[F(x_{(n)}) - F(x_{(1)})]^{n-2} f(x_{(1)}) f(x_{(n)})$$

for $x_{(1)} < x_{(n)}$ .... (20.9)

Applying the transformation

$$x_{(n)} - x_{(1)} = r, \; x_{(1)} = y$$

The joint p.d.f. of R and Y is

$$h(r, y) = n(n-1)\,[F(r+y) - F(y)]^{n-2} f(y) f(r+y) \qquad \dots (20.10)$$

The p.d.f. of R is

$$g(r) = n(n-1) \int_{\infty}^{-\infty} [F(r+y) - F(y)]^{n-2} f(y) f(r+y)\, dy \qquad \dots (20.11)$$

The distribution function of the range R is given by

$$G(r) = P(R \le r) = \int_0^r n(n-1) \int_{-\infty}^{\infty} [F(r+y) - F(y)]^{n-2} f(y) f(r+y)\, dy\, dr$$

$$= \int_{-\infty}^{\infty} [F(r+y) - F(y)]^{n-1} f(y)\, dy \quad \ldots$$

(20.12)

20.4.0. Quartiles

Assume that X is a continuous variate with strictly increasing cumulative distribution function $F(x)$. Then the equation

$$F(x) = p,\ 0 < p < 1$$

has a unique solution, say $x = \xi_p$, which is called the population quartile of order p. Thus $\xi_{\frac{1}{2}}$ is the median of the distribution. If $F(x)$ is not strictly increasing, $F(x) = p$ may hold in some interval in which case any point in the interval would serve as a quartile of order p.

When X is discrete variate ξ_p is defined by $P(x < \xi_p) \le p \le P(X \le \xi_p)$.

A sample quartile of order p $(0 < p < 1)$, denoted by Z_p is the order statistic $X_{(r)}$, where

$$r = \begin{cases} np \text{ if } np \text{ is an integer} \\ [np + 1], \text{ if } np \text{ is not an integer} \end{cases}$$

where $[x]$ is the largest integer $\le x$. If np is an integer, we can take any value between $X_{(np)}$ $X_{(np)+1}$ as the p-th sample quartile. Thus, if $P = \frac{1}{2}$ and n is even we can take any value between $X_{(\frac{n}{2})}$ and $X_{(\frac{n}{2})+1}$, the two midle values as the median. It is customary to take the average of these two as median. Thus the sample median is defined as

$$Z_{\frac{1}{2}} = \begin{cases} X_{(\frac{n+1}{2})} \text{ if } n \text{ is odd} \\ \dfrac{X_{(\frac{n}{2})} + X_{(\frac{n}{2})+1)}}{2} \quad \text{if } n \text{ is even} \end{cases}$$

Note that $\left[\frac{n}{2} + 1\right] = \left(\frac{n+1}{2}\right)$ if n is odd.

20.4.1. Distribution of Sample Median

Case I. n is odd, say $n = 2m + 1$.

The distribution of the sample median is the same as the distribution of $(m + 1)$th order statistic in a sample of size $(2m+1)$. The p.d.f. of $X_{(m+1)}$ is given by

$$g(z) = \frac{(2m+1)!}{(m!)^2} F^m(z)\,[1 - F(z)]^m f(z) \qquad \dots (20.13)$$

where F and f are the population distribution function and probability density function respectively.

Case II. n is even, say $n = 2m$, and the median is taken as the average of $X_{(m)}$ and $X_{(m+1)}$.

The joint p.d.f. of $X_{(m)}$ and X_{m+1} in a sample of size $2m$ is given by for $y > x$.

$$g(x, y) = \frac{(2m)!}{[(m-1)!]^2} F^{m-1}(x)\,[1 - F(y)]^{m-1} f(x) f(y).$$

To obtain the distribution of $Z = \dfrac{X_{(m)} + X_{(m+1)}}{2}$, we put

$$\begin{cases} \dfrac{x+y}{2} = z, \\ y = v \end{cases}$$

with $|J| = 2$.

The p.d.f. of Z and V is given by

$$h(z, v) = \frac{2\,(2m)!}{[(m-1)!]^2} F^{m-1}(2z - v)\,(1 - F(v))^{m-1} f(2z - v) f(v)$$

The p.d.f. of Z is given by

$$h(z) = \frac{2\,(2m)!}{[(m-1)!]^2} \int_z^{\infty} F^{m-1}(2z - v)\,[1 - F(v)]^{m-1} f(2z - v) f(v)\, dv$$

20.4.2. Asymptotic Distribution of Sample Quartile

The quartile of order p of the sample is the order statistic $X_{(r)}$, where r is the largest integer less than $(np + 1)$.

In considering the asymptotic distribution of the order statistic $X_{(r)}$, two cases may be distinguished:

(*i*) $\dfrac{r}{n} \to 0$ or 1, and

(*ii*) $\dfrac{r}{n} \to p,\ 0 < p < 1$ as $n \to \infty$

The first case will give us the asymptotic distributions of extremes. The second case will give the asymptotic distributions of quartiles.

Theorem. *If, in some neighbourhood of the population quartile ξ_p, the p.d.f. f is continuous and has a continuous derivatives and if*

$f(\xi_p) > 0$, then the distribution of the sample quartile Z_p is asymptotically normal with mean ξ_p and variance $p\ (1-p)/\ nf^2(\xi_p)$

Proof. Let $r = np$ or $[np]+1$ according as np is an integer or not. Then the sample quartite of order p is $Z_p = X_{(r)}$. The p.d.f. of Z_p is

$$g(z_p) = \frac{n!}{(r-1)!\,(n-r)!} F^{r-1}(z_p)\,[1-F(z_p)]^{n-r} f(z^p) \qquad \text{... (20.14)}$$

Let us make the transformation

$Y = \sqrt{\frac{n}{pq}}\ f(\xi_p)\ (Z_p - \xi_p)$, where $q = 1-p$

The p.d.f. of Y is

$$g_n(y) = \frac{n!}{(r-1)!\,(n-r)!} \frac{\sqrt{pq}}{n} \frac{1}{f(\xi_p)} \left[F\left(\xi_p + \sqrt{\frac{pq}{n}} \frac{y}{f(\xi_p)}\right)\right]^{r-1}$$

$$\times \left[1 - F\left(\xi_p + \sqrt{\frac{pq}{n}} \frac{y}{f(\xi_p)}\right)\right]^{n-r} f\left(\xi_p + \sqrt{\frac{pq}{n}} \frac{y}{f(\xi_p)}\right) \qquad \text{... (20.15)}$$

$$= \frac{1}{\sqrt{n}} \binom{n}{r} r\, p^{r-\frac{1}{2}} q^{n-r+\frac{1}{2}} \left[\frac{F\left(\xi_p + \sqrt{\frac{pq}{n}} \frac{y}{f(\xi_p)}\right)}{p}\right]^{r-1}$$

$$\times \left[\frac{1 - F\left(\xi_p + \sqrt{\frac{pq}{n}} \frac{y}{f(\xi_p)}\right)}{q}\right]^{n-r} \cdot \frac{F\left(\xi_p + \sqrt{\frac{pq}{n}} \frac{y}{f(\xi_p)}\right)}{f(\xi_p)}$$

$= A_1 . A_2\ A_3$, say. ...(20.16)

$$\text{Now } A_1 = \frac{r\binom{n}{r}}{\sqrt{n}} p^{r-\frac{1}{2}} q^{n-r+\frac{1}{2}}$$

$$= \frac{r\,(n)!}{r!\,(n-r)!} \left(\frac{r}{n}\right)^{r-\frac{1}{2}} \left(1 - \frac{r}{n}\right)^{n-r+\frac{1}{2}}$$

$\rightarrow \frac{1}{\sqrt{2\pi}}$, as $n \rightarrow \infty$ by applying Stirling's approximations.

or $\log A_1 \rightarrow -\log\sqrt{2\pi}$ ...(20.17)

$$A_3 = f\left(\xi_p + \sqrt{\frac{pq}{n}} \frac{y}{f(\xi_p)}\right)/ f(\xi_p) = f(\xi_p) + \sqrt{\frac{pq}{n}} \frac{y}{f(\xi_p)} f'(\xi_p) + 0\left(\frac{1}{n^{3/2}}\right) / f(\xi_p)$$

$\rightarrow 1$ as $n \rightarrow \infty$ .... (20.18)

or $\log A_3 \to 0$ as $n \to \infty$

Further,

$$\log A_2 = (r-1)\log\left[\frac{F\left(\xi_p + \sqrt{\frac{pq}{n}}\frac{y}{f(\xi_p)}\right)}{p}\right] + (n-r)\log\left[\frac{1 - F\left(\xi_p + \sqrt{\frac{pq}{n}}\frac{y}{f(\xi_p)}\right)}{q}\right] \qquad \text{.... (20.19)}$$

$\to -\frac{y^2}{2}$ as $n \to \infty$, using Taylor's theorem.

Hence as $n \to \infty$, the p.d.f. of y is

$$g(y) = \frac{1}{\sqrt{2\pi}} e^{-\frac{1}{2}y^2}$$

which is the p.d.f. of a standard normal variate.

Also A_1, A_2, A_3 are uniformly bounded in any interval (a, b). So Dominated convergence theorem, we have

$$P(a < y < b) \to \frac{1}{\sqrt{2\pi}} \int_a^b \exp\left(-\frac{1}{2}y^2\right) dy \qquad \text{... (20.20)}$$

i.e., the sample quartile Z_p is asymptotic normal with mean ξ_p and variance $pq/nf^2(\xi_p)$.

In particular, the sample median is asymptotically normal with the population median $\xi_{0.5}$ and $1/4nf^2(\xi_{0.5})$ as the variance.

20.4.3. Distribution-free confidence intervals for Quartiles.

Let ξ_p be the population quartile of order p $(0 < p < 1)$. first assume that X is continuous random variable with distribution function $F(x)$. We are to show that the interval $(X_{(r)}, X_{(s)})$ covers ξ_p with a probability which depends on r, s, n and p, but not on $F(x)$.

Note that the event $X_{(r)} \le \xi_p$ is the union of the disjoint compound events $X_{(r)} \le \xi_p$, $X_{(s)} \ge \xi_p$ and $X_{(r)} \le \xi_p, X_{(s)} < \xi_p$., Thus, since $X_{(s)} < \xi_p$ implies $X_{(r)} \le \xi_p$, we have whether X is continuous or not $P(X_{(r)} \le \xi_p) = P(X_{(r)} \le \xi_p \le X_{(s)}) + P(X_{(s)} < \xi_p)$.

or $P(X_{(r)} \le \xi_p \le X_{(s)}) = P(X_{(r)} \le \xi_p) - P(X_{(s)} < \xi_p)$

It follows that $(X_{(r)}, X_{(s)})$ covers ξ_p with probability $\Pi(r, s, n, p)$ given by

$$\Pi(r, s, n, p) = I_p(r, n-r+1) - I_p(s, n-s+1)$$

$$= \sum_{i=r}^{s-1} \binom{n}{i} p^i (i-p)^{n-i}$$

which does not depend on $F(x)$.

In the discrete case, $P(X < \xi_p) \leq p$ and $P(X, \leq \xi_p) \geq p$ and implies that

$$P(X_{(r)} < \xi_p) \leq I_p (r, n-r+1)$$

$$P(X_{(s)} \leq \xi_p) \geq I_p (s, n-s+1)$$

so that $P(X_{(r)} \leq \xi_p \leq X_{(s)}) \geq \Pi (r, s, n, p)$

and $P(X_{(r)} < \xi_p < X_{(s)}) \leq \Pi (r, s, n, p)$

Thus the lower and upper bounds are distribution-free.

20.5. Distribution-free Tolerance Intervels

A tolerance interval has random terminals, say L and V. The requirement of a tolerance interval (L, V) is that it contains atleast a proportion γ of the population with probability β, both β and γ being preassigned constants $(0 \leq \beta, \gamma \leq 1)$. Thus, if $f(x)$ is continuous we seek L, V such that

$$P\left(\int_L^V f(x)\,dx \geq \gamma)\right) = \beta$$

If we choose L and V, the r-th and s-th order statistics respectively, then we have

$$P\ (F(X_{(s)}) - F(X_{(r)}) \geq \gamma) = \beta$$

But $F(X_{(s)})$ and $F(X_{(r)})$ and just the order statistics $U_{(s)}$ and $U_{(r)}$ corresponding to a uniform distribution in $(0, 1)$.

Thus

$$P(U_{(s)} - U_{(r)} \geq \gamma) = 1 - I_\gamma (s-r, n-s+r+1) = \beta$$

which is independent of $F(x)$.

For $r = 1$, $s = n$, we have

$$1 - I_\gamma (n-1, 2) = \beta$$

or $$\frac{n!}{(n-2)!} \int_0^\gamma u^{n-2} (1-u)\ du = 1 - \beta$$

or $$n\gamma^{n-1} - (n-1)\gamma^n = \beta$$

For $\gamma = 0.95$, $\beta = 0.90$ we find $n = 77$.

Example 1. *Show that the sampling distribution of the largest observation $X_{(n)}$ from the rectangular distribution $R\ (0, \theta)$ has the probability density function given by $n\, x^{n-1\,(n)}/\theta^n$ in $(0, \theta)$ and 0, outside.*

Sol.

$$\text{Give } f(x) = \begin{cases} \frac{1}{\theta}, & 0 < x < \theta \\ 0, & \text{otherwise} \end{cases}$$

The distribution function of X is

$$F(x) = \int_0^x \frac{1}{\theta}\, dx = \frac{x}{\theta}$$

The distribution function of $X_{(n)}$ is

$$G(x_{(n)}) = [F(x_{(n)})]^n = \left(\frac{x_{(n)}}{\theta}\right)^n$$

The p.d.f. of $X_{(n)}$ is

$$g(x_{(n)}) = n\, \frac{x_{(n)}^{x-1}}{\theta^n},\ 0 < x_{(n)} < \theta$$

$$= 0, \text{ otherwise}$$

Example 2. *Show that for the uniform distribution over (0, 1) the correlation coefficient between* $X_{(r)}$ and $X_{(s)}$ *for* $r < s$ *is*

$$\left[\frac{r(n-s+1)}{s(n-r+1)}\right]^{\frac{1}{2}}$$

Hence show that the correlation coefficient between the mininum and maximum values in a sample of size n from a uniform distribution over (0, 1) is 1/n.

Sol. The joint p.d.f. of $X_{(r)}$ and $X_{(s)}$, $(r < s)$, when the sampling is done from uniform population over (0, 1) with $F(x) = \int_0^x dx = x$ is

$$g(x_{(r)}, x_{(s)}) = \begin{cases} \dfrac{n!}{(r-1)!\,(s-r-1)!\,(n-s)!}\, x_{(r)}^{r-1}\, (x_{(s)} - x_{(r)})^{s-r-1}\, (1 - x_{(s)})^{n-s}, \\ \qquad\qquad 0 < x_{(r)} < x_{(s)} < 1 \\ 0, \text{ otherwise} \end{cases}$$

The expected, value of $X_{(r)}$ is

$$E(X_{(r)}) = \int_0^1 x_{(r)} \cdot g_{(x_r)}\, dx_{(r)} = \frac{n!}{(r-1)!\,(n-r)!} \int_0^1 x \,.\, g(x_r)\, dx_r$$

$$= \frac{n!}{(r-1)!\,(n-r)!} \int_0^1 x_{(r)} \cdot x_{(r)}\, (1 - x_{(r)})^{n-r}\, dx\,(r)$$

$$= \frac{n!}{(r-1)!\ (n-r)!}\ B(r+1, n-r+1)$$

$$= \frac{n!}{(r-1)!\,(n-r)!}\ \frac{r!(n-r)!}{(n+1)!} = \frac{r}{n+1}$$

Similarly $E(X_{(s)}) = \dfrac{s}{n+1}$

The expected value of $X_{(r)}\ X_{(s)}$ is

$$E(X_{(r)} X_{(s)}) = \frac{n!}{(r-1)!\,(s-r-1)\ !\,(n-s)!} \int_0^1 \int_0^{x(s)} x_{(r)}\, x_{(s)}\, x_{(r)}^{r-1}\, (x_{(s)} - x_{(r)})^{s-r-1}$$

$$\times (1 - x_{(s)})^{n-s}\ dx_{(r)}\ dx_{(s)}$$

$$= \frac{n!}{(r-1)!\ (s-r-1)!\,(n-s)!} \cdot \frac{r!\,(s-r-1)!}{s!} \int_0^1 x_{(s)}^{s+1}\ (1 - x_{(s)})^{n-s}\ dx\ (s)$$

$$= \frac{n!}{(r-1)!\ (s-r-1)!\,(n-s)!} \cdot \frac{r!\,(s-r-1)!}{s!} \cdot \frac{(s+1)!\,(n-s)!}{(n+2)}$$

$$= \frac{r(s+1)}{(n+1)(n+2)}$$

Further, we have

$$Var(X_{(r)}) = \frac{r(n-r+1)}{(n+1)^2\,(n+2)}.$$

$$Var\ (X_{(s)}) = \frac{s(n-s+1)}{(n+1)^2\,(n+2)}.$$

The correlation coefficient between $X_{(r)}$ and $X_{(s)}$ is

$$P = \frac{Cov(X_{(r)}, X_{(s)})}{\sqrt{Var(X_{(r)})\,.\,Var(X_{(s)})}}$$

$$= \frac{\dfrac{r(s+1)}{(n+1)(n+2)} - \dfrac{r}{(n+1)} \cdot \dfrac{s}{(n+1)}}{\sqrt{\dfrac{r(n-r+1)\ s(n-s+1}{(n+1)^2\,(n+2)}}}$$

$$= \left[\frac{r(n-s+1)}{s(n-r+1)}\right]^{\frac{1}{2}}$$

The correlation coefficient between $X_{(1)}$ and $X_{(n)}$ is

$\rho = \dfrac{1}{n}$ for $r = 1$ and $s = n$.

Example 3. *Let $X_{(1)}, X_{(2)}, X_{(3)}$ be the order statistics of i.i.d. r.v.'s X_1, X_2, X_3 with common p.d.f.*

$$f(x) = \begin{cases} Be^{-x\beta}, \ x \geq 0, \ \beta > 0 \\ 0, \text{ otherwise} \end{cases}$$

Show that $X_{(3)} - X_{(2)}$ and $X_{(2)}$ are independent. Deduce the distribution of $Y_1 = X_{(3)} = X_{(2)}$.

Sol. The distribution function of X is given by

$$F(x) = \int_0^x \beta\, e^{-x\beta}\, dx = 1 - e^{-x\beta}$$

The joint p.d.f. of $X_{(2)}$ and $X_{(3)}$ is given by

$$g(x_{(i)}, x_{(3)}) = \begin{cases} \dfrac{3\,!}{1\,!\,01\,!} - (1 - e^{\beta x_{(2)}})\,.\, \beta\, e^{-\beta x_{(2)}}\, \beta\, e^{-\beta x_{(3)}}, \\ \qquad\qquad 0 < x_{(2)} < x_{(3)} < \infty \\ 0, \text{ otherwise} \end{cases}$$

$$= \begin{cases} 6\, \beta^2\, (1 - e^{-\beta x(2)})\ e^{-\beta (x_{(2)} + x_{(3)})} \\ \qquad\qquad , 0 < x_{(1)} < x_{(3)} < \infty \\ 0, \text{ otherwise} \end{cases}$$

Let us make the transformation

$$x_{(3)} - x_{(2)} = y_1,$$

$$x_{(2)} = y_2.$$

Clearly $|J| = 1$

The p.d.f. of Y_1 and Y_2 is

$$h(y_1, y_2) = 6\, \beta^2\, (1 - e^{-\beta y_2})\ e^{-\beta (y_1 + 2y_2)}$$

$$= 3\,!\, \beta\, e^{-2\beta y_2}\, (1 - e^{-\beta y_2})\, (\beta\, e^{-\beta y_1})\ ; \quad 0 < y_1, y_2 < \infty$$

It follows that Y_1 and Y_2 are independent.

Example 4. *Obtain the density function of the sample range from a sample of size n from the rectangular distribution having the p.d.f.*

$$f(x) = \begin{cases} \dfrac{1}{\theta}, \ \textbf{if } a - \dfrac{\theta}{2} < x < a + \dfrac{\theta}{2} \\ 0, \textbf{ otherwise} \end{cases}$$

The distribution function of X is given by

$$F(x) = \int_{a - \frac{\theta}{2}}^{x} \frac{1}{\theta}\, dx = \frac{1}{\theta}\left[x - (a - \frac{\theta}{2})\right]$$

The joint p.d.f. of $X_{(1)}$ and $X_{(n)}$ is given by

$$g(x_{(1)}, x_{(n)}) = n(n-1)\,\frac{[(X_{(n)} - (a - \frac{\theta}{2}) - (x_{(1)} + a - \frac{\theta}{2})]^{n-2}}{\theta^{n-2}} \cdot \frac{1}{\theta^2}$$

$$= \frac{n(n-1)}{\theta^n}\,(x_{(n)} - x_{(1)})^{n-2}$$

Let us make the transformation

$x_{(n)} - x_{(1)} = w, \quad x_{(1)} = y,$ with $|J| = 1$

The joint p.d.f. of W and Y is

$$h(w, y) = \frac{n(n-1)}{\theta^n}\,w^{n-2}$$

The p.d.f. of w is

$$h(w) = \frac{n(n-1}{\theta^n}\,w^{n-2}\int_{a-\frac{\theta}{2}}^{a+\frac{\theta}{2}-w} dy$$

$$= \frac{n(n-1)}{\theta^n}\,w^{n-2}(\theta - w), \; 0 < w < \theta.$$

Example 5. *Let $X_1, X_2 \ldots, X_n$ $(n = 2m)$ be a random sample from $U(0, 1)$ population. Obtain the distribution of the sample median.*

The joint p.d.f. of $X_{(m)}$ and $X_{(m+1)}$ is given by

$$g(x_1\, y) = \frac{(2m)!}{[(m-1)!]^2}\,x^{m-1}(1-y)^{m-1}, \; 0 < x < y < 1$$

Make the transformation

$$\frac{x+y}{2} = z,$$

$y = v.$

$|J| = 2$

$y > x \Rightarrow v > z$

$x > 0 \Rightarrow z > \frac{v}{2}$

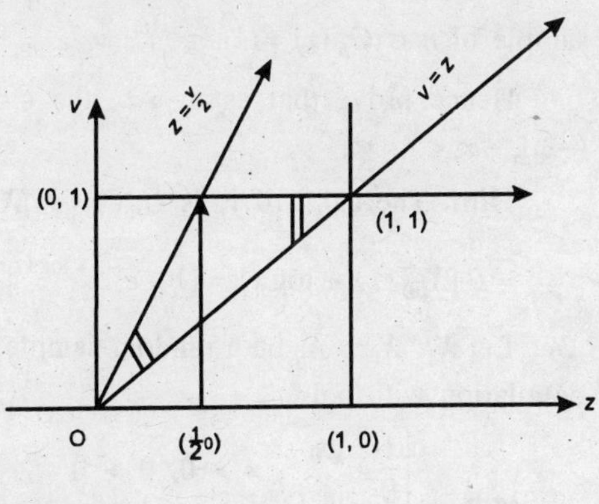

Thus

$$z > v < 2z \text{ if } 0 < z \le \frac{1}{2}$$

$$z < v < 1 \text{ if } \frac{1}{2} < z < 1$$

Thus the p.d.f. of the median of given by

$$\frac{2\,(2m)\,!}{[(m-1)\,!]^2}\int_z^{2z}(2z-v)^{m-1}\,(1-v)^{m-1}\,dv,\ 0<z\leq\frac{1}{2}$$

$$\text{and}\ \frac{2\,(2m)\,!}{[(m-1)\,!]^2}\int_z^{1}(2z-v)^{m-1}\,(1-v)^{m-1}\,dv\ \text{for}\ \frac{1}{2}<z<1$$

In particular, if $m=2$, the p.d.f. is

$$48\int_z^{2z}(2z-v)\,(1-v)\,dv=8z^2\,(3-4z),\ \text{if}\ 0<z\leq\frac{1}{2}$$

$$\text{and}\ 48\int_z^{1}(2z-v)\,(1-v)\,dv=8\,(6z-1-9z^2+4z^3),\ \text{if}\ \frac{1}{2}<z<1$$

Exercise 20

1. Let $X_1, X_2 ......, X_n$ be independent random variates, X_i having a geometric distribution with probability mass function with parameter p_i *i.e.*

$$P\,(X_i=x_i)\ =\ q_i^{x_i-1}\,p_i;\ x_i=1,2,...,q_i=1-p_i$$

Show that $X_{(1)}$ is distributed geometrically with parameter $1-q_1\,q_2\,....\,q_n$. **(M.Sc. Statistics,. Delhi, 1983)**

2. For the exponential distribution with p.d.f.

$$f(x)\ =\ \begin{cases} e^{-x}, x\geq 0 \\ 0\ \ x<0 \end{cases}$$

Show that the cumulative distribution function of $X_{(n)}$ in a random sample of n is $G_n\,(x)=(1-e^{-x})^n$.

Hence prove that as $n\to\infty$, the c.d.f. of $X_{(n)}-\log n\to\exp[-\exp(-x)]$, $-\infty<x<\infty$.

Hint. The c.d.f. of y_n is $G_n\,(\,)=P\,[X_{(n)}'-\log x\leq x]$

$$=P\,[X_{(n)}\leq x+\log x]=[1-e^{-(x+\log x)}]^n=(1-\frac{1}{n}\,e^{-x})\to\exp\,(-e^{-x}).$$

3. Let $X_1,\ X_i ...,\ X_n$ be a random sample of size n from the exponential population with p.d.f.

$$f(x)\ =\ \begin{cases} \dfrac{1}{\theta}\,e^{-x/\theta}\ \ ,\ x>0,\ \theta>0 \\ 0,\ \ x<0 \end{cases}$$

(a) Find the p.d.f. of $X_{(r)}$.

(b) Show that $X_{(r)}$ and $X_{(s)} - X_{(r)}$, $(s > r)$ are independently distributed.

(c) Show that the distribution of $X_{(r+1)} - X_{(r)}$ is exponential with parameter $(n - r)$.

(d) If $f(x) = 1/\theta_2, \theta_1 - \frac{1}{2}\theta_2 \le \theta_1 + \frac{1}{2}\theta_2$, show that

$$E\left(\frac{X_{(r)} - \theta_1}{\theta_2}\right) = \frac{r}{n+1} - \frac{1}{2}.$$

4. Show that the distribution function of the midrange

$M = \frac{1}{2}(X_{(1)} + X_{(n)})$ in random sample of n from a continuous population with distribution function $F(x)$ is

$$G(m) = P(M \le m) = n \int_{-\infty}^{m} [F(2m - x) - F(x)]^{n-1} f(x)\, dx$$

5. Show that the joint p.d.f. of range R and midrange M in random sample of size n from a uniform population over the interval $(-\frac{1}{2}, \frac{1}{2})$ is

$g(r, m) = n(n-1)\, r^{n-2}, 0 \le r \le 1 - 2\,|\,m\,| \le 1$

Hence show that the p.d.f. of M is

$g(m) = n\,[1 - 2\,|\,m\,|]^{n-1}, |\,m\,| \le 1.$

6. Let X_1, X_2, X_3 be a random sample of size 3 from a distribution having p.d.f. (*i*) $f(x) = 5x^4, 0 < x < 1$, zero elsewhere

(*ii*) $e^{-x/\theta}/\theta, x > 0, \theta > 0$, zero otherwise. Let Y be the largest item in the sample.

(*i*) Find the distribution function and p.d.f. of Y.

(*ii*) Show that $Y_1 = X_{(3)} - X_{(2)}$ and $Y_2 = X_{(2)}$ are independent.

7. Let $X_i, i = 1, 2, ..., n$ be i.i.d. non-negative r.v.'s of continuous type. If $M_n = X_{(n)} = Max(X_1, X_2, ..., X_n)$ and $E(|X|) < \infty$, then prove that.

$$E(M_n) \le E(M_{n-1}) + \int_0^{\infty} F^{n-1}(x)\,[1 - F(x)]\, dx.$$

(B.Sc. Stat. Hons. Delhi, 1990)

Hint. $E(M_n) = \lim_{a \to \infty} \int_0^a x f_x(x)\, dx$ where $f_n(x) = n\,[F(x)]^{n-1} . f(x)$

Intergrate by parts the R.H.S. and show that

$$E(M_x) = \lim_{a \to \infty} \left[\int_0^a (1 - F^n(x))\, dx - a\,(1 - F^n(a)) \right]$$

Now $E(X_{(r)}) < \infty$ if $E(x) < \infty$ since $E|X_{(r)}| = \int_0^\infty |x| f_r(x)\, dx$

$$= \int_0^\infty |x|, \frac{n!}{(r-1)!\,(n-r)!} f(x)\, F^{r-1}(x)\,(1 - F(x))^{n-r}\, dx.$$

$$\le n \binom{n-1}{r-1} \int_0^\infty |x| f(x)(dx \le n \binom{n-1}{r-1} E|X|;$$

Also it can be easily shown that $1 - F^n(a) \to 0$ as $a \to \infty$ so that

$$E(M_n) = \int_0^\infty 1 - F^n(x)\, dx = \int_0^\infty 1 - F^{n-1}(x).\,[1 - (1 - F(x)]\; dx$$

$$= E(M_{n-1}) + \int_0^\infty F^{n-1}(x)\;[1 - F(x)]dx$$

8. Show that for a random sample of size 2 from $N(0, \sigma^2)$ population, $E(X_{(1)}) = -\sigma/\sqrt{\pi}$ ***(M.Sc. Stat. Delhi, 1982, 88 Roorkee, 1992)***

Hint. For $n = 2$, $f_1(x) =$ p.d.f. of $X_{(1)} = \dfrac{1}{\beta(1, 2)}\,[1 - F(x)]\, f(x)$

$= 2\,[1 - F(x)]\, f(x)$ where $f(x) = e^{-x^2/2\sigma^2}/\sigma\sqrt{2\pi}$

$$\therefore E(X_{(1)}) = \int_{-\infty}^{+\infty} x f_1(x)\, dx = 2\int_{-\infty}^{+\infty} [1 - F(x)].\, x f(x)\, dx = 2\,(1 - F(x)](-\sigma^2 f(x)]_{-\infty}^{+\infty}$$

$$= -2\sigma^2 \int_{-\infty}^{+\infty} [f(x)]^2 = \frac{1}{\pi} \int_{-\infty}^{+\infty} e^{-x^2/\sigma^2}\, dx = \frac{-\sigma}{\sqrt{\pi}}$$

as $\log f_{(\lambda)}$ = const. $-x^2/2\sigma^2$ so that $f'(x)/f(x) = -x/\sigma^2 \Rightarrow \int x f(x)$ $= -\sigma^2 \int f'(x)\, dx = \ldots = -\sigma^2 f(x)$.

9. Let $X_{(1)} < X_{(2)} < X_{(3)} < X_{(4)} < X_{(5)}$ denote the order statistics of a random sample of size 5 from a continuous population. Compute: (*a*) $P(X_{(1)} < \xi_{0.5} < X_{(5)})$, (*b*) $P(X_{(1)} < \xi_{0.25} < X_{(3)})$.

(*c*) $P(X_{(4)} < \xi_{0.80} < X_{(5)})$.

10. Find the smallest value of n for which

$P(X_{(1)} < \xi_{0.5} < X_{(x)}) \ge 0.99$ where

$X_{(1)} < X_{(2)} .... < X_{(n)}$ are the order statistics of a random sample of size n from a continuous distribution.

11. Let $X_1, X_2...., X_n$ be r.v.s. with pdf

$$= \begin{cases} \frac{1}{\alpha} e^{-(x-\theta)/\alpha} & f(x), \text{ if } x \geq \theta \\ 0 & \text{if } x < \theta \end{cases}$$

Show that $X_{(1)}, X_{(2)} - (X_{(1)}, ..., X_{(n)} - X_{(n-1)})$ are independent.

Hint. Let $Y_1 = X_{(1)}, Y_2 = X_{(2)} - X_{(1)}, ..., Y_n = X_{(n)} - X_{(n-1)}$

$\therefore X_{(1)} = Y_1, X_{(2)} = Y_1 + Y_2, X_{(n)} = Y_1 + Y_2 + ... Y_n$

Hence the p.d.f. of $Y_1, Y_1, ..., Y_n$ is given by

$$g(y_1, y_2, ..., y_n) = n! \prod_{i=1}^{n} f(x_i) \mid J \mid$$

Where $J = \partial \begin{pmatrix} x, & x, & ...., & x \\ (1) & (2) & & (n) \end{pmatrix} \mid \delta (Y_1, Y_2, ... Y_n) \mid = 1$

$$= \frac{n}{\sigma} e^{-n(y-\theta)}/\sigma \cdot \frac{n-1}{\sigma} e^{-(x-1)y_2/\sigma} ..... \frac{2}{\sigma} e^{-2y_2/\sigma} \cdot \frac{1}{\sigma} e^{-y/\sigma}$$

Thus $Y_1, Y_2, ..., Y_n$ are independently distributed.

12. Let X_1 and X_2 be independent r.v. with common p.m.f.

$P(X = x) = p(1-p)^{x-1}, x = 1, 2, ..., 0 < p < 1$

show that $X_{(1)}$ and $X_{(2)} - X_{(1)}$ are independent.

13. Let $X_{(1)}, X_{(2)}, ..., X_{(n)}$ be the ordered variates from a population with p.d.f. $f(x) = 1$ if $0 < x < 1$ and $= 0$ otherwise. Show that $Y_1 = X_{(1)}/X_{(2)}, Y_2 = X_{(2)}/X_{(3)}$ ...... $Y_{n-1} = X_{(n-1)}/X_{(n)}$ and $Y_n = X_{(n)}$ are independent.

14. Let $X_1, X_2, ..., X_n$ be i.i.d with a distribution function $F(y) = y^{\alpha}; 0 < y < 1, \alpha > 0$. Show that $X_{(i)}/X_{(n)}, i = 1, 2, ..., n-1$ and $X_{(n)}$ are independent. ***(B.Sc. Stat. Hons. Delhi 1990)***

15. Let $x_{(1)}, x_{(2)}, ..., x_{(i)} = 1, 2, .., k$ be k random samples from rectangular population with range $(\frac{1}{2}, 1)$ Find the distribution of $x_{1(x)} \cdot x_{2(x)} ...... x_{k(x)}$ where $x_{i(x)}$ is the max of ith ample.

(M.Sc. Stat. Delhi, 1989)

16. Let X_1, X_2 be i.i.d r.v.'s from Posison distribution with parameter m. Let $M = \max(X_1, X_2)$ and $N = Min(X_1, Y_1)$. Find the marginal p.m.f.'s of M and N.

17. Show by means of an example that there may exist a r.v. X for which $E(X)$ does not exist but $E(X_{(r)})$ exists for some r.

Hint. Let $X_1, X_2, \ldots, X_n$ be a random sample from the population with p.d.f. $f(x) = 1/x^2, 1. < x < \infty,\ F(x) = 1 - 1/x$.

Clearly $E_{(X)}$ does not exist though $E(X_{(r)})$ exists some r, say $r \geq 2$.

18. Let X_1, X_2, X_3 and X_4 be independent random variables with distribution functions respectively $F_1(.), F_2(.), F_3(.)$ and $F_4(.)$ If w is defined by

$W = Min\ [Max\ (X_1, X_2), Max\ (X_3, X_4)]$,

Find the distribution of w.

Find also the joint distribution of Main (x_1, x_2) and $Min\ (x_2, x_3)$.

Miscellaneous Exercises

1. Let (X, Y) be a bivariate normal random variable with $EX = EY = 0$, Var X = Var $Y = 1$ and cov $(X, Y) = \rho$. Show that the random variable $Z = Y/X$ has a Cauchy distribution. ***(B.A. Hons. Mathematics Delhi, 1989)***

[**Hint.** Let $W = XY, X = \dfrac{W}{1+Z}, Y = \dfrac{WZ}{1+Z}, \left|\dfrac{\partial(x, y)}{\partial(W, Z)}\right| = \ldots = |w|/(1+Z)^2$

$$f_{xy}(x, y)\, dxdy = \frac{1}{2\pi\sqrt{1-\rho^2}} \exp.\left[-\tfrac{1}{2}(x^2 + y^2 - 2\rho xy)\right] dxdy = \text{etc.}$$

Now integrate for w from $-\infty$ to $+\infty$ to obtain the distribution of Z as $\dfrac{1}{\pi}\cdot\dfrac{1}{1+Z^2}$.

2. Let
$$f(x, y) = \tfrac{1}{2}\left[\frac{1}{2\pi\sqrt{1-\rho^2}} \exp\left(\frac{-1}{2(1-\rho^2)}(x^2 - 2\rho xy + y^2)\right)\right.$$
$$\left.+ \frac{1}{2\pi\sqrt{1-\rho^2}} \exp\left(\frac{-1}{2(1-\rho^2)}(x^2 + 2\rho x + y^2)\right)\right], -\infty < x, y < \infty,$$

then show that

(*i*) $f(x, y)$ is a joint *p.d.f.* such that both marginal densities are normal but $f(x, y)$ is not bivariate normal.

(*ii*) X and Y have zero correlation but X and Y are not independent.

(B.A. Hons. Statistics Delhi, 1989)

3. Let $X_1, X_2, \ldots, X_n$ be independent observations from a normal universe with mean μ and variance σ^2 and let $\overline{X}$ and S^2 be the sample mean and sum of squares of the deviations from the mean respectively. Let X' be one more observation, independent of previous ones. Show that

$$\left(\frac{X' - \overline{X}}{S}\right)\left(\frac{n(n-1)}{n+1}\right)^{1/2}$$

has a t-distribution with $(n-1)$ *d.f.* ***(B.A. Hons. Statistics, Delhi 1989)***

4. $F(X, Y)$ is a *c.d.* function, then

Is (*i*) $F(X, Y) = F(X) + F(Y)$?

(*ii*) $F(X, Y) = F(X)F(Y)$ ***(B.A. Hons. Math. Delhi, 1989)***

a joint *c.d.f.* ?

(*iii*) $P(X_1 < X_1 < X_2, Y_1 < Y \le Y_2) = P(x < x_2, y_1 < y \le y_{2)} - P(x \le x_1, y_1 < y \le y_2$

$$= F(x_1, y_2) - F(x_2, y_{1)} - F(x_1, y_2) + F(x_1, y_1)$$

5. Find the joint distributions of X and Y and $S^2 = \frac{1}{n-1}\sum_{i=1}^{n}(X_i - \overline{X})^2$.

Answer the following questions ?

(*i*) What is the distribution of $\sigma^2 [(k-1) S_k^{\ 2} + (n-k-1) S_{n-k}^{\ 2}]$?

(*ii*) What is the distribution of $S^2{}_k / S^2_{n-1}$?

(*iii*) What is the distribution of $\sqrt{n}\,(\overline{X} - \mu)/S$?

(B.A. Math. Hons. Delhi, 1989)

(*iv*) What is the distribution of $\sqrt{n}\,(X - \mu)/S$?

(B.A. Hons. Math. Delhi, 1899)

6. Let X be discrete random variable with density

$$f_X(x) = \frac{1}{8} T_{(-1)}(x) + \frac{6}{8} T_{(0)}(x) + \frac{1}{8} T_{(1)}(x)$$

For $k = 2$, evaluate $P[\,|X - EX| \geq k\sqrt{\text{Var}\, X}\,]$. **(B.A. Math. Hons. Delhi, 1989)**

(Ans. $\leq 1/4$**)**

7. If X has negative exponential distribution with parameter 1 and if $P(X \leq 1) = P(X > 1)$, then what is Var X ? **(B.A. Math. Hons. Delhi, 1989)**

8. Show that if $n_1 = n_2$, the median of the distribution of F with n_1, n_2 *d.f.* is at $F = 1$ and that the quartite Q_1 and Q_3 satisfy $Q_1 Q_3 = 1$.

(B.A. Hons. Statistics, Delhi, 1989)

9. Given that $P[F(10, 12)] > 2{\cdot}753] = 0{\cdot}05$ and $P[F(1, 12) > 4{\cdot}748] = 0{\cdot}05$ find

(*i*) $P\left[F(12, 10) > \frac{1}{2{\cdot}753}\right]$ (*ii*) $P[-\sqrt{4{\cdot}747} < t(12) < \sqrt{4{\cdot}747}] = 0{\cdot}05$

10. Show that for any $x > 0$. $1 - \Phi(x) \leq \frac{1}{x}\phi(x)$, **(B.Sc. Delhi, 1972)**

where $\Phi(x) = \int_{-\infty}^{x} \phi(t)dt$ and $\phi(t) = \frac{1}{\sqrt{2\pi}} e^{-\frac{1}{2}t^2}$

11. Show that the *p.d.f.* of a *r.v.* X defined over $(-1, 1)$ satisfies the equation

$$\frac{df}{dx} = C\,(1 - x^2)^{-1}, c \text{ being constant.}$$

12. A random variable X has a uniform distribution in $[0, 1]$. Find the *p.d.f.* of $-2 \log_e x$ and identify its distribution.

(B.A. Hons. Statistics Delhi, 1984)

13. If $F(x, y) = 1$ for $x + 2y \geq 1$

$= 0$ for $x + 2y < 0$,

then $F(x, y)$ can not be the joint distribution of the variates X and Y

14. Let X be a binomial variate with probability p of success. Truncate $x = 0$ and then find the limit of truncated binomial as $n \to \infty, p \to 0$ with np remaining finite.

15. Let X and Y have a bivariate normal distribution with parameters

$\mu_1 = 3, \mu_2 = 1, \sigma_1^2 = 16, \sigma_2^2 = 25$ and $\rho = \frac{3}{8}$. Determine the following probabilities ;

(*a*) $P(3 < Y < 8)$ (*b*) $P(3 < Y < 8 \mid x = 7)$
(*c*) $P(-3 < X < 3)$ (*d*) $P(-3 < X < 3 \mid y = -4)$

Ans. (*a*) 0·264 (*b*) 0·440 (*c*) 0·433 (*d*) 0·642

16. (*a*) Let $f(x, y) = (1/2\pi) \exp. [-\frac{1}{3}(x^2 + y^2)] \{1 + xy \exp. [-\frac{1}{3}(x^2 + y^2 - 2)]\}$,

$$-\infty < x, y < \infty$$

If $f(x, y)$ is a joint *p.d.f.*, it is not a normal bivariate *p.d.f.* Show that $f(x, y)$ actually is a joint *p.d.f.* and that each marginal *p.d.f.* is normal.

(*b*) Let X, Y and Z have the point *p.d.f.*

$$\left(\frac{1}{2\pi}\right)^{3/2} \exp[-(x^2 + y^2 + z^2)/2] \{1 + xyz \exp. [-(x^2 + y^2 + z^2)/2]\},$$

$-\infty < x, y < -\infty$. Show that X, Y and Z are pairwise stochastically independent and that each pair has a bivariate normal distribution.

17. Let X and Y have the joint *p.d.f.* $f(x, y) = e^y/[x! (y - x)!], y = 0, 1, 2, \ldots, x = 0, 1, \ldots, y$, zero elsewhere

(*a*) Find the *m.g.f.* $M(t_1, t_2)$ of the joint distribution

(*b*) Determine the conditional mean $E(x \mid y)$.

[**Hint.** Note that $\sum_{x=0}^{y} e^{'xy}/[x! (y - x)!] = [1 + e^{'1}]^y$

18. Let $f(x_1 \mid x_2) = c_1 x_1/x_2^2, 0 < x_1 < x_2, 0 < x_2 < 1$, zero elsewhere and $f_2(x_2) = c_2 x_2^4, 0 < x_2 < 1$, zero elsewhere be the conditional *p.d.f.* of X_1 ; given $X_2 = x_2$ and the marginal *p.d.f.* of X_2.

(*a*) Determine the constant c_1 and c_2 and also joint *p.d.f.* of X_1 and X_2

(*b*) Find $P\left(\frac{1}{4} < X_1 < \frac{1}{2} \,\middle|\, X_2 = \frac{3}{8}\right)$ and (*c*) $P\left(\frac{1}{4} < X_1 < \frac{1}{2}\right)$.

19. Let $f(X, Y) = 21 X^2 Y^3, 0 < X < Y < 1$, zero elsewhere, be the joint *p.d.f.* of X and Y. Find the conditional mean and conditional variance of X ; given $Y = y, 0 < y < 1$.

[**Hint.** Conditional means of $X = \int_0^y \frac{3}{y^3} x^3 \, dx$] **Ans.** $3y/4, 3y^2/80, 0 < y < 1$.

20. If $X_1, X_2, \ldots, X_n$ are mutually independent standard normal variates, then show that the mean value of $\min(|X_1|, |X_2| \ldots |X_n|)$ is

$$2^n \int_0^{\infty} \left\{1 - \phi(x)\right\}^n dx$$

where $\phi(x)$ denotes the standard normal distribution function.

$$\left[\textbf{Hint. } E\left[\min(|x_1|, |x_2| \ldots |x_n|)\right]\right.$$

$$= \int_{-\infty}^{+\infty}\int_{-\infty}^{+\infty} \ldots \int_{-\infty}^{+\infty} \min(|x_1| \ldots |x_n|) f(x_1, \ldots, x_n)\, dx_1 \ldots dx_n$$

$$= n\, 2^{n-1} \int_{-\infty}^{+\infty} \int_{|x_1|}^{\infty} \ldots \int_{|x_n|}^{\infty} |x_1|\, \phi(x_1) \ldots \phi(x_n) dx_1 \ldots dx_n$$

$$= n\, 2^n \int_0^{\infty}\int_{x_1}^{\infty} \ldots \int_{x_n}^{\infty} x_1\, \phi(x_1) \ldots \phi(x_n)\, dx_1 \ldots dx_n$$

$$\left.= n\, 2^n \int_0^{\infty} x_1\, \phi(x_1)\, [1 - \phi(x_1)]^{n-1}\, dx_1 = 2^n \int_0^{\infty} [1 - \phi(x)]^n\, dx\right]$$

21. The joint probability distribution of two r.v. X and Y is given by

$$p(x, y) = \frac{2}{n(n+1)} : x = 1, 2, \ldots n\, ;\, y = 1, 2, \ldots n.$$

Examine whether X and Y are independent.

Ans. No ; $p_{XY}(x, y) \neq p_X(x) \,.\, p_Y(y)$

[**Hint.** Write down the joint p.d. table along with the marginal distributions of X & Y.]

22. Show that with usual notation for all X, Y,

$F_X(x) + F_Y(y) - 1 \leq F_{XY}(x, y) \leq \sqrt{F_X(x)\, F_Y(y)}$ (***B.Sc. Delhi, Math. Hons. 1985***)

[**Hint.** Define the events

$$A = \{X \leq x\}\, ;\, B = \{Y \leq y\},\, A \cap B = \{X \leq x \cap Y \leq y\}.$$

$\therefore\ P(A) = F_X(x),\ P(B) = F_Y(y),\ P(A \cap B) = F_{XY}(x, y)$

Since $(A \cap B) \subset A\ \therefore\ F_{XY}(x, y) \leq F_X(x)$ .....(1) Similarly $F_{XY}(x, y) \leq F_Y(y)$ .....(2)

Multiply (1) and (2) and get $F_{XY}(x, y) \leq \sqrt{F_X(x)\, F_Y(y)}$.

Also $P(A \cup B) \leq 1 \Rightarrow P(A) + P(B) - P(A \cap B) \leq 1$

$\Rightarrow P(A) + P(B) - 1 \leq P(A \cap B)$.

23. If the joint distribution function of X and Y is given by

$$F(x, y) = 1 - e^{-x} - e^{-y} + e^{-(x+y)},\ x > 0\ y > 0\, ;\, = 0, \text{ elsewhere.}$$

(*a*) Find the marginal densities of X and Y,

(*b*) Are X and Y independent ?

(*c*) Find $P(X \leq 1 \cap Y \leq 1)$ and $P(X + Y \leq 1)$ (***I.A.S. 1996***)

Ans. e^{-x}, e^{-y}, yes, $(1 - e^{-1})^2$, $1 - 2e^{-1}$

[**Hint.** (*b*) $f_{XY}(x, y) = \dfrac{\partial F(x, y)}{\partial x, \partial y} = e^{-(x+y)}$, $x \ge 0, y \ge 0$; 0, elsewhere :

$$= e^{-x} \cdot e^{-y} = f_X(x) f_Y(y).]$$

(*c*) $P(X \le 1 \cap Y \le 1) = \int_0^1 \int_0^1 f(x, y)\, dx\, dy = (1 - e^{-1})^2 \cdot P(X + Y \le 1)$

$$= \int_0^1 \int_0^{1-x} \{f(x, y)\, dy\}\, dx > F$$

24. If the d.f. of X is $F(x)$ find the d.f. of (*i*) $Y = X - c$, (*ii*) $Y = cX$, (*iii*) $Y = X^3$ c having constant.

Ans. (*i*) $F(x + c)$ (*ii*) $F\left(\dfrac{x}{c}\right)$, $c > 0$; $1 - F\left(\dfrac{x}{c}\right)$, $c < 0$, (*iii*) $F(x^{1/3})$

25. The joint p.d. of two r.v. is given by $f(x, y) = e^{-(x+y)}$; $x > 0, y > 0$, 0 else where. Show that the p.d.f. of $U = \frac{1}{2}(X + Y)$ is $g(u) = 4ue^{-2u}$

(***B.Sc. Calicut, 1986***)

26. X and Y are two variates with finite expectations. Prove or disprove :

(*a*) $E[\text{Min}(X, Y)] \le \text{Min}[(E(X), E(Y)]$

(*b*) $E[\text{Max}(X, Y)] \ge \text{Max}[(E(X), E(Y)]$

(*c*) $E[\text{Min}(X, Y)] + \text{Max}[(X, Y)] = EX + EY$

(***B.A. Statistics Hons. Delhi, 1989***)

[**Hint.** Use $\text{Min}(X, Y) = \frac{1}{2}(X + Y) - |X - Y|$; $\text{Max}(X, Y) = \frac{1}{2}(X + Y) + |X - Y|$

Also $|E(X - Y)| \le E|X - Y|$ or $-E|X - Y| \Rightarrow -|E(X - Y)|$

$= -|E(X)| - E(Y)|$; Also $|E(X + Y)| \le E|X + Y|$

$\text{Min}(X, Y) + \text{Max}(X, Y) = X + Y$]

27. Let $f(x, y) = 8xy$, $0 < x < y < 1$; $= 0$, elsewhere. Find

(*i*) $E(Y \mid X = x)$ (*ii*) $E(XY \mid X = x)$ (*iii*) $E(Y^2 \mid X = x)$ (*iv*) $\text{Var}(Y \mid X = x)$

(***B.A. Maths. Hons. Delhi, 1990***)

Ans. (*i*) $E(Y \mid X = x) = \displaystyle\int_x^1 y \left(\frac{2y}{1 - x^2}\right) dy = \frac{2}{3} \cdot \frac{1 + x + x^2}{1 + x}$

(*ii*) $E(XY \mid X = x) = x\, E(Y \mid X = x) = \dfrac{2x}{3} \cdot \dfrac{1 + x + x^2}{1 + x}$

(*iii*) $E(Y^2 \mid X = x) = \displaystyle\int_x^1 y^2 \left(\frac{2y}{1 - x^2}\right) dy = \frac{1 + x^2}{2}$

(*iv*) $\text{Var}(Y \mid X = x) = E(Y^2 / X = x) - [E(Y \mid X = x)]^2 = \dfrac{1 + x^2}{2} - \dfrac{4}{9} \dfrac{(1 + x \pm x^2)^2}{(1 + x)^2}$

28. Let $F_n(x)$ be the d.f. defined by

$F_n(x) = 0$ for $x \le -n$; $\dfrac{x+n}{2n}$ for $-n < x < n$; $= 1$ for $x \ge n$; Show that

$\lim\limits_{n \to \infty} F_n(x) = \frac{1}{2} \pm 1.$

[**Hint.** C.F. $= \phi_n(t) = \int_{-n}^{+n} e^{itx} \frac{1}{2n} dx = \frac{1}{2n} \frac{e^{itn} - e^{-itn}}{it} = \frac{\sin nt}{nt}$

$$\left[\because f_n(x) = F_n'(x) = \frac{1}{2n} \right]$$

$\phi(t) = \lim\limits_{n \to \infty} \phi_n(t) = \lim\limits_{n \to \infty} \frac{\sin nt}{nt} = \begin{cases} 1 \text{ if } t = 0 \\ 0 \text{ if } t \ne 0 \end{cases}$ *i.e.* $\phi(t)$ is discontinuous at $t = 0$

$\lim\limits_{n \to \infty} F_n(x) = \frac{1}{2} \pm 1.$]

29. Examine if the WLLN holds for the sequence $\{X_n\}$ of i.i.d.r.v.s with $P(X = k)_i = (-1)^{k-1} k = {}^6/\pi^2 k^2$; $k = 1, 2, 3, \ldots\ldots$;

(B.Sc. Math Hons. Delhi 1990)

[**Hint.** $E(X_i) = \frac{6}{\pi^2} \sum_{k=1}^{\infty} (-1)^{k-1} \frac{k}{k^2} = \frac{6}{\pi^2} \left[1 - \frac{1}{2} + \frac{1}{3} - \frac{1}{4} + \ldots\ldots \right] = \frac{6}{\pi^2} \log_e 2$

and the series within bracket is convergent. Hence by Khinchin's theorem WLLN holds.]

30. Let X and Y be independent binomial variates, each with parametres n and p. Find $P(X - Y = k)$. **(B.Sc. Cal. 1993)**

[**Hint.** Let $Z = X - Y$. Then Z takes the values $-n, -(n-1). -, -1, 0, 1, \ldots, n$

Now $P(Z = P) = \sum_{r=0}^{n} P(X = k + r \cap Y = r)$

$$= \sum_{r=0}^{n} \binom{n}{k+r} \binom{n}{r} p^{2r+k} q^{2n-2r-k}, \quad q = 1 - p]$$

31. A drunk performs a random walk over positions 0, ... as follows : He starts at 0. He takes successive one unit steps, going to the right with probability p and to the left with probability $(1-p)$. His steps are independent. Let X denote his position after n steps. Find the distribution of $\frac{1}{2}(X + n)$ and find EX. **(B.A. Hons. Delhi, 1994)**

[**Hint.** With the step of the drunk, associate a variable X defined as follows : $X_i = 1$ if he takes step on the right ; $= -1$ if he takes step on the left.

Now $X = X_1 + X_2 + \cdots\cdots + X_n$ is the position of the drunk after n steps.

Let $Y_i = (X_i + 1)/2 \therefore Y_i = (1 + i)/2 = 1$ with prob. p

$= (-1 + 1)/2 = 0$ with probability $(1-p)$.

$$\sum_1^n Y_i = \sum_1^n \left(\frac{X_i + 1}{2}\right) = \frac{X + n}{2},\ EX = np.]$$

Ans. $EX = n\,(2p - 1)$

32. If X is a Poisson variate with parameter m prove that

$$m\left({}^rC_1\,\mu_{r-1} + {}^rC_2\,\mu_{r-2} \ldots\ldots + rc_r\,\mu_0\right) = \mu_{r+1}$$

[Hint. $\mu_{r+1} = E\,[X - E(X)]^{r+1} = \sum_{x=0}^{\infty} (x - m)^{r+1} \cdot \frac{e^{-m}\,m^x}{x\,!}$

$$= \sum_{y=0}^{\infty} \frac{(y - m + 1)^r\, e^{-m}\, m^{y+1}}{y\,!} - m\,\mu_r,\ x - 1 = y, \text{ etc.}]$$

33. If X has a P.D with parameter λ, show that the distribution function of X is given by

$$F(x) = \frac{1}{\lfloor (x+1)} \int_{\lambda}^{\infty} e^{-t}\, t^x \,.\, dt\ ;\ x = 0, 1, 2, \ldots\ldots$$

[Hint. $I_x = \frac{1}{x!}\int_{\lambda}^{\infty} e^{-t}\, t^x = \frac{e^{\lambda}\,\lambda^x}{x\,!} + \underset{x-1}{I} = \ldots\ldots = e^{-\lambda} + \lambda e^{-\lambda} + \frac{\lambda^2\, e^{-\lambda}}{2\,!} + \ldots + \frac{\lambda^x\, e^{-\lambda}}{x\,!}$

$= P\,(x = 0) + P\,(x = 1) + \ldots\ldots + P\,(X = x) = P\,(X \le x) = F(x)]$

34. (*a*) If $X \sim N(\mu, \sigma^2)$. Find the mean and variance of $Y = \frac{1}{2}\,[(X - \mu)/\sigma]^2$

Ans. $\frac{1}{2}, \frac{1}{2}$

(*b*) Show that for $1 - \le \alpha \le 1$,
$\underset{x,y}{fx}(x, y, d) = f_x(x)\, f(y)\, [1 + \alpha\, \{2\, F_x(x) - 1\}\ 2\, F_y(y) - 1\}]$ is a probability density function where $\underset{x}{F}(x)$ is the c.d.f. ***(B.A. Hons. Delhi, 1993)***

35. If $X \sim N(\mu, \sigma^2)$ and $Y \sim N(\mu, \sigma^2)$ are independent and if $P\,(X + 2Y \le 3) = P\,(2X - Y \ge 4)$; determine μ.

36. If $X \sim N(0, 1)$ and $Y \sim N(0, 1)$ prove that

$E\,|\,X - Y\,| \le \sqrt{\frac{8}{\pi}}$. ***(B.A. Hons. Stat. Delhi, 1988)***

[Hint. Use $|\,X - Y\,\} \le |\,X\,| + |\,Y\,|$, $E\,|\,X\,| = E\,|\,Y\,| = \sqrt{2/\pi}$, $X - Y \sim N(0, 1 + 1)$

$E\,|\,X - Y\,| = \sqrt{2/\pi}\ \sigma = \sqrt{4/\pi} < \sqrt{8/\pi}$.]

37. If $X \sim N(0, 1)$, find p.d.f. of $|\,X\,|$ and hence evaluate $E\,|\,X\,|$.

[Hint. $\underset{Y}{G}(y) = P\,(Y \le y) = P\,(|\,X\,| \le y) = P\,(-y \le X \le y)$

$= P\,(X \le Y) - P\,(X \le -y)$

$= F_X(y) - F_X(-y)]$

38. If $X \sim N(0, \sigma^2)$, obtain the distribution of e^X.

39. Let X and Y have joint p.d.f.

$$g(x, y) = e^{-(x+y)}\, x^3 y^4 / \lceil 4 \lceil 5, x > 0, y > 0 \,;\, 0, \text{ otherwise,}$$

(B.A. Hons. Stat., 1997)

Find (*i*) p.d.f. of $U = \dfrac{X}{X+Y}$ (*ii*) $E[U]$ and (*iii*) Var U **(B.Sc. Alld 1992)**

[**Hint.** Let $u = \dfrac{x}{x+y}$ and $v = x + y$

$\therefore \quad p(u, v) = \left(\dfrac{1}{\lceil 9} e^{-v} v^8\right)\left(\dfrac{\lceil 9}{\lceil 4 \lceil 5} u^3 (1-u)^4 \,;\, v > 0 \,;\, 0 < u < 1\right.$

$\therefore \quad U$ is a β-variate of first kind with parameters (4, 5)]

40. A random sample of size n is taken from a population with distribution

$$dP(x) = \frac{1}{\Gamma(v)} e^{-x/a} (x/a)^{v-1} \frac{dx}{a}, x > 0, a > 0, v > 0$$

Find the distribution of the mean $\overline{X}$.

[**Hint.** Use $M(t) = \left[M_X \left(\dfrac{t}{n}\right)\right]^n$ where $M_X(t) = E(e^{tx}) = $ etc.]

41. Let $X_i, i = 1, 2, ...n$ be independent *r.v.* with parameter θ_i, then $Z_i = \min(X_1, X_2, ..., X_n)$ has exponential distribution with parameter $\sum_{i=1}^{n} \theta_i$ **(B.Sc., Maths. Hons Delhi, 1986)**

[**Hint.** $G_Z(z) = P(Z \le z) = 1 - P(Z > z) = 1 - P[\min(X_1, X_2, ...X_n > z]$

$$= 1 - P[X > z] = 1 - \prod_{i=1}^{n} P(X_i > z)$$

$$= 1 - \prod_{i=1}^{n} [1 - P(X_i \le z)] = 1 - \prod_{i=1}^{n} [1 - F_{X_i}(z)]$$

$$= 1 - \prod_{i=1}^{n} \left[1 - (1 - e^{\theta_i z}) \right] = 1 - \exp[-\textstyle\sum_{\theta_i z}] \,;\, 0, \text{otherwise.}]$$

42. (*i*) Show that $Y = -(1/\lambda) \log F(X)$ is $E \exp(\lambda)$.

$\left[\textbf{Hint. } M_Y(t) = E(e^{ty}) = E \exp\left[-\dfrac{t}{x} \log F(x) \right] = E[F(x)^{-t/\lambda}] = E[Z^{-t/v}]\right]$

(*ii*) If X_1, X_2, X_3 and X_4 are *i.i.d* N (0, 1) variates, show that

$Y = X_1 X_2 - X_3 X_4$ has p.d.f.

$$f(y) = \frac{1}{2} \exp[-|y|], -\infty < y < \infty.$$

(I.C.S. 1984)

[**Hint.** Show that $\phi_y(t) = {}^1/(1 + t^2)$ where $\phi(t) = \int e^{ity} f(y)\, dy$

43. Let X_1, X_2 be i-i-d. r.v's with p.d.f.

$$P(X_i = x) = e^{-\lambda}\lambda^x/x!, x = 0, 1, 2, ..., i = 1, 2, \lambda > 0$$

Find the marginal d.f. of M and N where M and N are max. & min. of X_1, X_2.

44. (*i*) If X has Laplace distribution with parameters (ν, 10) and $Y = a + bx + cx^2$. Show that $\rho_{xy} = \sqrt{b^2 \nu^2 + 10 c^2}$

[Hint. $p(x) = \frac{\nu}{2} \cdot e^{-\nu|x|}, -\infty < x < +\infty$]

(*ii*) In a sample of n random observations from exponential distribution with parameter ν, the number of observations in $(0, 1/\nu)$ and $(1/\nu, 2/\nu)$, denoted by X and Y are noted, show that $P(X, Y) = (e-1)/\sqrt{e^2 - e - 1}$.

[**[Hint.** $p_1 = P(0 < x < 1/\nu) = \int_0^{1/\nu} \nu e^{-\nu x} = \frac{e-1}{e}, p_2 = P(1/\nu < Y < 2/\nu) = \frac{e-1}{e^2}$

$$p_3 = 1 - p_1 - p_2$$

$$\therefore \quad \rho(X, Y) = -\left[\frac{p_1 p_2}{(1-p_1)(1-p_2)}\right]^{1/2} = \text{etc.}$$]

45. Prove or disprove :

(*i*) $r(X, Y) = 0 \Rightarrow r(|X|, Y) = 0$, (*ii*) $r(X, Y) = 0, r(Y, Z) = 0 \Rightarrow r(X, Z) = 0$

[Ans. (*i*) False unless X & Y are independent (*ii*) Let $Z = X$ & X and Y be independent, then $r(X, Z) = r(X, X) = 1$; $r(X, Y) = 0$; $r(Y, Z) = 0$]

46. Given $f(x, y) = x e^{-(y+1)}, x \geq 0, y \geq 0$ find the regression curve of Y on X.

[Hint. Marginal p.d.f. of $X = f(y \mid x) = \frac{f(x, y)}{f_1(x)} = \frac{x e^{-x(y+1)}}{e^{-x}} = xe^{-xy}$

Regression curve of Y on $X = f_1(x) = \int_0^\infty f(x, y)\, dy = e^{-x}, x \geq 0$

$$y = E(Y \mid X = x) = \int_0^\infty y f(y \mid X = x)\, dy = \int_0^\infty yx e^{-xy}\, dy = \frac{1}{x} \Rightarrow xy = 1$$

Thus the regression of Y on X is not linear.]

47. Let X and Y denote the number of successes and failures respectively in n independent Bernoulli trials with p as the probability of success in each trial, show that

$$\frac{(X - np)^2}{np} + \frac{[Y - n(1-p)]^2}{n(1-p)}$$

can be approximated by a chi-square distribution with 1.d.f., when n is large.

(M.A. Eco. Delhi, 1992)

48. If $(X, Y) \sim$ B.VN (0, 0, 16, 9, 8), find the joint density of $X+Y$ and $X-Y$.

49. (*a*) Let (X, Y) be $BV \sim N(0, 0, 1, 1, p)$, show that $Z = Y/X$ has the Cauchy distribution.

Ans. $f(z) = \frac{1}{\pi}\left[\frac{\sqrt{1-p^2}}{(1-p^2)+(z-p)^2}\right], -\infty < z < \infty$

(*B.Sc. Math. Hons ; Delhi 1989*)

(*b*) If $(X, Y) \sim N(\mu_x, \mu_y, \sigma_x, \sigma_y, \rho)$, then

$$P(X > \mu_x \cap Y > \mu_y) = \frac{1}{4} + \frac{1}{2\pi}\sin^{-1}\rho$$ **(*M.Sc. Stat. Delhi, 1987*)**

[Hint. Put $X - \mu = r\cos\theta, Y - \mu = r\sin\theta, |J| = r.$]

(*c*) If $(X, Y) \sim \text{BVN}(0, 0, 1, 1, \rho)$, find $E[\max(X, Y)]$. **Ans.** $\left[\frac{1-\rho}{\pi}\right]^{1/2}$

[Hint. $\text{Max}(X, Y) = \frac{1}{2}(X+Y) + \frac{1}{2}|X-Y|, Z = X - Y \sim N[0, 2(1-\rho)]$]

50. (*a*) The joint distribution of *rv's* (X, Y) is given by

$$f(x, y) = \frac{1}{2\pi}.\exp\left[-\frac{1}{2}(x^2+y^2)\right].\left[1 + xy\exp(-(x^2+y^2-2)/2\right],$$
$$-\infty < (x, y) < \infty$$

Verify that (*i*) $f(x, y)$ is a p.d.f. (*ii*) X and Y are not independent and (*iii*) marginal distribution of each of X and Y is normal.

(*b*) Let X and Y be independent non-degenerate variates. Prove that

$$Var(XY) = Var\,X\,Var\,Y$$

iff $E(x) = 0 = E(Y)$. **(*B.Sc. Math. Hons. Delhi, 1989*)**

51. Let X_1 be a random sample of size one from a population with *p.d.f.* $f(x, \theta) = e^{-n/\theta}, x \geq 0\ \theta > 0$. Obtain (*i*) B.C.R. of size α for testing $H_0 : \theta = \theta_0$ against $H_1 : \theta = \theta_1$ and (*ii*) power of the test

(*B.Sc. Stat. Hons. Delhi, 1983*)

52. Suppose a random sample of size n is taken from the population $e^{-\lambda}\lambda^x/x!, x = 0, 1, 2, \ldots$. Give the most powerful critical region of size α for testing $H_0 : \lambda = \lambda_0$ against $H_1 : \lambda = \lambda_1 (> \lambda_0)$.

53. (*a*) Given a random sample $X_1, X_2, \ldots, X_n$ of size n from the distribution with *p.d.f.* $f(x, \theta) = \theta e^{-\theta x}, x > 0, 0 < \theta < \infty$. Show that UMP test for testing $H_0 : \theta = \theta_0$ against $H_1 : \theta < \theta_0$ is given by

$$\left\{x \mid \Sigma x_i \geq \frac{1}{2\theta_0}\chi^2_{d, 2n}\right.$$ **(*B.Sc. Stat. Hons. Delhi, 1988*)**

(*b*) Let $X_1, X_2, \ldots, X_n$ be a random sample from the distribution with *p.d.f.* $f(x \mid \theta) = \theta x^{0-1}, 0 < x < 1, 0 < \theta < \infty$. Show the MP test of level α for testing $H_0 = \theta = 1$ against $H_1 : \theta = 2$ is given by the critical region

$$\left\{ x \,\middle|\, \prod_{i=1}^{n} x_i > \exp\left(-\frac{1}{2} \chi^2_{1-\alpha, 2n} \right) \right\}$$

where $\chi^2_{1-\alpha, 2n}$ is the lower α-point of the χ^2-distribution with $2n$ *d.f.*

(*B.Sc. Stat. Hons. Delhi, 1987*)

54. (*a*) Let X have the distribution $f(x, \theta) = \theta^x (1-\theta)^{1-x}, x = 0, 1, 0 < \theta < 1$. For test $H_0 : \theta = \theta_0$ against $H_1 : \theta = \theta_1$, construct S.P.R.T. and obtain its A.S.N. and D.C. functions. **(*B.Sc. Stat. Hons. Delhi, 1985, 1993*)**

(*b*) Construct S.P.R.T. for testing $H : \theta = \theta_0$ against $H : \theta = \theta_0 \ (0 < \theta_0 < \theta_1)$, on the basis of a random sample drawn from pareto distribution with *p.d.f.* $f(x, \theta) = \theta a^\theta / x^{\theta + 1}, x \geq a$.

Also obtain its O.C. and A.S.N. functions. **(*B.Sc. Stat. Hons. Delhi, 1989*)**

(*c*) Let X have the p.d.f. $f(x, \theta = \theta e^{-\theta x}, x \geq 0, \theta > 0; 0.$ elsewhere. For testing $H_0 : \theta = \theta_0$ against $H_1 : \theta = \theta_1$. Construct S.P.R.T. and obtain its ASN and O.C. function. **(*B.Sc. Stat. Hons. Delhi, 1982, 1986*)**

55. Thirty observations as given below are obtained :

24, 35, 12, 50, 60, 70, 68, 49, 80, 25, 69, 28, 28, 11, 83,
31, 37, 34, 54, 75, 45, 95, 75, 26, 43, 57, 94, 48, 63, 45

Test their randomness by considering the sequence of positive and negative siezes. **(*B.Sc. Agra, 1989*)**

56. Let $X \sim \gamma(\lambda, a)$ and $Y \sim \gamma(\lambda, b)$ be independent gamma r.v.'s, show that

$$E\left(\frac{X}{X+Y}\right) = \left(\frac{E(X)}{E(X+Y)}\right)$$

Hint. $U = X/(x+y)$, $V = X + Y$ are independent.

Also $E(UV) = E(U)\,E(V) \Rightarrow E(X) = E\left(\frac{X}{X+Y}\right) E(X+Y)$

57. Suppose that the amount of money (in rupees) that a person has saved (negative saving representing debt) is found to be a random variable with a probability function specified by the distribution $F(x)$ as

$$F(X) = \frac{1}{2} e^{-(\frac{x}{50})^2}, x \leq 0$$

$$= 1 - \frac{1}{2} - e^{-(\frac{x}{50})^2}, \ x \geq 0$$

Show that the conditional probability that the amount of saving possessed by the person will be less than Rs 100, given that it is more than Rs 50, is $1 - e^{-3}$. **(*I.F.S., 1997*)**

Hint. $f(x) = \dfrac{x}{2500} e^{-\frac{x^2}{2500}}, \ x \geq 0$

Now $P(X < 100 \mid X 750) = P(50 < X < 100) / P(X > 750P)$

where $P(X 750) = \int_{50}^{\infty} \frac{x}{2500} e^{-(\frac{x}{50})^2} dx = \left[-\frac{1}{2} e^{-(\frac{x}{50})^2} \right]_{50}^{\infty}$

$= \frac{1}{2e}, P\ (50\ (X < 100) = \frac{1}{2}\ (e^{-1} . e^{-4})$

Ex. 1. Let $X_1\ X_2 ....,\ X_n$ be i.i.d. random variables. Then min $(X_1, X_2, ...X_n)$ has a distribution if and only if the common distribution of $X^{,s}$, is a Weibul distribution.

Hint. Let $Y = \min(X_1, X_2, ..., X_n)$. Then

$P(Y > y) = P[\min(X_1, X_2 ....., X_n) > y]$

$$= P\ [\bigcap_{i=1}^{n} X_i > y] = \prod_{i=1}^{n} p\ P(X_i > y)$$

$$= \{P(X_i > y)\}^n \text{ as } X_i^{,s} \text{ are i.i.d r.v.'}^s$$

Consider $P(X_i > y) = \int_y^{\infty} f(x)\, dx$

Where $F(x) = c\ \lambda^{-1} \left(\frac{x-\mu}{\alpha}\right)^{x-1} \quad \exp\left\{-\frac{(x-\mu)^l}{\alpha}\right\} dx$

is $P[X_i > y] = \int_{(\frac{y-\mu}{\alpha})}^{\alpha} -e^z\, dz$ where $z = \frac{(x-\mu)^c}{\alpha}$

is $P(Y > y) = \exp\left\{ -(\frac{y-\mu}{\alpha})^c \right\}$

$= \exp\left\{ -n\ (\frac{y-\mu}{\alpha})^c \right\}$

$= \exp\left[-\left\{ n^{1/x} (\frac{y-\mu}{\alpha})^c \right\} \right]$

This result shows that Y has the Weibul distribution with the parameters $\alpha\ n^{-1/c}$.

Ex. 2. If X_1, X_2 are independent X^2 variates each with n $d.f.$, show that $V = \frac{\sqrt{n}\,(X_1 - X_2)}{\sqrt{X_1 X_2}}$ is distributed as student's with n $d.f.$ independent of $V = x_1 + x_2$

Hint. Use. $\Gamma n = 2^{n-1}\ \Gamma n/2 \quad (\frac{n+1}{2})/\sqrt{n},$

$I = \frac{V}{2\sqrt{n}}\ (1 + \frac{u^2}{n})^{-3/2}$

APPENDIX I

Multivariate Normal Distribution

In chapters 7 and 10, we have studied in detail the normal distributions of one and two random variables. In this section, we shall derive the joint distribution of n random normal variables This distribution is known as the n-variate normal distribution. We shall assume that the n-variate normal density function is

$$^{*}f(x_1, x_2, \ldots\ldots, x_n) = k \exp\left[-\tfrac{1}{2}(x-b)' A (x-b)\right] \qquad \ldots(1)$$

where $k\,(>0)$ is chosen, so that, the integral of (1) over the entire n-dimensional Euclidean space of $x_1, x_2, \ldots, x_n$ is unity. A is a $n \times n$ real symmetric matrix, which is positive definite, b is $n \times 1$ matrix, such that, b', the transpose of b, is $b' = [b_1, b_2, \ldots\ldots, b_n]$, each b_1 is a real constant.

Further let x stand for $n \times 1$ matrix such that

$$x' = [x_1, x_2, \ldots\ldots, x_n].$$

To determine k, we have

$$\frac{1}{k} = \int_{-\infty}^{+\infty} \int_{-\infty}^{+\infty} \ldots\ldots \int_{-\infty}^{+\infty} e^{-\frac{1}{2}(x-b)' A (x-b)}\, dx_1\, dx_2 \ldots . dx_n$$

For simplicity, we shall re-write this as

$$\frac{1}{k} = \int_{-\infty}^{+\infty} e^{-\frac{1}{2}(x-b)' A (x-b)}\, dx \qquad \ldots(2)$$

Now since A is positive definite, there exists a non-singular matrix C such that

$$C' A C = 1, \qquad \ldots(3)$$

where

$$I = \begin{pmatrix} 1 & 0\ldots0 \\ 0 & 1\ldots0 \\ \ldots & \ldots \\ 0 & 0\ldots1 \end{pmatrix}, \text{ so that } I' = I. \qquad \ldots(4)$$

---

* The uni-variate normal density function can be written as

$$k\, e^{-\frac{1}{2} a (x-b)^2} \quad \text{or equivalently} \quad k\, e^{-\frac{1}{2}(x-b)\, a\, (x-b)}$$

where a is positive and $k\,(>0)$, is so chosen that the integral of the *p.d.f.* over the entire x-axis is unity.

Let $\quad x - b = Cy$ where

$$y' = (y_1\ y_2 \ldots. y_n),\ C' = \begin{pmatrix} c_{11} & c_{12} \ldots c_{1n} \\ \ldots\ldots\ldots\ldots \\ c_{n1} & c_{n2}\ c_{nn} \end{pmatrix}$$

and $$dx = \begin{vmatrix} \dfrac{\partial x_1}{\partial y_1} \ldots \dfrac{\partial x_n}{\partial y_1} \\ \ldots\ldots\ldots\ldots \\ \dfrac{\partial x_1}{\partial y_n} \ldots \dfrac{\partial x_n}{\partial y_n} \end{vmatrix} dy = mod \begin{vmatrix} c_{11} \ldots 0 \ldots c_{1n} \\ \ldots\ldots\ldots\ldots \\ c_{n1} \ldots 0 \ldots c_{nn} \end{vmatrix} dy = mod\,|\,C\,|\,dy$$

But from (3), $\quad |\,C'AC\,| = |\,I\,|$

$$\Rightarrow \qquad |\,C'\,||\,A\,||\,C\,| = 1 \qquad [\because\ |\,C'\,| = |\,C\,|]$$

in $$\text{mod}\,|\,C\,| = 1/\sqrt{|\,A\,|}$$

so that $$dx = 1/\sqrt{|\,A\,|} \cdot dy \qquad \ldots(5)$$

Thus (2) is transformed to

$$\frac{1}{k} = \frac{1}{\sqrt{|\,A\,|}} \int_{-\infty}^{+\infty} e^{-\frac{1}{2} y'C'ACy}\, dy$$

$$= \frac{1}{\sqrt{|\,A\,|}} \int_{-\infty}^{+\infty} e^{-\frac{1}{2} y'y}\, dy \text{ ; by (3)} = \frac{1}{\sqrt{|\,A\,|}} \prod_{i=1}^{n} \int_{-\infty}^{+\infty} e^{-\frac{1}{2} y_i^2}\, dy_i$$

$$= \frac{1}{\sqrt{|\,A\,|}} (2\pi)^{-\frac{n}{2}} \quad \Rightarrow k = \sqrt{|\,A\,|} \cdot (2\pi)^{-\frac{n}{2}}$$

Hence the n-variate normal density function is

$$f(x_1, x_2, \ldots\ldots, x_n) = \frac{\sqrt{|\,A\,|}}{(2\pi)^{n/2}} \cdot \exp\left[-\tfrac{1}{2}(x-b)' A (x-b)\right]. \qquad \ldots(1)$$

Cor. We shall show that :–

(*i*) $b = EX = \mu$

(*ii*) $\Sigma =$ Covariance matrix of $X = E(X-\mu)(X-\mu)' = A^{-1}$.

Proof. From (4) we have

$$\mu = EX = b + cEY$$

where $$EY_i = \frac{1}{(\sqrt{2\pi})^n} \int_{-\infty}^{+\infty} \cdots\cdots \int_{-\infty}^{+\infty} y_i \, e^{-\frac{1}{2} \cdot \sum_{i=1}^{n} y_i^2} \, dy_1 \cdots dy_i \cdots dy_n$$

$$= \frac{1}{\sqrt{2\pi}} \int_{-\infty}^{+\infty} y_i \, e^{-\frac{1}{2} y_i^2} \, dy_i \times \prod_{j=1}^{n} \left\{ \frac{1}{\sqrt{2\pi}} e^{-\frac{1}{2} y_j^2} \, dy_j \right\}$$

$$= \frac{1}{\sqrt{2\pi}} \int_{-\infty}^{+\infty} y_i \, e^{-\frac{1}{2} y_i^2} \, dy_i \times 0 = 0 \; \forall \; i$$

Now $$E\,Y = E \begin{pmatrix} y_1 \\ \vdots \\ y_n \end{pmatrix} = \begin{pmatrix} Ey_1 \\ \vdots \\ Ey_n \end{pmatrix} = \begin{pmatrix} 0 \\ \vdots \\ 0 \end{pmatrix} = 0$$

$\therefore$ From (6), $\mu = b + .0 = b \quad \Rightarrow b = \mu$.

Covariance matrix of X is defined as

$$E\,(X\,X') = E\,(X - \mu)\,(X - \mu)' = E\,(X_i - \mu_i)\,(X_j - \mu_j)$$

$$= E\,(X_i - \mu_i)^2 - E\,(X_i - \mu_i)\,(X_j - \mu_j) = EY_i^2 - CE\,(Y_i\,Y_j)C'$$

where $$EY_iY_j = \int_{-\infty}^{+\infty} \cdots\cdots \int_{-\infty}^{+\infty} y_i y_j \cdot \prod_{i=1}^{n} \frac{1}{\sqrt{2\pi}} e^{-\frac{1}{2} y_i^2} \, dy_1 \cdots dy_n$$

$$= \frac{1}{\sqrt{2\pi}} \int_{-\infty}^{+\infty} y_i \, e^{-\frac{1}{2} y_i^2} \, dy_i \cdot \frac{1}{\sqrt{2\pi}} \int_{-\infty}^{+\infty} y_j \, e^{-\frac{1}{2} y_j^2} \, dy_j \times \prod_{\substack{l=1 \\ l \neq i, j}}^{n} \left\{ \frac{1}{\sqrt{2\pi}} e^{-\frac{1}{2} y_l^2} \, dy_l \right\}$$

$$= 0$$

and $$Ey_i^2 = \frac{1}{\sqrt{2\pi}} \int_{-\infty}^{+\infty} y_i^2 \, e^{-\frac{1}{2} y_i^2} \, dy_i \prod_{\substack{l=1 \\ l \neq i}}^{n} \left\{ \int_{-\infty}^{+\infty} \frac{1}{\sqrt{2\pi}} e^{-\frac{1}{2} y_l^2 \, dy_l} \right\} = 1.$$

$$\therefore \quad E\,(X\,X') = C \cdot \begin{pmatrix} 1 & 0 \ldots\ldots 0 \\ 0 & 1 \ldots\ldots 0 \\ 0 & 0 \ldots\ldots 1 \end{pmatrix} C' = CC'$$

$$= A^{-1} \quad [\because \; CAC' = 1 \Rightarrow A = (C')^{-1} C^{-1} = (CC')^{-1}$$

Hence $\Sigma = A^{-1}$ and is positive definite.

Thus, if the density of a n-dimensional random vector X is given by (1), then the expected value of X is $b' = \mu$, and the covariance matrix is $A^{-1} (= \Sigma)$.

Conversely, given a vector μ and a positive definite matrix Σ, there is a n-variate normal density.

$$(2\pi)^{-\frac{n}{2}} |\Sigma|^{-\frac{1}{2}} e^{-\frac{1}{2}(x-\mu)' \Sigma^{-1} (x-\mu)}$$

Theorem. *Let $X = (x_1\, x_2 \ldots\ldots x_n)'$ be distributed as $N(\mu, \Sigma)$.*

then $Z = CX$ *is distributed as* $N(C\mu, C\Sigma C')$

where C is non-singular.

Proof. We have $x = C^{-1}z$, so that

$$dx = \text{mod}\, |C^{-1}|\, dz = \frac{1}{|\text{mod}\, |C|}\, dz = \frac{1}{\sqrt{|C|^2}}\, dz = \frac{|\Sigma|\, 1/2}{|CIC'|^{1/2}}\, dz$$

$$\text{Also } Q = (x-\mu)' \overset{-1}{\Sigma} (x-\mu) = (C^{-1}z - \mu)' \Sigma^{-1} (C^{-1}z - \mu)$$

$$= (C^{-1}Z - C^{-1}C\mu)' \Sigma^{-1} (C^{-1} Z - C^{-1} C\mu)$$

$$= [C^{-1} (Z - C\mu)]' \Sigma^{-1} c^{-1} (Z - C\mu)]$$

$$= (Z - C\mu)' C^{-1} \Sigma^{-1} C^{-1} (Z - C\mu)$$

$$= (Z - C\mu)' (C\Sigma C')^{-1} (Z - C\mu)$$

Thus Z is distributed as $N(C\mu, C\Sigma C')$.

Theorem. Let $X \sim N(\mu, \Sigma)$. *Also let* $X = \begin{pmatrix} X^{(1)} \\ X^{(2)} \end{pmatrix}$ *where*

$$X^{(1)} = (X_1 \cdots X_q)', X^{(2)} = (X_{q+1} \cdots X_n)', \mu = \begin{pmatrix} \mu^{(1)} \\ \mu^{(2)} \end{pmatrix} \text{ and } \Sigma = \begin{pmatrix} \Sigma_{11} & \Sigma_{12} \\ \Sigma_{21} & \Sigma_{22} \end{pmatrix}$$

we shall prove that

$X^{(1)}$ and $X^{(2)}$ *are independently normally distributed.*

Proof. We have $EX^{(1)} = \mu^{(1)}, EX^{(2)} = \mu^{(2)}$

$$E(X^{(1)} - \mu^{(1)})' = \Sigma_{11}, E(X^{(1)}) - \mu^{(1)}) (X^{(2)} - \mu^{(2)}) = \Sigma_{12} = 0$$

$$\therefore \quad \Sigma = \begin{pmatrix} \Sigma_{11} & 0 \\ 0 & \Sigma_{22} \end{pmatrix} \text{ and } \Sigma^{-1} = \begin{pmatrix} \Sigma_{11}^{-1} & 0 \\ 0 & \Sigma_{22}^{-1} \end{pmatrix}$$

$$\therefore \quad Q = (x - \mu)' \Sigma^{-1} (x - \mu)$$

$$= [(x^{(1)} - \mu^{(1)\prime}, x^{(2)} - \mu^{(2)}]' \cdot \begin{pmatrix} \Sigma_{11}^{-1} & 0 \\ 0 & \Sigma_{22}^{-1} \end{pmatrix} \begin{pmatrix} x^{(1)} - \mu^{(1)} \\ x^{(2)} - \mu^{(2)} \end{pmatrix}$$

$$= (x^{(1)} - \mu^{(1)})' \Sigma_{11}^{-1}, (x^{(2)} - \mu^{(2)})' \Sigma_{22}^{-1} \begin{pmatrix} x^{(1)} - \mu^{(1)} \\ x^{(2)} - \mu^{(2)} \end{pmatrix}$$

$$= (x^{(1)} - \mu^{(1)\prime} \Sigma_{11}^{-1} (x^{(1)} - \mu^{(1)}) + (x^{(2)} - \mu^{(2)})' \Sigma_{22}^{-1} (x^{(2)} - \mu^{(2)})$$

$$= Q_1 + Q_2, \text{ say}$$

wher$Q_1 = (x^{(1)} - \mu^{(1)})' \Sigma_{11}^{-1} (x^{(1)} - \mu^{(1)})$; similarly for Q_2.

Alsd $\Sigma \mid = \left| \begin{pmatrix} \Sigma_{11} & 0 \\ 0 & \Sigma_{22} \end{pmatrix} \right| = \left| \Sigma_{11} \right| \cdot \Sigma_{22} \mid$

$\therefore$ (1) can be written as $(2\pi)^{-n/2} \mid \Sigma \mid^{-\frac{1}{2}} e^{-\frac{1}{2} Q}$

$$= [(2\pi)^{-q/2} \mid \Sigma_{11} \mid^{-\frac{1}{2}} e^{-\frac{1}{2} Q_1}] \cdot [(2\pi)^{-(n-q)/2} \mid \Sigma_{22} \mid e^{-\frac{1}{2} Q_2}]$$

Thus $X^{(1)}$ and $X^{(2)}$ are independent normal variates and

$$X^{(1)} \sim N (\mu^{(1)}, \Sigma_{11}), X^{(2)} \sim N (\mu^{(2)}, \Sigma_{22})$$

Particular Case. We dedude bivariate normal distribution. Let $X^{(1)} = X_1, X^{(2)} = X_2, \mu^{(1)} = \mu_1, \mu^{(2)} = \mu_2, \Sigma_{11} = \sigma_{11} = \sigma_1^2,$

$\Sigma_{22} = \sigma_{22} = \sigma_2^2, \Sigma_{12} = \Sigma_{21} = \sigma_1 \sigma_2 \rho_{12}$. Thus

if X_1 and X_2 have a bivariate normal distribution, then X_1 and X_2 are independent iff $\rho_{12} = 0$ *i.e.* uncorrelated.

Exercises

1. Let $X \sim N(\mu, \Sigma)$ where $X = (X_1 X_2 X_3)', \mu = (000)'$ and $\Sigma = \begin{pmatrix} 7 & 3 & 2 \\ 3 & 4 & 1 \\ 2 & 1 & 2 \end{pmatrix}$. Obtain the distribution of $X_1 + 2X_2 - 3X_3$. **(*M.A., Delhi, 1980*)**

2. Find b and A if X and Y have the joint density given by

$$\frac{1}{2\pi} \exp\left[-\frac{1}{2} (x^2 + y^2 + 4x - 6y + 13) \right]$$

3. Show that the *m.g.f.* of the multivariate normal distribution (1) page 743, is exp $(t't\ V)$ where $t' = [t_1\ t_2 \ldots\ldots t_n]$ and V the positive definite covariance matrix.

(*b*) Let $n = 2$ and take $V = \begin{bmatrix} \sigma_1^2 & \rho\sigma_1\sigma_2 \\ \rho\sigma_1\sigma_2 & \sigma_2^2 \end{bmatrix}$

Determine the $|V|$, V^{-1} and $(x-\mu)'\ V^{-1}(x-\mu)$.

(*c*) Prove that the multivariate normal variates are mutually stochastically independent iff the covariance matrix is a diagonal matrix.

APPENDIX II

Theory of Errors

We know that the measurements of quantities leads to errors. They are of two types : 1. Determinate 2. Random.

Determinate errors are systematic and are due to the imperfections of our measuring instruments. Such instruments are either adjusted wrongly or a mistake is done by an observer while making a reading with such an instrument. The cause of these errors can always be traced and then proper corrections be made.

Random errors are indeterminate. They are due to unknown causes *i.e.*, causes which are under our control. However, correction can be made with the help of statistical methods.

Let X be the true value of a quantity. Let $X_1, X_2, \ldots\ldots, X_n$ be the values obtained when n equally reliable measurements of the given quantity are taken. Thus the corresponding *errors* in the measurements are given by

$$X_1 - X, X_2 - X, \ldots\ldots, X_n - X.$$

Let $a_i = X_i - X,\ i = 1, 2, \ldots\ldots, n$; so that errors are $a_1, a_2, \ldots.., a_n$

Let $\overline{X}$ (average of the whole data) be the *most* probable value.

Then the residuals of $X_1, X_2, \ldots\ldots, X_n$ are defined by $d_1, d_2, \ldots\ldots, d_n$

where $d_i = X_i - \overline{X}$.

The errors and residual are neither systematic nor fixed. They are assumed to be varying at random. We know that large errors generally do not occur. It is only the small errors which are common and occur more frequently. The errors obey a certain law, which is called the Gaussian Normal Law of Error. It states that the errors obey the probability law, given by $f(x) = \dfrac{1}{\sigma\sqrt{2\pi}} e^{-x^2/2\sigma^2}$

This expression can be reduced to a simpler form by putting $f(x) = N$ and $h^2 = (1/2\sigma^2)$. This form is $N = \dfrac{h}{\sqrt{\pi}} e^{-h^2x^2}$ ...(1)

The number h is known as the *precision index.*

The formula (1) enables us to known the relative number of measurements N which have an error x.

Gaussian Error Curve. It is obtained by plotting N and x on a graph paper and then joining these points by a smooth freehand drawing. We have plotted the Gaussian Error curve for the two particular values $h = 0{\cdot}5, h = 1$. Also it is clear that when h is large, N will be large for a small error x and will be small for the decreasing values of x. Whenever precision index h is high, it is concluded that a large number of measurements agree with the true value of quantity, while if the precision index is low, a lesser number of measurements agree with the true value of the quantity.

We can determine the probability of an error of a single measurement lying between the various limits. Thus if the limits are $\pm a$, then $P(-a \le x \le a) = \frac{h}{\sqrt{\pi}} \int_{-a}^{+a} e^{-h^2x^2} dx = \frac{2h}{\pi} \int_0^a e^{-h^2x^2} dx.$

The values of this integral have been tabulated for various values of a differing by 0·1. Further, the error functions is defined by

$$erf(t) = \frac{2}{\sqrt{\pi}} \int_0^t e^{-y^2} dy,$$

There are various types of errors, the important ones being—

1. *Probable error or Quartiles.* It was defined by Sir Francis Galton, an English Economist, as

$$pr.er.f\ (\alpha h) = \frac{1}{2} \quad \Rightarrow P(-\alpha \angle t \angle \alpha) = \frac{h}{\sqrt{\pi}} \int_{-\alpha}^{\alpha} e^{-h^2u^2} du = \frac{1}{2}$$

$$\therefore \quad \frac{2}{\sqrt{\pi}} \int_0^{h\alpha} e^{-x^2} dx = \frac{1}{2}, \ x = hu, dx = h\, du$$

From the table, we find that $\alpha h = 0{\cdot}676936$ or $\alpha = (0{\cdot}676936)/h$

Also $\quad h = 1/\sigma\sqrt{2\pi}$ $\therefore$ $\alpha/\sigma = 0{\cdot}676936$

Hence $\quad \alpha = 0{\cdot}6769\,\sigma = \frac{2}{3}\sigma$ (approx.)

2. *Laplacean Mean error.* It is defined by

$$\eta = \int_{-\infty}^{+\infty} |x| N\, dx$$

$$= \frac{h}{\sqrt{\pi}} \int_{-\infty}^{+\infty} |x| e^{-h^2x^2} dx = \frac{2h}{\sqrt{\pi}} \int_0^{\infty} x\, e^{-h^2x^2} dx = \frac{1}{h\sqrt{\pi}} = \int_0^{\infty} y\, e^{-y^2} dy$$

$$= \frac{1}{2h\sqrt{\pi}} = \sqrt{2\pi}\,\sigma = 0{\cdot}7979\,\sigma$$

Hence $\eta = \frac{4}{5}\sigma$ (approx.)

3. *Gaussian Mean error.* It is defined by $\alpha = \dfrac{1}{h\sqrt{2}}$

Arithmetic mean of all the measurements

Suppose that n independent measurements have been made and that the ith measurement has an error equal to x_i

Now the probability that $x_i, x_i + dx_i$ is given by $p_i = \dfrac{h}{\sqrt{\pi}} e^{-h^2 x_i^2} dx_i$

Since the measurements are independent, the probability that the errors x_i, $i = 1, 2, \ldots, n$ be in the intervals of length dx_i, is given by

$$p = \prod_{i=1}^{n} p_i = \left(\frac{h}{\sqrt{\pi}}\right)^n e^{-h^2 \sum_1^n x_i^2} \prod_{i=1}^{n} dx_i$$

Since h is fixed, p will be minimum when Σx_i^2 is maximum *i.e.*

$$S = \sum_{i=1}^{n} (X_i - X)^2 \text{ is minimum.} \quad \therefore \quad \frac{dS}{dX_i} = 0, \frac{d^2S}{dX_i^2} > 0$$

The first condition gives

$$\sum_{i=1}^{n} \left(X_i - X\right) = 0 \Rightarrow X = \frac{\sum_1^n X_i}{n} = \bar{X}$$

while second condition is clearly satisfied.

Hence $X = \bar{X} = A.M.$ of all the measurements.

APPENDIX III

Borel-Cantelli Lemma (Zero-one Law). Let $\{A_n\}$ be a sequence of independant events. Then $\{I(\frac{A}{n})\}$ is a sequence of independent r.v's. Hence $\lim I(A_n)$, lem $I(A_n)$ and $\lim I(A)$ (if existent are tail functions and hence a,s degenerate, but

$\lim I(A_n) = I(\text{Lim } A_n)$, $\lim I(A_n) = I$

$(\lim A_n)$, Lim $(I(A_n)\ 0 = I(\lim A_n)$

Hence $I(\lim A_n)$ is either zero or one, equivalently either $P(\lim A_n)$ is zero or one.

Similar P $(\lim A_n)$ is 0 or /. If $\lim A_n$ = A events, $P(A)$ is either 0 or /.

Theorem. (*i*) If $\Sigma P A_n < \infty$, then $P(\lim A_n) = 0$

(*ii*) If $\Sigma P A_n = \infty$ and A's are independent, then $P(\mathit{lim}\ A_{n)}\ P./.$

(*i*) First part of the theorem is true whether A_k are independent or not.

Since $P(U A_R) \leq \Sigma P A_r$ whether A_k are independent or not.

$$P(U_r^s A_k) \leq \sum_r^s P A_R \to 0.$$

as $r, s \to \infty$ if $\Sigma P A_n < \infty$. But this implies that

$\lim_r \lim_s P(U_r^s A_k) = P(\lim A_n) = 0$

(*iii*) We have $P(\lim_r A_n) \lim_r \lim_s [1 - P(A_r A_k^c)$

$$= \lim_r \lim_s [1 - \frac{\Pi^s}{k = r}(1 - PA_k) \text{ since } A_n^c \text{ are independent.}$$

But for $0 \leq P A_k \leq 1$ $\qquad 1 - P A_k \leq \exp(-P A_k)$

$$\Rightarrow 1 - \exp(-\sum_k PA_k) \leq 1 - \frac{\Pi}{k}(1 - P A_k) \frac{\Pi}{h}(1 - PA_k) \leq \exp(-\sum_R P A_k)$$

if $\Sigma p A_n = \infty$, then $\Sigma^\infty P A_k$ is infinite for all n.

$$\therefore\ k = r \lim \sum_n^\infty PA_k = \infty, \lim_r \lim_s \sum_r^s PA_k = \infty$$

$$\lim_r \lim_s \exp(-\sum_r^s P A_n) = 0 \Rightarrow P(\lim |A_n|) = 1$$

Hence the result.

Index

A

B

C

R

S

T

U

V

W

Y

LOGARITHMS

	0	1	2	3	4	5	6	7	8	9	Mean Differences.								
											1	2	3	4	5	6	7	8	9
10	00000	00432	00860	01284	01703						42	85	127	170	212	254	297	339	381
						02119	02531	02938	03342	03743	40	81	121	162	202	242	283	323	364
11	04139	04532	04922	05308	05690						37	77	116	154	193	232	270	309	348
						06070	06446	06819	07188	07555	37	74	111	148	185	222	259	296	333
12	07918	08279	08636	08991	09342						36	71	106	142	177	213	248	284	319
						09691	10037	10380	10721	11059	34	68	102	136	170	204	238	272	307
13	11394	11727	12057	12385	12710						33	66	98	131	164	197	229	262	295
						13033	13354	13672	13988	14301	32	63	95	126	158	190	221	253	284
14	14613	14922	15229	15534	15836						30	61	91	122	152	183	213	244	274
						16137	16435	16732	17026	17319	29	59	88	118	147	177	206	236	265
15	17609	17898	18184	18469	18752						28	57	85	114	142	171	199	228	256
						19033	19312	19590	19866	20140	28	55	83	110	138	165	193	221	248
16	20412	20683	20951	21219	21484						27	53	80	107	134	160	187	214	240
						21748	22011	22272	22531	22789	26	52	78	104	130	156	182	208	233
17	23045	23300	23553	23805	24055						26	50	76	101	126	151	176	201	227
						24304	24551	24797	25042	25285	25	49	73	98	122	147	171	196	220
18	25527	25768	26007	26245	26482						24	48	71	95	119	143	167	190	214
						26717	26951	27184	27416	27646	23	46	69	93	116	139	162	185	208
19	27875	28103	28330	28556	28780						23	45	68	90	113	135	158	180	203
						29003	29226	29447	29667	29885	22	44	66	88	110	132	154	176	198
20	30103	30320	30535	30750	30963	31175	31387	31597	31806	32015	21	43	64	85	106	127	148	170	190
21	32222	32428	32634	32838	33041	33244	33445	33646	33846	34044	20	41	61	81	101	121	141	162	182
22	34242	34439	34635	34830	35025	35218	35411	35603	35793	35984	20	39	58	77	97	116	135	154	174
23	36173	36361	36549	36736	36922	37107	37291	37475	37658	37840	19	37	56	74	93	111	130	148	167
24	38021	38202	38382	38561	38739	38917	39094	39270	39445	39620	18	35	53	71	89	106	124	142	159
25	39794	39967	40140	40312	40483	40654	40824	40993	41162	40330	17	34	51	68	85	102	119	136	153
26	41497	41664	41830	41996	42160	42325	42488	42651	42813	42975	16	33	49	66	82	98	115	131	148
27	43136	43297	43457	43616	43775	43933	44091	44248	44404	44560	16	32	47	63	79	95	111	126	142
28	44716	44871	45025	45179	45332	45484	45637	45788	45939	46090	15	30	46	61	76	91	107	122	137
29	46240	46389	46538	46687	46835	46982	47129	47276	47422	47567	15	29	44	59	74	88	103	118	132
30	47712	47857	48001	48144	48287	48430	48572	48714	48855	48996	14	29	43	57	72	86	100	114	129
31	49136	49276	49415	49554	49693	49831	49969	50106	50243	50379	14	28	41	55	69	83	97	110	124
32	50515	50650	50786	50920	51054	51188	51322	51455	51587	51720	13	27	40	54	67	80	94	107	121
33	51851	51983	52114	52244	52375	52504	52634	52763	52892	53020	13	26	39	52	65	78	91	104	117
34	53148	53275	53403	53529	53656	53782	53908	54033	54158	54283	13	25	38	50	63	76	88	101	113
35	54407	54531	54654	54777	54900	55023	55145	55267	55388	55509	12	24	37	49	61	73	85	98	110
36	55630	55751	55871	55991	56110	56229	56348	56467	56585	56703	12	24	36	48	60	71	83	95	107
37	56820	56937	57054	57171	57287	57403	57519	57634	57749	57864	12	23	35	46	58	70	81	93	104
38	57978	58092	58206	58320	58433	58546	58659	58771	58883	58995	11	23	34	45	57	68	79	90	102
39	59106	59218	59329	59439	59550	59660	59770	59879	59988	60097	11	22	33	44	55	66	77	88	99
40	60206	60314	60423	60531	60638	60746	60853	60959	61066	61172	11	21	32	43	54	64	75	86	97
41	61278	61384	61490	61595	61700	61805	61909	62014	62118	62221	10	21	31	42	53	63	74	84	95
42	62325	62428	62531	62634	62737	62839	62941	63043	63144	63246	10	20	31	41	51	61	71	82	92
43	63347	63448	63548	63649	63749	63849	63949	64048	64147	64246	10	20	30	40	50	60	70	80	90
44	64345	64444	64542	64640	64738	64836	64933	65031	65128	65225	10	20	29	39	49	59	68	78	88
45	65321	65418	65514	65610	65706	65801	65896	65992	66087	66181	10	19	29	38	48	57	67	76	86
46	66276	66370	66464	66558	66652	66745	66839	66932	67025	67117	9	19	28	37	47	56	65	74	84
47	67210	67302	67394	67486	67578	67669	67761	67852	67943	68034	9	18	27	36	46	55	64	73	82
48	68124	68215	68305	68395	68485	68574	68664	68753	68842	68931	9	18	27	36	45	53	63	72	81
49	69020	69108	69197	69285	69373	69461	69548	69636	69723	69810	9	18	26	35	44	53	62	70	79

LOGARITHMS. (Contd.)

	0	1	2	3	4	5	6	7	8	9	Mean Differences.								
											1	2	3	4	5	6	7	8	9
50	69897	69984	70070	70157	70243	70329	70415	70501	70586	70672	9	17	26	34	43	52	60	69	77
51	70757	70842	70927	71012	71096	71181	71265	71349	71433	71517	8	17	25	34	42	50	59	67	76
52	71600	71684	71767	71850	71933	72016	72099	72181	72263	72346	8	17	25	33	42	50	58	66	75
53	72428	72509	72591	72673	72754	72835	72916	72997	73078	73159	8	16	24	32	41	49	57	65	73
54	73239	73320	73400	73480	73560	73640	73719	73799	73878	73957	8	16	24	32	40	48	56	64	72
55	74036	74115	74194	74273	74351	74429	74507	74586	74663	74741	8	16	23	31	39	47	55	63	70
56	74819	74896	74974	75051	75128	75205	75282	75358	75435	75511	8	15	23	31	39	46	54	62	69
57	75587	75664	75740	75815	75891	75967	76042	76118	76193	76268	8	15	23	30	38	45	53	60	68
58	76343	76418	76492	76567	76641	76716	76790	76864	76938	77012	7	15	22	30	37	44	52	59	67
59	77085	77159	77232	77305	77379	77452	77525	77597	77670	77743	7	15	22	29	37	44	51	58	66
60	77815	77887	77960	78032	78104	78176	78247	78319	78390	78462	7	14	22	29	36	43	50	58	65
61	78533	78604	78675	78746	78817	78888	78958	79029	79099	79169	7	14	21	28	36	43	50	57	64
62	79239	79309	79379	79449	79518	79588	79657	79727	79796	79565	7	14	21	28	35	41	48	55	62
63	79934	80003	80072	80140	80209	80277	80346	80414	80482	80550	7	14	20	27	34	41	48	54	61
64	80618	80686	80754	80821	80889	80956	81023	81090	81158	81224	7	13	20	27	34	40	47	54	60
65	81291	81358	81425	81491	81558	81624	81690	81757	81823	81889	7	13	20	26	33	40	46	53	59
66	81954	82020	82086	82151	82217	82282	82347	82413	82478	82543	7	13	20	26	33	39	46	52	59
67	82607	82672	82737	82802	82866	82930	82995	83059	83123	83187	6	13	19	26	32	38	45	51	58
68	83251	83315	83378	83442	83506	83569	83632	83696	83759	83822	6	13	19	25	32	38	44	50	57
69	83885	83948	84011	84073	84136	84198	84261	84323	84386	84448	6	12	19	25	31	37	43	50	56
70	84510	84572	84634	84696	84757	84819	84880	84942	85003	85065	6	12	19	25	31	37	43	50	56
71	85126	85187	85248	85309	85370	85431	85491	85552	85612	85673	6	12	18	24	31	37	43	49	55
72	85733	85794	85854	85914	85974	86034	86094	86153	86213	86273	6	12	18	24	30	36	42	48	54
73	86332	86392	86451	86510	86570	86629	86688	86747	86806	86864	6	12	18	24	30	35	41	47	53
74	86923	86982	87040	87099	87157	87216	87274	87332	87390	87448	6	12	17	23	29	35	41	46	52
75	87506	87564	87622	87679	87737	87795	87852	87910	87967	88024	6	12	17	23	29	35	41	46	52
76	88081	88138	88195	88252	88309	88366	88423	88480	88536	88593	6	11	17	23	29	34	40	46	51
77	88649	88705	88762	88818	88874	88930	88986	89042	89098	89154	6	11	17	22	28	34	39	45	50
78	89209	89265	89321	89376	89432	89487	89542	89597	89653	89708	6	11	17	22	28	33	39	44	50
79	89763	89818	89873	89927	89982	90037	90091	90146	90200	90255	6	11	17	22	28	33	39	44	50
80	90309	90363	90417	90472	90526	90580	90634	90687	90741	90795	5	11	16	22	27	32	38	43	49
81	90848	90902	90956	91009	91062	91116	91169	91222	91275	91328	5	11	16	21	27	32	37	42	48
82	91381	91434	91487	91540	91593	91645	91698	91751	91803	91855	5	11	16	21	27	32	37	42	48
83	91908	91960	92012	92064	92117	92169	92221	92273	92324	92376	5	10	16	21	26	31	36	42	47
84	92428	92480	92531	92583	92634	92686	92737	92788	92840	92891	5	10	15	20	26	31	36	41	46
85	92942	92993	93044	93095	93146	93197	93247	93298	93349	93399	5	10	15	20	26	31	36	41	46
86	93450	93500	93551	93601	93651	93702	93752	93802	93852	93902	5	10	15	20	25	30	35	40	45
87	93952	94002	94052	94101	94151	94201	94250	94300	94349	94399	5	10	15	20	25	30	35	40	45
88	94448	94498	94547	94596	94645	94694	94743	94792	94841	94890	5	10	15	20	25	29	34	39	44
89	94939	94988	95036	95085	95134	95182	95231	95279	95328	95376	5	10	15	19	24	29	34	39	44
90	95424	95472	95521	95569	95617	95665	95713	95761	95809	95856	5	10	14	19	24	29	34	38	43
91	95904	95952	95999	96047	96095	96142	96190	96237	96284	96332	5	9	14	19	24	28	33	38	42
92	96379	96426	96473	96520	96567	96614	96661	96708	96755	96802	5	9	14	19	24	28	33	38	42
93	96848	96895	96942	96988	97035	97081	97128	97174	97220	97267	5	9	14	18	23	28	32	38	42
94	97313	97359	97405	97451	97497	97543	97589	97635	97681	97727	5	9	14	18	23	28	32	37	42
95	97772	97818	97864	97909	97955	98000	98046	98091	98137	98182	5	9	14	18	23	27	32	36	41
96	98227	98272	98318	98363	98408	98453	98498	98543	98588	98632	5	9	14	18	23	27	32	36	41
97	98677	98722	98767	98811	98856	98900	98945	98989	99034	99078	4	9	13	18	22	27	31	36	40
98	99123	99167	99211	99255	99300	99344	99388	99432	99476	99520	4	9	13	18	22	26	31	35	40
99	99564	99607	99651	99695	99739	99782	99826	99870	99913	99957	4	9	13	17	22	26	31	35	39

ANTILOGARITHMS

	0	1	2	3	4	5	6	7	8	9	Mean Differences. 1	2	3	4	5	6	7	8	9
.00	10000	10023	10046	10069	10093	10116	10139	10162	10186	10209	2	5	7	9	12	14	16	19	21
.01	10233	10257	10280	10304	10328	10351	10375	10399	10423	10447	2	5	7	10	12	14	17	19	21
.02	10471	10495	10520	10544	10568	10593	10617	10641	10666	10691	2	5	7	10	12	15	17	20	22
.03	10715	10740	10765	10789	10814	10839	10864	10889	10914	10940	3	5	8	10	13	15	18	20	23
.04	10965	10990	11015	11041	11066	11092	11117	11143	11169	11194	3	5	8	10	13	15	18	20	23
.05	11220	11246	11272	11298	11324	11350	11376	11402	11429	11455	3	5	8	11	13	16	18	21	24
.06	11482	11508	11535	11561	11588	11614	11641	11668	11695	11722	3	5	8	11	13	16	19	21	24
.07	11749	11776	11803	11830	11858	11885	11912	11940	11967	11995	3	5	8	11	14	16	19	22	25
.08	12023	12050	12078	12106	12134	12162	12190	12218	12246	12274	3	6	8	11	14	17	20	22	25
.09	12303	12331	12359	12388	12417	12445	12474	12503	12531	12560	3	6	9	11	14	17	20	23	26
.10	12589	12618	12647	12677	12706	12735	12764	12794	12823	12853	3	6	9	12	15	18	21	24	26
.11	12882	12912	12942	12972	13002	13032	13062	13092	13122	13152	3	6	9	12	15	18	21	24	27
.12	13183	13213	13243	13274	13305	13335	13366	13397	13428	13459	3	6	9	12	15	18	21	25	28
.13	13490	13521	13552	13583	13614	13646	13677	13709	13740	13772	3	6	9	13	16	19	22	25	28
.14	13804	13836	13868	13900	13932	13964	13996	14028	14060	14093	3	6	10	13	16	19	22	26	29
.15	14125	14158	14191	14223	14256	14289	14322	14355	14388	14421	3	7	10	13	16	20	23	26	30
.16	14454	14488	14521	14555	14588	14622	14655	14689	14723	14757	3	7	10	13	17	20	24	27	30
.17	14791	14825	14859	14894	14928	14962	14997	15031	15066	15101	3	7	10	14	17	21	24	28	31
.18	15136	15171	15205	15241	15276	15311	15346	15382	15417	15453	4	7	11	14	18	21	25	28	32
.19	15488	15524	15560	15596	15631	15668	15704	15740	15776	15812	4	7	11	14	18	22	25	29	32
.20	15849	15885	15922	15959	15996	16032	16069	16106	16144	16181	4	7	11	15	18	22	26	30	33
.21	16218	16255	16293	16331	16368	16406	16444	16482	16520	16558	4	8	11	15	19	23	26	30	34
.22	16596	16634	16672	16711	16749	16788	16827	16866	16904	16943	4	8	12	15	19	23	27	31	35
.23	16982	17022	17061	17100	17140	17179	17219	17258	17298	17338	4	8	12	16	20	24	28	32	36
.24	17378	17418	17458	17498	17539	17579	17620	17660	17701	17742	4	8	12	16	20	24	28	32	36
.25	17783	17824	17865	17906	17947	17989	18030	18072	18113	18155	4	8	12	17	21	25	29	33	37
.26	18197	18239	18281	18323	18365	18408	18450	18493	18535	18578	4	8	13	17	21	25	30	34	38
.27	18621	18664	18707	18750	18793	18836	18880	18923	18967	19011	4	9	13	17	22	26	30	35	39
.28	19055	19099	19143	19187	19231	19275	19320	19364	19409	19454	4	9	13	18	22	26	31	35	40
.29	19498	19543	19588	19634	19679	19724	19770	19815	19861	19907	5	9	14	18	23	27	32	36	41
.30	19953	19999	20045	20091	20137	20184	20230	20277	20324	20370	5	9	14	19	23	28	32	37	42
.31	20417	20464	20512	20559	20606	20654	20701	20749	20797	20845	5	10	14	19	24	29	33	38	43
.32	20893	20941	20989	21038	21086	21135	21184	21232	21281	21330	5	10	15	19	24	29	34	39	44
.33	21380	21429	21478	21528	21577	21627	21677	21727	21777	21827	5	10	15	20	25	30	35	40	45
.34	21878	21928	21979	22029	22080	22131	22182	22233	22284	22336	5	10	15	20	25	31	36	41	46
.35	22387	22439	22491	22542	22594	22646	22699	22751	22803	22856	5	10	16	21	26	31	37	42	47
.36	22909	22961	23014	23067	23121	23174	23227	23281	23336	23388	5	11	16	21	27	32	37	43	48
.37	23442	23496	23550	23605	23659	23714	23768	23823	23878	23933	5	11	16	22	27	33	38	44	49
.38	23988	24044	24099	24155	24210	24266	24322	24378	24434	24491	6	11	17	22	28	34	39	45	50
.39	24547	24604	24660	24717	24774	24831	24889	24946	25003	25061	6	11	17	23	29	34	40	46	51
.40	25119	25177	25236	25293	25351	25410	25468	25527	25586	25645	6	12	18	23	29	35	41	47	53
.41	25704	25763	25823	25882	25942	26002	26062	26122	26182	26242	6	12	18	24	30	36	42	48	54
.42	26303	26363	26424	26485	26546	26607	26669	26730	26792	26853	6	12	18	24	31	37	43	49	55
.43	26915	26977	27040	27102	27164	27227	27290	27353	27416	27479	6	13	19	25	31	38	44	50	56
.44	27542	27606	27669	27733	27797	27861	27925	27990	28054	28119	6	13	19	26	32	39	45	51	58
.45	28184	28249	28314	28379	28445	28510	28576	28642	28708	28774	7	13	20	26	33	39	46	52	59
.46	28840	28907	28973	29040	29107	29174	29242	29309	29376	29444	7	13	20	27	34	40	47	54	60
.47	29512	29580	29648	29717	29785	29854	29923	29992	30061	30130	7	14	21	28	34	41	48	55	62
.48	30200	30269	30339	30409	30479	30549	30620	30690	30761	30832	7	14	21	28	35	42	49	56	63
.49	30903	30974	31046	31117	31189	31261	31333	31405	31477	31550	7	14	22	29	36	43	50	58	65

ANTILOGARITHMS. (Contd.)

	0	1	2	3	4	5	6	7	8	9	Mean Differences.								
											1	2	3	4	5	6	7	8	9
.50	31623	31696	31769	31842	31916	31989	32063	32137	32211	32285	7	15	22	29	37	44	52	59	66
.51	32359	32434	32509	32584	32659	32735	32809	32885	32961	33037	8	15	23	30	38	45	53	60	68
.52	33113	33189	33266	33343	33420	33497	33574	33651	33729	33806	8	15	23	31	39	46	54	62	69
.53	33884	33963	34041	34119	34198	34277	34356	34435	34514	34594	8	16	24	32	40	47	55	63	71
.54	34674	34754	34834	34914	34995	35075	35156	35237	35318	35400	8	16	24	32	40	48	56	65	73
.55	35481	35563	35645	35727	35810	35892	35975	36058	36141	36224	8	16	25	33	41	50	58	66	74
.56	36308	36392	36475	36559	36644	36728	36813	36898	36983	37068	8	17	25	34	42	51	59	68	76
.57	37154	37239	37325	37411	37497	37584	37670	37757	37844	37931	9	17	26	35	43	52	61	69	78
.58	38019	38107	38194	38282	38371	38459	38548	38637	38726	38815	9	18	27	35	44	53	62	71	80
.59	38905	38994	39084	39174	39264	39355	39446	39537	39628	39719	9	18	27	36	45	54	63	72	82
.60	39811	39902	39994	40087	40179	40272	40365	40458	40551	40644	9	19	28	37	46	56	65	74	83
.61	40738	40832	40926	41020	41115	41210	41305	41400	41495	41591	9	19	28	38	47	57	66	76	85
.62	41687	41783	41879	41976	42073	42170	42267	42364	42462	42560	10	19	29	39	49	58	68	78	87
.63	42658	42756	42855	42954	43053	43152	43251	43351	43451	43551	10	20	30	40	50	60	70	80	89
.64	43652	43752	43853	43954	44055	44157	44259	44361	44463	44566	10	20	30	41	51	61	71	81	91
.65	44668	44771	44875	44978	45082	45186	45290	45394	45499	45604	10	21	31	42	52	62	73	83	94
.66	45709	45814	45920	46026	46132	46238	46345	46452	46559	46666	11	21	32	43	53	64	75	85	96
.67	46774	46881	46989	47098	47206	47315	47424	47534	47643	47753	11	22	33	44	54	65	76	87	98
.68	47863	47973	48084	48195	48306	48417	48529	48641	48753	48865	11	22	33	45	56	67	78	89	100
.69	48978	49091	49204	49317	49431	49545	49659	49774	49888	50003	11	23	34	46	57	68	80	91	103
.70	50119	50234	50350	50466	50582	50699	50816	50933	51050	51168	12	23	35	47	58	70	82	93	105
.71	51286	51404	51523	51642	51761	51880	52000	52119	52240	52360	12	24	36	48	60	72	84	96	108
.72	52481	52602	52723	52845	52966	53088	53211	53333	53456	53580	12	24	37	49	61	73	85	98	110
.73	53703	53827	53951	54075	54200	54325	54450	54576	54702	54828	13	25	38	50	63	75	88	100	113
.74	54954	55081	55208	55336	55463	55590	55719	55847	55976	56105	13	26	38	51	64	77	90	102	115
.75	56234	56364	56494	56624	56754	56885	57016	57148	57280	57412	13	26	39	52	66	79	92	105	118
.76	57544	57677	57810	57943	58076	58210	58345	58479	58614	58749	13	27	40	54	67	80	94	107	121
.77	58884	59020	59156	59293	59429	59566	59704	59841	59979	60117	14	27	41	55	69	82	96	110	123
.78	60256	60395	60534	60674	60814	60954	61094	61235	61376	61518	14	28	42	56	70	84	98	112	126
.79	61659	61802	61944	62087	62230	62373	62517	62661	62806	62951	14	29	43	58	72	86	101	115	130
.80	63096	63241	63387	63533	63680	63826	63973	64121	64269	64417	15	29	44	59	74	88	103	118	132
.81	64565	64714	64863	65013	65163	65313	65464	65615	65766	65917	15	30	45	60	75	90	105	120	135
.82	66069	66222	66374	66527	66681	66834	66988	67143	67298	67453	15	31	46	62	77	92	108	123	139
.83	67608	67764	67920	68077	68234	68391	68549	68707	68865	69024	16	32	47	63	79	95	110	126	142
.84	69183	69343	69503	69663	69823	69984	70146	70307	70469	70632	16	32	48	64	81	97	113	129	145
.85	70795	70958	71121	71285	71450	71614	71779	71945	72111	72277	17	33	50	66	83	99	116	132	149
.86	72444	72611	72778	72946	73114	73282	73451	73621	73790	73961	17	34	51	68	85	101	118	135	152
.87	74131	74302	74473	74645	74817	74989	75162	75336	75509	75683	17	35	52	69	87	104	121	138	156
.88	75858	76033	76208	76384	76560	76736	76913	77090	77268	77446	18	35	53	71	89	107	125	142	159
.89	77625	77804	77983	78163	78343	78524	78705	78886	79068	79250	18	36	54	72	91	109	127	145	163
.90	79433	79616	79799	79983	80168	80353	80538	80724	80910	81096	19	37	56	74	93	111	130	148	167
.91	81283	81470	81658	81846	82035	82224	82414	82604	82794	82985	19	38	57	76	95	113	132	151	170
.92	83176	83368	83560	83753	83946	84140	84333	84528	84723	84918	19	39	58	78	97	116	136	155	175
.93	85114	85310	85507	85704	85901	86099	86298	86497	86696	86896	20	40	60	79	99	119	139	158	178
.94	87096	87297	87498	87700	87902	88105	88308	88512	88716	88920	20	41	61	81	102	122	142	162	183
.95	89125	89331	89536	89743	89950	90157	90365	90573	90782	90991	21	42	62	83	104	125	146	166	187
.96	91201	91411	91622	91833	92045	92257	92470	92683	92897	93111	21	42	64	85	106	127	149	170	191
.97	93325	93541	93756	93972	94189	94406	94624	94842	95060	95280	22	43	65	87	109	130	152	174	195
.98	95499	95719	95940	96161	96383	96605	96828	97051	97275	97499	22	44	67	89	111	133	155	178	200
.99	97724	97949	98175	98401	98628	98855	99083	99312	99541	99770	23	46	68	91	114	137	160	182	205

NATURAL SINES

	0'	6'	12'	18'	24'	30'	36'	42'	48'	54'	Mean Difference				
	0°.0	0°.1	0°.2	0°.3	0°.4	0°.5	0°.6	0°.7	0°.8	0°.9	1	2	3	4	5
0	.00000	00175	00349	00524	00698	00873	01047	01222	01396	01571	29	58	87	117	145
1	.01745	01920	02094	02269	02443	02618	02792	02967	03141	03316	29	58	87	117	145
2	.03490	03664	03839	04013	04188	04362	04536	04711	04885	05059	29	58	87	116	145
3	.05234	05408	05582	05756	05931	06105	06279	06453	06627	06802	29	58	87	116	145
4	.06976	07150	07324	07498	07672	07846	08020	08194	08368	08542	29	58	87	116	145
5	.08716	08889	09063	09237	09411	09585	09758	09932	10106	10279	29	58	87	116	145
6	.10453	10626	10800	10973	11147	11320	11494	11667	11840	12014	29	58	87	116	145
7	.12187	12360	12533	12706	12880	13053	13226	13399	13572	13744	29	58	87	115	144
8	.13917	14090	14263	14436	14608	14781	14954	15126	15299	15471	29	58	86	115	144
9	.15643	15816	15988	16160	16333	16505	16677	16849	17021	17193	29	57	86	115	144
10	.17365	17537	17708	17880	18052	18224	18395	18567	18738	18910	29	57	86	114	143
11	.19081	19252	19423	19595	19766	19937	20108	20279	20450	20620	29	57	86	114	143
12	.20791	20962	21132	21303	21474	21644	21814	21985	22155	22325	28	57	85	114	142
13	.22495	22665	22835	23005	23175	23345	23514	23684	23853	24023	28	57	85	113	141
14	.24192	24362	24531	24700	24869	25038	25207	25376	25545	25713	28	56	85	113	141
15	.25882	26050	26219	26387	26556	26724	26892	27060	27228	27369	28	56	84	112	140
16	.27564	27731	27899	28067	28234	28402	28569	28736	28903	29070	28	56	84	112	139
17	.29237	29404	29571	29737	29904	30071	30237	30403	30570	30736	28	56	83	111	139
18	.30902	31068	31233	31399	31565	31730	31896	32061	32227	32392	28	55	83	110	138
19	.32557	32722	32887	33051	33216	33381	33545	33710	33874	34038	27	55	82	110	137
20	.34202	34366	34530	34694	34857	35021	35184	35347	35511	35674	27	55	82	109	136
21	.35837	36000	36162	36325	36488	36650	36812	36975	37137	37299	27	54	81	108	135
22	.37461	37622	37784	37946	38107	38268	38430	38591	38752	38912	27	54	81	107	134
23	.39073	39234	39394	39555	39715	39875	40035	40195	40355	40514	27	53	80	107	133
24	.40674	40833	40992	41151	41310	41469	41628	41787	41945	42104	27	53	79	106	132
25	.42262	42420	42578	42736	42894	43051	43209	43366	43523	43680	26	53	79	105	131
26	.43837	43994	44178	44307	44464	44620	44776	44932	45088	45243	26	52	78	104	130
27	.45399	45554	45710	45865	46020	46175	46330	46484	46639	46793	26	52	77	103	129
28	.46947	47101	47255	47409	47562	47716	47869	48022	48175	48328	26	51	77	102	128
29	.48481	48634	48786	48938	49090	49242	49394	49546	49697	49849	25	51	76	101	126
30	.50000	50151	50302	50453	50603	50754	50904	51054	51204	51354	25	50	75	100	125
31	.51504	51653	51803	51952	52101	52250	52399	52547	52696	52844	25	50	74	99	124
32	.52992	53140	53288	53435	53583	53730	53877	54024	54171	54317	25	49	74	98	123
33	54464	54610	54756	54902	55048	55194	55339	55484	55630	55775	24	49	73	97	121
34	.55919	56064	56208	56353	56497	56641	56784	56928	57071	57215	24	48	72	96	120
35	.57358	57501	57643	57786	57928	58070	58212	58354	58496	58637	24	48	71	95	118
36	.58778	58920	59061	59201	59342	59482	59622	59763	59902	60042	23	47	70	94	117
37	.60181	60321	60460	60599	60738	60876	61015	61153	61291	61429	23	46	69	92	115
38	.61566	61704	61841	61978	62115	62251	62388	62524	62660	62796	23	46	68	91	114
39	.62932	63068	63203	63338	63473	63608	63742	63877	64011	64145	23	45	67	90	112
40	.64279	64412	64546	64679	64812	64945	65077	65210	65342	65474	22	44	66	88	111
41	.65606	65738	65869	66000	66131	66262	66393	66523	66653	66783	22	44	65	87	109
42	.66913	67043	67172	67301	67430	67559	67688	67816	67944	68072	22	43	64	86	107
43	.68200	68327	68455	68582	68709	68835	68962	69088	69214	69340	21	42	63	84	106
44	.69466	69591	69717	69842	69966	70091	70215	70339	70463	70587	21	41	62	83	104

NATURAL SINES (Contd.)

	0' 0°.0	6' 0°.1	12' 0°.2	18' 0°.3	24' 0°.4	30' 0°.5	36' 0°.6	42' 0°.7	48' 0°.8	54' 0°.9	Mean Difference 1	2	3	4	5
45	.70711	70834	70957	71080	71203	71325	71447	71569	71691	71813	20	41	61	82	102
46	.71934	72055	72176	72297	72417	72537	72657	72777	72897	73016	20	40	60	80	100
47	.73135	73254	73373	73491	73610	73728	73846	73963	74080	74198	20	39	59	79	98
48	.74314	74431	74548	74664	74780	74896	75011	75126	75241	75356	20	39	58	77	96
49	.75471	75585	75700	75813	75927	76041	76154	76267	76380	76492	19	38	57	76	94
50	.76604	76717	76828	76940	77051	77162	77273	77384	77494	77605	19	37	56	74	93
51	.77715	77824	77934	78043	78152	78261	78369	78478	78586	78694	18	36	54	72	91
52	.78801	78908	79015	79122	79229	79335	79441	79547	79653	79758	18	35	53	71	89
53	.79864	79968	80073	80178	80282	80386	80489	80593	80696	80799	17	35	52	69	87
54	.80902	81004	81106	81208	81310	81412	81513	81614	81714	81815	17	34	51	68	84
55	.81915	82015	82115	82214	82314	82413	82511	82610	82708	82806	16	33	49	66	82
56	.82904	83001	83098	83195	83292	83389	83485	83581	83676	83772	16	32	48	64	80
57	.83867	83962	84057	84151	84245	84339	84433	84526	84619	84712	16	31	47	62	78
58	.84805	84897	84989	85081	85173	85264	85355	85446	85536	85627	15	31	46	61	76
59	.85717	85806	85896	85895	86074	86163	86251	86340	86427	86515	15	30	44	59	74
60	.86603	86690	86777	86863	86949	87036	87121	87207	87292	87377	14	29	43	57	72
61	.87462	87546	87631	87715	87798	87882	87965	88048	88130	88213	14	28	42	56	69
62	.88295	88377	88458	88539	88620	88701	88782	88862	88942	89021	14	27	40	54	67
63	.89101	89180	89259	89337	89415	89493	89571	89649	89726	89803	13	26	39	52	65
64	.89879	89956	90032	90108	90183	90259	90334	90408	90483	90557	13	25	38	50	63
65	.90631	90704	90778	90851	90924	90996	91068	91140	91212	91283	12	24	36	48	60
66	.91355	91425	91496	91566	91636	91706	91775	91845	91914	91982	12	23	35	46	58
67	.92050	92119	92186	92254	92321	92388	92455	92521	92587	92653	11	22	33	45	56
68	.92718	92784	92849	92913	92978	93042	93106	93169	93232	93295	11	21	32	43	54
69	.93358	93420	93483	93544	93606	93667	93728	93789	93849	93909	10	20	31	41	51
70	.93969	94029	94088	94147	94206	94264	94322	94380	94438	94495	10	19	29	39	49
71	.94552	94609	94665	94721	94777	94832	94888	94943	94997	95052	9	18	28	37	46
72	.95106	95159	95213	95266	95319	95372	95424	95476	95528	95579	9	17	26	35	44
73	.95630	95681	95732	95782	95832	95882	95931	95981	96029	96078	8	17	25	33	41
74	.96126	96174	96222	96269	96316	96363	96410	96456	96502	96547	8	16	23	31	39
75	.96593	96638	96682	96727	96771	96815	96958	96902	96945	96987	7	16	22	29	36
76	.97030	97072	97113	97155	97196	97237	97278	97318	97358	97398	7	14	20	27	34
77	.97437	97476	97515	97553	97592	97630	97667	97705	97742	97778	6	13	19	25	32
78	.97815	97851	97887	97922	97958	97992	98027	98061	98096	98129	6	12	17	23	29
79	.98163	98196	98229	98261	98294	98325	98357	98388	98420	98450	5	11	16	21	27
80	.98481	98511	98541	98570	98600	98629	98657	98686	98714	98741	5	10	14	19	24
81	.98769	98796	98823	98849	98876	98902	98927	98953	98978	99002	4	9	13	17	22
82	.99027	99051	99075	99098	99122	99144	99167	99189	99211	99233	4	8	11	15	19
83	.99255	99276	99297	99317	99337	99357	99377	99396	99415	99434	3	7	10	13	16
84	.99452	99470	99488	99506	99523	99540	99556	99572	99588	99604	3	6	8	11	14
85	.99619	99635	99649	99664	99678	99692	99705	99719	99731	99744	2	5	7	9	11
86	.99756	99768	99780	99792	99803	99813	99824	99834	99844	99854	2	4	5	7	9
87	.99863	99872	99881	99889	99897	99905	99912	99919	99926	99933	1	3	4	5	6
88	.99939	99945	99951	99956	99961	99966	99970	99974	99978	99982	1	2	2	3	4
89	.99985	99988	99990	99993	99995	99996	99998	99999	99999	1.0000	0	1	1	1	1
90	1.000														

NATURAL COSINES

Numbers in different columns to be subtracted, not added

	0'	6'	12'	18'	24'	30'	36'	42'	48'	54'	Mean Difference				
	0°.0	0°.1	0°.2	0°.3	0°.4	0°.5	0°.6	0°.7	0°.8	0°.9	1	2	3	4	5
0	1.000	99999	99999	99999	99998	99996	99995	99993	99990	99988	0	1	1	1	1
1	.99985	99982	99978	99974	99970	99966	99961	99956	99951	99945	1	2	2	3	4
2	.99939	99933	99926	99919	99912	99905	99897	99889	99881	99872	1	3	4	5	6
3	.99863	99854	99844	99834	99824	99813	99803	99792	99780	99768	2	4	5	7	9
4	.99756	99744	99731	99719	99705	99692	99678	99664	99649	99635	2	5	7	9	11
5	.99619	99604	99588	99572	99556	99540	99523	99506	99488	99470	3	6	8	11	14
6	.99452	99434	99415	99396	99377	99357	99337	99317	99297	99276	3	7	10	13	16
7	.99255	99233	99211	99189	99167	99144	99122	99098	99075	99051	4	8	11	15	19
8	.99027	99002	98978	98953	98927	98902	98876	98849	98823	98796	4	9	13	17	22
9	.98769	98741	98714	98686	98657	98629	98600	98570	98541	98511	5	10	14	19	24
10	.98481	98450	98420	98388	98357	98325	98294	98261	98229	98196	5	11	16	21	27
11	.98163	98129	98096	98061	98027	97992	97958	97922	97887	97851	6	12	17	23	29
12	.97815	97778	97742	97705	97667	97630	97592	97553	97515	97476	6	13	19	25	32
13	.97437	97398	97358	97318	97278	97237	97196	97155	97113	97072	7	14	20	27	34
14	.97030	96987	96945	96902	96858	96815	96771	96727	96682	96638	7	16	22	29	36
15	.96593	96547	96502	96456	96410	96363	96316	96269	96222	96174	8	16	23	31	39
16	.96126	96078	96029	95981	95931	95882	95832	95782	95732	95681	8	17	25	33	41
17	.95630	95579	95528	95476	95424	95372	95319	95266	95213	95159	9	17	26	35	44
18	.95106	95052	94997	94943	94888	94832	94777	94721	94665	94609	9	18	28	37	46
19	.94552	94495	94438	94380	94322	94264	94206	94147	94088	94029	10	19	29	39	49
20	.93969	93909	93849	93789	93728	93667	93606	93544	93483	93420	10	20	31	41	51
21	.93358	93295	93232	93169	93106	93042	92978	92913	92849	92784	10	21	32	43	54
22	.92718	92653	92587	92521	92455	92388	92321	92254	92186	92119	11	22	33	45	56
23	.92050	91982	91914	91845	91775	91706	91636	91566	91496	91425	11	23	35	46	58
24	.91355	91283	91212	91140	91068	90996	90924	90851	90778	90704	12	24	36	48	60
25	.90631	90557	90483	90408	90334	90259	90183	90108	90032	89956	13	25	38	50	63
26	.89879	89803	89726	89649	89571	89493	89415	89337	89259	89180	13	26	39	52	65
27	.89101	89021	88942	88862	88782	88701	88620	88539	88458	88377	13	27	40	54	67
28	.88295	88213	88130	88048	87965	87882	87798	87715	87631	87546	14	28	42	56	69
29	.87462	87377	87292	87207	87121	87036	86949	86863	86777	86690	14	29	43	57	72
30	.86603	86515	86427	86340	86251	86163	86074	85985	85896	85806	15	30	44	59	74
31	.85717	85627	85536	85446	85355	85264	85173	85081	84989	84897	16	30	46	61	76
32	.84805	84712	84619	84526	84433	84339	84245	84151	84057	83962	16	31	47	62	78
33	.83867	83772	83676	83581	83485	.83389	83292	83195	83098	83001	16	32	48	64	80
34	.82904	82806	82708	82610	82511	82413	82314	82214	82115	82015	16	33	49	66	82
35	.81915	81815	81714	81614	81513	81412	81310	81208	81106	81004	17	34	51	68	84
36	.80902	80799	80696	80593	80489	80386	80282	80178	80073	79968	17	35	52	69	87
37	.79864	79758	79653	79547	79441	79335	79229	79122	79015	78908	18	35	53	71	89
38	.78801	78693	78586	78478	78369	78261	78152	78043	77934	77824	18	36	54	72	91
39	.77715	77605	77494	77384	77273	77162	77051	76940	76828	76717	18	37	56	74	93
40	.76604	76492	76380	76267	76154	76041	75927	75813	75700	75585	19	38	57	76	94
41	.75471	75356	75241	75126	75011	74896	74780	74664	74548	74431	19	39	58	77	96
42	.74314	74198	74080	73963	73846	73728	73610	73491	73373	73254	20	39	59	79	98
43	.73135	73016	72897	72777	72657	72537	72417	72297	72176	72055	20	40	60	80	100
44	.71934	71813	71691	71569	71447	71325	71203	71080	70957	70834	20	41	61	82	102

NATURAL COSINES (Contd.)

Numbers in different columns to be subtracted, not added

	0'	6'	12'	18'	24'	30'	36'	42'	48'	54'	Mean Difference				
	0°.0	0°.1	0°.2	0°.3	0°.4	0°.5	0°:6	0°.7	0°.8	0°.9	**1**	**2**	**3**	**4**	**5**
45	.70711	70587	70463	70339	70215	70091	69966	69842	69717	69591	21	41	62	83	104
46	.69466	69340	69214	69088	68962	68835	68709	68582	68455	68327	21	42	63	84	106
47	.68200	68072	67944	67816	67688	67559	67430	67301	67172	67043	21	43	64	86	107
48	.66913	66783	66653	66523	66393	66262	66131	66000	65869	65738	22	44	65	87	109
49	.65606	65474	65342	65210	65077	64945	64812	64679	64546	64412	22	44	66	88	110
50	.64279	64145	64011	63877	63742	63608	63473	63338	63203	63068	22	45	67	90	112
51	.62932	62796	62660	62524	62388	62251	62115	61978	61841	61704	23	45	68	91	114
52	.61566	61429	61291	61153	61015	60876	60738	60599	60460	60321	23	46	69	92	115
53	.60181	60042	59902	59763	59622	59482	59342	59201	59061	58920	23	47	70	94	117
54	.58779	58637	58496	58354	58212	58070	57928	57786	57643	57501	24	47	71	95	118
55	.57358	57215	57071	56928	56784	56641	56497	56353	56208	56064	24	48	72	96	120
56	.55919	55775	55630	55484	55339	55194	55048	54902	54756	54610	24	49	73	97	121
57	.54464	54317	54171	54024	53877	53730	53583	53435	53288	53140	24	49	74	98	123
58	.52992	52844	52696	52547	52399	52250	52101	51952	51803	51653	25	50	74	99	124
59	.51504	51354	51204	51054	50904	50754	50603	50453	50302	50151	25	50	75	100	125
60	.5000	49849	49697	49546	49394	49242	49090	48938	48786	48634	25	51	76	101	126
61	.48481	48328	48175	48022	47869	47716	47562	47409	47255	47101	26	51	77	102	128
62	.46947	46793	46639	46484	46330	46175	46020	45865	45710	45554	26	52	77	103	129
63	.45399	45243	45088	44932	44776	44620	44464	44307	44151	43994	26	52	78	104	130
64	.43837	43680	43523	43366	43209	43051	42894	42736	42578	42420	26	53	79	105	131
65	.42262	42104	41945	41787	41628	41469	41310	41151	40992	40833	26	53	79	106	132
66	.40674	40514	40355	40195	40035	39875	39715	39555	39394	39234	27	53	80	107	133
67	.39073	38912	38752	38591	38430	38268	38107	37946	37784	37622	27	54	81	107	134
68	.37461	37299	37137	36975	36812	36650	36488	36325	36162	36000	27	54	81	108	135
69	.35837	35674	35511	35347	35184	35021	34857	34694	34530	34366	27	54	82	109	136
70	.34202	34038	33874	33710	33545	33381	33216	33051	32887	32722	27	55	82	110	137
71	.32557	32392	32227	32061	31896	31730	31565	31399	31233	31068	28	55	83	110	138
72	.30902	30736	30570	30403	30237	30071	29904	29737	29571	29404	28	56	83	111	139
73	.29237	29070	28903	28736	28569	28402	28234	28067	27899	27731	28	56	84	112	139
74	.27564	27396	27228	27060	26892	26724	26556	26387	26219	26050	28	56	84	112	140
75	.25882	25713	25545	25376	25207	25038	24869	24700	24531	24362	28	56	85	113	141
76	.24192	24023	23853	23684	23514	23345	23175	23005	22835	22665	28	57	85	113	141
77	.22495	22325	22155	21985	21814	21644	21474	21303	21132	20962	28	57	85	114	142
78	.20791	20620	20450	20279	20108	19937	19766	19595	19423	19252	29	57	86	114	143
79	.19081	18910	18738	18567	18395	18224	18052	17880	17708	17537	29	57	86	114	143
80	.17365	17193	17021	16849	16677	16505	16333	16160	15988	15816	29	57	86	115	144
81	.15643	15471	15299	15126	14954	14781	14608	14436	14263	14090	29	58	86	115	144
82	.13917	13744	13572	13399	13226	13053	12880	12706	12533	12360	29	58	87	115	144
83	.12187	12014	11840	11667	11494	11320	11147	10973	10800	10626	29	58	87	116	145
84	.10453	10279	10106	09932	09758	09585	09411	09237	09063	08889	29	58	87	116	145
85	.08716	08542	08368	08194	08020	07846	07672	07498	07324	07150	29	58	87	116	145
86	.06976	06802	06627	06453	06279	06105	05931	05756	05582	05408	29	58	87	116	145
87	.05234	05059	04885	04711	04536	04362	04188	04013	03839	03664	29	58	87	116	145
88	.03490	03316	03141	02967	02792	02618	02443	02269	02094	01920	29	58	87	117	145
89	.01745	01571	01396	01222	01047	00873	00698	00524	00349	00175	29	58	87	117	145
90	.0000														

STATISTICAL TABLES

Area under the Normal Curve

$\frac{x}{\sigma}$	0·00	0·01	0·02	0·03	0·04	0·05	0·06	0·07	0·08	0·09
0·0	·0000	·0040	·0080	·0120	·0159	·0199	·0239	·0279	·0319	·0359
0·1	·0398	·0438	·0478	·0517	·0557	·0596	·0636	·0675	·0714	·0753
0·2	·0793	·0832	·0871	·0910	·0948	·0987	·1026	·1064	·1103	·1141
0·3	·1179	·1217	·1255	·1293	·1331	·1368	·1406	·1443	·1480	·1517
0·4	·1554	·1591	·1628	·1664	·1700	·1736	·1772	·1808	·1844	·1879
0·5	·1915	·1950	·1985	·2019	·2054	·2088	·2123	·2157	·2190	·2224
0·6	·2257	·2291	·2324	·2357	·2389	·2422	·2454	·2486	·2518	·2549
0·7	·2580	·2611	·2642	·2673	·2704	·2734	·2764	·2794	·2823	·2852
0·8	·2881	·2910	·2939	·2967	·2995	·3023	·3051	·3078	·3106	·3133
0·9	·3159	·3186	·3212	·3238	·3264	·3289	·3315	·3340	·3365	·3389
1·0	·3413	·3438	·3461	·3485	·3508	·3531	·3554	·3577	·3599	·3621
1·1	·3643	·3665	·3686	·3708	·3729	·3749	·3770	·3790	·3910	·3830
1·2	·3849	·3869	·3888	·3907	·3925	·3944	·3962	·3980	·3997	·4015
1·3	·4032	·4049	·4066	·4082	·4099	·4115	·4131	·4147	·4162	·4177
1·4	·4192	·4207	·4222	·4236	·4251	·4265	·4279	·4292	·4306	·4319
1·5	·4332	·4345	·4357	·4370	·4382	·4394	·4406	·4418	·4430	·4441
1·6	·4452	·4463	·4474	·4485	·4495	·4505	·4515	·4525	·4535	·4545
1·7	·4554	·4564	·4573	·4582	·4591	·4599	·4608	·4616	·4625	·4633
1·8	·4641	·4649	·4656	·4664	·4671	·4678	·4686	·4693	·4699	·4706
1·9	·4713	·4719	·4726	·4732	·4738	·4744	·4750	·4756	·4762	·4767
2·0	·4772	·4778	·4783	·4788	·4793	·4798	·4803	·4808	·4812	·4817
2·1	·4821	·4826	·4830	·4834	·4838	·4842	·4846	·4850	·4854	·4857
2·2	·4861	·4865	·4868	·4871	·4875	·4878	·4881	·4884	·4887	·4890
2·3	·4893	·4896	·4898	·4901	·4904	·4906	·4909	·4911	·4913	·4916
2·4	·4918	·4920	·4922	·4925	·4927	·4929	·4931	·4932	·4934	·4936
2·5	·4938	·4940	·4941	·4943	·4945	·4946	·4948	·4949	·4951	·4952
2·5	·4938	·4940	·4941	·4943	·4945	·4946	·4948	·4949	·4951	·4952
2·6	·4953	·4955	·4956	·4957	·4959	·4960	·4961	·4962	·4963	·4964
2·7	·4965	·4966	·4967	·4968	·4969	·4970	·4971	·4972	·4973	·4974
2·8	·4974	·4975	·4976	·4977	·4977	·4978	·4979	·4980	·4980	·4981
2·9	·4981	·4982	·4983	·4983	·4984	·4984	·4985	·4985	·4986	·4986
3·0	·49865	·4987	·4987	·4988	·4988	·4989	·4989	·4989	·4990	·4990
3·1	·49903	·4991	·4991	·4991	·4992	·4992	·4992	·4992	·4993	·4993

Chi Square Test

ν \ P	0.99	0.95	0.50	0.30	0.20	0.10	0.05	0.01
1	0·0002	0·004	0·46	0·07	0·64	2·71	3·84	6·64
2	0·020	0·103	1·39	2·41	3·22	4·00	5·99	9·21
3	0·115	0·35	2·17	3·66	4·64	6·25	7·82	11·34
4	0·30	0·71	3·86	4·88	5·99	7·78	9·49	13·28
5	0·55	1·14	4·35	6·06	7·29	9·24	11·07	15·09
6	0·87	1·64	5·35	7·23	8·56	10·64	12·59	16·81
7	1·24	2·17	6·35	8·38	9·80	12·02	14·07	18·48
8	1·65	2·73	7·34	9·52	11·03	13·36	15·51	20·09
9	2·09	3·32	8·34	10·66	12·24	14·68	16·92	21·67
10	2·56	3·94	9·34	11·78	13·44	15·99	18·31	23·21
11	3·05	4·58	10·34	12·90	14·63	17·28	19·68	24·72
12	3·57	5·23	11·34	14·01	15·81	18·55	21·03	26·22
13	4·11	5·89	12·34	15·12	16·98	19·81	22·36	27·69
14	4·66	6·57	13·34	16·22	18·15	21·06	23·68	29·14
15	5·23	7·26	14·34	17·32	19·31	22·31	25·00	30·58
16	5·81	7·96	15·34	18·42	20·46	23·54	26·30	32·00
17	6·41	8·67	16·34	19·51	21·62	24·77	27·59	33·41
18	7·02	9·39	17·34	20·60	22·76	25·99	28·87	34·80
19	7·63	10·12	18·34	21·69	23·90	27·20	30·14	36·19
20	8·26	10·85	19·34	22·78	25·04	28·41	31·41	37·57
21	8·90	11·59	20·34	23·86	26·17	29·62	32·67	38·93
22	9·54	12·34	21·34	24·94	27·30	30·81	33·92	40·29
23	10·20	13·09	22·34	26·02	28·43	32·01	35·17	41·64
24	10·86	13·85	23·34	27·10	29·55	33·20	36·42	42·98
25	11·52	14·61	24·34	28·17	30·68	34·38	37·65	44·31
26	12·20	15·38	25·34	29·25	31·80	35·56	38·88	45·64
27	12·88	16·15	26·34	30·32	32·91	36·74	40·11	46·96
28	13·56	16·93	27·34	31·39	34·03	37·92	41·34	48·28
29	14·26	17·71	28·34	32·46	35·14	39·09	42·56	49·59
30	14·95	18·49	29·34	33·53	36·25	40·26	43·77	50·89

T-Table

ν = number of degrees of freedom

ν \ P	0·50	0·10	0·05	0·02	0·01
1	1·000	6·34	12·71	31·82	63·66
2	0·816	2·92	4·30	6·96	9·92
3	0·765	2·35	3·18	4·54	5·84
4	0·741	2·13	2·78	3·75	4·60
5	0·727	2·02	2·57	3·36	4·03
6	0·718	1·94	2·45	3·14	3·71
7	0·711	1·90	2·36	3·00	3·50
8	0·706	1·86	2·31	2·90	3·36
9	0·703	1·83	2·26	2·82	3·25
10	0·700	1·81	2·23	2·76	3·17
11	0·697	1·80	2·20	2·72	3·11
12	0·695	1·78	2·18	2·68	3·06
13	0·694	1·77	2·16	2·65	3·01
14	0·692	1·76	2·14	2·62	2·98
15	0·691	1·75	2·13	2·60	2·95
16	0·690	1·75	2·12	2·58	2·92
17	0·689	1·74	2·11	2·57	2·90
18	0·688	1·73	2·10	2·56	2·88
19	0·688	1·73	2·09	2·24	2·86
20	0·687	1·72	2·09	2·53	2·84
21	0·686	1·72	2·08	2·52	2·83
22	0·686	1·72	2·07	2·51	2·82
23	0·685	1·71	2·07	2·50	2·81
24	0·685	1·71	2·06	2·49	2·80
25	0·684	1·71	2·06	2·48	2·79
26	0·684	1·71	2·06	2·48	2·78
27	0·684	1·70	2·05	2·47	2·77
28	0·683	1·70	2·05	2·47	2·76
29	0·683	1·70	2·04	2·46	2·76
30	0·683	1·70	2·04	2·46	2·75
35	0·682	1·69	2·03	2·44	2·72
40	0·681	1·68	2·02	2·42	2·71
45	0·680	1·68	2·02	2·41	2·69
50	0·679	1·68	2·01	2·40	2·68
60	0·678	1·67	2·00	2·39	2·66
∞	0·674	1·64	1·96	2·33	2·58

F-Table

The variance ratio, 5 *and* 1% *'points' of F*

ν_1 is the number of degrees of freedom for the greater estimate of variance, and ν_2 for the smaller

ν_2 \ ν_1	1	2	3	4	5	6	8	12	24	∞
2	18·51	19·00	19·16	19·25	19·30	19·32	19·37	19·41	19·45	19·50
	98·49	99·00	99·17	99·25	99·30	99·33	99·36	99·42	99·46	99·50
3	10·13	9·55	9·28	9·12	9·01	8·94	8·84	8·74	8·64	8·63
	34·12	30·82	29·46	28·71	28·24	27·91	27·49	27·05	26·60	20·12
4	7·71	6·94	6·59	6·39	6·26	6·16	6·04	5·91	5·77	5·63
	21·20	18·00	16·69	15·98	15·52	15·21	14·80	14·37	13·93	13·46
5	6·61	5·79	5·41	5·19	5·05	4·95	4·82	4·68	4·53	4·36
	16·26	13·27	12·06	11·39	10·97	10·67	10·27	9·89	9·47	9·02
6	5·99	5·14	4·76	4·53	4·39	4·28	4·15	4·00	3·84	3·67
	13·74	10·92	9·78	9·15	8·75	8·47	8·10	7·72	7·31	6·88
7	5·59	4·74	4·35	4·12	3·97	3·87	3·73	3·57	3·41	3·23
	12·25	9·55	8·45	7·85	7·46	7·19	6·84	6·47	6·07	5·65
8	5·32	4·46	4·07	3·84	3·69	3·58	3·44	3·28	3·12	2·93
	11·26	8·65	7·59	7·01	6·63	6·37	6·03	5·67	5·28	4·86
9	5·12	4·26	3·86	3·63	3·48	3·37	3·23	3·07	2·90	2·71
	10·56	8·02	6·99	6·42	6·06	5·80	5·47	5·11	4·73	4·31
10	4·96	4·10	3·71	3·48	3·33	3·22	3·07	2·91	2·74	2·54
	10·04	7·56	6·55	5·99	5·64	5·39	5·06	4·71	4·33	3·91
12	4·75	3·88	3·49	3·26	3·11	3·00	2·85	2·69	2·50	2·30
	9·33	6·93	5·95	5·41	5·06	4·82	4·50	4·16	3·78	3·36
14	4·60	3·74	3·34	3·11	2·96	2·85	2·70	2·53	2·35	2·13
	8·86	6·51	5·56	5·03	4·69	4·46	4·14	3·80	3·43	3·00
16	4·49	3·63	3·24	3·01	2·85	2·74	2·59	2·42	2·24	2·01
	8·53	6·23	5·25	4·77	4·44	4·20	3·89	3·55	3·18	2·75
18	4·41	3·55	3·16	2·93	2·77	2·66	2·51	2·34	2·15	1·92
	8·28	6·01	6·09	4·58	4·25	4·01	3·71	3·37	3·01	2·57
20	4·35	3·49	3·10	2·87	2·71	2·60	2·45	2·28	2·08	1·84
	8·10	5·85	4·94	4·43	4·10	3·87	3·56	3·23	2·86	2·42
25	4·24	3·38	2·99	2·76	2·60	2·49	2·34	2·16	1·96	1·71
	7·77	5·57	4·68	4·18	3·86	3·63	3·32	2·99	2·62	2·17
30	4·17	3·32	2·92	2·69	2·53	2·42	2·27	2·09	1·89	1·62
	7·56	5·39	4·51	4·02	3·70	3·47	3·17	2·84	2·47	2·01
40	4·08	3·23	2·84	2·61	2·45	2·34	2·18	2·00	1·79	1·51
	7·31	5·18	4·31	3·83	3·51	3·29	2·99	2·66	2·29	1·81
60	4·00	3·15	2·76	2·52	2·37	2·25	2·10	1·92	1·70	1·39
	7·08	4·98	4·13	3·65	3·34	3·12	2·82	2·50	2·12	1·60
80	3·96	3·11	2·72	2·49	2·33	2·21	2·06	1·88	1·65	1·32
	6·96	1·88	4·04	3·56	3·25	3·04	2·74	2·41	2·03	1·49

Control Chart Factors

Factors for $\overline{X}$ charts			Factors for R charts		
Sample size n	$d_2 = \frac{R}{\sigma}$	$A_2 = \frac{3}{d_2 \sqrt{n}}$	$d_3 = \frac{\sigma_z}{\sigma}$	$D_3 = 1 - \frac{3ds}{d_2}$	$D_4 = 1 + \frac{3d_3}{d_2}$
2	1.128	1.881	0.853	0	3.269
3	1.693	1.023	0.888	0	2.574
4	2.059	0.729	0.880	0	2.282
5	2.326	0.577	0.864	0	2.114
6	2.534	0.483	0.848	0	2.004
7	2.704	0.419	0.833	0.076	1.924
8	2.847	0.373	0.820	0.136	1.864
9	2.970	0.337	0.808	0.184	1.816
10	3.078	0.308	0.797	0.223	1.777
11	3.173	0.285	0.787	0.256	1.744
12	3.258	0.266	0.779	0.283	1.717
13	3.336	0.249	0.770	0.308	1.692
14	3.407	0.235	0.763	0.328	1.672
15	3.472	0.223	0.756	0.347	1.653
16	3.532	0.212	0.750	0.363	1.637
17	3.588	0.203	0.744	0.378	1.622
18	3.640	0.194	0.739	0.391	1.609
19	3.689	0.187	0.734	0.403	1.597
20	3.735	0.180	0.729	0.414	1.586
21	3.778	0.173	0.724	0.425	1.575
22	3.819	0.167	0.720	0.434	1.566
23	3.858	0.162	0.716	0.443	1.557
24	3.895	0.157	0.712	0.452	1.548
25	3.931	0.153	0.708	0.460	1.540

Note: If $1 - \frac{3d_3}{d^2} < 0$, then $D_3 = 0$

Critical Values of D in Kolmogorov-Smirnov Goodness of Fit Test

Sample Size	*Level of significance for $D_n = Max \mid F_e - F_a \mid$*				
(n)	0.20	0.15	0.10	0.05	0.01
1	0.900	0.9	0.950	0.975	0.995
2	0.684		0.776	0.842	0.929
3	0.565		0.642	0.708	0.828
4	0.494		0.564	0.624	0.733
5	0.446		0.510	0.565	0.669
6	0.410		0.470	0.521	0.618
7	0.381		0.438	0.486	0.577
8	0.358		0.411	0.457	0.543
9	0.339		0.388	0.432	0.514
10	0.322		0.368	0.410	0.490
11	0.307		0.352	0.391	0.468
12	0.295		0.338	0.375	0.450
13	0.284		0.325	0.361	0.433
14	0.274		0.314	0.349	0.418
15	0.266		0.304	0.338	0.404
16	0.258		0.295	0.328	0.392
17	0.250		0.286	0.318	0.381
18	0.244		0.278	0.309	0.371
19	0.237		0.272	0.301	0.363
20	0.231		0.264	0.294	0.356
25	0.21		0.24	0.27	0.32
30	0.19		0.22	0.24	0.29
35	0.18		0.21	0.23	0.27
over 35	1.07		1.22	1.36	1.63
	$\sqrt{n}$		$\sqrt{n}$	$\sqrt{n}$	$\sqrt{n}$

VECTOR ALGEBRA

Contents : Vector Addition • Centroid • Vector equations of Straight Line and Plane • Scalar Product of Two Vectors • Vectors Product of two Vectors • Multiple Products • Vector Equations of Straight Line and Plane • Solution of Vector Euqations • Volume of Tetratedron • Vector Equation of Sphere • Applicatioon of Vectors tp Mechanics.

COORDINATE GEOMETRY

Contents : Coordinates • The Straight Line • Transformation of Co-ordinates • Pair of Straight Lines • Circles • System of Circles • Parabola • Ellopse • Hyperbola • The Polar Equation of a Conic • General Equation of the Second Degree • General Conic • System or Conic • Conlocal Conics • Circle of Curvature.

DIFFERENTIAL CALCULUS

Contents: Real Numbers • Limit and Continuity • Differentiation or Derivatives • Successive Differentiation • Indetrminate Forms • Tangents amd Normals • Some General Theorems and Exapansions of Fucntions • Maxima and Minima • Curvature • Asymptotes • Partial Differentiation • Points of Inflexion and Singular Points • Curve Tracing

IMPORTANT BOOKS ON MATHEMATICS

PROBABILITY THEORY AND RANDOM PROCESSES

S.P. Eugene Xavier

The purpose of this book is to present an introductory, yet comprehensive treatment of Probability Theory and Random Processes, with a strong emphasis on numerical examples. The author has attempted to explain these concepts and indicate their usefulness through discussion, examples and exercises.

CONTENTS: Probability Theory • Random Variables • Statistical Averages • Random Processes • Linear Systems and Random Noise Processes • Appendices • Bibliography.

14 369 **ISBN: 81-219-1395-0** **pp. 160**

A COURSE OF MATHEMATICAL ANALYSIS

Shanti Narayan

Designed for students who desire to proceed to a course preparatory to a serious study of the purely arithmetical theory of functions of real variables in its general aspects.

CONTENTS: Real Numbers • Bounded Sets, Open and Closed Sets • Real Sequences • Real Valued Functions of a Single Real Variable, Limit and Continuity • Real Valued Functions of a Single Real Variable • Derivability • Riemann Integrability • Sequences of Functions Point-wise and Uniform Convergence • Elementary Functions • Improper Integrals • Fourier Series • Euclidean Spaces • Open and Closed Sets • Compact Sets •

Real Valued Functions of Several Real Variables. • Limit • Continuity • Partial Derivatives • Invertible Functions Implicit Functions • Integrals as Functions of a Parameter • Integration in R^2 Line Integrals. Double Integrals • Curve Lengths. Surface Areas • Integration in R^3 Gauss's and Stoke's Theorems Mix-Exercises • Answers • Appendix.

14 045 **ISBN: 81-219-0472-2** **pp. 440**

MATHEMATICAL STATISTICS

J.N. Kapur & H.C. Saxena

This book is on the same model as books on calculus, differential equation, statics, dynamics, etc. A special feature is the large number of exercises especially constructed to illustrate the theory.

CONTENTS: Introduction • Frequency Distribution and Measures of Location • Measures of Dispersions, Skewness & Kurtosis, Moments of Frequency Distributions • Theory of Probability • Discrete Probability Distributions • Special Discrete Probability Distributions • Univariate Continuous Probability Distributions • Special Continuous Probability Distributions • Principle of Least Squares, Fittings of Curves & Orthagonal Polynomials • Correlation and Regression • Multiple and Partial Correlation • Theory of Sampling • Exact Sampling Distributions • Tests of Significance Based on the *T*, *F* and *Z* Distributions • Tests of Significance Based on the Chi-Square Distribution • Statistical Theory of Point Estimation • Testing of EHypotheses, Sequential Analysis Distribution-Free Methods, Statistical Decision Theory • Elements of Stochastic Processes • Appendices • Index • Log Tables.

14 046 **ISBN: 81-219-1246-6** **pp. 784**

ANALYTICAL SOLID GEOMETRY

Shanti Narayan

This book is intended as an introduction to Analytical Solid Geometry and is suitable for students of B.A., B.Sc., Pass and Honours courses. The book contains numerous exercises of varied types in a graded form.

CONTENTS: Co-ordinates • The Plane • Right Line • Interpretation of Equations-Loci • Transformation of Co-ordinates • The Sphere • Cones, and Cylinders • Homogeneous Cartesian Co-ordinates-Elements at infinity • The Conicoid • Plane Sections of Conicoids • Generating Lines of Conicoids • General Equation of the Second Degree: Reduction to Canonical forms • Appendix • Index.

14 0471 **ISBN:81-219-0473-0** **pp. 320**

A TEXTBOOK ON DYNAMICS

M. Ray & G.C. Sharma

A thoroughly revised edition where the chapters on Central Forces, Moment of Inertia and D'Alembert's Principle have been substantially retouched for the B.A. and B.Sc. students of Dynamics.

CONTENTS: Kinematics and Kinetics • Rectilinear Motion • Uniplanar Motion • Work Energy and Impulse • Impact • Circular and Harmonic Motions • Hodograph • Central Forces • Resisting Medium • Constrained Motion • Moment of Inertia • D'Almbert's Principle and the Equations of Motion.

14 048 **ISBN: 81-219-0342-4** **pp. 328**

DIFFERENTIAL CALCULUS

Shanti Narayan

The text is marked with unified and graphical presentation. The complicated issues are well illustrated.

CONTENTS: Real Numbers • Functions and Graphs-Elementary Functions • Continuity and Limit • Differentiation • Successive Differentiation • Tangents and Normals • Mean Value Theorem • Maxima and Minima • Indeterminate Forms • Partial Differentiation • Equality of Repeated Derivatives • Concavity and Points of Inflexion • Curvature & Evolutes Asymptotes • Singular Points • Curve Tracing • Envelopes • Partial Differentiation • Equality of Repeated Derivatives.

14 049 **ISBN: 81-219-0471-4** **pp. 400**

INTEGRAL CALCULUS

Shanti Narayan

Has a new chapter on `Centre of Gravity and Moment of Inertia'. Many illustrations are given so as to enlighten the subject-matter.

CONTENTS: Table of Standard Results • Methods of Integration • Integration of Algebraic Rational Functions • Integration of Trigonometric Functions • Integration of Irrational Functions • Definite Integral as the Limit of a Sum • Areas of Plane Regions • Rectification • Lengths of Plane Curves • Volumes and Surfaces of Revolution • Centre of Gravity: Moment of Inertia • Differential Equations of First Order and First Degree • Equations of First Order but not of the First Degree • Trajectories of a Family of Curves • Linear Equations • Appendices Mise. Exercises.

14 050 **ISBN: 81-219-0681-4** **pp. 360**

A TEXTBOOK OF VECTOR CALCULUS

Shanti Narayan & J.N. Kapur

Gives a systematic and self- contained treatment to the subject- matter. Good many illustrations are there to make understanding comprehensive.

CONTENTS: Vector Valued Functions of Scalar Variables • Differential Geometry I—Curves • Differential Geometry II—Surfaces • Mechanics I—Particles • Mechanics II—Rigid Body • Differential Operators—Point Functions • Integral Transfor-mations—Line, Surface, Volume Integrals • Hydrodynamics • Electromagnetic Theory.

14 051 **ISBN:81-219-0161-8** **pp. 328**

ORDINARY AND PARTIAL DIFFERENTIAL EQUATIONS

M.D. Raisinghania

Caters to the needs of the honours and postgraduate students. Part I deals with Elementary Differential Equations. Part II deals with ordinary differential equations, and Part III deals with partial differential equations.

CONTENTS: Part I: Elementary Differential Equations— Introduction • Equations of first order and first degree • Trajectories • Linear Equations with constant coefficients • Homogeneous linear equations or Cauchy-Euler equations • Equations of the first order but not of the first degree and singular solutions- Extraneous loci • Ordinary Simultaneous differential equations • **Part II: Ordinary Differential Equations**—Numerical Integration •Existence and uniqueness theorems • Independence of solutions of linear differential equations-Wronskian • Exact differential equations and equations of special forms • Linear Equations of second order • Simultaneous equations of the form dx/P = dy/Q = dz/R • Total differential equations • Riccati's equations • Series solutions of linear differential equations • Legendre Polynomials and Functions • Legendre functions of second kind • Bessel functions • Hermite Polynomials • Laguerre Polynomials • Hyper-geometric Function • **Part III:** Partial Differential Equations— Linear partial differential equations of order one • Non-linear partial-differential equations of Order one • Partial differential equation of the second order • Linear partial differential equations with constant coefficients • Monge's methods • Appendices.

14 282 ISBN:81-219-0892-2 pp. 576

EXAMPLES IN FINITE DIFFERENCES AND NUMERICAL ANALYSIS

H.C. Saxena

This book comprises of twelve chapters with an introduction to important concepts and the list of important formulae needed in the solution of the examples.

CONTENTS: Differences, Operators, Interpolation with Equal Intervals • Interpolation for Unequal Intervals of the Argument • Central Difference Interpolation Formulae • Numerical Differentiation • Inverse Interpolation • Numerical Quadrature • Summation of Series • Difference Equations • Roots of Algebraic and Transcendental Equations • Simultaneous Linear Algebraic Equations • Numerical solution of Ordinary Differential Equations • Linear Programming • Log Tables.

14 175 ISBN:81-219-0753-5 pp. 368

ADVANCED DIFFERENTIAL EQUATIONS
(Including Boundary Value Problems and Difference Equations)

M. D. Raisinghania

This book has been written for Honours, M.A., M.Sc. (Mathematics and Physics) and B.E. students of various universities. The students preparing for P.C.S., I.A.S., A.M.I.E. will also find it useful.

CONTENTS: Part I: Ordinary Differential Equations • Numerical Integration • Existence and Uniqueness Theorem • Independence of solutions of Linear Differential Equations – Wronskian • Exact Differential Equations and Equations of Special Forms • Linear Equations of Second Order • Simultaneous Equations of the form dx/P = dy/Q = dz/R • Total Differential Equations • Riccati's Equation • Series Solutions of Linear Differential Equations • Legendre Polynomials and Functions • Legendre Functions of Second Kind • Bessel Functions • Hermite Polynomials • Languerre Polynomials • Hypergeometric Function • • **Part II: Partial Differential Equations** • Linear Partial Differential Equations of Order One • Non-linear Partial Differential Equations of the Second Order • Linear Partial Differential Equations with Constant Coefficients • Monge's Methods • More an Partial Differential Equations • **Part III: Boundary Value Problems** • **Part IV: Difference Equations.**

14 271 ISBN:81-219-0893-0 pp. 648

A TEXTBOOK OF VECTOR ALGEBRA

Shanti Narayan

An attractively well-written book which deals rigorously and systematically with three dimensional Vector algebra and its application to Euclidean Geometry and Statics.

CONTENTS: Multiplication of Vectors by Scalars and Addition of Vectors • Geometry with Vectors Affine Geometry • Scalar Product • Applications to Metric Geometry • Vector Product and Scalar Triple Product • Application of Scalar Triple Products to Geometry • Geometry with Cartesian Co-ordinates • Some Miscellaneous Topics • Some Properties of Tetrahedra • Statics with Vectors • Answers.

14 064 **ISBN: 81-219-0952-X** **pp. 240**

FLUID DYNAMICS

M.D. Raisinghania

The present text has been well-organised and made up-to-date in the light of the latest syllabi. A detailed index has been provided at the end of the book and more solved examples have been added so that the readers may gain confidence in the technique of problem solving. Almost all the chapters have been rewritten so that in the present form, readers will easily understand the subject matter.

CONTENTS: Introduction • Kinematics of Fluids in Motion • Equations of Motion of Inviscid Fluid • One-Dimensional Inviscid Incompressible Flow (Bernoulli's equation and its applications) • Motion in Two-Dimensions and Sources and Sinks • General Theory of Irrotational Motion • Motion of Cylinders • Irrotational Motion in three Dimensions (Motion of a sphere) Stoke's Stream Function • Vortex Motion (Rectilinear vortices) • Waves • General Theory of Stress and Rate of Strain • The Navier-Stokes Equations and the Energy Equation • Dimensional Analysis • Dimensionless Constants (numbers) • Similarity of Flows • Laminar Flow of Viscious Incompressible Fluids • Boundary Layer Theory • Discontinuous Motion • Index.

14 349 **2nd Edn.** **Rep.1998**

ISBN: 81-219-0869-8 **pp. 872**

INTEGRAL TRANSFORMS

M.D. Raisinghania

This book is useful for Honours, M.A., M.Sc. (Mathematics) and Physics and B.E. students. It is also useful for students preparing for A.M.I.E., I.A.S. and other competitive examinations. Various types of problems have been fully illustrated by solved examples.

CONTENTS: Laplace Transform • Inverse Laplace Transform • Applications of Laplace Transforms • Fourier Transforms • Finite Fourier Transform • Hankel Transforms • The Finite Hankel Transforms • The Mellin Transform • Index.

14 283 **2nd Edn. 1995**

ISBN:81-219-0886-8 **pp. 296**

RAMANUJAN NUMBERS (Mathematical Thoughts and Ideas)

M. Meyyappan

This book is intended to provide a modest knowledge of mathematics with an excursion in the number world. The book has a lot of illustrations and through activities the students are helped to acquire basic concepts and facts.

The standard of students in the subject mathematics is gradually deteriorating in our country. Popular Science books may interest and help divert students' attention on to developing skills but, this book in its humble effort tries to rejuvenate the mathematical inclination of Indian students.

To remove the fear of mathematics at an early stage and to create strong interest in this vital subject, the mother of all Sciences, mathematics, may be introduced as fun and entertainment. In this book such an attempt has been made.

The book also presents algebra and problem solving in an exciting and amusing manner. On the whole the book endeavours to arouse the curiosity in the students and thereby develop interest.

CONTENTS: Ramanujan— Mathematician Extraordinary • Second Order Ramanujan Numbers • Third Order Ramanujan Numbers • Fourth Order Ramanujan Numbers • Appendix.

14 368 1st Edn. Rep. 1998 ISBN: 81-219-1272-5 pp. 128

ELEMENTS OF REAL ANALYSIS

Shanti Narayan

This book is an attempt to make presentation of Elements of Real Analysis more lucid. The book contains examples and exercises meant to help a proper understanding of the text.

CONTENTS : Sets and Statements • The Real Numbers • Limit Points of a Set • Sequences • Infinite Series with positive terms • Infinite Series with positive and negative terms • Real Functions—Limit and Continuity • Real Functions— The Derivative • Riemann Integrability.

14 052 6th Edn. Rep. 1998 ISBN: 81-219-0306-8 pp. 216

VECTOR ALGEBRA

M.D. Raisinghania

The present edition of the book has been thoroughly revised, and updated in the light of latest syllabi of various universities and institutions of India. More solved examples have been added and a detailed index has been provided at the end of the book inorder to make it more useful to students and teachers.

CONTENTS: Vector Addition • Centroid • Vector equations of straight line and plane (Parametric form) • Scalar products of two vectors • Vector product of two vectors • Product of three and four vectors (Multiple products) • Vector equations of straight line and plane (non-parametric forms) • Volume of tetrahedron • Solution of vector equation • Vector equation of sphere • Applications of vectors to mechanics (Dynamics and Statics) • Dynamics with vectors • statics with vectors **Section I: Concurrent Forces • Section II : Vector moment (or torque) and work • Section III : Parallel Forces • Section IV Couple • Section V : Coplaner non-concurrent forces acting on a rigid body • Section VI : Forces in three dimensions** • Index.

14 363 3rd Edn. 1997 ISBN : 81-219-4447-7 pp. 312

FINITE DIFFERENCES AND NUMERICAL ANALYSIS

H.C. Saxena

This thoroughly revised edition of the book completely covers the syllabi in the calculus of Finite Differences of various Indian universities. Examples given at the end of each chapter have been specially constructed, taken from university papers, and standard books.

CONTENTS : Part I—Calculus of Finite Differences — Finite Differences, Interpolation with Equal Intervals • Interpolation of Unequal Intervals of the Argument • Central Difference Interpolation Formulae • Inverse Interpolation • Numerical Differentiation • Numerical Quadrature • Summation of Series • Difference Equations • Generating Functions and Theorems • Bernoulli and Euler Polynomials • Remainder Terms in (or Errors associated with) • Interpolation Formulas **Part II —Numerical Analysis**— Eigen Value and Eigen Vector • Roots of Polynomial and Transcendental equation in one variable • Simultaneous Linear Algebraic Equations • Numerical Solution of Ordinary Differential Equations • Errors • Partial Differential Equations • Appendices • Answers • Index • Log Tables.

14 056 **14th Rev. Edn. 1998** **ISBN: 81-219-0339-4** **pp. 420**